# International Conference on Agriculture, Environment, and Food Security (AEFS 2017)



IOP Conference Series: Earth and Environmental Science
Volume 122

Medan, Indonesia
7 – 8 November 2017

ISBN: 978-1-5108-5920-3
ISSN: 1755-1307

**Printed from e-media with permission by:**

Curran Associates, Inc.
57 Morehouse Lane
Red Hook, NY 12571

**Some format issues inherent in the e-media version may also appear in this print version.**



Printed by Curran Associates, Inc. (2018)

For permission requests, please contact the Institute of Physics
at the address below.

Institute of Physics
Dirac House, Temple Back
Bristol BS1 6BE UK

Phone: 44 1 17 929 7481
Fax: 44 1 17 920 0979

techtracking@iop.org

**Additional copies of this publication are available from:**

Curran Associates, Inc.
57 Morehouse Lane
Red Hook, NY 12571 USA
Phone: 845-758-0400
Fax: 845-758-2633
Email: curran@proceedings.com
Web: www.proceedings.com

# TABLE OF CONTENTS

IOP Conf. Series: Earth and Environmental Science **122** (2018) 011001 doi:10.1088/1755-1315/122/1/011001

## Conference Information

| | |
|---|---|
| Dates | : November 7 - 8, 2017 |
| Organizer | : Faculty of Agriculture<br>Universitas Sumatera Utara |
| Venue | : Arya Duta Hotel Medan<br>Jalan Kapten Maulana Lubis No. 8 Medan, Sumatera Utara, Indonesia Phone: +62 61 4572999<br>Web: http://www.aryaduta.com |
| Official Language | : English |
| Secretariat | : Faculty of Agriculture Universitas Sumatera Utara<br>Jl. Dr. A. Sofyan No. 3 Kampus USU Medan Sumatera Utara, Indonesia |
| Email | : aefs@usu.ac.id<br>aefsfpusu2017@gmail.com |
| Conference Website | : https://ocs.usu.ac.id/AEFS/AEFS2017/index |

IOP Conf. Series: Earth and Environmental Science **122** (2018) 011001 doi:10.1088/1755-1315/122/1/011001

**Conference Committee**

| | |
|---|---|
| Chairman | :Dr. Ir. Tavi Supriana, M.Si |
| Co-chair | :Ir. Revandy Damanik, M.Sc, Ph.D |
| | Dr. Lisnawita, SP, M.Si |
| Secretary | :Riswanti Sigalingging, STP, M.Si, Ph.D |
| Honorary Board | :Prof. Dr. Runtung, SH, MHum |
| | Prof. Ing. Vladimir Jurcha, CSc |
| | Prof. Dr. Ing. Frantisek Kumhala |
| | Dr. Ir. Hasanuddin, MS |
| | Drs. Mahyuddin K M Nasution, M.I.T, Ph.D |
| Scientific Committee (Reviewer) | :Prof. Carolee T. Bull, Ph.D (USA) |
| | Dr. Ian T. Riley (Australia) |
| | Prof. Dr. Erman Munir, M.Sc (Indonesia) |
| | Prof. Dr. Abdul Razak Alimon (Malaysia) |
| | Prof. Dr. Purwiyatno Hariyadi (Indonesia) |
| | Ing. Abraham Kabutey, Ph.D (Czech Republic) |
| | Prof. Dr. David B. Weaver (USA) |
| | Dr. Rich Llewelyn (USA) |
| | Prof. Dr. Ir. Cece Sumantri, M.Agr Sc (Indonesia) |
| | Prof. Edison Purba, Ph.D (Indonesia) |
| | Riswanti Sigalingging, STP, M.Si, PhD (Indonesia) |

Local Committee :

1. Prof. Dr. Ir. Darma Bakti, MS
2. Dr. Eriyusni, M.Sc
3. Dr. Ir. Yaya Hasanah, M.Si
4. Dr. Ir. Elisa Julianti, M.Si
5. Dr. Ir. Ma'ruf Tafsin, M.Si
6. Dr. Khairunnisa, SP, MP
7. Irda Safni, SP, MCP, Ph.D
8. Ir. Diana Chalil, M.Si, Ph.D
9. Era Yusraini, S.TP, M.Si
10. Ir. Hotnida Sinaga, M.Phil, Ph.D
11. Rulianda Wibowo, SP, M.Sc, Ph.D
12. Lukman Adlin Harahap, STP, M.Si
13. Delima Lailan S. Nasution, S.TP, M.Sc
14. Nazif Ichwan, S.TP, M.Si
15. Achmad Sadeli , S.Pt, M.Sc
16. Fuad Hasan, S.Pt, M.Si
17. Tati Vidiana, S.Pt, M.S
18. Indra Elizar
19. Arung Buana
20. Antasari Malau
21. Kenzi
22. Hermanyanto Laia
23. Nisrina Hayati
24. Bobby Manullang
25. Stefani Limanto
26. Yerika
27. Sonia
28. Nina Unzila Angkat
29. Sylvia Bianca Sihombing

## Venue Map

**Arya Duta Hotel Medan**
Jalan Kapten Maulana Lubis No. 8 Medan, Sumatera Utara, Indonesia
Phone: +62 61 4572999

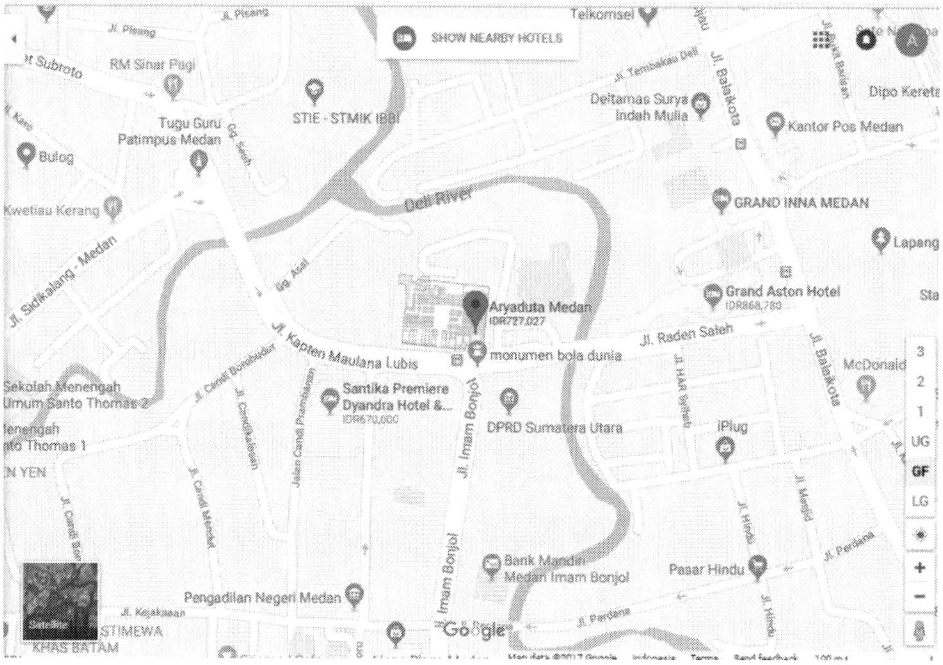

IOP Conf. Series: Earth and Environmental Science **122** (2018) 011002 doi:10.1088/1755-1315/122/1/011002

## Introduction

First of all let us praise and thank the presence of Allah Almighty, for the abundance of grace and the joy all of us can still gather in this place in good health.We would like to take this opportunity to tell you that this is our honour and privilege to welcome you here.

The honorable:

- Rector of Universitas Sumatera Utara, Prof. Dr. Runtung , MHum
- Dean of Faculty of Agriculture Universitas Sumatera Utara : Dr. Ir. Hasanuddin, M.S.
- Keynote Speakers:
    1. Prof. Dr. Chris Franco (University of Flinders, Australia)
    2. Prof. David Herak (Czech University of Life Science, Prague)
    3. Mirza Hasanuzzaman (Sher-e-Bangla Agricultural University, Bangladesh)
    4. Dr. Janice Sher Huay Lee (Nanyang Technology University, Singapore)
    5. Dr. Ir. Anton Apriyantono, MS (Former Minister of Agriculture, Indonesia)
- The distinguished guests from all around the world

I m greatly honored to welcome you to the first International Conference on Agriculture, Environment and Food Security (AEFS) 2017. AEFS conference aims to offer the opportunity for knowledge sharing, networking, and collaboration between engineers, scientists, and technologists as well as academician and researchers working in the specific areas of agriculture, social economics, biosystems engineering and food technology. For this year the committee has chosen "Agriculture, Plantation and Livestock for World Food Security" as the main theme, with 6 selected tracks including Agricultural Engineering, Agricultural Economics, Plant Science (Agronomy and Plantation, Plant Breeding, Biotechnology, Integrated Pest Management and Soil Science), Animal Science, Food Science and Technology, Marine and Fisheries Sciences. This conference is organized by The Faculty of Agriculture, University of Sumatera Utara (USU) as an annual event to celebrate the faculty anniversary and fully supported by Czech University of Life Sciences, Prague (CULS), The Institution of Engineers Indonesia (IEI/PII), Indonesian Association of Food Technologist (IAFT/PATPI), Indonesian Society of Agricultural Economics (ISAE/PERHEPI), Indonesian Association of Biochemistry and Biology Molecular (IABBM/PBBMI), The Indonesian Agronomy Association (IAA/PERAGI), Weed Science Society of Indonesia (WSSI/HIGI), Indonesian Phytopathological Society (IPS/PFI) and The Indonesian of Plant Breeding Society (IPBS/PERIPI). The AEFS 2017 program consists of the interactive presentation sessions, keynote speaking and social events including networking dinner and post-conference tour.

IOP Conf. Series: Earth and Environmental Science **122** (2018) 011002 doi:10.1088/1755-1315/122/1/011002

There are 220 papers that have been submitted to AEFS' commitee, but after the reviewing process there are 142 papers whichhave been approved. International seminar has been held for 2 days from $7^{th}$ – $8^{th}$ of November 2017 in Aryaduta Hotel with various important agenda.

I would like to express my appreciation to the presenters who are coming from the university in Indonesia, Czech Republic (Czech University of Life Science Prague), Australia (The University of Queensland), Ghana (CSIR - Council for Scientific and Industrial Research - Ghana · SARI), Japan (University of the Ryukyus, Nishihara), Turkey ( Ondokuz Mayis University, Samsun) and chairs of all the sessions. You deserve it and I think we would all agree that the quality of the presentations and the papers for this conference have been of a very high standard.

I hope you will have a pleasant post-conference tour and journey to your home countries. I look forward toseeing you on second ICAEFS 2018.

My personal respect and thanks goes out to all of you
Chair of the Organizing Committee of ICAEFS 2017
Dr. Ir. Tavi Supriana, MS

Scopus
NOVEMBER 7 – 9, 2017

AEFS
2017

NOVEMBER 7 – 9, 2017
ARYA DUTA HOTEL, MEDAN
MEDAN - INDONESIA
AEFS
2017

IOP Conf. Series: Earth and Environmental Science **122** (2018) 011002 doi:10.1088/1755-1315/122/1/011002

IOP Conf. Series: Earth and Environmental Science **122** (2018) 011002 doi:10.1088/1755-1315/122/1/011002

IOP Conf. Series: Earth and Environmental Science **122** (2018) 011003 doi:10.1088/1755-1315/122/1/011003

IOP Conf. Series: Earth and Environmental Science **122** (2018) 011004 doi:10.1088/1755-1315/122/1/011004

# KEYNOTE SPEAKER

**Prof. David Herak** (Biosystem Engineering)

*Czech University of Life Science, Prague*

**Prof. Dr. Chris Franco** (Microbiology and Biotechnology)

*University of Flinders, Australia*

**Dr. Ir. Anton Apriyantono, MS** (Food Science)

*Former Minister of Agriculture, Indonesia*

**Dr. Janice Ser Huay Lee** (Environment)

*Nanyang Technology University, Singapore*

**Mirza Hasanuzzaman, Ph.D** (Agronomy)

*Sher–e–Bangla Agricultural University, Bangladesh*

IOP Conf. Series: Earth and Environmental Science **122** (2018) 011004 doi:10.1088/1755-1315/122/1/011004

# SCIENTIFIC COMMITTEE AEFS 2017

Prof. Carolee T. Bull, Ph.D (USA)

Dr. Ian T. Riley (Australia)

Prof. Dr. Erman Munir, M.Sc (Indonesia)

Prof. Dr. Abdul Razak Alimon (Malaysia)

Prof. Dr. Purwiyatno Hariyadi (Indonesia)

Ing. Abraham Kabutey, Ph.D (Czech Republic)

Prof. Dr. David B. Weaver (USA)

Dr. Rich Llewelyn (USA)

Prof. Dr. Ir. Cece Sumantri, M.Agr Sc (Indonesia)

Prof. Edison Purba, Ph.D (Indonesia)

Riswanti Sigalingging, STP, M.Si, PhD (Indonesia)

Onrizal, S. Hut, M.Si, Ph.D (Indonesia)

# ORGANIZING COMITTEE

**Chairman** :Dr. Ir. Tavi Supriana, MS

**Secretary** : Riswanti Sigalingging, S.TP, M.Si, Ph.D

**Staff** :

1. Prof. Dr. Ir. Darma Bakti, MS
2. Prof. Ir. Edison Purba, Ph.D
3. Dr. Lisnawita, SP, M.Si
4. Dr. Eriyusni, M.Sc
5. Dr. Ir. Yaya Hasanah, M.Si
6. Dr. Ir. Elisa Julianti, M.Si
7. Dr. Ir. Ma'ruf Tafsin, M.Si
8. Dr. Khairunnisa, SP, MP
9. Ir. Revandi I.M. Damanik, M.Sc, Ph.D
10. Irda Safni, SP, MCP, Ph.D
11. Ir. Diana Chalil, M.Si, Ph.D
12. Era Yusraini, S.TP, M.Si
13. Ir. Hotnida Sinaga, M.Phil, Ph.D
14. Rulianda Wibowo, SP, M.Sc, Ph.D
15. Delima Lailan S. Nasution, S.TP, M.Sc
16. Nazif Ichwan, S.TP, M.Si
17. Achmad Sadeli , S.Pt, M.Sc
18. Fuad Hasan, S.Pt, M.Si
19. Tati Vidiana, S.Pt, M.S
20. Indra Elizar
21. Arung Buana
22. Antasari Malau
23. Kenzi
24. Hermanyanto Laia
25. Nisrina Hayati
26. Bobby Manullang
27. Stefani Limanto
28. Yerika
29. Sonia
30. Nina Unzila Angkat
31. Sylvia Bianca Sihombing

IOP Conf. Series: Earth and Environmental Science **122** (2018) 011004 doi:10.1088/1755-1315/122/1/011004

# SUMMARY KEYNOTE SPEAKERS

**Modern method sofdescription of mechanical behaviour of *Jatropha curcas*L. seeds undercompression loading**

**Prof. Ing. David Herák, Ph.D.Biosystem Engineering *Czech University of Life Science, Prague*.**

This presentation deals with the development of long time research related to the mechanical behaviour of Jatropha bulk seeds carried out at Czech University of Life Sciences Prague in cooperation with several Indonesian institutions. Presentationis divided into several chapter which logically follow. At a beginning of the presentation the mechanical behaviour of bulk seeds under compression loading, creep and relaxation are described. In the next chapterdifferent mathematical modelsas well ascomputer virtual models using finite element method are derived and presented. The connection of mathematical models with real environment of screw presses are also described in this presentation as well as their utilization in industrial engineering. Last but not least chapter is focused on the involvement of international projects and students cooperation in this research. The significant results of carried out research are an integral part of this presentation.

**Endophytic Actinobacteria – Beneficial partners for Crop production**

**Prof Christopher Franco, Medical Biotechnology, Flinders University, Adelaide, Australia**

Actinobacteria are filamentous bacteria that are prolific producers of antibiotics and other bioactive compounds. They are ubiquitous in soil but some have adapted to an endophytic life style.

The first part examines their isolation, taxonomy and their presence in the plants. This is done using a GFP-containing strain and confocal and electron microscopy (EM) to show that when added as a spore coating they can recolonise the intercellular spaces. It is also noted that they can enter the plant through crack in the roots especially at lateral root junctions. We have shown that they can be found in the leaves by scanning EM and growth using an antibiotic resistant marker.

We have found that endophytic Actinobacteria are ubiquitous in plants and some can be used to improve crop production.The next phase describes our Beneficial Microbes Program (BMP)supported by the Australian Grains Research and Development Corporation in which we have 5 streams.1. Nodulation increasing bacteria; 2. Control ofRhizoctonia on wheat; 3. Control Pythium on wheat; 4. Control Pythium on canola; 5. Canola tolerance to moisture stress.
For the control of fungal root diseases of wheat we have utilised a unique high throughput tube assay followed by the more rigorous replicated pot assays to confirm efficacy, reliability, stability on seed and determine the optimum amount of inoculum. The effect of farm chemicals such as fungicides and insecticides was also tested for compatibility.In addition, because Rhizoctonia manifests in patches a new form of field testing was employed so that the controls and treatments fall within a disease patch. We screened over 2000 strains for Pythium and Rhizoctonia root rot to give us 2 strains for control of each of the disease.

Of relevance to Indonesia is the microbiome of rice. This was doctoral research work done by Dr FitriWidiantini who showed that Australian rice plants harbour an abundance on *Microbispora* species even though the soils have a majority of *Streptomyces* species. A number of the isolates were found to be new on the basis of polyphasic taxonomy including DNA-DNA hybridisation. Further studies with different soil types confirmed the presence of *Microbispora* spp. as being the dominant genus in Australian rice plants.

Mechanisms of action are an important factor in progressing the use of microbial inoculants for use in agriculture. Here we report on our inoculants being able to induce systemic resistance using *Arabidopsis thaliana* as the model plant. The transcript analysis indicates that the inoculant primes the plant so that it can withstand infection either by *Fusarium oxysporum* or *Pectobacteriumcarotovorans*. A proteomics study showed that the priming effect is due to the inoculant being able to produce a boost of PR protein. These proteins appear to be able to counteract the ROS damage inflicted by the pathogen at levels that exceed the response to infection by the pathogen alone.

Finally we examined the association of actinobacteria with Rhizobia to increase biological N-fixation by 50-70% in legume plants grown in soils with low N content, compared to untreated controls. In crops like Lucerne the biomass yield increases by up to 50% and with grain legumes such as soya the dry bean yield increased by 20-50% when treated with specific actinobacteria.

Our work has served to illustrate that a few endophytes offer reliable sustainable solutions to modern agriculture.

*Prof Christopher Franco, Medical Biotechnology, College of Medicine and Public Health,*

*Flinders University, Adelaide, Australia.*

*Email: chris.franco@flinders.edu.au Tel# +618 7221 8554 ; Fax# +618 7221 8555*

# Ecosystem Management For Improving Oil Palm Agriculture & Ensuring Food Security

**Janice Ser Huay Lee**

**Assistant Professor at the Asian School of the Environment, Nanyang Technological University of Singapore**

In this talk, I will discuss the importance of ecosystem management for oil palm agriculture and highlight the ways in which plantation agriculture rely and is shaped by its ecosystem. Oil palm plantation agriculture cannot be seen as being isolated from the ecosystem as it has large impacts on the integrity of tropical ecosystems, where they are grown, and are equally affected when the ecosystem degrades and is in poor health. Well managed ecosystems have the potential to benefit oil palm plantation agriculture in the long-run, which will in turn benefit Indonesia's rural population and national economy.

Ecosystem management is an example of a coupled human and natural system where human decisions and policies govern the landscape and shape the outcomes of components of the landscape, including oil palm plantation agriculture. Firstly, I will discuss how to implement effective ecosystem management through the lens of existing government regulations for oil palm development, certification standards to reward well managed oil palm plantations, and jurisdictional approaches to include stakeholders which have yet to be convinced of the benefits of incorporating ecosystem health in the management of their oil palm plantations. Secondly, I will highlight how environmental changes pose significant risks to oil palm plantation landscapes and the ways in which ecosystem management can mitigate these risks. These include peatland management and restoration, riparian buffer development, and a moratorium on the development of relatively intact peatswamp forests, peat domes and mangrove forests. Lastly, I will talk about the use of palm oil in our diets and highlight how there is an increasing trend in the use of palm oil for processed food or junk food, which is nutritionally poor. I will present the extent of oil palm plantations required for palm oil consumption in junk foods and elaborate on the need to link nutritional values to agricultural land use.

**Contact Information:**

*Janice Ser Huay Lee (Dr.), Assistant Professor,Asian School of the Environment, Nanyang Technological University of Singapore,50 Nanyang Avenue, Block N2-01C-43, Singapore 639798,janicelee@ntu.edu.sg*

IOP Conf. Series: Earth and Environmental Science **122** (2018) 011004 doi:10.1088/1755-1315/122/1/011004

# Crop Responses And Adaptation To Environmental Stresses In The Era Of Climate Change

**Mirza Hasanuzzaman**[a,b]

[a]**Department of Agronomy, Sher-e-Bangla Agricultural University, Dhaka-1207, Bangladesh**
[b]**Center of Molecular Biosciences, University of the Ryukyus, Nishihara, Okinawa, Japan**
**E-mail: mhzsauag@yahoo.com**

'How will we feed the world in the next decades?' is one of the frequently asked questions among the agricultural scientists. The main evolution that will increase the demand for food is a continued rise in the number of people on our planet. While the world's population consists of 7.3 billion people today, the UN projects that the number will increase by 1.2 per cent annually, amounting to 8.5 billion people in 2030 and almost 10 billion people in 2050. Unfortunately, the population growth is high in developing countries that have the least area of arable lands per capita. Although plant scientists are largely successful in enhancing crop yield due to advanced technologies there is still crisis of food. Plants are the major providers of staple foods for human. But as a sessile organism plants cannot avoid environmental adversities. Due to the climate change different adverse environmental factors like salinity, drought, extreme temperatures, toxic metals/metalloids, waterlogging/flooding, ozone etc. are reducing plant productivity significantly. The complex nature of the environment, along with its unpredictable conditions and global climate change, are increasing gradually, which is creating a more adverse situation. Abiotic stress may reduce crop yield up to 70%. The episodes of drought and heat stress is very common now-a-days. In the last couple of years we have experienced record breaking high temperature. In contrary, both chilling and freezing are common in many temperate countries and also in other part of the world during winter. Some of the toxic metals/metalloids, e.g. cadmium, lead, arsenic etc. are exceeding their allowable limits which are also major concern for crop production and human health as well. Abiotic stresses modify plant metabolism leading to harmful effects on growth, development and productivity which ultimately affect food security.Therefore, understanding the molecular and physiological mechanisms of abiotic stress tolerance and to find the ways that would increase stress tolerance in plants are crucial in agriculture. Recently, scientists have explored the underlying mechanisms of stress-induced damages and the tolerance mechanisms. They also developed various techniques in conferring environmental stress tolerance in plants. Learning from the tolerant plants and transferring these traits to sensitive plants to gain productivity is now a major concern. With the advancement of molecular tools tailoring of stress responsive genes become possible for the development of tolerant genotypes. Overexpression of combinations of target genes in transgenic plants has been shown to have synergistic effects on stress tolerance. Therefore, increased emphasis is being placed on producing transgenic plants overexpressing genes – i.e., gene stacking – associated with more than one trait in order to achieve tolerance to multiple environmental stresses. Some eco-friendly approaches like phytoremediation has also drawn attention to researchers for the mitigation of metal/metalloids-induced damages. It is possible to minimize losses in agricultural production due to abiotic stresses by a judicious blend of knowledge in crop physiology and crop husbandry procedures. Since environmental stress is concerned with multidiscipline like agriculture, plant science, plant breeding, molecular biology, soil science, chemistry and many other field of biological science an integrated approaches with coordinated and sustainable research may attain the goal.

IOP Conf. Series: Earth and Environmental Science **122** (2018) 011004 doi:10.1088/1755-1315/122/1/011004

# Opening Speech From Rector of Universitas Sumatera Utara

## International Conference on Agriculture, Environment and Food Security (AEFS) 2017

07 November 2017 at Hotel Aryaduta, Medan

---

*Assalamu'alaikum warahmatullahi wabarakatuh,*

*Good morning and Peace be upon you all*

Excellencies, distinguished guest, Ladies and Gentlemen

It gives me great pleasure and privileged to extend to you all a very warm welcome on behalf of Universitas Sumatera Utara.We are very grateful to all our distinguished guests who accepted our invitation especially those who have travelled many hours to convene this International Conference on Agriculture, Environment and Food Security (AEFS) 2017.

AEFS 2017 is organized by The Faculty of Agriculture, Universitas Sumatera Utara asaannual event to celebrate the faculty anniversary and fully supported by other university and research institution partners. The cross cutting issue such as growing population, climate change, fair trade negotiation, food standard, environmental sustainability and globalization are putting pressure on future food security and environmental sustainability. These issues have motivated us to initiate this conference. The agricultural practices must rely on restoring the natural ecological balance by optimizing the production of food and the use of space. The challenge will be of how to produce enough food to meet all needs but still enhancing and preserving natural resources. This conference will not only give an opportunity to all of us to exchange idea and information but also to extend our corporation and discuss problems of mutual interest with academician and delegates not only from Indonesia but also from member countries of the region.

No matter how much we can do by ourselves on the local or national level, whether it is research or development action, it never enough to solve all problems in Agriculture, food security and environment. We must collaborate in an action-oriented effort to resolve the problems that trouble theagricultural development. I hope this conference can be a very fruitful first step to initiate our future collaboration. At the end of the conference, the knowledge from exchanging the information will give benefit to all of us with deeper understanding and give brighter idea of how to meet the future food need but with sustainable agricultural practice.

North Sumatera is famous known by its cultural heritage and having Danau Toba as one of the most popular tourism destination in Indonesia. Istana Maimun and Mesjid Raya are one of the famous cultural heritage which designed by Italian Architect. In addition, Danau Toba

IOP Conf. Series: Earth and Environmental Science **122** (2018) 011004 doi:10.1088/1755-1315/122/1/011004

is the largest volcanic lake in Indonesia which never fails to amaze any visitor. We hope that you may have the opportunity to visit any of these sites during your stay here.

I am looking forward to a productive discussion among all of the participants including both the speakers and the audience. I hope today's conference will be useful and interesting to all of us. With this, I proudly announce that on Tuesday, November 7th 2017, The International Conference on Agriculture, Environment and Food Security (AEFS) 2017 is officially open. Hope you have a good day and thank you very much for your attendance.

*Wassalamu'alaikum warahmatullahi wabarakatuh,*

Rector,

**Prof. Dr. Runtung, SH, M.Hum**

C_Humas-USU 17_Okt_2017

IOP Conf. Series: Earth and Environmental Science **122** (2018) 011004 doi:10.1088/1755-1315/122/1/011004

# Conference Schedule

| **Activity** | **Time** |
|---|---|
| **Tuesday, 07th November 2017** | |
| Registration of Participants | 07.30 – 08.30 |
| Welcome Address and Opening | 08.30 – 09.00 |
| **Keynote Speaches and Discussion Part I** | |
| Chris Franco | 09.0 -09.30 |
| David Herak | 09.30 – 10.00 |
| Mirza Hasanuzzaman | 10.00 – 10.30 |
| Discussion / Coffee Break | 10.30 – 11.00 |
| **Keynote Speaches and Discussion Part II** | |
| Janice Sher Huay Lee | 11.00 – 11.30 |
| Anton Apriyantono | 11.30 – 12.00 |
| Discussion | 12.00 – 12.20 |
| Lunch break | 12.20 – 13.00 |
| Parallel Session I | 13.30 – 15.00 |
| Coffee Break | 15.00 – 15.15 |
| Photo Session and Welcome Dinner at USU | 16.00 |
| **Wednesday, 08th November 2017** | |
| Paralel Session II | 08.00 – 09.30 |
| Coffee Break | 09.30 – 09.45 |
| Paralel Session III | 09.45 – 11.15 |
| Paralel Session IV | 11.15 – 12.30 |
| Lunch Break | 12.30 -13.30 |
| Paralel Session V | 13.30 – 15.00 |
| Coffee Break | 15.00 – 15.15 |
| Paralel Session VI | 15.15 – 16.45 |
| Closing Ceremony | 17.00 |

IOP Conf. Series: Earth and Environmental Science 122 (2018) 011004 doi:10.1088/1755-1315/122/1/011004

# Parallel Session Schedule

## CLASS A

**Session I 7/11/17**

| No | Paper ID | Author | Title | Time | Date | Track |
|---|---|---|---|---|---|---|
| A1 | 1515 | Witono Adiyoga | Farmers's Perceptions on Climate Change in Low Land and High Land Vegetable Production Centers of South Sulawesi | 13.30 – 13.45 | 11/07/2017 | AEC |
| A2 | 1524 | E.W. Riptanti, A. Qonita, Suprapti | Revitalization of Food Barns in Supporting Sustainable Food Security in Central Java | 13.45 – 14.00 | 11/07/2017 | AEC |
| A3 | 1625 | Diana Chalil, Riatri Barus | Questioning the Sustainable Palm Oil Demand : Case Study from French – Indonesia Supply Chain | 14.00 – 14.15 | 11/07/2017 | AEC |
| A4 | 1793 | S. Hariyanti, T. Supriana, R.S. Siregar | The Effect of Consumer Perception to the Satisfaction of Use of Traditional Drugs in Medan | 14.15 – 14.30 | 11/07/2017 | AEC |
| A5 | 1802 | A.F. Siregar, T. Supriana | Factors That Influence The Interest of Farmer in Shallots Farming at Cinta Damai Village of Simanindo Sub- District of Samosir District | 14.30 – 14.45 | 11/07/2017 | AEC |
| A6 | 1803 | Salsabila, T. Supriana | Strategies to increase the consumption of Traditional Medicine in Medan | 14.45 – 15.00 | 11/07/2017 | AEC |

**Session II 8/11/17**

| No | Paper ID | Author | Title | Time | Date | Track |
|---|---|---|---|---|---|---|
| A13 | 1571 | Dessy Adriani, Imron Zahri, Elisa Wildayana, Maryadi, Maryanah Hamzah, Yulius | Can tidal land prosper the farmers ? Lesson learn from migrant and local farmers in telang's integreted independent city, south sumatera | 08.00 - 08.15 | 11/08/2017 | AEC |
| A14 | 1766 | T. Supriana and Farida Yani | The analysis of consumption level and preferences of fresh shallots consumers in Medan | 08.15- 08.30 | 11/08/2017 | AEC |
| A15 | 1771 | M C Alam and T Supriana | Analysis of Supply Chain Management of Shallots at Medan | 08.30- 08.45 | 11/08/2017 | AEC |
| A16 | 1854 | IS Rianse, U Rianse, WG Abdullah, S Hartono, A | The Income Distribution and Contribution of Palm Sugar Producer in | 08.45- 09.00 | 11/08/2017 | AEC |

IOP Conf. Series: Earth and Environmental Science **122** (2018) 011004 doi:10.1088/1755-1315/122/1/011004

| | | | | | | |
|---|---|---|---|---|---|---|
| | | Suryantini, Jamhari, Muhidin | Increasing The Household Welfare of Palm Sugar Maker in Kolaka Southeast Sulawesi Indonesia | | | |
| A17 | 3032 | Rahmanta | Influence of Land Area and Capital Strenghtening Fund of Rural Economics Enterprise toward Corn Production in North Sumatera Province | 09.00-09.15 | 11/08/2017 | AEC |
| A18 | 1832 | Leni Handayani, Abdul Rauf, Rahmawaty, Tavi Supriana | The Strategy Of Sustainable Soybean Production To Increase Soybean Needs In North Sumatera | 09.15-09.30 | 11/08/2017 | AEC |

**Session III 8/11/17**

| No | Paper ID | Author | Title | Time | Date | Track |
|---|---|---|---|---|---|---|
| A19 | 3011 | Rulianda P Wibowo, Sumono, Noah Miller, Hasman Hashyim, Thompson Sebayang | Pricing behavior of USA exporter in wheat international market | 09.45 - 10.00 | 11/08/2017 | AEC |
| A20 | 1683 | Lindawati, Nunung Kusnadi, Sri Utami Kuntjoro, Dewa K.S. Swastika | The Impact Of Input And Output Prices On The Household Economic Behavior Of Farmers Rice-Livestock Integrated Farming System (Rlifs) And Non Rlifs In Westjava | 10.00 - 10.15 | 11/08/2017 | AEC |
| A21 | 1580 | Muhammad Buhari Sibuea, Faiz Ahmad Sibuea | Contribution of Village Cooperation Unit in Improving Farmers Incomes | 10.15 - 10.30 | 11/08/2017 | AEC |
| A22 | 1640 | Rina Br Bukit, Bode Haryanto, Paham Ginting | Environmental Performance, Profitability, Asset Utilization, Debt Monitoring and Firm Value | 10.30 - 10.45 | 11/08/2017 | GP |
| A23 | 1582 | Apri Heri Iswanto, Tito Sucipto, Elfrida Adlina | Passion fruit hulls particleboard: the effect of urea formaldehyde level on physical and mechanical properties | 10.45 - 11.00 | 11/08/2017 | GP |
| A24 | 1641 | Mohammad Ikbal Bahua, Dewa Oka Suparwata | Local of participation the community in land rehabilitation critical in das randangan pohuwato regency | 11.00 - 11.15 | 11/08/2017 | GP |

**Session IV 8/11/17**

| No | Paper ID | Author | Title | Time | Date | Track |
|---|---|---|---|---|---|---|
| A25 | 1831 | Siti Latifah, Yunus Afifuddin,Sri Widya | Analysis of Community Income on Suren (*Toona sureni* (Blume) Merr.) and Cacao Crops (*Theobroma cacao* L.) in Simalungun, North Sumatera-Indonesia | 11.15 – 11.30 | 11/08/2017 | GP |
| A26 | 1755 | Tito Sucipto, Rudi Hartono, Wahyu Dwianto | Determination of Wood Wettability Properties of Oil Palm Trunk, Shorea sp., and Paraserianthes falcataria by Contact Angle Method | 11.30 - 11.45 | 11/08/2017 | GP |
| A27 | 1643 | Z Noer, Hasanuddin, Lisnawita, D Suryanto | Pathotype profile of *Xanthomonas oryzae* pv. *oryzae* isolates from North Sumatera | 11.45 - 12.00 | 11/08/2017 | GP |
| B49 | 3030 | Amna Suresti [1*], Uyung Gatot S. Dinata [2*],Rahmi wati | Maturity Analysis of the Innovation System in the Livestock Industries of West Sumatra, Indonesia | 12.00 – 12.15 | 11/08/2017 | PS |
| B50 | 1584 | Herwenita and Yanter Hutapea | Swamp Land Optimization in Supporting Food Security and Enhancing Farmers Welfare in South Sumatra Indonesia | 12.15 – 12.30 | 11/08/2017 | PS |

**Session V 8/11/17**

| No | Paper ID | Author | Title | Time | Date | Track |
|---|---|---|---|---|---|---|
| D33 | 1510 | Jessica Natalina Manurung, Hasnudi, Tavi Supriana | Income Analysis Of Goat Farmers On The Farmers Group In District Of Serdang Bedagai | 13.30 - 13.45 | 11/08/2017 | AS |
| D34 | 3027 | Tafsin M, Y Khairani , ND Hanafi, Yunilas | In Vitro Digestibility Of Oil Palm Frond Treated By Local Microorganism (MOL) | 13.45 - 14.00 | 11/08/2017 | AS |
| D35 | 1631 | Obed Haba Nono, Ronny Natawidjaja, Burhan Arief, Dadi Suryadi, and Maximilian M. J. Kapa | The Impact Of Sharing Arrangement Institution On Beef Cattle Breeding Performance In Kupang District, East Nusa Tenggara Province, Indonesia | 14.00 - 14.15 | 11/08/2017 | AS |
| D36 | 1368 | Nevy Hanafi , Hadirin, Nini Rahmawati | Respone Of Forages By Administration Of Fermented Goat Urine | 14.15 - 14.30 | 11/08/2017 | AS |
| D37 | 1477 | David Yasin, Ari. Ashari Harahap, Muhammad Ary Syaputra, Randi Mulianda, Try Juli | Factors Affecting The Whole Beef Prices In Eight Provinces In Sumatera | 14.30 – 14.45 | 11/08/2017 | AS |

IOP Conf. Series: Earth and Environmental Science 122 (2018) 011004 doi:10.1088/1755-1315/122/1/011004

| | | | | | | |
|---|---|---|---|---|---|---|
| | | Adha, Ali Syahdana Harahap, Alfath Rusdhi , Musa Seno Ibarahim, Annur Rasyidah Siregar, Bobby Purwadi, Simon Petrus Ginting, Hasnudi, and Yusuf Leonard Henuk | | | | |

**Session VI 8/11/17**

| No | Paper ID | Author | Title | Time | Date | Track |
|---|---|---|---|---|---|---|
| B51 | 1826 | Sarifuddin; Zulkifli Nasution; Abdul Rauf And Budi Mulyanto | Changing Of Sumatra Backswamp Peat Properties By Seawater And Zeolite Application | 15.15 - 15.30 | 11/08/2017 | PS |
| B52 | 3036 | Virna Muhardina | Quality Evaluation Of Probiotic Capsule Prepared From Alginate, Carrageenan And Tofu Waste Flour Based On Bacterial Activity And Organoleptic Test | 15.30 - 15.45 | 11/08/2017 | PS |
| B53 | 3028 | Zulkifli Nasution | Some Ecologycal Factors Of Pine At Lake Toba Catchment Area | 15.45 - 16.00 | 11/08/2017 | GP |
| B54 | 989 | Achmad Siddik Thoha, Bambang Hero Saharjo, Rizaldi Boer, Muhammad Ardiansyah | Strengthening Community Participation In Reducing Ghg Emission From Forest And Peatland Fire | 16.00 - 16.15 | 11/08/2017 | PS |
| B55 | 3021 | Eva Sartini Bayu, Anggria Lestami, E. Harso Kardhinata And Rosemary | Phylogenetic Relationshipof Asam Gelugur (*Garcinia Atroviridis* Griff. Ex T. Anders) Based On Morphological Characters In Langkat And Serdang Berdagai, North Sumatra | 16.15 - 16.30 | 11/08/2017 | PS |
| A28 | 3037 | T. Supriana, T. Chairuna | The Influences Of Consumer Characteristics On The Amount Of Rice Consumption | 16.30 - 16.45 | 11/08/2017 | AEC |

IOP Conf. Series: Earth and Environmental Science **122** (2018) 011004 doi:10.1088/1755-1315/122/1/011004

## CLASS B

**Session I 7/11/17**

| No | Paper ID | Author | Title | Time | Date | Track |
|---|---|---|---|---|---|---|
| B1 | 1520 | Mujiyo, B.H. Sunarmingto, E. Hanudin, J. Widada, J. Syamsiah | The Effect Of Organic Paddy Field System To Soil Properties | 13.30 - 13.45 | 11/07/2017 | PS |
| B2 | 1538 | Alridiwirsah. E.M. Harahap, E.N.Aqoeb, Hamidah Hanum, Aisar Novita | Growth And Production Of Local Rice Varieties In The Shade Intensity | 13.45 - 14.00 | 11/07/2017 | PS |
| B3 | 3008 | AM Lubis, LAM Siregar, K Lubis, Lisnawita, I Safn , AR Tantawi | Root Morphology Of Several Potato Varieties – Infected *Meloidogyne* Spp. And Addition Of Organic Matters | 14.00 - 14.15 | 11/07/2017 | PS |
| B4 | 3019 | Mutia Rahmah, Diana Sofia Hanafiah, Luthfi Azis Mahmud Siregar | Selection Of Selected Individuals On Soybean Plant (*Glycine Max* L.Merrill) In Generation $M_5$ Based On High Quality Production And Tolerant Stem Base Rod Disease *Athelia Rolfsii* (Curzi) | 14.15 - 14.30 | 11/07/2017 | PS |
| B5 | 1844 | Yaya Hasanah, Mariani Sembiring | Effect On Foliar Application Of Chitosan And Salicylic Acid On The Growth Of Soybean (Glycine Max (L.).Merr.) Varieties | 14.30 - 14.45 | 11/07/2017 | PS |
| B6 | 1592 | Rita Noveriza, Jumsu Trisno, Haliatur Rahma, Sri Yuliani, Reflin and Martinius | Effectiveness Of Several Dosage Formula Of Oil And Nano Emulsion Of Citronella Against Vascular Streak Dieback (Vsd) Disease On Cocoa | 14.45 - 15.00 | 11/07/2017 | PS |

**Session II 8/11/17**

| No | Paper ID | Author | Title | Time | Date | Track |
|---|---|---|---|---|---|---|
| B17 | 1808 | Rahayu M , La Ode Santiaji Bande, Asmar Hasan, Agung Yuswana dan Fatma Rinambo | Contributionof Pod Borer Pests To Soybeancrop Production (Case In Pondidaha, Konawe District, Southeast Sulawesi) | 08.00 - 08.15 | 11/08/2017 | PS |
| B18 | 1843 | Ratna Rosanty Lahay, Srinidiyanti Misrun, Rosita Sipayung | The Storage Capacity Of Cocoa Seeds (Theobroma Cacao L.) Through Giving Polyethylene Glycol (PEG) In The Various Of Storage Container | 08.15 - 08.30 | 11/08/2017 | PS |
| B19 | 1648 | Nilahayati, | Genetic Variability And | 08.30 - | 11/08/2017 | PS |

IOP Conf. Series: Earth and Environmental Science **122** (2018) 011004 doi:10.1088/1755-1315/122/1/011004

| | | | | | | |
|---|---|---|---|---|---|---|
| | | Rosmayati, Diana Sofia Hanafiah, Fauziyah Harahap | Heritability In Kipas Putih Soybean Mutant Line Using Gamma Rays Irradiation In $M_3$ Generation | 08.45 | | |
| B20 | 1677 | M Basyuni, A Nuryawan, Yunasfi, and L A P Putri | Effect Of Long-Term Salinity On The Growth And Biomass Of Two Non-Secretors Mangrove Plants *Rhizophora Apiculata* And *Ceriops Tagal* | 08.45 - 09.00 | 11/08/2017 | PS |
| B21 | 1789 | Sakiah, Mariani Sembiring, Jamalluddin Hasibuan | Entisol Land Characteristics With And Without Cover Crop (*Mucuna Bracteata*) On Rubber Plantation | 09.00 - 09.15 | 11/08/2017 | PS |
| B22 | 1798 | M Hasibuan, I Safni, Lisnawita, K Lubis | Morphological Characterization Of Several Strains Of The Rice-Pathogenic Bacterium *Burkholderia Glumae* In North Sumatra | 09.15 - 09.30 | 11/08/2017 | PS |

**Session III 8/11/17**

| No | Paper ID | Author | Title | Time | Date | Track |
|---|---|---|---|---|---|---|
| B23 | 1842 | Posma Marbun,Zulkifli Nasution, Hamidah Hanum, Abubakar Karim | Classification Of Andisol Soil On Robusta Coffee Plantation In Silima Pungga - Pungga Sub District | 09.45 - 10.00 | 11/08/2017 | PS |
| B24 | 1761 | Ratna Rubiana, Araz Meilin | Effect Of Rice Husk Biochar Aplication To Soil Insect Diversity On Potato Cultivation | 10.00 - 10.15 | 11/08/2017 | PS |
| B25 | 1847 | La Ode Santiaji Bande, Mariadi1, Gusnawaty HS, Nuriadi, Lilis Trisulpa & Rahmania | Botanical Pesticides Effect From Shells Of Bean's Cashew Nut On Biological Agents Of*Trichoderma Sp.* And *Gliocladium Sp.* | 10.15 - 10.30 | 11/08/2017 | PS |
| B26 | 1838 | Rosmayati and D Bakti | Identification And Phylogenetic Analysis Of Local Yellow And Orange Sweet Potatoes Genotype In Sumatera Utara | 10.30 - 10.45 | 11/08/2017 | PS |
| B27 | 1858 | N Rahmawati and R I M Damanik | Effect Of Foliar Application Of A-Tocopherol On Vegetative Growth And Some Biochemical Constituents Of Two Soybean Genotypes Under Salt Stress | 10.45 - 11.00 | 11/08/2017 | PS |
| B28 | 1605 | Rini Sulistiani, Rosmayati, Luthfi | Differences In Morphology And Sugar | 11.00 - 11.15 | 11/08/2017 | PS |

IOP Conf. Series: Earth and Environmental Science **122** (2018) 011004 doi:10.1088/1755-1315/122/1/011004

| | | | | | | |
|---|---|---|---|---|---|---|
| | | A.M. Siregar and Fauziyah Harahap | Content Of Purple Sweetpotato (*Ipomoea Batatas*L.) With Potassium Treatment At Several Altitude | | | |

**Session IV 8/11/17**

| No | Paper ID | Author | Title | Time | Date | Track |
|---|---|---|---|---|---|---|
| B29 | 1853 | ChairaniHanum | Growth, yield and movement of phosphate nutrients in soybean on P fertilizer, straw mulch and difference of plant spacing | 11.15 – 11.30 | 11/08/2017 | PS |
| B30 | 3012 | Andi Khaeruni, EkoAprianto Johan, Teguh Wijayanto, Muhammad Taufik, Andi Abdul Rahman Syafar | Induction of Soybean Resistance to Bacterial Pustule Disease (*Xanthomonasaxonopodispvglycines*) byRhizobacteriaand Organic MaterialTreatment | 11.30 - 11.45 | 11/08/2017 | PS |
| B31 | 1855 | L Sitinjak and E Purba | Response to Growth and Production of Green Beans (*Vigna Radiata* L.) In Various Cropping Spots and Fertilizer Provision of Layer Chickens | 11.45 - 12.00 | 11/08/2017 | PS |
| B32 | 1811 | Revandy Damanik | Germinaton Performance of Selected Local Soybean Cultivars during Drought Stress Induced by Polyethylene Glycol (PEG) | 12.00 – 12.15 | 11/08/2017 | PS |
| B33 | 1861 | Yaya Hasanah, Tengku Chairun Nisa, Hapsoh, Hamidah - Hanum | Physiological Characters of Soybean Cultivars with Application of Nitrogen Sources Under Dry Land Conditions | 12.15 – 12.30 | 11/08/2017 | PS |
| B34 | 1810 | R I Damanik, B H Manurung and E S Bayu | Effect of Hypoxia Condition in Embryogenic Callus Growth of soybean cell culture | 12.30 - 12.45 | 11/08/2017 | PS |

**Session V 8/11/17**

| No | Paper ID | Author | Title | Time | Date | Track |
|---|---|---|---|---|---|---|
| B35 | 1647 | Nasruddin, E.M. Harahap, C. Hanum, L.A.M. Siregar | Growth and Yield of Patchouli (Pogostemon Cablin, Benth) due to Mulching and Method of Fertilizer on Rain/Fet Land | 13.30 - 13.45 | 11/08/2017 | PS |
| B36 | 1576 | Delvian Delvian, Ridahati Rambey | Decomposition Rate of Rhizopora Stylosa Litter in Tanjung Rejo Village, Deli Serdang Regency, North Sumatera Province | 13.45 - 14.00 | 11/08/2017 | PS |
| B37 | 1689 | M Basyuni, N Sulistyono, R Wati and R Hayati | Deforestation trend in North Sumatra over 1990-2015 | 14.00 - 14.15 | 11/08/2017 | PS |
| B38 | 1606 | A. Susilowati, A.H. Rachmad, | Genetic Diversity of Resin Yielder Pinus Merkusii from Weest Java | 14.15 - 14.30 | 11/08/2017 | PS |

| | | | | | | |
|---|---|---|---|---|---|---|
| | | I.Z. Siregar, Supriyanto | Indonesia Revealed by Microsatellites Market | | | |
| B39 | 1621 | C.R. Kholibrina, Aswandi, A.Susilowati | Flowering and Fruiting Phenology of Kemenyan Toba (*Styrax Sumaterana J.J.SM*) in Aek Nauli Forest, North Samatera | 14.30 - 14.45 | 11/08/2017 | PS |
| B40 | 1598 | Arida Susilowati, Cut Rizlani Kholibrina, Henti Hendalastuti Rahmad, muhtar Ardansah Munthe | Phylogeny of Kemenyan (Styrax Sp) From North Sumatra Based on Morphological Characters | 14.45 - 15.00 | 11/08/2017 | PS |

**Session VI 8/11/17**

| No | Paper ID | Author | Title | Time | Date | Track |
|---|---|---|---|---|---|---|
| B41 | 1809 | Hamidah Hanum, Lisnawita, Ahmad Rafiqi Tantawi | Agronomic Effect Of Empty Fruit Bunches Compost, Anorganic Fertilizer And Endophytic Microbes In Oil Palm Main Nursery Used Ganoderma Endemic Soil | 15.15 - 15.30 | 11/08/2017 | PS |
| B42 | 1805 | E Utari, Hasanuddin, Lisnawita, I Safni, K Lubis, AR Tantawi | The Potential Of Soil Fungi Associated With Potato Rhizosphere To Control Root Knot Nematode (*Meloidogyne* Spp.) On Potato". | 15.30 - 15.45 | 11/08/2017 | PS |
| B43 | 1804 | Desi Indah Permatasari, Lisnawita, Syahrial Oemri, Irda Safni, Khairunisa Lubis, Ahma Rafiki Tantawi | Use Of Organic Waste As Biofumigant For Controlling Root Knot Nematodes (Meloidogyne Spp.) On Potato | 15.45 - 16.00 | 11/08/2017 | PS |
| B44 | 3035 | S.A.A.Taridala, Suaib, S. Wahyuni, W.G. Abdullah, N. I. Wianti, M. Zani, L.O. Jabuddin, Rosmawaty, Yusria, H. Batoa | Understanding The Social And Economic Aspects Of Upland Rice Farming | 16.00 - 16.15 | 11/08/2017 | PS |
| B45 | 1814 | E Purba and D P Nasution | Planting Pattern And Weed Control Method Influence On Yield Production Of Corn (Zea Mays L.) | 16.15 - 16.30 | 11/08/2017 | PS |
| B46 | 3034 | I R Manurung, Rosmayati and N Rahmawati | Physiology Response Of Fourth Generation Saline Resistant Soybean (*Glycine Max* (L.) Merrill) With Application Of Several Types Of Antioxidants | 16.30 - 16.45 | 11/08/2017 | PS |

## CLASS C

**Session II 8/11/17**

| No | Paper ID | Author | Title | Time | Date | Track |
|---|---|---|---|---|---|---|
| C1 | 1522 | Maya Sharah, Farida hanum, Mushila Rizky, M.F. Hisham | Microwave Assisted Extraction Of Pectin From Cocoa Pin | 08.00 - 08.15 | 11/08/2017 | FST |
| C2 | 1774 | Zakwan, Pada Mulia Raja, Giyanto | The Combination Of Activated Natural Ziolit Bentonid To Reduce Fe, N, Cu, In Revined Bleach Palm Oil (RBPO) By Using Atomic Absorbstion Spectropotometer Method | 08.15- 08.30 | 11/08/2017 | FST |
| C3 | 1699 | Ika Ucha Pradifta Rangkuti, Elisa Julianti, Jenny Elisabeth | The Tocol Content Of Crude Palm Oil Based On The Level Ripeness And Their Relationship To The Quality And Their Stability | 08.30- 08.45 | 11/08/2017 | FST |
| C4 | 3024 | S Wahyuni, Holilah, Asranudin and Noviyanti | Estimation Of Shelf Life Of Wikau Maombo Brownies Cake Using Accelerated Shelf Life Testing (ASLT) Method With Arrhenius Model | 08.45- 09.00 | 11/08/2017 | FST |
| C5 | 1727 | Hotnida Sinaga, Hilton Deeth, Bhesh Bhandari | Effect Of Sodium Azide Addition And Aging Storage On Casein Micelle Size | 09.00- 09.15 | 11/08/2017 | FST |
| C6 | 1728 | Hotnida Sinaga, Hilton Deeth, Bhesh Bhandari | Effect Of Microfluidization On Casein Micelle Size Of Bovine Milk | 09.15- 09.30 | 11/08/2017 | FST |

**Session III 8/11/17**

| No | Paper ID | Author | Title | Time | Date | Track |
|---|---|---|---|---|---|---|
| C7 | 1251 | Irhamni, Setyati Pandia, Edison Purba, Wirsal Hasan | Heavy Metal Content in Leachate Garbage of Final Disposal Side (TPA) of Banda Aceh City | 09.45 - 10.00 | 11/08/2017 | AEN |
| C8 | 1508 | Sera Pita Loka, Sumono, D.L.S. Nasution | Revamping the Physical Properties of Entisol Soil with Compost Treatment | 10.00 - 10.15 | 11/08/2017 | AEN |
| C9 | 1574 | B Haryanto, M Sirait, M Azalea, Alvin, S E Cahyani | Ball Mill Tool for Crushing Coffee and Cocoa Beans Base on Fraction Size Sieving Results | 10.15 - 10.30 | 11/08/2017 | AEN |
| C10 | 1656 | R. Manurung, A. Syahputra, M.A. Alhamdi, W. Satria, E.M. Barus, R. Hasibuan, M.Z. Siswarni | Delignification and Hydrolisis Lignocellulosic of Bagasse in Colinegloride System | 10.30 - 10.45 | 11/08/2017 | AEN |

| | | | | | | |
|---|---|---|---|---|---|---|
| C11 | 1661 | B Haryanto, R Hasibuan, Alexander, M Ashari and M Ridha | Herbal Dryer: Drying of Ginger (*Zingiber officinale*) using Tray Dryer | 10.45 - 11.00 | 11/08/2017 | AEN |
| C12 | 1717 | GR Sadimantara, B Kadidaa, Suaib, LO Safuan, Muhidin | Growth Performance And Yield Stability Of Selected Local Upland Rice Genotypes In Buton Utara Of Southeast Sulawesi | 11.00 - 11.15 | 11/08/2017 | AEN |

**Session IV 8/11/17**

| No | Paper ID | Author | Title | Time | Date | Track |
|---|---|---|---|---|---|---|
| C13 | 1768 | Ulfani ikhwana, Taufik Rizaldi, Riswanti Sigalingging | The Design Of Dapog Rice Seeder Model For Laboratory Scale | 11.15 – 11.30 | 11/08/2017 | AEN |
| C14 | 3025 | Erni Misran, Okta Bani, Elfrida Margaretha Situmeang, Adelina Suciani Purba | Removal Efficiency Of Methylene Blue Using Activated Carbon From Waste Banana Stem: Study On Ph Influence | 11.30 - 11.45 | 11/08/2017 | AEN |
| C15 | 1815 | Sumono, Syarah Mulkan Parinduri. Nurul Huda, Nazif Ichwan | The Utilization Of Ultisol Soil For Horticulture Crops Cultivation | 11.45 - 12.00 | 11/08/2017 | AEN |
| C16 | 1545 | Kabutey | The Effects Of Heating Temperatures And Time On Deformation Energy And Oil Yield Of Sunflower Bulk Seeds In Compression Loading | 12.00 – 12.15 | 11/08/2017 | AEN |
| C17 | 2256 | A Kešner, R Chotěborský, M Linda and M Hromasová | Utilization Of FEM Model For Steel Microstructure Determination | 12.15 – 12.30 | 11/08/2017 | AEN |
| C18 | 1599 | R Sigalingging, Sumono and N Rahmansyah | Evapotranspiration And Crop Coefficient Of Oil Palm (Elaeis Guineensis Jacq.) On Main Nursery In A Greenhouse | 12.30 - 12.45 | 11/08/2017 | AEN |

**Session V 8/11/17**

| No | Paper ID | Author | Title | Time | Date | Track |
|---|---|---|---|---|---|---|
| C19 | 1849 | R Sigalingging, D Herak and A Kabutey | Mechanical Behaviour Of Arabica Coffee (*Coffea Arabica*) Beans Under Loading Compression | 13.30 - 13.45 | 11/08/2017 | AEN |
| C20 | 1859 | M.S. Hutasoit, E. Julianti, and Z. Lubis | Effectof Pretreatment On Purple-Fleshed Sweet Potato Flour For Cake Making | 13.45 - 14.00 | 11/08/2017 | FST |
| C21 | 3013 | Herla Rusmarilin, | Soy-Yamgurt Probiotic | 14.00 - | 11/08/2017 | FST |

| | | | | | | |
|---|---|---|---|---|---|---|
| | | Nurhasanah,RatnaYudythia Andayani | Drink As A Natural Potential Of Antioxidant | 14.15 | | |
| C22 | 1860 | E A Zebua, J Silalahi, dan E Julianti | Hypoglicemic Activity Of Gambier Drinks (*Uncaria Gambir* Robx.) In Alloxan-Induced Mice | 14.15 - 14.30 | 11/08/2017 | FST |
| C23 | 1851 | S. Alfat1 & La Ode Santiaji Bande | Numerical Analysis Heat And Mass Transfer On Drying Process Of Agricultural Product Using Finite Element Method | 14.30 - 14.45 | 11/08/2017 | FST |

**Session VI 8/11/17**

| No | Paper ID | Author | Title | Time | Date | Track |
|---|---|---|---|---|---|---|
| C24 | 1862 | Al Muzafri, E Julianti, H Rusmarilin | The Extractionof Antimicrobials Component Of Andaliman(*Zanthoxylu macanthopodium*dc.)An dit's Application On Catfish (*Pangasiussutchi*)Fillet | 15.15 - 15.30 | 11/08/2017 | FST |
| C25 | 1389 | Tri Marwati[1], Triahmadi Januarsyah[2], Purwoko[2] and Ridwansyah | Inhibitory Activity Of Bacteriocin Produced From *Lactobacillus* SCG 1223 Toward *L. Monocytogenes*, *S. Thypimurium* Ande. *Coli* | 15.30 - 15.45 | 11/08/2017 | FST |
| C26 | 1390 | Tri Marwati, Ukhdiah Tiara Astiati, Ridwansyah, Agus Budiyanto,Wahyudiono, Abdullah Bin Arif and Nur Richana | The Effect Of Alpha Amylase Enzyme On Quality Of Sweet Sorghum Juice For Chrystal Sugar | 15.45 - 16.00 | 11/08/2017 | FST |

## CLASS D

**Session II 8/11/17**

| No | Paper ID | Author | Title | Time | Date | Track |
|---|---|---|---|---|---|---|
| D1 | 1506 | H Wahyuningsih, A P Bangun, A Muhtadi | The Relation Of Sediment Texture To Macro- And Microplastic Abundance in Intertidal Zone | 08.00 - 08.15 | 11/08/2017 | MFS |
| D2 | 1087 | A P Bangun, H Wahyuningsih, A Muhtadi | Impacts of Macro - and Microplastic on Macrozoobenthos Abundance in Intertidal Zone | 08.15- 08.30 | 11/08/2017 | MFS |

| | | | | | | |
|---|---|---|---|---|---|---|
| D3 | 1523 | Darma Bakti , Tiurmaida Manullang, Rusdi Leidonald | Structure of Gastropod Communities at Mangrove Ecosystems in Lubuk Kertang Village, West Berandan District, Langkat Regency, North Sumatera Province | 08.30-08.45 | 11/08/2017 | MFS |
| D4 | 1525 | Jessica Tambun, Darma Bakti, Desrita | Growth and Exploitation Rate of Yellowstripe scad (*Selaroides leptolepis* Cuvier, 1833) In The Waters Belawan, Malacca Strait, North Sumatra | 08.45-09.00 | 11/08/2017 | MFS |
| D5 | 1527 | M Fauzan, D Bakti, I E Susetya, Desrita | Growth and Exploitation Rate of *Anadara gubernaculum* (Reeve, 1844) Family Arcidae in Asahan Aquatic of North Sumatra | 09.00-09.15 | 11/08/2017 | MFS |
| D6 | 1678 | M Basyuni, M S Lubis, and A Suryanti | Habitat characteristic of macrozoobenthos in Naborsahan River of Toba Lake, North Sumatra, Indonesia | 09.15-09.30 | 11/08/2017 | MFS |

**Session III 8/11/17**

| No | Paper ID | Author | Title | Time | Date | Track |
|---|---|---|---|---|---|---|
| D7 | 1526 | Muhammad Mulia Wisesa, Darma Bakti | Abundance Of Sea Cucumbers On The Ecosystem Of Seagrasses In Unggeh Island, Tapanuli Tengah Regency North Sumatera Province | 09.45 - 10.00 | 11/08/2017 | MFS |
| D8 | 1671 | Nurasiah Riza, Ani Suryanti | Length-Weight Relation and Condition Factor of White Shrimp Penaeus merguiensis Captured in Ecosystem Mangrove Waters of Bagan Asahan Village, Tanjungbalai District, Asahan Regency of North Sumatera Province | 10.00 - 10.15 | 11/08/2017 | MFS |
| D9 | 1679 | M Basyuni, P Yani and K S Hartini | Evaluation of mangrove management through community-based silvofishery in | 10.15 - 10.30 | 11/08/2017 | MFS |

| | | | | | | |
|---|---|---|---|---|---|---|
| | | | North Sumatra, Indonesia | | | |
| D10 | 1397 | Rita Rosmala Dewi , Desrita & Amanatul Fadhilla | The Prevalence of parasites in Ornamental Fish from Fish Market in Medan | 10.30 - 10.45 | 11/08/2017 | MFS |
| D11 | 1710 | Samadi, S. Wajijah, A.A. Munawar | Fast and Simultaneous Prediction of Animal Feed Nutritive Values Using Near Infra Red Reflectance Spectroscopie | 10.45 - 11.00 | 11/08/2017 | MFS |
| D12 | 3026 | Yunilas dan Edhy Mirwandhono | The Role Of Lactic Acid Bacteria (***Lactobacillus sp YEL133)***From Beef In Inhibiting Of Microbial Contaminants On Various Fillers of Starter Culture | 11.00 - 11.15 | 11/08/2017 | PS |

**Session IV 8/11/17**

| **No** | **Paper ID** | **Author** | **Title** | **Time** | **Date** | **Track** |
|---|---|---|---|---|---|---|
| D13 | 1409 | M. Tafsin, N.D. Hanafi, E. Kejora, E. Yusraini | Nutrition Quality Of Extraction Mannan Residu From Palm Kernel Cake On Broiler Chicken | 11.15 – 11.30 | 11/08/2017 | AS |
| D14 | 1594 | Nurhaita, Neli Devi Niati, Urip Santoso, Sahro Ali Akbar | The Effect of Fermented Cocoa Pod (Theobroma Cacao) Husk Supplemented with Mineral on In Vitro Digestibility Rumen Bacteria Population and Rumen Liquid Characteristic | 11.30 - 11.45 | 11/08/2017 | AS |
| D15 | 3023 | J. Gloria, M. Tafsin, ND. Hanafi[\], AH. Daulay | The Influence Of Aspergillus Niger Inoculum Dosage On Nutritive Value And Metabolizable Energy Of Apu-Apu Meal (Pistia Stratiotes L.) On Broiler Chicken | 11.45 - 12.00 | 11/08/2017 | AS |
| D16 | 1321 | E.Z.J.Nasution, M.Tafsin, N.D.Hanafi | The response of red ginger *(zinggiber officinalle var rubra)* with various processing in broilers were infected by *eimeria tenella* | 12.00 – 12.15 | 11/08/2017 | AS |
| D17 | 1339 | M M J Kapa, Y L Henuk, Hasnudi, Suyadi | Contribution of local Beef Catle Production on Farmers Income in the driyland farming of kupang regency | 12.15 – 12.30 | 11/08/2017 | AS |

| | | | | | | |
|---|---|---|---|---|---|---|
| | | | Indonesia | | | |
| D18 | 3003 | Hasnudi, Berutu I S, Daulay A H, N Ginting | Analysis of Cattle Breeder's Income in South Kualuh Sub-district of Labuhan Batu Utara Regency | 12.30 - 12.45 | 11/08/2017 | AS |

**Session V 8/11/17**

| No | Paper ID | Author | Title | Time | Date | Track |
|---|---|---|---|---|---|---|
| D19 | 1484 | Widihastuty·,MaryaniCyccu Tobing, Marheni, RetnaAstuti Kuswardani | Preypreference Of *Myopoponecastaneae*( Hymenoptera: Formicidae) Toward Larvae *Oryctes Rhinoceros*Linn (Coleoptera: Scarabidae) | 13.30 - 13.45 | 11/08/2017 | AS |
| D20 | 1507 | St.Y.F.G. Dillak, N.G.A Mulyantini, Geertruida M. Sipahelut, and UlrikusR.Lole | Carcass Yields Of Two Different Strains Of Ducks Raised In Different Altitude | 13.45 - 14.00 | 11/08/2017 | AS |
| D21 | 1509 | S.S.A.Purba, M.Tafsin, S.P Ginting,Y.Khairani | The Utilization Of Endopower B In Commercial Feed Which Contains Palm Kernel Expeller On Broiler | 14.00 - 14.15 | 11/08/2017 | AS |
| D22 | 1511 | Ronistra Ginting, Sayed Umar, Chairani Hanum | The Potential And Biological Test On Cloned Cassava Crop Remaining On Local Sheep | 14.15 - 14.30 | 11/08/2017 | AS |
| D23 | 1666 | Andhika Putra, S.Pt, M.Pt, drh.Mudhita Zikrullah Ritonga, M.Vet, | Effectiveness Duckweed (Lemna Minor) As An Alternative Native Chicken Feed Native Chicken (Gallus Domesticus) | 14.30 - 14.45 | 11/08/2017 | AS |
| D24 | 1684 | S A Sitepu, Zaituni U, Jaswandi and Hendri | Improved Quality Of Frozen Boer Goat Semen With The Addition Of Sweet Orange Essential Oil On Tris Yolk And Gentamicin Extender | 14.45 - 15.00 | 11/08/2017 | AS |

**Session VI 8/11/17**

| No | Paper ID | Author | Title | Time | Date | Track |
|---|---|---|---|---|---|---|
| D25 | 3004 | P. Siswoyo, M. Tafsin, R. Handarini | Potential Production And Response Of | 15.15 - 15.30 | 11/08/2017 | AS |

IOP Conf. Series: Earth and Environmental Science **122** (2018) 011004 doi:10.1088/1755-1315/122/1/011004

| | | | | | | |
|---|---|---|---|---|---|---|
| | | | Selenium And Zinc Mineral Supplementation On Quality Of Goat Samosir Semen | | | |
| D26 | 3015 | Mirwandono, E., S Mentari, Tri Hesti, Hasnudi, N Ginting, S Galih A | (Nutrition Quality Test Of Fermented Waste Vegetables By Bioactivator Local Microorganisms (Mol) And Effective Microorganism (Em4) | 15.30 - 15.45 | 11/08/2017 | AS |
| D27 | 1618 | N Ginting, Zuhri, Hasnudi, Mirwandono, S. Iskandar, D. Armyn Hakim | Financial Analysis Of Biogas Utilization : Input Cattle, Pig Feces And Coffee Waste In Karo Indonesia | 15.45 - 16.00 | 11/08/2017 | AS |
| D28 | 1619 | Tri Hesti. W, N Ginting,Yunilas,Hasnudi, Mirwandono E,Siregar GA, Sinaga IG | The Utilization Of Coconut Waste Ferment ated By Aspergillus Niger And Tape Yeast On Meat Quality Of Weaning Males Rex Rabbit | 16.00 - 16.15 | 11/08/2017 | AS |
| D29 | 1652 | Hamdan, Mirwandono, Hasnudi, U SAYed, N Ginting, Alwiyah, Saputra | Genetic Distance Estimates And Variable Factors Distinguishing Between Goat Kacay, Muara And Samosir Through | 16.15 - 16.30 | 11/08/2017 | AS |
| D30 | 1741/3 018 | Nevy Diana Hanafi, Ma'ruf Tafsin, Ulina Hutasuhut, Erifson Lubis | Anylisis Of Botanical Composition And Nutrient Content On Natural Pastures In Samosir Island Of Samosir Regency | 16.30 - 16.45 | 11/08/2017 | AS |

## CLASS E

**Session II 8/11/17**

| No | Paper ID | Author | Title | Time | Date | Track |
|---|---|---|---|---|---|---|
| A7 | 1807 | W.G. Abdullah, U. Rianse, Muhidin, W. Widayati, E.S. Mihrad, S.A.A. Taridala, IS. Rianse, W.K. Baka | Farmers Motivation in Aren Sugar Processing Bussiness | 08.00 - 08.15 | 8/11/2017 | AEC |
| A8 | 3009 | M. Afif Syahputra, Zahari Zein | Positive and Negative Impacts of Oil Palm Expantion in Indonesia and The Prospact to Achieve Sustainable Palm Oil | 08.15- 08.30 | 8/11/2017 | AEC |
| A9 | 3031 | Idriya Adhany, Diana Chalil, Rahmanta Ginting | Strategy to Incerase Barangan Banana Production in Kabupaten | 08.30- 08.45 | 8/11/2017 | AEC |

| | | | | | | |
|---|---|---|---|---|---|---|
| | | | Deli Serdang | | | |
| A10 | 1519 | Surya Abadi Sembiring | Analysis of Rice Policy Based on Presidential Instruction on Household Food Security | 08.45-09.00 | 8/11/2017 | AEC |
| A11 | 1642 | Sastro M Wantu, Usman Moonti and Asmun Wantu | Policy In Management Based On Corruption Growth Of Resources Agropolitan In The Gorontalo Province | 09.00-09.15 | 8/11/2017 | AEC |
| A12 | 3010 | Rulianda P Wibowo, Sumono, Yahayu Iddrisu, Mozart Darus, Luhut P Sihombing, Jufri | The Pricing Behavior Comparison Of Canada And Australia Exporter In Wheat International Market Using Pricing To Market (PTM) And Residual Demand Elasticity (RDE) | 09.15-09.30 | 8/11/2017 | AEC |

**Session III 8/11/17**

| No | Paper ID | Author | Title | Time | Date | Track |
|---|---|---|---|---|---|---|
| B7 | 1657 | Fachrina Wibowo, Armaniar | Physiological performance of the soybean (*glycine max* (l.) Merr.) Crosses in salinity stress | 09.45 - 10.00 | 8/11/2017 | PS |
| B8 | 1659 | Muhidin, Elkawakib Syam'un, Kaimuddin, Yunus Musa, Gusti Ray Sadimantara, Usman, Sitti Leomo and Tresjia C. Rakian | The Effect of Shade on Chlorophyl and Anthocyanin Content of Upland Red Rice | 10.00 - 10.15 | 8/11/2017 | PS |
| B9 | 1662 | Gusti Ayu Kade Sutariati, La Ode Santiaji Bande, Andi Khaeruni, Muhidin, La Mudi, Rian Maya Savitri | The Effectiveness of Preplant Seed Bio-Invigoration Techniques Using *Bacillus* sp. CKD061 to Improving Seed Viability and Vigor of Several Local Upland Rice Cultivars of Southeast Sulawesi | 10.15 - 10.30 | 8/11/2017 | PS |
| B10 | 1664 | Tresjia C. Rakian, La Karimuna, Muhammad Taufik, Gusti Ayu Kade Sutariati, Uli Fermin and Muhidin | The Effectiveness of Various Rhizobacteria Carriers to Improve the Shelf Life and the Stability of Rhizobacteria as Bioherbicide | 10.30 - 10.45 | 8/11/2017 | PS |
| B11 | 1665 | Ni Wayan Sri Suliartini, Teguh Wijayanto, Abdul Madiki, Dirvamena Boer, Muhidin dan Juniawan | Relationship of Some Upland Rice Genotype After Gamma Irradiation | 10.45 - 11.00 | 8/11/2017 | PS |
| B12 | 1669 | Sitti Leomo, Sahta Ginting, Laode Sabaruddin, M. | The Land Use Patterns for Soil Organic Carbon Conservation At Endanga | 11.00 - 11.15 | 8/11/2017 | PS |

IOP Conf. Series: Earth and Environmental Science **122** (2018) 011004 doi:10.1088/1755-1315/122/1/011004

| | | | | | | |
|---|---|---|---|---|---|---|
| | | Tufaila and Muhidin | Watershed Southeast Sulawesi Indonesia | | | |

**Session IV 8/11/17**

| No | Paper ID | Author | Title | Time | Date | Track |
|---|---|---|---|---|---|---|
| B13 | 3007 | Bintang, Supriadi, Eko Tampubolon | Evaluation Of Land Suitability For Onion (*Allium Ascalonicum*l.) And Lemon (Citrus Sp.) At Harian District Of Samosir Regency | 11.15 – 11.30 | 8/11/2017 | PS |
| B14 | 1667 | Aswandi Aswandi, Cut Rizlani Kholibrina | Growth And Yield Model For Non-Timber Forest Product Of Kemenyan Toba (Styrax Sumatrana J.J. Sm) In Tapanuli, North Sumatra | 11.30 - 11.45 | 8/11/2017 | PS |
| B15 | 3017 | Yudi Sudarno S., Asmarlaili Sahar Hanafiah and Mariani Sembiring | Study On Changes Of Soil Sulfuric Acidity And Plant Growth With Different Soil Water Condition Byapplication Of Sulphate Reducing Bacteria (SRB) | 11.45 – 12.00 | 8/11/2017 | PS |
| B16 | 3020 | Eva Sartini Bayu, Revandy I. M Damanik, Widia Febrianti | Flower Morphology Diversity In Some Accessions Of Asam Gelugur (*Garcinia Atroviridis*Griff, Ex T. Anders) In Several Districts Of Sumatera Utara | 12.00 – 12.15 | 8/11/2017 | PS |
| B47 | 1813 | Suprianto Sidik, Edison Purba, and Erwin Nyak Akub | Population Dynamics Of Weeds In Oil Palm (*Elaeisguineensis*jacq.) Circle Weeding Area Affected By Herbicide Application | 12.15 – 12.30 | 11/08/2017 | PS |

**Session V 8/11/17**

| No | Paper ID | Author | Title | Time | Date | Track |
|---|---|---|---|---|---|---|
| B48 | 3029 | Mariati Sinuraya, Diana Sofia Hanafiah, and Andreas Romulo, Asil Barus | Determination Quercetin Content, Antioxidant And Antimicrobial Activity Of Genotip Mutant Samosir Shallots Irradiated By Gamma Rays | 13.30 - 13.45 | 11/08/2017 | PS |
| D31 | 3006 | Angelia Utari Harahap | Effects Of Wheat Leaf Noni (*Morinda Citrifolia*) On Carcass And Production Quail Eggs (*Coturnix Coturnix Javonica)In The Different Level Concentrate* | 13.45 – 14.00 | 8/11/2017 | AS |

| | | | | | | |
|---|---|---|---|---|---|---|
| D32 | 1624 | Sukawaty Fattah, Yohanis Umbu Laiya Sobang, Fredeicus Dedy Samba, Erna Hartati, Maximilian Marthen J. Kapa, Yusuf Leonard Henuk | The Effect Of Feeding Bull Bali Cattle Kept In Extensive Husbandry System With Concentrates Contained Gliricidia Sepium Leaf Meal And Banana Starch Tuber Meal On Their Feed Consumption And Dried Organic Matter Digestibility | 14.00 – 14.15 | 8/11/2017 | AS |
| D38 | 3038 | Lollie Agustina P Putri, I E Setyo, M Basyuni, E S Bayu , H Setiado, N F Reynaldi, FATT Puteri D Arifiyanto, and I Syahputra | Molecular Performance of Commercial MTG Variety Oil Palm Based on RAPD Markers | 14.15-14.30 | 8/11/2017 | PS |
| D39 | 3039 | Jonatan Ginting and Carolina Permata Sari Simanjuntak | Application of Some Rice Varieties and Fertilizing NPK Against Increased Production | 14.30 – 14.45 | 8/11/17 | PS |

# Peer review statement

All papers published in this volume of *IOP Conference Series: Earth and Environmental Science* have been peer reviewed through processes administered by the proceedings Editors. Reviews were conducted by expert referees to the professional and scientific standards expected of a proceedings journal published by IOP Publishing.

# Farmers perceptions on climate change in lowland and highland vegetable production centers of South Sulawesi, Indonesia

**W Adiyoga**

Indonesian Vegetable Research Institute
Jl. Tangkuban Perahu 517 Lembang, Bandung-40391, West Java, Indonesia

adiyogawitono@gmail.com

**Abstract**. A survey was carried out in South Sulawesi, Indonesia interviewing 220 vegetable farmers. It was aimed at examining the vegetable farmers' perception of climate change and assessing the consistency of farmers' perception with available time series meteorological data. Results suggest that meteorological data analysis is in agreement with farmers' perception regarding faster start, longer ending, and longer duration of rainy season. Further data analysis supports the claim of most farmers who perceive the occurrence of increasing air temperature, changing or shifting of the hottest and coldest month. Most respondents also suggest that climate change has affected vegetable farm yield and profitability. Other respondents even predict that climate change may affect the quality of life of their future descendants. Meanwhile, significant number of farmers is quite optimistic that they can cope with climate change problems through adaptation strategy. However, the attitude of farmers towards climate change is mostly negative as compared to positive or neutral feeling. Informative and educational campaign should be continuously carried out to encourage farmers in developing positive attitude or positive thinking towards climate change. Positive attitude may eventually lead to constructive behavior in selecting and implementing adaptation options.

## 1. Introduction

Climate change and variability are serious threats to sustainable development because of their potential negative impacts to environment, human health, food security, economic activities, natural resources and physical infrastructures [1, 2]. Impacts of climate variability are manifested as floods, drought, uncertain rainfall, and extreme events that have consequences on increasing uncertainties/risks of agricultural and food security programs, especially in developing countries [3]. Some crops, especially those grown in lowland areas, have been allegedly cultivated in environmental conditions that are very close to their tolerable maximum heat. Hence, slight changes in atmospheric heat or rainfall may lead to substantial decrease of crop yield [4].

Climate change in Indonesia has received much attention due to its standing as the world's third largest emitter [5]. It is reported that the main sources of $CO_2$ emissions in Indonesia are deforestation, forest fires, and degradation of peat land. The impacts of climate change in Indonesia include, but are not limited to temperature increase, intense rainfall, a rise in sea-level, and a threat to food security [6]. Based on the simulation of climate data from 1971-2006, it is predicted that a combination of temperature increase and rainfall decrease will have significant impacts on the country's food balance by 2050 [7].

Climate change adaptation may only be carried out when (a) farmers perceive that climate change is really happening, (b) farmers are able to identify adaptation options available, and (c) farmers respond to climate change adaptation in accordance to their production circumstances [8]. Hence, a complete understanding of farmers' perception on climate change is very important since it may establish farmers' readiness to adapt to climate change by considering some adjustments to their cultivation practices [9]. Most climate change perception studies deal with temperature and rainfall that include amount, annual distribution, start and end dates [10, 11]. Time-series meteorological data are often used in studies for confirming farmers' perception or opinion regarding climate change [12, 13] or refuting them [11]. Other studies have dealt with perceptions of cropping seasonality, perceptions of production risks and threats related to climate variability [14, 15] and farmers' local knowledge in forecasting weather and adapting to climate [16]. This study was aimed at examining the vegetable farmers' perception of climate change and assessing the consistency of farmers' perception with available time series meteorological data.

## 2. Materials and Methods

The study was conducted in vegetable production centers of South Sulawesi province, Indonesia. Focus Group Discussion carried out during preparation stage suggested that farmer respondents should be selected not only based on different agro-ecosystems (lowland < 400 m and highland > 700m above sea level), but also on different season-pattern sectors (western, eastern and transitional sector). Western sector is a part of South Sulawesi that has rainy season during the period of October-March and dry season during the period of April-September. Eastern sector has the opposite seasonal pattern, i.e. rainy season in April-September and dry season in October-March. Transitional sector constitutes regions that are characterized to have both seasonal patterns. Simple random sampling procedure was employed in selecting a total sample of 220 respondents with the distribution as follows: (a) lowland-western sector – 55 respondents, (b) lowland–eastern sector – 55 respondents, (c) highland-transitional sector – 55 respondents, and (d) highland land-western sector – 55 respondents.

The selected respondents were interviewed by using a structured questionnaire. Data collected mostly focuses on socio-economic characteristics of the respondents and perception and knowledge on climatic variables or climate change in general. Various questions were designed based on their specific purpose. These might include questions that only needed "yes" or "no" answer (dichotomous), and questions that their responses were rated in several levels of measurement.

Descriptive statistics that include mean, percentages and frequency counts was used to describe socio-economic features of the respondents' perception about climate change. Chi-square test was used to analyze nominal data. Continuum-scale data, such as rainfall were analyzed by using simple regression. For statistical computation, ordinal-scale data were transformed to interval-scale data by using the method of successive interval (MSI) [17]. Meanwhile, the rank of importance of some factors was assessed by ranking method using multiple weighted score analysis [18].

## 3. Results and Discussions

### *3.1 Socio-economic characteristics of respondents*

Table 1 showed that 61.9% of respondents were within the age range of 30-49 years. It was reported that 30% of respondents had tertiary education and 7.7% had even a BS degree. The study also showed that 88.2% of respondents had engaged in vegetable production as long as 1-20 years. These three characteristics suggested that most of the respondents were at productive age, quite educated and experienced.

**Table 1**. Age, education and experience of the respondents.

| Age (years) | | | Age (years) | | | Experience (years) | | |
|---|---|---|---|---|---|---|---|---|
| | Σ | % | | Σ | % | | Σ | % |
| < 20 | 5 | 2.3 | No school | 3 | 1.4 | 1 – 10 | 145 | 65.9 |
| 20-29 | 32 | 14.5 | Primary | 75 | 34.1 | 11 – 20 | 49 | 22.3 |
| 30-39 | 78 | 35.5 | Secondary | 55 | 25.0 | 21 – 30 | 14 | 6.4 |
| 40-49 | 58 | 26.4 | Tertiary | 66 | 30.0 | 31 – 40 | 8 | 3.6 |
| 50-59 | 30 | 13.6 | Bachelor | 17 | 7.7 | 41 – 50 | 3 | 1.4 |
| 60-69 | 15 | 6.8 | Master | 1 | .5 | 51 – 60 | 1 | .5 |
| > 70 | 2 | .9 | Diploma | 3 | 1.4 | | | |
| | 220 | 100 | | 220 | 100 | | 220 | 100 |

*3.2 Farmers' perception on changes in rainy and dry season in the last five years*
Most respondents perceived that rainy season started earlier (65.9%), ended later (58.2%), and had longer duration (60%) (Table 2). However, 47.3% of respondents indicated the less frequent occurrence of heavy rains and more frequent occurrence of drought.

**Table 2**. Farmers' perception on changes in rainy and dry season.

| Variable | Σ | % | Σ | % | Σ | % |
|---|---|---|---|---|---|---|
| | earlier | | no change | | later | |
| Start of rainy season | 145 | 65.9 | 20 | 10.0 | 55 | 24.1 |
| | earlier | | no change | | later | |
| End of rainy season | 67 | 30.5 | 25 | 11.4 | 128 | 58.2 |
| | shorter | | no change | | longer | |
| Duration of rainy season | 59 | 26.8 | 29 | 13.2 | 132 | 60.0 |
| | less frequent | | no change | | more frequent | |
| Frequency of heavy rain occurrence | 104 | 47.3 | 40 | 18.2 | 76 | 34.5 |
| | less frequent | | no change | | more frequent | |
| Frequency of drought occurrence | 91 | 41.4 | 25 | 11.4 | 104 | 47.3 |

Since the p-value (0.000) was less than the significance level (0.05), then there was sufficient evidence to conclude that the proportion of respondents who expressed their perceptions based on three categorical choices was different (Table 3).

**Table 3**. Chi-squared test on changes in rainy and dry season.

| Indicator | Variable | | | | |
|---|---|---|---|---|---|
| | start of rainy season | end of rainy season | duration of rainy season | heavy rain occurrence | drought occurrence |
| Chi-Sq. (a,b) | 111.609 | 73.155 | 76.536 | 28.073 | 48.936 |
| df | 2 | 2 | 2 | 2 | 2 |
| Asymp. Sig. | .000 | .000 | .000 | .000 | .000 |

a. 0 cells (.0%) have expected frequencies less than 5. The minimum expected cell frequency is 73.3.
b. 0 cells (.0%) have expected frequencies less than 5. The minimum expected cell frequency is 110.0.

*3.3 Dynamics of rainfall in South Sulawesi based on meteorological data*
Rainfall data during the period of 2001-2010 showed a continuing annual increase of rainfall starting in 2007 with the exception in 2009. The highest rainfall occurred in 2007 that reached about 300% of rainfall in the previous years. Extremely high coefficient of variation in 2007 explained the high variability of monthly rainfall occurred during the year. On the contrary, there was a relatively stable monthly rainfall in 2010 with an ample monthly rainfall average. Mapping the rainfall time-series monthly data also indicated that there was a shift in months with highest rainfall or lowest rainfall from year to year. The highest rainfall in 2001 and 2001 occurred in January had shifted one month

faster to December 2003. January as the wettest month of the year recurred in 2004, 2005 and 2006, but the month with the highest rainfall had also shifted one month faster to December 2007. Similar shifting was also happened for months with the lowest rainfall.

Simple regression analysis of rainfall and months in Table 4 indicated that there was a significant positive relationship (37.287) between rainfall and months. A one-unit increase in explanatory variable (month) caused a one-unit increase (37.3 mm) of dependent variable (rainfall).

**Table 4**. Linear regression of rainfall and month, 2001-2010

| Model | | Un-standardized Coefficients | | Standardized Coefficients | t | Sig. |
|---|---|---|---|---|---|---|
| | | B | Std. Error | Beta | | |
| 1 | (Constant) | 2297.258 | 1015.925 | | 2.261 | .026 |
| | X_MONTH | 37.287 | 14.573 | .229 | 2.559 | .012 |

a Dependent Variable: Y_RAINFALL

### *3.4 Farmers' perception on changes temperature in the last five years*

Most respondents perceived that air temperature was getting hotter (69.5%), number of hot-days per month became higher (69.1%), and the number of cold-days per month became lower (66.8%). It was also perceived some changes in the hottest month (75%) and the coldest month (78.2%) of the year.

**Table 5**. Farmers' perception on changes in temperature.

| Variable | Σ | % | Σ | % | Σ | % |
|---|---|---|---|---|---|---|
| | colder | | no change | | hotter | |
| Air temperature | 36 | 16.4 | 31 | 14.1 | 153 | 69.5 |
| | less | | no change | | more | |
| Number of hot-days per month | 45 | 20.5 | 23 | 10.5 | 152 | 69.0 |
| | less | | no change | | more | |
| Number of cold-days per month | 147 | 66.8 | 29 | 13.2 | 44 | 20.0 |
| | changed | | | | do not change | |
| Hottest month | 165 | 75.0 | | | 55 | 25.0 |
| | changed | | | | do not change | |
| Coldest month | 172 | 78.2 | | | 48 | 21.8 |

The test results in Table 6 indicated that the p-value (0.000) was less than the significance level (0.05) implying that the null hypothesis could not be accepted. This meant that the proportion of respondents who expressed their perceptions based on three categorical choices was different.

**Table 6**. Chi-squared test on changes in air temperature

| Indicator | Variable | | | | |
|---|---|---|---|---|---|
| | air temperature | no. of hot-days/month | no. of cold-days/month | hottest month | coldest month |
| Chi-Sq. (a,b) | 129.991 | 129.882 | 112.536 | 55.000 | 69.891 |
| df | 2 | 2 | 2 | 1 | 1 |
| Asymp. Sig. | .000 | .000 | .000 | .000 | .000 |

a. 0 cells (.0%) have expected frequencies less than 5. The minimum expected cell frequency is 73.3.
b. 0 cells (.0%) have expected frequencies less than 5. The minimum expected cell frequency is 110.0.

### *3.5 Dynamics of air temperature in South Sulawesi based on meteorological data*

Temperature data during the period of 2002-2011 indicated that air temperature was increasing in 2005, decreasing in 2006, 2007 and 2008, and increasing again in 2009 and 2010. However, the highest maximum temperature (28.3$^0$ C) was actually occurred in 2006 and 2009. During those years, months with the lowest temperature were mostly occurred in January-February, at least 7 times (2001, 2003, 2007, 2008, 2009, 2010 and 2011). Meanwhile, months with the highest temperature were mostly happened in October-November, at least 7 times too (2002, 2003, 2006, 2007, 2008, 2009 and 20110. Shifts or changes of months with lowest or highest temperature did not show fixed pattern. Months with lowest temperature could occur 4-5 months later, while months with highest temperature could occur 1-5 months earlier.

A significant positive relationship between temperature and months was confirmed by the t-value of 2.146 at p-value of 0.034 (Table 7). A one-unit increase in explanatory variable (month) caused a one-unit increase (0.005 °C) of dependent variable (temperature).

**Table 7**. Linear regression of temperature and month, 2001-2010

| Model | | Unstandardized Coefficients | | Standardized Coefficients | t | Sig. |
|---|---|---|---|---|---|---|
| | | B | Std. Error | Beta | | |
| 1 | (Constant) | 1.421 | .004 | | 364.132 | .000 |
| | LOGX_MONTH | .005 | .002 | .185 | 2.146 | .034 |

a Dependent Variable: LOG Y_TEMPERATURE

### *3.6 Farmers' attitude towards climate change*

In the last 5 years, most respondents (95%) indicated that they had observed and experienced increasingly uncertain climate, weather and season. Most respondents also perceived that climate change had already affected farm yield or agricultural productivity (95%). About 93.6% of farmers even suggested that climate change had affected the profitability of their farms. High proportion of respondents (85.9%) stated their agreement suggesting concurrence among the farmers that climate change and variability might reduce the next generation quality of life. However, more than half of respondents (61.8%) inclined to have positive perception regarding farmers' ability to adapt.

Farmers' attitude to climate change becomes very important because it may affect their decision in taking actions and responding to challenges, incentives and rewards. About 19.1% of

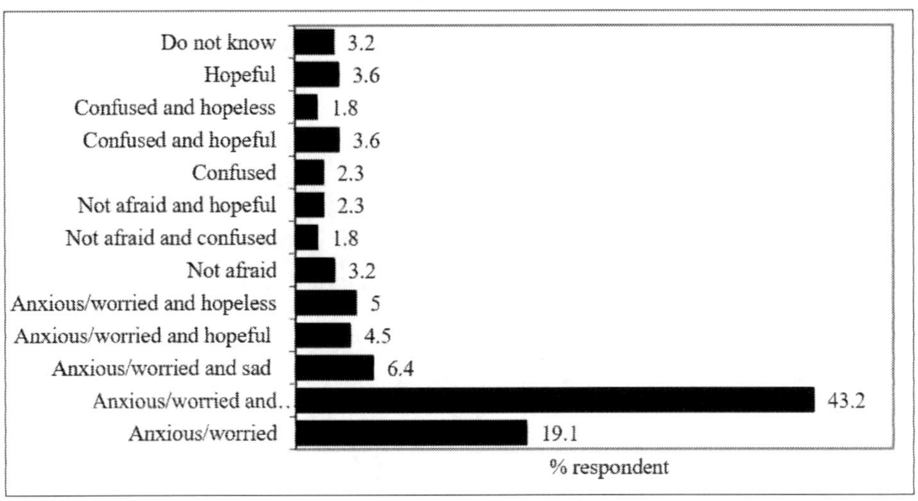


**Figure 1**. Farmers' attitude towards climate change

respondents were anxious or worried because climate change had created so many uncertainties. A significant proportion of respondents (43.2%) was not only anxious or worried, but also confused to deal with climate change. Figure 1 also showed some combinations of farmers' attitude towards climate change that could be grouped into (a) negative response – anxious/worried, sad, hopeless, (b) positive response – hopeful, and (c) neutral – not afraid, confused, do not know. Considering this grouping, it seemed that farmers' attitude towards climate change was more dominated by negative response. Previous studies eliciting attitudes toward climate change or global warming consistently indicated similar response: anxious or worried in specific negative outcomes such as ocean level rise, more frequent storms, and possible water shortages [19, 20].

## 4. Conclusion

Most respondents in lowland and highland vegetable areas perceived that rainy season was started early, ended later and had longer duration in the last five years. An increasing trend of temperature and changes in the hottest and coldest month during the studied period were also perceived to be occurred by most respondents in all seasonal patterns. In this study, the results of meteorological data analysis were consistent with most farmers' perceptions regarding some changes in climate. However, not all farmers' perceptions were consistent with data analysis. For example, rainfall data analysis did not support the perception of most farmers in all seasonal patterns claiming that the occurrence of heavy raining was becoming less frequent. The contradiction might also be possible since human perception of climate was strongly influenced by expectations, which might have little relationship to the true nature of climate as provided by the instrumental records.

Almost all the farmers interviewed were aware of climate change, and recognize its impacts. Most farmers perceived that climate change had already affected farm yield. Increased in temperature and changes in rainfall pattern could fasten the spread of fungus and diseases directly or indirectly affecting yield and farm profit. Some respondents even perceived that climate uncertainty and variability had the potential of reducing the next generation quality of life. Nonetheless, there was a quite significant proportion of respondents who still had positive perception suggesting that all farmers would always have the ability to adapt.

Most of the farmers in the study area tended to show negative attitudes towards climate change. Negative attitudes reflected lack of farmers' knowledge and information regarding climate change. This was actually a bit contradictive since most farmers had recognized extreme climate events such as floods and droughts due to climate change. To some extent, farmers were quite knowledgeable about climate change, but seemed to be in confusion as to the more technical or scientific aspects of climate change. Therefore, various informative and educational efforts should be carried out to encourage farmers for developing positive attitudes and showing constructive behavior and actions for climate change adaptation.

## References

[1] Rakib MA, Atiur Rahman MD, Akter MS and Bhuiyan MAH 2014 Climate change: Farmers perception and agricultural activities. *Herald J Geogr Rgnl Plann* 3 (3): 115-123.
[2] Rashid MH, Afroz S, Gaydon D, Muttaleb A, Poulton P, Roth C, and Abedin Z 2014 Climate change perception and adaptation options for agriculture in Southern Khulna of Bangladesh. *App Ecol Envir Sci* 2 (1): 25-31.
[3] Mengistu DK 2011 Farmers' perception and knowledge of climate change and their coping strategies to the related hazards: Case study from Adiha, central Tigray, Ethiopia. *Agric Sci* 2 (2): 138-145.
[4] Ayanwuyi E, Kuponiyi, Ogunlade FA and Oyetoro JO 2010 Farmers perception of impact of climate changes on food crop production in Ogbomosho agricultural zone of Oyo State, Nigeria. *Global J Hum Soc Sci* 10 (7): 33-39.
[5] Measey M 2010. Indonesia: A vulnerable country in the face of climate change. *Glob Maj E-J* 1 (1): 31-45.
[6] Case M, Ardiansyah F and Spector E 2007 Climate change in Indonesia: Implications for humans and nature (Gland, Switzerland: World Wide Fund for Nature (WWF)); available at: http://assets.panda.org/downloads/indonesian_climate_change_impacts_report_14nov07.pdf

[7] Syaukat Y 2011 The impact of climate change on food production and security and its adaptation programs in Indonesia. *J. ISSAAS* 17 (1): 40-51.

[8] Maddison D 2006 The perception of and adaptation to climate change in Africa. CEEPA Discussion Paper No. 10 Centre for Environmental Economics and Policy in Africa, University of Pretoria South Africa 47 pp.

[9] Bryan E, Deressa TT, Gbetibouo GA, and & Ringler C 2009 Adaptation to climate change in Ethiopia and South Africa: Options and constraints. *Envir Sci & Pol* 12:413-426.

[10] Deressa T, Hassan RM, Ringler C, Alemu T, and Yesuf M 2009 Determinants of farmers' choice of adaptation methods to climate change in the Nile Basin of Ethiopia.*Global Envir Change* 19:248-255.

[11] Fisher M, Chaudhury M, and McCusker B 2010 Do forests help rural households adapt to climate variability? Evidence from Southern Malawi. *World Dev* 38:1241-1250

[12] Vedwan N, and Rhoades RE 2001 Climate change in the Western Himalayas of India: A study of local perception and response. *Clim Res* 19:109-117.

[13] Orlove BS, Chiang JCH, and Cane MA 2000 Forecasting Andean rainfall and crop yield from the influenceof El Nino on Pleiades visibility. *Nature* 403:68-71.

[14] Grothmann T, and Patt A 2005 Adaptive capacity and human cognition: the process of individual adaptation to climate change. *Global Envir Change* 15:199-213

[15] Thomas DSG, Twyman C, Osbahr H, and Hewitson B 2007 Adaptation to climate change and variability: Farmer responses to intra-seasonal precipitation trends in South Africa. *Clim Change* 83:301-322.

[16] Saroar MM, and Routray JK 2012 Impacts of climatic disasters in coastal Bangladesh: Why does private adaptive capacity differ? *Reg Envir Change* 12:169-190.

[17] Waryanto B, and Millafati YA 2006 Transformasi data skala ordinal ke interval dengan menggunakan makro Minitab (Transforming ordinal data into interval data using Minitab macro). *Informatika Pertanian* 15: 881-895.

[18] Rahayu SH, Abdullah N, and Aziz A 2005 Ranking method using multiple weighted score analysis. School of Science and Technology, Universiti Malaysia, Kota Kinabalu, Sabah, Malaysia.

[19] Bord RJ, Fisher A, and O'Connor RE 1998 Public perceptions of global warming: United States and international perspectives. *Clim Res* 11 (1): 75–84

[20] Bord RJ, O'Connor RE, and Fisher A 2000 In what sense does the public need to understand global climate change? *Public Understand Sci* 9: 205–218.

IOP Conf. Series: Earth and Environmental Science **122** (2018) 012002 doi:10.1088/1755-1315/122/1/012002

# Revitalization of food barns in supporting sustainable food security in Central Java

**E W Riptanti**[1], **A Qonita**[1] **and Suprapti**[1]

[1]Agribusiness Department of Faculty of Agriculture Universitas Sebelas Maret
Corresponding Author Email: erlynawida@staff.uns.ac.id

**Abstract.** Rice barns have been developed in some areas in Central Java, but several problems seem to appear, leading to nonoptimal functions of nonactive food barns. The present article aims to examine revitalization of food barns through systematic, integrated, and sustainable empowerment. The research design is exploratory research to generate data that are in-depth qualitative and quantitative. Survey was applied in four regencies including Wonogiri, Purworejo, Temanggung, and Batang. Key informants comprise caretakers of food barns, village apparatus, public figures, and Food Security Office apparatus. The research results revealed that the food barns have not been managed in professional manners. Active roles of all members and caretakers, village government, and Food Security Agency are, therefore, required in revitalizing the food barns. In order to perform social functions well, the food barns should be profit-oriented to achieve sustainability.

## 1. Introduction

According to [1] pointed out that the world is encountering global climate change that leads to the decrease inworld's food production. By 2050, cereal production will have dropped by 1% and the world population will have increased by 1%. It is, therefore, crucial to establish food security. The evaluation results of Global Food Security Index released by Economist Intelligence Unit in July 2012 indicated that Indonesia ranked fifth in food security of 7 evaluated ASEAN countries. Top ranks were occupied by Malaysia, Thailand, Vietnam, and the Philippines. For Indonesia, food crisis is a paradox of an agricultural country [2].

On one hand, Central Java province serves as one of provinces with national buffer stocks, and therefore rice productivity is prioritized. In 2015, rice productivity reached 60.99 quintal per hectare with paddy field width of 1.80 million hectares and rice production of 11.00 million tons. Rice productivity in Sukoharjo regency was the highest of all (75.26 quintals per hectare), while that in Pekalongan was found to be the lowest (43.12 quintals per hectare) [3].

On the other hand, despite its position as a province with national buffer stocks, the number of poor population (population living below the poverty line) in March 2016 reached 4.507 million persons (13.27%), indicating an increase of 1.11 thousand persons if compared to the number of poor population in September 2015 which reached 4.506 million persons (13.32%). Although such fact revealed an absolute increase, the percentage of the number of poor population indeed showed a decline of 0.05%.

Food commodities play a greater role in dealing with poverty line than non-food commodities (such as housing, clothing, education, and health). In March 2016, food poverty line contributed 73.00% to poverty line, indicating less significant change compared to that in September 2015 (73.23%) [4].

According [5], poor households, which allocate 60% to 80% of their income to buy food, are more sensitive to food price fluctuation. They do not have any other choice to deal with the rise in food prices unless, except to reduce the quantities of food consumed. The food availability in poor households should be improved by establishing food reserves [6]. Central Java owns a food reserve system set up by either government or community. However, such system in the form of food barns can only be found in 35 regencies/ cities in Central Java. The table below gives an overview of the condition and the number of food barns in Central Java.

**Table 1**. The number of food barns in Central Java in 2013.

| Regency | The number of active government food barns | The number of inactive government food barns | The number of active community food barns | The number of inactive community food barns | The total number of food barns |
|---|---|---|---|---|---|
| Purworejo | 31 | 27 | 312 | 93 | 463 |
| Boyolali | 55 | 0 | 0 | 0 | 55 |
| Wonogiri | 48 | 234 | 50 | 19 | 351 |
| Karanganyar | 1 | 0 | 0 | 0 | 1 |
| Rembang | 26 | 240 | 0 | 0 | 266 |
| Temanggung | 9 | 0 | 29 | 167 | 205 |
| Batang | 1 | 61 | 19 | 0 | 81 |
| Tegal | 7 | 0 | 0 | 0 | 7 |
| Total | 178 | 562 | 410 | 279 | 1.429 |

Data source: Ditjen Pemberdayaan Masyarakat dan Desa[7], 2013.

The inactive government food barns were greater in number than the active ones. However, the number of active community food barns exceeded the number of the inactive ones. Such condition is of concern considering that the existence of food barns is marginal. Revitalizing food barns with regards to local conditions and development of an era is a significant way to realize a democratic and sustainable food system.The present tendency indicates that food is no longer regarded as stuff with socio-cultural values, and an important part of human rights, but rather it is a stock-in-trade to gain profits. A food barn is an institution and a symbol of solidarity of farmers and community to meet their own needs of food under the base of local resources [8]. According to the explanation above, the purpose of this study how the revitalization happened in the active and inactive food barn in Central Java.

## 2. Research Method

### *2.1. The Basic Method*

The design of this study is exploratory research, in which a research aims to obtain description or study on the condition of local food barns in Central Java. Explanatory research seeks explanations to the observes phenomena, problems, or behaviors [9]. Cross-sectional approach was applied, in which a study is carried out only at a certain time without any treatment or action to object in order to portrait situation at a certain period.

*2.2. Types of Data, Sources of Data, and Data Collection*
Data include primary and secondary data. Primary data consist of the condition of local food barns seen from the aspects of physical condition, institution and management, factors contributing to the absence of local food barns and the presence of local food barns. The data were obtained from in-depth interview using questionnaires with 42 caretakers of both government and community food barns which every food barn consist of 1 caretaker, as well as the relevant stakeholders and 3 respondents which chosen randomly from house-hold in every location of food barn. Meanwhile, secondary data were obtained from Central Bureau of Statistics, regional and provincial Food Security Agencies, and other relevant institutions.

*2.3. Research Location*
The research location consisting of regencies having either active or inactive food barns, managed either by government or community was selected using purposive sampling technique. The research was carried out from April to August 2016. Table 2 presents regencies selected with regards to their uniqueness of food barns and the potential to be role models for development of either active or inactive food barns. The research location covers Purworejo, Wonogiri, Temanggung, and Batang regencies.

**Table 2**. Data of samples of food barns in four selected regencies.

| No | Regency | The number of food barns (unit) | The number of inactive food barns (unit) | The number of active food barns (unit) |
|---|---|---|---|---|
| 1 | Wonogiri | 32 | 4 | 7 |
| 2 | Purworejo | 27 | 3 | 9 |
| 3 | Temanggung | 42 | 2 | 8 |
| 4 | Batang | 14 | 1 | 8 |
| | Total | 115 | 10 | 32 |

Data source: Secondary data of Food Security Agency, 2015

Key informants in this study consist of caretakers of both active and inactive food barns managed by government and community, village apparatus, Food Security Office apparatus, and farmers and households living nearby food barns.

*2.4. Data Analysis*
The data were analyzed using a method of descriptive analysis to obtain results of research which are able to convey an actual situation. Activities in analysis of qualitative data were performed in interactive and continuous manners until they finished reaching data saturation. Data saturation occurs at time when the added samples or respondents do not provide new information. Activities of data analysis include data reduction, data display, and conclusion drawing/ verification [10].

## 3. Result and Discussion

*3.1. Condition of Food Barns*
Food barns are closely related to food availability and food reserves. Community has traditionally established food reserve systems of villages and households, one of which is food barn institution. The

importance of food reserves at community level is stipulated in Regulation of Minister No. 30 of 2008 regarding Village Government Food Reserves [11]. Food barns in areas of research location have different conditions. Food barns fall into some categories according to their functions:

**Table 3.** Functions of food barns.

| No. | Category | The number of food barns in each regency | | Total | Percentage |
|---|---|---|---|---|---|
| | | Regency | Total | | |
| 1 | Unhulled rice saving and credit association | Batang | 6 | 23 | 7.188 |
| | | Purworejo | 7 | | |
| | | Temanggung | 7 | | |
| | | Wonogiri | 3 | | |
| 2 | Storage of foodstuff reserves | Batang | 2 | 7 | 21.875 |
| | | Purworejo | 0 | | |
| | | Temanggung | 1 | | |
| | | Wonogiri | 4 | | |
| 3 | Others | Batang | 0 | 2 | 6.25 |
| | | Purworejo | 2 | | |
| | | Temanggung | 0 | | |
| | | Wonogiri | 0 | | |
| | TOTAL | | 2 | 32 | 100 |

Source: Primary Data Analysis, 2016

Table 3 demonstrates that food barns mostly function as unhulled rice saving and credit association. Members of the association should return the borrowed unhulled rice by giving extra 10% and storing the unhulled rice after harvest. Inactive food barns have experienced a change in function; to store bric-a-brac such as tables, chairs, plates, and glass since most of the members did not return the unhulled rice. For them, food barns had already received supports from government, and therefore they did not need to return what they borrowed.

Not all of the active food barns deserves to be considered a place to store unhulled rice due to poor appearance of unhulled rice. Lusterless unhulled rice is resulted from over-drying or over-heating [12]. Most barns were so stuffy due to poor air circulation that many rats and insects made nests. For that reason, it is important for caretakers of food barns to repair air circulation and find ways to get rid of rats and insects [13].

When managing the function of food barns, approximately 68.75% of caretakers do not receive any incentives or wages. Poor managerial skills cause limitations of food barns to provide incentives/wages to the caretakers [14]. Only caretakers with strong motivation or those who are in urgent situations for nobody can replace their positions are willing to manage food barns.

Top-down dominates the mechanism of food barns, in which government regulate the caretakers and members. This mechanism is commonly not based on the desire of the people getting involved/having interest in food barns, and this, in turns, cause them to have poor relatedness compared to in bottom-up system. The aspect of public self-support is one of keys of successful food barn management. The working mechanisms of bottom-up and top-down are also obvious from the characteristics of society. If the society is compact and highly motivated in managing food barns, the government only need to guide them. However, if the society is not prepared to manage food barn or they require coaching, top-down mechanism should be applied.

One of ways to develop food barns is establishing cooperation with other parties, for instances, food barns in other villages, *BULOG* (the Indonesian Bureau of Logistics), or other concerned parties. There are only 4 out of 32 food barns cooperating with other parties. The business is maintained between Food Barns in Pijiharjo, Manyaran, Wonogiri and *BULOG* of Karangdowo. Other type of cooperative work is also maintained by food barns in Pijiharjo as rice and unhulled rice supplier for *BULOG* of Karangdowo. This cooperation develops food barns and facilitates the operation.

**Table 4**. The Whereabouts of Cooperation between Food Barn with Other Food Barn, BULOG, or Stakeholder.

| No. | Category | Total per regency | | Total | Percentage |
|---|---|---|---|---|---|
| | | Regency | Quantity | | |
| 1. | Maintain cooperative work | Batang | 0 | 4 | 12.5% |
| | | Purworejo | 1 | | |
| | | Temanggung | 0 | | |
| | | Wonogiri | 3 | | |
| 2. | Do not maintain cooperative work | Batang | 8 | 28 | 87.5% |
| | | Purworejo | 8 | | |
| | | Temanggung | 8 | | |
| | | Wonogiri | 4 | | |
| TOTAL | | | | 32 | 100% |

Sources: Primary Data Analysis, 2016

Some food barns do not have good mutual aid with other parties because the barns do not meet the requirements or criteria demanded by prospective partner institutions. Furthermore, inactivity of food barn caretakers in establishing cooperation with the third parties is due to relatively poor managerial or negotiating skills.

### *3.2. Sustainable Food Barn Revitalization*

According to [15], food barn needs to be developed and revitalized through systematic, integrated and continuing empowerment process. In the short-term, food barn revitalization needs to be directed to capacity improvement of public food security, among others are through strengthening farmer capital to trigger the development of farmer groups and providing privilege to farmers to set their business priority that will be run based on their own potential.Referring to the condition of food barns in Central Java, revitalization in supporting sustainable food security is carried out by:

#### *3.2.1 Improving institutional capacity building of food barns*

Food barn institution has to be reorganized particularly for the aspects of organizational structure, position filling, the article of association and bylaws, as well as management.Slim organizational structure is indeed able to perform all managerial function better than complex organizational structure[16], but it is difficult to operate. People who have certain skills in organizing food barn[17] fill positions for organizational structures. Dual position in organizational structure in a village should be avoided working performance. Food barn caretakers have to improve their technical and managerial skills through technical training or internship program. The article of association and bylaws has to be reviewed and adjusted to food barn visions and missions that are independent and continuing. Food barn flexibility is expanded by optimizing the potential of natural and human resources.

*3.2.2 Improving sense of belonging to food barn*

Socio-cultural condition today has made a change to members' behaviors. The growing slash system, savings and credit, and public economics which approaching industry cause the sense of belonging to start fading. Members' non-performed credits of rice and unhulled rice are resulted from their opinion that the rice is a governmental support that is free of charge. In fact, the rice is intended to stimulate food barns to perform their functions. Members of food barns and society need to improve their motivation to possess deeper relatedness to food barns. Efforts have to be made to revive society and members that food barns ought to be revitalized due to their vital functions in empowering food security.

*3.2.3 Improving the physical condition of food barns*

The physical condition of food barns is divided into permanent and non-permanent. Acording to the data below, there are 30 of food barns have permanent building. Food barns with permanent buildings are better able to protected unhulled rice.

**Table 5**. The physical condition of food barns

| No. | Category | Total per regency | | Total | Percentage |
|---|---|---|---|---|---|
| | | Regency | Quantity | | |
| 1. | Permanent | Batang | 8 | 30 | 93,75% |
| | | Purworejo | 9 | | |
| | | Temanggung | 8 | | |
| | | Wonogiri | 5 | | |
| 2. | Non-permanent | Batang | 0 | 2 | 6,25% |
| | | Purworejo | 0 | | |
| | | Temanggung | 0 | | |
| | | Wonogiri | 2 | | |
| | | TOTAL | | 32 | 100% |

Source : Primary Data Analysis, 2016

Research results show that the condition of some active food barns is not appropriate to store unhulled rice, and therefore, physical improvement has to be made. This effort requires active participation of all members, village officials of apparatus, and food security board. Adequate food barns can support all their activities and functions.

*3.2.4 Optimizing the functions of food barns*

Effective food barns have to be able to perform their functions continuously. There are three functions of food barns which have to be accomplished:

a. Social function

This function deals with the capability of food barn to manage public food problems when famine or food shortage happens [18]. The most common problem is that social function gets more priority than other functions so that more than half of the samplefood barn in research locations have savings do not continue to develop. Thus, balanced management is required to perform this social function.Strengthening rural food barns were implemented through the process of community development, consisting of: created social conditions conducive, strengthen economic potential, and protect people from his own weaknesses.[10]

b. Price control function

This function is developed through seasonal food storage system [19] and a food credit program [20]. In research locations about 6,25% of the food barns can be advanced by applying warehouse Receipt System. With this system, farmer or owner does not need to sell their commodities when the price drops and they still receive fund supports from food barns through Warehouse Receipt from storage [21].
c. Economic function
Sustainable food barns have the capabilities to generate profits such as selling seeds or other production facilities, and hence, they can finance operational activities and perform their social functions. The managerial skills of food barn maintainers are essential in performing this function because wrong decision made will cause loses to the food barns. The fact, most food barns in Central Java are used for savings and loans where un-hulled rice is stored to control rice availability in the market.

### *3.2.5 Improving Food barns' cooperation with other parties*
The real life, most offood barn in Central Java particularly in sample region doesn't have good patnership with any organization or company at every region. The condition is caused by the inability to communicate and negotiate well.Communication and negotiation skills play important roles in succeeding the partnership. Establishing cooperation is important to support the three functions of food barn. Cooperation can be built with farmers, rice barns in other regions, *Bulog*, grocers, food security office and other business partners. These activities will increase rice barns' capacities and roles in improving food security in certain areas.

## 4. Conclusion
Revitalization is a fundamental step in reviving and advancing food barns' capacities, both governmental and public ones. The efforts are accomplished through the aspects of institutional, human resources and physical condition of food barns. Revitalization entails support from all parties, including central government, local government, and other concerned parties.

## References
[1] Fisher, G 2009 *How do climate change and bioenergy after the long term outlook for food, agriculture and resources availability.* Food and Agriculture Organization of The United Nations Economics and Social Development Department.
[2] Krisdyatmiko 2012 Kemampuan desa dalam mewujudkan ketahanan pangan: Pembelajaran dari Indonesia Timur (Village ability in realizing food security: A study from East Indonesia). *WELFARE, J. Ilmu Kesejahteraan Sosial.* **1**
[3] BPS[a] 2016 *Provinsi Jawa Tengah dalam angka 2016* (Semarang: Central Bureau of Statisics of Central Java)
[4] BPS[b] 2016 Profil kemiskinan di provinsi Jawa Tengah Maret 2016 (Poverty profile in Jawa Tengah province March 2016) *Berita Resmi Statistik No. 49/07/33/Th. X, 18 Juli 2016* (Semarang: Central Bureau of Statisics of Central Java)
[5] Madina A A, Andzor H D, Hassan È T, Lomali A Shidaev and Eliza A.Shidaeva 2016 Global food security problems in the modern world economy. *Int. Jour. Of Env. & Sci. Ed.***11** 5320
[6] Guptal P, Kalyani S, Veenu S, Sidharth A and Pulkit M 2015 Coping strategies adopted by households to prevent food insecurity in urban slums of Delhi, India. *J. of Fo. Sec.***3** 6
[7] Ditjen Pemberdayaan Masyarakat dan Desa 2013 Data pokok Ditjen *PMD Ministry of Domestic Affairs of the Republic of Indonesia* Ditjen Pemberdayaan Masyarakat dan Desa www.pdm.kemendagri.go.id
[8] ErlynaW R, Suprapti dan AuliaQ[a] 2016 Laporan akhir revitalisasi lumbung pangan lokal dalam meningkatkan ketahanan pangan pada rumah tangga miskin di Jawa Tengah (Local food silo

revitalization on improving food security in rural households in Jawa Tegngah). *Lembaga Penelitian dan Pengabdian Kepada Masyarakat* (Surakarta: Universitas Sebelas Maret)

[9] Nueuman, W L 1994 *Social research methods: Qualitative and quantitative approach.* Allyn and Bacon.

[10] Muhammad A 2016 Farmers Community Empowerment through Institutional Strengthening of Rural Food Barn *Int. Jour. of Agr.e Sys.* **4** 161-167

[11] ErlynaW R, Suprapti dan AuliaQ[c] 2017 Barns society as a buffer stock (Case in Nglaris Village Bener District Purworejo Regency) *Int. Conf. F. Sci. and Eng. 2016* **193** 012051.

[12] Angelita M P, Melvin M D L, Francis T D Rio, Rey R M F and Darwin Z H 2015 Warehousing operation of national food authority in Batangas City *Int. J.of Aca. Res. in Eco. and Mana. Sci.***4** 1

[13] Luis A E, Chris A G, Randall G M and James F T 2014 Survey of rice storage facilities identifies research and education needs *Jour. Of California Agr.***68** 38-46

[14] ErlynaW R, Suprapti dan Aulia Q[b] 2017 The development of sustainable community food barn in Wonogiri Regency, Central Java, Indonesia. *Asi. J.of Appl. Sci.* **5** 274

[15] Rachmat M, Budhi G S, Supriyati and Sejati W K 2011 Lumbung pangan masyarakat Keberadaan dan perannya dalam penanggulangan kerawanan pangan (Local food silo: Existence and role in preventing food insecurity). Pusat Sosial Ekonomi dan Kebijakan Pertanian.

[16] Jannatul F 2016 OrganitationTheoris: From Clasical Perspective. *Int. Jour. of Bus., Eco. and L.* **92** 1-6

[17] Yayuk Y and Riyanti I 2016 The strategy to increase women farmer's participation in the program of village food barn in East Java *Int. J. Of Env. & Sci. Ed.* **11** 5728

[18] Ahmad M and Imron R 2015 *Model revitalisasi fungsi dan peran lumbung pangan desa untuk meningkatkan ketahanan pangan* (Function revitalization model and food silo role for food security). University Research Colloquium.

[19] Albinus M Makalle 2012 Post harvest storage as a rural household foods security strategy in Tanzania. *ARPN Journal Of Science And Technology.* **2** 814

[20] Karna B M W 2015 Evaluating seasonal food storage and credit program in east Indonesia *Jour. of Dev. Eco.* **115** 200

[21] Wahyudi S, Senator N B, Andi Cakravastia and Arisamadhi T M A 2012 A buffer stock model to ensure price stabilization and availability of seasonal staple foods under free trade considerations *J. Eng. Sci.* **44**

**Acknowledgements**

We would like to express our gratitude to the Ministry of Research, Technology and Higher Education for granting fund of *Penelitian Unggulan Perguruan Tinggi 2016.* Our sincere appreciation is extended to all parties who have supported us to complete this article.

IOP Conf. Series: Earth and Environmental Science **122** (2018) 012003 doi:10.1088/1755-1315/122/1/012003

# Questioning the sustainable palm oil demand: case study from French-Indonesia supply chain

**D Chalil and R Barus**

University of Sumatera Utara

chalildiana@gmail.com

**Abstract**. Sustainable palm oil has been widely debated. Consuming countries insist certified sustainable produces palm oil, but in fact the absorption of the certified palm oil is still less than 60%. This raise questions about the sustainable palm oil demand. In this study, such a condition will be analysed in French-Indonesia supply chain case. Using monthly and quarterly data from 2010 to 2016 with Autoregressive Distributed Lag (ARDL) approach and Error Correction Model, demand influencing factors and price integration in each market of the supply chain is estimated. Two scenarios namely re-export and direct export models are considered in the Error Correction Model. The results show that France Gross Domestic Product, prices of France palm oil import from Indonesia, Malaysia, and Germany, and price of France groundnut import significantly influence the France palm oil import volume from Indonesia. Prices in each market along palm oil re-export France-Indonesia supply chain are co-integrated and converge towards long-run equilibrium, but not in the direct export supply chain. This leads to a conclusion that France market preferences in specific and EU market preferences in general need to be considered by Indonesian palm oil decision makers.

## 1. Introduction

In Europe palm oil is the most imported vegetable oil, which is used in large volumes by the food, personal care and biofuel industries. Palm oil does not only have the lowest selling price, but could also replace various ingredients in all of these industries [1-2]. Since 2010-2016, the total European consumption of palm oil has increased at an annual rate of around 40% [3]. In general the increase can be related to the increase in the technical use of palm oil as biofuel. However, the consumption of edible palm oil has been decreasing in the last few years due to environmental issues impact in producing countries. The Ministry of Foreign Affair stated that opportunities for exporters in developing countries are likely to be found in sustainable certified palm oil [4].

Sustainable palm oil has been widely debated. On one side consuming countries insist certification as a guarantee of sustainable produced palm oil, while on the other side producing countries ask premium prices to cover additional costs for producing the certified palm oil. In this case consumers are likely referred to manufacturers or multinational companies as palm oil is not consumer product, but rather is used as ingredients of many further processed consumer products. Food manufacturers and large multinational companies are likely have more influences to the supply chain. As profit oriented business, manufacturers and companies still consider price as one of the important factors in selecting types of vegetable oils [4]. Most of these vegetable oils are continuously imported in processed form, in which one to another appears to be substituted as raw materials in manufacturers and processing industries [2 & 5].

Among the European countries, France appears to have the most stringent requirement on certified palm oil, despite its insignificant share as palm oil importer. However, France is recorded as the third largest main destination of palm oil imported from other European hubs such as The Netherland, Germany or Italy [4]. Moreover, France is recorded as the second main economy in Western European that might strongly relate to other European countries' economy [6]. With these considerations, growers from Indonesia and Malaysia made efforts to produce certified palm oil. In 2015, Indonesia and Malaysia produce 6,444,156 and 5,105,806 ton certified palm oil respectively. However, from 2008-2014 on average only 39.74% of the certified palm oil is traded with premium prices [7], raising questions about the sustainable palm oil demand. Using France market as the proxy of sustainable palm oil demand, this study is conducted to answer (i) the influencing factors of France palm oil import volume from Indonesia (ii) prices co-integration in the Indonesian-France palm oil supply chain and (ii) adjustments of the short run dynamics towards the long run equilibrium.

## 2. Material and Methods

In this paper, the sustainable palm oil demand will be analysed in French-Indonesia supply chain case using monthly and quarterly data from 2010 to 2016. The series include Indonesia palm oil import price in France ($P_M^{PPO.Ind}$), Malaysia palm oil import price in France ($P_M^{PPO.Mal}$), Netherland palm oil import price in France ($P_M^{PPO.Neth}$), Germany palm oil import price in France ($P_M^{PPO.Ger}$), ground nut import price in France ($P_M^{GNut.Sen}$), France gross domestic product (GDP), Indonesia-France trade balance (trade_balance), Palm oil import volume in France from Indonesia ($Q_M^{PPO}$), palm oil export (FOB) price in Indonesia ($P_X^{PPO.Ind}$), crude palm oil (CPO) local price in North Sumatra ($P_{Sumut}^{CPO}$), fresh fruit bunches (FFB) price in schemed SH ($P_{SSH}^{FFB}$), FFB price in independent SH ($P_{ISH}^{FFB}$), Exchange rate (EUR/USD) (ER).

France imports two main types of palm oil, which are crude palm oil (CPO) and processed palm oil (PPO) with HS Code 151110 (Palm oil and its fractions; whether or not refined, but not chemically modified) and 151190 (Vegetable oils; palm oil and its fractions, other than crude, whether or not refined, but not chemically modified), respectively. HS code 151190 (Crude palm oil for the manufacture of foodstuffs for human consumption) is chosen as it appears as the main imported palm oil with average share of 54.76% in 2010-2016 [8]. The data collected from European Statistics (Eurostat), International Trade Statistics Database (UN Comtrade), FAO, Statistics Indonesia (BPS), PASPI monitor, Estate Government Office (Dinas Perkebunan) North Sumatra Province, and St. Louis Fed's Economic Research Division. Consumers refer to the French importers as a number of EU processors stated that the additional premium price was not passed on end consumers. All the price data are measure in USD/kg, while the GDP is measure in billion USD, import volume is measure in thousand metric ton, and trade balance is measure in million USD.

Using monthly data from 2010-2016, stationarity of each variable is tested using the Augmented Dickey Fuller test by using the following equation.

$$\Delta Y_t = \propto + \rho Y_{t-1} + \sum_{j=1}^{n} \gamma Y_{t-j} + \mu_t \quad (1)$$

$Y_t$ is the series data, $\rho$ is the test coefficient and $j$ is the lag length chosen from ADF test. The null hypothesis is $Y_t$ has a unit root. Influencing factors of France palm oil import volume from Indonesia, co-integration and vector error correction model is tested with Autoregressive Distributed Lag Model (ARDL), which could be used if variable included in the equations are stationer at different level [9]

$$Y_t = \sum_{i=1}^{p} a_i Y_{t-i} + \sum_{i-1}^{n} c_i X_{t-1} + \varepsilon_t \quad (2)$$

Estimation of influencing factors of France palm oil import volume from Indonesia includes France gross domestic product (GDP), exchange rate (ER), prices of France palm oil import from Indonesia ($P_M^{PPO.Ind}$), Malaysia ($P_M^{PPO.Mal}$), The Netherlands ($P_M^{PPO.Neth}$), Germany ($P_M^{PPO.Ger}$) and price of

France groundnut import from Senegal ($P_M^{GNut.Sen}$). All of these countries are recorded as main exporters of France palm oil. The Netherlands accounts for over 60% of European re-exports of palm oil, while Germany follows with 24%. In addition, France is also recorded as the main importer of groundnut oil [4]. Except GDP, all data are in monthly base that are aggregated in quarterly data to run the following equation. The number of lag in price is selected by empirical consideration, which in fact producers' decision often influenced by the quarterly price information.

$$Q_M^{PPO} = f\left(GDP, ER, P_M^{PPO.Ind}, P_M^{PPO.Mal}, P_M^{PPO.Ger}, P_M^{PPO.Neth}, P_M^{GNut.Sen}\right) \quad (3)$$

Multicollinearity and linearity tests are conducted through Pearson Correlation and Ramsey Reset Test, respectively. Using ARDL approach, co-integration between prices along the supply chain is estimated for each level of the market namely (i) Netherland export price and France import price, (ii) France import price and Indonesia export price, (iii) Indonesia export price and Indonesia local mill price, (iv) Indonesia local mill price and Indonesia smallholder price. North Sumatra (Sumut) port (Belawan) FOB price is used as the representative of Indonesian export price. Smallholders' prices are differentiated between schemed and independent groups as in general the former receives Disbun prices, while the latter depends on their bargaining power [10-11].

Price integration along the Indonesia France supply chain us two possible scenarios namely (i) the re-export palm oil and (ii) the direct export palm oil. The former include (a) The Netherland or Germany palm oil (151190) export price to France, (b) France palm oil (151190) price imported from Indonesia, (c) Indonesia palm oil (151190) export price to France (Belawan FOB price), (d) Indonesia (Sumut) CPO local price, and (e) Indonesia (Sumut) schemed and independent Fresh Fruit Bunches (FFB) price. Using monthly data, the co-integrated prices in the Indonesia-France supply chain are then further estimated with VECM in order to analyse the short run dynamic and adjustment time needed to reach the long-run equilibrium. The VECM equations run separately for each of the co-integrated price. VECM is used to analyse the speed of adjustment of such integration.

$$\Delta Y_{it} = \alpha_i + \sum_{j=1}^{n} \beta_i \Delta Y_{jt-1} + \gamma_i \varepsilon_{it} + \varepsilon_t \quad (4)$$

$\varepsilon_{it}$is the residual series such that $\varepsilon_{it} = Y_{it-1} - \alpha_i - \beta_i Y_{jt-1}$ , and $\gamma_i$shows the speed of adjustment. Short-run dynamics from the non-stationer series is shown by $\beta_i$, which will adjusted towards the Long Run Equilibrium if $\gamma_i < 0$. The more negative $\gamma_i$ the faster the series converge to the Long Run Equilibrium.

## 3. Results and Discussion

The unit root test results show that the data are stationary at different levels, vary from I(0) to I(2).

**Table 1.** Unit Root Test Result

| No | Variabel | Quarterly | | | Monthly | | |
|---|---|---|---|---|---|---|---|
| | | t-Statistic | Prob.* | Ordo | t-Statistic | Prob.* | Ordo |
| 1 | Palm oil import price in France from Indonesia ($P_M^{PPO.Ind}$) | -5.29 | 0.00 | I(0) | -4.48 | 0.01 | I(1) |
| 2 | Palm oil import price in France from Malaysia ($P_M^{PPO.Mal}$) | -4.21 | 0.01 | I(2) | -4.02 | 0.01 | I(0) |
| 3 | Palm oil import price in France from Netherland ($P_M^{PPO.Neth}$) | -4.31 | 0.01 | I(2) | -8.46 | 0.00 | I(0) |
| 4 | Palm oil import price in France from Germany ($P_M^{PPO.Ger}$) | -4.88 | 0.00 | I(0) | -5.67 | 0.00 | I(0) |
| 5 | Ground nut import price in France ($P_M^{GNut.Sen}$),) | -4.68 | 0.03 | I(1) | | | |

| | | | | | | | |
|---|---|---|---|---|---|---|---|
| 6 | Palm oil import volume in France from Indonesia ($Q_M^{PPO}$) | -4.70 | 0.00 | I(0) | | | |
| 7 | France gross domestic product (GDP) | -4.06 | 0.02 | I(1) | | | |
| 8 | Indonesia-France trade balance (TB) | -4.69 | 0.00 | I(0) | | | |
| 9 | Exchange rate (USD/PKR) (ER) | -3.60 | 0.05 | I(0) | | | |
| 10 | Other palm oil export (FOB) price ($P_X^{PPO.Ind}$) | | | | -6.25 | 0.00 | I(1) |
| 11 | CPO local price in North Sumatera ($P_{Sumut}^{CPO}$) | | | | -8.26 | 0.00 | I(1) |
| 12 | FFB price in scheme SH ($P_{ISH}^{FFB}$) | | | | -7.87 | 0.00 | I(1) |
| 13 | FFB price in independent SH ($P_{SSH}^{FFB}$) | | | | -8.52 | 0.00 | I(1) |

Showing high Pearson Correlation (0.88) with GDP, exchange rate series are excluded from the equation estimations. Correlation of palm oil prices imported from The Netherland and Indonesia is 0.77, while that of palm oil prices imported from Malaysia and Indonesia is 0.72. However, all are still maintained, since literature suggests that as a rule of thumb correlation of two series is considered highly collinear if $R^2$ of the series exceeds 0.90. Moreover, since the OLS estimators are still BLUE as long as collinearity is not perfect, it is often suggested that the results of the fitted model is still acceptable [12]. In this case, considering the empirical importance information all of the prices are maintained including in the regression. Ramsey RESET result with F test (2.13) less than F stat (4.10) justifies the usage of linear specification in the ARDL model[1]. F-Bounds and t-Bounds Tests with 11.24 and -3.43 values, respectively, reject the "No levels relationship" null hypothesis. With this co-integrated result, influencing factors of France palm oil import from Indonesia can be estimated using the ADRL result.

**Table 2.** ADRL Test Result

| Variable | Coefficient | Std. Error | t-Statistic | Prob.* |
|---|---|---|---|---|
| $Q_M^{PPO}$ (-1) | 0.075 | 0.270 | 0.278 | 0.786 |
| GDP | 0.046 | 0.027 | 1.677 | 0.119 |
| GDP(-1) | -0.101 | 0.024 | -4.267 | 0.001 |
| $P_M^{PPO.Ind}$ | -15.079 | 8.016 | -1.881 | 0.084 |
| $P_M^{PPO.Ind}$(-1) | -21.518 | 9.217 | -2.335 | 0.038 |
| $P_M^{PPO.Mal}$ | 15.068 | 4.420 | 3.409 | 0.005 |
| $P_M^{PPO.Neth}$ | 9.162 | 5.231 | 1.751 | 0.105 |
| $P_M^{PPO.Ger}$ | 12.342 | 3.769 | 3.274 | 0.007 |
| $P_M^{GNut.Sen}$ | 7.304 | 2.545 | 2.870 | 0.014 |
| C | 31.734 | 10.359 | 3.063 | 0.010 |
| R-squared | 0.834 | | | |
| Adjusted R-squared | 0.709 | | | |
| F-statistic | 6.690 | | | |

The ADRL estimation results reveal that last quarter, France palm oil imported from Indonesia price, France palm oil imported from Malaysia price, France palm oil imported from Germany, France palm oil imported from Netherland, and France ground nut oil import price significantly influence the France palm oil import volume from Indonesia. All the coefficient signs are as expected. The prices of France palm oil imported from Indonesia and Malaysia have the greatest influence to France palm oil import volume from Indonesia. The magnitude of both have coefficients are similar although with different signs. Each dollar per kg decrease in the France palm oil imported from Indonesia price will increase 15 tons palm oil import per month from Indonesia, while vice versa of those from Malaysia.

[1] Correct specification is one of the classical linear regression model assumptions for OLS estimation approach [13]

Positive coefficients are also found for those from the Netherland and Germany, which are the main hubs for palm oil import to European countries. Therefore, all of palm oil import from these countries can be considered as substitutes of that imported from Indonesia. The groundnut oil price positive sign shows its role also as a substitute of palm oil. Each dollar per kg increase in soybean oil price will increase 7,30 ton per month France palm oil import from Indonesia. Last quarter France GDP negatively influences its palm oil import from Indonesia, indicating the status of palm oil as an inferior good. Among all vegetable oils use in France, palm oil has the lowest selling price. In general European countries are known as one of the most interesting markets for premium vegetable oil worldwide. There are special high-quality oils, new exotic oils or oils with a healthy composition. In fact, most of the processed imported palm oil (HS 151190) are used in the food industry such as margarine, ice cream, confectionary, filled milk and cocoa butter substitute. Crude palm oils or known as red palm oil (HS 151110) can be divided into lower-priced and standard quality and higher-priced and high quality, which can be purchased in ethnic and health shops, respectively [4]. Price integrations in each market along the Indonesia-France supply chain, both in re-export and direct export chain can be seen in the following table.

**Table 3.** Co-integration Test Result

| Price Series | Bounds Test | | CointEq(-1)* | Prob. |
|---|---|---|---|---|
| | F | t | | |
| $P_M^{PPO.Ind}$ - $P_M^{PPO.Neth}$ | 44.89*** | -9.58*** | -1.18 | 0.00 |
| $P_M^{PPO.Neth}$ - $P_X^{PPO.Ind}$ | 1.13 | -1.51 | -0.00 | 0.13 |
| $P_M^{PPO.Ind}$ - $P_X^{PPO.Ind}$ | 1.24 | 1.59 | 0.02 | 0.12 |
| $P_M^{PPO.Ind}$ - $P_M^{PPO.Ger}$ | 33.11*** | -8.21*** | -0.70 | 0.00 |
| $P_M^{PPO.Ger}$ - $P_X^{PPO.Ind}$ | 5.18** | -3.24** | -0.01 | 0.00 |
| $P_X^{PPO.Ind}$ - $P_{Sumut}^{CPO}$ | 7.48*** | -3.89 | -0.42 | 0.00 |
| $P_{Sumut}^{CPO}$ - $P_{SSH}^{FFB}$ | 2.05 | -2.04 | -0.08 | 0.05 |
| $P_{Sumut}^{CPO}$ - $P_{ISH}^{FFB}$ | 3.75 | 2.76* | -0.18 | 0.00 |

*, **, *** show significance at 10%, 5% and 1% respectively

F and t bound tests results reveal the co-integration between series in the Indonesia-France palm oil supply chain. All series, except The Netherland - Indonesia ($P_M^{PPO.Neth}$ - $P_X^{PPO.Ind}$) export prices, France palm oil import price from Indonesia and Indonesia export price ($P_M^{PPO.Ind} - P_X^{PPO.Ind}$), and local Indonesia and schemed smallholder FFB price ($P_{Sumut}^{CPO}$ - $P_{SSH}^{FFB}$), are co-integrated. In addition, all speed of adjustment coefficients (CointEq(-1)) of the co-integrated series are negative and significant, indicating the converge towards the long run equilibrium. The speed of adjustment values range from 0.01 to 0.42. Indonesia export price (Belawan FOB) and Indonesia local price ($P_{Sumut}^{CPO}$) has the highest speed towards the equilibrium. Each month the series adjust 42% towards the long-run equilibrium, thus need about 2.38 month to fully be adjusted. Germany export price and Indonesia export price has the slowest speed, with only 1 % of adjustment per month. Therefore, in can be concluded that prices in each market along palm oil re-export France-Indonesia supply chain are co-integrated and converge towards long-run equilibrium, but not in the direct export supply chain. This might be partly explained by the wide options for the France vegetable oils consumption. There are at least 6 other countries besides Indonesia that continuously export palm oil to France. In addition, there are at least 3 other vegetable oils that can be seen as substitutions for France palm oil consumption.

## 4. Conclusion

France is still an important market for Indonesia palm oil export. France can obtain the palm oil directly from producing countries such as Indonesia and Malaysia or from other European importers such as The Netherlands and Germany. Although prices along the direct export supply chain are not co-integrated, that of the re-export is co-integrated and converged towards the long-run equilibrium.

As the second main economy in Western European that might strongly relate to other European countries' economy, such a condition indicates that France market preferences in specific and EU market preferences in general need to be considered by Indonesian palm oil decision makers.

**References**

[1] Qiu Yin 2014 The substitution effect between soybean and palm oil and global carbon emissions Thesis Graduate School of Arts and Sciences of Georgetown University Washington DC
[2] Santeramo F G 2016 Cross-Price Elasticities for Oils and Fats in the US and the EU Report Italy
[3] European Union (EU-27) Palm Oil Domestic Consumption by Year 2017 https://www.indexmundi.com/agriculture/?country=eu&commodity=palm oil&graph=domestic-consumption
[4] CBI 2015 CBI Trade Statistics Ministry of Foreign Affairs Netherland
[5] Alias M H and Othman J 1998 Cointegration between Palm Oil Price and Soybean Oil Price: A Study on Market Integration Jurnal Ekonomi Malaysia 32 pp. 39-50
[6] ITPC (Indonesian Trade Promotion Centre) LYON 2016 Nutella Tax: Analisa Perdagangan Minyak Kelapa Sawit, Peluang dan Polemiknya di Perancis Market Intelligence (Nutella Tax: Business Analysis on Palm Oil, Opportunities and Polemics in France Market Intelligence) Kementerian Perdagangan Republik Indonesia
[7] GreenPalm 2015 Market volume and price charts http://greenpalm.org/the-market/market-overview/market-volume-and-price-charts
[8] European Statistics (eurostat) 2017 http://epp.eurostat.ec.europa.eu/newxtweb/setupdimselection.do
[9] Researchgate 2017 Cointegration testing? https://www.researchgate.net/post/Cointegration_testing
[10] Chalil, D. and Barus, R. 2013 Willingness to Pay and Benefit Cost Analysis in Implementing RSPO Principle and Criteria Research Report National Strategic Fund (Hibah Strategis Nasional).
[11] Chalil, D. and Barus, R. 2016 Crude Palm Oil Market Integration and the Impact on Smallholders' Income Research Report National Strategic Fund (Hibah Strategis Nasional).
[12] Gujarat D 2015 *Econometric by Example* Second Edition Palgrave Macmillan Education
[13] Gujarati D 1995 *Basic Econometrics* 3rd edition cGraw-Hill International, Inc.

# The effect of consumers' perception to the satisfaction of use of traditional medicines in Medan

**R S Siregar[1] , T Supriana[2] and S Haryanti[2]**

[1]Agribusiness of Universitas Muhammadiyah Sumatera Utara
[2]Agribusiness of Universitas Sumatera Utara
E-mail: tavihutasuhut@yahoo.co.id

**Abstract**. Consumption of chemical medicines fluctuated in 2009-2014, whereas the consumption of solid traditional medicine increased in 2009-2014. The purpose of this study is to analyse the influence of consumers's perception on the consumption of traditional medicinal plants. The data was analysed by using a binomial logit regression analysis. It is found that the consumers's perceptions affect customer satisfaction simultaneously are the health benefits variable, quality of traditional medicine variable, price of traditional medicine and available product; the health benefits variable and quality of traditional medicine variable partially have significant effects to customers's satisfaction simultaneously satisfaction partially; the health benefit variable and quality of traditional medicine is found to have a marginal effect of 7% and 4%, respectively.

## 1. Introduction

Indonesia has the fourth largest population in the world after the People's Republic of China, India and USA. The population of Indonesia according to BPS is 255,993,674 people. A large number of residents resulted in the need for medicines Indonesia continues to increase from year to year. 90% of Indonesian medicine needs is produced domestically, and 10% is obtained from imports. Knowledge and increasing income also leads to better public awareness of well-being. This causes increasing demands for medicines. Thus drug prices in Indonesia are still relatively expensive. Modern health services have grown in Indonesia, but the number of people using traditional medicine remains high.

The use of traditional medicines is also increasing because it has become a substitute for chemical drugs considering the side effects caused by chemical drugs. This increase in traditional medicine also follows the global trend of "back to nature". The tendency to return to nature shows rapid growth, including in Indonesia.

Traditional medicine has remained as the most affordable and easily accessible source of treatment in the primary health care system of resource poor communities. The local people have a long history of traditional plant usage for medicinal purposes. The medicinal use of plants is very old. The writings indicate that therapeutic use of plants is as old as 4000 - 5000 B.C. and Chinese used first the natural herbal preparations as medicines. [6]

Medicinal plants have played an essential role in the development of human culture. Medicinal plants are resources of traditional medicines and many of the modern medicines are produced indirectly from plants. It has been confirmed by WHO that herbal medicines serve the health needs of about 80 percent of the world's population; especially for millions of people in the vast rural areas of developing countries. Meanwhile, consumers in developed countries are becoming disillusioned with modern health care and are seeking alternatives[7]

Modern health services have grown rapidly in Indonesia, but the number of people using traditional medicine remains high. According to Susenas (2001), 31.7% of Indonesians use traditional medicine and 9.8% seek treatment in a traditional way to overcome their health problems [8]

Indonesian have known about traditional medicinal plants for a long time; besides, they also have used the medicinal plants as one of the efforts in solving health problems. The use of medicinal plants is partly based on their experiences and skills which are forwarded from generation to generation[5].

The data of traditional medicine profile in Province of North Sumatra shows that there are 2629 traditional medicines using medicinal plants, derived from self-made medicinal plants and traditional Indonesian medicines. The data Include the city of Medan which has 739 people of its population using taditional medicine [3] Based on the fact, a study to determine the cause of traditional medicine consumption in needed. The purpose of this study is to analyse the influence of consumers's perception on the consumption of traditional medicinal plants

## 2. Methods

*2.1 Method of Determining The Research Area*

The location of the research is determined purposively in Medan. According to Minister of Health are the people of Medan mostly use traditional medicinal plants compared to 7 cities and 25 other districts in North Sumatera. This is very contradictory when it is viewed from the activities of the city community which is much busier than other cities and districts. How the perception of traditional medicinal plants consumers affects the use of this traditional medicines and what other factors affect consumers are the question that need to be investigated more deeply.

Purposive sampling was used in determining the sample in this research, which were the consumers of traditional medicine. Sample size is determined based on [1] formula:

$$n = \left[\frac{Z\,{}^{a}/_{2}}{E}\right]^2$$

$$n = \left[\frac{1{,}96}{0{,}20}\right] \quad (1)$$

n = 96 respondents

The effect of consumers's perception to the satisfaction of use of traditional medicines in Medan analyzed by logistic regression model specified as follow :

$$ln\left\{\left[\frac{P}{1-pi}\right]\right\} = \beta_0 + \beta_1 X_1 + \beta_2 X_2 + \beta_3 X_3 + \beta_4 X_4 + \beta_5 X_5 \quad (2)$$

where :$ln\left\{\left[\frac{P}{1-pi}\right]\right\}$ is odd ratio where (p) explain the predicted probability of consumers are satisfaction of consume traditional medicine which is coded with 1 (interest), (1-p) is the predicted probability of the consumers are not satisfaction of consume traditional medicine which is coded with 0 (not interest).

$\beta_0$ = Intercept
$\beta_1$ = Coefficient of health benefits
$\beta_2$ = Coefficient of quality
$\beta_3$ = Coeffecient of price
$\beta_4$ = Coeffecient of product availability
$X_1$ = Health benefits
$X_2$ = Quality
$X_3$ = Price
$X_4$ = Product availability

*2.2 Parameter Testing*

*2.21Concurrent Testing*. It is performed to know the significance of β parameters as a whole or simultaneously. This recitation hypothesis is:

$H_0$ ; $\beta_0 = \beta_1 = \cdots = \beta_8 = \beta_p = 0$. With rejection: reject $H_0$ if value G > X_(α,1)^2 Where p is the number of statistical variables in the model or p- value < α.

a. Individual Test (Wald Test)

It is intended to examine the significance of individual β parameters. This recitation hypothesis is:

$H_0$ ; $\beta_j$ $H_1$ : $\beta_j \neq 0,\ j = 1, 2 \ldots, p$

With statistical tests:

W (Wald) = $\frac{\beta_j}{S.E\ (\beta_j)}$

Denial area : reject $H_0$ if $W_j^2 > X_{\alpha,1}^2$ atau p- value < α

*2.2.2 Hosmer and Lemeshow Test.* This test aims to compare the distribution of observation with the distribution of theory (model test). Criteria Test:
If sign < 0,1 then accept $H_1$ reject $H_0$
If sign > then accept $H_0$ reject H1 [2].

*2.2.3 Odd Ratio and calculation of marginal effect.* Odds ratio is the probability of results obtained by individuals with x = 1 defined $\pi(1)/[1-\pi(1)]$. Similarly, the likelihood of results being present between individuals with x = 0 defined as $\pi(0)/[1-\pi(0)]$. *Odds Ratio*Which is denoted by OR, Defined as the opportunity ratio for x = 1 And opportunities for x = 0. Which can be written in the following equation: $\left[\frac{\pi(1)/[1-\pi(1)]}{\pi(0)/[1-\pi(0)]}\right]$
Marginal Effect can be calculated by using the equation below:

$$Pi = \frac{e^{B_iX_i}}{1+e^{B_iX_i}} = (1+e^{-c\beta_1X_1})^{-1}$$
$$dp_i = \frac{(1+e^{-\beta_iX_i})^{-1}}{dX_i}$$
$$= -1(1+e^{-\beta_iX_i})^{-2}\frac{(1+e^{-c\beta_iX_i})^{-1}}{dX_i};(e^{u(X)}) = e^{u(X)}u(X) \quad (3)$$
$$=-\frac{1}{(1+e^{B_iX_i})^2}(0+e^{-c\beta_iX_i})(\beta_i)$$
$$\beta i\frac{1}{1+e^{-B_iX_i}}\frac{e^{-\beta_iX_i}}{1+e^{-B_iX_i}};1-p_i=\frac{e^{-\beta_iX_i}}{1+e^{-B_iX_i}}$$
$$=\hat{\beta}_1\hat{p}_1(1-\hat{p}_1)$$

**3. Result and Discussion**

The influence of perception factors on customer satisfaction can be seen in table 1.

**Table 1.** Panel Regression Result The influence of perception factors on customer satisfaction

| Variable | B | S.E. | Wald | Sig. | Exp(B) |
|---|---|---|---|---|---|
| Health Benefit | 0.298 | 0.143 | 4.35 | 0.037 | 1.348 |
| Quality | 0.171 | 0.076 | 5.090 | 0.024 | 1.187 |
| Price | 0.133 | 0.160 | 0.694 | 0.405 | 1.142 |
| Product availibility | 0.026 | 0.173 | 0.023 | 0.879 | 1.027 |
| Constant | 10.32 | 2.28 | 20.42 | 0.000 | 0.00 |

The estimated model can be write is:

$$Ln\frac{p}{1-p} = 0.00 + 1.348X_1 + 1.187\ X_2 + 1.142\ X_3 + 1.027\ X_4 \quad (4)$$

Interpretation of result in logistic regression model

*3.1 Wald Test*

Table 1 show that health and quality benefits influence the satisfaction of consumers of traditional medicinal plants partially. Where this is explained by the significance of the results obtained in each variable health benefits 0.037 and quality 0.024 with α amount 0,05. Price factor and product availability do not influence the satisfaction of consumer of traditional medicinal plants partially, where this is explained by the value of significance obtained are the price 0,405 and product availability 0.879 with α amount 0.05.

*3.2 Omnibus and Hosmer –Lemeshow Test*

**Table 2.** Omnibus and Hosmer&Lemeshow Consumer Perception Variables

| Test | Omnibus | Hosmer & Lemeshow |
|---|---|---|
| Chi Square | 46.548 | 8.505 |
| Sig | 0.00 | 0.386 |

Table 2 gave the result which indicated wheter the inclusion of variables contributed significantly to model fit. A p-value (sig) of less than 0.05. The null hypothesis was rejected. Therefore, the model is statistically significant.

The test results explain that there is no difference in the distribution of observation with the distribution of theory/model in accordance with the data held. Where this is explained by the value of significance obtained is equal to 0.386 with α amount 0.05 which means that was not statistically significant and therefore the model was quite a good fit.

*3.3 Interpretation of Category Variables Exp (B)*

The Exp (B) in the Table 1 presented the extent to which raising the corresponding measure by one unit influences the odds ratio. To find the value of the odd ratio using the marginal formula can be write is:

$$Pi = \frac{e^{B_i X_i}}{1 + e^{B_i X_i}}$$

$$Pi = \frac{1{,}348}{9.652{,}348}$$

$$Pi = 0.57$$

$$Ci = \beta Pi(1 - Pi)$$

Ci = (0.298) (0.57) (1-0.57)

Ci= 0.07 (5)

The odd ratio of health benefit has a value 0.07. The model predicts that the odds of perception to satisfaction consume traditional medicines are 7 % it can be explained that the benefits obtained are not proportional to the time spent to cure or increase the metabolism of a user using traditional medicinal plants.

The odd ratio of quality has a value 0.04. The model predicts that the odds of perception to satisfaction consume traditional medicines are 4 % Phenomenon that occur to the field are the taste, the smell, color and cleanliness get the lowest score from consumers. This can be explained because the smell, taste, color and hygiene of traditional medicinal plants are in homogeneous colors and can't be brighter after processing from traditional medicinal plants, the sense that consumers are expected can not be fulfilled, because the majority of the flavor generated by traditional medicinal plants is not delicious or bitter, the resulted smell stimulates the sensory nerves to refuse to consume the processing of traditional medicinal plants because in general the consumers have sensitive to the stinging smell and cleanliness that lacks standards while consumer tastes are heterogeneous meaning this will reduce the chances of consumers to be satisfied.

The odd ratio of price has a value 0.03. The model predicts that the odds of perception to satisfaction consume traditional medicines are 3 %. The phenomenon that occurs in the field is the price is not a major factor in determining the consumption of a health product so that the more variable price, the smaller of the influence on customer satisfaction opportunities.

The odd ratio of product availbilty has a value 0.006. The model predicts that the odds of perception to satisfaction consume traditional medicines are 0.6 %. The phenomenon that occurs to the field explains that the availability of traditional medicinal plants is sufficient, but there are several types of herbs from traditional medicinal plants that should be combined as traditional medicine with other traditional medicinal plants located in separate or different locations and have a long distance and some locations are difficult to be taken so as the result, the efforts to obtain the entire medicinal plant is low.

## 4. Conclusion and Recommendation

Based on the results of research conducted, it can be concluded that health benefits variable, quality, price and product availability have significant influence simultaneously to satisfaction of consumer of traditional medicine in Medan. Partially, health benefit and quality have significant effect on satisfaction of consumer of traditional medicine. The health and quality variables have a marginal effect of 7% and 4% on the satisfaction of consumers of traditional medicine.

Therefore, the researchers should increase investigation of variation of types of traditional medicine which has a good taste, practical and has a high quality. Then, provide education to consumers about the utility and how to use traditional medicine.

## References

[1] Arikunto, Suharsimi. 2002. Metodologi Penelitian (Research methods). Penerbit PT. Rineka Cipta. Jakarta.

[2] BPS (2009) Survei Sosial Ekonomi Nasional (Susenas) Tahun 2009 (National Social Economy Survey 2009), Jakarta:BPS.

[3] Depkes. Kepmenkes No.128/2004 tentang Kebijakan Dasar Pusat Kesehatan Masyarakat (Community health center basic policy). Jakarta Kementerian Kesehatan RI; 2013.

[4] Hosmer, David W and Stanley Lemeshow. 2000. Applied Logistic Regression, John Wiley and Sons, Inc, New York.

[5] Kotler, P.2000. Manajemen Pemasaran (Marketing management). Edisi Kesepuluh. Strategy Marketing for Educational Institution. Prenhallindo. Jakarta

[6] Prakash, P. and Gupta, N. 2005 Therapeutic Uses of Ocimum sanctum Linn (Tulsi) with a Note on Eugenol and Its Pharmacological Actions: A Review. Indian Journal of Physiology and Pharmacology, 49, 125-131.

[7] Saleh Hosseinzadeh and Azizollah,J.2015.The Aplication of Medicinal Plants in Traditional and Modren Medicine: A Review of *Tyhumus vulgaris*.International Journal of Clinical Medicine,6,635-642

[8] Sari Kumala, R, O, Lusia. 2006. Pemanfaatan Obat Tradisional Dengan Pertimbangan Manfaat dan Keamanannya (Traditional medicine utilization considering its benefit and safety). Majalah Ilmu Kefarmasian, **Vol:** III, No.1, April 2006.

IOP Conf. Series: Earth and Environmental Science **122** (2018) 012005 doi:10.1088/1755-1315/122/1/012005

# Factors that influence the interests of farmer in shallots farming at Cinta Dame village of Simanindo sub district of Samosir district

**A F Siregar[1] and T Supriana[2]**
[1]Master Program Agribusiness Alumnae of Universitas Sumatera Utara
[2]Lecturer of Universitas Sumatera Utara
tavihutasuhut@yahoo.co.id

**Abstract.** Shallots contains a lot of usefull ingredients for human life, especially as flavor to dishes by Indonesian.The need for shallots was increasing as increasing population. The increased demand of shallots caused the price to increase due to production in North Sumatera was low. The objective of this study is to analyze interest and factors that affect the interest of farmers in shallots farming and analyze the responses from each factors to the interest of farmers in shallots farming. The samples were 85 farmers in shallots farming. Binomial logit was used as data analysis method. The result of the study showed that the factors that influence the interest of farmers in shallots farming consist of land area, experience, income, supporting and trauma.. The opportunity of farmers in shallots farming increased 22% if the area of land increased by one acre. The probability variable with the supporting is higher 0,3 % compared without the supporting. While the probability variable without the trauma is higher 0,014 % compared with the trauma.

## 1. Introduction

Shallot has an important meaning for society. Both from its high economic value and its nutritional content. Shallot is not a basic necessity, but its needs can hardly be avoided by household consumers as a complement to dishes and useful for health.

One of the provinces producing shallots in Indonesia is North Sumatra Province. Shallots harvest area in North Sumatera in 2015 is 1,238 acre with production of 9,971 tons. As for the consumption of shallot in North Sumatra of 25,503 tons. There is a shortage of shallot production of 15,532 tons [1]

One of the shallot producing districts in North Sumatra is Samosir. It was famous for its local shallot since it used to be a national shallot producer. Samosir and surrounding areas include Samosir regency, Humbang Hasundutan regency, Simalungun regency, North Tapanuli regency is the main producer of shallots in North Sumatera with the name of local varieties Samosir.The characteristic of local shallots of Samosir has a more red color, low water content, has a more spicy taste. In addition, the price of local shallots has a high selling price in the market. At the time of harvest season, shallot farmers can also enjoy the proceeds from the sale of shallots which has become an economic source for farmers in Samosir.

Interest is closely related to attitude. Interest and attitude are also important in making decisions. Interest can cause a person to actively do something that has interest him. Interest can be a motivation that can provide benefits and satisfaction. Interest arises because of the feeling of being happy with something, having a positive nature and the desire and need to achieve goals.The use of interest as a key aspect of conformity between a person and a job makes it a reason to know what factors influence the interest of the farmers to plant shallots.

Simanido sub-district is one of the largest shallots producing areas in Samosir regency compared to other districts with productivity of 6.50 tons /acre. Therefore Cinta Dame village became the research destination area because the village is the biggest production center of shallot in Simanindo District with shallots harvest area in 2015 is 41.9 acre and production 270.9 ton.

In reality, the production produced by farmers in Cinta Dame has not been satisfy these days. The problems that occurred in recent years have caused farmers' constraints to increase shallots production. The main problems are pest attacks, expensive seed prices and unavailability of superior seeds. However, because the selling price of shallot and the high demand of shallot is expected by farmers can give hope again to plant shallots. Therefore, it can be a motivation for farmers to further increase the production of shallots. This study seek to empirically analysed interest and factors that affected interest and analysed responses from each factors in shallots farming.

## 2. Methodology

The location of the research is purposive in Cinta Dame Village, Simanindo Subdistrict, Samosir District. The reason of determining the location is because Samosir is one of production center of shallots in North Sumatera. While the determining of research location in the village of Cinta Dame with the reason that the village is the largest shallots production center area in Samosir district Simanindo with shallots crop area in 2016 is 41.9 acre and production 270.9 tons.

Population in this study were 109 sample of shallots farmer. The method of determining the sample size is determined by the Slovin method, sample size for each group based on the following formula:

$$n = \frac{N}{1+(N(\propto^2))} \quad (1)$$

Where :
N : Sample population
n : Sample size
$\propto$ : level of accuracy (level of precision 0.05) [2].

So the number of samples of shallot farmers is determined as many as 85 of the shallot farmer population. Samples were drawn randomly using simple random sampling.

In analyzing interest and factors that influence the interest of farmers in shallots farming and analyze the responses from each factors to the interest of farmers in shallots farming, we use logistic regression model. Our regression model will be predicting the logit, that is, the natural log of the odds of having made one or the other decision. Logistic regression analysis studies the association between a categorical dependent variable and a set of independent (explanatory) variables. The name logistic regression used when the dependent variable has only two values, such as 0 and 1.

### *2.1 Model Specification*

In analyzing the factors that influence the interest of farmers in shallots farming and analyze the responses from each factors to the interest of farmers in shallots farming, we specify a mathematical logistic regression model. The model is thus;

$$ln\left\{\left[\frac{P}{1-pi}\right]\right\} = \beta_0 + \beta_1 X_1 + \beta_2 X_2 + \beta_3 X_3 + \beta_4 X_4 + \beta_5 X_5 \quad (2)$$

Where : $ln\left\{\left[\frac{P}{1-pi}\right]\right\}$ is odd ratio where (p) explain the predicted probability of the event which is coded with 1 (interest), (1-p) is the predicted probability of the other decision which is coded with 0 (not interest).

$\beta_0$ = Intercept

$\beta_1, \beta_2, \beta_3, \beta_4, \beta_5$ = Coefficient of land area, experience, income, supporting from government and trauma

$X_1$ = Land area (acre)
$X_2$ = Experience (years)
$X_3$ = Income
(X=1 for income over the regional minimum wage)
(X=0 for income below the regional minimum wage)
$X_4$ = Supporting from government
(X=1 with supporting)
(X=0 without supporting)
$X_5$ = Trauma
(X=1 without trauma)
(X=0 for trauma)

*2.2 Parameter Testing*

2.2.1 *Concurrent Testing.* It is performed to know the significance of β parameters as a whole or simultaneously. This recitation hypothesis is:

$$H_0 ;\ \beta_0 = \beta_1 = \cdots = \beta_8 = \beta_p = 0. \tag{3}$$

With rejection: reject $H_0$ if value G > X_(α,1)^2 Where p is the number of statistical variables in the model or p- value < α..

*2.2.2 Individual Test (Wald Test).* It is intended to examine the significance of individual β parameters. This recitation hypothesis is:

$$H_0 ;\ \beta_j \qquad H_1 :\ \beta_j \neq 0,\ j = 1, 2 \ldots, p \tag{4}$$

With statistical tests:

$$W\ (\text{Wald}) = \frac{\beta_j}{S.E\ (\beta_j)} \tag{5}$$

Denial area : reject $H_0$ if $W_j^2 > \mathcal{X}^2_{\alpha,1}$ or p- value < α

*2.2.3 Hosmer and Lemeshow Test.* This test aims to compare the distribution of observation with the distribution of theory (model test).
Criteria Test :
If sign < 0,1 then accept $H_1$ reject $H_0$
If sign > then accept $H_0$ reject H1 [3]

*2.2.4 Odd Ratio and calculation of marjinal effect.* Odds ratio is a measure of risk or inclination to experience a certain incident between one category with another category is defined as the ratio of odds to = 1 to = 0. The value of odds ratio is used to interpret the relationship between independent variables with dependent variables that have differences with multiple linear regression with independent variables in the form of continuous (quantitative) [4].
Marginal Effect can be calculated by using the below equation:

$$p_{i=} \frac{1}{1 + e^{-\beta_i X_i}} = (1 + e^{-\beta_i X_i})^{-1}$$
$$\frac{dp_i}{dx_i} = \frac{d\,(1 + e^{-\beta_i X_i})^{-1}}{dx_i}$$
$$= -1(1 + e^{-\beta_i X_i})^{-2}\ \frac{d(1+e^{-\beta_i X_i})^{-1}}{dx_i};\ \left(e^{u\,(X)}\right) = e^{u\,(X)}\ U'(X) \tag{6}$$
$$= -\frac{1}{(1+e^{-\beta_i X_i})^2}\left(0 + e^{-\beta_i X_i}\right)(-\beta_1)$$

$\beta \frac{1}{1+e^{-\beta iXi}} \frac{e^{-\beta iXi}}{1+e^{-\beta iXi}}$ ;1 – pi = $\frac{e^{-\beta iXi}}{1+e^{-\beta iXi}}$

= $\beta_i\ P_i\ (1 - P_i\ )$ [5].

## 3. Result and Discussion

**Table 1.** Result of Factors that Influenced the Interest in Shallots Farming

| Variable | B | S.E. | Wald | Sig. | Exp(B) |
|---|---|---|---|---|---|
| Land area | 1.38 | 0.55 | 3.47 | 0.042 | 3.991 |
| Experience | 0.01 | 0.14 | 0.005 | 0.941 | 1.013 |
| Income | 0.73 | 1.16 | 0.15 | 0.699 | 2.092 |
| Supporting | 5.79 | 1.45 | 11.98 | 0.001 | 0.003 |
| Trauma | 4.21 | 1.20 | 7.831 | 0.005 | 0.015 |
| Constant | 0.18 | 1.84 | 0.005 | 0.944 | 1.204 |

The estimated model is thus:

$$Ln\frac{p}{1-p} = 1.204 + 3.991\ X_1 + 1.013\ X_2 + 2.092\ X_3 + 0.003\ X_4 + 0.015\ X_5 \quad (7)$$

Interpretation of result in logistic regression model :

### *3.1 Wald Test*

Table 1 show that land area, supporting and trauma contributed significantly influenced farmer interest in shallots farming but income and experience did not. The p-value for land area, supporting and trauma are 0.042, 0.001 and 0.005 (<0.05) while p-value for income and experience 0.699 and 0.941 (>0.005).

### *3.2 Omnibus and Hosmer –Lemeshow Test*

**Table 2.** Omnibus and Hosmer&Lemeshow Factors that Influence the Farmer Interest in Shallots Farming

| Test | Omnibus | Hosmer & Lemeshow |
|---|---|---|
| Chi Square | 93.959 | 1.760 |
| Sig | 0.00 | 0.972 |

Table 2 gave the result which indicated whether the inclusion of variables contributed significantly to model fit. A p-value (sig) of less than 0.05. The null hypothesis was rejected. Therefore, the model is statistically significant.

The Hosmer – Lemeshow tested the null hypothesis that predictions made by the model fit perfectly with observed group memberships. The Hosmer – Lemeshow statistic assumed sampling adequacy. Hosmer and Lemeshow statistic has a significance of 0.972 which means that was not statistically significant and therefore the model was quite a good fit.

### *3.3 Interpretation of Quantitative and Category Variables Exp (B)*

Quantitative variables consist of land area and experience. Marginal effect is used to explain the quantitative variables indicated by the value of $C_i$ where:

$$C_i = \beta_i\ P_i\ (1 - P_i\ ) \quad (8)$$

To calculate the value of $C_i$ , then first calculate the value of $P_i$ by antilog value of $ln\ \frac{P_i}{1-P_i} = \beta_i$ ($ln\ \frac{P_i}{1-P_i} = 1.38$). By entering value of $\beta_i$ then, $\frac{P_i}{1-P_i} = e^{1,38}$ the value is obtained $\frac{P_i}{1-P_i} = 3.991$. By solving the previous equation, the value will be obtained $P_i = 0,8$. After obtaining the value of $P_i$ then can be searched value of $C_i$ by entering value $\beta_i = 0,8$ into the equation of $C_i$ , $C_i = \beta_i\ P_i\ (1 - P_i\ )$. Then the value will be obtained $C_i = 0,22$.. Similarly, to calculate the marginal effect of experience variables by entering value of $\beta_i$. ($\frac{P_i}{1-P_i} = e^{0.013}$). The value is obtained $\frac{P_i}{1-P_i} = 1.013$ and then we can find the value will be obtained $P_i = 0.5$. By entering value of $P_i$ into the equation of $C_i = \beta_i\ P_i\ (1 - P_i\ )$. We can obtain the value of $C_i = 0.003$ for experience variable.

Marginal effects showed the change in probability when the predictor or independent variable increases by one unit. Interpretation marginal effect of land area has the value of marginal effect 0.22. It means that when land area is raised by one unit (one acre) so that opportunity the farmer to plant shallots increasing 22 %.

This is in accordance with previous research, the results of the study stated that the Odss Ratio of farmers' land area is 2.484, if the area of farmers increasing one acre then the possibility to participate increasing by 2.484. The land area affected farmers in participating in the certification of oil palm, the greater the farmers' land the more likely to participate in the sustainability of their businesses in the future [6].

Interest is essentially a result of experience. Interest develops as a result of an activity and will cause it to be used again in the same activity [7]. Interpretation marginal effect of experience has the value of marginal effect 0.003. It means that every increase of one unit (one year) will increase the opportunity of farmer to plant shallot 0,3 %.

Category variables consist of income, supporting and trauma. To interpret the category variable by specifying when $X_i = 1$ into this equation $ln\ (\frac{P_i}{1-P_i}) = \beta_0 + \beta_i X_i$ then $ln\ \left(\frac{P_i}{1-P_i}\right)$ increasing as $\beta_i$ or $\frac{P_i}{1-P_i}$ increasing as exp ($\beta_i$ ). The Exp (B) in the Table 1 presented the extent to which raising the corresponding measure by one unit influences the odds ratio.

Income variable consist of 2 variable $X_1 = 1$ and $X_2 = 0$ where $X_1 = 1$ show that income over the regional minimum wage and $X_2 = 0$ show that income below the regional minimum wage. The income variable is known $ln\ (\frac{P_i}{1-P_i}) = \beta_i X_i$ when income has $X_1 = 1$ then $ln\ (\frac{P_i}{1-P_i}) = \beta_i$ while $X_2 = 0$ then $ln\ \left(\frac{P_i}{1-P_i}\right) = 0$ . It means that value $ln\ \left(\frac{P_i}{1-P_i}\right)$ at income over the regional minimum wage is higher as $\beta$ compared by income below the regional minimum wage. The next step, we calculate $P_i$ for income variable by antilog $ln\ \frac{P_i}{1-P_i} = 0.738$ then, $\frac{P_i}{1-P_i} = e^{0.738}$. the value is obtained $\frac{P_i}{1-P_i} = 2.092$. By solving the previous equation, the value will be obtained $P_i = 0.67$. The results obtained that the probability variable of income over the regional minimum wage is higher 67% compared by income below the regional minimum wage.

This is accordance with the previous study, the regression analysis can be seen that the income gives a positive influence on the farmer's interest in Sendangan village and has a significant influence on the real level of $\alpha$ 0.1 percent, the increase of income has a positive effect on the interest of farmers as much as 0.001 rupiah by assuming that other factors are constant. [8].

Supporting variable consist of 2 variable $X_1 = 1$ and $X_2 = 0$ where $X_1 = 1$ show that with the supporting and $X_2 = 0$ show that without supporting. The supporting variable is known $ln\ (\frac{P_i}{1-P_i}) = \beta_i X_i$ when supporting has $X_1 = 1$ then $ln\ (\frac{P_i}{1-P_i}) = \beta_i$ while $X_2 = 0$ then $ln\ \left(\frac{P_i}{1-P_i}\right) = 0$ . It means that value $ln\ \left(\frac{P_i}{1-P_i}\right)$ with the supporting is higher as $\beta$ compared without the supporting. The next step, we calculate $P_i$ for supporting variable by antilog $ln\ \frac{P_i}{1-P_i} = 5.79$ then, $\frac{P_i}{1-P_i} = e^{5.79}$. the value is obtained $\frac{P_i}{1-P_i} = 0.003$. By solving the previous equation, the value will be obtained $P_i = 0{,}003$. The results obtained that the probability variable with the supporting from government is higher 0.3 % compared without the supporting from government.

Stimuli coming from the environment or scope that matches one's desires or needs will be easy to generate interest [9]. The supporting gained by each farmer, such as production and technology factors that can generate or increase production, will increase the interest of farmers higher and encourage farmers to keep working on paddy farming [10].

And the last variable is trauma. Trauma variable consist of 2 variable $X_1 = 1$ and $X_2 = 0$ where $X_1 = 1$ show that without the trauma and $X_2 = 0$ show that with the trauma. The trauma variable is known $ln\ (\frac{P_i}{1-P_i}) = \beta_i X_i$ when trauma has $X_1 = 1$ then $ln\ (\frac{P_i}{1-P_i}) = \beta_i$ while $X_2 = 0$ then $ln\ \left(\frac{P_i}{1-P_i}\right) = 0$. It means that value $ln\ \left(\frac{P_i}{1-P_i}\right)$ without the trauma is higher as $\beta$ compared with the trauma. The next step, we calculate $P_i$ for trauma variable by antilog $ln\ \frac{P_i}{1-P_i} = 4.21$ then, $\frac{P_i}{1-P_i} = e^{4.21}$. the value is obtained $\frac{P_i}{1-P_i} = 0.0015$. By solving the previous equation, the value will be obtained $P_i = 0.0014$. The results obtained that the probability variable without the trauma is higher 0.014 % compared with the trauma.

Changes in farmers' interest towards shallot are caused by pests attacking nearly 90% of the farmer in Cinta Dame Village. The infected shallots caused shallots were die and the crop to be failure. The virus attacks caused some trauma farmers to plant shallots again.

The results of this study in accordance with previous research, it explained that one of the factors that affect the low interest of farmers to apply environmentally friendly chili cultivation is pest and disease attack [11].

## 4. Conclusion and Recommendation

Based on the findings of this study, we conclude that land area, experience, income, supporting and trauma influenced the interest of farmer in shallots farming simultaneously. While land area, supporting and trauma influenced the interest of farmer in shallots farming partially.

Therefore, the government should increase facilities and infrastructure as a government's support to farmers in running of shallots farming. The government also should keep the stability of shallots's price in selling the products of shallots.

## References

[1] Central Agency on Statistics. 2014. Samosir in numbers, 2015
[2] Supriana, T. 2016. Socio-Economic Research Methods. USU press, Medan
[3] Hosmer, D.W., dan Lemeshow, S. 2000. Applied Logistic Regression. John Wiley dan Son. New York
[4] Gujarati, D. 2006. Fundamentals of Econometrics. Erlangga, Jakarta
[5] Chalil, D. 2014. Qualitative Data Analysis. USU press, Medan
[6] Emilia, R. 2014. Factors that Affect the Farmer Interest of People's Palm Oil in Product Certification in Kampar District. Agribusiness Department. Agriculture Faculty. Riau University
[7] Khairani , H. 2013. Psychologist Learning. National Library of Indonesia. Yogyakarta: Aswaja Persindo
[8] Muawiyah, S. 2014. Factors that Affect Rice Farmers' Interest in Sendongan Village, Kakas Subdistrict. Journal of Socio-Economic Research. Agriculture Faculty Sam Ratulangi University Manado. Manado
[9] Crow and crow. 1973. An outline of Pscicology (Translation Z. Kazijan) Surabaya : PT Bina Ilmu
[10] Widowati, S. 2007. Healthy with Low Glycemic Food Index. News Research and Development of Agriculture Vol. 29. No. 3
[11] Astuti, P. 2013. Factors - Causes of Low Interest in Farmers to Implement Green Chili Cultivation in South Lampung District. Agriculture Department of South Lampung Regency

IOP Conf. Series: Earth and Environmental Science **122** (2018) 012006 doi:10.1088/1755-1315/122/1/012006

# Strategies to increase the consumption of traditonal medicine in Medan

**Salsabila[1] and T Supriana[2]**
[1]Master Program Agribusiness Alumnae of Universitas Sumatera Utara
[2] Lecturer of Universitas Sumatera Utara
tavihutasuhut@yahoo.co.id

**Abstract**. As people improving their knowledge and increasing their income, it improves their awareness about healthy life. It affects of demanding of medicine highly including traditional medicine because of its using more safety and hereditary. The consumption of traditional medicine in Medan was still low comparing with modern medically. The objective of the study was to analyze the strategy of increasing traditional medicine consumption in Medan. The samples are the business actors of traditional medicine, the consumer of traditional medicine, the leader of Jamu's enterpreneur Association, and the supervisor of food and medicine in The National Agency of Drug and Food Control (NA-DFC). The analysis SWOT was used in this study. The Result of the study showed that the position of the strategy of increasing the using of herbal in Medan was in the first quadran. Commonly, it means that the strategy was in agresif level which showed the maximal probability. Therefore, jamu's entrepreneurs should implement the strategy of product development to increase the number and variation of types of traditional medicine which has a good taste and high quality and provide education to consumers about utility and how to use traditional medicine.

## 1. Introduction

Indonesia has biological resources and is one of the world's largest biodiversity nations, and ranks second position after Brazil. Herbal wealth of about 30,000 species of medicinal plants is a potential for the development of herbal products whose quality is equivalent to modern medicine and beneficial to the interests of society. However, the natural resources have not been utilized optimally, only about 1,200 species of medicinal plants are utilized and researched as traditional medicine [1].

Traditional medicine in Indonesia is part of the nation's culture and is widely used by the community, but in general its effectiveness and safety have not been fully supported by the research.

The use of traditional medicine is generally considered safer than the use of modern medicine. In addition to traditional medicine has relatively smaller side effects than modern medicine, the price of traditional medicine is also cheaper when compared with pharmaceutical drugs. Thus the role of traditional medicine in Indonesian society in particular is very important and felt the benefits.

Based on research data published in the informal meeting on Strategic Planning on Traditional Medicine in Manila, it was mentioned that the use of traditional medicine in Australia reached 48.50%, China 90%, Hongkong 60%, Japan 49%, Nauru 60%, Republic of Korea 69% , Philippines, 57.30%, Singapore 45% and Vietnam 50%, while Indonesian herbalists based on the last 15 years data as much as 5% of Indonesians often use herbs and 45.17% sometimes use them [2].

According to the Strategic Plan of BBPOM Medan based on data on drug consumption conducted in Medan, most people are still many who consume modern medicine compared with traditional medicine. Consumer modern medicine in 2015 reached 91.40%, while traditional medicine only as much as 24.33%. This indicates the low consumption of Medan city people against traditional medicine compared to modern medicine [3].

Traditional medicine producers are business actors who produce traditional herbal medicine products as an alternative treatment for most people, especially the people of Medan city. Traditional medicine entrepreneurs must compete with their competitors to attract customers in order to expand their business. The main competitor of traditional medicine is a modern medicine that contains chemicals that are easily available because it is available in every store.

This strategy is needed by traditional medicine business actors in order to increase the availability and quality and quality of traditional medicine, expand market share and increase the intensity of traditional medicine use.

Therefore, this study seek to empirically analyzed strategies to increase the consumption of traditional medicine in Medan. These strategies is expected to increase the consumption of community in Medan against traditional medicine.

## 2. Methodology

The location of the research is determined purposively in Medan because of the low consumption of the community against traditional medicine compared to modern medicine. In addition, there are several business actors that produce traditional medicines (herbs) that have received permits and supervision from The National Agency of Drug and Food Control (NA-DFC) in Medan

Purposive sampling was used in determining the sample in this study, which consist of 5 business actors of traditional medicine, 30 consumer of traditional medicine leader of Jamu's enterpreneur Association, and the supervisor of food and medicine in The National Agency of Drug and Food Control (NA-DFC).

Strategy is a tool to achieve goals. SWOT analysis is a suitable analytical tool to formulate the strategy. This analysis is based on logic that can maximize strength and opportunity, and can minimize weakness and threat [4]. The SWOT matrix is used to construct strategic factors. This matrix clearly illustrates how opportunities and threats can be adjusted to the strengths and weaknesses.

The first thing to do is identify the factors that influence the increase of traditional medicine consumption in Medan. From these factors are determined strategic factors that can be grouped on external factors and internal factors. Internal factors are factors that can be controlled by business actors, while internal factors are factors that can not be controlled by business actors.

Second, in analyzing the strategies we determined strength, weakness, opportunity and threat based on scoring. Scoring was identified among the internal factors (strengths and weaknesses) and external (opportunity and threats) through a questionnaire that has been compiled. Score values range from 1 - 4 from the lowest to highest assessment. Scores 1 and 2 show weaknesses and threats, while 3 and 4 show strengths and opportunity. Score values are determined based on predefined parameters obtained through interviews. The next step is to determine the arithmetic mean for each factor after obtaining the value for each factor of each respondent

Third, we determined weighting internal and external factors. Weighting is perfomed by pairwise comparison scale model with comparing factors that one with other factors in a hierarchy, so we can get the values of each of the factors.

Weighting is done by using pairwise comparison scale model with value of scale 1,2 and 3 so that obtained value of importance determined by ability of respondent to distinguish value between paired factors. Next step, we calculate the weighted score by multiplying the score of each factor by the weight obtained in each factor. The analysis results are made on the position matrix in cartesius diagram by calculating the difference of internal factors (strengths - weaknesses) for values on the horizontal axis and the difference of external factors (opportunity - threat) for values on the vertical axis. The last step, we can arrange strategies which consist of strength – weakness, weakness – opportunity, strength – threat and weakness – threats strategies.

## 3. Result and Discussion

Based on the observations obtained can be seen factors internal and external factors that affect strategies increased consumption of traditional medicine in Medan. Internal factors consist of strength and weakness. External factors consist of opportunity and weakness. Internal factors are factors that can be controlled by business actors. External factors can't be controlled by business actors.

*3.1 Description of Internal Factors*

*3.1.1 The Availability of the Raw Material for Traditional Medicine.* The production of traditional medicinal products depends on the herbal raw material which in fact depends on the nature, the availability of raw materials is an important concern to maintain the availability of sustainable, well cultivated so that the quality produced has good and high quality.

In General, traditional medicine is the herb ingredients or are made from plants, animals, minerals, or a combination of these materials are processed and traditionally has been used orally for the treatment. Raw materials used by the business actors are selected based on considerations by the criteria of having a good quality. The amount of raw materials used is also always meets the needs of the production and trade of traditional medicine.

*3.1.2 The Availability of Appropriate Labor.* According to National Agency of Drug and Food Control (NA-DFC) that the labor should have the knowledge, skills and competence appropriate to the duties and functions, and be available in sufficient quantities [5].

Based on the result of this research, the availability of labor in traditional medicine business consists of production area, pharmacist, packaging and marketing. Labors of production area in traditional medicine business in Medan consists of women and most of them are housewife and trained so that the work can be done quickly and regularly.

*3.1.3 The Availability of Buildings, Machines and Production Equipment Of Traditional Medicine.* Ideally the traditional small industries well and healthy should be in a location that is free from pollution. The building must also meet the requirements of sanitary and hygienic procedures.

The building of the traditional medicine business should have design rooms and the width of the building according to the shape,characteristic, quantity of the product, the number of using equipment and the number of employees employed.

Based on the the result of this research, traditional medicine business in Medan has adequate buildings, machines and production equipments which complete for business activity.

*3.1.4 Process of Traditional Medicine Production.* According to WHO that process of production is one of the key stages in which quality control is required to ensure the quality of natural medicines produced [6].

The traditional medicine production process based on the results of interviews with business actors consists of preparation of raw material for traditional medicine, production process, packaging and brands label and logos of herbal medicine. Every business actor of traditional medicine in all aspects of traditional medicine and traditional medicine-producing activitie based on Good Manufacturing Practice (GMP).

*3.1.5 Knowledge of Business Actor.* Knowledge owned by the business actor of traditional medicine including the knowledge of the truth of the raw materials used, regarding the accuracy of the dose, the precision processing of raw materials, use of timekeeping that is right and good in taking traditional medicine and the benefits of traditional medicine and knowledge of side effects.

*3.1.6 Information on Packaging.* Clarity and completeness of product information on the label of the package traditional medicine intended to communicate the product thoroughly so that it can affect consumers to increase consumption of traditional medicine.

*3.1.7 Marketting of Traditional Medicine.* In general they market their traditional medicine directly from the manufacturer to the customer directly, social media, distributor and salesman. Marketing on traditional medicine businesses generally use direct marketing. The goal is to obtain a response and direct response from consumers about the effects of the efficacy of herbal medicine consumed.

*3.1.8 The Number of Variations of Traditional Medicine.* According to National Agency of Drug and Food Control (NA-DFC) in strategic planning that the increasing number of people will increase the demand for drugs so that the supply of traditional medicine will also increase. A large market potential becomes an opportunity for traditional medicine producers to increase the number and variety of traditional medicine. Most of the traditional medicine producers in Medan only produce one or two types of traditional medicine with variations that are still a little bit.

*3.2 Description of external factor*

*3.2.1 Consumer Assessment of the Quality of Traditional Medicine.* Natural herbal medicine including herbal medicine produced by traditional medicine industry and small industry have the same requirement that is safe to use, useful and good quality [7].

The assessment of consumers in taking traditional medicine are side effects, rapid response to healing and efficacy felt by consumers after taking traditional medicine.The quality of traditional medicine in the city of Medan, according to consumers in Medan city included in the category of the good and safe for consumption.

*3.2.2 The Availability of Traditional Medicine in Medan.* One of the categories evaluated by consumers in the buying decision process is availability [8]. Based on the results of interviews with consumers of traditional medicine is obtained that most consumers did not trouble getting traditional medicine because it could be bought in pharmacies nearby, at drugstores, serves traditional medicine. Furthermore consumers argued that traditional medicine was always available at any time and can be purchased at any time consumers need it.

*3.2.3 The Role of the The National Agency of Drug and Food Control (NA-DFC).* The role of the National Agency of Drug and Food Control (NA-DFC) is to guarantee the protection of consumers from traditional drug abuse, ensure the security aspects, benefit and quality of traditional medicines that are circulating in Medan by testing a sampling of traditional medicine.

*3.2.4 The Role of the Jamu's Enterpreneur Association.* Based on the results of interviews with the leader of jamu's enterpreneur association in Medan that jamu's enterpreneur association supported towards the development of jamu's business in Medan.

*3.2.5 The Dangers of Chemical Medicine.* The consumer mentioned that taking modern medicine resulting in side effects directly or indirectly, that may be because the chemicals contained in modern medicine, in addition it can be addicted. Drugs containing chemicals are not good for the body if it is consumed continuous. The more routine taking of chemical drugs, the resistant body and disease will be more immune to fight treatment given.

*3.2.6 The Accuracy of the Consumer Taking Traditional Medicine.* Traditional medicine will be useful and safe if it is used with consideration of six aspects that consists of accuracy are appropriate dosage, timely and how to use, appropriate selection of materials and information review and in accordance with certain disease indications [9].

The results of interviews with consumers of traditional medicine that most people did not consume traditional medicine in accordance with the applicable dose and did not correspond with the time of the use of traditional medicine. That is because the consumers of traditional medicine is only trust the efficacy and benefits of traditional medicine without regard to the accuracy of the use of traditional medicine.

*3.2.7 The Opinion of Consumers about the Taste of Traditional Medicine.* The problem often encountered by the consumer at the time of taking traditional medicine traditional medicine is a less preferred because it tends to be more bitter. consequently some consumers complain about the traditional taste of medicine but because of the efficacy of the drug then consumers continue to consume.

*3.2.8 Consumer Confidence in Consuming Traditional Medicine.* When a consumer trusts a company, they will prefer to repurchase and share valuable personal information with the company [10]. Based on result of this research, the confidence of consumer with traditional medicine is quite high. This is evidenced by the results of interviews with 30 consumers of traditional medicine and the results obtained quite well.

*3.2.9 The Influence of Consumer Income.* According to the fact, the higher the income then the greater the public consumption of drugs that have standards and quality. Based on the results of interviews with consumers of traditional medicine that if consumer incomes increase, then consumers will be ncreasing the standard and quality of traditional medicines that they consumed. Some consumers say to prefer taking traditional medicine that has been clinically tested such as phytopharmaca products.

Evaluation of internal and external factor strategy is done by creating a matrix table. The steps are taken to make weighting, scoring and calculate weighted scores (score x weight). Based on the result of calculation weighted score, we obtained total score of strength is 3.21and weakness is 0.26.

Meanwhile difference of (strength – weakness) is 2.95. Total score of opportunity is 2.24 and threat is 0.66. Meanwhile difference of (opportunity – threat) is 1.58.

The position of the strategy consumption increase of traditional medicine in Medan uses the position matrix which be shown by the coordinate point (x, y). The value of x is obtained from the difference between the internal factor (strength - weakness) and the value of y obtained from the difference of external factors (opportunities - threats). The position of the x and y coordinate points can be seen in the cartesian diagram in Figure 1.

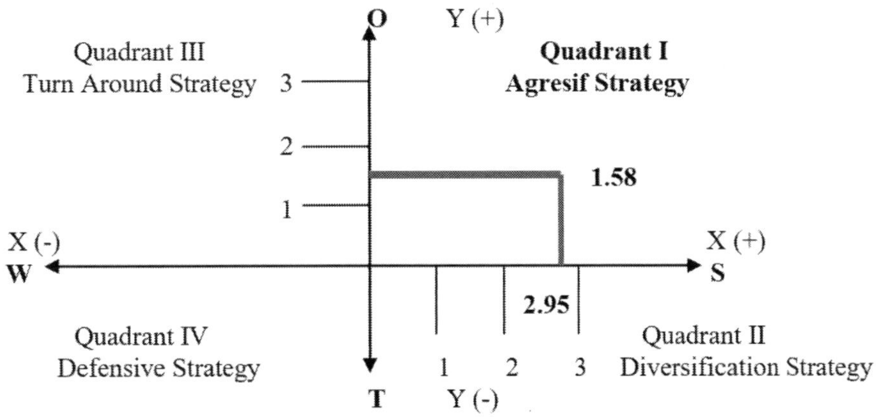


**Figure 1.** Position Matrix of Strategies Consumption Increase of Traditional Medicine

The position of the strategies to increase the consumption of traditional medicine in Medan is on the I quadrant means business actors have factors – factors of strength and great opportunities to improve the consumption of traditional medicine. The focus of the strategy is to maximize strengths by utilizing the opportunities. The strategy should be applied in this case is supportive of a growth oriented strategy. The I quadrant is the most profitable situation. It means that the company has opportunities and strength so that it can utilize the strengths and opportunities [11].

Based on determination of internal and external factors, then we evaluate strategies to increase the consumption of traditional medicine in Medan.

Strategies of Strength-Opportunity

1. Utilizing the availability of raw materials to increase the availability of traditional medicine
2. Utilizing the availability of labor to increase the availability of traditional medicine
3. Optimize the knowledge of business actor of traditional medicine to improve the quality of traditional medicine
4. Utilizing the clarity of information on traditional medicine packaging to increase consumer confidence in traditional medicine
5. Utilizing standar of the traditional medicine production process to improve the quality and to increase consumer confidence

Strategies of Weakness – Opportunity

1. Improve marketing network to boost the availability of traditional medicine.
2. Increase the number and variation of traditional medicine to claim the role of The National Agency of Drug and Food Control (NA-DFC) in the process of assessment and supervisor.

Strategies of Strength – Threat

1. Optimize the knowledge of business actor to provide education and guidance in raising consumer awareness in taking traditional medicine in accordance with the rules of the use of traditional medicine is good and true.
2. Optimize the knowledge of business actor and the availability of labor to create product innovations in order to change the perception of the consumer against a sense of traditional medicine that awful.

Strategies of Weakness – Threat

1. Improve the marketing of traditional medicine network by optimizing the combined role of jamu's enterpreneur association.

2. Increase the number and variation of traditional medicine by optimizing the combined role of jamu's enterpreneur association.

## 4. Conclusion and Recommendations

Based on the findings of this study, we conclude that factors affecting to increase the consumption of traditional medicine in Medan include 8 internal factors and 9 external factors. An alternative strategy that can be done that is aggressive strategies by maximizing the strength to take advantage of opportunities. In detail, there were 11 alternatives could be done that consist of strength-threat strategy, weakness-opportunity strategy, strength - threat strategy and weakness-threat strategy.

Therefore, jamu's enterpreneurs should implement the strategy of product development to increase the number and variation of types of traditional medicine which has a good taste, practical and has a high quality. Then, provide education to consumers about the utility and how to use traditional medicine.

## References

[1] Department of Health Indonesia. 2007. Traditional Medicine Policy. Department of Health of the Republic of Indonesia : Minister of Health Decree 381/ Minister of Health/ SK/ III/ 2007.
[2] Waspada Medan. 2016. Traditional Medicine Has Business Opportunities. Available at http://www.waspadamedan.com/
[3] NA-DFC. 2015. Regulation of the Head of the National Agency of Drug and Food Control of the Republic of Indonesia Number 2 about Strategic Planning of the National Agency of Drug and Food Control of the Republic of Indonesia, Medan.
[4] Rangkuti, F. 2003. SWOT Analysis : Dissecting Business Case Technique. Gramedia Pustaka Utama, Jakarta.
[5] NA-DFC Indonesia. 2005. Regulation of the Head of the National Agency of Drug and Food Control of the Republic of Indonesia Number HK 00.05.41.1384 about Criteria and Procedure of Registration of Traditional Medicines, Standarized Based Herbal Medicine and Clinical Based Herbal Medicine. Jakarta : Head of NA-DFC
[6] WHO. 2008. Traditional Medicine. Available at: www.who.int/mediacentre/factsheets/fs134/en/index.html
[7] Suharmiati dan Lestari. 2007. Medicinal Plants And Traditional Herbs To Overcome Dengue Hemorrhagic Fever. Jakarta: Agromedia pustaka
[8] Mullins, John W. dan Orville C. Walker, JR. 2010. Marketting Management : A Strategic Decision-Making Approach. New York: McGraw-Hill.
[9] Katno. 2008. Level Benefits, Safety and Effectiveness of Medicinal Plants and Traditional Medicines. Center for Research and Development of Medicinal Plants and Traditional Medicines, Central Java.
[10] Prasaranphanich. 2007. Consumer Behavior: Decision Model Analysis. Atma Jaya University Yogyakarta, Yogyakarta.
[11] Rangkuti, F. 2005. SWOT Analysis : Dissecting Business Case Technique. Gramedia Pustaka Utama, Jakarta.

# Farmer's Motivation in *Aren* Sugar Processing Business

**W G Abdullah[1], U Rianse[1], Muhidin[2], W Widayati[3], E S Mihrad[4], S A A Taridala[1], I S Rianse[1] and W K Baka[5]**

[1]Department of Agribusiness, Faculty of Agriculture, Universitas Halu Oleo, Kendari South East Sulawesi 93212 Indonesia
[2]Department of Agro technology, Faculty of Agriculture, Universitas Halu Oleo, Kendari South East Sulawesi 93212 Indonesia
[3]Department of Geography, Faculty of Technology and Earth Science, Universitas Halu Oleo, Kendari South East Sulawesi 93212 Indonesia
[4]Department of Agriculture Extension, Faculty of Agriculture, Universitas Halu Oleo, Kendari South East Sulawesi 93212 Indonesia
[5]Department of Culture Sciences, Faculty of Culture Science, Universitas Halu Oleo, Kendari South East Sulawesi 93212 Indonesia
Email : weka.gusmiarty@uho.ac.id/wkgusmiarty09@yahoo.com

**Abstract:** The objectives of the research were to analyze socio-economic characteristics and motivation of farmers in *aren* sugar processing business in Kolaka District, Southeast Sulawesi, Indonesia. The analysis used in this research was quantitative descriptive analysis. The results showed that average state of socio-economic characteristics of *aren* sugar farmer that were in the category of productive age, which was 46.12 years old, has fulfil basic education category (9 years education), low category of dependents as many as three people, the Average experience of *aren* sugar processing business during 18 years, and the average *aren* trees were tapped as much as seven trees, every day, (b) farmer's motivation in *aren* sugar processing business in low category, with the indicator is the motive of imitation, economic, security, affiliations, awards, and self-actualization.

## 1. Introduction

The gap between consumption and national sugar production has been filled with imported sugars. This means that there is a big chance for brown sugar agro-industry to thrive as an alternative source of sweetener Indonesian society. Great opportunities of brown sugar agro-industry as an alternative sweetener and raw materials of various kinds of food and beverages that also have advantages in health side requires high performance in order to capture these opportunities. The high performance is needed to improve the quantity and quality of sugar required by the market.

Indonesia is very rich in natural resources as a sweetener resources, primarily on palm group such as *aren* trees, coconut, *lontar, siwalan, sawit,* and *nipah*. Similarly in Southeast Sulawesi which has a land area of *aren* tree as a source of raw materials of brown sugar, as much as 3,477 ha [1]. Kolaka District was third for the *aren* land area (5.75%) after Bombana District (76.79%) and Muna District (8.66%). *Aren* land area in Muna District is not utilized as *aren* sugar production centers. District of Bombana and Kolaka also did not maximize the potential of its *aren* land area. It was interesting that the productivity of *aren* sugar in Kolaka District only 0.33 tons/ha/year. It was lower than the North

Kolaka District which its *aren* land area was lower. The central areas of *aren* sugar in Kolaka District was in Wolo Sub District, especially in the Village of Tolowe Ponre Waru and Langgomali. However, there has been decreasing the number of farmer in *aren* sugar processing business in both villages. It was very interesting to study. On the other hand, looks great opportunities for the development of *aren* sugar processing industry but it turned out there was a decrease in motivation of farmers to produce *aren* sugar. Decrease in the number of *aren* sugar processors indicates a decline in the motivation of farmers to process *aren* sap into *aren* sugar. Continuous decline in motivation can eliminate the great potential of *aren* sugar agroindustry as a source of food material and sweetener (sugar). Intrinsic and extrinsic motivations are often closely related to one's socioeconomic factors [2], as well as the palm sugar processors.

Based on the above description, the problem formulation in this study are as follows: (1) How socio-economic characteristics of farmer in *aren* sugar processing Business? (2) How is the motivation of farmer in *aren* sugar processing Business?

## 2. Material and Method

The population of this research was the farmer who was also *aren* trees owner who run *aren* sugar processing business as many as 149 people in the village of Talowe Ponre Waru and Langgomali, Wolo Sub District, Kolaka District, Southeast Sulawesi, Indonesia. Simple random sampling method was used to determine the sample. The number of samples was determined by *Slovin* formula ($\alpha$= 10%) in order to obtain the total sample of 60 people.

The analysis used in this research was quantitative descriptive analysis. Statistical techniques such as the value of maximum, minimum, and average, percentage, and the value of the interval were used to determine socio-economic characteristic (family dependents, business experience, and ownership *aren*) and the motivation criteria (high, medium, low). Socio-economic characteristic of age and education followed generally accepted category in Indonesia. There were three category of age, namely productive age yet (< 15 year), productive age (15-54 year), and not productive age (>54 year). Socio-economic characteristic of education contain six category, such as: no school, not completed primary school, elementary school, junior high school, high school, and bachelor degree. The indicator of motivation in this study, among others; motive of imitation, economic, security, affiliations, awards, and self-actualization. Formula of class interval as follows:

$$I = \frac{J}{K} \quad (1)$$

where:
I = class interval
J = Distance distribution (the highest score minus the lowest score)
K = class amount

## 3. Result and Discussion

### *3.1. Sosio-economic characteristic of farmers in aren sugar processing business*

Socio-economic characteristics of farmer in *aren* sugar processing business include age, education, family dependents, business experience, and the number of *aren* trees owned. Social-economic characteristics of respondents could be seen in Table 1.

Table 1 showed that *aren* sugar processing business was dominated by farmer in productive age (> 70%). It could be explained that *aren* sugar processing business practices required a lot of energy to do the preparation tapping, tapping, and sap cooking (boiling) into *aren* sugar, [5]. Tapping should be done twice a day. Moreover, the process of cooking (boiling) sap need very long time (4-8 hours) so it requires a lot of firewood. Activities to prepare firewood also requires a lot of energy. Respondents did all the activity by their selves. Respondents who were in productive age criteria will be willing to keep running the *aren* sugar processing business, because only requires their energy while the raw materials and auxiliary materials were still available in nature around their residence, abundantly. Good education is needed for existence and sustainability a business.

**Table 1.** Social-economic Characteristics of Respondents

| | Socio-economic Characteristics | | | | | | | | | | | | | | | | | |
|---|---|---|---|---|---|---|---|---|---|---|---|---|---|---|---|---|---|---|
| | age | | | Education | | | | | | Dependent | | | Experience | | | *Aren* tree tapped | | |
| | < 15 | 15-54 | >15 | NS | NCPS | ES | JHS | HS | BD | 0-3 | 4-7 | >7 | 0-15 | 15-30 | >30 | 0-7 | 8-15 | >15 |
| number | 0 | 43 | 17 | 3 | 6 | 21 | 19 | 11 | 0 | 37 | 22 | 1 | 28 | 22 | 10 | 39 | 19 | 2 |
| % | 0.00 | 71.70 | 28.30 | 5.0 | 10.0 | 35.0 | 31.7 | 18.3 | 0.0 | 61.7 | 36.7 | 1.7 | 46.7 | 36.7 | 16.7 | 65.0 | 31.7 | 3.3 |
| Highest | | 73.00 | | | | High school | | | | | 8.0 | | | 43 | | | 20.0 | |
| Lowest | | 18.00 | | | | No shool | | | | | 0.0 | | | 1 | | | 2.0 | |
| Average | | 46.12 | | | | | | | | | 3.0 | | | 18 | | | 7.0 | |

Note: NS= not school, NCPS= not complete primary school, ES= elementary school, JHS= junior high school, HS= high school, BD= bachelor degree

Base on Table 1 showed that 66.7% of respondents were on basic education category (elementary and junior high school). It could be explained that the *aren* sugar processing business could be done by anyone despite not having a high education. The low education level of respondents caused a lack of knowledge and the application of management concepts in the practice of *aren* sugar processing business. Next, their business scale was hardly ever increase since the beginning do.

The number of family dependent can be a motivation to run a business. A number of family dependents of respondents were varied There was the respondents who did not have dependents up to respondents who have 8 people of dependent. Table 1 also showed that the number of family dependents in small category (0-3 people) was dominated. The number of family dependents reflected their economic burden (clothing, food, house, and education) to be borne by respondent.

Almost respondents were in the category of small business experience (0-15 years). The average experience of respondents in *aren* sugar business were in middle category (age productive). It could be explain that *aren* sugar processing business was not a heritage business. The respondents obtained skills to process *aren* sugar from migrants from South Sulawesi.

Number of *aren* tree tapped showed economic side of respondent. The number of *aren* trees were tapped by almost respondents were in the small category (0-7 trees). An *aren* tree have abundant sap, ranging between 10-30 liters per day (twice tapping, morning and afternoon). Therefore the respondents considered it sufficient just to rely on wild *aren* tree with the average number of *aren* trees tapped only 7 tree per day.

### *3.2. Farmer's motivation*

Farming was a culture of Kolaka District community, especially in the plantation sector. However, they did not cultivate of *aren* plants. It was because nature has placed *aren* tree in abundance. *Aren* tree grows wild around the plantation community. *Aren* trees were a source of raw materials in the processing of *aren* sugar. The abundance of availability of *aren* trees in production central areas did not make 100% of the people did *aren* sugar processing business. It was varied motivation of people to join in *aren* sugar processing business. The indicator of motivation in this study, among others; motive of imitation, economic, security, affiliations, awards, and self-actualization. The motivation level of respondent and the social-economic characteristic for each motivation level could be seen in Table 2.

Table 2 showed that almost respondent (68.33%) have motivation in low category, and very few (3.33%) respondent with high motivation. It mean that the actual *aren* sugar processing business was not an option the main livelihood for the villagers in Kolaka District. They prefer plantations (such as, clove, cocoa, pepper), and produce *aren* sugar their leisure time.

There was the same average number of family dependents for the respondent in low and high motivation level. The average number of family dependents in the medium level of motivation have one person different with the other level. It mean that the dependent variations of respondent was relatively homogeneous, so it just have little contribution for motivation varied of respondents. Respondent with low motivation levels have the highest average for the characteristic of business experience and age. *Aren* sugar processing business requires high energy, while increasing age could reduce respondent's energy in conducting a series of activities to process *aren* sugar. It was as a deduction motivation of farmer in practice the *aren* sugar processing business. The low motivation of respondent was also shown with the least amount of *aren* trees which were tapped every day than respondents who have medium and high motivation. It mean respondent with low motivation levels did not devote more time and energy than respondent in medium and high level of motivation. Table 8 also showed that the average education of respondent in high motivation level was the highest compare than motivation level. Although, they only have basic education of 9 years. The low education of respondents caused the limited employment opportunities so that they rely on their own livelihoods in the field of plantation and processing of *aren* sugar.

**Table 2.** Characteristics of Respondents base on the Level of Motivation

| Level of Motivation | Average Social-economic Characteristics | | | | |
|---|---|---|---|---|---|
| | Age (year) | dependents (people) | education (year) | Business experience (year) | *Aren* tree taped (year) |
| Low (68.33%) | 47.63 | 3.00 | 7.60 | 19.10 | 6.50 |
| Medium (28%) | 42.71 | 2.00 | 7.53 | 15.59 | 8.41 |
| High (3.33%) | 44.00 | 3.00 | 9.00 | 17.50 | 8.50 |

The value of total motivation of respondents as in Table 3 was determined by six motive types underlying respondent did *aren* sugar processing business, including the motive of imitation, economic, security, affiliations, awards, and self-actualization. The six motive types of respondent clearly were be showed in the Table 3.

**Table 3.** Six Motive Types of Respondent

| | Criteria | Low | | Medium | | High | |
|---|---|---|---|---|---|---|---|
| | | Number (people) | Percent (%) | Number (people) | Percent (%) | Number (people) | Percent (%) |
| **Types of Motive** | Motive of Imitation | 25 | 41.67 | 32 | 53.33 | 3 | 5.00 |
| | Motive of Economic | 15 | 25.00 | 40 | 66.67 | 5 | 8.33 |
| | Motive of Security | 33 | 55.00 | 26 | 43.33 | 1 | 1.67 |
| | Motive of Affiliation | 44 | 73.33 | 16 | 26.67 | 0 | 0.00 |
| | Motive of Award | 40 | 66.67 | 17 | 28.33 | 3 | 5.00 |
| | Motive of Self-actualization | 48 | 80.00 | 11 | 18.33 | 1 | 1.67 |

The results of this study answers the phenomenon of decreasing the number of *aren* sugar processing business at the sites. Motivation of respondent that was varied in the category of high, medium and low could be caused by a variety of socio-economic characteristics.

*3.2.1 Motive of imitation.* Respondents produced *aren* sugar tend to be motivated by a desire to imitate people in the vicinity who had previously pursue the processing of *aren* sugar. Table 3 showed that the imitation motive of respondents was dominant on the medium criteria. Respondents more imitating *aren* sugar processing business conducted community around their homes (67.00%) than to imitate their p*aren*ts (13.33%). It could be explained that many p*aren*ts of respondents who were not background as *aren* sugar processor. *Aren* sugar processing was a new experience for the respondent. Firstly, they known it after seeing the community comes first in the Village of Talowe Ponrewaru and Langgomali. Almost respondents were immigrants from South Sulawesi. Respondents decided to run *aren* sugar processing business despite not having a heritage of expertise. Their decision were also due to the abundant availability of *aren* trees as a source of raw materials in their surroundings.

*3.2.2 Motive of economic.* The desire of respondents obtained economic benefits to meet the physical needs (food, drink, and home) was an encouragement for them in *aren* sugar processing business. Physical needs was the most fundamental thing in Maslow's hierarchy of needs. Table 3 showed that the almost motivation of respondents in *aren* sugar processing business have economic motive in the medium category. Only 8.33% of respondents who reasoned that *aren* sugar processing business provides high gain so that it became the main livelihood. It could be explained that the economic motive was more based on the grounds that the *aren* sugar processing business just requires small capital. It was because the main raw material source (*aren* sap) obtained free of charge from the natural resources around the respondents. The main auxiliary material for *aren* sugar processing such as firewood was for free (no fee), as well as labor because they used family labor (wife) only [4]. The benefit of *aren* sugar processing business could not be said to be high. However, regardless of *aren* sugar produced by respondents always sold out (95% of respondents). It supported also by the low risk category of *aren* sugar [6]. According to [8] palm sugar domestic industry was financially feasible. Therefore, *aren* sugar processing business was very helpful respondents to meet their daily basic needs. This was become a major economic motive to pursue the respondents in *aren* sugar processing business.

*3.2.3 Motive of security.* Not only the physical needs that was required by respondents but also security needs. It mean safety education for their children as well as savings for their unexpected needs. Security motive have become one of their reasons to pursue the *aren* sugar processing business. Table 3 showed that the almost security motive of respondents were in low category (55.00%). It could be explained that only 37% of respondents were motivated to produce *aren* sugar for reasons of profit that could be used as educate funds for their children. Thus, more respondents who did not feel safe enough with profit from *aren* sugar processing business to education needs of their children or for their unexpected needs. Respondents were often owed to an *aren* sugar broker. As a result, *aren* sugar produced was a form of debt payments so that they no longer accept cash. It caused the respondents could not feel the profit from their *aren* sugar processing business.

*3.2.4 Motive of affiliation.* Fulfillment of social needs (affiliates) were further needs that must be met in addition to physical needs and safety. The desire of respondents to interact and cooperate with other farmers was a major driving force to pursue the *aren* sugar of processing business. Table 3 showed that the affiliation motive of respondent mostly low category (73.33%) even none of the respondents were categorized as high (0.00%). However, there was 3.00% of respondents were motivated to practice *aren* sugar processing business that could be interconnected with other farmers, especially when sap tapping activity so as to tighten the brotherhood. There were also 7% of respondents who were motivated to practice *aren* sugar processing business in order to meet and cooperate with other farmers. In this condition they did *aren* sugar processing business together. This was usually done by the respondents with plantation land (cocoa, clove, pepper) adjacent.

*3.2.5 Motive of award.* A power to obtain the best results so as to create a sense of pride or valued by others was essential for the existence of the respondents pursued *aren* sugar processing business. Table 3 showed that the motive of awards of respondent mostly were in low category (66.67%). It mean that *aren* sugar processing business was not something to be proud of but it did not considered low. In other words that *aren* sugar processing business was not the main job but not also a marginal job. *Aren* sugar production was done every day by respondents, but just sidelines of their gardening activities.

*3.2.6 Motive of self- actualization.* Meeting of the highest needs base on popularity Need Theory is self-actualization needs. It could also motivate respondents in *aren* sugar processing business. Table 3 showed that mostly the self-actualization motive of respondents were at low category (80.00%). It mean *aren* sugar processing business could not be used as self-actualize place for the respondents. Only about 1.67% respondent who have high motive of self-actualization because of his awareness that *aren* sugar was a healthy sweetener and white sugar substitute. It was suitable with the result of [4] that palm plants that are free of artificial fertilizers and pesticides, the excellence nutrient content of *aren* sap and brown sugar and the uniqueness aroma of brown sugar.

*Aren* sugar processing business actually has great potential developed as a prestigious or competitive business as a container self-actualization for the respondents. In connection with that some things that can be done to improve business competitiveness is important to pay attention to customer behavior of *aren* sugar and or palm sugar. It has been investigated that the behavior of brown sugar customers was influenced by age, and income of customer, texture of brown sugar, and dummy for purchasing reason [7], [3].

## 4. Conclusion and Recommendation

Based on the results and discussion before, we could conclude some of the following (1) average state of socio-economic characteristics of *aren* sugar farmer that were in the category of productive age, which was 46.12 years old, has fulfill basic education category (9 years education), low category of dependents as many as three people, the average experience of *aren* sugar processing business during 18 years, and the average *aren* trees were tapped as much as seven trees, every day, (2) Farmer's motivation in *aren* sugar processing business in low category, with the indicator is the motive of imitation, economic, security, affiliations, awards, and self-actualization. Economic motive give main contribution for the total value of farmer's motivation.

Some things that could be recommended of the results of this study, namely (1) Farmers need to increase motivation in *aren* sugar processing business because of its healthy side, abundant available of sap from an *aren* tree, and its daily contribution, (2) Government need to increase attention to the increase in business scale, performance, and quality of *aren* sugar, also human resources quality in order to increase motivation for doing *aren* sugar processing business. It can be done by partnering with the universities and nongovernmental organizations.

## References

[1] BPS Sultra 2016 *Sulawesi Tenggara dalam Angka* (Southeast Sulawesi in number) (Kendari : Badan Pusat Statistik. Kendari)

[2] Luanmase, C.M., S. Nurtini, F.T, dan Haryadi 2011 Analisis Motivasi Beternak Sapi Potong bagi Peternak Lokal dan transmigran serta Pengaruhnya terhadap pendapatan di Kairatu, Kabupaten Seram Bagian Barat (Cow farm motivation analysis for local farmers and transmigrants and its effect on income in Kairatu, West District of Seram). *Buletin Peternakan.* **35(2)** 113-123

[3] Abdullah WG, Rianse U, Iswandi RM, Taridala SAA, Rianse IS, Zulfikar ZL, Widayati W, Baka LR, Abdi AL, Cahyono E 2014 Mathematical Model of Factors Influencing the Consumption of Brown Sugar in Southeast Sulawesi. 2014 *Proceedings of the 2th International Conference on Mathematical, Computational and Statistical Sciences (MCSS '14)* 421-428

[4] Abdullah WG, Rianse U, Iswandi RM, Taridala SAA, Rianse IS, Zulfikar LZ, Baka LR, La Abdi A, Cahyono E, Widayati W and Baka WK 2015 Potency of natural sweetener: Brown sugar. *WSEAS Transaction on on business and economics* **12** 99-110

[5] Abdullah WG, Rianse U, Iswandi RM, Taridala SAA, Rianse IS, Zulfikar ZL, Abdi AL, Widayati W, Baka WK, Cahyono E 2015 Economics of *Aren* Brown Sugar Production in Indonesia: Supply Analysis at Farm Level. *Proceedings of the 4th International Conference on Agricultural Science, Biotechnology, Food and Animal (ABIFA '15)* 332-339

[6] Abdullah WG, Rianse U, Iswandi RM, Taridala SAA, Yunus L,Fausayana I, Dirgantoro MA, Rianse IS 2016 Mapping and Preference of Marketing Risk of Eco Sweetener '*Aren* Sugar'. *International Journal of Environmental Science* **1** 133-139

[7] Baka WK, Rianse U, Sidu D, Widayati W, Cahyono E, Rianse IS, Zulfikar L, Abdi AL and Baka LR 2016 Customer Behaviour Model of Brown Sugar Commodity *Int. J. Business Innovation and Research* **11(3)** 444-60

[8] Rianse IS, Abdullah WG, Midi LO, Hartono S, Suryantini A, Jamhari, Widayati W. Financial, Economic and Environmental Feasibility Analysis of Palm Sugar Domestic Industry in Kolaka, Indonesia. *International Journal of Economics and Management Systems* **1** 104-109

# Positive and Negative Impacts of Oil Palm Expansion in Indonesia and the Prospect to Achieve Sustainable Palm Oil

**M A Shahputra[1] and Z Zen[2]**

[1]Department of Economics - State University of Medan , Indonesia
[2]School of Economics – STIE Harapan Medan, Indonesia
**E-mail:** afifchani@live.com, zahari.zen@gmail.com

**Abstract**. The aim of the study is to deepen understanding the role of palm oil on Indonesian economy, poverty elevation and to investigate the positive and negative impacts of oil palm expansion, due to the burden of GHG emissions; and prospect to be more sustainable palm oil industry. The statistics show that average rural poverty tends to be lower and Gross Regional Product tends to be higher in provinces which have greater levels of oil palm cultivation. Indonesian oil palm will grow from 10.6 in 2013 to 13.7 million ha by 2020. This will release 135.59 million tons of CO2 if nothing is done to mitigate BAU emissions. Unless there are sustained efforts to redirect development and expansion of oil palm, plantation growth will continue to encroach on intact forest and peat land.. In fact Indonesia has large areas of degraded land, an estimated total 19,144,000 ha is available for planting oil palm and other crops. A large-scale expansion program driven by estate companies needs to be accompanied by effective smallholder development program in order to achieve the best outcome for local farmers and avoid the conflicts.

## 1. Introduction

Several factors have led to rapid industry expansion: increasing returns associated with world demand for oil palm the suitability of Indonesia's land and climate for oil palm cultivation; and, in terms of land use and productivity, oil palm's higher productivity than other common oil seeds [19], allowing the price of palm oil to outcompete its competitors. Statistical data indicates that the area of oil palm and the production increase dramatically since 2003 until 2013.

In the last 8 years, Indonesian palm oil exports experienced a growth rate of 11.35% per year, from 11.88 million tons in 2007 to 20.58 million tons in 2013[3]. Likewise, the value of palm oil exports increased with a growth rate of 23.93% per annum. State revenue has also expanded, with export taxes now grossing Rp28.9 trillion in 2013 [3]. Given its contribution to Indonesia's balance of payments, Indonesia's international economy is currently very reliant on the export of palm oil [3].

In terms of regional economic development, Gross Regional Product (GRP tends to be higher in provinces which have greater levels of oil palm cultivation such as North Sumatera, Riau and Central Kalimantan. While issues of increasing inequality and distributional justice associated with oil palm are at times acute, nevertheless the statistics show that average rural poverty tends to be lower in the rural areas where oil palm is developed.

With global demand for palm oil expected to grow further, palm oil offers promising economic prospects for Indonesia, within the constraints of social responsibility and environmental sustainability. However, the pace of oil palm growth has led to an international debate regarding deforestation and climate change. There are concerns about serious unintended social, economic, and environmental consequences, as deforestation and land conversion for oil palm development now contributes 15% to 25% of whose Indonesia's carbon emissions [18].

This is the challenge of sustainable development: how we could find a settlement that accommodates the various stakeholders, Indonesia's environment, biodiversity, and the nation's next generations of people. Given this challenge, this research focuses on the question of how to accommodate the interests of various stakeholders without threatening the environment during future oil palm development.

## 2. Regional Development and Poverty Alleviation

In the centers of palm oil production, the level of GRP is higher than in the other provinces, were, contributing to poverty reduction; for example, rural poverty in Riau decreased from 21% to 10% in 5 years (2009-2013) [21]. It is evident that in gross terms, oil palm has significantly reduced the inequality of development between Java and Indonesia's outer islands.

Oil palm is the prevalent sector, average rural poverty, poverty depth and the poverty severity index are much lower than where the oil palm is less developed [8]. World Growth Report [25] argues that oil palm expansion has been a significant source of poverty alleviation for the rural poor.

## 3. The Government role in Economic Improvement of the local people

More than 3.9 million family farmers Palm oil production provides a reliable form of income for a large number of Indonesia's rural poor, about 3.9 million family farmers, employment generated from palm oil production in Indonesia could potentially reach over 6 million families by 2020 and take them out of poverty [25].

Productivity of independent smallholdings could also be improved, through use of high-yielding palm oil varieties and best agronomic practice. High-yielding varieties, under ideal climate conditions and good management, produce up to 5 tons per ha per year [25]; some suggestions are as high as 6-7 tons [13]. With per-hectare average annual production currently at only 2.31 tons for smallholders and 3.19 tons for government-owned and private plantations, there is potential for all holdings to greatly increase their output on existing land, without requiring additional land conversion [25].

Although Indonesia has an estimated 14.19 million ha of additional land which is suitable for oil palm development [8], given the social, environmental and biodiversity risks and associated national and international pressure (such as ISPO; RSPO), it will be important to develop sustainable palm oil practices, which focus on improving productivity on existing plantations while safeguarding Indonesia's remaining forests.

## 4. Debate of Negative impact of oil palm on Forests and Climate Change

Conversion of forests and peat lands for oil palm cultivation is considered by many to be the largest source of GHG emissions, due to changes in land cover; and oil palm plantations are estimated by Agus [1] to be responsible for a substantial and increasing of total carbon emissions in Indonesia, Malaysia and Papua New Guinea (up from 13% during 2000-2005 to 18% in 2006-2010). Since over 90% of conversion of land for palm oil has occurred in Sumatra and Kalimantan, this places Indonesia among the 10 biggest carbon emitters [5], with one 2012 report placing Indonesia third after China and the USA.

Recent data indicate that while some provinces with large expansions in oil palm plantations have had a concurrent increase in deforestation [6], others have not. For the latter, this indicates that plantation land is being converted from land previously used for food crops, including rice, which may have become less attractive due to poor irrigation infrastructure, falling term of trade for rice farmers, and concerns about climate impacts. Similarly, Fairhurst and McLaughlin [11] reports that many of

Kalimantan's oil palm plantations has been established on lands that were previously degraded forest. [22] emphasize that oil palm plantations store more carbon than alternative agricultural land uses. However, to increase oil palm on such land, we need better agronomic technics and required better field management. Due to unsustainable slash and burning farming, degraded land increased by 8% in last 5 years (2009 -2013) while planting oil palm increased by 38% [7].

Indonesia's land use activities are responsible for most of the nation's carbon emissions, and are estimated to continue to contribute over 50% of Indonesia's total carbon through 2030 – indicating that 2.5% of the world's GHG emission could be emitted from Land Use, Land-Use Change and Forestry (LULUCF) sector and peat land activities in Indonesia [14]. With regard to peat lands, although figures differ somewhat among researchers, some researchers estimate that around 1.7 million ha of peat land have already been planted for oil palm, representing about 20% of Indonesia's total oil palm area [1]. During peat land conversion, carbon emissions are caused not only by land clearing (with emissions estimates ranging from 26-100 Mg $CO_2$ p.a. for Indonesia; Rieley, J.O. and Jauhianen [20] but from illegal fire (446 M tons $CO_2$ p.a. for the period 2000-06; JICA 2014). Fires continue to destroy hundreds of thousands of ha of peat and forest in Indonesia per year. Importantly, it is estimated that if Indonesia is able to prevent peat fires, this will reduce the country's emissions by a massive 23-45% [16].

According to palm oil producers, GAPKI [12] cites findings that carbon emissions from oil palm on peat land are lower than emissions from intact peat forest, secondary peat forests and rice fields on peat land. Oil palm on peat land may also absorb more CO2 per year than tropical forests 64.5 versus 42.2 Ton per hectare per year, according to Henson 1999; although figures vary widely, with IOPRI [15] reporting that oil palm fixes only 18.7 tons $CO_2$ p.a. Other cited benefits of oil palm on peat, in comparison to other crops, include resilience against erosion, and lower fertilizer and pesticide requirements than other crops.

Given the environmental impacts of destroying Indonesia's remaining intact forest and peat lands, a key development strategy which Indonesia could adopt to support rural communities is to implement sustainable land use planning, involving expanding oil palm into degraded land mostly covered by *alang-alang* grass (*Imperata cylindrica*). Most provinces in Sumatera and Kalimantan contain large areas of degraded land; an estimated total 19,144, 000 ha is available – enough to expand not only oil palm, but also other agriculture crops [7,8].

## 5. Approach for Future of Oil Palm Expansion and Policy related to Climate Mitigation Actions

Recognizing both issues, ICCSR (2007) has developed a range of detailed policy guidance and mainstreaming tools for the sectorial and cross-sectorial development programs, in order to ensure climate change is considered in all aspects of national, regional and local development planning such as Climate change planning. Indonesia has promulgated Law No 17 of 2007 on long term national development plans (RPJPN), for the years 2005-2025. This legislation should serve as the basis for medium term 5 year development plans (RPJMN and RPJMD) for every province and district in Indonesia.

To follow up law No 17 of 2007, the Indonesian government promoted President Regulation no 61 of 2011, on the National Action Plan (RAN-GRK), which guides the implementation of reducing GHG at the national, province and municipality levels. President Regulation No.61 of 2011 states that Governors and Mayors throughout Indonesia should develop a local action plan (RAD-GRK) by October 2012, and this action plan should cover five areas: Forestry and Peat; Agriculture; Energy; Transportation and Waste Management. The RAD GRK is used to support local development for policy direction, strategy, programs and activities. Currently, 32 out of 33 provinces in Indonesia have set a RAD-GRK, has and these have been mainstreamed into RPJMD and into the local government budgets. One of the major targets to which the government is committed, through these RAN-GRK and RAD-GRK plans, is that Indonesia must reduce its carbon emissions by 26% of BAU by 2020 through its own efforts, and by 41% with international support. Emission reduction from 'land use, land use change and forestry' (LULUCF) is also featured in the national-level action plan (RAN-GRK) for greenhouse gas emission (GHG) reduction. The national RAN-GRK calls for reductions of GRK emissions of about 23% of Indonesia's total emission in 2020 [10].

On one side of the oil palm debate, industry expansion has real potential to further damage the environment and biodiversity, particularly with respect to contributing to climate change (which itself will have serious economic impacts); along with risks of conflict with local communities in association with environmental, social or economic concerns. On the other side, this research has discussed the important role of palm oil and benefits from a both a macro and micro economic perspective.

## 6. Scenario of oil palm expansion without harming the environment

To accommodate these complex issues, we developed assumptions to calculate acceptable area and GHG reduction to take into account in the development of new approaches as follow:

i. Oil palm must only be cultivated by converting secondary forests and degraded lands (including existing agricultural land), which have no ongoing conflicts with local communities.

ii. Oil Palm expansion will not occur on remaining peat lands and high carbon stock forest, in accordance with the following legislation and other requirements:

- Law No. 18 of 2013 on the prevention and eradication of forest destruction.
- Presidential Instruction No: 10/2011, amended to No. 6/2013, concerning the extension of the moratorium.
- Government Regulation No. 71/2014 on the Protection and Management of Peat lands
- Minister of Agriculture Regulation No.19 / Permentan / OT.140 / 3/2011 of March 29, 2011, regarding Guidelines for Sustainable Palm Oil in Indonesia.
- Market demand for plantation companies to hold RSPO certificates.
- Commit to no deforestation, respecting the rights of indigenous people and local community.
- Minister of Agriculture Regulation No.19 / Permentan / OT.140 / 3/2011 of March 29, 2011, regarding Guidelines for Sustainable Palm Oil in Indonesia.

iii. In accordance with government regulations No. 4 of 2001, regarding the control of pollution relating to forest and land fires, no further land clearing in Indonesia is conducted by burning.

iv. For our calculations, first we estimate trend projections of new palm cultivation through to 2020 (Figure1). We have assumed that the palm oil industries do nothing to mitigate or minimize GHG emission; i.e. Business as Usual (BAU). With that BAU, we calculate the amount of emissions released into the atmosphere.

v. We use an emission factor of 42.31 ton $CO_2$ per ha per year, which is based on Indonesian oil palm research conducted by the Indonesian Oil Palm Research Institute and representing estimates of emissions released from conversion of secondary forest and degraded lands [14].

vi. Therefore, in calculating the threshold, we do not take into account the emissions from peat and forest fires.

## 7. Interpretation of acceptable area subject to GHG mitigation actions

Based on estimated trends in oil palm areas using simple regression trend, we calculated that between the years 2013-2020 there would be additional new oil palm expansion areas in Indonesia equating to 13.67 million ha (Figure 1), which we estimated would release $CO_2$ emissions of 135.59 million tons, representing BAU emissions. Emissions from private estates and smallholdings is higher than from government-owned plantations, meaning that special attention should be given to improving mitigation efforts on private estates and smallholdings

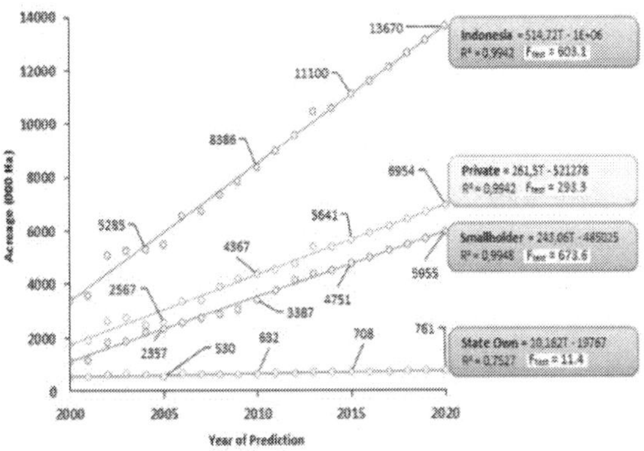


Figure 1. Indonesian oil palm area, and increase by holding type, 2014-2020

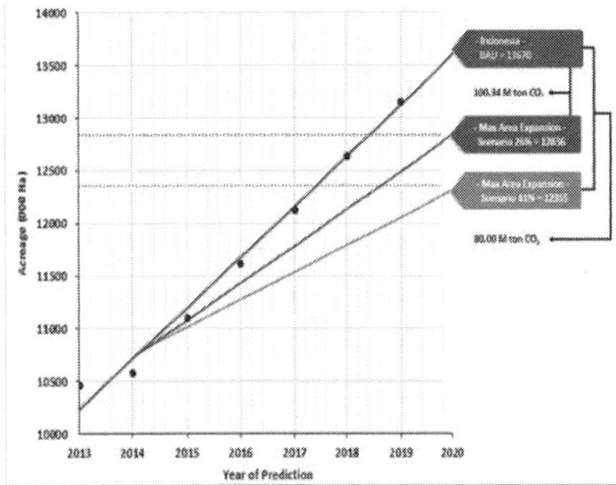


Figure 2. Mitigation actions scenario 26% and 41% of CO2 emission reduction

Given Indonesia's commitment to 26% reduction in emissions from present-day levels by 2020, and 41% reduction if international support is forthcoming, $CO_2$ emissions can be reduced by up to 35.25 Million ton and 55.6 Million ton $CO_2$ respectively. Therefore, total oil palm areas can only grow to a maximum of 12.84 Million ha (to meet the 26% emission reduction target, or 12.36 Million ha if meeting the 41% target (Figure 2).

The targets of emission reductions included in the RAN-GRK and RAD-GRK programs and Reduced Emissions from Deforestation and forest Degradation Programs (REDD+) include all activities associated with oil palm cultivation. Key actions of the programs to reduce emissions include reducing forest fire incidents, including through 'hot spot reduction programs' Other actions and activities of the programs include those associated with effective monitoring and law enforcement, community development, awareness-raising of communities, industries and local governments and intensification of smallholder plantations, not only to prevent forest encroachment but also to increase the productivity of oil palm cultivation.

To meet sustainable palm oil that expansion either by state own plantations, private plantations or smallholders could use three aspects of HCS+ implementation principles, include (1) No deforestation of High Conservation Value (HCS), fully implemented RSPO criterion 7.8, zero burning, national government regulation on ISPO and emission reduction program of RAN and RAD. (2) Protect peat areas means that no new development on peat land, regardless of depth, commit to applying best management practice in existing plantation on peat lands.(3) increase focus on driving socio-economic benefits for local people and transparently work with all parties to resolve verifiable complaints and conflicts [26].

## 8. Conclusion and Recommendation

Taking into account the challenges of economic issues, regional development, employment and poverty reduction objectives, as well as the government's target of reducing carbon emissions and settling land disputes, oil palm development could be supported and improved in the following ways:

- The goal of sustainable palm oil development cannot be separated from the goals of poverty reduction, conflict resolution, and reducing the impacts of global warming. Therefore A large-scale oil palm expansion program driven by estate companies needs to be accompanied by a well thought and effective smallholder development program. Companies need to bear in mind that they are developing profitable oil palm in areas where local communities may experience issues of poverty, poor levels of health, poor nutrition or food insecurity.

- Large oil palm companies can grow oil palm in degraded grass land. If they are able to use such lands, companies could double oil palm areas without having to open new forest land (intact forest) and peat lands. If land issues could be resolved and local landowners included, oil palm acreage could be increased up to two-fold, without having to convert additional new forest land and peat lands. With current agronomic technology and financing capabilities, large estates could remain profitable if this objective is pursued. In terms of the environment, planting oil palm on abandoned land can improve soil conditions and land coverage, providing GHG mitigation and increasing the accumulation of carbon stock.
- With international assistance for the 41% emission reduction target, the expansion of oil palm could extend to 12.3 million ha, with corresponding GHG emission reductions of 55.6 Million tons of CO2. Large estates could integrate their expansion program into the government's action plan of the RAN-GRK and RAD-GRK program. The New paradigm requires having empathy towards cultural values and the economy of surrounding communities.

**References**

[1] Agus, F., Gunarso, P., Sahardjo, B.H., Harris, N., van Noorrdwijk, M., Killeen, T.J. 2013. Historical CO2 emission from land use and land use change from the oil palm industry in Indonesia, Malaysia and Papua New Guinea, Report from the technical panels of the 2 nd Green House Gas working Group of the RSPO.

[2] Barani, Mangga A. 2014. Kelapa sawit berkelanjutan di Indonesia, presentation material "Associate Committee Sustainable Palm Oil Program SLP 20 March 2014. Forum Pengembangan Perkebunan Strategis Berkelnjutan (FP2SB).

[3] BPS Indonesia 2014. Indonesian Central Bureau of Statistics.

[4] Bank Indonesia 2014. The statistic of the Central Bank of the Republic of Indonesia.

[5] Carlson, K. M., Curran L. M., Asner, G. P., Pittman, A. M., Trigg, S. N., & Adeney J. M. 2013. Carbon Emissions from Forest Conversion by Kalimantan Oil Palm Plantations. Nature Climate Change (3), Macmillan Publisher Limited.

[6] Carlson K.M., Curran, L.M. Ratnasari, D., Pittman, A.M. Soares-Filho, B.S., Asner, G.P.Trigg, S.N., Gaveau, D.A., Lawrence D., and Rodrigues, H.O. 2012. Committed carbon emission, deforestation, and community land conversion from oil palm plantation expansion in West Kalimantan, Indonesia.

[7] KLHK, 2014, E-Book Statistik Kementerian Lingkungan Hidup dan Kehutanan.

[8] DITJENBUN 2014, Statistik Perkebunan Indonesia Komoditas Kelapa Sawit 2014.

[9] DNPI and JICA 2014. Kajian emisi gas rumah kaca (GRK) dari kebakaran hutan dan lahan: studi kasus di Indonesia, Laporan Kajian Awal.

[10] DNPI 2014. Policy Memo: Peat land Definition from Uncertainty to Certainty. Report on Peat land in Indonesia: Context and Issues.

[11] Fairhurst, T and McLaughlin, D. 2012, Sustainable Oil Palm Development on Degraded Land in Kalimantan. WWF publication.

[12] GAPKI (2013) Indonesia dan perkebunan kelapa sawit dalam isu lingkungan global, Gabungan Pengusaha Kelapa Sawit Indonesia (GAPKI)

[13] Gunadi, 2008 Perspective on Indonesian palm oil production. Paper presented on the International Food & Agricultural Trade Policy Council's Spring 2008 Meeting, 12 May 2008, Bogor, Indonesia.

[14] IOPRI, 2008. Carbon emission of oil palm plantation on peat soil. Report of Indonesian Oil Palm Research Institute.

[15] IOPRI 2009. CO2 emission on oil palm plantation: Field observation. Report of Indonesian Oil Palm Research Institute.

[16] JICA and DNPI 2014. Kajian emisi gas rumah kaca (GRK) dari kebakaran hutan dan lahan: Study kasus di Indonesia. Laporan Kajian Awal tentang Dampak kabut asap termasuk emisi Gas Rumah Kaca.

[17] Ministry of Forestry 2014. Ministry of Forestry Statistics 2013.
[18] Obidzinski, K. Andriani, R., Komarudin, H., and Andrianto, A. 2012. Environmental and social impacts of oil palm plantations and their implications for biofuel production in Indonesia, Ecology and Society 17 (1):25. E-Journal.
[19] Oil World 2013. Palm Oil Facts & Figures, publish by Sime Darby Plantation
[20] Rieley.J.O. and Jauhiainen, J. 2011. Review of peat surface greenhouse gas emissions from oil palm plantations in Southeast Asia. E-Journal
[21] Sipayung, T. 2014. Peranan industri persawitan dalam pertumbuhan ekonomi, pembangunan pedesaan, pengurangan kemiskinan dan pelestarian lingkungan hidup provinsi Riau. Presentation material from Palm Oil Agribusiness Strategic Policy Institute (PASPI).
[22] Sayer, J. ,Ghazoul, J., Nelson, P. and Boedhihartono, A.K. 2012. Oil palm expansion transforms tropical landscapes and livelihoods. Global Food Security. E-Journal.
[23] RAD-GRK 2013. The national action plan for reducing greenhouse gas emissions, Presidential Regulation of the Republic Indonesia Number 61 year 2011.
[24] World Growth (2009). The economic benefit of palm oil to Indonesia, Report Palm Oil Green Development Campaign
[25] World Growth (2011). The economic benefit of palm oil to Indonesia, Report Palm Oil Green Development Campaign.
[26] Zen, Z., McCarthy, J., Diana Chalil. D., Sitorus, H. and Shahputra, M. A. 2015. High Carbon stock (HCS) and the socio-economics of palm oil: Towards improving the sustainability of oil palm sector in Indonesia. Consulting Study 14.

**Acknowledgments**

Supported by High Carbon Stock (HCS) 2015 (Consulting Study 14: HCS and the Socio-Economics of Palm Oil –Towards Improving the Sustainability of the Palm Oil Sector in Indonesia).

IOP Conf. Series: Earth and Environmental Science **122** (2018) 012009 doi:10.1088/1755-1315/122/1/012009

# Strategy to increase Barangan Banana production in Kabupaten Deli Serdang

**I Adhany, D Chalil and R Ginting**

University of Sumatera Utara

agb.unusu@gmail.com

**Abstract**. This study was conducted to analyze internal and external factors in increasing Barangan Banana production in Kabupaten Deli Serdang. Samples were determined by snowball sampling technique and purposive sampling method. Using SWOT analysis method, this study found that there were 6 internal strategic factors and 9 external strategic factors. Among that strategic factors, support for production facilities appears as the most important internal strategic factor, while the demand for Barangan Banana.as the most important external strategic factor. Based on the importance and existing condition of these strategic factors, using support for production facilities and realization of supporting facilities with farming experience are the strategies covering strength-opportunity (SO), organizing mentoring to meet the demand for Barangan Banana are the strategies covering weakness-opportunity (WO), making use of funding support and subsidies to widen the land, using tissue culture seeds and facilities and infrastructures are the strategies covering strength-threat (ST), increas the funding support to widen the land, the use of tissue culture seeds and facilities and infrastructures are the strategies covering weakness-threat (WT) are discussed and proposed to increase Barangan Banana productivity in Kabupaten Deli Serdang.

## 1. Introduction

Barangan Banana (*Musa paradisiaca sapientum* L.) is one of local North Sumatera typical fruit. With sweeter flesh, yellow peel, redish yellow flesh, dry and distinctive flavored flesh, Barangan Banana is suitable as fresh consumption table fruit. The demand is stable and high, indicating by the stable distribution not only in North Sumatera but also to Jakarta, Bekasi, Riau, and Batam even overseas.In 2005-2015 the Bananas' export volume tends to increase, the highest was reached in 2014 at 26,694 tons and the lowest was in 2010, which only reached 14 tons [1] China was the main exporting country with export volume of 7,847 tonnes (35.17%) in 2015. The other countries was dominated by Middle-East region countries such as Kingdom of Saudi Arabia (3.499 tonnes), Japan (2,968 tonnes), Malaysia (2,847 tonnes), UAE (2,763 tonnes) and Kuwait (2,342 tonnes) [1]

In fact, the distribution has not fulfilled the whole demand to begin with. However, the Barangan Banana's production tends to decrease by years. In Kabupaten Deli Serdang, which notably centre of Barangan Banana production in North Sumatera Utara, In 2011, the banana production was still 1,750,604 quintals, yet only recorded 298,305 quintals in 2014 [2]

In 2016, land area of banana in Kabupaten Deli Serdang has reached 1,513,74 ha, with 3 main districts namely STM Hulu (674.5 ha), STM Hilir (260 ha), and Biru-Biru (129 ha). However, this might still be developed as Kabupaten Deli Serdang has 59,537 ha dry land. Therefore, Kabupaten Deli Serdang's Department of Agriculture has designed 100 ha Barangan Banana extensification program

[3] indicates that the internal factors which influence the banana comodity development are farmer's knowledge and cultivation of banana, farmer's motivation, communication facility, high production of banana and farming management, farmer's low purchasing power to production facility. Meanwhile, the extrenal factors are banana seed aid from government, banana's market prospect, government concern in order to boost farmer's skill.

[4] points that the excess factors are farmer's financial basis, low-waged labor, row material supply, production facility availability. Lackness factors are low-quality human resources either businessmen or labor. Limited financial basis (capital). Unstable market segmentation, limited market information, and often time unfulfilled buyer's demand.

[5] exhibits that production increment strategy in order to boost participation of farmer, can be achieved by intensive regulation, through setting the basic price guarantee, thus soy farming generates appropriate advantage to farmer. Besides, it requires farmer's empowerment and capacity development through capital supply support and socialization also revamping the trading mechanism through revitalization of Bulog as prior importer. Establishing long term elements, instances infrastructure development, irrigation system, research and development, strengthening of agricultural institutions, insentive system, macro economy regulation supports.

Relating to its overviews, thus it is needed an empirical research to be utilized as a basis on strategy to increase Barangan Banana's production. This research aims to analyze internal factors (strengths and weaknesses) and external factors (opportunities and threats) in term of Barangan Banana production's increment in Kabupaten Deli Serdang, to optimize strengths and opportunities, also to overcome the weaknesses and threats upon the strategy to increase Barangan Banana's production in Kabupaten Deli Serdang.

## 2. Research Method

This study was conducted in 3 districts namely STM Hulu, STM Hilir, and Biru-Biru, which are determined purposively as these were recorded as the main production centers of Barangan Banana in Kabupaten Deli Serdang but the production decreased significantly in 2011-2014 (Table 1).

**Table 1.** Banana Production (quintal) according to districs in Kabupaten Deli Serdang in 2011-2014

| District | 2011 | 2012 | 2013 | 2014 |
|---|---|---|---|---|
| Gunung Meriah | 34,300 | 19,275 | 9.662 | 30,550 |
| STM Hulu | 240,305 | 152,584 | 120,720 | 80,795 |
| Sibolangit | 744 | 1,157 | 726 | 351 |
| Kutalimbaru | 1,639 | 540 | 2,550 | 2,300 |
| Pancur Batu | 11,030 | 2,652 | 1,267 | 1.610 |
| Biru-Biru | 141,250 | 32,194 | 2,393 | 8,373 |
| STM Hilir | 124,700 | 179,479 | 201,321 | 151,250 |
| Jumlah | 1,676,268 | 387,881 | 338,630 | 275,229 |

Source : *Dinas Pertanian Deli Serdang, 2016.*

Respondents in this research consist of 4 groups namely KUPT Department of Agricultural staff, extension agent, head of farmer group, and the Barangan Banana farmers. Farmer samples are selected by snowball sampling technique, with a total of 60 farmers or 20 farmers from each village. Department of Agricultural staff, extension agent and head of the farmer group are selected by purposive sampling by considering their understanding about the whole strategic factors including in the analysis, with a total samples of 6 persons.

Barangan Banana production increment strategy formulation is analyzed by SWOT analysis. The strategy is expected to be implemented by the Agriculture Office in Kabupaten Deli Serdang. Based on previous studies and field observation, 15 (fifteen) strategic factors in increasing Barangan Banana production in Kabupaten Deli Serdang were identified namely farming experience, farmer's capability to control pest and disease, Barangan Banana selling price, farmer's financial basis (capital), land area,

Barangan Banana demand, labor availability, seed usage with plant tissue isolation method and facilities-infrastructures. These factors were then classified into (a) external factor, that is the factor which cannot be controlled by department of agriculture Kabupaten Deli Serdang and (b) internal factor, that is the factor which can be controlled by department of agriculture Kabupaten Deli Serdang. Each of the factors were scored and weighted, based on their existing condition and importance, respectively. The score is calculated based on the set indicators and parameters. In internal factors, score 1 and 2 are weakness while 3 and 4 are strengths. In external factors, 1 and 2 are threats while 3 and 4 are opportunities. The weighting is done by paired comparison technique with value 1, 2, and 3 [6].

The difference between the total weighted score of strength and weakness in internal factors and and opportunity and threat of external factors were then used to determine the matrix position. Strategy details of the matrix position were elaborated in matrix SWOT including (a) SO (Strength-Opportunity) Strategy, (b) ST (Strength-Threat) Strategy, (c) WO (Weakness-Opportunity) Strategy, and (d) WT (Weakness-Threat) Strategy. [7].

## 3. Results and Discussion

Based on score and weight calculation of each strategic factors, IFAS and EFAS Tables details can be arranged as follows.

**Table 2.** Table of IFAS

| Internal Factors | Score | Weight | Total |
|---|---|---|---|
| Strengths | | | |
| Support for production facilities | 4 | 0.30 | 1.20 |
| Realization of supporting equipment | 4 | 0.18 | 0.72 |
| Total | | | 1.92 |
| Weakneses | | | |
| Training | 1 | 0.10 | 0.10 |
| Mentoring | 2 | 0.10 | 0.20 |
| Extension agent | 1 | 0.11 | 0.11 |
| Funding support | 1 | 0.21 | 0.21 |
| Total | | 1.00 | 0.62 |
| Difference (Strengths - Weakness) | | | 1.30 |

Table 2 shows that support for production facilities is the most important factor in increasing the production of Barangan Banana, weighting 0.3. The condition of the existing fact it self is good enough, with the score 4. Therefore, support for production facilities can be optimized more for the production development of Barangan Banana in Kabupaten Deli Serdang. Furthermore, the realization of supporting equipment is likewise good and can be used. However, there are some weakness in internal factors even with relatively small weight.

Table 3 shows that selling price of Barangan Banana and demand for Barangan Banana are the most important factor in increasing the production of Barangan Banana, weighting 0.16. Table 3 also shows that all opportunities are in the good condition, by score of 4. Farmer has been planting Barangan Banana more than 15 years. This high experience causes the farmer is able to cope the pests and disease which strike the crops. Hence, the selling price of barangan is quite high by approximately > Rp. 10.000 per bunch. Score 4 for demand, showing that high and continous demand, meaning that regardless how much the production by the farmer, is invariably able to be absorbed by the market. It becomes an important factor to increase the production. Fund used by the farmers also points an excellent condition, by the score of 4, in which the farmer itself utilize the individual owning (financial supply) without the agricultural department or government's aids. However, it does exist some threats onto the external factors, instances of facilities and infrastructures which are notably not optimal yet. Furthermore, based on the calculation of weighted score difference of internal factors and

**Table 3.** Table of EFAS

| External Factors | Score | Weight | Total |
|---|---|---|---|
| Opportunities | | | |
| Farming Experience | 4 | 0.07 | 0.28 |
| The Ability of the Farmers to control pest and diseases | 4 | 0.07 | 0.28 |
| Selling Price of Barangan Banana | 4 | 0.16 | 0.64 |
| Fund used by the Farmers | 4 | 0.11 | 0.44 |
| Demand for Barangan Banana | 4 | 0.16 | 0.64 |
| Labor Availability | 4 | 0.07 | 0.28 |
| Total | | 0.64 | 2.56 |
| Threats | | | |
| Land area | 2 | 0.08 | 0.16 |
| The Use of Tissue Culture Seeds | 1 | 0.15 | 0.15 |
| Facilities and Infrastructures | 2 | 0.12 | 0.24 |
| Total | | 0.35 | 0.55 |
| Difference (Opportunities - Threats) | | | 2.01 |

external factors on Table 2. IFAS and Table 3. EFAS, the coordinate point (1.30 ; 2.01) in quadrant I is obtained.

Quadrant I shows that the production of Barangan Banana in Kabupaten Deli Serdang has relatively big opportunity and great strength. In such condition, the production increasement strategy should be more emphasized in increasing the condition that has been good. However, in reality, up until this study is done, the business condition is not yet optimized. The main problem lies on the Fusarium attack which reduces the planting area significantly. It is not effective to eradicate Fusarium disease with pesticide, and the impact is quite in a long time. Therefore, it is better not to plant the land that already got attacked anymore.

According [8], the most effective way is by using superior seeds from tissue culture. However, there are different views from interviewees in weighting. Therefore, the average geometric value of culture tissue is only 0.15. From the position of quadrant I strategy, the activities/programs that can be used to increase the production are arranged as follows.

**Table 4.** The Alternative Strategy to Increase Barangan Banana Production in Kabupaten Deli Serdang

| Internal<br><br>External | Strengths<br>1. Support for production facilities<br>2. Realization of supporting equipment | ***Weakness***<br>1. Training<br>2. Mentoring<br>3. Extension Agent<br>4. Funding support |
|---|---|---|
| Opportunities<br>1. Farming experience<br>2. The ability of the farmers to control pest and disease<br>3. Selling price of Barangan Banana<br>4. Fund used by the farmers<br>5. Demand for Barangan Banana | SO Strategies<br>1. Using support for production facilities and realization of supporting facilities with farming experience (S1, S2, O1).<br>2. Using support for production facilities to optimize labor availability and to utilize the selling price of Barangan Banana (S1,O3, O6). | WO Strategies<br>1. Organizing mentoring to meet the demand for Barangan Banana (W2, O5).<br>2. Organizing training to optimize farmers' experience. (W1, O1).<br>3. Using of extension agent to utilize the selling price of Barangan Banana (W3, O3).<br>4. Providing funding support to |

| | | |
|---|---|---|
| 6. Labor availability | 3. Optimizing funding support and production facilities to meet the demand for Barangan Banana (S1,S2, O5).<br>4. Using subsidies by making using of production facilities to control pest and disease (S1, S2,O2). | increase fund used by the farmers (W4, O4). |
| Threats<br>1. Land's wide<br>2. The use of tissue culture seeds<br>3. Facilities and infrastructures | ST Strategies<br>1. Using of funding support and subsidies to widen the land, using tissue culture seeds and facilities and infrastructures (S1,S2,T1, T2, T3). | WT Strategies<br>1. Increas the funding support to widen the land, the use of tissue culture seeds and facilities and infrastructures (W4,T1,T2, T3). |

## 4. Conclusion

Farming for Barangan Banana in Kabupaten Deli Serdang still has potential to be developed. This can be seen from the result of strategic factors analysis which shows bigger strength compared to the weakness and bigger opportunity compared to the threat. Therefore, if weakness and threat can be handled well, and the strength can be used well, the big opportunity can be potential to increase Barangan Banana production in Kabupaten Deli Serdang.

## References

[1] Pusat Data dan Sistem Informasi Kementrian Pertanian, 2017
[2] Dinas Pertanian Deli Serdang, 2016.
[3] Dini Kurnia Wardhani, 2011, Strategi Pengembangan Komoditi Pertanian di Kecamatan Baureno Kabupaten Bojonegoro (Agricultural commodity development strategy in Baureno sub-district Bojonegoro district) (Pendekatan Tipologi Klassen, SWOT, QSPM, FP. UNS, Surakarta
[4] Wahyuniarso Tri D. S., 2013 Strategi Pengembangan Industri Kecil Keripik Di Dusun Karangbolo Desa Lerep Kabupaten Semarang (Chips small industry development strategy in Lerep village Semarang District), FE Semarang
[5] Zakaria, Amar K. 2010. *Program Pengembangan Agribisnis Kedelai Dalam Peningkatan Produksi* dan *Pendapatan Petani (Soybean agribusiness development program in increasing farmers production and income)*. Jurnal Litbang Pertanian. Bogor.
[6] Saaty, Thomas L, 2008, Decision Making With The Analytic Hierarchy Process, Institute for Operations Reasearch and The Management Science, no 6, vol 24.
[7] David, Fred R. 2006. *Manajemen Strategis: Konsep, Edisi 10. Salemba Empat*. Jakarta.
[8] Agus Susanto, Edison HS dan Catur Hermanto, 2015, Pisang, Profil Komoditas Buah Nusantara, Prosiding Seminar Nasional Buah Nusantara, 2015 (Banana, National fruit commodity profile, national fruit seminar proceeding), Pusat Penelitian dan Pengembangan Hortikultura.

# Analysis of rice policy based on presidential instruction on household food security: simultaneous equation model

**S A Sembiring**
Catholic University of Saint Thomas North Sumatera

sabadi@ymail.com

**Abstract**. The objective of this research is to analyze the impacts of rice policy on the household food security. The research used cross section data, were collected from 74 respondent as determined by purposive sampling in Sei Rejo villages, the Sub District of Sei Rampah, Serdang Bedagai District in the Province of North Sumatera. Rice policy model specification uses the simultaneous equations consisting of 6 structural equations and 6 identity equations which was estimated using Two Stages Least Squares (2SLS) method. The results show that the effectiveness of government purchase price of dried harvest paddy gave a positive impact on paddy planted area and lead to an increase paddy production and an increase of the rice production gave a positive impact on household rice availability and household rice surplus, and the increase of household rice surplus gave the quantity of Raskin decrease, whereas the increase of fertilizers gave a negative impact on paddy planted area and lead to decrease paddy production and to decrease in rice production was followed by an decrease in household rice availability and household rice surplus, and the decrease of household rice surplus gave the quantity of Raskin increase.

## 1. Introduction

Through the Timbergen framework, policy instrument is an exogenous variable, while the policy objective is endogenous variables [1]. Based on this framework, food security on Presidential Instruction (Inpres) both No 8/2008 [2] and No7/2009 [3] is endogenous variables. Based on [4] it is shown that to achieve food security if government implemented government purchased on dried harvest paddy.

Although Indonesia has achieved food self-sufficiency, in certain areas are still found food insecurity. Food security at the national and provincial levels does not guarantee food security at the household level because in some provinces it is found that household food security is low [5]. Using BPS data, Susenas 1996-2002, shows that the percentage of households experiencing in Indonesia, the greatest food insecurity in low income groups. In 1999, the percentage of households facing food insecurity reached 34.0 percent in low-income groups. In contrast, food insecurity in high income groups is only 0.4 percent [6]. The lack of acces to food, 82 percent famer household in Sukajadi villages facing food insecurity [7]

The challenges facing food security in the future can be approached through supply side and demand side [8] This condition lead the fluctuation food production and food price changes. According to [9] show that the links fluctuation food production or changes in food prise as food insecurity. Once of the objective of government purchase price is to increase paddy production [10], on the other hand Government Purchase Rice is one way to give incentive to farmer [11].

According to [12, 13] used the food security indicators is rice production, community rice stock, Bulog rice stock, domestic rice stock and rice surplus. The limitations of the study [14,15] have included food security and farmer income in his study, although HPP policies effectiveness.
The formulation of the research problem is how the impact of government policy through Presidential Instruction on household food security? The purpose of this study is to analyze the impact of government policy through Presidential Instruction on household food security on farmers level.

## 2. Data and Methods

The research used cross section data were collected from respondent of 74 farmer which was collected in 2012 as determined by purposive sampling in Sei Rejo villages in the Sub District of Sei Rampah, Serdang Bedagai District in the Province of North Sumatera. Rice policy model specification uses the simultaneous equations consisting of 6 structural equations and 6 identity equations. Rice policy model specification using simultaneous equation consists of 6 structural equations and 6 identity equation.

System contains 12 equations (G = total number of equations) with 22 variables (K = number of total variables in the model). An equation paddy production containing 4 variables (M = number of variables, endogenous and exogenous, included in a particular equation ) and the counting rule (K-M) > (G-1) [16] gives ( 23 - 4) > (12-1), and paddy production equation is over-identified. Therefore other equation is must over-identified. When the all structural equation is over-identified it can be proved by 2SLS (Two Stage Least Squares). Definition of variables is as shown in Table 1.

**Table 1**. Definition of the Variables.

| Variable | Definition | Variable | Definition | Variable | Definition |
|---|---|---|---|---|---|
| LAIT | Paddy Planted Area (Ha) | QPIT | Paddy Production (Kg) | QBIP | Rice Production (Kg) |
| JUPT | The Quantity of Urea Fertilizers (Kg) | JPTT | The Quantity of SP-36 Fertilizers (Kg) | SRHH | Household Rice Available (Kg) |
| RAST | Rice for the poor people (Kg) | IFPT | Farmer Income (Rp) | TRUT | Total Revenue of Farmer (Rp) |
| TCUT | Total Cost of Farm (Rp) | EXRH | Household Rice Surplus (Kg) | HGRT | Price of Dried Paddy (Rp/Kg) |
| HPUT | Price uf Urea Fertlizers (Rp/Kg) | HPTT | Price of SP-36 Fertilizers (Rp/Kg) | KHPPE | The Effectiveness of Government Purchased Price |
| TVCP | Total Cost for Fertilzers (Rp) | TVCO | Total Cost for Pesticide (Rp) | TVCW | Total Cost for wage (Rp) |
| TVCS | Total Cost for Seed (Rp) | QCRT | Household Rice for Consumption (Kg) | D | Dummy; 1= Certificate Seed; 0 = non Certificate Seed |
| Fk | Factor Conversion: 0,63 | | | | |

## 3. Result and Discuss

According to Table 2 show that all the equation has the coefficient of determination $R^2$ than those for household rice available less than 0,70 (Table 2), Durbin-Watson (DW) value, Variance inflation factor (VIF) and F value. VIF value for all structural equation less than 10, it means there is no multicolenirity between exogenous variables [17]. All the equation has statistic F than those for quantity of SP-36 fertilizer are significant at the 1 per cent. It means that the explanatory variables of each equation together can explain the endogenous variables well.

According to 'as in equation (1)' paddy harvested area depend negatively and significantly on HPUT and HPTT but for the KHPPE variable is positive and significant. The estimates of the elasticity of LAIT with respect to the HPUT and HPTT are -4.482 and -1.930, respectively. It means, increasing HUPT by 1 per cent reduce LAIT by more than 4.482 per cent in short run and , increasing HPTT by 1 per cent reduce LAIT by more than 1,930 per cent in short run.

Based on ' as in equation (2)' paddy production depend negatively and insignificantly JPUT and JPTT, but for DS is positive not insignificant and LAIT is positive and significant. The coefficients JPUT and JPTT tending to contradict the hypothesis but the coefficients DS and LAIT agree with prior expectations.

**Table 2**. The estimated econometric model of rice policy on household food security

| The structural and identity equation | EN |
|---|---|
| LAIT = 1.299596 - 0,00133 HPUT* - 0,00051HPTT*** + 0,01018 KHPPE** , $R^2$ = 0.16465 | (1) |
| 1. (1.017/*-4.482*) 1.406/*-1.930*) (1.407/*1.104*) | |
| 2. DW = 0.3414, F = 4.60** | |
| QPIT = -268.961 – 268,961 JUPT – 10.4338 JPTT + 17.438 DS + 880749 LAIT*, $R^2$ = 0.45152 | (2) |
| 3. (3.196/*-8.265*) 1.816/*-0.259*) (1.018/*0,0017*) (3.160/*1.398*) | |
| 4. DW = 1.855, F =14.20* | |
| JUPT = 549.514 – 0.2268 HUPT** + 51.3570 LAIT, $R^2$ = 0.2031, DW = 1.2093 ; F = 9.05* | (3) |
| 5. (1.137/-3.953) ( 1.137/*0265*) | |
| JPTT = 94.110 – 0.0236 HPTT + 90.879 LAIT ** , R2 = 1.96; DW = 2.1461; F = 1.96 | (4) |
| (1.0054/- 0.571) ( 1.0054/ *0.580*) | |
| SRHH = 502.282 + 0.000142 IFPT*, $R^2$ = 0.749 ; DW = 0.7555; F = 215.91* | (5) |
| 6. (1.0000/0.0006) | |
| RAST = 27.129 – 0.00509 EXRH**, R2= 0.0539 ; DW = 2.1081 ; F = 4.11** | (6) |
| 7. (1.0000/- 0.748) | |
| QBIP = 0.63 x QPIT | (7) |
| IFPT = TRUT –TCUT | (8) |
| TRUT = HGRT * QPIT | (9) |
| TCVT = TVCP + TVCO + TVCW + TVCS | (10) |
| SRHH = QBIP + RAST | (11) |
| EXRH = SRHH – QCRT | (12) |

Note: * significant 1%, ** significant 5%, *** significant 10%. The number in parentheses are Variation Inflation Factor (VIF) and short run elasticity , and italic number is short run elasticity, Intriligator [9 ] ; the coefficient of determination $R^2$, Durbin-Watson statistics are given as DW, and F statistic value. EN = equations numbering.

The increase of quantity Urea fertilizers is to reduce paddy production in Sei Rejo village. The estimates of the elasticity of LAIT with respect to the JPUT and JPTT are -8.265 and -0.259, respectively.. It means, increasing JUPT by 1 per cent reduce LAIT by more than 8.265 per cent in short run and increasing JPTT by 1 per cent reduce LAIT by more than 0.259 per cent in short run. This result is consistent with the results of the previous study, where the urea and SP-36 fertilizer relationships were negative with paddy production [15]

On the cases Sei Rejo village, the government policy of raising HET of Urea ferlitizer is relevant to the soil conditions in the rice fields. According to [11] suggests that one of the advantages of increasing the price of subsidized urea fertilizer is to encourage farmers to reduce the excessive use of urea fertilizer. In fact, in Sei Rampah sub-district, the economic logic did not occur, due to the increase of urea fertilizer price followed by the decrease paddy production for rice farming activities. Based on [14] show that the implementation of seed current aid, subsidized of fertilizer and irrigation system is not effective whereas government purchased price (HPP) dried harvest paddy and dried paddy are effective.

Although dummy variable is unresponsive to rice production, farmers need to be encouraged to use certified seeds. On the other hand, the seed aid policy continues in the rainy season 2011/2012, which no farmers received seed aid from the government. In contrast, [12] shows that farmers in Sei Rampah sub-district receive certified seeds from the government

All coefficients sign in 'as in equation (3 and (4)' agree with prior expectations. Prices of Urea fertilizers depend negatively and significantly on the JPUT. The negative coefficient of HPUT indicates that the increase of HPUT to decrease JPUT. Prices of SP-36 fertilizers depend negatively and insignificantly on the JPTT. the estimates of the elasticity of JUPT with respect to the HPUT is -3,9532 , it indicates that, increasing HPUT by 1 per cent reduce JUPT by more than 3.9532 per cent in short run.

The result of this study is different with [15] that found that where the price of urea fertilizer is positively and insignificantly, although, the amount of urea fertilizer is responsive to the price of urea fertilizer, with its short-run elasticity 3.5465.

The effect of the government raise the highest retail price (HET) of urea fertilizer, lead to decrease the amount of fertilizer used for paddy farming. Whereas, the increase of quantity Urea fertilizers to reduce paddy production in Sei Rejo village. On the cases Sei Rejo villagest, the government policy of raising HET of Urea ferlitizer is relevant to the soil conditions in the rice fields. Based on [11] suggests that one of the advantages of increasing the price of subsidized urea fertilizer is to encourage farmers to reduce the excessive use of urea fertilizer

On the other hand, in the mind of farmers is urea fertilizer is the main fertilizer [18]. Therefore, it is necessary to study soil analysis in Sei Rejo villages, since the increase of 1% urea fertilizer causes the decrease of paddy production by 8.2652% in the short run. According to [19] stated that the use of urea fertilizers at the farm level exceeds the dosage. Based on [20], it was found that the use of urea fertilizer in Sei Rampah sub-district has not exceeded the dosage.

According to 'as in equation (5)' IFPT variable are statistically significant and IFPT coefficients agree with prior expectations. Income farmer depend positively and significantly on the SRHH. The positive sign of IFPT indicates that it leads to increase SRHH. As in 'equation (6)' it is shown EXRH variable are statistically significant and EXRH coefficients agree with prior expectations. The coefficient of EXRH is negative, it means that the increase of EXRH to decrease RAST.

Household rice the availability is derived from the sum of rice production and farmers with rice for the poor (Raskin). However, Raskin's contribution to the availability of household rice for small farmers is due to the fact that the farming families receive Raskin at 7 kg per month. The data of household rice availability in this study is season data, therefore the maximum number of Raskin received by household farmer is 28 kg, and note that not all households receive Raskin. Thus almost certainly, the source of household rice availability of farmers is the production of rice farmers.

Raskin implementation in West Java was not effective [21]. The study recommends that the government increase the control of Raskin programs at the village level and increase the monthly Raskin quota to the poor every month at affordable prices. The poor in the village of Sei Rejo, North Sumatra purchase Raskin for Rp. 2000/kg and receive 10 kg per month in 2010 and in 2011, the number of Raskin received every month is less than in 2010 (7 kg per month).

An increase of HPUT has a negative impact on the LAIT, and decrease of the LAIT an decrease of YPIT. The decrease in YPIT was followed by a decrease in QBIP. Decrease in QBIP leads SRHH to decrease. Decrease in SRHH leads EXRH to decrease and follow by increase in RAST. The result of an increase of HPTT is similarly an increase in HPUT. Whereas, an effectiveness of HPUT has a positive impact on the LAIT, and an increase of the LAIT lead to increase of YPIT. An increase in YPIT was followed by an increase in QBIP. An increase in QBIP leads to the SRHH to increase. An increasing in SRHH lead to the EXRH to increase and follow by decrease in RAST.

## 4. Conclusion

The effectiveness of government purchase price of dried harvest paddy gave a positive impact on paddy planted area, paddy production and rice production. An increase of the rice production gave a positive impact on household rice availability and household rice surplus, and the increase of household rice surplus gave the quantity of Raskin decrease.

An increase of price of Urea and SP-36 fertilizers has a negative impact on the paddy planted area, and a decrease of the paddy planted area and decrease of paddy production and rice production. The decrease in rice production was followed by an decrease in on household rice availability and household rice surplus, and the decrease of household rice surplus gave the quantity of Raskin increase.

*4.1 Policy Implication*

Government policy instrument to achieve household rice surplus based on output price policy does not based on input price policy

*4.2 The Limitation of Study*

This model needs to be refined by including the amount of seed use, household income from other farming activities, non-farm income, rainfall data, land conversion, and household expenditure. Therefore it is necessary to add or remove certain variables

**References**

[1] Ellis F 1992 Agricultural Policies in Developing Countries. (Cambridge University Press, Cambridge) P 18-21

[2] Sekretariat Kabinet 2008 Inpres No 8 Tahun 2008 tentang Kebijakan Perberasan (Rice Policy), Jakarta.

[3] Sekretariat Kabinet 2009 Inpres No 7 Tahun 2009 tentang Kebijakan Perberasan (Rice Policy), Jakarta.

[4] Sembiring S A, Harianto , H Siregar, B Saragih 2012 Dampak Kebijakan Pemerintah Melalui Instruksi Presiden Tahun 2005-2008 Tentang Kebijakan Perberasan terhadap Ketahanan Pangan (The Impact of Government Policy on Presidential Instruction in 2005-2008 for Rice Policy on Food Security). *Forum Pascasarjana*, Institut Pertanian Bogor. Bogor. **35**(1): 15-24

[5] Salim H P et al. 2001. Analisis Ketahanan Pangan Tingkat Rumahtangga dan Regional (Analysis of household and regional food security). Laporan Hasil Penelitian, Pusat Penelitian dan Pengembangan Sosial Ekonomi Pertanian, Bogor

[6] Ariani M, E Ariningsih, I K Kariyasa and M Maulana 2008 Empowerment on households facing food insecurity in the era of decentralization in Indonesia. In Rusastra I W, G Thompson, J W T Bottema, R Baldwin (Editor) Food Security and Poverty in the Era of Decentralization in Indonesia. CAPSA Working Paper No 102A reference

[7] Baliwaty Y F 2001. Model Evaluasi Ketahanan Pangan Rumah Tangga Petani: Desa Sukajadi, Kecamatan Ciomas, Kabupaten Bogor.( Evaluation model of household food security of farmers: Sukajadi Village, Ciomas Sub-district, Bogor District) Disertasi Doktor (Doctoral Dissertation). Program Pascasarjana, Institut Pertanian Bogor, Bogor.

[8] Suryana A 2014 Menuju Ketahanan Pangan Indonesia Berkelanjutan 2025: Tantangan dan Penangananya (Toward Sustainable Food Security Indonesia 2025: Challenges and Its response). *Forum Penelitian Agro Ekonomi*. **32** (2):123-135.

[9] Siamwalla A and A Valdes 1984 Food security in Developing Countries:Internasional Issue. In Eicher C K and J M Staatz (Editor). Agricultural Development in the Third World. (The John Hopkins-University Press, Baltimore and London).

[10] Maulana M and Rahman B 2011 Harga Pembelian Pemerintah (HPP) Gabah-Beras Tahun 2010: Efektifitas dan Implikasinya Terhadap Kualitas dan Pengadaan oleh Dolog.( Government Purchase Price (HPP) of Paddy-Rice 2010 : Effectiveness and its implications to quality and procurement by Dolog). *Analisis Kebijakan Pertanian*. **9** (4):331-347.

[11] Kariyasa K 2007 Usulan HET Pupuk Berdasarkan Tingkat Efektifitas Kebijakan Harga Pembelian Gabah (Proposed HET fertilizer based on effectiveness of grain price purchase policy level). *Analisis Kebijakan Pertanian*. **5** (1):72-85.

[12] Sembiring S A 2011 Analisis Ekonomi terhadap Instruksi Presiden tentang Kebijakan Perberasan Nasional Tahun 2005-2008.( Economic analysis of presidential instruction on National rice policy 2005-2008). Disertasi Doktor (Doctoral Dissertation). Sekolah Pascasarjana, Institut Pertanian Bogor, Bogor.

[13] Sembiring S A 2011 The Impacts of Exchange rate on National Food Security. Symp.Editor: Halimmatudahliana., Suryanto. D. (Medan: USU Press). P 273-278

[14] Sembiring S A, Harianto, H Siregar, B Saragih 2010 Implementasi Kebijakan Perberasan di Tingkat Petani: Kinerja dan Perspektif ke Depan ( Implementation of Rice Policy at Farmers Level: Performance and Future Perspective). Analisis Kebijakan Pertanian. **8** (4): 339-361
[15] Sembiring S A 2011 Model Ekonometrika Tentang Kebijakan Perberasan Pada Tingkat Usahatani (Econometrics model on rice policy at farming level. Proc.Reg.Symp on Regional Development of North Sumatera. P 174-183.
[16] Koutsoyiannis A 1977 *Theory of Econometrics: An Introductory Exposition of Econometric Methods*. (Harper and Row Publishers Inc, New York). P.352-356.
[17] Sitepu R and B M Sinaga 2006 Aplikasi Model Ekonometika: Estimasi, Simulasi dan Peramalan Menggunakan Program SAS ( Application of Econometric Model: Estimation, Simulation and Forecasting Using the SAS Program). (Program Studi Ilmu Ekonomi Pertanian. Sekolah Pascasarjana, Institut Pertanian Bogor, Bogor).P 111-116.
[18] Rachman B 2003 Evaluasi Kebijakan Sistem Distribusi dan Harga Pupuk di Tingkat Petani. (Evaluation of policy of distribution system and fertilizer price at farmer level). *Analisis Kebijakan Pertanian*. **1**(3): 221-237.
[19] Kariyasa K and Y Yusdja 2005 Evaluasi Kebijakan Sistem Distribusi Pupuk Urea di Indonesia: Kasus Propinsi Jawa Barat (Evaluation of system policy of urea fertilizer distribution in Indonesia: The Case of West Java Province). *Analisis Kebijakan Pertanian*. **3**(3): 201-216.
[20] Sembiring S A and R Simbolon 2011 Deskripsi Ketahanan Pangan Rumah Tangga Petani di Kecamatan Sei Rampah Kabupaten Serdang Bedagai Propinsi Sumatera Utara ( Description of Household farmers food security Farmers in ei Rampah District Serdang Bedagai, North Sumatra Province). Proc. Nat on association of Indonesian food technology. P 190-203
[21] Hutagaol M P and A Asmara 2008 Effectivity analysis of public policy for the interest of poor people: Case Study of the Implementation of rice for the poor program in West Java, 2007. *Agro Economic Journal*. **26**(2):145-165

**Acknowledgements**
The author is grateful to the Ministry of Education and Culture (Directorate General of Higher Education) for research funding support. Also to Hasan Basri, SP (PPL Desa Sei Rejo); Supianto, Jumadi and Suhery who have helped research activities in Sei Rejo Village, Sei Rampah Sub-District..

# Policy in management based on corruption growth of resources agropolitan in the Gorontalo Province

**S M Wantu[1], U Moonti[2] and A Wantu[3]**

[1]Faculty of Social Sciences, State University of Gorontalo, Indonesia
[2]Faculty of Economie and Bussines, State University of Gorontalo, Indonesia
[3]Faculty of Social Sciences, State University of Gorontalo, Indonesia
E-mail : sastrowantu@ung.ac.id

**Abstract.** Gorontalo was formed as a new province in Indonesia as a producer of national maize, but until now the agropolitan-based agricultural development condition is still inadequate and needs to be developed well. Thus, it is necessary to conduct a governance-related study on empowering farmers in order to increase their capacity in exploring the agricultural sector. This study aims to see the need for local government policies in managing human resources in the agricultural sector in this case the potential and contribution of farmers in creating new jobs in rural areas. This study uses a qualitative approach to the use of secondary data and primary data as a basis for tracing or researching local government management policies in empowering farmers to improve their welfare and also how this policy is expected to organize effective market mechanisms for farmers.

## 1. Introduction

Discussing the interrelationship between the dimensions of population resources and the dimensions of agricultural land is actually a proverbial term "symbiotic mutualism" in which both are interconnected and mutually beneficial. This relationship was once argued by Malthus in An essay on population that there is a strong trend of population growth to be faster than the growth of food supplies, this is because the land area is fixed. Closely related to Malthus's statement, the term for the carrying capacity of the land and the population. Population density is a measure of quantitative carrying capacity, while the qualitative carrying capacity can be measured by man land ratio. In the human ratio land quality aspects of land have been distinguished and calculated how much agricultural area (cropland), how much arable land (can be used as cropland). From this calculation this ratio will be obtained how many people can be considered every hectare of agricultural land.

Political strategies and policies relating to human resource issues will certainly have an impact on employment empowerment as regulated by law throughout Indonesia both central and local to protect every community while doing work including in the agricultural sector. The great role of the state accompanied by responsibility to the community or citizen has actually been mandated in the 1945 Constitution. The fourth amendment result concerning citizenship and the citizens of Article 27 paragraph 2 states that "Every citizen is entitled to decent work and livelihood for humanity".

When viewed from the same protection and giving opportunities in social life, in the context of local locus for Gorontalo province that is by looking at geographic dimension, social and economic structure that impact on agricultural development and positive feedback from farming society.

## 2. Methods And Data Resource

This research uses qualitative method, where researchers describe and find a phenomenon that has a unique character in the implementation of policies related to management based on resource empowerment of agropolitan corn farmers in the province of Gorontalo. In relation to this empowerment, farmers are regarded as a social phenomenon in the field of agricultural political development at the local level.

The data are collected in an integrative manner: First, through observation and interview activities in more depth conducted to generate sources of information about the existence and relationship with farmers resource management policy. Second, the source of data obtained through various documents obtained from local government, community institutions, media, literature, workshops and so on. Furthermore, data analysis using interactive model [1] consisting of three components of analysis, namely data reduction, data presentation and conclusion.

## 3. Result

### *3.1 Farmers' Resource Policy and Management Geographically*

Gorontalo has favorable conditions from the potential richness of natural resources including in the agricultural sector. Gorontalo's regional autonomy becomes a new province after the expansion of the province of North Sulawesi, the distance between regions with some other areas of the agricultural axis such as Central Sulawesi and North Sulawesi which became cross-economic area. In short, the superiority of Gorontalo province is due to its proximity to several other areas which become the pivot area for Sulawesi Island which is the agricultural economic area, so that geographically very profitable for developing the region and improvingthe community welfare . With the advantages in this geographical location, then the question that arises is why there is still poverty that reached a large enough number of data in 2016 is 17, 72%..

One of the triggers is based on observations in the field where there is a change of policy that initially agriculture including agropolitan maize as a priority area, but now is no longer a regional flagship, so that the corn farming factors experience policy discrimination and the result is corn is considered under the target Set. There are two triggers of this failure namely: (1). The transfer of policy on agricultural development has an impact on corn agropolitan agriculture, causing problems that lead to a continuous decline in production and not in accordance with the expected target (2). Lack of mobilization of attention concerning knowledge of peasant resources.

Based on the description and interview results, the obstacle that arises is that there is no definite data related to how big farm laborers in Gorontalo work in the agriculture sector, especially the maize plantation, although the data between 48.65% of the working population is not full on agricultural land and only 28 , 32% of the full work, the data is the whole picture of the farmers in all crops of rice, kedelei and so forth. The absence of definitive figures on maize farmers who earn from peasants and also the very poor data on how many corn laborers work in other sectors to provide for their family life. Farm laborers can not hang their livelihoods on only the corn crop sector, because these jobs are seasonal, even unexpected when the prolonged dry season in Gorontalo.

Although the arguments of corn farmers in Gorontalo have been largely successful since the time of the governor Fadel Muhammad's leadership, the fact remains that most of them are still traditional farmers, not modern farmers and struggling with poverty, where the poverty picture in Gorontalo is still staggering, compared to the national poverty rate Where the Gorontalo region is ranked nationally as the fourth poorest region in Indonesia. In the case of the local government from the time of the governor of Fadel to Rusli Habibie who was elected twice to 2022 made a policy of concern for the corn farmers even though only at the level of policy to stimulate and provide assistance to farmers such as aid equipment, seeding, medicine but not too much change Behavior of farmers. Even less touching on the issue of protection of peasant resources from their unfavorable policies such as price games and other protections.

With this argument with peoples power as peasants working in the agricultural sector, when properly organized and empowered, although most are still categorized as traditional farmers, these peasant communities can actually be a source of potential employment. Farmers community as a farmer resource capable of triggering in accelerating the progress of regional development of Gorontalo which automatically can give advantages in various things, especially for the community such as the farmers themselves in growing self-esteem to enjoy common prosperity. Priority attention to farmers is a major step towards resolving considerable employment issues in areas where unemployment is open until 2015 reaches 4.65 percent.

Although agricultural potential in Gorontalo province has a big share for the regional economic development in the effort to improve welfare and reduce poverty, agricultural development, especially from the agropolitan sector of maize as a traditional crop that has industrial export value still has some constraints. The problem surrounding the agrarian system policy in Gorontalo is agropolitan maize that is developed still in tarap only to meet the needs of local communities and national needs but not maximally to meet the needs of the industrial world in large quantities and high quality. The impact is very visible is the development has not shown the desired expectations, especially in the issue of productivity and competitiveness to meet national and international needs.

Other problems in agropolitan maize related to the price of products and exchange rates received by farmers are still too low, which is then coupled with minimal business capital, resulting in unattainable production that automatically competitiveness can not support maximum results. In addition to the maximum policy issues supporting the quality, quantity and competitiveness to land issues, the most decisive factor is also the resources of farmers who actually work in agriculture level of person capacity as farmers is still low, because the average does not pass primary school, so that With a very low level of education causes them not maximally to manage agricultural land properly. Yet it is undeniable that the agricultural sector, especially agropolitan maize as a potential source of local and very strategic in opening a field of workers who are very open to the village community.

*3.2 Contribution of Farmers' Resources in Agropolitan Maize*

Why agropolitan corn development is so important for the people of Gorontalo, there are several prologues that can be explained as follows: First, agricultural development that puts the core policy on corn agropolitan as the mainstay of Gorontalo people has a tremendous impact on society and regional development and at the same time easy To be done under conditions of agricultural land or plantation; Secondly, agricultural development has contributed at this time, especially with the condition of the economy that is quite considerate and has an impact on the living conditions of the community, so agriculture in Gorontalo society has a strategic meaning, especially as a source of job creation that is not difficult and requires administrative and normative procedures. So that agricultural development is considered as a land that is able to absorb worker resources, especially people who want to become farmers in large numbers, mostly located in rural areas; Third, agriculture development based on agropolitan maize as well as a source of life for the community, also has a substantial share for the inventory of the industrial world; Fourth, agriculture development based on agropolitan maize as a source of food security and sovereignty both locally and nationally; Fifth-based agricultural development Agropolitan corn has an influence on the stability of the political and social life at the local level, often agropolitan corn to the consumption of the political elite in the constellation of political life in the area.

Because that's a reason that is logical enough, from the perspective of local development which focuses on the problem of regional economic development based on the data above will not be separated from the problems of agriculture, especially maize. Even though this time has been pioneered other agricultural sectors such as rice and soybeans and so forth, but when compared with referring to information and field data that in terms of financing and land management, the corn is easier for farmers. This is not because the treatment is not difficult and in addition to the econmic value began to improve. Even corn for this decade become one of the priority for the needs of industry both domestically and abroad. Therefore, to see the living conditions of villagers who mostly work on agricultural areas, it would be nice agricultural development policy on locus of people living in rural areas done improvements where by arranging policies more follow the path associated with the farming population of corn.

To cite the parameters from various perspectives of agricultural development, there are several policy steps that must be taken by the local government in empowering and managing the resources of farmers in order to have a position and be a decisive and calculated part in determining the policy of regional development through the agricultural sector. Position empowerment can only be implemented if farmers as human resources in agriculture have bargaining positions in agricultural economic economic system, so that they gain high social and political energy in realizing future development. Citing the concept of national policy and the concept of the Ministry of Agriculture on parameters that have a perspective value for agriculture and farmers as described above, political bargaining only occurs when they have the power to place their position in the social, economic and Decisive politics like other social class societies in the region.

The policy measures that local governments should take in empowering and managing farmers' resources are directed so that they have a position and become a decisive and calculated part in determining regional development policy through the agricultural sector. Empowerment of corn farmers to gain a decisive position in agricultural development and this condition can only be implemented when corn farmers as rural resources have bargaining positions in the political economy system, so they have a share in the agricultural political circulation at the local level in realizing donations Empirical in the future development of the region.

## 4. Conclusion

Policies in human resources management in addition to facilities and infrastructure and institutions that support added value for farmers are also made efforts to empower the economy of farmers must be in line with the increase of farmers' resources in an effort to establish the peasant community so that they have the ability to change the structure and culture of traditional farmers become a modern farmer.

Such transitional status makes them able and independent in enlivening their families, even to improve their welfare in rural areas. In addition, empowering the farmers community through the improvement of human resources in the direction of farmers have a quality to competitiveness so as to bring productivity and production of corn in large quantities and quality at the local to national level.

## References

[1] Miles and Huberman 1992 *Qualitative Data Analysis* (Jakarta: UI Press)

[2] Nugroho and Riant 2008 *Public Policy: Policy Theory, Policy Analysis, Policy Process, Formulation, Implementation, Evaluation, Revised Risk Management In Public Policy: Policy As The Fifth Estate Policy Research Method* (Jakarta: Elex Media Komputindo Group Gramedia)

# The pricing behavior comparison of Canada and Australia exporter in wheat international market using Pricing to Market (PTM) and Residual Demand Elasticity (RDE)

**R P Wibowo[1], Sumono[1], Y Iddrisu[2], M Darus[1], L P Sihombing[1] and Jufri[1]**

[1]Department of Agribusiness, Faculty of Agriculture Universitas Sumatera Utara, Indonesia
[2]CSIR - Council for Scientific and Industrial Research - Ghana · SARI

E-mail: rulianda_wibowo@yahoo.com

**Abstract.** This paper try to identify and examined the degree of market power on wheat international market by 2 major exporting countries comprising Canada and Australia by using the Pricing to Market (PTM) method and Residual Demand Elasticity (RDE) method. The PTM method found that Canada impose noncompetitive strategy by applying price discrimination and apply market power to their importing. Different results come from Australian exporter as they are not using their market power to the importing. Conflicting result arise from estimation using RDE and PTM method suggest that the need to extend the theoretical model of both model by expand its economic and econometric model to have consistent expected result theoretically and empirically.

## 1. Introduction

Canada and Australia are two major wheat exporting countries. The export share of Canada is relatively declining but the market share of Australia is relatively stable over the years [1]. The change in export share of Canada could be because of the emerging of new wheat producing and exporting countries.

The wheat agriculture has been increasing over the years, shown by its increasing on wheat production and export [1]. The number of wheat producing countries has been increasing by two times from 1960 to 2011. However, the increase in world wheat production due to new wheat producing countries might change market competitive behavior in wheat international market. The new producing countries may reduce previous major exporting countries market share. In addition this development could result in more competitive wheat market. Whether the development in wheat agriculture may reduce uncompetitive behavior of major exporting countries is still a big question. The uncompetitive behavior is not also affected by crop production but also affected by other factors such as agreement, collusive behavior, etc.

Several agreements between importing and exporting countries such as FTA agreement between Australia and China may also give an effect to uncompetitive behavior. The Australian FTA with China may strengthen the market power of Australian exporter to wheat importing countries. Thus, it may result in uncompetitive pricing strategies from Australian exporter to other wheat importing countries. The same situation also may apply to other wheat major exporting countries.

All these facts are raising a question whether the wheat exporter countries have more competitive behavior upon world wheat production increasing. Though this topic is already being discussed by policy makers, however there are not many literatures comparing the pricing strategies between major wheat exporting countries. As far as I know the current literature in pricing strategies is more concentrated on one specific major exporting country without try to compare with other major exporting countries.

Exporter countries with market power may determine their price above the marginal cost, however this will not result to substantial losing in its market share even though the exporter countries has downward demand sloping. This raises the ability of exporter firm to generate positive economic profit in the long run. Market power of exporter country may be developed by its market share in output market, made the exporter country with big market share has the power to determine the price in the market become as price maker and other exporter country from smaller exporting countries become price follower and take the price as their baseline . The market power ability provides the ability to the exporter firm to engage in non-competitive and collusion behavior.

The purpose of this study is to measure market power of Australia and Canada exporter in world wheat market. There are two methods that are going to be used such as Pricing to Market Power (PTM) methods and the elasticity of residual demand (RDE). The PTM method is used to observe the competitiveness behavior of major exporter. The Residual demand elasticity (RDE) method employs the [2] model to measure exporter power to importing countries. This paper also will analyze whether the PTM method has a consistent result with result from RDE method.

## 2. Methods

This study uses annual data of wheat export for Canada and United Kingdom which are provided by [1] FAO (2013). The official exchange rate of importer currency against the exporter currency, GDP (gross domestic product), CPI (consumer price index) were acquired from World Bank. In addition, data related with PPI (producer price index) wheat is obtained from importer statistic center. Canada data set include 231 annual cross section time series observation from 1991-2011 for 11 wheat importing countries. Australia data set include 189 annual cross section time series observation from 1991-2011 for 9 wheat importing countries.

This paper adopt the two way fixed effect model PTM method proposed by [3] and [4] shown by:

$$\ln p_{i,t}^{x} = \sum \alpha_i A_i + \sum \theta_i t_i + \delta_i \ln\left(\frac{e_{i,t}^{x}}{CPI_{i,t}}\right) + \gamma_i \ln\left(\frac{GDP_t}{CPI_t}\right) + \beta_i Oil_t + \vartheta_i PPI_t^{x} \varepsilon_{i,t} \quad (1)$$

where $\propto_i$ is a importing effect, $\theta_i$ represents the specific time effect, $p_{i,t}^{x}$ is wheat export price in country exporter price to importing country i in period t. The parameter on exchange rate variable which represents the elasticity of the domestic currency export price with respect to the exchange rate is denoted by $\delta_i$. $e_{it}$ is the exchange rate in units of destination market currency per unit of the exporter's currency and $CPI_{i,t}$ is the consumer price index for importing country i. $GDP_t$ is the gross domestic product for exporting country, $CPI_t$ is the consumer price index for exporting country and $\varepsilon_{it}$ is the random disturbance with zero mean. $Oil_t$ is the oil price at time t and $PPI_t^{x}$ is the wheat producer price index (PPI) from exporting countries.

The RDE method is estimated using the SUR method and GMM non IV. The RDE model is being developed based on [2]. The econometric model of residual demand function can be written as formula below:

$$\ln p_{m,t}^{x} = \theta_m + \vartheta_m \ln Q_t^{m} + \rho_x \sum \ln\left(\frac{e_t^{m}}{e_t^{x}}\right) + \tau_x \sum \ln PPI_t^{x} + \varepsilon_t \quad (2)$$

$p_{m,t}^{x}$ is the price of export to importing country measured in domestic currency and divide by its CPI. $Q_t^{m}$ is total of wheat quantity demanded by the importer. We also use PPI of competitor which are for Canada and Australia. In this model we also try to add more new covariates such as paddy price and PPI for wheat for China and Korea as we assume that the paddy is substitute product for wheat and wheat domestic product is substitute product for wheat export product means that the new covariates is suitable to be used as demand shifter for wheat product. During the estimation we also try to see whether if there is conflicting result between model which we add new covariates with original model without new covariate or the result still the same with better estimation precision. If $\vartheta_i$ =0 then the demand schedule for importing countries is flat and the exporter cannot change the export price to importing by changing its quantity exported. The residual demand elasticity has downward slope if $\vartheta_i < 0$ indicate that exporter has market power over the importer countries. [2] mention that the highest the absolute value of $\vartheta_i$ means that the exporter has more market power over the importer as demand schedule of importer is becoming more inelastic.

## 3. Result and Analysis

Result for pricing behavior of Canada wheat exporter are presented in table 1 As can be shown in table 1 with respect to the country effect that Algeria, China, Republic of Korea and United Kingdom has a significant country effect . Canada wheat exporter amplify the exchange rate effect on local price currencies to China, Colombia, Indonesia, Italy, Japan, United Kingdom, USA and Venezuela as shown by its positive exchange rate effect. However Canada Wheat Exporter USA offset the exchange rate effect to Algeria which means they applying the LCPS to that market. Canada wheat exporters behave as a competitive supplier in Philippines. For that country we cannot reject $\propto_i$ and $\delta_i$ are zero means that we have competitive market without price discrimination across countries. There is also significant effect of oil price and wheat producer price index to the price export. The high R-squared value may imply that the Canada exporter may not apply the law of one price.

**Table 1.** PTM Estimation result for Canada wheat exporter 1991-2011

| Country | Fixed effect | Exchange Rate | Other Variable |
|---|---|---|---|
| Algeria | -1.152** | -0.470** | |
| China | 1.474*** | 0.954*** | GDP= 0.030 |
| Colombia | -0.314 | 0.615*** | |
| Indonesia | 0.00009 | 0.505** | |
| Italy | 0.0311 | 0.504* | Oil Price = -0.337*** |
| Japan | 0.120 | 0.769*** | |
| Philippines | -0.018 | 0.336 | |
| Republic of Korea | 0.776* | 0.136 | PPI=1.603*** |
| United Kingdom | 1.863** | 0.808*** | |
| USA | 0.617 | 0.598** | |
| Venezuela | 0.066 | 0.451** | $R^2$=0.9409 |

*10% significant, **5% significant, ***1% significant

Result for pricing behavior of Australia exporter are presented in table 2. Australian wheat exporters behave as a competitive supplier in all of his import destination such as China, Egypt, Indonesia, Japan, Malaysia, New Zealand, Philippines, Republic of Korea and Thailand . For these countries we cannot reject $\propto_i$ and $\delta i$ are zero means that we have competitive market without price discrimination across countries.

**Table 2.** PTM Estimation result for Australia wheat exporter 1991-2011

| Country | Fixed effect | Exchange Rate | Other Variable |
|---|---|---|---|
| China | 2.282 | 0.35 | GDP =0.032 |
| Egypt | --1.701 | -0.531 | |
| Indonesia | -0.279 | -0.519 | Oil Price=-0.124 |
| Japan | -1.119 | -0.583 | |
| Malaysia | -1.762 | -0.560 | PPI= 1.019 |
| New Zealand | -0.535 | -0.224 | |
| Philippines | -2.128 | -1.237 | $R^2$=0.4059 |
| Republic of Korea | -1.585 | -0.099 | |
| Thailand | -1.221 | -0.514 | |

*10% significant, **5% significant, ***1% significant

The RDE is estimated using SUR and GMM non IV with the assumption that there is no endogeneity problem in the model. There are 2 model using to estimate comprising model 1 which include new covariate PPI of paddy and wheat and model 2 which not include those covariates.

**Table 3.** Coefficient estimate of variable log quantity in RDE method

| Country | Method | Model with PPI Paddy and Wheat | | Model without PPI Paddy and Wheat | |
|---|---|---|---|---|---|
| | | China | Republic of Korea | China | Republic of Korea |
| Canada | SUR | -0.019 | -0.007 | -0.041** | -0.0202 |
| | GMM non IV | -0.014 | 0.009 | -0.032** | -0.013 |
| Australia | SUR | -0.044*** | -0.896 | -0.023 | -0.903*** |
| | GMM non IV | -0.044*** | -0.871*** | -0.019 | -0.876*** |

*10% significant, **5% significant, ***1% significant

From table 3 we may see that there is consistent result between estimated parameter of log quantity using both SUR and GMM non IV for Canada case. However there is not consistent result for Australian export to Republic of Korea using model 2. There is consistent result for Canada export as we found that Canada have market power over China importer but do not have market power over Republic of Korea importer. The parameters of log quantity for Canada exporter to China importer are shown to be significantly negative but it is not significantly different than zero for Republic of Korean importer.

Interesting result come from Australian case as when we use SUR, we found that model 1 shows Australia exporter has market power over China importer but do not have market power over Republic of Korean importer. However different result come from when we use model 2 as we found that Australia has market power over Republic of Korean importer but do not have market power over China importer. When we use GMM non IV we found that in model 1, Australia exporter has market power over both countries but in model 2 it has only market power over Republic of Korea. These results also contradict with what we found using PTM as we expect that Australia do not have market power over those two countries (see table 2).

Contradicting result between PTM model result and RDE model result is likely to be happened as similar result also found in market power of German beer industry [5]. [5] suggest that the theoretical PTM and RDE model need to be extended to consistently match the observed market solution.

In table 4 we found that in model 1, cost shifter for Canada export to China importer is statistically significant using both SUR and GMM non IV for variable; GDP, time trend, both importer currency to

IOP Conf. Series: Earth and Environmental Science **122** (2018) 012012 doi:10.1088/1755-1315/122/1/012012

competitor, both PPI competitor and PPI wheat China. Similar result also occur in Republic of Korean import, the difference is only it has not significant result in PPI wheat Korea variable using SUR model but the other cost shifter variable is significant. Not significant result found in PPI paddy Korean variable using GMM non IV model but the other cost shifter variable is statistically significant. Positive significant result of PPI competitor means that if Australia reduce the cost then Canada must decrease the export price. Positive significant result on PPI on exchange rate of china currency to Australia, following by positive PPI Australia imply that Australia plays important role for restricting Canada market power to China and Korean market.

**Table 4.** Estimation of RDE for Canada using SUR and GMM IV

| Variable | SUR | | GMM non IV | |
|---|---|---|---|---|
| | China | Republic of Korea | China | Republic of Korea |
| LnQ | -0.019 | -0.007 | -0.014 | 0.009 |
| LnGDP | 0.684** | -1.307** | 0.718*** | -1.204*** |
| Time trend | -0.067** | 0.039** | -0.064*** | 0.038*** |
| ln (Importer currency/Australia) | 1.361*** | 0.428* | 1.454*** | 0.409* |
| ln PPI Australia | 0.989*** | 0.570*** | 1.058*** | 0.616*** |
| ln PPI Paddy | 0.082 | 0.701*** | 0.044 | 0.675*** |
| ln PPI Wheat | -0.528*** | -0.455 | -0.542*** | -0.468 |
| Constant | 135.651 | -66.356** | 129.508*** | -64.57*** |
| Breusch-Pagan Statistic | 5.095, Pr=0.024 | | 2,216 Pr=0.1366 | |

According to table 5 we found statistically significant result in model 1 for Australia using SUR method to all cost shifter variables in China import market. However in Korean import market only GDP, PPI paddy Korean and PPI Wheat Korean is statistically significant. Using GMM no IV method to model 1 in China import market give us statistically significant result to all cost shifter variable except GDP. Table 5 gives different result to Korean import market using GMM non IV on model 1 as we have more significant variable compare to estimation result using SUR. According to SUR and GMM non IV estimation result we found that Canada has restrictive power to influence Australia market power in China and Korean Market.

**Table 5.** Estimation of RDE for Australia using SUR and GMM IV

| Variable | SUR | | GMM IV | |
|---|---|---|---|---|
| | China | Republic of Korea | China | Republic of Korea |
| LnQ | -0.044*** | -0.896 | -0.044*** | -0.871*** |
| LnGDP | 0.184 | -3.565* | 0.103 | -2.959*** |
| Time trend | -0.059* | -0.0474 | -0.053*** | -0.074*** |
| ln (Importer currency/Canada) | 2.791*** | 0.0378 | 2.754*** | 0.434 |
| ln PPI Canada | 1.234*** | 0.009 | 1.222*** | 0.018 |
| ln PPI Paddy | 0.235*** | 1.393*** | 0.269*** | 1.419*** |
| ln PPI Wheat | -0.278* | 4.379*** | -0.330** | 4.411*** |
| Constant | 121.442** | 129.982 | 109.418*** | 175.654*** |
| Breusch-Pagan Statistic | 4.512, Prob = 0.0337 | | 4.812, Prob=0.0283 | |

## 4. Conclusion

This paper tries to identify and examined the degree of market power on wheat international market by 2 major exporting countries comprising Canada and Australia. Canada wheat exporter amplify the exchange rate effect on local price currencies to China, Colombia, Indonesia, Italy, Japan, United Kingdom, USA and Venezuela. However Canada Wheat Exporter offset the exchange rate effect to Algeria. Surprisingly the PTM method shows Australian wheat exporters behave as a competitive supplier in his entire import destination. I assume this behavior is due to its smaller market share in world wheat market compare to USA and Canada.

The RDE result shows consistent result for Canada export as we found that Canada have market power over China importer but do not have market power over Republic of Korea importer. Not consistent result found in Australian export using SUR model as we found that model 1 shows Australia exporter has market power over China importer but do not have market power over Republic of Korean importer. However different result come from when we use model 2 as we found that Australia has market power over Republic of Korean importer but do not have market power over China importer. The GMM non IV estimation shows that in model 1, Australia exporter has market power over both countries but in model 3 its only has market power over Republic of Korea. This result also contradict with what we found using PTM as we expect that Australia do not have market power over those two countries. The Contradict result between PTM and RDE method suggest that we need to extend the theoretical model of both model by expand its economic and econometric model to have consistent expected result theoretically and empirically.

## References

[1] FAO (2013). Statistic. www.fao.org.
[2] Goldberg, P.K. and M.M. Knetter (1999): Measuring the intensity of competition in export markets. Jounal of International Economics, 47: 27-60.
[3] Knetter, M.M. (1989): Price discimination by U.S. and German exporters. The American Economic Review, 79, 1: 198-210.
[4] Falk, M. and R. Falk (1998). *Pricing to Market by German Exporters: Evidence from Panel Data, Centre for European Economic Research.* Discussion Paper No. 28. Mannheim.
[5] Glauben, Thomas. and Loy, Jens Peter (1999): Market Power of the German Beer Industry on Export Markets - An Empirical Study -, FE Working paper / Universität Kiel, Department of Food Economics and Consumption Studies, No. 99.

IOP Conf. Series: Earth and Environmental Science 122 (2018) 012013 doi:10.1088/1755-1315/122/1/012013

# Farmer's welfare in Telang's integrated independent city: lesson learned from migrant and local farmers in tidal land, South Sumatera

**D Adriani[1], I Zahri[1], E Wildayana[1], Maryadi[1], M Hamzah[1] and Yulius[1]**

[1)]Universitas Sriwijaya

Corresponding Author Email: dessyadriani@fp.unsri.ac.id

**Abstract.** Telang's Independent Integrated City development is a model of tidal agricultural development through technological innovation to improve farmers' welfare. In this area, the diversity of origin of population is also suspected to have an impact on the achievement of success. The purpose of this study is to analyze and prove the hypothesis that farmers are able to prosper in tidal land by Telang's Independent Integrated City Program based on migrant and local farmers disaggregation. The research was conducted at Tanjung Lago District Banyuasin Regency, South Sumatra. The research method is survey with stratified simple random sampling. Data is processed by mathematics and statistics. Telang's Independent Integrated City is an area of rice production center development and rice industry. Since 2008, farmers have implemented rice cultivation innovations by the twice-cropping index. This program by utilization of tidal land proved that farmers have a great chance of success. Farmers have succeeded in breaking the myth that has been growing in the food agriculture sector, that farmers cannot prosper only with food crops. Both Farmer's income has been above the necessities of life. Judging by the diversity of origin of population, the income of migrant is higher than local farmers.

## 1. Introduction

One of the rural areas management models developed by the government in Indonesia is The Independent Integrated City Program. In Particularly, this program is intended to develop rural areas into integrated agricultural areas through exogenous and endogenous approaches. Exogenous approach or industrial modernization is implemented by introducing external investments or resources, requiring prerequisites for high quality labor guarantees, the adequacy of the cluster industry, the development of infrastructure (especially transportation facilities) and the appeal of useful industrial policies. While endogenous approach is implemented by utilizing the full potential of resources and capital, as well as paying attention to the environmental balance (Asriati [1]).

Independent Integrated City is a village or a region that grows and develops as a centre of collection, processing, distribution and services of the Transmigrant Development are designed as a structured development direction of transmigration settlement units and surrounding villages within a single infrastructure network and regional as shown in Fig.1. The objective of Independent Integrated City development is to improve the ease of meeting the basic needs that enable the opening of socio-economic growth opportunities of the transmigrant areas and to create centres of business activities that attract investors as an effort to foster the economic activities of transmigrants and surrounding communities. Function of Independent Integrated City as a place to live, as a place of agricultural activities to meet the needs of city dwellers (On Farm) and as a place of post-harvest business activities and service activities (Off Farm). The Centre of WPT is called Integrated City with the hierarchy of Order II City is the highest order in the scope of WPT Nas, [8], Chiara [2], Emilya K and T W Caesariadi [5].

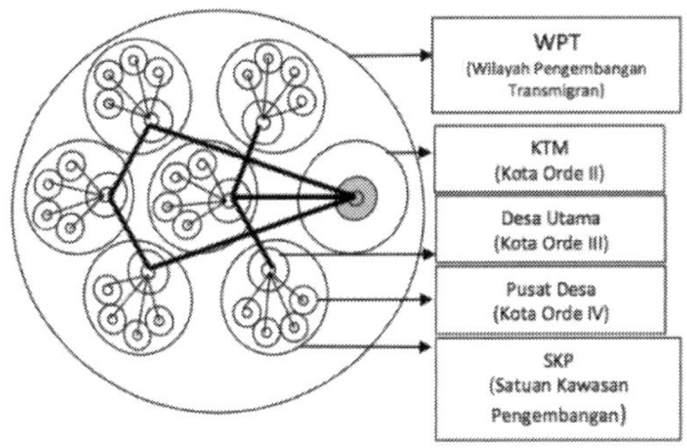


**Figure 1.** Hierarchy of Region Relations (Nas, [8])

Transmigration Development Policy is essentially an integral part of national and regional development in Indonesia, to accelerate agriculture development and to improve the welfare of the migrants and the surrounding communities especially in a remote area as well as. In fact, not all transmigration settlement units progressed well. Various problems occured and impacted the level of transmigrants' level of welfare. Transmigration has many challenges and obstacles in various sectors in several residential units, so in its development the government takes a strategic step in supporting and developing transmigration programs, by implementing a program called New Urban Area/ Independent Integrated City. Independent Integrated City is a holistic and comprehensive program designed to build an urban transmigration area, as an acceleration of rural economic development and the realization of an independent Transmigration Area. This new development paradigm of Transmigration Program is shaping the transmigration area to be the center of new growth so as to involve all stakeholders to become more participative, and sustainable as stated by Emilya K and T W Caesariadi, 2016 [5].

In South Sumatera Province, since 2008 the government has developed Telang's Independent Integrated City. Telang consists of 4 sub-districts, that is Muara Telang Sub-district, Tanjung Lago Sub-district, Sumber Marga Telang Sub-district, and Makarti Jaya Sub-district with 113,783 Ha of total Area. For the area of each sub-district included in the Independent Integrated City (KTM) Telang: (a) Tanjung Lago Sub-district 61.223 Ha or equivalent to 53, 81% (area with the largest area), (b) Muara Telang Sub-district 27,609 Ha or equivalent to 24.26%, (c) Sumber Marga Telang Sub-district 22,297 Ha or equal to 19,60%, and (d) Makarti Jaya Sub-district 2,654 Ha or equal to 2,33% (area with smallest area).

In Telang's Independent Integrated City, there are 34 village units, with 24 ex-transmigration settlements, which are devided 24 definitive villages and 10 local villages. Thus, there are differences in socio-cultural characteristics between the two sites groups. Several research results indicate differences in socioeconomic characteristics could affect the difference in achievement in welfare. Therefore, it is expected that the differences in farmers' the characteristics could affect their responses, although the aims to improve their welfare is the same.

Every development program, physical or non-physical, aims to increase the welfare of society, as well as the development of an Independent Integrated City. Nevertheless, several problems also are found within the execution of program, such as the location of transmigration or with the participants themselves. While the environment and society undergoes change, a review of program impact must be done in order to achieve maximum results. The synergy of change brings a variety of effects such as the segregation of culture, environmental order and a new rule of order.

During this time, the tidal area of Telang's Independent Integrated City is an tidal area with the characteristics of land that is highly dependent on nature. Farming activities are not optimal to do. Therefore, Independent Integrated City program is expected to address these issues. From research aspect, research on the socio-economic condition of the community in this region has been done a lot. However, studies on the impact of the program in improving Telang's Independent Integrated City socio-economic conditions, both local and migrant farmers are still limited. This study analyzes the welfare of farmers on the of both local and migrant farmers, especially related to their activity on the tidal area of Telang's Independent Integrated City.

## 2. Research Method

This research was carried out in Tanjung Lago District Banyuasin Regency as shown in Fig.2. This research was carried out in February 2017. The research method was survey methods, in which the sample are (1) local from Tanjung Lago Village and (2) migrant farmers from Telang Sari Village. Sampling methods were simple random sampling. In total, 120 farmers were selected at two sample categories. Data collection is done through interviews using questionnaire method. In addition, data collection was conducted with a quick survey for migrant data collection, in-depth interviews in interviewing informants; community leaders, village officials, village forum members and officers. Data obtained in the field processed in tabulations, calculated mathematically and descriptive quantitatively. To analyzes the welfare of farmers on the of both local and migrant farmers, especially related to their activity on the tidal area of Telang's Independent Integrated City, we calculate the structure of income source on both of them. We divided structure of income source into (1) on-farm income (rice and non-rice income) and (2) off farm income. To analyze household welfare, we compared their income with the Decend Living Standards of South Sumatera Province in 2016. If their income covered the Decend Living Standards, we concluded that the household's farmers are welfare.

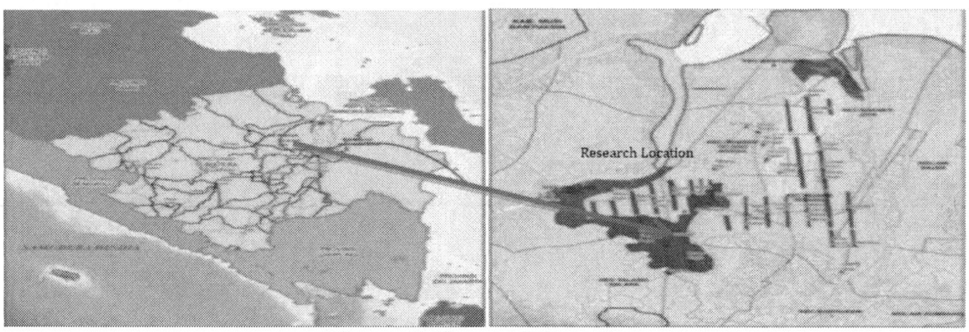


**Figure 2.** Research Location

## 3. Results and Discussion

Central Bureau of Statistics data of 2016 shows the number of population in Telang's Independent Integrated City scattered in 4 (four) District of 94.232 person (24,666 KK) consisting of residents in transmigration settlements area of 68,754 person (72.96%) with 18,692 households (75.78%) and local villagers as many as 25,478 person (27.04%) with 5,974 families (24.22%). Details for population structure can be seen in Fig. 3.

Farmers in the Telang Area are mostly transmigration residents who come from Java Island, but there are also Bugis tribe and local people of South Sumatra. The migrant's population were mostly occupied as rice farmers, while the occupation of the Bugis were farmers (especially coconut plants) and traders. The occupation of local people is generally traders, and only a little occupied as farmers and fishermen.

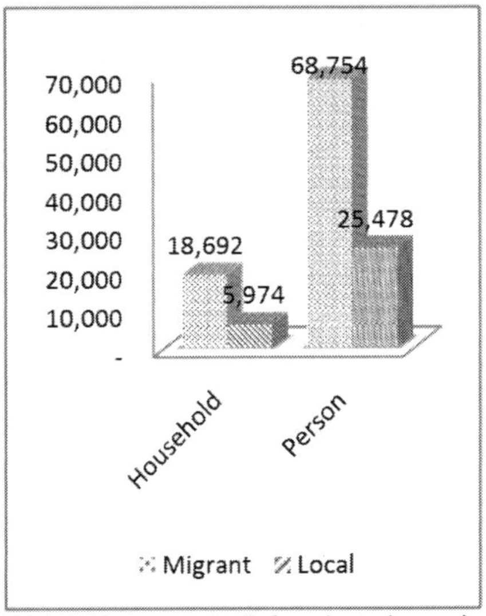


**Figure 3.** Population Structure at Telang's Independent Integrated City, 2016

For migrant farmers, rice from on-farm income is the biggest income compared to the income of non-rice (corn, beans, vegetables livestock) and off-farm. This is because rice farming is done on vast land with an average of more than 2 ha although only one season planting. As for the corn farming is lower than rice farming when planted on the same land area, because corn farming needs much attention itself very much. In addition, the production and price of corn is lower than rice farming. For off-farm activities of farming actually has lowest level of income, but because not all farmers who seek outside activities farming as seen in Table 1.

**Table 1.** Income structure for migrant and local farmers in Telang's Independent Integrated City, 2017

| No | Income Structure | Income (IDR/year) | Percentage (%) |
|---|---|---|---|
| | MIGRANT FARMERS | | |
| 1. | On-farm (rice) | 34.087.121 | 50.10 |
| 2. | On-farm (corns, beans, vegetables,livestocks) | 9.406.252 | 13.80 |
| 3. | Off-Farm | 24.480.556 | 36.00 |
| | **Total** | 67.973.929 | 100.00 |
| | LOCAL FARMERS | | |
| 1 | On-farm (rice) | 12,320,818 | 30.98 |
| 2 | On-farm (corns, beans, vegetables,livestocks) | 1,439,999 | 3.62 |
| 3 | Off-Farm | 26,014,667 | 65.40 |
| | **Total** | 39,775,484 | 100.00 |

Table 1 showed the income structure for migrant and local farmers in Telang's Independent Integrated City. Based on Table 1 can be concluded that the contribution of rice farming to the total income of migrant farmers is the largest contribution of 50.10 percent. For the total income of corns, beans, vegetables, livestocks farming has the smallest contribution to the three types of sources of income, which is 13.80 percent and for off-farm income has a contribution of 36.00 percent.

The opposite occurs in the structure income of local farmers. For local farmers, the largest income is derived from the off-farm sector (such as traders, drivers, and laborers). Then followed by income from rice and non-rice farming. For non-rice farming, almost all local farmers raise livestock and do not plant other commodities. Based on Table 1 can be concluded that the contribution of rice farming to the total income of farm households is only 30.98 percent. For the total income of corns, beans, vegetables, and livestocks farming has the smallest contribution to the three types of sources of income, which is 3.63 percent and off-farm income has the biggest contribution of 65.40 percent.

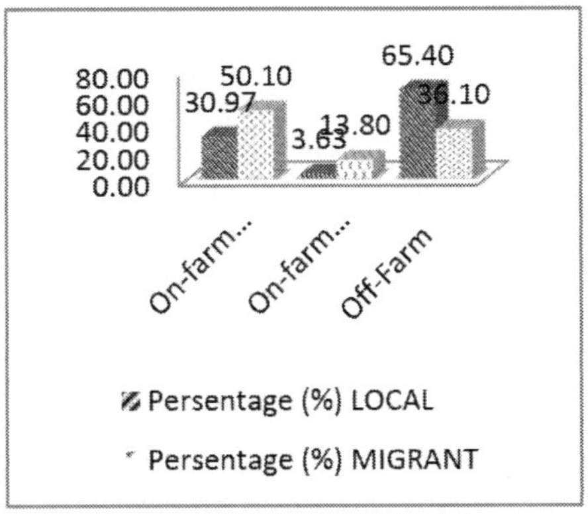


**Figure 4.** Comparison of Income Structure between migrant and local farmers in Telang's Independent Integrated City, 2017

If it is related to the ability of farmers to covered the decent living needs of South Sumatera Province in 2016 amounting to IDR 2,206,000 / month, it can be concluded that the income of both local and migrant farmers has been able to covered their decend living needs. In other words, both local and migrant farmers are welfare, especially related to their activity on the tidal area of Telang's Independent Integrated City. So, this research result is same with Murti (2012)[9], there was positive impact from this program. The independent integrated city program has boosted household economic growth in this area. In general, this research result indicated a positive change that indicates there will be an economic acceleration after at all [3].

There are some new finding in this research. As we know, the South Sumatera Government developed Telang's Independent Integrated City with agriculture as the main sector. Develop agriculture as the main sector in Telang's Independent Integrated City impacted on the main occupation in this area, that is as farmers. But, there some anomaly finding in this research result. Not both of local and migrant farmers have main occupation in agriculture now. The results of the analysis indicate that there has been a change in the structure of occupation for local farmers, who initially worked on-farm turned into off-farm. Limitations of on-farm activities are actually related to the smaller farming area owned. Local farmers only have 1.37 hectares of land, while farmers have 2.4 hectares. With a small land area, rice farming is not a lucrative option for local farmers in on-farm activities. This result is in accordance with the research findings in Micklethwait & Wooldridge [7], human life and other species will experience difficulties due to contamination of various sources of life, thus disrupting the balance of nature and human life. In situations like this there will be a change of occupation or transition of work to the peasant community for the sake of survival.

The transformation of livelihoods pattern of local farmers is marked by very small agricultural income, the narrower land, the large production cost is not proportional to the sale value of agricultural products, the non-agricultural business opportunities are more open with considerable income, the younger local generation and the farmers are unwilling continue the agricultural work of his parents, and there is mobility of the population that can generate transactions of work between farmers as a party in need of employment with migrant communities as the party in need of labor. In addition there is also a perception among the younger local generation that agricultural work is considered less prestigious, impressed dirty, shabby, poor, less educated and provide no guarantee of a clear future. In the opinion of local farmers in the region, agricultural work is not the only type of income-generating job. Although agricultural work is dominant in rural areas, the facts show that the socio-economic life of farmers' societies is determined not only by the agricultural sector. Along with the increasing opening of information and changes to the nature of life among the community, the farmers also experienced changes in the way of learning and economic behavior in everyday life. The fulfillment of the necessities of life is not enough to be satisfied only by one type of work, but also to be supported by other work and a more productive social role.

Based on the results of this study, there has been social change, especially for local communities. If this social change continues to be left, in the future more and more groups will leave the agricultural sector. A number of suggestions were proposed by Hardika (2011) [6], dan Qodir [10] regarding the transformation of livelihood patterns of farming communities in the transition areas, namely (1) the need to identify the needs and sources of as a material for the formulation of socio-economic empowerment programs of the community, and (2) it is necessary to develop non-farm business models relevant to the characteristics.

## 4. Conclusion

Telang's Independent Integrated City is an area of rice production center development and rice industry. Telang's Independent Integrated City Program in tidal land proved to have a great chance of success due to various technological innovations that are available. Farmer's income has been able to covered their decend living needs, so the household are welfare. Judging by local and migrants farmers, the income of migrant farmers is higher than local farmers. From the income structure side, migrant farmers rely more on the livelihoods of the on-farm sector, while local farmers rely more on the livelihoods of the off-farm sector.

## References

[1] Asriati N 2015 Pengembangan kawasan terpadu mandiri dengan pendekatan model one village one product (OVOP) daerah transmigrasi rasau jaya (Independent and integrated region development with one village one product (OVOP) approach: transmigration region rasau jaya). *Prosiding Seminar Nasional 9 Mei 2015 Kementerian Transmigrasi dan Ketenagakerjaan.* 705-721

[2] Chiara J D 1982 *Urban Planning and Design Criteria* 3rd Edition (New York: Van No strand Reinhold)

[3] Danarti 2011 Akselerasi perekonomian masyarakat transmigrasi di hinterland kota terpadu mandiri telang (Economy acceleration transmigration society in hinterland in integrated and independent city Telang). J. Ketransmigrasian. **28** 13-24.

[4] Dirjen Pembinaan Pengembangan Masyarakat dan Kawasan Transmigrasi 2006 Kota Terpadu Mandiri (Independent and Integrated City) (Jakarta: Dirjen Pembinaan Pengembangan Masyarakat dan Kawasan Transmigrasi)

[5] Emilya K and Caesariadi T W 2016 Konsep permukiman kota terpadu mandiri (Independent and integrated city settlement concept) J. Langkau Betang. **3** 12-24.

[6] Hardika 2011 Transformasi pola mata pencaharian petani: Strategi dan perilaku belajar petani di kawasan transisi dalam mengembangkan kehidupan (Farmers' livelihood transformation: Strategy and study behaviour of farmers in transition region in developing life) Jur. Pend. dan Pemb. **18** 90-97.

[7] Micklethwait, J and Wooldridge A 2000 A future perfect *The Challenge and Hidden Promise of Globalization* (New York: Crown Publishers)

[8] Nas, P J M D 1979 Kota di dunia ketiga *Pengantar Sosiologi Kota dalam Tiga Bagian* (City in the third world: *Introducing city in three parts*) (Jakarta: Bhratara Karya Aksara)

[9] Murti, S 2012 Mitigasi sosial di kota terpadu mandiri (ktm) Labangka Sumbawa Besar NTB (Social mitigation in integrated and independent Labangka Sumbawa Besar NTB). Jur. Pen. Hum. **17** 61-78

IOP Conf. Series: Earth and Environmental Science **122** (2018) 012014 doi:10.1088/1755-1315/122/1/012014

# The analysis of consumption level and preferences of freshshallots consumers in Medan

**F Yani[1] andT Supriana[2]**
[1] Program Agribusiness of Universitas Sumatera Utara
[2]Lecturer of Universitas Sumatera Utara
tavihutasuhut@yahoo.co.id

**Abstract.** Comprehension about consumer preferences gives a very important contribution to the consumer's decision making process, which eventually affects the level of consumer consumption. This study aims to analyze the level of consumer consumption, consumer preferences, as well as the combination of the most preferred attributes by the fresh shallots consumers in Medan. The number of respondents in this study is 100 households. Consumer preferences were identified using conjoint analysis. Attributes of the product that were observed are the size of the root, aroma, price, and dryness. The result of this study shows that the average level of fresh shallots consumption in Medan is 11.67 grams/capita/day, which is higher compared to the 7 grams/capita/day average of national consumption. Consumers assume that the dryness is a very high importance value in judging and deciding whether to buy the fresh shallots, followed by the attribute of aroma, price, and size of the root. Meanwhile, the combination of attributes that is most preferred are fresh shallots with bulb diameter of 2.5 cm, strong aroma, low price ranging below Rp 35000/kg and dryness level in medium category which means the products are in a rather moist/dry condition, and has a little skin waste.

## 1. Introduction

Fresh shallot is one of the leading vegetable commodities that has long been cultivated intensively by farmers. This vegetable commodity belongs to a group of unsubstituted herbs that serve as food seasonings as well as traditional medicinal ingredients[1]. Fresh shallot is also a nutrient-dense food which means it contains low calorie and high nutrients and is beneficial as vitamins, minerals and antioxidants [1]. The needs or consumption of shallots in North Sumatra continues to fluctuate, and the production of shallots has not been able to meet the consumption of shallots in North Sumatra.Medan as an administrative city which has a population of 2,135,156 people, with population growth rate of 0.60%, is very dependent on food consumption, including shallots[2].This is caused the characteristics of people in Medan is very fond of spicy cuisine with the use of shallots as a natural flavoring seasoning.

The diversity of characteristics attached to fresh shallots products raises a variety of preferences from consumers as the ultimate decision-maker to buy. Differences in social status, culture, environmental influences, purchasing power, motivation and lifestyle create different consumer behavior regarding consuming fresh shallots.

This condition simultaneously creates opportunities for producers to provide a wide variety of fresh shallots products with differences in price, moisture/dryness and varying sizes according to the needs and desires of consumers and market segments targeted by traders. With the results, that the level of consumer consumption and preferences with the most preferred combination of shallots attributes, especially in Medan.

Consumer preferences in determining the choice of a particular product are reflected in consumer's attitude towards the product.According to Kotler and Armstrong[3], attitude is an evaluation, feeling, and tendency of someone who consistently likes or dislikes an object or idea.

The needs for fresh shallots are very high. Almost all cuisine use shallots as a flavoring herb. Vegetable commodities are included in the group of spice vegetables that are used to add taste and pleasure to the cuisine. It's also used as a traditional medicine[7].

## 2. Methodology

*2.1 Method of Determining the Research Location*
The research was conducted in two districts, namely Medan Deli and Medan Marelan, which were deliberately chosen because they can represent and have the largest of households in Medan. Every household in Medan would use shallots for cooking. Medan Deli District has 40054 household and Medan Marelan has 34423 household[5].This type of descriptive research can describe and analyze consumer consumption level, consumer preferences and the combination of the most desirable fresh shallot attributes.

*2.2 Method of Determining Sample*
The nonprobability method was employed to determine the sample. The research population consists of 100 households, assuming all households consumed fresh shallot as a spice. Respondents were chosen by using accidental sampling amount 100 respondents, that is a sampling technique where the person encountered by chance or anyone whom the researcher considers suitable is selected as the data source

Respondents in this study are the final consumers who decided to buy fresh shallots and the majority are women who are housewives. Respondents were found in a condition of buying fresh shallots in traditional markets as well as vegetable stalls in research areas and were found by chance, that the respondents were considered suitable as the source of data. Data used in this study was primary data and secondary data.

*2.3 Method of Data Analysis*
The first questionnaire was used to identify the attributes and levels desired by the respondents, which will then be used to form various combinations of attributes and levels that have been obtained from the first questionnaire. The combinations created were offered back to the consumer by giving out the second questionnaire to find out the level of attributes that are most desired by consumers. The second stage questionnaire is the result of stimuli from several attribute levels that were assessed using Likert scale (1 = very unlikely the combinations of attribute of shallots, 2 = unlikely the combinations of attribute of shallots, 3= normal, 4= like the combinations of attribute of shallots,and 5 = very like).

The level of fresh shallot consumption is the volume of fresh shallots consumed by consumers in units per time.To determine the level of shallot consumption in the research area, descriptive analysis was used by tabulating the primary data of fresh shallots consumer consumption from 100 households in the research area. Then, it was compared with the secondary data of the average shallot consumption at the national level, the level of North Sumatra Province and Medan City so that it will be known whether the average consumption of fresh shallot per gram/capita/day falls into the category of low, medium or high consumption level.

The next analysis tool used in this research was conjoint analysis. This analysis was used to determine consumer preferences on the decision to buy fresh shallots. The results obtained from this analysis are the order of preferences formulation desired by the respondents and the level considered important. Conjoint analysis is very useful in formulating marketing strategies, as well as in determining prices, predicting sales and finding out the characteristics of the market.

*2.4 Model Specification and Parameter Testing*
Utility is a conceptual basis for measuring value in a conjoint analysis, which is an assessment of subjective preferences that is unique for each individual. Products with higher utility values have a higher preference and a higher chance of being selected.

Characteristics of fresh shallot consumers are reflected by gender, age, occupation, consumption level and income level. Meanwhile, consumer preferences are reflected by several attributes of fresh shallot, which are the size of the bulb, aroma, price and moisture/dryness level, of which the combination is then scored using the Likert scale.

The main output of a conjoint analysis is a series of part-worth (utility) interval scale from each level for each attribute. From the merging of these utilities, predicted preference from each level for each attribute of the product would be obtained.

$Y = X1 + X2 + X3 + .......... + Xn + Constants$

Where:

The independent variables are (X1, X2, X3, ....Xn) the characteristic of fresh shallots. The independent variable is in the form of non-metric data, while the dependent variable (Y) is the overall preference of the respondent to the level of each factor of a product. This dependent variable also includes consumer ratings of the importance of factors to the attributes of a product

*Total utility = utility X1 (attribute level 1 to-i) +utilityX2 (attribute level 2 to-i) + utility X3 (attribute level 3 to-i) + utilityX4 (attribute level 4 to-i).....+ constants* (1)

*Where :*

Y : total utility
X1 : utility value of the size attribute
X2 : utility value of the aroma attribute
X3 : utility value of the price attribute
X4 : utility value of the moisture/dryness attribute
Constant: the constant value in the analysis

According to Green and Krieger **Error! Reference source not found.**,conjoint analysis can be utilized for several purposes as follows:

1. Pricing
   Predict the level of sale or use of the product (market share), test the concept of new products.
2. Preference segmentation
3. Designing a promotional strategy

Conjoint analysis describes the nature of consumer preferences based on consumer trade-off analysis of various product attributes. Attributes can be defined as the characteristics that differentiate products from other brands or products, or may also be the factors consumers consider in making decisions about purchasing a brand or product category attached to a product or product part

A brief description concerning the level of fresh shallot consumer consumption in Medan that has diverse preferences and the role of the composition of product attributes combination will provide a clearer picture of its contribution in designing the market strategy of fresh shallot products which are most favored by consumers.

## 3. Result and Discussion

### *3.1 Characteristics of Respondents*

The characteristics of the shallot respondents (combined from two districts, namely Medan Deli and Medan Marelan) were obtained from a total of 100 respondents. From the data, it was found that by gender, 78 respondents are female and 22 are male. They are who make decision in buying the shallots.

Respondents aged between 31-40 years amounted to 46%. In aggregate, most fresh shallot respondents, that is 67%, have more than 3 family members, and as many as 40% of the respondents are working in private institutions.Most of the respondents is earn around Rp.2 million-Rp.5 million/month, whereas respondents with an average consumption level of 50 grams per day comprise 63% of the total respondents, with the assumption that it is for the consumption of one household, family size considered with amount of family members.

### *3.2 Attribute Combination and Attribute Level Orthogonal Design*

To avoid overburdening consumers when choosing a combination of attributes, only a few shallot attributes were selected based on the result of the first stage questionnaire. Only the attributes that were considered essential and easy for respondents to look at and make choices on were chosen. Therefore, by using SPSS orthogonal procedure, a combination of attributes was obtained using the Orthogonal Design.

IOP Conf. Series: Earth and Environmental Science **122** (2018) 012014 doi:10.1088/1755-1315/122/1/012014

**Table 1.** Attribute Combination and Attribute Level

| No. | Attribute Combination and Attribute Level | | | |
|---|---|---|---|---|
| | Size | Aroma | Price (Rp/Kg) | Moisture/Dryness |
| 1 | Large | Weak Scent | > 40000 | Bad |
| 2 | Large | Strong Scent | ≤ 35000 | Medium |
| 3 | Medium | No Scent | > 40000 | Medium |
| 4 | Medium | Strong Scent | 35000-40000 | Bad |
| 5 | Medium | Weak Scent | ≤ 35000 | Good |
| 6 | Small | Strong Scent | > 40000 | Good |
| 7 | Small | No Scent | ≤35000 | Bad |
| 8 | Large | No Scent | 35000-40000 | Good |
| 9 | Small | Weak Scent | 35000-40000 | Medium |

Based on the formation of stimuli, the combination of attributes obtained amounted to 81 (eighty-one) combinations. With orthogonal design, the number of combinations of attributes can be simplified into 9 (nine) combinations (see Table 1), to facilitate the consumers to describe their preferences of fresh shallot.

For utility estimates, the greatest positive value from the consumer preferences indicates the most preferred attribute level by the consumer, whereas the smallest negative value indicates the least favorable attribute level by the consumer. Meanwhile, the total value of utility with the largest positive value shows the most preferred combination of attributes.

**Table 2.**Attribute and Attribute Level ofShallot

| No. | Attribute | Level of Attribute | Description |
|---|---|---|---|
| 1 | Size | 1 | Small |
| | | 2 | Medium |
| | | 3 | Large |
| 2 | Aroma | 1 | No Scent |
| | | 2 | Weak Scent |
| | | 3 | Strong Scent |
| 3 | Price (Rp/Kg) | 1 | >Rp.40000/kg |
| | | 2 | Rp.35000/kg-Rp.40000/kg |
| | | 3 | ≤Rp.35000/kg |
| 4 | Moisture/Dryness | 1 | Bad |
| | | 2 | Medium |
| | | 3 | Good |

Based on Table 2, there are 3x3x3x3 = 81 combinations of attributes. It tends to be less practical and consumers would find it difficult to rank the 81 attribute combinations and levels. To overcome this, the fractional factorial design was used instead of the full-factor design. By using this design, a part of the entire combinations of the product that really affects the main effect and design, known as the orthogonal array, was selected. The formation of attribute combinations that satisfy the orthogonal properties is a crucial initial step.

From Table 3, it is found that in general, the most preferred attribute is the attribute of moisture and dryness with good attribute level (0.507) and the least favored attribute is also the attribute of moisture/dryness that has a bad attribute level (-0.877). For the characteristic of aroma, it is found thatconsumers prefer fresh shallots with a strong scent/aroma (0.460), while the least popular one is a weak scented fresh shallot (-0.637).

**Table 3.**Utility Value of each Attribute Level (consumers in general)

| | Level of Attribute | Utility Value |
|---|---|---|
| Attribute 1 (Size) | Small | -0.397 |
| | Medium | -0.020 |
| | Big | 0.417 |
| Attribute 2 (Aroma) | No Scent | -0.637 |
| | Weak Scent | 0.177 |
| | Strong Scent | 0.460 |
| Attribute 3 (Price (Rp/kg)) | > 40000 | -0.577 |
| | 35000-40000 | 0.303 |
| | ≤ 35000 | 0.273 |
| Attribute 4(Moisture/Dryness) | Bad | -0.877 |
| | Medium | 0.370 |
| | Good | 0.507 |
| (Constant) | | 3.053 |
| Pearson's R = 0.985 | | Significance = 0.000 |
| Kendall's tau =0.915 | | Significance = 0.000 |

Where in general most of the consumers like fresh shallots with price ranging from Rp.35000/kg to Rp.40000/kg, and below Rp.35000/kg, and they do not like shallots with price attribute level above Rp.40000/kg.Furthermore, for the shallot bulb size, consumers, in general, are very fond of large bulb size (0.417) and dislike shallots with a small size. It can be seen from its utility value of -0.397.

**Table 4.**The most preferred combination of Respondents (Maximum Utility Rule)

| Number of preferred combination of respondent | Size | Aroma | Price (Rp/Kg) | Moisture/Dryness | Total | Constant | Utilities |
|---|---|---|---|---|---|---|---|
| 2 | 0.147 | 0.460 | 0.273 | 0.370 | 1.520 | 3.053 | 4.573 |
| 5 | -0.020 | 0.177 | 0.273 | 0.507 | 0.937 | 3.053 | 3.990 |
| 8 | 0.417 | -0.637 | 0.303 | 0.507 | 0.590 | 3.053 | 3.643 |
| 9 | -0.397 | 0.177 | 0.303 | 0.370 | 0.453 | 3.053 | 3.506 |
| 7 | -0.397 | -0.637 | 0.273 | 0.877 | 0.116 | 3.053 | 3.169 |
| 6 | -0.397 | 0.460 | -0.577 | 0.507 | -0.007 | 3.053 | 3.046 |
| 4 | -0.020 | 0.460 | 0.303 | -0.877 | -0.134 | 3.053 | 2.919 |
| 1 | 0.417 | 0.177 | -0.577 | -0.877 | -0.860 | 3.053 | 2.193 |
| 3 | -0.020 | -0.637 | -0.577 | 0.370 | -0.864 | 3.053 | 2.189 |

The reality found in the research area is that at the level of fresh shallots consumers, the respondents' choices are still dominant with the attribute of low price. Therefore, low price attribute is still used as the benchmark that is considered essential for prospective consumers to decide whether to buy and consume the fresh shallots.From the total value of utility for combination number two, that is 1.52, the average number of consumers who like this combination composition is found to be 28% (28 respondents).

The results of this research will be the right basis for producers/traders to arrange a strategy in marketing fresh shallot. It can also be used by the government as an input in designing market strategy, making shallot export policy, arranging planting season, and arranging patterns distribution of local shallot, as well as stimulating local farmers, especially in North Sumatra to increase the production and quality of fresh shallots, which has a strong aroma, good level of moisture/dryness, competitive price and not inferior to the quality of other local shallots, such as Bima varieties from Brebes, Central Java.

*3.3 Average ShallotConsumption per gram/capita/day*

The level of shallot consumption in the study area is determined by the amount of shallot consumption at the time of the research conducted in 2016. The amount of shallot consumption can be known from the questions in the questionnaire. Then, the results obtained are compared with three indicators, namely:

| Level of Consumption: | kg/cap/ year | kg/cap/ month | gram/ cap/day |
|---|---|---|---|
| National: | 2.51 | 0.21 | 7 |
| Province of North Sumatera: | 2.59 | 0.21 | 7 |
| Medan City: | 3 | 0.25 | 8.33 |

From the result of the research on 100 respondents, the average of household consumption was 46.7 grams/RT/day, divided by the average number of household members, the average of fresh shallots consumption is 11.67 grams/cap/day. When compared with these three indicators, the level of fresh shallot consumer consumption per gram/ capita/day in the research area is still relatively higher than the consumption level of the three indicators.

In relation to the reality faced by consumers today, many varieties of shallots are offered. However, the results of this study indicate that the emphasis on loyalty to fresh local shallot consumption is still high.This is seen from the conclusion. From the results of research generally local or domestic shallotshas the characteristics of the level of attributes with medium size, sharp aroma of the price ≤Rp.35000/kg, humidity / moderate drought. and this study was conducted when the price of fresh onion is very high ranging from Rp.40000/kg- Rp.35000/kg.

## 4. Conclusion and Recommendation

Average fresh shallot consumption is 11.67 grams/cap/day. It indicates that fresh shallot consumption in the research area is still relatively higher when compared with the national average consumption level (7 grams/cap/day), consumption level of North Sumatera Province (7 grams/cap/day), and consumption level of Medan City (8.33 grams/cap/day).

Combination model preferred by fresh shallot consumers in general is the combination number two, that is fresh shallot with combination composition of large shallot bulb size, strong scent, price ranging below Rp.≤35000/kg and has medium moisture/dryness level, which means it is moist/slightly dry and has a little skin waste.

## References

[1] BadanLitbangPertanian. 2006. *ProspekdanArahPengembanganAgribisnisBawangMerah.* (Prospect and Direction of Agribusiness Development of Shallots).BadanPenelitiandanPengembanganPertanianKementerianPertanian, (Ministry of Research and Development for Agriculture) ,Jakarta.

[2] BKP Kota Medan. 2014. *PublikasiNeracaBahanMakanan 2014.*(Condensed Balance Sheet Foodstuffs) Medan.

[3] Kotler, Philip &Amstrong,Gary. 2008. *Prinsip – prinsipPemasaran* ($12^{th}$)(the principle of Marketing). PenerbitErlangga, Jakarta.

[4] Green, Paul E. & Krieger, Abba M. 1991. Segmenting Market with Conjoint Analysis.*Journal or Marketing*, Vol. 55

[5] BPS Medan. 2016.*MedandalamAngka*. Medan.

[6] Green, Paul E. & Krieger, Abba M. 1991. Segmenting Market with Conjoint Analysis.*Journal or Marketing*, Vol. 55

[7] Simamora, B. 2004. *PanduanRisetPerilakuKonsumen*. (The Guide of Consumer Behavior Research) PT GramediaPustakaUtama, Jakarta.

[8] Supriana, T. 2016. *MetodePenelitianSosialEkonomi*. USU Press, Medan.

# Analysis of supply chain management of shallots in Medan

**M C Alam and T Supriana**

Department of Agribusiness Universitas Sumatera Utara, Medan, Indonesia
tavihutasuhut@yahoo.co.id; mega@xtremenitro.org

**Abstract.** Supply chain is important for business. One of supply chain that needs to be studied is the shallots supply chain. Medan have high demand while the supply of shallots is limited. This study aims to analyze the flow of shallots supply chain distribution in Medan. The method used was survey by using questionnaires to shallots producers, collecting traders, distributors, traders as well as government involved in shallots supply chain. Descriptive analysis was used to explain the shallots supply chain distribution flow. The results showed that there are two shallots supply chain model in Medan that was local shallots model and imported shallots model. Local shallots model could be distinguished based on three producer area, those were models of Medan Marelan, Samosir, and Simalungun. Medan Marelan and Simalungun models have seven supply chains, while the Samosir Model has eight supply chains. This condition indicates that the local shallots supply chain management in Medan was not efficient because of the length of the distribution channel. Supply chain imported shallots was more efficient because it had a shorter distribution flow with five supply chains.

## 1. Introduction

Shallot is a horticultural commodity belonging to spice vegetables. Problem nowadays is Medan's Central Market needs a supply of 100 tons/day of shallots to complement the needs of consumers. However, the amount of production in North Sumatra aimed at the Medan's Central Market is not sufficient to meet these needs. This can lead to scarcity and increasing the price of the shallots. Reported on Info Medan [1] the price of shallots reaches Rp 40,000.00/kg in Medan and shallots supply comes from Samosir and Java.

According Pujawan [2], components in supply chain management is divided into three main stream, namely: (1) The product component contains the flow of goods from the supplier to the consumer; (2) Component information contains order delivery and review of delivery status; (3) Financial component consists of credit line, payment and payment schedule, delivery accuracy and identity of owner.

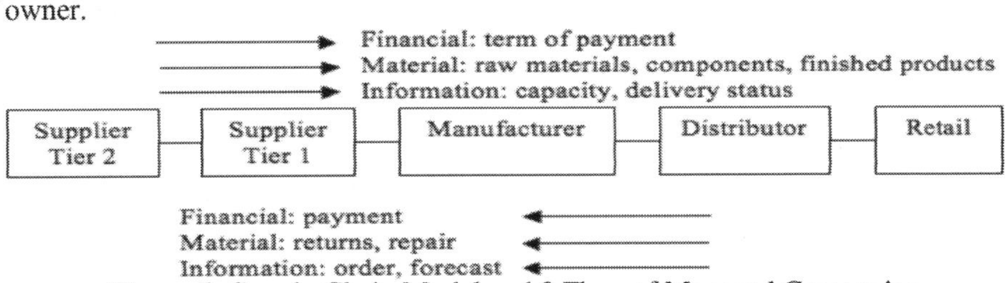


**Figure 1.** Supply Chain Model and 3 Flow of Managed Companies

Supply chain model and 3 kinds of flow managed by company as shown in Figure 1 [2]. Supply chain management is a process of regulating the flow of goods or products from a supply chain flow.

Supply chain management model applies how a network of production and distribution activities of a company can work together to meet consumer needs. The main goal of supply chain management is to deliver products on time to satisfy consumers, reduce costs, improve all outcomes of the entire supply chain, reduce time, centralize planning and distribution activities [3].

The structure of the supply chain of agricultural products according to Marimin and Maghfiroh [4] has a uniqueness because it does not always follow the order of the chain above. Farmers can directly sell their agricultural products directly to the retail market, thus cutting off the chain of middlemen, manufacturers and distributors. Manufacturers also do not have to supply products through their distributors to retail, but can go directly to customers. Customers here are usually big customers like restaurants, hospitals, or hotels. Manufacturers also use many services of exporters as distributors to market their products to international customers. The structure of the agricultural supply chain is shown in Figure 2 [4].

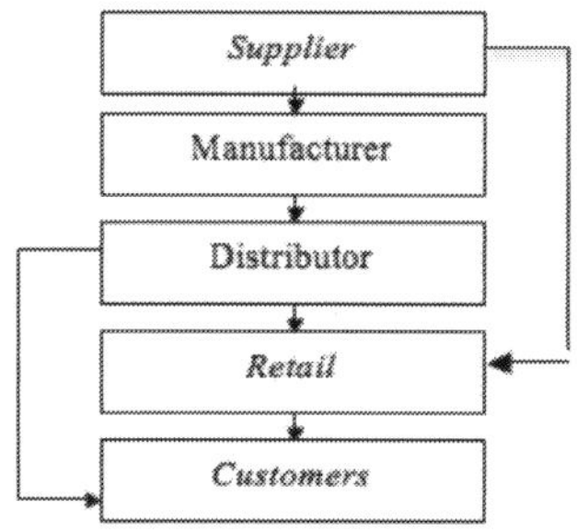


**Figure 2.** Agricultural Supply Chain Structure

This case applies to the supply of shallots in Medan. The supply of shallots that pass through the supply chain is too long can hamper the way of distribution and cause of high price of shallots received by consumers. So supply chain needs cuts that can reduce the price of shallots and that commodities can enter the Medan central market on time and there is no scarcity.

Based on the implementation of the supply chain management theory, it is expected that shallots supply in Medan can be supplied evenly in the central market or distributor so that the price of shallots does not bounce up due to the low supply of shallots and this research purpose is analyzing the supply chain distribution flow at Medan.

## 2. Material and Method

### *2.1 Experimental design*

The selected research location was Medan Central Market with the consideration that the market is the biggest consumers and distributors of the biggest shallots in Medan.

The sample of chains start from farmers is determined by snowball sampling method. Irianto and Mardikanto [5] stated that the snowball method is a sample selection technique by first defining a key person, then subsequent sample selection depends on the first informant, so on the longer rolling like a snowball. Thus it is expected that shallots supply chain at Medan can be identified.

For the farmers sample, at Medan there are 4 groups of shallots farmers are concentrated in the district of Medan Marelan. As for the area of Simalungun and Samosir which is the main supplier of shallots at Medan, joined 15 farmer groups. Based on the snowball sampling technique, there were 36 samples of producers (farmers) to be informants who have high enough shallot productivity scattered in some areas.

### *2.2 Data Analysis*

Method of data retrieval is a process of activity needed in a research. The process will generate data. The data were divided into two sources, namely Primary and Secondary Data. The way of collecting primary data as follows (a) Field Observation; (b) Survey; (c) Documentation; (d) Questionnaire.

The way of collecting secondary data obtained from the department or institution associated with this research. Sources from these secondary data were obtained from the Department of Agriculture Medan and the Central Bureau of Statistics (BPS) of Medan. These data were the general state of the research area, number of traders, distributors, and producers and other data relating to this research.

Data analysis used in this research was descriptive analysis. Qualitative descriptive analysis is an analysis that describes the distribution and supply chain of shallots at Medan. In addition, this analysis also describes the implementation of shallots supply chain management at Medan.

## 3. Results

Result of research on distribution of shallots commodities at Medan there are two model of supply chain based on its variety. The first model for the shallots supply chain of Medan and Brebes, and the second model for imported shallots supply chains. In Medan and Brebes shallots supply chain, can be differentiated based on the producer's shallot. The results of the study of supply chain model for the Medan and Brebes shallots displays in Figures 3 to 6.

1) Supply Chain Model of Shalllots Medan and Brebes from Medan Marelan

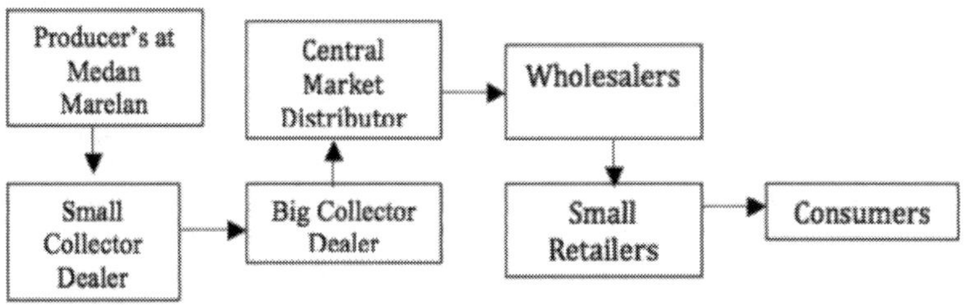


**Figure 3.** Model Supply Chain Shallots Medan and Brebes from Marelan District Entering into the Central Market Medan

Based on the study of primary data of supply chain actors and comparison of Indrajit and Djokopranoto [6] theory, can be found of supply chain of Shallots Medan and Brebes from Marelan District is having 7 (seven) supply chains. The first chain is the producer in Marelan, the second is a small collector dealers, the third chain is a big collector dealer, the fourth chain is the distributor of the central market, the fifth chain is the wholesaler, the sixth chain is the small retailers, and the last chain is the consumer.

According to the results obtained in the field, supply chain Shallots Medan and Brebes originating from the producers Marelan not efficient because of the long supply chain that is in the supply chain model of the shallot. This is similar to the theory expressed by Pujawan [2] that the longer the supply chain, the supply chain becomes more inefficient, due to the higher costs incurred.

2) Model of Supply Chain of Shallots Medan and Brebes from Samosir Regency

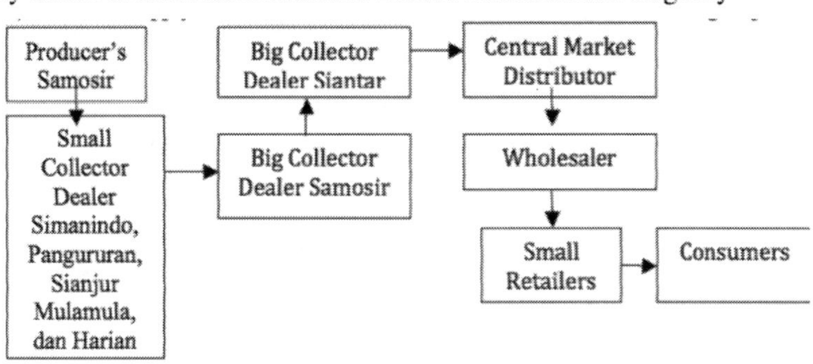


**Figure 4.** Model *Supply Chain* Shallots Medan dan Brebes from Samosir Regency entering to Central Market in Medan

Based on the study of supply chain primary data and comparison of Indrajit and Djokopranoto [6] theory, it is found that supply chain of Shallots Medan and Brebes from Samosir Producers is having 8 (eight) supply chains. The first chain is a producers in Samosir, the second is a small collector dealer spread in each subdistrict of shallots production center, the third chain is Samosir big collection dealer, the fourth chain is Siantarbig collector dealer, the fifth chain is the distributor of the central market, the sixth chain is the wholesalers, and the seventh chain is the small retailer and the last chain is the consumer.

The results obtained in the field, supply chain shallots Medan and Brebes from Samosir Producers not efficient because of the length of supply chain that is in the supply chain model of the shallot. This is similar to the theory expressed by Pujawan [2] that the longer the supply chain, the supply chain becomes more inefficient, due to the higher costs incurred.

3) Model Supply Chain Shallots Medan and Brebes from Simalungun Regency

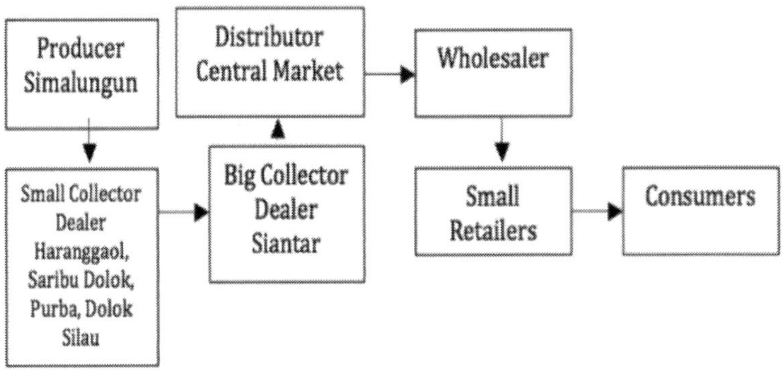


**Figure 5.** Model *Supply Chain* Shallots Medan dan Brebes from Simalungun which Entering to Central Market at Medan

Based on the primary data supply chain actors and comparison of the theory of Indrajit and Djokopranoto [6] found supply of Shallots Medan and Brebes from Simalungun is have 7 (seven) supply chain. The first chain is a producers in Simalungun, the second is a small collector dealer in each sub-district of shallots production center, the third chain is Siantar big collector dealers, the fourth chain is the distributor of the central market, the fifth chain is the wholesaler, and the sixth chain is a small retailer and the last chain is the consumer.

The results obtained in the field, supply chain shallots Medan and Brebes from Simalungun have not been efficient because of the length of the supply chain that is in the supply chain model of the shallots. This is similar to the theory expressed by Pujawan [2] that the longer the supply chain, the supply chain becomes more inefficient, due to the higher costs incurred.

4) Model Supply Chain of Imported Shallots

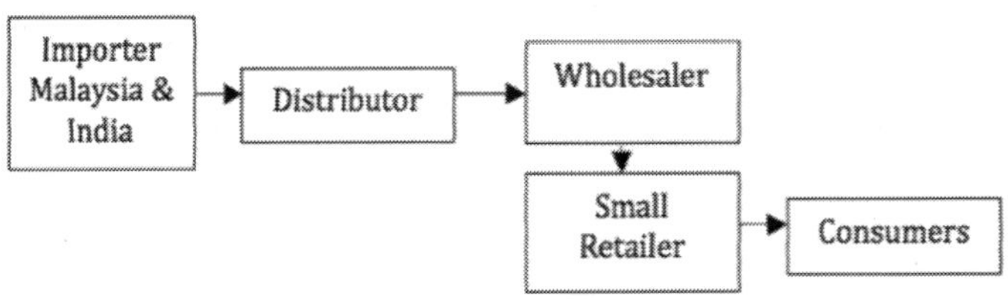


**Figure 6.** Model of Supply Chain of Imported Shallots

Based on the primary data supply chain actors and comparison of the theory Indrajit and Djokopranoto [6] found the offender supply of Bawang Merah Medan and Brebes from Imported is have 5 (five) supply chain. The first chain is the importer, the second is the distributor of the central market, the third chain is the wholesaler, the fourth chain is a small retailer, and the fifth chain is the consumer.

Results obtained based on interviews with Head of Sarana and Infrastructure of North Sumatera Province and Deputy Head of Central Market Medan, imported shallots supply chain in Medan including efficient supply chain. Because the chain is not too long, then the selling price of imported shallots tends to be cheap. However, imported shallots in the central market is indicated the illegal shallots because the policy of shallots import has been closed by the government.

Based on the theory of Pujawan [2] there are 3 (three) streams managed in the supply chain. First is the flow of goods or commodities that flow from upstream to downstream, the second is the financial/money flow from downstream to the upstream, and the third is the flow of information that can flow from upstream to downstream or instead.

This study also obtained an overview of the three streams managed in the shallots supply chain at Medan. This description was obtained based on interviews of respondents and comparisons of the Pujawan's theory [2]. Figure 8 shows the flow pattern in the shallots supply chain at Medan.

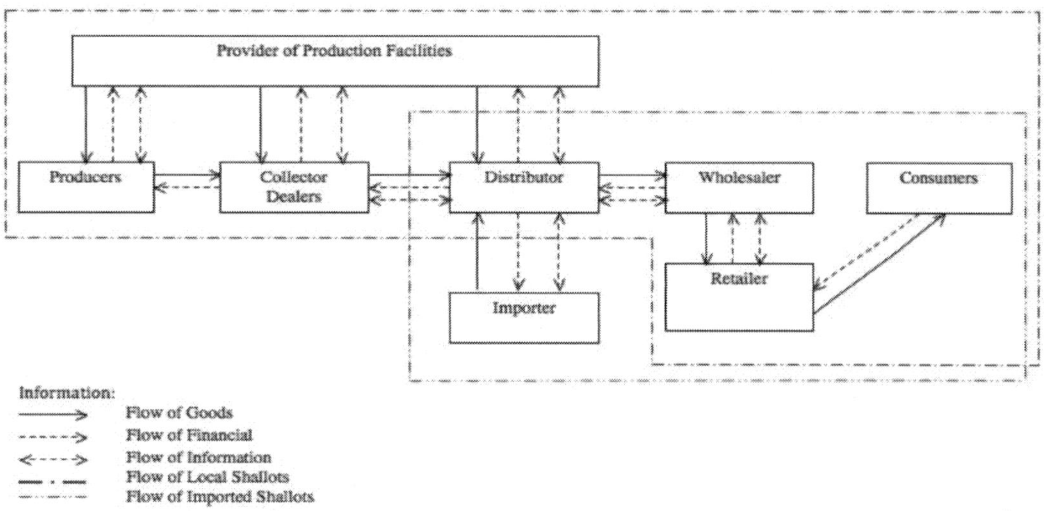


**Figure 7.** Flow in the Supply Chain of Shallots at Medan

## 4. Conclusion and Suggestions

Based on the results of the analysis that has been done in the study, it can be obtained the following conclusions:

The local shallots supply chain distribution flow in Medan has not been efficient, due to the length of supply chain from producer to consumer. For imported shallots, the supply chain distribution flow is efficient because of the short supply chain from importer to consumer. The supply chain distribution flow is as follows: Local shallots supply chain distribution pattern (Medan and Brebes) based on production centre of Medan Marelan was Producers Marelan → Small Collector Dealers → Big Collector Dealers → Distributor → Wholesalers → Retailer → Consumer; Local shallots supply chain distribution pattern (Medan and Brebes) based on Samosir production center was Producers Samosir → Small Collector Dealer Simanindo, Pangururan, Sianjur Mulamula, Harian → Big Collector Samosir → Big Collector Dealer Siantar → Distributor → Wholesale → Retailer → Consumer; Local shallots supply chain distribution pattern (Medan and Brebes) based on Simalungun production center was Producers Simalungun → Small Collector Dealers Haranggaol, Saribu Dolok, Purba, Dolok Silau → Big Collector Dealers Siantar → Distributor → Wholesaler→ Retailer → Consumer; Distribution pattern of imported shallots supply chain was Importer → Distributor → Wholesalers → Retailer → Consumer.

Recommendations proposed to establish efficient supply chain management, need support from all supply chain actors in order to be optimally implemented. The establishment of efficient supply chain management is able to optimize the speed of service time, creating prosperity among business actors throughout the supply chain. This can be realized through collaborative planning those are production and sales planning undertaken jointly by the parties that collaborate.

**References**

[1] Info Medan, 2017. *Harga Bawang Merah di Pusat Pasar Melambung Tinggi.* (Price of Shallots in the Market Centre was Soared). Available at http://www.dimedan.co/medan/harga-bawang-merah-di-pusat-pasar-medan-melambung-tinggi/

[2] Pujawan, I Nyoman. 2010. *Supply Chain Management.* Second Edition. Guna Widya. Surabaya.

[3] Chopra, S., and Meindl, P. 2001. *Supply Chain Management: Strategy, Planning, and Operations.* New Jersey – Prentice Hall. Jersey: Prentice Hall

[4] Marimin, M. dan Magfiroh, 2011. *Aplikasi Teknik Pengambilan Keputusan Dalam Manajemen Rantai Pasok.* (Application of Decision Making Technique in Supply Chain Management). Bogor: IPB Press.

[5] Irianto, H dan Mardikanto, T. 2011. *Metoda Penelitian dan Evaluasi Agribisnis.* (Research Method and Evaluation of Agribusiness). Jurusan/Program Studi Agribisnis Fakultas Pertanian UNS (Major Courses Agribusiness Agriculture Faculty UNS).

[6] Indrajit dan Djokopranoto, 2005. *Strategi Managemen Pembelian dan Supply Chain.* (Purchasing Management Strategy and Supply Chain). Grasindo. Jakarta.

# The income distribution and contribution of palm sugar producer in increasing the household welfare of palm sugar maker in Kolaka Southeast Sulawesi Indonesia

**I S Rianse[1], U Rianse[1], W G Abdullah[1], S Hartono[2], A Suryantini[2], Jamhari[2] and Muhidin[3]**

[1]Department of Agribussines, Faculty of Agriculture, Universitas Halu Oleo, Kendari South East Sulawesi 93212 Indonesia
[2]Department of Agribussines, Faculty of Agriculture, Universitas Gadjah Mada Jl. Flora, Bulaksumur, Yogyakarta, 55281 Indonesia
[3]Department of Agrotechnology, Faculty of Agriculture, Universitas Halu Oleo, Kendari South East Sulawesi 93212 Indonesia

Email : Ilma_lov3@yahoo.co.id

**Abstract.** The objective of the research were to analyze socio-economic characteristics and motivation of farmers in aren sugar processing business in Kolaka District, Southeast Sulawesi, Indonesia. The analysis used in this research was quantitative descriptive analysis. The results showed that average state of socio-economic characteristics of aren sugar farmer that were in the category of productive age, which was 46.12 years old, has fulfill basic education category (9 years education), low category of dependents as many as three people, the Average experience of aren sugar processing business during 18 years, and the Average aren trees were tapped as much as seven trees, every day, (b) farmer's motivation in aren sugar processing business in low category, with the indicator is the motive of imitation, economic, security, affiliations, awards, and self-actualization

## 1. Introduction

The target of agricultural development is directed to increase the income and welfare of farmers through the empowerment of farming communities. In achieving the development goals, the policy and strategy of agricultural development is done through the development and system of agribusiness efforts as a whole and integrated. Agribusiness is not only agriculture, but includes industries that produce agricultural production facilities and agricultural processing industry (agroindustry) [1].

*Aren* sugar agroindustry is an effort to diversifying sugar and increase the utilization of *aren* sugar economically and stimulate the rural community to participate actively in small scale or household industry. *Aren* agroindustry is also a container to obtain or increase family income even though the processing is still very simple (process and its equipment), with limited human resources [2-5]. Discussion about income distribution concerning poverty and inequality was released in the 1970s in both developed and developing countries [6]. Assessment of the welfare problem is also widely done in Indonesia, including the distribution problem of land, farm income, and household incomes.

Aren sugar has been produced in several districts in Southeast Sulawesi, namely, Bombana, Kolaka, and Muna. The most popular brown sugar in Kendari City (the capital of Southeast Sulawesi) is *aren* sugar from Kolaka District. Therefore, Objectives of this research were to find out how the

IOP Conf. Series: Earth and Environmental Science **122** (2018) 012016 doi:10.1088/1755-1315/122/1/012016

income distribution of *aren* sugar producer in household industry scale, to know the income contribution of *aren* sugar household industry to the total income of *aren* sugar producer house hold, and to know the role of *aren* sugar house hold industry to the welfare level of sugar producer in Kolaka District.

## 2. Material and Method

### *2.1. Sampling method*

The population of this research was people who tapped *aren* sap and produce *aren* sap into *aren* sugar in house hold industry scale in the Wolo Sub District, Kolaka District, as much as 72 people. Sampling method for aren sugar producers was purposive random sampling method. Sample was taken using purposive random sampling method.

### *2.2. Data analysis*

*2.2.1. Income distribution.* The measurement of the income distribution used the drawing method with the Lorenz Curve and the number method with the Gini coefficient. Gini coefficient was calculated by using the formula that is:

$$G = 1 - \sum_{i=1}^{k} fi\,(Y_i + Y_{i-1}) \tag{1}$$

Where:

G : Gini indeks
k : number of group
$f_i$ : The proportion of the number of households in the classi,i = 1
$Y_i$ : The cumulative proportion of income from the number of sample households to the it's grade
$Y_{i-1}$ : The cumulative proportion of the previous amount of income

The criteria used to determine the level of equally or imbalances of *aren* sugar producer's income distribution were:

a. Imbalance is low when Gini coefficient is between 0.20 – 0.35
b. Imbalance is medium when Gini coefficient is between 0.36 – 0.49
c. Imbalance is high when Gini coefficient is between 0.50 – 0.70

*2.2.2. Income contribution of aren sugar household industry to the total income of aren sugar producer household.* The amount of income contribution of *aren* sugar household industry to the aren sugar producer household could be obtained by calculating the percentage of income of *aren* sugar household industry to the total household income, by using formula:

$$\text{income contribution} = \frac{\text{income of } \textit{aren} \text{ sugar household industry}}{\text{total household income}} \times 100\% \tag{2}$$

The criteria of income contribution [7] as follows:

- Income contribution is very low if its value < 25%.
- Income contribution is low if its value about 25 - 49%.
- Income contribution is very high if its value about 50 - 75%.
- Income contribution is very high if its value < 75%.

### *2.3. Analysis of household welfare*

Welfare was analysed by GSR (Good Service Ratio), ie compare the consumption expenditure of (primary needs) to services and services (secondary needs).

$$GSR = \frac{\textit{the expenditure of primary needs}}{\textit{the expenditure of secondary needs}} \tag{3}$$

Welfare criteria:
GSR > 1 means; *aren* sugar producer is less prosperous
GSR = 1 means; *aren* sugar producer is prosperous
GSR < 1 *aren* sugar producer is more prosperous

## 3. Result and Discussion

### *3.1. Household income of aren sugar producer*

The income of *aren* sugar producer household comes from various sources. It was divided into two kind source, namely income from in and out of *aren* sugar house hold industry.

**Table1.**Distribution of *aren* sugar producer by household income, Kolaka District, 2014

| No | Income of aren sugar household industry (Rp year$^{-1}$) | Number (people) | Percentage (%) |
|---|---|---|---|
| 1 | < 50,402,685.19 | 50 | 69.44 |
| 2 | 50,402,685.19 – 93,409,787.04 | 20 | 27.78 |
| 3 | > 93,409,787.04 | 2 | 2.78 |
| | Total | 72 | 100.00 |
| Average | | | 39,684,523.15 |
| Max | | | 136,416,888.89 |
| Min | | | 7,395,583.33 |
| No | Household income without income of aren sugar household industry (Rp year$^{-1}$) | Number (people) | Percentage (%) |
| 1 | < 41,911,500 | 69 | 95.83 |
| 2 | 41,911,500 – 83,823,000 | 1 | 1.39 |
| 3 | > 83,823,000 | 2 | 2.78 |
| | Total | 72 | 100.00 |
| Average | | | 11,107,720.15 |
| Max | | | 125,734,500.00 |
| Min | | | 0.00 |

| No | Total household income of *aren* sugar producer (Rp/year) | Number (people) | Percent (%) |
|---|---|---|---|
| 1 | <64,820,055.57 | 54 | 75.00 |
| 2 | 64,820,055.57 – 122,244,527.78 | 14 | 19.44 |
| 3 | >122,244,527.78 | 4 | 5.56 |
| | Total | 72 | 100.00 |
| Average | | | 50,792,245.29 |
| Max | | | 179,669,000.00 |
| Min | | | 7,395,583.33 |

Other income sources of *aren* sugar producer come from farming and services. The farming income came from food crops and or plantation, such as rice field, cocoa, clove, *langsat, durian,* and others. The incomes from services, among others carpenters, retailer of nine basic materials of daily needs, and merchant collectors, and other.

### *3.2 Income distribution*

*3.2.1. Gini index method (gini ratio).* Based on the analysis of Gini Index, household income of *aren*sugar producer without income of aren sugar household industry and income of aren sugar household industry both were in the category of low imbalance with Gini Index each other of 0.35, and 0.20. As well as the total household income of aren sugar producer was in the low imbalance category with Gini Index of 0.22

Gini index of the income of aren sugar household industry lower than household income of *aren* sugar producer without income of aren sugar household industry. It means that the income of *aren* sugar household industry made the income distribution of *aren* sugar producer more balance. The analysis result of income distribution by using Gini Index could be seen in Table 2

**Table 2.** Gini index value of household income index of aren sugar producer, Kolaka District, 2014

| No | Income source | Gini index value | Imbalance category |
|---|---|---|---|
| 1 | Aren sugar household industry | 0.20 | Low |
| 2 | Total household income without industry | 0.35 | Low |
| 3 | Total Income | 0.22 | Low |

*3.2.2. Lorenz curve.* The Lorenz curve provide an overview of the balance level and quantitative relationship between the percentage of income recipient and the percentage of income received. Here was a figure of Lorenz curve calculated from the Gini Index on Table 2.

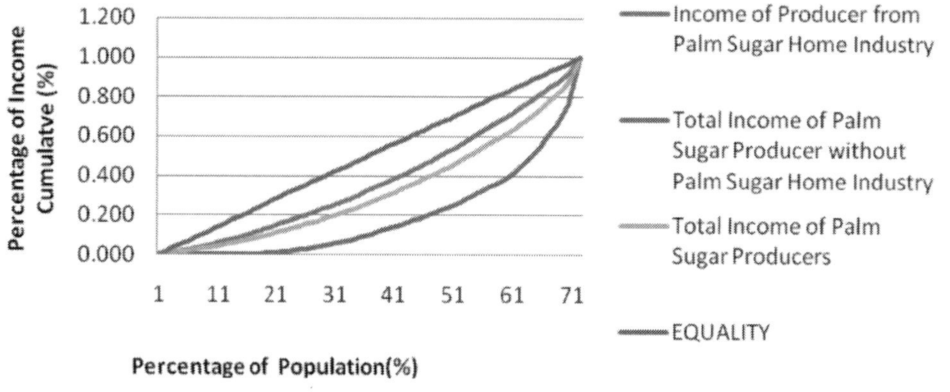


Figure1. Lorenz curve of income distribution of aren sugar producer in Kolaka District

Figure 1 showed that the Lorenz curve of total income distribution without the *aren* sugar household industry lies somewhat away from the perfect balance line, whereas Lorenz curve of income of *aren* sugar households industry was closest to the balance line, and the *Lorenz curve* of the total income distribution was between both. It could be explained that the income distribution of *aren* sugar household industry was more balance than others. Therefore, the income of *aren* sugar household industry played a role in improving the income distribution of aren sugar producer more balance.

*3.2.3. Income contribution of aren sugar household industry to total household income.* The income contribution of *aren* sugar house hold industry to the household's income as follows:

$$\text{income contribution} = \frac{\text{income of } aren \text{ sugar household industry}}{\text{total household income}} \times 100\% \quad (4)$$

IC = (39,684,523.15/50,792,245,29) x 100% = 78.13%

The income contribution of *aren* sugar household industry to the total household income of aren sugar producerwas 78.13 percent. This showed that the *aren*sugarhousehold industry give high contribute in supporting the household economy of aren sugar producer. It also showed by the result of one sample t-test, where the value of t-count was greater than t-table at error level α 0.05 (10.529> 2.646). The result of one sample t-Test of income contribution of aren sugar household industry could be seen *in Table 3.*

**Table 3.** The result analysis of one sample t-Test of income contribution, aren sugar household industry in Kolaka District, 2014

| Variable | Mean | Standar Deviation | Mean Error Standard | Criteria | t-count |
|---|---|---|---|---|---|
| Income contribution | 77.4899 | 22.15420 | 2.61090 | ≥ 50% | 10,529 |

α =0.05; N= 72; df = 71; t-table = 2.646

*3.2. 4.The welfare level of aren sugar producer*
*a. Criteria of good service ratio (GSR) welfare.*
The average expenditure of *aren* sugar producer household for primary needs amounted to Rp 1,449,861.00 while for secondary needs amounted to Rp 2,148,905.00. Based on that value, GSR (Good Service Ratio) was calculatedas follows:

$$GSR = \frac{\textit{the expenditure of primary needs}}{\textit{the expenditure of secondary needs}}$$

$$GSR = 1{,}449{,}861/2{,}148{,}905 \tag{5}$$

$$GSR = 0{,}67$$

Based on the results of GSR analysis, the ratio of household consumption expenditure for primary and secondary needs was amounted 0.67(GSR < 1). It means that the aren sugar producer household was more prosperous. It also showed by the result of one sample t-test that could be seen in Table 4.

**Table 4.** Result analysis one sample t-Test good service ratio, craftsman household industry sugar Aren in Kolaka District, Year 2014

| Variable | Mean | Standar Deviation | Mean Error Standard | Criteria | t-count |
|---|---|---|---|---|---|
| GSR | 0.8315 | 0.4896 | 0.0577 | < 1 | -2.920 |

α =0.05%; N= 72; df = 71; t-tabel = 2.646

Table 4 showed that t-count value was smaller than t-table at error level α 0.05 (-2,920 <2,646). This means that the null hypothesis is rejected and the alternative hypothesis is accepted, therefore, it could be concluded that aren sugar producer was economically prosperous.

## 4. Conclusion

Based on the research concluded that (1) the income of aren sugar household industry made the income distribution of aren sugar producer more balance, (2) income contribution of aren sugar household industry to the total income of aren sugar producer household, (3) the aren sugar household industry have important role to the welfare level of aren sugar producer in Kolaka District.

**References**

[1] Buwono HX 2001 *Otonomi Pembangunan Pertanian dalam Pembangunan Pertanian Pada Otonomi Daerah* (Agriculture Development Autonomy in Agriculture Development on Region Autonomy) (Yogyakarta : Edisi revisi LP2KP Pustaka Karya Yogyakarta)

[2] Rianse IS, Hartono S, Any S and Jamhari 2016. Analysis of Aren sugar Agroindustry Value Added in Kolaka Regency Indonesia *International Journal of Business and Management Studies* **05(01)** 13–22

[3] Rianse IS 2016 *Analisis Keberlanjutan Usaha Industri Rumah Tangga Gula Aren di Kabupaten Kolaka (*Jogjakarta : Disertasi. Universitas Gadjah Mada)

[4] Abdullah WG, Rianse U, Iswandi M, Taridala SAA, Rianse IS, Zulfikar LZ, Baka LR, La Abdi A, Cahyono E, Widayati W and Baka WK 2015 Potency of natural sweetener: Brown sugar. *WSEAS Transaction on on business and economics* **12** 99-110.

[5] Baka WK, Rianse U, Sidu D, Widayati W, Cahyono E, Rianse IS, Zulfikar L, Abdi AL and Baka LR 2016 Customer Behaviour Model of Brown Sugar Commodity *Int. J. Business Innovation and Research* **11(3)** 444-60

[6] Todaro, Michael P and Smith SC 2003 *Economic Development: 8th Edition* [Boston : Addison Wesley)

[7] Patty Z 2010 Kontribusi Komoditi Kopra Terhadap Pendapatan Rumah Tangga Tani di Kabupaten Halmahera Utara (Copra commodity contribution in farmer households in North Halmahera District) *Jurnal Agroforesty* **3(3)** 213-20

IOP Conf. Series: Earth and Environmental Science 122 (2018) 012017 doi:10.1088/1755-1315/122/1/012017

# Influence of land area and capital strengthening fund of rural economic enterprises toward corn production in North Sumatera province

**Rahmanta**

Lecturer of Agribusiness Study Program, Faculty of Agriculture, University of North Sumatra. Jl. Prof. A Sofyan No. 3 Kampus USU Medan Email: gintingrahmanta@gmail.com, Phone: 081319035250

**Abstract.** Corn is one of the staple food crops. Corn can also be processed into various foods and also as animal feed. The need for corn will continue to increase from year to year so it is necessary to increase production. The government has targeted corn crop self-sufficiency to achieve the corn production standards required by the animal feed industry. The purpose of this study is to analyze the effect of land area and capital strengthening funds to rural economic enterprises on corn production. This study uses secondary data obtained from the Central Statistical Agency of North Sumatra Province. The research method used is panel regression method. The result shows that the area of land has a significant effect on corn production and the capital strengthening fund to the rural economy institution has an insignificant effect on corn production in North Sumatera Province.

## 1. Introduction

Corn is the second most important food commodity in Indonesia after rice but corn is not a major product in the agricultural sector. Corn is one of the staple crops consumed by most residents other than rice, cassava, sweet potatoes, and sago. In addition, corn can also be processed into various foods that are a source of calories and also as animal feed. As a product between rice cultivation, corn is also produced intensively in some areas in Indonesia which is a corn producer.

The need for corn will continue to increase from year to year in line with the improvement and progress of corn-based industry, so it is necessary to increase production through human resources and natural resources, land availability and yield and technology potential. This condition makes corn cultivation has a very promising prospect, both in terms of demand and selling price. The government has targeted self-sufficiency of corn cultivation to achieve the corn production standards required by the animal feed industry. To realize this, the government made several efforts, among others, to cooperate with private parties engaged in the field of animal feed and food industries that use corn as raw materials. So the government in the effort to develop corn crop will be developed in areas known as corn production center with the district system. These areas include northern Sumatera, southern Sumatera, central Java, eastern Java, and Sulawesi.

The extent of agricultural land cultivation is something that is very important in the production process or farming and agricultural business. In farming, for example, the ownership or control of narrow land is definitely less efficient than the wider land. The narrower the business land, the less efficient the farming is. Unless farming is done in good order and good administration and appropriate technology. The level of efficiency actually lies in the application of technology. Because of the narrower extent, the application of technology tends to be exaggerated (this is closely related to land-to-hectare conversion) and makes business inefficient.

Farm land is said to be productive if the agricultural land can produce the production in the field of farming is satisfactory. To increase agricultural production, every farmer is increasingly dependent on the resources of his wider environment. Various information on the possibilities of land use and the limitation of the physical environmental factors is very important in discussing land use planning and patterns. In addition, information on social, economic, and community factors in the land itself is also needed as a support for consideration in planning and land use patterns.

In order to cope with the fall in corn prices that often occur during harvest time, the government through the food security agency - the department of agriculture since 2003 has made a policy through the distribution of Bridging Fund, which is also called Capital Strengthening Funds Rural Economic Enterprises (DPM- LUEP) to rural economic enterprises to improve their ability to buy grain from farmers especially at harvest time with reference to government purchasing price (HPP).

The Capital Strengthening (DPM) Funds Channeled to Rural Economic Enterprises (LUEP) in 2016 are APBD funds allocated to maize production centers as interest-free bailouts that can be used repeatedly by LUEP to buy farmers' grain and the deadline for utilization of DPM- LUEP which is a maximum of 2 (two) years from the date of signing of cooperation in accordance with the established mechanism.

Based on the description above, the problem in this research is how the influence of land area and capital strengthening fund to a rural economy business institution to corn production in North Sumatera Province. The purpose of this study is to analyze the effect of land area and strengthening of capital to rural economic enterprises on corn production in North Sumatera Province.

## 2. Research Methods

### *2.1. Data Analysis Method*

The free variables in this study are the area of land, capital strengthening funds channeled to rural economic enterprises. The dependent variable in this research is corn production in North Sumatera Province. So the equation model in this research is as follows.

$$PRD_{it} = a_o + a_1LH_{it} + a_2DPM_{it} + e_{it} \quad (1)$$

Where:

$PRD_{it}$ : Corn production in North Sumatera Province
LHit : Land corn area
$DPM_{it}$ : Fund strengthening capital to rural economic enterprises
eit : Error term
$a_o$ : Constants
$a_1$, $a_2$ : Regression coefficients
i : 7 cross-section units (Deli Serdang, Serdang Bedagai, Langkat, Karo, Simalungun, Dairi, and Toba samosir)
t : Time series of 5 (2011-2015)
n x t : Number of panel data (7x5 = 35)

## 3. Results and Discussion

### *3.1 Selection of Panel Data Estimation Model*

To find out whether the PLS or FEM model to be selected for data estimation is done F-test/Chow Test. To be more clear about the model PLS or FEM to be used in the estimation can be tested Chow.

IOP Conf. Series: Earth and Environmental Science **122** (2018) 012017 doi:10.1088/1755-1315/122/1/012017

*3.1.1 Chow Test*

**Table 1.** Chow Test

Redundant Fixed Effects Tests
Pool: POOL01
Test cross-section fixed effects

| Effects Test | Statistic | d.f. | Prob. |
|---|---|---|---|
| Cross-section F | 16.239793 | (6,26) | 0.0000 |
| Cross-section Chi-square | 54.517702 | 6 | 0.0000 |

Cross-section fixed effects test equation:
Dependent Variable: PRD?
Method: Panel Least Squares
Date: 09/11/17 Time: 14:18
Sample: 2011 2015
Included observations: 5
Cross-sections included: 7
Total pool (balanced) observations: 35

| Variable | Coefficient | Std. Error | t-Statistic | Prob. |
|---|---|---|---|---|
| C | -50.25181 | 12.14507 | -4.137632 | 0.0002 |
| LH? | 6.534269 | 0.289414 | 22.57757 | 0.0000 |
| DPM? | 0.040587 | 0.037835 | 1.072751 | 0.2914 |

| | | | |
|---|---|---|---|
| R-squared | 0.946913 | Mean dependent var | 169.8837 |
| Adjusted R-squared | 0.943595 | S.D. dependent var | 165.7939 |
| S.E. of regression | 39.37556 | Akaike info criterion | 10.26598 |
| Sum squared resid | 49613.91 | Schwarz criterion | 10.39930 |
| Log likelihood | -176.6547 | Hannan-Quinn criter. | 10.31200 |
| F-statistic | 285.3927 | Durbin-Watson stat | 0.400502 |
| Prob(F-statistic) | 0.000000 | | |

H0 : Common Effect Model
H1 : Fixed Effect Model (H1 diterima FEM Model Terbaik)

From the table 1, it can be seen that the value of prob cross section chi-square is 0.000 whose value is smaller than the alpha value of 0.10. This shows that the null hypothesis (H0) is rejected means the appropriate model to be used in this equation is the Fixed effect model (FEM).

*3.1.2 Hausman Test.* With the result of better estimation between PLS with FEM, the next step is to determine a more efficient model whether to use fixed effect model or random effect model with Hausman test. To be more clear about the FEM or REM model that will be used in the estimation it can be done Hausman Test.

IOP Conf. Series: Earth and Environmental Science **122** (2018) 012017 doi:10.1088/1755-1315/122/1/012017

**Table 2.** Hausman Test

Correlated Random Effects - Hausman Test
Pool: POOL01
Test cross-section random effects

| Test Summary | Chi-Sq. Statistic | Chi-Sq. d.f. | Prob. |
|---|---|---|---|
| Cross-section random | 4.820674 | 2 | 0.0898 |

Cross-section random effects test comparisons:

| Variable | Fixed | Random | Var(Diff.) | Prob. |
|---|---|---|---|---|
| LH? | 5.016771 | 5.961076 | 0.269355 | 0.0688 |
| DPM? | -0.008466 | -0.005182 | 0.000009 | 0.2748 |

Cross-section random effects test equation:
Dependent Variable: PRD?
Method: Panel Least Squares
Date: 09/11/17 Time: 14:26
Sample: 2011 2015
Included observations: 5
Cross-sections included: 7
Total pool (balanced) observations: 35

| Variable | Coefficient | Std. Error | t-Statistic | Prob. |
|---|---|---|---|---|
| C | 8.580397 | 22.86662 | 0.375237 | 0.7105 |
| LH? | 5.016771 | 0.694106 | 7.227674 | 0.0000 |
| DPM? | -0.008466 | 0.022241 | -0.380664 | 0.7065 |

Effects Specification

Cross-section fixed (dummy variables)

| | | | |
|---|---|---|---|
| R-squared | 0.988818 | Mean dependent var | 169.8837 |
| Adjusted R-squared | 0.985378 | S.D. dependent var | 165.7939 |
| S.E. of regression | 20.04823 | Akaike info criterion | 9.051193 |
| Sum squared resid | 10450.22 | Schwarz criterion | 9.451139 |
| Log likelihood | -149.3959 | Hannan-Quinn criter. | 9.189254 |
| F-statistic | 287.4026 | Durbin-Watson stat | 1.324508 |
| Prob(F-statistic) | 0.000000 | | |

H0 : Random Effect Model
H1 : Fixed Effect Model (H1 accepted with α 10% FEM as the best model)

From Table 2, it can be seen that the probability of cross section random is 0.0898 whose value is more than the alpha value of 0.10. This shows that the null hypothesis (H0) is accepted that the appropriate model for use in this equation is the fixed effect model (FEM).

This research uses panel data and after done data processing using fixed effect model (FEM), then the result of fixed effect approach can be seen estimation result in the following table.

**Table 3.** Fixed Effects Model

Dependent Variable: PRD?
Method: Pooled Least Squares
Date: 09/11/17 Time: 14:15
Sample: 2011 2015
Included observations: 5
Cross-sections included: 7
Total pool (balanced) observations: 35

| Variable | Coefficient | Std. Error | t-Statistic | Prob. |
|---|---|---|---|---|
| C | 8.580397 | 22.86662 | 0.375237 | 0.7105 |
| LH? | 5.016771 | 0.694106 | 7.227674 | 0.0000 |
| DPM? | 0.008466 | 0.022241 | 0.380664 | 0.7065 |
| Fixed Effects (Cross) | | | | |
| _DS—C | -68.62331 | | | |
| _SB—C | -28.53563 | | | |
| _LKT—C | -10.99563 | | | |
| _KR—C | 107.7711 | | | |
| _SML—C | 33.17984 | | | |
| _DRI—C | -22.73709 | | | |
| _TS—C | -10.05926 | | | |

Effects Specification

Cross-section fixed (dummy variables)

| | | | |
|---|---|---|---|
| R-squared | 0.988818 | Mean dependent var | 169.8837 |
| Adjusted R-squared | 0.985378 | S.D. dependent var | 165.7939 |
| S.E. of regression | 20.04823 | Akaike info criterion | 9.051193 |
| Sum squared resid | 10450.22 | Schwarz criterion | 9.451139 |
| Log likelihood | -149.3959 | Hannan-Quinn criter. | 9.189254 |
| F-statistic | 287.4026 | Durbin-Watson stat | 1.324508 |
| Prob(F-statistic) | 0.000000 | | |

### *3.2 Coefficient of Determination ($R^2$)*

Determination Coefficient value ($R^2$) is used to determine the strength of independent variables (independent variable) describes the variation of the dependent variable (dependent variable). From the estimation result using fixed effect model obtained coefficient of determination $R^2$ equal to 0.9888 which means that variation of independent variable that is land area and capital strengthening fund to rural economy enterprise able to explain corn production variable in North Sumatera province as dependent variable equal to 98,88 percent and the rest 1.12 percent is explained by other variables not included in the estimation model.

*3.3 Test Together/F-Test*

Assessment of hypotheses together is done by using test F or with prob. If F-arithmetic > F-table or prob score < alpha (0.10) then the null hypothesis must be rejected and the alternative hypothesis must be accepted. Which means that together independent variables affect the dependent variable. From probability test obtained is equal to 0.000000 < alpha (0.05) then H0 is rejected and Ha accepted so that together independent variable (land area and capital strengthening fund to rural economic enterprise) have a significant effect on the dependent variable of corn production in the province of North Sumatra at a 90 percent confidence level.

*3.4 Influence of Land Area to Corn Production in North Sumatera Province*

Land area has the positive and significant effect on corn production in North Sumatera Province. From the results of the processed data presented in the table above can be seen that obtained the value of prob 0.000 <α = 0.05 consequently H0 rejected and Ha accepted so that there is a positive and significant influence between the land area of corn production in North Sumatra Province. The value of land area coefficient of 5.016 this explains that if the variable of the land area increased by one hectare then the maize production will increase by 5,0016 tons. In general, the more land (cultivated), the greater the amount of production produced by the land. The effect of the land area not only on the efficiency of farming, but also has an impact on the transfer and application of technology in agricultural development. If land ownership is more plentiful with a narrow mastery area, agricultural development efforts will be difficult. But if the tenure is wide enough, then the process of technology transfer will be easier to do.

*3.5 Influence of Capital Strengthening Funds to Rural Economic Enterprises on Corn Production in North Sumatera Province*

Capital strengthening funds to rural economic enterprises have a positive and insignificant effect on corn production in North Sumatra Province. From the results of the processed data presented in the table above can be seen that obtained the value of prob 0.706> α = 0.05 consequently H0 accepted and Ha rejected so that there is a positive and insignificant influence between capital strengthening funds to rural economic enterprises on corn production in the Province North Sumatra. The coefficient value of capital strengthening fund to rural economy enterprises amounted to 0.008. This explains that if the variable of capital strengthening fund to rural economy business institution increased by one million rupiahs then corn production will increase by 0,008 ton. Capital is the most important factor in agriculture especially related to production input and labor cost. In other words, the existence of capital very determines the level or kind of technology applied. Lack of capital causes a lack of input given so as to cause a risk of failure or low yields that will be accepted by farmers.

## 4. Conclusion

Based on the results of research and discussion that has been described, it can be concluded as follows: (a) Partially, the area of land has a positive and significant effect on corn production, and the capital strengthening fund to the rural economic enterprise has a positive and insignificant effect on corn production in North Sumatera Province, (b) Simultaneously, the area of land and capital strengthening funds to rural economic enterprises have a significant effect on corn production in North Sumatra Province.

## References

[1] Central Statistics Agency, 2016. North Sumatra In Figures 2016.
[2] Gujarati, D.2006. Basic Econometrics.
[3] Ghozali, I. 2016. Multivariate Analysis Application with IBM SPSS 23 Program.
[4] Mubyarto. 1989. Introduction of Agriculutural Economic.
[5] Mudrajad Kuncoro.2001. Quantitative Research Methods.

[6] Nachrowi, D.N. dkk.2005. Use of Econometrics Technik.
[7] Pratomo, W.A and Hidayat. 2010. Use of Eviews in Econometrics.
[8] Soekartawi.1996. Agricultural Development.
[9] Suratiyah.2015. Science Self Employed.
[10] Widarjono, A. 2013. Introduction Econometrics and Applications.

# The strategy of sustainable soybean development to increase soybean needs in North Sumatera

**L Handayani[1], A Rauf[2], Rahmawaty[3] and T Supriana[4]**

[1]AGraduate Student of Agricultural Science, Universitas Sumatera Utara, Indonesia
[2]Department of Agriculture Agroecotechnology, Universitas Sumatera Utara, Indonesia
[3]Departemen Agriculture Forestry, Universitas Sumatera Utara, Indonesia
[4]Department of Agriculture Agribusiness, Universitas Sumatera Utara, Indonesian
E-mail: tavihutasuhut@yahoo.co.id

**Abstract:** The objective of the research was to analyze both internal and external factors influencing the strategy of sustainable soybean development to increase soybean needs in North Sumatera. SWOT analysis was used as the method of the research through identifying internal factors in the development of sustainable soybean the strategy to increase soybean production in research area is aggressive strategy or strategy of SO (Strengths - Oppurtunities) that is using force to exploit existing opportunity with activities as follows: (1). Use certified seeds in accordance with government regulations and policies. (2). Utilizing the level of soil fertility and cropping patterns to be able to meet the demand for soybeans. (3). Utilizing human resources by becoming a member of farmer groups.

## 1. Introduction

Soybeans are the most important food crops after rice and maize. In addition, soy is also a commodity palawija rich in protein. Fresh soybeans are much needed in the food industry and soybean meal is needed for the feed industry. Soybean plays a role as an important vegetable source in order to increase the nutrition of the community because in addition to safe for health is also relatively cheap compared to the source of animal protein [1]

National soybean production has not been able to meet the needs, because the actual harvested area is still inadequate and the productivity is still low. Productivity at the farm level averaged 1.3 tons / ha, while the production potential reached 2.0 to 2.5 tons /ha. The large gap is partly due to the application of technology at the farmer level is still low, the use of limited seed and human resources still weak. Every year the industry knows tempe requires at least 1.85 million tons of soybeans, as well as the soy and tauco industry of around 325.220 tons. In addition, the seed industry also needs 25.843 tons of soybean, as well as for feeding about 8.319 tons of soybeans. To meet the shortage of needs each year, soybean imports reached 1.3 million tons. In fact, Indonesia is one of the countries that depend on imports of soybeans from the United States (US).

Availability of food is essential for the stability of a country. The ability to be self-sufficient can save foreign exchange that can be utilized for other strategic purposes. In fact, to date in Indonesia import needs especially in the sectors of food needs are always increasing because the demand exceeds the supply of available food. healthy lifestyle changes and progress in the field of agro-industry and farming sectors. Conditions such as these that ultimately trigger a shortcut to import soybeans to meet the needs of the public soy consumption. in addition, an increase in the volume of i mports was also caused by increased demand for import soybean because it has cheaper prices than local soybean [2]

Currently, soybean cultivation in paddy fields only reaches 65% of the total harvest area. The opportunity to increase soybean production is wide open due to the availability of potential area of about 800 thousand ha spread over 10 provinces. But the problem faced is to instill trust and interest of farmers to want to plant soybeans aback. Altemeier, K and T. Bottema[3] Argue that price protection is an effective way of encouraging increased production, adoption of fertilizer technology and labor absorption compared to fertilizer price subsidies. Acid Soils amelioration technology with high Al content is generally well known, that is by liming, giving organic matter and P fertilization from natural phosphate[4]. The soybean cultivation that reaches over 20 million ha in Brazil is almost entirely done on acid soils with high productivity after land amelioration is applied [5].

Conditions in the field also show that the application of soybean production technology components by farmers is not complete. Implementation of drainage channels, such as the width and depth of the channel, the number and location of the channel) is not in accordance with the recommendations. The soybean production technology package in NTB area shows the diversity of production level achieved by farmers between planting time, location and season. Application of true and exact technology package able to obtain result > 2.0 ton/ha [6]

There are some environmental factors related to the lower productivity of soybean. Drought and floodingas results of climatic anomaly and climate change are the main causal factors. Many sectors are affected by climate changes, and agriculture is the most susceptible sector related to extreme climate change. Ecosystem of rice and other food crops such as soybean are the common examples that impacted by the extreme climate changes [7].

Soybean (*Glycine max* L) is a highly nutritious food commodities as a source of vegetable protein an d low cholesterol at an affordable price. Soybeans also an important food commodity after rice and maize. Soy consumption in the form of fresh or in processed form can improve nutrition. Soybean plants can grow well in areas with rainfall around 100-400 mm / month with temperatures bet ween 21-34 degrees C and at a height of not more than 500 m above sea. In Indonesia, many proces sed soybeans for various foodstuffs, such as tempeh, soy milk, tofu, bean curd, soy sauce, oncom, tauco, soybean cake, ice cream, edible oil, and soy flour. In addition, it is also widely used as an animal feed ingredient [8].

From the above problems, it needs to be studied further about The Strategy of Sustainable Soybean Development to Increase Soybean Needs in Nort Sumatera.

## 2. Research Method

The data collected in this research consist of primary data and secondary data. Primary data is in the form of questionnaires obtained from the results of interviews as many as 120 soybean farmers consisting of 4 districts in North Sumatra namely Langkat, Deli Serdang, Serdang Bedagai and Simalungun using a list of questions that have been prepared previously. Secondary data are supporting data obtained from related institutions such as Central Bureau of Statistics (BPS) of North Sumatra and other related institutions. The method used in this research is SWOT analysis and position matrix to determine strategy in soybean development. This matrix produces four sets of alternative strategic possibilities, as illustrated in the Table 1.

Prior to the data analysis as above then first performed data collection using the matrix model internal strategy factor, external factor matrix as as illustrated in the Table 1.

**Table 1.** Category Assessment of Internal and External Factors

| Rating | Category | Internal Factors | External Factors |
|---|---|---|---|
| 4 | Very Good | Strength | Oppurtunity |
| 3 | Good | Strength | Oppurtunity |
| 2 | Pretty Good | Strength | Oppurtunity |
| 1 | Not Good | Strength | Oppurtunity |
| -4 | Very Good | Weakness | Threats |
| -3 | Good | Weakness | Threats |
| -2 | Pretty Good | Weakness | Threats |
| -1 | Not Good | Weakness | Threats |
| Total Score | | | |

Each internal factor of strength and external factors of opportunity is given very good category until it is not good and given a 4 start rating for category very good up to 1 for not good category. whereas any internal factors of weakness and external threat factors are categorized as excellent until not good and given a rating ranging from - 4 very good category up to - 1 for not good category.

## 3. Results And Discussions

### *3.1. Sustainable Soy Development Strategy*

Every farming certainly faces problems in the production process. But the problems in dealing with these goals must be able to determine the right development strategy of production in order to be able to position them selves in a favorable position. Based on the of data processing obtained from soybean farmers in the study area, the result is as follow.

**Table 2.** Merging of Internal and External Strategy Factor Factoring Matrix for Sustainable Soybean Development

| Strategic Factors | Rating | Weight | Score |
|---|---|---|---|
| Internal Strategy Factor | | | |
| *Strength* | | | |
| 1. Human Resources in Producing Soybean | 3 | 10 | 30 |
| 2. The capital used by farmers | | | |
| 3. Use of Certified Seeds | 2 | 6.67 | 13.34 |
| 4. Improving Cropping Patterns and Utilizing | 4 | 13.33 | 53.32 |
| Potential Land | 4 | 13.33 | 53.32 |
| 5. The Level of Land Fertility | 2 | 6.67 | 13.34 |
| Total Strength Score | 15 | 50 | 163.32 |
| *Weakness* | | | |
| 1. Utilization of Natural Potential | -3 | 11.54 | 34.62 |
| 2. Technology used Farmers | -3 | 11.54 | 34.62 |
| 3. Management System in farming | -3 | 11.54 | 34.62 |
| 4. The area of land cultivated | -2 | 7.69 | 15.38 |
| 5. Use of production facilities | -2 | 7.69 | 15.38 |
| Total Weakness Score | 13 | 50 | 134.62 |
| Difference (Strength - Weakness) | | | 28.70 |
| External Strategic Factors | | | |
| Oppurtunity | | | |
| 1. Selling Price of Soybean | 2 | 5.89 | 11.78 |
| 2. The existence of Soybean Processing Industry | 4 | 11.76 | 47.04 |
| 3. Organization of Farmer Groups | 3 | 8.83 | 26.49 |
| 4. Demand for Soybeans | 4 | 11.76 | 47.04 |

| | | | |
|---|---|---|---|
| 5. Regulations and government policies | 4 | 11.76 | 47.04 |
| Total Opportunity Score | 17 | 50 | 152.90 |
| *Threats* | | | |
| 1. Entry of imported soybeans | -3 | 10.71 | 32.13 |
| 2. Lack of Counseling | -2 | 7.14 | 14.28 |
| 3. Climate Change and Weather | -2 | 7.14 | 14.28 |
| 4. Development of Information and Communication | -3 | 10.71 | 32.13 |
| Technology | -3 | 10.71 | 32.13 |
| 5. Availability kiosks Means of Production | -1 | 3.57 | 3.57 |
| Threats Total Score | 14 | 50 | 128.52 |
| Range (Opportunity – Threats) | | | 24.38 |

*Source: Primary Data Analysis*

The data above shows that the difference of internal strategy factor (strength-weakness) is equal to 28.70 which means the influence of strength is greater than the influence of weakness in soybean development in research area. While the difference of external strategy factor (opportunity-threat) is 24.38 which means the influence of opportunity is greater than influence of the threat on soybean development in the research area.

**Table. 3.** SWOT Matrix

| INTERNAL / EXTERNAL | STRENGTH (S) | WEAKNESS (W) |
|---|---|---|
| | 1. Human Resources in Producing Soybean (S1)<br>2. The capital used by farmers (S2)<br>3. Use of Certified Seeds (S3)<br>4. Improving Cropping Patterns and Utilizing Potential Land (S4)<br>5. The Level of Land Fertility (S5) | 1. Utilization of Natural Potential (W1)<br>2. Technology used Farmers (W2)<br>3. Management System in farming (W3)<br>4. The area of land cultivated (W4)<br>5. Use of production facilities (W5) |
| OPPURTUNITY (O)<br>1. Selling Price of Soybean (O1)<br>2. The existence of Soybean Processing Industry (O2)<br>3. Organization of Farmer Groups (O3)<br>4. Demand for Soybeans (O4)<br>5. Regulations and government policies (O5) | STRATEGY SO<br>1. Utilizing human resources by being a member of farmer group (S1, O3)<br>2. Using certified seeds according to government regulations and policies (S4, O5)<br>3. Utilizing land fertility and cropping patterns to meet soybean demand (S5, S4, O1, O2, O4) | STRATEGY WO<br>1. Utilizing demand and selling price of soybean by expanding farming land and managing natural potency (W1, W4, O1, O3)<br>2. Following training to improve skills in the use of technology (W2, O2, O3, O5)<br>3. Encourage the improvement of farmer group activities in the form of farm management training (W3, O3) |
| THREATS (T)<br>1. Entry of imported soybeans (T1)<br>2. Lack of Counseling (T2)<br>3. Climate Change and Weather (T3) | STRATEGY ST<br>1. Using the land fertility level and improved cropping patterns to overcome the attacks of pests and diseases (S5, S4, T4)<br>2. Using capital to obtain the | STRATEGY WT<br>1. Looking for information that can give hope for a better farming (W3, T5)<br>2. Utilizing existing counseling for soybean farming on |

| | | |
|---|---|---|
| 4. Development of Information and Communication Technology (T4)<br>5. Availability kiosks Means of Production (T5) | required production facilities (S2, T6) | available land (W4, T2, T3, T6) |

From the above strategy results can be poured policies, programs and activities of soybean development as follows:
Strategy (S1,O3) :
Utilizing human resources by becoming a member of farmer groups
Effort :
Appropriate human resource upgrading so as to ensure the sustainability of a good so that the production can be maintained its quality, through the organization of farmer' groups, the farmers' experience will increase
Program :
Increased production, productivity and quality of sustainable soybean
Activities :
Agricultural institutions provide training to farmers
Strategy (S4,O5) :
Using seeds certified in accordance with government regulations and policies
Effort :
Increase the production of soybean with the use of certified seed to guarantee quality yields the good one
Program :
Prevention use of seeds that less quality
Activities :
Certified seeds obtained from agricultural services provided free of charge through farmer groups designated as seed breeders of F2 and F3 derivatives
Strategy (S5, S4, O1, O2, O4) :
Utilizing the land fertility and cropping patterns to meet the demand for soybeans
Effort :
Increasing production by planting soybeans on suitable land for soybean crops and the utilization of vacant land (fallow) post harvest rice can keep the soybean stock for industry.
Program :
The dependence of farmers on inorganic fertilizers (chemical) that make agricultural land damaged, the use of organic fertilizer can improve soil structure to be better
Activities :
Provision of subsidized fertilizer by the government to soybean farmers.

## 4. Conclusion

1. The strategy to increase soybean production in the research area is aggressive strategy or SO (Strengths - Oppurtunities) strategy that uses the power to exploit the existing opportunities
2. with uses the power to exploit the existing opportunities with the following activities : (1). Use certified seeds in accordance with government regulations and policies. (2). Utilizing the level of soil fertility and cropping patterns to be able to meet the demand for soybeans. (3). Utilizing human resources by becoming a member of a farmer group.

**References**

[1] Swastika D K, Nuryanti S S dan Sawit M H 2007 *Kedudukan Indonesia dalam Perdagangan Internasional Kedelai Dalam: Kedelai: Teknik Produksi dan Pengembangan (Indonesia Position In Soybean International Trade In Soybean Production Tecnique and Development)* Agriculture Research and Development Centre pp 20-44

[2] Atman 2009 Strategy to increase soybean production in Indonesia *Tambua Scientific Journals* **8**:39-45

[3] Altemeier K and Bottema T 1991 Agricultural diversification in Indonesia : price responses and linkganes in the food crop sektor, 1969-1988 on outlook to 2000 *Working Paper Series* No. 11. CGPRT Centre Bogor p 21

[4] Maid F X 2000 Soil fertility management for crop production on acid solis, Experiences from Brazil.p. 85 – 88. Dalam : L.W Gunawan et al, (Eds). *Penelitian dan Pengembangan Produksi Kedelai di Indonesia*. Kerjasama BPPT, Badan Litbang Pertanian, Kementrian Pendidikan, Sains Riset dan Teknologi Jerman Dit. Teknologi Lingkungan BPPT, Jakarta (Research and Development Production Soybean in Indonesia. Teamwork BPPT, Research and Development institusion Agriculture. Ministry of education, Science Research and Technology Jerman. Dit. Technology environment BPPT, Jakarta

[5] Maidl F X 1996 *Soil Fertility Management for Crop Production on Acid Soils*. Experiment from Brazil

[6] Adisarwanto T, Santoso B S and Sumarno 1998 Technology packing for soybean after wetland rice in Indonesia *Proc. World Soybean Research Conference V. (Eds)* Banpot Napampeth Bangkok Thailand pp 498-499

[7] Kaimuddin K, Kamaluddin A dan Sasmono M S 2013 Analisis tingkat kerentanan dan adaptasi terhadap perubahan iklim berbasis ekosistem padi di provinsi Sulawesi Selatan [Analysis of vulnerability and adaptation to climate change with rice-based ecosystem in South Sulawesi]. Retrieved from http://balitbangda.sulselprov. go.id/artikel-analisis-tingkat kerentanan dan adaptasi terhadap perubahan iklim berbasis ekosistem padi

[8] Fatah A, Satna and Andi 2012 *Soybean Cultivation Technology on Wetland* Institute for Agricultural Technology (BPTP) South Sulawesi

**Acknowledgements**

We thank the Faculty of Agriculture, Universitas Sumatera Utara for the implementation of the International on Conference on Agriculture, Environment, An Food Security, November, 7, 2017

# Pricing behavior of USA exporter in wheat international market

**R P Wibowo[1], Sumono[1], Y Iddrisu[2], M Darus[1], L P Sihombing[1] and Jufri[1]**

[1]Department of Agribusiness, Faculty of Agriculture Universitas Sumatera Utara, Indonesia
[2]CSIR - Council for Scientific and Industrial Research - Ghana · SARI

E-mail: rulianda_wibowo@yahoo.com

**Abstract.** The number of wheat producing countries is changing over time. It is expected the change in wheat supply will lead world wheat market become more competitive and reduce market power of major exporter country. This paper tries to identify and examined the degree of market power on wheat international market for USA by using the Pricing to Market (PTM) method. USA is the biggest producer and exporter in wheat market. The PTM method found that USA impose noncompetitive strategy by applying price discrimination and apply market power to their importer country.

## 1. Introduction

The number of countries who are producing wheat is increasing over time. Total number of wheat producing countries in 2011 is more than two times of total number of wheat producing countries in 1960 [1]. The total of wheat product in world market is constantly increasing and market share of major wheat exporting countries such as USA, Canada and Australia is steadily decreasing. All these changes give indication that the global wheat market will become more competitive in the near future. However due to increasing world population lead to higher demand and collusive behavior between major exporting countries, then more exporter and producer which we expect lead to more competitive market will not always come to reality.

There is a steady increase in export and import from 1961 to 2010 as can be shown in figure 1 and 2. Both the import and export has increased more than three times from 1960 to 2010. The major cause of increasing export is because there is a sharp increase in area harvested as there are many countries increased their wheat plantation such as Kazakhstan, Iran, Ukraine, Pakistan, India, etc. The increase in wheat plantation is also derived by high demand of wheat which is induced by the increase in population and consumption trend. Thus, it results in steady increase of import over the year. The increase in export and production also being enhanced by innovation and technology of wheat plantation as wheat farm productivity is increasing over the year significantly especially in major exporting country. The increasing of area harvested by countries especially from country who are not wheat major exporter may contribute to more competitive market as market importer may have more option in exporting wheat. However whether the development in wheat international market will lead to more competitive market is still a big question.

The market power ability provides the ability to the exporter firm to engage in non-competitive and collusion behavior. However even the exporter firm has the market power, but the degree in which they can impose their market power also depend on convexity of importer demand schedule. The application of market power can be traced by price discrimination imposed by exporter firm to importer. [2] noted that the demand characteristic and market competitiveness will determine the degree of price discrimination. The variation of degree of competition for each market depends on inside and outside competition [3].

**Figure 1.** World export by Tonnes

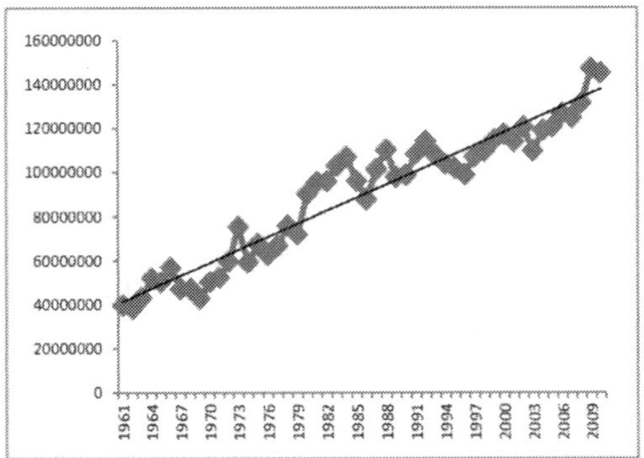


Source: FAO (2013)

**Figure 2.** World Import by Tonnes

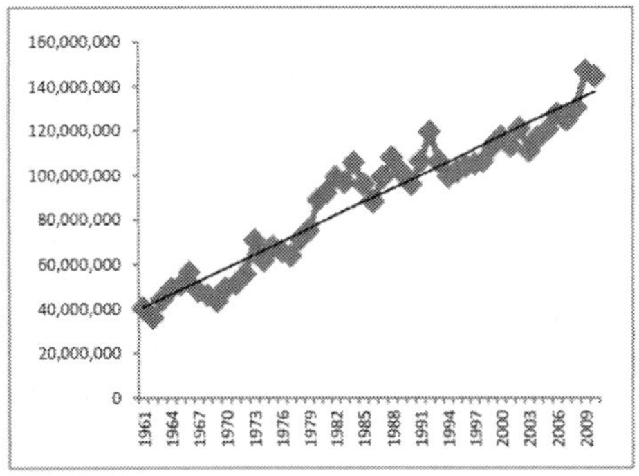


Source: FAO (2013)

[4] using USA wheat quarterly export data from 1978-1988 to eight destination market found that USA apply price discrimination to importer countries. Similar result from [5] also found that USA exporter apply PTM to six from seven importer between 1978-1988. This previous study suggests that the USA exporter seems to imposed price discrimination across different importer countries. There are two possible PTM method that the exporter may apply which are local-currency price stability (LCPS) and amplify the exchange rate effects. [6] found evidence that USA exporter tend to apply LCPS between period 1980-1998.

The purpose of this study is to measure market power of USA exporter in world wheat market after USSR breakup. The method that is going to be used is Pricing to Market Power (PTM) methods. The PTM method try to analyze the pricing behavior of producer when the exchange rate changes, a producer may choose to past the cost shock fully to its selling price or to absorb the shock to keep its selling price unchanged or combination between those two [7]. The PTM method is used to examine the competitiveness behavior of exporter. PTM occurs when an exporter holds his domestic currency export price steady or raises it for an importer who has realized a domestic currency appreciation [8]. This has the effect of allowing the importer's domestic currency price either to fall or to remain stable.

## 2. Data and Model

We use annual data on USA wheat export, PPI (producer price index) which are provided by FAO. The official exchange rate of importer currency against the exporter currency, CPI (consumer price index), GDP (gross domestic product) was obtained from World Bank. Some data related with PPI wheat is obtained from each importer statistic center. USA data set include 252 annual cross section time series observation from 1991-2011 for 12 wheat importing countries.

This paper adopts two way fixed effect model for of export prices across destination for PTM method proposed by [3].

$$lnP_{it} = \theta_t + \alpha_i + \beta ilne_{it} + \varepsilon_{it} \quad (1)$$

By using two way fixed effect model suggest by [9] we make commodity group to analyze how exporter apply price discrimination between export market. The PTM model become:

$$\ln p^x_{i,t} = \sum \alpha_i A_i + \sum \theta_i t_i + \delta_i \ln\left(\frac{e^x_{i,t}}{CPI_{i,t}}\right) + \gamma_i \ln GDP_t + \beta_i Oil_t + \vartheta_i PPI^x_{i,t}\varepsilon_{i,t} \quad (2)$$

where $p^x_{i,t}$ is wheat export price in country exporter price to importing country i in period t, $\alpha_i$ is a importer country effect, $\theta_i$ represents the specific time effect, $\delta_i$ denote the parameter on exchange rate variable which represents the elasticity of the domestic currency export price with respect to the exchange rate, $e_{it}$ is the exchange rate in units of destination market currency per unit of the exporter's currency and $CPI_{i,t}$ is the consumer price index for importing country i and $\varepsilon_{it}$ is the random disturbance with zero mean. Comparing to the original model we add more covariate by adding real GDP, oil price and wheat producer price index (PPI) from exporting countries. The reason that we add oil price and producer price index as the trend in price export may greatly effect by the variation in those covariate. We also estimate the PTM method with model not using Oil price and PPI. We compare between model using oil price and PPI with model which not using these covariates, surprisingly we have the same result for estimated parameter and variance of variable.

There are three possible form of market behavior that can be interpreted from the estimated parameter. The first is the competitive market behavior without price discrimination where export price are all the same across destination and price equal marginal cost. In this scenario we will find that $\propto_i$ and $\delta_i$ will be zero. The second is there is price discrimination across importer countries with constant elasticity of demand for each export market i.e $\propto_i \neq 0$, $\delta_i$=0. The last is there is price discrimination and elasticity of demand is not constant ie i.e $\propto_i \neq 0$, $\delta_i \neq 0$. Positive expected sign of $\delta_i i$ shows that the exporter amplify the exchange rate effect and negative sign of $\delta_i i$ imply that exporter try to stabilize the price in buyer currency and not pursuing the constant markup policy.

## 3. Result and analysis

Result for pricing behavior of US wheat exporter is presented in table 1. The F test and R-squared are highly significant indicate that the wheat export market for USA is not having a competitive market in which they are not applying the law of one price. There is strong belief that the price of wheat export are differ across country export destination. As can be shown in table 1 that China, Colombia, Guatemala, Japan, Peru, Philippines, Thailand and Venezuela has a significant country effect which means that USA exporter applies price discrimination to those countries.

There is a significant positive exchange rate effect to Colombia, Guatemala, Indonesia, Japan, Peru, Philippines, Republic of Korea, Thailand and Venezuela which means that the USA wheat exporters amplify the exchange rate effect on local price currencies. Different situation happen with China as it has significant negative exchange rate effect which means that USA exporter apply local-currency price stability (LCPS). The application of LCPS to China is may due to China is the largest importer of USA wheat product. In addition, China market has big potential import market for USA wheat product where USA faces strong outside competition from other exporter, i.e. Australia and Canada (see table 2). USA wheat exporters behave as a competitive supplier in Egypt and United Kingdom. For these countries we cannot reject $\propto_i$ and $\delta_i$ are zero means that we have competitive market without price discrimination across countries .

**Table 1.** PTM Estimation result for USA wheat exporter 1991-2011

| Country | Fixed effect | Exchange Rate | Other Variable |
|---|---|---|---|
| China | -2.499*** | -0.407*** | |
| Colombia | 1.117*** | 0.345*** | GDP = 0.029 |
| Egypt | 0.206 | 0.192 | |
| Guatemala | 1.159*** | 0.470*** | |
| Indonesia | 0.809 | 0.445*** | Oil Price = -0.366** |
| Japan | 0.982** | 0.356** | |
| Peru | 1.389** | 0.539*** | |
| Philippines | 0.962*** | 0.364*** | |
| Republic of Korea | 0.727 | 0.491*** | PPI = 1.687*** |
| Thailand | 1.115*** | 0.451*** | |
| United Kingdom | -0.325 | 0.155 | $R^2$=0.9575 |

*10% significant, **5% significant, ***1% significant

USA has more market power compare to its respective competitor. One of potential reason is USA has bigger market share in world wheat market (see table 2) so that they have more ability to impose noncompetitive behavior in the market and set up the price to gain more profit.

**Table 2.** USA, Canada and Australia share of world wheat total export 1991-2011

| Year/Country | Australia | Canada | United States of America |
|---|---|---|---|
| 1991 | 11.021% | 21.464% | 28.687% |
| 1992 | 7.097% | 20.725% | 29.694% |
| 1993 | 8.824% | 16.936% | 33.171% |
| 1994 | 12.277% | 20.617% | 29.482% |
| 1995 | 7.690% | 16.683% | 31.891% |
| 1996 | 14.740% | 16.715% | 31.517% |
| 1997 | 18.106% | 17.620% | 24.076% |
| 1998 | 13.915% | 16.173% | 24.671% |
| 1999 | 14.447% | 14.113% | 24.845% |
| 2000 | 15.124% | 16.018% | 23.748% |
| 2001 | 13.664% | 15.524% | 22.666% |
| 2002 | 12.207% | 10.135% | 20.138% |
| 2003 | 8.672% | 10.679% | 23.204% |
| 2004 | 15.513% | 12.712% | 26.553% |
| 2005 | 11.550% | 11.559% | 22.561% |
| 2006 | 11.844% | 14.630% | 18.489% |
| 2007 | 5.420% | 14.081% | 26.433% |
| 2008 | 6.311% | 12.031% | 22.942% |
| 2009 | 10.204% | 13.118% | 14.930% |
| 2010 | 10.945% | 12.672% | 19.034% |
| 2011 | 11.909% | 11.017% | 22.115% |

Interesting result come from exchange rate effect of China from USA as USA applies LCPS. If we see table 3 then we may found that USA has the biggest market share in import market for China and it is relatively stable which possible induced by LCPS strategy apply by USA exporter. However different condition may apply to other competitive exporter such as Canada with less market share as its import share in China is steadily declining and its market share is being captured by other competitive exporter, Australia, who is not applying market power to China makes its wheat more attractive to China importer.

**Table 3.** China Import Share 1991-2011

| Year | Australia | Canada | United States of America | Rest of the World |
|---|---|---|---|---|
| 1991 | 10.6089% | 35.4583% | 39.0626% | 14.8701% |
| 1992 | 2.0478% | 51.6170% | 34.1016% | 12.2336% |
| 1993 | 9.1810% | 44.4272% | 44.1137% | 2.2781% |
| 1994 | 19.0888% | 45.7806% | 34.7476% | 0.3830% |
| 1995 | 3.7217% | 40.3641% | 35.7419% | 20.1723% |
| 1996 | 26.2493% | 41.8122% | 29.5555% | 2.3830% |
| 1997 | 11.1621% | 57.5430% | 27.3435% | 3.9513% |
| 1998 | 11.5026% | 48.0029% | 40.2142% | 0.2804% |
| 1999 | 14.0731% | 15.0251% | 67.4848% | 3.4170% |
| 2000 | 8.3802% | 43.9404% | 47.6792% | 0.0002% |
| 2001 | 5.6199% | 33.7255% | 59.8612% | 0.7934% |
| 2002 | 10.3397% | 32.2142% | 54.5573% | 2.8887% |
| 2003 | 3.4489% | 19.6372% | 69.5988% | 7.3152% |
| 2004 | 23.4220% | 32.5401% | 42.2563% | 1.7816% |

| | | | | |
|---|---|---|---|---|
| 2005 | 26.2187% | 34.8721% | 25.0539% | 13.8553% |
| 2006 | 36.0240% | 8.5336% | 55.1853% | 0.2571% |
| 2007 | 26.7265% | 50.9297% | 22.2993% | 0.0445% |
| 2009 | 30.5040% | 9.4597% | 56.2612% | 3.7752% |
| 2010 | 45.5400% | 13.2800% | 38.1100% | 3.0700% |
| 2011 | 35.5800% | 6.7200% | 55.9100% | 1.7900% |

## 4. Conclusion

The increase in wheat production and produces is not necessarily will result in less anti- competitive behavior in world wheat market. This paper tries to identify and examined the degree of market power on wheat international market by world wheat biggest producer and exporter, USA. We found that USA exporter applies price discrimination to China, Colombia, Guatemala, Japan, Peru, Philippines, Thailand and Venezuela. The USA wheat exporters amplify the exchange rate effect on local price currencies to Colombia, Guatemala, Indonesia, Japan, Peru, Philippines, Republic of Korea and Thailand but applying local price stabilization strategy (LCPS) to China. One of the main reasons why USA exporter applies LCPS to China could be because of to maintain the market share in China wheat market. Algeria, China, Republic of Korea and United Kingdom have a significant country effect.

## References

[1] FAO (2013). Statistic. www.fao.org.

[2] Goldberg, P.K. and M.M. Knetter (1997): Goods Prices and Exchange Rates: What Have We Learned?. Journal of Economic Literature, 35, 3:1243-1272

[3] Goldberg, P.K. and M.M. Knetter (1999): Measuring the intensity of competition in export markets. Jounal of International Economics, 47: 27-60.

[4] Pick, D.H., Park, T.A. (1991): The competitive structure of U.S. agricultural exports. Am. J. Agric. Econ. 73(1), 133–141.

[5] Pick, D.H., Carter, C.A.( 1994): Pricing to marketwith transactions denominated in a common currency. Am. J. Agric. Econ. 76(1), 55–60.

[6] Carew, R., Florkowski, W.J. (2003). Pricing to market behavior by Canadian and U.S. agri-food exporters: Evidence from wheat, pulse and apples. Can. J. Agric. Econ. 51(2), 139–159.

[7] Gaulier, G., Lahrèche-Révil, A., & Méjean, I. (2008). Exchange-rate pass-through at the product level. Canadian Journal of Economics/Revue canadienne d'économique, 41(2), 425-449.

[8] Tasdogan, C., Τσακιρίδου, Ε., & Μάττας, Κ. (2015). Country market power in EU olive oil trade. South-Eastern Europe Journal of Economics, 3(2).

[9] Falk, M. and R. Falk (1998). Pricing to Market by German Exporters: Evidence from Panel Data, Centre for European Economic Research. Discussion Paper No. 28. Mannheim.

# The Impact of Input and Output Prices on The Household Economic Behavior of Rice- Livestock Integrated Farming System (Rlifs) and Non Rlifs Farmers

**L Lindawati[1], N Kusnadi[2], S U Kuntjoro[3] and D K S Swastika[4]**

[1]North Sumatera Islam University, Indonesia
[2]Bogor Agricultural University, Indonesia
[3]Bogor Agricultural University, Indonesia
[4]Indonesian Centre for Agricultural Socio Economics and Policy Studies, Indonesia

lindatan84@fp.uisu.ac.id;nunungkusnadi@yahoo.com sriutamikuntjoro@yahoo.com
sadradewa@yahoo.com

**Abstract.** Integrated farming system is a system that emphasized linkages and synergism of farming units waste utilization. The objective of the study was to analyze the impact of input and output prices on both Rice Livestock Integrated Farming System (RLIFS) and non RLIFS farmers. The study used econometric model in the form of a simultaneous equations system consisted of 36 equations (18 behavior and 18 identity equations). The impact of changes in some variables was obtained through simulation of input and output prices on simultaneous equations. The results showed that the price increasing of the seed, SP-36, urea, medication/vitamins, manure, bran, straw had negative impact on production of the rice, cow, manure, bran, straw and household income. The decrease in the rice and cow production, production input usage, allocation of family labor, rice and cow business income was greater in RLIFS than non RLIFS farmers. The impact of rising rice and cow cattle prices in the two groups of farmers was not too much different because (1) farming waste wasn't used effectively (2) manure and straw had small proportion of production costs. The increase of input and output price didn't have impact on production costs and household expenditures on RLIFS.

## 1. Introduction

### *1.1. Background an Issues*

The contraction of the rice field area requires an agricultural technology that is able to encourage the production increasing without having to increase land area. Intensification of agricultural systems is appropriate to resolve the issue. The problems arise when the use of chemical inputs is disproportionate that causes degradation of soil fertility and declined the income of farmers (Ashby, 2001). Soil fertility improvement requires a systematic and gradual effort in reducing the use of high external input with low external input. One of the efforts in addressing these problems was the implementation of an integrated farming system. Integrated farming system is a system that emphasized the existence of linkages and synergism several units of farming (crops, livestock, fisheries and plantations) through the utilization of farming waste from each unit of farming that aims to increase production (Maudi and Kusnadi, 2011). Integrated farming system is capable of restoring the fertility of the soil and stabilize the incomes of farming (Lightfoot and Minnick, 1991). Rice- livestock integrated farming system (RLIFS) is an environmentally friendly technology and appropriately done for several reasons; (1) support the nature which the integration used the concept of zero waste that minimize waste by recycling process, (2) minimize the use of chemical inputs.

Integrated farming system is one of the farming system that has been developed particularly in the developing countries that the people still rely on agriculture as a source of household income. The implementation of an integrated farming system generally has a positive and negative impact. Some studies showed the positive impact of integrated farming system and applied in several countries such as Indonesia, Philippine, Thailand, Vietnam, Nigeria, Bangladesh and India. Integrated farming system gave positive impact and was able to increase production and income significantly (Channabasavana et al, 2007; Priyanti, 2007; Prayitno, 2009; Nageswaran, 2009; Jayanthi et al, 2009; Ugwumba et al, 2010). Integrated farming system gave negative impact which the efficiency of labor and capital usage on the Minapadi farming system was lower than the farming of rice monoculture (Dwiyana and Mendoza, 2006). The study by Handayani (2009) showed that the integrated farming system positively impact like rice-cocoa-livestock program at Donggala, Indonesia. This is due to the availability of cocoa rind as the raw material was very low caused by the pests stricken and the difficulty of obtaining probiotics for the making of animal feed (Handayani, 2009). The study at Zhujian River delta, China suggested that the integrated farming system of sugar cane farming, fisheries, mulberry leaves and silkworm gave negative impact on farmers' profit. The low availability of input was low, causing high production costs so the profit became smaller (Ruddle and Zhong, 1988).

Based on the varying results from integrated farming research in some areas that requires a detailed analysis about what kind of factors that affect rice-livestock integrated farming system in particular the farmers' household economic behavior and then analyze the impact of input and output prices on both Rice Livestock Integrated Farming System (RLIFS) and non RLIFS farmers.

## 2. Methodology

### 2.1. *Location and Method of Samples Determination*

The research conducted in West Java Province as one of the provinces that ever implemented rice-livestock integrated farming system development program from Indonesia government. Based on surveys and information from a related department, there were selected three districts consisted of Subang, Sumedang and Tasikmalaya which were representative enough particularly of the characteristics of respondents. The sub-district and village samples were determined using a purposive method based on (1) rice production centers, (2) livestock population centers. The sample of this study was the Rice-Livestock Integrated Farming System (RLIFS) and non RLIFS farmers. Based on the sampling frame, the sample households were determined using a simple random technique. The number of samples was 199 farmers consisted of 134 RLIFS farmers and 65 non RLIFS farmers

### 2.2. *Types of Data*

The types of data were cross sectional and time series data, while the data source were primary and secondary data. Primary data were obtained through direct interviews with respondents, whereas secondary data were obtained from the Bureau of Statistics Center (BPS) in West Java, Department of agriculture and animal husbandry, district or sub district as well as the village, the results of the study, related publications and references.

### 2.3. *Analysis Methods*

The analysis method used econometric model in the form of a simultaneous equations system consisted of 36 equations (18 behavior equations and 18 identity equations). The impact of changes in some variables was obtained through simulation of input and output prices increases by 10%. Before the model was simulated, validation was performed to check whether the alleged model could reflect reality well. Validation aims to determine the extent to which the estimated value of the model approaches the actual situation. Size used in validation of this research was Root Mean Square Percent Error (RMSPE) and U-Theil statistics. The smaller the error, the better built model. If the statistical value is close to zero then the model simulation is considered to explain the actual data (Sitepu and Sinaga, 2006).

## 3. Result And Discussions

### 3.1. *Spesification Model of The Simultaneous Equation*

The simultaneous equation model is a model specification of a problem as a system of equations. Model specification is an important step so it needs to be done precisely and carefully in order to obtain an appropriate model before the model estimation. The simultaneous equation model must also be checked in advance of the model identification requirements. The requirement is also called the order conditions obtained by calculating the variables in one particular equation by the following procedure (Koutsoyiannis, 1977):

Underidentified: $(K - M) < (G - 1)$

Exactly identified: (K - M) = (G - 1)

Overidentified: (K - M)> (G - 1)

The total number of variables in the model (K) of 77 variables consisted of 36 endogenous and exogenous variables of 41. The number of endogenous and exogenous variables belonging to a given equation in model (M) of 6 and the total equations in model (G) Endogenous variables in the model of 36 (Appendix 1). So the structural equation based on the order condition was required to be overidentified and therefore the parameter estimation method using the 2SLS method. The use of the 2SLS method avoided the simultaneous equation of bias, which basically suspects a system of simultaneous equations by forecasting any partial structural equations (Koutsoyiannis 1977). Analysis in this research using Statistical Analysis System / Econometric Time Series (SAS / ETS) 9.1 for Windows.

*3.2. Validation of Farmer Household Economy Model*

Validation is the stage that must be done before the simulation. Validation aims to determine the extent to which the estimated value of the model approaches the actual situation. Validation results in RMSPE size (Appendix 2) showed good results because of 22 endogenous variables in both groups of farmers yielding values below 100%. However, this quantity has not been able to provide guidance in the use of the model because there were still other criteria that was often used in validation i.e. the coefficient U-Theil. The validation results using U-Theil in this study using the minimum and maximum quantities and benchmarks of certain numbers, namely $\leq 0.50$ (Kusnadi, 2005). There were 36 endogenous variables measured, 26 endogenous variables have a coefficient of U-Theil $\leq 0.50$. The coefficient value of U-Theil ranges from 0.00 to 1.00 so that the validation results showed the goodness of the model was quite large

Decomposition of U-Theil values showed more detailed validation results as presented (Appendix 3) Both groups of farmers who had UM $\leq 0.10$ as many as 24 and 21 endogenous variables so that in general the model is quite good. The US criteria (variance proportions) indicate the model's ability to replace variations of the dependent variable. The result of model validation evaluation using US criterion showed that in household economic model from both farmers there were only 7 and 11 endogenous variables that met the standard $\leq 0.10$. This was because this research used cross section data, so the data variation did not have a certain pattern. Then the UC criteria (covariance proportion) of the model indicated to both farmers there are 17 and 15 of 36 equations close to 1 so the model was said to be not good enough because the simulation error was less randomly fluctuated. Based on the criteria developed, it could be assessed that the overall model validation results were less satisfactory, mainly in US and UC criteria. Nevertheless, there were some fairly good assessment criteria, such as the UM coefficient, RMSPE and U-Theil coefficients.

*3.3. The Impact of Input and Output Price Changes on Household Economics*

The simulation was done to see the impact of changes that occur when there are fluctuations in market prices and by the government. The simulation of rising input and output prices and wage increases is as done in 2 parts ie simulation on input price output price. In integrated cattle-paddy farming system, the use of production input included the use of rice seed, the price of SP-36 fertilizer, urea, bran, straw, manure, medications and livestock vitamin. The input price simulations were performed with two scenarios, namely: (1) Increase in input price of rice-cow cattle production included the price of rice seed, SP-36 fertilizer, urea, medications and livestock vitamin, manure by 10%. (2) The wage increase of 10%. The output in integrated farming consisted of rice and cow cattle production so that at the output price was simulated with two scenarios, namely: (3) Increase in grain price by 10%, (4) increase in cow cattle price by 10%

*3.3.1.* The Impact of Input Price Increase

The policy of input price increase affected the economic activities of integrated households of rice-livestock farming ie production activities, input use, labor, and income. The increase of input price of rice-livestock production included the price of rice seed, SP-36 fertilizer, urea, medications and vitamins for livestock, manure, rice bran and straw by 10% (scenario 1) impacted the decrease of all production on integrated farming (Table 1) . The decrease of rice production was greater in non RLIFS farmers by 0.05% compared with non RLIFS farmers by 0.02%

The policy of input price increase had a decreasing effect on the manure produced by 4.32% on the RLIFS farmers. Farmers kept doing manure processing as usual because of low dependence on external inputs. Meanwhile, the non RLIFS farmers, the increase of input price had no impact on the production of manure. The non RLIFS farmers did not produce manure due to several things, namely
(1) labor shortage to process the cow crap into manure, (2) difficulty in transporting manure rice fields, and (3) the availability of chemical fertilizers was quite easy to obtain.

**Table 1** The Impact of Input Price and Wage Increase

| Endogen Variables | Scenario 1 | | Scenario 2 | |
|---|---|---|---|---|
| | RLIFS Farmers | non RLIFS Farmers | RLIFS Farmers | non RLIFS Farmers |
| PRUP | | | | |
| POSA | -0.0352 | -0.0714 | 0.1409 | 0.1429 |
| PRPK | -4.3181 | . | 11.2117 | . |
| PRJS | -0.0284 | -0.0547 | 0.1949 | 0.1977 |
| JLBP | -12.1326 | -13.2392 | 0.000 | 0.000 |
| JLPU | -11.0356 | -12.3631 | 0.000 | 0.000 |
| JLDK | 0.0000 | -0.0255 | 0.0255 | 0.0255 |
| JLOT | -0.0861 | -0.1233 | 0.000 | 0.000 |
| JLJS | 0.0082 | 0.0205 | -0.0659 | -0.0758 |
| TPDP | -1.7300 | -1.7321 | 1.6513 | 1.6510 |
| TWDP | -1.0765 | -1.0838 | 17.1384 | 17.4618 |
| TPDS | 0.4969 | 0.4720 | 2.4845 | 2.3599 |
| TWDS | -0.0060 | -0.0123 | 74.5835 | 73.2225 |
| TPLP | -0.6582 | -0.7073 | -4.1325 | -4.4277 |
| TWLP | -2.3194 | -1.9973 | -12.8755 | -11.0094 |
| JTDP | -1.4235 | -1.4304 | 8.9143 | 8.9852 |
| JTLP | -1.6243 | -1.5255 | -9.2155 | -8.6078 |
| JTUS | 0.4459 | 0.4785 | 9.8662 | 9.4099 |
| JTUP | -1.4749 | -1.4563 | 4.2581 | 4.2065 |
| BYSP | 0.000 | 0.000 | 0.000 | 0.0000 |
| BYSS | 0.000 | 0.000 | 0.000 | 0.0000 |
| PNUP | -0.0263 | -0.0575 | 0.1931 | 0.2017 |
| PDUP | -0.0203 | -0.0456 | 0.1491 | 0.1599 |
| PNUS | -0.0207 | -0.0435 | 0.1514 | 0.1527 |
| PDUS | -0.0748 | -0.1336 | 0.5482 | 0.4691 |
| PUTS | 0.000 | 0.000 | 0.000 | 0.000 |
| PDTR | -3.0531 | -2.7828 | 4.7387 | 4.6906 |
| PDTD | 0.000 | 0.000 | 0.000 | 0.000 |
| KSBS | 0.000 | 0.000 | 0.000 | 0.000 |
| KSPG | 0.000 | 0.000 | 0.000 | 0.000 |
| KSNP | 0.000 | 0.000 | 0.000 | 0.000 |
| IVSD | 0.000 | 0.000 | 0.000 | 0.000 |
| TABN | 0.000 | 0.000 | 0.000 | 0.000 |
| KSTL | 0.000 | 0.000 | 0.000 | 0.000 |
| IVTL | 0.000 | 0.000 | 0.000 | 0.000 |
| PGTR | 0.000 | 0.000 | 0.000 | 0.000 |

The use of inputs of rice and cow cattle production has decreased due to the increase in input prices of production in scenario 1. The use of production inputs such as rice seeds, SP-36 fertilizer, straw, grass and the use of livestock medications and vitamins decreased greater for the non RLIFS farmers than the RLIFS farmers. The larger decrease was seen in the use of rice seed and urea fertilizer ie 13.2% and 12.4% for the non RLIFS farmers compared to the RLIFS farmers that was 12.1% and 11%.

Increased input prices such as urea fertilizers had less responsive impact on the RLIFS farmers (11%) with reduced use of urea as compared to the non RLIFS farmers (12%). This was because the RLIFS farmers had an alternative use of manure as a result of their own processing. Increase in input prices such as urea fertilizer gave a greater impact to the non RLIFS farmers than the RLIFS farmers.

Increased input prices such as rice seeds had a less responsive impact on the RLIFS farmers (12%) with reduced use of rice seed compared with the non RLIFS farmers (13%). The RLIFS farmers are generally open minded farmers and always receive innovation and apply new technology. This could be seen from the application of integrated farming system conducted by the RLIFS farmers. The use of new technology in agriculture with less use of rice seeds in integrated farming system caused less responsive farmers in responding to rising rice seed prices.

The increase in input prices, especially those related to inputs that were the integration factors such as the manure, bran and fresh straw had no impact on production costs of rice and cow cattle farming. The use of straw as animal feed did not have a good impact on the RLIFS and non RLIFS farmers. The inputs of the straw, manure and bran were inputs derived from household wastes which obtained at low prices and generally were also obtained freely, so that the increase in input prices had no impact on production costs of farming. But the increase in input prices had an impact on the decline in medications and vitamin livestock use. This decrease was quite small in both RLIFS and non RLIFS groups, which was 0.086% and 0.12% respectively. The non RLIFS farmers had relatively larger numbers of livestock than RLIFS farmers, so the increase in input prices had a significant impact on the decline in in medications and vitamin livestock use.

The increase of input price includes the price of rice seed, SP-36 fertilizer, urea fertilizer, medications and vitamin livestock in general became obstacle in increasing farmer's income. The increase of input prices had a decreasing impact on the income of rice and cow cattle farming in both groups of the RLIFS and non RLIFS farmers. The decrease in income of both rice and cow cattle farming was responded greater by the non RLIFS farmers 0.05% and 0.13% while the RLIFS farmers responded by 0.02% and 0.08% respectively. Decrease in income was not too large between the two farmer groups. This was because both groups of farmers had a land area that was not too much different that was 0.190 and 0.165 Ha and the number of cattle production 1.058 and 1.057. Meanwhile, input price increase (scenario 1) did not affect the production cost in the two groups of farmers due to differentiation factors ie integrated rice farming inputs such as straw and manure could be obtained at a low price.

### *3.3.2. The Impact of Wage Increase*

The production input in scenario 2 was the labor which included labor in rice and cow cattle farming. The increase in the wage of labor affected the decision of the farm household in the allocation of family labor (internal labor) and the demand for outside family labor (external labor). The wage increases had a positive impact on both groups of both RLIFS and non RLIFS farmers. The negative impacts were seen in the use of outside labor and the amount of fresh straw used as animal feed.

The wage increases had a positive impact on increasing rice production, production of manure and the amount of fresh straw produced by farmers in each farmer group. Both farmers gave almost equal response values to both farmer groups, but the non RLIFS farmers were slightly more responsive than the RLIFS farmers. This was because the non RLIFS farmers had a smaller land area so that the wage increases were responded by reducing the use of agricultural inputs (production inputs) and external labor. Table 1 showed the labor cost of the non RLIFS farmers was relatively lower than the RLIFS farmers. The production cost efficiency lead to a positive and more responsive wage increase in rice income for the non RLIFS farmers. The wage increases have a positive and more responsive impact on the cow cattle farming income for the RLIFS farmer. But for the total household income, the positive and more responsive impacts were shown by the RLIFS farmers.

The wage increase by 10% caused farm households to increase their labor time outpouring in both cow cattle farming and other farming businesses and reduce their outpouring to non-farm businesses. The wage increases also encouraged the farmers to reduce the demand for external labor. The behavior of farm households in demand for external labor for other farms, behaved differently when there was a change in wages. In the case of an increase in wages, the demand for the first off-shore workforce was female labor. This was because the number of female labor greater than men in a type of work on the farming. The Household reactions are always rational, that was when wages increased then households would add to the internal labor outpouring for farming, and reduce the demand for external labor. The wage increase by 10% did not have an impact on all expenditures. The farm households in the study area were farmers with an average age of 49 years which they had a small number of family dependents so that farmers' expenditures were not too large for consumption. This caused wage increases to have no impact on spending.

### *3.3.3. The Impact of Rice Price Increase*

There were two main products that produce in rice-livestock integrated farming system. They were grain production and cow cattle production. In this section simulations were done with two scenarios, namely: (3) Increase in the grain price by 10% and (4) Increase in the cow price by 10%. Furthermore, the increase in the price of grain and cow were called scenario 3 and 4.

The policy of increase in the grain price (scenario 3) affected the economic activities of integrated households of rice-cow cattle farming ie the production, the use of inputs and the use of labor and income. The increase in the price of grain had a positive impact on increasing the cow cattle production, the manure production, the demand for bran, The cow cattle farming labor and the income. But the increase in grain prices had a negative impact on the rice production, the fresh straw production, the demand and use of external labor and total household income.

**Table 2.** The Impact of Output Prices Increase

| Endogen Variables | Scenario 3 | | Scenario 4 | |
|---|---|---|---|---|
| | RLIFS Farmers | non RLIFS Farmers | RLIFS Farmers | non RLIFS Farmers |
| PRUP | -0.0871 | -0.0870 | 0.1216 | 0.1249 |
| POSA | 0.0032 | 0.0032 | -0.0046 | -0.0046 |
| PRPK | 0.0027 | 0.0023 | -0.0038 | -0.0033 |
| PRJS | -0.0870 | -0.0870 | 0.1217 | 0.1248 |
| JLBP | 0.000 | 0.000 | 0.000 | 0.000 |
| JLPU | 0.000 | 0.000 | 0.000 | 0.000 |
| JLDK | 0.0005 | 0.0005 | -0.0008 | -0.0010 |
| JLOT | 0.000 | 0.000 | 0.000 | 0.000 |
| JLJS | 0.0295 | 0.0326 | -0.0412 | -0.0469 |
| TPDP | 0.0000 | 0.0000 | 0.0001 | 0.0001 |
| TWDP | 0.000 | 0.000 | 0.000 | 0.000 |
| TPDS | 0.0006 | 0.0006 | -0.0006 | -0.0006 |
| TWDS | 0.0010 | 0.0009 | -0.0014 | -0.0013 |
| TPLP | -0.0001 | 0.0000 | 0.0001 | 0.0001 |
| TWLP | 0.000 | 0.000 | 0.000 | 0.000 |
| JTDP | 0.000 | 0.000 | 0.000 | 0.000 |
| JTLP | 0.000 | 0.000 | 0.000 | 0.000 |
| JTUS | 0.0006 | 0.0005 | -0.0011 | -0.0005 |
| JTUP | 0.000 | 0.000 | 0.000 | 0.000 |
| BYSP | 0.000 | 0.000 | 0.000 | 0.000 |
| BYSS | 0.000 | 0.000 | 0.000 | 0.000 |
| PNUP | 0.0042 | 0.0043 | 0.1210 | 0.1252 |
| PDUP | 0.0033 | 0.0034 | 0.0934 | 0.0993 |
| PNUS | 0.0033 | 0.0033 | 0.0949 | 0.0948 |
| PDUS | 0.0120 | 0.0100 | 0.3434 | 0.2913 |
| PUTS | 0.000 | 0.000 | 0.000 | 0.000 |
| PDTR | -0.0602 | -0.0578 | 0.0843 | 0.0830 |
| PDTD | 0.000 | 0.000 | 0.000 | 0.000 |
| KSBS | 0.000 | 0.000 | 0.000 | 0.000 |
| KSPG | 0.000 | 0.000 | 0.000 | 0.000 |
| KSNP | 0.000 | 0.000 | 0.000 | 0.000 |
| IVSD | 0.000 | 0.000 | 0.000 | 0.000 |
| TABN | 0.000 | 0.000 | 0.000 | 0.000 |
| KSTL | 0.000 | 0.000 | 0.000 | 0.000 |
| IVTL | 0.000 | 0.000 | 0.000 | 0.000 |
| PGTR | 0.000 | 0.000 | 0.000 | 0.000 |

The increase in grain prices had a negative impact on rice production. Increasing rice production in the research area was difficult due to: (1) the average land area of rice cow cattle farming under 0.2 Ha (subsistence), (2) location of rice farming in hilly topography area and (3) planting season in certain area that can only one time planting season in subdistrict area like Cisalak (Subang), Tanjungmedar and Tanjungkerta (Sumedang). So that the increase of grain price had a negative impact on both RLIFS and non RLIFS farmer groups. The impact of decreased rice production was followed by decreased straw production. This straw decline eventually encouraged the farmers to collect straw from outside their own farming as it was needed as a feed for livestock.

The increase in grain prices had no impact on demand for rice seed and urea fertilizer. The development of rice cultivation technology is primarily the use of the right superior seed, reduce the use of inorganic fertilizers and maximize the use of organic fertilizer. The negative impact due to increase in grain price was the decrease of labor usage in cow cattle farming. Decreasing rice production made farmers maximize the outpouring of internal labor on rice farming and reducing labor outpours in livestock business. Reducing external labor the family was done to reduce production costs

The increase in grain prices hasd no impact on farmers' expenditure. The average farmer in the study area was the less productive farmers with the age range of 49 years and even more. The need for consumption and investment was quite small that could be seen on the average number of school children i.e. 0.8 people and 3.5 family members. In addition, the small land area did not make the farmers could not enjoy a considerable profit if the increase in grain prices in the range of 10%. Many factors influence: farmers are subsistence, grain quality and harvest time.

The increase in the price of grain had a positive impact on the income of rice farming and cow cattle farming. The authors agreed on several aspects of Heatubun's (2001) research that was integrated/multicommodity farming system could increase farm income according to local agro ecosystem. The scale of business and production of each business is inelastic to the price variable. Food crops were less market-oriented and more subsistent. However, the research results of Lindawati in West Java province in 2013 (Subang Regency, Sumedang, Tasikmalaya) that integrated farming system of paddy-cattle had not been able to increase expenditure which is an indicator of welfare. Several aspects of other authors were also approved, among others, Sawit (1993) states that the increase in rice prices would increase family income, employment, and the amount of rice sold in the market. Absorption of labor in integrated farming system of rice-cow cattle was derived from the labor of cow cattle farmi ng in farmer's household.

The policy of raising the price of live cattle (scenario 4) had a negative impact on cattle production, production of manure, the amount of bran demand, the labor of cow cattle business and income. But the increase of cow cattle prices had a positive impact on rice production, fresh straw production, demand and use of external labor and total household income.

The increase in the prices of cow cattle had a negative impact on cow cattle production. The increase in the prices of cow cattle was generally enjoyed for large investors. The conditions of the control of the farmers' resources in the study sites were (1) the average ownership of rice fields below 0.2 Ha (subsistence farmers), (2) the average cattle ownership of 2.35 animals (equivalent to 1.42 livestock units). The conditions of the location of the study area were (1) some farmers had two planting seasons and the others were only one planting season due to the hilly topography condition, (2) the cattle business was done to sustain the need if needed sometime. This condition caused most farmers selling their cows in times of need with unprofitable price situations. Thus, the increase of cow cattle prices was rarely enjoyed by subsistence farmers of integrated rice-cow cattle farming. The cow maintenance was relatively simple because it was done independently by farmers such as feed that only rely on straw, grass and bran causing cow cattle did not have the maximum weight. The impact of declining cattle production is followed by declining production of manure.

The negative impact of the increase in cow cattle prices was the decline in the use of labor in cow cattle business. The decreased cattle production enabled farmers to maximize the outpouring of internal labor on rice farming and reducing labor outpouring of livestock business. This causing the production of rice and straw increased

The increase in the price of grain had a positive impact on the income of rice farming and cow cattle farming/business and the total income of households. Table 2 could be seen that the impact of 10% grain price increase on both groups of farmers influenced some economic variables of production and income blocks. Most of the impact of this increase in cow cattle price approves some aspects of Kusnadi (2005) which stated that the effect of the increase in output prices in general leads to an increase almost all household economic variables on farming activities.

## 4. Conclusions And Policy Implications

### *4.1. Conclusions*

1. The increase in price of seed input, SP-36 fertilizer price, price of urea fertilizer and livestock / vitamins, price of manure, bran price and straw price negatively affect rice production, cattle production, manure production, straw production, and income
2. The wage increases had a positive impact on manure, bran, straw on the allocation of cow cattle and labor labor in the family of rice farming, total income and expenditure of integrated cattle-paddy farming. Wage increases have a negative impact on the allocation of labor outside the family of rice farming.
3. The increase in rice paddy price has a positive impact on cow production, manure production, bran amount, straw amount, internal labor, rice farming income and cow cattle business income.

   The increase in rice prices had a negative impact on rice production, straw production and the allocation of male external labor of rice farming and total household income.
4. The increase in cow cattle prices had a positive impact on rice production, straw, the labor allocation of internal and external of rice farming and total household income. The increase in cow cattle prices negatively impacted cow cattle production, production of manure, the amount of bran demand, the amount of straw used and the labor allocation of cow cattle business
5. The increase in production inputs, wages, rice paddy prices and live cattle prices by 10% did not have an impact on production costs and household expenditures on rice- cow cattle integrated farming system.

### *4.2. Policy Implications and Advanced Research*

1. The results showed that the production of manure was carried out independently and has not been managed properly and optimally. Production of manure was used to meet the needs of fertilizer itself or subsistence. Need further research on institutional integrated farming. Preliminary observations indicate that the production of manure was quite good and the number was quite large if managed in a farmer group.
2. An integrated farming system of rice-livestock can be an alternative government policy to be developed in the countryside but a comprehensive concern is needed regarding the influence of external factors affecting the sustainability of the integrated farming system.

## 4. Appendices

- **Appendix A.** Endogenous Variables and Examples of validation programs Economic Model of Integrated Farming of Rice-Cow Cattle, prosedur PROC SIMNLIN, SAS versi 9.0
- **Appendix B.** RMSPE and U-Theil coefficients of household economic model RLIFS Farmers and non RLIFS Farmers
- **Appendix C**. Model validation (UM, US and UC) Economy Farm households

## 5. References

[1] Ashby J A 2001 Integrating research on food and the environment: An exit strategy from the rational fool syndrome in agricultural science *Ecol. Soc.* **5**.

[2] Channabasavanna A S and Biradar D P 2007 Relative Performance of Different Rice-Fish-Poultry Integrated Farming System Models with Respect to System Productivity and Economics *Karnataka J. Agric. Sci.* **20**(4): (706-709)

[3] Dwiyana E and Mendoza T C 2006 Comparative Productivity, Profitability, and Efficiency of Rice Monoculture *Journal of Sustainable Agriculture* vol.**29**(1): 145-160.

[4] Handayani S 2009 *Model Integrasi Tanaman-Ternak di Kabupaten Donggala Provinsi Sulawesi Tengah: pendekatan optimasi program linier* [Tesis]. Bogor : InstitutPertanian Bogor.

[5] Haryanto B, Inounu I, Arsana I G M B dan Diwyanto K 2002 *Panduan Teknis Sistem Integrasi Padi-Ternak* Departemen Pertanian Jakarta.

IOP Conf. Series: Earth and Environmental Science **122** (2018) 012020 doi:10.1088/1755-1315/122/1/012020


[6] Jayanthi C, Vennila C, Nalini K and Chandrasekaran B 2009 *Sustainable integrated management of crop with allied enterprises- Ensuring livelihood security of small and marginal farmers*. Special Feature: Sustainable Agriculture.

[7] Koutsoyiannis A 1977 *Theory of Econometrics: An Introductory Exposition of Econometric Methods* Second Edition The Macmillan Press Ltd, London

[8] Kusnadi N 2005 Perilaku ekonomi rumah tangga petani dalam pasar persaingan tidak sempurna di beberapa provinsi di Indonesia [*Disertasi*] Bogor : IPB.

[9] Lightfoot C and Minnick D R 1991 Farmer-first qualitative methods: Farmers diagrams for improving methods of experimental design in integrated farming systems *Journal for Farming Systems Research and Extension*, **2**:11-34.

[10] Lindawati 2015 Analisis Faktor yang Memengaruhi Perilaku Ekonomi dan Kesejahteraan Rumah Tangga Petani Usahatani Terpadu Padi-Sapi di Provinsi Jawa Barat [*Disertasi*] Bogor : IPB.

[11] Mamun S, Nusrat F, and Debi M R, 2011 Integrated Farming System: Prospects in Bangladesh *J. Environ. Sci. & Natural Resources* **4**(2): 127-136, 2011

[12] Maudi F, dan Kusnadi N 2011 Model Usahatani Terpadu Sayuran Organik-Hewan Ternak (Studi Kasus:Gapoktan Pandan Wangi, Desa Karehkel, Kecamatan Leuwiliang, Kabupaten Bogor) *Forum Agribisnis*, vol.**1** (1): 76-94.

[13] Nageswaran, M., Selvaganapathy, E., Vijay, R. and Sudha, N. (2009). *Demonstration and replication of integrated farming systems at Chidambaram* M. S. Swaminathan Research Foundation, Chennai.

[14] Pindyck R, and Rubinfeld D 1998 *Econometric Models and Economic Forecast*. 3 rd Edition. Singapore: Mc Graw Hill International Edition.

[15] Priyanti A 2007 *Dampak program sistem integrasi tanaman-ternak terhadap alokasi waktu kerja, pendapatan dan pengeluaran rumah tangga* [Disertasi] Bogor (ID): Institut Pertanian Bogor.

[16] Prajitno D 2009 Sistem usahatani terpadu sebagai model pembangunan pertanian berkelanjutan di tingkat petani *Materi Pidato* Pengukuhan Jabatan Guru Besar pada Fakultas Pertanian. Yogyakarta (ID): Universitas Gadjah Mada.

[17] Ravisankar N, Pramanik S C, Rai R B, Nawaz S, Biswas T K, and Bibi N 2007 Study on integrated farming system in hilly upland areas of Bay Islands *Indian Journal of Agronomy* **52**(1):7–10

[18] Ruddle K and Zhong G 1988 *Integrated Agriculture-Aquaculture in South China: The Dike- Pond System of The Zhujian Delta* Cambridge University Press.

[19] Sawit M H 1993 A farm household model for rural households of West Java Indonesia [Ph.D.*Dissertation*] Wollongong: The University of Wollongong.

[20] Sitepu RK, Sinaga BM 2006 *Aplikasi Model Ekonometrika. Estimasi, Simulasi dan Peramalan Menggunakan Program SAS* Institut Pertanian Bogor.

# Contribution of village cooperation unit in improving farmers incomes

**M B Sibuea[1] and F A Sibuea[2]**

[1] Agribusiness Department University of Muhammadiyah Sumatera Utara
[2] A student of Magister Agribusiness of Institut Pertanian Bogor
Email: buchari65@yahoo.com, Tel: +628126588264

**Abstract**. One of the government and private efforts to improve people's welfare particularly to improve farmer's income is to activate the Village Cooperation Unit (KUD). The objective of research was to know the efficiency level of farming organized by farmers together with cooperative. Theoretically some social economic variables have been known influences the rate of farmer's income, therefore three social variables, such us the level of cooperation services, members participation and friendship among farmers with cooperation were studied. List of questions divided into forty family's leader members of KUD which become samples. Analysis models were used production function of Cob-Douglass and Output Input Ratio models. It was concluded that level of participation and friendship partially were significantly to the income's rate meanwhile variable of cooperation services level were not significant. Simultaneously, three factors gave very significant contribution, where R-square was 0.97 so that very significant. It's also concluded that the biggest contribution given by the friendship level. Efficiency level or farming efforts of the farmers is very well and feasible with the average of OIR rate 19.23. This research recommended that this effort could be improved from friendship process among institution since the contribution was significantly improving the farmer's income.

## 1. Introduction

The cooperative holds an important role in the Indonesian economy which is a representation of the constitutional form of business for the Indonesian nation as stated in Article 33 Paragraph 1 of the 1945 Constitution where is stated that the economy is arranged as a joint effort on the principle of kinship. Logical consequence is that in every dynamics of national development must always give greater space and motion to the cooperatives development in order to be more instrumental because it is considered a form of the existence of a populist economy. On the implementation, the cooperative is gone through Law Number 25 of 1992 about Cooperatives and Presidential Instruction Number 18 year 1998 about the cooperative improvement and development [1].

Village Cooperative Unit (KUD) is one of the most famous and most widely found type of cooperatives in Indonesia, especially service of various economic activities in rural area teaching farmers have a distribution function of production facilities such as fertilizers, pesticides, seeds and various equipment Farming, distribute of daily necessities by fit prices, provide low interest credits and easy rules, do counselling, processing and marketing of products and other fields in accordance with the capabilities and circumstances of the cooperative [2].

KUD is a vital business entity for rural communities that can excite and improve the performance farmers so that the produced production will increase both quantitatively and qualitatively. KUD is a collection of people who work together to fill one or more economic need or cooperation to do business. It is different from other types of business and usually has the orientation to capital and profit that will be accepted. Cooperatives prefer the human factor and work on the basis of the economy for the welfare of its members. Nevertheless, the situation but the cooperative is not a charity [3].

The key of cooperative advancement is ability on accumulating savings from members and its business have surplus. With stronger capital and the ability to serve better members, agricultural cooperatives will certainly gain wider trust in the midst of society so that in turn it will increase the number of members of the cooperative. His stronger ability will enable the cooperative to accommodate the aspirations of other community members so that the scope of the cooperative will be able to include more diverse activities. In this context, it is hoped that the cooperative can become a forum for activities as well as a teacher of the rural economy [4].

The agricultural sector in general is one of the main drivers aimed at increasing national production, improving the living standards of the large majority of people working in agriculture and to establish a strong foundation for the gradual execution of development. In the context of implementing these objectives, from the very beginning of nation-building, the development of the agricultural sector has been placed as the first priority gradually directed to economic development [5].

In the development activities of the agricultural sector, so that the implementation process goes well, many elements or supporting tools that really determine progress or not the development of agriculture. One of the most prominent is the continuity KUD in researching referred to as agricultural cooperatives [6].

The objective condition found in the research area is felt less optimal cooperation between farmers with KUD in terms of business management, shelter/distribution, palm oil marketing and still lack of participation level of farmers to the existence of cooperatives and otherwise cooperative services that are still not good so that not yet fully touching to all member layers. Therefore, this research is intended to know how far the influence of cooperative services, the level of farmer participation to the existence of cooperatives and farmer-cooperative partnership partially and simultaneously to increase farmer's income and also to know the efficiency level farmer-managed farmer.

Through this research is expected to provide an alternative problem solving in empowering the economy of the villagers. The model of social economy which is reflected by the large participation of the lower community / farmers in an effort to change their life for example through a farming cooperative should be able to provide welfare to its members that can be measured from the efficiency level of farming.

To further direct and focus the research, hypothesis is proposed (1) the real influence between the level of cooperative service to the members, the level of participation of members on the existence of cooperatives and the level of partnership between farmers and cooperatives to increase farmers' income either simultaneously or partially and (2) Agricultural enterprises managed by farmers members of agricultural cooperatives are feasible.

## 2. Material and methods

This research is done by case study so that it is expected that the result of this research will be able to provide an alternative solution as well as comparative or comparison material for other relevant research cases. This research was conducted in working area KUD Harta located in Selayang Village, Selesai District, Langkat Regency covering work area in 5 villages. The villages selected to be the object of research are Selayang Village and Ujung Bandar Village because as many as 75% of the members of the cooperative are from both villages. This agricultural cooperative is one of the largest in Langkat Regency so it is very representative as a research location.

The reason for choosing this KUD as research location is because this cooperative is the biggest in Langkat Regency in terms of number of members (reaching 1169 members of cooperative) managerial aspect as well as from profit aspect of their business. The data collected is in the form of primary and secondary data. Primary data obtained direct interviews with respondents consisting of farmers, farmer groups and administrators KUD own. Secondary data is obtained from Cooperative Office, other relevant Government Offices and Private Offices and references relevant to this research.

The population in this study is the palm oil cooperative members where the main livelihood is derived from the oil palm business. The research unit is the head of the farm household which is taken at random. This can be done because the area, type and pattern of farming business are relatively homogeneous. Sampling was done by purposive sampling method. The sample size of 40 farmers with the details as shown in table below

**Table 1**. Size of research area

| No | Village | Number of members | Percentage (%) | Number of samples |
|---|---|---|---|---|
| 1 | Selayang | 467 | 53.25 | 21 |
| 2 | Ujung Bandar | 410 | 46.75 | 19 |
| | Total | 877 | 100.00 | 40 |

This research use survey method where to facilitate data acquisition is done through interview which divided into 2 (two) part that is structured interview and field observation technique. Data analysis is done by using model of Coob Douglass production function where to more easy per-count, first was logarithm become form of linear function that is as follows:
Model Cob Douglass:

$$Y = aL^{b1}T^{b2}M^{b3} \quad (1)$$

Linear function model:

$$\text{Log } Y = \text{Log } a + b1 \text{ Log } L + b2 \text{ Log } T + b3 \text{ Log } M \quad (2)$$

Where :
Y = income level of farmers members of cooperatives;
L = variable level of cooperative service to member;
T = the variable level of member participation to the cooperative;
M = the degree of partnership between farmers and cooperatives;
a = intercept;
b1, b2, b3 = the level of elasticity or contribution of each independent variable.

To find out the predictor contribution (X) to the dependent variable (Y) used the concept of elasticity and for hypothesis testing used t and F test [7].

While to test the second hypothesis used Output-Input Ratio (OIR) formula that is:
OIR = Output (Result) / Input (Cost)
Where if:
OIR > 1 the business has been efficient
OIR = 1 the business is in a state of breakeven point (BEP)
OIR < 1 the business is inefficient [8].

## 3. Result and discussion

### *3.1. Partnership service, partnership and farmer cooperative*

The level of service given by the cooperative in this case is the readiness / alertness of the provision of production facilities (tools) required by the farmers in their farming efforts, the speed and accuracy in collecting the farmers harvest, the marketing of agricultural products, the distribution of daily necessities, And speed of credit giving to farmers, electricity bill services and agricultural extension services. From the result of the research, it can be seen that the level of service provided by the cooperative to the member farmers with the criteria used is very good 75% -100%, both 50% -75%, less 25% - 50% and 0% -25% bad as shown in table of the following:

**Table 2**. Level of service given by KUD to member farmers

| Number | Cooperative type services to members | Rating (%) |
|---|---|---|
| 1 | Tools provision | 66.7 |
| 2 | Result collection | 65.8 |
| 3 | Result marketing | 63.3 |
| 4 | Daily goods production | 65.0 |
| 5 | Credit giving | 60.0 |
| 6 | Electrical account service | 81.6 |
| 7 | Counseling | 66.7 |

From table it is clear that the level of service given by the cooperative to members is look relatively good where the electricity bill payment service is the highest level of service although the service is actually not directly related to the farming activities. Therefore, the levels of agricultural cooperative services in essence still have a very big opportunity to be improved especially services that are directly related to the production process of farming.

While the level of participation contributed by the farmers in the business of agricultural cooperatives, especially in terms of purchasing the needs of production facilities associated with farming activities, the accuracy of harvesting the results of farming, willingness in buying daily needs, accuracy and paying credits that have matured, accuracy in paying electricity bills and activitation in following the activities of cooperatives and the idea of the development of cooperatives, more clearly can be seen in table below:

**Table 3.** Participation rate given by farmers to cooperatives

| Number | Cooperatives services type given farmers to cooperatives | Rating (%) |
|---|---|---|
| 1 | Buying tools production | 73.3 |
| 2 | Accurate harvest | 62.5 |
| 3 | Selling harvest | 67.5 |
| 4 | Buying daily needs | 84.2 |
| 5 | Accurate buying credits | 55.0 |
| 6 | Accurate buying electricty accounts | 80.0 |
| 7 | Giving ideas | 46.7 |

According the table, the level of participants on the context of empowerment cooperative has been relatively well although it is still to improve continually. It looks that the level of member participants concerning direct something about live or die a cooperative institute such as giving ideas, buying accurate credit are still low. It shows that actually, the member participants concerning existence cooperative are still more improved especially things be an example cooperatives for farmers to the development his business agriculture.

Partnerships or partnerships between farmers and cooperatives include the provision and purchase of production facilities for farmers, the accuracy of harvesting and harvesting, marketing of crops, credit and accuracy in paying due credits, provision of daily necessities, In terms of electricity bill payments for members of the cooperative, agricultural extension and distribution of ideas/aspirations from members to cooperatives. The results of the research can be seen in table below. According to the table, the relationships between farmer and cooperative is relatively good although there are some components partnership researched to look the bad condtion. Variable indicating the bad level is counseling and aspiration channel. It is essential factor for cooperative life future. This fact also provides a strong indication that the aspects directly related to the empowerment of cooperatives precisely provide an unfavorable value while the aspects that are actually less related to the essence of the presence of a cooperative proved to provide good value (e.g. provision of daily needs). Therefore, there is still a need to increase the quality of partnership between farmers and cooperatives so that it will give more benefit to both parties that cooperate.

**Table 4.** The Level of Partnerships Between Farmers and Cooperatives

| Number | Type of done partnership | Rating (%) |
|---|---|---|
| 1 | Provision and tools purchase | 75.0 |
| 2 | Accurate harvest and result collection | 75.0 |
| 3 | Marketing and purchasing of result | 67.5 |
| 4 | Purchasing of credits and purchasing accurately | 71.7 |
| 5 | Daily needs Provision | 72.5 |
| 6 | Consultation and aspiration channel | 49.2 |

### *3.2. Influence analysis of service level, participation and partnership concerning farmers members farmers income level*

To see the extent of simultaneous or simultaneous influence of the three variables studied on farmer member farmer's income, an analysis is performed with Coob-Douglass production function model and can be seen in table below:

**Table 5**. The results of analyze of influence level service, participants and partnership concerning level of incoming the cooperative member

| Variable | Regression coefficient | Error standard | t-count (df=36) | Probability | $r^2$ partial |
|---|---|---|---|---|---|
| Service logarithm | 1.0621 | 0.9228 | 1.151 | 0.2602 | 0.0485 |
| Participant logarithm | 1.7543 | 1.2303 | 1.826 | 0.1658 | 0.0725 |
| Partnership logarithm | 2.8475 | 1.0422 | 4.732 | 0.0111 | 0.2231 |
| Constant | -1.3209 | | | | |
| Std error of est | 0.0280 | | | | |
| adj. R-squared | 0.9694 | | | | |
| R-squared | 0.9726 | | | | |
| Multiple R | 0.9862 | | | | |
| F-Count | 30.7328 | | | | |

Based on the results of test from table can be formulated linear function of the analysis are:
$\text{Log } Y = -1.3209 + 1.0621\log L + 1.7543\log T + 2.8475\log M$
So if converted to Coob Douglass rank function becomes:
$Y = 0.05\, L^{1.06}T^{1.75}M^{2.85}$

Based on the results of hypothesis testing above it is seen that simultaneously the variables studied can explain the change in income of 97.26% indicated by the value of R-squared or coefficient of determination of 0.9726. While the rest is explained by other variables that are not researched.

Simultaneously, the three independent variables are the level of real cooperative service to the income level of the farmers of the cooperative members so that if these variables are proportionally increased, it will give the farmers income increase at 95% confidence level. This is also evidenced by the existence of a very high relationship between independent variables with dependent variable. While the influence of each independent variable to the dependent variable partially or individually can be explained follows:

*3.2.1. Influence the level of service concerning the level of incoming.* Partially, this variable has positive effect to income because it has positive elasticity value that is equal to 1.06 which means if service level added 1% then it will raise income level equal to 1.06% where other variables are considered fixed (ceteris paribus). While based on t test obtained that the value of t arithmetic smaller than t table which means the influence of service is not significant at 95% confidence level. This means that although the elasticity is positive but it does not show any real effect. The above indicates that the level of cooperative services to members is still not good and optimal so it needs to be improved because based on the results of interviews with respondents in the field looks strong impression as if the board of cooperatives put forward the interests of a group of people, especially the management of the cooperative itself. It is precisely this that makes agricultural cooperatives less desirable by some farmers.

*3.2.2. Influence the level of participants concerning the level of incoming.* Partially, participants level variable have positive elasticity value 1.75 that mean increasing participants level are amount 1 %. It will give influence concerning increasing amount 1.75%. Moreover, according to t-amount, it is got t-amount bigger t-table so that there is real influence or significant from member participants level concerning cooperative in increasing the member farmer incoming on the truth level 95%. The objective condition in field shows that actually, farmers are relatively response about cooperative attendance especially thing of tools provision so that it must look for other alternative to finish it. Therefore, farmers must be insisted and cared about the agriculture cooperative existence.

*3.2.3. Influence the level of partnership concerning the level of incoming.* Partially, the variable of partnership between farmers and cooperatives has a very real effect on the increase of income at 99% confidence level. This partnership variable is what should be further developed if there is intention to increase farmer's income. Based on the value of elasticity as positive as 2.85 then this means with the addition of 1% level of partnership will increase the income level by 2.85%. If it is related to the situation in the field it is seen that if the cooperative really shows its identity as the extension of farmer's hand in this case marketing of agricultural products, then real farmers are helped and apart from the trick of price manipulation that has been a scourge for farmers. From the interview results obtained a resume that farmers really hope that the concept of this partnership can be developed and expanded in more integrated forms again.

*3.2.4. Analyze of the efficiency business farm level.* Based on the results of research obtained data that the average level of farming efficiency managed by cooperative member farmers is very good because from the calculation results using Output-Input Ratio method obtained the average level of efficiency of palm oil farming from members of the cooperative was very high of 19.23 or 1923% means that it goes far beyond the 1 or 100% feasibility. Thus economically, the farming that is held in partnership between farmer and cooperative institution give benefit which multiply to both side although in essence managerial handling has not been optimally implemented yet.

## 4. Conclusion

Hence, the research result that the services level given by cooperative to the member farmers are relatively good although it is still good enough opportunity/potential to improve [4,9]. The participant's level from members farmer cooperative concerning cooperative existence are relatively good although it is not total to show to be back. It happens because member farmer's cooperative ideas are limited. The partnerships level between member farmer and cooperative institution also show relatively good indicator.

Simultaneously, there is influence or real contribution and positive from variables researched. It is the service level, the participant and partnership concerning the incoming level member farmer cooperative. Partially, variable of the participant level and partnership shows significant contribution but variable of the service level is not significant.

Farmer business of the member cooperative has very good or comfortable the efficiency level.

## 5. Suggestion

Looking for big contribution showed by the partnership level between farmer and cooperative so suggested to both partner to increasingly focus on improving cooperation with other institutions or agencies that have more relevance to the oil palm farming. And suggested to farmer in order to use more technology well in handling farming so that can improve harvest productivity because the efficiency business level is very high or comfortable.

## References

[1] Reksohadiprodjo S 2005 Cooperative management Yogyakarta
[2] Aziz M A 2004 Cooperatives and agroindustry Jakarta
[3] Kartasapoetra, G., et al 2011 Management of cooperative practices Jakarta
[4] Sibuea M B 2016 *International journal of business, economics and law* Modeling of cooperation to improve rural economic in Langkat Vol. **9(5)** pp. 74-81.
[5] Baharsjah S 1995 Fifty years building Indonesian modern agriculture: Reflection, challenges and prospects scientific address for the 32nd Anniversary of IPB Bogor
[6] Mubyarto and Daryanti 2008 Plantation crops: Socio-economic review Jakarta.
[7] Soekartawi 2011 Agribusiness, theory and application Jakarta
[8] Soekartawi 2003 Risks and uncertainties in agribusiness Jakarta
[9] Sibuea M B 2011 Analisis peranan koperasi dalam kegiatan usaha tani (Cooperative role analyses in agricultural activities) *Sosial dan ekonomi pertanian* Vol. **8(1)** pp. 169-190.
[10] Sibuea M B 2016 Analisis faktor-faktor penyebab kemiskinan petani padi sawah (Factors of paddy farmers poverty analyses) *Jurnal ilmu pertanian Agrium* Vol. **20(1)** pp. 344-348.

# The influences of consumer characteristics on the amount of rice consumption

**T Supriana[1] and TC Pane[2]**

[1]Universitas Sumatera Utara
[2]Universitas Islam Sumatera Utara

Email: tavihutasuhut@yahoo.co.id, tasyacpane@yahoo.com

**Abstract**. This study aimed to analyze the characteristics of rice consumers and the influences of consumer characteristics on the amount of rice consumption. The research areas were determined purposively in the sub-districts with the most significant population in Medan City. The analytical methods used were descriptive and multiple linear regression analysis. The results showed that consumers in the study areas have various characteristics, concerning age, income, family size, health, and education. Simultaneously, characteristics of rice consumers have the significant effect on the amount of rice consumed. Partially, age and the number of family members have the significant effect on the amount of rice consumed. The implications of this research are, need different policies toward consumers of rice based on their income strata. Rice policies cannot be generalized.

## 1. Introduction

Rice is one of the food commodities which plays an important role both from the side of producers, consumers, government, as well as society and the environment in general. From the producer side, 34 percent of Indonesians work in the agricultural sector. While from the consumer side, more than 90 percent of Indonesians consume rice as staple food, even 30 percent of the total expenditure of low-income households is used to buy rice [2].

Rice consumers in North Sumatra can be distinguished as consumers living in rural and urban areas. The need for rice by the population living in rural areas is also higher when compared to the population living in urban areas. It is due to differences in income levels and the lack of variety of rice substitute foods sold in rural areas compared to urban areas [9]. Table 1 shows that the rate of food grains consumption per capita of North Sumatra population is still very high, both in rural and urban areas, where consumption of food grains in rural areas is higher than in urban areas.

The behavior of consumers in the purchase of foodstuffs continues to develop. The increase in public incomes results in demands on quality. Changes in the demographic structures such as educational level, knowledge, lifestyle, technology, transportation, and communication affect consumer preferences and satisfaction.

The average rice consumption in North Sumatra in 2009 was 134.13, while in 2013 it was 131.46. The average per capita rice consumption in Medan City was 134 kg/capita, even higher than the national average rice consumption that was 114 kg/capita [4].

**Table 1.** Average Daily Calorie Consumption Per Capita by Types of Food and Urban/Rural Areas (Kcal) Year 2013 – 2014 [3]

| No | Details | Urban | | Rural | | Urban+Rural | |
|---|---|---|---|---|---|---|---|
| | | 2013 | 2014 | 2013 | 2014 | 2013 | 2014 |
| 1. | Grains | 798.03 | 805.55 | 1,080.53 | 1080.25 | 941.53 | 944.95 |
| 2. | Tubers | 13.43 | 15.79 | 24.67 | 26.23 | 19.14 | 21.09 |
| 3. | Fishes | 66.30 | 67.87 | 73.91 | 73.20 | 70.16 | 70.57 |
| 4. | Meats | 42.35 | 48.63 | 28.45 | 32.09 | 35.29 | 40.23 |
| 5. | Eggs and Dairies | 65.41 | 83.61 | 42.82 | 45.35 | 53.93 | 64.19 |
| 6. | Vegetables | 33.28 | 35.56 | 39.55 | 40.10 | 36.47 | 37.87 |
| 7. | Nuts | 30.70 | 44.33 | 25.40 | 25.53 | 28.01 | 34.79 |
| 8. | Fruits | 40.33 | 37.67 | 37.76 | 38.86 | 39.02 | 38.27 |
| 9. | Oils and Fats | 282.82 | 311.50 | 295.41 | 297.76 | 289.22 | 304.53 |
| 10. | Beverages | 90.51 | 91.87 | 91.70 | 90.27 | 91.11 | 91.05 |
| 11. | Seasonings | 11.25 | 10.93 | 11.44 | 10.31 | 11.35 | 10.62 |
| 12. | Other Consumptions | 34.90 | 34.61 | 33.14 | 35.43 | 34.00 | 35.03 |
| 13. | Finished Foods | 238.90 | 214.10 | 161.48 | 167.82 | 199.57 | 190.62 |
| | Total Foods | **1,748.21** | **1,802.03** | **1,946.26** | **1,963.18** | **1,848.80** | **1,883.81** |

The number of consumer consumption and preferences for rice varies. These differences are influenced by consumer characteristics and attributes of rice to be purchased by consumers. Consumer characteristics will be assessed in this study comprehensively. This study divides consumers into various characteristics by age, income, the number of family members, health condition, and educational level. In general, if the income level rises, then the amount and types of food tend to improve as well. At low-income levels, the primary energy sources are obtained from grains, tubers, and vegetables.

Low-income families will more easily meet their food needs if the number of family members who need to be fed a little. The food available for a large family may be sufficient for the family which has the half of the family members, but not enough to prevent malnutrition in the large family.

Concerning age, the age differences will lead to differences in tastes and preferences for the product. Thus, the age differences will also affect the differences in the amount of consumption. So also with health conditions, consumption of rice is closely related to diabetes mellitus and cholesterol, changes in health condition lead to changes in the consumption of rice.

The levels of education affect the amount of consumer consumption. It is because the higher the educational level of a person then will be the better the type of work, and it will affect the level of income. The levels of income itself will also affect the purchasing power, tastes, and the amount of consumption. Besides, the levels of education also affect consumer knowledge about rice products.

This study aimed to analyze the characteristics of rice consumers and the influences of consumer characteristics on the amount of rice consumption in Medan City. It was expected that consumer characteristics by age, income, the number of family members, health condition, and educational level have the significant effect on the amount of rice consumption.

## 2. Material and Methods

The research areas were determined purposively in the sub-districts with the most significant population in Medan City. Medan City was chosen because it is the capital of North Sumatra Province. The study population was all households of rice consumer in Medan City. The number of households in Medan City in 2015 was 507,205 households [4]. The sample size was determined using the Slovin Method, by the following equation [17].

$$n = \frac{N}{1 + \epsilon^2 N} \quad (1)$$

Where $n$ is Sample size, $N$ is Population size, $\in$ is Error tolerance of 10%. So the size of samples to be examined was:

$$n = \frac{507{,}205}{1 + 0{,}1^2 x 507{,}205} = 100\ Households \quad (2)$$

The sampling method used was non-probability sampling, the combination of accidental sampling and purposive sampling. The samples were determined purposively based on income levels. At first, accidental sampling was done, the samples encountered would be interviewed, so the income levels were known. Once the sample size for specific income categories was known, then the samples for the other income categories would be searched. The sample size was taken based on each income category as many as 20 people. Sample size by categories of income is presented in Table 2.

**Table 2.** Sample Size by Income Categories

| No | Income Categories | Sample Size (People) |
|---|---|---|
| 1 | < Rp 2,500,000 | 20 |
| 2 | Rp 2,500,000 – Rp 6,000,000 | 20 |
| 3 | >Rp 6,000,000 – Rp 10,000,000 | 20 |
| 4 | > Rp 10,000,000 – Rp 15,000,000 | 20 |
| 5 | > Rp 15,000,000 | 20 |
| | **Total** | **100** |

The data used in this study were primary data and secondary data. The primary data collection technique was conducted by interviewing rice consumers using structured questionnaires, while the secondary data was obtained from BPS Medan. The data were analyzed qualitatively and quantitatively. Qualitative methods were presented by interpreting and describing the data obtained, while the quantitative data obtained would be tabulated based on the activities.

The analytical methods used in this research were descriptive and multiple linear regression analysis. The descriptive analysis method was used to study the characteristics of rice consumers based on their socioeconomic conditions. The multiple-linear regression method was used to analyze the influences of consumer characteristics on the amount of rice consumption. The analysis was performed using SPSS 17 software.

Multiple-linear regression is a regression model used to analyze the effects of several independent variables on one dependent variable. The Model of the influences of consumer socioeconomic characteristics, those are age, income, the number of family members, health condition, and educational level, to the amount of rice consumption, systematically can be written as follows [16]:

$$Y = a + b1X1 + b2X2 + b3X3 + b4X4 + b5K + e \quad (3)$$

Where:
Y = The amount of rice consumption (kg/month)
a = Constant
b1-b5 = Regression coefficients
e = Error
X1 = Age (year)
X2 = Income (Rp/month)
X3 = The number of family member (person)
X4 = Educational level (level)
K = Health condition,
K = 1; cholesterol, other diseases, and healthy
K = 0; diabetes

## 3. Results and Discussion

The age of the samples ranged from 26 - 70 years. No group dominated the age of the samples, but most of the samples were aged 40 - 49 years those were 34 samples. Followed by the samples were aged 50 - 59 years those were 26 samples. Furthermore, the samples were aged 30 - 39 years those were 23 samples, 60 - 69 years those were 11 samples, 20 - 29 years those were 5 samples, and 70 - 79 years those were 1 sample. Characteristics of the sample by age groups are shown in Table 3.

**Table 3.** Characteristic of Sample by Age Groups

| Ages | Total (Person) | Percentage (%) |
|---|---|---|
| 20 – 29 years | 5 | 5.00 |
| 30 – 39 years | 23 | 23.00 |
| 40 – 49 years | 34 | 34.00 |
| 50 – 59 years | 26 | 26.00 |
| 60 – 69 years | 11 | 11.00 |
| 70 – 79 years | 1 | 1.00 |
| **Total** | **100** | **100.00** |

The highest sample income level was Rp 51,000,000, while the lowest was Rp 1,000,000. The highest income group was >Rp 15,000,000 and the lowest income group was <Rp 2,500,000.

The largest number of family members of sample households was 9 persons. The lowest number of family members of sample households was 1 person, while the average number of family members of sample households was 2 persons. There were 31 sample households with 0 - 3 family members, 66 sample households with 4 - 6 family members, and 3 sample households with 7 - 9 family members. Table 4 shows the number of family members of sample households by groups that had been set.

**Table 4.** Characteristic of Sample by The Number of Family Members Groups

| Number of Family Members | Total (Person) | Percentage (%) |
|---|---|---|
| 0 – 3 person | 31 | 31.00 |
| 4 – 6 person | 66 | 66.00 |
| 7 – 9 person | 3 | 3.00 |
| **Total** | **100** | **100.00** |

The health condition of the sample was seen from the presence or absence of the illnesses. The diseases were diabetes, cholesterol, and other diseases. If the sample did not suffer from any illnesses, then the sample was declared healthy. Of the 100 samples, 20 samples were suffering from diabetes, 6 samples suffering from cholesterol, 2 samples suffering from other diseases, and the remaining 72 samples stated themselves healthy. Table 5 shows the health conditions of the sample.

**Table 5.** Characteristic of Sample by Health Conditions

| Health Conditions | Total (Person) | Percentage (%) |
|---|---|---|
| Diabetes | 20 | 20.00 |
| Cholesterol | 6 | 6.00 |
| Other diseases | 2 | 2.00 |
| Healthy | 72 | 72.00 |
| **Total** | **100** | **100.00** |

The most samples had Senior High School Education those were 62 samples. Followed by the samples had S1 education those were 15 samples. The samples had Elementary School education those were 5 samples, Junior High School education those were 13 samples, Diploma education those were 2 samples, S2 education those were 2 samples, and S3 education that was 1 sample. Table 6 shows the educational levels of the samples.

**Table 6.** Characteristic of Sample by Educational Levels

| Educational Levels | Total (Person) | Percentage (%) |
|---|---|---|
| Elementary School | 5 | 5.00 |
| Junior High School | 13 | 13.00 |
| Senior High School | 62 | 62.00 |
| Diploma | 2 | 2.00 |
| S1 | 15 | 15.00 |
| S2 | 2 | 2.00 |
| S3 | 1 | 1.00 |
| **Total** | **100** | **100.00** |

Before the estimation of the analysis of the influences of consumer characteristics on the amount of rice consumption, it was necessary to test the assumption of Ordinary Least Square as follows.

a. Test of Normality

After passing the test of data normality with Kolmogorov Smirnov Test, the significance value obtained was 0.364> 0.05 which means the residual data of model was distributed normally.

b. Test of Multicollinearity Symptoms

After performing the test of multicollinearity, the VIF values obtained for each variable were < 10 and Tolerance values obtained for each variable were > 0.1 which means there were no multicollinearity symptoms in the model.

c. Test of Heteroscedasticity Symptoms

After performing the test of heteroscedasticity with the graph method and Park Test, the form of the graph did not show the specific patterns, and the significance values of Park Test for each variable obtained were > 0.05 which means there were no heteroscedasticity symptoms in the model.

d. Autocorrelation Test

After performing the test of autocorrelation with Durbin-Watson Test, the Durbin-Watson value (d) obtained was $du \leq d \leq 4 - du$ or $1.80 \leq 1.990 \leq 2.20$ which means there was no positive or negative autocorrelation in the model or the trend was $\rho = 0$.

So after testing the OLS assumption, the final result of the estimation obtained showed the influence of consumer characteristics on the amount of rice consumption as follows.

**Table 7.** The Results of The Influences of Consumer Characteristics on The Amount of Rice Consumption in Medan City

| Variables | Coefficient | Std. Error | t | Sig |
|---|---|---|---|---|
| Constant | -8.010 | 5.259 | -1.523 | 0.131 |
| Age | 0.195 | 0.072 | 2.716 | 0.008 |
| Income | -0.153 | 0.000 | -1.600 | 0.113 |
| Number of Family Member | 5.422 | 0.549 | 9.882 | 0.000 |
| Educational Level | -0.516 | 0.765 | -0.675 | 0.501 |
| Health Condition | 1.724 | 1.776 | 0.972 | 0.334 |
| $R^2 = 0.589$ | | | t - table = 1.671 | |
| F = 26.989 | | | | |
| F- table = 2.769 | | | | |

Based on Table 7, obtained the following equation:

$$\hat{Y} = -8.010 + 0.195\,X1 - 0.153\,X2 + 5.422\,X3 - 0.516\,X4 + 1.724\,K$$

Based on the model, the coefficient of determination obtained was 0.589. It showed that 58.9% of the variation of the amount of rice consumption had been explained by age, income, number of family members, health condition, and educational level, while the remaining 41.1% was explained by other variables that had not been included in the model. The F value obtained was 26.989 > F-table = 2.769 with the significance value obtained was 0.000 < 0.05, which means that H1 was accepted. It means, simultaneously, age, income, number of family members, health condition, and educational level significantly affect the amount of rice consumption at 95% level.

Partially, consumer characteristics that influence the amount of rice consumption are described as follows.

a. The regression coefficient of the age effect on the amount of rice consumption was 0.195 and had the positive sign, which means that if the age increases by 1 year, then the amount of rice consumption will increase by 0.195 kg/month assuming other variables are considered fixed. Based on t-test, t values of the age obtained was 2.716 > t-table = 1.671 with significance value obtained was 0.008 < 0.05, which means that H1 was accepted. It means, partially, the age significantly affects the amount of rice consumption at 95% level.
b. The regression coefficient of the income effect on the amount of rice consumption was -0.153 and had the negative sign, which means that if the income increases by 1 million Rupiah, then the amount of rice consumption will decrease by 0.153 kg/month assuming other variables are considered fixed. Based on t-test, t values of the income obtained was 1.600 < t-table = 1.671 with significance value obtained was 0.113 > 0.05, which means that H0 was accepted. It means, partially, the income does not significantly affect the amount of rice consumption at 95% level.
c. The regression coefficient of the number of family members effect on the amount of rice consumption was 5.422 and had the positive sign, which means that if the number of family members increases by 1 person, then the amount of rice consumption will increase by 5.422 kg/month assuming other variables are considered fixed. Based on t-test, t values of the number of family members obtained was 9.882 > t-table = 1.671 with significance value obtained was 0.000 < 0.05, which means that H1 was accepted. It means, partially, the number of family members significantly affect the amount of rice consumption at 95% level.
d. The regression coefficient of the educational level effect on the amount of rice consumption was -0.516 and had the negative sign, which means that if the educational level increases by 1 level, then the amount of rice consumption will decrease by 0.516 kg/month assuming other variables are considered fixed. Based on t-test, t values of the educational level obtained was 0.675 < t-table = 1.671 with significance value obtained was 0.501 > 0.05, which means that H0 was accepted. It means, partially, the educational level does not significantly affect the amount of rice consumption at 95% level.
e. The regression coefficient of the health condition effect on the amount of rice consumption was 1.724 and had the positive sign, which means that the amount of rice consumed by consumers who have cholesterol, other diseases, or in healthy condition is higher by 1.724 kg/month than the amount of rice consumed by consumers who have diabetes. Based on t-test, t values of the health condition obtained was 0.972 < t-table = 1.671 with significance value obtained was 0.334 > 0.05, which means that H0 was accepted. It means, partially, the health condition does not significantly affect the amount of rice consumption at 95% level.

## 4. Conclusion

Simultaneously, characteristics of rice consumers have the significant effect on the amount of rice consumed. Partially, age and the number of family members have the positive and the most significant effect on the amount of rice consumed, income and educational level have the negative effect on the amount of rice consumed, and diabetes lowers rice consumption. The implications of this research are, need different policies toward consumers of rice based on their income strata. Rice policies cannot be generalized.

## References

[1] Astuti. 2008. Analysis of Consumer Preference and Satisfaction on Rice in Mulyorejo Sub-district Surabaya East Java [*Analisis Preferensi dan Kepuasan Konsumen terhadap Beras di Kecamatan Mulyorejo Surabaya Jawa Timur*]. Undergraduate-Thesis. Study Programme of Agricultural Management, Faculty of Agriculture, Bogor Agricultural University. Bogor.

[2] BPS. 2015. Average Monthly Expenditure Per Capita by Commodity Groups (Rupiah) Year 2013 - 2014 [*Rata-rata Pengeluaran per Kapita Sebulan menurut Kelompok Barang (Rupiah) Tahun 2013 - 2014*]. Downloaded December 30th 2015. http://www.bps.go.id.

[3] BPS Sumatera Utara, 2016. Average Daily Calorie Consumption Per Capita by Types of Food and Urban/Rural Areas (Kcal) Year 2013 - 2014 [*Rata-Rata Konsumsi Kalori Perkapita Sehari Menurut Jenis Bahan Makanan dan Daerah Perkotaan/Perdesaan (Kkal) Tahun 2013 - 2014*]. Downloaded February 21st 2016. http://www.bps.go.id.

[4] BPS Sumatera Utara, 2016. North Sumatera Province in Figures [*Provinsi Sumatera Utara dalam Angka*]. Downloaded February 21st 2016. http://www.bps.go.id.

[5] Damardjati, D.S. 1995. Characteristics of Properties and Standardization of Rice Quality as the Platform of Agribusiness and Agro-industry Development of Rice in Indonesia [*Karakteristik Sifat dan Standarisasi Mutu Beras sebagai Landasan Pengembangan Agribisnis dan Agroindustri Padi di Indonesia*]. Agency for Agricultural Research and Development, Department of Agriculture. Jakarta.

[6] Engel, J.F., R.D. Blackwell, and P.W. Miniard. 1994. Consumer Behavior [*Perilaku Konsumen*]. Binarupa Aksara. Jakarta.

[7] Frankilawati, D.A.M. 2013. Correlation Between Diet Patterns, Genetics, and Sports Habits against Type II Diabetes Mellitus Occurrence in the Work Area of Puskesmas Nusukan, Surakarta [*Hubungan Antara Pola Makan, Genetik dan Kebiasaan Olahraga Terhadap Kejadian Diabetes Melitus Tipe II di Wilayah Kerja Puskesmas Nusukan, Surakarta*]. Faculty of Health Sciences. University of Muhammadiyah Surakarta. Surakarta.

[8] Kotler, P. 2000. Marketing Management [*Manajemen Pemasaran*]. Volume I. Millenium Edition. Prenhallindo. Jakarta.

[9] Lastry, Y. 2006. Analysis of Rice Consumption Patterns of Households in Bogor City [*Analisis Pola Konsumsi Beras Rumah Tangga di Kota Bogor*]. Undergraduate-Thesis. Study Programme of Agricultural Management, Faculty of Agriculture, Bogor Agricultural University. Bogor.

[10] Nazir, M. 2005. Research Methods [*Metode Penelitian*]. Ghalia Indonesia. Jakarta.

[11] Selamet, R. 2003. Analysis of Consumer Decision Processes in Rice Research and Rice Marketing Strategy [*Analisis Proses Keputusan Konsumen dalam Penelitian Beras dan Strategi Pemasaran Beras*]. Undergraduate-Thesis. Study Programme of Agribusiness, Department of Agricultural Socio Economic Sciences, Faculty of Agriculture, Bogor Agricultural University. Bogor.

[12] Silalahi, N.E., Salmiah, and Jufri. 2013. Levels and Patterns of Rice Consumption of Medan City Society [*Tingkat Konsumsi dan Pola Konsumsi Beras Masyarakat Kota Medan*]. Journal on Social Economic of Agriculture and Agribusiness. Vol 2, No 9 (2013). University of Sumatera Utara. Medan.

[13] Simamora, B. 2004. Consumer Behavior Research Guide [*Panduan Riset Perilaku Komsumen*]. Gramedia Pustaka Utama. Jakarta.

[14] Simamora, B. 2005. Marketing Multivariate Analysis [*Analisis Multivariat Pemasaran*]. Gramedia Pustaka Utama. Jakarta.

[15] Sumarwan U. 2004. Consumer Behavior, Its Theory and Application in Marketing [*Perilaku Konsumen, Teori dan Penerapannya dalam Pemasaran*]. Ghalia Indonesia. Bogor.

[16] Supriana, T. 2013. Introduction to Econometrics [*Pengantar Ekonometrika*]. USU Press. Medan.

[17] Supriana, T. 2016. Socio-Economic Research Methods [*Metode Penelitian Sosial Ekonomi*]. USU Press. Medan.

[18] Supriana, T. and Riantri Barus. 2013. Nonparametric Statistics, Applications in Agricultural Socio Economics [*Statistik Nonparametrik, Aplikasi dalam Bidang Sosial Ekonomi Pertanian*]. USU Press. Medan.

# The effect of organic paddy field system to soil properties

**Mujiyo[1*], B H Sunarminto[2], E Hanudin[2], J Widada[3] and J Syamsiyah[1]**

[1]Department of Soil Science, Faculty of Agriculture, Universitas Sebelas Maret, Surakarta, Indonesia;
[2]Department of Soil Science and[3]Department of Microbiology, Faculty of Agriculture, Universitas Gadjah Mada, Yogyakarta,Indonesia.
E-mail: mujiyo@staff.uns.ac.id

**Abstract**. This study aims to compare the soil properties among organic paddy field and other cultivation systems. Soil profile description was done to determine the soil horizons. The characteristics of organic paddy soil were better than others as it has C-organic in horizons I and II which were 2.09% and 2.00%, respectively; black soil color in horizon I (10 YR 2/1); microbial biomass C in horizon I and II which were 98.39 µg C/g soil and 418.01 µg C/g soil respectively; and microbial biomass N in horizon I and II which were 72.59 µg N/g soil and 59.11 µg N/g soil respectively. However, the unfavorable qualities of organic paddy soil including the clay leaching from horizon I (clay 24.56%) to horizon II (clay 40.16%) and high production potential of methane ($CH_4$) (1.81 µg $CH_4$/kg soil/day and 1.67 µg $CH_4$/kg soil/day, respectively in horizon I and II) need to be anticipated. The high amount of C-organic in organic paddy field affects the soil color was darker, and the increasing of clay leaching, microbial biomass C and N, and potential of $CH_4$ production.

## 1. Introduction

Many studies investigated the good effects of organic paddy rice systems in soil properties. The organic paddy field is characterized by organic fertilizers exertion in each growing season. It has been widely reported that organic fertilizers significantly increase organic matter content in the soil. Benbi [1] stated that long-term applications of farmyard manure and rice straw resulted in a build-up of not only labile but also recalcitrant soil organic carbon pool. The presence of organic matter in the soil has an impact on the improvement of physical, chemical and biological properties of the soil.

But, until now, no study has predicted how the long-term disadvantage impact of organic farming systems on the soil properties. The long-term organic matter existence in the soil is related to the complex organic-clay formation, which affects to the mass mobilization and immobilization. The organic acids lead the improvement of clay fraction dispersion, so it is easily leached to the below horizon [2]. Some other studies indicate that the use of organic fertilizers in paddy field appears to increase the $CH_4$ production. The series decomposition process of organic matter through the oxidation-reduction process will end with the formation of $CO_2$ and $CH_4$ [3, 4]

Over the last years, the effect of agricultural management practices on the soil properties and on crop yield has been widely studied in diverse types of agricultural soils [5]. Based on that problem, it is important to determine soil properties in paddy soils with diverse management practices. The objective of this study is to compare soil properties between organic paddy field and other cultivation systems, and also to find what the disadvantage impact to soil properties are in the organic paddy field.

## 2. Material and Methods

All site of soil profile samples were taken from paddy fields in Sukorejo Village, Sragen, Indonesia. The soil profiles are organic (P1), semi-organic (P2), conventional (P3) paddy field, and the teak forest land-use (P4) as a control. The criteria for organic paddy field selection is the use of organic fertilizers; for semi-organic is the use of chemical and organic fertilizers; for conventional is the use of most types of fertilizers (including both chemical and only occasionally organic fertilizers); and for the teak forest as a control is the land-use type of teak forest [6].

A soil profile is made at each site sampling with a size of width, length and depth; 1.5 m, 2.0 m and 1.5 m, respectively. The horizon was determined based on Soil Taxonomy USDA [7]. All observed soil and land character parameters were included on the boardlist with data and information referring to the Guidelines for Soil Description [8] and the Land Observation Technical Manual by Hidayat [9]. The soil texture and C-organic were analyzed based on Eviati and Sulaeman [10], soil colour were used Munsell Soil Color Charts [237], microbial biomass C dan N were determined based on Mueller [11], and $CH_4$ production potential was adopted from Susilowati [12] with the modification of adding acetylene as an inhibitor of methane oxidation [13, 14].

The differences soil properties between all paddy field systems were determined using one-way analysis of variance (ANOVA) with the Duncan's Multiple Range test. The correlations between soil properties were determined using the correlation analysis [15] and the Pearson's correlation coefficient was calculated as well. All statistical analysis were performed with SPSS Statistics 17.0 software.

## 3. Results and Discussion

The description of the soil profile shows that P1 and P2 consist of four horizons; P3 consists of five horizons; and P4 consists of six horizons. Table 1 shows horizon P1-P4 and selected soil properties. There are some significant differences among the organic, semi-organic, conventional paddy field and the teak forest. In this paper is presented the diagnostic of soil properties that can distinguish among the soil samples, which is the content of C-organic, clay, colour, microbial biomass C and N, and methane production potential. The soil properties of C-organic, microbial biomass C and methane production potential have been explained by Mujiyo [6].

There was very significant difference in the content of C-organic between all soil profiles observed (one-way ANOVA $F = 8.89$, $P < 0.01$). C-organic on conventional paddy field was the lowest among the others because of the continuous rice cultivation without the using of organic fertilizer. This practice affected C-organic content is lower than its original soil (teak forests, as a control). Nakhro and Dkhar [16] stated the use of inorganic fertilizers in the paddy field resulted in low organic carbon content of the soil. Komatsuzaki and Syuaib [17] also stated that conventional farming leads the soil with lower soil carbon storage capacity than organic farming. Long term practice of organic paddy field with the use of organic fertilizer in each growing season became the primary factor that causes the organic paddy has higher amount of C-organic than the others.

Clay content on the horizon II in organic paddy field is the highest (40.16%). The long-term organic matter existence in the soil is related to complex organic-clay formation, which affects to mass mobilization and immobilization. Organic compounds interact with clay minerals on the external and internal surfaces through adsorption, and other mechanisms [18]. Organic acids improves clay fraction dispersion, so it is easily leached to the below horizon [2]. Dispersion of a clay particle from micro-aggregates are caused by adsorption of organic acid complex that increase negative charge of clay [19, 20], and it causes a certain clay particles release [2]. Finally, clay on horizon I was leaching and then was acumulating in the horizon II.

The significant differences occurred is the colour in horizon I at organic paddy field. It's matrix colour was black (10 YR 2/1), which was the darkest among the others. It was caused by C-orgin organic paddy field is higher. The value from bright to dark has commonly become an indicator for estimating the C-organic content in the soil. The high value (dark colours) generally indicates high content of C-organic in the soil [21]. C-organic had an important role in the correlation of Munsell colour value [22].

**Table 1**. Soil horizon of profile P1-P4 and selected soil properties

| Soil Profile | Horizon (Notation) | C-organic (%) [6] | Clay (%) | Colour (Hue Value/Chroma) | MBC (µg C/g soil) [6] | MBN (µg N/g soil) | $CH_4$ PP (µg $CH_4$/kg soil/day) [6] |
|---|---|---|---|---|---|---|---|
| P1 | I (Apg) | 2.09 | 24.56 | Black (10 YR 2/1) | 498.39 | 72.59 | 1.81 |
| | II (Bt) | 2.00 | 40.16 | Dark Yellowish Brown (10 YR 3/4) | 418.01 | 59.11 | 1.67 |
| | III (Bw1) | 1.31 | 31.58 | Dark Yellowish Brown (10 YR 4/6) | 84.98 | 30.07 | 1.05 |
| | IV(Bw2) | 1.30 | 37.62 | Yellowish Brown (10 YR 5/6) | 110.24 | 30.59 | 0.92 |
| P2 | I (Apg) | 1.78 | 29.63 | Dark Yellowish Brown (10 YR 3/4) | 475.42 | 59.37 | 1.23 |
| | II (Bw1) | 1.65 | 28.07 | Yellowish Brown (10 YR 5/6) | 199.82 | 56.00 | 0.98 |
| | III (Bw2) | 1.56 | 30.41 | Yellowish Brown (10 YR 5/6) | 172.26 | 38.89 | 0.82 |
| | IV (Bw3) | 1.48 | 9.36 | Yellowish Brown (10 YR 5/6) | 321.54 | 33.70 | 0.66 |
| P3 | I (Apg) | 1.08 | 37.82 | Dark Yellowish Brown (10 YR 4/6) | 367.48 | 49.26 | 0.84 |
| | II (Bw1) | 1.00 | 33.92 | Dark Yellowish Brown (10 YR 4/4) | 257.23 | 20.74 | 0.59 |
| | III (Bw2) | 1.01 | 24.17 | Grayish Red (2.5 YR 5/2) | 303.17 | 27.48 | 0.63 |
| | IV (Bw3) | 0.68 | 33.14 | Reddish Brown (2.5 YR 4/6) | 261.83 | 27.22 | 0.66 |
| | V (Bw4) | 0.78 | 26.90 | Reddish Brown (2.5 YR 5/3) | 266.42 | 16.33 | 0.53 |
| P4 | I (Ap) | 1.83 | 21.44 | Dark Brown (7.5 YR 3/3) | 468.53 | 66.11 | 1.21 |
| | II (Bw1) | 1.88 | 23.39 | Brown (7.5 YR 4/4) | 443.27 | 57.04 | 1.12 |
| | III (Bw2) | 1.63 | 26.12 | Bright Brown (7.5 YR 5/6) | 271.02 | 45.89 | 0.80 |
| | IV (Bw3) | 1.52 | 28.07 | BrightBrown (7.5 YR 5/8) | 282.50 | 36.30 | 0.67 |
| | V (Bw4) | 1.35 | 15.79 | Bright Brown (7.5 YR 5/6) | 197.52 | 22.30 | 0.73 |
| | VI (Bw5) | 1.24 | 22.22 | Orange (7.5 YR 6/8) | 287.09 | 27.48 | 0.61 |

P1 – organic; P2 – semi-organic; P3 – conventional paddy field; P4 – teak forest
Clay – relative percent with content of sand and silt to determine the soil texture, data not shown
MBC – microbial biomass C; MBN – microbial biomass N; $CH_4$ PP – methane production potential

The microbial biomass C and N were describing the microbial population in the soil. The correlation between C-organic with microbial biomass C is significant ($r = 0.52$, $P < 0.05$, $n = 19$) [6], and with microbial biomass N is very significant ($r = 0.84$, $P < 0.01$, $n = 19$). Soil with high substrate of C-organic also has a high microbial population. C-organic in soil contains both living and non-living components. Living components consist of various soil microbial biomass [23]. Soil with high substrate of C-organic is able to support life and activity of different types of microbial metabolism with different types of morphology and physiology.

There was significant difference in methane production potential between all soil profile observed (one-way ANOVA $F = 5.19$, $P < 0.01$), which is on organic paddy field is the highest among the others. The most obvious soil properties which has close relationship with the $CH_4$ production potential is the C-organic content ($r = 0.80$, $P < 0.01$, $n = 19$) [6]. The increasing content of C-organic soil will raise the $CH_4$ production potential. Organic paddy soil has the highest average C-organic and $CH_4$ production potential among other soils [6]. The soil has high amount of C-organic will be followed by the high amount of $CH_4$ production [24, 25, 26]. Organic matter appears to increase the methane production [3, 4] by providing C sources and decreasing oxidation-reduction potential (Eh) [27, 28].

The $CH_4$ production potential was also significantly correlated with microbial biomass C ($r = 0.56$, $P < 0.05$, $n = 19$) and N microbial biomass ($r = 0.84$, $P < 0.01$, $n = 19$). Microbial biomass C and N are significantly correlated with C-organic ($r = 0.52$, $P < 0.05$, $n = 19$; $r = 0.84$, $P < 0.01$, $n = 19$, respectively). This correlation reinforces conjecture on soils containing high C-organic and microbial populations (including microbial methanogens forming $CH_4$) so also leads the high amount of the $CH_4$ production potential. The high amount of C-organic soil supports the development of methanogen populations to produce $CH_4$ from the reduction of $CO_2$ and $H_2$, $CH_3OH$ and $CH_3COOH$ [4, 29, 30].

## 4. Conclusion

The organic paddy system has led to better changes in soil properties. The long term practice of organic paddy field with the use of organic fertilizer in each growing season become the main factor that causes higher amounts of C-organic than semi-organic paddy field, conventional paddy field and teak forest. The soil in organic paddy field, especially in the horizon I, has highest of C-organic among the others, as the color of soil matrix in the horizon I was the darkest. The high amount of C-organic in organic paddy field affects the soil color was darker and also affects in the increasing of microbial biomass C and N. But, the unfavorable soil properties in the organic paddy field were clay leaching and high of methane production potential. The clay leaching from horizon I to horizon II needs to be anticipated as the clay is a key characteristic of soil fertility. The high production potential of methane also needs to be mitigated in order methane emissions from organic paddy field has to remain low.

## References

[1] Benbi D K, Brar K, Toor A S and Sharma S 2015 Sensitivity of labile soil organic carbon pools to long-term fertilizer, straw and manure management in rice-wheat system *Pedosphere* **25**(4): 534-545.

[2] Wilding L P, Smeck N E and Hall G F 1983 *Pedogenesis and Soil Taxonomy* Elsevier Science Publishers. B. V. Amsterdam, The Netherlands p 303

[3] Le Mer J and Roger P 2001 Production, oxidation, emission and consumption of methane by soils: A review *European Journal of Soil Biology* **37**: 25-50

[4] Nieder R and Benbi D K 2008 *Carbon and Nitrogen in the Terrestrial Environment* Berlin, Springer Science + Bussines Media B.V. p 430

[5] Lopes A R, Faria C, Fernández A P, Cepeda C T, Manaia C M and Nunes O C 2011 Comparative study of the microbial diversity of bulk paddy soil of two rice fields subjected to organic and conventional farming *Soil Biology and Biochemistry* **43**(1): 115-125.

[6] Mujiyo M, Sunarminto B H, Hanudin E, Widada J and Syamsiyah J 2017 Methane production potential of soil profile in organic paddy field. Online first articles in *Soil and Water Research* (SWR). Available at http://www.agriculturejournals.cz/web/swr.htm?type=article&id=58_2016-SWR (accessed July 14th, 2017)

[7] Soil Survey Staff 2014 Keys to soil taxonomy, 12th Ed. Washington DC: USDA – Natural Resources Conservation Service p 360

[8] FAO 2006 Guidelines for soil description. 4th Edition. Food and Agriculture Organization of The United Nations Rome p 97

[9] Hidayat A, Djaenudin D, Suhardjo H and Subardja D 2004 Technical manual: Soil Description. Bogor, Indonesian Soil Research Institute, Research Center for Agricultural Land Resources, Indonesian Agency for Agricultural Research and Development, Ministry of Agriculture. (in Indonesian)

[10] Eviati and Sulaeman 2009 Chemical analysis of soil, plant, water, and fertilizer. Technical manual. 2nd Ed. Bogor, Indonesian Soil Research Institute, Research Center for Agricultural Land Resources, Indonesian Agency for Agricultural Research and Development, Ministry of Agriculture. (in Indonesian)

[11] Mueller T, Joergensen R G and Meyer B 1992 Estimation of soil microbial biomass C in the presence of fresh roots by fumigation extraction. *Soil Biology and Biochemistry*, **24**: 179-181

[12] Susilowati H L 2007 Measurement of Gas Production Potential of $CH_4$, $CO_2$ and $N_2O$ by Soil Incubation Technique. Jakenan Pati, Indonesian Agricultural Environment Research

Institute, Indonesian Agency for Agricultural Research and Development, Ministry of Agriculture. (in Indonesian)

[13] Watanabe I, Hashimoto T and Shimoyama A 1997 Methane-oxidizing activities and methanotrophic populations associated with wetland rice plants *Biology and Fertility of Soils* **24**: 261–265

[14] Watanabe I, Takada G, Hashimoto T and Inubushi K 1995 Evaluation of alternative substrates for determining methane-oxidizing activities and methanotrophic populations in soils *Biology and Fertility of Soils* **20**: 101–106

[15] Steel R G D and Torrie J H 1980 Principles and procedures of statistics, a biometrical approach. 2nd Edition. New York, McGraw Hill. p 633

[16] Nakhro N and Dkhar M S 2010 Impact of organic and inorganic fertilizers on microbial populations and biomass carbon in paddy field soil *Journal of Agronomy* **9**(3):102-110

[17] Komatsuzaki M and Syuaib M F 2010 Comparison of the farming system and carbon sequestration between conventional and organic rice production in West Java Indonesia *Sustainability* **2**: 833-843S

[18] Lagaly G, Ogawa M, and Dékány I 2013 Clay mineral–organic interactions *Developments in Clay Science* **5**: 435-505

[19] Durgin P B, and Chaney J G 1984 Dispersion of kaolinite by dissolved organic from Douglas-FIR roots. *Canadian Journal of Soil Science*, **64**: 445-455

[20] Oades J M 1984 Soil organic matter and structural stability: Mechanisms and implications for management. *Plant and Soil* **76**: 319-337

[21] Zelenak E M 1995 Relationship between Munsell color value and organic carbon content in Montana soils. A thesis submitted in partial fulfillment of the requirements for the degree of Master of Science in Soils Montana State University p 98p

[22] Pereira P, Úbeda X, Solera J M, Oliva M and Novara A 2014 Short-term changes in soil Munsell colour value, organic matter content and soil water repellency after a spring grassland fire in Lithuania. *Solid Earth* **5**: 209–225

[23] Dalal R C, Allen D E, Chan K Y and Singh B P 2011 Soil organic matter, soil health and climate change. In: Singh B.P., Cowie A.L., Chan K.Y. (eds): *Soil Health and Climate Change. Soil Biology Series*, Berlin, Heidelberg, Springer- Verlag: 87-106

[24] Joulian C, Olliver B, Neu H U, and Roger P A 1996 Microbiological aspects of methane emission by a ricefield soil from the Camargue (France): 1. Methanogenesis and related microflora *European Journal of Soil Biology* **32**: 61-70

[25] Oelbermann M and Schiff S L 2008 Quantifying carbon dioxide and methane emissions and carbon dynamics from flooded boreal forest soil. *Journal of Environmental Quality*, **37**: 2037–2047

[26] Liu D Y, Ding W X, Jia Z J and Cai Z C 2011 Relation between methanogenic archaea and methane production potential in selected natural wetland ecosystems across China. *Biogeosciences* **8**: 329-338

[27] Hou A X, Chen G X, Wang Z P, Van Cleemput O and Patrick W H Jr 2000 Methane and nitrous oxide emissions from a rice field in relation to soil redox and microbiological processes *Soil Science Society of America Journal* **64**: 2180-2186

[28] Li C 2007 Quantifying greenhouse gas emission from soils: Scientific basis and modeling approach. *Soil Science and Plant Nutrition* **53** 344-352

[29] Reddy K R and DeLaune R D 2008 Biogeochemistry of wetlands: science and applications. Boca Raton, London, New York, CRC Press, Taylor & Francis Group p 774

[30] Thauer R K, Kaster A K, Seedorf H, Buckel W and Hedderich R 2008 Methanogenic archaea: Ecologically relevant differences in energy conservation. *Nature Reviews Microbiology* **6**: 579–591

**Acknowledgements**
This research study was supported by the Ministry of Education and Culture of the Republic of Indonesia under the research grant for doctoral program 2012–2014, and Sebelas Maret University Surakarta under the research grant PNBP UNS 2017. We would like to thank Citra Recha Sari for her participation in the paper preparation.

# Growth and production of new superior rice varieties in the shade intensity

**Alridiwirsah[1], E M Harahap[2], E N Akoeb[2]and H Hanum[2]**

[1]Department of Agrotechnology, Faculty of Agriculture, University of Muhammadiyah Sumatera Utara, Jl. Muhtar Basri No.3, Medan,Indonesia
[2]Department of Agrotechnology, Faculty of Agriculture, Universitas Sumatera Utara, Jl. Dr. A. Sofyan No.3, Medan, 20155,Indonesia
E-mail: alridiwirsah@gmail.com

**Abstract**. Shade intensity is one of the most important requirements for plant growth, affecting growth, development, survival, and crop productivity. This study aims to evaluate the growth and productiom of New Superior Rice Varieties In The shade Intensity. This study was conducted in BalaiPengkajianTeknologiPertanian, Pagar Merbau, Deli Serdang, North Sumatra. The research used completely randomized design with twofactors. The shade intensity (N) were 25%, 50% and no shade intensity as a control. Whereas new superior rice varieties were $V_1$: Inpara 2, $V_2$: Suluttan Unsrat 2, $V_3$: Inpari Mugibat, $V_4$: Inpari Sidenuk, $V_5$: Mekongga, $V_6$: Ciherang, $V_7$:Inpari 10, $V_8$: Inpari 3, $V_9$: Inpari 4, $V_{10}$: Inpari 30, dan $V_{11}$: Cibogo. The result indicated that new superior rice varietiesshowedsignificant effectonthe growth and productionvariablesuch as leaf area, where Inpari Sidenuk variety was the highest among the varieties. Total chorophyll, the highest was found on Inpari variety. Number of tillers and plant height where the highest was found on Ciherang variety. The shade intensity showed significant effect on leaf area, where 25% shade intensity was the highest. Total chlorophyll, the highest was found on 50% shade intensity, number of tillers, the highest was found on no shade intensity.

## 1. Introduction

Rice is an important staple food in Asian populations. World's major rice producing countries are Thailand, China and Vietnam, while Indonesia is the main consumer of rice. Future rice demand is expected to increase, in line with the population growth. Therefore there is a need to increase rice production by at least the rate of the increase in population growth, in order to achieve a sustainable rice production for local consumption. Rice compliance challenges ahead, especially in Indonesia, will be more difficult due to the conversion of agricultural land, especially irrigated land to non-agricultural interests [1]. Rice (*Oryza sativa* L.) is the world's second most important cereal crop, belonging to the family Gramineae. Rice provides 75% calories and 55% proteins to the average daily diet of consumers and contributes tremendously to the economies of many nations [2]. It is estimated that demand for rice in 2025 will be increased by 60% compared to today in order to meet the food needs. Indonesia, which has a population of 237 million people (2010 census) with main meals majority (95%) of the population is rice. Rice consumption is about 137 kg / capita / year, the predicted demand for rice in 2020 to the people of Indonesia reached 35.97 million tons / year. Where the vast rice fields in Indonesia reached 8.061 million hectares consisted of 4.896 million hectares of irrigated rice and non-irrigated rice fields reached 3.16 million hectares has the potential to increase rice yields [3].

The role of environmental factors in rice production cannot be overemphasized. There are diverse environmental factors that affect rice production; growth and yield, among them are sunlight, drought, and temperature. Growth of autotropic plants is directly and dramatically influenced by light intensity (i.e. quantum flux density) which is the driving force of photosynthesis and provides nearly all of the carbon and chemical energy needed for plant growth. In addition, opined that increased light intensity improved rice yield till plant reaches its light saturation. Furthermore, light intensity is among important requirements for plant growth, development, survival, and crop productivity. Because of the difficulty of controlling light intensity, researchers have evaluated the effects of variation in light regimes on morphological characteristics, physiological characteristics, yield, and quality of agricultural crops [2].

Wang *et al.*[4] examined that the effects of shading on starch content and starch pasting viscosity in rice genotypes. The research results may lay a theoretical foundation for the selection of shade-tolerant varieties of rice and the improvement of cultivation technologies. Therefore, the development of rice varieties are required to test rice variety productivity. In plant photosynthesis, chlorophyll is the most important photosynthetic pigment, and shading also affected the chlorophyll content of plants. Shading altered light-use efficiency by increasing leaf chlorophyll a, chlorophyll b, and chlorophyll a+b, and decreasing chlorophyll a/b ratios [5].

Effect of shade on upland rice varieties decreases the number of tillers, number of panicle, number of productive grains, production of grains per hill of plants and total sugar content of plants [6].Light intensity is one of the most important requirements for plant growth, affecting growth, development, survival, and crop productivity. Because of the difficulty of controlling light intensity [4], researchers have evaluated the effects of shading on morphological characteristics, physiological characteristics, yield, and quality of agricultural crops. Multiple studies [5] have shown that the morphological changes resulting from shading included increases in leaf width, length, and area index, and decreases in leaf thickness due to the reduction of palisade layer number, palisade cells, and spongy parenchyma length. On the rice plant, shade can reduce the number of tillers, dry weight of the canopy, leaf area index, grain yield and the efficiency of solar radiation use [7].

There are many new superior rice varieties produced by such research institutes as Agricultural Research and Development Agency, universities, and private research institutions, and BATAN. Those varieties, however, are unknown by local farmers regarding their superiority. In addition, fertilization conducted by farmers is not in accordance with recommendation of location specific. Thus, identification of new superior rice varieties will be essential in developing high-yielding varieties that can survive in the shade intensity. Therefore, this present study was carried out in order to examine growth and production of new superior rice varieties in the shade intensity.

## 2. Materials and Methods

In order to study productivity test of new superior rice varieties in the shade intensity an experiment was conducted in Balai Pengkajian Teknologi Pertanian, Pagar Merbau, Deli Serdang, North Sumatra. The experiment was conducted from November 2014 until March 2015. The experiment was designed as a completely randomized design with two factors. The shade intensity (N) were 25%, 50% and no shade intensity as a control. Whereas new superior rice varieties were $V_1$: Inpara 2, $V_2$: Suluttan Unsrat 2, $V_3$: Inpari Mugibat, $V_4$: Inpari Sidenuk, $V_5$: Mekongga, $V_6$: Ciherang, $V_7$:Inpari 10, $V_8$: Inpari 3, $V_9$: Inpari 4, $V_{10}$: Inpari 30, dan $V_{11}$: Cibogo. If the effect of different treatments on the real variance, then tested further by Duncan's multiple range test.

Land preparation, Tillage was done by using a tractor plow and harrow. Plowed soil until conditions crumb processed layer 15-25 cm deep. Experimental Plots, Created with a length of 200 cm and width of 100 cm by the number of beds 45 plot. The experiments were replicates three times. The distance between replications of 200 cm, 50 cm distance between plots. Shade Manufactures, Shade was made of bamboo poles and the roof put on paranet 25% and 50%. Appropriate treatment shade 10 m long and 4.5 m wide shade. Tall bamboo from the ground level was 2 m. Seed preparation, Seeds cleaned then soaked for 24 hours in saltwater. Washed with clean water and then drained and ripened for 24 hours. Cultivation, Seedlings planted at 18 days after sowed. Row spacing 25 cm x 25 cm ,1 seed per hole.Regulate Irrigation,Seeds are planted in water-saturated soil conditions and mapped fields in circulated again after 3-4 days. Water rotation must be checked every 3 days. Fertilization,Fertilizer used was 14.49 kg urea fertilizer with 3 times applications when the plant was 10 days, 30 days and 45 days, 9.66 TSP twice applications when the plant was 30 days and 45 days, and KCL 9.66 kg with once application at age of plant 30 days. Pest and Disease Control, Pests controlled by kenfas 100 EC (Alfa sipermethria 100 g/l), walang sangit pest controlled by baycarb (BPMC (2-Ci-methyl propy) Phenyl Methyl carbamate), snails pest controlled mechanically (citations) and spraying molluscicides debestan (besnoid 400 gr), and stem rot disease was controlled byfungicides (Difenokonazol 250 g/l).Harvest, Rice was harvested when 95% of panicles have yellowed.

Parameters observed: 1)leaf Area ($cm^2$), which the measurement was done before harvest calculated using the leaf area meter. The leaves were observed which leaves the largest, medium and smallest unfolded. 2) Number of tillers. The number of tillers was calculated starting from 30 days after planting. Puppies were calculated by counting the number of tillers that grow from the main stem, once in 15 days until 60 days after planting. If the clump of rice plants was 20 stems, then the number of tillers was 19 straws, because the rod rest was rice plant stem [8]. 3) Plant height (cm). Plant height measurements started when the plant was 10 days after planting until flowering, once every two weeks. 4) Total of Chlorophyll (mg/g), Chlorophyll content was calculated by spektrotofometerUvis following the method proposed by Henry and Grime [9]. The extraction of chlorophyll done with acetone 80%, weighed 0.1 g leaves, crushed in a mortar, plus 10 ml acetone. Then filtered with Whatman filter paper. The filtrate was measured absorbance at 645nm and 663nm. The calculation of the levels of chlorophyll as follows:

Chlorophyll a mg/g weight of leaf

$$= (12{,}7 \times A663 - 2{,}69 \times A645)10^{-1}$$

Chlorophyll b mg/g weight of leaf

$$= (22{,}9 \times A645 - 4{,}68 \times A663) \times 10^{-1}$$

Chlorophyll total mg/g weight of leaf

$$= (8{,}02 \times A663 + 20{,}2 \times A645) \times 10^{-1}$$

## 3. Results and Discussion

In this research, New superior rice varieties showed significant effect on the growth and production variable such as leaf area, where Inpari Sidenuk variety was the highest among the varieties. Total chorophyll, the highest was found on Inpari variety. Number of tillers and plant height where the highest was found on Ciherang variety. The shade intensity showed significant effect on leaf area, where 25% shade intensity was the highest. Total chlorophyll, the highest was found on 50% shade intensity, number of tillers, the highest was found on no shade intensity.

There were significant effect on leaf area amongnew superior rice varieties (Table), where Inpari Sidenuk variety was the highest among the varieties, followed by Inpara and Inpari as the lowest. The shade intensity showed significant effect on leaf area, where 25% shade intensity was the highest then followed by 50% shade intensity and no shade intensity was the lowest. The rice plants respond and adapt to various types, timing and duration of flooding in different ways which include elongation for deepwater rice varieties and submergence tolerance for lowland varieties exposed to transient flash floods.[15] stated that the contribution of superior varieties on the increase in national production is 56%. Research and Development Agency has produced 233 superior varieties consist of 144 Inbrida superior rice varieties, 35 hybrid superior rice varieties, 30 superior upland rice varieties, and 24 wet rice varieties. [4] reported that shading at different growth stages and in different rice varieties all affected the starch pasting characteristics of rice. The effects of shading on starch pasting viscosity at middle and later growth stages were greater than those at earlier stages. Shading enhanced breakdown but reduced hold viscosity and setback at tillering-elongation stage. Most pasting parameters changed significantly with shading after elongation stage. Furthermore, the responses of different varieties to shading differed markedly. The change scope of starch pasting viscosity in Dexiang 4103 was rather small after heading, while that in IIyou 498 and Gangyou 906 was small before heading. We observed clear tendencies in peak viscosity, breakdown, and pasting temperature of the five rice varieties with shading in 2010 and 2011. Correlation analysis indicated that the rice amylose content was negatively correlated with breakdown, but was positively correlated with setback. Based on our results, IIyou 498, Gangyou 906, and Dexiang 4103 had higher shade endurance, making these varieties most suitable for high-quality rice cultivation in low-light regions. Thats why in their research reported that to be successful staple crops, crops need to be resistant to varying growing conditions, providing consistent yield and quality under a range of environmental conditions. While many previous studies focused on shading effects on rice morphology, physiology, and yield, the responses of starch quality to shading in indica hybrid rice are unclear. Therefore, they examined the effects of shading on starch content and starch pasting viscosity in rice genotypes. Their research results may lay a theoretical foundation for the selection of shade-tolerant varieties of rice and the improvement of cultivation technologies.

New suoerior rice varieties showed significant effect on number of tillers 4 weeks after planting where the highest was found on Ciherang variety then followed by inpari variety and Suluttan Unsrat was the lowest. the shade intensity showed significant effect on number of tillers 4weeks after planting, where no shade intensity was the highest followed by shade intensity (25%) then shade intensity (50%) was the lowest (Table). It was caused by factor of shade is one of the issues that need to be addressed in the development of lowland rice cultivation as intercrops. Rice productivity comes under some influencing indices which are often interdependent. The present study revealed that total number of tillers; and more importantly, number of effective tillers increased with increasing light intensity; this agreed with the findings of [10], who reported that the numbers of tiller increased in parallel with increases in the level of light intensity. In the same vein, the quantity of harvest is directly related to the number of tillers that produced panicles provided there were no disease or pest incidence. Effect of shade on upland rice varieties decreases the number of tillers, number of panicle, number of productive grains, production of grains per hill of plants and total sugar content of plants [6]. There are varieties of red rice that can adapt to low and high light intensities [11]. Wang *et al.* [4] examined the effects of shading on starch content and starch pasting viscosity in rice genotypes. The research results may lay a theoretical foundation for the selection of shade-tolerant varieties of rice and the improvement of cultivation technologies. Therefore, the development of rice varieties are required to test rice variety productivity. However, this factor can interfere with the development of plant morphological characteristics and metabolic processes resulting in the ability of the plant. Light availability is the main constraint to grow upland red rice in shading trees. Vacant land between crops and forestry tree that is not utilized is the potential for the development of upland rice. The land area in Indonesia is estimated at 11.5 million hectares every year and 3-4% of them rejuvenated. Deficit light on upland rice caused disruption of metabolic processes that have implications for decreasing the rate of photosynthetic and carbohydrate synthesis, The main ecological constraint in the utilization of land under plantation crops and forest stands is the low intensity of solar radiation, so that the plant will light stress, which will result in the decrease of growth and production. On the rice plant, shade can reduce the number of tillers as reported earlier by Cruz [7] and shade can reduce dry weight of the canopy, leaf area index, grain yield and the efficiency of solar radiation use.

There were significant differences in plant height among the new superior rice varieties at 14 weeks after planting, the highest was obtained in Ciherang variety followed by Inpari Sidenuk and the lowest was obtained in Suluttan Unsrat. The result of this study indicated there are several indices in the cultivation and management of the new superior rice varieties that could be improved. Emmanuel and Mary [2] reported that plant height decreased with increasing light intensity; this is encouraging because tall rice stems make them susceptible to lodging consequently heavy harvest losses. Also, plants grown in the dark or in very weak light have delicate, soft and slender stem since they continue in rapid growth and elongate, looking for light until they reach their maximum dimension and bend toward light direction. A condition generally referred to as etiolation. Short erect plant type is also an impetus for direct seeding which is often cheaper than transplanting using machines.

**Table 1**. Average plant height, number of tillers, leaf area, total chlorophyll on new superior rice varieties and shade intensity

| Treatment | Leaf Area | Number of Tillers 4 WAP | Plant Height 14 WAP | Total Chlorohyll |
|---|---|---|---|---|
| New Superior Rice Variety | | | | |
| $V_1$ | 47.74 ab | 7.97 cd | 115.30 bc | 4.43 c |
| $V_2$ | 41.51 c | 6.08 d | 103.99 c | 5.30 abc |
| $V_3$ | 42.49 bc | 10.14 b | 111.26 bc | 6.30 a |
| $V_4$ | 49.05 a | 10.31 ab | 119.98 ab | 5.96 ab |
| $V_5$ | 45.42 abc | 10.24 ab | 112.37 bc | 5.20 abc |
| $V_6$ | 41.33 c | 12.31 a | 128.87 a | 5.64 ab |
| $V_7$ | 43.07 bc | 10.30 ab | 112.97 bc | 5.60 abc |
| $V_8$ | 41.81 c | 6.90 d | 107.48 bc | 5.23 abc |
| $V_9$ | 40.40 c | 10.36 ab | 111.98 bc | 5.49 abc |
| $V_{10}$ | 45.34 abc | 9.60 bc | 113.47 bc | 5.59 abc |
| $V_{11}$ | 45.79 abc | 6.23 d | 108.12 bc | 5.04 bc |
| Shade Intensity | | | | |
| $N_0$ | 40.91 b | 13.35 a | 110.85 | 4.83 b |
| $N_1$ | 46.97 a | 7.63 b | 115.88 | 5.59 a |
| $N_2$ | 44.11 a | 6.42 c | 113.03 | 5.88 a |
| New Superior Rice Variety and Shade Intensity | | | | |
| $V_1N_0$ | 46.10 | 12.97 bc | 121,43 | 3.15 |
| $V_1N_1$ | 47.87 | 7.23 defghi | 113.73 | 5.31 |
| $V_1N_2$ | 49.25 | 3.70 i | 110.73 | 4.84 |
| $V_2N_0$ | 37.67 | 7.23 defghi | 105.06 | 4.27 |
| $V_2N_1$ | 44.59 | 5.70 efghi | 101.16 | 6.20 |
| $V_2N_2$ | 42.28 | 5.30 fghi | 105.73 | 5.43 |
| $V_3N_0$ | 38.07 | 16.10 ab | 107.9 | 5.31 |
| $V_3N_1$ | 48.17 | 6.90 defghi | 110.56 | 6.02 |
| $V_3N_2$ | 41.24 | 7.43 defghi | 115.3 | 7.58 |
| $V_4N_0$ | 45.84 | 14.40 ab | 115.86 | 5.86 |
| $V_4N_1$ | 48.83 | 8.50 defgh | 122.6 | 5.82 |
| $V_4N_2$ | 52.49 | 8.03 defgh | 121.46 | 6.19 |
| $V_5N_0$ | 41.48 | 15.03 ab | 112.3 | 4.51 |
| $V_5N_1$ | 51.16 | 8.67 defg | 111.63 | 5.49 |
| $V_5N_2$ | 43.63 | 7.03 defghi | 113.16 | 5.61 |
| $V_6N_0$ | 40.85 | 17.33 a | 109.76 | 5.32 |
| $V_6N_1$ | 43.39 | 10.17 cd | 155.8 | 5.20 |
| $V_6N_2$ | 39.76 | 9.43 de | 121.03 | 6.39 |
| $V_7N_0$ | 39.61 | 16.17 ab | 165.33 | 5.17 |
| $V_7N_1$ | 50.20 | 6.67 defghi | 158.86 | 5.76 |
| $V_7N_2$ | 39.40 | 8.07 defgh | 170.76 | 5.88 |
| $V_8N_0$ | 38.64 | 9.17 def | 108.5 | 5.09 |
| $V_8N_1$ | 41.16 | 6.93 defghi | 115.16 | 4.27 |
| $V_8N_2$ | 45.64 | 4.60 hi | 115.23 | 6.34 |
| $V_9N_0$ | 38.99 | 14.50 ab | 109,53 | 5.07 |
| $V_9N_1$ | 43.68 | 8.93 defg | 105.03 | 5.84 |
| $V_9N_2$ | 38.54 | 7.63 defghi | 107.86 | 5.56 |
| $V_{10}N_0$ | 36.47 | 15.53 ab | 111.7 | 4.60 |
| $V_{10}N_1$ | 50.85 | 8.33 defgh | 113.26 | 5.88 |
| $V_{10}N_2$ | 48.69 | 4.93 ghi | 110.96 | 6.30 |
| $V_{11}N_0$ | 46.28 | 8.37 defgh | 108,26 | 4.74 |
| $V_{11}N_1$ | 46.77 | 5.87 efghi | 114.9 | 5.77 |
| $V_{11}N_2$ | 44.32 | 4.47 hi | 117.23 | 4.61 |

Means followed by the same letters in the same columns and rows are not significantly different at 5% level of probability by DMRT

Some research indicates that the level of shade more than 50%, can decline production. The rate of decline rice production by shade is depending on the level of tolerance and growth phase of each variety. A process in plants that may be affected by shade is photosynthesis, transpiration, respiration, nitrate reduction, protein synthesis, hormone production and photosyntate translocation, root growth and mineral absorption [12]. Rice leaf responds to different levels of light received. For high light intensity, the ratio of a/b chlorophyll will be the highest from plants that can adapt to high light. Whereas in low-light conditions, the a/b chlorophyll ratio was highest in plants that was adapted to the low light, ± 20% of light in [11].

In this research, the new superior rice variety showed significant effect on total chlorophyl, where the highest was found on Inpari Mugibat then followed by Inpari Sidenuk variety and Inpara variety was the lowest (Table). Rice leaf responds to different levels of light received. There are varieties of red rice that can adapt to low and high light intensities. For high light intensity, the ratio of a/b chlorophyll will be the highest from plants that can adapt to high light. Whereas in low-light conditions, the *a/b* chlorophyll ratio was highest in plants that was adapted to the low light,± 20% of light in. Watanabe *et al.* [11] reported that leaf morphology and anatomy indicate changes in correlated with shaded conditions. Research to determine the morphological characteristics of plants that are good for full light have been carried out. While knowledge of the morphological characteristics of plants that are good at low light intensity is still limited. Based on the pattern of assimilation rate and the linear erect leaves to changes in light intensity and pattern of flat leaf assimilation rate reaches a maximum at low light intensity which can be presumed that the arrangement of the leaves is good for low-light leaf leaves are erect leaves on top and flat leave on bottom.

Furthermore, shade will affect plant morphology changes will also affect the quality of upland rice grain obtained. Leaf morphology and anatomy indicate changes in correlated with shaded conditions. The leaves tend to be thinner and wider [13].In 2014, Agricultural Research and Development Agency has launched new superior varieties (VUB) in irrigated land (Inpari 1 – Inpari 33), superior upland rice varieties (Inpago 4-10), and for marsh ecosystem of Inpara 1-7 varieties [14], continues to play a key role in rice farming in many parts of Asia.

Wang *et al.* [4]examined the effects of shading on starch content and starch pasting viscosity in rice genotypes. The research results may lay a theoretical foundation for the selection of shade-tolerant varieties of rice and the improvement of cultivation technologies. Therefore, the development of rice varieties are required to test rice variety productivity. In plant photosynthesis, chlorophyll is the most important photosynthetic pigment, and shading also affected the chlorophyll content of plants. Shading altered light-use efficiency by increasing leaf chlorophyll a, chlorophyll b, and chlorophyll a+b, and decreasing chlorophyll a/b ratios [5].

## 4.Conclusions

New superior rice varieties showed significant effectonthe growth and production variable such as leaf area, where Inpari Sidenuk variety was the highest among the varieties. Total chorophyll, the highest was found on Inpari variety. Number of tillers and plant height where the highest was found on Ciherang variety. The shade intensity showed significant effect on leaf area, where 25% shade intensity was the highest. Total chlorophyll, the highest was found on 50% shade intensity, number of tillers, the highest was found on no shade intensity. New superiorrice vareties should be recommended. thus, enhancing the yield of new superior variety like ciherang variety which performs better in the shade intensity.

## References

[1] Muhidin, Jusoff K, Elkawakib S U, Yunus M, Kaimuddin, Meisanti, Ray S Gand Rianda B L 2013 The Development of Upland Red Rice under Shade Trees *World Applied Sciences Journal* **24 (1)**: 23-30

[2] Emmanuel G A and Mary D M 2014 Effect of Light Intensity on Growth and Yield of a

Nigerian Local Rice Variety-Ofada *International Journal of Plant Research* **4(4)**: 89-94

[3] BPS Sumut (Badan Pusat Statistik Sumatera Utara) 2013 *Sumatera Utara 5 besar Swasembada Beras*http://setkab.go.id

[4] Wang L, Deng F, Ren W J and Yang W Y 2013 Effects of Shading on Starch Pasting Characteristics of Indica Hybrid Rice (*Oryza sativa* L.) *PLoS ONE* **8(7)**: e68220. doi:10.1371/journal.pone.0068220

[5] Gregoriou K, Pontikis K and Vemmos S 2007 Effects of reduced irradiance on leaf morphology, photosynthetic capacity, and fruit yield in olive (Olea europaea L.) *Photosynthetica* **45(2)**: 172–181

[6] Ginting J, Damanik B S J, Sitanggang J M and Muluk C 2015 Effect of Shade, Organic Materials and Varieties on Growth and Production of Upland Rice *International Journal of Scientific & Technology Research* Volume 4

[7] Cruz P 1997 Effect of Shade on Growth and Mineral Nutrition of a C4 Perenial Grass under Field Condition *Plant and Soil* **188**: 227-237

[8] Kaderi H 2004 *Pengamatan Percobaan Bahan Organik TerhadapTanaman Padi Di Rumah Kaca* Balai Penelitian Pertanian Lahan Rawa (Balittra) Banjarbaru

[9] Hendry GAF and Grime J P 1993 *Methods on Comparative Plant Ecology* A Laboratory Manual London : Chapman and Hill

[10] Aumonde T Z, Peds T, Borella J, Amarante L, and Villela F A 2013 Seed vigor, antioxidant metabolism and initial growth characteristics of red rice seedlings under different light intensities *Acta.Bot. Bras.* **27 (2)**, http://dx.doi.org/10.1590/S0102-33062013000200007

[11] Watanabe N, Fujii C, Shirota M and Furuta Y 1993 Changes in Chlorophyll, Thylakoid Proteins and Photosynthetic Adaptation to Sun and Shade Environments in Diploid and Tetraploid OryzapunctuataKotschy and Diploid Oryzaeichingeri Peter Plant *Physiol BiochemParis*, **31(4)**: 469-474

[12] Marschner H 1995 Mineral Nutrition of Higher Plants *Academics Press Inc. San Diego* pp: 889

[13] Fitter AH and HayRKM 1991 *Environmental Physiology of Plants* 3 Edition Academic PressINC Waltham MA USA pp: 102

[14] Mejaya M J, Satoto, Sasmita P, Baliadi Y, Guswara A and Suharna 2014 Deskripsi Varietas Unggul Baru Padi *Badan Penelitian dan Pengembangan Pertanian* Kementerian Pertanian 73 hal.

[15] Las I, Supriharto B, Daradjat A A, Suwarno, Abdullah B and Satoto 2004 Inovasi Teknologi Varietas Unggul Padi: Perkembangan, Arah dan Strategi ke Depan dalam Ekonomi Padi dan Beras Indonesia. *Badan Penelitian dan Pengembangan Pertanian*. Jakarta.

# Root morphology of several potato varieties – infected *Meloidogyne* spp. and addition of organic matters

**K Lubis[1*], A M Lubis[1], L A M Siregar[1], Lisnawita[1], I Safni[1] and A R Tantawi[2]**

[1] Department of Agrotechnology, Faculty of Agriculture, Universitas Sumatera Utara, Jl. Prof. A. Sofyan No. 3, Medan 20155, Indonesia
[2] Departement of Agrotechnology, Faculty of Agriculture, University of Medan Area, Jl. Kolam No. 1 Medan Estate, Medan 20233, Indonesia
[*] E-mail: nisalbz@gmail.com

**Abstract.** This research was aimed to determine root morphology of several potato varieties which were applied by organic materials into the planting medium inoculated nematodes. The research was conducted at Research Station of Horticulture in Berastagi, Sumatera Utara on May to November 2016. The randomized block design was used with two factors ; the first factor was K1 = Positive control ( no use compost / inoculation of nematodes ) K2 = Negative control (no use compost / no inoculation of nematodes) K3 = Using compost mucuna and inoculation of nematode, K4 = Using compost peanuts and inoculation of nematodes and the second factor was potato varieties (Tenggo, Maglia, and Margahayu). The results showed that organic matters increased the shoot fresh weight, the root fresh weight, the tubers weight and the number of tubers, root diameter, root lenght. However, organic matters also increased the number of nematodes. Varieties of Tenggo and Maglia showed significant affect to all observed characters. The interaction of the two treatments had significant affect to the shoot fresh weight, the number of root-knot, and the number of tubers, root lenght. However, no significant affect was observed in root wet weight, and tuber weight.

## 1. Introduction

Potato (*Solanum tuberosum* L.) is an important crop in Indonesia. In terms of the value of its nutrition value, is one type of potato tubers that can serve as a potential source of nutrition. The nutritional substances contained in Potato tubers, among others, carbohydrates, minerals, iron, magnesium, phosporus , sodium, calcium and potassium, protein and vitamins especially vitamin C and vitamin B1. In addition, the potatoes also contain fat in relatively small amounts, i.e., of 1.0-1.5 percent [1].

Based on the data of the Central Bureau Of Statistics in district of Berastagi in 2015. Potato production in 2014 of 2,333 tons with an area 147 Ha harvest, while in 2013 with a vast harvest 104 Ha was able to produce production 2,754 tons. Based on the above it can be concluded that there has been a significant decline in spite of the addition of a vast harvest [2].

One of the causes of the decline in crop production is the presence of plant pest organism, one of which is root-knot nematodes (*Meloidogyne spp.*). In addition to attacking the potato nematodes also attacked crops of tomatoes, cucumbers, carrots and others [3].

Control efforts have been done in an attempt to suppress the nematode population density in the field. One of these by using nematicide. Different types of nematicide that can be used to take control of *Meloidogyne spp.* such as Furadan, Carbofuran, Fenamilos and others. The effectiveness of nematicide depends on the dose and mode of application [4].

Nematode control with use of chemical nematicide still play an important role. This occurs because other means of control have not been able to provide satisfactory results. But controlling nematodes with nematisidea may cause negative effects for the environment and the organism was not target. This is because the nematicidecan be toxic to humans and pets. Additionally nematicide can be persistent in the soil, causing soil to water pollution, as well as a slew of other organism they are not natural enemies include target nematodes such as fungi, bacteria and other microorganisms. In an attempt to maintain environmental sustainability, control of nematodes is directed at controlling biological basis like using antagonistic microorganisms (natural enemies),organic materials, crop plants, and plants that are nutritious as pesticides [5]. Legumes can serve as an amendement to the soil and plants cover the soil to reduce the population of nematodes. Plant legumes can produce organic material such as a Lectin, Rotenone Deguelin, and Tephrosin are used as pesticides. The use of nuts as organic amendement to pest control may not yet be widely practice except in the developing countries, but the practice is likely to be one of the alternatives in the future. Most species examined to start constituent legumes nematicidal [6].

One of the techniques of breeding resistance disease that using varieties resistant. Resistant varieties are varieties of plant that have the ability to reject or avoid, recover back from the attack of pests or disease. Resistant varieties are varieties that are able to produce more and better compared to other varieties of pest populations at the level of the same [7].

## 2. Material and methods

The research was conducted at Research Station of Horticulture in Berastagi, Sumatera Utara on May to November 2016. The randomized block design was used with two factors ; the first factor was organic materials (mucuna and peanut compost) and the second factor was potato varieties (Tenggo, Maglia, and Margahayu). Againts a significant variety of prints, continued follow-up analysis by using Duncan Multiple Range Test at a 5% level [8]. The materials used were plant seed potato varieties (Tenggo, Margahyu, Maglia), waste plant, Leguminoceae (peanuts, mucuna), top soil, the roots-knot of tomato plants.

### 2.1. *Estimation of genetic parameter*

Prediction component of genetic diversity, variance of interaction genotype and environment , environmental diversity and variance of phenotypes. Performed in the following way :

$$\sigma^2_e = M1 \quad (1)$$

$$\sigma^2_g = \frac{M3-M2}{rl} \quad (2)$$

$$\sigma^2_p = \sigma^2_g + \frac{\sigma^2_{AB}}{l} + \frac{\sigma^2_e}{rl} \quad (3)$$

Description : $\sigma^2_e$= error variance ; $\sigma^2_g$ = genetic variance ; $\sigma^2_p$= phenotype variance$\sigma^2_{AB}$= interaction genetic and environment variance

The value of heritability ($h^2$) in the broad sense is :

$$\text{Heritability } h^2 \text{ mean base} = \frac{\sigma^2_g}{\sigma^2_p} \times 100\% \quad (4)$$

Broad or narrow of genetic diversity value of a character depends on genetic variance. Coefficient of genetic variance predictable by knight equation

$$\text{Coefficient genetic variance} = CGV = (\sqrt{\sigma^2_g} / x) \times 100\% \quad (5)$$

Description : $\sigma^2_g$genetic variance

Crietria: narrow coef genetic diversity (0-10%) moderate (10-20%) and wide (>20%).

*2.2. Correlation*

Analysis correlation was used to determine character related to main characters, to improve the follow response in the implementation of indirect selection. Analysis of correlation calculated by Gaspersz

$$r_{xy} = \frac{n\, \Sigma\, xiyi - (\Sigma xi)\, (\Sigma yi)}{\sqrt{(n\, \Sigma xi^2 - (\Sigma xi)^2)(n\Sigma yi^2 - (\Sigma yi)^2)}} \quad (6)$$

Description : $r_{xy}$ = correlation of variable x and y; n = number of observations object ; x = value of variable x; and y = value of variable y. This analysis can be used to find out which morphology characters correlated with outcomes, so it can be made into character selection.

**3. Results and Discussions**

The results (Table 1,2,4,5,6,7) in some varieties of potatoes with the applied of the organic material showed that the applied of the organic material in the form of compost peanut and mucuna would increased all parameters ,because with the applied of the organic material will increase the availability of nutrient elements in soil, improve the structure of the soil and improve soil in water binding, so plants can grow and thrive [9].

**Table 1.** The shoot fresh weight on each treatment organic materials and potato varieties

| Organic materials | Varieties | | | Mean |
|---|---|---|---|---|
| | V1 | V2 | V3 | |
| | .......................g..................... | | | |
| K1 | 5.67cd | 8.93c | 1e | 5.2c |
| K2 | 3.36de | 8.18c | 0.92e | 4.15c |
| K3 | 19.97b | 35.99a | 7.39cd | 21.12a |
| K4 | 20.19b | 17.23b | 7.49cd | 14.97b |
| Mean | 12.3b | 17.58a | 4.2c | |

Description : K1 = Positive control ( no use compost / inoculation of nematodes )K2 = Negative control (no use compost / no inoculation of nematodes)K3 = Using compost mucuna and inoculation of nematode. K4 = Using compost peanuts and inoculation of nematodes. V1 = Tenggo variety. V2 = Maglia variety. V3 = Margahayu variety. The figures followed a different letter notation is different based on the significant test of multiple distance duncan at a 5% level.

The results (Table 3) in some varieties of potatoes with the applied of the organic material showed that the applied of the organic material in the form of compost peanut and compost mucuna can not suppress the growth and development of nematodes.

However. the granting of organic compost and compost peanut mucuna shows the result of a higher number of root-knot than positive control (no use compost / inoculation of nematodes). This is due to the applied of the organic material in the form of composting and compost soil mucuna beans are able to improve the physical properties of the soil. keeping the water availability in the soil pore spaces of the soil as well as add that makes soil conditions supporting the growth and development of nematode[9]. Yet despite the number of root-knot administering more organic matter in the form of composting and compost soil mucuna beans is capable of supporting the growth and development of the plant so that plant potatoes stay afloat against attack nematodes.

**Table 2**. The root fresh weight on each treatment organic materials and potato varieties

| Organic Materials | Varieties | | | Mean |
|---|---|---|---|---|
| | V1 | V2 | V3 | |
| | ........................g........................ | | | |
| K1 | 2.04 | 2.76 | 0.22 | 1.68b |
| K2 | 0.66 | 3.08 | 0.47 | 1.40b |
| K3 | 5.93 | 8.52 | 2.3 | 5.58a |
| K4 | 3.07 | 8.56 | 1.59 | 4.41a |
| Mean | 2.93b | 5.73a | 1.15c | |

Description : K1 = Positive control ( no use compost / inoculation of nematodes )K2 = Negative control (no use compost / no inoculation of nematodes)K3 = Using compost mucuna and inoculation of nematode. K4 = Using compost peanuts and inoculation of nematodes. V1 = Tenggo variety. V2 = Maglia variety. V3 = Margahayu variety. The figures followed a different letter notation is different based on the significant test of multiple distance duncan at a 5% level.

**Table3.** Number of root-knot on each treatment organic materials and potato varieties

| Organic Materials | Varieties | | | Mean |
|---|---|---|---|---|
| | V1 | V2 | V3 | |
| | ........................root-knot..................... | | | |
| K1 | 169.44cd | 232.89c | 5.89e | 136.07c |
| K2 | 1.89e | 2.22e | 1.22e | 1.78b |
| K3 | 384.22b | 406.89b | 93.33d | 294.81a |
| K4 | 358b | 501a | 154.78cd | 337.93a |
| Mean | 228.39a | 285.75a | 63.81b | |

Description : K1 = Positive control ( no use compost / inoculation of nematodes )K2 = Negative control (no use compost / no inoculation of nematodes)K3 = Using compost mucuna and inoculation of nematode. K4 = Using compost peanuts and inoculation of nematodes. V1 = Tenggo variety. V2 = Maglia variety. V3 = Margahayu variety. The figures followed a different letter notation is different based on the significant test of multiple distance duncan at a 5% level.

**Table 4**. The root diameter on each treatment organic materials and potato varieties

| Organic Materials | Varieties | | | Mean |
|---|---|---|---|---|
| | V1 | V2 | V3 | |
| | ........................cm..................... | | | |
| K1 | 5.26 | 6.12 | 3.19 | 4.86bc |
| K2 | 4.17 | 5.42 | 2.95 | 4.18c |
| K3 | 5.83 | 6.74 | 5.23 | 5.94ab |
| K4 | 4.93 | 8.78 | 5.37 | 6.36a |
| Mean | 5.05b | 6.77a | 4.18b | |

Description : K1 = Positive control ( no use compost / inoculation of nematodes )K2 = Negative control (no use compost / no inoculation of nematodes)K3 = Using compost mucuna and inoculation of nematode. K4 = Using compost peanuts and inoculation of nematodes. V1 = Tenggo variety. V2 = Maglia variety. V3 = Margahayu variety. The figures followed a different letter notation is different based on the significant test of multiple distance duncan at a 5% level.

Based on the results of (Table 1-7) Tenggo varieties and maglia varieties showed better result on all parameters than margahayu varieties. It is alleged to be caused by the nature of the distinction or excelence of each variety complies with its own genotype.

**Table 5.** The root length on each treatment organic materials and potato varieties

| Organic Materials | Varieties | | | Mean |
|---|---|---|---|---|
| | V1 | V2 | V3 | |
| | ...................cm.................. | | | |
| K1 | 25.78b | 24.06bc | 8.72e | 19.52b |
| K2 | 22.44bc | 35.28a | 22.94bc | 26.89a |
| K3 | 30.89a | 34.22a | 20.00c | 28.37a |
| K4 | 33.06a | 35.00a | 15.39d | 27.81a |
| Mean | 28.04b | 32.14a | 16.76c | |

Description : K1 = Positive control ( no use compost / inoculation of nematodes )K2 = Negative control (no use compost / no inoculation of nematodes)K3 = Using compost mucuna and inoculation of nematode. K4 = Using compost peanuts and inoculation of nematodes. V1 = Tenggo variety. V2 = Maglia variety. V3 = Margahayu variety. The figures followed a different letter notation is different based on the significant test of multiple distance duncan at a 5% level.

**Table 6.** The tubers weight on each treatment organic materials and potato varieties

| Organic Materilas | Varieties | | | Mean |
|---|---|---|---|---|
| | V1 | V2 | V3 | |
| | ........................g......................... | | | |
| K1 | 9.12 | 13.19 | 3.85 | 8.72b |
| K2 | 15.11 | 12.40 | 5.33 | 10.94b |
| K3 | 48.97 | 51.88 | 23.25 | 41.37a |
| K4 | 42.10 | 44.35 | 21.70 | 36.05a |
| Mean | 28.82a | 30.45a | 13.53b | |

Description : K1 = Positive control ( no use compost / inoculation of nematodes )K2 = Negative control (no use compost / no inoculation of nematodes)K3 = Using compost mucuna and inoculation of nematode. K4 = Using compost peanuts and inoculation of nematodes. V1 = Tenggo variety. V2 = Maglia variety. V3 = Margahayu variety. The figures followed a different letter notation is different based on the significant test of multiple distance duncan at a 5% level.

**Table 7.** The number of tubers on each treatment oranic materials and potato varieties

| Organic Materilas | Varieties | | | Mean |
|---|---|---|---|---|
| | V1 | V2 | V3 | |
| | .........................tuber......................... | | | |
| K1 | 3.78de | 3.00ef | 1.67g | 2.81b |
| K2 | 2.56fg | 1.89g | 2.00g | 2.15b |
| K3 | 10.56a | 7.78b | 4.67cd | 7.67a |
| K4 | 9.22a | 7.11b | 5.11c | 7.15a |
| Mean | 6.53a | 4.94b | 3.36c | |

Description : K1 = Positive control ( no use compost / inoculation of nematodes )K2 = Negative control (no use compost / no inoculation of nematodes)K3 = Using compost mucuna and inoculation of nematode. K4 = Using compost peanuts and inoculation of nematodes. V1 = Tenggo variety. V2 = Maglia variety. V3 = Margahayu variety. The figures followed a different letter notation is different based on the significant test of multiple distance duncan at a 5% level.

The value of the heritabilty some of the varieties tested showed high criteria (Table 8). This show that some varieties are tested against genetic diversity contributes to phenotype in some oganic material tested.

**Table 8.** The value of the expect variety genotype and heritability broad sense as well as coefficients of genetic diversity

| Characters | $\sigma^2G$ | $h^2_{bs}$ | Criteria | CGD(%) | Criteria |
|---|---|---|---|---|---|
| Shoot fresh weight | 5373 | 0.97 | High | 2.50 | High |
| Root fresh weight | 700.08 | 0.99 | High | 1.06 | High |
| Root length | 8414.76 | 0.99 | High | 3.57 | High |
| Root diameter | 225.96 | 0.99 | High | 2.82 | High |
| Number of root-knot | 1653651 | 0.98 | High | 146.96 | High |
| Number of tubers | 311.28 | 0.98 | High | 0.08 | High |
| Tubers weight | 11317.56 | 0.99 | High | 02.23 | High |

Description : $\sigma^2G$ = Genotype variety. $h^2_{bs}$ = heritability broad sense. CGD = Coefficient of genetic diversity

Based on the correlation of (Table 9) can be seen that all the parameters are correlated positively indicated by the value of the correlationthat is very significant. This shows that every parameter is correlated with each other.

**Table 9.** Corelation between morphological Characters with the Character of production and the number of root-knot potatoes

| | RL | TW | NT | SFW | RFW | NRK |
|---|---|---|---|---|---|---|
| TW | 0.565** | | | | | |
| NT | 0.482** | 0.819** | | | | |
| SFW | 0.615** | 0.792** | 0.749** | | | |
| RFW | 0.624** | 0.772** | 0.577** | 0.797** | | |
| NRK | 0.554** | 0.745** | 0.787** | 0.777** | 0.790** | |
| RD | 0.557** | 0.486** | 0.437** | 0.557** | 0.722** | 0.661** |

Description : RL =root length. RD = root diameter. TW = tubers weight. RFW = root fresh weight. NRK = number of root-knot. SFW = shoot fresh weight. NT = number of tubers. * = corellate the significant on the extent 0.05; ** = corellate the significant on the extent 0.01.

## 4. Conclusions

1. Varieties Tenggo and Maglia shows growth and better production compared with margahayu varieties based on The shoot fresh weight. root fresh weight. tubers weight and number of tuber.

IOP Conf. Series: Earth and Environmental Science **122** (2018) 012025 doi:10.1088/1755-1315/122/1/012025

2. Applied organic materials compost peanut and compost mucuna can increase the growth and production of potato plants on medium inoculated nematodes
3. Highly heritability and correlation show on shoot fresh weight. root fresh weight. root length. root diameter. number of root-knot. number of tubers. and tubers weight. All parameters shown significant corellation. Thats mean all parameters : shoot fresh weight. root fresh weight. root length. root diameter. number of root-knot. number of tubers. and tubers weight potentially used for character selection to early selected resistance varieties to nematode

**References**

[1] Talburt W F and O Smith 1987 Potato Processing (New York: AVI Book Published by Van Nostrand Reinhold. CO.)
[2] The Central Bureau Of Statistics 2015 Berastagi In Numbers 2015 The Central Bureau Of Statistics. Karo Regency..
[3] Sherf A F and A A Macnab 1986 Vegetables diseases and their control (New York: John Wiley and Sons. New York. 728
[4] Marwoto B 1994 Root-Knot Nematode Control (Meloidogyne spp) Integrated On Tomato Plants. Lembang Horticultural Research Hall. Bandung. 96-97.
[5] Mustika I 1992 Introduction to The Namatologi Plants. Spices and Plants Research Hall The Medicine of Bogor.
[6] Morris J B and Walker J T 2002 Non-Traditional Legumes as Potential Soil Amendments for Nematode Control. Journal of Nematology **34**(4): 358– 361
[7] Samsudin 2008 Pest Control With a Botanical Insecticide Agricultural Institutions Healthy. www.pertaniansehat.or.id. Retrieved May 2017
[8] Bangun M K 1991 *The Design Of The Experiment* The Faculty Of Agriculture. USU Medan.
[9] Leiwakabessy F M. Wahjudin U M and Suwarno 2003 Soil fertility IPB Bogor

**Acknowledgements**

This Research is part of The National Competitif Research on Scheme National Strategys (Stranas) (2016-2017) from The Ministry of Research. Technology. and Higher Education

IOP Conf. Series: Earth and Environmental Science **122** (2018) 012026 doi:10.1088/1755-1315/122/1/012026

# Selection individual on mutant genotype of soybean (*Glycine max*l.merrill) in $m_5$ generation based on resistance of stem rot disease *Athelia rolfsii* (curzi)

**M Rahmah[1], D S Hanafiah[1*], L A M Siregar[1] and I Safni[1]**

[1]Faculty of Agriculture, Universitas Sumatera Utara, Indonesia
*E-mail: dedek.hanafiah@yahoo.co.id

**Abstract.** This study was aimed to obtain selected individuals on soybean plant *Glycine max* L. (Merrill) in M5 generation based on high production character and tolerance of stem rot disease *Athelia rolfsii* (Curzi). This research was conducted in Plant Disease Laboratory and experimental field Faculty of Agriculture Universitas Sumatera Utara Medan, Indonesia. This research was conducted from December 2016 to June 2017. The treatments were 15 mutant lines genotypes and Anjasmoro variety. The results showed that some lines mutant genotypes can gave the good agronomic appearance character than Anjasmoro variety on inoculation treatment of stem rot disease. Selection performed on population $M_5$ producesselected individuals with tolerance of stem rot disease from 100 and 200 Gy population.

## 1. Introduction

Oktara [1] stated that soybean production in 2015 is known to 963.10 thousand tons, an increase of 8.10 thousand tons compared to the previous year. The increase outside Java Island was 30.41 thousand tons, but in Java decreased 22.31 thousand tons. The increase in soybean production was due to a productivity increase of 0.18 quintal per hectare or 1.16 percent. However, it is known that the harvested area has decreased by 1.80 thousand hectares or about 0.29 percent. The decline in harvested area is also related to global climate change affecting agriculture.

Global climate change affects schedule and cropping pattern in which climate and weather are the determining factors as well as limiting the success of the business to farmers. Climate change is causing several threats in agriculture, one of which is the increase of pests and plant diseases. This will affect the national soybean production.

In Indonesia, soybean (*Glycine max* L. Merr.) is one of the leading commodities of legumes, however itis susceptible to plant pests and diseases. Some harmful diseases in soybeans mainly due to fungal pathogens include stem rot disease of *A. rolfsii*. Blum and Rodriguez [2] reported that *A. rolfsii* infections cause low percentage of seeds to germinate, thus reducing plant populations in soybean .

The study of *rolfsii* from Pudjihartati *et al.* [3] showed that *A. rolfsii* (formerly *Sclerotium rolfsii*) is a soil pathogenic fungus that is necrotropic, and is the cause of collar rot in peanut crops. *A. rolfsii* in humid environment conditions, also infects branches and leaves near the surface of the soil, and can be a bridge spreading the growth of mycelium to other parts of the plant. *A. rolfsii* infection in susceptible pea plants in the field can decrease pod yields up to 74%.

The study from Sibarani *et al.*[4] stated that the productivity of soybean crops can be improved through the use of technology. One of the technological components is the use of gamma ray irradiation to obtain new high yielding varieties, and to change the morphological appearance of plants aimed at increasing crop productivity .

The results of this study are expected to reveal various useful information in the development of soybean varieties through low-dose gamma ray irradiation to produce mutant genotypes resistance to stem rot disease of *A. rolfsii* Curzi and high yielding in M5 generation.

## 2. Materials and methods

This research was conducted at Plant Disease Laboratory and research area of Faculty of Agriculture University of Sumatera Utara, Medan. This research was conducted in December 2016 until July 2017. Soybean seeds were planted mutant M5 genotypes obtained from the selection of M4 generation pedigrees. The control plants were Anjasmoro variety as the control and parent plant. There were 7 groups of genotypes used in plastic housing including M5 (100)-A-25 (180 plants), M5 (100)-A-6 (48 plants), M5 (100)-A-12 (100 plants), M5 (100)-A-17 (65 plants), M5 (100)-A-11 (122 plants), M5 (100)-A-8 (38 plants) and M5 (300)-A-6 (45 plants).

Each genotypes were planting in plastic housing (inoculation with *A. rolfsii* fungus). Data analysis was done by calculating the mean value of each character observed then the mean value of each population is tested by using t test. The parameters were plant height, flowering age, harvest age, number of primary branches, number of pods per plant, number of seeds per pods, seed weight per plant and disease incidence.

## 3. Results and Discussion

The results showed that the average of several genotypes with agronomic characters showed different results for each character. The results of $M_5$ (100 Gy) showed that flowering age and plant height characters gave very significant value while the number of primary branches, the number of pods per plant, the age of harvest, the number of seeds per plant, and the weight of seeds per plant did not give significant value. The mean value of agronomic characters on $M_5$ (100 Gy) with Anjasmoro can be seen in Table 1.

**Table 1.** The mean value of agronomic characters on Population $M_5$ (100 Gy) compared to Anjasmoro on inoculated media with *Athelia rolfsii* (Curzi)

| Characters | Mean | | t value |
|---|---|---|---|
| | Anjasmoro | $M_{5(100)}$ | |
| Flowering age (days) | 32.880 | 34.627 | 0,007** |
| Plant height (cm) | 73.578 | 85.468 | 0,006** |
| Number of primary branches | 2.560 | 2.902 | 0,185 |
| Number of pods per plant | 69.760 | 68.045 | 0,773 |
| Harvest age (days) | 101.080 | 100.586 | 0,099 |
| Number of seeds per plant | 145.520 | 142.850 | 0,821 |
| Weight seeds per plant (g) | 23.592 | 24.821 | 0,552 |

Note : * = Significantly different from the Anjasmoro population at the 5% level based on t tes
** = Very significantly different from the Anjasmoro population at the level of 1% based on t test

The result of t-test analysis on $M_5$ (200 Gy) compared to Anjasmoro on the inoculated medium of fungus *Athelia rolfsii* (Curzi) showed not significant value to all observed characters. The average value of agronomic characters on $M_5$ (200 Gy) can be seen in Table 2.

**Table 2.** The mean value of agronomic characters on population $M_5$ (200 Gy) compared to Anjasmoro on inoculated media .

| Characters | Mean | | t value |
|---|---|---|---|
| | Anjasmoro | $M_{5(200)}$ | |
| Flowering age (days) | 32.880 | 33.544 | 0.139 |
| Plant height (cm) | 73.568 | 77.483 | 0.327 |
| Number of primary branches | 2.560 | 2.694 | 0.590 |
| Number of pods per plant | 69.760 | 68.408 | 0.818 |
| Harvest age (days) | 101.080 | 100.639 | 0.139 |
| Number of seeds per plant | 145.520 | 142.537 | 0.797 |
| Weight seeds per plant (g) | 23.592 | 25.841 | 0.292 |

Note : * = Significantly different from the Anjasmoro population at the 5% level based on t test
** = Very significantly different from the Anjasmoro population at the level of 1% based on t test

The result of t test on M5 (300 Gy) compared to Anjasmoro parent in the inoculated medium of *A. rolfsii* (Curzi) showed a significant difference in the number of primary branches and the harvest age gave very significantly difference value compared to Anjasmoro. Characters of flowering age. plant height. number of pods per plant. number of seeds per plant. weight of seeds per plant. and weight of 100 seeds showed not significantly difference compared to Anjasmoro. The average value of agronomic characters on $M_5$(300 Gy) with Anjasmoro can be seen in Table 3.

**Table 3.** The mean value of agronomic characters M5(300 Gy) generation compared to Anjasmoro on Inoculated Media

| Characters | Mean | | t value |
|---|---|---|---|
| | Anjasmoro | $M_{5(300)}$ | |
| Flowering age (days) | 32.880 | 34.476 | 0.133 |
| Plant height (cm) | 73.568 | 71.448 | 0.745 |
| Number of primary branches | 2.560 | 3.476 | 0.023* |
| Number of pods per plant | 69.760 | 82.095 | 0.239 |
| Harvest age (days) | 101.080 | 102.524 | 0.001** |
| Number of seeds per plant | 145.520 | 169.286 | 0.328 |
| Weight seeds per plant (g) | 23.592 | 33.148 | 1.000 |

Note : * = Significantly different from the Anjasmoro population at the 5% level based on t test
** = Very significantly different from the Anjasmoro population at the level of 1% based on t test

The results showed that the appearance of agronomic characters on the inoculated medium of *A. rolfsii* (Curzi) causes stem rot disease is lower than planting under optimum conditions without inoculation of fungi in population M5. This is because *A. rolfsii* is a soilborne pathogen that infect the base of the stem in the soybean crop and forms a white sclerotium that eventually becomes brown. Ferreira and Boley[5] stated that *A. rolfsii* . infects some parts of the plant under appropriate environmental conditions including roots. petioles. leaves and flowers. however it infects stem at the early stage of disease cycle

Based on the selection of inoculated media of *A. rolfsii* (Curzi) obtained 62 selected individuals from each genotype with the tolerant character of stem rot disease *A. rolfsii* (Curzi). The number of selected individuals is more generated from the population of 100 Gy and 200 Gy irradiation. Harsanti and Yulidar [6] stated that the advantages of mutation techniques can alter and improve one character of a variety without changing the other character. This change creates a new trait that is not owned by its parent.

The result of observation diagram of incidence of 24 days after inoculation disease can be seen in Figure 1. The results showed that the affected plant from plot 1 were G M100-A-25 (2/7) by 73.1%. G M100-A-25 (3/4) by 77.1%.and G M100-A-25 (3/7) by 76.9%. G M300-A-8 (37/2) by 100%.G M300-A-8 (35/7) by 87.5%. G M300-A-8 (33/8) by 75.0%.G M300-A-8 (33/3) by 100% and Anjasmoro by 80%. The number of affected plant from plot 2 were G M200-A-12 (6/5) by 88.8%. G M200-A-11 (39/7) by 81.4%. G M200-A-11 (32/3) by 67.4% and Anjasmoro by 87.5%. The number of affected plant fromplot 3 were G M100-A-6 (30/2) by 84.0%.G M100-A-6 (31/1) by 88.8%. G M200-A-17 (18/5) by 80.6%.G M200-A-17 (13/6) by 76.9% G M100-A-25 (5/3) by 91.7% and Anjasmoro by100%.

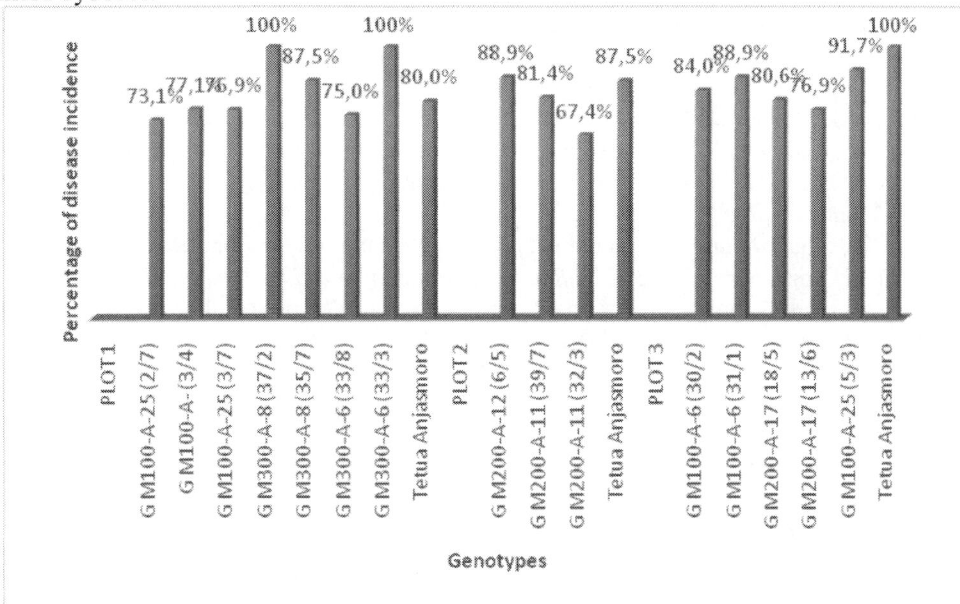


*Note : DAI : days after inoculation*

**Figure 1.** Diagram of disease incidence at 24 days after inoculation (DAI)

The highest percentage of disease incidence was on G M300-A-8 (37/2). G M300-A-8 (33/3) and Anjasmoro (100%). Stem rot disease can decrease the production of soybean crops. the varieties used also depend on the sensitivity or absence of disease infection. This is in accordance with Rahayu[7] who states that Anjasmoro variety in humid environment such as in Genteng- Banyuwangi. infected with *A. rolfsii* with a high incidence of disease. which reached 23%. Symptoms of malformation include root rot and the base of the stem. fallen seed (damping-off). wilted. dead plants. and pod rot. Stem rot disease caused by *A. rolfsii* in the field resulted in dying suddenly after entering the generative phase. the initial symptoms at the base of the stem are surrounded by white mycelium and wilted leaves then plants are blackened like burning. The progression of stem rot disease spread very rapidly due to the humid and wet environment conditions in the planting area when the soybean was planted during the rainy season in March. Sclerotium of fungi can spread through water. *A.rolfsii* is one of the soil pathogens in legumes. Xie[8] reported that *A. rolfsii* likes warm conditions. wet weather and continues to spread when hyphae and sclerotia reach the vulnerable plant tissue so that this condition is advantageous to survive.

## 4. Conclusions

The results showed the appearance of agronomic characters on the inoculated medium of *A. rolfsii*more lower than treatment without inoculation of *A. rolfsii* in population M5. Selection performed on population $M_5$ produces selected individuals with tolerance of stem rot disease from 100 and 200 Gy population.

**References**

[1] Oktara D 2016 Central Bureau of Statistics: Corn Production Up 3.17 percent. Soybean 0.85 percent. www.tempo.com
[2] Blum L E B and Rodríguez-Kábana R 2004 Effect of organic amendments on sclerotial germination. mycelial growth. and *Sclerotium rolfsii* induced diseases. Fitopatol. Bras.Vol.29(1).
[3] Pudjihartati E. Ilyas S. Sudarsono 2016 Oxidative Burst. Peroxidase Activity. and Lignin Content of *Sclerotium rolfsii* Infected Peanut Tissue. Hayati. Vol.13 (4) : 166-172.
[4] Sibarani I B. Ratna R L and Diana S H 2015 Soybean (*Glycine max* (L.) Merrill) morphology response on Anjasmoro variety at several levels of gamma ray irradiation. University of Sumatra Utara. *Online Journal of Agroecotechnology*. Vol.3. (2) : 515- 526.
[5] Ferriera S A and Boley R A 2006 *Sclerotium rolfsii*. http:/www.extento.edu.
[6] Harsanti L dan Yulidar 2015 Effect of gamma ray on early growth of soybean [*Glycine max* (L.) Merill] on Denna 1 variety. Isotope and Radiation Application Center - National Nuclear Power Agency
[7] Rahayu M 2015 *Sclerotium rolfsii* stem rot disease in various beans. Research Institute of Peanut and Tubers Crops. *http:// balitkabi. litbang. pertanian .go. id/ info- teknologi/ 2069.*
[8] Xie C 2012 Mycelial compatibility and pathogenic diversity among *Sclerotium rolfsii* isolates in southeastern United States. University of Florida. USA.

**Acknowledgments**

The authors wish to thank Rector of the University of Sumatera Utara and ministry of Research. Technology and Higher Education. Republic of Indonesia that has supportedand funded this research.

IOP Conf. Series: Earth and Environmental Science **122** (2018) 012027 doi:10.1088/1755-1315/122/1/012027

# Effect of foliar application of chitosan and salicylic acid on the growth of soybean (*Glycine max* (L.) Merr.) varieties

**Y Hasanah*and M Sembiring**

Faculty of Agriculture, Universitas Sumatera Utara, Jl. Prof. A. Sofyan No. 3 Kampus USU, Padang Bulan Medan 20155, Indonesia.
*E-mail: yaya@usu.ac.id

**Abstract.** Elicitors such as chitosan and salicylicacid could be used not only to increase isoflavone concentration of soybean seeds, but also to increase the growth and seed yield. The objective of the present study was to determine the effectsof foliar application of elicitor compounds (i.e. chitosan, and salicylic acid)on the growth of two soybean varieties under dry land conditions.Experimental design was a randomized block design with 2 factors and 3 replications. The first factor was soybean varieties (Wilis and Devon). The second factor was foliar application of elicitors consisted of without elicitor ; chitosan at V4 (four trifoliate leaves are fully developed) ; chitosan at R3 (early podding) ; chitosan at V4 and R3 ; salicylic acid at V4 ; salicylic acid at R3 and salicylicacid at V4 and R3. Parameters observed was plant height at 2-7 week after planting (WAP), shoot dry weight and root dry weight. The results suggest that the Wilis variety had higher plant height 7 WAP than Devon. The foliar application of chitosan increased the plant height at 7 WAP, shoot dry weight and root dry weight. The foliar application of chitosan at V4 and R3 on Devon variety increased shoot dry weight.

## 1. Introduction

Interest in soybeans has greatly increased in recent years because of the positive effects of soybean on human health, particularly due to isoflavone content in soybean. Isoflavones as a secondary metabolite compound in soybean plants are phytoestrogens that act as antioxidants for chemopreventive agents and estrogenic compounds that can inhibit cell proliferation, as well as to prevent osteoporosis and heart damage, reduce menopausal syndrome and anti-tumor to prevent degenerative diseases such as arterosclerosis, prostate cancer and colon cancer and diabetes mellitus [1, 2, 3, 4].

The accumulation of isoflavones in soybean is determined by genetic (internal) and environmental (external) factors. Internal factors that affect the accumulation of isoflavones are related to gene activity. Influencing environmental factors consist of biotic factor, such as wounding, nodulation and pathogen attack, and abiotic elements such as temperature, water regime, UV light, soil nutrient content, carbon dioxide and also elicitor [5, 6, 7, 8, 9, 10, 11, 12].

Increased accumulation of isoflavones in soybeans can be done by inducing soybeans with biotic or abiotic elicitors that will stimulate the formation of phytoalexin in soybeans. Elicitation is an efficient strategy that by means of compounds or treatments induces plants to synthesize phytoalexins at enhanced levels [13]. Several studies have shown that the elicitation method may increase the phytoalexin content and secondary metabolites in soybean crops. The usefulness of elicitors is as a strategy in inducing and promoting the formation of secondary metabolites and can increase the activity of specific enzymes associated with the formation of secondary metabolites [14]. Previously research by authorreported that the treatment of chitosan and salicylic acid elicitors in the screen house experiments resulted in the highest levels of genistein, daidzein and isoflavone content compared to other treatments [15].

Elicitors such as chitosan and salicylicacid could be used not only to increase isoflavone concentration of soybean seeds, but also to increase the growth and seed yield. Several studies carried out under laboratory or field conditions strongly suggest that salicylic acid play an important role in many biological responses in plants. The effect of these substances on the physiology of the plants is variable, promoting some processes and inhibiting others [16]. Salicylic acid and its regulatory role in plant physiology include inhibiting ethylene biosynthesis, interfering with membrane depolarisation, blocking wound responses, and an increase in photosynthetic rate and chlorophyll content in soybeans [17]. The results from the literature show that, chitosan can increase the yield when used in plants, [18], reduce transpiration [19] and induce a range of metabolic changes as a result of which, plants become more resistant to viral, bacterial and fungal infections [20].

Based on the background, the objective of the present study was to determine the effectsof foliar application of elicitor compounds (i.e. chitosan, and salicylic acid)on the growth of two soybean varieties under dry land conditions.

## 2. Materials And Methods

### *2.1. Study area*

The field experiment was conducted at the field grown of Medan Selayang, Medan (Indonesia). The soil content of Nitrogen was low (0.26%), organic matter was 1.02%, with a pH of 4.5.

### *2.2. Procedures*

Treatments were arranged in a Randomized Block Design with two factors and three replications. The first factor was soybean varieties (Wilis and Devon). The second factor was foliar application of elicitors consisted of without elicitor ; chitosan at V4 (four trifoliate leaves are fully developed) ; chitosan at R3 (early podding) ; chitosan at V4 and R3 ; salicylic acid at V4 ; salicylic acid at R3 and salicylicacid at V4 and R3. Preparation of elicitor refered to standard procedures. Chitosan and salicylic acid is a product of Sigma Aldrich. Autoclaved stock solution of 120°C for 20 minutes, and sterile distilled water to obtain a final concentration of chitosan solution 0.5 mg/mL. Salicylic acid dissolved in distilled water and diluted to concentrations (0.5 mM). The determination of the concentration of salicylic acid refers to previous research by Al Hetar *et. al* [21].

Soybean seeds that have been inoculated by *Bradyrhizobium japonicum* are planted with plant spacing 40 cm x 20 cm. N fertilizer application at dose of 50 kg $ha^{-1}$ Urea was given half the dose of N fertilizer at planting time and the rest at 4 WAP. At planting time, phosphor and potassium fertilizer application at a dose of 150 kg $ha^{-1}$ $P_2O_5$ and 75 kg $K_2O$ $ha^{-1}$ for all soybean crops. Weeding was done manually by removing the weeds in accordance with the conditions of the field.
Parameters observed was plant height at 2-7 week after planting (WAP), shoot dry weight and root dry weight.

### *2.3. Data analysis*

Data were subjected to analysis of variance (ANOVA) for comparison of means. Means were separated using Duncan's Multiple Range Test at the 0.05 probability level.

## 3. Results and Discussion

### *3.1. Plant height*

Based on Table 1. it can be seen that the treatment of varieties has significant effect on plant height at 2, 3, 4 and 7 WAP. The elicitor treatment had significant effect on plant height at 4 WAP. Devon variety have plant height at 2-4 WAP that is significantly higher than Wilis, while at plant height 7 WAP the Wilis variety significantly higher than Devon. It is assumed that the difference in plant height between varieties is due to plant genetic factors and is associated with faster Devon plant growth rate compared to Wilis at the beginning of the vegetative growth phase. At the age of 7 WAP, the combination of Wilis and elicitor of chitosan at R3 increased plant height. This correlated with increases in stomatal conductance and transpiration rate. Chitosan foliar application did not have any effect on the intercellular $CO_2$ concentration. Previously authors reported that the observed effect on the net photosynthetic rate is, in general, common in maize and soybean after foliar application of high molecular weight chitosan [22]

**Table 1.** Effect of foliar application of chitosan and salicylic acid on plant height 2-7 WAP of two soybean varieties

| WAP | Variety | Elicitor E0 | E1 | E2 | E3 | E4 | E5 | E6 | Mean |
|---|---|---|---|---|---|---|---|---|---|
| | | ……… | ……… | ……… | cm | ……… | ……… | ……… | |
| 2 | V1 (Wilis) | 9.28 | 8.83 | 9.25 | 9.49 | 9.02 | 9.35 | 8.55 | 9.11b |
| | V2 (Devon) | 13.18 | 12.39 | 11.87 | 12.88 | 12.17 | 11.49 | 12.20 | 12.31a |
| | Mean | 11.23 | 10.61 | 10.56 | 11.19 | 10.59 | 10.42 | 10.37 | |
| 3 | V1 (Wilis) | 12.94 | 12.14 | 14.06 | 13.16 | 13.00 | 13.50 | 13.13 | 13.13b |
| | V2 (Devon) | 16.19 | 17.09 | 16.86 | 17.34 | 16.92 | 15.25 | 17.01 | 16.67a |
| | Mean | 14.57 | 14.61 | 15.46 | 15.25 | 14.96 | 14.38 | 15.07 | |
| 4 | V1 (Wilis) | 25.67 | 21.62 | 28.20 | 25.37 | 25.90 | 23.73 | 25.40 | 25.13b |
| | V2 (Devon) | 26.87 | 28.43 | 29.22 | 29.33 | 28.58 | 24.20 | 29.47 | 28.01a |
| | Mean | 26.27cd | 25.03de | 28.71a | 27.35bc | 27.24bc | 23.97e | 27.43b | |
| 5 | V1 (Wilis) | 34.37 | 30.40 | 37.83 | 34.97 | 34.87 | 35.07 | 34.70 | 34.60 |
| | V2 (Devon) | 35.43 | 36.69 | 38.67 | 36.92 | 38.77 | 32.00 | 35.07 | 36.22 |
| | Mean | 34.90 | 33.54 | 38.25 | 35.94 | 36.82 | 33.53 | 34.88 | |
| 6 | V1 (Wilis) | 52.13 | 44.80 | 54.93 | 51.73 | 49.53 | 48.53 | 52.10 | 50.54 |
| | V2 (Devon) | 47.20 | 49.83 | 52.73 | 51.10 | 54.10 | 45.07 | 50.67 | 50.10 |
| | Mean | 49.67 | 47.32 | 53.83 | 51.42 | 51.82 | 46.80 | 51.38 | |
| 7 | V1 (Wilis) | 65.17 | 55.82 | 67.99 | 64.93 | 61.03 | 61.11 | 63.17 | 62.74a |
| | V2 (Devon) | 55.64 | 58.83 | 61.69 | 59.16 | 60.10 | 53.69 | 57.56 | 58.09b |
| | Mean | 60.40 | 57.32 | 64.84 | 62.04 | 60.56 | 57.40 | 60.36 | |

Note : E0 =withoutelicitor ; E1 = Chitosan at V4 ; E2 = Chitosan at R3 ; E3 = Chitosan at V4 dan R3 ; E4 = Salicylicacid at V4 ; E6 = Salicylicacid at R3 dan E6 =Salicylicacid at V4 dan R3. Means followed by the same letters at the same coloum, row and time of abservation not significantly differ base on Duncan's Multiple Range test (p=0.05). This note also used for Table 2 and 3.

*3. 2. Root dry weight*

The result showed that Devon variety have higher root dry weight than Wilis. The chitosan treatment on V4 (E2) yielded higher root dry weight than other treatments. The combination of Devon treatment and chitosan elicitor on V4 increased root dry weight (Table 2). This suggests that chitosan plays a role in increasing root growth which is indicated by increasing root dry weight. Foliar application of chitosan at V4 increases root dry weight. In line with previously research reported that foliar application of chitosan stimulates increased vegetative growth of plants and protects plants against oxidative stress [23, 24, 25].

**Table 2.** Effect of foliar application of elicitor (chitosan and salicylic acid) on root dry weight of two soybean varieties

| Variety | Elicitor E0 | E1 | E2 | E3 | E4 | E5 | E6 | Mean |
|---|---|---|---|---|---|---|---|---|
| | ……… | ……… | g | ……… | ……… | ……… | | |
| V1 (Wilis) | 1.76 | 1.11 | 1.70 | 1.51 | 1.68 | 1.49 | 1.39 | 1.52 |
| V2 (Devon) | 1.70 | 1.60 | 2.09 | 1.48 | 1.37 | 1.20 | 1.54 | 1.57 |
| Mean | 1.73 | 1.36 | 1.89 | 1.49 | 1.53 | 1.35 | 1.46 | |

### *3. 3. Shoot dry weight*

Devon varieties have higher shoot dry weight than Wilis. The chitosan treatment on V4 (E3) yielded higher shoot dry weight than other treatments. The combination of Devon treatment and chitosan elicitor on V4 increased shoot dry weight (Table 3). This suggests each variety has different growth rates and chitosan application on V4 enhances the growth rate that indicated with increased shoot dry weight. Previously research reported that chitosan play a positive role in plant growth associated with its effects in increased nutrient uptake such as N, P and K which play an important role in the biosynthesis and translocation of carbohydrates and stimulates cell division, cell turgor and DNA and RNA formation [26]. The other authors reported that the increase in *Phaseolus*yield due to chitosan application may be due to its effects in stimulating physiological processes, improving vegetative growth, followed by active translocation of photoassimilates from source to sink tissues. The increases in plant biomass may be due to improving photosynthetic machinery, and chitosan promoted the plant and root growth have suggested that the growth promotion might be a nitrogen effect because chitosan contains about 8.7% N [27]

**Table 3.** Effect of foliar application of elicitor (chitosan and salicylic acid) on shoot dry weight of two soybean varieties

| Variety | Elicitor | | | | | | | Mean |
|---|---|---|---|---|---|---|---|---|
| | E0 | E1 | E2 | E3 | E4 | E5 | E6 | |
| | ....................g.................... | | | | | | | |
| V1 (Wilis) | 13.47 | 3.63 | 13.37 | 9.52 | 15.19 | 11.02 | 8.72 | 10.70 |
| V2 (Devon) | 11.74 | 12.79 | 7.61 | 17.55 | 8.71 | 10.03 | 10.19 | 11.23 |
| Mean | 12.61 | 8.21 | 10.49 | 13.53 | 11.95 | 10.53 | 9.46 | |

## 4. Conclusion

Wilis variety had higher plant height 7 WAP than Devon. The foliar application of chitosan increased the plant height at 7 WAP, shoot dry weight and root dry weight. The foliar application of chitosan at V4 and R3 on Devon variety increased shoot dry weight.

## References

[1] Barnes S 2001 Soy isoflavones and cancer p. 49–53. In:K. Descheemaeker and I. Debruyne (ed.) Soy and health. Garant. Leuven. Belgium.

[2] Bennett JO, OYu, LG Heatherly and HB Krishnan 2004 Accumulation of genistein and daidzein soybean isoflavones implicated in promoting human health. is significantly elevated by irrigation. J Agric. Food Chem. 52: 7574 – 7579.

[3] Shinde AN, N Malpathak, DP Fulzele2009 Enhanced production of phytoestrogenic isoflavones from hairy root cultures of *Psoralea corylifolia* L. using elicitation and precursor feeding Biotechnology and Bioprocess Engineering 14, 288–294.

[4] Tsukamoto C, Shimada S, Igita K, Kudou S, Kokubun M, Okubo K and Kitamura K 1995 Factors affecting isoflavones content in soybean seeds: changes in isoflavones, saponins, and composition of fatty acids at different temperatures during seed development J. Agric. Food. Chem. 43: 1184 -1192.

[5] Nelson RL, Lozovaya V, Lygin A and Widholm J 2001 Variation in isoflavones in seeds of domestic and exotic soybean germplasma In: 2001 Agronomy Abstracts (CD-ROM). ASA. CSSA and SSSA. Madison. WI.

[6] Ososki ALand Kennelly EJ2003. Phytoestrogens: a review of the present state of research. Phytother. Res. 17: 845 - 869.

[7] Vyn T J, Yin X, Bruulsema TW, Jackson CC, Rajcan I and Brouder SM 2002 Potassium fertilization effects on isoflavone concentrations in soybean (*Glycine max* (L.) Merr.) J. Agric. Food. Chem. 50: 3501–3506.

[8] Lozovaya VV, Lygin AV, Ulanov AV, Nelson RL, Dayde Jand Widholm AM 2005 Effect of temperature and soil moisture status during seed development on soybean seed isoflavone concentration and composition. Crop Sci. 45 : 1934 - 1940 .

[9] SubramanianS, G Stacey, O Yu 2005 Endogenous isoflavones are essential for the establishment of symbiosis between soybean and *Bradyrhizobium japonicum*. Plant J.48:261–273.

[10]Zhang B Hettiarachchy N Chen PHorax R CorneliousB ZhuD 2006 Influence of The Application of Three Different Elicitors on Soybean Plants on The Concentrations of Several Isoflavones in Soybean Seeds. Journal Of Agricultural and Food Chemistry 54, 5548–5554.

[11]Naoumkina M Farag MA Sumner LW Tang Y Liu CJand Dixon RA 2007 Different mechanisms for phytoalexin induction by pathogen and wound signals in *Medicago truncatula*. Proc. Natl Acad. Sci. USA 104 : 17909 - 17915.

[12]Phommalth S JeongYS Kim YH Dhakal KH Hwang YH 2008 Effects of light treatment on isoflavone content of germinated soybean seeds. Journal of Agriculture and Food Chemistry 56, 10123–10128.

[13]Gonzalez JJ SK Guttikonda LS Phan Tran DL Aldrich R Zhong OYu HT Nguyen and DA Sleper 2010Differential expression of isoflavone biosynthetic genes in soybean during water deficits. Plant and Cell Physiol.51(6):936-948.

[14]Rao RS Ravishankar GA 2002 Plant Cell Cultures: Chemical Factories Of Secondary Metabolites. Biotechnology Advances 20, 101–153.

[15]Roberts SC 2005 Plant metabolic engineering for pharmaceutical production. www. metabolic engineering .gov/me2005/Roberts.pdf.

[16]Hasanah Y, LAM Siregar, L Mawarni 2016Final Report of Competitive Grant Research Year "The Role of Elicitors in Increasing Soybean Isoflavones Content as the Functional Food". Universitas Sumatera Utara. Medan.

[17]Raskin I 1992 Role of salicylic acid in plants, Annu. Rev. Plant Physiol. Plant Mol. Biol. 43 : 439-463.

[18]Leslie CA, Romani RJ 1988 Inhibition of ethylen biosynthesis by salicylic acid, Plant Physiol. 88 : 833- 837.

[19]Mondal MMA, Malek MA, Puteh AB, Ismail MR, Ashrafuzzaman M, Naher L 2012 Effect of foliar application of chitosan on growth and yield in okra. Aust. J. Crop, 6: 918-921

[20]Dzung NA, Khanh VTP, Dzung TT 2011 Research on impact of chitosan oligomers on biophysical characteristics, growth, development and drought resistance of coffee. Carbohydr. Polym., 84: 751-755.

[21]Al-Hetar MY, Zainal Abidin MA, Sariah M, Wong MY 2011 Antifungal activity of chitosan against Fusarium oxysporum f. sp. cubense J. Appl. Polym. Sci., 120: 2434-2439.

[22]Saini RK, M Akithadevi, Parvatamgiridhar, G Ravishankar 2013 Augmentation of major isoflavones in *Glycine max* L. Acta Bot. Croat. 72 (2), 311–322.

[23]El Hadrami ALR Adam I El Hadrami and D Daayf 2010 Chitosan in Plant Protection Mar Drugs 8(4) : 968-987.

[24]Guan Ya-jing, Jin Hu, Xian-ju Wang, Chen-xia Shao 2009 Seed priming with chitosan improves maize germination and seedling growth in relation to physiological changes under low temperature stress J. of Zhejiang University Science B 10(6):427-433.

[25]Farouk S Mosa AA Taha AA Ibrahim Heba M EL-Gahmery AM 2011 Protective effect of humic acid and chitosan on radish (Raphanus sativus L. var. sativus) plants subjected to cadmium stress. Journal of Stress Physiology and Biochemistry 7(2):99-116

[26]Ghoname AA EL-Nemr MA Abdel-Mawgoud AMR El-Tohamy WA 2010 Enhancement of sweet pepper crop growth and production by application of biological, organic and nutritional solutions. Research Journal of Agriculture and Biological Science 6(7): 349-355.

[27]Farouk S and AR Amany 2012 Improving growth and yield of cowpea by foliar application of chitosan under water stress Egyptian J. of Biology 14:14-26.

[28]Khan MH, Singha KLB, Panda SK 2002 Changes in antioxidant levels in *Oryza sativa* L. roots subjected to NaCl salinity stress Acta Physiol. Plantarum 24:145–148.

**Acknowledgements**

The authors would like to thank Research Institution, Universitas Sumatera Utara that have funded this research in accordance with Research Contract TALENTA Universitas Sumatera Utara, Fiscal Year 2017 Number: 5338 / UN5.I.R / PPM / 2017 dated May 22, 2017. The authors also thank Rijalul Afkar for his assistance in the research.

# Effectiveness of several dosage formula of oil and nano emulsion of citronella against vascular streak dieback (VSD) disease on cocoa

**R Noveriza[1], J Trisno[2], H Rahma[2], S Yuliani[3], Reflin[2] and Martinius[2]**

[1] Indonesian Spice and Medicinal Crops Research Institute, Bogor, Indonesia
[2] Faculty of Agriculture. Andalas University,Padang, Indonesia
[3] Indonesian Center for Agricultural Post Harvest Research and Development, Bogor, Indonesia
E-mail : rita_noveriza2000@yahoo.com

**Abstract.** The disease of *Vascular streak dieback* (VSD) is a deadly disease of cocoa plants, because it attacks the vascular tissue of cocoa at growing point of the plant. In West Sumatra the disease was first reported in 2015 with an incidence of disease range 58.82% - 100% and an intensity of disease range 24.29% - 44.7%. The purpose of this study was to examine the effectiveness of dosage application of oil formula and nano emulsion of citronella formula against *Vascular streak dieback* (VSD) disease on cocoa plants in West Sumatra (in Padang Pariaman District and Limapuluh Kota District). The results showed that the percentage of VSD disease attacks in both testing sites was 100% .The oil and nano emulsion of citronella formulas can reduce the intensity of VSD disease on cocoa plants in West Sumatra, particularly in Padang Pariaman District and Limapuluh Kota District. The reduction of VSD intensity in Padang Pariaman district ranged from 8.32 to 21.13%; while in Limapuluh Kota district ranged from 4.33 to 11.80%. The nano emulsion of citronella formulation is effective to suppress the intensity of VSD disease on cocoa plants at doses 0.1% ($\geq$ 30% of effectiveness level).

## 1. Introduction

VSD (*Vascular streak dieback*) disease is a deadly disease of cocoa plants, because it attacks the vascular tissue at growing point. The development and spread of the disease in Indonesia is very rapid, because VSD disease was first discovered in 1983 in Sebatik island (East Kalimantan), and in 2013 almost all cocoa plantations are infected with the disease [1]. In West Sumatra the disease was first reported in 2015 with disease incidence are 58.82% - 100% and an intensity of disease are 24.29% - 44.7% [2]. Losses due to VSD disease worldwide can reach 30,000 tons per year equivalent to US $ 28 million [3].

In Indonesia, especially in West Sumatra the potential for loss of results has not been reported. However, on the basis of farmers' surveys and interviews many farmers' gardens have been destroyed and replaced with other crops, because the diseases cause deciduous leaves, barren plants and stop production. On the other hand, poor garden maintenance can accelerate the spread of disease in the field. Cocoa cultivation in Indonesia is generally a smallholder plantation, where in the management and maintenance has not done well. Various efforts to control the disease have been applied, but not yet effective. Wahap and Sulle [4] reported that VSD disease can be controlled by (1) placing the seedlings in isolated and quarantine locations for 6 months, (2) pruning the sick branch once a month and good drainage, (3) resistant clone cultivation and (4) the use of propiconazole and biloxazole fungicides, [5] reported that the disease is difficult to control because it is in the vessel network. Therefore it is necessary to find a technology that utilize the natural plant potential as pesticide that may be able to control it.

Citronella is known to control several pests. Natural pesticides from citronella plants can be utilized to overcome the spread of mold spores causing plant diseases. The citronella plant is reportedly one of the most potential plant-based pesticides to control plant pathogens [6]. The use of clove oil and potential citronella extracts for control the VSD disease that it can suppress disease progression up to 38.6% and 31.6% [5]. The potential of active ingredients on biopesticides can be improved by using nano technologies such as nano emulsion and nano encapsulation.

Nanoemulsion is a transparent, translucent emulsion system and is a water oil dispersion stabilized by a film coating of a surfactant or surfactant molecule, having a droplet size of 50 - 500 nm [7]. Small size nanoemulsion droplets make nanoemulsically stable kinetically to prevent sedimentation and kriming during storage [8]. In addition, nanoemulsion with an oil-in-water emulsion system (o/w) is an alternative to increase the solubility and stability of the bioactive components present in oil [2].

Nanotechnologies, which can reduce particle size to nano size ($10^{-9}$ m), are expected to increase the efficiency and effectiveness of the active ingredients contained in such materials on a broader scale or wider scale test. An improved perfumed oil formulation using spontaneous emulsification (nanotechnology) and greenhouse test results showed biopesticide nano emulsion at doses of 1-1.5% suppressed viral progression up to 82.5%; While the oil formula of citronella is only 65-70% [9][10].

The purpose of this study was to examine the effectiveness of dosage application of oil formula and nano emulsion of citronella against *Vascular streak dieback* (VSD) disease in cocoa plants in West Sumatra.

## 2. Materials and method

This research was conducted at Indonesian Spice and Medicinal Crop Research Institute (Plant Protection Laboratory) and Nano Technology of Post Harvest Technology, Bogor (for preparation of citronella formula) and cocoa farmers' land of VSD disease in Padang Pariman and Limapuluh Kota District of West Sumatra Province from March to December 2016.

*2.1. Preparation of Citronella Formula.* Citronella plant oils obtained from Indonesian Spice and Medicinal Crop Research Institute of Manoko experimental garden in Lembang, Bandung. The oil is distilled from the leaves of 3 year-old citronella by steaming. Citronella oil formula was prepared according to the method of Wang and Liu [11] and nano-emulsion formula citronella oil is made from the modified method of Bouchemal *et al.* [12].

*2.2. Application of biopesticides in the field.* The study used a randomized block design with 2 levels of treatment (dosage of biopesticide formula and type of formula (oil biopesticide and nano emulsion of citronella oil biopesticide), each with 6 replicates. The treatment was (1) the formula of citronella oil at a dose 0.1%, 2) the formula of citronella oil at a dose 0.5%, (3) the formula of citronella oil at a dose 1%, (4) the formula of citronella oil at a dose 1.5%, (5) nano emulsion formula of citronella oil at a dose 0.1% 6) nano emulsion formula of citronella oil at a dose 0.5%, (7) nano emulsion formula of citronella oil at a dose 1%, (8) nano emulsion formula of citronella oil at a dose 1.5%, (9) synthetic fungicide with mankozeb active site, and (10) control (without treatment). Fifteen cocoa trees was observed for each treatment as a samples.

*2.3. Technique application of biopesticide formula.* Applications biopesticide formula is done by spraying using sprayer in all parts of the plant with a total volume of 500-1000 ml per plant (because the plant already aged over 2 years). Spraying is done every single month for 6 months application times. Cultivation of cocoa plants follows the standard operational procedures carryout out by farmers.

*2.4. Observation parameters.* Observations were made on disease symptoms, disease progression by counting the percentage of diseases, the intensity of diseases each month, and the degree of efficacy of the formula at the end of the observation.

Observations percentage VSD disease using the formula of Strange [13]:

$$\text{Percentage of diseases} = \frac{\text{Number of plants affected by VSD}}{\text{Number of all the plants observed}} \text{ x } 100\% \quad (1)$$

Observation of disease intensity was done on each tree by counting the number of leaves and branches affected by the disease based on the attack category in Table 1. The intensity of disease was calculated by the formula of Strange [13]:

$$\text{I} = \left(\frac{\Sigma(\text{ni x vi})}{\text{Z x N}}\right) x\ 100\% \quad (2)$$

I = intensity of disease; ni = number of plants in each attack category; vi = scale values of each attack category; Z = the scale value of the highest attack category; N = number of plants observed.

**Table 1**. Scoring of VSD symptoms in cocoa plants.

| Scoring | Attack category | Symptom |
|---|---|---|
| 0 | Healthy | 0% are infected. |
| 1 | Light Mild | 1-10% leaves are infected. |
| 2 | Mild | 11-50% leaves are infected, chlorosis, necrosis, deciduous leaves, existing lenticell swelling. |
| 3 | Light Heavy | 51-75% leaves are infected, chlorosis, necrosis, deciduous leaves, lenticell swelling, there is a fruit body. |
| 4 | Heavy | >75% leaves are infected, chlorosis, necrosis, deciduous leaves, lenticell swelling, there is a fruit body and thre are dead branches. |

*Source*: Susilo and Anita-Sari [14] (modified).

The level of efficacy (EI) of biopesticide is calculated by comparing the intensity of the disease in the treated plots with the untreated plots, using the formula:

$$EI = \left(\frac{Ca-Ta}{Ca}\right) X\ 100\% \quad (3)$$

EI = the level of efficacy; Ca = the intensity of the disease in the control (untreated pesticide plant); Ta = Intensity of disease attack on the treatment of natural pesticides. The formula studied can be considered effective if the value of efficacy level (EI) $\geq$ 30%.

## 3. Results and discussion

The result of VSD disease survey on cocoa plant in West Sumatera based on symptoms showed that the severity of VSD disease in Limapuluh Kota district was higher than the severity of VSD disease in Padang Pariaman district (figures 1 and 2). The percentage of VSD disease attacks at both sites reached 100%. Both sites are used as research sites for the application of biopesticide citronella oil and nano emulsion of citronella formulation.

**Figure 1**. VSD disease symptoms on the cocoa crop in the district of Limapuluh Kota (a. Vascular tissue infected with VSD, b and c. Necrosis on leaves).

**Figure 2**. Symptoms of VSD disease attack on cocoa plants in Padang Pariaman District (a & b, leaf necrosis symptoms, c. VSD infected tissue vessels) and plant conditions after application of nano emulsion of citronella formula (d)

In the observation of disease intensity of VSD that has been done every month in Limapuluh Kota district, some treatments still showed an increasing trend and it was decreasing trend of disease compared to with the control without treatment (figure 3). Similarly, with the cocoa plant in Padang Pariaman district (figure 4), the downward trend is more evident when compared to the disease intensity of VSD before application (0 month observation) with disease intensity of VSD after application (observation at 1 to 5 months). This suggests that both tested formulas were able to decrease the progression of VSD disease.

Citronella oil has been reported to be effective against some pathogenic fungi such as *Aspergillus* sp. and *Penicillium* sp.[15], as well as *Phytophthora palmivora* cause of fruit rot of cocoa [1][16][17][18].

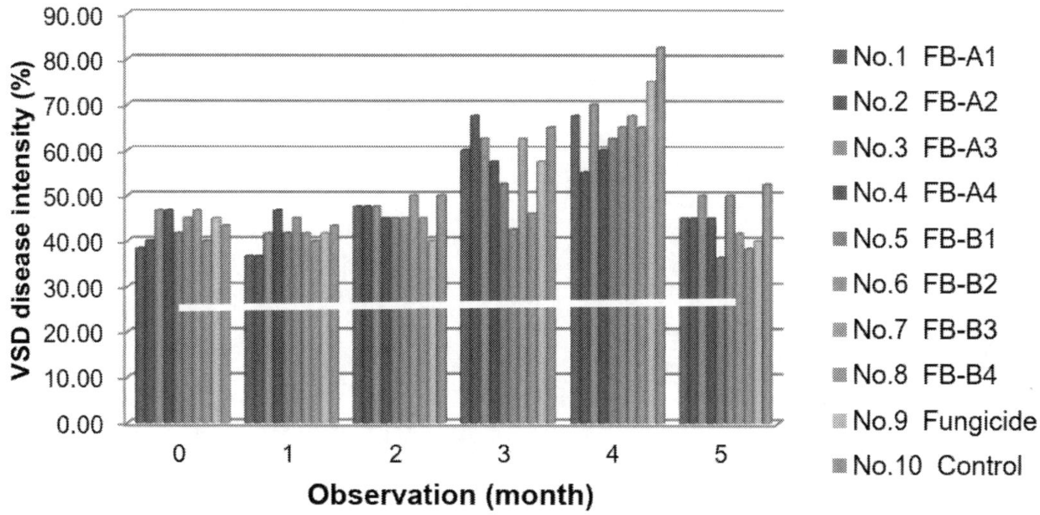


**Figure 3**. Percentage disease intensity of *Vascular streak dieback* (VSD) on cocoa plants in Limapuluh Kota district before and after application of natural biopesticide. *Description*: No.1 = nano emulsion of citronella oil at a dose 0.1%. No.2 = nano emulsion of citronella oil at a dose 0.5%. No.3 = nano emulsion of citronella oil at a dose 1%. No.4 = nano emulsion of citronella oil at a dose 1.5%. No.5 = citronella oil at a dose 0.1%. No.6 = citronella oil at a dose 0.5%. No.7 = citronella oil at a dose 1%. No.8 = citronella oil at a dose 1.5%. No.9 = no treatment as a comparison for control. No.10 = Fungicide with active ingredient mankozeb.

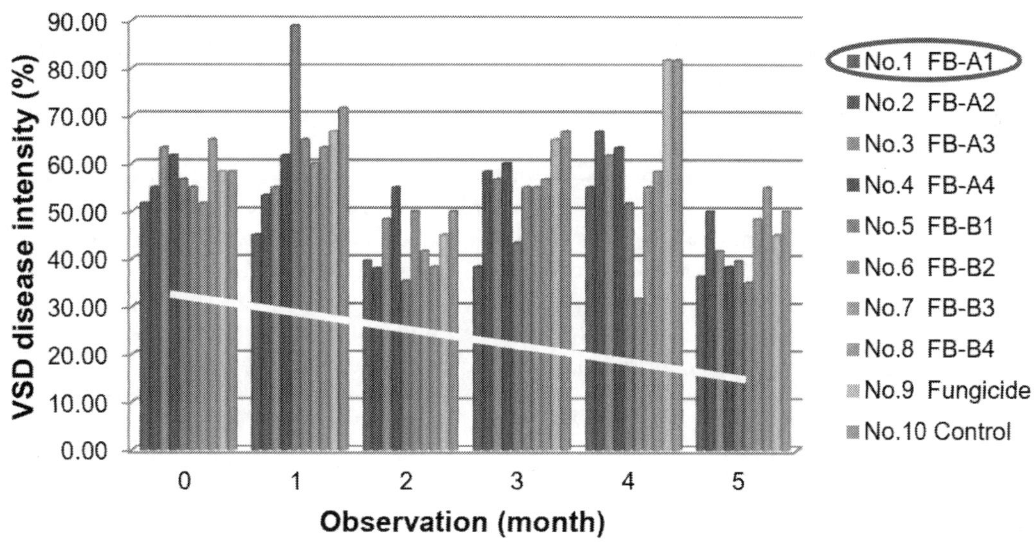


**Figure 4**. Percentage intensity of *Vascular streak dieback* (VSD) disease in cocoa plants in Padang Pariaman District before and after application of biopesticide. *Description*: No.1 = nano emulsion of citronella oil at a dose 0.1%. No.2 = nano emulsion of citronella oil at a dose 0.5%. No.3 = nano emulsion of citronella oil at a dose 1%. No.4 = nano emulsion of citronella oil at a dose 1.5%. No.5 = citronella oil at a dose 0.1%. No.6 = citronella oil at a dose 0.5%. No.7 = citronella oil at a dose 1%. No.8 = citronella oil at a dose 1.5%. No.9 = no treatment as a comparison for control. No.10 = Fungicide with active ingredient mankozeb.

**Table 2**. Climate data (average of air temperature, average of relative humidity and rainfall) from June to December 2016 in Padang Pariaman and Limapuluh Kota district.

| Month *(Bulan)* | Average of air temperature (°C) *(Temperatur Udara Rata-rata)* | Average of relative humidity (%) *(Kelembaban Udara Rata-rata)* | Rainfall (mm) *(Curah hujan)* |
|---|---|---|---|
| Parit Malintang-Padang Pariaman district | | | |
| June | 25.7 | 89 | 157 |
| July | 26.0 | 84 | 324 |
| August | 25.6 | 84 | 385 |
| September | 25.5 | 86 | 317 |
| October | 29.2 | 87 | 559 |
| November | 28.0 | 86 | 288 |
| December | 25.3 | 88 | 341 |
| Limapuluh Kota district | | | |
| June | 23.2 | 81 | 91 |
| July | 22.0 | 83 | 124 |
| August | 21.9 | 82 | 60 |
| September | 21.7 | 84 | 61 |
| October | 21.6 | 86 | 34 |
| November | 26.0 | 87 | 308 |
| December | 21.5 | 86 | 149 |

On the basis of analysis of the measurement of Area Under Disease Progress Curve (AUDPC) in Padang Pariaman district, the lowest AUDPC was found in the treatment of nano emulsion formula with a dose of 0.1% (FB-A1) compared with control. While in the area of Limapuluh Kota district the lowest AUDPC was found in the treatment of citronella oil formula at a dose 1.5% (FB-B4) (Table 3). This suggests that the two biopesticide formulations can decrease the development of VSD disease at both test sites. The best activity in controlling VSD disease using nano emulsion formula at a dose 0.1% with decrease of disease intensity up to 21.13%, while formula of citronella oil at a dose 1.5% with decrease of disease intensity up to 11,80%. From several studies it is revealed that citronella at doses of 12 ppm could decrease the intensity of *Colletotrichum* sp that infected onion plants to 20% [19], citronella oil with concentration 0.2-1.5 µl/ml could control *Alternaria alternata* in tomato plant [20] and with concentration 0.5% can kill the conidia of *Aspergillus niger* [21].

**Table 3**. Reduced VSD disease intensity (%) of VSD and efficacy level (%) after treated with oil and nano emulsion biopesticide formulations for 6 months of application time.

| | | Padang Pariaman district | | | | Limapuluh Kota district | | | |
|---|---|---|---|---|---|---|---|---|---|
| **No Treatment** | **Treatment Bio-pesticide** | **AUDPC** | Decrease in intensity of disease (%) | Efficacy level (%) | Assessment of efficacy level | **AUDPC** | Decrease in intensity of disease (%) | Efficacy level (%) | Assessment of efficacy level |
| 1 | FB-A1 | 221.93 | 21.13 | 32.08 | Effective | 253.34 | 7.33 | 12.11 | Not Effective |
| 2 | FB-A2 | 268.83 | 10.72 | 16.08 | Not Effective | 249.17 | 8.33 | 12.83 | Not Effective |
| 3 | FB-A3 | 274.17 | 11.32 | 16.53 | Not Effective | 270.01 | 4.33 | 6.52 | Not Effective |
| 4 | FB-A4 | 289.99 | 8.32 | 11.94 | Not Effective | 255.01 | 7.83 | 11.08 | Not Effective |
| 5 | FB-B1 | 267.38 | 12.22 | 19.54 | Not Effective | 240.69 | 11.06 | 17. | Not |

| | | | | | | | | | |
|---|---|---|---|---|---|---|---|---|---|
| | | | | | | | | 61 | Effective |
| 6 | FB-B2 | 246.67 | 16.65 | 23.59 | Not Effective | 245.00 | 9.17 | 13.35 | Not Effective |
| 7 | FB-B3 | 261.67 | 11.99 | 17.27 | Not Effective | 265.84 | 6.00 | 9.30 | Not Effective |
| 8 | FB-B4 | 276.66 | 9.66 | 13.7 | Not Effective | 235.17 | 11.80 | 19.02 | Not Effective |
| 9 | Fungicide | 310.01 | 3.32 | 5.88 | Not Effective | 256.67 | 7.83 | 13.65 | Not Effective |
| 10 | Control | 324.11 | - | - | | 288.75 | - | - | |

*Description*: No.1 = nano emulsion of citronella oil at a dose of 0.1%. No.2 = nano emulsion of citronella at a dose of 0.5%. No.3 = nano emulsion of citronella oil at dose of 1%. No.4 = nano emulsion of citronella oil at a dose of 1.5%. No.5 = citronella oil at a dose of 0.1%. No.6 = citronella oil at a dose of 0.5%. No.7 = citronella oil at a dose of 1%. No.8 = citronella oil at a dose of 1.5%. No.9 = no treatment as a comparison for control. No.10 = fungicide with active ingredient mankozeb.

The observations also show that formula of nano emulsion of citronella oil was more effective and efficient in suppressing the development of VSD disease than the formula of citronella oil. Improving the formula of citronella oil using nanotechnology can improve the effectiveness of the biopesticide, because of its small particle size (nanometer in size) so it can directly reach the target microorganism. Moreover, citronella oil is more stable in the field. According to [22], nanoparticles can improve the stability of neem products (biopesticides from *Azadirachta indica*) against ultraviolet radiation and increase dispersion in the aqueous phase.

## 4. Conclusion

The percentage of VSD disease attacks in Padang Pariaman and Limapuluh Kota district was 100%. Citronella oil and nano emulsion formulation can reduce the intensity of *Vascular streak dieback* (VSD) disease in cocoa plants in West Sumatra, especially in Padang Pariaman and Lima Puluh Kota districts. The decrease in intensity of VSD disease in Padang Pariaman district ranged from 8.32 to 21.13%; while in the district of Lima Puluh Kota ranged from 4.33 to 11.80%. The nano emulsion formula at a dose of 0.1% is effective and efficient in suppressing the intensity of VSD attacks on cocoa plants compared to the formula of citronella oil itself.

## References

[1] Harni R dan Khaerati 2013 Potensi beberapa fungisida nabati untuk mengendalikan penyakit vascular streak dieback (VSD) pada bibit kakao *Paper presented at* Seminar dan Kongres PFI Padang, 8–10 Oktober 2013: Universitas Andalas

[2] Trisno J, Reflin dan Martinius 2016 *Vascular Streak Dieback* (VSD) Penyakit Baru Tanaman Kakao di Sumatera Barat. *J. Fitopatol. Indo.* **12(4):**142-147

[3] World Cocoa Association. 2001. *Estimated annual reduction in potential cocoa production by major diseases*. Retrieved from http:// www.papillonsartpalace.com/ impactofchocolate.htm

[4] Wahab A and Sulle A 2008 Penyakit *Vascular streak dieback* (VSD) dan pengendaliannya pada tanaman kakao *Buletin Teknologi dan Informasi Pertanian* pp 14-20

[5] Harni R dan Baharuddin 2014 Kefektifan minyak cengkeh, seraiwangi dan ekstrak bawang putih terhadap penyakit Vascular streak dieback (*Ceratobasidium theobromae*) pada kakao. *J.TIDP* **1(3)**: 167-174

[6] Mariana M dan Noveriza R 2013 Potensi minyak atsiri untuk mengendalikan *Potyvirus* pada Tanaman Nilam *Jurnal Fitopatologi Indonesia* **9(1)**: 53-58. DOI: 10.14692/jfi.9.2.53

[7] Yuliasari S dan Hamdan 2012 Karakterisasi nanoemulsi minyak sawit merah yang disiapkan dengan high pressure homogenizer *Prosiding InSiNas* 25-28

[9] Noveriza R, Mariana M dan Yuliani S 2017 Keefektifan formula nanoemulsi minyak seraiwangi terhadap *Potyvirus* penyebab penyakit mosaik pada tanaman nilam *Bul. Littro* **28(1)**:47-56

[10] Noveriza R, Mardiningsih, Lestari T, Miftakhurohmah and Mariana M 2016 *Antiviral Effect of Clove Oil Combined with Citronella Oil to Control Mosaic Disease and its Vector on Patchouli Plant* In: Djiwanti,S.R. et al. {eds.) lnnovation on Biotic ond Abiotic Stress Management to Maintain Productivity ol Spice Crops in lndonesia IAARD Press pp 91-96

[11] Nakahara K, Alzoreky N S, Yoshihashi T, Nguyen H T T and Trakoontivakom G 2003 Chemical composition and antifungal activity of essential oil from *Cymbopogon nardus JARQ 37*(**4**): 249–252

[12] Bouchemal K, Briancon S, Perrier E and Fessi H 2004 Nano-emulsion formulation using spontaneous emulsification: solvent, oil and surfactant optimization *International J. Pharmaceutics* **280**: 241-251

[13] Strange R N 2008 *Introduction to plant pathology* (New York (US): John Wiley and Sons Ltd.)

[14] Susilo A W and Sari I A 2011 Respon ketahanan beberapa hibrida kakao (*Theobroma cacao* L.) terhadap serangan penyakit pembuluh kayu (*Vascular streak dieback*) *Pelita Perkebunan* **27(2)**: 77–84

[15] Wang and Liung 2007 Foliar uptake of pesticides present status and future challenge *Pesticide Biochemistry and Physiology* **87**: 1–8

[16] Harni R, Amaria W dan Supriadi 2013 Keefektifan beberapa formula fungisida nabati eugenol dan sitronella terhadap *Phythopthora palmivora* Bult asal kakao *Buletin Riset Tanaman Rempah dan Aneka Tanaman Industri* **4(1)**:11–18

[18] Nurmansyah. 2010. Efektivitas minyak serai wangi dan fraksi sitronellal terhadap pertumbuhan jamur *Phytophthora palmivora* penyebab penyakit busuk buah kakao *Bul. Littro.*, **21(1)**: 43–52

[19] Kurniasih R, Djauhari S, Muhibuddin A dan Utomo E P 2014 Pengaruh sitronelal seraiwangi (*Cymbopogon winterianus* Linn) terhadap penekanan serangan *Colletotrichum* sp. Pada tanaman bawang daun (*Allium fistulosum* L.) *Jurnal HPT* **2(4)**: 11-21

[20] Chen Q, Xu S, Wu T, Guo J, Sha S, Zeng X and Yu T 2014 Effect of cittronella essential oil on the inhibition of post harvest *Alternaria alterna* in cherry tomato *J. Sci. Food Agric.* **94(12)**: 2441–2447

[21] Li W R, Shi Q S, Ouyang Y S, Chen Y B and Duan S S 2013 Antifungal effect of citronella oil against *Aspergillus niger* ATCC 16404 *App. Microbiol Biotechnol*. **97**:7483–7492

[22] Forim M R, Costa E S, da Silva M F G F, Fermandes J B, Mondego J M and Junior A L B 2013 Development of a new method to prepare nano/microparticles loaded with extract of *Azadirachta indica*, their characterization and use in controlling *Plutella xylostella J. Agric. Food Chem.* **61**: 9131-9139

**Acknowledgements**
The author would like to thank Siti Nuryanih, Zuhisnain, Rahil A. Rifqah and Selviana Anggraini who have helped in the implementation of field research. The research was funded by the SMARTD project of IAARD 2016.

# Physiological performance of the soybean crosses in salinity stress

**F Wibowo and Armaniar**

Universitas Pembangunan Panca Budi, Gatot Subroto Km 4,5 Medan, 061-30106067, Indonesia
E-mail : fachrinawibowo@dosen.pancabudi.ac.id

**Abstract.** Plants grown in saline soils will experience salinity stress. Salinity stresses, one of which causes oxidative stress, that cause an imbalance in the production ROS compounds (Reactive Oxygen Species), antioxidants and chlorophyll. Where the reaction of this compound can affect plant growth and plant production. This study aims to inform performance and action gene to soybean physiological character that potential to tolerant from salinity soil that characterized by the presence of SOD and POD antioxidant compounds and chlorophyll. This research used a destructive analysis from crossbred (AxN) and (GxN). A = Anjasmoro varieties and G = Grobogan varieties as female elders and N = Grobogan varieties as male elders ($N_1$,$N_2$,$N_3$,$N_4$,$N_5$) that have been through the stage of saline soil selection. Research result can be concluded that GxN cross is more potential for Inheritance of the offspring. This can be seen from the observed skewness of character SOD, POD compounds, Chlorophyll a and chlorophyll b.

## 1. Introduction

Soybean is a favorite crop food in Indonesia because of its high nutritional value. [1] BPS recorded soybean production in 2012 of 843.153 tons, in 2013 the production decreased by 779.992 tons. The next year 2014 soybean production increased by 954.997 tons and in 2015 the production is 963.099 tons. Despite increased by production but still not in line with the needs and the amount of public consumption. Therefore in Indonesia, this plant needs special attention.

In Indonesia, salinity-exploited agricultural land may be expanding, due to rising sea levels, sea water intrusion, waste pollution and excessive land exploitation, especially coastal areas [2]. [3] It is estimated that the area of agricultural land affected by salinity is 440.300 ha. The challenge is how the area of saline soils is used for a variety of competitive crops [4]. Salinity is an abiotic factor that limits the production of soybean crops worldwide, which is a limiting factor inhibiting growth and yield potential. Salinity causes depressed plants, where subsequent plants undergo structural changes that include changes in morphological, physiological and biochemical mechanisms.

The mechanisms of plant resistance to salinity depend on species and varieties, start from the sensive to the level of tolelan. Response of plant to salt resistance is avoidance mechanism and those who do adapt mechanisms to be tolerant. One particularly effective way for plant breeders to overcome the decline in soybean production and saline expansion is through genetic improvement efforts by developing varieties of saline-resistant crops through crosses and adaptation selection. The response of plants to salinity is of course important in efforts to improve the potential of this plant. This study aims to inform performance and action gene to soybean physiological character that potential to tolerant from salinity soil that characterized by the presence of SOD, POD compounds, Chlorophyll a and chlorophyll b.

## 2. Materials and Methods

This research was conducted at the screen house of the Faculty of Agriculture, University of North Sumatra and Physiology Laboratory of Sungai Putih. This research used a destructive analysis from crossbred seeds (AxN) and (GxN). A = Anjasmoro varieties and G = Grobogan varieties as female elders and N = Grobogan varieties as male elders (N1,N2,N3,N4,N5) that have been through the stage of saline soil selection.

Research methods that used is statistic descriptive analysis, where normality test data distribution and genotype frequency do to SOD, POD, chlorophyll a and chlorophyll b. Graph form for skewness and kurtosis value calculated with pearson theory, furthermore result obtained interpreted and characterized as follow (Table 1).

**Table 1.** Estimation of the action of gene value of skewness and kurtosis

| Test Data Normality | Graphic Form | Information |
|---|---|---|
| Skewness = 0 | Normal Distribution | Action gene additive |
| Skewness < 0 | Abnormal Distribution | Action of additive gene with the effect of duplicate epistasis |
| Skewness > 0 | Abnormal Distribution | Action of additive gene with complementary epistatic effect |
| Kurtosis = 3 | Mesokurtik | |
| Kurtosis < 3 | Platykurtik | The characters are controlled by many genes |
| Kurtosis > 3 | Leptokurtik | Characters are controlled by few genes |

## 3. Results and Discussion

### *3.1 Superoxide Dismutase Activity (SOD)*

The observed Superoxide dismutase activity AxN crossover (figure 1.a.) shows result not normally distributed. The result of AxN crossing shows average activity of SOD 11.910 shows crossover result higher from elder N and lower than form elder A. The highest SOD actifity is 25.272 and lowest is 1.731. Action gene expected value for SOD activity (Table 2) skewness explain additive action gene with epistatic complementer influence and kurtosis with playkurtik graph form explain character is controlled by many genes.

The observed Superoxide dismutase activity GxN crossover (figure 1. b.) shows result not normally distributed. The result of GxN crossing shows average activity of SOD 16.777 shows crossover result higher from elder N and G. The highest SOD actifity is 40.246 and lowest is 6.726. Action gene expected value for SOD activity (Table 2) skewness explain additive action gene with epistatic complementer influence and kurtosis with playkurtik graph form explain character is controlled by many genes.

In the beginning phase of salinity stress, osmotic stress attributed by physiologist alteration, like membrane interference, unbalanced nutrition, disturb ability to oxygen reactive species (ROS) detoxification. First step during oxygen reduction is superoxide ($O_{2}$.) establishment that can be converted to hydrogen peroxide ($H_2O_2$) with superoxide dismutation (SODs) action [5]. [6] Sopandie explained that plant develop detoxification system through enzymatic or non-enzymatic to face up ROS, so be able to protect cell from oxidative stress. SOD formed in plant very influential to some physiology phenomenon, including $H_2O_2$ lost, toxic reductant oxidation, biosynthesis and lignin degradation in cell wall, auxin catabolism, self defense response against injury and some respiration process.

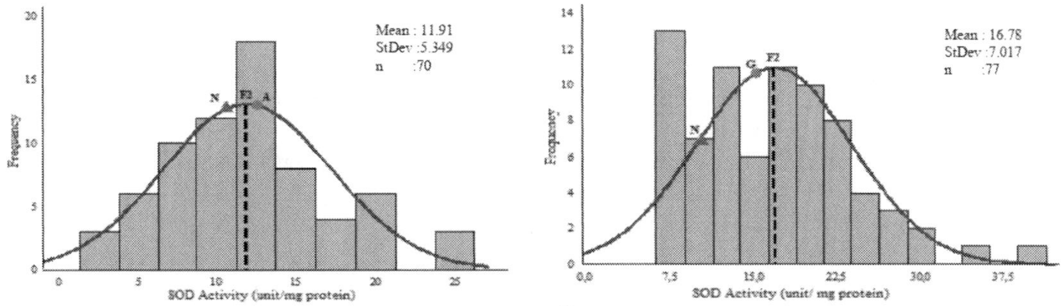


a) b)

**Figure1**.Grap of SOD activity distribution (unit/ mg protein) (a), AxN (b) and GxN crossover

*3.2 Peroxide Activity (POD)*

The observed peroxide activity AxN crossover (figure 2.a.) shows result not normally distributed. The result of AxN crossing shows average activity of POD 0.663 shows crossover result higher from elder N and lower than form elder A. The highest POD actifity is 1.643 and lowest is 0.094. Action gene expected value for POD activity (Table 2) skewness explain additive action gene with epistatic complementer influence and kurtosis with playkurtik graph form explain character is controlled by many genes.

The observed peroxide activity GxN crossover (figure 2. b) shows result not normally distributed. The result of GxN crossing shows average activity of POD 0.627 shows crossover result higher from elder N and G. The highest POD actifity is 0.469 and lowest is 0.111. Action gene expected value for POD activity (Table 2) skewness explain additive action gene with epistatic complementer influence and kurtosis with playkurtik graph form explain character is controlled by many genes.

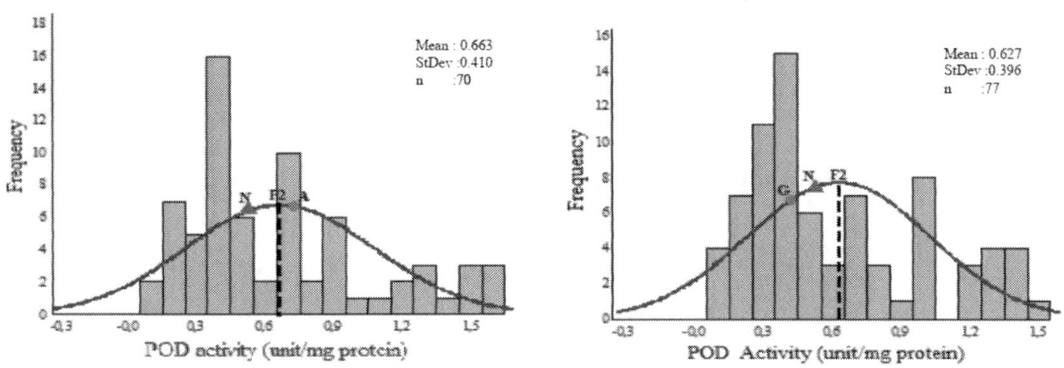


a) b)

**Figure2.** Grap of POD activity distribution (unit/mg protein) (a), AxN (b) and GxN crossover

Observed parameter result peroxide activity (figure 2.a,b) highest peroxide activity by AxN crosses result. Peroxide is antioxidant compound that can parry Hydrogen Peroxide ($H_2O_2$) compound, peroxide boost oxidation of some peroxide compound, especially $H_2O_2$ that reduced to form water, enzyme that have active character, when in peroxide plant can help to minimize damage caused by stress factor [7]. Peroxide activity enhancement will increasing toxin product to pathogens so withstand infection. [8] Peroxide is a member of the reductase enzyme, isoenzyme that play a role in grow, differentation and defense. Peroxide catalyze $H_2O_2$ to $H_2O$ and $O_2$ substrate feniline diamin compound like 3-amino-9etil karbazole will oxidized with oxygen from reduction, that can be marked with formed a brownish red colored precipitate. [9] Other research mention that high enzyme peroxide activity can increase plant endurance to decease. In this research POD activity take effect to production, when POD enhancement parallel with number of plant escalation.

### 3.3 Chlorophyll a

The observed chlorophyll a AxN crossover (figure 3.a.) shows result not normally distributed. The result of AxN crossing shows average chlorophyll a 3.098 shows crossover result higher from elder n and lower than form elder a. the highest chlorophyll a value is 3.870 and lowest is 1.755. Action gene expected chlorophyll a (Table 2) skewness explain additive action gene with epistatic duplicate influence and kurtosis with playkurtik graph form explain character is controlled by many genes.

The observed chlorophyll a GxN crossover (figure 3.b.) shows result not normally distributed. The result of GxN crossing shows average chlorophyll a 1.652 shows crossover result lower from elder N and G. The highest chlorophyll a value is 3.492 and lowest is 0.525. Action gene expected chlorophyll a (Table 2) skewness explain additive action gene with epistatic complementer influence and kurtosis with playkurtik graph form explain character is controlled by many genes.

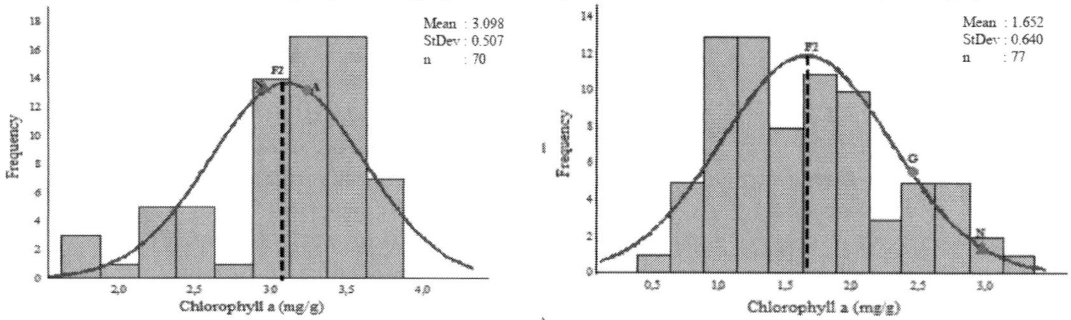


a) b)

**Figure 3.**Grap of chlorophyll a distribution (mg/g) (a), AxN (b) and GxN crossover

### 3.4 Chlorophyll b

The observed chlorophyll b AxN crossover (figure 4. a) shows result not normally distributed. The result of AxN crossing shows average Chlorophyll b 2.910 shows crossover result higher from elder A and lower than form elder N . The highest chlorophyll b value is 3.786 and lowest is 1.851. Action gene expected chlorophyll b (Table 2) skewness explain additive action gene with epistatic duplicate influence and kurtosis with playkurtik graph form explain character is controlled by many genes.

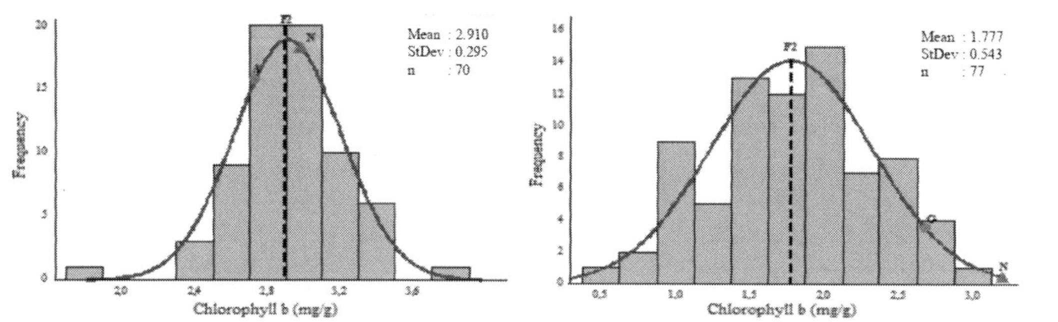


a) b)

**Figure 4.**Grap of Chlorophyll b distribution (mg/g) (a), AxN dan (b) GxN crossover

The observed chlorophyll b GxN crossover (figure 4. b) shows result not normally distributed. The result of GxN crossing shows average Chlorophyll b 1.777 shows crossover result lower from elder

N and G. The highest Chlorophyll b value is 2.943 and lowest is 0.554. Action gene expected chlorophyll b (Table 2) skewness explain additive action gene with epistatic complementer influence and kurtosis with playkurtik graph form explain character is controlled by many genes.

Observed parameter result chlorophyll a and chlorophyll b (figure 3.a,b and 4.a,b) that potentially developed in salinity soil based highest chlorophyll value is result of crosses AxN. Chlorophyll is the main component of chloroplasts and chlorophyll positive correlated with photosynthesis rate. High content of chlorophyll in a plant shows that a plant potentially grow in salinity soil. Genotypically from observed result can be seen that chlorophyll a decreased if compared by chlorophyll b in salinity condition. According to Mane *et al.*[10] that content of chlorophyll a can be replaced by chlorophyll b, that have the same trend, from observed result obtained that in salinity condition chlorophyll a is more sensitive if compared by chlorophyll b. Chlorophyll decreased may be caused by obstacle of accumulation from the salts in biosynthesis chlorophyll process. [11] Salinity influence the strength from protein complex pigment liquid in chloroplasts structure. Chlorophyll decrease caused by photosynthesis decrease and cause decreased crop production, although plant tolerance level not only determined by one plant attribute. In this research prove that plant with high adaptation can defend the level of chlorophyll compared by sensitive plant or cant be able to adapt well. Other research [12] mention salinity stress inhibit chlorophyll syntetic on leaves as a result of decreased rate of photosynthesis and ocurrence temperature enhancement and transpiration that cause chlorophyll disentegration.

**Table 2**. Action gene expected value for SOD and POD activity (unit/mg protein) through skewness and kurtosis value

| Crosses | SOD Activity | | POD Activity | | Chlorophyll a | | Chlorophyll b | |
|---|---|---|---|---|---|---|---|---|
| | Skewness | Kurtosis | Skewness | Kurtosis | Skewness | Kurtosis | Skewness | Kurtosis |
| Ax N | 0.548 | 0.296 | 0.898 | -0.020 | -1.107 | 0.488 | -0.388 | 2.200 |
| GxN | 0.766 | 0.640 | 1.654 | 2.522 | 0.518 | -0.452 | 0.035 | -0.712 |

Observed parameter SOD and POD activity, chlorophyll a and chlorophyll b have skewness > 0 means additives gene with epistatic complement influence, positive marked skewness means a character can be inherited to the offspring, epistatic complement influence means dominant gene occurs interaction that complete each other so it raises that character. Kurtosis < 3 platikurtik graph, means character is controlled by many genes. When character is controlled by many genes then possibility to inherit to the offspring became smaller. Skewness describes the slope level in distribution, the tendency to deviate a curve based on data centered relationship concept between average count, mode and median [13]. [14] Skewness also called slope measure can show symmetric curve form of a distribution. If frequency curve extends to the right means a possitive value distribution pattern, if frequency extends to the left means a negative value distribution pattern, if symmetric data inclanation means value of mode and mean are the same. Enlightenment of skewness and kurtosis normal distribution explain gene interaction of Mendel law deviation. [15] Skewness analysis potentially to inherit to the offspring is skewness with positive value that have action gene epistatic complementer with positis kurtosis indicated by leptokurtik graph where only few gene witch are involved in form a phenotype character[16]. [17] Between population, expected value gene action in all of peanut F2 population show different result between population, there are character that controlled by many gene and few gene with additive gene action, domination, duplicate epistatic as well complementer.

## 4. Conclusion

Research result can be concluded that Salinity can influence growth and plant production, plant response to salinity environment is generally cause by plant metabolism change. Physiology result from SOD, POD, Chlorophyll a and Chlorophyll b character dominated by GxN crosses are more potential in inherit nature of descendants, it seen from observation result in distribution chart and skewness value, however to ensure soy genotype that potentially saline tolerant and high produce required more research and much more crops to development.

**References**

[1] Badan Pusat Statistik (BPS) 2016 Produksi kedelai menurut propinsi (ton) 1993-2015 (Production of soybean by province (ton) 1993-2015). (https://www.bps.go.id/linkTableDinamis/view/id/871

[2] Taufiq A, Kristiono A dan Harnoto D 2015 Respon varietas unggul kacang tanah terhadap cekaman salinitas (response of peanuts superior variety to salinity stress) balai penelitian tanaman aneka kacang dan umbi *penelitian pertanian tanaman pangan* **34**(2)

[3] Rachman A, Subiksa I G M dan Wahyunto 2007 Perluasan areal tanaman kedelai kelahan suboptimal (Expansion of soybean area to suboptimal land) Kedelai: Teknik Produksi dan Pengembangan Puslitbang, Bogor

[4] Atman 2009 Strategi peningkatan produksi kedelai di Indonesia (Strategy for increasing soybean production in Indonesia) *Tambua* Vol **VIII** No(1): 39-45

[5] Bose J A, Moreno R and Shabala S 2013 Review paper ROS homeostasis in halophytes in the context of salinity stress tolerance. *Experimental Botany* doi:10:109. http://jxb.oxfordjournals.org/

[6] Sopandie D 2014 Fisiologi adaptasi tanaman terhadap cekaman abiotik pada agroekosistem (Physiology of plant adaptation to abiotic stress in agroecosystem) *Tropika*. IPB Press, Bogor, Indonesia

[7] Yulianti E 2009 *Klasifikasi Enzim (Enzyme Classification)* Universitas Unsyiah SyahKuala. Fakultas MIPA

[8] Cahyarini R D, Yunus A dan Purwanto E 2004 Identifikasi keragaman genetik beberapa varietas lokal kedelai di jawa berdasarkan analisis isozim (Genetic variability identification of several local varieties of soybean in java based on isozyme analysis) *agrosains* **6**(2) : 79-83

[9] Syukur M, Sujiprihati S, Koswara J dan Widodo 2009 Ketahanan Terhadap antaroksa yang disebabkan oleh colletotriclum acutatum pada beberapa genotipe cabai (*Capsicum annum L.*) dan korelasinya dengan kandungan kapsaicin dan peroksidase. *Agronomi Indonesia.* **37**(3) 233-239

[10] Mane A V, Saratale G D, Karadge B A dan Samant J A 2011 Studies on the effects of salinity on growth, polyphenol content and photosynthetic response in *Vetiveria zizanioides* (L.) Nash.

[11] Aini N, Sumiya W D Y, Syekhfani, Dyah R P dan Setiawan A 2014 Kajian pertumbuhan, kandungan klorofil dan hasil beberapa genotipe tanaman kedelai (*Glycine max L.*) pada kondisi salinitas *Prosiding Seminar Nasional Lahan Sub Optimal; Palembang* 26-27 September 2014 Palembang ISBN : 979-587-529-9: 591-597

[12] Ai N S dan Banyo Y 2011 Konsentrasi klorofil daun sebagai indikator kekurangan air pada tanaman *Ilmiah Sains* FMIPA Universitas Sam Ratulangi Manado **11**(2)

[13] Fauziah I 2016 Tingkat kemiringan dan tingkat keruncingan kurva (*Skewness* dan *Kurtosis*) *Academia.*http://www.academia.edu/9325652.

[14] Kartika D A 2015 *Skewness* dan *Kurtosis* (Statistika) (Skewness and Kurtosis (Statistics). DocSlide. http://documents.tips/documents/*skewness*-dan-*kurtosis*-statistika.html.

[15] Sobir dan Syukur M 2015 *Genetika Tanaman (Plants Genetica)* IPB Press, Bogor, Indonesia

[16] Jambornias E 2014 *Analisis Genetik dan Segregasi Transgresif Berbasis Informasi Kekerabatan untuk Potensi Hasil dan Panen Serempak Kacang Hijau* [Disertasi] Bogor: Institut Pertanian Bogor Program Pasca Sarjana

[17] Inayah I 2014 Analisis parameter genetik dan deteksi segregan transgresif pada dua populasi f2 persilangan kacang tanah *(Arachis hypogaea L.)* genetic parameter analysis and transgressive segregan detection in two populations f2 peanut crosses [Skripsi] Bogor: Institut Pertanian Bogor

IOP Conf. Series: Earth and Environmental Science **122** (2018) 012030 doi:10.1088/1755-1315/122/1/012030

# The effect of shade on chlorophyll and anthocyanin content of upland red rice

**Muhidin[1], E Syam'un[2], Kaimuddin[2], Y Musa[2], G R Sadimantara[1], Usman[1], S Leomo[1] and T C Rakian[1]**

[1]Department of Agrotechnology, Faculty of Agriculture, Universitas Halu Oleo, Kendari South East Sulawesi 93212 Indonesia
[2]Department of Agronomy, Faculty of Agriculture, Universitas Hasanuddin University, Makassar South Sulawesi, 90245 Indonesia
Email : muhidinunhalu@gmail.com

**Abstract.** Upland red rice (*Oryza sativa*) is a staple food and contains anthocyanin, which can act as antioxidants, plays an important role both for the plant itself and for human health. Levels of antioxidants in rice can be affected by the availability of light. The results showed that the difference of shade, cultivar, and interaction both significantly affect the content of chlorophyll *a*, chlorophyll *b* and total chlorophyll. The results also showed that shade could increase chlorophyll in all cultivars tested. The highest levels of chlorophyll *a* were present in the moderate shade level ($n_2$), then decreased at the shelter level ($n_3$) and increased again at high levels ($n_4$). While on chlorophyll content *b*, it appears that shade increased chlorophyll *b* in all cultivars tested and this increase was linear to the increase of shade. The shade treatment may increase the anthocyanin content and the increase depending on the type of cultivar. Increased levels of anthocyanin highest due to shade occurred on Jangkobembe cultivar. The original level of anthocyanin on *Jangkobembe* cultivar averaged 0.096 mg $g^{-1}$ increased to 2.487 mg $g^{-1}$ or increased 26 fold. It is concluded that the shade had a significant effect on the chlorophyll and anthocyanin content.

## 1. Introduction

Red rice (*Oryza sativa*) is a staple food of high health value. In addition to carbohydrates, fats, proteins, fiber, and minerals, brown rice also contains anthocyanins [1]. Anthocyanins are phenolic compounds that enter the flavonoid group and act as antioxidants, plays an important role both for the plant itself and for human health. The role of antioxidants for human health to prevent liver disease (*hepatitis*), colon cancer, stroke, diabetes, is essential for brain function and reduce the effects of brain aging.

Increased red rice production has associated with efforts to increase rice production, especially the upland rice production. Red rice derived from upland rice and the efforts to increase upland rice production with local cultivars are therefore always associated with increase red rice production [2]. Indonesia actually is major rice producing countries together with China, India, Bangladesh and Vietnam [3-4], but at the same time, Indonesia also the main consumer of rice [5]. The Indonesian government has two police to reach rice self-sufficiency [4,6], through the increasing rice production and trying to curb rice consumption, while promoting consumption of other source staple foods, such as sago [7].

One of the alternatives to increase the national rice production is through the development of upland rice [8] as an insertion plant under stands of plantation crops/forestry or crops intercropping in agroforestry patterns [2,5]. Upland rice cultivation can be planted between the rows of plantation or young forestry crops [9] when the leaf canopy has not shaded each other [10]. The plantation area in 2013 reaches 22 million hectares [11] and every year there is approximately two million ha of plantation reforestation land that can be cultivated upland rice as intercrops.

The potential of the forestry sector to support food security are also very large by making the forest for food production [2,5]. Development of upland rice production in forest areas can be done through the provision of food forests in forest areas, community forests and other types of forest exploitation such as social forestry. Until 2013 there are potentials of more than 10 million hectares of forest land, which can be used for the development of upland rice, both as insertions in agroforestry and food reserves in forest areas.

The main ecological constraints of land use under stands of plantation crops and forestry are the low intensity of solar radiation, and the plants will be light stress, which will result in the process of growth and production. The decrease in the intensity of solar radiation will affect other micro climate elements such as air temperature, relative humidity, and soil temperature.

Shade can affect on growth and production, and reduce the efficiency of solar radiation [12-14]. Some research results showed that at a shade level of more than 50%, there is a drastic reduction in production. The magnitude of the decline in rice production due to shade, is depending on the level of tolerance and growth phase of each strain. Improved and moderate upland rice cultivar is recommended for planting upland rice between plantation stands and forestry. However, the shade or low light intensity, can affect the increase of anthocyanin levels. Therefore, although shade can reduce the production of upland rice but the anthocyanin content has increased.

Development of red rice will have a broad economic impact for farmers because the types or varieties of rice planted can produce rice with high selling value and the pattern of planting has integrated with existing farming, especially mixed farming in the form of plantation or forestry. Under these conditions, the choice is to plant shade of red rice that is tolerant to shade, so that it can be planted under the stands of a plantation tree or a young forestry.

## 2. Material and Method

The research design using split plot design. The main plot was the different of shae treatment consisting of 4 levels of shading and the second factor is different of cultivar. The main plot is the difference of shading as follows: $n_1$=shade level < 25 %, $n_2$=shade level 25-50 %, $n_3$ = shade level 50-75 % and $n_4$= shade level > 75%. While in subplot is the difference of upland red rice cultivar i.e., ($V_1$) = *La Bandiri*, ($V_2$) = *Jangkobembe*, ($V_3$) = *Ranggohitam*, ($V_4$) = *Paedara*.

## 3. Results and Discussion

### *3.1. Chlorophyll content*

The result showed that the difference of shade, cultivar, and interaction both have a significant effect on the content of chlorophyll a, chlorophyll b and total chlorophyll of upland red rice. Based on the cross-tabulation between cultivars and shade as in Table 1 shows that shade can increase chlorophyll content in all cultivars tested.

The highest levels of chlorophyll a are present at moderate shade level ($n_2$), then decrease at the medium shade ($n_3$) and increase again at heavy shade levels ($n_4$). While on chlorophyll b content, it appears that shade increases chlorophyll b in all cultivars tested and this increase is linear, where the higher the shade level, the higher the chlorophyll b content.

A similar pattern is also found in total chlorophyll content, which increases as a result of shade. At low shade (n2) and medium shade (n3) levels, total chlorophyll content increases, but at a heavy shade (n4) the total chlorophyll content decreases. Based on this pattern it appears that plants in the shaded condition will adapt by increasing the content of chlorophyll, both for the content of chlorophyll a, chlorophyll b and total chlorophyll. In general, chlorophyll content increases with increasing shade. Further test results of shade and cultivar interaction on chlorophyll a content as shown in Table 2, it appears that in Labandiri cultivars, there is no significant difference in the content of chlorophyll a due to shade.

**Table 1.** Leaf chlorophyll content of various rice cultivars of upland red rice at different levels of shade [2]

| Shade Treatment | Chlorophyll Content (mg $100g^{-1}$) | | | | |
|---|---|---|---|---|---|
| | Chlorophyll a | | | | |
| | $v_1$ | $v_2$ | $v_3$ | $v_4$ | Average |
| $n_1$ | 4.12 | 4.12 | 4.12 | 4.12 | 4.12 |
| $n_2$ | 4.56 | 4.56 | 4.56 | 4.56 | 4.56 |
| $n_3$ | 4.18 | 4.18 | 4.18 | 4.18 | 4.18 |
| $n_4$ | 4.46 | 4.46 | 4.46 | 4.46 | 4.46 |
| Average | 4.33 | 4.33 | 4.33 | 4.33 | 4.33 |
| | Chlorophyll b | | | | |
| | $v_1$ | $v_2$ | $v_3$ | $v_4$ | Average |
| $n_1$ | 44.22 | 44.22 | 44.22 | 44.22 | 44.22 |
| $n_2$ | 51.47 | 51.47 | 51.47 | 51.47 | 51.47 |
| $n_3$ | 54.30 | 54.30 | 54.30 | 54.30 | 54.30 |
| $n_4$ | 59.19 | 59.19 | 59.19 | 59.19 | 59.19 |
| Average | 9.1 | 9.1 | 10.1 | 9.4 | 9.4 |
| | Total Chlorophyll | | | | |
| | $v_1$ | $v_2$ | $v_3$ | $v_4$ | Average |
| $n_1$ | 21.96 | 21.96 | 21.96 | 21.96 | 21.96 |
| $n_2$ | 32.68 | 32.68 | 32.68 | 32.68 | 32.68 |
| $n_3$ | 33.71 | 33.71 | 33.71 | 33.71 | 33.71 |
| $n_4$ | 32.34 | 32.34 | 32.34 | 32.34 | 32.34 |
| Average | 30.17 | 30.17 | 30.17 | 30.17 | 30.17 |

Rem : $n_1$ = Shade level < 25 %, $v_1$ = Cultivar *Labandiri*
$n_2$ = Shade level 25-50 % $v_2$ = Cultivar *Jangkobembe*
$n_3$ = Shade level 50-75 % $v_3$ = Cultivar *Ranggohitam*
$n_4$ = Shade level > 75% $v_4$ = Cultivar *Paedara*

### *3.2 Anthocyanin content*

Shade treatment may increase anthocyanin levels of upland red rice, and the increasing depends on the type of cultivars tested as shown in Table 3. The highest increase of anthocyanin levels due to shade stress occurs in Jangkobembe cultivar.

The anthocyanin content of the original *Jangkobembe* cultivar averaged 0.096 mg $g^{-1}$ increased to 2.487 mg $g^{-1}$ or increased 26 fold. In the other three cultivars also increased levels of anthocyanin from the shade, but the increase was not as large as the *Jangkobembe* cultivar. The subsequent increase in the highest levels of anthocyanin occurred in the *Labandiri* cultivars, followed by *Ranggohitam* and *Paedara* cultivars.

In general, shade stress plants have an increased amount of chlorophyll (chlorophyll a, chlorophyll b and total chlorophyll). This increase in chlorophyll content is in line with the higher level of shade (Table 1) and occurs for the four cultivars tested. Increasing the chlorophyll content is one of the plant's efforts to adjust to low light intensity conditions. The decrease in light intensity to some extent can increase the chlorophyll content. If the available light is no longer able to support the plant for photosynthesis, the rate of chlorophyll increase also decreases. Adaptation mechanism for low light intensity in soybean through increasing the chlorophyll content occurring in all shading-treated genotypes [15]. The content of chlorophyll a and b are both increased, but the proportion of increase occurs more in chlorophyll b.

**Table 2.** Interaction between shade and cultivar on chlorophyll a content [2]

| Shade Treatment | Chlorophyll a content (mg 100g$^{-1}$) | | | | | | | | DMRT 0.05 |
|---|---|---|---|---|---|---|---|---|---|
| | $v_1$ | | $v_2$ | | $v_3$ | | $v_4$ | | Between v |
| $n_1$ | 4.12 | a | 1.40 | c | 2.30 | d | 1.36 | b | |
| | p | | r | | q | | r | | 2= 0.71 |
| $n_2$ | 4.57 | a | 2.91 | ab | 3.40 | c | 1.00 | b | |
| | p | | q | | q | | r | | 3= 0.75 |
| $n_3$ | 4.19 | a | 2.47 | b | 4.45 | b | 3.01 | a | |
| | p | | Q | | P | | q | | 4= 0.78 |
| $n_4$ | 4.46 | a | 3.37 | a | 5.69 | a | 2.63 | a | |
| | q | | r | | p | | s | | |

Remarks : Number followed by the same index in the same row, are not significantly different at Duncan's multiple Range Test (DMRT).

**Table 3.** Efect of shade on anthocyanin content of upland red rice [2]

| Shading treatment | Anthocyanin content (mg g$^{-1}$) | | | | | | | |
|---|---|---|---|---|---|---|---|---|
| | $v_1$ | | $v_2$ | | $v_3$ | | $v_4$ | |
| Low shade ($n_1$) | 0.127 | a | 0.096 | a | 3.832 | a | 0.153 | a |
| | r | | P | | Q | | p | |
| Heavy shade ($n_3$) | 0.336 | b | 2.487 | b | 6.226 | b | 0.218 | b |
| | q | | R | | Q | | p | |
| Increasing level | 2.64 | | 26.03 | | 1.62 | | 1.42 | |

Remark : Number followed by the same index in the same row, are not significantly different at Duncan's multiple Range Test (DMRT).

The results showed that shade could increase the anthocyanin level of red rice, even can reach 26-fold as happened in Ranggohitam cultivars. While in the other three varieties (Labandiri, *Jangkobembe and Paedara*) there was also an increase in anthocyanin levels up to nearly threefold (Table 3).

Anthocyanins are phenolic compounds belonging to the flavonoid group. Flavonoids are potent antioxidants that prevent the formation of free radicals. Light has a very significant effect on anthocyanin content of red rice gogo rice. An increase in anthocyanin content is due to the leaves exposed to full light, anthocyanins found only in epidermal cell vacuoles (Generally treated plants generally have increased chlorophyll (chlorophyll a, chlorophyll b and total chlorophyll).

This increase in chlorophyll content is in line with the higher level of shade and occurs for the four cultivars tested. Increasing the chlorophyll content is one of the plant's efforts to adjust to low light intensity conditions. Increasing the formation of anthocyanins in low light will reduce the asymilat formed to reduce the efficiency of light use [16], but the increase of anthocyanin formation at low light intensity is a consequence of decreasing carotene content at low light intensity [17].

## 4. Conclusion

The results also showed that shade could increase chlorophyll content in all cultivars tested. The shade treatment may increase the anthocyanin content and the increase depending on the type of cultivar. It is concluded that the shade had a significant effect on the chlorophyll and anthocyanin content.

## Refferences

[1] Suliartini N W S, Kuswanto, Basuki N and Soegianto A 2016 Superior lines candidates evaluation of two local red rice Southeast Sulawesi cultivars (Indonesia) derived from gamma rays irradiation techniques *International Journal of Plant Biology* **7 : 6475**

[2] Muhidin 2015 *Morphological Characterisation and the Quality of Red Rice Irradiated with Gamma Rays in Various Kinds of Shade and the Application of Bacterial Fixation in Dry Land in Southeast Sulawesi* (Makassar : Disertation on Postgradaute Program Hasanuddin University Makassar)

[3] FAOSTAT 2014 *Food and Agricultural commodities production / Countries by commodity* (Rome : Food and Agriculture Organization of the United Nations Statistics Division)

[4] Sutariati G A K, Khaeruni A, Pasolon Y B, Muhidin and Mudi L 2016 The effect of seed bio-invigoration using indigenous rhizobacteria to improve viability and vigor of upland rice (*Oryza sativa* L.) seeds *Internasional Journal PharmTech Research* **9(12)** 565-73

[5] Muhidin, Jusoff K, Elkawakib S, Musa Y, Kaimuddin, Meisanti, Sadimantara G R and Baka L R 2013 The development of upland red rice under shade trees *World Applied Sciences Journal* **24(1)** 23-30

[6] Kadidaa B, Sadimantara G R, Suaib, Safuan L O and Muhidin 2017 Genetic diversity of local upland rice (*Oryza sativa* L) genotypes based on agronomic traits and yield potential in marginal land of North Buton Indonesia *Asian Journal of Crop Science* **9(4)** 109-117

[7] Muhidin, Leomo S, Alam S and Wijayanto T 2016 Comparative studies on different agroecosystem base on soil physicochemical properties to development of sago palm on dryland. International *Journal of ChemTech Research* **9(8)** 511-18

[8] Syaiful S A, Syam'un E, Dachlan A, Jusoff K and Haerani N 2013 The effect of inoculating nitrogen fixing bacteria on production of rice *World Applied Sciences Journal* **26(26)** 94-99

[9] Toha H M, Permadi K, Prayitno and Julairdi I 2005 *Peningkatan produksi padi gogo melalui pendekatan model pengelolaan tanaman dan sumber daya terpadu PTT* Seminar Pusat Penelitian dan Pengembangan Pangan. Juli 2006. p17

[10] Harahap E and Lubis Z 1995 *Pengembangan Tanaman Padi Gogo Sebagai Tanaman Sela Di Daerah Perkebunan. Prosiding Diskusi Pengembangan Teknologi Tepat Guna Di Lahan Kering Untuk Mendukung Pertanian Berkelanjutan* (Bogor : Jurusan Budidaya Pertanian. Fakultas Pertanian IPB)

[11] Kementan 2013 *Statistik Pertanian Indonesia Tahun 2012* (Jakarta : Kementerian Pertanian Republik Indonesia) p 178

[12] Las I 1982 *Efisiensi Radiasi Surya dan Pengaruh Naungan Fisik terhadap Padi Gogo (*Bogor : Tesis Fakultas Pasca Sarjana IPB)

[13] Marler T L 1994 Developmental light level effect of growth, morphology and leaf physiology of young carambola trees *J. Amer. Soc. Hort. Sci.* **199(4)** 711-18.

[14] Cruz P 1997 Effect of shade on growth and mineral nutrition of a C4 perenial grass underfield condition *Plant and Soil* **188** 227-37

[15] Khumaida N 2002 *Studies on Adaptability of Soybean and Upland Rice to Shade Stress* (Tokyo : PhD Thesis The University of Tokyo)

[16] Levit J 1980 *Response of Plant to Environmental Stress* (New York : Academic Press) p570

[17] Buchanan B B, Gruissem W and Jones R L 2000 *Biochemistry and Molecular Biology of Plants* (Maryland : American Society of Plant Physiologists)

# The effectiveness of preplant seed bio-invigoration techniques using *Bacillus* sp. CKD061 to improving seed viability and vigor of several local upland rice cultivars of Southeast Sulawesi

**G A K Sutariati[1*], L O S Bande[1], A Khaeruni[1], Muhidin[1], L Mudi[1] and R M Savitri[2]**

[1]Department of Agrotechnology, Faculty of Agriculture, Universitas Halu Oleo, Kendari Southeast Sulawesi 93212 Indonesia
[2]Graduated Student at Department of Agrotechnology, Faculty of Agriculture, Universitas Halu Oleo, Kendari Southeast Sulawesi 93212 Indonesia
Email : sutariati69@yahoo.co.id

**Abstract**. Research was aimed to evaluate the bio-invigoration techniques using *Bacillus* sp. CKD061 in improving seed viability and vigor of local upland rice. The research is arranged in factorial with completely randomized design (CRD). The different upland rice cultivars as first factor that consists of 11 cultivars, namely: *Pae Tinangge, Pae Rowu, Pae Uwa, Pae Tanta, Pae Waburi-Buri, Pae Mornene, Pae Indalibana, Pae Lawarangka, Pae Huko, Pae Wagamba* and *Pae Momea*. The second factor is the seed bio-invigoration technique, consists of 5 treatments, namely: without seed bio-invigoration ($B_0$), NaCl + Bacillus sp. CKD061 ($B_1$), $KNO_3$ + Bacillus sp. CKD061 ($B_2$), Ground burned-rice husk + Bacillus sp. CKD061 ($B_3$), and Ground brick + Bacillus sp. CKD061 ($B_4$). The results showed that seed bio-invigoration using Bacillus sp. CKD061 gave effect on the seed viability and vigor. Interaction of the seed bio-invigoration and upland rice cultivars were able to improve seed viability and vigor. Seed bio-invigoration ttreatment using ground brick + Bacillus sp. CKD061 was the best treatment, which could improve the viability and vigor of *Pae Waburi-Buri, Pae Mornene* and *Pae Indalibana*. The treatment increased vigor index by 133% in *Pae Waburi-Buri* and 127% in *Pae Mornene*, and *Pae Indalibana* compared with control.

## 1. Introduction

Rice (*Oryza sativa* L.) is a very important food crop because rice is still the main staple food in Indonesian. The demand of rice is increase every year. Various efforts have been done to increase rice production such as through the increasing productivity, breeding new variety [1] and development rice under the shade [2]. Another program is by the decreasing the level of rice consumption and promoting a local source of staple foods such sago [3], cassava or corn [4], but the results are not yet optimal. One effort to rice production is through the development of upland rice in dry land [4].

Constraints encountered include limited input of cultivation technology, especially in terms of the use of quality seeds and control techniques of plant-disturbing organisms. In addition, the decline in rice production is also attributed to the decrease in productive wetland due to the transfer of functions to the interests of industry, housing and other non-agricultural land use. The development of upland rice in dry land can be a solution to increase production. However, upland rice received less attention because of its low productivity. Development of upland rice (especially local upland rice) in Southeast

Sulawesi is still limited, in addition to marginal land issues, as well as the implementation of cultivation techniques of upland rice, especially in the use of quality seeds. The use of high quality seeds is an important prerequisite for generating economically profitable crop production. Therefore, preparation and treatment of seeds to improve the quality is very important to do, especially with the physiological dormancy (after ripening) problem in post-harvest upland rice seed in the field.

The alternative to overcome these problems is through seed invigoration technology integrated with biological agents of rhizobacterial groups, microorganisms capable of acting as biofertilizers and biopestisides [5]. Invigoration is a way to improve the physiological quality of seeds, especially seed vigor, through physical or chemical treatment. High-vigorous seeds are able to demonstrate good performance in germination processes under diverse environmental conditions [6]. Seed invigoration is physiological and biochemical improvements associated with synchronous germination, velocity, and increased seed germination using low-potency matrix solids or low osmotic potential solutions. This treatment is known as seed matriconditioning or osmoconditioning treatments can be integrated with rhizobacterial applications, called bio-matriconditioning [7-8] or bio-osmoconditioning. This treatment aims to improve the viability and vigor of seeds, growth and yield of plants [5,9], also proved able to protect the seed from seedborne and soilborne fungi at an important phase at the beginning of its growth [10]. Treatment can be recommended as a growth promoting of local rice crops of Southeast Sulawesi.

## 2. Materials and Methods

The research was conducted in Agrotechnology Laboratory, Agriculture Faculty of Halu Oleo University, from September 2015 to March 2016. The research was arranged based on factorial in completely randomized design (CRD). The first factor is the local upland rice cultivar consisting of 11 local upland rice cultivar of Southeast Sulawesi: V1=*Pae Tinangge*, V2=*Pae Rowu*, V3=*Pae Uwa*, V4= *Pae Tanta*, V5=*Pae Waburi-Buri*, V6=*Pae Mornene*, V7=*Pae Indalibana*, V8=*Pae Lawarangka*, V9=*Pae Huko*, V10 =*Pae Wagamba* dan V11=*Pae Momea*. The second factor is bio-invigoration treatment with *Bacillus* sp. CKD061 Consists of 5 treatments namely: $B_0$ = without bio-invigoration treatment (as control), $B_1$ = NaCl + *Bacillus* sp. CKD061, $B_2$ = $KNO_3$ + *Bacillus* sp. CKD061, $B_3$ = ground burned-rice husk + *Bacillus* sp. CKD061 dan $B_4$ = ground brick + *Bacillus* sp. CKD061. Each treatment was replicated 3 times, therefore, overall there were 165 experimental units.

The effects of seed bio-invigoration on the seed viability and vigor were evaluated by measuring their germination percentage, vigor index, relative growth rate, and growth uniformity.

1. Germination percentage (GP), depicting seed potential viability [11], was measured based on the percentage of normal seedlings (NS) during the first (i.e. 5 days after planting (dap)) and the second (i.e. 7 dap) observation by using the following formula:

$$GP = \frac{\Sigma NS \text{ at observation 1} + \Sigma NS \text{ observation 2}}{\Sigma \text{ seeds planted}} \text{ X } 100\%$$

2. Relative growth rate (RG-r), depicting seed vigor, is the ratio of KCT to maximum RG-r. The maximum RG itself was obtained from the assumption that at the first observation, normal seedlings had reached 100%.

3. Seed uniformity, depicting seed vigor, was measured based on the percentage of normal seedlings (NS) on the day between the first count (5 dap) and second (7 dap) i.e. at 6 dap.

4. Vigor index (VI), depicting the growth rate vigor [10], was measured based on percentage of normal seedlings at the first observation (i.e. 5 dap):

$$VI = \frac{\Sigma NS \text{ at observation 1}}{\Sigma \text{ seeds planted}} x \text{ } 100\%$$

## 3. Results and discussion

### *3.1. Results*

#### *3.1.1.The effects of seed bio-invigoration treatment on germination percentage on local upland.*

There is an interaction effect between seed bio-invigoration treatment and cultivars on the germination percentage (Table 1). Different cultivars show different responses to the seed bio-invigoration treatment. In Pae Uwa cultivar the best interaction is in the treatment of $KNO_3$ + *Bacillus* sp. CKD061 and ground burned rice husk + *Bacillus* sp. CKD061 with the value of each germination percentage is 94.67% and 93.33%. In *Pae Waburi-buri* cultivar the best interaction is in the treatment of NaCl + *Bacillus* sp. CKD061, $KNO_3$ + *Bacillus* sp. CKD061, ground burned rice husk + *Bacillus* sp. CKD061, ground brick + *Bacillus* sp. CKD061 with the value of each germination percentage is 93.33%, 93.33%, 94.67% and 93.33%.

**Table 1.** The effects of seed bio-invigoration treatment on germination percentage of several local upland rice cultivars of Southeast Sulawesi

| Local upland rice cultivar | Seed Bio-invigoration Treatment | | | | | | | | | |
|---|---|---|---|---|---|---|---|---|---|---|
| | B0 | | B1 | | B2 | | B3 | | B4 | |
| V1 | 65.33 | abQ | 64.67 | bQ | 81.33 | abP | 72.00 | bPQ | 78.67 | abPQ |
| V2 | 50.67 | bQ | 83.33 | abP | 54.67 | cQ | 84.00 | abP | 62.67 | bQ |
| V3 | 69.33 | aQ | 81.33 | abPQ | 94.67 | aP | 93.33 | aP | 81.33 | abPQ |
| V4 | 13.33 | cR | 84.00 | abP | 61.33 | bcQ | 69.33 | bQ | 67.33 | bQ |
| V5 | 65.33 | abQ | 93.33 | aP | 93.33 | aP | 94.67 | aP | 93.33 | aP |
| V6 | 61.33 | abQ | 60.00 | bQ | 80.00 | abPQ | 73.33 | bQ | 90.67 | aP |
| V7 | 61.33 | abQ | 60.00 | bQ | 80.00 | abPQ | 73.33 | bQ | 90.67 | aP |
| V8 | 65.33 | abQ | 78.67 | abPQ | 77.33 | bPQ | 73.33 | bPQ | 84.00 | aP |
| V9 | 57.33 | abQ | 78.67 | abP | 69.33 | bcPQ | 78.67 | abP | 64.00 | bPQ |
| V10 | 73.33 | aP | 80.00 | abP | 86.67 | abP | 89.33 | aP | 88.00 | aP |
| V11 | 61.33 | abQ | 73.33 | bPQ | 77.33 | bP | 73.33 | bPQ | 84.00 | aP |

Means in the same column (a-c) or in the same line (P-Q) suffixed with different letters are different at 5% levels of significance according to DMRT. V1=*Pae Tinangge*, V2=*Pae Rowu*, V3=*Pae Uwa*, V4=*Pae Tanta*, V5=*Pae Waburi-Buri*, V6=*Pae Mornene*, V7=*Pae Indalibana*, V8=*Pae Lawarangka*, V9=*Pae Huko*, V10 =*Pae Wagamba*, V11=*Pae Momea*, B0 = control, B1 = NaCl + *Bacillus* sp. CKD061, B2 = KNO3 + *Bacillus* sp. CKD061, B3 = ground burned rice husk + *Bacillus* sp. CKD061, B4 = ground brick + *Bacillus* sp. CKD061

In *Pae Wagamba* cultivar the best interaction is in the treatment of ground burned rice husk + *Bacillus* sp. CKD061, ground brick + *Bacillus* sp. CKD061 with the value of each germination percentage is 89.33% dan 88.00%. Meanwhile, *Pae Mornene*, *Pae Indalibana*, Pae *Lawarangka*, and *Pae Momea*, provides the best response to the treatment of ground brick + *Bacillus* sp. CKD061 with the value of each germination percentage is 90.67%, 90.67, 84.00% and 84.00%.

#### *3.1.2. The effects of seed bio-invigoration treatment on germination growth rate on local upland rice.*

*Pae Waburi-Buri*, *Pae Mornene*, and *Pae Indalibana* cultivars provide a better response to the treatment of ground brick + *Bacillus* sp. CKD061 with the value of each germination percentage is 93.33%, 90.67% dan 90.67%. Among all of seed bio-invigoration treatments tested, *Pae Waburi-Buri* cultivar was able to respond all of seed bio-invigoration treatments provided. Among all of seed bio-invigoration treatments tested, only NaCl + *Bacillus* sp. CKD061 was able to solve the dormancy of Pae Tanta cultivars by increasing the relative growth rate by 565% (Table 2).

**Table 2.** The effects of seed bio-invigoration treatment on relative growth rate of several local upland rice cultivars of Southeast Sulawesi

| Local upland rice cultivar | Seed Bio-invigoration Treatment | | | | | | | | | |
|---|---|---|---|---|---|---|---|---|---|---|
| | B0 | | B1 | | B2 | | B3 | | B4 | |
| V1 | 59.90 | abQ | 64.35 | bPQ | 78.22 | abP | 67.56 | bPQ | 66.67 | bcPQ |
| V2 | 43.11 | bQ | 81.27 | aP | 47.08 | cQ | 74.03 | bP | 54.70 | cQ |
| V3 | 67.17 | abQ | 79.68 | abPQ | 91.68 | aP | 88.92 | abP | 77.46 | bPQ |
| V4 | 12.29 | cR | 81.75 | aP | 59.52 | cQ | 68.35 | bPQ | 65.56 | bcQ |
| V5 | 58.51 | abQ | 86.54 | aP | 92.95 | ,aP | 94.29 | aP | 93.33 | aP |
| V6 | 57.40 | abQ | 60.41 | bQ | 80.00 | abPQ | 70.06 | bQ | 90.67 | aP |
| V7 | 57.40 | abQ | 60.41 | bQ | 80.00 | abPQ | 70.06 | bQ | 90.67 | aP |
| V8 | 62.05 | abQ | 76.38 | abP | 76.79 | bP | 72.38 | bPQ | 80.83 | abP |
| V9 | 49.56 | bQ | 76.48 | abP | 62.86 | bcPQ | 76.29 | bP | 63.17 | bcPQ |
| V10 | 70.67 | aQ | 77.59 | abPQ | 82.51 | abPQ | 88.73 | abP | 85.68 | abP |
| V11 | 60.19 | abQ | 75.24 | abP | 74.79 | bP | 72.38 | bPQ | 80.83 | abP |

Notes: Means in the same column (a-c) or in the same line (P-Q) suffixed with different letters are different at 5% levels of significance according to DMRT.

*3.1.3. The effects of seed bio-invigoration treatment on vigor index of several local upland rice cultivars of Southeast Sulawesi.*

Cultivars of *Pae Waburi-Buri, Pae Mornene,* and *Pae Indalibana* provide a better response to the treatment of ground brick + *Bacillus* sp. CKD061 with value vigor index respectively 93.33%, 90.67% and 90.67%. Consistently, seed bio-invigoration treatment using NaCl + *Bacillus* sp. CKD061 was able to solve the dormancy of *Pae Tanta* cultivars by increasing the vigor index by 817% (Table 3).

**Table 3.** Effect of interaction bio-invigoration technique interaction on seeds vigor several cultivars of local rice seedlings of Southeast Sulawesi

| Local upland rice cultivar | Bio-invigoration | | | | | | | | | |
|---|---|---|---|---|---|---|---|---|---|---|
| | B0 | | B1 | | B2 | | B3 | | B4 | |
| V1 | 41.33 | aQ | 41.33 | bQ | 69.33 | bP | 45.33 | cQ | 40.00 | cQ |
| V2 | 18.67 | bQ | 60.00 | aP | 32.00 | cQ | 34.67 | cQ | 32.00 | cQ |
| V3 | 57.33 | aQ | 66.67 | aPQ | 78.67 | abP | 70.67 | bPQ | 60.00 | bQ |
| V4 | 8.00 | bR | 73.33 | aP | 53.33 | bcQ | 65.33 | bPQ | 60.00 | bPQ |
| V5 | 38.67 | aR | 64.00 | aQ | 92.00 | aP | 93.33 | aP | 93.33 | aP |
| V6 | 45.33 | aQ | 58.67 | abQ | 80.00 | abP | 61.33 | bcQ | 90.67 | aP |
| V7 | 45.33 | aQ | 58.67 | abQ | 80.00 | abP | 61.33 | bcQ | 90.67 | aP |
| V8 | 53.33 | aQ | 70.00 | aPQ | 76.00 | abP | 66.67 | bPQ | 60.00 | bPQ |
| V9 | 17.33 | bR | 69.33 | aP | 40.00 | cQ | 65.33 | bP | 60.00 | bP |
| V10 | 57.33 | aQ | 62.67 | aQ | 69.33 | bPQ | 86.67 | abP | 76.00 | abPQ |
| V11 | 57.33 | aQ | 70.67 | aQ | 75.33 | abP | 66.67 | bPQ | 70.67 | bPQ |

Notes: Means in the same column (a-c) or in the same line (P-R) suffixed with different letters are different at 5% levels of significance according to DMRT.

### *3.2. Discussion*

Seed bio-invigoration treatment using *Bacillus* sp. CKD061 integrated with ground burned rice husk or ground brick or $KNO_3$ gives better results in enhancing the viability and vigor of local upland rice seed compared with NaCl and control. *Bacillus* spp. is a group of PGPR (Plant Growth Promoting Rhizobacteria) which has been shown to be effective in increasing plant growth and yield [12]. The role of PGPR in increasing plant growth and production was presumably caused by the ability of rhizobacteria to produce IAA [9], gibberellins [13] and to dissolve phosphate [5,14]. In general, the utilization of *Bacillus* sp.CKD061 integrated with matriconditioning of ground burned rice husk or ground brick resulted in more effective in increasing the viability and vigor of local upland rice seed

## 4. Conclusions

It is concluded that seeds bio-invigoration treatment using *Bacillus* sp. CKD061 which is integrated with a medium of ground burned rice husk or ground brick or $KNO_3$ solution, is better able to increase the viability and vigor of local upland rice seed compared to other treatments and controls. Cultivars *Pae Waburi-Buri, Pae Mornene*, and *Pae Indalibana* responded better to the treatment of ground brick + *Bacillus* sp. CKD061 through increased vigor index respectively 131%, 100% and 100%. Integration of bio-invigoration seed treatment using NaCl + *Bacillus* sp. CKD061 was able to overcome the dormancy of local upland rice seed cv. *Pae Tanta.*

## References

[1] Sadimantara G R, Muhidin M, Ginting S and Suliartini N W S 2016 The potential yield of some superior breeding lines of upland rice of southeast sulawesi indonesia *Biosci. Biotech Res. Asia* **13(4)** 1867-70

[2] Muhidin, Jusoff K, Syam'un E, Musa Y, Kaimuddin, Meisanti, Sadimantara G R and Baka L R 2013 The development of upland red rice under shade trees *World Applied Sciences Journal* **24(1)** 23-30

[3] Muhidin, Leomo S, Alam S and Wijayanto T 2016 Comparative studies on different agroecosystem base on soil physicochemical properties to development of sago palm on dryland *International Journal of ChemTech Research* **9(8)** 511-18

[4] Sutariati G A K, Khaeruni A, Pasolon Y B, Muhidin and Mudi L 2016. The effect of seed bio-invigoration using indigenous rhizobacteria to improve viability and vigor of upland rice (*oryza sativa* L.) Seeds *International Journal of PharmTech Research* **9(12)** 565-73

[5] Sutariati G A K, Jusoff K, Sadimantara G R, Khaeruni A, Muhidin and Meisanti 2013 Effectiveness of bio-invigoration technologies on seed viability and vigor of cocoa (*Theobroma cacao* L.) *World Applied Sciences Journal* **26 (26)** 31-36.

[6] ISTA 2007 *International Rules for Seed Testing Edition 2007* (Switzerland : International Seed Testing Association)

[7] Ilyas S, Sutariati G A K, Suwarno F C and Sudarsono 2002 Matriconditioning improved quality and protein level of medium vigor hot pepper seed *Seed Technol.* **24** 65-75

[8] Moradi A and Younesi O 2009 Effects of osmo and hydro-priming on seed parameters of grain sorghum (*Sorghum bicolor* L.) *Australian Journal of Basic and Applied Sciences* **3(3)** 1696-700

[9] Ilyas S, Asie K V, Sutariati G A K and Sudarsono S 2014 Biomatriconditioning or biopriming with biofungicides or biological agents applied on hot pepper (*Capsicum annuum* L.) seeds reduced seedborne *Colletotrichum capsici* and increased seed quality and yield *ISHS Acta Horticulturae* **1105** 89-96

[10] Sutariati G A K and Wahab A 2010 Isolasi dan uji kemampuan rizobakteri indigenous sebagai agens pengendali hayati penyakit pada tanaman cabai *J. Hort* **20(1)** 86-95

[11] Sadjad S 1999 *Kuantifikasi Metabolisme Benih* (Jakarta : Gramedia Widiasarana Indonesia)

[12] Datta M, Palit R, Sengupta C, Pandit M K and Banerjee S 2011 Plant growth promoting rhizobacteria enhance growthand yield of chilli (*Capsicum annuum* L.) under field conditions

*Australian Journal of Crop Sciences* **5(5)** 531-36

[13] Ambawade S and Pathade G R 2015 Production of gibberellic acid by *Bacillus siamensis* BE 76 isolated from banana plant (*Musa* spp) *International Journal of Science and Research* **4(7)** 394-98

[14] Mujahid T Y, Subhan S A, Wahab A, Masnoon J, Ahmed N and Abbas T 2015 Effects of different physical and chemical parameters on phosphate solubilization activity of plant growth promoting bacteria isolated from indigenous soil *Journal of Pharmacy and Nutrition Sciences* **5** 64-70

**Acknowledgements**

The authors extend their gratitude to the Directorate General of Higher Education, Ministry of Research, Technology and Higher Education of the Republic of Indonesia for providing research grant under *Hibah Kompetensi* in the fiscal year 2016 to support this study.

IOP Conf. Series: Earth and Environmental Science **122** (2018) 012032 doi:10.1088/1755-1315/122/1/012032

# The effectiveness of various Rhizobacteria carriers to improve the shelf life and the stability of Rhizobacteria as Bioherbicide

**T C Rakian[1], L Karimuna[1], M Taufik[2], G A K Sutariati[1], Muhidin[1] and U Fermin[1]**

[1] Department of Agrotechnology, Faculty of Agriculture, Universitas Halu Oleo, Kendari, Southeast Sulawesi 93212, Indonesia

[2]Department of Plant Protection, Faculty of Agriculture, Universitas Halu Oleo, Kendari, Southeast Sulawesi 93212 Indonesia

Email : tresjia_junus@yahoo.com

**Abstract**. Deleterius Rhizobacteria (DRB) has a potential to control of weed and act as a bioherbicide. Developing a method to weed control that environmentally sound friendly has been increasingly studied. Rhizobacteria can form colonies on weed rooting and synthesize the secondary metabolite compounds. The effectiveness of rhizobacteria as bioherbicide is determined by its survival to be stored for a long time. The objective of this study is to obtain the type of carrier which effectively maintains the life and stability of DRB. Therefore it is necessary to do in vivo and in-vitro research. This study consists of two stages of testing the effectiveness of the carrier in increasing the shelf life of rhizobacteria and testing the effectiveness stability as a bioherbicide on *Ageratum conyzoides* weed after storage for 20 weeks. Research was conducted in Agronomy Lab, Agriculture Faculty, Halu Oleo University Kendari, Since August to December 2016. Research found that the talc powder and chaff charcoal powder were effective as a carrier of rhizobacteria and able to maintain the viability of rhizobacteria *Bacillus lentus* A05 and *Pseudomonas aeruginosa* A08 for five months and also able to maintain the stability of rhizobacteria as bioherbicide.

## 1. Introduction

Weed control is generally done conventionally using synthetic chemicals that is herbicide. The use of herbicides is considered as the most appropriate and fast method, because it is very effective to suppress the growth of weeds. However, continuous use of synthetic herbicides will have a negative effect on the environment, human health, and can cause weeds to become resistant to herbicides [1-3]. To reduce the negative effects of using synthetic herbicides in weed control is by developing biological controls using microorganisms or biological agents to regulate weed population density [4].

Microorganisms are used as bioherbicides because they do not have a negative impact on the environment, are effective during plant lifetimes and some agents can produce compounds that function as growth hormones thereby providing multiple benefits to the plants [5-6]. Rhizobacteria that have the ability to inhibit the growth of weeds are called deleterious rhizobacteria (DRB), which is a non parasitic bacteria, capable of forming colonies on weed roots and synthesizing secondary metabolite compounds [7].

The effectiveness of rhizobacterial applications as bioherbicides in the field is determined by the survival of rhizobacteria both in storage and in the field. Application of field rhizobacteria in the form of suspension will be ineffective due to unsupportive environmental factors, such as very high solar radiation, rainfall, and nutrients for rhizobacteria for survival in both storages and field.

Therefore, rhizobacteria applications should be packed in a carrier material that does not contain toxins and does not affect the ability of these rhizobacteria as the bioherbicide. The formulations of *P. putida* strain 30 and 180 in the talc carrier can still live in observation for six months with a density of colonies greater than 1 x $10^8$ cfu $g^{-1}$ [8]. In addition to *P. putida*, research suggests that *B. subtilis* in a talk formulation can still live with a density of 1.0 x $10^6$ cfu $g^{-1}$ colonies on observation 45 days after application [9].

## 2. Materials and Methods

The study consisted of two stages of testing the effectiveness of the carrier in increasing the shelf life of rhizobacteria as the first stage. Furthermore, in the second stage of rhizobacteria originating from storage for 20 weeks, it was taken and tested its stability as a bioherbicide on *A. conyzoides* weed. Both stages of this study were conducted on a laboratory scale. Research was conducted in Lab of Agronomy Faculty of Agriculture Halu Oleo University Kendari, from August to December 2016.

### *2.1. Effectiveness of carrier in the increasing shelf life of rhizobacteria*

The test was performed in a formulation, prepared by adding a carrier material (talc, bentonite, kaolin and ground burned rice husk) of 50 g each into 50 ml of a bacterial suspension with a density of $10^9$ $ml^{-1}$, then mixed evenly for 10 minutes. After that, it is inserted into a bottle of formulation and stored at room temperature ($22^0$C- $24^0$C). Determination of life bacteria in the formulation material is done every 14 days with spread method. The bacterial formulation of 1 g was diluted with sterile aquadest until $10^{-10}$ dilution. Furthermore, bacterial cultures of each formulation dilution of the formulation material are spread on TSA media. At the first 14 days observation, the dilution was carried out to $10^{-10}$ then dispersed at dilution levels of $10^{-9}$ and $10^{-10}$. It aims to see if bacteria still grow on the dilution. Each further observation remains dilution until level $10^{-10}$, except in the dilution there is no growing colony, the next observation is dilution at one level below ($10^{-9}$, $10^{-8}$, and so on). The observation of rhizobacterial life ability in each formulation was done until the age of 5 months in storage.

### *2.2. Stability of rhizobacteria in carrier as bioherbicide*

Rizobacterial isolates that have been stored in various carrier materials for 20 weeks are taken and dispersed in a petri dish. After 24 hours, rhizobacteria were grown using ose needle, then rizobacteria was reproduced by growing in TSA medium and incubated for 48 hours, then made in a liquid suspension with a cell density of $10^9 ml^{-1}$. Test of the ability of rizobacterial isolates as bioherbicides was performed by seed biopriming treatment on *A. conyzoides* seeds, disinfected seeds were inserted into rizobacterial isolate suspension and incubated for 24 hours using shaker. Weed seeds *A. conyzoides* which have been incubated in suspension for 24 hours Is added in a plastic plate which is dampened with a dampened sterile CD in an oven with $121^{\circ}$C for 24 hours, each cup planted with 25 weed seeds then placed in a germinator.

## 3. Results and discussion

### *3.1. Effectiveness of carrier in the increasing shelf life of rhizobacteria*

The growth patterns of rhizobacterial isolates of *B. lentus* A05 and *P. aeruginosa* A08 on various carrier types for 20 WAA (weeks after application) showed that the growth of DRB isolate up to the shelf life of 4 WAA experienced a logarithmic phase of a rapidly increasing population of DRB isolates. Furthermore, in the shelf life of 6 WAA, the growth of DRB isolates tended to have decreased the number of bacterial colonies up to the shelf life of 20 WAA. Observations on the shelf life of 2 to 20 WAA indicate that the isolate of B. *lentus* A05 can live on various carriers such as talc, bentonite, kaolin and ground burned rice husk (figure 1). As for the isolate rhizobacteria *P. aeruginosa* A08, able to live for 20 MSA in talc carrier material, bentonite and ground burned rice husk, but on kaolin carrier with a shelf life of 10 WAA have no live bacteria.

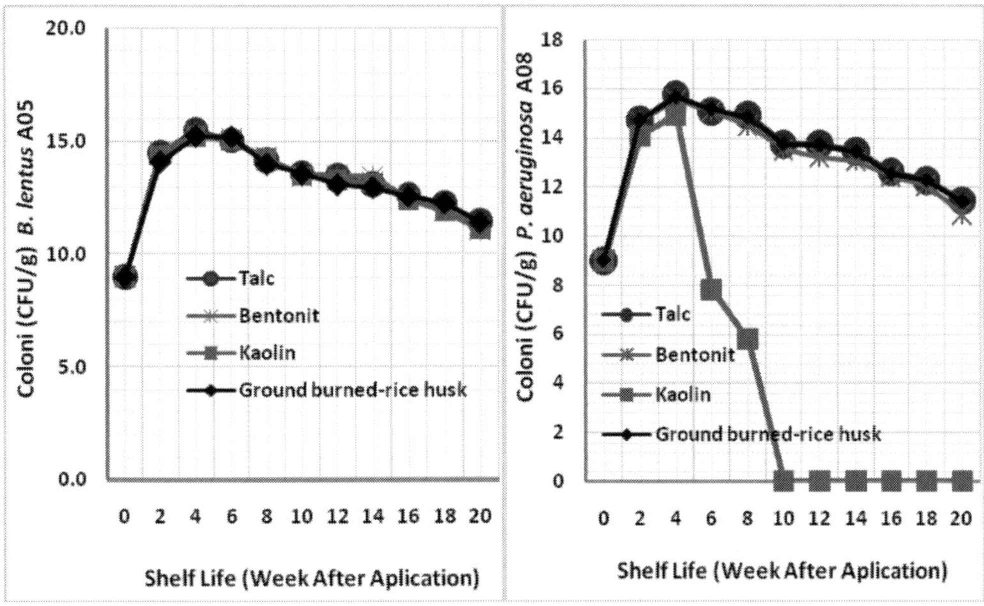


**Figure 1.** The growth model of *Bacillus lentus* A05 and *Pseudomonas aeruginosa* A08 in various rhizobacterial carriers.

The talc carrier material is called magnesium silicate ($Mg_3Si_4O_{10}(OH)_2$) and is available in powder form [10]. Chemical composition of $Mg_3SiO_{10}(OH)_2$ is magnesium 26.28%, silicon 10.10%, oxygen 63.36% and 0.36% hydrogen so that talc is a carrier that can provide nutrients for rhizobacterial viability, especially the fulfillment of nutrients O and Mg. Rhizobacteria need nutrients of C, O, N, H, P, and S, for element O required about 20%. Besides it also needed micro elements such as Mg [11]. The talc carrier is an appropriate material compared to tapioca starch, chitosan and corn flour to *Bacillus* sp [12].

The best carriers for *B. lentus* A05 and *P. aeruginosa* are also shown by ground burned rice husk. Analyzed of ground burned rice husk showed that the moisture content contained in it was 7.40%, C-organic was 7.51%, N total was 0.49%, $P_2O_5$ was 0.07%, $K_2O$ 0.08% with CEC 88.08 cmol $kg^{-1}$ and the pH of $H^2O$ was 6.73 [13]. The content contained in ground burned rice husk is very suitable with the nutritional needs of rhizobacteria both *B. lentus* A05 and *P. aeruginosa*. Microbes need a source of key nutrients such as carbon, protein, water, vitamins, and minerals to support its growth [11].

### *3.2. Stability of rhizobacteria in carrier as bioherbicide*

The carrier of rhizobacteria (talk, bentonite, and ground burned rice husk) in addition to improving the survival of rhizobacteria in storage for five months can also maintain the stability of rhizobacteria as bioherbicides. The results showed that *P. aeruginosa* A08 and *B. lentus* A05 derived from the carrier were still capable of suppressing the weed of *A. conyzoides* when compared with the control. However, the best treatment was demonstrated by the isolate *P. aeruginosa* A08 using talc and ground burned rice husk powder as the carrier. *P. aeruginosa* derived from the carrier of talc and ground burned rice husk, can inhibit germination of *A. conyzoides*, 77.92% (figure 2), inhibit root length of 79.53% and 81.89% respectively (figure 3). The talc and ground burned rice husk can maintain the stability of *P. aeruginosa* as bioherbicide.

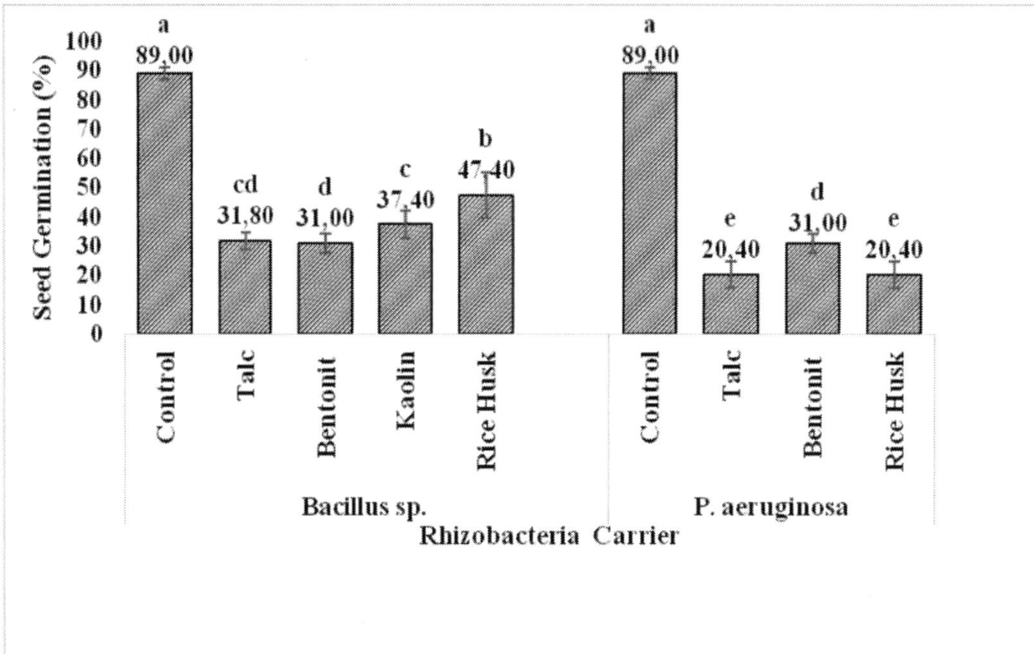


**Figure 2.** Effect of rhizobacterial isolate *Bacillus lentus* A05 and *Pseudomonas aeruginosa* A08 on vigor of *Ageratum conyzoides*

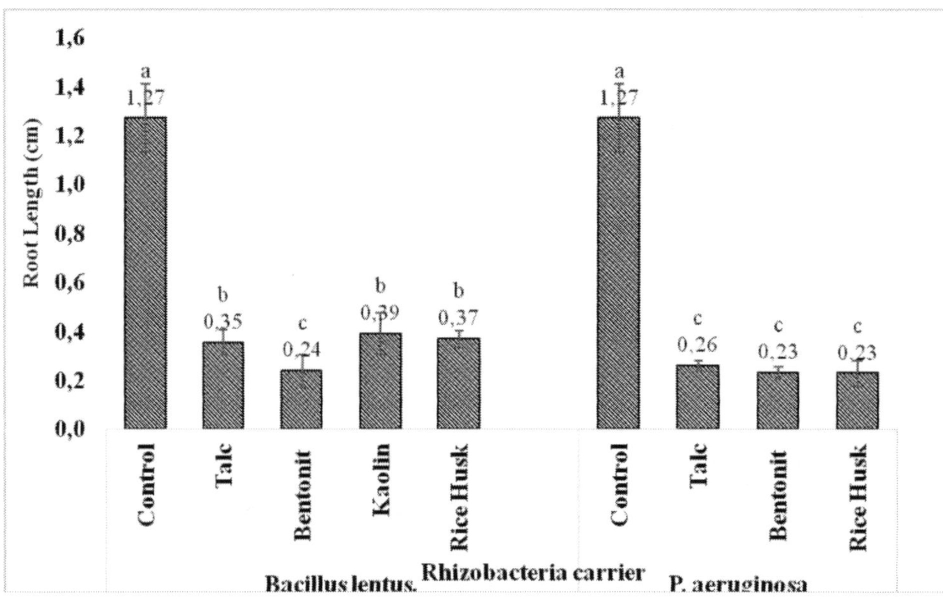


**Figure 3.** Effect of rhizobacterial isolate *Pseudomonas aeruginosa* on length root of *Ageratum conyzoides*

*P. aeruginosa* A08 has ability to inhibit germination and growth of *A. conyzoides* weeds. *P. aeruginosa* A08 isolate has a biochemical characteristic that produces HCN (*hydrogen cyanide*) with 5.26 ppm concentration [14-15]. Another research showed that *P. putida* produces HCN compounds that have capable to inhibit the growth of weeds *Abutilon theophrasti* [16]. *P. fluorescens* also can inhibit the growth of downy brome weed (*Bromus tectorum* L. Brote) to 86%, because *P. fluorescens* rhizobacteria produce hydrogen cyanide and *indole-3-acetic acid* [17].

Cyanide is a growth inhibitor that has been shown to play a role in plant respiration, carbohydrate metabolism, $CO_2$ and nitrate assimilation. HCN produced by DRB isolates in weed seed rhizosphere can cause phytotoxicity similar to that of herbicide. The isolate B. *lentus* A05 can also inhibit the germination and length root of *A. conyzoides* and produce high amounts of IAA compounds [14-15]. The mechanism inhibition of downy brome weeds by *P. fluorescens,* in addition to producing HCN is also due to the production of IAA (*indole-3-acetic acid*) with high concentrations [12].

## 4. Conclusions

The best carriers for rhizobacterial isolates potentially as bioherbicides were talc and ground burned rice husk; this is because the carrier can retain the shelf life of *B. lentus* A05 and *P. aeruginosa* A08 to 20 WAA. In addition to maintaining shelf life, the carrier can maintain the stability of rhizobacteria as bioherbicide, especially in *A. conyzoides.*

## References

[1] Kughur P G 2012 The effects of herbicides on crop production and environment in makurdi local government area of Benue State Nigeria *Journal of Sustainable Development in Africa* **14(4)** 206-216
[2] Mada D, Duniya N and Adam I G 2013 Effect of continuous application of herbicide on soil and with crop protection machinery in Southern Adamawa State. *International Refereed Journal of Engineering and Science* **2(6)** 44-49
[3] Khan M S, Chaudhry P, Wani P A and Zaidi A 2006 Biotoxic effects of the herbicides on growth, seed yield, and grain protein of greengram *J. Appl. Sci.Environ. Mgt.* **10(3)** 40-46
[4] Sembel D T 2010. *Pengendalian Hayati Hama-hama Serangga Tropis dan Gulma*. Penerbit Andi Yogyakarta. p282.
[5] Silva H S A, Romeiro R S R, Macagnan D, Vieira B A H, Pereira M C B and Mounteer A 2004 Rhizobacterial induction of systemic resistance in tomato plants: non-specific protection and increase in enzyme activities *Biol Control* **29** 288-295
[6] Yan Z, Ryu C M, McInroy J, Reddy M S, Woods F, Wilson M and Kloepper J W 2002. Induction of systemic resistance againts tomato late blight by PGPR *Phytopathology* **92(12)** 1329-33
[7] Kremer R J and Kennedy A C 1996 Rhizobacteria as biocontrol agent of weed *Weed Technol. USA* **10** 601-609
[8] Bora T, Ozaktan H, Gore E and Aslan E 2004 Biological control of *Fusariumoxysporumf.* sp. Melonis by wettable powder formulations of the two strains of *Pseudomonas putida. J. Phytopathology* **1529 (7)** 471-75
[9] Amer G A and and Utkhede R S 2000 Development of formulations of biological agents for management of root rot of lettuce and cucumber *Can. J. Microbiol* **46(1)** 809-16
[10] Nakkeeran S, Fernando W G D and Siddiqui Z A 2005 *PGPR: Biocontrol and Biofertilization.* Chapter 10. Plant Growth Promoting Rhizobacteria Formulations and Its Scope in Commercialization for the Management of Pests and Diseases. (Netherlands) pp: 257-296.
[11] Rahayu W P and Nurwitri C C 2012. *Mikrobiologi Pangan*. (Bogor : Penerbit IPB Press) 132p
[12] Devi HS, Munir E and Priyani N 2015 *Viabilitas Bacillus sp. BK17 pada Berbagai Bahan Pembawa* (Medan : Fakultas MIPA Universitas Sumatera Utara Medan)
[13] Nurbaity A, Setiawan A and Mulyani O 2011 Efektivitas arang sekam sebagai bahan pembawa pupuk hayati mikoriza arbuskula pada produksi sorgum. *Agrinimal* **1(1)** 1-6
[14] Rakian T C, Karimuna L, Taufik M, Sutariati G A K, Muhidin, Pasolon Y B 2015 The effectivesness of indigenous rhizobacteria as bioherbicide to control of weed *Australian Journal of Basic and Applied Sciences* **9(31)** 707-11
[15] Rakian T C 2015 *The Development of Deleterius Rhizobacteria as A Biological Control of*

*Agent of Weed (Bioherbicide) on Soybean Plant* (*Glycine max* L Merril) (Kendari : Dissertation on Halu Oleo University Kendari)

[16] Biswas M M, Zdor R and Miller C 2009 Using Gene Fusions to Study Cyanogenesis in a Weed Deleterious Rhizobacterium *Proceedings IXth International Bioherbicide Group Workshop* p20-22

[17] Kennedy C and Johnson B N 2001. Host range of a deleterious rhizobacterium for biological control of downy brome *Weed Science* **49** 792–97

# Relationship of some upland rice genotype after gamma irradiation

**N W S Suliartini[1], T Wijayanto[1], A Madiki[1], D Boer[1], Muhidin[1] and Juniawan[2]**

[1]Department of Agrotechnology, Faculty of Agriculture, Universitas Halu Oleo Kendari 93232 Southeast Sulawesi Indonesia;
[2]Indonesian Agriculture Training Center, Ketindan Malang, East Java Indonesia
Email : sri.suliartini@gmail.com

**Abstract.** The objective of the research was to group local upland rice genotypes after being treated with gamma irradiation. The research materials were upland rice genotypes resulted from mutation of the second generation and two parents: *Pae Loilo* (K3D0) and *Pae Pongasi* (K2D0) Cultivars. The research was conducted at the Indonesian Sweetener and Fiber Crops Research Institute, Malang Regency, and used the augmented design method. Research data were analyzed with R Program. Eight hundred and seventy one genotypes were selected with the selection criteria were based on yields on the average parents added 1.5 standard deviation. Based on the selection, eighty genotypes were analyzed with cluster analyses. Nine observation variables were used to develop cluster dendrogram using average linked method. Genetic distance was measured by euclidean distance. The results of cluster dendrogram showed that tested genotypes were divided into eight groups. Group 1, 2, 7, and 8 each had one genotype, group 3 and 6 each had two genotypes, group 4 had 25 genotypes, and group 5 had 51 genotypes. Check genotypes formed a separate group. Group 6 had the highest yield per plant of 126.11 gram, followed by groups 5 and 4 of 97.63 and 94.08 gram, respectively.

## 1. Introduction

Indonesia has 50 million hectares of dry land but only 35 million hectares that can be utilized for cultivation [1]. The problem of dry land is low soil fertility. Working with cassava plants showed that there was a yield loss of 6.2 tons ha$^{-1}$ due to low soil fertility [2]. One alternative solution to solve the problem of dry land is by cultivation of plants tolerant with the condition of the land, such as upland rice [3,4]. Upland rice (*Oryza sativa* L.) is rice cultivated on dry land. Upland rice has become an alternative in increasing rice production and has been widely consumed by the community. Upland rice has nutrient content [5] that is able to meet 63% of total energy sufficiency and 37% protein. Policies of Government of Indonesia through the increasing rice production [4] and also promote local food [6] as an alternative food.

The production of upland rice is still far below the production of wetland rice. Productivity of wetland rice was 4.79 tons ha$^{-1}$ [7], while upland rice was only 2.69 tons ha$^{-1}$. The low production of upland rice was due to the decrease of harvested area caused by land conversion, varied soil and climate conditions, high pest and disease rate, and not optimal application of cultivation technology, especially in the use of superior varieties. The most crucial problem faced by farmers in the cultivation of upland rice was the lack of availability of varieties and superior seeds [8]. In general, farmers cultivate local varieties which have good taste, tolerance to marginal land, resistant to several types of pests and diseases, require low input of fertilizer and easy and simple maintenance, although the production is still considered low.

Improvement of rice varieties continues to be done by researchers through plant breeding [4,5], both conventional and unconventional methods. Conventional assembly of varieties takes a long time and more than 5 years, more over by using a variety of varieties or parents that have desirable traits or the available population has low diversity for traits will be fixed. The method of mutation is one method to accelerate the assembly of new superior varieties. Induction of mutations is one way to increase plant diversity. Mutations are sudden and random changes in genetic material. The purpose of induction of mutation in plants is to create a population with a high level of genetic diversity [9], including the diversity of production supporting factors such as number of productive tillers, number of grains per panicle and 100 grain weight. Genetic diversity is very important in a plant breeding program to assemble a new superior variety [10]. New genotypes from mutation breeding can then be analyzed for their kinship relationship to facilitate further breeding programs [11].

Cluster analysis is useful to determine kinship and similarity between mutant genotypes. Genotypes in one group have a high degree of similarity, so if researchers want to select genotypes, they can choose one genotype per group depending on superiority of each group. Analyzing kinship relationships among local crops has been done, among others, on local rice in Central Java [12], local red rice from Central Kalimantan [13], taro [14]. However, similarity information between the parents and their derived mutants is still very limited. The cluster results will allow researchers to select which genotypes are re-selected, where the genotypes have almost the same characters, or close genetic distance, so one of them is selected for plant breeding further. This will improve research efficiency as it saves time, land and energy.

## 2. Materials and methods

The research materials were the second generation of upland rice mutants, *Pae Loilo* cultivar and *Pae Pongasi* cultivar as parent and check test, *Inpago* 7 and *Inpago Unram* as check tests. The research design used was Augmented Design which was divided into 6 blocks. The number of mutant genotypes was 18 genotypes. Genotype that already existed on one block was not repeated in the next blocks except check tests. Each genotype was planted as many as 50 plants, numbered according to the genotype code of the parent and the sequence of plants in the row. Observations were done to several parameters such as plant high, the number of productive tillers, the length of panicle, harvest age, the number of empty grains, the number of full grain, the weight of 1000 grains, and grain number per clump.

Research data were analyzed with R Program. There were 871 genotypes analyzed for selection. Selection was done based on production above 1.5 standard deviation plus the average of both parents. Based on the results of the selection, 84 genotypes were analyzed by cluster analysis to determine the relationship between genotypes. Genetic distance between genotypes was tested by the euclidean distance method.

## 3. Results and discussion

The kinship relationship or similarity relationship can be used as a foundation for cross breeding to obtain new superior varieties. The crosses will be better if it done between cultivars of distant kinship [14]. Kinship relationships can also be used to streamline the selection process. Similar genotypes or close kinship relationships are assumed to have similar genotypes and phenotypes that can be selected as a further breeding material.

A total of 871 genotypes were analyzed for selection. Selection was based on production above 1.5 standard deviation plus the mean of both parents ((yield = $\geq$ 60.694 + (1.5 * 18.009)). The selection resulted in 84 genotypes being analyzed to know the kinship relationship between upland rice genotypes.

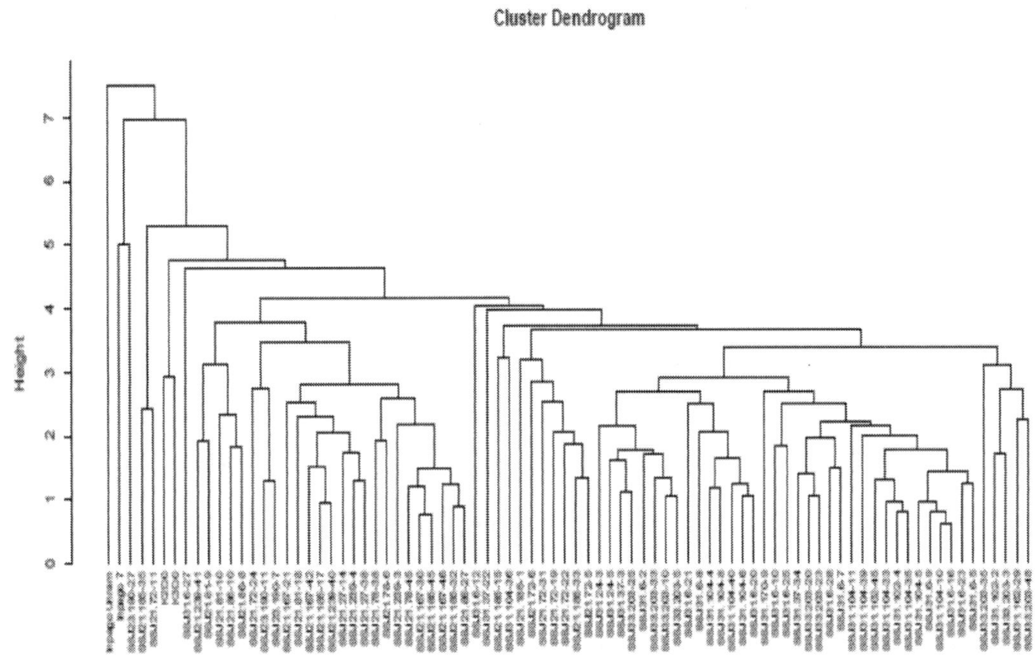


**Figure 1.** Dendrogram of upland rice genotypes.

Dendrogram as a result of cluster analysis is shown in Figure 1. Similarity analysis of 84 mutant upland rice genotypes on quantitative characters resulted in the division of 8 genotype groups reflecting similarities within groups or subgroups and dissimilarities between groups or subgroups. *Inpago Unram* (check variety) has the highest genetic distance (> 7) compared to other genotypes.

The high genetic distance indicates the greatest dissimilarity. Another check genotype, *Inpago* 7, has a higher similarity with SSJ 23.190-27 compared to other mutant genotypes. The genetic distance between the two genotypes namely *Inpago* 7 is the only high-yielding varieties of upland red rice released by the Indonesian government in 2011. As with *Inpago 7*, all mutant genotypes and parent are an upland red rice, contain anthocyanins from the group sianidin [15].

There were nine genotypes of upland rice in Southeast Sulawesi (*Paebiu Tamalaki, Paebiu Kolopua, Pae Tanta Mohalo, PaebiuTamalaki Mataiwoi, Paebiu Sitoro, Paebiu Tamalaki Pewutaa, Paebiu Kolopua Kosebu, Ranggo Hitam, and Hitam Lapodidi*) which had high anthocyanin content [16].

Group 1, 2, 7 and 8 consisted of one genotype (Figure 1; Table 1). This was due to the considerable dissimilarities between genotypes in each group with other group genotypes. *Inpago* 7 (group 1) and *Inpago Unram* (group 2) genotypes were superior varieties of upland rice that served as comparative plants. The separation of *Inpago Unram* into a separate subgroup with the highest genetic distance was because the genotype had a distinct difference compared to other genotypes. *Inpago Unram* was the shortest plant (125.2474 cm) and the shortest harvest age (111.59 day) compared to the other 7 groups. Group 7 (SSJ23.190-27) had the highest total number of grains but with the highest grain of hollow (90,000 grains) compared to the other groups. Group 8 (SSJ31.6-27) was the highest plant (193.00 cm) with the longest harvest age (138.00 day), the least productive tiller (16.00 tillers).

The advantage of this genotype was the highest 100 grain weight (4.25 g) (Table 2). The two parents (K2D0 and K3D0) formed a separate group from mutant genotypes. The genetic distance was 3. This suggested that gamma irradiation successfully altered the production and production support characteristics of the mutant genotypes. This also indicated a fairly high incidence between the genotypes of the parents and the mutants. In group 4, the highest dissimilarity was between genotype SSJ21.239-41 and genotype SSJ21.86-27. The group was subdivided into 2 subgroups i.e.4 subgroup 1 with the highest dissimilarity between genotype SSJ21.239-41 and genotype SSJ21.86-8 and 4 sub group 2 between genotype SSJ21.72-24 with genotype SSJ21.86-27. Subgroup 2 consisted of 2 sub-sub groups with a sufficiently large number of members (17 genotypes). The highest dissimilarity in sub-subgroup 1 occurred between the genotype SSJ21.167-21 and genotype SSJ21.27-38. The highest dissimilarity in sub-sub group 2 occurred between genotypes SSJ21.78-38 and genotype SSJ21.86-27.

**Table 1.** Name of genotype of upland rice in each group

| Group | Genotypes | Number of group members |
|---|---|---|
| 1 | Inpago 7 | 1 |
| 2 | Inpago Unram | 1 |
| 3 | K2D0; K3D0 | 2 |
| 4 | SSJ21.239-41 SSJ21.81-9 SSJ21.81-10 SSJ21.86-10 SSJ21.86-8 SSJ21.72-24 SSJ23.190-11 SSJ23.190-7 SSJ21.167-21 SSJ21.81-18 SSJ21.167-42 SSJ21.185-17 SSJ21.239-40 SSJ21.27-14 SSJ21.239-4 SSJ21.27-38 SSJ21.78-38 SSJ21.78-6 SSJ21.239-3 SSJ21.78-45 SSJ21.167-30 SSJ21.185-45 SSJ21.167-46 SSJ21.185-32 SSJ21.86-27 | 25 |
| 5 | SSJ31.6-12 SSJ31.37-22 SSJ21.185-15 SSJ31.104-36 SSJ21.185-1 SSJ21.72-6 SSJ21.72-31 SSJ21.72-19 SSJ21.72-22 SSJ21.185-33 SSJ21.72-5 SSJ31.24-3 SSJ31.24-5 SSJ31.37-3 SSJ33.203-25 SSJ31.6-2 SSJ33.203-33 SSJ33.203-10 SSJ33.203-5 SSJ31.6-21 SSJ31.6-8 SSJ31.104-4 SSJ31.104-8 SSJ31.104-40 SSJ31.104-6 SSJ31.6-20 SSJ31.170-9 SSJ31.6-10 SSJ31.6-25 SSJ31.37-34 SSJ33.203-20 SSJ33.203-23 SSJ31.6-28 SSJ31.6-7 SSJ31.104-11 SSJ31.104-39 SSJ31.102-45 SSJ31.104-33 SSJ31.102-4 SSJ31.104-35 SSJ31.104-5 SSJ31.6-9 SSJ31.104-10 SSJ31.6-16 SSJ31.6-23 SSJ31.6-5 SSJ33.203-35 SSJ31.102-2 SSJ33.203-3 SSJ31.162-29 SSJ33.203-48 | 51 |
| 6 | SSJ21.185-35; SSJ21.72-11 | 2 |
| 7 | SSJ23.190-27 | 1 |
| 8 | SSJ31.6-27 | 1 |

Group 5 had the largest number of member namely 51 genotypes (Figure 1; Table 1). This group was divided into two sub groups namely 5 sub group 1 consisting of one mutant genotype and 5 sub group 2 consisting of 50 mutant genotypes. The highest dissimilarity occurred between genotype SSJ31.6-12 and genotype SSJ33.203-48. The highest similarity existed between the genotype SSJ31.104-10 and SSJ31.6-16.

Group 6 consisted of 2 genotypes i.e.genotype SSJ21.185-35 and SSJ21.72-11, had the lowest number of empty grain (7.00 grains) and the highest number of filled out grains (248.00 grains). This caused the groups to have the highest yield (126.115 gclump$^{-1}$) (Table 2). This result was higher than 8 genotypes from crosses of upland rice with wetland rice obtained by that was 35.92 -61.61g clump$^{-1}$ [17].

**Table 2.** Characters of production and production support in each group

| No. Group | Harvest age (day) | Plant height (cm) | Productive tillers (tiller) | Number of filled out grains (grain) | Number of empty grains (grain) | Number of total grains (grain) | Long panicle (cm) | Yield (g clump$^{-1}$) | 100 grain weight (g) |
|---|---|---|---|---|---|---|---|---|---|
| 1 | 131.2727 | 151.80 | 22.34667 | 145.0865 | 55.95870 | 201.0452 | 27.27 | 66.91 | 2.58 |
| 2 | 111.5918 | 125.25 | 24.85537 | 130.2697 | 17.38421 | 147.6539 | 22.90 | 66.72 | 2.49 |
| 3 | 137.7500 | 178.32 | 19.08606 | 162.7965 | 16.36257 | 179.1590 | 24.37 | 61.24 | 3.08 |
| 4 | 132.6000 | 184.29 | 21.32000 | 208.7200 | 11.56000 | 220.2800 | 27.45 | 94.08 | 2.99 |
| 5 | 136.0196 | 176.53 | 25.66667 | 147.6275 | 11.41176 | 159.0392 | 26.83 | 97.63 | 3.67 |
| 6 | 129.0000 | 188.50 | 27.00000 | 248.0000 | 7.50000 | 255.5000 | 29.78 | 126.11 | 2.83 |
| 7 | 132.0000 | 181.30 | 22.00000 | 186.0000 | 90.00000 | 276.0000 | 24.70 | 91.59 | 2.91 |
| 8 | 138.0000 | 193.00 | 16.00000 | 197.0000 | 13.00000 | 210.0000 | 31.90 | 88.42 | 4.25 |

## 4. Conclusions

The results of cluster dendrogram showed that tested genotypes were divided into eight groups. Group 1, 2, 7, and 8 each had 1 genotype, group 3 and 6 each had 2 genotypes, group 4 had 25 genotypes, and group 5 had 51 genotypes.Check genotypes formed a separate group. Group 6 had the highest yield per plant of 126.11 gram, followed by groups 5 and 4 of 97.63 and 94.08 gram, respectively

## References

[1] Agus W and Suwardi 2015 *Status of Indonesian Soil Resources*. Asian Soil Partnership Consultation Wor on Sustainable Management and Protection of Soil Resources 13-15 May 2015, Bangkok, Thailand. 31p

[2] Kintche K, Hauser S, Mahungu N M, Ndonda A, Lukombo S, Nhamo N, Uzokwe V N E, Yomeni M, Ngamitshara J, Ekoko B, Mbala M, Akem C, Pypers P, Matungulu K P, Kehbila A and Vanlauwe 2017 Cassava yield loss in farmer fields was mainly caused by low soil fertility and suboptimal management practices in two provinces of the Democratic Republic of Congo *European Journal of Agronomy* **89** 107-23

[3] Muhidin, Jusoff K, Elkawakib S, Musa Y, Kaimuddin, Meisanti, Sadimantara G R, Baka L R 2013 The development of upland red rice under shade trees *World Applied Sciences Journal* **24(1)** 23-30

[4] Kadidaa B, Sadimantara G R, Suaib, Safuan L O, Muhidin 2017 Genetic diversity of local upland rice (*Oryza sativa* L) genotypes based on agronomic traits and yield potential in marginal land of North Buton Indonesia *Asian Journal of Crop Science* **9(4)** 109-117

[5] Suliartini N W S, Kuswanto, Basuki N, Soegianto A 2016 Superior lines candidates evaluation of two local red rice Southeast Sulawesi cultivars (Indonesia) derived from gamma rays irradiation techniques *International Journal of Plant Biology* **7** : 6475

[6] Muhidin, Leomo S, Alam S and Wijayanto T 2016 Comparative studies on different agroecosystem base on soil physicochemical properties to development of sago palm on dryland. International *Journal of ChemTech Research* **9(8)** 511-18

[7] BPS Sultra 2015 *Produksi Tanaman Padi dan Palawija di Sulawesi Tenggara* (Kendari : BPS Provinsi Sulawesi Tenggara) 72p

[8] Soerjandono N B and Robi'in 2012 Teknik pengujian galur harapan padi gogo *Buletin Teknik Pertanian* **17(1)** 7-9

[9] Poespodarsono S 1988 *Dasar-dasar Ilmu Pemuliaan Tanaman* (Bogor : Pusat Antar Universitas, Institut Pertanian Bogor)

[10] Mangoendidjojo W 2003 *Dasar-dasar Pemuliaan Tanaman* (Yogyakarta : Penerbit Kanisius Yogyakarta)

[11] Makmur A 1992 *Pengantar Pemuliaan Tanaman* (Jakarta : Penerbit Rineka Cipta Jakarta Cetakan 3)

[12] Susila A, Rustini S, Rohman E, Cempaka I G and Prasetya V E 2015 *Kekerabatan Kultivar Padi Lokal Jawa Tengah Berdasarkan Karakter Agronomi Dan Morfologi* Prosiding Seminar Nasional Sumber Daya Genetik Pertanian "Pengelolaan Sumber Daya Genetik Lokal sebagai Sumber Pertumbuhan Ekonomi Daerah" 2015 p 80-89

[13] Subekti A 2015 *Karakteristik Dan Pola Kekerabatan Plasma Nutfah Padi Beras Merah Di Kalimantan Barat. Prosiding Seminar Nasional Sumber Daya Genetik Pertanian* Prosiding Seminar Nasional Sumber Daya Genetik Pertanian "Pengelolaan Sumber Daya Genetik Lokal sebagai Sumber Pertumbuhan Ekonomi Daerah" 2015.p 118-124

[14] Hafsah, Hidayat T and Kusdianti 2014 The phenetic relationship among taro cultivar *(Colocasia esculenta)* based on vegetative morphological characters *Jurnal Bioslogos* **4(1)** 17-25

[15] Reddy A R 1996 *Genetic and molecular analysis of anthocyanin pigmentation pathway in rice* Proceedings of the third international rice genetics symposium. 16-20 Oct 1995. IRRI. Manila.Phillipines. p 16-20

[16] Suliartini N W S, Sadimantara G R, Wijayanto T and Muhidin 2011 Pengujian kadar antosianin padi gogo beras merah hasil koleksi plasma nutfah Sulawesi Tenggara *Jurnal Crop Agro* **4 (2)**:43-48

[17] Sadimantara G R, Muhidin, Ginting S and Suliartini N W S 2016 The Potential Yield of Some Superior Breeding Lines of Upland Rice of Southeast Sulawesi Indonesia. *Biosciences Biotechnology Research Asia* **13(4)** 1867-70

# The land use patterns for soil organic carbon conservation at Endanga watershed Southeast Sulawesi Indonesia

**S Leomo[1], S Ginting[1], L Sabaruddin[2], M Tufaila[1] and Muhidin[2]**

[1]Department of Soil Science, Faculty of Agriculture, Halu Oleo University, Kendari Southeast Sulawesi 93212 Indonesia
[2]Department of Agrotechnology, Faculty of Agriculture, Halu Oleo University, Kendari Southeast Sulawesi 93212 Indonesia
E-mail : sittileomo@yahoo.com

**Abstract**. The Endanga basin is one part of the Konaweeha watershed located in South Konawe, Southeast Sulawesi Province, covering an area of 1,353.67 hectares. The land use patterns in Endanga Watershed contained forests, shrubs, oil palm plantations, pepper fields, and cultivated fields of field rice, corn monoculture and intercropping of peanuts and corn. This watershed needs serious attention because most of its territory is on slope of 15-40%, with erosion hazard levels (EHL) varying from mild erosion to severe erosion. The loss of organic carbon (C-organic) soil is measured from the soil carried along with the surface stream and into the reservoir on various land uses. The result measurement of C-organic soil loss on forest land use is 14.02 kg $ha^{-1}$, shrubs land 22.71 kg $ha^{-1}$, oil palm 151.32 kg $ha^{-1}$, pepper garden 93.69 kg $ha^{-1}$, field rice 313.80 kg.$ha^{-1}$, monoculture of maize 142.44 kg $ha^{-1}$, intercropped maize and corn 51.10 kg $ha^{-1}$ and open land 1,909.16 kg $ha^{-1}$. The forest land and shrubs is best in conserving soil C-organic, but economically unfavorable for the community, so land use pattern for intercropping and pepper plantation can be used for soil C-organic conservation

## 1. Introduction

Development progress has increased the conversion of forest to agricultural or non-agricultural land. This conversion is unavoidable as development progress increases the demand of better quality of life. The conversion seems rational from economical point of view with direct short-term benefit that can be obtained, yet this does not consider long-term environmental impact from the loss of forest or protected area [1-2]. Conversion of forest or grassland into farms causes the loss of soil organic carbon. Grassland and forest suffer 20–50% loss of their soil organic carbon after developed for 40–50 years [3].

Organic materials are essential in soil ability to support plants, thus the decrease of soil organic will result in the degradation of soil productivity. Soil degradation caused by decrease of organic materials is a frequent problem in developing countries, since the amount and intensity of degraded soil increase as the development is progressing. Soil degradation in on-site area of a watershed will reduce soil productivity as it will decrease soil fertility and its water retention ability, which subsequently affect production, ecological, and hydrology functions of the watershed [1, 4-6].

Endanga watershed is a part of Konaweha watershed located in South Konawe Region, Southeast Sulawesi Province, which covers ± 1,353.67 ha area. The land use in Endanga watershed includes forest, shrubs, intercropping farming, upland rice, and sago palm plantation [7]. Endanga watershed is a safety and preservation buffer of Konaweha watershed that serves as buffer zones for protecting water quality and flood control of Konaweha River. The Endanga watershed requires serious attention as most of its area is located in 15-40% slope with erosion hazard varied from low to high level [8].

Erosion causes loss of top soil that contains organic materials and nutrients. It also has better physical and biological properties than lower soil. Indonesia has high soil erosion rate of agricultural land with slope 3-15%, with the number reaches 97.5–423.6 t $ha^{-1}$ [9]. In reality, there are many agricultures in Indonesia located in area with slope more than 15%, thus their soil erosion rate is likely to be significantly higher. Hence, land use need to consider soil and water conservation to prevent or reduce erosion that will degrade soil productivity.

This study was aimed to measure the number of soil organic carbon (C-organic) transported/lost by erosion, calculate economic value of loss of soil organic material, review efficiency of land use based on soil organic components that are preserved by Endanga watershed.

## 2. Materials and Methods

This study was conducted in Endanga watershed is a part of Konaweha watershed located in South Konawe Region, Southeast Sulawesi Province, which covers $\pm$ 1,353.67 ha area. Geographically, it is located in $04^0$5'20"- $04^0$9'46" South and $122^0$14'49" – $122^0$16'42" East. Soil analysis was conducted in Soil Laboratory, Faculty of Agriculture, Halu Oleo University and Laboratory of Testing, Soil Research Center, Agricultural Land Resources Research Center, Bogor.

Approximately 84% (1,131.24 ha) of Endanga watershed was located in 15–40 % slope. Dominant types of soil are Ustult (84.42%) and Tropept (15.58%), formed from various parent materials such as conglomerate, sandstone, phyllite, and quartz. Average annual rainfall for 10 years (2005–2014) was 2,388.96 mm, the lowest average monthly rainfall was 62.93 mm on September and the highest average was 305.58 mm on March [10].

Soil samples for C-organic measurement were taken from erosion sedimentation, which was directly conducted in erosion plot (sized 22.1 x 2 m, 9% slope). A "soil collector" or a sedimentary reservoir was placed in the lower end of the plot. Erosion plots were used to simulate land use of forest, shrubs, oil palms, pepper, upland rice, corn monoculture, intercropping of corn and peanuts, and a clearing with no vegetation. Observation was conducted every day after rain, which causing run-off, occurred.

Soil C-organic concentration was stated in percentage based on Walkley and Black method on every type of cover crop. Soil C-organic data was multiplied by the amount of eroded soil and converted into organic materials (C-organic x 1.724). Soil C-organic data was then tested statistically by ANOVA followed by BNT test with $\alpha = 0.05$. Economic value of loss of soil C-organic was calculated by the number of C-organic transported by eroded soil converted into organic material, then it was used to calculate the cost of loss of soil C-organic, by multiplying the number of C-organic loss with the price of organic fertilizer used by farmers (Rp 3,000 $kg^{-1}$).

## 3. Results and discussion

### *3.1. Erosion and soil C-organic concentration*

Erosion causes the transportation of soil particles, nutrients, and organic materials. The number of soil C-organic lost depended on the amount of erosion the concentration of C-organic in the sediment. The result of erosion and soil C-organic concentration calculation in Endanga watershed is shown in Table 1. Table 1 shows the amount of sedimentation in various types of land use varied from 0.86 ton $ha^{-1}y^{-1}$ to 328.46 ton $ha^{-1}y^{-1}$. Forest and shrubs had the lowest amount of erosion, because they are considered as dense vegetation with litter in soil surface, which covers soil surface from raindrop impact, thus, soil is protected from destruction of soil aggregates.

This in turns prevent or reduce the number of run-off and erosion. Forest conversion to oil palms plantation or monoculture of upland rice and corns increased the amount of eroded soil. Lower and denser vegetation is more effective in protecting the soil from erosion. Vegetation effectiveness in reducing the number of run-off and erosion is influenced by canopy height, canopy area, vegetation density, and root density [1,11]. Vegetation is essential factor influencing the amount of erosion, and this factor can be easily changed by humans. Vegetation is able to cover soil surface from raindrop impact, to prevent soil dispersion or destruction of soil aggregates.

**Table 1.** The amount of erosion and percentage of soil C-organic lost in various types of land uses in Endanga watershed

| No. | Land Use | Erosion (ton $ha^{-1}$ $y^{-1}$) | Soil C-organic (%) | Soil organic material (%) |
|---|---|---|---|---|
| 1 | Forest | $0.86^{e}$ | 1.63 | 2.81 |
| 2 | Bush | $3.39^{e}$ | 0.67 | 1.15 |
| 3 | Oil palms | $18.64^{c}$ | 0.81 | 1.40 |
| 4 | Pepper | $6.25^{de}$ | 1.50 | 2.58 |
| 5 | Paddy fields | $48.27^{b}$ | 0.65 | 1.12 |
| 6 | Corn monoculture | $14.58^{cd}$ | 0.98 | 1.68 |
| 7 | Intercropping of corn and peanuts | $8.88^{de}$ | 0.58 | 0.99 |
| 8 | No vegetation | $323.46^{a}$ | 0.59 | 1.02 |

*Source: Calculation and laboratory analysis results in 2015*
Note: Numbers followed by similar letter did not have real difference in BNT test with $\alpha = 0.05$

The highest number of soil C-organic and organic materials concentration in erosion sedimentation was found in plots with forest and pepper, with the numbers for other types of land use were low. The high number of C-organic lost in forest caused by organic materials from litter on soil surface. Most of organic materials in soil surface are decomposed and mineralized, thus the number of organic materials in soil surface is high. In land with pepper plantation, the high number of organic carbon is caused by the addition of organic fertilizer in flowering phase.

### *3.2. Economic value of loss of soil c-organic*

Economic value of loss of soil C-organic is translated as an indirect value of one type of land use. The estimation of economic value was aimed to measure how much the loss of C-organic cost in Indonesian Rupiah. To obtain economic value of loss of soil C-organic, the number of C-organic lost was converted to organic material or organic fertilizer with specific price in order to calculate economic value of each type of land use. The price of organic fertilizer in area of Endanga watershed is Rp 3,000 $kg^{-1}$. The number of soil C-organic lost was directly proportional to the amount of erosion (Table 2).

The economic value of loss of soil C-organic due to erosion per hectare is shown in Table 2. As can be seen in Table 2, the highest loss of soil C-organic occurred in land with no vegetation, while in contrast, the lowest number was in forest. Meanwhile, in farmland, the highest loss occurred in upland rice while the lowest occurred in intercropping of corn and peanuts.

Forest conversion to oil palms plantation or monoculture of paddy fields and corns yielded the high number of soil C-organic lost. On the contrary, land covered with intercropping plants reduced the number of soil C-organic lost. A land with no vegetation had 3,291.39 kg $ha^{-1}$ loss of soil C-organic, or economically, the loss cost Rp. 9,874,160.- per hectare. Meanwhile, the loss of organic materials in forest was 24.17 kg $ha^{-1}$, with Rp. 72,510.- economic value.

In farmland, land with upland rice had the highest economic loss of soil C-organic (Rp. 1,622,970) compared to oil palms, pepper, corn monoculture, and intercropping of corn and peanuts. In contrast, land with intercropping plants had low economic loss of soil C-organic (Rp. 264,290.-), this number was nearly as low as the economic cost of land with bushes. The low economic cost was influenced by the cover of soil surface from dense vegetation, which protects the soil from raindrop impact, and subsequently reduces erosion.

**Table 2.** Economic value of loss of soil C-organic due to erosion per hectare in various types of land use in Endanga watershed

| No. | Land Use | Soil C-organic (kg ha$^{-1}$) | Soil organic material (kg ha$^{-1}$) | Economic value (Rp ha$^{-1}$) |
|---|---|---|---|---|
| 1 | Forest | 14.02$^{d}$ | 24.17$^{d}$ | 72,510 |
| 2 | Shrubs | 22.71$^{d}$ | 39.15$^{d}$ | 117,460 |
| 3 | Oil palms | 151.32$^{c}$ | 260.88$^{c}$ | 782,630 |
| 4 | Pepper | 93.69$^{cd}$ | 161.52$^{cd}$ | 484,560 |
| 5 | Upland rice | 313.80$^{b}$ | 540.99$^{b}$ | 1,622,970 |
| 6 | Corn monoculture | 142.44$^{c}$ | 245.57$^{c}$ | 736,700 |
| 7 | Intercropping of corn and peanuts | 51.10$^{cd}$ | 88.10$^{cd}$ | 264,290 |
| 8 | No vegetation | 1,909.16$^{a}$ | 3,291.39$^{a}$ | 9,874,180 |

Note: Numbers followed by similar letter did not have real difference in BNT test with $\alpha = 0.05$

### *3.3. Efficiency of land use in soil c-organic retention*

The loss of soil C-organic (organic materials) due to erosion can be minimized by proper cropping pattern, since it will protect soil surface from raindrop impact, which in turns reduces run-off and erosion, and subsequently minimizes the loss of soil organic material. This means that organic material retention ability of the soil and its productivity is maintained. The low number of soil organic material lost means that farmers could reduce organic fertilizer costs, which will bring economic benefits for them. The economic value of organic material retained by soil with various types of land use in Endanga watershed is displayed in Table 3.

**Table 3.** Economic value of retained organic materials in Endanga watershed

| No | Land use | Number of retained organic materials (kg ha$^{-1}$) | Economic value of retained organic materials (Rp ha$^{-1}$) |
|---|---|---|---|
| 1 | Forest | 3,267.22 | 9,801,660 |
| 2 | Shrubs | 3,252.23 | 9,756,690 |
| 3 | Oil palms | 3,030.51 | 9,091,530 |
| 4 | Pepper | 3,129.87 | 9,389,610 |
| 5 | Upland rice | 2,750.41 | 8,251,230 |
| 6 | Corn monoculture | 3,045.82 | 9,137,460 |
| 7 | Intercropping corn and peanuts | 3,203.30 | 9,609,900 |
| 8 | No vegetation | 0.00 | 0.00 |

Note : Unit cost of organic material = Rp. 3,000/kg

Table 3 shows forest is type of vegetation with the highest efficiency in retaining organic materials, with the number reached 3,267.22 kg ha$^{-1}$ or its economic value was Rp. 9,801,660.-. Land with paddy field was ineffective in retaining organic materials, thus from economic point of view, paddy field is had low benefit as farmers need to spend more money for fertilizer. The number of retained organic materials in paddy field was 2,750.41 kg ha$^{-1}$ with economic value of Rp. 8,251,230.-.

Based on the economic value, forest and shrubs had high economic value from soil and water conservation point of view. However, as population grows, it is impossible for a watershed area to be maintained as forest and bush. Therefore, land with intercropping plants can be a choice for agricultural crop as it showed the highest ability in retaining soil C-organic materials or low cost of soil C-organic lost. The use of land for upland rice, oil palms, pepper, and corn plantation should be intercropped with other cover crops to reduce erosion and subsequently minimizing cost of loss of soil organic materials.

## 4. Conclusions

It is concluded that (1) The number of soil C-organic lost in various types of land use in Endanga watershed in ascending order was 14.02 kg ha$^{-1}$ (forest), 22.71 kg ha$^{-1}$ (shrubs), 51.10 kg ha$^{-1}$ (intercropping of corn and peanuts), 93.68 kg ha$^{-1}$ (pepper), 142.44 kg ha$^{-1}$ (corn), 151.32 kg ha$^{-1}$ (oil palms), 313.80 kg ha$^{-1}$ (upland rice), and 1,909.16 kg ha$^{-1}$ (no vegetation); (2). The number of economic value of C-organic lost (per hectare) due to erosion in Endanga watershed in ascending order was Rp 72,510.- (forest), Rp. 117,480.- (shrubs), Rp. 264,270.- (intercropping of corn and peanuts), Rp. 484,560.- (pepper), Rp. 736,710.- (corn), Rp. 782,640.- (oil palms), Rp. 1,622,940.- (upland rice), and Rp. 9,874,170.- (no vegetation); (3). Types of land use with efficient ability of retaining soil C-organic were forest, bush, intercropping plants, and pepper plantation.

## References

[1] Jiang X, Huang C H and Ruan F 2008 Impacts of land cover changes on runoff and sediment in the Cedar Creek Watershed, St, Joseph River, Indiana, United States, *Journal of Mountain Science-England* **5** 113–21
[2] Solaimani K, Modallaldoust S and Lotfi S 2009 Investigation of land use changes on soil erosion process using geographical information system Int. J. Environ Sci. Tech. 6 415-24.
[3] Bruce J P, Frome M, Haites E, Janzen H, Lal R and Paustian K 1999 Carbon Sequestration in Soils *J. Soil and Water Conservation* **54** 382–89
[4] Gao J 2008 Detection of changes in land degradation in northeast China from landsat TM and Aster Data, The International Archives of the Photogrammetry *Remote Sensing and Spatial Information Sciences* **37** 1533-38
[5] Meusburger K, Banninger D and Alewell C 2010 Estimating vegetation parameter for soil erosion assessment in an alpine catchment by means of QuickBird imagery *International Journal of Applied Earth Observation and Geoinformation* **12** 201-07
[6] Kovacs A S, Fulop B and Honti M 2012 Detection of hot spots on soil erosion and reservoir siltation in ungauged Mediterranean catchments *Energy Procedia* **18** 934-43
[7] Balai Pengelolaan Daerah Aliran Sungai Sampara 2012 *Pola Rehabilitasi Lahan dan Konservasi Tanah* (Kendari : DAS Sampara)
[8] Leomo S, Ginting S, Sabaruddin L and Tufaila M 2016 Estimation of erosion hazard level using universal soil loss equation (USLE) method in Endanga Watershed Southeast Sulawesi Indonesia *Advances in Environmental Biology* **10(1)** 101-106
[9] Banuwa IS 2013 *Erosi* (Jakarta : Edisi Pertama, Kencana Prenada Media Group, Jakarta)
[10] Leomo S 2016 *Model Tindakan Konservasi Tanah dan Air pada DAS Endanga Sulawesi Tenggara (*Kendari : Disertasi Program Pascasarjana Universitas Halu Oleo Kendari)
[11] Arsyad S 2010 *Konservasi Tanah dan Air* (Bogor : Edisi Kedua IPB Press) Bogor

IOP Conf. Series: Earth and Environmental Science **122** (2018) 012035 doi:10.1088/1755-1315/122/1/012035

# Evaluation of land suitability for shallot (*A. ascalonicum* L.) and orange (*Orange sp.)* at Harian District of Samosir Regency

**Bintang, Supriadi and E Tampubolon**

Departement of Agroecotechnology, Agriculture Faculty, USU, Medan 20155
E-mail: bintangsitorus@ymail.com

**Abstract.** Evaluation of land suitability for shallot *(Allium ascalonicum* L.) and lemon (*Citrus sp.*) aimed to determine the level of suitability to the plants that would be cultivated at Harian District, Samosir Regency. Operation of soil type, altitude map and slope maps with scale of 1:50000 obtained nine Land Map Units namely LMU1 and 2 – SampurToba village; LMU3 - JanjiMartahan village; LMU4 – Turpuk Limbong village; LMU5 – Sosor Dolok village; LMU6,7,8 and 9 in PartungkoNaGinjang village. Land suitability assessment criteria based on Bogor Soil Research Center Staff 1983 and the limit method by Djaenuddin 2011.The results showed that shallot crop, there is only on LMU 2 (Sampur Toba) that has barier that can be overcome, such as actual class S3 (n, r); S2 (eh) had an inhibiting factor pH and slope that can be managed with liming and terracing. Orange crop on LMU 1 can be overcome by giving organic material; it might be overcome with terracing on LMU 3; and managed by liming and terracing on LMU 8. On the other LMU (4, 5, 6, 7, 9) are not suggested for orange cultivation.

## 1. Introduction

Land evaluation is the process of estimating the potential of land for alternative kinds of utilization. The process of land classification can basically be carried out with two approaches or methods, namely limiting factor method and parametric method. In the limiting factor method, each land characteristic or land quality is arranged sequentially from the best (with the lowest limit) to the worst (top obstacle). Each classes are organized in the criteria table in such a way in terms of articular use that the smallest limiting factor for the best class and the biggest one for the worst class [1]

Land suitability assessment distinguishes the actual land suitability carried out under conditions of current land use without input and the potential land suitability performed under condition after given improvements such as fertilizer increase, irrigation, terraces managment depending on the type of limiting factors. Assessment of land suitability is done by matching land quality and land characteristics (physical and chemical properties of the land) as a parameter to the criteria of the land suitability classes which have been prepared based on terms of use or growing crops or evaluated agricultural commodities [2,3] .

Shallot *(Allium ascalonicum* L.) and Orange *(Orange sp.)* have quite high sale value or market price. Most people grow other varieties namely leek and batak shallot (chieves) but their market values are lower than shallot's because they are sold in freshness and not shelf-resistant. People want to grow shallots with which larger capital so as will not take the risk to farming [4]. Orange is currently popular as consumption for a healthy life and has a good market price compared to the price of orange fruits as the table dishes. Farmers aim to work on it but wish the certain parties would provide information on whether the plant can be cultivated. Based on the matters, the research was conducted in an effort to evaluate the suitability of land at Harian Subdistrict for shallot and orange. Land suitability evaluation was performed by matching soil property and the data of the research area with the grow requirements of the plants and the prominet limiting factors.

Harian District with its capital Harian Boho, is one of 9 districts in Samosir Regency. Harian District at coordinates 2°30'-3°45' LU and 98°30'-98°49' BT with the altitude of 900-1.847.5 meters above sea level is the largest district in Samosir Regency with the area of 56,045 Ha (38.81% of Samosir Regency). Most people subsist from agriculture and forest products. Common types of farming are potato, cabbage, coffee, chili and other crops. Dry land area ranges from 494.5 ha while paddy field area is around 833.0 ha. Topographical conditions are hilly and bumpy to the mountain, the soil structure is unstable and is in the path of tectonic and volcanic earthquakes [5]

The objective of the research is to determine the level of land suitability at Harian District of Samosir Regency to shallot and orange and to supply alternative practical management in an effort to increase the production of both crops.

## 2. Materials and Methods

Place and time of research held was in Harian District of Samosir Regency, Laboratory of Chemistry and Soil Fertily and Laboratory of Research and Technology at Agriculture Faculty of University of Sumatera Utara.

Land suitability assessment based on the criteria of Bogor Soil Research Center in 1983 and the limit method referring to the limiting factors on land characteristics based on [2] . Implementation begun with determination of Land Map Unit (LMP) from overlay of land type map, slope altitude map 1:50000. On each Land Map Unit drilling was performed as representative of the land character. The data of physical environmet, physical and chemicl properties of the soil included: soil texture, length and slope, soil drainage, crude material, soil / solum depth, dominant vegetation, crop management, soil conservation efforts, flood / inundation hazards, surface rocks and rock outcrops .

## 3. Results and Discussion

**Table 1.** Evaluation of Land Suitability for Shallot *(Allium ascalonicum* L.)

| Usage Requirements/ Land Characteristics | Land Map Unit | | | | | | | | |
|---|---|---|---|---|---|---|---|---|---|
| | 1 | 2 | 3 | 4 | 5 | 6 | 7 | 8 | 9 |
| | ST | | JM | TL | SD | | PNG | | |
| Temp (tc) | | | | | | | | | |
| The average temperature ($^0$C) | S1 | S1 | S2 | S1 | S1 | S2 | S2 | S3 | S3 |
| Oxygen avaibility (oa) | | | | | | | | | |
| Drainage | S1 | S1 | S1 | S2 | S1 | S1 | S1 | S1 | S1 |
| Rooting channel (rc) | | | | | | | | | |
| Texture | S3 | S1 | S3 | S3 | N | n | S3 | S3 | S3 |
| Base Saturation | S1 | S1 | S1 | S1 | S1 | S3 | S1 | S2 | S1 |
| Crude material (%) | S1 | S1 | S1 | S2 | S2 | S1 | S1 | S1 | S1 |
| Soil depth (cm) | S1 | S1 | S1 | S1 | S1 | S1 | S1 | S1 | S1 |
| Nutrient Retention (nr) | | | | | | | | | |
| CEC clay (cmol) | S2 | S1 | S1 | S2 | S1 | S1 | S1 | S1 | S1 |
| Base saturtion (%) | S1 | S1 | S1 | S1 | S1 | S3 | S1 | S2 | S1 |
| pH $H_2O$ | S3 | S3 | S1 | S3 | S2 | S3 | S3 | S3 | S3 |
| C-organic (%) | S3 | S2 | S1 | S3 | S1 | S1 | S1 | S1 | S1 |

| | | | | | | | | | |
|---|---|---|---|---|---|---|---|---|---|
| Toxicity (xc) | | | | | | | | | |
| Salinity (dS/m) | S1 | S1 | S1 | S1 | S1 | S1 | S1 | S1 | S1 |
| Sodicity (xn) | | | | | | | | | |
| Alkalinity /ESP (%) | S1 | S1 | S1 | S1 | S1 | S1 | S1 | S1 | S1 |
| Erosion hazard (eh) | | | | | | | | | |
| Slope (%) | S1 | S2 | S2 | n | S2 | S2 | n | S2 | n |
| Erosion hazard | S1 | S1 | S1 | S2 | S2 | S1 | S1 | S1 | S1 |
| Flood Hazard (fh) | | | | | | | | | |
| Puddle | S1 | S1 | S1 | S1 | S1 | S1 | S1 | S1 | S1 |
| Land Preparation (lp) | | | | | | | | | |
| Rocks on surface (%) | S1 | S1 | S1 | S1 | S1 | S1 | S1 | S1 | S1 |
| Rock outcrop (%) | S1 | S1 | S1 | S1 | S1 | S1 | S1 | S1 | S1 |

Explanation :
ST : Sampuran Toba village ; JM : Janji Martahan village ; TL : Turpuk Limbong village ;
SD : Sohor Dolok village ; PNG : Partungko Na Ginjang village

**Table 2.** Evaluation of Land Suitability for Orange Crop *(Orange sp.)*

| Usage Requirements/ Land Characteristics | Land Map Unit | | | | | | | | |
|---|---|---|---|---|---|---|---|---|---|
| | 1 | 2 | 3 | 4 | 5 | 6 | 7 | 8 | 9 |
| | ST | | JM | TL | SD | | PNG | | |
| Temp (tc) | | | | | | | | | |
| The average temperature ($^0$C) | S1 | S2 | S1 | S1 | S1 | S1 | S1 | S1 | S2 |
| Oxygen avaibility (oa) | | | | | | | | | |
| Drainage | S1 | S1 | S1 | S1 | S1 | S1 | S1 | S1 | S1 |
| Rooting channel (rc) | | | | | | | | | |
| Texture | S1 | S1 | S1 | S1 | N | N | S1 | S1 | S1 |
| Base Saturation | S1 | S1 | S1 | S1 | S1 | S2 | S1 | S1 | S1 |
| Crude material (%) | S1 | S1 | S1 | S2 | S2 | S1 | S1 | S1 | S1 |
| Soil depth (cm) | S1 | S1 | S1 | S2 | S1 | S1 | S1 | S1 | S1 |
| Nutrient Retention (nr) | | | | | | | | | |
| CEC clay (cmol) | S2 | S1 | S1 | S1 | S1 | S1 | S1 | S1 | S1 |
| Base saturtion (%) | S1 | S1 | S1 | S1 | S1 | S2 | S1 | S1 | S1 |
| pH $H_2O$ | S1 | S1 | S1 | S1 | S2 | S3 | S2 | S2 | S3 |
| C-organic (%) | S3 | S2 | S1 | S3 | S1 | S1 | S1 | S1 | S1 |
| Toxicity (xc) | | | | | | | | | |
| Salinity (dS/m) | S1 | S1 | S1 | S1 | S1 | S1 | S1 | S1 | S1 |
| Sodicity (xn) | | | | | | | | | |
| Alkalinity /ESP (%) | S1 | S1 | S1 | S1 | S1 | S1 | S1 | S1 | S1 |
| Erosion hazard (eh) | | | | | | | | | |
| Slope (%) | S1 | S2 | S2 | N | S2 | S2 | n | S2 | N |
| Erosion hazard | S1 | S1 | S1 | S2 | S2 | S2 | S2 | S1 | S1 |
| Flood Hazard (fh) | | | | | | | | | |
| Puddle | S1 | S1 | S1 | S1 | S1 | S1 | S1 | S1 | S1 |
| Land Preparation (lp) | | | | | | | | | |
| Rocks on surface (%) | S1 | S1 | S1 | S1 | S1 | S1 | S1 | S1 | S1 |
| Rock outcrop (%) | S1 | S1 | S1 | S1 | S1 | S1 | S1 | S1 | S1 |

Explanation:

ST : Sampuran Toba Village ; JM : Janji Martahan Village ; TL : Turpuk Limbong Village ;
SD : Sohor Dolok Village ; PNG : Partungko Na Ginjang Village

The evaluation of shallot crops showed that at some land map units there were the severe limiting factor (LMU 8 & 9) the inappropriate soil texture or marginal (S3 at LMU 1, 3, 4, 7, 8, 9 and N at LMU 3). The barrier which can be resolved (n, r) reaching S3. However, soil pH and organic matter can be managed by liming and applying solid or liquid compost. The obstacle of slope can be handled by constructing terrace (S2 at LMU 2, 3, 5, 6 and 8) but the slope is not appropriate for LMU 4, 8 and 9 because it reaches more 30%. Thus, only at Sampur Toba village LMU 2 has the barrier which can be solved.

Orange has the inappropriate barrier to soil temperature at LMU 2 and 9 (S2 tc) which is very tough to change; soil texture (Nr, c) for LMU 6 and 7 and slope more 30% (Neh) at LMU 4, 7 and 9. The blockade which be able to deal with to increase the land potential found at LMU 1, 3 and 8 wherein each LMU the obstacle of nr can be controlled with appropriate input and for the obstacle of slope by terrace.

## 4. Conclusions

The results showed that the shallot crop is only on LMU 2 at Sampur Toba village that has barrier that can be overcome, such as actual class S3 (n, r); S2 (eh) had an inhibiting factor pH and slope that can be managed with liming and terracing.

On orange crop, class S3 (nr) on LMU 1 can be overcome by giving organic material; class S2 (eh) on LMU 3 might be overcome with terracing; and class S2 (nr, eh) on LMU 8 can be managed by liming and terracing. On the other LMU (4, 5, 6, 7, 9) is not suggested for orange cultivation.

Oranges has the smallest constraint on LMU 1- Sampuran Toba Village, it can be overcome by organic treatment. LMU 3 – Janji Martahan village, the potensial suitability of the land might be overcome with terracing; LMU 8 in Partungko Na Ginjang village be managed by liming and terracing. On the other LMU (4,5,6,7 and 9) are not suggested for cultivation.

## References

[1] Hakim, N., M.Y. Nyakpa, A.M. Lubis, S.G. Nugroho, M.A. Diha, G.B. Hong *dan* H.H. Baailey, 1986. Dasar-Dasar Ilmu Tanah., Universitas Lampung Press, Lampung.

[2] Djaenudin, D., Marwah., S. Subagjo., 2011. Petunjuk Teknis Evaluasi Lahan untuk Komoditas Pertanian. Balai Besar Penelitian dan Pengembangan Tanah, Bogor.

[3] Ritung, S. Wahyunto, F., dan H. Hidayat, 2007. Panduan Evaluasi Kesesuaian Lahan Balai Penelitian Tanah Bogor

[4] Deptan, 2003. Pengembangan Usaha Agribisnis Bawang merahTerpadu, Direktorat Tanaman Sayuran, Hias dan AnekaTanaman. Direktorat Jenderal Bina Produksi Hortikultura. Deartemen Pertanian Jakarta.

[5] Badan Pusat Statistik Kecamatan Harian, 2014. Statistik Daerah Kecamatan Harian. Badan Pusat Statistika Kabupaten Samosir

## Appendix

**Appendix 1.** Characteristics of Land Suitability for Shallot Crop (Allium cepa L)

| Usage Requirements/ Land Characteristics | Land Suitability Classes | | | |
|---|---|---|---|---|
| | S1 | S2 | S3 | N |
| Temp (tc) | | | | |
| The average temperature ($^0C$) | 20-25 | 18-20 / 25-30 | 15-18 /30-35 | <15 / >35 |

| | | | | |
|---|---|---|---|---|
| Water avaibility (wa) | | | | |
| Rainfall (mm/tahun) | 350-600 | 300-350 / 600-800 | 250-300 / 800-1600 | <250/>1600 |
| Length of dry period (month) | | | | |
| Humidity (%) | | | | |
| Oxygen avaibility (oa) | Good. Slightly | Middle | Hamperd | quite hamperd |
| Drainage | hampered | | | Fast |
| Rooting channel (rc) | | | | |
| Texture | fine. slightly fine | midle | slighly rough | rough |
| | < 15 | 15-35 | 35 - 55 | > 55 |
| | > 50 | 30 – 50 | 20 – 30 | < 20 |
| Crude material (%) | | | | |
| Soil depth (cm) | | | | |
| Nutrient Retention (nr) | | | | |
| CEC clay (cmol) | > 16 | ≤ 16 | - | - |
| Base saturtion (%) | > 35 | 20 - 35 | < 20 | - |
| pH $H_2O$ | 6.0 – 7.8 | (5.8 - 6.0/ 7.8 - 8.0) | < 5.8 > 8.0 | - |
| C-organic (%) | > 1.2 | 0.8 – 1.2 | < 0.8 | - |
| Toxicity (xc) | | | | |
| Salinity (dS/m) | < 2 | 2 - 3 | 3 – 5 | > 5 |
| Sodicity (xn) | | | | |
| Alkalinity /ESP (%) | < 20 | 20 - 35 | 35- 50 | > 50 |
| Erosion hazard (eh) | | | | |
| Slope (%) | < 8 | 8-16 | 16 - 30 | > 30 |
| Erosion hazard | Lowest | low middle | severe | very severe |
| Flood Banjir (fh) | | | | |
| Puddle | F0 | - | - | >F1 |
| Land Preparation (lp) | | | | |
| Rocks on surface (%) | < 5 | 5 - 15 | 15 - 40 | > 40 |
| Rock outcrop (%) | < 5 | 5 - 15 | 15 – 25 | > 25 |

**Appendix 2**. Characteristics of Land Suitability for Orange Crop (*Orange sp*)

| Usage Requirements/ Land Characteristics | Land Suitability Classes | | | |
|---|---|---|---|---|
| | S1 | S2 | S3 | N |
| Temp (tc) | | | | |
| The average temperature ($^0$C) | 19 - 33 | 33 – 36<br>16 – 19 | 36 – 39<br>13 – 16 | < 39<br>> 13 |
| Water avaibility (wa) | | | | |
| Rainfall (mm/tahun) | 1.200 - 3.000 | 1.000 - 1.200<br>3.000 - 3.500 | 800 - 1.000<br>3000 - 4000 | < 800<br>> 4.000 |
| Length of dry period (month) | 2.5 – 4<br>50 – 90 | 4-5<br><50; >90 | 5 - 6 | > 6 |
| Humidity (%) | | | | |

IOP Conf. Series: Earth and Environmental Science **122** (2018) 012035 doi:10.1088/1755-1315/122/1/012035

| | | | | |
|---|---|---|---|---|
| Oxygen avaibility (oa) | | | | |
| Drainage | Good. somewhat | slightly obstructed | obstructed; slightly fast | very obstructe. fast |
| Rooting channel (rc) | | | | |
| Texture | smooth. slightly smooth. | - | slightly smooth | rough |
| Crude material (%) | medium. | 15-35 | 35 - 55 | > 55 |
| Soil depth (cm) | slightly rough < 15 > 100 | 75 – 100 | 50 – 70 | < 50 |
| Nutrient Retention (nr) | | | | |
| CEC clay (cmol) | > 16 | ≤ 16 | - | - |
| Base saturtion (%) | > 35 | < 35 | - | - |
| pH $H_2O$ | 5.6 – 7.0 | 5.2 - 5.6 7.0 - 8.0 | < 5.2 > 8.0 | - |
| C-organic (%) | > 1.2 | 0.8 – 1.2 | < 0.8 | - |
| Toxicity (xc) | | | | |
| Salinity (dS/m) | < 3 | 3 – 2 | 5 – 6 | > 6 |
| Sodicity (xn) | | | | |
| Alkalinity /ESP (%) | < 25 | 25 – 35 | 35- 45 | > 45 |
| Erosion hazard (eh) | | | | |
| Slope (%) | < 8 | 8-16 | 16 - 30 | > 30 |
| Erosion hazard | lowest | low. middle | severe | quite severe |
| Flood Banjir (fh) | | | | |
| Puddle | F0 | - | - | >F0 |
| Land Preparation (lp) | | | | |
| Rocks on surface (%) | < 5 | 5 - 15 | 15 - 40 | > 40 |
| Rock outcrop (%) | < 5 | 5 – 15 | 15 – 25 | > 25 |

*Source : Djaenudin. et al.. 2011*

# Growth and yield model for non-timber forest product of kemenyan (*Styrax sumatrana* J.J. Sm) in Tapanuli, North Sumatra

**Aswandi and C R Kholibrina**

Forestry and Environment Research Development Institute of Aek Nauli, Kampus Kehutanan Aek Nauli. Jl Raya Parapat Km 10.5 Sibaganding, Parapat Simalungun North Sumatra, Indonesia. 21174.
E-mail: andiasw@yahoo.com

**Abstract**. Kemenyan is *Styrax* tree resin, the main of non-timber forest product commodity in Lake Toba catchment area, North Sumatra since hundreds years ago. However, there are lack of information about the growth and yield prediction for this tree species. The objective of study is to construct the growth and yield models for *Styrax sumatrana* in Tapanuli region, North Sumatra. Measurement data from 20 temporary plots were used to formulate stand diameter and height equations, and to project the incense production. The highest Current Annual Increment (*CAI*) of diameter occurs in the stand's age 21 to 25 years (1.00 cm/year). The growth of diameter declines significantly to 0.48 cm/year in age 46 to 50 years, and decrease to 0.26 cm/year at age 50 years up. The intersection of *CAI* and *MAI* curves occur in stand age 31 to 35 years. It shows that the optimal growth occurs in this period. The average of incenses production was 318.59 g/tree/year. The optimum incense production was achieved when the diameter growth was maximal and tapping scars accumulation was limited.

## 1. Introduction

Yield regulation is an important aspect in sustainable forest management. Estimation of growth, yield projection, and harvest intensity are some of regulatory tools that influence the silviculture options, length of management cycle, and effectiveness of forest management activities. These estimations are generally conducted by model approach. In timber management, the models are built on the mathematical relationships of stand parameters consist of stand density, age, and site productivity [1]. The growth model provides a quantitative description of forest stand development due to the influence of management actions which applied over a time period [2].

Over the last few decades, community forest management with non-timber forest products as the main commodity have not supported by the proper yield regulation [3, 4]. One of NTPFs management that have a long history in the Lake Toba catchment area, North Sumatra is the Kemenyan (*Styrax*) forests [5]. Kemenyan or incense is resin from *Styrax* tree. The resin contains compounds that used for industrial raw materials, medicines, cosmetics, natural insecticides, and food and beverage preservatives [5, 6,7]. It have cultivated by most communities in Tapanuli since hundred years ago. However, there are lack of information about the growth and yield prediction for this tree species.

The objective of study is to construct the growth and yield models for *Styrax sumatrana* in Tapanuli region, North Sumatra. The models can be applied projection of resin production and to determine an appropriate silviculture options.

## 2. Materials and Methods

### *2.1. Study sites*

The research was conducted on *Styrax* forests in North Tapanuli and Humbang Hasundutan district, North Sumatra. These community forests are commonly grow in areas with hilly topography with varying altitude ranging from 933 to 1,488 m above sea level. The average of annual rainfall is 3200 mm, daily temperature during the day 22.2 to 30.3°C and relative humidity 55.59 to 82.90%. The development of measurement plots were conducted in May until November 2012.

### *2.2. Data collection*

Growth data were obtained by measurement of diameter and height on 20 temporary plots. Plots 0.1 ha (20 m x 50 m) were constructed in several stand conditions with stand age range from 8 to 65 years old. All data were calculated as one population. There has been limited standardized calculation for NTFPs productivity [3, 4, 8]. In this study, the productivity was calculated by collecting the incenses from all tapping points from sample trees in a harvest period. The variables measured were diameter, height, number of tapping points and scars.

### *2.3. Data analysis*

The age classes (*AC*) of stand is set by 5-year interval from *AC I* (<10 years) to *AC X* (> 50 years). The determination of *AC I* which start at tree age <10 years were due to the average of trees began to produce a productive incense at 8 years old. This interval refers to forest management of *Pinus merkusii* in Indonesia, and also applied by Calama *et al.* [8] to study ecological processes in measuring the NTPFs productivity such as flowering, leaf growth, and exudates production. In this study, the incense production ($W_{TIG}$) is calculated by summing the incenses from all the tapping points in a harvest period.

$$W_{TIG} = \sum_{i=1}^{n} Wig \quad (1)$$

Remarks: $W_{TIG}$= incense production (g/tree); $Wig$ = incense production per tapping point (g)

Current Annual Increments (*CAI*) and Mean Annual Increments (*MAI*) of diameter, height and incense production are calculated by the formula:

$$CAI = Y_{t+1} - Y_t \quad (2)$$

$$MAI = \frac{Yt}{t} \quad (3)$$

Remarks: $Yt$ = diameter, height, and incense production at time $t$, $t$ = age.

The stand models were based on the mathematical relationship between diameter and height of stand with age using the equation ln $Y$ = a + b 1/$X$ with nested regression analysis.

$$\ln D = a + b\ 1/A \quad (4)$$

$$\ln H = a + b\ 1/A \quad (5)$$

Remarks: $D$ = Dbh (cm); $H$ = total height (m); $A$ = Age (years).

Referring to various studies [9,10], the models were constructed base on the regression between resin productivity with stand's parameters such as diameter, height and harvest intensity. The intensity of harvesting is a function of the number of tapping points measured and the accumulation of scars due to tapping.

$$W_{TIG} = f\,(Dbh, N_{NT}, N_{BL}) \quad (6)$$

Remarks: $D$ = Dbh (cm); $N_{BL}$ = Number of scars tapping (N); and $N_{NT}$ = Number of new tapping point.

Regression analysis was performed using the MINITAB 14 for Windows program. The built models were tested for their reliability through the variance of regression and coefficient of determination. Model validation performed on the Aggregate of Difference-*AgD* with criteria less than 1% and the Average of Percentage Deviation (*AvR*) was less than 10%.

## 3. Result and Discussion

### *3.1. Data distribution*

The average of stand diameter and height of 463 sample trees derived from 20 temporary measurements plots are listed in Table 1. The mean of diameter (*Dbh*) is 21.4 cm, while the mean of trees height ($H_{tot}$) is 15.8 m.

**Table 1**. Descriptive statistics of some stand variables

| Age Class | | Number of trees | Average of Dbh (cm) | Average of Total Height (m) |
|---|---|---|---|---|
| KU I | < 10 years | 22 | 08.16 | 7.5 |
| KU II | 11-15 years | 75 | 12.35 | 12.9 |
| KU III | 16-20 years | 111 | 16.55 | 14.9 |
| KU IV | 21-25 years | 88 | 21.56 | 15.9 |
| KU V | 26-30 years | 86 | 26.55 | 16.9 |
| KU VI | 31-35 years | 45 | 31.11 | 18.0 |
| KU VII | 36-40 years | 23 | 35.23 | 20.3 |
| KU VIII | 41-45 years | 8 | 39.23 | 21.9 |
| KU IX | 46-50 years | 4 | 41.64 | 25.0 |
| KU X | > 50 years | 1 | 42.93 | 26.0 |
| Total average | | | 21.35 | 15.8 |

The distribution of stand diameter, height, age, incenses productivity and the relationship between each stand parameter are shown in Figure 1.

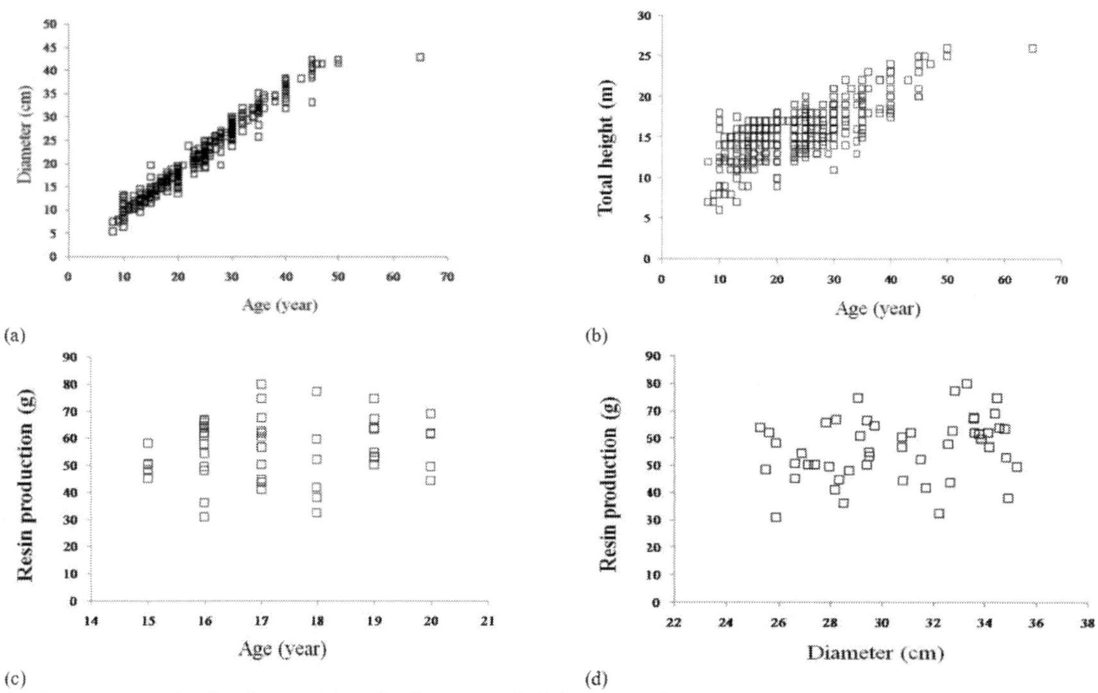


**Figure 1**. distribution of stand diameter, height, age, incenses productivity and the relationship between each stand parameter. (a) *Dbh* distribution by age; (b) $H_{tot}$ distribution by ages; (c) resin production by age; and (d) resin production by *Dbh*.

The relationship between diameter and stand age tend sigmoid, whereas stand height on age tends to logistics. It illustrates the principle of developmental biology of stand where growth would tend to increase in initial period and slowed down after a certain time period or when it reaches a certain growth point [2]. According to figure 1, the incense production tends to spread to stand age and diameter. These suggest that age (*A*) and diameter (*Dbh*) have a poor ability to explain the variance of incense production ($W_{TIG}$). The difficulty to understanding the behavior of natural ecosystems, especially NTFPs lead to the prediction are often imprecise and difficult [10,11].

### *3.2. Growth of Diameter and Height*

The standgrowth is shown in table 2 and figure 2. The highest *CAI* diameter occurs in *CA IV*at 1.00 cm/year, decreasing significantly in *AC IX* (46-50 years) at 0.48 cm/year, and the lowest at 0.26 cm/year in *AC X* (> 50 years). The *MAI* diameter ranging from 0.82 to 0.95 cm/year, thus the tree with diameter> 20 cm is achieved at age over 21 years. The *MAI* height ranges from 0.51 to 0.99 m/year.

**Table 2.**The stand growth for *S. sumatrana* in Tapanuli, North Sumatera

| Age Class (year) | Diameter at breast height (Dbh) | | | Total Height (m) | | |
|---|---|---|---|---|---|---|
| | Average *Dbh* (cm) | *CAI* (cm/year) | *MAI* (cm/year) | Average of $H_{tot}$ (m) | *CAI* (m/year) | *MAI* (m/year) |
| AC I | 08.16 | 0.82 | 0.82 | 7.50 | 0.80 | 0.80 |
| AC II | 12.35 | 0.84 | 0.95 | 12.93 | 1.09 | 0.99 |
| AC III | 16.55 | 0.84 | 0.92 | 14.96 | 0.41 | 0.83 |
| AC IV | 21.56 | 1.00 | 0.94 | 15.88 | 0.18 | 0.69 |
| AC V | 26.55 | 1.00 | 0.95 | 16.91 | 0.21 | 0.60 |
| AC VI | 31.11 | 0.91 | 0.94 | 17.99 | 0.22 | 0.54 |
| AC VII | 35.23 | 0.82 | 0.93 | 20.28 | 0.46 | 0.53 |
| AC VIII | 39.23 | 0.80 | 0.91 | 21.88 | 0.32 | 0.51 |
| AC IX | 41.64 | 0.48 | 0.87 | 25.00 | 0.63 | 0.52 |
| AC X | 42.93 | 0.26 | 0.86 | 26.00 | 0.20 | 0.52 |

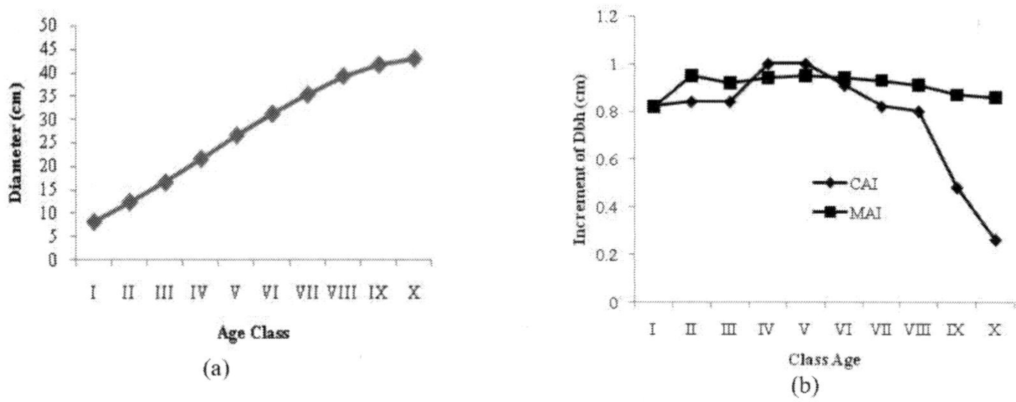


**Figure 2**. Curve of diameter growth (a) and CAI and MAI curves (b).

The diameter growth curve (figure 2a) illustrates the sigmoid pattern. The growth tends to be slow at the beginning of growth, increasing linearly to exponential at the mid-period, and slow down in late period. This pattern is also indicated by most of forest trees growth in various parts of the world. Furthermore, Figure 2b shows the intersection of *CAI* and *MAI* curves on 31 to 35 years old. In timber regulation, the intersection of these curves has important roles in determining the forest management rotation. This point represents the optimal growth, so that rotation is usually determined at that period. The incense productivity is influenced by several factors such as the intensity of harvesting (the number of tapping points and tapping scars). Therefore, the area of productive tapping is affected by the tree diameter, thus intersection of *CAI* and *MAI* curves remains a concern. The decline in incense productivity may be thought to be due to a decrease in the growth rate.

### *3.3. Growth and Yield Model*

Based on the regression analysis were obtained prediction model results as follows:

a. Equation of diameter estimation:

$$\ln D = 3.88 - 18.0\ 1/A \qquad (R^2_{adj} = 90.3) \qquad (7)$$

a. Equation of height estiamation

$$\ln H_{ot} = 3.11 - 7.64\ 1/A \qquad (R^2_{adj} = 49.9) \qquad (8)$$

$$H_{tot} = 9.23 + 0.307\ D \qquad (R^2_{adj} = 55.7) \qquad (9)$$

b. Incense production estimation

$$W_{TIG} = 23.4 + 0.895\ Dbh + 11.4\ N_{NNT} - 0.604\ N_{BL} \qquad (R^2_{adj} = 63.6) \qquad (10)$$

Remarks: $D$ = dbh (cm); $H$ = Total height (m); $A$ = Age (year); $W_{TIG}$ = incense production (g/tree/year); $N_{BL}$ = Number of scars tapping (N); and $N_{NT}$ = Number of new tapping point.

The equation (9) is considered reliable in predicting the growth and projection the diameter at a certain age. Regards$R^2_{adj}$ value = 90.3, the agecan explain variance of diameter with a confidence level 90.3%. The models have also met the validated model criterias with *AgD*<1% and *AvR*<10%.

**Table 3**.Statistics precision for growth and incenses production equations

| No. | Equation | $R^2_{adj}$ | *AgD* (%) | *AvR* (%) |
|---|---|---|---|---|
| 1 | $\ln D = 3.88 - 18.0\ 1/A$ | 90.3 | 0.91 | 9.28 |
| 2 | $\ln H_{tot} = 3.11 - 7.64\ 1/A$ | 49.9 | 1.31 | 11.90 |
| 3 | $H_{tot} = 9.23 + 0.307\ D$ | 55.7 | (-)0.03 | 11.50 |
| 4 | $W_{TIG} = 23.4 + 0.895\ Dbh + 10.4\ N_{NT} - 0.604\ N_{BL}$ | 63.6 | (-)0.21 | 11.76 |

According to $R^2_{adj}$ = 49.9, *AgD*> 1% and *AvR*> 10%, equation (8) has a relatively low reliability. The stand age can only explain 49.9% of the total height, whereas the remaining 51.1% of height can be affected by other variables. The $R^2_{adj}$ = 55.7 on the equation (9), it consequences that height variance is only 57.7% influenced by diameter. Different to diameter, stand height is not affected by silviculture treatments such as stand density. These low values reflect variations in the quality of growing sites.

The equation (10) is considered quite reliable in estimating the incense production ($R^2_{adj}$ = 63.6). Although it has an *AvR* value 11.76%, higher than validation criteria, considering the peculiarities of the incense characteristics as NTPFs and there are limited model for this commodity, the equation can be suggested as a reliable model. The equation indicates that tree diameter (*Dbh*) and harvesting intensity ($N_{NT}$) are directly proportional to incense productivity. However, increasing the unproductive surface area to the addition of tapping scars ($N_{BL}$) can be a limiting the productivity.

The optimal production is achieved at maximum diameter growth and minimal tapping scars accumulation. These conditions also have shown by the incense productivity of *Boswellia papyrifera* in Africa [10]. Increasing the tapping intensity in the short term will increase the resin productivity, especially in larger trees. However, Rijkers *et al.* [12] explains that high harvesting intensity will disrupt the vitality, reproductive system and stand productivity in long-term.

The highest productivity is found on larger diameter trees with the smallest number of tapping scars. In larger trees with little tapping scars, an increase in harvesting intensity by adding new tapping points will increase the resin productivity. But, It will decrease the proportion of productive tapping fields, and as the consequence it will disrupt the growth and productivity in the long term.

The average of incense production is 159.28 g/tree/harvest period. The harvesting carried out every six month or twice a year, where the average of resin productivity is 318.59 g/tree/year. The new tapping point is 5 to 10 points/harvest, where incense production is 11.37 g/point/harvest. The productivity is not vary with the previous studies. Sasmuko [13] obtained an average yield at 12.11

g/tapping point, while Waluyo [14] obtained a lower average at 9.54 g. This lower production is thought to be due to the smaller sample trees diameter (13 to 23 cm).

The low correlation between incense productivity to diameter (*r*=0.27) also were showed by Waluyo (1993) for *S. benzoin* in North Tapanuli (*r*=0.35). The prediction of *Pinus helepensis* resin production in Greece also shown a relatively low correlation (*r*=0.56) [9]. According to Purnomo [11], the reliable models for natural forest are limited due to their ecosystem behavior cannot be fully understood so that the predictions can be inappropriate and difficult.

## 4. Conclusion

The highest *CAI* occurs in 21-25 age class at 1.00 cm/year and decreases significantly to 0.26 cm/year in age class over 50 years. The intersection of *CAI* and *MAI* curves occurs in stand age 31-35 year old age. This indicates optimal diameter growth occurs over that time period. The stand age can describe the variance of diameter with high level of accuracy. Conversely, stand height cannot be explained by age. Accumulation of tapping scars becomes a limiting the incense productivity. The optimal incense production is achieved at maximum growth of diameter and minimal tapping scars accumulation. The average incense productivity is 159.28 g/tree/harvest period, and the average incense production is 318.59 g/tree/year. The average incense productivity is 11.37 g/tapping point/harvest.

## References

[1] Pretzsch H, Dieler J, and Rőtzer T, 2012. *Growth and Defense in Plants*. Ecological Studies 220. R. Matyssek *et al.* (eds). Principles of Growth Partitioning Between Trees in Forest Stand Under Stress. Springer-Verlag Berlin Heidelberg.

[2] Rahman MM and Ahmad IU, 2000. Growth and Yield Model of Gamar (*Gmelina arborea*). *Journal of Tropical Forest Science,* 12(2):276-286.

[3] Lemenih M and Kassa H, 2011. *Opportunities and challenges for sustainable production and marketing of gums and resins in Ethiopia*. CIFOR, Bogor, Indonesia.

[4] do Valle DR, Staudhammer CL, Cooper Jr WP, 2007. Simulating Nontimber Forest Product Management in Tropical Mixed Forests. *Journal of Forestry,* 105(6):301-306.

[5] Susilowati A, Hendalastuti H, Kholibrina CR, Ramadhani R. 2017. Weak delineation of *Styrax* species growing in North Sumatra, Indonesia by *matK* + *rbcL* gene. Biodiversitas, 18(3): 2085-4722. DOI: 10.13057/biodiv/d180353.

[6] Heyne K, 1987. *Useful Plants of Indonesia* Volume I, Forestry Research Center, Jakarta.

[7] Aswandi, Kholibrina CR. *Restoring Lake Toba Ecosystem*. Bina Media Perintis, Medan.

[8] Calama R, Tomé M, Sánchez-González M, Miina J, Spanos K, and Palahí M, 2010. Modelling non-wood forest products in Europe: a review. *Forest Systems*, 19(SI): 69-85.

[9] Spanos KA, Gaitanis D, Spanos I, 2009. *Proceedings of the 14th Pan-hellenic Forestry Conference*. Resin production in natural Aleppo pine stands in N. Evia-Greece and possibilities for increasing income under sustainable forest management. (Helenic Forestry Society. November, 1-4, 2009, Patra, Greece).

[10] Lemenih M, Feleke S, TadesseW, 2007. Factors constraining the production and marketing of frankincense by local people in Metema district, North-Western Ethiopia. *Journal of Arid Environments*, 71: 393–403.

[11] Purnomo H, 2004. *Theory of Complex Systems, Modeling and Resource Management*. Forest Management Department Faculty of Forestry Bogor Agricultural University.

[12] Rijkers T, Ogbazghi W, Wessel M, Bongers F, 2006. The effect of tapping for frankincense on sexual reproduction in *Boswellia papyrifera*. *J. of Appl. Ecology,* 43: 1188–1195.

[13] Sasmuko SA, 1996. The effect of tapping method on Incense resin production (*Styrax sumatrana* J.J. SM). *Buletin Penelitian Kehutanan*, 12(2), October 1996. Pematangsiantar.

[14] Waluyo TK, 1993. The relationship between the tree diameter and resin production of incense tapping. *Buletin Penelitian Kehutanan*, 9(4) December 1993. Pematangsiantar.

# The soil sulphate effect and maize plant (*Zea mays* L.) growth of sulphate reducing bacteria (SRB) inoculation in acid sulfate soils with the different soil water condition

**Asmarlaili S, A Rauf, D S Hanafiah, Y Sudarno and P Abdi**

Faculty of Agriculture, Universitas Sumatera Utara, Medan 20155
E-mail: assaharhanafiah@yahoo.com

**Abstract.** The objective of the study was to determine the potential application of sulphate reducing bacteria on acid sulfate soil with different water content in the green house. The research was carried out in the Laboratory and Green House, Faculty of Agriculture, Universitas Sumatera Utara. This research used Randomized Block Design with two treatments factors, ie sulphate reducing bacteria (SRB) isolate (control, LK4, LK6, TSM4, TSM3, AP4, AP3, LK4 + TSM3, LK4 + AP4, LK4 + AP3, LK6 + TSM3, LK6 + AP4, LK6 + AP3, TSM4 + TSM3, TSM4 + AP4, TSM4 + AP3) and water condition (100% field capacity and 110% field capacity). The results showed that application of isolate LK4 + AP4 with water condition 110% field capacity decreased the soil sulphate content (27.38 ppm) significantly after 6 weeks. Application of isolate LK4 + AP3 with water condition 110% field capacity increased soil pH (5.58) after-week efficacy 6. Application of isolate LK4 with water condition 110% field capacity increased plant growth (140 cm; 25.74 g) significantly after week 6. The best treatment was application isolate LK4 with water condition 110% field Capacity (SRB population $2.5x10^8$; soil sulphate content 29.10ppm; soil acidity 4.78; plant height 140cm; plant weight 25.74g).

## 1. Introduction

Based on Bos [1], there are about 12 million ha of acid sulphate soil worldwide and 1.5 million ha of which are found in Indonesia. Adhi and Alihamsyah [2] also reported if combined this soil association with peat and saline soils of acid sulphate in Indonesia to 6.7 million ha . Sludge sulphate is widely used for rice crop (232 ha), annual crops (59,237 ha) and annual crops / plantations (4.0 million ha) but faces many obstacles, including high soil acidity, increased solubility of toxic elements such as Al and Fe and low nutrient availability.

Konsten dan Sarwani [3] explained that the pyrite ($FeS_2$) which is widely contained in acid sulphate soils is stable if it is in a reductive condition, but if the sulfuric soil is drained the pyrite will oxidize, causing the formation of $H_2SO_4$ compounds which can increase soil acidity, in this condition soil pH can reach Less than 3.5 (pH <3.5). The reaction equation that describes the oxidation of pyrite and causes soil acidification is as the following:

$$2FeS_2 + 15/2O_2 + 4H_2O \longrightarrow Fe_2O_3 + 4SO_4^{-2} + 8H^+$$

Widyati [4] stated that arrangement of land and water system in accordance with the characteristics of the land, selection of appropriate commodities and varieties, and the application of appropriate amelioration and fertilization technology is a comprehensive effort that can be done to ensure the successful management of sulphate sulphate into productive agricultural land. The results showed that if sulphate sludge is properly managed, almost all food commodities, horticulture, and plantations are well developed and can produce well . High acidity has a negative impact on chemical properties and soil microbial activity because not all soil microbes are able to survive in very acid conditions. Therefore, a technology package is required to improve and improve the productivity of the land. Sulphate Reducing Bacteria (SRB) is a bacteria that can survive in acidic conditions. The bacteria can be utilized to increase the productivity of acid sulphate soil and SRB likes acid conditions then the bacteria can also be applied to sulfur contaminated environments such as post-coal mining land, drainage water and sludge paper waste disposal sites.

Hanafiah et al [5] stated that sulphate reducing bacteria are anaerobic obligate bacteria that use $H_2$ as an electron donor (chemolithotroph). SRB can reduce sulphate under anaerobic conditions into sulfides, then the resulting $H_2S$ can precipitate toxic metals (Cu, Zn, Cd) as sulphide metal. SRB requires organic substrates derived from short-chain organic acids such as pyruvic acid. Under natural conditions, the acid is generated by other anaerobic activities.

Widyati [4] explained that in sulphate reduction process, SRB uses sulphate as an energy source that is an electron acceptor and uses organic matter as a carbon source (C). Decreasing sulphate concentration will increase soil pH. This occurs because of several interrelated processes, namely due to inundation, addition of organic matter and SRB activities . The sulphate reduction reaction by SRB is as follows:

$$SO_4^{2-} + 4H_2 + 2H^+ \rightarrow H_2S + 4H_2O$$

While the reaction of sulphate reduction by water is as follows:

$$SO_4^{2-} + H_2O + 2e^- \longrightarrow SO_3^{2-} + 2OH^-$$

$$SO_3^{2-} + H_2O + 6e^- \longrightarrow S^{2-} + 6OH^-$$

Based on Sitinjak's [7] research also explained that sulphate reducing bacteria can be found in virtually every environment on earth including sour soil sulphate, paper sludge and sulfur hot water. Isolation, potential test, and benchmarking of sulphate reducing bacteria from paper waste, sulfur hot water and acid sulfuric soil have been found in the same growth medium in laboratory scale. Based on that, it is necessary to test the potential of sulphate reducing bacteria with different isolate sources on acid sulphate soil media with indicator of maize plant.

## 2. Materials and Methods

This research was conducted in Laboratory of Soil Biology and in the Green house at Faculty of Agriculture, University of Sumatera Utara. The study was conducted from April 2016 to April 2017.The materials used in this research were SRB collection of Soil Biology Laboratory (LK4, LK6, TSM4, TSM3, AP4, AP6), Postgate E (Postgate, 1984) consisting of (g / L): $KH_2PO_4$ (0.5); $NH_4Cl$ (1.0); $Na_2SO_4$ (1.0); $CaCl_2.6H_2O$ (1.0); $MgCl_2.7H_2O$ (2.0); Yeast extract (1.0); Ascorbic acid (1.0); $FeSO_4.7H_2O$ (0.5); In order (18.0); Sodium lactate (8 ml); Thioglycollic acid (0.76 ml) for growth of SRB, anaerobic kit as source of SRB $CO_2$, label as marker of each sample, sample plastic for sample place, acid sulfuric soil that has been oxidized source PT. Mopoli Raya, Kebun Paya Rambe Aceh Tamiang, polybag 5 kg, plastic hose as a watering hose, BISI hybrid crown seeds as planting material, straw compost as bacteria carrier media, NPK fertilizer as basic fertilizer, and other supporting research materials.

Equipments used included anaerobic jar, incubator, test tube, concrete nail, soil drill, ice box, petridish, inoculation needle, analytical scale, pipette scale, pH meter, microscope, laminar air flow, oven, autoclave, bottle, hoe, ground sieves, 10 kg-scale, gauges, cameras, and other glass tools used for laboratory analysis.

This research used Factorial Random Block Design with two treatments and 2 replication factors, which the treatment, factor I was soil water content with 2 treatment levels; T1 = 100% Field Capacity; T2 = 110% Field Capacity and treatment factor II was SRB isolates with 16 different Isolate; B0 = Without giving inoculum (control); B1 = Isolate LK4; B2 =: Isolate LK6; B3 = Isolate TSM4; B4 = Isolate TSM3; B5 = Isolate AP4; B6 = Isolate AP6; B7 = Isolate LK4 + Isolate TSM3; B8 = Isolate LK4 + Isolate AP4; B9 = Isolate LK4 + Isolate AP3; B10 = Isolate LK6 + Isolate TSM3; B11 = Isolate LK6 + Isolate AP4; B12 = Isolate LK6 + Isolate AP6; B13 = Isolate TSM4 + Isolate TSM3; B14 = Isolate TSM4 + Isolate AP4; B15 = Isolate TSM4 + Isolate AP3.

- LK4 (SRB isolate from waste paper sludge of PT. Toba Pulp Lestari code 4)
- LK6 (SRB isolate from waste paper sludge of PT. Toba Pulp Lestari code 6)
- TSM4 (SRB isolate from acid sulfate soil of PT. Mopoli Raya code 4)
- TSM3 (SRB isolate from acid sulfate soil of PT. Mopoli Raya code 3)
- AP4 (SRB isolate from hot water sulfur in Sidebukdebuk code 4)
- AP6 (SRB isolate from hot water sulfur in Sidebukdebuk code 6)

Data was analyzed by variance based on linear model as follows:
$Yijk = \mu + \alpha i + \beta j + (\alpha\beta)\ ij + \gamma k + \varepsilon ijk$
The data was statistically tested using variance analysis at 5% level,

## 3. Results and Discussion

### *3.1. Total Population of Sulphate Reducing Bacteria*

Based on Table 1, it appeared that the population of SRB was not much different in each treatment, but there is an increase in population with the added moisture content given. In the average population of sulphate predominant bacteria with soil moisture treatment, the treated water content of 110% Field Capacity (2,07 x $10^8$) was higher than the treatment of 100% Field Capacity (2,01 x $10^8$). This is because all the macro and micro pores on the planting medium of the treatment of 110% Field Capacity was filled with full of water, therefore it created anaerob condition which is preferred by SRB. The population of bacteria with treatments of groundwater level 100% of field capacity and 110% of field capacity ws not statistically significant.

There are also several groups of sulphate reducing bacteria that can live in aerobic conditions. This is supported by Yusron *et al.*[8], which states that sulphate reducing bacteria is an obligate anaerobic bacteria but there are several groups of sulphate reducing bacteria capable of growing under aerobic conditions.

In the treatment of addition of sulphate reducing bacteria into soil, there was no significantly different population increase compared to control. This is caused by sulphate reducing bacteria requiring time to adapt to very low acid acid soil pH conditions where the initial soil pH conditions are 2,17 and soil media conditions that are inconsistent with environmental conditions of isolate sources. Yusron *et al.*[8], explains that the growth and activity of sulphate reducing bacteria is influenced by pH. At low pH it will inhibit the bacterial enzymatic system resulting in inhibition of energy formation, and at a neutral pH of faster energy formation.

**Table 1.** Total Population of Sulphate Reducing Bacteria

| Treatments | Population of SRB (CFU) | | |
|---|---|---|---|
| | Water Condition (Field Capacity) | | Average |
| | T1 (100%) | T2 (110%) | |
| B0 (Control) | 2.5 x $10^8$ | 2.5 x $10^8$ | 2.5 x $10^8$ |
| B1 (LK4) | 1.475 x $10^8$ | 2.5 x $10^8$ | 1.98 x $10^8$ |
| B2 (LK6) | 2.5 x $10^8$ | 1.47 x $10^8$ | 1.98 x $10^8$ |
| B3 (TSM4) | 1.29 x $10^8$ | 9.5 x $10^7$ | 1.12 x $10^8$ |
| B4 (TSM3) | 1.47 x $10^8$ | 2.5 x $10^8$ | 1.98 x $10^8$ |
| B5 (AP4) | 2.5 x $10^8$ | 2.5 x $10^8$ | 2.5 x $10^8$ |
| B6 (AP3) | 2.5 x $10^8$ | 2.5 x $10^8$ | 2.5 x $10^8$ |
| B7 (LK4+TSM3) | 1.72 x $10^8$ | 1.72 x $10^8$ | 1.72 x $10^8$ |
| B8 (LK4+AP4) | 2.5 x $10^8$ | 2.5 x $10^8$ | 2.5 x $10^8$ |
| B9 (LK4+AP3) | 2.5 x $10^8$ | 1.72 x $10^8$ | 2.11 x $10^8$ |
| B10 (LK6+TSM3) | 1.72 x $10^8$ | 2.5 x $10^8$ | 2.11 x $10^8$ |
| B11 (LK6+AP4) | 2.5 x $10^8$ | 2.5 x $10^8$ | 2.5 x $10^8$ |
| B12 (LK6+AP3) | 1.29 x $10^8$ | 5.2 x $10^7$ | 9.1 x $10^7$ |
| B13 (TSM4+TSM3) | 2.5 x $10^8$ | 2.5 x $10^8$ | 2.5 x $10^8$ |
| B14 (TSM4+AP4) | 7 x $10^7$ | 2.5 x $10^8$ | 1.6 x $10^8$ |
| B15 (TSM4+AP3) | 2.5 x $10^8$ | 1.72 x $10^8$ | 2.11 x $10^8$ |
| Average | 2.01 x $10^8$ | 2.07 x $10^8$ | |

Description: The numbers followed by the same letter are not significantly different at the 5% level according to the DMRT test

*3.2. Soil Sulphate*

The highest soil sulphate content was shown by treatment without SRB isolate and 100% soil water condition (B0T1) condition was 988,76 ppm and the lowest was on treatment of SRB LK4 + AP4 isolate and 110% field capacity on soil water condition (B8T2) that is 27.38 ppm (Table 2). The B8T2 treatment was able to significantly decrease the soil sulphate content statistically to 97.2% lower than B0T1. Treatment of B8T2 showed decrease of sulphate concentration to low category (40-99 ppm), which this category also showed by treatment of B1T2 (29.10 ppm), B9T2 (46 ppm), B10T2 (58,41 ppm), B7T2 (56 ppm), B6T2 (67,03 ppm), B14T2 (79,10 ppm), B5T2 (83,24 ppm), B15T2 (83,24 ppm), and B12T2 (84,97 ppm).

The provision of SRB was able to decrease sulphate concentration in acid sulphate soil to achieve the characteristic of medium level by giving isolates LK4 (B1), TSM4 (B4), AP4 (B5), LK4 + AP4 (B8) and isolate LK6 + AP3 (B12) The value of isolates were respectively 92.55 ppm, 98.07 ppm, 72.72 ppm, 67.21 ppm and 86 ppm, and capable of decreasing by 83%, 82%, 87%, 88% and 84% better than without provision of CPM (552.38 ppm). The decrease in sulphate content is influenced by good capability of SRB in reducing sulphate.

**Table 2.** Soil Sulphate (ppm) by treatment with SRB isolates and soil water condition

| Soil Sulphate (ppm) | | | |
|---|---|---|---|
| Bacteria | Water Condition (Field Capacity) | | Average |
| | T1 (100%) | T2 (110%) | |
| B0 (Control) | 988.76a | 116.00c | 552.38 |
| B1 (LK4) | 156.00c | 29.10c | 92.55 |
| B2 (LK6) | 115.66c | 117.03c | 116.34 |
| B3 (TSM4) | 320.83c | 130.14 c | 225.48 |
| B4 (TSM3) | 108.41 c | 87.72c | 98.07 |
| B5 (AP4) | 62.21 c | 83.24 c | 72.72 |
| B6 (AP3) | 163.24 c | 67.03 c | 115.14 |
| B7 (LK4+TSM3) | 779.79ab | 56.00 c | 417.90 |
| B8 (LK4+AP4) | 107.03 c | 27.38 c | 67.21 |
| B9 (LK4+AP3) | 189.79 c | 46.00 c | 117.90 |
| B10 (LK6+TSM3) | 197.38 c | 58.41 c | 127.90 |
| B11 (LK6+AP4) | 226.34 c | 65.31 c | 145.83 |
| B12 (LK6+AP3) | 87.03 c | 84.97 c | 86.00 |
| B13 (TSM4+TSM3) | 169.10c | 129.79 c | 149.45 |
| B14 (TSM4+AP4) | 616.34b | 79.10 c | 347.72 |
| B15 (TSM4+AP3) | 123.93c | 83.24 c | 103.59 |
| Average | 275.74 | 78.78 | |

Description: The numbers followed by the same letter are not significantly different at the 5% level according to the DMRT test

In soil moisture content 110% of field capacity (T2) was able to decrease (Table 2). This is supported by Nenny [9], which states that in sulphate reduction reactions, not only $H_2S$ is released but also hydroxyl ions ($OH^-$). Nenny [9], also states that the more reduced sulphate ions, the more OH-ions are produced so the pH will increase.

Treatment of groundwater content of 110% field capacity (T2) or slightly stagnant showed a 15% increase in pH value with an average pH of 4.39 compared to groundwater level 100% field capacity with a pH of 3.82. This is caused by the condition of 110% of the capacity of the field created conditions of reduction or anaerob while in 100% of the field capacity there is still some macro pore space that is not filled by water, thus it creates aerobic conditions. In the aerobic conditions of acid sulfate soils become more stable. Groudev *et al.* [10] states that the saturation of water causes the soil to become anaerobic because the oxygen that fills the soil pores is pushed and replaced by water. When sulphate accepts electrons from organic matter it will undergo reduction to form sulphide compounds as described by Foth [11].

In the study there were several treatments given by SRB with 100% groundwater content showed a lower pH value than without being given SRB with the same groundwater conditions. This might be caused by the death or decline in the ability of SRB due to the presence of oxygen or the presence of bacteria that are antagonistic to SRB isolate. Posgate [12] explained that there is a common occurrence that inhibits microbial growth especially anaerobic types: 1) the presence of oxygen 2) the presence of other bacteria that inhibit the growth of SRB.

### *3.3. Plant height*

The highest plant height due to the application of sulphate reducing bacteria isolate and different soil conditions was indicated by treatment with LK4 bacteria and 110% field capacity water level (B1T2) of 103.25 cm and the lowest in treatment with LK4 + TSM3 bacteria with 100% field capacity (B7T1) is 41 cm. The treatment of B1T2 was not significantly different from the B9T2 treatment and was significantly different from B0T1, B0T2, B1T1, B2T1, B2T2, B3T1, B3T2, B4T1, B4T2, B5T1, B5T2, B6T1, B6T2, B7T1, B7T2, B8T1, B8T2, B9T1, B10T1, B10T2, B11T1, B11T2, B12T1, B12T2, B13T1, B13T2, B14T1, B14T2, B15T1, B15T2, B0T1, and B0T2.

**Table 3**. Plant Height (cm) by treatment with SRB isolates and soil water condition

| | Plant height (cm) | | |
|---|---|---|---|
| | Water Condition (Field Capacity) | | |
| | T1 (100%) | T2 (110%) | Average |
| B0 (Control) | 70defgh | 50fgh | 60 |
| B1 (LK4) | 66.5defgh | 140a | 103.25 |
| B2 (LK6) | 73.5cdefgh | 92.5bcde | 83 |
| B3 (TSM4) | 83cdefg | 99bcd | 91 |
| B4 (TSM3) | 76.5cdefg | 106.5bc | 91.50 |
| B5 (AP4) | 70defgh | 92.5bcde | 81.25 |
| B6 (AP3) | 50fh | 72cdefgh | 61 |
| B7 (LK4+TSM3) | 41h | 89bcde | 65 |
| B8 (LK4+AP4) | 68defgh | 90.5bcde | 79.25 |
| B9 (LK4+AP3) | 85cdef | 123ab | 104 |
| B10 (LK6+TSM3) | 83cdefg | 88.5bcde | 85.75 |
| B11 (LK6+AP4) | 52.50fgh | 77.5cdefg | 65 |
| B12 (LK6+AP3) | 49gh | 82.5cdefg | 65.75 |
| B13 (TSM4+TSM3) | 89bcde | 83cdefg | 86 |
| B14 (TSM4+AP4) | 59efgh | 82.5cdefg | 70.75 |
| B15 (TSM4+AP3) | 98bcd | 83cdefg | 90.50 |
| Average | 69.63 | 90.75 | |

Description: The numbers followed by the same letter are not significantly different at the 5% level by the DMRT test

### *3.4. Dry Shoot Weight*

The increase of dry shoot weight by giving of sulphate reducing bacteria and difference of soil water condition compared with control showed a varied increase.The highest dry shoot weight of plant where treatment with LK4 isolate with the condition of soil water 110% field capacity (B1T2) is 25.74 g and the lowest is on treatment of LK6 isolate plus AP4 with 100% field capacity (B11T1) that is 0.69 g (Table 4). In the research obtained the highest plant weight on treatment with LK4 bacteria isolate with water content 110% field capacity (B1T2) is 25.774 g where the pH of the treatment is 4.78 . The sulfate content of 29.10 ppm is better than the treatment without sulfate reducing bacteria isolate with 100% soil water level. Low soil pH and high sulfate content in the treatment of B0T1 inhibit the growth of corn crops.

The provision of sulphate reducing bacteria will decrease the sulfate content and soil pH increases. An increase in pH leads to a reduction in heavy metals that can interfere plant growth. This is in accordance with Widyati [13], which states that with the provision of SRB into acid sulphate soil will result in reduced sulfate so that the soil pH increases and the solubility of heavy metals will decrease so as not to disrupt the growth of plants or other soil microbes.

**Table 4.** Dry Shoot Weight of Corn Plant

| Treatments | Dry Shoot Weight<br>Water Condition (Field Capacity) | | Average |
|---|---|---|---|
| | T1 (100%) | T2 (110%) | |
| B0 (Control) | 4.14def | 4.56def | 4.35 |
| B1 (LK4) | 3.20def | 25.74a | 14.47 |
| B2 (LK6) | 3.38def | 7.37bcdef | 5.37 |
| B3 (TSM4) | 4.96def | 7.79bcdef | 6.38 |
| B4 (TSM3) | 5.25def | 12.95bc | 9.10 |
| B5 (AP4) | 1.63ef | 1.69ef | 1.66 |
| B6 (AP3) | 2.84def | 4.00def | 3.42 |
| B7 (LK4+TSM3) | 2.12ef | 9.95bcd | 6.04 |
| B8 (LK4+AP4) | 2.06ef | 6.60bcdef | 4.33 |
| B9 (LK4+AP3) | 6.52cdef | 14.08b | 10.30 |
| B10 (LK6+TSM3) | 6.50cdef | 6.29cdef | 6.39 |
| B11 (LK6+AP4) | 0.69f | 2.25def | 1.47 |
| B12 (LK6+AP3) | 2.48def | 5.63cdef | 4.05 |
| B13 (TSM4+TSM3) | 6.29cdef | 5.16def | 5.72 |
| B14 (TSM4+AP4) | 2.95def | 7.32bcdef | 5.13 |
| B15 (TSM4+AP3) | 8.99bcde | 6.62bcdef | 7.80 |
| Average | 4.00 | 8.00 | |

Description: The numbers followed by the same letter are not significantly different at the 5% level by the DMRT test

Plant height growth is the result of a long process in plant metabolism from the absorption of nutrients, carbon dioxide and sunlight. In some plants will form a certain metabolic system in a less favorable situation. When viewed on the terms of growth of maize crop included into the indicator plants that have broad growing requirements. Havlin *et al.* [14] states that plant growth is the result of a complex process through plants synthesizing solar energy, carbon dioxide, water and nutrients from the soil.

## 4. Conclusions

Sulphate reducing bacteria isolates was able to increase soil sulfuric acid pH and increase the growth of plant with the best SRB isolate types to increase soil acid saturation pH was LK4 isolate. Increased soil moisture content could also increase soil sulphate acid pH and increased the growth of crown crops. The best groundwater conditions to increase acid soil pH was 110% field capacity. The best interaction in reducing acid acid sulphate acidity and increasing the growth of crown plants was shown by LK4 isolates with groundwater content of 110% KL (population of SRB $2.5 \times 10^8$, soil sulphate 29.10 ppm, soil pH 4.78, plant height 140 cm, canopy dry weight 25.74 g).

## References

[1] Bos L 1990 Pengantar Virologi Tumbuhan. Gadjah Mada University Press, Yogyakarta.
[2] Adhi W dan Alihamsyah T 1998 Pengembangan Lahan Pasang Surut : Potensi, Prospek dan Kendala Serta Teknologi Pengelolaannya untuk Pertanin, *Dalam* Prosiding Seminar Himpunan Ilmu Tanah Jawa Timur, Malang, 18 Desember 1998
[3] Konsten C J M and Sarwani M 1994. Actual and potential acidity and related chemical charactericties of acid sulfate soil in Pulau Petak Kalimantan. Workshop on acid sulfate soil in the Humid Tropics, 20- 22 November, Bogor Indonesia. AARD and LAWOO, Bogor,Indonesia.
[4] Widyati E 2006 Tanah Bekas tambang Batubara dengan Sludge Industri Kertas untuk Memacu Revegetasi Lahan.Disertasi.IPB. Bogor
[5] Hanafiah A S, Sabrina T dan Guchi 2009 Ekologi dan Biologi Tanah. USU Press. Medan.
[6] Sitinjak M. S., 2016. Isolasi dan Uji Potensi Beberapa Isolat Bakteri Pereduksi Sulfat Terhadap Perubahan Kemasaman Media Tumbuh.Skripsi, USU e-Repository.
[7] Yusron M, Lay B W, Fauzi A M dan Santosa D W 2009 Isolasi Dan Identifikasi Bakteri Pereduksi Sulfat Pada Area Pertambangan Batu Bara Muara Enim, Sumatera Selatan.*Jurnal Matematika, Sains dan Teknologi* 9 (1) : 26-35.
[8] Nenny Andriyetni 2006 Dinamika Populasi Mikrob dalam Campuran Tanah Bekas Tambang Batu Bara dengan Sludge selama Proses Bioremediasi. Skripsi Prodi Ilmu Tanah Fakultas Pertanian.Institut Pertanian Bogor
[9] Groudev S N , Komnitsas K, Spasova I I and Paspaliaris I 2001 Treatment of AMD by a natural wetland. Minerals Engineering 12: 261-270.
[10] Foth HD 1990 Fundamentals of Soil Science. 8th ed. John Willey&son. New York.
[11] Postgate J R 1984 The Sulfate Reducing Bacteria. Cambridge University Press,Cambridge.
[12] Widyati E 2007 Pemanfaatan Bakteri Pereduksi Sulfat untuk Bioremediasi Tanah Bekas Tambang Batubara.*Biodiversitas* 8 (3): 283-286.
[13] Havlin J L, Beaton J B, Tisdale SL and Nelson W L 1999 Soil Fertility and Fertilizers. An Introduction to Nutrient Management.Prentice Hall. New Jersey.

# Flower morphology diversity of asam gelugur (*Garcinia atroviridis* griff, ex t. anders) accessions in several districts of North Sumatera, Indonesian

**E S Bayu*, W Febrianti and R I M Damanik**

Agrotechnology, Faculty of Agriculture, Universitas Sumatera Utara, Medan 20155
Email : evagirsang61@gmail.com; tinigirsang@yahoo.com

**Abstract.** The research aims to study of flower morphology of *G.atroviridis* plants in several Districts of North Sumatera. This study was conducted in Asahan, Batubara and Serdang Berdagai districts in February to April 2017,exploration using survey method with IPGRI descriptors guide,and accidental sampling. The research results showed that in three Districts of Sumatera Utara, there are 26 accessions of Asam Gelugur characterized based on morphology of flowers. Based on clustering analysis using the NTSYS-pc Program,all flower morphology data resulted in dendogram with similar coeffisien range from 0.34 to 0.75 into four groups . The first group (I) consists of 16 accessions i.e Asahan, Batubara and Serdang Bedagai districts . The second group (II) in Asahan, Batubara and Serdang Bedagai districts there are 7 accessions The third group (III) consists of two accessions, i.e. Asahan and Batubara Districts . The fourth group (IV) consists of one accession in Serdang Bedagai District is the only outgroup access.

## 1. Introduction

Asam Gelugur is a plant that spread in tropical area of Asia. The plant is divided into two widely known species, namely *Garcinia cambogia* commonly found in southern India, while other species, namely *Garcinia atroviridis* (Asam Gelugur) are commonly found in the Malay Peninsula region [1]. Asam Gelugur (*Garcinia atroviridis* Griff ex T. Anders) is a tropical and subtropical annual tree originating from South Asia and Southeast Asia and is widely grown Sumatera island [2]. The spreading area of *G.atroviridis* in Indonesia is from Aceh to South Sumatra. Asam Gelugur so far is categorized into two kinds of fruit by Sumatera Utara citizen, namely *G.atroviridis* rock type and water type. Rock type is smaller, while water type larger and contain more water.

One commodity that has the potential to become an export commodity in North Sumatera is *G.atroviridis*. Actually this plant has already existed in this area for a long time, but the utilization is only for simple purposes. *G.atroviridis* in North Sumatera used by the community as a food ingredient. Generally this fruit is cut into thin, then dried and after that it's used as a mixture for food. *G.atroviridis* also can be processed into sweets and tasty jams, can be prepared as a substitute for Asam Jawa (*Tamarinus Indica*), also can be used for seasoning, fish preservatives, syrups and acidic ingredients for latex processing[3].

Based on [4], *G.atroviridis* fruit contains strong antioxidants because of the content of hydroxysitic acid compounds. *G.atroviridis* the bioactivity of methanol-DMSO extracts from this plant which gave the result that the extract was antibacterial (root), antifungal (fruit and leaves), antioxidants (roots, fruit and stems), and antitumor (leaves, fruits and root).

Crop characteristics of each cultivation area has not been identified and the utilization is still very modest due to the lack of information in terms of characteristics [5]. In particular, knowledge of its sexual and breeding systems, important for designing suitable breeding strategies in order to increase fruit production, is limited [6].

Based on [7] literature, stated that flowers on Asam Gelugur plant is hermaphrodite or mono-sexual, so it has the ability to breed with flowers derived from the same tree and called plant with two houses (*Diecious*). The flowers have four dark red petals and on the edge of each petal are yellow. The sexual system of *Garcinia atroviridis* is evaluated regarding of basic structural specialization and reproductive character in natural conditions. This species is included as *gynodioecious* with female flowers (the tree produces female flowers/stigma), but also hermaphrodite flowers (trees that produce perfect flower) [8].

The hermaphrodite flowers have more flowers than the female flowers, but fall gradually before they bear fruit. In contrast, female flowers have relatively larger ovules on each flower, resulting in larger fruits and more seeds in fruit than the hermaphrodite flowersIn diecious plants, the sex identification can see based on the flower morphology, so needed the information for characteristics the flower morphology and the plant can be distinguished before fruiting, whether hermaphrodite or female.

## 2. Materials And Methods

The materials used in this research are 26 accessions of *G.atroviridis* in three districts in Sumatera Utara i.e Asahan include the Sub-district of Air Joman, Silo Laut, Setia Janji, Pulau Bandring, Batubara include Sub-district of air Putih. Lima Puluh, Talawi, and Serdang Bedagai district include Sipispis, Pantai Cermin Kiri, Perbaungan, Sei Rempah and Tebing Tinggi Sub-districts. The tools used in this research are camera (Nikon D5300), meter, term slide, scales, white cloth, label, GPS, questionnaires and stationery. Determining the location of research based on information from the community for the area of G. atroviridis plants.

Parameter observed of flower morphology include position of flower, flower size, petal colour and sepal colour.

To know phenotypic diversity and kinship relationship between accession of gelugur acid, morphological data of each accession of gelugur acid was processed using cluster analysis and dendogram using UPGMA method (Unweighted Pair Group Method Arithmetic) method using NTSYS (Numerical Taxonomy and Multivariate System) version 2.02 [9].

## 3. Results And Discussion

The survey results in three Sumatera Utara's districts; Asahan (P1-P9), Batubara (P10-P20) and Serdang Bedagai (P21-P26) showed that *G.atroviridis* accession was spread in several villages. Other data found in the survey is plant's age from interview of farmer, altitude and coordinat using gps.

### *3.1. Asam Gelugur Flower Morphological Characteristics*

Based on morphological identification of Asam Gelugur in three districts in North Sumatera that has been done, from 26 Asam Gelugur accession found in various villages by the three districts. There are some dominant morphological characters that can distinguish the 26 accessions of Asam Gelugur. One of which is the morphology of flowers.

In flower size parameter there are three variations that is small, medium flower size found in Asahan districts, and big flower in Asahan district, Batubara and Serdang Bedagai districts. On the parameter of interest position there is only one variation that is at the end contained in 26 accessions of *G.atroviridis* identified in Asahan, Batubara and Serdang Bedagai districts. In the color parameter of petals there are only five variations on 26 accessions of gelugur acid which is identified as green yellow in Asahan district two accessions and Serdang Bedagai district there is one accession, green petal color is found in Batubara district there is one accession and Serdang Bedagai there is one accession, red There are six accessions, yellow with red fringes contained in three districts of Asahan, Batubara, Serdang Bedagai districts, yellow green with red margin in Asahan, Batubara and Serdang

Bedagai districts. In the color parameter of the petal flowers there are three variations on 26 accessions of G.atroviridis that is identified is green yellow contained in Batubara District with one accession, yellow with red margin is in District of Asahan with two accession, district of Batubara with three accession and Serdang bedagai there are two accessions , Red of petal colour in Asahan districts there are eight accession, five district of Batubara accession and in district of Batubara there are four accession.

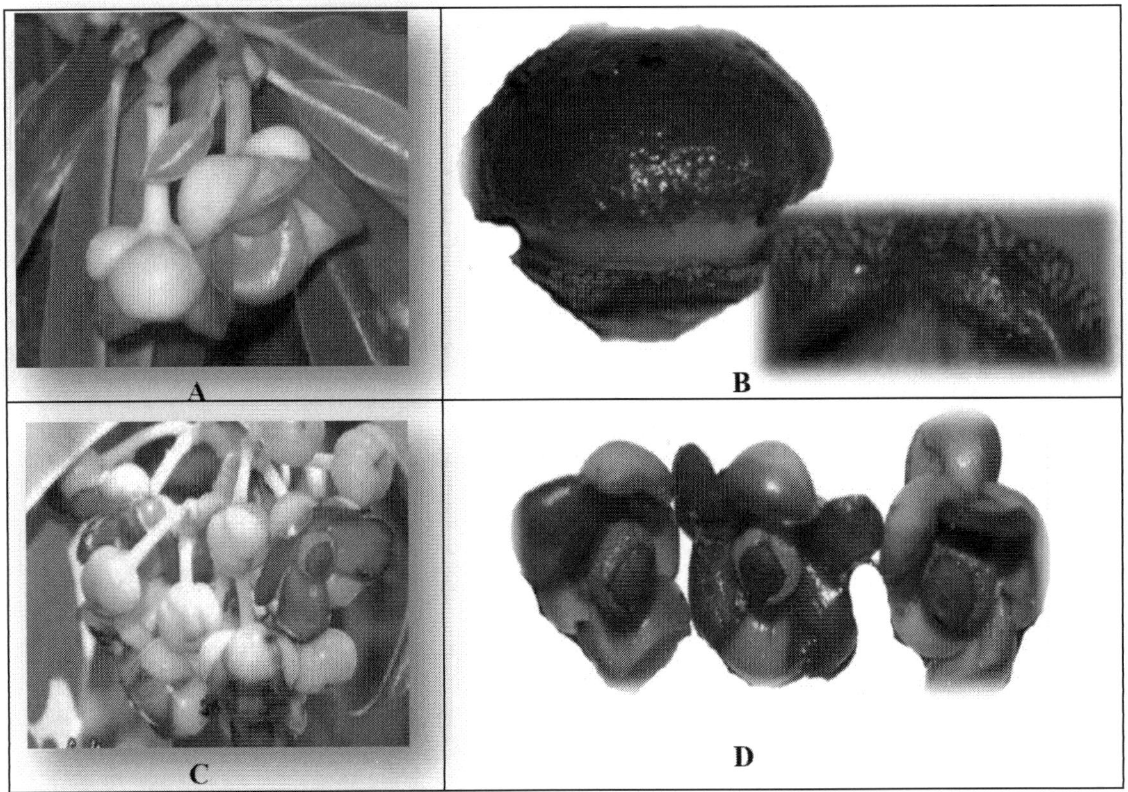


**Figure 1.** Difference of female and hermaphrodite flowers (A) from typical female flower is solitery (B) Female flowers produces less pollen (C) From typical hermaphrodite flowers is cluster (D) Hermaphrodite flowers produce less pollen has a relatively long filament and produces abundant and fertile pollen

**Figure 2.** the position of female flowers become fruit (A) pistillate flower and buds at terminal stalk, (B) pistillate flower and immature fruits, (C) fruits of *G.atroviridis*

There were significant morphological and anatomical differences between female flowers and hermaphrodite flowers (figure 1). The hermaphrodite flower has a relatively long filament and produces abundant and fertile pollen, whereas females produce less pollen. They also differ significantly in reproductive traits. The hermaphrodite flowers have more flowers than the female flowers, but fall gradually before they bear fruit. In contrast, female trees have relatively larger ovules on each flower, resulting in larger fruits and more seeds per fruit than the hermaphrodite tree [10]. Following the position of female flowers become fruit (figure 2).

The Characterization of observed flower morphology include flower position of female and hermaphrodite, the size of flowers (small medium large), sepal and petal colours of flowers . The diversity of flower morphological character in three districts in Sumatera Utara; Asahan, Serdang Bedagai and Batubara are: (figure 3 as follows).

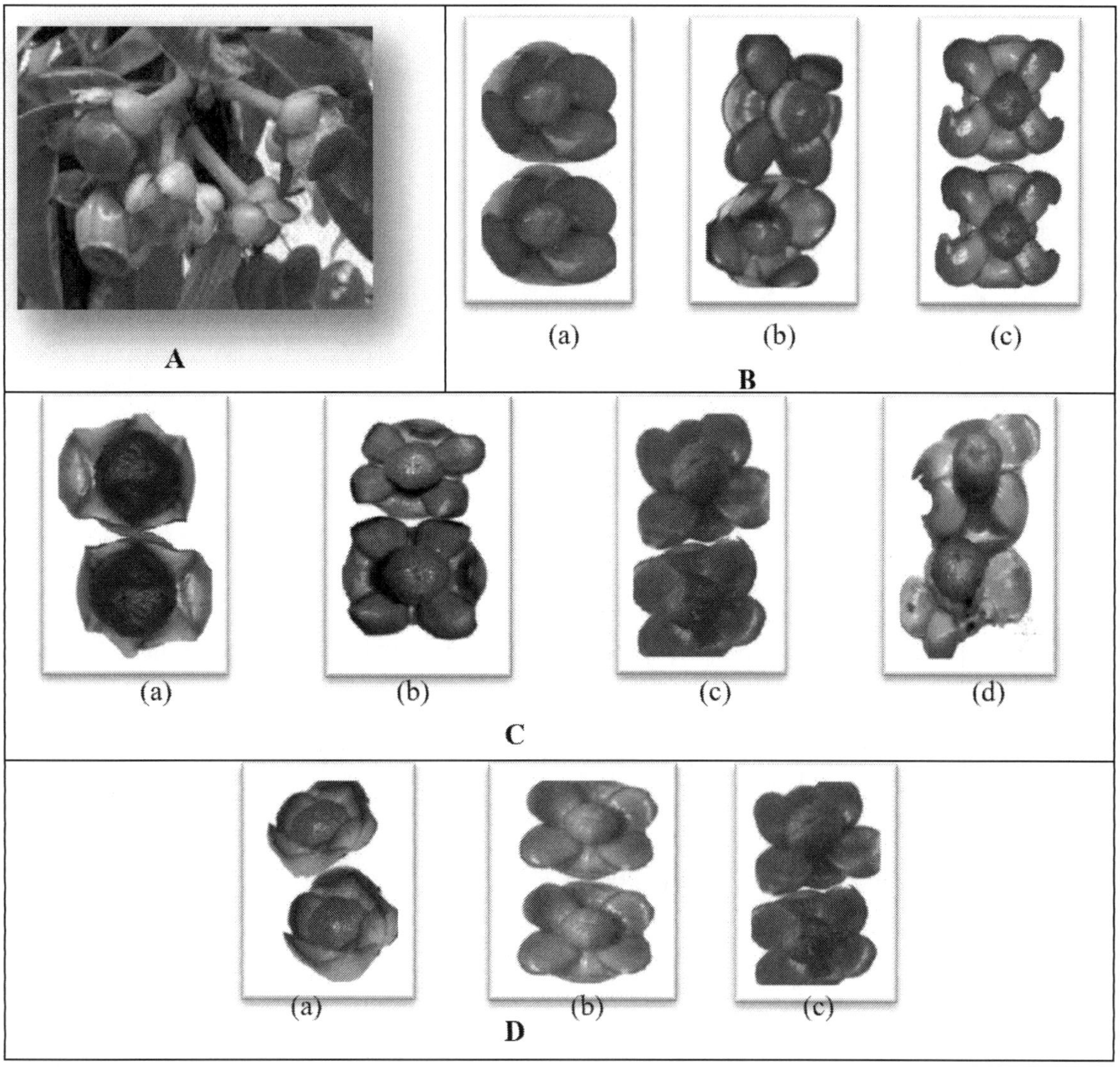


**Figure 3.** A. position of flowers terminal of the stalk, B. Flowers siza are (a) small (b) medium (c) large, C. Petal colours are (a) yellow green (b) yellow with red margin (c) red (d) yellow with red margin, D. Sepal colours are (a) green with red margin (b) yellow with red margin (c) red

### *3.2. Phenotypic Diversity between Individuals*

The UPGMA analysis of 26 Asam Gelugur accessions based on 4 characters of flower morphology is presented in Figure 4. Based on the similarity of characteristics that exist among several accessions of Asam Gelugur in some North Sumatera districts can be seen in the picture below:

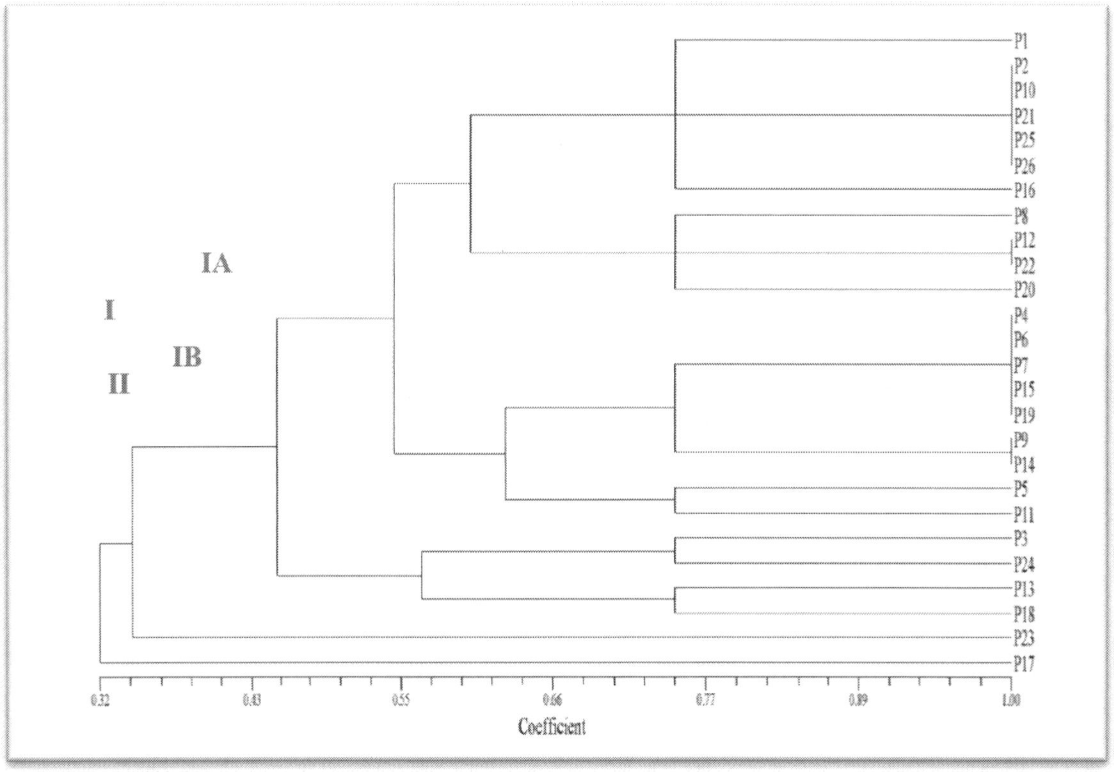


Figure 4. Dendogram from UPGMA analysis of 26 Asam Gelugur accessions in Asahan, Batubara and Serdang Bedagai districts.

From cluster analysis of flower morphology resulted in dendogram obtain two main group, the first group (I) consist of 25 accessions in Asahan, Batubara and Serdang Bedagai districts with coefficient 75%, which is generally the petal colour is red, sepal colour is yellow with red margin. The second group (II) in Batubara district is the only outgroup access because the petal colour in yellow green to distinguish from other accessions with coefficient 32% (figure 5). the first group (I) is divided into two subgroups where the first group (IA) consists of 24 accessions with coefficient 75%, which is generally the petal colour is red, sepal colour is yellow with red margin. The second group (IB) in Serdang Bedagai district is the only outgroup access because medium size flower as well petal and sepal colours is yellow with red margin with coefficient 35% (Figure 5).

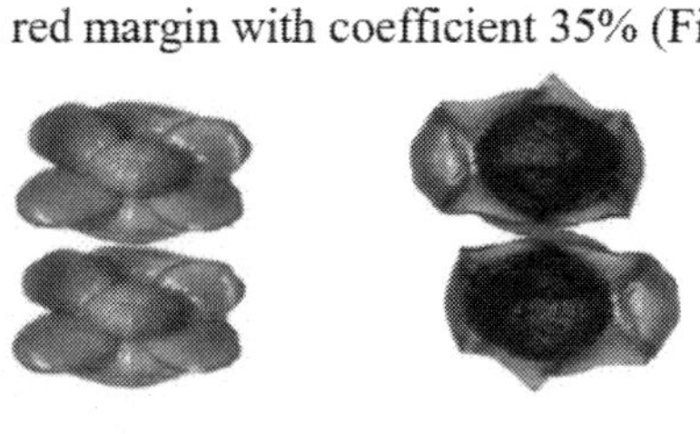

P23 P17

Figure 5. Two accessions out group are P23 and P17

Based on the phylogenetic relation analysis of Asam Gelugur in several districts in North Sumatera, i.e, Asahan, Batubara and Serdang Bedagai districts 26 accessions were obtained that each Asam Gelugur accession in three Sumatera Utara districts have relatively close relation that is equal to 56%.

## 4. Conclusion

Based on cluster analysis on flower morphological characters of 26 accessions in three Districts of North Sumatera using the NTSYS-pc Program,all flower morphology data resulted in dendogram with similar coeffisien range from 0.32 to 0.75. The characterization of flower morphology 26 accessions from position of flowers terminal of the stalk, various such as petal colour are yellow green, red and yellow with red margin, various such as sepal colour are green with red margin, red and yellow with red margin

## References

[1] Te-chato S 2007 Floral and fruit morphology of some species in Garcinia spp. Faculty Of Natural Resource, Prince Of Songkla University, Hat Yai, Thailand *Songklanakarin J.Sci. Technol.* Vol. **29** No. 2 Mar-Apr 2007

[2] Nazre M, Clyde M M and Latiff A phylogenetic relationships of locally cultivated *Garcinia* species with some wild relatives. Faculty Of Science and Technology, Universty Kebangsaan Malaysia **36** (1) : 31-40

[3] Bermawie N 2005 *Karacterization of Germplasm Plasma* Manual for Plant Germplasm Management, Plantation Research and Development Center Bogor Vol. 38-52

[4] Uji, T. 2007. Diversity of Indonesian Original Fruits and Their Potential. Botany Field, Center for Biological Research, Indonesian Institute of Sciences (LIPI) Bogor Vol. **8** No. 2 Hal. 157-167, ISSN: 1412-033X

[5] Mansyah E dan Edison H S 2012 *Estimation of Diversity of Garcinia sp and Nephelium sp in West Sumatera and Jambi Provinces as well as Potential of Utilization in Bioindustry Agriculture* Indonesian Tropical Fruits Research Institute, Solok.

[6] Mackeen M M, Ali A M, Lajis N H, Kawazu K, Hassan Z, Amran M, Habsah M, Mooi L Y, Mohamed S M 2000 Antimicrobial, antioxidant, antitumour-promoting and cytotoxic activities of different plant part extracts of *Garcinia atroviridis* Griff. ex T. Anders, Journal of Ethnopharmacology Vol. **72**(3): 395-402

[7] Pangsuban S, Noparat B, Kamnoon K dan Charassri N 2007 An Evaluation Of The Sexual System Of *Garcinia atroviridis* Griff. ex T. Anders (Clusiaceae), Based On Reproductive Features. Departmen Of Plant Science, Faculty Of Natural Resource, Prince Of Songkla Univetsity, Hat Yai Thailand Vol **29** No 6 Vov-Dec 2007

[8] Mansyah E dan Edison H S 2012 Estimation of Diversity of Garcinia sp and Nephelium sp in West Sumatera and Jambi Provinces as well as Potential of Utilization in Bioindustry Agriculture. Indonesian Tropical Fruits Research Institute Solok

[9] Rolhf F J 2000 NTSYSpc. Numerical taxonomy and multivariate analysis system. Version 2.1. Departement of Ecology and Evolution State of NewYork. USA

[10] Pangsuban S, Bamroongrusa N, Kamnoon N and Charassri N 2009 Facultative Apomixis in *Garcinia atroviridis* (Clusiaceae) and Effects of Different Pollination Regimes on Reproductive Success Faculty of Natural Resources Prince of Songkla University Thailand

# Contribution of pod borer pests to soybean crop production (case in Pondidaha, Konawe District, Southeast Sulawesi)

**M Rahayu[1], LOS Bande[1], A Hasan[1], A Yuswana[1] and F Rinambo[2]**

[1]Department of Plant Protection, Faculty of Agriculture, University of Halu Oleo, Kendari, 93232, Southeast Sulawesi, Indonesia
[2]Agricultural Quarantine Centre for Class II Kendari, Southeast Sulawesi, Indonesia
*E-mail: yayukmallarangeng@yahoo.com

**Abstract**. Soybean (*Glycine max* L.) is one of the most important crops which production continues to be improved in all areas of soybean cultivation centers in an effort to maintain the availability of soybean foods, including Southeast Sulawesi. The purpose of this study was to analyze the contribution of pod borer pests to soybean crop production. Methods of direct observation were made on observed variables, including species and population of pest pod borer, intensity, and crop production. The result that found four types of pod borer pests are *Nezara viridula, Riptortus linearis, Etiella zinckenella,* and *Leptocorisa acuta*, each with a different population and contribution to the intensity of pod damage. The result of path analysis showed that directly population of *N. viridula* (61.14) and *E. zinckenella* (66.44) gave positive contribution in increasing pod damage, by 0.332 and 0.502 respectively, while the negative contribution was shown by population of *R. linearis* and *L. acuta*. Damage of the pod causes increased production of low soybean is only about 0.202, therefore required appropriate control techniques to control pod borer pests populations in soybean crops.

## 1. Introduction

Soybean (*Glycine max)* is a strategic commodity and plays an important role for the life of society and economy of Indonesia. Soybean is the raw material for making tofu, tempe, soybean sauce, and tauco as a processed food product that is always consumed almost by all layers of society in Indonesia, both in the village and in the city. The productivity of soybean crop in Indonesia is about 1.2 ton ha$^{-1}$, still far below the potential yield that can reach 2.0-2.5 ton ha$^{-1}$ [1]. In addition to the growing planting area and the lack of interest of farmers to plant soybean, weather disturbances such as *La Nina*, low productivity and loss of yield due to pest and disease damage are the main causes of low soybean production in Indonesia.

Damage to pests and diseases in soybean plants is a stress biotic environment that greatly affects the response of plants to fertilization is the absorption of nutrients to be not optimal. As a result, soybean production becomes low due to loss of yield due to pest damage. In the production process, the pest problem cannot be ignored, because it will affect the production qualitatively and quantitatively and able to reduce production by 20.7%, even causing crop failure, if not control.

Main pests commonly found and able to cause crops are *Ophyomia phaseoli* [2], *Spodoptera litura* [3], *Phaedonia inclusa*, *Etiella zinckenella* [4], *Riptortus linearis* and *Nezara viridula* [5-6]. Data and information on how big pest destructive pests soybean contribute to the loss of production is still limited, therefore research on the amount of pest the destructive contribution to soybean crop production becomes important to do. Related to this matter, this research aims to analyze the contribution of pest destructive pods to soybean plant production.

## 2. Materials and methods

This research was carried out on farm land owned at Belutu Village, Pondidaha Sub-districts, Konawe Districts, Southeast Sulawesi. The study used the method of direct observation on soybean crop fields. The research sampling plots in observations plot are determined diagonally as many as five plots with the size of each plot is 1.5 m × 1.5 m, so that in one observation plot there are 12 soybean plants as sample plants. Observations were made since the plants were two weeks after planting (WAP) to 9 WAP with one-week of observation interval.

The observed parameters, includes (a) Types of insect pests: calculated all types of pests that damage pods countain in the plant sample. (b) Inscet pest population: calculated all pest populations found in sampel plant. (c) Iintensity of pod damage. To calculate the intensity of damage use the formula [7]

$$I = \frac{n}{N} \times 100\%$$

I = intensity of damage
n = the number of pods that stricken
N = total number of pods observed

## 3. Results and discussion

### *3.1.Results*

#### *3.1.1. Types of pest populations, the intensity of damage to pods, and crop production.*

Based on the results of the study found three types of destructive pest insect are *Nezara viridula, Riptortus linearis, Leptocorisa acuta* as pods sucking pest and *Etiella zinckenella* as pod borer pests. The intensity of pod damage caused by four types of pest insects varies as shown in Figure 1.

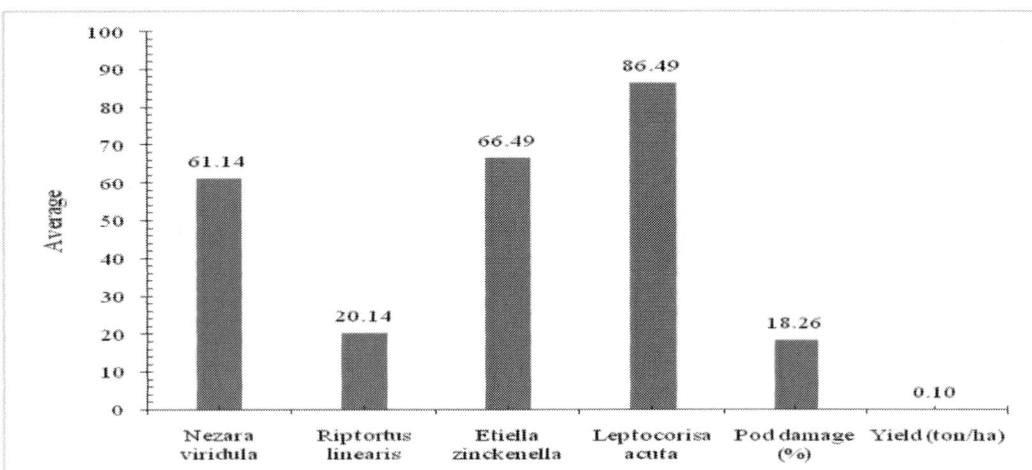


**Figure 1.** Population, the intensity of damage, and soybean production

Figure 1 shows that the average population of *N. viridula* as 61.14 individuals, *R. linearis* 20.14 individuals, *E. zinckenella* 66.49 individuals, and *L. acuta* 86.49 individuals. It is seen that the population of *L. acuta*, is highes , while *R. linearis* is the lowst. Four pest destructive pods, the attack caused the average pod damage of 18.26%. The average soybean production is quite low, at 0.10 ton $ha^{-1}$.

*3.1.2. Influences of pest population against the intensity of damage and production.*

Naturally in the field there was a relationship between insect pest destructive pod populations with the level of pod damage and soybean crop production. The third component of the relationship can be explained through the path diagram of the SEM-PLS (structural equation modelling-partial least square) whose results are shown in Figure 2.

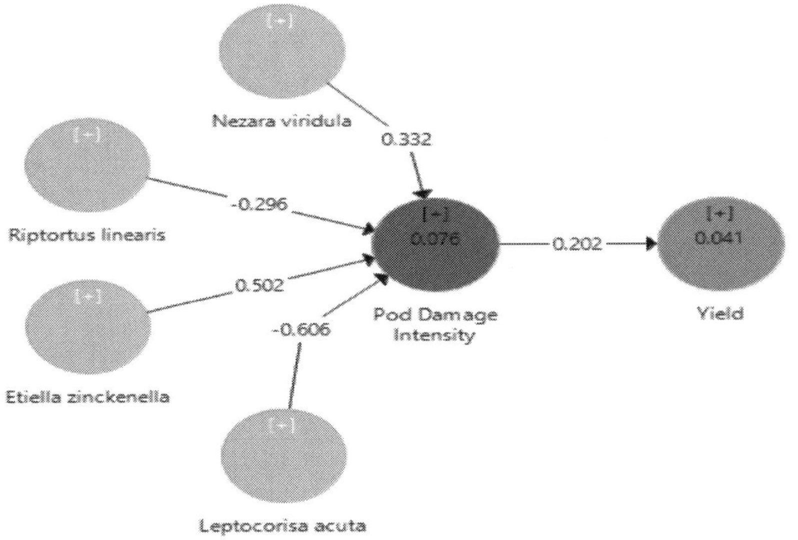


**Figure 2.** Diagram of the path analysis (path diagram) using the PLS-SEM (SmartPLS Student v. 3.2.3)

Figure 2 shows that the intensity of pod damage in the field as influenced by pest destructive population *N. viridula, R. linearis ,E. zinckenella), L. acuta* of 7.6% (R2 = 0.076), while the remaining 92.4% influenced by other factors in the field. This illustrates that the large extent of the pod damage in the field is not only influenced by pest destructive population, but can also be influenced by other factors such as climate factors, the control measures under taken car by farmers, natural enemy conditions, and plant resistance level.

The intensity of damage to pods may affect crop production amounted to 4.1% (R2 = 0.041), while the rest of 95.9% is affected by other factors, meaning damage to pods of the plant are not the main cause of high to low crop production of soybean. Crop production was also much influenced by the potential of genetic engineering, plant cultivation, soil fertility, and soil water availability.

Based on the results of path analysis it was found that from four types destructive pest found in the field, only the population of *N. viridula and E. zinckenella* could potentially can increase the intensity of pod damage directly. Every increase of 1% of the population of both the pest, then the extent of the damage potentially pods have elevated each of 0.332 and 0.502%. The results of path analysis also indicated that soybean production only increased by 0.202% caused by the destruction of soybean pods. Indirectly, this is caused by the activity of the fourth type of pest destruction pods. Plants basically still tolerate pod damage by *R. linearis* and *L. acuta* attacks because only *N. viridula* and *E. zinckenella* alone have the potential to increas pod damages so that can still produce despite small increases.

*3.2. Discussion*

Based on research results, the highest population was *L.acuta*, followed by *E. zinckenella*, *N. viridula*, and lowest *R. linearis*. The four types of pest destruction pods, the attack only causes the average pod damage of 18.26%. The population of *N. viridula* and *E. zinckenella* contributed positively to the intensity of pod damage, meaning that both pests have the potential to increase the intensity of pod damage, while L. acuta and *R. linearis* contribute negatively to the intensity of pod damage, meaning that it does not contribute to increased intensity pod damage. Although the population of *L. acuta* is the highest, its contribution to low pod damage is due to soybean crop is not the main host. *Leptocorisa oratorius* Fabricius, (Hemiptera: Alydidae), syn *Leptocorisa acuta*) is an insect that becomes an important pest on cultivated plants, especially rice [8]. This suggests that pest insect populations will not always cause great damage.

The high damage of plants by plant pest organisms is influenced by the population as well as the destructive ability of the organism. *R.linearis* is an important pest of soybean crop, but its population is quite low compared to other pests, this is thought to be caused by the predatory population of *Dolichoderus* sp. is abundant enough to eat eggs *R. linearis*. *Dolichoderus* sp., *Solenopsis geminata*, and *Paederus* sp. is a predator of *R.linearis* eggs [9]. Some types of insects can cause very serious damage even though the population is slightly natural [10].

The population of *E. zinckenella* 66.49 individuals contributed to the highest damage of pods of 0.502 compared with the pest destruction of *N. viridula* pod 61.14 individuals which was only 0.33. Soybean is the main host for *E. zinckenella* pest, whereas *N. viridula* has many alternative hosts and soybean is not its main host. The main host plant of *E. zinckenella* besides soybean was the *Crotalaria* sp.[11]. *N. viridula* has a wide range of host plants, in addition to attacking food crops also attack fruit crops, vegetables, and some weeds [12]. Damage of soybean pod is basically not solely caused by insect pest activity alone although it is known that insect pest is the main cause of crop damage, but the stress of plant environment also contribute significantly to insect pests and plants.

## 4. Conclusions

Base on the research concluded that (1) four pest destructive pests were found, each with the following population averages: *N. viridula* 61.14, *R. linearis* 20.14, *E. zinckenella* 66.49, and *L. acuta* 86.49 individuals. (2) Intensity of pod damage in field was 18.26% influenced by pest destructive population (*N. viridula, R. linearis, E. zinckenella, L. acuta*) 7.6% (R2 = 0.076), while the rest 92.4% is influenced by other factors in the field. (3) Intensity of pod damage can affect crop production by 4.1% (R2 = 0.041), while the rest of 95.9% is influenced by other factors

## References


[1] Saleh N 2004 Strategi pengendalian penyakit tanaman kedelai (Strategy to control Soybean's diseases) balai proteksi tanaman Kacang-kacangan dan umbi-umbian. *Buletin Palawija* **7&8** 51–60

[2] Biswas GC and Islam R 2012 Infestation and management of the lef roller (*Lamprosema indicata* Fab.) in soybean (*Glycine max* L.) *Bangladesh J. Agril. Res.* **37(1)** 19-25

[3] Biswas G C 2013 Insect pest of soybean (*Glycine max* L.), their nature of damage and succession with the crop stages *J. Asiat. Soc. Bangladesh Sci.* **39(1)** 1-8

[4] Berg H V D, Shepared B M and Nasikin 2010 Damage incidence by *Etiella zinckenella* soybean in East Java, Indonesia *Int. J. Pest Manage* **44(3)** 153-9

[5] Marwoto 2007 Dukungan pengendalian hama terpadu dalam program bangkit *kedelai Iptek Tan. Pang.* **2(1)** 79-92

[6] Marwoto dan Suharsono 2008 Strategi dan komponen teknologi pengendalian ulat grayak (Spodoptera litura) pada tanaman kedelai *J. Litbang Pertanian* **27(4)** 131-6

[7] Suharsono 2009 Hubungan kerapatan trikoma dengan intensitas kerusakan penggerek polong kedelai. *J. Penelitian Pertanian Tanaman Pangan* **28(3)** 176-82

[8] Suharsono and Trisnaningsih 2000. Komposisi jenis penggerek polong *Etiella* spp. di Jawa Timur. Participatory of Agriculture Departement Technology Project (PAATP), 1999/2000

*[9]* Van den Berg H, Bagus A, Hassan K, Muhammad A and Zega S 1995 Predation and parasitism on eggs of two pod-sucking bugs, *Nezara viridula* and *Piezodorus hybneri in*

*soybean International Journal of Pest Management **41(3)** 134-42*
[10] Sembel D T 2010 *Pengendalian Hayati Hama-Hama Serangga Tropis dan Gulma (Biological Control of Tropical Insects and Weeds)*. Yogyakarta : Andi Offset. p282
[11] Kalshoven L G E 1981 The Pest of Crops in Indonesia Revised by P.A Van der Laan (Jakarta : PT. Ichtiar Baru-Van Hoeve) p701
[12] Panizze A R, Macpherson J E, James D G, Javahery M and Mcpherson R M 2000 Stink bug (Pentatomidae) In: Schaefer, C.W., A. R. Panizzi (eds.) Heteroptera of Economy Importance. Boca Raton, Florida, USA: CRC Press. p: 432-434.

# The storage capacity of cocoa seeds (*Theobroma cacao* L.) through giving Polyethylene Glycol (PEG) in the various of storage container

**R R Lahay, S Misrun and R Sipayung**

Department of Agrotechnology, Faculty of Agriculture, Universitas Sumatera Utara, Medan, Indonesia
E-mail: ratna.rlahay@gmail.com

**Abstract.** Cocoa is plant which it's seed character is recalcitrant. Giving PEG and using various of storage containers was hoped to increase storage capacity of cocoa seeds as long as period of saving. The reseach was aimed to identify the storage capacity of cocoa seeds through giving PEG in the various of storage containers. Research took place in Hataram Jawa II, Kabupaten Simalungun, Propinsi Sumatera Utara, Indonesia. The method of this research is spit-split plot design with 3 replication. Storage period was put on main plot which was consisted of 4 level, PEG concentration was put on sub plot, consisted of 4 level and storage container was put on the sub sub plot consisted of 3 types. The results showed that until 4 days at storage with 45 % PEG concentration at all storage container, percentage of seed germination at storage can be decreased to be 2.90 %, and can be defensed until 16 days with 45 % PEG concentration at perforated plastic storage container. Percentage of molded seeds and seed moisture content were increased with added period of storage but seed moisture content was increased until 12 days at storage and was decreased at 16 days in storage.

## 1. Introduction

Recalcitrant seeds originated from tropical and subtropical plant species are characterized by the absence of maturation drying, and most of them have a high water content and active metabolism when they are shed from the mother plant. They are sensitive to drying and low temperatures, and quickly lose viability during storage [1,2]. Seeds of many important economical plants are recalcitrant, including many important tropical plantation crop species such as rubber and cocoa, tropical fruit crops such as mango, lychee and longan, and tropical timber species which belong to the families Dipterocarpaceae and Araucariaceae, etc. Seed germination during storage can be inhibited by using growth inhibitors such as Polyethylene glycol (PEG) which maintains the osmotic potential of cells that can be used to limit changes in water content and $O_2$ in germination or storage medium so that PEG molecules that are outside the seed cell membrane will forming a thin layer that protects the seeds and serves as a buffer of seed water and oxygen inlet [3]. Based on the description above, we interested to conduct research to know the provision of PEG-6000 to the cocoa seed with various storage container so that later beneficial in terms of delivery of seeds outside the area without any seeds germinated during the trips.

## 2.Materials and Methods

This research was conducted at farm in Jl. Hataran Jawa II Simpang Pure Kabupaten Simalungun, and Tissue Culture Laboratory of Faculty of Agriculture, University of Sumatera Utara, Medan, Indonesia.

### *2.1.Statistical analysis*

This study uses Split Split Plot Design with three treatments. Main plot is storage period (T) with four levels : 4,8, 12, and 16 days. Sub plot is concentration of PEG-6000 (P) with four levels : 0, 15, 30, and 45 %. Sub-sub plot is storage container (S) with three types : perforated plastic, plastic + sawdust, and plastic + hush ash. Data are presented as means- standard deviation of three replicates. The statistical differences were tested using an analysis of variance (ANOVA) and means were compared used Duncan Multiple Range - test. Figures show significant differences using letter designations. Data points with different letters show significant differences ($P < 0.05$).

### *2.2.Pulp cleansing and fungicidal seed treatment*

The material used in this research is cocoa seed clone DR 38. Cocoa fruit is split with a batter, then taken seeds in the middle of the fruit, and cleaned the pulp of cocoa beans with ash. After clean pulp, washed with water and soaked for 5 minutes in a Delsene MX-80 WP fungicide solution with a concentration of 2 g / l of water then dried for 1 hour.

### *2.3.Treatment of PEG-6000 concentration*

Seeds that have been soaked with fungicide are then incorporated into PEG-6000 (gr / vol) solution for 10 minutes according to the treatment concentration level and then dried for 5 hours.

### *2.4.Seed storage*

The treated seeds were stored in the medium according to the storage container treatment, where the sawdust and the husk ash to be used as storage container, soaked with Delsene MX 200 fungicide then dried and then stored into room temperature for 4, 8, 12, 16 days.

### *2.5.Seed germination*

Germination media is 10-15 cm thick layer sand. The seeds are sowed with radicles at the bottom with a seed spacing of 2 cm x 3 cm and a 1 cm of planting depth. Watering done twice a day ie in the morning and afternoon with handsprayer.

## 3. Results and Discussion

### *3.1.Percentage of seeds germinated in storage*

Storage duration (T), PEG concentration (P), storage container (S), interaction of T x P; T x S; P x S and interaction of T x P x S significantly affect the percentage of seed germinated in storage. Data percentage of seeds germinated in storage are presented in Table 1.

The lowest of percentage seed germinated in storage (2.90%) was in the seed treatment with PEG 45% in perforated plastic containers for 12 days which was not significantly different with the treatment of seeds with PEG 45% in 4 or 8 day in perforated plastic containers as well were not significantly different from seed treatment at all PEG concentrations (0, 15, 30, 45%) on plastic containers + sawdust and plastic + ash husk stored for 4 days. For each use of storage container, it is increasing PEG concentrations at storage time of 4, 8 and 12 days will prevent the percentage of seeds germinated during storage.

PEG in some concentrations has demonstrated its role in suppressing water absorption. This suggests that the higher concentrations of PEG used may decrease the number of seeds germinated during storage although it can not prevent seed germinate by 0 %. PEG has certain osmotic values equal to the osmotic value of cocoa seed, is expected to prevent imbibition so that the seed germination process will not occur because the imbibition is the main condition of germination [3].

**Table 1.** Percentage of seeds germinated in storage in the interaction of storage period, PEG concentration and storage container

| Storage period | PEG Concentration | Storage container | | |
|---|---|---|---|---|
| | | Perforated plastic | Plastic + sawdust | Plastic + husk ash |
| | | ............................ % ............................ | | |
| 4 days | 0 % | 49.63 ef | 5.42 j | 13.49 j |
| | 15 % | 37.07 g | 5.42 j | 6.93 j |
| | 30 % | 26.50 hi | 2.90 j | 6.93 j |
| | 45% | 2.90 j | 2.90 j | 2.90 j |
| 8 days | 0 % | 73,18 ab | 46.26 f | 84.58 a |
| | 15 % | 68.31 bc | 37.77 g | 80.55 a |
| | 30 % | 53.72 def | 36.11 g | 79.39 ab |
| | 45% | 10.60 j | 20.42 hij | 65.60 bc |
| 12 days | 0 % | 57.66 cde | 83.05 a | 82.03 a |
| | 15 % | 26.57 h | 60.77 cd | 87.13 a |
| | 30 % | 7.94 j | 49.58 ef | 81.94 a |
| | 45% | 2.90 j | 48.83 ef | 79.39 ab |
| 16 days | 0 % | 14.11 ij | 2.90 j | 7.94 j |
| | 15 % | 6.93 j | 2.90 j | 2.90 j |
| | 30 % | 2.90 j | 2.90 j | 2.90 j |
| | 45% | 2.90 j | 5.42 j | 2.90 j |

*3.2.Percentage of moldy seeds in storage*

Storage period (T), storage container (S), interaction of T x S; P x S had a significant effect on the percentage of moldy seeds in storage, but PEG concentration (P), interaction T x P; and interaction of T x P x S had no significant effect on percentage of moldy seeds in storage. Data percentage of moldy seeds in storage are presented in Table 2 and Table 3.

**Table 2.** Percentage of moldy seeds in storage in the interaction of storage period with storage container

| Storage Period | Storage container | | |
|---|---|---|---|
| | Perforated plastic | Plastic + sawdust | Plastic + husk ash |
| | ............................ % ............................ | | |
| 4 days | 2.90 c | 2.90 c | 2.90 c |
| 8 days | 3.53 c | 2.90 c | 2.90 c |
| 12 days | 13.14 b | 8.06 c | 4.54 c |
| 16 days | 24.70 a | 15.50 b | 4.79 c |

The longer of storage period for all types of storage container, the percentage of moldy seeds in storage is increasing. The lowest percentage of moldy seeds (2.90%) was obtained at seed storage up to 8 days in plastic container + sawdust or plastic container + husk ash which was not significantly different with 4 days seed storage in each type of storage container (perforated plastic, plastic +sawdust, and plastic + ash husk), also not significantly different with 12 day storage using plastic + sawdust or plastic + husk ash and 16 day storage using plastic + husk ash.

High levels moisture content of recalcitrant seed during storage have several risks, namely seed germination and growth of fungus in storage. According to [4] stated that cocoa seeds including recalcitrant seeds have the unstable properties stored at moisture content of less than 12% and at temperatures. Too low, while the fungus can grow and develop well at high seed moisture content and temperature levels so that cocoa seeds are susceptible to fungal attack during storage. The longer the seeds are stored, the fungus will continue to grow so as to decrease the power of seed sprouts.

**Table 3.** Percentage of moldy seeds in storage (%) in the interaction of PEG concentration with storage container

| PEG Concentration | Storage container | | |
|---|---|---|---|
| | Perforated plastic | Plastic + sawdust | Plastic + husk ash |
| | ……………… | % | ……………… |
| 0 % | 7.72 bcd | 4.66 cd | 4.16 d |
| 15 % | 9.90 abc | 12.12 ab | 2.90 d |
| 30 % | 12.89 ab | 6.46 cd | 4.16 d |
| 45 % | 13.76 a | 6.12 cd | 3.91 d |

In the use of perforated plastic containers, increasing PEG concentration will increase the percentage of moldy seeds. When using a storage container in the form of plastic + husk ash, then the increase of PEG concentration in the seeds has no significant effect on the percentage of moldy seeds. The lowest percentage of moldy seeds (2.90%) was found in seed treatment with 15% PEG stored in plastic + husk ash. It can be seen that perforated plastic container is unable to suppress the growth of the fungus during storage because by giving the aeration hole, oxygen and free air out and into the seed so that the fungus can be carried by air and attached to the seed and spread to the other seed. In the plastic + hush ash container, the percentage of moldy seeds in storage is relatively low when compared with other storage containers. This is because in the container store the ash husk has passed through the combustion process so that the pathogen microbes have died during the burning process.

*3.3.Seed moisture content after storage*

Storage period (T), PEG concentration (P), storage container (S), interaction of T x P; T x S had significant effect on seed moisture content, and interaction of P x S, and interaction of T x P x S not significant effect on seed moisture content. Data seed moisture content after storage are presented in Table 4.

**Table 4.** Seed moisture content in the interaction of storage treatment with storage container

| Storage Period | Storage container | | |
|---|---|---|---|
| | Perforated plastic | Plastic + sawdust | Plastic + husk ash |
| | ……………… | % | ……………… |
| 4 days | 45.31 bc | 35.88 e | 38.39 d |
| 8 days | 43.80 c | 43.52 c | 47.76 b |
| 12 days | 33.11 f | 45.85 bc | 53.30 a |
| 16 days | 25.91 g | 12.12 h | 24.25 g |

When using perforated plastic storage container, the longer the seeds are stored, the moisture content of the seeds decreases. But when using a storage container in the form of plastic + sawdust and plastic container + husk ash, then the water content of the seeds will increase until the storage time of 12 days, then the water seed will decrease after the seeds are stored 16 days. The highest seed water content (53.30%) was found in seeds stored in plastic + husk ash for 12 days which was significantly different from other treatment combinations.

According [5] that the high moisture content of seed during storage due to additional water from the process of respiration seed. Increased moisture content of this seed is in line with the reduced carbohydrate content in the seeds. The reducing sugar content is closely related to the respiration rate of the seeds in which the higher the sugar content of the respiration rate of the seed respiration rate will increase. As a result the water content in the seeds will increase because in the process of respiration will be released water. Therefore seeds stored for 12 days have high water content but will experience decrease when stored longer due to decline of seed viability. Recalcitrant seeds can not stand stored for long periods of time and there will be a process of declining the seed during storage.

Storage containers affect changes in moisture content after storage. In perforated plastic container , oxygen ($O_2$) and other gases can come out and enter into the container that causes the seed to do the imbibision. Sawdust has a hygroscopic property so that the water vapor attached to sawdust will be absorbed by the seed so that the determination of the container is very crucial since the cocoa seed is the recalcitrant seed.

### *3.4.Percentage of seeds germination in nursery*

Storage period (T), storage container (S), interaction of T x S, significantly effect on percentage of seed germination, while PEG concentration (P), interaction of Tx P; P x S, an interaction of T x P x S had not significant effect on the percentage of seed germination in nursery. Data percentage of seeds germination in nursery are presented in Table 5.

**Table 5.** Percentage of seeds germination in nursery in the interaction of storage period with storage container

| Storage Period | Storage container | | |
|---|---|---|---|
| | Perforated plastic | Plastic + sawdust | Plastic + husk ash |
| | ................................... % ........................................... | | |
| 4 days | 84.96 a | 86.41 a | 84.96 a |
| 8 days | 87.13 a | 87.13 a | 84.56 b |
| 12 days | 87.13 a | 87.13 a | 87.13 a |
| 16 days | 85.28 a | 7.22 c | 81.67 b |

The seed storage up to 12 days with different containers resulted in a high percentage of seeds germination in nursery. Especially when using perforated plastic containers, storage up to 16 days can still maintain the percentage of seeds germinated in the nursery is high enough. Meanwhile, when using a plastic + sawdust and plastic + husk ash stored up to 16 days will decrease percentage of seeds germination in nursery.

Storage containers have a great influence on seed conditions during storage. Environmental factors such as water and oxygen have an effect on seed germination. The moisture content of the seeds is too low causing decreased power growing seeds and cocoa seeds belonging to recalcitrant seeds are not able to withstand excessive dehydration and will soon lose its viability at relatively low humidity. According to [6] states that in general the recalcitrant seeds are high moisture content and the moisture content is correlated with its viability.

## 4. Conclusion

Treatment of cocoa seeds with PEG 45% stored in perforated plastic container up to 12 days can minimize the percentage of seeds germinated during storage up to 2.90%. The longer of storage period for all types of storage container, the percentage of moldy seeds in storage is increasing. The lowest percentage of moldy seeds (2.90%) was found in seed treatment with 15% PEG stored in plastic + husk ash. The highest seed water content (53.30%) was found in seeds stored in plastic + husk ash for 12 days. The seed storage up to 12 days using different storage containers can still maintain a high percentage of germination seeds in nursery.

## References

[1] Roberts, E H 1973 Predicting the storage life of seeds Seed Sci. Technol. **1** pp. 499 - 514.

[2] Vertucci C W and Farrant J M 1995 Acquisition and loss of desiccation tolerance In Kigel J, Galili G (Eds.) Seed Development and Germination. Marcel Dekker Inc. New York. pp. 237- 271

[3] Rahardjo P 1986 Penggunaan Polyethylene Glikol (PEG) sebagai medium penyimpanan benih kakao (Theobroma cacao L.) Pelita Perkebunan **II (3)** pp. 103 – 108

[4] Pusat Penelitian Kopi dan Kakao Indonesia 2004 Panduan lengkap budidaya kakao Agromedia Pustaka Jakarta

[5] Purwaningsih O 2001 Kajian fisiologis dan biokimiawi benih rambutan (Nephelium lappaceum L.) selama penyimpanan dengan perlakuan ABA dan $GA_3$. Ilmu Pertanian Vol. **8 (2)** pp. 66-75

[6] Sutopo L 1985 *Teknologi Benih (Seed Techonology)* Rajawali Jakarta

# Genetic variability and heritability on Kipas Putih soybean mutant lines using gamma rays irradiation (M3 generation)

**Nilahayati[1], Rosmayati[2], D S Hanafiah[2] and F Harahap[3]**

[1]Lecturer of Agroecotechnology, Agriculture Faculty, Universitas Malikussaleh, Aceh-Indonesia 24355/Doctoral Student of Agriculture Science at Agriculture Faculty, Universitas Sumatera Utara (USU), Medan 20155.
[2]Lecturer of Agroecotechnology, Agriculture Faculty, USU, Medan 20155.
[3]Lecturer of Biological Science, UNIMED, Medan
E-mail : nilahayati@unimal.ac.id

**Abstrack.** The objective of the research was to determine the selection criteria in Kipas Putih Soybean with gamma rays irradiation in M3 generation. In this study there are four populations namely population 0 Gray (control), 100 Gray, 200 Gray and 300 Gray. The results showed that high genotypic coefficient of variation (GCV) was obtained on the number of pods and seed weight per plant in population 200 Gy, number of branches on 100 Gy and 300 Gy, number of pods and seed weight per plants 100 and 200 Gy, moderate GCV in number of branches on 100 and 300 Gy while other characters such as plant height, numbers of branches on 200 Gy, flowering and harvest age have narrow GCV criteria. High heritability values are found in 300 Gy on plant height, number of branches 300 Gy, number of pods and seed weight per plant 200 Gy, days to flower 200 Gy and 300 Gy and days to harvest of all doses. Based on the genetic variabiliy and heritability, characters that can be used as selection criteria in this study is number of pods and seeds weight per plant, days to flower and days to harvest.

## 1. Introduction

The mutation using gamma rays irradiation is the one way to create genetic variation in new varieties with better characteristics. The application of gamma rays in soybean has produced various varieties of early maturity soybeans. [1] used local soybean material (Tidar variety), then it was irradiated with 200 Gy gamma irradiation dose. Furthermore, selection of early maturity mutants were undertaken from M2 generation and selected to M7 generation. Their result of the research was 2 mutants showing significant differences in productivity. The Q-298 and 4-Psj mutants had 66 days and 68 days to harvest with high productivity (2.41 and 2.42 tons / ha). Now, these mutants have released as new superior varieties of soybean (better known as Gamasugen 1 and Gamasugen 2).

Selection is the important process in plant breeding for character improvement. Selection activities are largely determined by the availability of wide genetic variability with high heritability. In plant population, only the phenotype values can be observed and measured directly. The observed variance of phenotypes of population shows the degree of phenotypic differences between individual groups arising from the presence of genetic and/or environmental variability [2]. High heritability values indicate a high degree of relationship between phenotype and genotype, and in this case genetic factors have a greater influence than environmental influences on the appearance of the character or phenotype. This suggests that the action of the additive gene is dominant (Malik *et al.*, 2006)

The previous research used gamma ray irradiation of local Aceh soybean (Kipas Putih) to obtain the early maturity and high yielding mutants in M1 generation [3]. To find out whether the desired characters are inherited or not, the information about heritability and genetic diversity of the desired characters is required. This study aims to determine the heritability and genetic variability values of Kipas Putih soybean mutant using gamma ray irradiation in M3 generation.

## 2. Materials and Methode

This research was conducted at Reuleut Timu, North Aceh-Indonesia. In M2 generation, each plant at each dose of gamma ray irradiation treatment was harvested and all of the seeds were grown as $M_3$ generation. $M_3$ seeds at each dose treatment were planted with spacing 40 x 20 cm. All technical cultivation such as fertilizing, watering, weeding and pest control were carried out during the period of plant growth. The plant was selected based on morphological appearance, early maturity and high yielding mutant.

The observation was done on plant height, number of branches, number of pods, number of seeds per plant, seed weight per plant, 100 seed weight, flowering and harvesting age. The data analysis was done by calculating the mean of each character observed, then the mean value of each population was tested by using t test. The data analysis was followed by various phenotypes and environmental variations and estimates of genotypes, heritability and genetic variability coefficient values of potential mutants selected for each population in accordance with irradiated doses. Differential selection was obtained from the difference between the average of selected population and the avarage of basic population. Genetic variation in $M_3$ generation was calculated with the formula:

$$\sigma^2 = \frac{(\sum x^2) - [(\sum x)^2 / n]}{n-1}$$

$\sigma^2 M_3 = \sigma^2 p$
$\sigma^2 p = \sigma^2 g + \sigma^2 e$
$\sigma^2 g = \sigma^2 p - \sigma^2 e = \sigma^2 M_3 . \sigma^2 M_0$
$\sigma^2$ = variance
n = number of population
$\sigma^2 p$ = phenotipic variance
$\sigma^2 g$ = genotypic variance
$\sigma^2 e$ = environmental variance
$\sigma^2 M_2$ = M3 population variance
$\sigma^2 M_0$ = M0 poplation variance

Heritability values were calculated by using the following formula:

$$h^2 = \frac{\sigma_g^2}{\sigma_p^2}$$

$\sigma^2 p = \sigma^2 g + \sigma^2 e$

$$h^2 = \frac{\sigma_g^2}{\sigma_g^2 + \sigma 2e}$$

Genetic variation was determined based on the coeffisient of genetic variation (CGV) with the formula:

$$CGV = \frac{\sigma_g}{\bar{X}} \text{ x } 100\%$$

$\sigma_g$ = genotypic standard deviation

$\overline{X}$ = character means

## 3. Result and Discussion

Genotype selection is a basis for improving plants in obtaining the new superior varieties. The role of genotypic variation aiming to assembly the superior varieties is important because high genotypic coefficient of variation provides an opportunity to obtain the source of genes for the improved character [4]. In $M_3$ generation, genotypic coefficient of variation values ranged from 2.43-32.06 with narrow to wide criteria. High CGV values were obtained on number of pods (32.06) and seed weight per plant (31.91) in 200 Gy population, moderate CGV criteria were found at number of branches 100 Gy and 300 Gy, number of pods and seed weight per plant at 100 Gy and 200 Gy, while other characters such as plant height, number of branches 200 Gy, days to flower and days to harvest had narrow CGV criteria.

**Table 1.** Genetic variation of $M_3$ plants at different gamma irradiation doses in Kipas Putih soybean

| No. | Characters | Irradiation doses | | | |
|---|---|---|---|---|---|
| | | P 0 Gy | P 100 Gy | P 200 Gy | P300 Gy |
| 1. | Plant height | 77.70 | 68.20 | 66.47 | 67.44 |
| | $\sigma^2$p (fenotipe variance) | 95.88 | 100.19 | 93.27 | 61.40 |
| | $\sigma^2$g (genotipe variance) | | 4.31 | 2.61 | 34.47 |
| | CGV (coeffisien of genotypic variance) | | 3.04 | 2.43 | 8.70 |
| | CGV criteria | | Narrow | Narrow | Narrow |
| 2. | Number of branches | 5.46 | 5.05 | 6.14 | 5.44 |
| | $\sigma^2$p | 2.62 | 2.03 | 2.80 | 1.51 |
| | $\sigma^2$g | | 0.58 | 0.18 | 1.10 |
| | CGV | | 15.06 | 7.06 | 19.29 |
| | CGV criteria | | moderate | Narrow | Moderate |
| 3. | Number of pods | 241.47 | 186.97 | 253.37 | 218.60 |
| | $\sigma^2$p | 5126.45 | 4577.76 | 11726.55 | 4372.24 |
| | $\sigma^2$g | | 548.75 | 6600.09 | 754.21 |
| | CGV | | 12.52 | 32.06 | 12.56 |
| | CGV criteria | | moderate | High | Moderate |
| 4. | Seed weight per plant | 51.88 | 41.32 | 52.71 | 44.47 |
| | $\sigma^2$p | 260.93 | 203.97 | 544.08 | 213.19 |
| | $\sigma^2$g | | 56.96 | 283.14 | 47.74 |
| | CGV | | 18.26 | 31.91 | 15.53 |
| | CGV criteria | | moderate | High | Moderate |
| 5. | Days to flower | 40.20 | 39.13 | 39.28 | 39.67 |
| | $\sigma^2$p | 0.42 | 0.78 | 1.38 | 0.89 |
| | $\sigma^2$g | | 0.35 | 0.96 | 0.46 |
| | CGV | | 2.50 | 2.49 | 2.47 |
| | CGV criteria | | Narrow | Narrow | Narrow |
| 6. | Days to harvest | 89.97 | 88.99 | 90.00 | 89.01 |
| | $\sigma^2$p | 0.05 | 5.26 | 13.17 | 5.17 |
| | $\sigma^2$g | | 5.21 | 13.11 | 5.12 |
| | CGV | | 4.06 | 4.02 | 4.06 |
| | CGV criteria | | Narrow | Narrow | Narrow |

Characters that had wide CGV values were number of pods and seed weight per plant in 200 Gy population. [5] found high CGV values on the number of pods per plant (29,59) Mungbean variety

K851 in $M_3$ generation. Other characters such as 50% days to flower (2.76), days to harvest (7.42), seed yield per plant (15.73), plant height (12.74) had low to moderate CGV values. Their research also obtained a high value of heritability on all characters of mungbean mutant.

The heritability values observed in this $M_3$ generation population ranged from 0.02 - 0.99 (Table 2). High heritability values were found in plant height 300 Gy (0,56), number of branches 300 Gy (0,72), number of pods (0,56) and seed weight per plant (0,52) 200 Gy, days to flower 200 Gy and 300 Gy (0.69 and 0.52) and days to harvest of all doses (0.98, 0.99 and 0.98). The population of 200 Gy had high heritability values except for the plant height and number of branches. The population of 100 Gy had low to moderate heritability values in all observed characters except for the days to harvest. The population of 300 Gy had high heritability values for all observed characters except for the number of pods and seed weight per plant.

Previous study conducted by [6] observed high heritability in greenbean with ranged from 41,30-88,96%. This result of the study indicated that the inheritance of characters observed was large enough and the selection activities for the improvement of these properties were easy to perform. The character of days to flower, days to harvest and number of pods had heritability of 68.81, 61.79 and 63.29 respectively.

[7] states that heritability values are needed to predict whether the character is much influenced by the environment or genetics. High heritability values indicate that genetic factors influence is greater than phenotypic appearance compared to environmental influences. High heritabilility values play an important role in improving the effectiveness of selection. [8] reports that heritability values indicates whether a character is controlled by genetic or environmental factors, which in turn it can be known to what extent characters can be inherited to the next generation. If the heritability values are high, most phenotypic variations are caused by genetic variation, then the selection will gain genetic progress [9].

**Table 2.** Heritability values in $M_3$ generation of Kipas Putih soybean using gamma irradiation

| No | Characters | Heritability | | |
|---|---|---|---|---|
| | | P 100 | P 200 | P 300 |
| 1. | Plant height | 0.04 | 0.02 | 0.56 |
| 2. | Number of branches | 0.28 | 0.06 | 0.72 |
| 3. | Number of pods | 0.11 | 0.56 | 0.17 |
| 4. | Seed weight per plant | 0.27 | 0.52 | 0.22 |
| 5. | Days to the first flower | 0.45 | 0.69 | 0.52 |
| 6. | Days to harvest | 0.98 | 0.99 | 0.98 |

The variability in F3 wheat family can be caused by genetic and environmental factors. In order that selection provides the expected genetic progress, it must be determined in advance the appropriate selection characters. Character selection for a population in a particular environment is chosen based on heritability values. Characters that can be used to select the genotype are characters that have high heritability [10].

The heritability values are one of the genetic parameters considered for selecting the characters in selection [7][11]. Days to flower and days to harvest characters had high heritability values. This indicated that the appearance of the character was influenced a little by environment. Selection might be more effective in characters with high heritability because the environmental impact was very small. Based on the high heritability values, the character that can be considered to select the best family in the next stage of this research were days to flower, days to harvest, number of pods and seed weight per plant.

The results showed that high CGV with high heritability values were number of pods and seed weight per plant characters in 200 Gy population, while days to flower and days to harvest had high heritability values but low CGV values. [12] stated that the high values of genetic coefficients are generally followed by high heritability and genetic variability, so it can be used for selection. The results of this study indicated that days to flower and days to harvest had low CGV but showed high heritability. [13] reports that the days to harvest character in wheat strain is one of character that has high heritability, so that this character can be used as the character of selection in the early generation in the process of assembling varieties.

Selection to obtain early maturity soybean was relatively easy to do because it has a high heritability values in days to flower and days to harvest. Meanwhile, for high yielding improvement, number of pods and seed weight per plant characters can be used as selection criteria. According to[14], the number of pods per plant has high heritability and directly affected the production.

## 4. Conclusions

In summary, it can be concluded that in 200 Gy population was identified high CGV and heritability in number of pods and seed weight per plant characters. Days to flower and days to harvest characters had high heritability with narrow CGV. Based on the genetic variabiliy and heritability, the characters that can be used as selection criteria in this study were number of pods, seeds weight per plant, days to flower and days to harvest.

## References

[1] Arwin, Mulyana, H I, Tarmizi, Masrizal, Fauzi K and Mukhlis A 2012 Galur mutan harapan kedelai super genjah Q298 dan 4 PJS. *Jurnal ilmiah aplikasi isotop dan radiasi* **8 (2)** 107-116.

[2] Barmawi M, A Yushardi and N. Sa'diyah 2013 Daya waris dan harapan kemajuan seleksi karakter agronomi kedelai F2 hasil persilangan antara Yellow Bean dan Taiching. *J. Agrotek Tropika* **1** 20-24

[3] Falconer O S, Mackay T F C 1996 *Introduction to quantitative genetic*, Fourth edition. Longman.

[4] Lestari A D,W. Dewi W, Qosim W A, Rahardja M, Rostini N and Setiamihardja 2006 Variabilitas genetik dan heritabilitas Karakter Komponen Hasil dan Hasil Lima Belas Genotip Cabai Merah *Zuriat* **17(1)** 94-102

[5] Martono B 2009 Keragaman genetik, heritabilitas dan korelasi antar karakter kuantitatif nilam (*Pogostemon sp*) hasil fusi protoplasma *J. Littri* **15** 9-15

[6] Nilahayati, Rosmayati, D S Hanafiah and F Harahap *2015 Proceedings of The 1th Almuslim International Conference on Science, Technology and Society (AICSTS)* Induction of genetic variability in Kipas Putih soybean with gamma ray irradiation (M1 generation). Bireuen, Indonesia p178-183

[7] Nur A, Trikoesoemaningtyas, Khumaida N and Yahya S 2012 Evaluasi dan keragaman genetik 12 galur gandum introduksi di lingkungan tropika basah *J. Agrivigor* **11** 230-243

[8] Rao Ch M, Rao Y K and Reddy M 2006 Genetic variability and path analysis in Mungbean *Legume Res.* **29(3)**:216-218

[9] Rathore J, Lavanaya G R, Kumar N and Khaedar O P 2014 Evaluation of gamma rays induced genetic variability and divergence in M3 generation og Mungbean. *International Journal of Innovation and Research (IJAIR)* **3(2)** 2319-1473

[10] Suprapto and N Kairudin 2007 Variasi genetik, heritabilitas, tindak gen dan kemajuan genetik kedelai (*Glycine max* Merrill) pada Ultisol *Jurnal Ilmu-ilmu Pertanian Indonesia* **9(2)**:183-190

[11] Syukur M, Sujiprihati S, Yunianti S and Nida K 2010 Pendugaan komponen ragam, heritabilitas dan korelasi untuk menentukan kriteria seleksi cabai (*Capsicum annuum* L.) populasi F5 *J. Hort.Indonesia* **1(3)**: 74-80

[12] Sulistyo, Yuliasti A 2012 Nilai duga heritabilitas galur-galur mutan kacang hijau (*Vigna radiata*) Seminar nasional pangan UPN Veteran Yogyakarta p 13-16

[13] Wardani, Wirnas D S and Wahyu Y 2015 Seleksi segregan gandum (Triticum aestivum L) pada dataran tinggi *J. Agron. Indonesia*. **43(1)**:45-51

[14] Yunianti, Sastrosumarjo R S, Sujiprihati S, Surahman M and Hidayat S H 2010 Kriteria seleksi untuk perakitan varietas cabai tahan *Phytopthora capsici* Leonian. *J. Agron. Indonesia* **38**:122-129

# Effect of long-term salinity on the growth and biomass of two non-secretors mangrove plants *Rhizophora apiculata* and *Ceriops tagal*

**M Basyuni[1*], A Nuryawan[1], Yunasfi[1], L A P Putri[2] and S Baba[3]**

[1] Department of Forestry, Faculty of Forestry, Universitas Sumatera Utara, Jl. Tri Dharma Ujung No. 1 Medan, North Sumatera 20155, Indonesia
[2] Department of Agrotechnology, Faculty of Agriculture, Universitas Sumatera Utara, Medan, North Sumatera 20155, Indonesia
[3] International Society for Mangrove Ecosystems (ISME), Faculty of Agriculture, University of the Ryukyus, Nishihara, Okinawa 903-0213, Japan
[*]E-mail: m.basyuni@usu.ac.id

**Abstract.** The present study describes the effect of long-term salinity on morphological character and biomass content of two non-secretors mangrove plants *Rhizophora apiculata* and *Ceriops tagal*. Two mangrove seedlings were grown for six months in 0%, 0.5%, 1.5%, 2.0% and 3.0% salt concentration. The growth of *R. apiculata* was significantly enhanced by salt with maximal stimulus at 1.5% (equal to 50% natural seawater), and this increase appeared to be attenuated by increasing the salinity concentration above 1.5%. By contrast, the growth of *C. tagal* thrived up to 0.5% salt concentration. Our findings, therefore, suggested that within the range of treatments used, 1.5% and 0.5%, respectively were the optimal salinity of *R. apiculata* and *C. tagal* for growth. The highest leaf area of *C. tagal* was obtained at 1.5% salinity concentrations and, on the other hand, *R. apiculata* showed much greater extent. The wet and dry weight of the two seedlings was changed in the same manner as the height of plants upon salt treatment. Our results indicated that *R. apiculata* was more salt tolerant than *C. tagal*, which may provide valuable information for mangrove rehabilitation in North Sumatra, Indonesia.

## 1. Introduction

Mangrove plants comprise a heterogeneous group of independently derived lineages that are denoted ecologically by their location in inter-tidal zones of tropical and sub-tropical areas and physiologically by their ability to withstand high concentrations of salt or low levels of soil aeration [1]. Indonesia is one of the world's great mangrove nations, covering over 300,000 square kilometers, which is 22.6% of the global total [2]. According to their morphological characteristics in salt management, mangrove plants fell into two broad groups [3]. The first group is the salt-secreting species that have either salt glands or salt hairs to remove excess salt. The second is non-secreting species that do not have such morphological features for excretion of excess salt [3-4]. *Rhizophora apiculata* and *Ceriops tagal* (Rhizophoraceae) are common mangrove species in Pulau Sembilan, North Sumatra, Indonesia and belonging to the latter group.

Mangroves are one of the most threatened ecosystems all over the world today due to direct and indirect deforestation and degradation [5]. Conservation of mangrove implicates not only to protect the coastal areas and communities from seawater intrusion and potential changes in sea level rise but also to ensure the availability of resources for future use through adaptation to changing environments.

Studies on the salt tolerance of North Sumatran mangroves is rare, and this information is vital to the success of rehabilitation and conservation effort in the region. Our previous studies shed light on the triterpenoid content and gene expressions of triterpenoid synthases in salt-treated and re-adapted mangrove plants [6-8]. It is important to obtain more insight into the morphological aspects of mangrove seedlings to cope with salinity, which may experience temporary periods of exposure to high salt concentrations. The present study, therefore, aimed to investigate the response of two non-secretors mangrove species, *R. apiculata* and *C. tagal* under various salt concentrations with the emphasis on seedling growth and biomass.

## 2. Materials and methods

### 2.1. *Salt tolerant experiment*

Viviparous mature and healthy propagules of two mangrove species *Rhizophora apiculata* BL. and *Ceriops tagal* (Perr.) C.B. Robinson (Rhizophoraceae) were collected from Sembilan Island, Langkat, North Sumatra, Indonesia, and planted in bottle pots with exposure to natural sunlight in a glass house, Faculty of Agriculture, University of Sumatera Utara. A seawater solution was prepared by dissolving commercial salt powder (Tropical Marine Reef Sea Salt, Thailand) to produce a salt concentration of 0%, 0.5%, 1.5%, 2.0% and 3% (equal to the level in sea water) by the manufacturer's protocol. Salinity in this study was defined as the mass of salt powder/weight of solution [9]. Salt concentration in the pots for each treatment was checked once weekly during the experiments using a salinity Refractometer S/Mill-E (Atago Co. Ltd., Tokyo, Japan), and adjusted accordingly. Nine to twelve plants per treatment grew for six months. After six months of cultivation, the two mangrove plants were harvested, washed, and the leaves and roots were stored at –4 °C for further analysis.

### 2.2. *Growth measurement*

The growth of *R. apiculata* and *C. tagal* seedlings was determined under varying salt concentrations. The stem heights and diameter of *R. apiculata* and *C. tagal* after six months of cultivation were the indices of growth used in this study. Measurement of the leaf area, the leaves were digitally scanned to quantify leaf area using ImageJ software [10].

### 2.3. *Biomass measurement*

Mangrove seedlings were separated into root, shoot, and leaf components to weight before drying. Dry weights of leaves, shoots, and roots of each plant were determined separately on balance, after drying in an oven at 75 °C for 48 h. From these measurements, we calculated the wet and dry weight of leaves, shoots, and roots of individual seedling of both species.

### 2.4. *Statistical analysis*

Data were analyzed by one-way analysis of variance (ANOVA) followed by Dunnett's test for comparisons of all treatments against the control. The value of $P < 0.05$ was selected as a limit of statistical significance. All statistical analyses were performed using the SAS 9.1 statistical software program (SAS Institute Inc. Cary, NC, USA).

## 3. Results and discussion

### 3.1. *Effect of salinity on the seedling growth*

Seedling growth was measured by the height and diameter of the plants as described in Materials and Method. The growth of *R. apiculata* enhanced in the presence of salt with maximal stimulation at 1.5%. This increase appeared to be attenuated when the salinity was elevated more than 1.5% (Figure 1A). By contrast, the growth of *C. tagal* thrived up to 0.5% salt concentration (Figure 1A).

The salinity affected the secondary growth of diameter seedlings significantly in 0.5-2.0% salt concentration of *C. tagal* and only 1.5% in *R. apiculata* (Figure 1B). Effect of salinity on the number of leaves and leaves area was shown in Figure 1 (C and D). Salinity influenced the number of leaves of both species (Figure 1C) and leaves area only in *C. tagal* (Figure 1D). No significant change was noted in the leaves are of *R. apiculata* (Figure 1D).

Mangrove species often show growth stimulation at low salinity (25% seawater/0.5% salt concentration) and moderate salinity (50% seawater/1.5% salinity) and then a decline in growth with further elevates in salinity [11-15]. However, the range of salinity in which the plant can survive varies according to the species [16]. This study suggested that mangrove seedlings survived to adapt either in saline or fresh water tolerance.

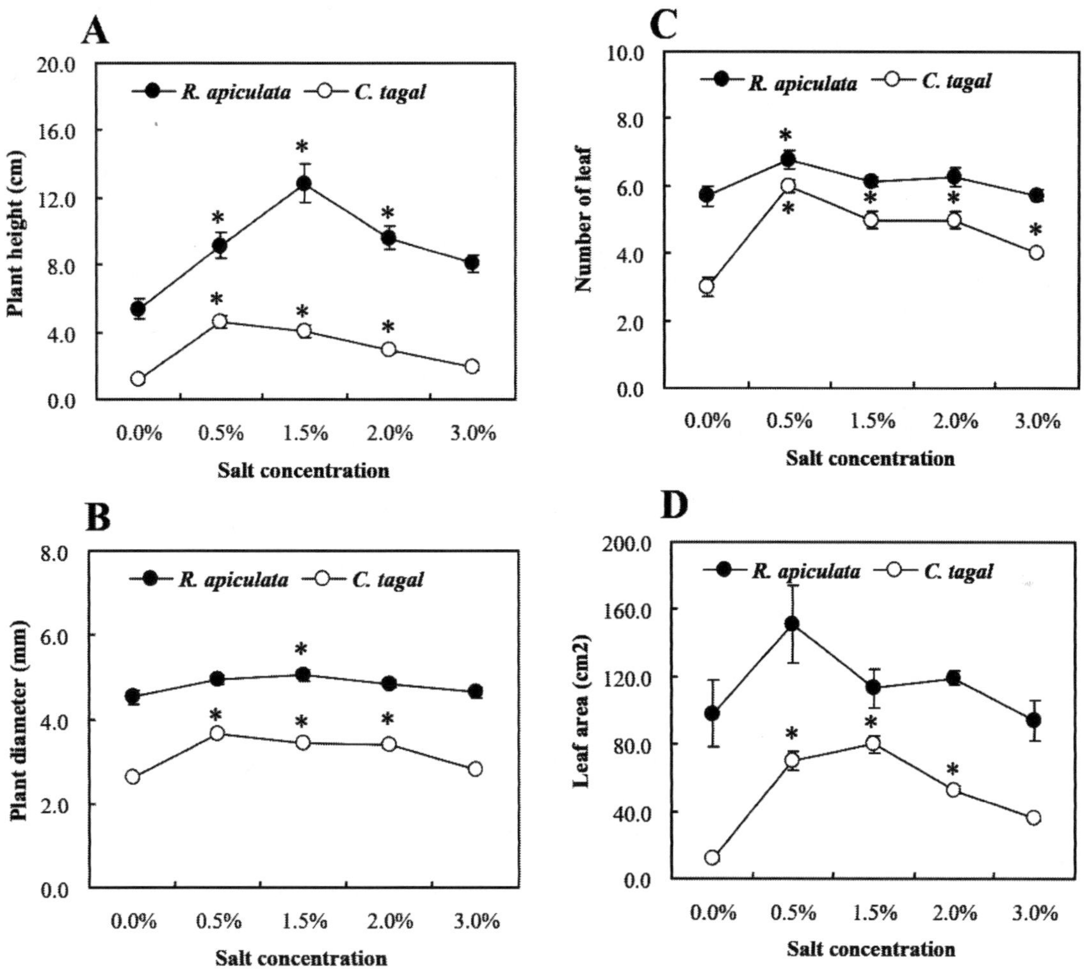


**Figure 1**. Effect of salinity on plant height (A), diameter (B), some of the leaves (C), and leaf area (D) in the leaves in *R. apiculata* (•) and *C. tagal* (○) seedlings. Data presented are the mean ± SE ($n$=9-12) and ($n$=5) for growth measurements and many leaves and leaf area, respectively. The asterisk indicates a statistically significant difference from 0% at $P< 0.05$ using Dunnett's test.

Our current results also well agree with our previous reports on the optimal growth at 0.5% salinity in *Kandelia candel* and *Bruguiera gymnorrhiza*, respectively [6]. The development of both species slightly increased after removal to salinity [8]. Furthermore, the optimum growth at 1.5% salt concentration found in *Avicennia marina* and *Rhizophora stylosa* grown in a glass house [7].

Patel *et al.* [15] studied that *C. tagal* is a salt-tolerant mangrove with the high ability to develop in hyper saline and poorly inundated sites. It has been reported that *C. tagal* from Pakistan showed an optimal growth at 50% seawater salinity [17]. Growth and survival of halophytes are reliant on the high levels of ion accumulation in its tissue for the maintenance of turgor and osmotic alteration [18]. Seedling growth and establishment would extend mangrove distribution and increase mangrove rehabilitation as well as the land creation.

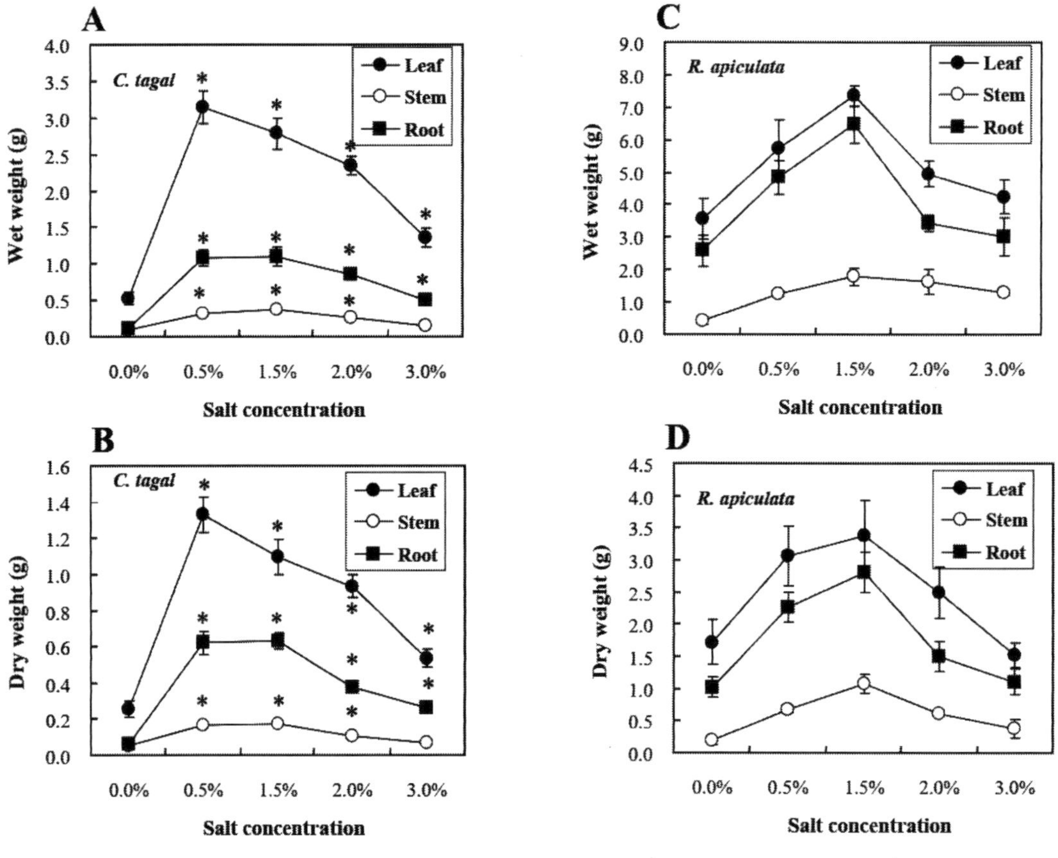


**Figure 2**. Effect of salinity on wet (A and C) and dry weight (B and D) in *C. tagal* and *R. apiculata* seedlings, respectively. Data presented are the mean ± SE ($n$=5) in biomass measurements. The asterisk indicates a statistically significant difference from 0% at $P < 0.05$ using Dunnett's test.

### *3.2. Effect of salinity on the seedling growth*

Seedling biomass was studied concerning to different salt concentration in two non-secretors species *R. apiculata* and *C. tagal*. As shown in Figure 2, salinity changed the wet and dry weight of both seedlings. No statistical difference was noted in wet and dry weight of *R. apiculata* (Figure 2 C and D), on the other hand, statistically different was found only in *C. tagal* seedling (Figure 2 A and B).

Biomass production of *C. tagal* and *R. apiculata* was stimulated at 0.5 and 1.5% salt concentration. Similar results were obtained in *Avicennia marina, Ceriops tagal* and *Rhizophora mucronata* [16]. Our study indicates that *C. tagal* and *R. apiculata* growing along the aridest region of East Sumatra coast are more tolerant to salinity.

Our present findings supported our previous results on the salt tolerance of four mangrove seedlings studied in our laboratory increases in the order of *B. gymnorrhiza<K. candel<C. tagal<R. apiculata<R. stylosa<A. marina*, which is in agreement with their habitat zonation. *A. marina* grows closest to sea, whereas *B. gymnorrhiza* distributed rather inland than in the coastal area compared to *C. tagal, R. apiculata, R. stylosa* and *K. candel*, indicating *B. gymnorrhiza* was less tolerant to salt stress [1,7].

## 4. Conclusions

The current study apparently showed that within the range of treatments used, 1.5% and 0.5% were the optimal salinity by *R. apiculata* and *C. tagal* for growth, providing necessary information for mangrove rehabilitation in North Sumatra. Furthermore, our study indicated that the salt tolerance of the two non-secretors mangrove seedlings increased *C. tagal<R. apiculata*, which is accordance with habitat zonation of mangrove.

## References


[1] Basyuni M, Oku H, Tsujimoto E, Kinjo K, Baba S and Takara K 2007 Triterpene synthases from the Okinawan mangrove tribe, Rhizophoraceae *FEBS J.* **274** 5028–5042.

[2] Giri C, Ochieng E, Tieszen L L, Zhu Z, Singh A, Loveland T, Masek J and Duke N 2011 Status and distribution of mangrove forests of the world using earth observation satellite data *Global Ecol. Biogeogr.***20** 154–159.

[3] Scholander P F, Hammel H T, Hemmingsen E and Garey W 1962 Salt balance in mangroves *Plant Physiol.* **37** 722–729.

[4] Tomlinson P B 1996 *the Botany of Mangroves* Cambridge University Press London.

[5] Duke NC, Meynecke J O, Dittmann S, Ellison A M, Anger K, Berger U, Cannicci S, Diele K, Ewel K C, Field C D, Koedam N, Lee SY, Marchand C, Nordhaus I and Dahdouh-Guebas F 2007 A world without mangroves? *Science* **317** 41–42.

[6] Basyuni M, Baba S, Inafuku M, Iwasaki H, Kinjo K and Oku H 2009 Expression of terpenoid synthase mRNA and terpenoid content in salt-stressed mangrove *J. Plant Physiol.***166**1786–1800.

[7] Basyuni M, Baba S, Kinjo Y and Oku H 2012 Salinity increases the triterpenoid content of a salt secretor and a non-salt secretor mangrove *Aquat. Bot.***97** 17–23.

[8] Basyuni M, Baba S, Kinjo Y, Putri LAP, Hakim L and Oku H 2012 Salt-dependent increase in triterpenoids is reversible upon transfer to fresh water in mangrove plants *Kandelia candel* and *Bruguiera gymnorrhiza J. Plant Physiol.***169** 1903–1908.

[9] Fofonoff NP and Lewis EL 1979 A practical salinity scale *J. Oceanogr.***35** 63–64

[10] Schneider C A, Rasband W S, Eliceiri K W 2012 NIH image to ImageJ: 25 years of image analysis *Nature Methods* **9** 671–675

[11] Clough B F 1984 Growth and salt balance of the mangroves *Avicennia marina* (Forssk.) Vierh. and *Rhizophora stylosa* Griff. in relation to salinity*J. Plant Physiol.***11** 419–430.

[12] Downton W J S 1982 Growth and osmotic relations of the mangrove *Avicennia marina*, as influenced by salinity *Aust. J. Plant Physiol.* **9** 519–528.

[13] Naidoo G 1987 Effects of salinity and nitrogen on growth and plant water relations in the mangrove *Avicennia marina* (Forssk.) Vierh*New Phytol.***107** 317–326.

[14] Burchett M D, Clarke C J, Field C D and Pulkownik A 1989 Growth and respiration in two mangrove species at a range of salinities *Physiol. Plant.***75** 299–303.

[15] Patel NT, Gupta A and Pandey AN 2010 Strong positive growth responses to salinity by *Ceriops tagal*, a commonly occurring mangrove of the Gujarat coast of India*AoB PLANTS***2010** plq011.

[16] Ball MC 1988 Salinity tolerance in the mangroves, *Aegiceras corniculatum* and *Avicennia marina*. Water use in relation to growth,carbon partitioning and salt balance*Aust. J. Plant Physiol.* **15** 447–464.

[17] Khan MA and Aziz I 2001 2001 Salinity tolerance in some mangrove species from Pakistan*Wetl. Ecol. Manag.* 9 219–223.
[18] Flowers TJ, Troke PF and Yeo AR 1977 The mechanism of salt tolerance in halophytes*Annu. Rev. Plant Physiol.* **28** 89–121.

**Acknowledgment**
This work was supported in part by a Competence Grant (No. 017/SP2H/LT/DRPM/II/2016 to MB) from the Directorate for Research and Community Service, Ministry of Research, Technology and Higher Education, Republic of Indonesia. A Grant-in-Aid for a Community Service Research-Based of BOPTN grant (No.453C/UN5.2.3.2.1/PPM/2015 to MB) from University of Sumatera Utara 2015 is acknowledged.

# Entisol land characteristics with and without cover crop (*Mucuna bracteata*) on rubber plantation

**Sakiah, M Sembiring and J Hasibuan**

Plantation agriculture, STIPER-Agribusiness Estate (STIPAP), Indonesia
E-mail: sakiah2@gmail.com

**Abstract**. Optimal nutrient delivery is one way to improve the quality and quantity of crop production. This is because the crops needs for nutrient is quite high, while the soil capacity in providing nutrients is limited. In addition to fertilization, nutrients can be given in the form of added organic material or planted as cover crop. The research took place from April to August 2016 in Bandar Pinang, Bandar Sumatera Indonesia Ltd. (SIPEF Group) plantation, with survey method. Soil samples were taken based on: Topography (flat and slope 15-30%), cover crop (with or without *Mucuna bracteata*) and plant age (seedling periods 1, 2 and 3). The soil sample is taken composite by zig zag method. The observed parameters were organic matter, N total, soil texture, bulk density and infiltration rate. *Mucuna bracteata* planting increased the contain of soil organic matter by 30.43% in flat area and 53.33% in hilly area, amount of N total soil by 27.27% in flat area and 7.69% at hilly area, bulk density 3.73 % In flat area and 0.41% in hilly area, soil infiltration by 48.88% with sandy clay dominant soil texture.

## 1. Introduction

Entisol has a loose consistency, low aggregation rate, is sensitive to erosion and low availability of nutrient content. The soil potential which derived from this volcanic ash is rich in nutrients but not available yet, corrosion will be accelerated when there is sufficient organic matter activity as a provider of organic acids [1]

In the management of rubber plantations, the policy to plant legume cover crop is carried out on young rubber cultivation. The planting of legume cover crop are aimed to overcome surface erosion and soil nutrient leaching, enrich organic materials, enrich soil N-nutrients, improve soil structure, maintain soil moisture and suppress weed growth.

The covering rate of *Mucuna bracteata* can reach 2-3 $m^2$ in a month. The plantations area will perfectly covered by entering the $2^{nd}$ year of planting, with 40 to 100 cm vegetation thickness. In the 3rd year the dry matter produced can reach 8-10 tons $ha^{-1}$, while other legume cover crop are 4.4 tons $ha^{-1}$. Besides biomass, soil fertility rates also increase because these legume cover crop also have root nodules that can fixate free nitrogen from the air into the $NH4^+$ form that is available for plants. [2]

Seeing the rapid used of *Mucuna bracteata* as cover crops on various land uses, hence this research is done to know the role of *Mucuna bracteata* at rubber plantation phase of immature plant in improving the properties of Entisol in flat and slope land topography.

## 2. Research Methods

The research was conducted in Bandar Pinang, Bandar Sumatera Indonesia Inc. (SIPEF Group) plantation from April to August 2016. This research used survey method; the soil was taken compositely at a depth of 20 cm in accordance with the specified sampling point. The soil sampling point was taken based on: topography (flat and slope), cover crop (with and without *Mucuna bracteata*) and plant age (Seedling period 1, 2 and 3).

The soil samples were dried, sieved with screen of 10 mesh, then analyzed in the laboratory. The observed parameters were; organic matter (*Walkley & Black* methode), N-total level (*Kjeldahl* methode), soil texture (hydrometer methode), Bulk Density (BD) (gravimetry methode) and soil infiltration (infiltrometer)

**Table 1.** Observation results of organic matter content, N-total, texture, bulk density and soil infiltration

| No | Treatment | Organic matter (%) | N-total (%) | Texture | Bulk density (gr $cm^{-3}$) | Infiltration (mm $Hour^{-1}$) |
|---|---|---|---|---|---|---|
| 1 | Flat & without MB Seedling Period 1 | 2.08 | 0.16 | sandy clay loam | 1.30 | 142.2 |
| 2 | Flat & without MB Seedling Period 2 | 2.08 | 0.09 | sandy clay loam | 1.27 | 40.8 |
| 3 | Flat & without MB Seedling Period 3 | 1.77 | 0.08 | sandy clay loam | 1.21 | 31.8 |
| 4 | Flat & with MB Seedling Period 1 | 2.58 | 0.15 | sandy clay loam | 1.27 | 75.0 |
| 5 | Flat & with MB Seedling Period 2 | 2.82 | 0.14 | sandy clay loam | 1.20 | 24.0 |
| 6 | Flat & with MB Seedling Period 3 | 2.35 | 0.13 | sandy clay loam | 1.17 | 10.8 |
| 7 | Slop & without MB Seedling Period 1 | 0.97 | 0.07 | sandy clay loam | 1.25 | 55.2 |
| 8 | Slop & without MB Seedling Period 2 | 1.94 | 0.09 | sandy clay loam | 1.19 | 34.8 |
| 9 | Slop & without MB Seedling Period 3 | 1.41 | 0.10 | sandy clay loam | 1.25 | 62.4 |
| 10 | Slop & with MB Seedling Period 1 | 1.98 | 0.10 | sandy loam | 1.27 | 60.0 |
| 11 | Slop & with MB Seedling Period 2 | 1.98 | 0.07 | sandy clay loam | 1.25 | 75.0 |
| 12 | Slop & with MB Seedling Period 3 | 2.68 | 0.11 | sandy clay | 1.16 | 199.8 |

## 3. Results and Discussions

### *3.1 Organic Matter*

Organic materials play an important role in building sustainable agriculture and preserving nature. Organic materials can improve the physical, chemical and biological properties of soil.

In this research, organic matter was derived from leaf litter produced by *Mucuna bracteata.* Average soil organic matter content in flat area without *Mucuna bracteata* is 1.98% and with *Mucuna bracteata* is 2.58%. Average soil organic matter content in slop area without *Mucuna bracteata* is 1.44%, and with *Mucuna bracteata* is 2.22%. In flat area planted with *Mucuna bracteata* increased the organic matter content by 30.43% while in slope area by 53.53%.

*Mucuna bracteata* has higher organic content than other beans for cover crop. *Mucuna bracteata* planting is able to contribute wet leaf litter up to 5.23 ton $ha^{-1}$, much higher compared to conventional LCC with only 4.41 ton $ha^{-1}$ [3]

### *3.2 N-total Content*

Nitrogen nutrient is one of the essential elements needed by plants. Nitrogen can be obtained from the air through N fixation with the help of microorganisms.

The research results showed; the average of N total on flat area without *Mucuna bracteata* is 0.11%, with *Mucuna bracteata* is 0.14%. The average of N total on slop area without *Mucuna bracteata* is 0.08%, with *Mucuna bracteata* is 0.09%. In flat area planted with Mucuna bracteata, soil N-total content increased by 27.27% while in the slope area increased by 7.69%.

Organic matter is the main source of Nitrogen in soil [4]. *Mucuna bracteata* is a cover crop which produces quite a lot leaf litter so that it will directly affect the levels of Nitrogen in the soil apart from the influence of *Rhizobium* bacteria found in the root nodule. Entisol chemical property in organic farming system shows that entisol which cultivated with organic crops have a higher of N-total content, organic farming system obviously improve soil chemical properties [5].

*3.3 Soil Texture*

The relative size of soil particles is expressed in terms of texture, which refers to the fineness or roughness of the soil. The rate and extent of various important physical and chemical reactions in plant growth is governed by texture, since this texture determines the number of surfaces at which the reaction occurs [6]. Soil texture contributes to soil structure, water system, air system and soil temperature.

Soil texture at Bandar Pinang Estate is dominant with sandy clay loam, other than that is also found sandy loam and sandy clay soil texture. In accordance with the type of soil at the location of research i.e. Entisol, one of its characteristics is rough-textured. Rough textured soil is difficult to retain water, consequently the rate of soil infiltration is relatively fast, and this is evidenced from the observation of infiltration which on average is quite fast.

*3.4 Bulk Density*

Bulk density is an indication of soil density. The denser a soil is, the higher it bulk density, which means it is increasingly difficult to continue the water or the plants root to penetrate. Bulk density is important to calculate the need for fertilizer or water for each hectare of land based on soil weight/hectare [4].

The average level of bulk density on flat area without *Mucuna bracteata* is 1.26 gr $cm^{-3}$ and with *Mucuna bracteata* is 1.21 gr $cm^{-3}$. The average level of bulk density on slop area without *Mucuna bracteata* is 1.23 gr $cm^{-3}$ and with *Mucuna bracteata* is 1.22 gr $cm^{-3}$. In flat area planted with *Mucuna bracteata* can fix bulk density by 3.73% while on slope area by 0.41%. The improvement of bulk density on the slope area planted with *Mucuna bracteata* tends to be smaller due to the surface flow that transports soil particles.

The ideal of bulk density level in soil is around 1.1 - 1.6 gr/$cm^3$ [4]. The research results obtained, showed that bulk density on each sample of observation is between 1.1 - 1.3 gr/$cm^3$. This result illustrated that the soil did not experience compaction, thus the development of the rubber plants roots and the entry of water into the soil runs well.

Previous research on palm oil, the planting of *Mucuna bracteata* in palm oil plantations was able to improve some physical and chemical soil properties such as fixing bulk density by 12.64% on flat area and 6.04% on slope area [7]

*3.5 Infiltration*

One of the important soil physical properties to be known is the infiltration rate, which is the maximum velocity of water entering vertically into the soil.

Average infiltration rate in flat area without *Mucuna bracteata* is 71.6 cm $hour^{-1}$, mean while in area planted with *Mucuna bracteata* is 36.6 cm $hour^{-1}$. In a flat area planted with *Mucuna bracteata* can improve the soil infiltration rate by 48.8%. The average infiltration rate in the slope area without *Mucuna bracteata* is 50.8 cm $hour^{-1}$, mean while with *Mucuna bracteata* is 111.6 cm $hour^{-1}$. Soil infiltration class is belonged to the very quick class; this is influenced by sand-dominated soil particles.

Cover crops can neutralize the destructive power of raindrops and suppress runoff; which can inhibit erosion and nutrient leaching. This is reflected by the positive effect of cover crops on soil physical properties [8]

## 4. Conclusions

The leaf litter produced by *Mucuna bracteata* is the main important source in improving the entisol properties through the formation of organic matter. *Mucuna bracteata* planting on rubber plantation is able to repair around 3.74% of bulk density in flat area and 0.41% in slope area, 48.88% of soil infiltration in flat area, the content of soil organic matter around 30.43% in flat area and 53.33% in slope area, and soil nitrogen content around 27.27% in flat area and 7.69% in slope area.

**References**

[1] Tan K H 1986 *Degradation of Soil Minerals by Organic Acid.* (SSSA Publ) p 1-25

[2] Harahap I Y, Hidayat T C, Simangungsong G, Sutarta E S, Pangaribuan Y, Listia E and Rahutomo S 2011 *Mucuna bracteata* Development and utilization in oil palm plantations (Pusat Penelitian Kelapa Sawit : Medan) p 2

[3] Mathews J 1998 The introduction and establishment of a new leguminose cover *Mucuna bracteata* under oil palm in Malaysia *Planter* **74 : 868** p 359-368

[4] Hardjowigeno S 1987 Soil science (Mediyatama Sarana Perkasa : Jakarta) p 49 – 73

[5] Utami S N and Handayani S 2003 Entisol chemistry in organic farming system *Ilmu Pertanian* **Vol 10 : 2** p 63 – 69

[6] Foth H D 1984 *Fundamentals of soil science 7th edition* (John Wiley & sons Inc: United States of America) p 21

[7] Sakiah, Sembiring M and Novitasari 2017 Physical and chemical characteristics of oil palm plantation (*Elaeis guineensis* Jacq) with cover crop *Mucuna bracteata Jurnal Agro Estate* **Vol 1 no 1** p 36 – 45

[8] Hanafiah K A 2010 Fundamental of soil science (Raja Grafindo Persada : Jakarta) p 179

**Acknowledgements**

Our thanks go to the academic community of STIPER - Agribusiness Estate (STIPAP) and management of SIPEF Group Bandar Pinang Estate which has facilitated the author in conducting this research.

# Morphological characterization of several strains of the rice-pathogenic bacterium *Burkholderia glumae* in North Sumatra

**M Hasibuan[1], I Safni[1*], Lisnawita[1] and K Lubis[1]**

[1]Department of Agrotechnology, Faculty of Agriculture, Universitas Sumatera Utara, Padang Bulan, Medan, Indonesia 20155
*Email:irda@usu.ac.id

**Abstract.** *Burkholderia glumae* is a quarantine seed-borne bacterial pathogen causing panicle blight disease on rice. This pathogen has been detected in some locations in Java, and recently, farmers in North Sumatra have reported rice yield loss with symptoms similar with those on rice infeced by the rice-pathogenic bacterium *B. glumae*. This research was aimed to isolate several bacterial strains from several rice varieties in various locations in North Sumatra and characterize the morphology of the strains to detect and identify the unknown bacterial strains presumably *B. glumae*. Several rice seed varieties were collected from Medan and Deli Serdang Districts. The seed samples were extracted, isolated and purified, then grown in semi-selective media PPGA. The morphological characteristics of the bacterial strains were determined including Gram staining, bacterial colony's and bacterial cell's morphology. The results showed that of eleven strains isolated, two strains were Gram negative and nine strains were Gram positive. On the basis of colony morphology, all strains had circular form, flat elevation and cream colour while the colony margin varied, i.e.entire and undulate. Most strains had bacillus/rod shape (8 strains) and only 3 strains were coccus.

## 1. Introduction

Rice (*Oryza sativa* L) is the main staple food for more than two million people worldwide, mostly live in Asia including in Indonesia. Rice contains high carbohydrate for human's source of energy. Of the total rice production in the world, i.e. 490,6 million tons [1], more than 90% is produced and consumed in Asia. Rice consumption of Indonesian people has ranked as the first position in the world which resulted in high rice demand [2].

Rice production in Indonesia is forecast to decline in 2016 [1]. In North Sumatra, rice production has been fluctuated with 4,06 million tons in 2015, which is slightly increase 11,40% compared to 2014 [3].

A number of factors affect the decrease of rice production in Indonesia including un-remunerative prices, unseasonable weather, lack of good quality of seed and pest and dieases [1]. Bacterial panicle blight (BPB) on rice was first reported in Japan in 1950s and the disease has been one of serious plant diseases in the world [4]. BPB has been reported in rice producing countries including Southern and Middle America (Dominique Republic, Venezuela, Equador, Brazil, Panama, Columbia, Nicaragua and Costarica), Southern Africa and Tanzania, Asia (Japan, Korea, Vietnam, Phillipines, India, Indonesia, Malaysia, Thailand and China) [5]. Rice panicle blight can cause yield losses up to 75% in several states in the USA and more than 40% crop losses in Panama and several tropical and sub-tropical countries have considered BPB as potential risk disease on rice [6,7].

Recent report from Dr. Suryo Wiyono, a lecturer at Bogor Agricultural Institute, confirmed the occurance pathogen of *B. glumae* in hybrid rice seeds in Tegal (Middle Java) and Blitar (East Java), although Ministry of Agriculture of Indonesia denied his statement [8]. Similarly, BPB has been detected in Southern Sulawesi [9]. In Indonesia, panicle blight was a minor rice disease since 1987, however no report causing significant crop losses after that [10]. Indonesian government has listed *B. glumae* as a quarantine pathogen (Group I, Category of A2), which cannot be removed from the carrier through quarantine treatments [11]. This disease has become an emergence disease that affects rice production in several locations in Indonesia since 2015 [12]. The infected rice plants produce empty panicle which causes crop failure. Recently, similar symptoms have been found in several rice fields in North Sumatera and have caused serious concerns to the farmers. As there is no information and research about this disease in North Sumatera, therefore research on detection and identification of this disease is urgently required. This research was aimed to characterize the morphology of several bacterial strains isolated from several symptom and symptomless rice seeds in North Sumatera. Bacterial morphology is one aspect of bacterial identification.

## 2. Materials and methods

### *2.1 Rice seed samples*

Rice seed samples were collected from several places in North Sumatra. 90 g of the seeds were placed in a polyetylene plastic bag and labelled, then brought to the Laboratory of Plant Disease, Faculty of Agriculture, University of Sumatera Utara for further analysis.

### *2.2. Seed extraction, isolation and purification*

The seeds sample were extracted, isolated and purificated to obtain pure cultures on semi-selective media PPGA and King's B media according to Safni *et al.* [13] and IRRI [14] with modification. The seeds were surface-sterilization with NaClO 0.5% for three minutes to sterilize the pathogens on the seed surface.

### *2.3. Morphology test*

#### *2.3.1. Cellular morphology*

Cellular morphology of the bacteria included the cell shape which was observed under the microscope and Gram-staining. Gram-staining test was performed according to Schaad *et al.* [15]. Bacterial cell shape and color was observed using microscope with 1,000 times magnification. Gram-negative bacteria is red or pink color while Gram-positive bacteria expose blue or purple color.

#### *2.3.2. Colony morphology*

The colony morphology was determined after 48 hours of incubation on the basis of colony's color, shape, elevation and margin according to Bergey's Manual of Systematic Bacteriology [16].

## 3. Results and discussion

### *3.1. Rice seed samples*

From four different locations, eleven bacterial strains were collected (Table 1). Eight strains were isolated from Deli Serdang and three strains were isolated from Medan.

**Table1.** Bacterial strains from several rice varieties isolated from two locations

| No | Bacterial strains | Location |
|---|---|---|
| 1 | IR64 M | Medan |
| 2 | CH MD | Medan |
| 3 | CH M | Medan |
| 4 | IP30 | Deli Serdang |
| 5 | IP32 | Deli Serdang |
| 6 | ME | Deli Serdang |
| 7 | SIB | Deli Serdang |
| 8 | CIB | Deli Serdang |
| 9 | CID | Deli Serdang |
| 10 | IR64 SR | Deli Serdang |
| 11 | CH SR | Deli Serdang |

### *3.2. Morphology test*

Morphology results of cell shape, Gram staining and colony's morphology tests are displayed in Table 2. It showed that only 5 strains were Gram negative, IR64 M isolated from Medan and ME, SIB, IR64 SR and CH SR isolated from Deli Serdang. The rest strains (6) were confirmed as Gram positive bacteria. The cell shape majority was rods/bacillus (8 strains) and only 3 strains were coccus. The colony's color, shape and elevation were similar for all strains, i.e. cream, circular and flat respectively. For colony's margin varied with entire margin for 5 strains and undulate margin for 6 strains.

According to Brenner *et al.* [16] in Bergey's Manual of Distriminative Bacteriology, *B. glumae* is characterized as rods/bacillus, 0.5-0.7 x 1.5-2.5 µm, Gram negative. On the basis of these characteristics, the possible bacterial strains of *B. glumae* are strains of IR64 M isolated from Medan, and SIB isolated from Deli Serdang.

For further work, biochemical characteristics as well as molecular essay are required for bacterial identification.

**Table 2.** Morphology of bacterial strains isolated from rice seeds

| No | Bacterial strains | Cell shape/ Arrangement | Size | Gram-staining | Colony's morphology | | | |
|---|---|---|---|---|---|---|---|---|
| | | | | | Color | Shape | Elevation | Margin |
| 1 | IR64 M | Bacillus | 0.69 µm x 0.59 µm | - | Cream | Circular | Flat | Undulate |
| 2 | CH MD | Bacillus | 3.24 µm x 0.74 µm | + | Cream | Circular | Flat | Undulate |
| 3 | CH M | Bacillus | 2.29 µm x 0.67 µm | + | Cream | Circular | Flat | Undulate |
| 4 | IP30 | Bacillus | 2.53 µm x 0.61 µm | + | Cream | Circular | Flat | Entire |
| 5 | IP32 | Bacillus | 3.13 µm x 0.78 µm | + | Cream | Circular | Flat | Entire |
| 6 | ME | Coccus | 0.85 µm x 0.45 µm | - | Cream | Circular | Flat | Entire |
| 7 | SIB | Bacillus | 1.076 µm x 0.48 µm | - | Cream | Circular | Flat | Entire |
| 8 | CIB | Bacillus | 3.24 µm x 0.66 µm | + | Cream | Circular | Flat | Undulate |
| 9 | CID | Bacillus | 2.43 µm x 0.86 µm | + | Cream | Circular | Flat | Undulate |
| 10 | IR64 SR | Coccus | 1.03 µm x 0.39 µm | - | Cream | Circular | Flat | Entire |
| 11 | CH SR | Coccus | 0.93 µm x 0.42 µm | - | Cream | Circular | Flat | Undulate |

## 4.Conclusions

It is concluded that on the basis of morphological characteristics of eleven bacterial strains isolated from several rice varieties and two districts in North Sumatra, one strain isolated from Medan (IR64 M) and one strain isolated from Deli Serdang (SIB) are presumably confirmed as *B. glumae*, the causal bacterial pathogen of rice panicle blight.

## References

[1] Food and Agriculture Organization of the United Nations 2016 Rice market monitor Volume **19** **(2)** 1-37

[2] Risnawaty R 2012 Identifikasi Cendawan Terbawa Benih Pada Padi Lokal Aromatik Pulu Mandoti, Pulu Pinjan, dan Pare Lambau Asal Kabupaten Enrekang, Sulawesi Selatan Artikel Publikasi Universitas Hasanuddin.

[3] BPS Provinsi Sumatera Utara 2016 Produksi Padi dan Palawija Sumatera Utara (Angka Tetap Tahun 2015). Badan Pusat Statistik Provinsi Sumatera Utara, Medan

[4] Xie G L, Luo J Y, Li B 2003 *Bacterial Panicle Blight*: A Rice Dangerous Diseases and its Identification. *Plant Prot.* **29** 47–49

[5] Nandakumar R, Rush M, Shahjahan A, O'Reilly K, Groth D 2005 Bacterial Panicle Blight of Rice in The Southern United States Caused by *Burkholderia glumae* and *B. gladioli*. *Phytopathology* **95** S73

[6] Trung H M, Van N V, Vien N V, Lam, D T and Lien, M 1993 Occurrence of Rice Grain Rot Disease in Vietnam. *Int Rice Res Notes* **18(3)** 30

[7] Ham J H, Groth D E 2011 Bacterial Panicle Blight, an Emerging Rice Disease. Baton Rouge, Louisiana, USA: Louisiana State University Agricultural Center pp. 16–17

[8] http://bbpadi.litbang.pertanian.go.id/index.php/berita/info-aktual/content/414-pernyataan-pakar-ipb-terkait-padi-hibrida-mengandung-bakteri-keliru

[9] Baharuddin, Harniati R., Faisal F, Yani A, Suparmi, Hamid H, Kuswinanti T, and Jahuddin R 2017 Keberadaan Penyakit Busuk Bulir (*Burkholderia glumae*) pada Tanaman Padi di Sulawesi Selatan. Prosiding Simposium Nasional Fitopatologi Indonesia 2017: Kemunculan penyakit baru dan impor benih pp. 19-26.

[10] http://tabloidsinartani.com/content/read/tak-perlu-kuatir-bakteri-burkholderia-tak-ganggu-produksi-padi/

[11] Badan Karantina Pertanian 2015 Lampiran Keputusan Menteri Pertanian (Permentan) No 51/Permentan/KR010/9/2015 tentang Jenis-jenis Organisme Pengganggu Tumbuhan Karantina (OPTK) Kategori A1 dan A2.

[12] Joko, T 2017 *Burkholderia glumae* sebagai *Emerging Pathogen*: Status, Potensi Kerusakan, dan Strategi Pengendalian. Prosiding Simposium Nasional Fitopatologi Indonesia 2017: Kemunculan penyakit baru dan impor benih pp. 27-35.

[13] Safni I, Sepulveda L R, Goldman, P H, Derie, M J, du Toit, l J, Koike, S T, Stockwell, V O and Bull C T 2016 Genetic diversity of *Pseudomonas syringae* causing bacterial leaf spot on table beet (*Beta vulgaris*) and Swiss chard (*Beta vulgaris* subsp. *cicla*). Proceedings of 2016 American Phytopathological Society Annual Meeting pp.74.

[14] International Rice Research Institute 1994 A Manual of Rice Seed Health Testing (Manila: *International Rice Research Institute)*

[15] Schaad N W, Jones J B, and Chun W 2001 *Laboratory Guide for Identifications of Plant Pathogenic Bacteria* 3rd Edition (Minnesota: APS Press)

[16] Brenner D J, Krieg N R, Staley J T 2015 *Bergey's Manual of Systematic Bacteriology* 2nd edition (USA: Springer)

## Acknowledgement

The authors gratefully acknowledge that this research was supported by Grant of TALENTA 2017, Universitas Sumatera Utara, Indonesia (Contract number 285/UN5.2.3.1/PPM/KP-TALENTA USU/2017).

# Classification of andisol soil on robusta coffee plantation in Silima Pungga - Pungga District

**P Marbun[1], Z Nasution[1], H Hanum[1] and A Karim[2]**

[1]Faculty of Agriculture, Universitas North Sumatera, Indonesia
[2]Faculty of Agriculture, Universitas Syiah Kuala, Indonesia
E-mail: posma.marbun@usu.ac.id

**Abstract.** The survey study aims to classify the Inceptisol soil on Robusta coffee plantation in Silima Pugga-Pungga District, from Order level to Sub Group level. The study was conducted on location of sample soil profiles which were determined based on Soil Map Unit (SMU) with the main Andisol Order, i.e. SMU 12, SMU 15 and SMU 17 of 18 existing SMU. The soil profiles were described to determine the morphological characteristics of the soil, while the physical and chemical properties were done by laboratory analysis. The soil samples were taken from each horizon in each profile and analyzed in the laboratory in the form of soil texture, bulk density, pH $H_2O$, pH KCl, pH NaF, C-organic, exchangeable bases ($Ca^{2+}$, $Mg^{2+}$, $K^+$, $Na^+$), ZPC (*zero point charge*), base saturation, cation exchange capasity (CEC), P-retention, Al-Oxalate (Al-O) and Si-Oxalate (Si-O). The results showed that the classification of Andisol soil based on Soil Taxonomy only has one Sub Group namely *Typic Hapludand*. It is expected that the results of this study can provide information for more appropriate land management in order to increase the production of Robusta coffee plant in Silima Pungga-Pungga Sub district.

## 1. Introduction

Andisol soil is one of orders from eleven orders of soil in soil taxonomy. In this world, there are around 100 million hectare or 0.76% of the total land in the world [1]. There are 6.4 million hectare of Andisol soils in Indonesia and 200 thousand hectare in North Sumatera Province which is spread in wet and dry climate which has various characteristic.

The order of this Andisol soil is the newest order, formerly this soil is called Andosol or Andept. The change in the classification of Andept sub-order into Andisol order because some irregularities were found in the definition of sub-order in the soil taxonomy.

According to Soil Research Center [2] and Center for Agricultural Land Resources Research and Development Agency for Agricultural Research and Development [3] define Andosol soil as a soil that has a molic horizon, or A umbrik horizon and may be above the B kambik horizon; or A okhrik and B kambik horizon that have one or both of the following conditions: (a) bulk density (on water content of 1/3 bar and fine earth (< 2mm)) less than 0.85 g /$cm^3$ and the exchange complexes are dominated by amorphous materials, (b) 60% or more are vitric ash, or pyroclastic materials in the fraction of dust, sand, and gravel.

According to Soil Survey System [4] in soil survey system, the soil which is known as Andosol above is equivalent to Andisol order, land that has Andik soil properties ≥ 60% of soil thickness. Physically, the soil has a low bulk density (< 0.90 g/$cm^3$), high water content, water availability for plants form medium to low, high water holding capacity and high total porosity.

Regional shape or topography is one of the factors that form the soil that greatly affect the process of land formation and its management. Seen from the shape of its land area Andisol spread in choppy, bumpy, hilly, and mountainous areas. However, most of the land lies in hilly to mountainous areas.

In North Sumatra Province there are several districts that have coffee, one of which is Dairi Regency for Robusta coffee. Silima Pungga Pungga District is the region with highest production and productivity of Robusta coffee in Dairi Regency: 467 ton and 610.46 ton / hectare / year, but this level of productivity is still low. The low productivity of Robusta coffee is caused by improper land management.

Proper land management based on its characteristics and potential will provide optimal productivity. One effort that can be done is to know the nature and specific characteristics of a soil type, commonly called the soil classification [3]. The classification of soil in this study, as did also by Marbun [5] in this study refers to the land classification system which was developed by the USDA in 1975, known as Soil Taxonomy. Until now the classification system has grown until 2014 with many additional contents in the granting of land nomenclature categories.

This study aims to determine the soil classification of Andisol to sub group based on Keys To Soil Taxonomy 2014 [4]. It is expected that the results of this study will be able to provide more appropriate land management information in order to increase the production of Robusta coffee in Silima Pungga Pungga District.

## 2. Material and methods

This research was conducted on Robusta coffee plantation in District of Silima Pungga-Pungga, Dairi Regency ($2^0$80' - $2^0$88' NL and $98^0$04' – $98^0$17' EL), with a height of 400 to > 800 meters above sea level. Soil analysis was conducted at the Faculty of Agriculture's Research & Technology laboratory, University of Sumatera Utara.

Tools and materials which was used in this research are; administration map, soil types map, land slope map and land altitude map with scale of 1 : 25,000 for each map, rainfall data and air temperature data. Soil samples taken from each horizon on each soil profile are made in each SMU. Tools which are used are GPS (Global Positioning System), earth drill, tape measure, ring sample, Munsell Soil Colour Chart book, plastic bag, labels, cleaver, knife, camera, marker, etc.

This research used survey method to determine the morphology and characteristics of the soil, in order to classify the soil from Order to Sub-group level according to Keys to Soil Taxonomy. The procedures of this research were devided into five stages, namely: secondary data collection, field observation, soil sampling, soil analysis and report compilation.

Determination of coordinate points and location of the profile is determined based on secondary data on SMU 12, 15 and 17. Preparation of the profile begins with the drilling at several points representing each SMU, after considered representative then soil profile point which want to be observed can be determined.

Field observations were carried out on soil profiles of 1 m x 1 m x 1.5 m and depicted according to soil horizons for soil characterization which shows soil morphological properties and characteristics to be observed include: soil horizon limit, soil color, soil texture, soil structure, soil consistency and effective depth.

Soil samplings, undisturb soil and disturbed soil, were performed on each soil horizon for analysis in the laboratory. At the time of sampling the soil is also recorded data from the study area, including vegetation, physiography, drainage, altitude, slope, geographical location and land use. The soil properties analyzed and the method of analysis are presented in Table 1.

Data of field research and laboratory are then used for land classification based on "Keys To Soil Taxonomy 2014" [5]. The stages of classification of land are as follows: (1) Determination of the main horizon symbol and sub horizon, (2) Determination of the horizon of the identifiers, (3) Determination of the horizon under the identifiers, (4) Determination of other identifiers, (5) Determination of the Land Order, (6) Determination of Sub Order, (7) Determination of Great Group and (8) Determination of Sub Group.

**Table 1**. Observed variable factors and analysis methods

| No. | Component Analysis | Method/Tool |
|---|---|---|
| | I. Environmental factors | |
| 1. | Climate | Secondary data (meteorology climatolog and geophysics agency) |
| 2. | Vegetation | Field Observation |
| 3. | Physiography | See the form of environment |
| 4. | External Drainage | Predicted from the slopes |
| 5. | Internal Drainage | Characteristic shown by rust/gley |
| 6. | Surface Rock | Observation of rock distribution |
| | II. Physical Properties in the Laboratory | |
| 1. | Texture 3 fractions | Pipette |
| 2. | Content weight | Ring Sample |
| 3. | Water Content | Gravimetry |
| | III. Chemical Properties in the Laboratory: | |
| 1. | pH $H_2O$, pH KCL and pH NaF | pH meter |
| 2. | C- Organik | Walkey dan Black |
| 3. | Cation Exchange Capacity (CEC) | Extract of $NH_4Oac$ 1 N pH 7 |
| 4. | Bases exchangeable (K,Ca,Mg,Na) | Extract of $NH_4Oac$ 1 N pH 7 |
| 5. | N-total (%) | Kjedhal |
| 6. | P-available (ppm) | Bray II |
| 7. | P-total | Extract of HCL |
| 8. | K-total | Extract of HCL |
| 9. | EC | Platina Elektroda |
| 10. | P-retention | Blackmore |
| 11. | Al-amorf | $NH_4$Oksalat |
| 12. | Fe- amorf | $NH_4$Oksalat |

## 3. Result and discussion

The Andisol land order is found in SMU 12, 15 and 17 of 18 (eighteen) SMU in District of Silima Pungga-Pungga. Order of land, altitude, slope, villages' name and area of each SMU can be seen in Table 2 below.

**Table 2**. Order of land, altitude, slope, villages' name and area SMU 12, SMU 15 and SMU 17

| SMU | Soil | Altitude (m asl) | Slope (%) | Village's Name | Area (ha) |
|---|---|---|---|---|---|
| SMU 12 | Andisol | 400 - 500 | 0 – 4 | Polding Anak – anak | 605.92 |
| SMU 15 | Andisol | 500 - 600 | 0 – 4 | Bakal Gajah | 916.42 |
| SMU 17 | Andisol | 500 - 600 | 8 - 16 | Bakal Gajah | 720.89 |

Based on the observation in the field, there are similarities in these three soil profiles so that only one of the three soil profiles are discussed namely SMU 12 because its profile represents the SMU 15 and SMU 17 profiles.

### *3.1. Soil morphology*

The soil color in the SMU 12 profile ranges from grayish black (2.5 Y 3/2) to faded yellow (2.5 Y 6/6). The more down the soil colors become brighter because of the reduced of organic material content in the lower horizon. Soil organic material affects various chemical and physical properties and increases soil biological activity as well as crop productivity. The dark color of humus on the surface horizon (black or dark brown) with crumb structure, loose consistency, high organic matter content as well as smeary is one of the most decisive properties in Andisol soil.

The soil structure on the surface horizon is generally glob. According to Prasetyo [6] Andisol land has a soil structure that reflects the high level of soil material in the form of non-crystalline minerals and soil organic material. These ground-level horizons generally have grain structures and lumpy structures. Agricultural cultivation tends to cause grain structure changes to become globules.

The consistency of the soil in the field is examined in moist and wet conditions. In general the consistency of the soil is generally loose, while in wet conditions are generally somewhat adhesive. As is known Andisol soil contains many non-crystalline mineral materials that affect consistency. Soil consistency is usually less adhesive and less plastic, but a low C-organic content with increased clay content results in consistency of being sticky and plastic.

Soil texture on SMU 12 varies greatly, but the horizon on this profile generally has a sandy clay loam texture (Table 2). Andisol soil has a very varied texture from sandy loam to sandy clay, it depends on the type and size of particles released during eruption.

### *3.2. Physical properties of soil*

Physical properties of soil observed in the field include color, structure and consistency, while the soil physical properties analyzed in the laboratory are soil texture and bulk density. The physical properties of soil SMU 12 can be seen in Table 3 below.

**Table 3**. Morphology and analyses soil physical properties at SMU 12, SMU 15 and SMU 17 profile

| Soil Profile | Horizon | Depth (cm) | Color | Structure | Consistency | | Particle Size Distribution (%) | | | Texture | Bulk Density (g/cm$^3$) |
|---|---|---|---|---|---|---|---|---|---|---|---|
| | | | | | Moist | Wet | Sand | Silt | Clay | | |
| | A | 0 - 14 | 2,5 Y 3/2 | sab | friable | ss | 47.84 | 19.28 | 32.88 | scl | 0,85 |
| SMU 12 | Bw$_1$ | 14 - 35 | 2,5 Y 4/4 | sab | friable | ss | 54.56 | 18.56 | 26.88 | sl | 0,70 |
| | Bw$_2$ | 35 - 150 | 2,5 Y 6/6 | sab | friable | ss | 48.56 | 16.56 | 34.88 | scl | 0,75 |
| | Ap | 0 - 45 | 2,5 Y 3/2 | sab | friable | s | 49.84 | 23.28 | 26.88 | sl | 0,60 |
| SMU 15 | Bw$_1$ | 45 - 90 | 2,5 Y 4/4 | sab | friable | s | 54.56 | 20.56 | 24.88 | scl | 0,75 |
| | Bw$_2$ | 90 - 150 | 2,5 Y 6/6 | ab | firm | vs | 36.56 | 15.84 | 47.60 | sl | 0,84 |
| | Ap | 0 - 40 | 2,5 Y 3/2 | cr | friable | ss | 49.84 | 31.28 | 18.88 | l | 0,80 |
| SMU 17 | A | 40 - 75 | 2,5 Y 4/4 | cr | friable | ss | 54.56 | 26.56 | 18.88 | sl | 0,76 |
| | Bw | 75 - 150 | 2,5 Y 3/2 | b | firm | s | 34.56 | 17.84 | 47.60 | c | 0,87 |

Notes: structure: ab = angular blocky; sab = sub angular blocky; cr = crumb
consistency: ss = slightly sticky; s = sticky; vs = very sticky
texture: scl = sandy clay loam; sl = sandy loam ; l = loam; c = clay

### ***3.3. Chemical properties of soil***

Chemical Properties of Soil observed in the laboratory are: pH of $H_2O$, pH of KCl, pH NaF, Bases exchange (Ca, Mg, K and Na), Cation Exchange Capacity (CEC), Base saturation, C-organic, Al-O, and Fe-O can be seen in Table 4 below.

**Table 4.** The chemical properties of soil data analysis result in soil profile of SMU 12, SMU 15 and SMU 17 robusta coffee's field in sub district silima pungga pungga

| Soil Profile | Horizon | Depth (cm) | pH | | | CEC (me/100g) | BS (%) | Ca-exch | Mg-exch | K-exch | Na-exch | C-org (%) | Ret-P (%) | AlO (%) | Fe0 (%) | Al + ½ Fe (%) |
|---|---|---|---|---|---|---|---|---|---|---|---|---|---|---|---|---|
| | | | $H_2O$ | KCl | NaF | | | (me/100g) | | | | | | | | |
| | A | 0 – 14 | 4.97 | 3.50 | 9.81 | 10.21 | 27.11 | 1.406 | 0.923 | 0.210 | 0.229 | 1.10 | 90.89 | 1.67 | 3.43 | 3.38 |
| SMU 12 | $Bw_1$ | 14 – 35 | 4.72 | 3.61 | 9.62 | 12.10 | 29.88 | 1.058 | 1.386 | 0.258 | 0.913 | 1.07 | 87.89 | 1.63 | 4.41 | 3.83 |
| | $Bw_2$ | 35 – 150 | 4.58 | 3.61 | 10.21 | 21.09 | 17.69 | 0.659 | 0.735 | 0.079 | 0.182 | 0.63 | 88.89 | 1.28 | 3.38 | 2.97 |
| | Ap | 0 – 45 | 4.76 | 3.62 | 9.46 | 13.88 | 13.61 | 0.675 | 0.798 | 0.170 | 0.246 | 1.14 | 87.74 | 3.16 | 1.07 | 3.69 |
| SMU 15 | Bw1 | 45 – 90 | 4.88 | 3.74 | 9.47 | 12.10 | 13.74 | 0.578 | 0.796 | 0.056 | 0.232 | 0.99 | 90.80 | 6.49 | 0.66 | 6.82 |
| | Bw2 | 90 – 150 | 4.87 | 3.58 | 9.85 | 9.39 | 25.43 | 0.577 | 1.526 | 0.148 | 0.137 | 0.74 | 85.46 | 8.04 | 1.95 | 8.51 |
| | Ap | 0 – 40 | 4.99 | 4.16 | 9.21 | 14.43 | 26.66 | 0.973 | 2.237 | 0.432 | 0.205 | 1.77 | 90.67 | 0.70 | 3.45 | 2.42 |
| SMU 17 | A | 40 – 75 | 4.76 | 3.85 | 9.70 | 13.10 | 8.71 | 0.574 | 0.295 | 0.041 | 0.231 | 1.58 | 93.33 | 1.30 | 3.41 | 3.00 |
| | Bw | 75 – 150 | 5.01 | 3.61 | 9.72 | 11.96 | 25.21 | 0.595 | 2.073 | 0.231 | 0.116 | 0.33 | 95.56 | 2.15 | 4.55 | 4.42 |

### *3.4. Soil classification*
Based on data obtained from both climate data, field observation and soil sample data from the laboratory, soil classification by using Keys Soil Taxonomy 2014 [4] is done. The first step is to determine the surface horizon identifiers (epipedon), the sub surface horizon identifiers as well as the other identifiers. In addition, the determination of orders, sub-orders, great groups and sub groups.

### *3.5. Determination of surface horizon identifier/epipedon*
Epipedon soil identifier on soil surface for SMU 12, SMU 15 and SMU 17 is epipedon umbrik. Terms for epipedon umbrik are the position located above the soil, has a color value and chroma of 3 or less (humid), base saturation < 50% and moist soil in a state of more than 3 months.

### *3.6. Determination of sub surface horizon identifier*
Kambik horizon is found in SMU 12, SMU 15 and SMU 17, because it has a very fine sand texture called sandy loam, horizon tickness is more than 15 cm, the absence of clay illuviasi process and is not part of the Ap horizon, also did not experience the aquatic condition.

### *3.7. Determination of other identifier*
SMU 12, SMU 15 and SMU 17 have a moisture regime udik, because the soil never dries within 90 days (cumulative), more than 90 days or from the data obtained by the average rainfall in the wet season range from 7 - 10 months per year or 210 days to 300 days (cumulative), and has soil temperature regime isohipertermik, because the hottest and coldest is smaller than $6^0$C, it is $1.71^0$ dan average annual soil temperature bigger than $22^0$C, it is $22.5^0$C. This soil is also have the characteristic of Andik because it has Phosfat retention which is bigger than 85%, bulk density is smaller than 0.90 g/cm3 and Al+1/2Fe (extract of ammonium oxalate) is bigger than or same as 2%.

### *3.8. Order determination*
SMU 12, SMU 15 and SMU 17 belong to *Andisol* order because they have the characterictics of Andik soil, bigger than or same as 60% of the tickness.

### *3.9. Sub order determination*
SMU 12, SMU 15 and SMU 17 belong to *Udand* because they have soil moisture regime udik.

### *3.10. Great group determination*
SMU 12, SMU 15 and SMU 17 belong to great group *Hapludand* because they have udand with the other characterictics.

### *3.11. Sub group determination*

SMU 12, SMU 15 and SMU 17 belong to *Typic Hapludand* sub group because they have another characteristic of hapludand.

## 4. Conclusion

Classification soil of SMU 12, SMU 15 and SMU 17 are order: *Andisol*, sub order: *Udand*, great group; *Hapludand* and sub group: *Typic Hapludand*.

## References

[1] Dinas Perkebunan Provinsi Sumatera Utara [Plantation Office of North Sumatra Province] 2009 Data Statistik Perkebunan Sumatera Utara Tahun 2008 [North Sumatera Plantation Statistics Data of 2008]

[2] Pusat Penelitian Tanah [Soil Research Center] 1983 *Balai Penelitian dan Pengembangan Pertanian Departemen Pertanian Bogor [Indonesian Center for Agricultural Research and Development Agriculture Department Bogor]* Kriteria Penilaian Data Sifat Analisis Kimia Tanah [Criteria Assessment Data Properties of Soil Chemistry Analysis]

[3] Subardja D S, Ritung S, Anda M, Sukarman, Suryani E dan Subandiono R E 2014 *Balai Besar Penelitian dan Pengembangan Sumberdaya Lahan Pertanian Badan Penelitian dan Pengembangan Pertanian Bogor [Center for Agricultural Land Resources Research and Development Agency for Agricultural Research and Development Bogor].* Petunjuk Teknis Klasifikasi Tanah Nasional [National Soil Classification Technical Guidelines] pp 22

[4] Soil Survey Staff 2014 United States Departement of Agriculture-Natural Resources Conservation Service Washington DC. Keys To Soil Taxonomy Twelfth Edition

[5] Marbun P, Nasution Z, Hanum H, Karim A 2017 *International Journal of Sciences Basic and Applied Research.* Classification of Inceptisol Soil on Robusta Coffee Plantation in Silima Pungga - Pungga District. Vol. **30** (5) pp 20-27

[6] Prasetyo B H, Adiningsih J S, Subagyono K dan Simanungkalit R D M 2005 *Jurnal Sumberdaya Lahan [Journal of Land Resources]* Andisol, Karakteristik dan Pengelolaannya untuk Pertanian di Indonesia [Andisol, Characteristics and Management for Agriculture in Indonesia]. Vol. **1** (1) pp 1-9

IOP Conf. Series: Earth and Environmental Science **122** (2018) 012046 doi:10.1088/1755-1315/122/1/012046

# Effect of rice husk biochar application to soil insect diversity on potato cultivation

**A Meilin and R Rubiana**

Jambi Assessment Institute for Agricultural Technology, Jalan Samarinda Paal Lima, Kotabaru, Jambi 36128, Indonesia.
E-mail: araz_meilin@yahoo.com

**Abstract**. High intensity of disease infection and the intensive use of fertilizers and pesticidescause saturated fertilizer and pesticide to the land. Remediation using biochar rice husk is one of the technology to decrease fertilizer and pesticide residue. The diversity of soil insects can be used as bioindicators because of their existence dependsg on soil structure and condition. This study was aimed to study the diversity and structure communities of soil insect in potatoes on difference husk rice biochar application. The sampling of soil insects was done on potato farmer's land with four treatments i.e control (farmers' technique), trichokompos without biochar, trichokompos + biochar with dose 1 ton/ha, and trichokompos + biochar with dose 2 ton / ha. At each point a single pitfall trap was installed for two nights and then it was taken for identification. The results showed that biochar application had significant effect on the number of soil insect species ($P = 0.037$). The soil insect species composition pattern also showed significant differences between the four treatments (R: 0.2306, Pvalue = 0.001). This mean that the application of biochar affects the number of insects species and plays a role in the formation of soil insect diversity beta patterns.

## 1. Introduction

The high incidence of leaf and stem diseases and the use of seedlings that have been infected by pathogens encourage potato farmers in Kerinci Jambi district to use fertilizers and pesticides on the recommended dosage [1]. For that matter, in the soil potato cultivation especially in Kayu Aro sub-district which is a vegetable center experiencing saturated fertilizer and pesticide. Excessive use of N-chemical fertilizers, damaging the environment and interfering with the absorption of other nutrients [2]. Pesticides applied to agricultural systems can be detected in air, surface and ground water, sediment, soil, vegetables, and to some extent in food [3]. Pesticides applied in soil for soil-borne pest control and pathogens will also produce accumulated residues and metabolites in soil at very high levels [4].

Various technologies are developed to deal with pollution of hazardous materials. Utilization of microorganisms (bacteria and fungi) is known to be a detoxifying agent of pesticides. Treatment of microbial and compost addition on pesticide contaminated soil resulted in decreasing percentage of chlorpyrifos, ethoprophos and carbofuran residues after 6 weeks of treatment were 86.35%; 83.91% and 82.32% [5]. Bioremediation with the use of compost is a good technique for solving environmental problems such as on contaminated soil. Microorganisms are very important part of the process [6]. Biochar from paper mill waste has proved beneficial to ferrous soil at a dose of 10 ton/ha which provides improvements to soil quality and affects plant growth [7].

Bioremediation technology with biochar utilizing agricultural waste has also been widely developedfor the restoration of polluted soil contaminated by microbial activity as well as the enzymes it produces to transform harmful ingredients into non-hazardous materials. Agricultural wastes processed into biochar include corn cob, rice husk, coconut shell. Biochar is the result of degradation of organic matter in the absence of air that can be utilized to improve soil fertility and ecosystem services provided by soil organisms [8]. Biochar derived from rice husks showed higher ash content than biochar derived from tree branches, allowing interactions between organic and inorganic so as to decrease soil density, allowing the exchange of Al, dissolve Fe and increase porosity [9].

Improved land function can also affect the presence of soil insects because insects can show sensitivity to changes in their environment. Research shows that soils with high organic matter and good soil fertility show less abundance of herbivorous insects [10]. On the other hand, intensive farming practices with excessive use of synthetic fertilizers can lead to nutrient imbalances resulting in low pest resistance. The presence of soil insect populations may indicate a link to healthy or polluted farmland due to environmental regulation and nutrient availability allowing suitable living spaces for soil insects. So research on the diversity of insects can be used to find out more of the application role of these technologies in decreasing residual and increasing the production of potatoes indirectly.

This study was aimed to: 1) study the diversity of soil insects in potatoes in four different biochar rice husk treatments, 2) study the structure of the insect community.

## 2. Materials and Methods

The research was conducted in Sangir Tengah Village, KayuArosub-district, KerinciDistrict, Jambi Province on farmer's land. From the previous observation, it is known that the land has an organosulfate residue. The research location is a potato centre. The study was conducted from April to August 2016.

Soil processing was done by making 20 bunds on peasant potato area of 1 ha. The technology of pesticide residual decrease studied consisted of four treatments: 1) Control (using the farmer's way that only with basic fertilizer before planting); 2) Trichokompos 10 tons / ha; 3) Trichokompos 10 tons / ha + Biochar / rice husk charcoal 1 ton / ha; 4) Trichokompos 10 ton / ha + Biochar / rice husk charcoal 2 ton / ha.Biochar was produced in a traditional way made from raw rice husks. Rice husks were inserted into a closed policator drum. At the bottom of the drum was given a fire. The husks were left to charcoal. Next the rice husk that has become charcoal, issued through the bottom door of the drum and cooled [11].Trichokompos is an organic material composted by a decomposer of self-produced Trichoderma biological agents. Cattle manure, biochar rice husk, trichoderma mixed with a ratio of 400: 40: 1 stirred until blended and moistened with enough water then covered with black plastic / sack and incubation 7-10 days [12].

Each treatment was applied to 5 bunds before planting. Furthermore on each planting hole with a depth of 10 cm diameter 25 mm planted one seed potato local varieties Granola. Spacing used is 60 cm between rows and 30 cm between seeds in a single bundle.Harvest was done after the plant was 90-100 HST, with the criteria of stems and leaves had dried and potato tubers when rubbed with each other tuber skin was not released.

### 2.1. *Sampling and insects identification*

In each treatment was mounted 10 traps (pitfall trap) so that on each bundle there were 2 points of sampling. A glass trap with a diameter of 13 cm contained a 70% alcohol solution and a soapy liquid deposited 10 cm deep in the soil [13].Traps were installed for the next two nights then were taken for identification [14].

The insect identification data was tabulated to the database in Microsoft Excel 2010 format. The views of the entire insects obtained were depicted in the species accumulation curve to estimate the existing insect population from the estimated value of S (observation)by randomized the number of species acquired at each sub-plot point of type of land use 50 times using the software EstimateS

IOP Conf. Series: Earth and Environmental Science **122** (2018) 012046 doi:10.1088/1755-1315/122/1/012046

version 9.1.0 [15].To determine the diversity of insects that focus on species richness or the Shannon-Wiener Diversity Index was calculated using the equations [16].

$$H' = \sum_{i=1}^{\infty} (Pi\ Ln\ Pi) \tag{1}$$

H'= Shannon-wiener Diversity
$P_i$= proportion of insect species i (ni) towards total of individu (N) ; (ni/N)
The criteria used to interpret Shannon-Wiener's diversity are H '<1 = Low Diversity; 1 <H <3 = Medium Diversity; And H> 3 = High Diversity [16].

The effect of treatment on the number of insect species (S) and the index of insect diversity (H ') was analyzed using ANOVA [17]. Furthermore, to know the similarity of soil insect structure at different treatment was by ANOSIM test (Analysis of Similarity) to get the difference coefficient statistic value. The analysis was performed in the R program. The composition of insect species present in each treatment based on the Bray-Curtis index was made in a non-metric multidimensional scaling (NMDS) ordinance using multidimensional scaling (MDS) analysis [18]. The analysis was performed using R Statistic software: Vegan package [19-20]. The difference of composition based on the role of ants between types of land use is shown in the pie chart.

## 3. Results and Discussion

Species Accumulation Curves (SAC) in the form of S (observation) estimation values are obtained from data showing the total number of species from all sample points. The SAC showed an increase and the absence of asymptot sampling which means that the list of soil insect species on the land is incomplete or not yet optimal the number of collected soil insects species (Figure 1). However, for some taxa such as Arthropods especially in tropical habitats, asymptotes are highly unattainable [21]. So the curve is used to compare the species of soil insects obtained from the soil by different treatments.

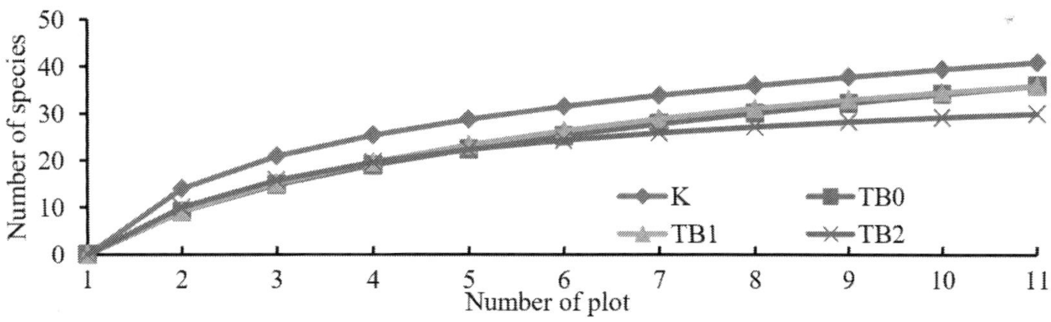


**Figure 1.**Soil accumulation curve accumulation at different treatments K = Control (farmer), TB0 = trichocompos only, TB1 = trichocompos + biochar 1 ton / ha; TB2 = trichokompos + biochar 2 ton / ha.

Based on the data obtained, there were 4,201 individuals belonging to 55 types of soil insects & the other Arthropods which belong to 11 orders (Table 1). Results of calculation of the highest number of species in the control and lowest land on the land with the giving of trichokompos + biochar 2 ton / ha. The provision of biochar significantly decreased the number of soil insect species in the field ($F_{3,34}$ = 2.739, P = 0.037). The high number of species and index H 'in the control area is due to the presence of herbivorous insects such as Hemiptera (Aphididae) and the other Athropods (Acarina) which are sheltered in weed plants. This is in line with Altieri and Nicholls's research on soils with high organic materials generally exhibit fewer abundances than some herbivorous insects [10]. The decrease in herbivorous insects is caused by lower nitrogen content.

**Table1.** Diversity of insects on potato fields

| Treatment[a] | Ordo | Number of species | Insect Diversity Index[b] | |
|---|---|---|---|---|
| | | | Abundance | H' |
| K | 11 | 41 | 1077 | 2,54 |
| TB0 | 10 | 36 | 876 | 1,97 |
| TB1 | 11 | 37 | 1258 | 1,79 |
| TB2 | 10 | 30 | 990 | 1,97 |
| Total | 11 | 55 | 4201 | |

[a]Research plot with the treatment
K = Control (farmer),
TB0 = only trichokompos,
B1 = trichokompos+biochar1 ton/ha;
TB2 = trichokompos+biochar 2 ton/ha.

The results of the Shannon-Wiener (H ') soil insect diversity index calculation on all four biochar treatments were moderate. Treatment of biochar did not affect H 'index of diversity of insects ($F_{3,34}$ = 2.138, P = 0.114). H 'index was obtained with insect type parameters and the proportion of abundance of each species in a habitat so that the field with trichocompos + biochar 1 kg / ha treatment had the lowest H' index because the abundance of insects at this treatment was highest with the large number of Collembola findings. The increasing association of Collembola's abundance due to biochar was also performed in Reibe's research [22] which is further attributed to the morphological mechanism, biomass, the volume of roots of wheat crops. The results of NMDS analysis showed that different doses of biochar administration affected the structure of soil insect communities (ANOSIM statistic R: 0.2306, Significance: 0.001) (Figure 2).

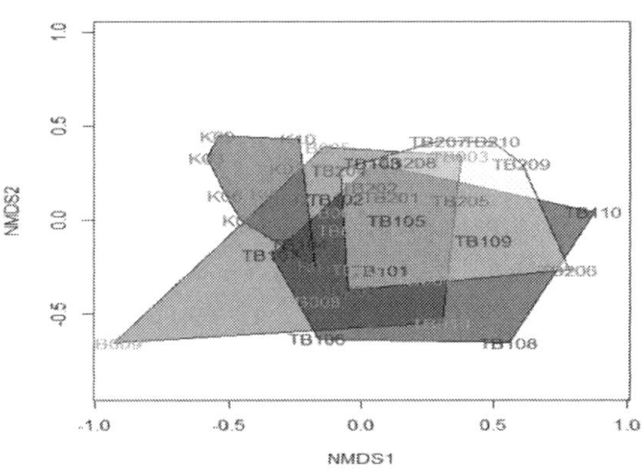


**Figure 2.**NMDS of insect composition based on Bray-Curtis inflation index in different treatments K = Control (farmer / pink field), TB0 = trichocompos only / green area, TB1 = trichokompos + biochar 1 ton / ha or colored field blue; TB2 = trichokompo + biochar 2 ton / ha or yellow field; The numbers (01 - 10) show the sample point as a repeat.

The formation of soil insect compositions provides more information than just the ratio of species. The composition of soil insects is indicated by the separation of plane between the treatment groups. It appears that the interplanes overlap, indicating that the insects in each treatment are similar in species.

The dimension of the control field has a field that is not too broad compared to other treatment dimensions because each sample point shows a high similarity. This is due to the discovery of lice (Aphididae) are abundant in the control field, while predators are found to specialize. A series of habitat characteristics may contribute to the residence preferences of most species of insects trapped in traps [23]. The preferred habitat of insects with detritivor role is on Trichokompos land of 10 tons / ha + Biochar / rice husk rice 1 ton / ha.

While ants (Hymenoptera: Formicidae), which is one of the cosmopolitan social insects group which is often used as bioindicator most found in Trichokompos land 10 ton / ha + Biochar / rice husk rice 2 ton / ha that is 6 point from 15 point of encounter. The family showed that densely fed biochar fields contain more diversity than controls. The ants belong to the genus Myrmicinae of small ants such as Pheidole and large ants such as Tetraponera. Pheidole is known to be found in litter which is included in generalized myrmicinae group functions are cosmopolitan and sub-dominant ants occur in most habitats, have the ability to survive and quickly adapt to collect food sources.

On Trichokompos land of 10 ton / ha + Biochar / rice husk husk 0 ton / ha it appears that the NMDS field is wider. This is due to the discovery of the group of kutu-kutuan, spiders (Araneae), mites (Acarina) are varied because generally predators found are active predators foraging.

The results of the analysis of the types of insects found in different treatment sites are shown in Figure. The composition of insect species based on their role indicates differences between treatments. Insects with a particular role dominate on certain types of land use. In the field of trichocompos + biochar treatment of 1 ton / kg with trichocompos + biochar 2ton / kg area showed relatively few insect species when compared to the type of insects in the control field. The soil possesses fertile soil properties so that herbivorous insects are not found.

Herbivorous insects that dominate the field of observation sites were *Gryllotalpa hirsuta*, *Phyllophaga javana*, *Brachytrypes portentosus*, and ground worms (*Agrotis ipsilon*). These insects become pests, because they attack the stems and roots which are also reported as the cause of the low production of potatoes and are able to attack all the growing phases [24]. In addition to the control area, aphids that attack the plant by sucking leaf liquid were also found, therefore the attacked plant leaves appeared to be wrinkled and kinky. The abaxial surface of the leaves was silvery because the inside of the hollow leaves after the liquid is inhaled.

Herbivorous insects can be categorized as pests if they result in losses that exceed the economic threshold. However, farmers often do not want to take the risk, so often apply insecticides as soon as possible when they find some insects. Often also encountered, farmers have not been able to distinguish which insects are acting as pests or natural enemies that are important for pest control.

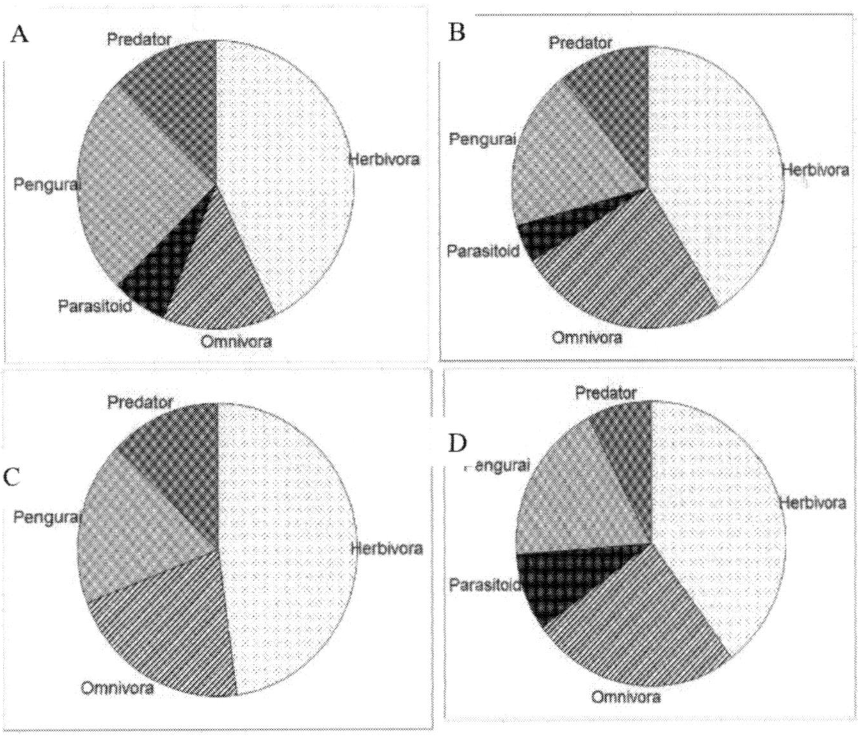


**Figure 3.** The role and composition of insects in four treatments (A) Controls; (B) Trichokompos 10 tons / ha; (C) Trichokompos 10 ton / ha + Biochar / rice husk charcoal 1 ton / ha; (D) Trichokompos 10 ton / ha + Biochar / rice husk charcoal 2 ton / ha.

Some fauna besides insects that are also found in the four differently treated fields belonging to the macrofauna group of soil (size> 2 mm) consist of milipides, isopods, earthworms. The role of fauna besides insects is decomposition, carbon flow, nutrient redistribution, nutrient cycle, and soil structure formation. Earthworm biomass has been known to be a good bioindicator for detecting pH change, presence of organic horizon, soil moisture and humus quality.

In the land treated with Trichokompos 10 tons / ha + Biochar / rice husk charcoal 2 tons / ha of parasitoid and predator insects more than other treatment areas because the land is dominated by predators and omnivorous ants. Habitat transformation is very influential on ant behavior such as competitive interaction, avoidance of predators, and parasitism. The dominance of ant groups in a habitat indicates that the habitat is relatively more stable because it is associated with the presence of prey [25].

## 4. Conclusion

The application of rice husks biochar affected the number of soil insects. The higher rice husk biochar dosage was showed fewer abundances than some herbivorous insects species. The structure of insect community showed the composition of soil insects dominated by herbivorous insects species at control, and trichokompos + rice husk biochar treatments showed soil insect composition dominated by insects omnivorous, detrivor, parasitoid and predator role.

## References

[1] Henny H, Mahbub I A and Arzita 2013 *Bioplantae* **2**(2): 70–81
[2] Galloway J N, Aber J D, ErismanJ W, SeitzingerS P, HowarthR W, Cowling E B and B J Cosby 2003 *American Institute of Biological Sciences* **53**(4): 341
[3] Sutanto and Rachman 2001 *Jurnal Perlindungan Tanaman Indonesia* **7**(1): 9–15
[4] BaileyK L and Lazarovits G 2003*Soil and Tillage Research* **72**(2): 169–80
[5] Shalaby, Shehata and Abdou G 2010 *Journal of Plant Protection Research* **50** (1): 1–7
[6] Soccol C R, Luciana P S, Woiciechowski A L, Soccol V T, Correia C T and Pandey A 2003 *Indian Journal of Experimental Biology* **41**(9): 1030–45
[7] Zwietenn L V, Kimber S, Morris S, Chan K Y, Downie A, Rust J, Joseph S and Cowie A 2010 *Plant and Soil* **327**(1): 235–46
[8] Beesley L, Moreno-Jimenez E, Gomez-Eyles J L, Harris E, Robinson B and Sizmur T 2011 *Environmental Pollution* **159**(12)
[9] Masulili A 2010 *Journal of Agricultural Science* **2**(1): 38–47
[10] Altieri M A Clara I N 2003 *Soil and Tillage Research* **72**(2): 203–11
[11] Nurida N L, Rachman A and Sutono S 2015 *IAARD Press*
[12] BPTP Sumbar 2001 *Teknologi Pengompos Cepat Menggunakan Trichoderma Harzianum*
[13] Bestelmeyer B T, Agosti D, AlonsoL E, Brandão C R F, Brown Jr.W L, Delabie J H C, and Silvestre R 2000"Field Techniques for the Study of Ground-Dwelling Ants." In *Ants: Standard Methods for Measuring and Monitoring Biodiversity*, 122–44
[14] Naumann I D, P B Came, J F Lawrence, E S Nielsen, J P Spradbery, TaylorR W, WhittenM J, and Littlejohn M J (eds.) 1991 *The Insects of Australia. A Textbook for Students and Research Workers* Melbourne: Melbourne University Press
[15] Colwell R K and Johanna E E 2014 *Ecography* **37**(6): 609–13
[16] Magurran A E and Brian J M 2011 *Challenges* 368
[17] Faraway J J 2002 *Reproduction* **21**(July): 212
[18] Clarke K R 1993 *Australian Journal of Ecology* **18**(1988): 117–43
[19] R Development Core Team 2016 *R Foundation for Statistical Computing Vienna Austria*
[20] Oksanen J 2015 *R Documentation*, **43**
[21] Gotelli N J and Colwell R K 2001 *Ecology Letters* **4**(4): 379–91
[22] Reibe K, Klaus Peter Götz K P, Roß C L, Döring T F,Ellmer F and Liliane R 2015 *Soil Biology and Biochemistry* **83** (February): 84–87

[23] Lassau S, Hochuli D F,Cassis G and M R Chris 2005 *Diversity & Distributions* **11**(1): 73–82.
[24] Setiawati W, Hudayya A and Jayanti H 2014*Jurnal Hortikultura* **24**(1): 65–75
[25] Kaspari, Michael, Yuan M and Alonso L 2003 *The American Naturalist* **161**(3): 459–77

**Acknowledgment**
This research was funded by Ministry of Agriculture. We would like to thank Jon Hendri, Sigid Handoko who helped during sampling of soil and insects in the field. We are grateful Prof. Dr. Gono Semiadi and Prof. Rosichon Ubaidillah as mentor who has provided input for the improvement of the manuscriptand a further anonymous reviewer for their comments on our manuscript.

# Botanical pesticides effect from shells of bean's cashew nut on biological agents of *trichoderma* sp. and *gliocladium* sp.

**L O S Bande**[1], **Mariadi**[1], **Gusnawaty HS**[1], **Nuriadi**[1], **L Trisulpa**[2] **& Rahmania**[2]

[1] Plant Protection Department, Agricultural Faculty, Halu Oleo University, Kendari

[2] Undergraduate of Agrotechnology Department, Agricultural Faculty, Halu Oleo University, Kendari

E-mail: Ls.bande@yahoo.co.id

**Abstract.** A shell of cashew nut (*Anacardium occidentanle*) has contained Cashew Nut Shell Liquid (CNSL) that is used as botanical pesticides. CNSL oil consists of active substance such as anacardat acid, cardol and cardanol. Utilization of the pesticides from shells of cashew nut to control pests and diseases of plants would be affected on biological agents. The objective of this research was to investigate pesticides inhibition on the increase of mycelium *Trichoderma* sp. and *Gliocladium* sp. by in vitro method. The tested concentration sample consisted of 0.0% (control), 2.5%, 7.5% and 10.0% in PDA media. The results of this research showed that 2.5% botanical pesticides concentration could minimize mycelium of *Trichoderma* sp. and *Gliocladium* sp. 22.73% and 21.04% respectively and also the increase shells of cashew extract could be affected the increase of mycelium inhibition. The extract with 2.5% concentration was the recommended concentration to control of fruit rot diseases and if concentration was 10.0% then its inhibition become 54.98% and 49.35%, respectively. The results proved that uncontrolled utilization of the pesticides could be affected on decrease of *Trichoderma* sp. and *Gliocladium* sp. growth.

## 1. Introduction

Current, utilization of the chemical pesticides is very massive on agriculture area, therefore it affects on the decrease of agricultural products. Utilization of pesticides in cocoa farming area has been used by farmers for cultivation, land preparation, plant maintenance and harvest. Excessive use of pesticides will causehealth problems and environmental pollution [1], decreasing of soil quality [2] and pesticides residue on cocoa beans [3]. There were many methods to reduce the use of chemical pesticides, such as utilization of biological agents, botanical pesticides, resistant varieties, setting cropping pattern and habitat management (4,5)

In Southeast Sulawesi, extract of cashew nut shells is one of potential source of botanical pesticides. Cashew nut shells contain cashew nut shell liquid (CNSL) oil, which has fungicidal effect [8] and insecticidal effect [9]. Active ingredients on CNSL are anacardat acid, cardol and cardanol [10,11,12]. Laboratory testing showed that 3.5% botanical pesticide concentration could control *Phytophthora palmivora* growth, pod rot disease in cocoa plants by 87.67% [6] and 2.5% of botanical pesticide concentration could be decreasing pod rot diseases in cocoa plants by 64.16-71.02% [7].

*Trichoderma* sp. and *Gliocladium* sp. are biological control agents for controlling diseases that are widely used [13, 14]. Utilization of uncontroled chemical pesticides can be affected on decreasing of pathogen control ability [14]. On the other hand, utilization of botanical pesticides can also be decreasing the pathogen. Application of botanical pesticides from extract of cashew nut shells to control pest and disease will minimize of *Trichoderma* sp. and *Gliocladium* sp. on plant roots.

The recommendation concentration of botanical pesticides from cashew nut shells to control of fruit rot disease in cocoa plants is 2.5% [17]. Study of biological safety against using biological pesticides of cashew nut shells should be studied further, this is to avoid the negative effects of these vegetable pesticides. The objective of this research was to observe the inhibition of botanical pesticides from cashew nut shells extract against mycelium growth of *Trichoderma* sp. and *Gliocladium* sp. by *in vitro* method.

## 2. Materials and methods

This research was conducted in Plant Protection Laboratory, Agriculture Faculty, Halu Oleo University from March to May 2017. There werefive different treatments that were arranged randomly with three replications. The treatments were $H_0$ (0.0%), $H_1$ (2.5%), $H_2$ (5.0%), $H_3$ (7.5%), and $H_4$ (10.0%). This research used a botanical agents of *Trichoderma* sp. and *Gliocladium* sp., which werethe colllection of Plant Protection Laboratory, Agriculture Faculty, Halu Oleo University. *Trichoderma* sp. and *Gliocladium* sp. are reproduced through Potato Dekstrosa Agar (PDA) medium. As for the botanical pesticides from extract of cashew nut shell is produced by Ir. Mariadi, M.S that contain active acid *anakardat*.

The experiment of botanical pesticides to inhibit *Trichoderma* sp. and *Gliocladium* sp. used toxic medium method, which was the mixture of botanical pesticides based on concentration on PDA medium [6]. After mixing process, the botanical pesticides was sterilized at 121 °C and 151 psi for 20 minutes. After that, the medium was cooled to temperature of 65 °C and 10 ml of the steriled mixture pesticide medium wass poured into a petridish and cooled. Inoculum of *Trichoderma* sp. and *Gliocladium* sp. was cultured on the center of petridish and incubated at room temperature separately.

The parameters were dry weight of mycelium and inhibitory of botanical pesticides against dry weight of mycelium. Observation of the mycelium was started $\pm$ 3-5 days after inoculation or if biological agents colony fulfilled a petridish. Meanwhile, the fungal mycelium was harvested by adding sterile aquadest on the medium sufficiently and then heated. The suspension was filtered, therefore only fungal mycellium was obtained. The fungal mycelium was dried in the oven at 50 °C until its weight is stable or the weight is not changing again. The inhibition of botanical pesticides was calculated by the equation as [6] as follows:

$$IB = \frac{DWC_1 - DWB_2}{DWC_1} x100\%$$

IB : The inhibit of botanical pesticides (%)
$DWC_1$: The dry weight of mycelium without botanical pesticides (control)
$DWB_2$: The dry weight of mycelium with botanical pesticides

Data was analyzed and if it has a significant affect then continue with Duncan's multiple-range test in 95% level.

## 3. Results and Discussion

The result of this study showed that the higher stimulation of botanical pesticides from cashew nut shells extract by in vitro method could be affected on decreasing of *Trichoderma* sp. and *Gliocladium* sp. biomass (Tables 1 and 2). Data analyzed of biomass weight showed that botanical pesticide concentration had a significant effect against biomass weight of *Trichoderma sp*. and *Gliocladium sp*. Clearly, this research implied that botanical pesticide was one of some alternatives to control pets and also to minimize the use of chemical pesticides. This result similar with some other research that mention about about advantages of botanical pesticides such as; environmentally friendly, easily degraded, without toxic, and according to the concept of sustainable agriculture [15,16].

The other result of the research (Table 1) show that that the higher concentration could be decreased biomass weight *Trichoderma* sp. This is in line with the result of previous resarch wich mentions that utilization dose or concentration of pesticides should be larger to get maximum results.

In the another hand, the Table 1 given information that utilization of botanical pesticides should be precise, so negative effect is neglected. Inhibition of 2.5% of botanical pesticides against *Trichoderma sp.* was 22.73% while the use of 5.0% botanical pesticides was not significantly different to 2.5% concentration (30.52%). If the concentration become 10.0%, it could be affected on increasing of inhibition of botanical pesticides by 54.98%. In this condition, *Trichoderma sp.* only growth 45.20% normally. The utilization of combination of *Trichoderma* sp. and botanical pesticides from cashew nut shells extract showed a good interaction if both of them will use to control pest simultaneously. Studied that *Trichoderma* sp. and botanical pesticides from clove extract had effect to control ginger rhizome rotten disease [18].

**Table 1.** Investigation data of biomass *Trichoderma* sp. and inhibition of botanical pesticides against biomass *Trichoderma* sp.

| Treatments | Biomass weight (g) | Inhibition of botanical pesticides (%) |
|---|---|---|
| Concentration 0.0% ($H_0$) | 0.13 d | 0.00 a |
| Concentration 2.5% ($H_1$) | 0.10 c | 22.73 b |
| Concentration 5.0% ($H_2$) | 0.09 bc | 30.52 bc |
| Concentration 7.5% ($H_3$) | 0.07 ab | 44.81 cd |
| Concentration 10.0% ($H_4$) | 0.06 a | 54.98 d |

Notes: the number is followed by alphabeth symbol is not different and the degree of trust is 95%

Biomass weight of *Gliocladium* sp. decreased by the increasing of botanical pesticides of cashew nut shells extract (Table 2). Higher biomass occurred on control treatment (without botanical pesticides) and it differed significantly to other treatments. Biomass weight of *Gliocladium* sp. on 2.5% as many as 0.12 g is not different with biomass weight on 5.0% as many as 0.10 g. This data showed that utilization of 5.0% botanical pesticides from cashew nut shells extract had affect on the increase of *Gliocladium* sp., and this is almost same with the 2.5% concentration of botanical concentration.

**Table 2.** Investigation data of biomass *Gliocladium* sp. and inhibition of botanical pesticides against biomass *Gliocladium* sp.

| Treatments | Biomass weight (g) | Inhibition of botanical pesticides (%) |
|---|---|---|
| Concentration 0.0% ($H_0$) | 0.15 c | 0.00 a |
| Concentration 2.5% ($H_1$) | 0.12 b | 21.04 ab |
| Concentration 5.0% ($H_2$) | 0.10 ab | 32.48 bc |
| Concentration 7.5% ($H_3$) | 0.07 a | 49.35 bc |
| Concentration 10.0% ($H_4$) | 0.07 a | 49.35 c |

Notes: the number is followed by alphabet symbol is not different and the degree of trust is 95%

Inhibition of 2.5% botanical pesticides against *Gliocladium* sp. did not differ significantly with control treatment, which means that the ability of botanical pesticides still normal on this concentration. This study proved that botanical pesticides from cashew nut shells extract could be applied with *Gliocladium* sp. to control pest simultaneously. This data is also important as information to control pest of plants. Supriadi [19] said that one of efforts to minimize of chemical pesticides was optimization of botanical pesticides and biological control that are compatible simultaneously.

## 4. Conclusions

In summary utilization of botanical pesticides from shells of cashew nut by in vitro method could decrease biomass weight of *Trichoderma* sp. and *Gliocladium* sp. There are decreasing of weight's *Trichoderma* sp. and *Gliocladium* sp. against inhibition of botanical pesticides using extract of cashew nut shells. The amount of inhibition using 2.5% extract of cashew nut shells against biomass weight of *Trichoderma sp.* and *Gliocladium sp.* were 22.73% and 21.04% respectively. Meanwhile, 10.0% concentration of botanical pesticides could inhibit by *Trichoderma* sp. and *Gliocladium* sp. by 54.98% and 49.35% respectively.

## References

[1] Sande D., J. Mullen, M. Wetzstein and J. Houston 2011 Environmental impacts from pesticide use: a case study of soil fumigation in florida tomato production. *Int. J. Environ. Res. Public Health* **8**: 4649-61

[2] Chowdhury A., Pradhan S., Saha M. and N. Sanyal 2008. Impact of pestisides on soil mikrobiological parameters and possible bioremediation strategies. *Indian J. Microbial* **48**: 114-27

[3] Wiryadiputra S 2013. Pesticide residue on Indonesian cacao beans and its varaiant products, and the efforts to solve the problem. *Rev. Penel. Kopi dan Kakao* **1(1):** 40-62.

[4] Birch A.N.E., G.S. Begg and G. R. Squire 2011 How agro-ecological research helps to address food security issues under new IPM and pesticide reduction policies for global crop production systems. *J. of Exp. Botany* **62(10):** 3251-61

[5] Harni, R., E. Taufiq and W. Amaria 2014 The effect of clove and citronella oil as botanical fungicides formula to control black pod disease. *J. TIDP.* **1(1)**: 41-48

[6] Bande L.O.S., Gusnawaty HS., Mariadi, Nuriadi dan Sukamawati 2017 Daya hambat pestisida nabati dari kulit biji jambu mete terhadap *Phytophthora palmivora* penyebab penyakit busuk buah kakao secara in vitro. *Prosiding Seminar Nasional Pengendalian Penyakit pada Tanaman Pertanian Ramah Lingkungan II.* pp: 314-21. PFI Komda Yogyakarta, Solo dan Semarang.

[7] Bande L.O.S, Gusnawaty HS., Mariadi, dan Nuriadi 2016 Formulasi pestisida nabati untuk pengendalian hama penggerek buah dan penyakit busuk buah kakao *Laporan Penelitian PUPT.* Lembaga Penelitian dan Pengabdian Kepada Masyarakat Universitas Halu Oleo. Kendari.

[8] Gassa A., S. Mulia, Yumarto, and M. Junaid 2016 Phythophthora black pod disease of cocoa caused by *Phytophthora palmivora*: development of bio-fungicidal package in controlling the disease and the vector by food bait. *J. Chem. Pharm. Res.* **8(6)**:129-35

[9] Harlita dan Muzayyinah,2004. Serbuk kulit biji jambu mete (*Anacardium occidentale* Linn) sebagai moluskisida untuk menghambat perkembangbiakan keong emas (*Pomacea* sp.). *Biosmart.* **6(1)**: 29-32

[10] Santos M. L. D. and D. C. D. Magalhaes 1999 Utilisation of cashew nut shell liquid from *Anacardium occidentale* as strating material for organic synthesis: a novel route to lasiodiplodin from cardols. *J.Braz. Chem. Soc.* **10 (1)** : 13-20

[11] Simpen I. N 2008 Isolasi cashew nut shell liquid dari kulit biji jambu mete *(Anacardium occidentale L.)* dan kajian beberapa sifat fisiko-kimianya. *J. Kim.* **2(2)**: 71-6

[12] Ola A.R. B. Z. Ikawati, Sisimindari E.D. Meye and B.D. Tawo 2008 Molecular identification and anticancer activity of alkylphenol from cashew nut shell oil (*Anacardium occidentale*) grown In Timor Island. *M. Farm. Indonesia.* **19(3)**: 137-44

[13] Blaszczyk L., M. Siwulaski, K. Sobieralski, J. Lisiecka and M. Jedryczka 2014 *Trichoderma* sp. application and prospects for use in organic farming and industry. *J. P. Protec. Res.* **54(4):** 309-17

[14] Brimner T.A and G.J. Boland 2003 A review of the non-target effects of fungi used to biologically control plant diseases. *Agric. Eco. and Envir.* **100**:3-16

[15] Hartati, S.Y. 2012 Prospec of essential oils delepode as botanical pesticides. *Prespektif.* 11(1): 45-58

[16] Dalimunthe C.I, and A. Rachmawan 2017 The prospek of plant secondary metabolite as botanical

pesticide againts pathogen on rubber. *W. Perkar*. **31(1):** 15-28
[17] Supriadi, 2015. Research innovation to support the commersialization of biopestisides in Indonesia. *Prespektif.* **14(1)**: 15-25
[18] Soessanto L., Sudarmono, N. Prihatiningsih, A. Manan, E. Iriani, and J. Pranomo 2015 Potency of biological and botanical agents in controling ginger rhizoama rot. *J. HPT.* **5(1)**: 50:57
[19] Supriadi 2013 Optimal use of various kins of pesticides to control plant pests and disease. *J. Litbang Pert.* **32(1):** 1-9

IOP Conf. Series: Earth and Environmental Science **122** (2018) 012048 doi:10.1088/1755-1315/122/1/012048

# Identification and phylogenetic analysis of local yellow and orange sweet potatoes genotypes in Sumatera Utara

**Rosmayati and D Bakti**

Faculty of Agriculture, Universitas Sumatera Utara, Padang Bulan, Medan 20155, Indonesia
E-mail : rosmayati@usu.ac.id

**Abstract**. The objectives of this research was to identify and create database about the diversity of local yellow and orange sweet potatoes accessions in Sumatera Utara, have diversity accession local sweet potatoes genotype in Sumatera Utara selection for classifying populations get high production and good fruit quality. The experiment was conducted in areas of production centers of sweet potatoes in the exploration survey methods in 2 districts in Sumatera Utara, which is in the Kabupaten Simalungun and Dairi. The study was conducted on June to July 2017. Observations were made based on the identification and characterization Description List of International Board for Plant Genetic Resources standard and purposive random sampling technique. The result of this research indicate there 15 genotype of sweet potato yellow and orange in KabupatenSimalungun consistedof KecamatanPurba (G3, G4 and G7), Silimakuta (G5, G6 and G14), and Pamatang Silimahuta (G15) in Kabupaten Dairi consists of Kecamatan Parbuluan (G1, G2, G8 and G9), Sidikalang (G10 and G13), Sumbul (G11), and Sitinjo (G12) with nearest relationship is G13 and G15 with a coefficient similarity 23.908 and farthest relationship is G2 and G7 with a coefficient similarity 140.029.

## 1. Introduction

Sweet potato (*Ipomoea batatas* (L.) Lam.) from the family Convolvulaceae, is an important food crop, which is widely grown in tropical, subtropical and warm temperate regions [1]. The Sweet potato was originally domesticated at least 5000 years ago in Tropical America [2; 3]. Based on the analysis of key morphological characters of sweet potato and the wild Ipomoea species, Austin [2] postulated that sweet potato originated in the region between the Yucatan Peninsula of Mexico and the Orinoco River in Venezuela. Using molecular markers the highest diversity was found in Central America, supporting the hypothesis that Central America is the primary center of diversity and most likely the center of origin of sweet potato [4; 5].

Sweet potato are rich in starch, sugar, vitamin C, β-carotene, iron, and several other minerals [6; 7]. The root reported to usually have higher protein content than other roots and tubers, such as cassava and yams. In addition, some varieties of sweet potatoes contain colored pigments, such as β-carotene, anthocyanin, and phenolic compounds. These pigments form the basis for classifying the foods as nutraceuticals [7]. Orange flesh sweet potatoes are high in carotenoids and beta carotene [8]. Consumption of orange flesh sweet potato roots can provide sustainable vitamin A, which plays a major role in preventing night blindness [9].

Germplasm of the world's sweet potato plants is estimated to number over 1000 species, but only 142 species are identified by the researchers. In the 1960s sweet potato planting had spread to all provinces in Indonesia. Production center of sweet potato is West Java Province, Central Java, East Java, Irian Jaya, and North Sumatra [10].

Sumatera Utara is one of sweet potato production center in Indonesia. According to Central Agency on Statistics (BPS) (2016), the area of sweet potato harvest in North Sumatera in 2015 is 8,952 ha with sweet potato production of 122,362 tons and productivity 136.69 Ku / ha. The largest production center of sweet potato in Sumatera Utara is in Simalungun Regency (35,080 tons) and Dairi Regency (27,818 tons).

Morphological characterization has been used extensively on various crop plants diversity assessments in many places of the world [11; 12; 13; 14; 15]. Despite the environmental influences on plant morphology, this direct inexpensive and easy to use method of estimations was perceived as the strongest determinant of the agronomic value and taxonomic classification of plants [14] and the first step in the assessment of plant diversity. On sweet potato, this tool has been used successfully to analyse genetic diversity necessary for the germplasm conservation, to reduce accession number by identification and elimination of duplicates and to enhance crop breeding[14; 16; 17; 18; 19; 20].

Based on the above description, the authors intend to conduct research on morphological identification and phylogentic relationship of sweet potato located in Sumatera Utara namely Simalungun and Dairi regencies, especially yellow and orange sweet potato containing β-carotene which is higher than other genotypes of sweet potatoes.

## 2. Materials and methods

The research was conducted in Simalungun Regency and Dairi Regency. Starting from June 2017 to July 2017. Determination of research location is determined based on the highest production of sweet potato data located in Sumatera Utara namely Simalungun Regency and Dairi Regency. There were 15 accessions were grown in these regencies (Table 1).

The materials used in this study were sweet potato crops with yellow and orange tuber color grow in Simalungun Regency and Dairi Regency.Observations were made based on the identification and characterization Description List of International Board for Plant Genetic Resources standard and purposive random sampling technique. The location of the study is determined based on the area of harvested area and production data from the Central Agency of Statistics (BPS), then searching the location of the study by accidental sampling. Qualitative and quantitative data were scored and processed using SPSS 21 program with cluster analysis to determine the level of relationship between genotypes. Cluster analysis is used to construct dendogram used to assess the diversity patterns of survey data. The technique used is Agglomerative Hierarchical Clustering, the method of average linkage (between group) with euclidien distance as follows :

$$d_{(i,j)} = \sqrt{\sum_{k=1}^{p} (x_{ik} - x_{jk})^2}$$

$d_{i,j}$ =the distance between object i and object k

$x_{ik}$ = object value $i$ on the variable$k$

$x_{jk}$ = object value$j$ on the variable$k$

$p$ = the number of cluster variables[21]

The parameters observed were: 1) winding, 2) stem morphology: stem type, stem diameter, stem length, main stem color, bar hair, 3) leaf morphology: leaf general shape, lobe type, lobe number, leaf size, Leaf repeating color, old leaf color, shoot leaf color, long petiole, leaf color,4) tuber morphology: tuber formation, tuber length, tuber shape, tuber defect, cortical thickness, tuber skin color, skin color intensity, secondary skin of tuber bulbs, main color of tuber meat, secondary color of tuber meat, secondary color distribution Meat, sap rate, oxidation reaction, average number of tubers, and average tuber weight.

**Table 1.** List of accessions collected in Sumatera Utara the varieties introduced used for the characterization.

| Code | Regency | District | Village | Folk name |
|---|---|---|---|---|
| G1 | Dairi | Parbuluan | Parbuluan 1 | Ubi Jepang |
| G2 | Dairi | Parbuluan | Parbuluan 1 | Ubi Sibolu |
| G3 | Simalungun | Purba | Nagori Ritongah | Ubi Jepang |
| G4 | Simalungun | Purba | Nagori Ritongah | Ubi Taiwan |
| G5 | Simalungun | Silimakuta | SaribuDolok | Ubi Sibolu |
| G6 | Simalungun | Silimakuta | Kasemak | Ubi Sibolu |
| G7 | Simalungun | Purba | TigaRunggu | Ubi Taiwan |
| G8 | Dairi | Parbuluan | Simallopuk | Ubi Jepang |
| G9 | Dairi | Parbuluan | Simallopuk | Ubi Sibolu |
| G10 | Dairi | Sidikalang | BatangBeruh | Ubi Sibolu |
| G11 | Dairi | Sumbul | LaeSiboban | Ubi Jepang |
| G12 | Dairi | Sitinjo | Sitinjo | Ubi Sibolu |
| G13 | Dairi | Sidikalang | LaeMungkur | Ubi Jepang |
| G14 | Simalungun | DolokSilou | Perasmian | Ubi Sibolu |
| G14 | Simalungun | PematangSilimahuta | Mardinding | Ubi Sibolu |

**3.Results and discussion**

A comprehensive analysis of the extent and distribution of the genetic variation in sweet potato is essential for sound genetic conservation strategies. Conservation and sustainable use of genetic resources is essential to meet the demand for future food security. Successful conservation of any given gene pool is largely dependent on understanding the diversity and its distribution in a given region [22]. In this study, high level of genetic diversity was exhibited in sweet potato. Morphological characters were highly variable among the genotypes studied. The morphological characterization of accessions in the collection is essential not only to have a description of each accession but also to use this information to identify duplicate accessions. The genotypes exhibited high morphological diversity in aerial and storage tuber characters.

The results of surveys conducted in Simalungun and Dairi regencies have found 15 genotypes of yellow and orange tuber plant consisting of 3 types of sweet potato (Ubi Jepang, Sibolu and Taiwan). The morphological characteristics of sweet potato plants are observed visually. Each of the sweet potato genotypes is present in several locations that have different altitudes from different types of sweet potatoes that have different morphological characteristics. Based on morphological character of sweet potato in Regency of Simalungun and Regency of Dairi obtained value of kinship relation which can be seen in table 2.

Table 2 shows if the value of coefficient smaller between variables one with other variables, then the relationship between the two variables are closer or the greater the similarity and vice versa. So it can be seen that the highest level of similarity or nearest kinship on 33 qualitative and quantitative characters is found in G13 and G15 is 23.908, whereas the lowest similarity level or the furthest kinship relationship is found in G2 and G4 is 140.029. Genetic diversity study is necessary to understand the genetic relationships among populations and to assign accessions to specific heterogeneous groups that help to identify parental lines for best combinations with maximum genetic potential for further selection [23]

Based on the morphological character of sweet potato in Simalungun and Dairi districts, it is found that all the genotypes of sweet potato can be grouped into two groups, three groups and four groups of members from each group (Table 3)

**Table 2.** Kinship relationship of 15 sweet potatoes genotypes in Simalungun and Dairi Districts Based on Dissimilarity Matrix

| No. | Kinship relationship | | Coefficient Value | No. | Kinship relationship | | Coefficient Value | No. | Kinship relationship | | Coefficient Value |
|---|---|---|---|---|---|---|---|---|---|---|---|
| 1. | G13 | G15 | 23.908 | 13. | G1 | G12 | 38.812 | 25. | G4 | G5 | 98.062 |
| 2. | G5 | G15 | 26.694 | 14. | G8 | G10 | 39.254 | 26. | G2 | G7 | 98.428 |
| 3. | G5 | G6 | 27.741 | 15. | G8 | G11 | 40.083 | 27. | G4 | G7 | 98.521 |
| 4. | G11 | G12 | 29.066 | 16. | G11 | G14 | 40.404 | 28. | G3 | G13 | 100.663 |
| 5. | G12 | G14 | 29.105 | 17. | G11 | G15 | 42.131 | 29. | G1 | G8 | 101.032 |
| 6. | G5 | G14 | 34.765 | 18. | G1 | G2 | 88.315 | 30. | G4 | G9 | 104.256 |
| 7. | G14 | G15 | 35.487 | 19. | G1 | G13 | 89.159 | 31. | G4 | G8 | 106.368 |
| 8. | G8 | G15 | 36.502 | 20. | G2 | G12 | 89.339 | 32. | G4 | G10 | 109.660 |
| 9. | G6 | G11 | 37.207 | 21. | G4 | G6 | 90.679 | 33. | G4 | G13 | 132.067 |
| 10 | G5 | G13 | 37.989 | 22. | G4 | G14 | 91.054 | 34. | G2 | G4 | 140.029 |
| 11. | G9 | G12 | 38.184 | 23. | G2 | G3 | 94.960 | | | | |
| 12. | G5 | G8 | 38.783 | 24. | G4 | G15 | 98.033 | | | | |

**Table 3.** The grouping of sweet potato genotypes in 33characters in Simalungun and Dairi districts

| Code | 4 Group | 3 Group | 2 Group |
|---|---|---|---|
| G1 | 1 | 1 | 1 |
| G2 | 2 | 2 | 1 |
| G3 | 3 | 1 | 1 |
| G4 | 4 | 3 | 2 |
| G5 | 1 | 1 | 1 |
| G6 | 1 | 1 | 1 |
| G7 | 1 | 1 | 1 |
| G8 | 1 | 1 | 1 |
| G9 | 1 | 1 | 1 |
| G10 | 1 | 1 | 1 |
| G11 | 1 | 1 | 1 |
| G12 | 1 | 1 | 1 |
| G13 | 1 | 1 | 1 |
| G14 | 1 | 1 | 1 |
| G15 | 1 | 1 | 1 |

The identified variations are expected to form a pattern of genetic relationship useable in grouping accessions. In grouping of accessions, cluster analysis has ability to identify crop accessions with highest level of similarity using the dendrogram [24] (Aliyu et al., 2000). Result of survey of sweet potato in Regency of Simalungun and Dairi Regency on 33 characters obtained dendogram kinship relationship which can be seen in Figure 1.

Analysis of kinship relationships based on 33 characters on a 25-degree scale formed into two groups. Where the first group (I) consists of G1, G2, G3, G5, G6, G7, G8, G9, G10, G11, G12, G13, G14, and G15 while the second group (II) consists of an accession of G4. In the first group (I) is divided into two sub-groups namely the subgroups IA and IB, the first group (IA) consists of G1, G3, G5, G6, G11, G12, G13, G14, And G15 while the second group (IB) consists only of G2.

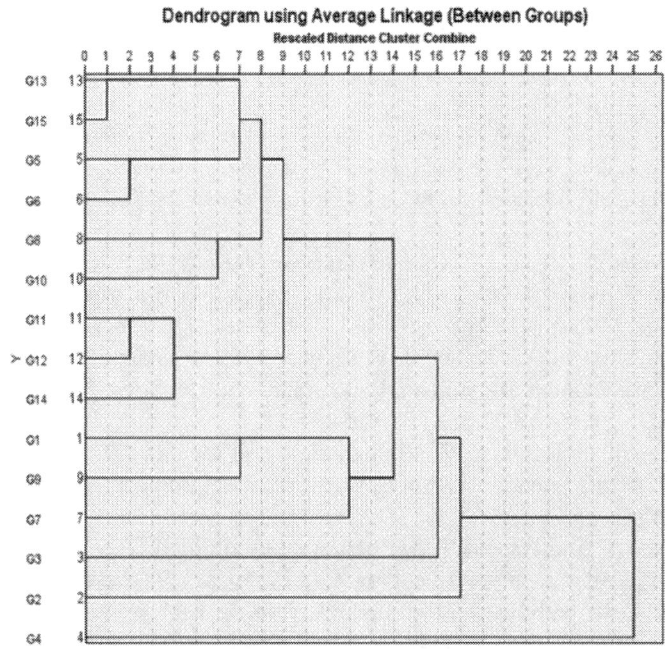


**Figure 1.** Dendrogram of the 15 sweet potato accessions revealed by average linkage cluster analysis based on the 33 discriminant phenotypic characters.

The further the kinship relationship will be the higher the level of diversity and the lower the level of uniformity, and vice versa [25]. The influence of environmental factors such as climate, temperature, soil type, altitude and humidity will cause the variation of plant morphology. This is not the case when the environmental factor conditions are the same [26]. Survey showed the planting area of sweet potato in Simalungun ranged from 1.160 – 1.407 m above sea level, while the area of sweet potato planting in Dairi ranged from 1.120 – 1.718 m above sea level. In Simalungun, in June the average temperature was 19, rainfall 190 mm, relative humidity 75.33%, in July average temperature 18.60, rainfall 176 mm and relative humidity 75%. While In Dairi, in June the average temperature was 18.30, rainfall 153.5 mm and relative humidity 82.5, in July average temperature 17.80, rainfall 23,5 mm and relative humidity 76,5%.

## 4..Conclusions

Yellow and orange sweet potatoes that have been found 15 genotypes are found in Simalungun District consisting of PurbaSubdistrict (G3, G4 and G7), Silimakuta (G5 and G6), DolokSilou (G14) and PamatangSilimahuta (G15) at Dairi Regency consists of Parbuluan sub-districts (G1, G2, G8 and G9), Sidikalang (G10 and G13), Sumbul (G11), and Sitinjo (G12) with nearest kinship values obtained at G13 with G15 of 23.908 and kinship value With the farthest distance obtained at G2 and G4 is 140.029.

## References

[1] Sihachakr D, Haı ¨cour R, Cavalcante-Alves J M, Tizroutine S, Allot M, Mussio I, Servaes A, Nozhge D, Ducreux G 1995 Embryogene `se somatique chez la patate douce (*Ipomoea batatas* (L.) Lam.): Caracte ´risation et re ´ge ´ne ´ration des plantes, in: Dubois J, Demarly Y-Aupelf-Uref (Eds.), Quel avenir pour l'ame ´lioration des plantes, John Libbey eurotext pp 251–261

[2] Austin D F 1988 The taxonomy, evolution and genetic diversity of sweet potatoes and related wild species IGregory P (ed.) *Exploration, maintenance and utilization of sweet potato genetic resources* CIP, Lima, Peru pp 27–60

[3] Yen DE 1982 Sweet potato in historical perspective. In Villareal R L and Griggs T D(eds.) Sweet Potato *Proceedings of the First International Symposium* AVRDC, Tainan, Taiwan pp 17–30

[4] Huang, J C and Sun M. 2000 Genetic diversity and relationships of sweetpotato and its wild relatives in Ipomoea series batatas (Convolvulaceae) as revealed by inter-simple sequence repeat (ISSR) and restriction analysis of chloroplast DNA *Theor. Appl. Genet.* Vol. **100** pp 1050–1060

[5] Zhang D, Cervantes J, Huaman E, Carey E. and Ghislain M 2000 Assessing genetic diversity of sweet potato (*Ipomoesa batatas* (L.) Lam.) cultivars from tropical America using AFLP *Genet. Resour. Crop* Vol. **47** pp. 659–665.

[6] Laurie SM, van Jaarsveld PJ, Faber M, Philpott MF, Labuschagne MT 2012 Trans-b-carotene, selected mineral content and potential nutritional contribution of 12 sweet potato varieties *J. Food Compost. Anal.* Vol **27** pp. 151-159

[7] Oloo BO, Shitandi A, Mahungu S, Malinga J B and Ogata B R 2014 Effects of lactic acid fermentation on the retention of betacarotene content in orange fleshed sweet potatoes *Int. J. Food Stud.* Vol **3** pp 13-33

[8] Jakahata Y, Noda T, and Nagata T 1999 Determination of β carotene content in sweet potato cultivars and its relationship with colour value *Japan Jrl of Breed.* Vol **43** pp 421-427

[9] Ndirigue, J 2004 Adaptability and acceptability of orange and yellow fleshed sweet potato genotypes in Rwanda BSc (Agri) Hons IFA Yagambi, DRC

[10] Deputi Menegristek 2008 Ubi Jalar/Ketela Rambat (*Ipomoea batatas*). Kantor Deputi Menegrestik Bidang Pendayagunaan dan Pemasyarakatan Ilmu Pengetahuan dan Teknologi MIG Corp. http://warintek.ristek.go.id.

[11] Bos HJ, Vos J, Strulk PC 2000 Morphological analysis of plant density effects on early leaf area growth in maize *Netherlands J. Agric.Sci.* Vol **48** pps 199-212

[12] Kaplan DR 2001 The science of plant morphology: definition, history, and role in modern biology *Am. J. Bot.* Vol **88**(10) pp 1711-1741

[13] Lacroix C, Jeune B, Barabe D 2005 Encasement in plant morphology: an integrative approach from genes to organisms. Can. J. Bot. Vol. 83 pp.1207-1221

[14] Li P, Wang Y, Sun X and Han J 2009 Using microsatellite (SSR) and morphological markers to assess the genetic diversity of 12 falcata (*Medicagosativa spp.* falcata) populations from Eurasia *Afr. J. Biotechnol.* Vol b(10) pp. 2102-2108

[15] K'Opondo FB 2011 Morphological characterization of selected spiderplant (*Cleome gynandra*L.) types from western Kenya *Ann. Biol. Res.* Vol **2**(2) pp.54-64.

[16] Huaman Z 1992) Morphological identification of duplicates in collections of *Ipomoea batatas*. CIP Research Guide

[17] Mok IG and Schmiediche P 1998 Collecting, Characterizing and maintaining sweet potatogermplasm in Indonesia. International Potato Center

[18] Tairo F, Mneney E, Kullaya A 2008 Morphological and agronomical characterization of sweet potato germplasm from Tanzania *Afr. J. Plant Sci.* Vol **2** pp 77-85

[19] Karuri H W, Ateka E M, Amata R, Nyende A B, Muigai A W T, Mwasame E and Gichuki S T 2010 Evaluating diversity among Kenyan sweet potato genotypes using morphological and SSR markers *Int. J. Agric. Biol.* Vol **12** pp 33-38

[20] Yada B, Tukamuhabwa P, Alajo A, Mwanga ROM 2010 Morphological Characterization of Ugandan Sweetpotato Germplasm *Crop Sci* Vol **50** pp 2364–2371.

[21] Sholihah A 2015 Perbandingan Analisis Klaster Menggunakan Metode Single Linkage, Complete Linkage, dan K-Means Untuk Pengelompokan Kecamatan Berdasarkan Variabel Jenis Ternak Di Kabupaten Semarang. Universitas Negeri Semarang, Semarang

[22] Zhang DP, Carbajulca D, Ojeda L, Rossel G, Milla S, Herrera C and Ghilsan M 1999 Microsatellite Analysis of Genetic Diversity in Sweet Potato Varieties from Latin America. Program report 1999-2000. CIP, International Potato Centre, Lima, Peru
[23] Mohammed W, Ali S, Shimelis B, and Burga S 2015 Genetic diversity of local and introduced sweet potato (*Ipomoea batatas* (l.) lam.) collections for agro physicochemical attributes in Ethiopia *Science, Technology and Arts Research Joutnal* Vol **4** (1) pp 9 – 19. DOI http://dx.doi.org/10.4314/star.v4i1.2
[24] Aliyu B and Fawole N G N Q 2000 Inheritance of Pubescence in crosses between Vigna unguiculata and V. rhomboidea. Nigeria Journal of Genetics Vol **15** pp 9-14
[25] Suratman, Dwi P and Ahmad D S 2000 analisis keragaman genus Ipomoea berdasarkan karakter morfologi Biodiversitas Vol **1**(2) pp 72-79
[26] Sitompul S M and Guritno B 1995 *Analisis Pertumbuhan Tanaman* Gadjah Mada University Press

**Acknowledgments**
The research was supported by Universitas Sumatera Utara under the scheme of Penelitian TALENTA USU 2017 with contract number : 5538/UN5.1.R/PPM/2017, 22 May 2017. The Authors would like to thank University of Sumatera Utara for financing and supporting this study.

# Effect of foliar application of α-tocopherol on vegetative growth and some biochemical constituents of two soybean genotypes under salt stress

**N Rahmawati and R I M Damanik**

Faculty of Agriculture, Universitas Sumatera Utara, Padang Bulan, Medan 20155, Indonesia
E-mail : nini@usu.ac.id

**Abstract**. Foliar spray of plant growth regulating compounds including antioxidants is an effective strategy to overcome the adverse effects of environmental constraints on different plants. A field experiment was conducted on May – July 2017 at the experimental farm in Paluh Merbau Village Deli Serdang (EC 6 – 7 dS/m). The aim was to study the effects of foliar spray of α-tocopherol (0, 250, 500, 500 ppm) on vegetative growth and some chemical constituents of 2 soybean genotypes (Grobogan x Grobogan and Grobogan x Anjasmoro) under salt stress (EC 6 – 7 dS/m). Most of morphological and biochemical parameters were significantly affected by application of α-tocopherol. The α-tocopherol at 500 ppm recorded the best value of root fresh weight, shoot and root dry weight, number of leaves, chlorophyll b, and soluble protein content. There was significant difference found between plants treated with α-tocopherol in terms of number of branch, shoot fresh weight, and chlorophyll a. Soybean genotypes showed diverse morphology and physiological responses to salt stress. Grobogan x Anjasmoro genotype was salt-sensitive based on all variable, while Grobogan x Grobogan genotype was more tolerant based on morphological and biochemical characters.

## 1. Introduction

Soybean is a strategic crop plant grown to obtain edible oil and forage. Nevertheless, soybean production may be limited by environmental stresses such as soil salinity [1]. Soybean is classified as a moderately salt tolerant crop and the final yield of soybean will be reduced when soil salinity exceeds 5 dS/m-1 [2]. High salt imposes damages in the whole life cycle of soybean. The difference in a plant response to a given level of salinity is dependent on the concentration and composition of ions in solution as well as the genotype that is exposed to the salinity [3]

The degree of salt tolerance of soybean germplasms varies among different developmental stages [4]. The essential processes leading to plant adaption to salt stress include control of water loss through stomata, metabolic adjustment, toxic ion homeostasis, and osmotic adjustment [5;6]. Soybean germplasms display a spectrum of salt tolerance capability from high to low [4].

Application of antioxidants including vitamins has gained considerable attention for alleviating the negative impact of water and salinity stress on plants in terms of plant growth and yield quantity and quality [7].

Alpha-tocopherol is a small molecule that is synthesized in the plants, mainly concentrated in plastids, and is one of the most effective single-oxygen quenchers [8]. Tolerance to salt stress, chilling stress, UV-B stress and pollutant stress is partly correlated with tocopherol content [9]. Alpha-tocopherol is a strong antioxidant that assists in maintaining membrane stability [9], intracellular signaling, and transport of electrons in the photosystem-II system [10]) and is photoprotective in nature.

Foliar spray with α-tocopherol on faba bean plants induced increase in growth parameters, yield components, chlorophyll a, b and carotenoids content [11]. Tocopherols play a role in a range of different physiological phenomena including plant growth and development, senescence, preventing lipid peroxidation and to interact with the signal cascade that convey abiotic and biotic signals [12; [13].

The aim was to study the effects of foliar spray of α-tocopherol (0, 250, 500, 500 ppm) on vegetative growth and some chemical constituents of 2 soybean genotypes (Grobogan x Grobogan and Grobogan x Anjasmoro) under salt stress (EC 6 – 7 dS/m)

## 2. Materials and methods

Field experiment was conducted on May – July 2017 at the experimental farm in Paluh Merbau Village Deli Serdang (EC 6 – 7 dS/m). The experiment was arranged in a randomized block design with two factors, the first, factor is the genotype of soybean (Grobogan x Anjasmoro and Grobogan x Grobogan) and the second factor is the concentration of α - tocopherol (0, 250, 500 and 750 ppm) with three replicates per each treatment with sampling. The research started from land preparation, planting, application of exogenous antioxidant, maintenance, fertilizing and analysis morphological and biochemical parameters.

In application of exogenous antioxidant treatment plots, the whole shoots of plants were foliar application with different concentration of exogenous antioxidant. For the control treatment, plants were sprayed with water. Foliar application of exogenous antioxidant was applied once a week, started from second week after planting.

The study of morphology characters were began when the plants were 14 days. The observation on number of branch and number of leaves were measured every week began 2 week after planting (WAP) until 5 WAP. Shoot and root fresh weight, shoot and root dry weight were the plants were 6 WAP. The observation on biochemical constituents (chlorophyll a, chlorophyll b and soluble protein content) were the plants were 6 WAP.

The data were analyzed statistically using F-test and then following by Duncan Multiple Range Test (DMRT) at 5 % level.

## 3. Results and discussion

The foliar application of α-tocopherol significantly increased root fresh weight, shoot dry weight, and root dry weight compared with application of 250 ppm α-tocopherol and untreated plants (Table 1). The previous study reported that application of α-tocopherol increased fresh weight of shoots and roots in *Hibiscus rosasineses* L. Plants [14].

**Table 1.** Effect of application α-tocopherol on shoot and root weight in soybean genotype under salt stress

| Treatment | Shoot fresh weight (g) | Root fresh weight (g) | Shoot dry weight (g) | Root dry weight (g) |
|---|---|---|---|---|
| Control | 12.06 | 5,39 b | 3.56 b | 1.33 b |
| α-tocopherol 250 ppm | 12.22 | 5.71 ab | 3.70 ab | 1.37 b |
| α-tocopherol 500 ppm | 12.89 | 6.17 a | 4.20 a | 1.69 a |
| α-tocopherol 750 ppm | 12.17 | 6.04 ab | 3.62 ab | 1.37 b |

Note: Mean values by the same letter in same column do not significantly differ based on DMRT at 5% level

Shoot and root dry weight in grobogan x grobogan genotype significantly different from grobogan x anjasmoro genotype (Table 2). Maximum leaf shoot dry weight and root dry weight obtained in grobogan x grobogan genotype lead to increasing 15.43% and 11,76%, respectively. The growth of root and shoot is the most important parameter for salt tolerance because roots are in direct contact with the soil and absorb water from soil and shoot supply it to the rest of the plant. For this reason, root and shoot length provides an important clue to the response of plants to salt stress [15]. Suppression of plant growth under saline conditions may either be due to the decreasing availability of water or to the increasing toxicity of NaCl associated with increasing salinity [16].

**Table 2.** Root fresh weight, root dry weight and root volume in two soybean genotype under salt stress

| Genotype | Shoot fresh weight (g) | Root fresh weight (g) | Shoot dry weight (g) | Root dry weight (g) |
|---|---|---|---|---|
| Grobogan x Grobogan | 12.77 | 5.87 | 4.04 a | 1.52 a |
| Grobogan x Anjasmoro | 11.90 | 5.79 | 3.50 b | 1.36 b |

Note: Mean values by the same letter in same column do not significantly differ based on DMRT at 5% level

Data in Table 3 indicated exogenous α-tocopherol application has significant effect on number of leaves. Although α-tocopherol treatment at a concentration of 500 ppm gives the highest increase in number of leaves compared with others treatment. Neverless, application of α-tocopherol has no significant effect on number of branch. None of different genotype led to significant changes in the number of branch and leaves. Meanwhile, data presented in Table 4 showed that number of branch and leaves increased in grobogan x grobogan genotype. Leaf injury and death is probably due to the high salt load in the leaf that exceeds the capacity of salt compartmentation in the vacuoles causing salt to build up in the cytoplasm to toxic levels [16;17]. Ludlow and Muchow [18] also reported that reduced leaf growth and accelerated leaf senescence are the common responses to water stress. The reduction in leaf number under salinity and water stress resulted from premature leaf senescence and defoliation. The other study reported, the reduction in number of internodes and number of branches could severely be affected by salinity [19] (Change et al., 1994).

**Table 3.** Effect of application α-tocopherol on number of branch and number of leaves in soybean genotype under salt stress

| Treatment | Number of branch | Number of leaves |
|---|---|---|
| Control | 2.00 | 6.33 b |
| α-tocopherol 250 ppm | 2.17 | 6.67 b |
| α-tocopherol 500 ppm | 2.50 | 11.33 a |
| α-tocopherol 750 ppm | 2.33 | 6.67 b |

Note: Mean values by the same letter in same column do not significantly differ based on DMRT at 5% level

**Table 4.** Number of branch and number of leaves in two soybean genotype under salt stress

| Genotype | Number of branch | Number of leaves |
|---|---|---|
| Grobogan x Grobogan | 2.33 | 8.17 |
| Grobogan x Anjasmoro | 2.27 | 7.33 |

Note: Mean values by the same letter in same column do not significantly differ based on DMRT at 5% level

Chlorophyll content can be considered as one of the indices to screen plant growth. Application of α-tocopherol at a concentration of 500 ppm gives the highest increase in all biochemical characters (Table 5). Foliar appplication of α-tocopherol has significant effect on chlorophyll b and soluble protein content, but no significant effect on chlorophyll a. In plants, tocopherols are believed to protect chloroplast membranes from photooxidation and help to provide an optimal environment for the photosynthetic machinery [9]. El-Quesni et al. [14] reported that foliar application of α-tocopherol increased chl a, chl b and carotenoids content compared to untreated plants. El-Bassiouny et al. [11]

reported that foliar spray of α-tocopherol on faba bean plants also increased chl a, b and carotenoids content. In the other study Orabi and Abdelhamid [20] showed that although salinity (sea water) induced considerable suppression in total proteins in the seeds of faba bean, foliar application of α-tocopherol at the rate of 50 or 100 mg L–1 caused a significant increase in protein content but in both stressed and nonstressed plants.

**Table 5.** Effect of application α-tocopherol on chlorophyll a,b and soluble protein content in soybean genotype under salt stress

| Treatment | Chlorophyll a | Chlorophyll b | Soluble protein |
|---|---|---|---|
| Control | 1.04 | 0.40 b | 0.26 b |
| α-tocopherol 250 ppm | 1.40 | 0.62 b | 0.29 ab |
| α-tocopherol 500 ppm | 1.51 | 1.13 a | 0.36 a |
| α-tocopherol 750 ppm | 1.37 | 0.48 b | 0.26 b |

Note: Mean values by the same letter in same column do not significantly differ based on DMRT at 5% level

**Table 6.** Chlorophyll a,b and soluble protein content in two soybean genotype under salt stress

| Genotype | Chlorophyll a | Chlorophyll b | Soluble protein |
|---|---|---|---|
| Grobogan x Grobogan | 1.44 | 0.75 | 0.30 |
| Grobogan x Anjasmoro | 1.22 | 0.57 | 0.28 |

Note: Mean values by the same letter in same column do not significantly differ based on DMRT at 5% level

There wereno significantly different in two genotype. Generally, biochemical constituents increased in grobogan x grobogan genotype compared than grobogan x grobogan genotype. Chlorophyll a, b and soluble protein increased 18.03%, 31.58%, and 7.14 respectively (Table 6). The decrease in chlorophyll content under salinity conditions is reported by Yasar et al. [21], Kusvuran [22], and Nazarbeygi et al. [23]. In the salt tolerant genotypes the chlorophyll content was protected probably because of the high antioxidant enzyme activities that prevented degradation of leaf chlorophyll. It is well known that soluble protein content is a significant indicator of physiological status of plants. Golezani and Noori [24] reported that protein content of three soybean cultivars was decreased by increased salinity levels and they attributed that to the disturbance in nitrogen metabolism. Also, the decrease in protein content by salt stress may be due to the effects of sodium chloride on protein synthesis [25].

## 4. Conclusions

Evaluation of morphophy and biochemical parameters indicated that application of exogenous α-tocopherol with different concentration induced the salt tolerance and increased vegetative growth of salt resistance soybean genotype. The highest growth of soybean genotype under salt stress was with the application of α-tocopherol 500 ppm. Grobogan x Anjasmoro genotype was salt-sensitive based on all variable, while Grobogan x Grobogan genotype was more tolerant based on morphological and biochemical characters.

The research was supported by Universitas Sumatera Utara under the scheme of Penelitian TALENTA USU 2017 with contract number : 5538/UN5.1.R/PPM/2017, 22 May 2017

## References

[1] Ghassemi-Golezani K., Ghanehpoor S, Dabbagh Mohammadi-Nasab A 2009 Effects of water limitation on growth and grain filling of faba bean cultivars *Journal of Food Agriculture and Environment* Vol **7** pp 442–447

[2] Ashraf M 1994 Breeding for salinity tolerance in plants *Crit. Rev. Plant Sci.* Vol.**13** pp 17-42

[3] Cramer G R 1992 Response of a Naexcluding cultivar and a Na- including cultivar to varying Na/Ca *J. Exp. Bot.* Vol. **43** pp 857–864

[4] Phang T H, Shao G and Lam H M 2008 Salt tolerance in soybean *Journal of Integrative Plant Biology* Vol. **50** (10) pp 1196-1212

[5] Hasegawa P M, Bressan R A, Zhu J K and Bohnert H J 2000 Plant cellular and molecular responses to high salinity *Ann Rev Plant Biol.* Vol. **51** pp 463-99

[6] Munns R and Tester M 2008 Mechanisms of salinity tolerance *Ann Rev Plant Biol.* Vol. 59 pp 651-81

[7] Sadak M S and Dawood M G 2014 Role of ascorbic acid and α tocopherol in alleviating salinity stress on flax plant (*Linum usitatissimum* L.) *J Stress Physiol Biochem* Vol **10** pp 93-111

[8] Fryer M J 1992 The antioxidant effects of thylakoid vitamin E (α-tocopherol) *Plant Cell Environment* Vol **15** pp 381-392

[9] Munné-Bosch S and Alegre L 2002 The function of tocopherols and tocotrienols in plants *Critical Reviews in Plant Scinces* Vol **21** pp 31-57

[10] Munné-Bosch S and Falk J 2004 New insights into the function of tocopherols in plants *Planta* Vol **218** pp 323-326

[11] El-Bassiouny H M S, Gobarah M E and Ramadan A A 2005 Effect of antioxidants on growth, yield, savism causative agents in seeds of *Vicia faba* L. plants grown under reclaimed sandy soils *J. Agr. Pak.* Vol **7**(4) pp 653-659

[12] Sattler S E, Gilliland L U, Magallanes-Lundback M, Pollard M and Penna D D 2004 Vitamin E is essential for seed longevity and for preventing lipid peroxidation during germination *Plant Cell* Vol 16 pp 1419-1432

[13] Baffel S O and Ibrahim M M 2008 Antioxidants and accumulation of αtocopherol induce chilling tolerance in *Medicago sativa*. *Int. J. Agric. Biol.* Vol **10**(6) pp 593-598

[14] EL-Quesni F E, Abd E L-Aziz N and Maga M K 2009 Some studies on the effect of Ascorbic Acid and α-tocopherol on the growth and some chemical composition of *Hibiscus rosasinensis* L. at Nurbaria *Ozean J. Appl. Sci.* Vol. **2**(2) pp 159-167

[15] Jamil M and Rha E S 2004 The effect of salinity (NaCl) on the germination and seedling growth of sugar beet (*Beta vulgaris* L.) and cabbage (*Brassica oleracea* capitata L.) *Korean J. plant es.* Vol.b pps 226–232

[16] Munns R 2003 Comparitive physiology of salt and water stress *Plant Cell Environ.* Vol. 25 pp. 239–250

[17] Munns R, James R A and Läuchli A 2006 Approaches to increasing the salt tolerance of wheat and other cereals *J. Exp. Bot.* Vol 57 pp1025–1043

[18] Ludlow M M and Muchow R C 1990 A critical evaluation of traits for improved crop yields in water-limited environments *Adv. Agron.* Vol **43** pp 107-153

[19] Chang R, Chen Y W, Shao G H and Wan C W 1994 Effect of salt stress on agronomic characters and chemical quality of seeds in soybean *Soybean Science* Vol **13** pp 101-105

[20] Orabi SA and Abdelhamid MT 2014 Protective role of α-tocopherol on two Vicia faba cultivars against seawater-induced lipid peroxidation by enhancing capacity of anti-oxidative system *J. Saudi Soc Agric Sci* Vol **15** pp 145-154

[21] Yasar F, Ellialtioglu S and Yildiz K 2008 Effect of salt stress on antioxidant defense systems, lipid peroxidation, and chlorophyll content in green bean *Russian J. Plant Physiol.* Vol **55**(6) pp 782-786

[22] Kusvuran S, Ellialtioglu S, Yasar F and Abak K 2007 Effects of salt stress on ıon accumulations and some of the antioxidant enzymes activities in melon (Cucumis melo L.) *Inter. J. Food Agric. Environ.* Vol **2**(5) pp 351-354

[23] Nazarbeygi E, Yazdi H L, Naseri R and Soleimani R 2011 The effects of different levels of salinity on proline and A-, B- chlorophylls in canola *Amer-Eurasian J. Agric. Environ. Sci.* Vol **10**(1) pp 70-74

[24] Golezani K G and Noori M T 2011 Soybean Performance under Salinity Stress Soybean - *Biochem Chem Physiol.* Vol b pp 631-642

[25] Omar M S, Yousifn D P, Al-Jibouri A J M, Al-Rawi M S and Hameed M K 1993 Effects of gamma rays and sodium chloride on growth and cellular constituents of sunflower

(Helianthus annuus L.) callus cultures *J. Islamic Acad. Sci* Vol. **6** (1) pp 69-72

# Differences in morphology and sugar content of purple sweet potato (*Ipomoea batatas* L.) with potassium treatment at several altitudes

**R Sulistiani[1], Rosmayati[2], L A M Siregar[2] and F Harahap[3]**

[1]Student of Agricultural Science Study Program, Post Graduate Program Faculty of Agriculture Universitas Sumatera Utara, Medan, Indonesia.
[2]Postgraduate Program, Department of Agrotechnology. Faculty of Agriculture, Universitas Sumatera Utara, Medan, Indonesia.
[3]Postgraduate Program, Department of Biology Education. Faculty of Biology, Universitas Negeri Medan, Sumatera Utara, Indonesia.
e-mail: tanjungrosmayati@yahoo.co.id

**Abstract**. This research was conducted at three locations in low, medium and high plains. This study was aimed to determine the morphological changes and sugar content of sweet potato caused by potassium dose treatment and climate change. Data was analyzed by factorial randomized block design in time series with two factors. The first factor was Altitude: $A_1$ (50 meter above sea level (MASL)), $A_2$ (750 MASL) and $A_3$ (1450 MASL). The second factor was Potassium: $K_0$ (0 kg/ha), $K_1$ (50 kg/ha), $K_2$ (100 kg/ha) and $K_3$ (150 kg/ha). The data of plant morphology change and sugar content was descriptively analized, while agronomic and harvest component data analysis by F test and continued with Duncan Multiple Range Test. The results of morphological observations showed different types of plant growth, which in the lowlands and medium plant growth types spread, twisted, more branches, wider leaf area. However, the plateau of plant growth was relatively dwarf, erect, no twist, huddled and short rods and small leaf area. The tuber weight per plant, number of tuber yield and tuber weight per plot in high altitude were significantly higher than lowland and medium. Similarly, increased altitude will be increased the sugar content significantly.

## 1. Introduction

Sweet potato as a crop of carbohydrate producing tubers is needed to support sustainability of food security as a complementary of rice. Indonesia needs rice up to 31 million tons annually. Meanwhile, the national rice production per year is only 30 million tons, therefore we must import about 500 thousand to 1 million tons of rice. Opportunities to develop tubers as alternative food are still open widely especially sweet potatoes. Similarly, utilization of purple sweet potatoes as functional food ingredients is increasing with public awareness in efforts to maintain natural health and potential sources of food coloring [1] and as high quality natural food dye [2].

Sweet potatoes are unique as vegetables [3,4], it has high nutrients and their physiological activities serve as treatments such as reducing risk of death of ischemic heart attack, stroke, anticancer, antidiabetic, antiinflamatory, antimicrobial, antioxidant [5] and as basic ingredients of the food industry.

Determination of suitable location needs to get maximum tuber production considering sweet potato crops adaptation widely. It can be planted from altitude 0-3.000 meters above sea level (MASL). Tuber is formed maximally at 25°C at night [6], therefore the effects of latitude, altitude and environmental factors such as light and temperature are important to consider.

Nitrogen, phosphorus and potassium is an essential nutrient, especially potassium that affects most of the biochemical and physiological processes in plant growth and metabolism, primarily contribute to accelerate photosynthesis process [7]. The ability of tuber production is closely related to availability of N, P and especially K elements to help tubers formation and sugar levels, where the starch synthase enzyme is activated by K [8], to form tuber flesh sweeter.

## 2. Materials and Methods

The research was conducted using top shoot cuttings of local variety, potassium chloride, urea, TSP, sweet potato flesh, ethanol, alcohol, 25% HCl and 45% NaOH, aquadest, Luff solution, phenolphthalein, KI 30%, Sodium Thiosulfate 0.1 N and starch indicator. Tools used were hoes, machete tripe, string, label, plastic, pipette, Erlenmeyer, filter paper and volumetric flask.

The research was conducted in 3 locations by Completely Randomized Block Design with Time Series Design for multi-site combined analysis [9]. The first factor was Altitude (A): $A_1$ = 50 MASL (Medan Johor). $A_2$ = 750 MASL (Sibolangit). $A_3$ = 1.450 MASL (Dolat Rakyat). The second factor was Potassium (P): $P_0$= 0 kg $K_2O$/ha, $P_1$= 50 kg $K_2O$/ha, $P_2$= 100 kg $K_2O$/ha and $P_3$= 150 kg $K_2O$/ha [10].

The parameters were primary stem length, vein internode, number of branch, leaf area, tuber weight/plant, number of tubers/plant, tuber weight/ha, total sugar content (%) and sugar/ha production. Data of observation result were analyzed by F test, while in stastistical data test obtained significant result furthermore test continued by DMRT test (Duncan's Multiples Range Test). The data was descriptively analyzed for data obtained based on visual observation.

## 3. Results and discussion

### 3.1. *Differences of plant morphology*

On the basis of visual observation, there was a difference of plant morphology at three locations tested. These differences include growth type, main stem length, length of internode, number of branches and leaf area (Table 1). Plant growth type spread with main stem and vein internode longer in the lowlands than highlands where type of plant growth was tend to be erect without twining and branch shape of huddled. This condition was due to considerable temperature difference where in the lowlands average temperature of the day was 30°C and 28°C at night, whereas on the plateau average temperature of the day was 23°C and 17°C at night [11]. According to Ravi *et al.* [12] and Gajanayake *et al.* [13], both of air and soil temperatures control the development of canopies and roots of sweet potato storage. Increased air temperature plays an important role in stem extension and canopy development as well as biomass accumulation [14]. The optimum temperature for stem extension and leaf area expansion was approximately 29-30°C.

Table 1 shows the differentt length of main stem, length of internode, number of branches and leaf area sweet potato at different altitudes above sea level. The length of main stem spread longer and very twining in the lowlands such as Medan Johor at altitude 50 MASL than type of plant in Sibolangit. In highland plant was shorter, huddled without twisted in Dolat Rakyat. Length of internode was not significantly different in Medan Johor with Sibolangit, but it was shorter significantly in Dolat Rakyat. Numbers of branch in Sibolangit was significantly less than Medan Johor and Dolat Rakyat. The leaf area showed no difference in three areas studied. The shape of tubers in lowlands tend to be round up to ovate, in medium plains tend to obovate and long oblong to long irregular in highland.

**Table 1.** Differences of morphology sweet potato at several altitude.

| | Treatment | length of main stem (cm) | length of internode (cm) | number of branches (pieces) | Leaf Area ($cm^2$) |
|---|---|---|---|---|---|
| 1. | $A_1$ (50 MASL) | 129.71a | 3.20a | 11.46a | 42.68a |
| 2. | $A_2$ (750 MASL) | 96.13b | 3.01a | 9.04b | 40.26a |
| 3. | $A_3$ (1450 MASL) | 51.25c | 2.57b | 10.54a | 37.49a |

Note: The numbers followed different letter in the same column is significantly difference at $\alpha = 5\%$ by Duncan's Multiple Range Test.

Difference of temperature, visually show prominently influence growth type and tuber shape compared to length of internode, number of branches and leaf area. In the summer, due to high temperatures around roots, high temperatures in air and both of its combination encourages canopy growth. Under temperature stress conditions will cause delays in initiation and tuber formation [15,16]. Beside temperature difference, possibility of different intensities of solar irradiance and air humidity causes of different types of growth. Reduction of radiation intensity will promote branch growth and reduce tuber weight [17].

### *3.2. Harvesting Components*

Based on data analysis of crop yield component, there are difference in number of tuber per plant, tuber weight per plant and tuber weight per hectare due to difference of altitude of plant growth, which can be seen in Table 2.

**Table 2.** Differences of numbers and weight of tuber in various altitudes.

| | Treatment | Amount of tuber/ Plant (pieces) | Weight of Tuber/ Plant (g) | Weight of Tuber/ ha (ton) |
|---|---|---|---|---|
| 1. | $A_1$ (50 masl) | 3.75a | 352.46b | 22.09b |
| 2. | $A_2$ (750 masl) | 2.42c | 268.15c | 13.73c |
| 3. | $A_3$ (1450 masl) | 2.88b | 593.02a | 30.67a |

Note: The numbers followed different letter in the same column is significantly difference at $\alpha = 5\%$ by Duncan's Multiple Range Test.

The number of tubers per plant is significantly different in each area where the most number of tubers in lowland (Medan-Johor) followed by Dolat Rakyat and Sibolangit. The maximum number of storage roots will increase quadratically as temperature rises to 25°C and decrease as temperature increases [18].

Tuber weight per plant and tuber weight per hectare were significantly heavier at Dolat Rakyat, followed by Medan-Johor and Sibolangit. In high altitude, intensity of solar irradiance is reduced, whereas high air humidity leads to reduced evaporation so the photosynthetic activity is more widely used for tuber formation than shoot growth. According to Haynes [17] the duration of irradiation of 56 cal/$m^2$ /day will encourage formation of tubers and the decrease of radiation intensity will reduce weight of tubers. Temperatures around the roots are critical for root and tuber initiation, canopy and plant growth [15].

Combination of treatments between potassium and altitude caused a significantly difference in response due to the interaction both of them, where amount of tuber was in Medan-Johor with $K_2O$ 50 kg/ha treatment and the lowest amount of tuber in Sibolangit with $K_2O$ 50 kg/ha treatment. However in same location with different levels of potassium treatment did not show difference amount of tubers per plant. The function of potassium [18] will increase production and yield quality as well as increase the translocation of sugar and starch required in large quantities for the process of plant growth.

The result of combination treatment between potassium and altitude caused differences in the number of tubers per plant are listed in Table 3.

**Table 3.** Differences of nmount of tuber per plant caused by interaction altitudes and Potassium.

| Treatment | Number of tuber/plant (pieces) |
|---|---|
| $A_1P_0$ | 3.33abc |
| $A_1P_1$ | 4.33a |
| $A_1P_2$ | 3.92ab |
| $A_1P_3$ | 3.42abc |
| $A_2P_0$ | 3.08bcd |
| $A_2P_1$ | 2.08e |
| $A_2P_2$ | 2.25de |
| $A_2P_3$ | 2.25de |
| $A_3P_0$ | 2.50cde |
| $A_3P_1$ | 2.83cde |
| $A_3P_2$ | 3.17bcd |
| $A_3P_3$ | 3.00bcd |

Note: The numbers followed different letter in the same column is significantly difference at $\alpha = 5\%$ by Duncan's Multiple Range Test.

Role of single potassium was not significant for each level treatment at certain same location, but after get interaction with environment it causes significant differences in number of tubers. The genotype and environmental interactions significantly affect to growth and production of sweet potatoes [8].

### *3.3. Components of Biochemical Test*

Test of total sugar content and sugar production per hectare at three different locations based on altitude can be seen in Table 4.

**Table 4.** Differences of total sugar content (%) and sugar production/ha at various altitudes

| | Treatments | Total sugar content (%) | Sugar production/ha (ton) |
|---|---|---|---|
| 1. | $A_1$ (50 MASL) | 5.21b | 1.13b |
| 2. | $A_2$ (750 MASL) | 7.14b | 0.99b |
| 3. | $A_3$ (1450 MASL) | 10.57a | 3.24a |

Note: The numbers followed different letter in the same column is significantly difference at $\alpha = 5\%$ by Duncan's Multiple Range Test.

Based on statistical analysis of sugar content and sugar production per hectare was higher significantly in the highlands, whereas in low and medium plains did not show any significant difference. Although sugar content was higher in medium plains than lowland, but sugar production was low because tuber production per hectare in medium plains was lower than lowland but not significant. The environment will affect to formation of sugar, where the altitude increases, the temperature decreases, it will increase sugar content. This is consistent with the research that differences in the amount of sugar content feasible to occur due to differences in varieties, endogenous amylase activity and environmental conditions [20].

## 4. Conclusions

The highest production of tubers and sugar content were found in the highlands at Dolat Rakyat District (an average of 30.67 tons/ha tuber production) and sugar production (3.24 tons/ha). The dosage of fertilization did not effect in each region, whereas altitude caused different production of tubers and sugar significantly in all areas studied. Combination of altitude treatmenst and potassium doses caused significant tuber numbers and resulted the largest number of tubers was $A_1P_1$ with an average tuber of 4.33 pieces/plant.

It is suggested to cultivate local genotypes in highlands to obtain tuber yields and the highest sugar production even without potassium fertilization.

## References

[1] Truong V D, Deighton N, Thompson R T, McFeeters R F, Dean Lo, Pecota K V and Yencho G C 2009 Characterization of anthocyanins and anthocyanidins in purple-fleshed sweetpotatoes by HPLC-DAD/ESI-MS/MS *J. Agr. Food Chem.* **58**: 404-410.

[2] Mervat M M El Far and Hanan A A Taie 2009 Antioxidant activities, total anthocyanins, phenolics and flavonoids contents of some sweetpotato genotypes under stress of different concentrations of sucrose and sorbitol *Australian Journal of Basic and Applied Sciences*, **3(4)**: 3609-3616, 2009 ISSN 1991-8178 © 2009, INSInet Publication.

[3] Mohanraj R and Sivasankar S 2014 Sweet potato (*Ipomoea batatas* [L.] Lam) a valuable medicinal food: a review. *J. Med. Food* 2014 Jul:**17(7)**:733-41.

[4] Tennant D R, Davidson J and Day A J 2014 Phytonutrient intakes in relation to european fruit and vegetable consumption patterns observed in different food surveys. *Br. J. Nutr.* **112**:1214–1225.

[5] Kumar P and Chandra R 2006 Decolourisation and detoxification of synthetic molasses melanoidin by individual and mixed cultures of *Bacillus* spp. *Biores.Technol.*7.2096-2102.

[6] Food and Agriculture Organization 2008 The State of Food and Agriculture 2008. Biofuels: prospects, risks and opportunities *Food and Agriculture Organization of The United Nations.* ISBN 978-92-5-105980-7. *http://www.fao.org*. Accessed 20 June 2016.

[7] Wang M, Qingsong Z, Qirong S and Shiwei G 2013 The critical role of potassium in plant stress response. *Int J Mol Sci.* 2013 Apr; **14(4)**: 7370–7390.

[8] Better Crops 1998 Functions of potassium in plants International plant nutrition institute (Inc) (US) *Better Crops* Vol. **82 (3)** 1998. *https://www.ipni.net/ ppiweb/bcrops.nsf.*

[9] Gomez K A and Gomez A A 1995 *Statistical procedures for agricultural research.* UI-Press Indonesia.

[10] Putra S and Karsidi P 2011 The effect of potassium fertilizer on the improvement of sweetpotato results of narutokintoki variety in wetland area. *J. Agrin* Vol. **15**, No. **2**, October 2011. ISSN: 1410-0029.

[11] Data.org. 2017 Climate data for cities around the world. *https://id.climate-data.org*. Accessed 20 June 2017.

[12] Ravi V, Naskar S, Makeshkumar T, Babu B and Krishnan BSP 2009 Molecular physiology of storage root formation and development in sweet potato (*Ipomoea batatas* (L.) Lam.). *J Root Crops*. **35**: 1-27.

[13] Gajanayake B, Reddy K R, Shankle M W *and* Arancibia R A 2013 Early season soil moisture reduces sweet potato storage root initiation and development. *HortScience* **48**:1457–1462.

[14] Patel D and Franklin K A 2009 Temperature-regulation of plant architecture. *Plant Signal Behav*. **4**:577–579.

[15] Chang D C, Jeong J C and Lee Y B 2006 Effect of root zone cooling on growth responses and

tuberization of hydroponically grown superior potato (Solanum tuberosum) in summer *Journal of Bio-Environment Control* **15**: 340–345.

[16] Struik P C 2007 Response of the potato plant to high temperatures. In potato biology and biotechnology: Advances and perspectives, ed. D Vreugdenhil, J Bradshaw, C Gebhardt, F Govers, D Mackerron, M A Taylor and H A Ross. Oxford: *Elsevier*.

[17] Haynes K G, Haynes F L and Swallow W H 1988 Temperature and photoperiod effect on tuber production and specific gravity in diploid potatoes *HortScience* **23**:562-565.

[18] Gajanayake B, Reddy K R, Mark W Shankle, Ramon A Arancibia and Arthur O Villordon 2014 Quantifying storage root initiation, growth and developmental responses of sweetpotato to early season temperature *Agronomy Journal* Volume **106**, Issue **5**, 2014.

[19] Kivuva B M, Githiri S M, Yencho G C and Sibiya J 2014 Genotype x environment interaction for storage root yield in sweetpotato under managed drought stress conditions *J. Agric. Sci.* **6**:41–56.

[20] Adebisola J A, Kolawole O Falade, John O Akingbala and Pathlene T 2009 Physicochemical properties of twenty-one caribbean sweetpotato cultivars *International Journal of Food Science and Technology*, Vol. **44 (9)** 2009 pp. 1696-1704.

**Acknowledgments**
Authors gratefully acknowledge to Universitas Sumatera Utara, Research Laboratory, phyitochemical Laboratory and Seeds Centers of Horticulture of Tongkoh for providing the facilities and technical support for this experiment and all contribution to this work.

# Growth, yield and movement of phosphate nutrients in soybean on P fertilizer, straw mulch and difference of plant spacing

**C Hanum**

Agrotechnology, Faculty of Agriculture, Universitas Sumatera Utara. Jalan Prof A Sofyan No. 3 Kampus USU Padang Bulan Medan.
E-mail: chairani_as@yahoo.com

**Abstract**. Soybean is one of the plants that require much amounts of phosphate. P nutrient, microclimate modification and plant spacing arrangement is the efforts to improve grain yield. The objective of the research was to study the effect of P fertilization, mulching straw and plant spacing on growth, yield and movement of P nutrient on soybean. The study was conducted at Cengkeh Turi Binjai using factorial randomized block design with 3 factors. The first factors was P fertilizer 0, 100, and 200 kg/ha, the second factor was thickness of rice straw mulch 0 and 5 cm, and third factors was plant spacing 30 cm x 15 cm, 40 cm x 20 cm, and 50 cm x 25 cm. The results of the research showed that phosphate fertilizer (200 kg/ha) significantly increased levels of phosphate in the shoot. Plant spacing (50 cm x 25 cm) increased root volume. The interaction of phosphate fertilizer (200 kg/ha) and spacing (50 cm x 25 cm) increased the phosphate level by 93.33% in shoot. Plant spacing (50 cm x 25 cm) produced the largest of 100 grains weight as compared to other plant spacing.

## 1. Introduction

Soybean yield potential is genetically determined. The agronomically important traits affect the enhancement of grain quantity and quality. However, the majority of agronomic and genetic traits such as complex metabolic, biological, and pharmaceutical pathways are polygenetic traits and are produced in a complex pathway [1], but actual yield potential depends on environmental conditions and management practices. Higher levels of soybean production may require higher levels of some nutrients appropriate environmental conditions to achieve yield goals. Obtaining maximum yield is only possible when the plant's nutritional requirements are met and environmental stress is limited. Maintaining a fertile growing environment is a risk management strategy that produces higher yield and healthier plants that are often better suited to withstand the yield-robbing effects of biotic and abiotic stress. Soybean respond to fertile soils [2]. A well planned soil fertility program is a management strategy that leads to profitable soybean production.

Soybean is one of the leguminous plants that require more P than others. Plants require adequate P from the very early stages of growth for optimum crop production [3]. Restricted early-season P supply frequently limits crop production, and P fertilizer is commonly applied to ensure that sufficient P is available to optimize crop yield and maturity. Therefore it is important that P management balances the goal of providing sufficient P to the crop to optimize crop yield with the goal of avoiding excess P and environmental risk. Management of the cropping system to improve the availability of P

IOP Conf. Series: Earth and Environmental Science **122** (2018) 012051 doi:10.1088/1755-1315/122/1/012051

to the crop early in the growing season may improve P nutrition while reducing the potential for excess accumulation of P in soils and risk of movement of P into water systems.

Some studies, Fonseka et al. [4], Ifenkwe and Allen [5], in which the relation between plant spacing and growth were examined, the results showed an increase in plant spacing to be accompanied by an increased stem length. The increased branching at the wider spacing did not compensate for fewer plants/m$^2$. They attributed increased branching at wider spacing to the availability of more space at lower plant densities. More space meant that plants were able to exploit the available nutrients in the soil and the photosynthetic active radiation for growth than plants at close spacing. In other words, the growth rate was increased. Setting plant spacing per unit area will reduce the effect of plant competition that will occur.

Several studies on plant spacing showed that the closer the spacing, it will cause competition plant height, number of branches and leaf area. Crop productivity could be greatly influenced by even a small change in soil water storage [6].

Mulches are essential components for up land agriculture, limit the negative effects of soil cultivation and increase the stability of natural environment Owing to a protective layer produced by these mulches the development of weeds is restrained and the water resistance of soil aggregates is improved [7].

Management of soybean cultivation practices through phosphor nutrient management, plant spacing, and mulching, is believed to increase soybean production and not damage the environment.In this context, this study analyzed the effect of different plant spacing, phosphor fertilizer, and mulching on growth, development and yield of soybean.

## 2. Material and Methods

The experiment was conducted in Cengkeh Turi - Binjai. This study used randomized block design (RBD) with three factors. The first factor is Phosphate fertilizer which consists of three levels (0, 100, and 200 kg TSP/ha). The second factor is Straw Mulch, consists of two levels of thickness 0, 5 cm. The third factor is Plant spacing which consists of three levels were: (30x15cm; 40x20cm; 50x25cm). The soybean cultivar used was Grobogan, a genotype which have large seed, adapted and widely grown in Sumatera Utara climate. Data were subjected to analysis of variance and means were compared by *Duncan Multiple Range Test* ($p<0.05$) using the SAS statistical software.

## 3. Results and Discussion

The result of statistic analysis showed that the application of P fertilizer and straw mulch did not affect on root volume, except the plant spacing. The increase in plant spacing will increase soybean root volume (Table 1).

**Table 1.** Root volume of soybean on plant spacing and straw mulch treatment

| Plant spacing | Thickness of Straw Mulch (cm) | | Average |
|---|---|---|---|
| | 0 | 5 | |
| 30 cm x 15 cm | 2.83 | 3.67 | 3.25c |
| 40 cm x 20 cm | 4.22 | 4.94 | 4.58b |
| 50 cm x 25 cm | 7.00 | 7.78 | 7.39a |
| Average | 4.69 | 5.46 | |

The number followed by the same letters on the same row are not significantly different according to Duncan Multiple Range Test (DMRT) at 5% level.

Individual plant growth and development is directly influenced by the space available to plants. Plant spacing would be good for root growth. Roots will grow and expand in all directions. In addition, with a more extensive root growth, competition between plant roots decreases thus expanding the water and nutrient uptake. Nutrient and water availability will increase the volume of plant roots. Regular spacing will provide space to grow plants that are uniform so that the process of making food by the plant will be the same. The effect of different plant spacing on growth, development and yield of stems and roots of commercial cassava grown in a subtropical environment also found by N.A. Streck et al [8].

Although statistically mulching does not effect on root volume parameter, however, mulching with a thickness of 5cm increases root volume. The yield differences were caused by the effects of the mulches on the temperature and water content of the soil surface. Ground that given mulch straw has a higher soil moisture than without mulch. Observations during the vegetative growth period obtained an average moisture content of 82% in mulched soil and 75% of non-mulch. The mean temperature also has a difference, 28 on mulch and 31 on non-mulch. Differences in temperature and humidity on mulch-weighted soil were also found Mbagwu [9]. Maximum daily soil temperature was lowest on the straw plots (32·5°C), about equal on the bare and white polythene treatments (38·6°C vs 39·8°C) and highest on the black polythene plots (45·0°C). Maximum daily soil temperature was lowest on the straw plots. (32·5°C), about equal on the bare and white polythene treatments (38·6°C vs 39·8°C) and highest on the black polythene plots (45·0°C). And the research of Bunna *et al* [10] found that The positive effect of straw mulch on grain yield was related to increased crop establishment and reduced weed biomass, particularly under weedy conditions. The effect of mulch on yield may be considered through its effect on improved crop establishment and reduced weed biomass, and also its effect on mungbean growth directly through favourable water and perhaps temperature conditions, or indirectly through increased crop establishment and reduced weed growth. Their contribution to yield is analysed below.

The high moisture of the soil on mulched straw will help the availability of water for nutrient and root growth. The relatively constant temperature on the ground given mulch will also help the process of nutrient uptake.

**Table 2.** Soybean root dry weight on phosphate fertilizer, straw mulch and plant spacing treatment

| Phosphate Fertilizer (kg/ha) | Thickness of Straw Mulch (cm) | Plant Spacing | | | Average |
|---|---|---|---|---|---|
| | | 30 cm x 15 cm | 40 cm x 20 cm | 50 cm x 25 cm | |
| 0 | 0 | 1.10 | 1.13 | 0.90 | 1.04 |
| | 5 | 0.90 | 1.13 | 0.93 | 0.99 |
| 100 | 0 | 1.32 | 1.23 | 0.97 | 1.18 |
| | 5 | 0.83 | 0.87 | 1.30 | 1.00 |
| 200 | 0 | 1.03 | 1.03 | 1.23 | 1.10 |
| | 5 | 1.27 | 1.20 | 1.33 | 1.27 |
| Average | | 1.06 | 1.10 | 1.11 | |

Phosphate fertilization, straw mulching and plant spacing did not significant effect on root (Table 3) and shoot growth (Table 4).

The results of this study indicate the root growth pattern has decreased at larger plant spacing (Table 2), different from the shoot growth, the extension of the plant spacing will increase the shoot growth, although statistically not significant effect (Table 3). roots and shoots are two sinks that compete during the plant life cycle [11]. Data of weight per plant and weight per root in table 4 can be explained by the hypothesis of Aguiar et al. [12] who found that, under conditions of lower densities (wide spacing), the sink of roots exceeds that of shoot tips.

The wide plant spacing will increase the weight of 100 soybean seeds, while the application of P fertilizer and straw mulch does not affect (Table 4). Narrow spacing (30x15cm), which produce lower shoot dry weight than in wide spacing. This lower shoot growth will result in lower assimilate accumulation, so the biomass for seed filling is lower, and produce smaller seed yields.

**Table 3**. Shoot dry weight of soybean on phosphate fertilizer, rice straw mulch and spacing treatment

| Phosphate Fertilizer (kg/ha) | Thickness of Straw Mulch (cm) | Plant spacing | | | Average |
|---|---|---|---|---|---|
| | | 30 cm x 15 cm | 40 cm x 20 cm | 50 cm x 25 cm | |
| 0 | 0 | 10.10 | 15.07 | 12.40 | 12.52 |
| | 5 | 14.60 | 12.00 | 10.77 | 12.46 |
| 100 | 0 | 9.50 | 14.27 | 14.27 | 12.68 |
| | 5 | 10.93 | 12.43 | 10.70 | 11.36 |
| 200 | 0 | 12.97 | 14.27 | 13.43 | 13.56 |
| | 5 | 13.90 | 14.20 | 16.03 | 14.71 |
| Average | | 12.00 | 13.71 | 12.93 | |

**Table 4.** Weight of 100 seeds of soybean on the of plant spacing and phosphate fertilizer treatment

| Phosphate Fertilizer (kg/ha) | Plant Spacing | | | Average |
|---|---|---|---|---|
| | 30 cm x 15 cm | 40 cm x 20 cm | 50 cm x 25 cm | |
| 0 | 19.97 | 21.09 | 22.08 | 21.05 |
| 100 | 17.76 | 20.34 | 22.40 | 20.17 |
| 200 | 20.99 | 22.28 | 22.79 | 22.02 |
| Average | 19.57b | 21.24ab | 22.42a | |

The number followed by the same letters on the same column are not significantly different according to Duncan Multiple Range Test (DMRT) at 5% level.

This result confirms the individual advantage for plants with larger area for exploration and is according to Barros et al. [13] who reported that the yield per cassava plant is greater in plants with wider plant spacing. And research of Onat, *et al* [14] found that The average pod yield was significantly affected by the row distance and plant spaces. Increasing the plant spaces from 5 cm to 25 cm, the pod yield per plant was increased from 24.5 g to 96.2 g in a two year average. In this equation, the pod yield per plant was decreased when the plant density increased, whereas plant number per hectare was increased when the plant density increased.

The increase of phosphate fertilizer 200 kg/ha at plant spacing of 50 x 25cm resulted in the highest P content on shoot, while the lowest rate was obtained on soybean without P fertilizer with narrow spacing (Table 5). P fertilization may improve P nutrition and crop yield potential [3].

**Table 5.** P content of soybean shoot on phosphate fertilizer and plant spacing treatment

| Phosphate Fertilizer (g/plot) | Plant Spacing | | | Average |
|---|---|---|---|---|
| | 30 cm x 15 cm | 40 cm x 20 cm | 50 cm x 25 cm | |
| 0 | 0.15h | 0.15h | 0.16g | 0.15 |
| 10 | 0.19f | 0.22e | 0.24d | 0.21 |
| 20 | 0.26c | 0.28b | 0.29a | 0.28 |
| Average | 0.20 | 0.22 | 0.23 | |

The number followed by the same letters on the same column and row are not significantly different according to Duncan Multiple Range Test (DMRT) at 5% level.

The results of this study indicate the highest root volume obtained in plants that have a wide spacing. In this condition the plant has the potential to have a higher root proliferation. An increase in P fertilizer dosage will cause an increase in total P of soil that affects its availability. This condition is suspected to cause why soybean with fertilization of phosphat 200 kg/ha at wide plant spacing yields higher P levels. When compared with no fertilizer P, the fertilized plant will increase P levels to 93.3%. As described by Deressegn and Telele [15] that Plant spacing have effect on the crop growth, yield components and yield of soybean due to competition effect of the plant such as light, nutrients and water, the rate and extent of vegetative growth and development of crops particularly that of leaf area index, plant height, root length and density, yield components and yield the major problems in the country. With the rise of density, the dry matter produced in both sole cropping (monocultures) and intercropping increased. High plant density particularly as to forage crops, creates a suitable microclimate and results in the rise of total dry matter yield.

## 4. Conclusion

Phosphate fertilizer (200 kg/ha) significantly increased levels of phosphate in the shoot. Plant spacing (50 cm x 25 cm) increases root volume. The interaction of phosphate fertilizer (200 kg/ha) and spacing (50 cm x 25 cm) increased the phosphate level by 93.33% in shoot. Plant spacing (50 cm x 25 cm) produces the largest of 100 grains weight as compared to other plant spacing.

## References

[1] Lee, H, S Y Park and Z J Zhang 2012 *An Overview of Genetic Transformation of Soybean* INTECH. Chapter 23 p 489-506

[2] Grath, C M, D Wright, A P Mallarino and A Lenssen 2013 *Soybean Nutrient Needs* (Iowaa: Iowa State University of Science and Technology)

[3] Grant, C A, Flaten, D N, Tomasiewicz D J and Sheppard S C 2001 The importance of early season P nutrition. Can *J. Plant Sci.* **81** 211–224

[4] Fonseka H D, Asanuma K I, Kusutani A, Ghosh A K, Ueda K 1996 Growth and yield of potato cultivars in spring cropping: I. Plant morphology, growth, assimilate partitioning and yield under two planting densities Jpn. *J. Crop Sci.* **65** 269-276

[5] Ifenkwe O P and Allen E J 1978 Effect of row width and planting density on growth and yield of two maincrop potato varieties *J. Agric. Sci. Camb.* **91** 279-289

[6] Liu Y, Li S Q, Chen F, Yang S J, Chen X P 2010 Soil water dynamics and water use efficiency in spring maize (Zea maysL.) fields subjected to different water management practices on the Loess Plateau, China *Agricultural Water Management* **97** 769–775

[7] Mangan, F, Degregorio R, Schonbeck M, Herbert S, Guillard K, Hazzard R, Sedeman E, Litchfild G 1995 Cover cropping systems for brassicas in the Northeastern United States: 2. Weed, insect and slug incidence *J. Sustainable Agric.* **5** 15-36.

[8] Streck, N A, Diego G P, Alencar J Z, Luana F G, Thiago S M R, André T S, Michel R S 2014 Effect of plant spacing on growth, development and yield of cassava in a subtropical environment *Bragantia* **73** 407-410

[9] Mbagwu J S C 1991 Influence of different mulch materials on soil temperature, soil water content and yield of three cassava cultivars *Science Food and Agriculture Journal* **54** 569–577

[10] Bunna, S, P Sinath, O Makara, J Mitchell, and S Fukai 2011 Effect of straw mulch on mungbean yield in rice fields with strongly compacted soils *Field Crop Research. Science Direct.* **124** 295-301

[11] Alves, A A C Fisiologia da mandioca In: EmpresaBrasileira de PesquisaAgropecuária – EMBRAPA Mandioca e Fruticultura Tropical. Aspectossocioeconômicos e agronômicos da mandioca. Cruz das Almas: EMBRAPA, **2006** p138-169

[12] Aguiar, E B, Valle T L, Lorenzi J O, Kanthack R A D, Miranda F H dan Granja 2011 N P Efeito da densidadepopulacional e época de colheitanaprodução de raízes de mandioca de mesa. *Bragantia* **70**561-569

[13] Barros, R S, Merces W C dan Alvim R 1978 Sink strength and cassava productivity *Hort Science* **13** 474-475

[14] Onat, B, H Bakal, L Gulluoglu and H Arioglu 2017 The effects of row spacing and plant density on yield and yield components of peanut grown as a double crop in mediterranean environment in Turkey *Turkish Journal of Field Crops* **22** 71-80

[15] Deressegn, F and T Telele 2017 Review on effect of inter and intra row spacing on yield and tield components of soybean (*Glycine max* (L.) Merrill) in Ethiopia *Journal of Biology, Agriculture and Healtcare* **7** 53-59

# Induction of soybean resistance to bacterial pustule disease (*Xanthomonas axonopodis pv. glycines*) by rhizobacteria and organic material treatment

**A Khaeruni[1], E A Johan[1], T Wijayanto[2], M Taufik[1], A A R Syafar[3] and G A Kade Sutariati[2]**

[1]Department of Plant Protection, Faculty of Agriculture, Universitas Halu Oleo, Kendari, Indonesia
[2]Department of Agrotechnology, Faculty of Agriculture, Universitas Halu Oleo, Kendari, Indonesia
[3]Department of Agrotechnology, Faculty of Agriculture, Universitas Islam Makassar, Makassar, Indonesia
E-mail: andikhaeruni.uho@gmail.com

**Abstract.** This study aimed to evaluate the role of different formulations and types of organic matter in improving yield and resistance of soybean plants to bacterial pustule disease. The study was prepared based on a randomized block design with a factorial pattern. The first factor was the application of rhizobacterial formulation (biofresh), ie F0 = without the application of rhizobacteria, F1 = application of biofresh in solid formulation, and F2 = application of biofresh in liquid formulation. The second factor was the application of organic materials, namely B1 = compost of soybean litter + cow dung, B2 = compost of rice straw + cow dung, B3 = compost of soybean litter + rice straw + cow dung. Observation of disease severity and soybean yield was conducted on five sample plants in each treatment. The results showed that the treatment of biological agent biofresh in solid formulation combined with compos of soybean litter, was the best treatment in increasing plant resistance to bacterial pustule disease and seed weight. Plant resistance induction occurred systemically characterized by salicylic acid increase of 0.3 mg and peroxidase increase of 0.07 unit / mL in the sample plants.

## 1. Introduction

Pustule bacterial disease *(Xanthomonas axonopodis pv. glycines)* is an important disease in soybean crop, and has been spread in soybean crop centers in Indonesia such as: West Java, Central Java, Yogyakarta, Lampung, South Sulawesi, Southeast Sulawesi and West Sumatera [1, 2]. The pathogen spreads through seeds [3], water and wind [4]. In humid environment conditions, the disease may decrease yields by about 21-40% [5]. Chemical disease control has been reported as one of the causes of environmental damage or resistance of pathogenic bacteria [6], while the supply of bacterial pustule resistant varieties is still very limited.

To minimize the use of chemicals in the control of plant diseases, the use of biological agents such as rhizobacteria is one of the alternative pathogen-friendly controls. Currently, biological fertilizers have been developed with three types of rhizobacteria contents, namely Bacillus cereus ST21b, B. subtilis ST21e, *Serratia* sp SS29a. This third mixture of rhizobacteria isolates produced a biological agent formulation called "Biological agents biofresh". Biofresh has the ability as a biological antagonist and plant growth promoter to reduce the use of synthetic pesticides and chemical fertilizer [7].

The survival of biological agents is very important because it is associated with its ability to suppress pathogens and stimulate plant growth. Application of organic materials can improve microbial life and activity, since organic matter is a source of energy and microbial food in the soil [8]. In addition to organic contents in the soil, the materials and formulations of biological agents also affect the survival of a biological agent in the field [9]. The objective of this research was to evaluate the role of biofresh biodegradable formulation form and different organic materials in increasing plant resistance to bacterial pustule disease and soybean crop yield.

## 2. Material and methods

### *2.1 Materials*

Soybean seeds of Gema variety were obtained from Indonesian Legumes and Tuber Crops Research Institute, Indonesian Agency for Agricultural Research and Development, Malang, Indonesia. *Bacillus cereus* ST21b, *B. subtilis* ST21e, *Serratia sp* SS29a, and *X. axonopodis* pv. *glycines* were obtained from Phytopathology Laboratory, Plant Protection Department of Halu Oleo University, Kendari.

### *2.2 Experimental design*

This study used factorial randomized complete block design consisting of two factors. The first factor was a biological agent biofresh formulation which consisted of three levels i.e.: F0 = without biofresh (control), F1 = solid formulations, F2 = liquid formulations.The second factor was organic matter compost consisting of four levels i.e.: B1 = compost of soybean litter waste, B2= compost of rice straw, B3 = compost of soybean litter waste + rice straw. Overall, there were 9treatment combinations, each treatment was repeated 3 times so that there were 27treatment units and each treatment consisted of 5 units of the plant samples. Data were analyzed using the SAS 9.1.3 version statistical software. The difference of treatment means were separated by Duncan multiple range test at 5% level.

### *2.3 Preparation of growing media and cultivation of soybean seeds*

The growing media used in this research was ultisols that has been sterilized with vapor sterilization. Soil thoroughly mixed with straw compost, soybeans compost and cow dung according to treatments, with a ratio of 1:7 [10]. The mixture were then placed in a polybag sized of 30 × 40 cm and placed on open land at Experimental Farm, of Halu Oleo University, Kendari according to the research design. Three soybean seeds were cultivated at depth of 5 cm from the surface of growing media. Two weeks after cultivation, two plants were selected and were retained as sample plants, while the rest was removed and discarded.

### *2.4 Formulation and application of biofresh*

The rhizobacteria (*Bacillus cereus* ST21b, *B. subtilis* ST21e and *Serratia* sp SS29a) were used as raw materials of biofresh formulation. Before mixing with the formulation, the rhizobacteria were propagated on solid TSA medium separately and were incubated for 48 h. Bacteria colonies growth was suspended in sterile distilled water until it reached population density of $10^{10}$cfu/mL and was used in two formulations of biofresh.

Solid formulation was prepared by mixing the third rhizobacteria suspension with carrier materials such as peat organic matter and animal manure with certain ratio. Furthermore, the mixture of biological agent and material formulation was dried for 48 h, was packed in plastic bags to 5 kg and was ready for application. For the liquid formulation, it was prepared by mixing the third rhizobacteria suspension with a carrier material in the form of a mixture of water, coconut water and Tryptic Soy Broth with certain ratio, in a 1 L plastic container, the mixture was then incubated for 48 h and was ready for application [10].

Applications of biological agent biofresh was conducted during planting process. Four week after planting (WAP), biofresh was sprinkled with liquid formulation around the planting hole as much as 10 mL and solid formulations for covering planting hole during planting and was spreaded around 10 grams per planting hole at the age of 4 WAP.

*2.5 Preparation and inoculation of Xanthomonas axonopodis* pv. *glycines*

*X. axonopodis* pv *glycines* purified on yeast extract dextrose agar (YDA) media for 4 days were suspended in aquadest solution until cell concentration reached $10^{10}$-$10^{11}$ cells/mL. Inoculation was performed on 3-week-old plants after planting (WAP) by spraying the pathogen suspension on the entire leaf surface using a 1 L sprayer. After inoculation, the plant was enclosed for 24 hours with transparent plates that had been hollowed both sides.

*2.6 Salicylic acid and* peroxidase *enzymes content analysis*

Analysis of salicylic acid and peroxidase enzymes was performed on the control treatment (F0B1) and best treatment (F1B1) through leaf composite, one day prior to pathogen inoculation and one week after pathogen inoculation. Analysis of salicylic acid levels was performed according to Warrier*et al.* [11], while the peroxidase enzyme activity analysis used Patra and Mishra method [12].

*2.7 Observation variables*

Observations were carried out on five plant samples of each treatment unit. The observed variables were:

(1) Disease severity.Disease severity was calculated using scoring method through assessment score of the plant leaves based on pustule symptoms formed in plant samples [10]. Scoring result was then used to calculate the severity of the disease by using the formula:

$$DS = \sum_{i=0}^{n} \left(\frac{n.V}{Z.N}\right) \text{x } 100\ \% \quad (1)$$

DS: disease severity (%); n: the number of leaves attacked in each category; N: the number of observed leaves; V: value scale of each category of that were attacked and Z: the highest scale value of attacked category. In this method, we measure disease severity(yi) collected at various times(ti). Besides disease severity values, the value of area under disease progress curve (AUDPC) was also calculated to observe the disease progress. Values was calculated based on the formula AUDPC of Van der Plank [13]

$$\text{AUDPC} = \sum_{i=1}^{n-1} \left(\frac{y_i + y_{i+1}}{2}\right) \text{x } (t_{i+1} - t_i) \quad (2)$$

(2) Measurement of salicylic acid content and peroxidase enzyme activity

Salicylic acid content was determined by measurement of absorbance value using UV-Vis spectrophotometer at 525 nm wavelength, while the determination of peroxidase enzyme activity was done by measuring absorbance at 420 nm wavelength in the same tool.

(3) Crop yield: Crop yield was observed when the plants were in harvested stage which was characterized by yellowing of the leaves as much as 80%. Observed crop yield variables included pods and crop seed weight.

## 3. Results and Discussion

*3.1 Disease Severity of Pustule Bacterial Disease*

The treatments significantly affected the severity of bacterial pustule disease at age 2, 3, 4, 5 and 6 weeks after inoculation. The rate of development of disease severity in each treatment combination is presented in Table 1.

**Table 1.** Combination effects of biofresh formulation and organic matter on disease severity of pustule bacterial disease on soybean

| Treatments | Disease Severity (%) on *n* | | | | | AUDPC |
|---|---|---|---|---|---|---|
| | 2 WAI | 3 WAI | 4 WAI | 5 WAI | 6 WAI | (unit/week) |
| F0B1 | 12 a | 25 a | 33 a | 58 a | 58 a | 10.54 |
| F0B2 | 11a | 23 a | 34 a | 60 a | 60 a | 10.73 |
| F0B3 | 12a | 26 a | 31 a | 58 a | 59 a | 10.49 |
| F1B1 | 6b | 12 c | 24 b | 46 b | 46 b | 7.59 |
| F1B2 | 9ab | 19ab | 33 a | 51 a | 52 b | 9.30 |
| F1B3 | 9ab | 17ab | 30 a | 52 a | 53 b | 9.11 |
| F2B1 | 7b | 19ab | 33 a | 56 a | 56 b | 9.79 |
| F2B2 | 7b | 18ab | 28 a | 52 a | 53 b | 8.99 |
| F3B3 | 8ab | 18ab | 28 a | 55 a | 55 b | 9.26 |

Note : WAI = week after inoculation. Numbers followed by same letters at same column showed no significant different at α 5%.

The best treatment combination in reducing bacterial pustule infections was the treatment of solid formulation with the addition of soybeancompos (F1B1), with the lowest disease severity and AUDPC, ie 46% and 7.59 units/week, respectively. This indicated that biofresh biological agents applications in the form of solid formulations can reduce the development of bacterial pustule disease in soybean crops.

### *3.2 Salicylic Acid Content and Peroxidase Enzyme Activity*

The results showed that there was a higher salicylic acid content and peroxidase enzyme activity in the biodegradation biofresh formulation (F1B1) treatment than non the biofresh treatments (F0B1), as presented in Table 2.

**Table 2.** Comparison of salicylic acid increases and peroxidase activity between biofresh treatment and compos soy litter with control

| Treatments | Content of Salicylate acid (mg) | | |
|---|---|---|---|
| | Before inoculation | After inoculation | Increase |
| F0B1 | 0.75 | 0.98 | 0.24 |
| F1B1 | 1.10 | 1.40 | 0.30 |
| | Peroxidase Activity (unit/mL) | | |
| | Before inoculation | After inoculation | Increase |
| F0B1 | 1.30 | 1.54 | 0.23 |
| F1B1 | 2.22 | 3.29 | 1.07 |

Increased accumulation of salicylic acid was a rapid reaction of plants to fight pathogenic bacterial infections. Salicylic acid is an important compound for plant defense against pathogens. This was reported by Zhang *et al.* [14], that the content of endogenous salicylic acid in tobacco plant sprouts treated with three strains of rhizobacteria increased significantly in the first week after treatment.

In other observations, peroxidase increased in the test plant of 1.07 units / mL. The peroxidase function is to strengthen the cell wall against the degradation of enzymes produced by pathogens through the formation of structural proteins in cell walls [15]. The use of PGPR as a biological induction agent has been widely reported. Application of B. megaterium in cucumber can inhibit the incidence of damping-off disease caused by Pythium aphanidermatum of 75.78% [16]. Increased levels of salicylic acid and peroxidase enzyme activation in plants treated with biodegradable biofresh agents indicated that the resistance of the soybean plant to bacterial pustule disease was related to the mechanism of systemic resistance induction.

*3.3 Crop yield*

The results of observation on soybean crops given biofresh biofertilizer and different organic materials significantly affect the weight of dry pod and seed weight. The highest dry pod weight and seed was found in biodegradation treatment of biofresh in solid formulation and soy litter organic matter (F1B1), 64.17 g / plant and 50.03 g / plant, respectively (Table 3). This indicated that the application of biofresh in a solid formulation with the addition of organic compos soy litter can provide high yields.

**Table 3.** Combination effects of Biofresh formulation and organic matter on crop yield

| Treatment | Pod Weight (g/plant) | Seed Weight (g/plant) |
|---|---|---|
| F0B1 | 55.83 b | 39.10 b |
| F0B2 | 51.90 ab | 39.07 b |
| F0B3 | 50.03 ab | 38.17 b |
| F1B1 | 64.17 c | 50.03 c |
| F1B2 | 49.97 ab | 36.13 ab |
| F1B3 | 49.10 ab | 34.67 a |
| F2B1 | 64.07 c | 46.40 c |
| F2B2 | 38.77 a | 37.43 ab |
| F3B3 | 52.27 ab | 37.13 ab |

Note : WAI = week after inoculation. Numbers followed by same letters at same column showed no significant different at α 5%.

## 4. Conclusion

Application of biological agent biofresh in a solid formulation combined with compos of soybean litter was the best treatment in increasing plant resistance to bacterial pustule disease and soybean crop yield. Induction of plant resistance occurred systemically, characterized by salicylic acid increase of 0.3 mg and peroxidase increase of 0.07 unit/mL in sample plants.

## References

[1] Dirmawati S R 2009 Jurnal Hama dan Penyakit Tanaman Tropika (Journal of Tropical Plant Pests and Diseases) ( 9(1) 54-57

[2] Habazar T 1989 Inventarisasi Penyakit-penyakit Bakteri pada Tanaman Kedelai (*Glycine max*) (Inventory of Bacterial Diseases on Soybean Plants). Laporan Penelitian Pusat Penelitian Universitas Andalas Padang

[3] Khaeruni A, Tjahjono B, Suwanto A, Sinaga MS 2007 Hayati *J. Bioscineces* 14(2) 76-80

[4] Goradia L, Hartman GL, Daniel S 2004 *J. Biological Sciences* 1(2):115-123

[5] Rahayu M 2007. *Tanggapan Varietas Kedelai terhadap Penyakit Pustul* Xanthomonas axonopodis pv. glycines *dan Potensi Ekstrak Nabati untuk Pengendaliannya. Di dalam Inovasi teknologi kacang-kacangan dan umbi-umbian mendukung kemandirian pangan dan kecukupan energi. Prosiding Seminar Nasional Balai Penelitian Tanaman Kacang-kacangan dan Umbi-umbian Malang,19 November2007*

[6] Sivan A and Chet I 1993 Microbial Control of Plant Diseases, In R.Mitchell (ed.), Environmental Microbiology New York: John Wiley and Son, Inc

[7] Khaeruni A, Sutariati G A K and Wahyuni S 2010 Jurnal Hama dan Penyakit Tanaman Tropika 10(2)123-130

[8] Syukur A 2005 *J. Ilmu Tanah dan Lingkungan* **5**(1)30-38

[9] Soekarno B P W, Surono and Susanti 2014 *J. Fitopatol Indones* **10**(5)153-159

IOP Conf. Series: Earth and Environmental Science **122** (2018) 012052 doi:10.1088/1755-1315/122/1/012052

[10] Khaeruni A, Satriana D, Wijayanto T, Harnowo D, Syafar AAR and Wahab A 2016 International Journal of Biosciences 8(5) 136-145
[11] Warrier RR, Paul M, Vineetha MV 2013 *Genetics and Plant Physiology* 3(12)90–97
[12] Patra HK, Mishra H 1979 *Plant Physiol.* 63:318-32
[13] Cooke BM.1998 *Disease bassessment and yield loss* In: Jones DG, Ed. *The Epidemiology of Plant Diseases*. London, UK: Kluwer 42-47
[14] Zhang S, Reddy M S and Klopper J W 2002 *Biol Control.* 23:79-86
[15] Vance C P, Kirk T K and Sherwood R T 1980 *Annu Rev Phytopathol* **18**:259-288
[16] Liang J, Tao R, Hao Z, WangL, Zhang X 2011 *African Journal of Biotechnology* **10**(36) 6920-6927

**Acknowledgment**
The authors gratefully acknowledge the financial support from Indonesian Agency for Agricultural Research and Development Ministry of Agriculture, Indonesia, under "KKP3N 2015" research

IOP Conf. Series: Earth and Environmental Science **122** (2018) 012053 doi:10.1088/1755-1315/122/1/012053

# Response to growth and production of green beans (*Vigna radiata* L.) in various cropping spots and fertilizer provision of layer chickens

**L Sitinjak[1] and E Purba[2]**

[1]Universitas Gunung Leuser, Aceh Tenggara,
[2]Universitas Sumatera Utara, Medan.
E-mail: epurba@yahoo.com

**Abstract.** Agroecology affecting plant growth can be influenced by factors such as plant spacing and growing media. This study aims to determine the effect of plant spacing and chicken manure on the growth and production of green beans. Plants were planted at three spacing plants (20cm x 20cm, 20cm x 30cm, and 20cm x 40cm) while manure was applied at 3.75 ,7.50 and 11.25 ton/ha and without manure as a comparison. The treatments were arranged in a randomized complete block design with three replicates. The result showed that the highest diameter of stem was resulted from the application of 11.25 ton/ha of manure combined with the planting space of 20cm x 40 cm. Similarly, the highest number of branch of stem we found at the plot where chicken manure of 11.25 ton/ha with planting space of 20cm x 30cm. The highest production (4,944.4 kg/ha) was resulted from 7.50 ton/ha manure with combined with planting space of 20cm x 40cm. There was an interaction between the treatment of manure and the planting space of soybean. The production of green bean was 24% higher in 7.50 ton/ha manure combined with 20cm x 40cm planting space compared to no manure (control).

## 1. Introduction

Green beans (*Vigna radiata*. L) is a food crop and belong to the family of nuts which have long been cultivated in Indonesia. This crop is the third most widely grown nuts plant after soybeans and peanuts [1]. This crop is known as one of the important leguminoceae plants, but until now the attention of the community on the cultivation of green beans is still lacking. This lack of attention is caused by the harvest of green beans must be done several times so that farmers think that planting green beans is considered less effective [2].

Rukmana [3] stated that when compared with other nuts, green beans have advantages such as mature, more tolerant to drought, can be planted less fertile and at the same time can fertilize the soil because it can simbiosis with rhizobium bacteria, easy cultivation and pests that attack Relatively little. However, the production of green beans in Indonesia is still low and has not been able to meet domestic needs. One cause of low production of a plant is the low level of soil fertility. One effort that can be made to improve soil fertility is the supply of nutrients through fertilization.

One of the efforts to increase the production of green beans is by improving the quality of cultivated land that can be done by improving the quality of the soil which decreases due to cultivation practices through fertilization. Fertilization is one of the most important factors for plant growth to determine the success of crop production. Fertilization is intended to meet the need for nutrients in sufficient quantities and balanced with the hope to support vegetative growth and generative plant wich leads to high production and good quality.

Fertilizers are all ingredients supplied to the soil in order to improve the physical, chemical and biological properties of the soil. Fertilization is done to provide nutrients needed by plants in each period of growth.

Organic materials ranks first in a series of cultivation plants because this material is used as a basic fertilizer so that the application is done early and in large quantities. Organic compounds or elements are the main ingredients of this fertilizer can be utilized after the process of decomposition in the soil [4]. Organic fertilizer is also called natural fertilizer, because all or most of this fertilizer comes from nature. Animal waste, crop residues, household waste, and stones are the basic ingredients of organic fertilizer [5]. [6] in his research found the organic fertilizer chicken manure that best affect the corn crop is on the treatment of organic fertilizer chicken poo 10 tons / ha).

Manure comes from animal waste that has undergone the process of decomposition or the influence of microorganisms. The signs of mature manure are not sharp (ammonia smell), dark brown, dry looking, not hot when held and loose when squeezed [7]. Melati and Andriyani [8], in his research showed that giving 10 tons of chicken manure / ha can increase vegetative growth and organic soybean production.

Organic fertilizer is a biological buffer that has a function in improving the physical, chemical and biological properties of the soil, so the soil can provide nutrients in a balanced amount. Organic fertilizers are widely used for food crops and animal feed plants are generally of animal waste, among others are chicken manure. From result of Kartika research [9] known that chicken manure of chicken manure give best result to growth and yield of grassy plant of *Brachiaria humidicola* when compared with cow manure and cow manure. Provision of chicken manure can improve the soil structure that is very short of organic elements and can fertilize spinach plants. That is why the giving of organic fertilizer into the soil is necessary so that the plants that grow on the soil can grow well [10].

Poultry manure is a solid fertilizer that contains lots of water and mucus. Chicken manure includes cold fertilizer because the change from the material contained in the fertilizer becomes available in the soil, takes place slowly. Based on the results of research that has been done [11], it was found that the provision of chicken manure on the observation of the growing kompenen highest average crops and production components produce the highest average weighted berangkasan. Sri Susanti Ningsih in his research found that the provision of chicken manure manure affect the growth and production of mustard plants.

Different plant spacing may affect the efficiency of radiation energy capture for plant growth. At a dense spacing, the transmission of solar radiation on the ground surface is smaller than the more loose spacing. [12] the arrangement of plant spacing of a plant can be more optimal. [13] in his research found that the more distance spacing in the nutrient balance test on elephant grass showed higher growth and production (100 x80).

The distribution of light on the plant canopy becomes an important factor affecting the efficiency of the use of sunlight by plants. One of the factors affecting the distribution of light in the canopy is the canopy structure. Spread of light on the upright leaf canopy is better than the horizontal leaf canopy [14]. At a wide spacing the growth is faster than at a narrow spacing. This is due to the happening of nutrients in the soil and sunlight in the process of photosynthesis.

Setting spacing is expected to affect the increase in green bean production. Use of plant spacing of 30 cm x 20 cm with 2 seeds per hole resulted in optimal production / dry seed ha [15]. [16] in his research found that the spacing of 15 x 30 gave the highest yield of 0.85 tons / ha in green beans.

## 2. Materials and methods

The treatments in this study were arranged in randomized complete block design (RCBD) with two factors namely 1. fertilizer application with four levels (P0) control, (P1) 1.5 kg/plot = 3.75 ton/ha, (P2) 3.0 kg/plot = 7.50 ton/ha and (P3) 4.5 kg/plot = 11.25 ton/ha, and the second factor namely planting space with three levels i.e (J1) 20cm x 20 cm, (J2) 20cm x 30cm and (J3) 20cm X 40cm.

Therefore there were 12 treatment combinations and each of them was repeated three times. In accordance with the treatment of plant spacing, the number of plant population in the plot varies for (J1) 100 plants or equal to 250.000 plants per hectare , (J2) 70 plants 166.667 plants per hectare and (J3) 50 plants 125.000 plants per hectare. The field was cultivated in accordance two times, a week before planting has done the application of chicken manure laying in accordance with the treatment. By way of placing 2-3 seeds of green beans the hole with the a depth of 3-5cm soil, then covered with loose soil. After plant growth is done the determination of the sample plants.

## 3. Results and Discussion

From the observations made on the three observation parameters obtained the following results:

**Table 1**. Average stem diameter (mm) at the end of observation (6 MST)

| Treatments | J1 | J2 | J3 | Total | Average |
|---|---|---|---|---|---|
| P0 | 15.92 | 18.54 | 18.67 | 53.13 | 17.71 |
| P1 | 15.58 | 19.11 | 19.34 | 54.03 | 18.01 |
| P2 | 17.16 | 20.34 | 20.99 | 58.49 | 19.50 |
| P3 | 16.75 | 20.67 | 21.33 | 58.75 | 19.58 |
| Total | 65.41 | 78.66 | 80.33 | | |
| Average | 16.35 | 19.67 | 20.08 | | |

The highest dimater ( 21,33 mm) was obtained on the application of manure as much as 11.5 ton/ha with spacing 20cm x 40xm (P3J3) with the stem diameter 21.3 mm, when compared with no chicken manure (P0J3). By applying chicken manure it can supply adequate nutrient content in soil required by plants. Prihmantoro (2003) stated that manure contains macro and micro nutrients complete although with a small amount. [17] manure has natural properties and does not damage the soil, providing macro and micro elements. In addition, manure functions to increase water holding capacity, soil microbiology activities, cation exchange rate and improve soil structure. The state of planting media that strongly supports the growth of green bean plants is seen in vegetative growth by showing the diameter of the stem that grows well/large and this makes the system of nutrient transgrtasi and photosynthesis results in green beans plant is good. [18] in her research found that the application of chicken manure 40 ton/ha gave better results on the growth and production of *Brachiaria humidicola* grass.

**Table 2**. Average number of branches at six weeks after planting

| | | Planting space | | | |
|---|---|---|---|---|---|
| Treatments | J1 | J2 | J3 | Total | Average |
| P0 | 6.30 | 7.00 | 7.00 | 20.30 | 6.77 |
| P1 | 6.30 | 7.70 | 7.00 | 21.00 | 7.00 |
| P2 | 7.30 | 8.00 | 8.00 | 23.30 | 7.77 |
| P3 | 7.30 | 8.90 | 8.00 | 24.20 | 8.07 |
| Total | 27.20 | 31.60 | 30.00 | | |
| Average | 6.80 | 7.90 | 7.50 | | |

Table 2 shows that most branching is in the application of laying chicken manure 11.50 ton/ha with spacing of 20cm x 30cm (P3J2) with the number of branching as much as 8.9 branches when compared with the treatment without the application of laying chicken fertilizer with spacing (20cm x

30cm) in the treatment (P0J2). Chicken manure and plant spacing affected the number of branches on green bean plants. The emergence of flowers that will later develop into a pod that will contain green bean seeds. The more the number of branching there are green bean plants the more the number of pods and seeds. The The height of branching in the application chicken manure as much as 11:5 tons/ha caused by the plants sufficient supply of nutrients and nutrients other nutrients provided by laying chicken manure. Ability of laying chicken breeders to improve the structure of chemically, physics and biologi, make the root system to be good, so it can absorb water and nutrients well, which is very influential dalm support vegetative growth of plants included in the formation of branches. [8] showed that 10 tons of chicken manure/ha increased vegetative growth and organic soybean production.

The optimum spacing in support of green bean planting is at 20cm x 30cm. Where at this planting distance, jumah branching more. Treatment of plant spacing is closely related to competition among plants in obtaining growing space such as water, nutrients and sunlight that are essential for photosynthesis. The distribution of light on the plant canopy becomes an important factor affecting the efficiency of the use of sunlight by plants. One of the factors affecting the distribution of light in the canopy is the canopy structure. Spread of light on the upright leaf canopy is better than the horizontal leaf canopy [14]. At a wide plant spacing (optimum) growth faster than at a narrow spacing. This growth includes mula formation of leaves and stems which in this case also includes the process of branching formation. When taaman gets a suitable growing space where the nutrients in the motherland, and enough sunlight, will support the process of photosynthesis, which is needed to produce food for plants (photosynthesis) used by plants for their vegetative and generative growth processes.

**Table 3**. Average number of grain production (kg / ha)

| Treatments | J1 | J2 | J3 | Total | Average |
|---|---|---|---|---|---|
| P0 | 2,805.56 | 3,728.40 | 3,986.11 | 10,520.07 | 3,506.69 |
| P1 | 3,458.33 | 2,765.44 | 4,305.56 | 10,529.33 | 3,509.78 |
| P2 | 3,916.67 | 3,851.86 | 4,944.44 | 12,712.97 | 4,237.66 |
| P3 | 4,277.78 | 4,296.30 | 4,000.00 | 12,574.08 | 4,191.36 |
| Total | 14,458.34 | 14,642.00 | 17,236.11 | | |
| Average | 3,614.59 | 3,660.50 | 4,309.03 | | |

From the result of observation, it was found that the highest production was in the application of laying chicken manure of 7.5 tons/ha with spacing 20cm x 40cm (P2J3) obtained by the production amount 4,944.4 kg /ha. Compared to no chicken manure with the same planting space (P0J3) is increased by 24%. Thus, in order to obtain the production of green bean plants, the optimum layer chicken fertilizer is at 7.5 tons/ha with spacing of 20cm x 40cm. There is an interaction between the application of laying chicken manure with the spacing to increase the production of green bean plants. From the results of [19], showed that the increasing of organic fertilizer from 1 ton/ha and the growing spacing of 0.75 x 25 cm in corn plant did not give better result. So it does not mean if if the provision of layers of chicken fertilizer required crop should be given as much as possible, but there is an optimum limit where at doses of 7.5 tons / ha not on the treatment of 11.5 tons/ha, laying chicken manure has supported the growth and production of plants Green beans, if more than the optimum limit, the available nutrient suplay will be wasted.

Different plant spacing may affect the efficiency of radiation energy capture for plant growth. At a dense spacing, the transmission of solar radiation on the ground surface is smaller than the more loose spacing. According to [12], the arrangement of plant spacing of a plant can be more optimal. I.J. Abadi, H.T. Sebayang, E.Widaryanto, 2013, in his research showed that the increasingly loose spacing of 75 x 30 showed higher growth and production in sweet potato plants. The right spacing makes the plants do not shade each other, the growth of the canopy is free so that the competition in obtaining sunlight is very small. This supports the process of photosynthesis is good, so the production of dry matter (carbohydrate) as raw material for the production takes place well and sufficiently available in the plant body.

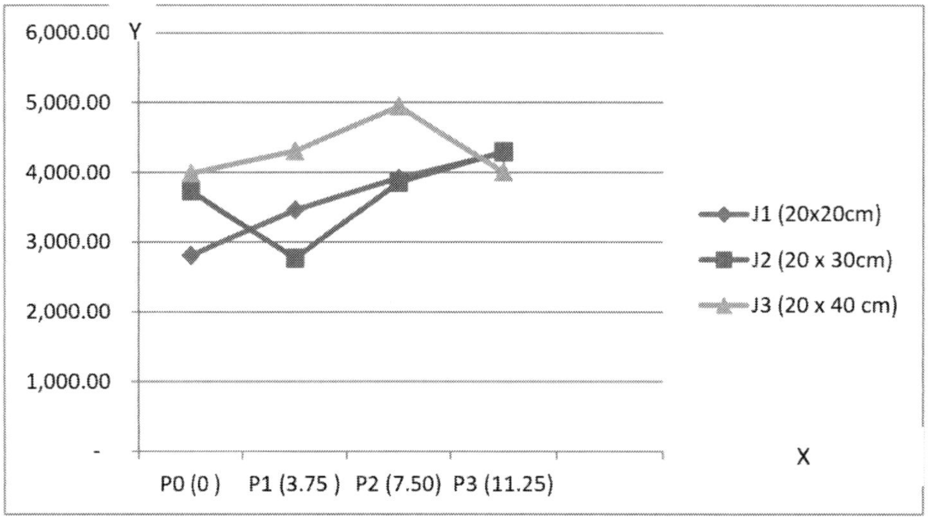


**Figure 1**. The increase of production result due to the application of chicken manure and spacing can be seen in the following graph

## 4. Conclusions

This study concluded that :

- Laying chicken fertilizer can increase the production of green kaccang plants up to 24% at the optimum dose of 7.5 tons/ha when compared to that without the application of chicken manure.
- The looser spacing (20cm x 40cm) between plants and supports green bean plants for the body and production.
- There is an interaction between the application of laying chicken manure with plant spacing on increasing the production of green bean plants.

## References

[1] Purnomo dan Hartono 2005 *Kacang Hijau (Green Beans)* (Jakarta: Penebar Swadaya)
[2] Soeprapto 2000 *Bertanam Kacang Hijau (Green Beans Cultivation)* (Jakarta: Penebar Swadaya) p 33
[3] Rukmana R 2006 *Kacang hijau, budidaya dan pasca panen (Green Beans, Cultivation and Post Harvesting)* (Yogyakarta: Kanisius) p 68
[4] Marsono dan Sigit P 2001 *Pupuk akar, Jenis dan Aplikasi (Root Fertilizer, Types and Aplication)* (Jakarta: Penebar swadaya)
[5] Hendarsin dan Srijono 2001 *Pupuk Organik (Organic Fertilizer)* (Jakarta: Kanisius)
[6] Ningsih S S 2013 Pengaruh pemberian pupuk kandang ayam dan pupuk (ZA) terhadap pertumbuhan serta produksi tanaman sawi (*Brasiaca juncea L*) (Effect of chicken manure and ZA fertilizer to mustard growth and production) *Jurnal Penelitian Pertanian BERNAS* **9** **(1)** pp 1-6
[7] Musnamar E 2003 *Pupuk Organik (Organic Fertilizer)* (Jakarta: Penebar Swadaya)
[8] Melati M dan Andriyani W 2005 The effect of chicken manure and green manure calopogonium mucunoides on growth and production of vegetable soybean under organic farming system *Bul. Agron.* **(33)** 2 pp 8 – 15

[9] Kartika Y 2013 *Pengaruh 3 Jenis Pupuk Kotoran Ternak (SapiI, Ayam dan Kambing) Terhadap Pertumbuhan dan Produsi Rumput Brachiaria Humidicola (Effect of 3 Kinds of animal composts (Cow, Chicken and Goat) to Brachiaria humidicola Growth and Production)* Skripsi Fakultas Peternakan Universitas Kristen Palangka Raya

[10] Subroto 2009 *Kesuburan dan Pemupukan Tanah Pertanian (Fertility and Fertilization of Agricultural Land)* (Bandung: Pustaka Buana)

[11] Latuamury N 2015 *Pengaruh Tiga Jenis Pupuk Kandang Terhadap Pertumbuhan dan Produksi Tanaman Kacang Hijau (Vigna radiate L.)(Effects of 3 types of manures to Green Bans Growth and Production)* Program Studi Agroteknologi Universitas Nani Bili Nusantara – Sorong ISSN : 1907-7556 *Jurnal Agroforestri* X Nomor **3** September 2015

[12] Janick J 1963 *Horticulture science* ( San Fransisco: W H Freeman and Company)

[13] Sandiah N, Pasolon Y B dan Sabaruddin L O 2011 Uji keseimbangan hara dan variasi jarak tanam terhadap pertumbuhan dan produksi rumpt gajah (*Pennisetum purpureum var. Hawai)(Balance Testing of nutrients and distance variation to Pennisetum purpureum var. Hawai Growth and Production) AGRIPLUS* Volume **21** (02) Mei 2011 ISSN 0854-0128

[14] Newton J E and Blacman G E 1969 *The Penetration of solar radiation through leaf canopies of difference structure* (New York: Academic Press)

[15] Soeprapto H S dan Sutarman T 1986 *Bertanam Kacang Hijau (Green Beans Cultivation)* (Bandung: PT. Penebar Swadaya) p 38

[16] Marsiwi T, Purwanti S, Prajitno D 2015 The effect of distance planting and rates of NPK ferrtilizer on the growth and results of mung bean seed *(Vigna radiate L. Wlcjek) Vegetalika* Vol. **4** No. 2 pp 124-132

[17] Syekhfani 2000 *Sifat dan Fungsi Pupuk Kandang (Characters and Fuction of Manure)* http://etd.eprints.ums.ac.id/ 14422/2/BAB_I.pdf. [ 28 April 2012]

[18] Kusuma M E 2015 The advance effect of dosage chicken manure to the growth and production brachiaria humidicola at the secon defoliation Fakultas Peternakan Universitas Kristen Palangka Raya *Jurnal Ilmu Hewani Tropika* Vol **4**. No. 2. Desember 2015 ISSN : 2301-7783

[19] Wahyudin A, Ruminta D and Bachtiar C 2015 Effect of different plant spacing with various dosages of organic fertilizer on growth and yield of maize hybrid P-12 in Jatinangor *Jurnal Kultivasi* Vol. **14**(1) Maret 2015

IOP Conf. Series: Earth and Environmental Science **122** (2018) 012054 doi:10.1088/1755-1315/122/1/012054

# Germinaton performance of selected local soybean (*Glycine max* (L.) Merrills) cultivars during drought stress induced by Polyethylene Glycol (PEG)

**R F Pane[1], R I Damanik[1]* and E H Khardinata[1]**

[1]Department of Agrotechnology, Faculty of Agriculture,
Universitas Sumatera Utara, Jl. Prof A. Sofyan No. 3, Medan 20155, Indonesia
*Email: d_revandy@hotmail.com

**Abstract.** Drought stress is one of the factors that can decreased growth and production, so that required a variety that has the ability to sustain cellular metabolism, and growth during the stress. This research was aimed to investigated the involvement of germination performance *invitro* of five local soybean cultivars, Grobogan, Kaba, Anjasmoro, Argomulyo, and Dering to drought stress induced by polyethylene glycol (PEG) 6000 (0%, 2%, 4%, and 6%). The measurable seedling traits as the day appearance of shoots and roots, total of leaves, shoot length, root length, fresh plant weight, dry plant weight, fresh root weight, and dry root weight under control as well as water stress condition were recorded. The experiment units were arranged in factorial completely randomized design with four replications. The result showed that the value for most parameters was recorded highest for Argomulyo cultivar compared with Dering cultivar which is known to be tolerant to drought. In terms of roots performance, Grobogan and Argomulyo cultivars produced the longest and heaviest of roots, while Grobogan cultivar had no significant different for root length compared with control. In conclusion, the root length and fresh weight root parameters can be used as quick criteria for drought tolerance.

## 1. Introduction

Drought or any other abiotic stresses constraining plant growth and comprehensive crop production seriously, also current global climate change has made this condition more serious [1]. Drought is a complex physical-chemical process, which in many biological macromolecules and small molecules are implicated, such as nucleic acids (DNA, RNA, microRNA), proteins, carbohydrates, lipids, hormones, ions, free radicals and mineral elements. Drought is connected with almost all aspects of biology needed of plants to increase the uptake of water, which is usually more available deep down in the soil [2]. At this time, drought study has been one of the main commands in global plant biology and biological breeding related to the fertile root system, ability to maintain stomatal opening and growth within their capacity at low levels of water potential.

Most of the agricultural plants, including soybean, rice, maize, barley, oats, sorghum, and wheat, tolerate even the most transient water stress. Stronger seed germination and seedling growth under drought conditions for requires above-normal tolerance of the low-level water conditions for the improvement of direct seeding cultivation. In rain fed agricultural conditions, low-level water conditions may prove to be a critical constraint to crop growth in the soil detrimental. The amount of water used by a crop is closely associated with the photosynthetic activity, dry matter production and yield in many species [3].

The use of polyethylene glycol (PEG) as a screening agent has been widely used for plantsdrought screening [4], such as Wheat [5], Sorghum [6], Maize [7], Tomato [8], and Sunflower [9]. PEG is a chemical drought-induced molecules, can be used to reducethe osmotic potential of water and nutrient solution culture without causing physiological damage and maintain the condition during the experiment periods. According to Sirait [10], *in vitro* screening genotypes for abiotic stress (drought) resistance properties has several advantages, such as shorter selection times, requires less space, ease of handling, less laborious, and is not limited by seasons. Furthermore selection for drought tolerance at early stage of seedling using PEG traits is a potentially cheaper method than conventional planting, Seedling period is often considered to be less consistent when facing the low-level water effect in the dry soil areas during the dry season, also for the selected genotypes to study in detail on water stress on plant germination indices [11].

The present investigation was carried out to find out appropriate criteria for screen out of soybean cultivars that have higher drought resistant and to study the morpho-physiological changes of soybean seed germination *in vitro* under osmotic stress.

## 2. Materials and methods

The study was conducted using five selected soybean genotypes on the basis of their tolerance of drought. All those genotypes were provided by Indonesian Agricultural Research Centre, Malang, Indonesia. Drought stress was simulated by polyethylene glycol (PEG) with the molecular weight (MW) 6000 (Merck Schuchardt OHG, Germany). Mature Seeds were dehusked and surface washed with 70% (v/v) ethanol for 2 min followed by 40% (v/v) sodium hypochlorite for another 20 min shaking. After rinsing five times with sterile distilled water, the sterilized seeds were placed and allowed to germinate on *in vitro* solidified basal Murashige and Skoog (MS) medium [12]. Osmotic stress *in vitro* was simulated by adding PEG to MS medium in 2% w/v, 4% w/v, and 6% w/v concentrations. The pH of the media was set to 5.8 prior to autoclaving at $121^0$C for 25 min. The cultures were incubated at $25^0C \pm 2^0C$ under artificial lightening to have photoperiod of 16 h.

About 7 days seed germinating, the day appearance of shoot and root were recorded. Seeds were considered as germinated if they exhibited radicle extension by more than 3 mm. The experiment was terminated by harvesting seedlings after 4 weeks and amount of leaves (AL), shoot length (SL), root length (RL), fresh weight shoot (FWS), fresh weight root (FWR), dry weight shoot (DWS), and dry weight root (DWR) of five soybean cultivars in four replication were recorded experimentally. Fresh shoot or root weight was measured on digital analytical balance (Sartorius, Germany), while dry shoot or root weight was obtained after drying the sample by putting shoots and roots in paper bags separately for 72h in the oven at $70^0$C for constant dry weight.

## 3. Results and discussion

Seedling development under laboratory conditions has been accepted as suitable growth stages for testing drought tolerance in soybean. The effect of osmotic stress induced by polyethylene glycol (PEG) on the day appearance root and shoot of germinated seeds of five selected soybean cultivars is presented in Fig. 1A and B. Most of the cultivars had longer time progressively for their appearance root or shoot compared to control condition, except for Grobogan, Kaba and Argomulyo cultivarswhich showed no significant different for the appearance shoot of their germinated seeds (Fig. 1B).Usually, the drought tolerance genotype has the highest germination rate and better survival. In this experiment, Grobogan cultivar seemed to have the fastest germination rate. The faster germination rate (in term of the day of appearance root and shoot) may be due to the capability of the seed to the increased diffusivity of seed coat to water even under osmotic stress condition. Turk *et al.* [13] reported that drought stress at germination stage delayed or hinder the emergence of the radicle and further germination completely. However, once the grain attains a critical level of hydration it will lead to full germination. The earlier seed forms the roots/shoot, so it can be provided energy from photosynthesis for growth and root formation at later stages [14].

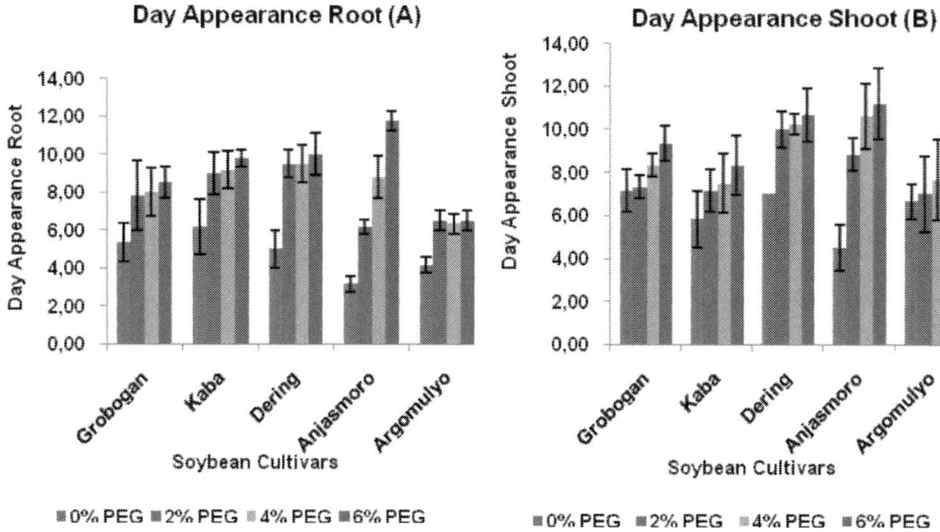


**Figure 1.** The day appearance root (A), the day appearance shoot (B) of soybean cultivars at different osmotic conditions on MS medium. Values are mean ± standard deviations based on four replications

Total of leaves was also measured, which considered as the fully open leaves. A number of leaves during osmotic stress showed different profiles depending on cultivars (Fig. 2).Cultivar grobogan at osmotic condition maintained the higher amount of leaves until 6% PEG when compared with control condition (0% PEG). A marked increased in total of leaves was observed in Argomulyo cultivar in 2% PEG, and Anjasmoro cultivar in 4% PEG compared to control (Fig. 2). Reduction in leaves perhaps due to less water absorption and decreased in external osmotic potential by PEG [15]. Grobogan cultivar showed maintain higher total of leaves that maybe it could preserve their higher leaf water-potential in the stressed environments.

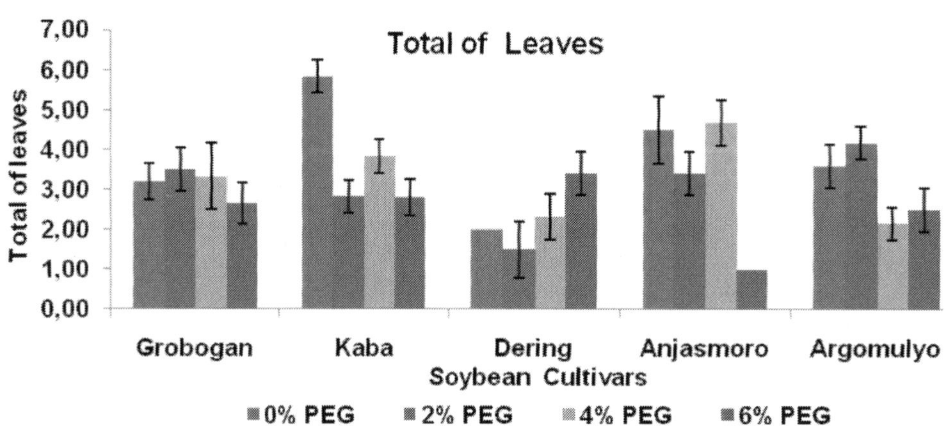


**Figure 2.** Total of leaves of soybean cultivars at different osmotic conditions on MS medium. Values are mean ± standard deviations based on four replications

The highest percentage reduction on shoot length (SL) was noted at Grobogan cultivar (32.5%), followed by Anjasmoro (31.7%), Kaba (27.0%), Dering (26.5%), and Argomulyo (5.7%) cultivars due to drought stress (2% PEG) *in vitro* MS medium (Fig 3.). The results suggested that Argomulyo cultivar was least affected by drought stress compared with other cultivars, and showed no significant different between control and osmotic stress in 2% PEG, and after that, the SL was decreased in 4% and 6% PEG.

Root length (RL) also decreased with increasing of PEG concentration, except for Grobogan cultivar (Fig. 5). At 2%, 4%, and 6% of PEG, the RL in Grobogan cultivar was slightly similar with the control (0% PEG). RL was decreased at 2% PEG, and after that the RL was increased at 4% PEG for Anjasmoro and Argomulyo cultivars.

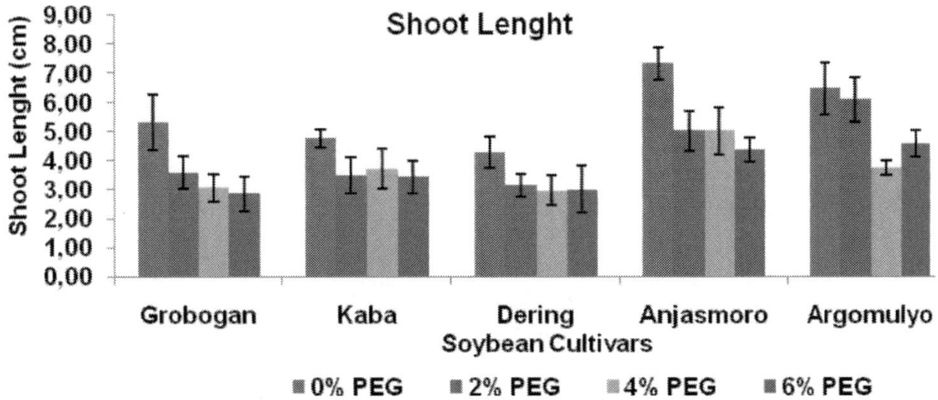


**Figure 3.** Shoot length of soybean cultivars at different osmotic conditions on MS medium. Values are mean ± standard deviations based on four replications

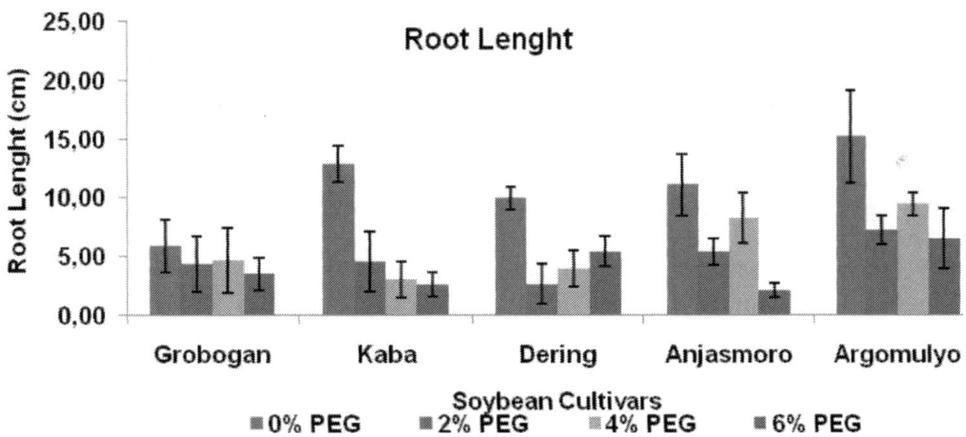


**Figure 4.** Root length of soybean cultivars at different osmotic conditions on MS medium. Values are mean ± standard deviations based on four replications

All fresh shoot weight (FSW) of the soybean cultivars were decreased with the increasing concentration of PEG. Osmotic stress caused significant reduction in FSW for all cultivars, with the highest decreased at Dering cultivar (56.3%), followed by Kaba cultivar (38.2%), Argomulyo cultivar (37.9%), Anjasmoro cultivar (32.6%), and Grobogan cultivar (14.3%) (Fig. 5A).

There was no significant decrease in dry shoot weight (DSW) detected for Grobogan, Kaba, and Argomulyo cultivars compared to control at 2-6% of PEG treatment (Fig. 5B). However, Dering cultivar which is known as one of tolerant for drought stress, had significant different in DSW at 2% PEG. Osmotic stress negatively affects photosynthesis and carbohydrate production, resulting in growth disturbances [16].The decrease of dry matter was noted in many plant species under drought conditions [17]. Some authors believe thatdry matter increases in plants during water stress may

IOP Conf. Series: Earth and Environmental Science **122** (2018) 012054 doi:10.1088/1755-1315/122/1/012054

indicate tolerance to drought. In the presented experiments, DSW of Grobogan, Kaba, and Argomulyo cultivars were no significant different with control (Fig. 5B). Our results revealed that those of cultivars improved to persistent and use the limitation of their surroundings such as water, nutrients supply, to be alive under stress.

Fresh root weight (FRW) in osmotic condition had shown significant differences decreased for all soybean cultivars, except for Grobogan cultivar relative to non osmotic condition control (Fig. 6A). Kaba, Argomulyo, and Anjasmoro cultivars had decreased of their FRW at 2% PEG by 83.4%, 78.7%, and 75.9%, respectively, relative to control.

There were significantly decreased in dry root weight (DRW) for all cultivar in osmotic conditions (Fig. 6B). Among soybean cultivars, Argomulyo has shown low reduction in DRW. Fresh and dry root weight were decreased due to osmotic stress in soybean cultivars, except for Grobogan cultivar. Similar results were reported by Savitri [18]. Water uptake by the root is a complex parameter that depends on root anatomy and the pattern by which different parts of the root contribute to overall water transport [19].

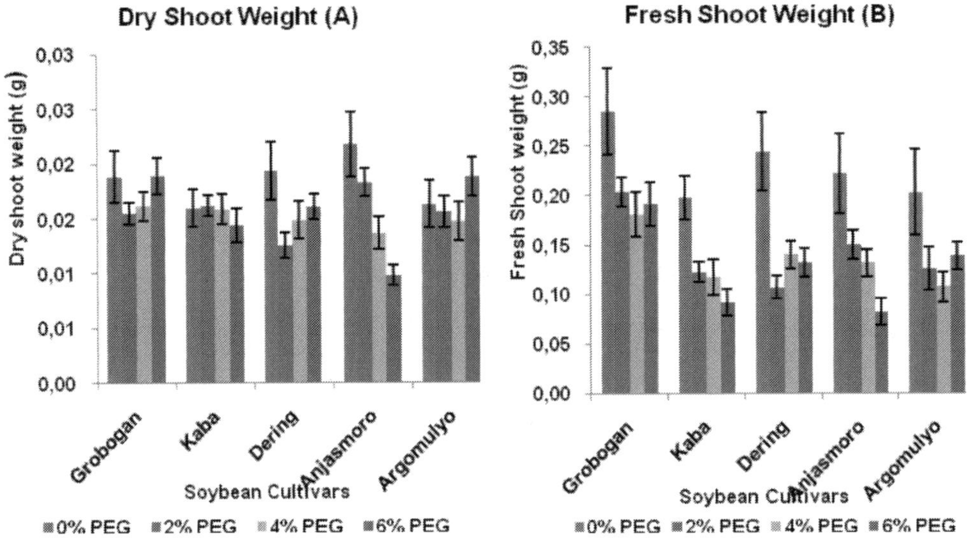


**Figure** Dryshoot weight (A), fresh shoot weight (B) of soybean cultivars at different osmotic conditions on MS medium. Values are mean ± standard deviations based on four replications

The cultivar which is showing better performance can be considered as drought tolerant. The current study showed that cultivars grobogan and Argomulyo found better growth than others, such as root length, fresh weight root and dry weight root (Fig. 7). In conclusion, these three characteristics of *in vitro* development are possible candidates for criteria in developing models and quick criteria for preliminary screening of drought tolerance. In addition to other physiological components involved in the seed germination process under drought stress conditions, may be considered for breeding purposes.

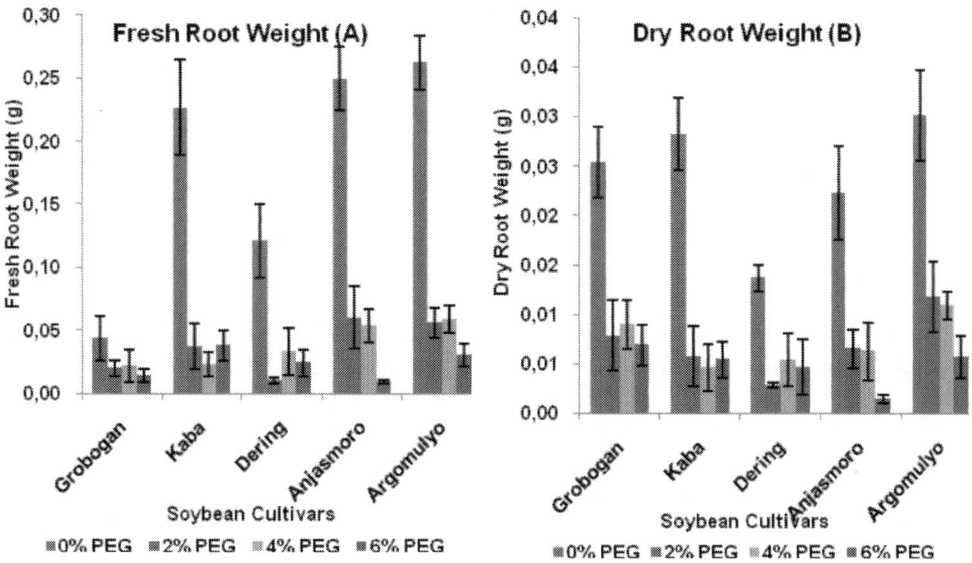


**Figure 6.** Fresh root weight (A), dry root weight (B) of soybean cultivars at different osmotic conditions on MS medium. Values are mean ± standard deviations based on four replications

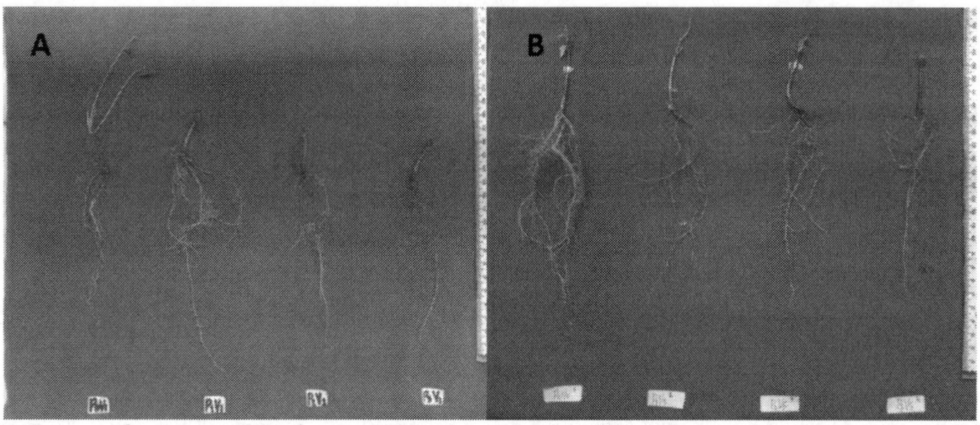


**Figure 7.** Root performance of Grobogan (A) and Argomulyo (B) soybean cultivars *in vitro* condition on MS medium subjected to 0, 2%, 4%, and 6% PEG

## 4. Conclusions

Root length, fresh weight root and dry weight root are possible candidates for criteria in developing models and quick criteria for preliminary screening of drought tolerance. In addition to other physiological components involved in the seed germination process under drought stress conditions, may be considered for breeding purposes.

## References

[1] Pan XY, Wang YF, Wang, G X, Cao QD, and Wang, J 2002 Relationship between growth redundancy and size inequality in spring wheat populations mulched with clear plastic film *Acta Phytoecol. Sinica* **26** 177-184

[2] Xiong L, Wang R, Mao G, and Koczan J M 2006 Identification of drought tolerance determinants by genetic analysis of root response to drought stress and abscisic acid *Plant Physiology* **142** 1065-1074

[3] Qing Z M, Jing L G, Kai CR 2001 Photosynthesis characteristics in eleven cultivars of sugarcane and their responses to water stress during the elongation stage *Proc. ISSCT* **24** 642-643.
[4] Widoretno W, and Sudarsono 2004 Evaluation of soybean varieties from somaclonal variants *in vitro* selection on drought stress *Hayati* **11** 11-20
[5] Dhanda S S, Sethi G S, and Behl R K 2004 Indices of drought tolerance in wheat genotypes at early stages of plant growth *J. Agron. Crop Sci.* **190** 1-6
[6] Bibi A, Sadaqat H A, Akram H M, and Mohammed M I 2010 Physiological markers for screening sorghum (*Sorghum bicolor*) germplasm under water stress condition *Int. J. Agric. Biol.* **12** 451–455
[7] Khayatnezhad M, Gholamin R, Jamaatie-Somarin S H, Zabihi-Mahmoodabad, R 2010 Effects of PEG stress on corn cultivars (*Zea mays* L.) at germination stage *World Appl. Sci. J.* **11(5)** 504-506
[8] Kulkarni M, and Deshpande U 2007 *In Vitro* screening of tomato genotypes for drought resistance using polyethylene glycol *Afr. J. Biotech.* **6(6)** 691-696
[9] Rauf S, Sadaqat H A, and Khan I A 2008 Effect of moisture regimes on combining ability variations of seedling traits in sunflower (*Helianthus annuus* L.) *Canadian J. Plant Sci.* **88** 323-329
[10] Sirait B 2001 Evaluation of morphophysiological character and production of soybean strains (*Glycine max* (L) Merr) aluminum tolerant selected *in vitro Thesis* Postgraduate Programme of IPB, Indonesia
[11] Sakthivelu G, Devi M K A, Giridhar P, Rajasekaran T, Ravishankar G A, Nedev T, and Kosturkova G 2008 Drought- induced alterations in growth, osmotic potential and *in vitro* regeneration of soybean cultivars *Genet. Appl. Plant Physiol.* **34** 103-112
[12] Murashige T, and Skoog F 1962 A revised medium for rapid growth and bioassay with tobacco tissue culture *Physiol Plant* **15** 473-479
[13] Turk M A, Rahman A, Tawaha M, and Lee K D 2004 Seed germination and seedling growth of three lentil cultivars under moisture stress *Asian J. Plant Sci.* **3** 394-397
[14] Bayoumi T Y, Eid, M H, and Metwali EM 2008 Application of physiological and biochemical indices as a screening technique for drought tolerance in wheat genotypes *Afr. J. Biotechnol.* **7** 2341-2352
[15] Kaydan D, and Yagmur M 2008 Germination, seedling growth and relative water content of shoot in different seed sizes of triticale under osmotic stress of water and NaCl *Afr. J. Biotech.* **7 (16)** 2862-2868
[16] Farooq M, Wahid A, Kobayashi N, Fujita D, and Basra, S M A 2009 Plant drought stress: effects, mechanism and management *Agr. Sustain.Dev.* **29** 185-212
[17] Zeid I M, and Shedeed, Z A 2006 Response of alfafa to putrescine treatment under drought stress *Biol. Plantarum* **50 (4)** 635-640
[18] Savitri E 2010 *In Vitro* Tests on Multiple Soybeav Varieties (*Glycine max* L. Merr) Tolerant Drought Using Polyethylene Glycol (PEG) 6000 on Solid and Liquid Media Faculty of Science and Technology UIN Maliki Malang **Vol.1** 9-13
[19] Cruz R T, Jordan W R, and Drew M C 1992 Structural changes and associated reduction of hydraulic conductance in roots of *Sorghum- bicolor* L. following exposure to water deficit *Plant Physiol.* **99** 203–212

# Physiological characters of soybean cultivars with application of nitrogen sources under dry land conditions

**Y Hasanah[1]*, T C Nisa[1], Hapsoh[2] and H Hanum[1]**

[1] Faculty of Agriculture, Universitas Sumatera Utara, Jl. Prof. A. Sofyan No. 3 Kampus USU, Padang Bulan Medan 20155, Indonesia
[2] Faculty of Agriculture, Universitas Riau, Pekanbaru, Indonesia.
*E-mail: yaya@usu.ac.id

**Abstract.** The objective of this study was to evaluate the influence of nutrient N management on physiological characteristics of three different soybean cultivars under dry land conditions. The study was conducted under dry lands of Desa Sambirejo (Langkat Regency) in the dry season. The study was conducted with a Randomize Block Design with two factors and three replication. The research was used a randomized block design with 2 factors and 3 replications. The first factor was soybean cultivars (Anjasmoro, Wilis, Sinabung). The second factor was N source, with Urea (50 kg/ha), *Bradyrhizobium* sp., farmyard manure (10 ton/ha), a combination of *Bradyrhizobium* sp. + farmyard manure (5 ton/ha) and a control with no N. The parameter observed in this study was the content of root N, shoot Nitrogen, shoot Phosphor, shoot Potassium and total of chlorophyll content. The results suggest that Anjasmoro and Sinabung cultivars had higher physiological characteristics (root N, shoot P and shoot K) compared to Wilis. Nitrogen source of Urea gave a higher physiological characteristics (content of root N, shoot Phosphor and shoot Potassium) compared to different treatment of N source in this study. The interaction between Anjasmoro cultivar and Urea gave the highest of content of shoot Phosphor and shoot Potassium, otherwise the interaction between Sinabung cultivar and *Bradyrhizobium* sp. gave the highest of content of shoot Nitrogen.

## 1. Introduction

Soybean [*Glycine max* (L.) Merril], as an important source of vegetable and oil for human and animal consumption [1]. As one of the most important legumes species, soybean contains about 40% highly nutritious protein with all essential amino acids, rich in magnesium, phospholipids, vitamins and minerals [2-3] and also contains secondary metabolites such as isoflavones [4], saponins, phyticacid, oligosaccharides [4] and phytoestrogens [5] which are beneficial to health. In addition to its nutritional values, soybean is also used as important nitrogen ($N_2$)-fixing crop throughout the world for the restoration and maintenance of soil fertility in a sustainable way and consequently the improvement of crop yields [6-7]

Demand for soybean in Indonesia continues to increase, along with increasing public knowledge of the benefits of soy as a functional food. Production of soybean is not keeping up with demand, so that efforts are necessary to improve national soybean production [8]. This is done through the approach of increasing productivity, increasing the intensity of soybean cultivation and expansion of planting soybean area to marginal land (sub optimal) such as dry land. Dry land has many problems such as drought stress during dry season, soil pH less suitable for soybean cultivation (pH 5.0), low macro nutrient N, P, K content. Based on these problems, dry land management for increasing soybean production can be done with two main approaches namely choosing adaptive soybean varieties of dry land and improving soil fertility through the management of nitrogen [9]

Nitrogen is one of the essential nutrients for plants, the basic components of protein formation, nucleic acids, amino acids and protoplasm, required in the synthesis of chlorophyll, the formation and growth of vegetative parts of plants (leaves, stems, and roots). Application of N nutrient under dry land conditions will influence on physiological characteristics of soybean varieties. Therefore, the objective of this research was to evaluate the influence of nutrient N management on physiological characteristics of three different soybean cultivars under dry land conditions.

## 2. Materials and methods

Research was conducted in Sambirejo Village, Binjei District, Langkat, Sumatra Utara (Indonesia), a dry land area, June to September 2012. Location determination refers to soybean production center in North Sumatera and according to Agroecological Zone map, FSZ North Sumatra. The soil texture of the experimental site was a sandy clay loam which had 11% coarse sand, 38% fine sand, 29% silt and 22% clay. Nitrogen content was low (0.14%), organic matter was 1.02%, with a pH of 5.0 [8]

### *2.1. Experimental design and crop management*

The research conducted by using a Factorial Randomized Block Design with 2 factors and 3 replications. The first factor is soybean cultivars are Anjasmoro (V1), Wilis (V2) and Sinabung (V3). As the second factor is the source of N consisted of without N application (N0) ; Urea 50 kg/ha (N1) ; *Bradyrhizobium* sp. (N2) ; manure of 10 tons/ha (N3) and combination *Bradyrhizobium* sp. + manure 5 tons/ha (N4).

The land used consists of 45 plot with the size of plot 2 m x 2 m. Among the plots are limited by drainage channels. Liming with dolomite 500 kg/ha was done during the second tillage by spread evenly 2 weeks before planting. Application of P and K fertilizer when planting is done at each time planting with the same dose (recommended dose of P and K fertilizer for soybean crop is 150 kg TSP/ha and 75 kg KCl/ha, respectivelly).

Three seeds per planting hole were planted at the study. Thinning is done 1 week after planting by cutting plants that grow poorly and leaving 1 plant per planting hole. Isolate *Bradyrhizobium* sp. is inoculated according to treatment. Isolate is mixed with soybean seeds, done in the morning just before planting in the shade. Soybean seeds that have been mixed isolates *Bradyrhizobium* sp.

Application of Urea and farmyard manure according to treatment done at planting time. Urea fertilizer is given half the dose at the time of planting and half the remaining dose at 30 HST. Application of cow manure is done by mixing cow manure with planting media evenly.

Harvesting is done when soybean have shown the criteria of harvest that the pod has brownish color and the stems and leaves have dried up.

### *2.2. Analysis of shoot N, P, K content and total chlorophyll*

Analysis of shoot N, P, K content was conducted at Asian Agri Laboratory, while the total chlorophyll was analyzed at Integrated Laboratory, Faculty of Agriculture, Universitas Sumatera Utara. Analysis of shoot N content were determined using method from Kjeldahl, shoot P content were determined using Spechtrophotometry method, shoot K content were determined using AAS method and total chlorophyll were determined using method from Arnon.

### *2.3. Statistical analysis*
Data were subjected to analysis of variance (ANOVA) for comparison of means. Means were separated using Duncan's Mutiple Range Test at the 0.05 probability level.

## 3. Results and Discussion
### *3.1. N shoot content*
Based on Table 1 it can be seen that the cultivars of Anjasmoro and Sinabung have significantly higher N content of shoot than Wilis. Source N *Bradyrhizobium* sp. has significantly higher N content of shoot (3.78%) than with other treatments, whereas the treatment of without N and combinations of *Bradyrhizobium* sp. + farmyard manure (5 tons / ha) has the lowest N shoot content. In the three varieties, the treatment of *Bradyrhizobium* sp. increased the N shoot content of Anjasmoro and also Sinabung varieties compared with other treatments.

### *3.2. N root content*
The cultivars of Anjasmoro and Sinabung have N content of roots significantly higher than Wilis. Sources of N Urea has significantly higher N root content (1.29%) than *Bradyrhizobium* sp. + farmyard manure (5 tons/ha) and without N application, but not significantly difference from *Bradyrhizobium* sp. and farmyard manure. The interaction between the varieties of Sinabung and Urea has a significantly higher N root content than other treatments, whereas the interaction of Wilis treatment and without N administration and the interaction of Wilis and farmyard manure treatment has the lowest N root content (Table 1)

### *3.3. P shoot content*
Anjasmoro and Sinabung cultivars have significantly higher P content than Wilis. Sources of N Urea have significantly higher P shoot content than other treatments. The interaction between Anjasmoro variety and Urea has the highest P content (0.39%), while the interaction between Anjasmoro variety and farmyard manure (10 tons/ha) and interaction between Wilis and combination of *Bradyrhizobium* sp. + farmyard manure (5 tons/ha) have the lowest P shoot content (Table 1).

### *3.4. K shoot content*
The cultivars of Anjasmoro and Sinabung have significantly higher K shoot content than Wilis. Sources of N Urea and *Bradyrhizobium* sp. have significantly higher K shoot content than other N treatments. The interaction between Anjasmoro variety and Urea has the highest K shoot content, whereas *Bradyrhizobium* sp. + farmyard manure (5 ton / ha) or manure on Wilis varieties give the lowest K content (Table 1)

### *3.5. N uptake of shoot*
N uptake of shoot in Anjasmoro cultivar tend to be higher than Wilis and Sinabung. Source N *Bradyrhizobium* sp. has significantly higher N uptake of shoot compared with other N source treatments, whereas treatment without N application has the lowest of N uptake of shoot. The interaction between the Anjasmoro variety and *Bradyrhizobium* sp. tend to be the highest N uptake of shoot (Table 1)

### *3.6. Total of Chlorophyll*
Anjasmoro cultivar tend to have the lower total chlorophyll content than Wilis and Sinabung. Treatment without N application tend to produce the lowest total chlorophyll compared with other N source treatments. Treatment of *Bradyrhizobium* sp. tend to produce the highest total chlorophyll. Urea treatment on Sinabung variety tend to increase the total leaf chlorophyll (5.68 mg/g wet weight of leaf).

**Table 1.** Physiological characters of three soybean cultivars with application of Nitrogen sources under dry Land conditions on June to September growing season.

| Treatment | Content of (%) | | | | N uptake of shoot (g/plant) | Total of chlorophyll (mg/g ww)[a] |
|---|---|---|---|---|---|---|
| | Shoot N | Root N | Shoot P | Shoot K | | |
| Cultivar (V) | | | | | | |
| $V_1$ = Anjasmoro | 3.40a | 1.17a | 0.31b | 2.06a | 16.87 | 4.93 |
| $V_2$ = Wilis | 3.19b | 0.98b | 0.34a | 1.53b | 13.14 | 5.03 |
| $V_3$ = Sinabung | 3.40a | 1.28a | 0.35a | 1.86a | 14.05 | 5.06 |
| N source (N) | | | | | | |
| $N_0$ = Without N application | 3.09c | 1.05b | 0.32b | 1.76b | 11.40c | 4.87 |
| $N_1$ = Urea | 3.46b | 1.29a | 0.38a | 2.00a | 13.58b | 5.07 |
| $N_2$ = *Bradyrhizobium sp.* | 3.78a | 1.14ab | 0.34ab | 1.88a | 19.44a | 5.26 |
| $N_3$ = Farmyard manure | 3.33b | 1.16ab | 0.31b | 1.72b | 14.97b | 4.98 |
| $N_4$ = *Bradyrhizobium sp.* + farmyard manure | 2.98c | 1.10b | 0.32b | 1.73b | 14.05b | 4.85 |
| Interaction V x N | | | | | | |
| $V_1 N_0$ | 3.07fg | 1.00def | 0.32c | 2.08bcd | 11.01 | 5.15 |
| $V_1 N_1$ | 3.66bc | 1.18bcd | 0.39a | 2.49a | 15.16 | 4.70 |
| $V_1 N_2$ | 3.97a | 1.18bcd | 0.33c | 2.22bc | 22.66 | 5.58 |
| $V_1 N_3$ | 3.41cde | 1.35b | 0.26d | 2.01cde | 15.61 | 5.31 |
| $V_1 N_4$ | 2.87gh | 1.16cd | 0.38ab | 1.48ij | 19.92 | 4.93 |
| $V_2 N_0$ | 3.29ef | 0.86f | 0.31c | 1.58ghij | 11.23 | 5.02 |
| $V_2 N_1$ | 2.87gh | 0.94ef | 0.38ab | 1.76efg | 11.44 | 4.84 |
| $V_2 N_2$ | 3.43cde | 1.10de | 0.34bc | 1.51hij | 20.26 | 5.26 |
| $V_2 N_3$ | 3.00g | 0.90f | 0.35abc | 1.43j | 10.66 | 4.54 |
| $V_2 N_4$ | 3.35de | 1.12cd | 0.25d | 1.39j | 12.11 | 5.48 |
| $V_3 N_0$ | 2.91gh | 1.29bc | 0.33c | 1.62ghij | 11.97 | 4.44 |
| $V_3 N_1$ | 3.85ab | 1.74a | 0.36abc | 1.74fgh | 14.13 | 5.68 |
| $V_3 N_2$ | 3.94a | 1.13cd | 0.35abc | 1.90def | 15.40 | 4.94 |
| $V_3 N_3$ | 3.59bcd | 1.02def | 0.32c | 1.71fghi | 18.65 | 5.09 |
| $V_3 N_4$ | 2.71h | 1.02def | 0.34bc | 2.33ab | 10.10 | 5.16 |

Note : Different letters at the same of group treatment and colomn represent significant differences at Duncan's Multiple Range Test ($p < 0.05$).
[a]) ww = wet weight

## 3. Discussion

Inoculation *Bradyrhizobium* sp. on Anjasmoro cultivar has higher shoot N content and N uptake of shoot compared with other N source treatments (Table 1). Anjasmoro cultivar required higher N nutrients for growth compared to Wilis and Sinabung, probably because Anjasmoro has higher plant growth rate and nett assimilation rate, so that N nutrients are indispensable for plant growth. N nutrient is obtained by mutualistic symbiotic between soybean and *Bradyrhizobium* sp. through the mechanism of assimilation of N in the form of direct ammonium from the nodule. In addition, the compatibility between Anjasmoro variety and *Bradyrhizobium* sp. determine the success rate in N absorption as indicated by an increase in shoot N content and N uptake of shoot (Table 1).

The high content of P and K canopy due to the urea treatment on the Anjasmoro cultivar showed that the Anjasmoro cultivar is more responsive to Urea fertilization compared to Wilis and Sinabung. Plant response to nutrient availability is highly dependent on nutrient uptake mechanisms by plants. Absorption of Urea containing 45% N results in increased availability of N so that vegetative growth is well and the ability of plant to absorb P increases. As a result, the ability of plants in the formation of seeds is also increased because P is very instrumental in the formation of primordia flowers, plant reproductive organs and as a key substrate in energy metabolism [10]

The high content of K due to Urea treatment on Anjasmoro cultivar is suspected to be related to the synergy between N and K nutrients. The increase in N uptake due to Urea fertilization results in increased K content, because K functions as an activator of enzymes in the process of plant metabolism and translocation of photosynthetic carbohydrates [11] so that the metabolic processes of plants is well done.

The role of *Bradyrhizobium* sp. in fixing N is seen from the increasing trend of total chlorophyll by the treatment of *Bradyrhizzobium* sp. in this study. Total chlorophyll of leaf is important to see the contribution of biological nitrogen fixation to the fulfillment of plant nitrogen needs. As it is known that nitrogen is one of the constituents of the porphyrin ring in chlorophyll (Salisbury and Ross, 1995). The fact that nitrogen is one of the constituents of chlorophyll, while chlorophyll itself is the engine for the process of photosynthesis, making N also a factor affecting the rate of photosynthesis. If nitrogen supply is limited, chlorophyll may not form and will eventually decrease the rate of photosynthesis (Yoshiki *et al.*, 2013).

## 4. Conclusions

Anjasmoro and Sinabung cultivars had a higher physiological characteristics (root N, shoot P and shoot K) compared to Wilis cultivar. Nitrogen source of *Urea* gave a higher physiological characteristics (content of root N, shoot Phosphor and shoot Potassium) compared to different treatment of N source in this study. The interaction between Anjasmoro cultivar and Urea gave the highest of content of shoot Phosphor and shoot Potassium, otherwise the interaction between Sinabung cultivar and *Bradyrhizobium* sp. gave the highest of content of shoot Nitrogen.

## References

[1] Couto C, Silva L R, Valentão P, Velázquez E, Peix A and Andrade PB 2011 Effects induced by the nodulation with *Bradyrhizobium japonicum* on *Glycine max* (soybean) metabolism and antioxidant potential *J. Food Chem.* 127 1487–1495

[2] Raghuvanshi R S, Bisht K 2010 *Uses of Soybean : Products and Preparation.* In: Singh G (Ed) The Soybean: Botany, Production and Uses (USA : CAB International) pp 345-374

[3] Ali N 2010 *Soybean Processing and Utilization.* In Singh G (ed,) The Soybean : Botany, Production and Uses, (USA : CAB International) pp 345-374

[4] Sakai T and Kogiso M 2008 Soy isoflavones and immunity. *J. Med. Invest.* **55** 167-173

[5] Ososki A L and Kennelly E J 2003 Phytoestrogens: a review of the present state of research *Phytother. Res.* **17** 845

[6] Smaling E M A, Roscoe R, Lesschen J P, Bouwman A F and Comunello E 2008 From forest to waste : Assessment of the Brazilian soybean chain, using nitrogen as a marker. *Agriculture, Ecosystems and Environment* **128** 185-197

[7] Charlotte C Z, Houngnandan P, Houinsou D and Fatiou T 2015 Determinants of Soybean [*Glycine max* (L.) Merrill] Production System in Benin. *J. of Exp. Biol. and Agric. Sci.* **3.**

[8] Hasanah Y, Nisa T C, Hapsoh A and Hanum H 2015 Isoflavone content of soybean [*Glycine max* (L). Merr.] cultivars with different nitrogen sources and growing season under dry land conditions *J. of Agric. and Env. for International Development* **109** 5 – 17

[9] Hasanah Y, Nisa T C, Hapsoh and Hanum H 2014 Optimasi Produksi dan Kandungan Isoflavone Kedelai (Glycine max (L.) Merril) pada Kondisi Cekaman Air Melalui Pengelolaan Hara N Disertasi (Medan : Universitas Sumatera Utara)

[10] Roper T J, Davenport C, DeMoranville S, Marchand A, Poole J, Hart and Patten K 2004 Phosphorus for bearing cranberries in North America

[11] Taiz L and Zeiger E 2002 *Plant Physiology* (Tokyo :The Benyamin/Cumming Publishing Co. Inc.) p 219-247

[12] Zhao, D, K R Reddy, V G Kakani and V R Reddy 2005 Nitrogen deficiency effects on plant growth, leaf photosynthesis, and hyperspectral reflectance properties of sorghum. *Europ. J. Agronomy* **22** 391-403

[13] Hua C, Feng Z S, Qin W Z, Ming W X and Chang Y Z 2008 Effect of Nitrogen and Phosphorus on the amino acids in root exudates and grains of rice during grain filling *Acta Agronomica Sinica* **34** 612-618

[14] Farhad , I S M, Islam M N, Hoque S and Bhuiy M S I 2010 Role of potassium and sulphur on the growth, yield and oil content of soybean (*Glycine max* L.). *An academic Journal of Plant Sciences* **3** 99-103

**Acknowledgments**

The authors would like to thank Irma Afriyanti and Yuda P. Surbakti for their assistance in conducting this research.

# Effects of hypoxia condition in embryogenic callus growth of soybean cell culture

**R I Damanik*, B H Manurung and E S Bayu**

Agrotechnology Department, Faculty of Agriculture, Universitas Sumatera Utara, 20155 Medan, Indonesia
*E-mail: revandy@hotmail.com

**Abstract.** The study was performed at Tissue Culture Laboratory, Agrotechnology Department, University of Sumatera Utara, to investigate the effect of plant growth regulator (PGR) and embryogenic callus performance soybean cultivars on hypoxia condition. This research had two stages, induction of embryogenic callus and analysis metabolism of callus after hypoxic condition with T-test. The analysis was used factorial Completely Randomized Design with two factors. The first factors were cultivars of soybean (Baluran, Gepak Kuning, and Grobogan) and the second factors were combinations of PGR (5 mg/l 2,4-D + 1 mg/l BAP, 10 mg/l 2,4-D + 1.5 mg/l BAP, and 15 mg/l 2,4-D + 2 mg/l BAP). The result showed the cultivars, combination of PGR, and interaction between cultivars and PGR gave significant effect to weight callus. The result of T-test showed that in hypoxic condition, POD enzyme exercise on Gepak Kuning's callus in 5 mg/l 2,4-D + 1 mg/l BAP was different before and after hypoxic condition.

## 1. Introduction

The soybean is one of the most important food commodity in Indonesia. In year 2015, soybean production increased 4.5% based on Central Statistics Agency (BPS) which reached 998.870 tons of soybean dry seeds. The increasing production is supported by the addition of harvest area of soybean plants [1]. Soybean plants susceptible to flooding. Puddles cause premature aging so that the leaf chlorosis, necrosis, fall and stunted plant growth, which in turn decreases the results. Loss of the vegetative phase is smaller than in the reproductive phase, ie 17-43% in the vegetative phase and 50-56% in the reproductive phase [2]. Plants that were flooded in a short time will have the condition of hypoxia (lack of $O_2$) [3].

Selecting cultivars for tolerant to flooding can be done with some experiment in the field or in the laboratory [2]. With *in vitro,* is expected to provide solutions resistant cultivars, tolerant or sensitive to flooding and is expected to maintain its integrity and grow into fully plants [4]. Somatic embryogenesis offers an in vitro method for mass propagation of important plant species. In addition, somatic embryogenesis enables to produce large numbers of plantlets [5] and also suitable for germplasm conservation, the selection of genetic variants for desirable characters, the generation of somaclonal variants and for performing cellular genetic manipulations [6].

In vitro culture techniques provide a means of rapid plant propagation and a tool for crop improvement. The success of tissue culture techniques is determined by the type and concentration of plant growth regulator used. Growth regulator auxin and cytokinin are needed to stimulate cell division and formation of callus [7]. Auxin has a role on the growth of cells, the apical dominance and callus formation. Cytokinins are growth regulators a role in regulating cell division and differentiation shoots influence on callus tissue [8]. By mixing some combination of growth regulators 2,4-Dichlorophenoxyasetat (2,4-D) and Benzylaminopurine (BAP) to the culture medium is able to stimulate the formation of callus on the explants. Based on [9], in young cotyledons soybean explants have started callus at the age of 1-2 weeks after explant callus culture with percentage 75%, in the treatment of 10 mg/l 2,4-D + 10 mg/l NAA. Research conducted by [10] on the culture *in vitro* of fruit Napier stated that BAP 1.5 mg/l is the optimum concentration in macassar fruit seed growth *invitro* for the purpose of propagation.

One of the alternatives that can be done to produce hypoxic-tolerant soybean with techniques *in vitro* in embryogenesis using combination of plant growth regulator (PGR) 2,4-D (2,4 Dichlorophenoxyacetic acid) and BAP (6-Benzylaminopurine) to induce embryogenic callus of selected soybean cultivars.

## 2. Materials and Methods

This study was conducted at Tissue Culture Laboratory of the Faculty of Agriculture, University of Sumatera Utara, Medan, with two phases of activity that embryogenic callus formation using a completely randomized design factorial and test the metabolic activity of callus on hypoxic conditions using t-test analysis Minitab 16. The study was consists of two factors, which as first factor, soybean cultivars (Baluran/$V_1$, Gepak Kuning/$V_2$, and Grobogan/$V_3$) and second factor, plant growth regulator (PGR) treatment with a combination of 5 mg/l 2,4-D + 1 mg/l BAP ($Z_1$), 10 mg/l 2,4-D + 1.5 mg/l BAP ($Z_2$), and 15 mg/l 2,4-D + 2 mg/l BAP ($Z_3$).

Mature seeds of the three selected soybean cultivars were dehusked and surface sterilized with 70% (v/v) ethanol for 2 min followed by 40% (v/v) sodium hypochlorite for another 20 min shaking. After rinsing five times with sterile distilled water, the sterilized seeds were used for callus induction. Soybean callus induction was following the method of [11]. The mature seeds were placed on MS medium [12] supplemented with 2,4 D + BAP, sucrose (30 g $l^{-1}$), casein hydrolysate (0.4 g $l^{-1}$) and gelrite (2.75 g $l^{-1}$). The pH of the media was set to 5.8 and added the PGR prior to autoclaving at $121^0$C for 25 min. The cultures were incubated at $25^0C \pm 2^0C$ under dark condition. After 2 weeks of culture, the primary callus was removed from the endosperm parts, and subcultured every 2 weeks in the same media for up to 8 weeks.

Planting was carried out in the laminar air flow (LAF), the matured seed that had been sprayed with 70% alcohol are planted soybean as cotyledon parts. The seeds were placed on petridish, then separated the embryo and endosperm part by using tweezers and a scalpel. Then the endosperms were planted into the media culture flask which has been prepared according to treatment. Each bottle consists of 1-2 explant culture and after planting is completed culture bottle rack placed in the culture of light such as light fixtures.

The bottles have been planted explant cultures were sprayed with 70% alcohol every day to be free from microorganisms (bacteria and fungi) that cause contamination. The temperature of the culture that is used is 18-22 $^{\circ}$C and light intensity of 2000 lux.

## 3. Results and Discussion

### *3.1. Growth of Callus*

The percentage of callus growth has shown at Fig. 1. It showed that callus growth of Baluran cultivar was (0%), Gepak Kuning (100%) and Grobogan (100%). The percentage of callus growth was observed at 4 weeks after planting. Baluran cultivar stagnated (no growth) because of contaminated and therefore can't be forwarded to the advanced stage.

The results showed that the cultivar of Baluran not induced callus. Explants of cultivar baluran stagnated which started from planting until a certain time is not dead but is not growing. There are several factors that influence the success of tissue culture one of which is the explants. Callus growth is the result of a very complex interaction between explant, medium composition and environmental conditions during the incubation period.

Grobogan cultivar capable of forming callus at 1 week and Gepak kuning capable of forming callus on 2 weeks. Explants callus marked by the growth of cells in which tissue explant growth and raised lumps cell mass was yellowish white. The callus that formed were in the injured area on the explants. Callus formed generally yellowish white color and textured friable. Media fastest induce callus in this study is 15 mg/l 2,4-D + 2 mg/l BAP.

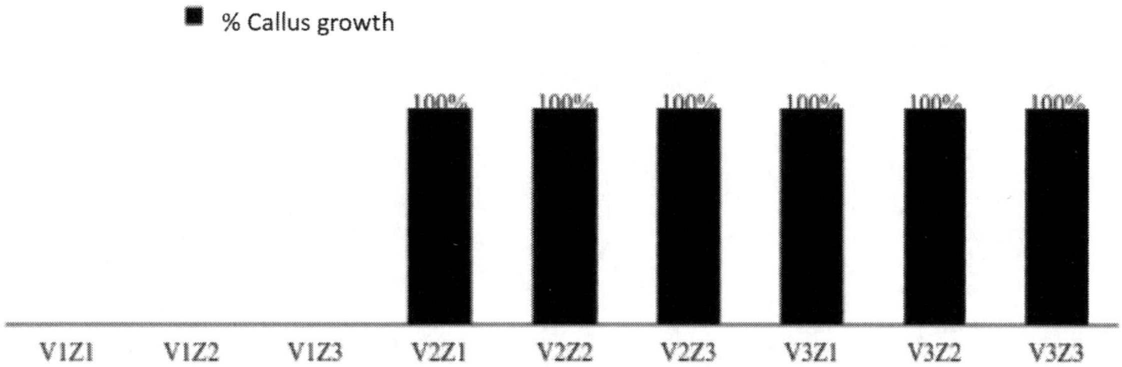


**Figure 1.** Histogram percentage growth of callus with different combinations of plant growth regulator at the age of 4 weeks

Plant growth regulators auxin and cytokinin stimulates cell division and the formation of callus. This is consistent with Khumaida and Handayani [9] reported that young cotyledons explants callus soybean crop began at the age of 1-2 weeks after explant percentage callus culture with 75% in the treatment of 10 mg/l 2,4-D + 10 mg/l NAA and the media 40 mg/l 2,4-D. Another research conducted by Manurung [10] on the culture *in vitro* of Makassar fruit (local fruit in South Sulawesi, Indonesia), giving the BAP cytokinin and auxin 2,4-D with different concentration level has a different response to the growth of fruit seeds makasar explants. The higher the concentration of BAP and 2,4-D, the higher the percentage of callus formation. BAP 1.5 mg/l is the optimum concentration in Makasar fruit seed growth *invitro* for the purpose of propagation.

### *3.2.Shape and Color of Callus*

Figure 2 showed that the cultivar of Gepak Kuning produce callus denser texture than Grobogan cultivar and callus brighter colors. Callus was produced had a type of globular callus growth stage. Generally embryogenic callus can be induced using a plant growth regulator auxin or in combination with cytokines. Increased use of 2,4-D and BAP provide a good influence on the resulting structure. The higher the concentration of 2,4-D and BAP band is used then the friable callus produced.

Callus color tends to be yellowish white and bright yellow callus. Callus color formed indicates the activity of cell division that occurs in the callus. Callus is an collection of cells amorphous that occur from tissue cells that divide continually. This is consistent based on Lizawati [13] stated that the progress indicator on the cultivation of explants *in vitro* in the form callus visual representations that can be known that the callus which formed the cells are still actively dividing or die. Callus formed from an explant usually raises a different color.

Callus culture on every media appearance of callus induction showed almost the same. The callus that formed yellowish white, friable callus texture, there are nodules, and transparent. In relation to the embryogenic callus formed callus is embryogenic callus because of the color of a yellowish white callus and nodules. On visual observation is the embryonic stage known globular embryo stage. The size of callus induced from explants of Grobogan larger than Gepak Kuning cultivars. Yelnititis and Komar [14] stated that the embryogenic callus has friable structure, nodular and creamy. According to [15] embryogenic callus initiation occurs in response to the stress caused by the effect of auxin concentration is relatively high. Callus is a disorganized mass of cells.

IOP Conf. Series: Earth and Environmental Science **122** (2018) 012056 doi:10.1088/1755-1315/122/1/012056
***** Formatting Issue - Best Available Paper/Graphic *****

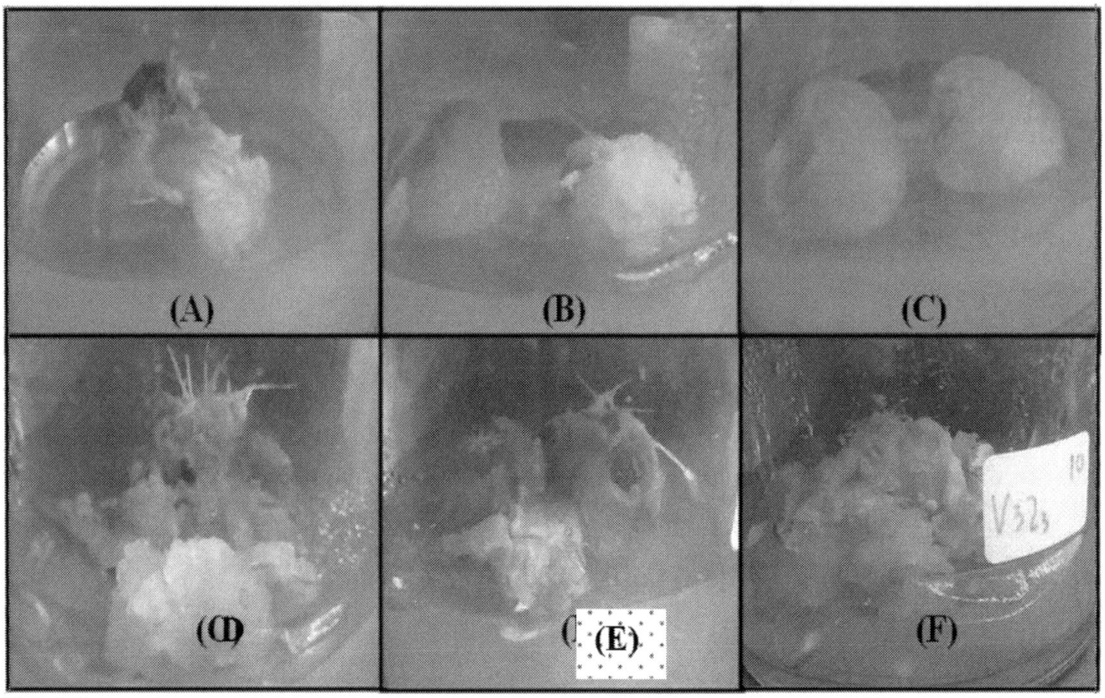


Descriptions :
(A). Treatment $V_2Z_1$ produces yellowish white callus with texture semi friable and types of development globular root
(B). Treatment$V_2Z_2$ produces yellowish white callus with texture semi globular friable and types of development.
(C). Treatment $V_2Z_3$ produce callus cream-colored with texture semi friable and types of development globular + root
(D). treatment $V_3Z_1$ produce callus yellowish bright with texture friable and types of development globular + root
(E). treatment $V_3Z_2$ produce callus yellowish bright with texture friable and types of development globular root+
(F). treatment$V_3Z_3$ produce callus yellowish with texture friable and types of development globular

**Figure 2.** Visualization Callus

### *3.3.weight of callus*

The fresh weight obtained by weighing the callus using an analytical balance. The results obtained are presented in Table 1.

**Table 1.** Fresh weight of callus from Gepak Kuning and Baluran cultivars with different combinations of plant growth regulator (PGR) at age 6 Weeks

| Cultivars | Growing Media | | | Mean |
|---|---|---|---|---|
| | $Z_1$ | $Z_2$ | $Z_3$ | |
| | ..................... | g | .............................. | |
| $V_1$(Baluran) | 0.00 b | 0.00 b | 0, 00 b | 0,00 a |
| $V_2$(Gepak Kuning) | 2,60 a | 4,13 a | 3,10 a | 3,28 b |
| $V_3$(Grobogan) | 0.00 b | 0.00 b | 3.33 a | 1,11 c |
| Mean | 0,87 a | 1,38 ab | 2,14 a | 1 , 46 |

Description: the numbers followed by the same letters in a row and the same column shows the real effect on DMRT at the level of 5%

Based on Table 1, the fresh weight of callus and results of variance showed treatment cultivars, and some combination of PGR and their interaction are both significant different at 6 weeks. The highest callus fresh weight on interactions obtained in combination Gepak Kuning and media 10 mg/l 2,4-D +

1.5 mg/l BAP.

Observations of variable weights callus observations indicate that genotype and media influence on the magnitude of the weight of the callus. Media interaction and the best cultivar found in the Gepak Kuning variety with a combination of growth regulators media 10 mg/l 2,4-D + 1.5 mg/l BAP. Growth regulator auxin and cytokinin plays a role in cell division and elongation so that the activity of both cause callus cell growth continuously. The better the callus growth then the resulting higher weight, greater callus clumps characterized by a growing callus mass. Indrianto [15] which states that the granting of cytokinin and auxin in the culture medium can stimulate cell division and morphogenesis. Based on research Khumaida and Handayani [9], the young cotyledons explants callus were best for soybean explant in the treatment of 10 mg/l 2,4-D + 10 mg/l NAA and also supported by Manurung [8] on the in vitro culture of Makassar fruit stating that BAP 1.5 mg/l is the optimum concentration in makassar fruit seed growth in tissue culture.

*3.4.Protein concentration*

Based on the results from Table 2, showed that the treatment inundation effect which was not significantly different to the increase or decrease in the protein concentration. On the average results of analysis of the protein concentration is known that the Gepak Kuning and Grobogan cultivars on the media 15 mg/l 2,4-D + 2 mg/l BAP to increase the protein concentration while two other treatment decreased protein concentration. Gepak Kuning and Grobogan cultivars showed metabolic response proteins in hypoxic conditions. The response to the waterlogged conditions caused changes to the process leading to the formation of protein. The use of cytokines to increase the rate of protein synthesis so that the faster the cell division process. In tissue cultures of nitrogen are building blocks of nucleic acids, proteins, and other compounds containing N. Dennis *et al.* [3] suggested that in conditions of hypoxia and anoxia tissues synthesize more soluble rice protein. In the second phase (4-24 hours) stages of the process plant responses to an oxygen-deficient condition is metabolic adaptation process. At this stage lasts induction of glycolysis and fermentation genes that are important for sustaining energy excretion. Metabolic response at this stage is more complex than expected because it involves changes in nitrogen metabolism.

Based on the analysis made known that the induced callus Grobogan cultivar of media 15 mg/l 2,4-D + 2 mg/l BAP have an increased metabolic activity after the hypoxic condition is the observation variable protein concentration and enzyme activity peroksidase (POD) and superoxide dismutase enzyme (SOD).

**Table 2.** Mean t-test against the protein concentration of soybean callus Gepak Kuning and Grobogan cultivars in different media combinations PGR

| Treatment | Mean | | t-Value |
|---|---|---|---|
| | Before hypoxic | After hypoxic | |
| $V_2Z_1$ | 1.265 | 1.133 | 0.070 |
| $V_2Z_2$ | 1.217 | 1.205 | 0.759 |
| $V_2Z_3$ | 0.853 | 1.228 | 0.390 |
| $V_3Z_3$ | 0.084 | 0.476 | 0.408 |

## 4. Conclusion

Gepak Kuning and Grobogan cultivars are more responsive to the growth of embryogenic callus by percentage growth of 100%. Generally, the resulting callus creamy texture Interaction friable cultivar and the best growing media for embryogenic callus induction based on this research that Gepak Kuning cultivar on the media 10 mg/l 2,4-D + 1.5 mg/l BAP the fresh weight callus higher (4.13 g). In hypoxic conditions, the protein concentration callus Gepak Kuning and Grobogan cultivars induced by the media 15 mg/l 2,4-D + 2 mg/l BAP increased after hypoxic condition.

**References**

[1] Central Bureau of Statistics 2015 *Production of Food Crops* I Forecast Figures 2014 http://www.bps.go.id Accessed on March 30, 2016

[2] Hapsari R T and Adie M M 2010 Assembly and development opportunities tolerant soybean puddle *J of Agri Research and Develop* **29**(2): 50

[3] Dennis E S, Dolferus R, Ellis M, Rahman M, Wu Y, Hoeren F U, Grover A, Ismond K P, Good A G and Peacock W J 2000 Molecular strategies for improving water-logging tolerance in plants *J.*Exp.Bot. **51**:89-97

[4] Yelnititis and Komar M 2010 The formation of callus from explants leaves crumb ramin *(Gonystylusbancanus* (Miq) Kurtz) *Research Institute of Forest Biotechnology and Plant Breeding* Indonesia

[5] Martin K P 2004 Plant regeneration through somatic embryogenesis in medicinally important *Centella asiatica* L. *In Vitro Cell Dev Biol* **40**:586-591

[6] Vasil I K 1988 Progress in the regulation and genetic manipulation of cereal crops *Biotechnol* **6**:397-402

[7] Yusnita 2003 *Tissue Culture: How to Increase Plant Efficiently* (Jakarta: Agromedia Book)

[8] Zulkarnain H 2009 *Tissue Culture* (Jakarta: Earth Literacy)

[9] Khumaida N and Handayani T 2010 Induction and proliferation of embryogenic callus on some soybean genotypes *J. Agron. Indonesia* **38**(1): 19-24

[10] Manurung L Y S 2007 Effect of auxin (2,4-D) and cytokinin (BAP) In *Invitro* Culture of Fruit Makasar (*Brucea javanica* (L.) Merr.). Agriculture department. Makasar, Indonesia.

[11] Damanik R I, Ismail M R, Shamsuddin Z, Othman S, Zain A M, Maziah M 2012 Response of antioxidant systems in oxygen deprived suspension cultures of rice (*Oryza sativa* L.) *Plant Growth Regul.* **67**:83-92

[12] Murashige T and Skoog F 1962 A revised medium for rapid growth and bioassays with tobacco cultures *Physiol Plant* **15**: 473-497

[13] Lizawati 2012 Induction of embryogenic callus from explants of Jatropha Plant Apical buds *(jatropacurcas* L.) with of 2,4-D and TDZ *Faculty of Agriculture* Edinburgh University

[14] Yunita R 2009 Utilization of somaclonal variation and selection *in vitro* in assembly plant abiotic stress tolerance *J of Agr Research* **28**(4): 143-144

[15] Indrianto A 2003 *Plant Tissue Culture* (Yogyakarta: Faculty of Biology) p 134

IOP Conf. Series: Earth and Environmental Science 122 (2018) 012057 doi:10.1088/1755-1315/122/1/012057

# Growth and yield of patchouli (*Pogostemon cablin*, Benth) due to mulching and method of fertilizer on rain-fed land

**Nasruddin[1,3,4], E M Harahap[2], C Hanum[2] and L A M Siregar[2]**

[1]Student of Doctoral Program of Agricultural Science, Universitas Sumatera Utara, Medan - Indonesia 20155
[2] Lecturer, Faculty of Agriculture, Universitas Sumatera Utara, Medan – Indonesia 20155
[3] Lecturer, Faculty of Agriculture, Universitas Malikussaleh, Aceh – Indonesia 24355
E-mail: Nasruddin.fp@unimal.ac.id

**Abstract.** The drought stress that occurs during growth results in a drastic reduction in growth and yield. This study was aimed to study the effect of mulching and method of fertilizer application in reducing the impact of drought stress on patchouli plants. The experiment was conducted from July to December 2016 using a split plot design into three replications with two treatment factors. The first factor was mulch factor with three levels, i.e. $M_0$ (without mulch), $M_1$ (rice straw mulch) and $M_2$ (silver black plastic mulch). The second factor was the method of fertilizer application consisting of three stages: $C_1$ (once), $C_2$ (twice), $C_3$ (three times). The parameters included plant height, number of branches, number of leaves, root length, wet weight of plant, root canopy ratio, total of chlorophyll, soil temperature and soil moisture content. The results showed the use of straw mulch reduce the impact of drought stress on patchouli plants. Two times fertilizer application gave better growth and yield. The use of straw mulch produced lower temperature degrees and maintained soil moisture content.

## 1. Introduction

Patchouli (*Pogostemon cablin*, Benth) is one of the essential oil-producing plants of the Labiatae family. Living in the subtropics, growing well at altitudes of up to 1,200 m asl in warm and wet climates with rainfall between 1,500 mm to 3,000 mm annually evenly throughout the year, 70-90% humidity, 24-28°C temperatures. Good soil types are regosol, latosol and alluvial. Sandy clay texture or clay dusty, have a good absorption and not inundated during the rainy season with soil pH 5.5-7.0 [1].

Patchouli plants have shallow roots that are less resistant to drought. Such rooting morphological characteristics result in patchouli sensitivity to soil moisture deficits. Generally patchouli are cultivated on dry land with irrigation from rainfall only, and no fertilization with the dose and the right time in accordance with the needs of patchouli plants.

In areas with low rainfall and uneven distribution, such as in North Aceh with rainfall of 1,478 mm/year the need for soil and water conservation through mulching. The use of mulch has an important role in reducing excess water evaporation, thus helping to maintain soil moisture [2], modify the soil temperature [3].

Patchouli plants need a sufficient amount of nutrients and their giving at the right time. Fertilization is one way for patchouli plants to grow fertile, leaves and high production. Provision of fertilizer inappropriate doses of plants is expected to stimulate vegetative growth and improve the process of plant assimilation and oil production.

The use of mulch and fertilizer with the right dose and time is expected to reduce the impact of water shortage on patchouli plants. This study aims to study the effect of mulch and the appropriate way of fertilizing the patchouli plants in an effort to reduce the impact of drought stress in North Aceh.

## 2. Materials and methods

The research was conducted in Reuleut Timu Village Muara Batu District, North Aceh Regency at altitude of ± 8 m above sea level (asl) from July to December 2016. The materials used are Tapaktuan variety seedlings, cow manure, mulch, urea fertilizer, SP-36, KCl and Mg, Furadan 3 G, Sevin 85 SP and Dithane M 45. The instrument used is Traceable ® Digital ground thermometer (range -50°C - 300°C), oven, digital scales, poly bag, sprayer, plastic strap, ruler.

The experiment was conducted by using a split plot design 3 replications. The main plot of mulch (M), consisting of $M_0$ (without mulch), $M_1$ (rice straw mulch) and $M_2$ (silver black plastic mulch), while subplot is a method of fertilizer (C), consisting of $C_1$ (once, all doses at planting time), $C_2$ (twice, ½ dose at plant + ½ dose age 3 months) and $C_3$ (three times (1/3 dose at plant + 1/3 dose of plant age 2 months + 1/3 dose of plant age 4 months). The observational data were analyzed by anova, if there was a significant followed by a further 5% BNT advanced test.

Plant material derived from shoots cuttings grown in polybag that contain a mixture of soil and cow manure. Cuttings kept for three weeks, then planted with a distance of 60 cm x 40 cm. Giving mulch straw done after planting, while silver black plastic mulch given before planting.

Watering plants done until the age of 30 days after planting (dap). Furthermore, water needs of plants only from rainfall only. The amount of rainfall is measured by an observatory measuring device located at the experimental site.

Observations of growth and yield were performed on plant height, number of branches, number of leaves, root length, wet weight of plant, root canopy ratio, chlorophyll count, soil temperature and soil moisture content. Soil temperature observations carried out at 07:30, 13:30 and 17:30 every 15 days at a depth of 0 cm, 5 cm and 10 cm. Measurement of soil water content is done by the tensiometer every 7 days at 07:30 and 17:30.

## 3. Results and discussion

The analysis of variance indicates that single mulching does not show any significant in average plant height, average number of branches, average number of leaves, average root of length, average wet weight of plants, average root canopy ratio and total of chlorophyll, but instead a highly significant on the average soil temperature and the average soil water content. The method of single fertilizer use shows significant to the average of plant height. There was no significant interaction between mulch treatment and fertilizer method on all observed parameters (Table 1).

Although the single did not show significantly, the level of rice straw mulch treatment ($M_1$) showed higher growth and yield compared without mulch ($M_0$) and black silver plastic mulch ($M_2$). As indicated on the larger average plant height parameter of 2.66% of $M_0$ and 9. 21% of $M_2$. the average number of branches greater than 0.18% of $M_0$ and 28.80 of $M_2$, the average number of leaves is 34.38% greater than $M_0$ and 37.69% of $M_2$, average root length greater than 5.35% of $M_0$ and 8.12% of $M_2$, average wet weight of the larger plant 14.45% of $M_0$ and 11.41% of $M_2$, and root canopy ratio 49.62% greater than $M_0$ and 30.95% of $M_2$.

The use of rice straw mulch is able to influence better growth of plant height, number of branches, number of leaves, root length, wet weight of the plant, the ratio of plant root of patchouli compared with without mulch and black silver plastic mulch. It is suspected that rice straw mulch suppresses lower soil temperatures, maintaining higher soil water levels so that nutrient uptake, water, and metabolism processes can occur well (Table 2). Mulch reduces the negative impact of water pressures on plant growth and fruit yield and increases the availability of N [4], reducing the amount of water provided, improving crop yields and plant chlorophyll content [5].

Rice straw mulch that has weathering can increase the nutrients for plants, as expressed [2] that the use of rice straw mulch can increase the organic carbon content and soil nutrient availability, while also improving the soil structure in the long term. Rice straw mulch also has other benefits such as improving the micro climate and soil texture, moisture conservation, soil fertility, weed control, plant pest control [6].

**Table 1**. Summary analysis of variance with 95% confidence in all observation parameters.

| No | Parameters | F critis value | | | | | |
|---|---|---|---|---|---|---|---|
| | | Mulch (M) | | Fertilization method (C) | | Interaction (M x C) | |
| 1. | Plant height | 2.57 | ns | 4.79 | * | 0.16 | Ns |
| 2. | Number of branches | 0.35 | ns | 1.72 | ns | 1.44 | Ns |
| 3. | Number of leaves | 3.82 | ns | 2.46 | ns | 0.49 | Ns |
| 4. | Root of length | 1.36 | ns | 1.27 | ns | 0.69 | Ns |
| 5. | Wet weight of plants | 0.56 | ns | 0.38 | ns | 0.49 | Ns |
| 6. | Root canopy ratio | 1.81 | ns | 2.62 | ns | 0.27 | Ns |
| 7. | Chlorophyll | 0.48 | ns | 3.48 | ns | 1.52 | Ns |
| 8. | Soil temperature | 27.77 | ** | - | - | - | - |
| 9. | Soil water content | 26.07 | ** | - | - | - | - |

ns = non significant, * = significant ** = highly significant

Level treatment without mulch according to observational data have a lower water content than rice straw mulch and mulch black plastic silver, this resulted in an effect on the vegetative growth of the plants. In soil with decreased soil water content (low moisture content) will experience the effects of drought stress so as to decrease plant growth such as plant height [7, 8], number of leaves, number of branches, wet weight of patchouli plant [9], root growth [10]. Sufficient water content is available, resulting in high and increasing number of plant leaves [11].

Unlike the case with underground growth, the roots are less inhibited than the top canopy growth due to lack of water. This is the same that [12] which reveals that some roots continue to extend to a low groundwater potential and inhibit shoot growth.

Mulch applications have not been able to improve the yield and production of patchouli oil than without mulch. The absence of significant to the yield and production of patchouli oil is believed to be the availability of adequate amount of moisture content for growth and production of patchouli oil. On average the difference in the amount of groundwater content is only ± 10% between mulch without mulch.

Treatment level $C_2$ showed higher growth of patchouli plants and significantly from $C_1$, which was indicated by the plant height parameter of 48.66 cm and larger 9.06% of $C_1$ and 1.77% of $C_3$ (Table 2).

The way fertilizer twice affects the increase of plant growth and patchouli yield. It is suspected that the fertilization effect twice gives enough nutrients at the right time. Increased cell division and synthesis of chlorophyll, protein and amino acids related to fertilization with the right dose and time [13].

IOP Conf. Series: Earth and Environmental Science **122** (2018) 012057 doi:10.1088/1755-1315/122/1/012057

Patchouli responds well to optimal fertilizer application at intervals twice that increase the production of fresh leaves [14] increase plant height, plant wet weight and production of essential oil of [15], increase the dry weight of patchouli leaf [16].

Application of fertilizers with appropriate doses and timing affects the growth and yield of patchouli plants [9]. In addition, fertilization affects vegetative growth resulting in increased crop production [13].

Although there was no significant, the amount of chlorophyll of patchouli leaf was higher at $C_3$ level (three times the fertilizer application). This is related to the time of chlorophyll measurements taken several days after fertilizer application where nutrients have been absorbed by plants so that the leaves of plants are greener. Absorption of nutrients from the soil in the root plays an important role in the synthesis of chlorophyll by plants [17]. Plants containing enough nutrients will show the color of dark green leaves which means high chlorophyll content in the leaves.

**Table 2.** Summary of the mean value test for the average parameters of plant height (PH), average soil temperature (ST) and average soil moisture content (SMC) due to mulch treatment and fertilizer application.

| Parameters | Average value | | |
|---|---|---|---|
| | PH | ST | SMC |
| | Mulch (M) | | |
| $M_0$ | 47.54 | 34.73 a | 85.14 c |
| $M_1$ | 48.84 | 31.70 c | 96.46 a |
| $M_2$ | 44.34 | 33.54 b | 92.02 b |
| BNT 0,05 | Ns | ** | ** |
| | Fertilization method (C) | | |
| $C_1$ | 44.25 b | - | - |
| $C_2$ | 48.66 a | - | - |
| $C_3$ | 47.80 a | - | - |
| BNT 0,05 | * | | |

ns = non significant, * = significant ** = highly significant

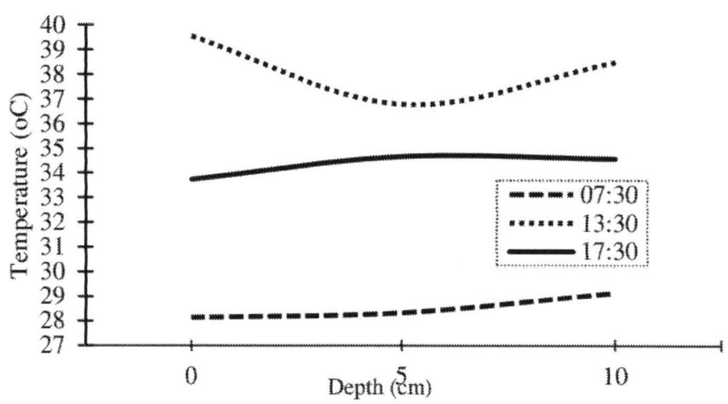


**Figure 1.** Soil temperature profile with depth at various times

IOP Conf. Series: Earth and Environmental Science **122** (2018) 012057 doi:10.1088/1755-1315/122/1/012057

Mulch treatment showed a highly significant against the average daily of soil temperature. The highest average soil temperature was achieved in without mulch ($M_0$) that was 34.73°C followed by $M_2$ 33.54°C and $M_1$ 31.70°C. This suggests that the use of rice straw mulch provides a lower temperature than the use of silver black plastic mulch and without the use of mulch (Table 2).

The average daily soil temperature according to depth indicates the temperature decreases by increase in depth. This decrease occurs both on $M_0$, $M_1$ and $M_2$. Soil temperature profile in the morning, according to the depth indicates that the lower soil temperature at a depth of 0 cm is 28.13°C, increased by 0.19°C at a depth of 5 cm and increased again by 0.82°C at a depth of 10 cm. Profile soil temperature at noon at the depth shows that higher soil temperature at a depth of 0 cm is 39.55°C, decreased by 2.76°C to a depth of 5 cm and decreased again by 1.70°C at a depth of 10 cm. Soil temperature profile in the afternoon, according to the depth shows that, soil temperature is lower at a depth of 0 cm ie 33.73°C, increased by 0.93°C at a depth of 5 cm and decreased again by 0.09°C at a depth of 10 cm (Figure 1).

Straw mulch is able to maintain a lower soil temperature compared to silver black plastic mulch and without mulching. The same is obtained in [18] studies using various types of mulch revealing that plastic mulch provides a higher maximum ground temperature compared to hay mulch. The degree of soil temperature indicated in rice straw mulch was at the patchouli environmental temperature between 24-32°C and did not experience large fluctuations.

The level of rice straw mulch treatment showed highly significant with the level of treatment without mulch and black silver plastic mulch on soil moisture content. The highest average soil moisture content was reached by a rice straw mulch level of 96.46%, followed by silver black plastic mulch of 92.02% and without mulch of 85%. The amount of soil water content in the mulch treatment is influenced by the ability of rice straw mulch in infiltrating more water when it rains and to maintain lower soil temperatures (31.70°C) compared without mulching (34.73°C) and silver black plastic mulch (33.54°C). Higher soil temperatures affect larger evapotranspiration rates in soil and plants (Table 2 and Figure 2).

Plastic and straw mulch significantly reduces evaporation from the soil surface, thus higher groundwater content is always present in the topsoil of the plot without mulch usage. Humidity found in the topsoil has an important role for plants to absorb nutrients to carry on their lives. Research shows that mulch provides many benefits to crop production through soil and water conservation. According to [19] the provision of organic mulch yields higher canopy and oil. Straw mulch and plastic mulch increase plant yield and leaf chlorophyll content. This shows that mulch reduces the negative impact of drought stress on plant growth [20].

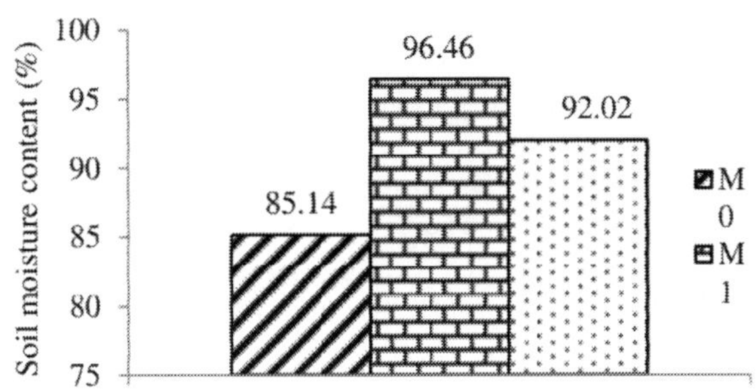


**Figure 2.** The water content of the soil (%) due to the mulch treatment (M)

## 4. Conclusions

The treatment of rice straw mulch has a better effect on the growth of patchouli plants, resulting in lower temperatures and maintaining higher soil moisture content than without mulch and black silver plastic mulch. Fertilization twice gave a significant to the height of patchouli plants, as well as affecting better growth in leaf number parameters, root length and wet weight of patchouli plants.

## References

[1] Ramya H, Palanimuthu V and Rachna S 2013 *Agricultural Engineering International: CIGR Journal.* An Introduction to Patchouli (Pogostemon cablin Benth.)–A Medicinal and Aromatic Plant: It's Importance to Mankind. Vol **15 (2)**:243-50

[2] Ramakrishna A, Tam H M, Wani S P, Long T D. 2006 *Field Crops Research.* Effect of mulch on soil temperature, moisture, weed infestation and yield of groundnut in northern Vietnam. Vol **95 (2)**:115-25

[3] Qin W, Hu C, Oenema O. 2015 *Sci Rep.* Soil mulching significantly enhances yields and water and nitrogen use efficiencies of maize and wheat: a meta-analysis. Vol **5**.

[4] Kirnak H, Kaya C, Higgs D, Tas I. 2003 *Journal of Plant Nutrition.* Responses of Drip Irrigated Bell Pepper to Water Stress and Different Nitrogen Levels with or without Mulch Cover. Vol **26 (2)**:263-77

[5] Kirnak H, Demirtas M N. 2006 *Journal of Plant Nutrition.* Effects of Different Irrigation Regimes and Mulches on Yield and Macronutrition Levels of Drip-Irrigated Cucumber Under Open Field Conditions. Vol **29 (9)**:1675-90

[6] Hu W, Duan S, Sui Q. 1995 *High-yield technology for groundnut*: ICRISAT

[7] Amiri R, Nikbakht A, Rahimmalek M, Hosseini H. 2017 *J Plant Growth Regul.* Variation in the Essential Oil Composition, Antioxidant Capacity, and Physiological Characteristics of Pelargonium graveolens L. Inoculated with Two Species of Mycorrhizal Fungi Under Water Deficit Conditions. Vol:**1-14**

[8] Zhao T-J, Sun S, Liu Y, Liu J-M, Liu Q, Yan Y-B, Zhou H-M. 2006 *Journal of Biological Chemistry.* Regulating the drought-responsive element (DRE)-mediated signaling pathway by synergic functions of trans-active and trans-inactive DRE binding factors in Brassica napus. Vol **281 (16)**:10752-9

[9] Sumathi M, Shashekala S, Shankaraiah N, Ravi Kumar P, Kavitha V. 2012 *Intern J Adv Bio Res.* Effect of nitrogen and VAM levels on herbage and oil yield of Patchouli. Vol **2 (3)**:403-11

[10] Aslam M, Zamir M S I, Afzal I, Yaseen M, Mubeen M, Shoaib A. 2013 *Cercetari Agronomice in Maldova.* Drought stress, its effect on maize production and development of drought tolerance trought potasium aplication. Vol **XLVI (2)**:154

[11] Taweesak V, Lee Abdullah T, Hassan S A, Kamarulzaman N H, Wan Yusoff W A. 2014 *The Scientific World Journal.* Growth and flowering responses of cut chrysanthemum grown under restricted root volume to irrigation frequency. Vol **2014**

[12] Kirnak H, Kaya C, Tas I, Higgs D. 2001 *Bulg J Plant Physiol.* The influence of water deficit on vegetative growth, physiology, fruit yield and quality in eggplants. Vol **27 (3-4)**:34-46

[13] Saha P, Chaudhuri A, Yadav G S, Babu S. 2014 *Annals of Agricultural Research.* Effect of nitrogen levels and planting geometry on growth, herbage and oil yields of patchouli (Pogostemon cablin, Benth.) under humid sub-tropical climate of North Eastern Himalayas. Vol **35 (3)**

[14] Swamy M K, Sinniah U R. 2016 *Industrial Crops and Products.* Patchouli (Pogostemon cablin, Benth.): Botany, agrotechnology and biothechnological aspects. Vol **87**:161-76

[15] Singh M, Rao R S G. 2008 *Industrial Crops and Products*. Influence of sources and doses of N and K on herbage, oil yield and nutrient uptake of patchouli [Pogostemon cablin (Blanco) Benth.] in semi-arid tropics Vol **29 (1)**:229-34
[16] Costa A G, Deschamps C, Cocco L C, Scheer A d P. 2014 *Bioscience Journal*. Vegetative growth, yield and essential oil of patchouli subjected to different levels of nitrogen fertilizer in the planting and maintenance. Vol **30 (2)**:387-92
[17] Rahmatzadeh S, Kazemitabar S K. 2013 *International Journal of Agriculture and Crop Sciences*. Biochemical and antioxidant changes in regenerated periwinkle plantlets due to mycorrhizal colonization during acclimatization. Vol **5 (14)**:1535
[18] El-Shaikh A, Fouda T. 2008 *Misr J Ag Eng*. Effect of different mulching types on soil temperature and cucumber production under Libyan conditions. Vol **25 (1)**:160-75
[19] Singh M, Sharma S, Ramesh S. 2002 *Industrial Crops and Products*. Herbage, oil yield and oil quality of patchouli [Pogostemon cablin (Blanco) Benth.] influenced by irrigation, organic mulch and nitrogen application in semi-arid tropical climate. Vol **16 (2)**:101-7
[20] Kirnak H, Kaya C, Higgs D, Gercek S. 2001 *Australian Journal of Agricultural Research*. A long-term experiment to study the role of mulches in the physiology and macro-nutrition of strawberry grown under water stress. Vol **52 (9)**:937-43

**Acknowledgements**
The author would like to thank the Directorate of Research and Community Service (DRPM) Ministry of Research, Technology and Higher Education, For the grant of research grant for a doctoral dissertation in 2017.

# Decomposition rate of *Rhizopora stylosa* litter in Tanjung Rejo Village, Deli Serdang Regency, North Sumatera Province

**R Rambey[1*], Delvian[1] and S D Sianturi[2]**

[1]Department of Forestry,Faculty of Forestry, Universitas Sumatera Utara, Jl. Tri Dharma Ujung No.1 Medan, North Sumatera 20155, Indonesia
[2]Student Faculty of Forestry Universitas Sumatera Utara
*Email:ridahatirambey@yahoo.com

**Abstract.** Research on the decomposition rate of *Rhizopora stylosa* litter in Tanjung Rejo village, Deli Serdang Regency, North Sumatera Province was conducted from September 2016 to May 2017. The objectives of this research were (1) to measure the decomposition rate of *Rhizophora stylosa* litter and (2) to determine the type of functional fungi in decomposition of litter. *R. stylosa* litter decomposition is characterized by a reduction in litter weight per observation period. Decomposition rate tended to increase every week, which was from 0.238 in the seventh day and reached 0.302 on the fiftysixthth day. The decomposition rate of *R. stylosa* litter of leaf was high with the value of k per day > 0,01 caused by macrobentos and fungi, and also the decomposition of *R. stylosa* litter conducted in the pond area which is classified far from the coast.Therefore, to enable the high population of fungi which affect the decomposition rate of the litter. The types of fungi decomposers were: *Aspergillus* sp.-1, *Aspergillus* sp.-2, *Aspergillus* sp.-3, *Rhizophus* sp.-1., *Rhizophus* sp.-2, *Penicillium* sp., *Syncephalastrum* sp. and *Fusarium* sp.

## 1. Introduction

Litter production includes the vegetative and reproductive litterfall of *R. stylosa*/mangove trees caused by aging, death, and damage of the whole plant by climate (rain and wind) [1]. The mangove litter produced in the form of leaf plays the most important litter role compared to other organs as it is a source of nutrition for the organism. The higher litter production also causes the higher productivity of mangove forest. The difference of the total litter of each organ for each mangove varies due to be the environmental conditions and biological characteristics. Biological characteristics include the small size of leaves and round fruit. Other litter components, due to the shape and size of the leaves are wide and thin so it is easily lost by wind and rain [2].

Decomposition is the process of destruction or decay of organic matter by biological agents and physical changes into mineral materials and organic colloidal humus. Therefore, the decomposition of organic matter is also often called the process of mineralization. This process is a microbial process (decomposer) in obtaining energy for the regeneration process. The factors that affect the decomposition of organic materials from the side of decomposer are temperature, humidity, salinity, and pH. This process has a very essential role in the energy and food chain cycles in the mangove ecosystem [3]. This decomposition process is very important because it converts non-digestible mangove leaf fibers into more easily digestible fibers. Rotting mangove litter will bedegadated into smaller pieces and digested by crabs and other invertebrates. These pieces are known as POM (particular Organic Matter). Once digested, organic particles are formed smaller and then used by food filtering organism (filter feeder) [4].

The objectives of this study were (1) to measure the decomposition rate of leaf litter of *R. stylosa* and (2) to know the type of fungi that play a role in litter decomposition.

## 2. Materials and Methods

### *2.1.Time and Place*

The research was conducted from September 2016 to May 2017. The sampling of *R. stylosa* leaf litter was conducted in Tanjung Rejo village, Percut Sei Tuan subdistrict, Deli Serdang district, North Sumatera. While the isolation of litter decomposer fungi was done in Soil Biology Laboratory, Faculty of Agiculture, Universitas Sumatera Utara.

### *2.2. Experiment of Litter Decomposition Rate*

The litter was weighed as much as 50 g and then put into a litter bag and tied up. The bags were then placed on the surface of the mangove substrate randomly and tied to the root or base of the mangove plant stems so as not to drift away at high tide. Each week for two months experiment, one sample was taken randomly. Then each bag was purged from mud using distilled water or clean water and dried with 75°C drying oven for 1 × 24 hours then weighed using an analytic scale. The same activity was done for next week's litter sampling.

### *2.3. Data analysis of litter decomposition*

The estimation of litter decomposition rate was performed according to the following Olson equation:

$$Xt / Xo = e^{-kt}$$

Where:

Xt = Weight of litter after the observationperiod
X0 = Initial weight of litter
e = Natural logarithm number (2.72)
t = Observation period
k = Decomposition rate

### *2.4. Isolation of litter decomposer fungi*

The tools used were firstly sterilized using an oven with a temperature of ± 80 °C for 3 hours. Media used for culture of fungi is Potato Dextrose Agar (PDA) medium and added with Chloramphenicol. The medium was heated and sterilized using an autoclave.

Leaf litter taken was first cleaned then cut aseptically with a knife into pieces measuring approximately 1cm × 1cm and placed directly on the surface of PDA in a petri dish. Then incubated in the incubator. After incubated for 2 -7 days at the appropriate temperature (28 °C), fungi colonies that gow apart or gow singly were observed and are immediately transferred aseptically to another petri dish with a PDA medium [5].

### *2.5. Identification of litter decomposer fungi*

a. Macroscopic identification
   Each type of fungi obtained, cultured singly on PDA media and identified microscopically by observing the color of spores, top surface, bottom surface and colony diameter..
b. Macroscopic identification

The identification was performed by observation of hyphae, conidia, spore shape and spore color performed under the light microscope. From macroscopic and microscopic observation. The results were identified and matched using fungi identification books [5].

*2.6. Calculation of fungi population*

A total of 10 g of litter sample of *R.stylosa* has gounded with mortar and pestle aseptically. Then the smoothed litter was suspended with 90 ml destilled water and diluted into $10^{-1}$ to $10^{-3}$. One milliliter of each dilution was inserted into a petri dish containing PDA medium with the spread plate method then incubated for 5-8 days. The number of growing fungi was counted by the formula as follows: *Total number of fungi = Number of colonies appeared x dilution factor* [6].

## 3. Results and discussion

Decomposition of *R. stylosa* litter was characterized by a reduction in litter weight per observation period. The weight of residual litter at each observation time is shown in Table 1.

**Table 1**. Decomposition rate of *Rhizophora stylosa* litter

| Day observation | Initial weight (g) | Residual litter (g) | Decomposition litter (g/day) |
|---|---|---|---|
| 7 | 50 | 10.564 | 0.029 |
| 14 | 50 | 10.021 | 0.061 |
| 21 | 50 | 9.671 | 0.094 |
| 28 | 50 | 8.887 | 0.132 |
| 35 | 50 | 8.098 | 0.174 |
| 42 | 50 | 8.20 | 0.208 |
| 49 | 50 | 7.545 | 0.253 |
| 56 | 50 | 6.962 | 0.302 |

Table 1 showed that the decomposition of *R. stylosa* is relatively fast. It refers to Graca *et al.* [7] stating that the decomposition rate is slow if the value of k is <0.005 per day (<1.8 per year), the rate is moderate if k = 0.005 - 0.01 per day (1.8 - 3.6 per year) and the rapid rate if k> 0.01 per day > 3.6 per year). It is thought to be influenced by a low salinity level of 0-10 ppt. In high salinity little microorganisms and livestock decomposers macrobentos can survive. Kurniawan [8] stated that the number of fungi types at the salinity level of 0-10 ppt and 10-20 ppt geater than the salinity level of 20-30 ppt and> 30 ppt further explained Damanik [9] that one of the responses of microorganisms to salinity is not Can tolerate and will die under high salinity conditions. From Table 2 it is found that the residual litter tends to decrease, whereas the residual litter on the 7th day is 10.564 g and the residual litter at 56th day is 6.962 g, while the decomposition rate tends to increase from 0.029 g / day on 7th day and reached 0.302 g/day on day 56. The above shows that the litter decomposition rate can be said to be inversely proportional to the remaining litter at each observation time, can be seen in (figure 1). The more residual litter left behind, meaning slow decomposition of litter. Conversely, the less residual litter left behind, the decomposed litter can be said to be short-lived. The figure below shows the weight and decomposition rate of *R. stylosa* leaf litter at each observation time (figure 1.)

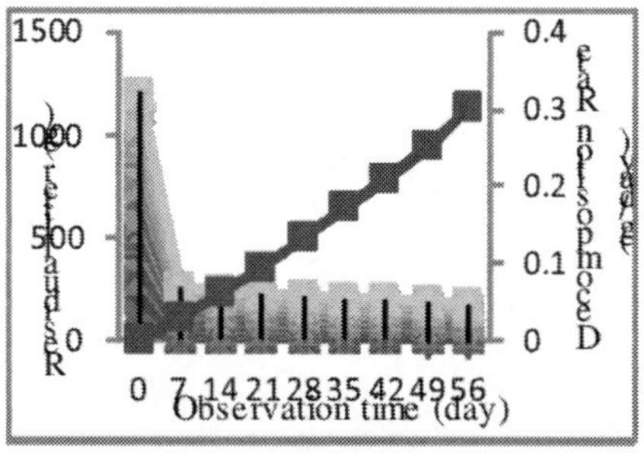


**Fig. 1.** Reduction and Decomposition *Rhizophora stylosa* litter

From figure 1 it can be seen that the biggest weight reduction occurred at first observation on the 7th day from 50 g to 10.564 g. The remainder of the second observation litter on the 14th day was 10.021 g. The difference in cistern litter was on the first observation on the seventh day and the second observation on the 14th day and then very different. The decrease in residual litter at the beginning of the observation is much geater than subsequent observations. According to Leo [10] decomposition of leaf litter in each week is different, where the weight loss of the litter will initially be high. This is possible because in the new litter there is still a lot of supply of elements that are food for organism decomposers, so litter quickly destroyed. Arief (2003) in Kurniawan [8] said that the nutrients contained by mangove leaves are carbon, nitrogen, phosphorus, potassium, calcium and magnesium.

The element is reduced until only the element that is not required by the decomposer. One such decomposer, decomposer is a worm. On the 14th day and the 28th day found worms in the litter bag (Appendix 1). The worms affect the increase in decomposition rate. Kurniawan [8] stated that macrobentos is one of the earliest decomposers that chopped the remains of leaves which were then released back as dirt after which was continued by bacteria and fungi to decompose organic matter. Yahya (2014) says that the most number of bacteria are found in environments with salinity levels of 10-20 ppt. Bacteria will be more easily isolated from the soil with a certain depth. Nurrochman [11] stated that the most populations of bacteria found in mangove forest land with a depth of 20 cm, while the fungi can be isolated from leaf litter. Therefore, in this research the isolated decomposer microorganisms are fungi.

### *3.1.Fungi as Decomposer*

***Types of fungal decomposer.*** The results of fungal decomposer isolation from *Rhizophora stylosa* leaf litter obtained 8 types of fungal isolates. Macroscopic observation could be seen through the gowth of litter fungi decomposers colonies, while microscopic observation could be done by using a microscope. These types of fungi are as follows:

**Isolate 1.** Macroscopically, small round shape colonies like the original white columns, then turn yellowish geen, then within 4 days turn brown to black on days 5-7. Microscopically, conidia are round with tapered bulges. This is consistent with Gandjar's [5] statement which stated that *Aspergillus* sp. is characterizedby a black conidial head, round in shape, and tends to break into columns in old colonies.Conidia are round to semi-round in shape, brown, has ornamentation in the form of bulge and uneven spines.Based on the explanation, isolate 1 is classified as *Aspergillus sp*-1. The figures below are macroscopic and microscopic shapes of *Aspergillus sp.*

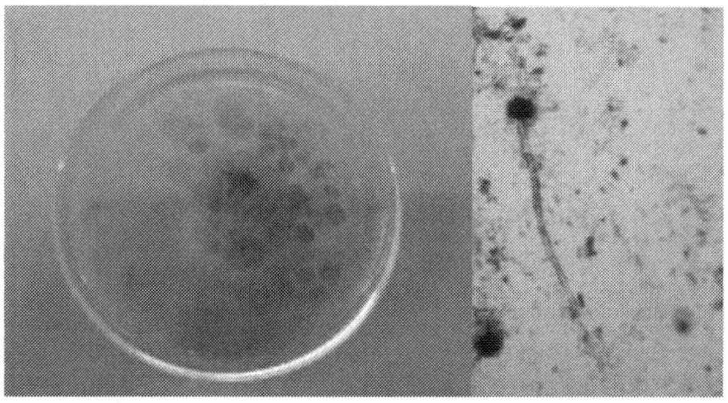

(a) (b)

**Figure 2.** Macroscopic (a) and microscopic (b) shapes of *Aspergillus* sp.-1 isolated from *Rhizophora stylosa* leaf litter.

**Isolate 2.** If it is observed macroscopically, the first colonies are yellowish geen with the white base become black within 7-10 days, The colony fills the cup, the diameter of the colony can reach 5 cm in 7 days. Microscopically, it is characterized by a smooth conidiophore. This is in line with Gandjar's [5] statement that *Aspergillus* sp. colony is characterized by a colony that reaches 4-5 cm in diameter in 7 days and consists of a compact basal layer of white to yellow and conidiophore layer that is dark brown to black. Conidiophore walled smooth, hyaline color, but can also brown. Based on the explanation, it is concluded that isolate 2 is *Aspergillus* sp.-2. It is in details can be seen in Figure below which is the macroscopic and microscopic shapes of *Aspergillus* sp.-2.

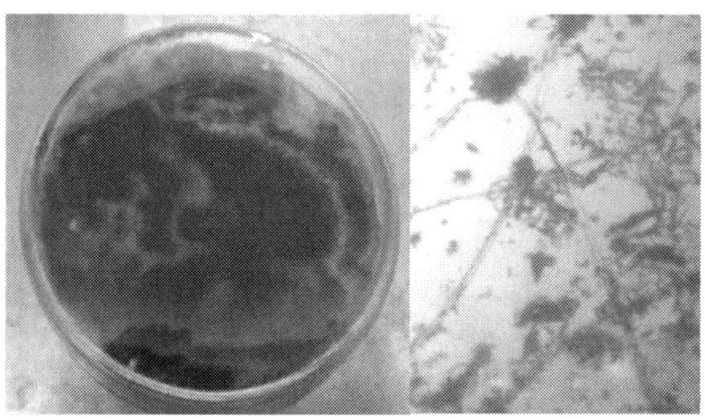

(a) (b)

**Figure 3.** Macroscopic (a) and microscopic (b) shapes of *Aspergillus* sp.-2 isolated from *Rhizophora stylosa* leaf litter.

**Isolate 3**. Macroscopically, isolate 3 is characterized by brown and dispersed colonies. Microscopically, conidia tend to be elliptical. Referring to Gandjar [5] stated that *Aspergillus* sp. colonies consist of a solid layer formed by darker brownish conidiophores that darken with increasing age of colonies. Based on the explanation, this isolate is a fungi type *Aspergillus* sp.-3. It can be seen in the figures below which are macroscopic and microscopic shapes of *Aspergillus* sp.-3.

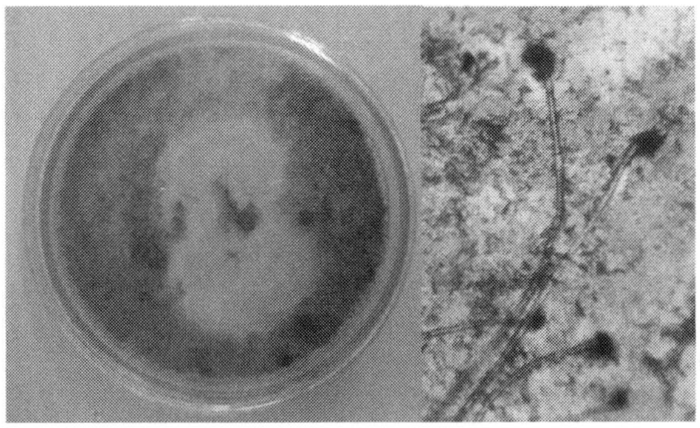

(a) (b)

**Figure 4.** Macroscopic (a) and microscopic (b) shapes of *Aspergillus* sp.3 isolated from *Rhizophora stylosa* leaf litter.

**Isolate 4.** Macroscopically, this isolate waslooked like white cotton and brownish gay to black in the middle with a quite slow gowth. The diameter is 1-2 cm within 3-5 days. Microscopically, the isolates have sporangiophore and at the bottom are shaped like roots, referring to Dwidjoseputro [12] which stated that the mycelium of *Rhizophus* sp. looks like a goup of cotton, over the time the colony became blackish due to a large number of sporangium and spores. The mycelium of *Rhizophus* is divided into stolons which produce tools like roots (Rhizoid) and sporangiophore.Based on these explanations, then this isolate is *Rhizopus* sp.-1.The figure below is macroscopic and microscopic shapes of *Rhizopus* sp.-1.

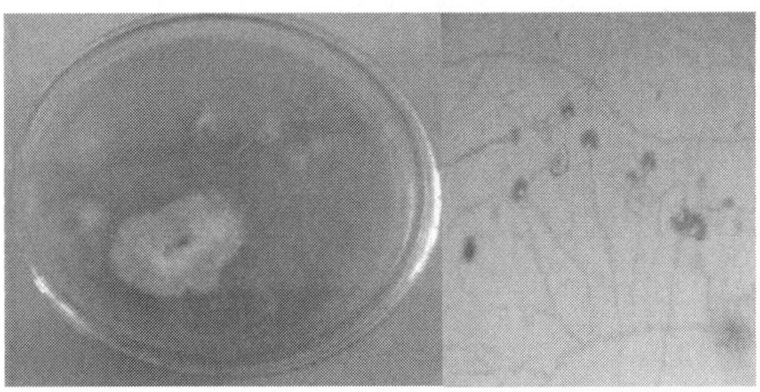

(a) (b)

**Figure 5.** Macroscopic (a) and microscopic (b) shapes of *Rhizopus*sp.-1 isolated from *Rhizophora stylosa* leaf litter.

**Isolate 5.** The macroscopic observation shows that the isolated colony is white and diffuse with slower gowth. While microscopically, this isolate has sporangia with branched sporangiophore which divided into rhizoids. This is consistent with the Gandjar's statement [5] that rhizoid of *Rhizopus* sp. has a brownish color, bifurcated in opposite direction with its sporangiophore which is single or gouped up to 5 and sometimes forming structures such as fork branching. Thus, it can be concluded that isolate 5 is *Rhizopus* sp.-2. The figures below are macroscopic and microscopic shapes of *Rhizopus*sp.-1.

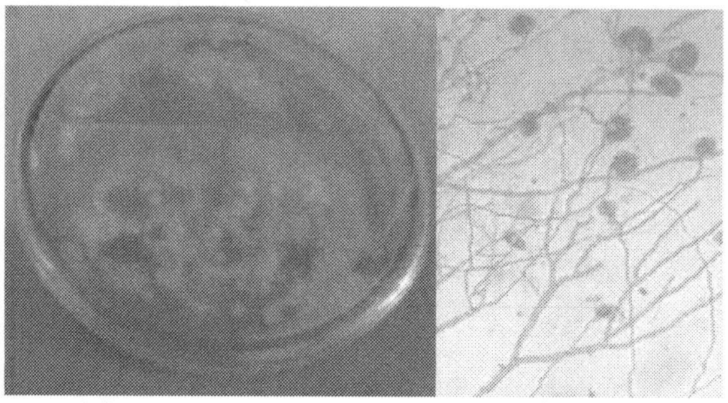

(a) (b)

**Figure 6.** Macroscopic (a) and microscopic (b) shapes of *Rhizopus* sp.-2 isolated from *Rhizophora stylosa* leaf litter.

**Isolate 6.** Macroscopically, this colony is brownish geen and velvety. While microscopically, the conidia are shaped like an elongated radiate chain. Referring to the book of Gandjar [5] which says that *Penicilliumsp*. colony has a surface like velvet and sometimes like cotton and yellow to brownish geen. Meanwhile, according to Kurnia (2011) who conidiophore of *Penicillium* sp. has a length of 5-50μm. Conidia radiate with 80μm chain length. So according to the literature, it can be concluded that isolate 6 is a fungi type of *Penicillium* sp. In more details, it can be seen in figures below which are macroscopic and microscopic shapes of *Penicillium* sp.

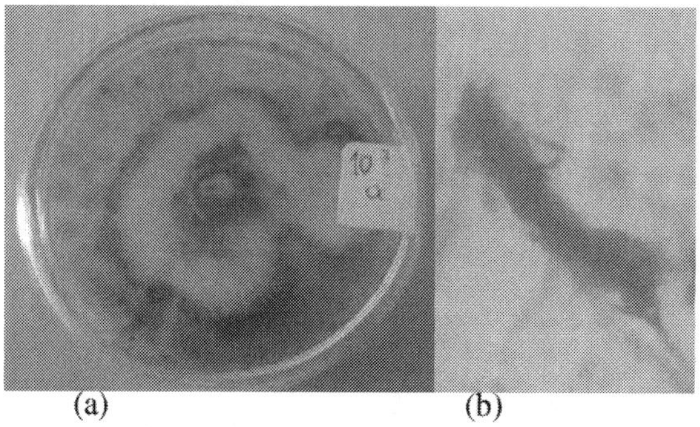

(a) (b)

**Figure 7.** Macroscopic (a) and microscopic (b) shapes of *Rhizopus* sp. isolated from *Rhizophora stylosa* leaf litter.

**Isolate 7.** Macroscopically, the colony initially was a white colony then turn to gay and furtherly to brown. Microscopically, the sporangia head is in the form of a round with such a rhizoid place to gow and branch. This is in line with the statement of Gilman (1957) which stated that the colony of *Syncephalastrumsp* is white at the first, then turn to gay. Moreover, its Konidiopor is strong, initially not branched, then form a laterally curved branch and connected to the short rhizoid that becomes the place to gow. Based on the explanation, the isolate 7 is *Syncephalastrum* sp. The figures below show the macroscopic and microscopic shapes of *Penicillium* sp.

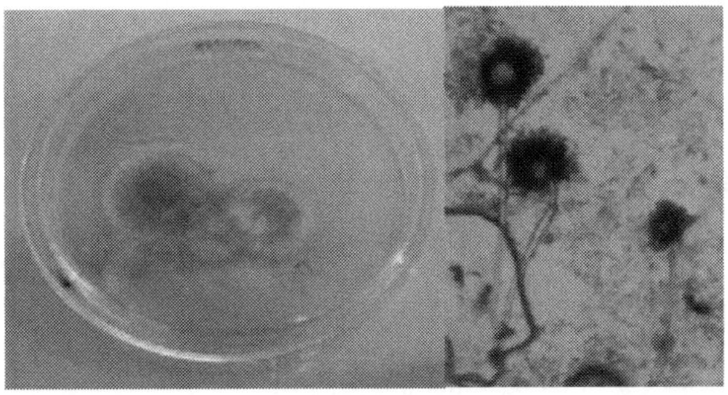

(a) (b)

**Figure 8.** Macroscopic (a) and microscopic (b) shapes of *Syncephalastrum* sp. isolated from *Rhizophora stylosa* leaf litter.

**Isolate 8.** Macroscopically, colonies are white like cotton and spread. Microscopically, conidia are macro-lengthwise. Referring to the Gandjar's literature [5] which stated that *Fusarium* sp. has mycelia like cotton, then becomes velvety. Macroconidia shaped almost like a sickle, slightly straight, to slightly bent. Based on these explanations, isolate 8 can be classified as *Fusarium* sp. The figures below are macroscopic and microscopic shapes of *Fusarium* sp.

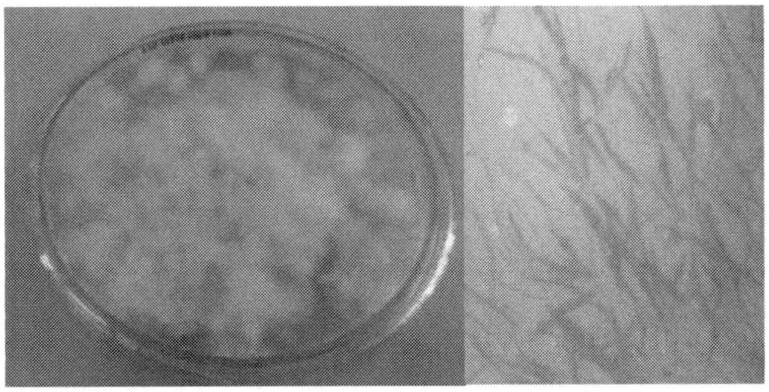

(a) (b)

**Figure 9.** Macroscopic (a) and microscopic (b) shapes of *Fusarium* sp. isolated from *Rhizophora stylosa* leaf litter.

Based on the observation on the isolates, there were 8 types of decomposer fungi. The types of fungi were from the genus of *Aspergillus, Rhizopus, Penicillium, Syncephalastrum, and Fusarium.* Compared with the study [14] by Ratna's [14] on the fungal diversity on the decomposition processes of *Ceriopstagal* litter at various salinity levels in Kampung Nipah Sei Nagalawan North Sumaterathey found 5 genus of decomposer fungi from *Ceriopstagal* leaves including *Aspergillus* sp.1, *Aspergillus* sp.2, *Aspergillus* sp.3, *Aspergillus* sp.4, *Aspergillus* sp.5, *Aspergillus* sp.6, *Aspergillus* sp.7, *Aspergillus* sp.8, *Aspergillus* sp.9, *Syncephalastrum*sp.1, *Syncephalastrum*sp.2, *Syncephalastrum*sp.3, *Syncephalastrum*sp.4, *Tricodherma*sp. *and Penicillium* sp.

In *R. stylosa* litter, the decomposer of the genus *Aspergillus* is the largest of the three types. The result of the observations showed that *Aspergillus* is the easiest type of fungi to gow on the petri dish. This is in line with Gandjar's [5] statement that *Aspergillus* cosmopolites in the tropics and subtropics, and easily isolated from soil, air, water, and leaf litter. Ratna [14] also reported that nine types of *Aspergillus* sp. played important role in litter decomposition process of Ceriopstagal mangoves (family *Rhizoporaceae*).

The second largest type of fungus is *Rhizopus* sp. Soeroyo [13] stated that the types of *Rhizopus* from Sonneratia let a slightly salt enter their root system through the roots of which the very small entering salts are excreted and the excess salt may be associated with leaf storage.

The type of *Penicillium* obtained was one type. Gandjar [5] stated that the species is cosmopolitan, and common in the tropics area. This species is easily isolated from air, cereals, spices, litter, vegetables, pulp, and paper. The next type of fungus is *Syncephalastrum* sp. Ratna [14] reported that at the salinity level of 21-30 ppt, the highest colony is *Syncephalastrumsp*that is 66.6%. This shows that this type of fungus is able to compete with other cosmopolite fungi types in litter decomposition. The last type is *Fusarium* sp. Gandjar [5] stated that *Fusarium* sp is generally found only in parts of plants that have died.

The types of fungi above are microorganisms that help the fertility of the gowing place and gowth of *R. stylosa* as well as provide natural nutrients for biota in the ecosystem.

## 4.Conclusions

Decomposition rate of *R. stylosa* litter on salinity level 0-10 ppt is classified as fast decomposition with decomposition value>0,01 g/day. Eight types of fungi were successfully identified to play important role in the decomposition process in leaf litter *R. stylosa* at salinity level 0-10 ppt. The fungi were *Aspergillus* sp.-1, *Aspergillus* sp.-2, *Aspergillus* sp.-3, *Rhizopus* sp.-1, *Rhizopus* sp.-2, *Penicillium* sp., *Syncephalastrum* sp. and *Fusarium* sp.

## References

[1] Fitriani F 2016 Production and decomposition of mangove leaf litter *Rhizophora stylosa* in island village nine sub districts milk districts Langkat North Sumatra Aquatic Resource Study Program Faculty of Agiculture University of Northern Sumatra

[2] Eka W D 2016 Leaf decomposition rate at Kelong Island mangove ecosystem,Bintan Regency. *Journal of Oceanogaphy and Limnology in Indonesia* **1(1)** 1-10

[3] Andrianto F, Bintoro A, and Budi S Y 2015 Production and decomposition rate of mangrove litter in Durian Village and Batu Menyan Village Padang Cermin District Pesawaran Subdistrict *Jurnal Lestari* **3(4)** 9-20

[4] Silitonga E L 2009 The types of fungi present in *Rhizophora mucronata* leaf litter are decomposed at Sylva various levels of salinity [Thesis]. Program Study of Forestry, Agiculture Faculty, University of North Sumatra

[5] Gandjar I, Samson R A,van den Tweel Vermeulen K,Oetari A, and Santoso I 1999 Introduction of common tropical molds Yayasan Obor Indonesia Jakarta.

[6] Deni and Delvian 2011 Guide of Soil Biology Practicum. Progam Study of Forestry, Faculty of Agiculture, University of North Sumatra.

[7] Graca M A S, Barlocher F, Gessner M. O, 2005 Methods to Study Litter Decomposition a Practical Guide Springer New York

[8] Kurniawan W P 2011 Lumaceous Decomposition Rate of *R. mucronata* After Application of Fungi *Penicillium* sp., *Aspergillus* sp., and *Curvularia* sp. at Different Levels of Salinity. [Thesis]. Forestry Program Study of Forestry, Faculty of Agiculture, University of North Sumatra

[9] Damanik A Y 2010 The types of fungi that are associated with the decomposition process of *Avicenia marina* leaf (Forsk) vierh after application of *Aspergillus* sp fungi, *Curvullaria* sp, *Penicillium* sp, at various levels of salinity in Sicanang Belawan village USU Medan

IOP Conf. Series: Earth and Environmental Science **122** (2018) 012058 doi:10.1088/1755-1315/122/1/012058

[10] Leo F D 2017 Estimation of Leaf Decomposition Rate Leaf Shorea Balangeran (Korth.) Burck Dan Hopea Bancana (Boerl.) Van Slooten in Dramaga Research Forest, Bogor, West Java [Thesis] Department of Silviculture, Faculty of Forestry, Bogor Agicultural Institution
[11] Nurrochman F 2015 Exploration of Selulotitic Bacteria from Mangrove Forest Land Baros Kretek Bantul Yogyakarta. Program Study of Biology Education, Faculty of Teacher Training and Education, Muhammadiyah University of Surakarta
[12] Dwidjoseputro 1978 Basic Microbiology Djambatan Jakarta
[13] Soeroyo 1992 Nature, Function and Role of Mangove Forest. Center for Oceanogaphic Research and Development-LIPI. Jakarta
[14] Ratna, S. 2016. Diversity of Fungi Types In Licensed *Ceriops tagal* Decomposition Processes at Various Salinity Levels in Nipah Sei Nagalawan Village of North Sumatra. Faculty of Forestry, University of North Sumatra.

# Deforestation trend in North Sumatra over 1990-2015

**M Basyuni[1*], N Sulistiyono[1], R Wati[1] and R Hayati[1]**

[1] Department of Forestry, Faculty of Forestry, Universitas Sumatera Utara, Jl. Tri Dharma Ujung No. 1 Medan, North Sumatera 20155, Indonesia
*Email: m.basyuni@usu.ac.id

**Abstract.** Deforestation and forest degradation have been previously reported to contributing greenhouse gas emission, the primary driver of global warming. The present paper studies deforestation and reforestation trend in North Sumatra, Indonesia using land-use/land-cover change from 1990-2015. The land-use consists of three classes derived from forest land (primary and secondary dry land forest, primary and secondary swamp forest, primary and secondary mangrove forest). Non-Forest (shrub, oil palm plantation, forest plantation, settlement, barren land, swamp shrub, dry land farming, mixed dry land farming, paddy field, aquaculture, airport, transmigration, and mining), and water body (water and swamp). Results showed that from 33 regencies/city in North Sumatra, among them, 25 districts deforested, which was the highest deforestation rate in Labuhanbatu and South Labuhanbatu (2,238.08 and 1,652.55 ha/year, respectively), only one area reforested, and seven districts showed no deforestation or reforestation. During 25 years observed, the forest has been deforested 22.92%, while nonforest has been increased 11.33% of land-use. The significant increasing loss of North Sumatran forest implies conservation efforts and developing sustainable forest management.

## 1. Introduction

Indonesian tropical rain forest contributes the highest annual deforestation among the tropical countries [1]. Deforestation denotes as the permanent conversion of forest to another land-use while degradation shows that forests remain forests but lose their ability to support ecosystem services [2]. The deforestation led to increased green house gas, biodiversity loss, and net biomass carbon loss [1-4]. The deforestation accounts about 90% net emission including shifting cultivation and degradation for about 10%, mostly derived from logging (with regrowth) for industrial and fuel wood [2]. Conversion forest to cropland, pasture, the draining and burning of peat land, have become the forces driving of deforestation in the tropic [2].

Based on Landsat data Indonesia has been reported to increase in forest loss (1021 $km^2.year^{-1}$), with a little of under 10,000 $km^2.year^{-1}$ from 2000-2003 and a high of over 20,000 $km^2.year^{-1}$ in 2011 to 2012 [5]. Peat swamp forests converted to oil palm plantation resulted to biodiversity reduces of 1% in Borneo (4 species), 3.4% in Sumatra (16 species), and 12.1% in Peninsula Malaysia (46 species) [4]. Forest conservations as well as sustainable forest management, therefore, are needed to maintain the presence of primary and secondary dry land, mangrove, and peat land forest and to increase reforestation programs.

Deforestation was globally or regionally well-documented [1,2,4,5], nonetheless, deforestation and reforestation as well from land-use and land-cover changes have not been previously reported primarily at regency level in Indonesia. To get more insight into a better understanding of spatial and temporal variation in the land uses that in displacing forests, the present study aimed to describe the drivers of forest loss and reforested from land-use and land-cover changes from 1990, 2000, 2009 and 2015 in 23 regencies of North Sumatran forest, Indonesia.

## 2. Materials and methods

### *2.1. Study area and dataset*

The study was carried out in North Sumatran forest, Indonesia, where the land-use and land cover consist of three classes derived from forest land (primary and secondary dry land forest, primary and secondary swamp forest, primary and secondary mangrove forest). Non-forest (shrub, oil palm plantation, forest plantation, settlement, barren land, swamp bush, dry land farming, mixed dry land farming, paddy field, aquaculture, airport, transmigration, and mining), and water body (water and swamp). The land-use distributed in 23 Regencies/cities of North Sumatra province, namely Asahan, Batubara, Dairi, Deli Serdang, Gunung Sitoli, Humbang Hasundutan, Karo, Binjai city, Medan city, Tanjungbalai city, Labuhanbatu, South Labuhanbatu, North Labuhanbatu. Next are Langkat, Mandailing Natal, Nias, West Nias, South Nias, North Nias, Padang Lawas, North Padang Lawas, Padangsidimpuan, West Pakpak, Pematangsiantar, Samosir, Serdang Bedagai, Sibolga, Simalungun, South Tapanuli, Central Tapanuli, North Tapanuli, Tebingtinggi and Toba Samosir (Figure 1). Land-use and land-cover changes data period 1990-2015 was obtained from Ministry of Forestry, Government of Indonesia. Landsat 7 Enhanced Thematic Mapper Plus (ETM+) satellite image was acquired from USGS (http://govis.usgs.gov/).

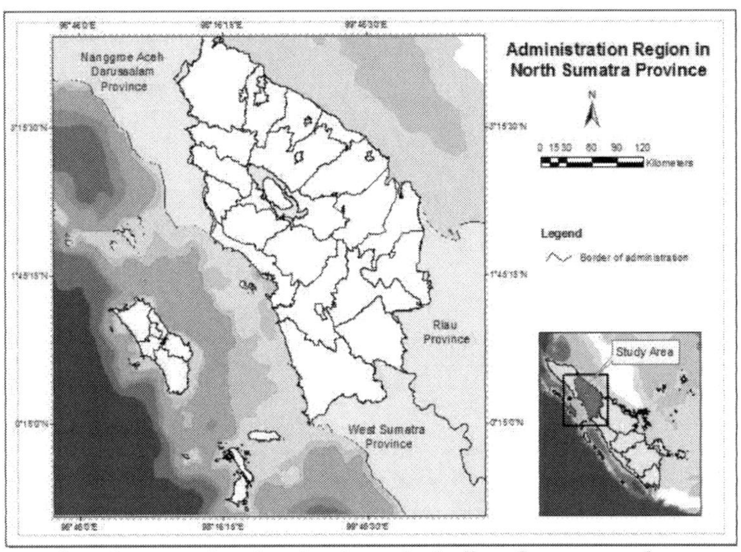


**Figure 1**. Location of study area showing land-use including forest, non-forest, and water body in North Sumatra

### *2.2. Analysis of land-use and land-cover changes*

Analysis of the Landsat images was carried out by applying supervised classification with maximum likelihood as previously reported [3,6]. Image pre-processing, the process of image interpretation, image classification and change detections were done by ArcGIS 9.3.1 and ArcView 3.3as previously reported [3,6]. The information was used as guidance for image geometric correction and image rectification.

### *2.3. Analysis of deforestation rate of North Sumatran mangrove*

Three classes of land-use/land-cover changes namely forest land, nonforest land, and water body existed in 33 regencies/cities in North Sumatra as previously described [3]. The term of plantation/regrowth land-use class, such as plantation forest was not considerationas a forest land cover class as previously reported [6-7]. Deforestation rate was measured from total forest loss (primary and secondary dry land, mangrove, and swamp) from the year 1990 to the year 2015. From this measurement, annual deforestation, deforestation proportion and rate, and class of deforestation were determined.

## 3. Results and discussion

### *3.1. Land-use/land-cover changes between 1990 and 2015*

Figure 2 shows the deforestation trend in North Sumatra over 1990-2015. Forest land occupied 2,322,770.86 ha in 1990 but/and in 2015 has been lost 1,790,425.30 ha. On the other hand, forest cover was 32.79% in 1990 to be 25.27 % in 2015 (Table1). In contrast to this observation, Non-forest land increased significantly from 4,716,416.42 ha in 1990 to 5,250,676.66 ha in 2015 covered 74.11% of land-use in North Sumatra.

Deforestation remains huge in North Sumatra with more than 595,220.57 ha primary and secondary forest (dry land, mangrove, and swamp) lost between 1990 and 2015 (Figure 2, Table 2). Furthermore, reforestation occurred 34,347.06 ha, less than 10% of deforestation area during 25 years observed. The nonforest land-use tend to increase with time, no change of water body land-use, however, forest land-use seems to decrease during 1990-2015 in North Sumatra (Figure 2). Water body might play an important role in the reforestation activity in North Sumatra, conversion from water body to be secondary mangrove forest [3].

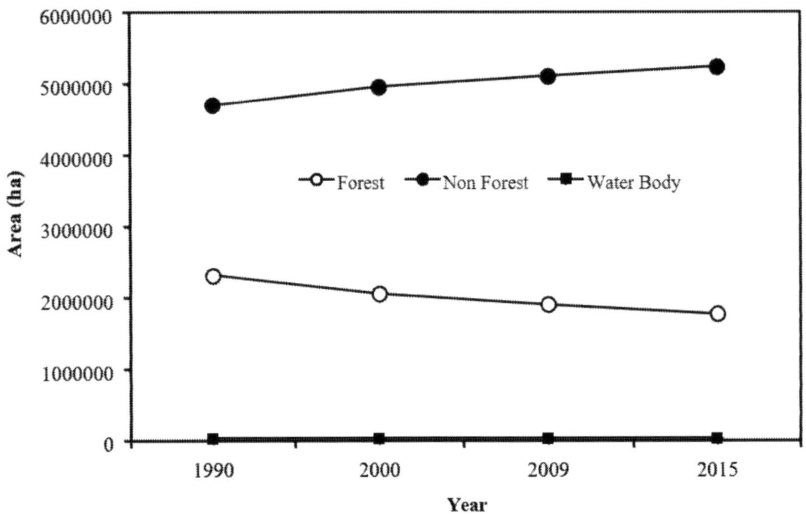


**Figure 2.** Trend of deforestation in North Sumatran forest, non-forest, and water body over 1990-2015

Several studies have reported drivers of deforestation in Indonesia [7-8]. Fiber plantation and logging concessions are responsible for the largest forest loss (~1.9 million ha and ~1.8Mha, respectively) in Kalimantan, Sumatra, Papua, Sulawesi, and the Moluccas during 2000-2010 [7]. Oil palm industry was ranked third as driver deforestation (~1 Mha) in this study, contributing the second leading source of carbondioxide emissions (~1,300 – 2,350 Mt $CO_2$) [7]. Oil palm plantation in Kalimantan alone has been reported to release 18-22% equal to 0.12-0.15 GtC/year of Indonesian's 2020 carbon dioxide equivalents emissions [8].

Table 2 depicts the deforestation and reforestation from land-cover changes 1990-2015 in North Sumatra. The largest deforestation has displayed the regency of North Sumatra province: South Labuhanbatu, Labuhanbatu, North Padang Lawas, Serdang Bedagai, and Asahan. In South Labuhanbatu, the main land-use changes occurred in forest lands: secondary dry land forest was 38,575.79 ha in 1990 decreased to 1,570.79 ha in 2015, whereas secondary swamp forest 4,308.67 ha in 1990 reduced to 0 ha in 2015. In contrast, barren land was 1,945.11 ha in 1990 significantly enhanced 100,693.66 ha in 2015. Main driver deforestation in South Labuhanbatu derived from the degraded land of barren land.

In Labuhan Batu Regency source of deforestation from land-use changes were secondary dry land forest, secondary swamp forest, and primary mangrove forest. The secondary dry land forest had been deforested from 15,634.20 ha in 1990 to 0 ha in 2015. Similarly, secondary swamp forest was 42,063.06 ha in 1990 had lost to 1,606.14 ha in 2015. Labuhanbatu was also lost primary mangrove forest 52.63 ha during 1990-2015. Besides, deforestation in North Padang Lawas was mainly from secondary dry land forest: 84,274.78 ha in 1990 to 21,691.18 ha in 2015 (Table 2).

Furthermore, the source of deforestation in Serdang Bedagai district came from land-use changes of secondary dry land forest and secondary mangrove forest to other land-use. Both forest lands have been deforested significantly, land-use of secondary dry land forest 1,839.79 ha in 1990 lost to 633.42 ha in 2015. Secondary mangrove forest-cover decreased in 1990 of 299.18 ha to 16.26 ha in 2015. In Asahan district, the decreasing secondary dry land forest from 34,333.75 ha to 32,619.06 ha during 1990-2015. It is noteworthy that oil palm plantation increased significantly in North Padang Lawas, Langkat, and Asahan districts. In North Padang Lawas, oil palm plantation area was 79,464.64 ha in 1990, enhanced to 92,426.36 ha in 2015. Similar results in Langkat and Asahan, oil palm industry, increased significantly from 97,408.63 and 114,599.44 ha to 152,302.46 and 154,250.28 ha, respectively during 25 years observed.

**Table 1.** Percentage of Land-use changeperiod 1990, 2000, 2009, and 2015

| Land-use | 1990 | 2000 | 2009 | 2015 |
|---|---|---|---|---|
| Forest | 32.79 | 29.22 | 27.13 | 25.27 |
| Non-forest | 66.57 | 70.13 | 72.25 | 74.11 |
| Water body | 0.64 | 0.65 | 0.62 | 0.62 |
| Total | 100.00 | 100.00 | 100.00 | 100.00 |

Forest land: primary and secondary dry land forest, primary and secondary swamp forest, primary and secondary mangrove forest). Non-forest: shrub, oil palm plantation, forest plantation, settlement, barren land, swamp shrub, dry land farming, mixed dry land farming, paddy field, aquaculture, airport, transmigration, and mining). Water body: water and swamp.

### *3.2. Deforestation and reforestation analysis*

Table 2 displays deforestation and reforestation from land cover change 1990-2015. Figure 3 shows the percentage of mangrove deforestation in North Sumatra between 1990 and 2015 was classified into six groups, the first group with negative deforestation rate occurred in only in Humbang Hasundutan. The second group without deforestation due to no forest land covered, such as in Binjai city, West Nias, Tanjung Balai city, Padang Sidempuan, Pematang Siantar, Sibolga, and Tebingtinggi. The third group contained less than 1% deforestation rate, occurred in majority study sites, Batubara, Dairi, Deli Serdang, Gunung Sitoli, Karo, Medan, Langkat, South Nias, West Pakpak, Simalungun, South Tapanuli, North Tapanuli, Simalungun, and Toba Samosir. The fourth group of 1-2 %, were deforested in Asahan, North Labuhanbatu, Mandailing Natal, Nias, North Nias, Central Tapanuli, Padang Lawas, and Samosir. The fifth group of 2-3% deforestation rate was in North Padang Lawas and Serdang Badagai. The last group with more than 3% deforestation rate was in Labuhanbatu and South Labuhan Batu.

**Table 2.** Deforestation and reforestation from land cover changes 1990-2015

| Regency/city | Land-cover changes 1990-2015 (ha) | | | | | Def (year/ha) | Def rate (%) |
|---|---|---|---|---|---|---|---|
| | Def | Ref | Forest | NF | WB | | |
| Asahan | 30,567.21 | - | 38,019.33 | 306,193.69 | 2531.09 | 1,222.69 | 1.78 |
| Batubara | 16.06 | - | 676.27 | 89,885.91 | 538.01 | 0.64 | 0.09 |
| Dairi | 7605.35 | - | 51,301.19 | 139,154.16 | 67.26 | 304.21 | 0.52 |
| Deli Serdang | 2,060.93 | 56.93 | 20,475.35 | 234,696.57 | 1,031.88 | 80.16 | 0.36 |
| Gunungsitoli | 15.02 | - | 135.56 | 30,886.54 | 46.03 | 0.60 | 0.40 |
| Humbang Hasundutan | 15,513.15 | 24,897.37 | 74,047.99 | 138,230.57 | 269.02 | -357.37 | -0.42 |
| Karo | 1,444.09 | 0.00 | 59,674.25 | 152,830.88 | 171.17 | 57.76 | 0.09 |
| Binjai | - | - | - | 9,413.74 | - | - | - |
| Medan | 114.77 | - | 589.75 | 27,150.24 | 419.21 | 4.59 | 0.65 |
| Tanjungbalai | - | - | - | 5,664.49 | 409.03 | - | - |
| Labutanbatu | 55,951.91 | - | 4,534.99 | 206,647.33 | 5,325.48 | 2,238.08 | 3.70 |
| South Labuhanbatu | 41,313.33 | - | 1,570.79 | 262,959.77 | 1,231.38 | 1,652.55 | 3.85 |
| North Labuhanbatu | 35,023.33 | - | 40,124.37 | 274,281.52 | 4,733.88 | 1,400.93 | 1.86 |
| Langkat | 32,932.58 | 505.63 | 214,323.19 | 367,090.50 | 6,406.46 | 1,292.01 | 0.52 |
| Mandailing Natal | 120,831.78 | 1,954.93 | 289,161.40 | 236,275.73 | 1,559.26 | 4,755.07 | 1.16 |
| Nias | 1,937.77 | - | 4,315.87 | 77,572.79 | 22.13 | 77.51 | 1.24 |
| West Nias | - | - | 0.01 | 47,297.60 | 129.26 | - | - |
| South Nias | 6,544.18 | - | 100,090.41 | 140,563.65 | 675.59 | 261.77 | 0.25 |
| North Nias | 2,231.25 | 0.00 | 6,025.82 | 111,168.02 | 201.03 | 89.25 | 1.08 |
| Padang Lawas | 81,142.50 | 439.97 | 91,373.03 | 200,054.29 | 513.76 | 3,228.10 | 1.87 |
| North Padang Lawas | 65,439.31 | 2,425.65 | 35,427.26 | 291,744.64 | 635.28 | 2,520.55 | 2.50 |
| Pd Sidempuan | - | - | - | 13,461.96 | - | - | - |
| West Pakpak | 4,844.84 | - | 90,186.48 | 37,257.61 | 22.21 | 193.79 | 0.20 |
| Pmsiantar | - | - | - | 7,684.02 | - | - | - |
| Samosir | 15,047.05 | 651.90 | 17,888.32 | 89,290.72 | 750.71 | 575.81 | 1.75 |
| Serdang Bedagai | 1,314.01 | - | 856.45 | 189,842.94 | 652.38 | 52.56 | 2.42 |
| Sibolga | - | - | - | 1,061.18 | - | - | - |
| Simalungun | 8,163.94 | - | 42,984.47 | 386,835.86 | 2,710.28 | 326.56 | 0.65 |
| South Tapanuli | 35,080.83 | 3,414.68 | 153,276.78 | 218,749.23 | 1,087.61 | 1,266.65 | 0.67 |
| Central Tapanuli | 23,860.75 | - | 61,077.06 | 147,827.59 | 722.59 | 954.43 | 1.13 |
| North Tapanuli | 3,138.58 | - | 161,821.09 | 204,591.00 | 1,050.95 | 125.54 | 0.08 |
| Tebingtinggi | - | - | - | 3,894.09 | 95.84 | - | - |
| Toba Samosir | 3,085.66 | - | 62,726.59 | 147,913.17 | 678.16 | - | 0.19 |
| Total | 595,220.57 | 34,347.06 | 1,621,684.08 | 4,798,768.65 | 34,686.92 | | |

Def: Deforestation, Ref: Reforestation, NF: Non-forest, WB: Water body, -: not available

IOP Conf. Series: Earth and Environmental Science **122** (2018) 012059 doi:10.1088/1755-1315/122/1/012059

Furthermore, during 1990-2015 the deforestation occurred in North Sumatra was 23,808.82 ha.year$^{-1}$ with percentage rate was 1.06 %. The highest deforestation was in Mandailing Natal 4,755.07 ha.year$^{-1}$, however, in this regency reforested 1,954.93 ha.year$^{-1}$ period 1990-2015 and grouped into the fourth category. By contrast, Labuhanbatu and South Labuhanbatu showed lower deforestation 2,238.08 and 1,652.55 ha, respectively. However, during study observed, no reforestation occurred (Table 2). The high deforestation in this province implies to increase conservation efforts and reforestation thoroughly the North Sumatra province.

The present study indicated the increasing plantation estates were from 999,596.12 ha in 1990 to 1,233,929.56 ha in 2015. The developing area of estate land was 9,373.33 ha.year$^{-1}$. In this context, oil palm industry contributed to deforestation in North Sumatra was 39.37%. Major increased area of oil palm industry derived from the conversion of forest land (Table 2).

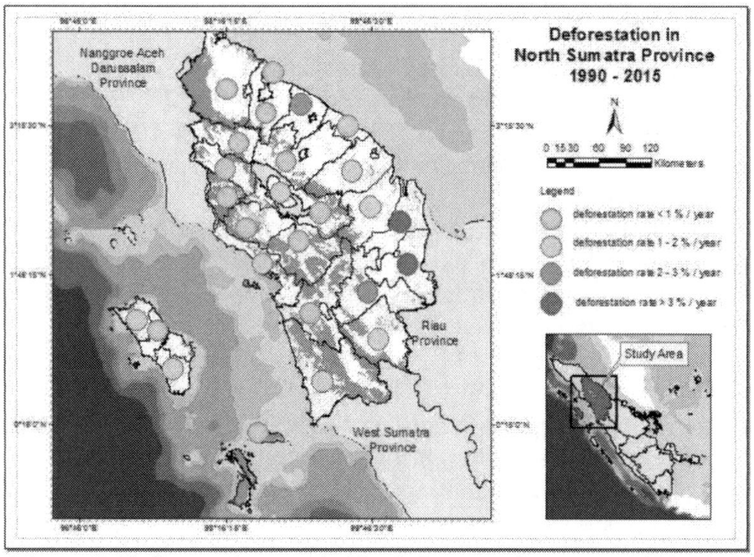


**Figure 3.** Deforestation rate in North Sumatra forestover 1990-2015

Oil palm plantation also contributed to mangrove deforestation in North Sumatra [3]. The threat of oil palm to mangrove is likely to increase in future [9]. Furthermore, it has been reported that mangrove deforestation between 2000 and 2012 in the east coast of Sumatra around 500-1000 ha [9]. Indonesian mangrove, however, has percentage mangrove loss was 1.7% [9], slightly similar to this study (1.06%). Mangrove conversion to aquaculture also has been reported in Kalimantan and Sulawesi [9].

Reforestation activity was observed in this study and occurred in ten regencies for 34,347.06 ha during observation, while the deforestation occurred 595,220.57 ha for 25 years (Table 2). Humbang Hasundutan was the only regency had negative reforestation where reforested 24,897.37 ha compared to deforestation 15,513.15 ha. The rehabilitation program is proposed by the involvement of the local communities dependent on the mangrove ecosystem for sustenance [10]. Protecting logged forests lead greater carbon emissions reductions (21%) than protecting intact forests alone (9%) and is vital for mitigating carbon dioxide emissions [8]

## 4. Conclusions

Major driver of North Sumatra deforestation, especially from the secondary and primary forest (dry land, mangrove, and peat land) was derived from plantation estate (oil palm plantation). The increasing loss of North Sumatra forest and increasing non-forest lands implies to conservation efforts and developing sustainable forest management.

## References

[1] Margono BA, Potapov P V, Turubanova S, Stolle F and Hansen M C 2014 Primary forest cover loss in Indonesia over 2000-2012 *Nat. Clim. Change.* **4** 730–735.

[2] Houghton R A 2012 Carbon emissions and the drivers of deforestation and forest degradation in the tropics. *Curr. Opin. Environ. Sustain.* **4** 597–603.

[3] Basyuni M, Putri L A P and Murni M B 2015 Implication of land-use and land-cover changes into carbon dioxide emission in Karang Gading and Langkat Timur Laut Wildlife Reserve, North Sumatra, Indonesia *J. Man. Hut. Trop.***21** 25–35.

[4] Koh L P, Miettinen J, Liew S C and Ghazoul J 2011 Remotely sensed evidence of tropical peatland conversion to oil palm*Proc. Natl. Acad. Sci. U.S.A.* **108** 5127–5132.

[5] Hansen M C, Potapov P V, Moore R, Hancher M, Turubanova S A, Tyukavina A, Thau D, Stehman S V, Goetz S J, Loveland T R and Kommareddy A 2013 High-resolution global maps of 21st-century forest cover change *Science* **342** 850–853.

[6] Sulistiyono N, Jaya I N, Prasetyo L B and Tiryana T 2015 Spatial model of deforestation in Sumatra Islands using typological approach *J. Man. Hut. Trop.* **21** 99–109.

[7] Abood S A, Lee J S, Burivalova Z, Garcia-Ulloa J and Koh L P 2015 Relative contributions of the logging, fiber, oil palm, and mining industries to forest loss in Indonesia *Conserv. Lett.* **8**58–67.

[8] Carlson K M, Curran L M, Ratnasari D, Pittman A M, Soares-Filho B S, Asner G P, Trigg S N, Gaveau D A, Lawrence D and Rodrigues H O 2012 Committed carbon emissions, deforestation, and community land conversion from oil palm plantation expansion in West Kalimantan, Indonesia *Proc. Natl. Acad. Sci. U.S.A.* **109** 7559–7564.

[9] Richards D R and Friess D A 2012 Rates and drivers of mangrove deforestation in Southeast Asia, 2000-2012 *Proc. Natl. Acad. Sci. USA* **113** 344–3349.

[10] Basyuni M, Rouf R A, Saragih M, Asbi A M and Yuriswan W 2017 Local wisdom and mitigation action to maintain secondary mangrove forest: a case study of Jaring Halus village in Langkat, North Sumatra, Indonesia *Adv. Soc. Sci. Educ. Hum. Res.* **81** 551–555.

## Acknowledgment

This study was partly supported by an International Research Collaboration and Scientific Publication Grant 2017 (No. 003/SP2H/LT/DRPM/IV/2017 to MB) from the Directorate for Research and Community Service, Ministry of Research, Technology and Higher Education, Republic of Indonesia.

# Genetic diversity of resin yielder *Pinus merkusii* from West Java - Indonesia revealed by microsatellites marker

**A Susilowati[1], H H Rachmat[2], I Z Siregar[3] and Supriyanto[3]**

[1]Faculty of Forestry, Universitas Sumatera Utara, Jln. Tridharma Ujung No 1, Kampus USU Medan
[2] FORDA, Ministry of Environment and Forestry, Jln Raya Gunung Batu. Bogor
[3] Faculty of Forestry Bogor Agricultural University, Jl. Lingkar Kampus Faculty of Forestry, Bogor Agricultural University, Bogor 16680
Email:arida_iswanto@yahoo.co.id

**Abstract**. Phenotypic observation of resin yielder *Pinus merkusii* showed higher value of genetic variation and *narrow sense* heritability values for resin production trait. This result indicated that genetic factor played as dominant aspect. However, further observation using molecular marker would still be needed to overcome the weakness of phenotypic observation. This study was carried out in order to characterize the genetic diversity and genetic differentiation of resin yielder genotype candidate *P.merkusii* using microsatellite markers and to characterize the genetic structure in the resin yielder populations. Seventy needle and inner bark samples were collected from resin yielder in Cijambu *Seedling Seed Orchard* (SSO) Sumedang, West Java and further divided into two genotype candidates (lower and high resin yielder). Seven microsatellites loci (*pm01, pm04, pm05,pm07,pm08, pm09a, pm12, pde5* and *SPAC 11.6*) were used for detection of genetic diversity. Results showed that genetic diversity in higher resin candidates was (0.551), slightly different compared lower candidates (0.545). However, cluster analysis determined that higher resin yielder grouped with lower one. Molecular variation was found to be low among populations (21%) and high among individuals within the populations (79%). Private alleles were detected both in higher yielder and also normal population.

## 1. Introduction

*Pinus merkusii* is the only member of genus *Pinus* that crosses the equator [1] and become one of the considerable species for machining wood, gum resin production, reforestation and land rehabilitation. One of valuable product from resin curing is *gondorukem* (gum resin). Gum resin was prospective export commodities in Indonesia because it can contribute about US$ 50 million/year [2].

Breeding programs for resin yielder have been started since 2006 through research on phenotypic observation, progenies and descendants test. High yielder trees producing 50 g/3 days/tree or more were successfully identified. Later on, Susilowati *et al.* [3] through their phenotypic study found that higher value of genetic variation and *narrow sense* heritability values indicated that genetic factor was dominant for resin production trait. However, the relationship between genetic diversity and the quantity of resin yield remain unclear. Thus, collecting the information about the relationship between genetic diversity and resin production is important to carry out [4].

Genetic diversity indicates differences among individual genotypes within and among the population. Genetic diversity is known to be a prerequisite for selection successes of desirable traits.

Identification of high yielder resin trees has been performed using morphological traits such as leave shape, resin process system, bark structure and resin production measurement. Recently, identification of these high yielder resin trees gains its more advanced technique by using a molecular marker.

A molecular marker, such microsatellite, can be better used than those of morphological characteristics because it is a dominant expression, having high polymorphisms and shows random distribution throughout the genome. This hyper-variability makes them informative and excellent not only for genetic diversity analysis, but also for a broad range of applications, including high-density genetic mapping, molecular tagging of genes, genotype identification, paternity exclusion, and quantitative identification and mapping. Molecular-based research has been conducted to quantify genetic diversity of *P. merkusii* [5], to detect reproductive system [6], genetic and strengthen the study of pine tree [7], and to determine the genetic diversity of offspring using microsatellite marker [8]. So far, there has no record of research which focuses on assessing the genetic diversity of resin yielder genotype candidate using microsatellites marker as a tool for verification of its phenotypic research.This study was carried out in order to: i ) characterize the genetic diversity and genetic differentiation of resin yielder genotype candidate *P.merkusii* using microsatellite markers and ii) characterize the genetic structure in the resin yielder populations.

## 2 Materials and Methods

### *2.1.Material*

Samples were collected from SSO Cijambu-Sumedang and Bogor (Table 1), both location are in West Java-Indonesia. Seventy needles and inner barks samples from seventy different individuals were collected randomly then they were clustered into normal producer (≤ 21g/3 days), lower yielder (≤ 50 g/3 day) candidates and higher yielder ((≥150 g/3 days). All samples stored in plastic bags with silica gel and kept at room temperature until DNA extraction.

**Table 1**. Sampling sites for resin yielder *Pinus merkusii*

| Location | Genotype | Geographic position | Elevation (m.asl) | Temperature ($^0$c) |
|---|---|---|---|---|
| Sumedang | High yielder | 107°45' N -6°.52'E | 900-1300 | 17-22 |
| Sumedang | Lower yielder | 107°45' N - 6°.52'E | 900-1300 | 17-22 |
| Bogor | Normal yielder | 106' 48' N-6° 26'E | 330 | 23-32 |

### *2.2. DNA extraction and microsatellite analysis*

Seven microsatellites loci developed for *P.merkusii* were used, those including: *pm*01, *pm*04, *pm*05, *pm*07, *pm*08, *pm*09a, *pm*12 (Table 2). Cross-amplification loci from different species *pde5* and *SPAC* 11.6 also used in this research. Needles and wood DNA samples were extracted using modified CTAB method (*Cetyl Trimethyl Ammonium Bromide*) according to Murray and Thompson [9]. PCR amplification was performed using a PCR Thermal Cycler (PTC-100), in reaction mixture 13.5 μL containing 2 μL DNA template, 2 μL nuclease free water, 7.5 green Go Taq and 2 μL loci. The PCR cycling conditions were as follows: pre-denaturation at 95 $^0$ C for 2 min, denaturation at 95 $^0$ C for 1 min, annealing for 2 minutes with temperature following each loci procedure, extension at 72 $^0$ C and the final extension at 72 °C for 5 minutes.

PCR products was run on 1 % polyacrylamide gel and followed by staining process using silver staining method. Band pattern then scored, and the scoring result further analyzed using POPGENE ver 1.31, NTSYS 2.02 and Genalex 6.3.

### *2.3. Measuring of variability*

The parameters for genetic diversity were determined for a percentage of polymorphic loci (*PLP*), observed number of alleles (*na*), the number of effective alleles *(ne),* expected heterozygosity (*He*), and genetic differentiation ($F_{ST}$). Principal Component Analysis (PCA) and Analysis of Molecular Variance (AMOVA) were also calculated to detect clustering pattern.

**Table 2**. Seven microsatellite primer used for genetic analysis

| Primer | Repeatable | Sequence | Ta (°c) | Gene bank accession |
|---|---|---|---|---|
| 1. *pm*01 | (TG)12 | F: AGAGAAGGCACGATTTTGTC<br>R: TCCCACTAATCACTTTGAAAG | 56 | AB201535 |
| 2. *pm*05 | (TG)9 | F: GAGTCTAATTGCAAACCCCA<br>R: TGGAGATCTACCACTTTTTC | 52 | AB201537 |
| 3. *pm*07 | (AC)8(AT)4 | F: GAATCTAAGCATATGAAATGAG<br>R: CTTGTTAATGCTACTAGTTATG | 52 | AB201538 |
| 4. *pm*09a | (AT)5(GT)1<br>8(AT)2 | F: CCTTCTCATTTCGATATGCAC<br>R: ATTAAAGGTTATATGGGGCT | 52 | AB201540 |
| 5. *pm*12 | (GT)5CT(GT<br>)5(AT)5 | F: GAACAATCATTGCGGGTCCCG<br>R: ATGCTGCGTTTATATGTATAAGTGTC | 59 | AB201541 |
| 6. *pde*5 | $(CA)_5C(CA)_5$<br>1 | F: AACGCACCTTTCTCAATGCAC<br>R: ATAAAGAGGCTACATGGTCCC | 56 | ABO38258 |
| 7. *SPAC* 11.6 | $(CA)_{29}(TA)_7$ | F: CTTCACAGGACTGATGTTCA<br>R: TTACAGCGGTTGGTAAATG | 56 | AJ223767 |

## 3. Results and discussion

Application of microsatellite loci on conifers was performed for the first time for *P.radiata* [10]. Compared with other Gymnospermae genus, microsatellites transfer for pine species is more difficult. For example, in *loblolly pine,* there was only 60% of microsatellite loci amplified. In this research among 9 loci studied, 7 showed polymorphic patterns (*pm*01, *pm*05, *pm*07, *pm*09a, *pm*12, *Pde*5 and *SPAC* 11.6) while two loci (*pm*04 and *pm*08) were monomorphic and excluded for further analysis. The specific allele for normal yielder was found in 117 bp, while for higher yielder found in 118 bp. Allele for higher and lower yielder were found in 88 bp and 132bp (Table 3).

**Table 3**. Amplification pattern of 7 microsatellites loci

| | Allele/ Lenght of fragment (bp) | | | | | | | | | | | | | | | | | | | | | | |
|---|---|---|---|---|---|---|---|---|---|---|---|---|---|---|---|---|---|---|---|---|---|---|---|
| Loci | 85 | 88 | 95 | 99 | 108 | 111 | 115 | 116 | 117 | 118 | 127 | 128 | 131 | 132 | 135 | 138 | 140 | 150 | 197 | 232 | 236 | 260 | 280 |
| *pm*01 | - | - | - | - | - | + | + | - | N | - | - | - | - | - | - | - | - | - | - | - | - | - | - |
| *pm*05 | - | - | - | - | + | - | - | + | - | T | - | - | - | - | - | - | - | - | - | - | - | - | - |
| *pm*07 | - | - | - | - | - | - | - | - | - | - | - | - | - | - | - | - | - | - | - | + | - | + | + |
| *pm*09a | + | T<br>R | + | + | _ | _ | _ | _ | _ | _ | _ | _ | _ | _ | _ | _ | _ | _ | _ | _ | _ | _ | _ |
| *pm*12 | _ | _ | _ | _ | _ | _ | _ | _ | _ | _ | _ | + | + | T | + | _ | _ | _ | _ | _ | _ | _ | _ |
| *Pde*5 | _ | _ | _ | _ | _ | _ | _ | _ | _ | _ | _ | _ | _ | _ | _ | + | _ | + | + | _ | _ | _ | _ |
| *SPAC* 1.6 | _ | _ | _ | _ | _ | _ | _ | _ | _ | _ | + | _ | + | _ | + | _ | + | _ | _ | _ | _ | _ | _ |

Note: +found in all population; -:no amplification; N:found only in normal yielder;
T: found only in higher yielder, R:found only in lower yielder.

*3.1. Genetic variation within populations*

The number of polymorphic loci in this genetic diversity analysis determines the level of diversity in a population. Table 4 showed data obtained from DNA band scoring analyzed using POPGENE version 3.2

**Table 4**. Genetic variability within resin yielder population

| Population | *PLP* | *Na* | *ne* | *He* |
|---|---|---|---|---|
| High resin yielder | 100% | 3.286 | 2.315 | 0.551 |
| Lower resin yielder | 100% | 3.143 | 2.244 | 0.545 |
| Normal | 100% | 3.000 | 2.4031 | 0.566 |

note: *P LP = Percentage of Polymorphic Loci; na = Observed number of alleles; ne = Effective number of alleles;* $H_e$*= Nei's (1973) gene diversity*

Based on genetic diversity value (*He*), high resin yielder candidates have a lower value (0.5511) compared with normal (0.566) but higher compared with lower yielder (0.545). These results are slightly different from previous research which determined that *He* for parent population and saplings in SSO Sumedang, Baturaden and Sempolan was 0.501, 0. 468, and 0.528 [8]. However, the *He* value in this study was in common range of previous results (He = 0.50–0.80) according to [10];[11].

Resin yielder Cijambu SSO and normal pine were characterized by high genetic diversity; this might be an indication that both groups would have the upper adaptability to environmental changes. As previously understood that microsatellites detection in quantitative characters determined by occurred selection, mutation process and selection pressure [12]. These reasons are likely to be one of the causes for differences in genetic variation of resin yielder and normal pine.

*3.2.Genetic variation between populations*

Genetic diversity between populations can be identified using genetic distance parameter. Genetic distance measures the differences of genetic structure between two populations at a particular gene locus. In general, the genetic difference between two or more population was analyzed by using genetic distance matrix in pairwise combination [13]. A pairwise genetic distance from this study was presented at Table 5.

**Table 5**. A pairwise genetic distance matrix

| Population | Higher yielder | Lower yielder | Normal |
|---|---|---|---|
| Higher resin yielder | **** | | |
| Lower resin yielder | 0. 114 | **** | |
| Normal | 0. 483 | 0. 322 | **** |

This study resulted in a different pattern of genetic distance and dendrogram analysis. Higher resin yielder clustered into a different group of that normal pine, while the genetic distance between high and low yielder was relatively small (0.114). Based solely on this result, higher yielder and low yielder population are genetically closely related but not for higher resin yielder and normal population (0.480). Further analysis by NTSYS *software*2.02 and GenAlex as shown in figure 1.

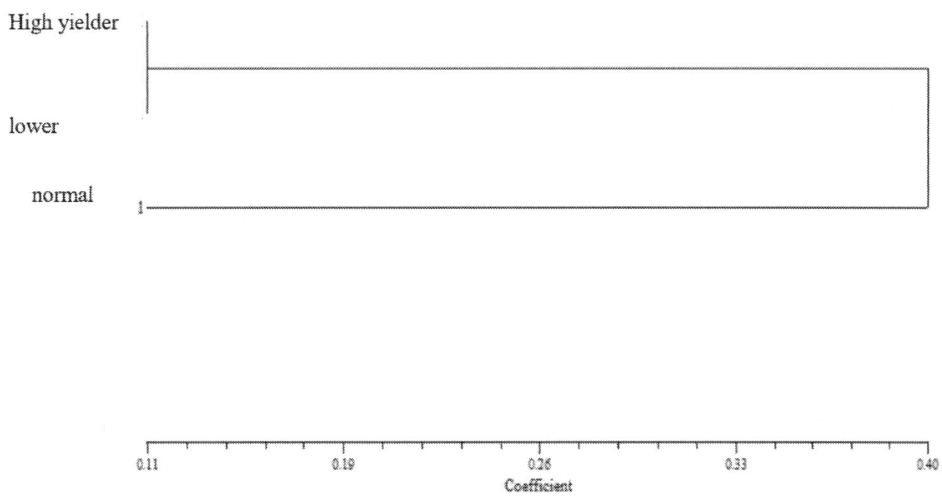


**Figure 1.** Dendrogram of genetic distance

*3.3. Source of genetic diversity P.merkusii based on microsatellite*

Based on Analysis of Molecular Variance (AMOVA) it was determined that 79% of the diversity resided within a population while variation between population only counts for 21%. This result was in accordance by previous study [14].

**Table 6.** Results from analysis of molecular variance (AMOVA)

| Sources | df | Sum of square | Mean of Square | Percentage of Variety | Fixation Index |
|---|---|---|---|---|---|
| Between populations | 2 | 25,590 | 12,795 | 21% | F*st*: 0.119 |
| In populations | 67th | 105067 | 3283 | 79% | F*is*: -0.241 |
| Total | 69th | 130657 | | 100% | |

Population differentiation in this study was higher (F*st* = 0.1191) compared to that of relatively similar study (He = 0.03) in the same area [8]. However, both values, in general were categorized as low population structuring. The low population structuring at SSO was recognizable as SSO were derived from materials with limited genetic diversity [15]. While Fst determined the population structuring, Fixation index or FIS described as inbreeding level occurred in population. Fis (–) indicated a high percentage of heterozygosity as a result from lower inbreeding while Fis (+) indicated a high level of homozygosity which means the occurrence of high inbreeding level within a population. From Table 4 it is known that Fis value in Cijambu SSO was negative which indicated lower inbreeding level within the population and this condition is promising for further improvement.

Genetic diversity is one of the key factors in effectiveness conservation strategies formulation and management germplasm. Generally, coniferous trees species have high the genetic diversity compared with other tree species. For selection series of resin yielder, information about the genetic diversity of high resin yielder candidate was needed to identify superior genotypes to conservation and breeding purposes in the future. Knowledge about genetic diversity in a population level also very important for conservation activities due to high genetic diversity will help populations adapt to surrounding environment changes and biodiversity preservation. Furthermore Namkoong *et al.* [16] states that degree of genetic diversity was one of sustainable forest management Indicators.

Results showed that even planted trees in Cijambu SSO were genetically related, but have high genetic diversity. This indicates that selection and maintenance activities are undertaken able to maintain genetic diversity.

## 4. Conclusions

Genetic diversity of high resin yielder in Cijambu SSO classified as high value, it was indicated by the high value of heterozygosity (He) of 0.551 (higher resin yielder) and 0.545 (lower resin yielder) slightly different from normal pine (0.565). High capacity microsatellite marker amplification indicated by the high *percentage of polymorphic loci* (100%) in all populations tested. The source of variability in resin yielder occurred by 79% within the population, while 21% occurred between, and based on genetic differentiation (FST) showed measurement genetic uniformity between populations.

Cluster analysis showed high resin yielder candidates separately clustered with normal pine, it indicates a difference in genetic structure between resin yielder candidates and normal, even genetic material originated from same seed source. Probably this fact caused by selection and thinning influence that has been done.

## References

[1] Mirov NT 1967 *The genus of Pinus* New York: Roland Press Company 602

[2] Fachrodji A, Sumarwan U, Suhendang E, Harianto 2009 Perbandingan daya saing gondorukem di pasar international *Jurnal Manajemen dan Agribisnis* **6 (2)** 140-151

[3] Susilowati A, Siregar I Z, Supriyanto, Wahyudi I, Corryanti 2013 Genetic variation, heritability and correlation between resin production character of *Pinus merkusii* high resin yielder (Hry) *Biotropia* **20 (2)** 122-133

[4] Falconer D S, Mackay TFC 1996 *Introduction To Quantitative Genetic* Essex: Addison Wesley Longman

[5] Kartikawati N K 1998 *Study of genetic variation of Tusam* (*Pinus merkusii* Jungh. et de Vriese) *natural forest in Aceh and plantation forest in Java using isozyme analysis method.* Master Thesis, Gadjah Mada University, Indonesia

[6] Siregar I Z, Hattemer H H 2004 Pattern of genetic structure and variation of merkus pine (*Pinus merkusii*) in Indonesia *J Tropical Forest Science* **16** 160-122

[7] Soeseno O H 2001 Current Status of Tree Improvement in Indonesia. In situ and Ex situ Conservation of Commercial Tropical Timber. *IUFRO Conf.* Yogyakarta:11-13 June 2001

[8] Nurtjahjaningsih I G, Saito Y, Tsuda Y, Ide Y 2007 Genetic diversity of parental and offspring population in a *Pinus merkusii* seedling seed orchard detected by mikrosatelit marker *Bull Tokyo Univ For* **118** 1-14

[9] Murray M G, Thomson WF Rapid isolation of high molecular weight plant DNA Nucleic Acids Res. 1980 Oct 10; 8(19): 4321–4325

[10]Smith D N, Devey M E 1994 Occurrence and inheritance of microsatellites in *Pinus radiata* genome *Genome* **37 (6)** 977-983

[11]Keys R N, Autino A, Edwards J, Fady B, Pichot C, Vendramin G G 2000 Characterization of nuclear microsatellites in *Pinus halapensis* Mill and their inheritance in *P. halapensis* and *Pinus brutia* Ten *Journal of Molecular Ecology* **9 (12)** *2157-2159*

[12]Karhu A 2001 *Evolution and Application of Pine Microsatellite.* Dissertation University of Oulu. Finladia

[13]Finkeldey R 2005 *Pengantar Genetika Hutan Tropis.* E. Jamhuri , I Z Siregar, U J Siregar dan AW Kertadikara, penerjemah (Gottingen : Institute of Forest Genetics and Forest Tree Breeding Georg-August-Univerity-Göttingen)

[14]Wu H x 1999 Study of early selection in tree breeding: Advantage of early selection through shortening the breeding cycle *Silvae genetica* **48** 78-93

[15]Hardiyanto E B, Danarto S 2000 *Second-generation breeding strategy for Pinus merkusiiJungh.et de Vriese in Java, Indonesia.* (In: Proceeding of Forest Genetics for the Next Millennium. Durban, South Africa, 8-13 October 2000). pp. 133-135.

[16] Namkoong, Barnes R D, Burley J 1980 A philosophy of breeding strategy for tropical forest trees. *Commonw For Rev* **59**

IOP Conf. Series: Earth and Environmental Science **122** (2018) 012061 doi:10.1088/1755-1315/122/1/012061

# Flowering and fruiting phenology of Kemenyan toba (*Styrax sumatrana* J.J.Sm.) in AekNauli forest, North Sumatra

**C R Kholibrina[1], Aswandi[1] and A Susilowati[2]**

[1]Forestry and Environment Research Development Institute of Aek Nauli
Jl Raya Parapat Km 10.5 Sibaganding Parapat, Simalungun North Sumatra, 21174
[2]Faculty of Forestry, Universitas Sumatera Utara, Jl. Tridarma Ujung No. 1,
Kampus USU, Medan North Sumatra
E-mail: rizlanicut@gmail.com

**Abstract**. The observation on flowering, fruiting phenology and germination of Kemenyan toba (*Styrax sumatrana*) has not been widely reported. It isrequired to support the breeding activities for this tree improvement, the most Non-Timber Forest Product commodity in Lake Toba Catchment Area, North Sumatra. The objectives of the research were to identify the development of flowering, fruiting andto calculate the number of fruits that germinate for *S. sumatrana* in certain cycle period. The flowering and fruiting observation were conducted on ten sample trees in Aek Nauli forest from July 2012 to February 2013. The seeds viability was observed from January to November 2014 in the greenhouse. The study showed that the flowering development occurred for 30 to 152 days, began from the growing of generative buds, the flower's shoots and bursts were developed, and young fruits were matured. All of processes proceeded for 30 to 152 days. The average percentage of flowering is 53.5%, and 72.8% for flowering to fruiting, and 47.3% for young to mature fruit. The percentage of mature fruit to germinate was 89.3%.

## 1. Introduction

The kemenyan trees are closely related to the development of civilization in Tapanuli, North Sumatra. In the mountainous region around Lake Toba, these trees cultivation have been going on since the 17th century or longer [1, 2]. The main NTFPs commodities in the region contribute substantially to the household economy of farmers, reaching 70% to 75% [3]. Kemenyan resin contains benzoin compounds that used for industrial raw materials, cigarette adders, medicines, cosmetics, insect and mosquitoes repellents (natural insecticides), pharmaceuticals, and food and beverage preservatives [1, 2, 3]. There are two species of Kemenyan that cultivated by most of communities in Tapanuli, i.e. Kemenyan durame (*Styrax benzoin*) and Kemenyan toba (*S. sumatrana*). However, the latter species is preferred because it has a better and more expensive resin quality [2, 4].

Indisputable, the incense productivity from the kemenyan forests have been decreased [5]. Some factors are identified as the cause of this decline, such as forest and land conversion, illegal logging, and limited cultivation techniques, especially application the higher quality seeds. The seeds production, especially from superior mother trees which higher quality and quantity resin is still limited. The incense tapping were conducted when flowering, thus interfering the fruiting process. This condition triggers a decline in seeds production.

Fulfillment of higher quality seeds would be obtained by managing the superior mother tree as a source of seeds. The management efforts start from flowering to fruit ripening and harvesting.

Generally, seed production is influenced by various factors, i.e. flower initiation, flower fallen, periodicity, dichogamy (protandry and protoginous), pollen viability and stigma conditions, and environmental conditions (climate, pests or diseases attack) [6].These information haven't been much explored in flowering of Kemenyan toba before. But, this is needed as a basis in an effort to improve the flowering process.

The objectives of the research were to identify the development of flowering, fruiting and to calculate the number of fruits that germinate for *S. sumatrana* in certain cycle period. Observation the flower development cycle until the fruits maturation can also be used as a tool in predicting the right time for seed harvesting. Measurement of germination ability can also to describe flowering efficiency. These are important information in establishing the proper pollination time in seed orchard management.

## 2. Methods

### *2.1.Study sites*

The flowering and fruiting observation were conducted on ten sample trees in Aek Nauli forest from July 2012 to February 2013. Furthermore, the seeds viability was investigated during January until November 2014 in greenhouse managed by Forestry Environment Research Development Institute of Aek Nauli. The trees grow at +/- 1200 m above sea level with the monthly average temperature of 23-24 °C and daily humidity 84 mm Hg. Rainfall in this region ranges from 2199 - 2452 mm/year so the area is included in type A according to Smith and Ferguson.

### *2.2. Data collection*

The observations were conducted in two stages: flowering phase and seed germination from observed flower. Flowering observation was conducted on ten sampling trees. The investigation were conducted on the morphology of flowers and development stages ranging from initiation, differentiation, generative shoot generation, flower buds and flowers bloom [7]. The flowering structure changes were observed and recorded (date and period) both shapes and colors. This investigation was applied in four times a week (intermittent day) during three months. The seed germination was applied on the collected seeds. The seeds were collected from previously observed flowers. The germination abilities were calculated as the percentage of seed germinating.

## 3. Results and discussion

### *3.1. Flower of kemenyan toba*

Kemenyan Toba flowers have a compound structure. Itare arranged in a bunch or panicle (inflorescence) with 5 to 12 flowers in a bunch (figure 1). The panicle of flower is at the tip of the branch with an upright position. The development of flowers from buds form to matured fruits is not the same period in one bunch [7]. There are a tendency that flowers in the top position is bloom first and then continued by the flowers at the bottom. The flower character and its structure are closely related to the type of pollination either by animals or wind [5].

**Figure 1.** The type of panicles and flowers of *Styrax sumatrana*

According to flower types, shapes and colors, the pollination were identified occurs with insect vectors, naturally. Bees and butterflies were observed flying around the flowers at 08.00-11.15 WIB [7] or 01.00-04.15 GMT. The flowers have white or cream color, irregular or tubular shapes, consisting of several parts, there are places of landing and flowers frequented by bees and moths [8].

The genital expression is hermaphrodite, where ♂ and ♀ consisted in a same flower, with anther and stigma close together, allowing for self-pollination. There were identified nine anthers that surrounded, with sticky pollen. There were indicated from the observed anther have been opening during investigation. Another tree species with sticky pollen is Teak (*Tectona grandis*) with medium tricolpatc form [9]. An anther and stigma position adjacent to each other is shown in Figure 2. The nature and shape of pollen will affects the controlled pollination patterns.

The anther and stigma positions are close together, but the compatibility is unknown. This condition raises the possibility of self pollination (autogamy) and cross (allogamy). The proportion of self and cross-pollination in a population are influenced by self-sterility, flowering behavior and the presence of pollinating vectors [10, 11].Generally, flowers have a natural mechanism to reduce their self-pollination both temporal and spatially. In this case, temporarily, the ripening of male and female flowers occurs at different times, whereas male flowers are spatially separated from female.

Petal is white and it is a part of flower to attract insects to pollinate. White pistil surrounded by orange stamens, light green sepals attached to bottom of the ovary. Diameter of flowers reaches 18 mm at the bud form, when blooming it reach 91 mm [7]. The panicle length is 65.4 to 75.0 mm, and the length of the stylus is 15 mm. The stamen length is 11 mm and it has 9 anthers. The stylus is longer than stamen. This is a defense mechanism to minimize the self-pollination. The anther is slightly hairy with a bright orange color when ripe and the stigma (pistil) is slightly slimy with a bright greenish white color when receptive condition.

### *3.2. Flowering periods*

The flowering development starts from the emergence of generative shoots that come out from the leaves armpits in the form of small bends, then developed into a flower bud (designated panicles). Shoots of panicles will develop into a clearer flower arrangement with petal still closed. Flowers on the buds develop to a complete flower structure with petal still in bud form. Further, individual flowers bloom with light green sepal color and white pistil parts (receptive flower conditions). If pollination occurs, the flowers will abort the petals and ovule starts to swell [7].

The designated flowers or generative shoots begin to appear at the end of June to July, then flower will appear and elongate at the end of August. In August to September the flower buds on the panicle grow and emerge a white petal. Generally, the flower blooms from August to September. In late September to early October, the flowers are fallen, and fertilization process has begun.

The flowering development occurs for 30 to 152 days, begin from the growing of generative buds, the flower's shoots and bursts isdeveloped, and young fruits ismatured. The developmental period of flowering of *S. sumatrana* is shown in Table 1. It is illustrates that the flowering cycle of this species incidence relatively longer than *S. benzoin*. The latter takes between 35 and 62 days [12].

**Table 1**. Flowering period of *Styrax sumatrana* in Aek Nauli forest, North Sumatra

| No | Sequence of reproductive organs | Period | Times | Remarks |
|---|---|---|---|---|
| 1. | Flowering initiation | 1day | June | |
| 2. | Generative shoots | 4-5 days | End June – July | |
| 3. | Shoots of the panicles | 7-8 days | 1st week August | |
| 4. | Designated panicles open, individual flowers buds | 4-5 days | 2nd – 3th weeksAugust | |
| 5. | Panicles developed, flower buds enlarged | 10-14 days | LateAugust | The flower on the lower part of the stalk is open but not yet blooming |
| 6. | Individual flowers bloom | 4-5 days | Late August – early September | Most flowers bloom on panicles |
| 7. | Receptive flower | 1-2 days | Early – mid September | Characterized by the slimy stigma and opening anthers |
| 8. | Petal fallen | 4-5 days | Late September– early October | |

Source: [7]

The flowers appear after the budding period. The number of *Styrax* flowers in one panicle varies in one tree as well as compared to other trees and the wind direction. Variations in the number of flowers often reported in tree plantations such as variations of flower production among clones, canopy and season in the seed orchard of *Pinus sylvetris*[13].

Environmental factors such as the adequacy of sunlight and soil nutrients affect the flowering. The sunlight reception is related to photosynthesis rate as a energy source for flowering process. The observations were showed that branching in the East has a higher proportion of flowers than in the West direction [7]. This is related to the intensity of sunlight received. Conversely, the less intensity of sunlight affects the flowering inhibition in reverse direction. While the availability of soil nutrients associated with energy supply and building materials for formation and development of flowers. The effects of competition among individual trees also determine the flowering [13].

### *3.3. Flowering effectiveness*

The results of observing the panicles, flowers, young and ripe fruits, and seed germination are shown in Table 2.The flower has an average 7 panicles with a range of 5 to 10 panicles on the observed branch. There are average 54 flowers in each panicle (range 30 - 79 flowers). These flowers will develop into young fruits with a percentage of 72.8% (Table 2). The young fruit will grow and matured. The percentage of young fruits that developed to mature fruits reached 47.3%. Furthermore, the matured fruit is germinated. The percentage of mature fruit to germinate was 89.3%.

**Table 2**. Number of panicles, fruits, flowers and germination of *Styrax sumatrana* seeds

| No | Number of panicles | Number of flowers | Number of young fruits | % flower developed to fruits | Number of mature fruits | % young fruits developed to mature fruits | Amount of germination | % of germination |
|---|---|---|---|---|---|---|---|---|
| 1 | 7 | 58 | 40 | 68.9 | 28 | 70.0 | 25 | 89.3 |
| 2 | 5 | 43 | 31 | 72.1 | 20 | 64.5 | 15 | 75.0 |
| 3 | 6 | 30 | 23 | 76.7 | 19 | 82.6 | 18 | 94.7 |
| 4 | 5 | 38 | 30 | 78.9 | 14 | 46.7 | 13 | 92.9 |
| 5 | 5 | 41 | 22 | 53.7 | 9 | 40.9 | 8 | 88.9 |
| 6 | 8 | 64 | 54 | 84.4 | 8 | 14.8 | 8 | 100.0 |
| 7 | 8 | 57 | 40 | 70.2 | 12 | 30.0 | 11 | 91.7 |
| 8 | 9 | 70 | 43 | 61.4 | 13 | 30.2 | 11 | 84.6 |
| 9 | 10 | 79 | 58 | 73.4 | 29 | 50.0 | 25 | 86.2 |
| 10 | 7 | 55 | 49 | 89.1 | 21 | 42.9 | 19 | 90.5 |
| Average | 7 | 54 | 39 | 72.8 | | 47.3 | | 89.3 |

## 4. Conclusions

The genital expression of *S.sumatrana* flower is hermaphrodite, where the ♂ and ♀ consisted in a same flower, with the anther and stigma close together, allowing for self-pollination. The flowering development occurs for 30 to 152 days, begin from the growing of generative buds, the flower's shoots and bursts isdeveloped, and young fruits is matured. The average percentage of flowering was 53.5%, and 72.8% for flowering to fruiting, and 47.3% for young to mature fruit. The percentage of matured fruit to germinate was 89.3%. This preliminary study should be continued by examining the pollination vector by investigating the pollen attached to body part of vector agent. To identify the grain pollen, microscopic observation and grain pollen germination should be necessary.

## References

[1] Heyne K 1987 Useful Plants of Indonesia Volume I Forestry Research Center Jakarta.

[2] Kholibrina C R 2012 Strategi Pemuliaan dan Peningkatan Teknik Budidaya Kemenyan Toba (*Styrax sumatrana* J.J.Sm. sinonim *S. paralleloneurum*) *Prosiding Ekspose Hasil-hasil Penelitian Kehutanan* Medan, 3 Juli 2012 Pusat Penelitian dan Pengembangan Konservasi dan Rehabilitasi Kementerian Kehutanan.

[3] Sianipar H and Simanjuntak B 2000 Isolation and Identification of Sinamat Acid from Kemenyan Sumatrana *Media Farmasi* **4(1)** 22-28

[4] Susilowati A, Hendalastuti H, Kholibrina C R, and Ramadhani R 2017 Weak delineation of *Styrax* species growing in North Sumatra, Indonesia by *matK* + *rbcL* gene Biodiversitas **18(3)** 2085-4722. DOI: 10.13057/biodiv/d180353.

[5] Central Bureau of Statistics of North Sumatra Province 2008 *North Sumatra in Figures 2008* Medan: BPS Provinsi Sumatera Utara

[6] Sedgley M and Griffin A R 1989 *Sexual Reproduction of Tree Crops* Academic Press.

[7] Kholibrina C R 2017 Pembungaan PohonKemenyan Toba (*Styrax sumatrana* J.J.Sm.) di Arboretum Aek Nauli,Sumatera Utara *Prosiding Seminar Hasil Penelitian Balai Penelitian Pengembangan Lingkungan Hidup dan Kehutanan Aek Nauli*. Pusat Litbang Hasil Hutan. Badan Litbang dan Inovasi.

[8] Palupi E R 2001 *Pollination, fertilization and seed development* The 3rd Training on Seed Biology. IFSP-BTP. Badan Litbang Kehutanan. Departemen Kehutanan dan Perkebunan, Bogor

[9] Owens J N, Sornsathapornkul P, Tangmitchareon S 1996 *Studying Flowering and Seed Ontogeny inTropical Forest Trees*. ASEAN-Canada Forest tree Seed Centre. Muak Lek, Saraburi. Thailand.

[10] Kittelson P M and Maron J L 2000 Outcrossing rate and inbreeding depression in the perennial

yellow bush lupine, *Lupinus arboreus* (Fabaceae) *American Journal of Botany* **87**652-660
[11] Satifah and Darjanto 1984 *Pengetahuan Dasar Biologi Bunga dan Teknik Penyerbukan Silang Buatan*. Gramedia, Jakarta.
[12] Syamsuwida D, Aminah A, Nurochman N, Sumarni EB dan Ginting J 2014 Flowering and Fruiting Development Cycle and Fruit Set of Kemenyan (*Styrax benzoin*) at Aek Nauli *Jurnal Penelitian Hutan Tanaman* **11(2)** 89-98
[13] Burczyk J and Lewandowski C 1998 Resin production of scots pine trees maybe associated with multilocus allozyme variation *Arboretum kdrnickie* **43** 79-83

# Phylogeny of kemenyan (*Styrax* sp.) from North Sumatra based on morphological characters

**A Susilowati[1], C R Kholibrina[2], H H Rachmat[3] and M A Munthe[1]**

[1]Faculty of Forestry, Universitas Sumatera Utara, Jl. Tridharma Ujung No.1, Kampus USU, Medan
[2] Aeknauli Forest Research Agency, Jl. Raya Parapat Km 10.5, Parapat, Sumatera Utara
[3] Forest Research Devolepment Agency, Ministry of Environment and Forestry, Jalan Raya Gn. Batu. Bogor
E-mail: arida_iswanto@yahoo.co.id

**Abstract.** Kemenyan is the most famous local tree species from North Sumatra. Kemenyan is known as rosin producer that very valuable for pharmacheutical, cosmetic, food preservatives and vernis. Based on its history, there were only two species of kemenyan those were kemenyan durame and toba, but in its the natural distribution we also found others species showing different characteristics with previously known ones. The objectives of this research were:The objectives of this research were: (1). To determine the morphological diversity of kemenyan in North Sumatra and (2). To determine phylogeny clustering based on the morphological characters. Data was collected from direct observation and morphological characterization, based on purposive sampling technique to those samples trees atPakpak Bharat, North Sumatra. Morphological characters were examined using descriptive analysis, phenotypic variability using standard deviation, and cluster analysis. The result showed that there was a difference between 4 species kemenyen (batak, minyak, durame and toba) according to 75 observed characters including flower, fruits, leaf, stem, bark, crown type, wood and the resin. Analysis and both quantitative and qualitative characters kemenyan clustered into two groups. In which, kemenyan toba separated with other clusters.

## 1. Introduction

*Styrax*is a genus of about 130 species of large shrubs or small trees in the family Styracaceae. The genus of *Styrax* known as benzoin rosin producer in the world and widespread but disjunctive distribution occurring in the Americas, eastern Asia, and the Mediterranean region [1]. However, different species of *Styrax* could also be found in some South East Asia countries such as Indonesia, Laos, Vietnam and Thailand. The province of North Sumatra was reported as the largest distribution site of kemenyanin Indonesia [2]. Although some experts suggest that these species are also found in South Sumatra and Papua, the accurate study about its existence has not yet been done.

The production of kemenyan in North Sumatra was 4.460 tons every year [3], which distributed to 5 districts: North Tapanuli, Humbang Hasundutan, Pakpak Bharat, Toba Samosir, and South Tapanuli. The existence of these species is important for forest communities in North Sumatra because it contributes 70-80% of farmers' income.

The information about kemenyan species in Indonesia has not been determined yet, especially in North Sumatra. Prevoius research conducted by Jayusman [4] found two species that distributed in North Sumatra, i.e. kemenyan toba (*Styrax sumatrana* J.J.SM) and kamenyan durame (*Styrax benzoin*).Kemenyan bulu (*Styrax paralleloneurum*) is also found in North Sumatra, but this species is rarely cultivated and has still not widely known until now. This species was referred by Stenis [5] as *S. benzoine* var hiliferum. In North Tapanuli farmer also found different species assumed as kemenyan dairi [6]. In Pakpak Bharat, based on farmer information there were kemenyan toba, kemenyan durame, kemenyan minyak and kemenyan batak, but there is no recent information about the species. Unclear information and community preference to cultivate only certain species (toba and durame) were the triggers to potential loss of species. The cultivation preference was chosen based on abundance of certain natural regeneration and limited knowledge of other species [7]. Comprehensive knowledge and database about kemenyan species could be a baseline in breeding activities in order to obtain superior characters and give easier guidance for farmer identification. The knowledge could also prevent species extinction due to species preferences.Therefore, we suggest specific study to differentiate the kemenyan species, particularly in North Sumatra.

Species identification of kemenyan could be done by morphological characterization. Morphological characterization played important role in systematic biology because despite of many approaches used to compile the classification system, all was originated from morphological characterization [8]. Morphological characterization was one activity to avoid gene erosion and also as a guide in the empowerment of genetic resources in plant breeding programs [9]. Therefore study about morphological diversity and phylogenetic of kemenyan species is needed to obtain precise information on the identity and species clustering in North Sumatra.

## 2. Materials dan Methods

### *2.1. Materials*

Direct observation method using generative and vegetative morphological characterization measurements were used in this research. The vegetative character observation and measurement were done to habitus, stem, canopy, and leaf. The generative character observation was done to flower, petal, flower crown, fruit, and seed. We also observed the rosin character. The research was conducted at PakpakBharat district on North Sumatra due to the fact it has the largest local kemenyan species varieties compared to other districts.

The species studied were kemenyan batak, kemenyan durame, kemenyan bunga and kemenyan minyak. Identification and determination referred to Steenis [5] and Tjitrosoepomo [10]. The tools used for observation were a magnifier, razor blade, tweezers, knife, plant scissor, plastic bag, stationary, ruler, stereo microscope, petri dish and a book titled Munsell Colour for Plant Tissue for color identification.

### *2.2. Methods*

Living plant speciments were taken both for generative (flowers, fruits, and seeds) and vegetative (roots, stems, leaves) organs. Direct observation and measurement were conducted into detailed specific part of the speciment. The quantitative and qualitative data of morphological characters observed were transformed into binary score data and formed into data matrix using Microsoft Excel program. The data matrix was used to analyze sample clustering based on SAHN (*Sequential Agglomerative Hirearchical and Nested Clustering*) using UPGMA (*Unweighted Pair Group Method Using Average*) method with NTSYS PC program to obtain phylogenetic dendrogram. The results were compared descriptively.

## 3. Results and Discussion

### *3.1. Morphological Characters*

Similar observation to leaf, stem, flower, fruit, seed and rosin characters showed morphological differences of all four kemenyan species. The morphological observation result of rosin, generative and vegetative part is presented in Table 1.

IOP Conf. Series: Earth and Environmental Science **122** (2018) 012062 doi:10.1088/1755-1315/122/1/012062

**Table 1.**Morphological characters of four species of kemenyans in North Sumatra

| No | Part | Characters | Observation Result | | | |
|---|---|---|---|---|---|---|
| | | | Batak | Minyak | Durame | Toba/Bunga |
| 1 | Canopy | a.Canopy shape | rounded | rounded | rounded | rounded |
| | | b. Canopy width | 4.4 m | 3.10 m | 3.7 m | 2.94 m |
| | | c. Canopy height | 11 m | 24.1 m | 15.6 m | 12.6 m |
| 2 | Leaf | a.Leaf shape | elliptical | lanceolate | lanceolate | Lanceolate |
| | | b.Leaf margin | entire | entire | entire | entire |
| | | c. Leaf tip shape | acuminate | acuminate | acuminate | acuminate |
| | | d. Leaf base shape | obtuse | obtuse | obtuse | obtuse |
| | | e. Leaf area | 47.48 cm | 25.60 cm | 46.28 cm | 22.63 cm |
| | | f. Lamina length | 10.7 | 10 | 12.8 cm | acute |
| | | g. Number of lamina | 14 | 12 | 16 | 16 |
| | | h. Upper surface color | green | dark green6/6 2.5YG | Green | 4/8 5YG |
| | | i. Lower surface color | whitist green | silvery white7/4 | whitist | whitist |
| | | j. Leaf arrangement | alternate | alternate | alternate | alternate |
| | | k. Venation | pinnate | pinnate | pinnate | pinnate |
| | | l. Leaf surface | smooth | smooth | smooth | smooth |
| | | m. Leaf composition | simple | simple | simple | simple |
| | | n. Leaf bud | naked bud | naked bud | naked bud | naked bud |
| 3 | Stem | a.Stem shape | cylindrical | cylindrical/with rosin | cylindrical | cylindrical |
| | | b.Outer bark color | reddish yellow | reddish brown | reddish brown | reddish yellow |
| | | c. Inner bark color | yellow | yellow | yellow | yellow |
| | | d. Bark thickness | 0.6 cm | 0.6 cm | 1 cm | 0.5 cm |
| | | e. Wood color | creamy white | creamy white | creamy white | creamy white |
| | | f. clear bole | 4.3 m | 7.7 m | 7.8 m | 7.4 m |
| | | g. Diameter | 17.1 cm | 28.7 cm | 50 cm | 16.3 cm |
| | | h. stem character | monopodial | monopodial | monopodial | monopodial |
| | | i. Branch construction | orthotropic | orthotropic | orthotropic | orthotropic |
| | | j. Branch position | continuous | continuous | continuous | continuous |
| 4 | Flower | a. Flower position | axillary | axillary | axillary | axillary |
| | | b. Number of petal | 4-5 | 4-5 | 4-5 | 4-5 |
| | | c. Flower color | white | white | white | white |
| | | d. Number of stamen | 10 | 8-10 | 10 | 10 |
| | | e. Stalk color | green | green | Green | green |
| 5 | Fruit | a. Fruit shape | flat round | flat round | flat round | oval round |
| | | b. Fruit color | deeply green | deeply green | green | green |
| | | c. Skin thickness | 0.4 cm | 0.4 cm | 0.4 cm | 0.5 cm |
| | | d. Fruit length | 1.9 | 2.6 cm | 3.2 | 2.8 |
| | | e. Fruit width | 2.3 | 2.9 cm | 2.4 | 2.5 |
| 6 | Seed | a. Seed shape | flat round | flat round | round | round |
| | | b. Length | 1.2 | 1.7 cm | 1.2 cm | 1.6cm |
| | | c. Width | 1.5 | 2 cm | 1.5 cm | 1.5cm |
| | | d. Seed color | brownish yellow | brownish yellow | brownish white | pitch black |
| 7 | Rosin | a. Rosin color | Creamy white | yellowish white | yellow | Creamy white |
| | | b. Rosin scent | scented | resin | aromatic | Aromatic |
| | | c. Rosin type | hard | scented/aromatic soft/liquid | soft/liquid | Crystal /hard |

Kemenyan's canopy consisted of monopodial branching with orthotropic branching and continuous branch position. All four kemenyan have rounded shaped canopies with variations in size and height, light branching, clear bole 4.5-7.8 cm. The canopy of kemenyan batak tended to be wider (4.4 m) but also the shortest compared to kemenyan minyak, durame and toba. Kemenyan had a cylindrical stem with up to 7.8 meters branch-free bole and monopodial branching. The bark was chapped and grooved; the outer bark was yellow in colour. The colour of inner bark varied from reddish yellow to reddish brown. Bark thickness varied between 0.5-1 cm.

Kemenyan had simple leaf in alternate arrangement, 12-16 pinnate laminas, smooth leaf surface. Kemenyan bunga, minyak and durame had lanceted leaves, but kemenyan batak had ovatus leaves. The leaf had entire margin, acuminate tip and obtuse base. The upper surface was moderate yellow green to deep yellow green. The lower surface colour was varied from silvery white to white. The fruit of kemenyandurame was flat round to oval in shape, hard shell, color from pale green (6/10) to green (8/6). The skin thickness was varied from 0.4 cm-0.5 cm. The diameter was varied from 1.9-3.2 cm. The fruit of kemenyan minyak had the largest size and it tended to have flat round in shape compared to other species. The seed was round to flat round, diameter 1.2-1.7 cm and the color varied from brownish yellow to pitch black. Kemenyan had compound flower. The flower was inflorescence and composed of 5-12 florets. It was axilary flower with 4-5 petals, the color varied from creamy white (9/4 YR) to white (9/6 YR). There were 8-10 stamens and the flower stalk was green (7/4).

Rosin was commonly used by farmers to identify kemenyan types in North Sumatra. The observation of rosin characteristics showed that kemenyan minyak had different color than other types (yellowish white). The rosin of kemenyan minyak was liquid, while kemenyan batak and bunga were hard.

### *3.2. Phylogeny of kemenyan in North Sumatra*

Phylogeny could be used to estimate the degree of similarity between species or population,in contrast to the diversity coefficients which was used to estimate the degree of inter species differences or population on chosen characters. From this relationship, we can conclude that species with far relatedness distance have higher level of diversity and lower level of uniformity. This relatedness in kemenyan species is shown in phylogenetic dendrogram (figure 1).

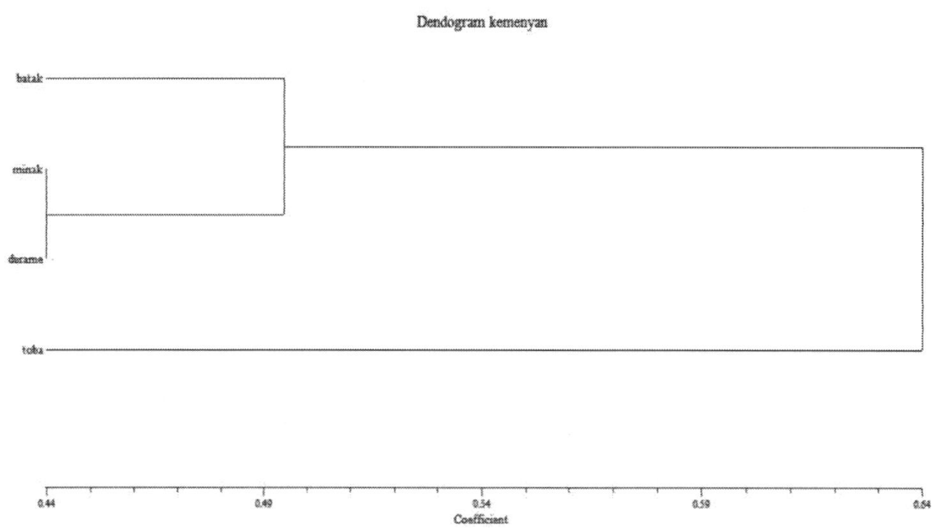


**Figure 1.** Relatedness between kemenyan species

The phylogenetic dendrogram that determined using dissimilarity index according to morphological characters, showed that kemenyan durame and kemenyan minyak were clustered in a small group with genetic distance 0.44, this cluster later joined with kemenyan batak in a larger group and had genetic distance as 0.49. Kemenyan toba was separated from the others and had genetic distance 0.64. We concluded that kemenyan durame and kemenyan minyak had the highest relatedness level and the largest morphological character uniformity, but lower diversity compared to kemenyan toba and kemenyan minyak.

However, the difference in character determination could result in different dendrogram pattern and classification model due to the difference of concept review and subjectivity between observers, although in generalthe characterare still similar. The more characters are analyzed, the higher validity, the higher a character's ability as distinguishing characteristic, then the ability to know relatedness level is more obvious [11]. Therefore, we suggest further study with the help of molecular markers to ensure species identification.

## 4. Conclussions

Phylogenic relationship based to morphological characteristics on four types of kemenyan showed that kemenyan minyak, durame and batak formed into a cluster according to kemenyan characteristic similarities of stem, leaf, flower, fruit, seed, canopy and rosin. Kemenyan toba was separated from the others due to its characteristic differences, in particular rosin characteristic, outer and inner bark color, the number of stamens and fruit size.

## References

[1] Fritsch P W 2001 Phylogeny and biogeography of the flowering plant genus Styrax (Styracaceae) based on chloroplast DNA restriction sites and DNA sequences of the internal transcribed spacer region *Molecular Phylogenetics and Evolution Journal* **19 (3)** 387-401

[2] Susilowati A, Hendalastuti H, Kholibrina C R, Ramadhani R 2017 Weak delineation of *Styrax* species growing in North Sumatra, Indonesia by *matK+ rbcL*gene *Biodiversitas***18 (3)** 1270-1274

[3] Silalahi J, Sukmana A, Antoko B S, Sunandar A D, Barus J A, Manik W S, Sanjaya H 2013 *Buku Kecil: Kemenyan Getah Berharga Tano Batak* (Parapat : Balai Penelitian Kehutanan Aek Nauli).

[4] Jayusman 1997 *Kajian Sistem Pemasaran Getah Kemenyan (Styrax* sp.*) Bul Penelitian Kehutanan Balai Penelitian Kehutanan Pematang Siantar* **13** (3-18)

[5] Steenis V 1953. *Styracaceae*. Flora Malesiana Seri I Vol. **42**.

[6] Iswanto A H, Susilowati A, Azhar I, Riswan, Supriyanto, Tarigan J E, Fatriasari W 2016 Physical and mechanical properties of local styrax woods from North Tapanuli in indonesia *J. Korean Wood Sci. Technol* **44 (4)** 539-550

[7] Susilowati A, Hartini K S, Rahmat H H, Alvaroby M 2017 *IOP Conf. Series :Materials Science and Engineering Propagation of Valuable North Sumatera Benzoin Trees(Styrax Sp) Using Macrocutting Technique* **180** 012046

[8] Rahayu S E, Handayani S 2008 Keanekaragaman morfologi dan anatomi *Pandanus* (Pandanaceae) di Jawa Barat *Vis Vitalis Journal* **1(20)**29- 44.

[9] Hershey G H 1995. Cassava Germplasm Resources In CIAT Cassava Breeding,*Multidiscplinary Review*. p. 1-24.

[10] Tjitrosoepomo G 1990 *Morphologi Tumbuhan*. Yogyakarta: Gajah Mada University Press.

[11] Suratman, Dwi P, Setyawan A D 2000 Analisis keragaman genus *Ipomoea* berdasarkan karakter morfologi *Biodiversitas***1 (2)** 72-79

## Acknowledgement

We would like to express my sincerely thanks to SEAMEO-BIOTROP for supporting our research fund with *Joint research grant*. My sincere appreciation also goes to Farmer in Pakpak Bharat North Sumatra for the access to their kemenyan forest and providing samples as well as for technical assistance during fieldwork

# Agronomic effect of empty fruit bunches compost, anorganic fertilizer and endophytic microbes in oil palm main nursery used *Ganoderma* endemic soil

**H Hanum[1*], Lisnawita[1] and A R Tantawi[1]**

[1]Department of Agrotechnology, Faculty of Agriculture, Universitas Sumatera Utara, Padang Bulan Medan 20155 Indonesia
[2]Department of Agrotechnology, Faculty of Agriculture, Universitas Medan Area, Jl. Medan Estate No. I. Medan 20223 Indonesia
*E-mail : hamidah1@usu.ac.id

**Abstract**. Using of *Ganoderma* endemic soil in oil palm main nursery is not recomended because produce bad quality seedling. The application of organic and anorganic fertilizer and endophytic microbes are the alternative for solving the problem. The objective of this research is to evaluate the effect of empty fruit bunches compost, anorganic fertilizer and endophytic microbes on growth of oil palm seedling in main nursery. This research used factorial randomized block design. The first factor was combination of empty fruit bunches compost and anorganic fertilizer, The second factor was endophytic microbes consisting of *Trichoderma* and *Aspergillus*. The results showed that interaction effect of the both treatment factor used increased growth of seedling in third and fourth month after application. The best growth of seedling was on the treatment of empty fruit bunches compost combined with anorganic fertilizer 150% recommended dosage and *Trichoderma viride*.

## 1. Introduction

One of the significant plantations commodity in Indonesia is oil palm (*Elaeis guineensis* Jacq.) because its contribution in increasing foreign exchange and the supply of vegetable oil. The oil palm plantation area in Indonesia and the average production of crude palm oil was 8.91 million hectares in 2011 [1] and around 31 million ton per year [2] respectively.

The main problem as that causes major damage to oil palm plantations especially the oil palm plantation of the people in the area is *Ganoderma*. *Ganoderma* caused the basal stem rot disease. It is happened in many properties land. The pathogen does not only attack the produced plant but also attack plants that have not produced even in plants seedling nursery. On land which has high sand content (>50%), basal stem rot disease has spread with the level of disease incidence between 0.71% to 50% in vary years old oil palm. The symptoms of the basal stem rot disease are chlorotic leaves, the appearance of fruiting body, collapsed plants, and the existence of holes on the basal stem [3]

On the plants that are 15 years of age which replanted with underplanting technique, the basal stem rot can reach 33% and reduce the oil palm production to 35% [4].

Commonly, the infection rate of *Ganoderma* in the soil which endemic *Ganoderma* especially textured sandy is high [5] and poor nutrition [6]. Although soil has many risks but many farmers in community's plantations such as in Bukit Kijang Village in Sumatera Utara Province use the soil as nursery planting media. Generally the seedling quality is very poor.

The seedling growth planted use *Ganoderma* endemic soil can be improve by minimizing the infection rate of *Ganoderma*. The improvement of the endemic soil with *Ganoderma* can be done through the integrated methods namely the use of the combinations of organic and anorganic fertilizers and microbial antagonists. The organic fertilizer will improve soil properties and increase nutrition absorption from anorganic fertilizer. Microbial antagonists as endophytic fungus such as *Trichoderma* sp, *Aspergillus* sp. and *Penicillium* sp. showed antagonistic reaction with *Ganoderma* [7], [8]. The effectivity of microbe determined themicrobe types.

Generally, farmer used compound fertilizer that contain N, P, K and Mg in ratio 15-15-6-4 or 12-12-7-2. It is difficult to supply nutrition need because nutrition composition determined. So, beside that, it is too used single fertilizer. Compost application as organic matter can be reduced the problem. Compost is very important material to give in seedling growth media Empty fruit bunches contain 42.8 % C, 2.90% $K_2O$, 0.80% N, 0.22% $P_2O_5$, 0.3% MgO 10 ppm B, 23 ppm Cu, 51 ppm Zn [9]. The objective of this research is to compare the effect combination of empty fruit bunches compost and anorganic fertilizer and endophytic microbes on oil palm growth in main nursery seedling

## 2. Material and methods

### *2.1. Determination of the location*

The research was done in 'community' oil palm plantations in Desa Bukit Kijang, Gunung Melayu, in Sumatera Utara, Indonesia. The location is at N 02°42°34.6°E 099°30°38.2°, plant number is 350, plant age is seventeen years old and disease incidence is 50%.

### *2.2. Material: Endophytic Microbes, Compost, Soil Sample*

The research use pure cultures of three isolates of Endophytic Microbial namely *Trichoderma* koningii, *Trichoderma* viride, *Aspergillus* sp as collection Phytophatology Laboratory of Universitas Sumatera Utara isolated from oil palm trees in the village of Gunung Melayu, Asahan, Sumatera Utara-Indonesia. Propagation endophytic fungus carried on the carrier grits. After that reproduced by using a zeolite with a ratio of 1: 9 (fungus endophyte in corn flour: zeolite). Soil sample is taken from location determined which textured sandy loam (64.82% sand, 21.47% silt and 13.71% clay) and pH 5.31. Compost of empty fruit bunches is taken from PT. Socfindo.

### *2.3. Analysis of empty fruit bunches compost and soil sample*

The result of empty fruit bunches compost and soil analysis is showed in Table 1.

**Table 1.** Properties of empty fruit bunches compost and soil sample

| No | Properties | Empty fruit bunches compost | Soil sample s |
|---|---|---|---|
| 1 | C-Organic (%) | 17.79 | 2.26 |
| 2 | N-Total (%) | 1.27 | 1.54 |
| 3 | $P_2O_5$ (%) | 0.02 | - |
| 4 | P-available (ppm) | - | 13.96 |
| 5 | $K_2O$ (%) | 0.286 | - |
| 6 | K-exch (me/100) | - | 0.628 |
| 7 | MgO (%) | 0.529 | - |
| 8 | Mg-exch (me/100) | - | 0.438 |

### *2.4. Experimental design*

This research used factorial randomized block design. The first factor was the combination of empty fruit bunches compost and anorganic fertilizer. The treatment consisted of compost, 100% dosage of NPKMg fertilizer, compost and 50% dosage of NPKMg fertilizer, compost and 100% dosage of NPKMg fertilizer, compost and 150% dosage of NPKMg fertilizer. The second factor is type of endophytic microbes that consist of control (no endophytic microbes), *T.koningii*, *T. viride* and *Aspergillus sp.*

### *2.5. Procedure of compost, fertilizer and endophytic microbes application*

Each polybag contained 5 kg of soil sample taken from the location. The dosage of compost and endophytic microbes respectively were 600 g of compost in each polybag and 30 g of endophytic microbes inoculant. The application of compost and inoculation of endophytic microbes adapted to the treatment that has been compiled. It is done at transplanting time of seedling four months years old. The anorganic fertilizer also adapted to the treatment that has been compiled used PT Socfindo recommendation (Table 2). The fertilizer is applicated around the stem in 10 cm space

**Table 2.** The dosage of NPKMg compound fertilizer for seedling since 4 to 7 months years old

| Old of the seedling (month) | Application Time (weeks) | NPKMg (g/polybag) | | |
|---|---|---|---|---|
| | | 50% | 100% | 150% |
| 4 | 17 | 2 | 4 | 6 |
| | 19 | 3,75 | 7,5 | 12,5 |
| 5 | 21 | 3,75 | 7,5 | 12,5 |
| | 23 | 3,75 | 7,5 | 12,5 |
| 6 | 25 | 3,75 | 7,5 | 12,5 |
| | 27 | 3,75 | 7,5 | 12,5 |
| 7 | 29 | 5 | 10 | 15 |
| | 31 | 5 | 10 | 15 |

### *2.6. Leaf- Sampling*

Leaf sample was taken from each unit of the treatment. It is done at seedling 8 month years old. The sample is dried in oven at $70^0$C for two days until the weight is constant.

## 3. Results and discussion

The compost treatment enriched endophytic fungi at various doses of NPKMg fertilizer did not affect plant growth until 2 months after application but the real influence started to occur at 3 month age crop. The main effects of endophytic fungi and the interaction effect between fertilizer and endophytic fungus did not affect plant height, but significantly affected the stem circumference and number of leaves. The diversity of plant growth is shown in Table 3, Table 4 and Table 5.

Table 3 showed that generally in the treatment of each type of endophytic fungus, the amount of midrib does not differ between treatments of fertilizer combination. Similarly, in each treatment of fertilizers, the effect of endophytic fungi on the number of stem is also the same, At the age of 8 months of plants, the highest number of barks is found in the treatment of 100% inorganic fertilizer recommendations. The effect was no different from the treatment that received compost input and various recommendation doses and also was inoculated with endophytic microbes.

Table 4 showed that the main effects of endocrine fungal application of *T.viride* increase the highest stem circumference compared to *T. koningii* and *Aspergillus*. Furthermore, in 8-month-old plants, from the interaction effect it is known that the highest stem circumference was found in combination of *T.viride* and compost treatment with 150% recommended inorganic fertilizer.

**Table 3.** Effect of combination empty fruit bunches compost and anorganic fertilizer and kind of endophytic microbe on number of leaf

| Treatment | Compost | 100% NPKMg | Compost + 50% NPKMg | Compost +100% NPKMg | Compost +150% NPKMg | Average |
|---|---|---|---|---|---|---|
| 3 month after fertilizer application (7 months years old) | | | | | | |
| | ......................................the number of leaf................................................ | | | | | |
| Control | 5.67 e | 7.67 abcd | 8.00 abc | 7.67 abcd | 6.67 cde | 7.13 y |
| *T. viride* | 8.00 abc | 7.67 abcd | 8.33 abc | 8.33 abc | 8.00 abc | 8.07 x |
| *T.koningii* | 7.67 abcd | 6.00de | 7.33 bcde | 8.67 ab | 8.33 abc | 7.60 xy |
| *Aspergillus* | 9.00 a | 8.00 abc | 7.33 bcde | 8.33 abc | 8.67 ab | 8.27 x |
| Average | 7.58 | 7.33 | 7.75 | 8.25 | 7.92 | |
| 4 month after fertilizer application (8 months years old) | | | | | | |
| Control | 7.67 d | 10.00 abc | 9.33 abcd | 10.00 abc | 8.67 bcd | 9.13 y |
| *T. viride* | 9.67 abc | 10.67 a | 10.33 ab | 9.67 abc | 10.33 ab | 10.13 x |
| *T.koningii* | 9.67 abc | 8.33 cd | 9.33 abcd | 10.00 abc | 10.00 abc | 9.47 xy |
| *Aspergillus* | 10.33 ab | 10.00 abc | 9.33 abcd | 8.67 bcd | 10.00 abc | 9.67 xy |
| Average | 9.33 | 9.75 | 9.58 | 9.58 | 9.75 | |

Note: The number in same rows followed the same notation represent not significant differences based on Duncan's Multiple Range Test (p = 0.05)

**Table 4.** Effect of combination empty fruit bunches compost and anorganic fertilizer and kind of endophytic microbe on stem circle

| Treatment | Compost | 100% NPKMg | Compost + 50% NPKMg | Compost + 100% NPKMg | Compost + 150% NPKMg | Average |
|---|---|---|---|---|---|---|
| 3 month after fertilizer application (7 months years old) | | | | | | |
| | ........................................the stem circle (cm)................................ | | | | | |
| Control | 4.43 | 4.93 | 4.93 | 5.10 | 4.70 | 4.82 c |
| *T. viride* | 5.23 | 5.63 | 5.63 | 6.57 | 6.63 | 5.94 a |
| *T.koningii* | 5.70 | 4.50 | 4.90 | 5.27 | 5.77 | 5.23 bc |
| *Aspergillus* | 6.13 | 5.90 | 4.93 | 5.57 | 6.10 | 5.73 ab |
| Average | 5.38 | 5.24 | 5.10 | 5.63 | 5.80 | |
| 4 month after fertilizer application (8 months years old) | | | | | | |
| Control | 5.27 d | 8.13 ab | 7.17 abcd | 6.93 abcd | 6.33 bcd | 6.77 x |
| *T. viride* | 6.97 bcd | 7.70 abc | 8.33 ab | 8.37 ab | 9.20 a | 8.11 y |
| *T.koningii* | 7.20 abcd | 6.23 bcd | 6.53 bcd | 7.07 abcd | 7.50 abc | 6.91 y |
| *Aspergillus* | 8.00 ab | 8.23 ab | 5.73 cd | 6.83 bcd | 7.57 abc | 7.27 xy |
| Average | 6.86 | 7.58 | 6.94 | 7.30 | 7.65 | |

Note: The number in same rows or same colomsfollowed the same notation notation represent not significant differences based on Duncan's Multiple Range Test (p = 0.05)

There is no interaction effect between the combination of compost - inorganic fertilizer and endophytic microbe to shoot dry weight (Table 5) . Conversely, the dry weight of the shoot is influenced by major factors of endophycytic microbial inoculation and compost and inorganic fertilizer combination. The highest shoot dry weight is affected by the *T. koningii* treatment. Meanwhile, in the treatment of fertilizers, the highest shoot dry weight is affected by combination of organic fertilizer and inorganic fertilizer dose 100% recommend.

**Table 5.** Effect of combination empty fruit bunches compost and anorganic fertilizer and kind of endophytic microbe on shoot dry weight

| Treatment | Compost | 100% NPKMg | Compost+ 50% NPKMg | Compost + 100% NPKMg | Compost + 150% NPKMg | Average |
|---|---|---|---|---|---|---|
| | ........................................ shoot dry weight (g)........................................ | | | | | |
| Control | 67.62 | 70.47 | 71.72 | 79.17 | 63.57 | 70.51 c |
| *T. viride* | 62.13 | 67.84 | 63.15 | 77.49 | 77.84 | 69.69 c |
| *T.koningii* | 73.69 | 83.50 | 77.47 | 82.19 | 81.27 | 79.62 a |
| *Aspergillus* | 78.58 | 79.17 | 68.19 | 82.53 | 74.84 | 76.66 b |
| Average | 70.50 c | 75.24b | 70.13 c | 80.34 a | 74.38 b | |

Note: The number in same row or same colom followed the same notation notation represent not significant differences based on Duncan's Multiple Range Test (p = 0.05)

The results showed that compost and fertilizer compound fertilizer NPKMg compound cause better growth response than compost only or microbes only. Even when viewed from the circumference of the stem and the amount of midrib is known to have an interaction effect between the combination of compost and fertilizer as well as microbial endophytic. This is because compound fertilizers have a certain nutrient composition while the soil used also has a low nutrient status when the plants need nutrients that are higher than those available in the soil and from the fertilizer. The positive effects of compost given are related to the quality of the compost. This means that the compost that is given has good quality, among others, from the initial analysis, it is known that the C / N ratio is about 14 and contains high N, K and Mg nutrients. The addition of organic materials with low C / N will increase the fertility of the soil more quickly because the organic matter is more easily decomposed [10]

This study showed that the interaction effect of treatment factor on plant growth is seen after the plants are 8 months or 4 months after application. This is in line with Riniarti's study [11] which showedagricultural waste such as bagases and empty fruit bunches that have been more biodegradable, at incubation 3 weeks. On the other hand, growth indicators. Plant stem diameter and dry weight significantly influenced by combination treatment of compound fertilizer and compost of empty fruit bunches. This is because the compost used has been ripe so as not to cause nutrient immobilization in the soil.

Based on this research it is known that to increase plant growth on endemic soil require compost and organic fertilizer according to recommendation even up to 1.5 times recommendation. This means that the requirement of plant nutrients in biotic stress conditions is higher than the ideal state as one of the mechanisms to increase disease resistance.

The results also showed that endophytic microbial administration can improve plant growth and there are differences in effect between microbial types. The presence of endophytic microbes in addition to increasing plant resistance to Ganoderma is also associated with improved nutrient status for plants [12] thereby increasing the growth of stem circumference and number of plant leaflets [13]

The quality of seeds from the results of this study is still below the standard. It is known that the standard 8-month-old palm seedlings have 11.5 cm in height and 64.3 cm in diameter and 3.6 cm in diameter [14]. This indicates that additional ameliorants such as lime are still needed because the soil pH used in this research is classified as acid. Lime can increase soil pH and increase the nutrients Ca and Mg.

## 4. Conclusions

The interaction effect of the both treatment factor used increase growth of seedling in third and fourth month after application. The best growth of seedling is on the treatment of empty fruit bunches compost combined anorganic fertilizer 150% recommended dosage and *Trichoderma viride*.

## References

[1] http://www.deptan.go.id/kelapa-sawit-informasi-ringkas-komoditas-perkebunan (retrieved on: 13 December 2013)
[2] http://www.indonesia-investments.com. (retrieved on: 20 August 2015)
[3] Lisnawita, Hanum H and Tantawi A R 2016 2nd International Conference on Agricultural and Biological Sciences (ABS 2016) IOP Publishing. IOP Conf. Series: *Earth and Environmental Science* **41** 012007
[4] Khairuddin H 1990 *Agriculture Science Thesis*: *Basal stem rot of oil palm: incidence, etiology and control* (Universiti Pertanian Malaysia, Selangor, Malaysia)
[5] Susanto A, Prasetyo A E, and Wening S 2013 *Jurnal Fitopatologi Indonesia* **1 9** 39-46
[6] Hanum H, Lisnawita and Tantawi A R 2014 *Proc The 3rd Int Conf of Multidisciplinary Research* ( Islamic North Sumatera University Medan) pp71-74
[7] Lisnawita, Hanum H, Tantawi A R 2014 *Proc.Nat. Conf. Pengendalian Penyakit pada Tanaman Pertanian Ramah Lingkungan* (Universitas Gajah Mada Yogyakarta) pp 244-252
[8] Herliyana E N, Taniwiryono D, Minarsih H, Firmansyah M A, Dendang B 2011 *Jurnal lmu Pertanian Indonesia* **16** 14-27.
[9] Darmosarkoro W and Winarna 2010 *Lahan dan Pemupukan Kelapa Sawit* (Pusat Penelitian Kelapa Sawit)
[10] Stevenson J *1984 Humus Chemistry; Genesis, Composition, Reactions* ( John Willey and Sons Inc New York) 496 pp.
[11] Riniarti D, Kusumatuty A and Tahir M 2013 *Jurnal Penelitian Pertanian Terapan* **13** 123-130
[12] Hanum H, Lisnawita and Tantawi A R 2015 Proc Int The 5th Int Conf (AIC 2015) in conjunction with The 8th Chemical Engineering on Science and Application (Syiah Kuala University Banda Aceh) pp 55-61
[13] Gusmawartati, Sampurna and Sitorus M 2013 *Jurnal Teknobiologi* **4** 99–103
[14] http://ditjenbun.pertanian.go.id/bbpptp medan [retrieved on: 4 Mei 2016]

**Acknowledgment**
This Research is part of The National Competitive Research on Scheme MP3EI (2015) from The Ministry of Research, Technology, and Higher Education of Republic of Indonesia with project number DIPA-023.04.1.673453/2015.

IOP Conf. Series: Earth and Environmental Science **122** (2018) 012064 doi:10.1088/1755-1315/122/1/012064

# The potential of soil fungi associated with potato rhizosphere to control root knot nematode (*Meloidogyne* spp.) on potato

**E Utari[1], Lisnawita[1*], I Safni[1], K Lubis[1], AR Tantawi[2] and Hasanuddin[1]**

[1] Department of Agrotechnology, Faculty of Agriculture, Universitas Sumatera Utara, Jl. Prof. A. Sofyan No. 3, Medan 20155, Indonesia
[2]Department of Agrotechnology, Faculty of Agriculture, University of Medan Area, Jl. Kolam No. 1 Medan Estate, 20223, Indonesia
*Email : lisnawita@usu.ac.id

**Abstract.** The root knot nematode (*Meloidogyne* spp.) is one of important pathogens on potato crops in North Sumatra, Indonesia. This nematode causes significant crop losses on potatoes directly and indirectly. The effect of fungal isolates (*Trichoderma* sp. 1, *Mucor* sp.1, *Aspergillus* sp. 2, *Mucor* sp. 2) that were isolated from rhizosphere of potato in North Sumatra were studied in green house experiments on the growth of potato and the reproduction of the nematode (*Meloidogyne* spp). The results showed that *Trichoderma* sp. 1 caused a significant gall reduction, while *Mucor* sp.1 and *Mucor* sp.2 could improve the growth of potato.

## 1. Introduction

North Sumatra is one of the potato production centres in Indonesia. The data showed that potato productions in North Sumatera in 2012 was 128,965 tons, more than the production in 2011, which was 123,078 tons [1]. However, the results of previous survey and research found that most locations of the potato farming in North Sumatera have been infested by root knot nematode (*Meloidogyne spp.*/RKN). Moreover, more than one RKN species is found in one location in several places [2].Three RKN species, namely *Meloidogye arenaria, Meloidogyne incognita,* and *Meloidogyne javanica,* were found at the potato farming in Karo District, North Sumatera [2].

Some strategies to control nematode have been reported, including cultural practices, plant resistence, and nematicide, but they have not given the satisfying result yet. The use of nematicide for controlling nematode gives the negative impact to the quality of environment, the balance of ecosystem and the human health. The use of nematode-resistant crop varieties are effective and economical way in controlling the plant parasite nematodes. Besides it can reduce the nematode population and the application is easily implemented on the field, this technique can also reduce the cost production to the lowest level. Unfortunately, potato resistance to *Meloidogyne spp.* is not available yet at the agricultural stores in North Sumatera.

Biological control with fungi which associates with potato plant rhizosphere, to control the parasite nematode especially the root knot nematode is the potential alternative technology to be developed. This is because fungi are the organisms that are naturally existed in the nature and have the same habitat with the plant parasite nematode, not dangerous to environment, easily multiplied on artificial medias with low cost, easily applied and will develop naturally and can survive when there is no host nematode, so it will become saprophyte inside the soil [3]. The aim of this study was to observe the fungi potential of the potato plant rhizosphere in controlling the population of root knot nematode in the green house.

## 2. Material and methods

### *2.1. The evaluation in the greenhouse*

The evaluation was conducted in the greenhouse of Balai Penelitian Tanaman Buah (BPTB), Tongkoh, Brastagi district by using potato G0 cultivar Granola. Each potato tuber was planted in polybag with the size of 25 cm x 25 cm that contained the mixture of topsoil, sand and fertilizer (2 : 1 : 1 / w / v) (5 kg / polybag). The polybags are arranged based *on* Complete Randomized Design in the green house. The plants were taken care of for 12 weeks. The moisture of sand inside each polybag was kept with daily watering.

### *2.2. The evaluation of antagonistic fungal isolates*

Based on *in vitro* screening, four antagonistic fungi isolates were selected and used for the evaluation in the greenhouse. They were *Trichoderma* sp1. (A1), *Mucor* sp1. (A2), *Aspergillus* sp2. (A3), and *Mucor* sp2. (A4). All fungi isolates were first cultured on Potato Dextrose Agar (PDA) on petri plates. The plates were incubated at 24°C for 14 days. The produced conidia were collected from cooked rice. 30 gr/polybag from each fungi ($10^6$ spores/g cooked rice [4], which had been counted with hemocytometer [5], were applied separately into each polybag. At the same time 10 ml suspension that contained 500 juveniles *Meloidogyne* spp. were applied into each polybag. The control treatments, which were the plants that were inoculated with only *Meloidogyne* spp. (A0), and only nematicide (Carbofuran 3%) (A5). Each treatment was conducted with 4 replications. Plants were fertilized by using NPK fertilizer (1 g/1 liter water) if it was needed. The observation was conducted to plant height at 1 – 12 weeks after application, tuber weight, root fresh weight and the number of root knot [6] at the end of evaluation (12 weeks after the application).

### *2.3 Statistical analysis*

Data was statistically analyzed according to standard analysis of variance by a one way ANOVA with SPSS software. The comparison between means was carried out by Duncan's Multiple Range Test [7] with level of 5 %.

## 3. Results and discussion

The results showed that four fungal isolates, gave the significat influence in increasing the addition of height of potato plants compared with the control treatment and nematicide (Carbofuran 3%) (Table 1). The highest plant height was at the treatment of *Mucor* sp 1 (19.48 cm). This result was not significantly different with *Muco*r sp.2 (19.10 cm), *Trichoderma* sp.1 (18.31 cm), and *Aspergillus* sp.2 (17.94 cm). Similar results were obtained on the observations of tuber weight and root fresh weight. All 4 fungal isolates gave the better result compared with control and the giving of nematicide (Carbofuran 3%) (Table 2).

The evaluation result indicated that the antagonistic fungal application including *Trichoderma, Mucor* dan *Aspergillus* could increase the plant growth like the plant height, root fresh weight, tuber weight and reduce the number of root knot. This result proved that the secondary metabolite activity produced by antagonistic fungi had nematicidal effect, therefore it affected the existence of nematode inside the root. Nematode could not develop well inside the root [8, 9, 10] (Figure 1). *Trichoderma* sp. produces volatile compound, i.e. 6-pentyl-a-pyrone (6-PAP), 1 ßvinylcyclopentane-1a, 6-pentyl-2H-pyran2-one, and 4-(2-hydroxyethyl) phenol that had nematicidal effect [11]. Furthermore, the attacked root would die and caused the plant growth inhabit [11]. The response of the plant to the root knot nematode was the response from all parts of the plant and response from the cells of the plant, all parts of the plant gave response to infection, and the decrease rate of photosynthesis, growth and production.

In contrast, the application of nematicide (Carbofuran 3%), did not give the significant difference with the control treatment. On this treatment, the plant height, root fresh weight, lower tuber weight were obtained, meanwhile the number of root knot were higher compared with the treated plants with the antagonistic fungi. This result indicated that nematicide (Carbofuran 3%) was resistant to control root knot nematode. This type of nematicide (Carbofuran 3%) is usualy applied by the farmers in controlling nematode.

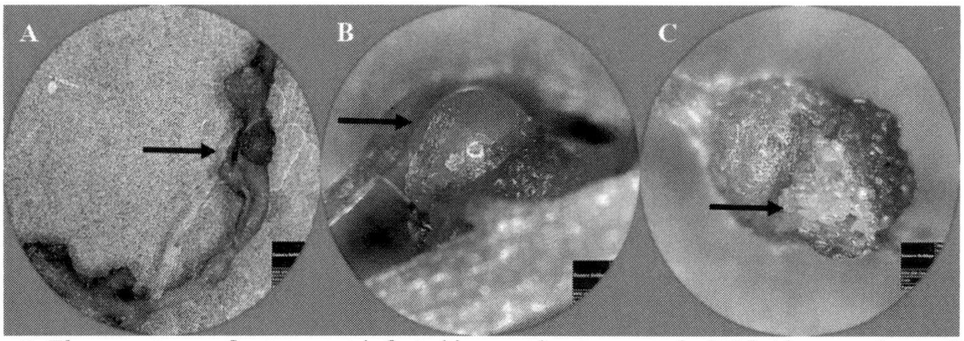


**Figure 1.** The symptoms of potato root infected by root knot nematode (*Meloidogne* sp.) : A. Galls on potato root, B. Female nematode, C. Nematode egg mass (arrows)

**Table 1.** The effects of antagonistic fungal application to potato plant

| | Plant height (cm) | | | | | | | | | | | |
|---|---|---|---|---|---|---|---|---|---|---|---|---|
| | Time of observation (weeks after application) | | | | | | | | | | | |
| Treatments | 1 | 2 | 3 | 4 | 5 | 6 | 7 | 8 | 9 | 10 | 11 | 12 |
| Control (A0) | 0.00 | 0.00 | 0.38 | 0.79 | 3.49 | 4.28 a | 6.80 a | 9.44 a | 10.54 a | 11.69 a | 11.76 a | 11.76 a |
| *Trichoderma* sp. 1 (A1) | 0.00 | 0.00 | 0.00 | 1.61 | 5.10 | 9.06 b | 12.50 b | 16.35 b | 16.68 b | 18.31 b | 18.31 b | 18.31 b |
| *Mucor* sp. 1 (A2) | 0.00 | 0.61 | 0.43 | 1.86 | 6.68 | 9.90 b | 12.63 b | 16.44 b | 16.69 b | 19.44 b | 19.44 b | 19.48 b |
| *Aspergillus* sp. 2 (A3) | 0.00 | 0.13 | 0.00 | 0.63 | 5.75 | 8.98 b | 12.54 b | 15.06 b | 16.00 b | 17.94 b | 17.94 b | 17.94 b |
| *Mucor* sp. 2 (A4) | 0.00 | 0.68 | 0.38 | 1.81 | 6.61 | 9.46 b | 12.55 b | 16.43 b | 16.46 b | 19.06 b | 19.06 b | 19.10 b |
| Nematicide (Carbofuran 3%) (A5) | 0.00 | 0.00 | 0.37 | 0.94 | 4.63 | 6.18 ab | 7.64 a | 10.29 a | 11.56 ab | 12.38 a | 12.38 a | 12.38 a |

Notes : The numbers followed by the same letter notations in the same table are not significantly different on Duncan's multiple range test with level of 5 %

**Table 2.** The effects of antagonistic fungi application to tuber weight, root fresh weight, and number of root knot on potato

| Treatments | Tuber weight (g) | Root fresh weight (g) | Number of galls |
|---|---|---|---|
| Control (A0) | 79.6 a | 0.19 b | 41.75 b |
| *Trichoderma* sp. 1 (A1) | 24.31 bc | 1.61 a | 19.13 a |
| *Mucor* sp. 1 (A2) | 30.98 c | 1.88 b | 36.00 b |
| *Aspergillus* sp. 2 (A3) | 20.35 bc | 2.09 b | 38.50 b |
| *Mucor* sp. 2 (A4) | 31.09 c | 2.82 b | 54.13 bc |
| Nematicide (Karbofuran 3%) (A5) | 13.29 ab | 0.35 a | 67.25 c |

Notes : The numbers followed by the same letter notations in the same table are not significantly different on Duncan's multiple range test with level of 5%

These results convince us that antagonistic fungi give the direct or indirect effects to the reproduction of nematode (number of root knot) and the response of host (plant growth). All antagonistic fungi that are obtained from the rhizosphere of potato plants that are used on this research have potentials as the biocontrol agents and as the biological fertilizer to control *Meloidogyne* spp. on potato plants under the condition inside the greenhouse.

## 4. Conclusions

In greenhouse trial, *Trichoderma* sp. 1 gave significat effect as antagonistic agents, which caused lower number of gallscompared with than the other fungi. *Mucor* sp. 1 and *Mucor* sp. 2 gave effect as the biological fertilizer because both isolated increased the plant height, tuber weight and root fresh weight.

## References

[1] Badan Pusat Statistik (BPS) 2015 Luas panen, produksi dan produktivitas kentang, 2009-2013. http://www.pbs.go.id/ (Dikunjungi 23 Februari 2016 )

[2] Lisnawita dan Hasanuddin 2014 Karakterisasi morfologi dan molekuler nematoda puru akar (*Meloidogyne* spp.) asal kentang di Sumatera Utara. Laporan Penelitian Hibah Fundamental Universitas Sumatera Utara

[3] Winarto dan Trizelia 2010 Aktivitas Antagonistik dan Karakterisasi Jamur yang Berasosiasi dengan Nematoda Bengkak Akar (*Meloidogyne* spp.) pada Tanah dan Akar Tanaman Tomat. Universitas Andalas, Padang. http://repository.unand.ac.id/6460/1/artikel.pdf

[4] Lisnawita, Tantawi A R and Pinem M I 2013 Aplikasi cendawan endofit terhadap perkembangan nematoda *Radopholus similis* pada pisang Barangan *J. Fitopatol Indonesia* **9 (5)** 133-138

[5] Booth C1971 Introduction to general methods. In: Booth, C. (Ed.), In: Methods in Microbiology, vol 4 Academic press Inc. (London) Limited, UK pp. 1–49

[6] Hussey R S and Barker K.R 1973 A comparison of methods of collecting inocula for *Meloidogyne* spp. including a new technique *Plant Dis. Rep* **57** 1025–1028

[7] Duncan B 1955 Multiple ranges and multiple F. test. Biometriex **11** 1-42

[8] Radwan M A 2007 Comparative effects of culture filtrate of soil-borne fungi on mortality and infectivity of juveniles of *Meloidogyne incognita Indian J. Nematol* **37**109-114

[9] Xalxo P.C, Karkun D and Poddar A N 2013 Rhizosperic fungal association of root knot nematode infested cucurbits : *in vitro* assessment of their nematicidal potential *J. Microbiol* **8 (2)** 81-91

[10] Santoso A, Toto H dna Hagus T 2014 Pengaruh filtrat biakan *Trichoderma* spp. terhadap penetasan telur nematoda puru akar (*Meloidogyne* sp..) *J Hama Penyakit Tumbuhan* **2 (4)** 85-91

[11] Yang Z, Yu Z, Lei L, Xia Z, Shao L, Zhang K dan Li G 2012. Nematicidal effect of volatiles produced by *Trichoderma* sp. *J. Asia Pasific Entomol* **15** 647-650

[12] Starr JL, R Cook and Bridge J 2002. Plant Resistance to Parasitic Nematodes CABI Publishing

## Acknowledgements

This Research is part of The National Competitive Research on Scheme National Strategy (Stranas) (2016-2017) from The Ministry of Research, Technology, and Higher Education of Republic of Indonesia.

IOP Conf. Series: Earth and Environmental Science **122** (2018) 012065 doi:10.1088/1755-1315/122/1/012065

# Use of organic waste as biofumigant for controlling root knot nematodes (*Meloidogyne* spp.) on potato

**D I P Sari [1], Lisnawita [1*], S Oemry [1], I Safni [1], K Lubis [1] and A R Tantawi [2]**

[1]Department of Agrotechnology, Faculty of Agriculture, Universitas Sumatera Utara, Jl. Prof. A. Sofyan No. 3, Medan 20155, Indonesia
[2]Department of Agrotechnology, Faculty of Agriculture, University of Medan Area, Jl. Kolam No. 1 Medan Estate, 20223, Indonesia
[*]E-mail : lisnawita@usu.ac.id

**Abstract.** Root knot nematode (*Meloidogyne* spp.) is one of the important pathogens that causes big impact on potato crop yields. One of the control strategies for controlling this nematode is the use of biofumigants. Biofumigants are *volatile toxic compound* derived from plants, and have biocide properties against insects and plant pathogens. Organic waste such as *Brassicaceae*, *Leguminoceae*, and *Solanaceae* can be used as biofumigant sources. This research was conducted to determine the effectiveness of *Brassicaceae*, *Leguminoceae*, and *Solanaceae* as biofumigants against *Meloidogyne* spp. The experiment was set in a *completely randomized design* (CRD) with the treatments were organic wastes including *Brassicaceae*, *Leguminoceae*, and *Solanaceae*, both single and combinations, and 2 controls (positive and negative controls) with 3 replications. Each of the biofumigant treatments was prepared and stored for 2 weeks. Potato tubers were transplanted 15 days after germination into polybag inoculated with 1,000 *Meloidogyne* spp. J2s. The results showed that *Brassicaceae* + *Solanaceae* were effective in decreasing the number of galls in potato plants, however only *Solanaceae* improved plant growth.

## 1. Introduction

Potato (*Solanum tuberosum* L.) the fourth most important food crop in the world after corn, rice and wheat. Potatoes and other plants use these tubers to store sugars, called carbohydrates, as energy and nutrition sources they need to reproduce. These carbohydrates and nutrients are also beneficial to the human diet . One factor that reduce the potato production is the attack of nematodes [1], which one of them is root knot nematode (*Meloidogyne* spp.) [2]. The loss that is caused by the attack of root knot nematode could happen on the field, so it could reduce the quality and the quantity of the product. This nematode attack could decrease the crop yields around 15% on potato plants [3]. *Meloidogyne* spp. larvae infect plant *roots*, *causing* the development of *root*-knot *galls*. This condition causes the limitation of water absorption and even the essential fertilizer to the plants, hence it would harm the growth and the development of the plants and the loss could not be avoided [4].

During the last 50 years, the nematode control using chemical nematicide has been very significant role, but this control causes negative impact on living organisms and environment including polluting land and waterland causing death for non-targeted organisms, such as nematode natural enemies such as fungi and bacteria. Therefore, looking for an alternative environmental friendly control strategies is urgently needed. One of those alternative methods is the use of biofumigants.

Daulay [5] reported that the addition of the organic material such as the residue of cabbage, broccoli, mustard and radish could be made use as biofumigants to control nematodes and the other soil borne pathogens through the biofumigant process. Biofumigants are *volatile toxic compound* derived from plants, and have biocide properties against insects and plant pathogens. Horticultural biofumigant is based on the organic molecular exploitation derived from the nature that possesses ability as biocide, which the organic moleculs are contained in a number of plants species such as from *Brassicaceae* [6]. There are several plant species that could be used as biofumigants [5]. The purpose of this research is to study the effectiveness of *Brassicaceae*, *Leguminoceae*, and *Solanaceae* as single or combination as biofumigants to control the root knot nematode (*Meloidogyne* spp.) on potato plants.

## 2. Material and methods

### *2.1 The supply of biofumigant*

The organic wastes *Leguminoceae, Brassicaceae,* and *Solanaceae* were cleaned and cut into small pieces with the size around 1-3 cm for *Leguminoceae, Brassicaceae* and around 5 – 7 cm for *Solanaceae*. After being cut,all organic wastes were ready to be used.

### *2.2 Extraction of root knot nematode*

The potato roots that were infested by root knot nematode from the potato plantations in Karo District were collected and put into a plastic bag before brought to Plant Disease Laboratory, Faculty of Agriculture, Universitas Sumatera Utara. The roots were washed under the running water. The egg masses of nematode were collected under a stereo microscope. The egg masses were soaked in the sterile water, and incubated for 24 hours at $22^0$ C. Juvenil 2 (j2) that germinated from the eggs were used for the evaluation.

### *2.3 Screenhouse studies*

The experiment was conducted at Balai Penelitian Tanaman Buah (BPTB), Tongkoh, Karo District. The experiment was conducted with the *Completely Randomized Design* (CRD) with 3 organic waste treatments, namely *Brassicaceae*, *Leguminoceae*, and *Solanaceae,* single and combinations, and 2 controls (positive control which was nematicide with carbofuran (Furadan 3G) with the suggested dosage, and negative control which was treatment without biofumigation) with three replications. The fumigational treatment were conducted by mixing 300 g biofumigant / 3 kg soil (1 : 10) in polybags and the polybags were tightly closed for two weeks. The polybags were unbound one day before the potato tubers were planted. The 15 day old- potato seedlings of Granola G0 cultivar were planted in polybags. The nematode inoculation was conducted during the planting of potatoes. Each plant was inoculated more or less 1000 J2 nematodes.

The observation was conducted toward the plants growth such as : tuber weight (g), root length (cm), shoot fresh weight (g), shoot dryweight (g), and the nematodes growth including the number of galls, and the final nematode population.

### *2.4 Statistical analysis*

Data were statistically analyzed according to standard analysis of variance by a one way ANOVA with SPSS software. The comparison between means was carried out by Duncan's Multiple Range Test [7] with level of 5%.

## 3. Results and Discussions

The results showed that the treatment of application of the agricultural waste as biofumigant gave the better influences to the plants growth such as : tuber weight, shoot fresh weight, shoot dry weight, root length and diameter of root distribution compared to the negative control plants (K0) and the application of nematicide (K8) (Table 1). On the positive control treatment (K8) and the negative control treatment (K0), seen the nematode attack that retarded the root growth. This occurs because of the ability of nematode to use the significant elements such as phosphor and carbon needed by the plants. Nematodes transfered the elements for its importance to proceed its life cycle, thefore roots lacked of nutrition and could not grow well. According to [8] *Meloidogyne* spp. could transfer 10 % of total phosphor compound and carbon derived from shoot to the root part of the plants for the importance of nematode activities in accomplishing its life cycle.

**Table 1.** The impact of organic wastes *Brassicaceae*, *Leguminoceae*, and *Solanaceae* as the single and combination of biofumigant toward the parameter of potato growth infected by *Meloidogyne* spp.

| Treatment | Tuber weight (g) | Shoot fresh weight (g) | Shoot dry weight (g) | Root length (cm) | Diameter of root distribution(cm) |
|---|---|---|---|---|---|
| K0 | 3.48 d | 1.84 c | 0.33 d | 3.65 c | 4.61 b |
| K1 | 32.42 bc | 3.70 abc | 1.75 abc | 5.19 ab | 11.1 a |
| K2 | 16.38 cd | 3.24 abc | 1.51 bcd | 4.98 ab | 9.58 a |
| K3 | 39.04 abc | 5.42 a | 3.30 a | 5.74 a | 12.3 a |
| K4 | 46.90 ab | 3.95 abc | 2.11 ab | 5.21 ab | 8.75 a |
| K5 | 43.32 ab | 4.45 ab | 2.13 ab | 5.28 ab | 10.9 a |
| K6 | 51.18 ab | 4.65 a | 2.72 ab | 5.76 a | 12.6 a |
| K7 | 56.09 a | 4.83 a | 2.40 ab | 5.66 a | 11.4 a |
| K8 | 8.80 d | 2.48 bc | 0.50 cd | 4.47 bc | 7.76 ab |

Note : K0 (negative control), K1 (*Brassicaceae*), K2 (*Leguminoceae*), K3 (*Solanaceae*), K4 (*Brassicaceae+Leguminoceae*), K5 (*Brassicaceae+Solanaceae*), K6 (*Leguminoceae+Solanaceae*), K7 (*Brassicaceae+Leguminoceae+ Solanaceae*), K8 (carbofuran = positive control)

By increasing the growth of potato that were applied by biofumigants *Brassicaceae, Leguminoceae,* and *Solanaceae*, these organic wastes did not only have roles as biofumigants but also had a role as biofertilizer. The application of organic waste improve the process of organic material decomposition. The addition of organic materials into the soil can improve the chemical and physical conditions of the soil and therefore can increase the microbes which can balance the natural environment underneath the soils. *Brassicaceae, Leguminoceae,* and *Solanaceae* contain fertilizer such as N, P, K which are important requirement for plants for their growth. Similarly, *Brassicaceae* produces nitrogen for the plant growth as well as P, K, Mg and C-Organic. This condition is also similar to the wastes *Leguminoceae* and *Solanaceae*. Besides *Brassicaceae* can reduce the nematode population, it can also improve the soil structure, increase fertilizer elements, and increase the useful microbe activities [9].

It can be seen from Table 2 that the impact of organic wastes *Brassicaceae*, *Leguminoceae*, and *Solanaceae* as the single and combinations of biofumigants toward the number of galls and the nematode population potato plants infected by *Meloidogyne* spp. showed that the application of agricultural wastes could reduce the number of galls and nematode population compared to control treatment.

The agricultural waste application as the biofumigants on potato plants infected by *Meloidogyne* spp. gave lower average of galls number, compared to the plants that were not applied by biofumigants (negative control / K0) (36 galls/10 g root) and the application of nematicides with active carbofuran material (positive control / K8) (52.33 galls/10 g root). The lowest average galls number was the treatment *Brassicaceae+Solanaceae* (K5), which was 9.16 galls/10 g root. The highest final nematode population from roots and from the soil was potatoes treated with carbofuran (949 nematodes) comparing to control (K0) which was 616.5 nematodes. The successfull treatment of combination of *Brassicaceae* and *Solanaceae* because potato is grouped as Solanaceae and and *Brassicaceae* waste has been proven to be able to control nematodes through biofumigant process. Brassica produces *Glukosinolat* that can be converted into the chemical material (*isothiosianat*) with biofumigant activities [8].

**Table 2.** The impact of organic wastes *Brassicaceae*, *Leguminoceae*, and *Solanaceae* as the single biofumigants and combinations toward the number of galls and the nematode population on potato plants infected by *Meloidogyne* spp.

| Treatment | The number of galls/10 g root | Nematode population (10 g root + 100 g soil) |
|---|---|---|
| K0 | 36.00 ab | 616.5 abc |
| K1 | 18.00 bc | 461.0 c |
| K2 | 9.33c | 507.6 bc |
| K3 | 10.16 c | 853.6 ab |
| K4 | 10.50 c | 542.5 bc |
| K5 | 9.16c | 378.3 c |
| K6 | 15.16 bc | 497.8 bc |
| K7 | 18.50 bc | 510.5 bc |
| K8 | 52.33 a | 949.0 a |

Note : K0 (negative control), K1 (*Brassicaceae)*, K2 (*Leguminoceae)*, K3 (*Solanaceae)*, K4 (*Brassicaceae+Leguminoceae)*, K5 (*Brassicaceae+Solanaceae)*, K6 (*Leguminoceae+Solanaceae)*, K7 (*Brassicaceae+Leguminoceae+ Solanaceae)*, K8 (carbofuran = positive control

## 4. Conclusions

The results showed that combination treatment of *Brassicaceae* + *Solanaceae* were effective in decreasing the number of galls in potato plants, however only *Solanaceae* improved plant growth.

## References

[1] Hafez S L & Sundararaj P 2005 Potato nematode management tactics with biological and chemichal strategis idaho potato *Conference*

[2] Supramana A dan Gede S 2015 *Meloidogyne incognita* penyebab umbi berbintil pada kentang di beberapa sentra produksi kentang di Jawa (*Meloidogyne incognita* the cause of potato tubers berries in some potato production centers in Java) *J Fitopatologi Indonesia* **11**(5) : 143-149

[3] Mustika I 2010 Konsepsi dan strategi pengendalian nematoda parasit tanaman di Indonesia (Conception and strategy to control plant parasitic nematodes in Indonesia) *Pengembangan Inovasi Pertanian* **3**(2) : 81-101

[4] Lamovsek J, Gregor U and Stanislav T 2013 Biological control of root-knot Nematodes (*Meloidogyne* spp.): microbes against the pests. *Acta Agriculturae Slovenica* **101**(2): 263-275

[5] Daulay N S 2013 Sisa tanaman cruciferae sebagai biofumigan untuk mengendalikan nematoda puru akar (*Meloidogyne* spp.) (Cruciferae plant waste as biofumigant to control the root nematodes) *Departemen Proteksi Tanaman Institut Pertanian Bogor*, Bogor

[6] http://www.agwest.sk.ca/events 2001 Biofumigation. Agroscience (Diakses 02 Maret 2017).

[7] Duncan B 1955 Multiple ranges and multiple F. test. *Biometriex* **11**: 1-42

[8] Wardhiany C K, Made S and Ketut A Y 2014 Studi uji ekstrak beberapa jenis gulma dalam menekan nematoda puru akar *Meloidogyne* spp. pada tanaman tomat (*Licopersicum esculentum* Mill) (Test of several extraxt types of weeds to suppressing root nematodes *Meloidogyne spp*. on tomato plants) *Jurnal Agroekoteknologi Tropika* **3**(1)

[9] Nur JN. Supramana. Abdul M 2016 Keefektifan limbah tanaman *Brassicaceae* untuk pengendalian nematoda puru akar (*Meloidogyne spp.)*pada mikroplot di lapangan *Jurnal HPT Tropika* **16** (2) : 99-106

**Acknowledgements**
This Research is part of The National Competitive Research on Scheme National Strategys (Stranas) (2016-2017) from The Ministry of Research, Technology, and Higher Education of Republic of Indonesia.

# Understanding the social and economic aspects of upland rice farming

**S A A Taridala[1], W G Abdullah[1], Suaib[2], S Wahyuni[3], N I Wianti[4], M Zani[1], L O Jabuddin[1], W O Yusria[1], M A Limi[1] and A S Ekaputri[5]**

[1]Department of Agribusiness, Faculty of Agriculture, Universitas Halu Oleo, Kendari, Southeast Sulawesi 93212, Indonesia
[2]Department of Agrotechnology, Faculty of Agriculture, Universitas Halu Oleo, Kendari, Southeast Sulawesi 93212, Indonesia
[3]Department of Food Science and Technology, Faculty of Technology and Industry, Universitas Halu Oleo, Kendari, Southeast Sulawesi 93212, Indonesia
[4]Department of Extension, Faculty of Agriculture, Universitas Halu Oleo, Kendari, Southeast Sulawesi 93212, Indonesia
[5]Alumnus of Marketing Department, University of Birmingham, UK
Email: sitti.taridala@uho.ac.id; aidataridala@yahoo.com

**Abstract.** The objectives of this study were to (1) analyze the socio-economic characteristics of upland rice farmers, and (2) to analyze the productivity of farming in South Konawe Regency of Southeast Sulawesi Province, Indonesia. The analysis used in this research was combine economic research through quantitative and qualitative analysis. This research was conducted by survey method. The results showed that (1) farmers are generally in productive age, dominated by men, with low formal education level, and moderate family members, (2) upland rice farming is cultivated in medium land area, with fixed costs higher than variable cost, productivity that has been increased but still lower than rice paddy, and the price of rice production is relatively higher than rice paddy production price, and (3) feasible to cultivate dryland rice, and has a high efficiency value.

## 1. Introduction

As the world's population increases, food provision is crucial to every country and it is a debate on how to ensure global food security [1]. As in some countries in Asia [2] and Africa [3, 4], rice also plays an important role in the economy and is a staple food for most Indonesians [5]. Therefore, the provision of sufficient quantities and affordable prices is a top priority of national development. Every ruling leader will make every effort to meet the needs of his people. In the 2015 was born Special Effort (UPSUS) program, which establishes sustainable self-sufficiency for rice, corn, and soybeans that can be achieved in 2018 [6]. It is a fact that so far the mainstay of national rice production has always focused on irrigated wetland rice fields, especially in Java. While the contribution of dry land (where to grow upland rice, or gogo rice) that its availability is spread in various islands in Indonesia is not utilized optimally. Upland rice that produces a product of red or black rice, is underestimated so that the productivity achieved is low [7].

Similarly, in South Konawe Regency of Southeast Sulawesi Province, the development of upland rice did not show a significant increase, whereas this area is quite potential, both in terms of supply and demand side. In terms of supply, there is dryland, high biodiversity, suitable climate, and farmers have been used to cultivate upland rice. From this aspect of supply, one way to increase production is by optimizing the utilization of various biodiversity potentials [5]. From the demand side, the demand for brown rice tends to increase, because the community considers its benefits as functional food [7, 8]. Approximately 80 percent of consumers of red or brown rice (local rice) in Kendari buy rice from other regions because it is more profitable. Where there is a price difference of IDR 18,500 $kg^{-1}$. The average buying price is IDR 9,000 $kg^{-1}$. If you buy brown rice sold in mini market or super market in Kendari City, the price is very expensive that reaches IDR 27,500 $kg^{-1}$. The price continues to increase. The latest survey shows that the price of brown rice in Kendari has reached IDR 33,500 $kg^{-1}$ [9]. The price of brown rice is relatively very expensive, because it is generally imported from outside the province of Southeast Sulawesi, namely from the island of Java.

One of the main agendas to fulfill rice needs is to encourage better, effective and efficient production systems [10]. A good, effective, and efficient production system is one of the aspects that must be met to achieve a sustainable livelihood for the fulfillment of various needs of farmers and their families. As a result of low productivity, which only reached 412.55 kg $Ha^{-1}$ per planting season [9], lower than the productivity figures released by the Central Bureau of Statistics, which reached 3.12 tonnes $Ha^{-1}$ [11]. The upland rice farmers in South Konawe can not rely on income from upland rice farming. Dryland rice farmers can be called small farmers (peasant). Because they are farmers who do not fully gain market access and because the household functions both as a unit of production and consumption. Therefore, this research needs to be done because to obtain information about the socio-economic condition of farmers and their families, as well as the state of production, productivity, income, and efficiency of farming. Because during this time, the production of upland rice is obtained from land that has been planted several times with traditional cultivation practices, the productivity is so low that it affects the welfare of the farmer's family [9]. Information on socio-economic aspects and aspects related to the production of upland rice is very beneficial for the government to reduce the gap, especially for dryland rice farmers.

## 2. Material and Method

The population of this research is rice farmer in 2 (two) districts, namely Wolasi and Baito in South Konawe Regency, Southeast Sulawesi Province, Indonesia. As is known from previous studies [9], that the dryland rice farmers in Wolasi Sub-district are all Tolaki ethnic, while in Baito sub-district is a mixture of ethnic Tolaki and Java. The number of selected samples is 81 people representing farmers' households. Determination of the sample using simple random sampling, by taking 25 percent of the total population. This research was conducted by survey method.

Primary data is the data collected by (1) conducting interviews at the group level to the individual scope in the household, (2) observation at the location of cultivation, and (3) application of cultivation technology, especially the introduction of technological packages of increasing productivity of dryland rice. The analytical tool used descriptive statistics such as maximum, minimum, and average values, as well as percentages. Also used income analysis and R-C ratio. For income analysis, using the profit formula [12]: Profit = TVP – TC = TVP – TVC – TFC

$$= Py\,Y - Px\,X - TFC \qquad (1)$$

In accordance with the purpose of this study, adjustments are made to (1) to:

$$I = TR - TC \qquad (2)$$

$$TR = Pq * Q \qquad (3)$$

$$TC = Pxi * Xi + TFC \qquad (4)$$

Where: I = Income, IDR/Year

TR = Total Revenue, IDR/SP (IDR/SP); 1 Season Planting about 4.5 month

TC = Total Cost, IDR/PS; TFC = Total Fixed Cost, IDR/PS

Pq = Price of Brown Rice, IDR $Kg^{-1}$; Q = Quantity of Brown Rice, $Kg^{-1}$

Pxi = Price input (IDR per unit); Xi = Amount of input (unit)

The formula for assessing farm efficiency is R-C *ratio*: R/C = TR/TC

In Indonesia, for socio-economic aspect, there were three category of age, namely productive age yet (< 15 year), productive age (15-54 year), and not productive age (>54 year). For socio-economic aspect of education contain at least five category, such as: no school, elementary school, junior high school, high school, and bachelor degree.

## 3. Result and Discussion

### *3.1. An overview of the practice of upland rice cultivation*

The cultivation system practiced by the upland rice farmers in South Konawe Regency is still very simple. In many aspects, they still follow the habits of their predecessor parents. In fact, in some aspects they are still very strong in holding these habits. For example, should not kill pests such as rats, pigs, and monkeys. Because it will cause the animals are angry and will increasingly increase his attacks on the plant. The use of modern technology in gogo rice farming is very limited, both equipment technology to facilitate various work in the process of cultivation to harvest, and technology such as superior seeds, fertilizer, and the eradication of weeds, pests and diseases. The results showed that the practice of upland rice cultivation by farmers is still done in 2016 [9], except in the case of the use of liquid organic fertilizer. The results showed that farmers have used organic fertilizers in the form of ZPT, NPK, and POC, both in soaking rice seed and to increase the availability of nutrients for plants. Activities in the upland rice farming are carried out independently by farming families (land clearing, weeding, fertilizing), and some are carried out in mutual assistance with farmers or other families (in cultivation, and harvesting).

### *3.2. Social characteristics of farmers*

The socio-economic characteristics of the upland rice farmers is important to illustrate the efforts of the farmers. So it can be understood what and how the actions they take in managing dryland rice farming. Description of the social characteristics of the upland rice farmers is presented in Table 1.

**Table 1.** Social Characteristics of Farmers

| Description | Tipes of Social Characteristics | | | | | | | | | | | |
|---|---|---|---|---|---|---|---|---|---|---|---|---|
| | Age (Year) | | Gender | | Education | | | | | Family Size (Person) | | |
| | 25-54 | >54 | M | W | NS | ES | JHS | HS | BD | 1-3 | 4-6 | >6 |
| Number of Respondent (n) | 62 | 19 | 67 | 14 | 5 | 48 | 13 | 12 | 3 | 19 | 50 | 13 |
| Persentage (%) | 76.54 | 23.46 | 82.72 | 17.28 | 6.17 | 59.27 | 16.05 | 14.81 | 3.70 | 23.50 | 60.50 | 16.00 |
| Total (n, %) | 81 (100%) | | 81 (100%) | | 81 (100%) | | | | | 81 (100%) | | |

Note: n = number of samples; M = Man; W = Woman; NS = No school; ES = Elementary school
JHS= Junior high school; HS = High school; BD = Bachelor degree

Age dimension is a reflection of maturity level of one's thinking, also shows the level of rational consideration of a person in making decisions related to the problems encountered. Besides, the age of a person will determine the physical ability in doing a job. The results showed that the majority of respondents in South Konawe Regency were in the age range 25 to 54 years with the percentage of 76.54%, while the remaining 23.46% were in the age range more than 54 years.

The concept of gender in this paper is not only interpreted as a physical distinction or segregation, but also related to the consequences of such segregation socially. Table 1 informs that upland rice farmers in South Konawe Regency are dominated by male respondents (82.72%), while female respondents are 17.28%. It caused in managing rice farming gogo (upland rice farming) required a strong physical ability in performing stages of production ranging from land processing to post-harvest.

There are 6.17% of respondents who have never received formal education, 59.27% have elementary school education, and 3.70% have gone through higher education. Thus it can be said that most of the farmers have a formal education that is low. Therefore, to improve the knowledge, skill, way of thinking and creativity in running upland rice farming needs to be supported by non formal education.

The number of family members owned can be a source of manpower to assist in the management of farming, so as to increase the production and income of the family. Therefore, the number of family members in a household, describes the availability of labor [13]. With the availability of sufficient labor in a household, then in a farming activity is not required rental labor from outside the family.

*3.3. Characteristics of farming*

Characteristics of farming can vary between place, time, and culture of the surrounding community. This difference will have implications on many aspects, including on production, productivity, production costs, revenues, profits, and efficiency of cultivated upland rice. In this section will be described on the area of arable land, production costs, productivity, and price of upland rice. Description of the characteristics of upland rice farming is presented in Table 2.

**Table 2.** Characteristics of Gogo Rice Farming

| Description | Tipes of farming characteristics | | | | | | | | | |
|---|---|---|---|---|---|---|---|---|---|---|
| | Land area | | Production cost (IDR) | | Productivity (Kg Ha$^{-1}$) | | | Production price (IDR Kg$^{-1}$) | | |
| | >0.5 Ha | 0.6-2 Ha | VC | FC | Min | Max | Average | Min | Max | Average |
| Amount | 19 | 62 | 94938 | 274993 | 120 | 2128 | 476 | 12000 | 13000 | 12500 |
| Persentase (%) | 23.46 | 76.54 | 25.67 | 74.33 | | | | | | |
| Total (n, %) | 81 (100%) | | 369.931,- (100%) | | | | | | | |

Note: VC= Variable cost; FC = Fixed cost

Land is a very important asset for a farmer. The size of the land determines the scale of the farm, affects the amount of use of other factors of production, and will ultimately determine the level of production and income of farmers. The results showed that 19 respondents or 23.46% manage the land with the width of <0.5 Ha (narrow land) and 62 respondents or 76.54% of the total respondents manage the land of 0.5 - 2 Ha (medium category). This result confirms that based on the area of land managed, generally farmers are categorized as moderate land owners (average 1.02 Ha). In Indonesia, researchers often use farmers' categorization based on the area of land under control, ie narrow land area (<0.5 ha), medium land area (0.5 - 2 ha) and large area (> 2 ha) [14]. The extent of land tenure is also a measure of one's social status, which of course will have different economic impacts.

The cost of production is the value of all input in conducting a production process of farming. The cost of production of upland rice farming in this study consists of variable costs and fixed costs. The results showed that the production cost incurred by the upland rice farmers in South Konawe Regency consisted of wage labor (as variable cost) that is average of IDR 94,938 per planting season or 25.67% of total production cost. The fixed costs consisting of land and equipment depreciation amounted to IDR 1,444 (0.39%) and IDR 273,549 per planting season (73.94%), with an average total production cost of IDR 369,931 per planting season. The cost is much lower than the production cost of paddy field farming which can reach 3 to 6 million per planting season.

Based on the results obtained, it is known that the productivity of upland rice farming ranges from 120-2,128 Kg Ha$^{-1}$ rice with an average of 476 Kg Ha$^{-1}$ rice per planting season. The productivity of upland rice is much lower than the productivity of lowland rice farming which can reach 2,400 - 3,000 Kg Ha$^{-1}$ per planting season. The low productivity of upland rice farming system is generally caused by the lack of application of good and proper cultivation technology in the production process. One of the most prominent causes is the absence of nutrients into the soil, so that the soil structure becomes hard. Also the area of ubiquitous rice cultivation that is still overgrown with grass, planted with agroforestry and intercropping systems, and with irregular spacing. One proven way to increase productivity of upland rice is by bokashi application [5] as well as the technology of varieties and terraces [15].

The nowdays phenomenon is rarely encountered is the main motive of the production process of gogo rice farmers. Their main goal is to meet the food needs of his family. The grain or rice they produce is primarily intended to meet the staple food needs of all family members. Price information shown in Table 2 confirms that the price of upland rice ranges from IDR 12,000 to IDR 13,000 $Kg^{-1}$, with an average price of IDR 12,500 $Kg^{-1}$. This indicates that the price of gogo rice is higher than rice price of wetland rice which only reach IDR 8,000 to IDR 10,000 $Kg^{-1}$. The higher price of red rice from upland rice is a common phenomenon for organic products [7]. Thus, upland rice has good prospects to be developed in the future. Because it is not only a healthy product [16], but also economically expensive.

### *3.4. Revenue, income, and efficiency of farming*

Table 2 shows that the amount of income from farmers of respondents from upland rice farming ranges from IDR 960,000- IDR 25,536,000 with an average of IDR 5,961,798 per planting season. Ownership of cultivated land for an average of 1.02 hectares of paddy field is obtained per hectare amount of IDR 5,844,891 per planting season.

The result of the analysis shows that the activity of upland rice farming in South Konawe Regency gets income/profit from economic aspect. The farmers' income amounts to IDR 725,000-IDR 24,892,667 per planting season, average income of IDR 5,601,743 per planting season. If the value of such income is converted into income per month, the value of IDR 1,400,436 is obtained. If the value is divided by the average number of household members for 5 people, then per capita income per month is IDR 280,087. The amount of this income is lower than the standard of the rural poverty line of rural Southeast Sulawesi Province of IDR 317,881 per capita per month [17]. Based on this condition, if gogo paddy farmers rely solely on income from working in gogo rice farming without improving cultivation technology, then the income earned is very small. Thus, the households of upland rice farmers in South Konawe Regency are still classified as poor.

**Table 3.** Revenue, Income and Efficiency of Upland Rice Farming

| Description | Total revenue (IDR/Season) | Income (IDR/Season) | Farming Efficiency (R/C) |
|---|---|---|---|
| Minimum | 960,000 | 72,5000 | 2.0 |
| Maximum | 25,536,000 | 2,489,2667 | 77.7 |
| Average | 5,961,798 | 5,601,743 | 27.3 |

Income earned, not the primary measure to assess the success of a farm. By him it is necessary to know the efficiency of the production process undertaken, to determine the sustainability of upland rice farming. This is done so that farmers can know how much the level of profits obtained due to the cost that has been sacrificed during the production process. Analysis of the efficiency of upland rice farming in this study is proxyed based on the return cost ratio (R/C) approach. The results showed that the efficiency of upland rice farming in South Konawe Regency was very good (efficient), with an average value of 27.3. The results of the study in Aceh also corroborate the findings of this study, that the gogo rice farming is financially feasible [18].

When compared to the value of farming efficiency gained previously [9], this result is higher and the value improves. The increased efficiency of farming can be attributed to the improved production practices practiced by farmers, which in previous studies [9] show some inefficient gogo rice farming. This condition also implies that there is an increase in income obtained by upland rice farmers. A good income condition shows that the upland rice farmer has been able to sovereignly meet his daily food needs. Nevertheless, the income condition of these dryland rice farmers needs to be harmonized with the social factors of other farmers' households. This is for the formulation of specific policies related to the adoption of organic agricultural technology in sustainable rice farming. This should be done considering that upland rice has important importance [18] is a people's plant, where most of the people's agriculture is dry land without irrigation.

## 4. Conclusion

Some conclusions can be drawn: (1) Gogo farmers are generally in productive age, dominated by men, with low formal education level, and moderate family members, (2) upland rice farming is cultivated in medium land area, and the price of rice production is relatively higher than wetland rice production price, and (3) in South Konawe Regency is feasible to cultivate dryland rice, and has a high efficiency value.

The policy implication that can be recommended to local government of South Konawe Regency is to encourage the introduction of good and proper dryland rice cultivation technology (environmentally friendly), the use of liquid organic fertilizer, and the improvement of the human resources of farmers through the strengthening of community institutions. In addition, the government also needs to prepare professional extension workers, in order to assist the community in carrying out the cultivation of upland rice.

## References

[1] Frimawaty E, Basukriadi A, Syamsu A J, and Soesilo Budhi T E 2013 Sustainability of Rice Farming Based on Eco-Farming to Face Food Security and Climate Change: Case Study in Jambi Province, Indonesia. *Procedia Environmental Sciences* **17(2013)** 53-59

[2] Nwinya C E, Obienusi E A, and Onuoha D C 2014 Comparative Economic Analysis of Upland and Lowland Rice Production in Izzi Local Government Area of Ebonyi State. *Journal of Economics and Sustainable Development* **5(17)** 144-160

[3] Ayambila N S, Kwadzo M-T G and Brempong-Asuming S 2008 Economics of Rice Production an Economic Analysis of Rice Production Systems in the Upper East Region of Ghana. *Ghana Journal of Development Studies* **5(1)** 95 - 108

[4] Chen S, Zheng X, Wang D, Chen L, Xu C and Zhang X 2012 Effect of Long-Term Paddy-Upland Yearly Rotations on Rice *(Oryza sativa)* Yield, Soil Properties, and Bacteria Community Diversity. *The Scientific World Journal* **2012** 1-11

[5] Saputra E, Tabrani G and Zuhry E 2016 The Experiment of Bokashi and Ferinsa Composition on Upland Red Rice *(Oryza nivara L.)* Inpago 7 Variety *JOM FAPERTA* **3(2)** 1-13

[6] Departemen Pertanian RI 2015 *Peraturan Menteri Pertanian Republik Indonesia Nomor 03/Pementan/OT.140/2/2015 tentang Pedoman Upaya Khusus (UPSUS) Peningkatan Produksi Padi, Jagung dan Kedelai Melalui Program Perbaikan Jaringan Irigasi dan Sarana Pendukungnya TA 2015*. Dirjen Prasarana dan Sarana Pertanian, Departemen Pertanian RI. Jakarta. http://psp.pertanian.go.id, accessed on 5 August, 2015.

[7] Lalel D J H, Abidin Z and Jutomo L 2009 The Physico-chemical Properties of Local Ende High Land Brown Rice. **XX (2)** 109-116

[8] Taridala S A A 2014 *Preferensi Konsumen Beras 'Daerah' di Kota Kendari*. Research Report. Un-published

[9] Taridala S A A, Wianti N I, Abdullah W G 2016 Pemetaan Sosial Ekonomi Rumahtangga Petani Padi Gogo sebagai Dasar Kebijakan Menuju Kedaulatan Pangan di Kabupaten Konawe Selatan. *Parnership Program between Universitas Halu Oleo and Bappeda Konawe Selatan. LPPM-UHO* Kendari

[10] Subejo 2014 *Ekonomi Perberasan Indonesia*. Perhimpunan Ekonomi Pertanian Indonesia. Bogor

[11] BPS Konawe Selatan 2014 *Konawe Selatan dalam Angka*. BPS Konawe Selatan. Andoolo.

[12] Doll J P and Orazem F 1984 *Production Economics: Theory with Applications*. Second Editon. John Wiley & Sons. New York.

[13] Hernanto F 1989 *Ilmu Usahatani*. Penebar Swadaya. Jakarta.

[14] Soekartawi 2002 *Prinsip Dasar Ekonomi Pertanian: Teori dan Aplikasinya*. PT. Raja Grafindo Persada. Jakarta.

[15] Wang H, Pandey S, Hu F., Xu P, Zhou J, Li J, Deng X, Feng L, Wen L, Li J, Li Y, Velasco L E, Ding S and Tao D 2010 Farmers Adoption of Improved Upland Rice Technologies for Sustainable Mountain Development in Southern Yunnan. Mountain *Research and Development* **30(4)**:373-380

[16] Santoso N K, G Hartono and Nuswantara B 2012 Analisis Komparasi Usahatani Padi Organik dan Anorganik di Kecamatan Sambirejo Kabupaten Sragen. *Agric* **24 (1)** 63-80

[17] BPS Sulawesi Tenggara 2015 *Garis Kemiskinan Penduduk Provinsi Sulawesi Tenggara*. BPS Sulawesi Tenggara. Kendari

[18] Fitria E and Ali N M 2014 The Feasibility of Upland Rice Farming with Integrated Crop Management (ICM) In Aceh Besar, Aceh. *Widyariset* **17(3)** 425-434

# Planting pattern and weed control method influence on yield production of corn (*Zea mays* L.)

**E Purba[1] and D P Nasution[2]**

[1] Department of Agrotechnology, Faculty of Agriculture, Universitas Sumaera Utara, Medan, Indonesia 20155
[2] Former student, Department of Agrotechnology, Faculty of Agriculture, Universitas Sumaera Utara, Medan, Indonesia 20155
E-mail: epurba@yahoo.com

**Abstract.** Field experiment was carried out to evaluate the influence of planting patterns and weed control methods on the growth and yield of corn. The effect of the planting pattern and weed control method was studied in a split plot design. The main plots were that of planting pattern single row (25cm x 60cm), double row (25cm x 25cm x 60cm) and triangle row ( 25cm x 25cm x 25cm). Subplot was that of weed control method consisted five methods namely weed free throughout the growing season, hand weeding, sprayed with glyphosate, sprayed with paraquat, and no weeding.. Result showed that both planting pattern and weed control method did not affect the growth of corn. However, planting pattern and weed control method significantly affected yield production. Yield resulted from double row and triangle planting pattern was 14% and 41% higher, consecutively, than that of single row pattern. The triangle planting pattern combined with any weed control method produced the highest yield production of corn.

## 1. Introduction

Various efforts to improve agricultural technology such as improvements to varieties and how to grow crops are continually developed in order to meet the needs of food for global consmption. Research on varieties continues to be done to improve the resistance to pests and diseases, response to fertilization, shorter harvest age and so forth. While improvement of cultivation techniques include the amount and type of fertilizer, time and weeding method, plant point pattern, the number of plants per hectare, water stress or drought and others.

The area of maize in North Sumatra Province and Indonesia in 2015 is 3787367 ha and 243770 ha with production of 15,19407 and 19,612,435 tons respectively [1]. The productivity of maize in Indonesia (5.07 tons / ha) in 2015 is above the average production per hectare of ASEAN (4.29 tons / ha) [2].

Increased maize production can be carried out with increasing population through plant spacing system. The commonly used spacing system is one-row pattern, and as the demand for maize increases, two-row pattern begins to be applied for greater yield [3]. The triangle pattern is also a concern for farmers to increase production per unit of land. The larger population on the triangle line increased production by 8.98% over one-row and 4.59% with two-row [4].

The presence of weeds can significantly reduce the growth and production of crops as a competitor in the fight for nutrients and sunlight, thus reducing production by 48% [5]. Weed control method has been adopted varies such as mechanical control by pulling or clearing, burning, flooding, using mulch, natural enemies, crop rotation and spraying herbicides [6]. Each of these methods has advantages and disadvantages, but farmers often implement manual weeding and the use of herbicides. The objective of the study was to evaluate the system of plant-spacing pattern and appropriatel weed control methods on the growth and production of maize (*Zea mays* L.)

## 2. Materials and Methods

The study investigating two factors, firstly, planting pattern consist of three patterns: one-row (60 cm x 25 cm), two-row ((25 cm x 25 cm) x 60 cm), and triangle ((25 cm x 25 cmx 25 cm) x 60 cm) whilst the second factor is weed control method: manual weeding, spraying with glyphosate at 4 weeks after planting (WAP), sparaying with paraquat 4 WAP, weed-free along the growing season, and unweeded for a comparison. All treatments were arranged in a split plot design with three replicates.

On plot size of 275 cm x 275 cm, seeds of corn DK3 variety were planted one seed per planting point according tp the planting patterns. The corn was fertilized with nitrogen (135 kg N/ha), P2O5 (36 kg /ha) and K2O (25 kg K2O/ha) (Warisno,1998)

Weed was controlled by using herbicides, glyphosate or paraquat, as much of 960 g ,a,i. and 600 g a.i consecutively at 4 WAP but for manual weeding at 2 WAP. One treatment was kept weed free as long as growing season and the remaining treatment, unweeding, let weed grew as long as growing season.

## 3. Results and Discussion

Plant heights at different weed control were shown in Table 1. At 6 and 8 WAP, corn height where the weed was controlled was signifintly higher compared to those of no-weed control. No difference of corn height when the weed was controlled by herbicides, glyphosate or paraquat (Table 1). On the other hand, corn height grown in one-row pattern (267.4cm) was higher than those in two-row and triangle planting patterns.

**Table 1.** Average of plant height on different planting pattern and weed strategy management 2, 4, 6, and 8 weeks after planting

| Time of measurement (WAS) | Weed Management Strategy | Planting patterna[a)] | | | Average |
|---|---|---|---|---|---|
| | | One row | Two-row | Triangle | |
| | | ---------- cm ---------- | | | |
| 2 | Unweeded | 54,96 | 54,43 | 55,05 | 54,81 |
| | Weed free | 55,04 | 55,85 | 55,25 | 55,38 |
| | Manual | 54,99 | 56,67 | 54,37 | 55,34 |
| | Glyphosate | 54,14 | 55,25 | 54,13 | 54,51 |
| | Paraquat | 54,68 | 55,57 | 55,86 | 55,37 |
| | *Average* | 54,76 | 55,56 | 54,93 | |
| 4 | Unweeded | 108,09 | 106,88 | 107,00 | 107,32b |
| | Weed free | 112,25 | 109,98 | 110,27 | 110,83a |
| | Manual | 109,65 | 107,97 | 108,50 | 108,71ab |
| | Glyphosate | 108,32 | 106,98 | 106,29 | 107,20b |
| | Paraquat | 108,40 | 106,35 | 106,15 | 106,97b |
| | *Average* | 109,34 | 107,63 | 107,64 | |
| 6 | Unweeded | 173,07 | 172,00 | 173,00 | 172,69b |
| | Weed free | 186,63 | 186,43 | 186,03 | 186,37a |
| | Manual | 186,47 | 185,40 | 185,90 | 185,92a |
| | Glyphosate | 185,87 | 184,03 | 184,07 | 184,66a |
| | Paraquat | 185,30 | 183,77 | 183,63 | 184,23a |

| | | | | | |
|---|---|---|---|---|---|
| | *Average* | 183,47 | 182,33 | 182,53 | |
| 8 | Unweeded | 262,26 | 259,43 | 259,6 | 260,43b |
| | Weed free | 269,80 | 262,36 | 262,13 | 264,76a |
| | Manual | 269,73 | 261,90 | 261,63 | 264,42a |
| | Glyphosate | 267,80 | 261,60 | 260,30 | 263,23ab |
| | Paraquat | 267,23 | 261,60 | 259,70 | 262,94ab |
| | *Average* | 267,36 a | 261,90 b | 260,67 b | |

[a]Numbers followed by the same letters on the same column and treatment category are not significantly different at the level of 5% based on the least significant difference

Grain production per plant of corn on weed-free is not difference than that of manual weeding, ranging from 190,5 to 197,1 g per plant. The grain per plant of corn where the weed was controlled with glyphosate is similar with those of plant using paraquat which is ranging from 174,4 to178.9 g /plant. The lowest grain production per plant (164 g) was found in unweeded area.

Grain production per plant of corn with one-row planting pattern (190,9 g/plant) is not different compared to that of triangle planting pattern (184,1 g/plant) but higher than that of two-row planting pattern (160,2 g/plant). (Table 2).

Grain production of corn per hectare planting in one-row, two-row, and triangle patterns with different weed control methods is shown on Table 3.

**Table 2.** Dry weight of grain production per plant from one-row, two-row and triangle planting pattern of corn

| Weed Management Strategy | Planting Pattern[b] | | | Average |
|---|---|---|---|---|
| | One-row | Two-row | Triangle | |
| | ---------- g/ plant ---------- | | | |
| Unweeded | 170,20 | 160,65 | 162,42 | 164,42 c |
| Weed free | 205,80 | 190,27 | 195,19 | 197,09 a |
| Manual | 200,75 | 180,46 | 190,31 | 190,51 a |
| Glyphosate | 193,12 | 157,00 | 186,58 | 178,90 b |
| Paraquat | 184,72 | 152,43 | 185,90 | 174,35 b |
| *Average* | 190,92 a | 160,16 b | 184,08 a | |

[b]Numbers followed by the same letters on the same column and treatment category

**Table 3.** Dry weight of grain production per hectare per plant from one-row, two-row and triangle planting pattern of corn

| Weed Manageme nt Strategy | Planting Pattern[c] | | | Average |
|---|---|---|---|---|
| | One-row | Two-row | Triangle | |
| | ---------- ton/ha ---------- | | | |
| Unweeded | 11,35 cx (G) | 15,04 by(D) | 15,83 bz(CD) | 14,07 c |
| Weed free | 13,72 ax(EF) | 17,81 ay(AB) | 19,03 az(A) | 16,85 a |
| Manual | 13,38 abx(FG) | 16,89 ay(BC) | 18,55 az(A) | 16,28 a |
| Glyphosate | 12,87 abx(FG) | 14,69 by(D) | 18,19 az(AB) | 15,25 b |
| Paraquat | 12,31 bcx(FG) | 14,27 by(DE) | 18,12 az(AB) | 14,90 b |
| *Average* | 12,73 c | 15,74 b | 17,94 a | |

[c]Numbers followed by the same letters on the same column and treatment category are not significantly different at the level of 5% based on least significant difference. Letters followed by x, y, and z are compared in the same main plot whilst letters in brackets are compared in general.

are not significantly different at the level of 5% based on the least significant difference
Dry weight of grain production of corn per hectare is presented on Table 3. The highest grain production (17,9 ton/ha) is found in triangle planting pattern followed by two-row planting pattern (15.7 ton/ha) and single-row planting pattern (12.7 ton/ha). In other words there is 40.9% and 23.6% increase in triangle and two-row planting pattern consecutively compared to one-row planting pattern. Grain production per hectare from different weed management strategies is presented on Table 3. The highest production (16.3 to 16.8 ton/ha) was found in area where the area is weed free or manually controlled. Grain production of corn where weed control was spraying with glyphosate (16.3 ton/ha) is no difference from that of spraying with paraquat (15.3 ton/ha). On the top of all, corn planted in triangle pattern combined with any of those weed control method tested produce the highest grain production compared to any other treatments (Table 3).

## 4. Conclusion

The triangle planting pattern produced the highest corn grain production followed by two-row and one-row planting patterns. The triangle pattern increased grain production as much of 40.9% and 14.0% compared to those of one-row planting pattern and two-row planting pattern consecutively. Weed control by spraying with herbicides, glyphosate or paraquat, increased the corn production as much of 5.9 to 8.4% whereas by manual weeding about 19.4% in triangle planting pattern.

## References

[1] https://www.bps.go.id (accessed 15 Aug 2017)
[2] http://www.pertanian.go.id. *Produksi Jagung Indonesia 2016 Diperkirakan Tertinggi di Asia Tenggara* (Accessed 10 October 2016)
[3] Stalcup L 2008 *Twin Rows Help Boost Yields: Stil, The Jury's Out on Whether Twin Rows are Always Profitable* Corn and Soybean Digest ; Jan 2008; 68,1; ABI/Inform Trade and Industry p 6
[4] Cox W J, Cherney D R and Hanchar J J 2006 Row spacing, hybrid, and plant density effects on corn silage yield and quality *J. Prod. Agric.* **11**:128-134. (In Row Spacing, Plant Density and Hybrids Effects on Corn Grain Yield and Moisture 2001 *Agron. J.* **93**:1049-105)3
[5] Tanveer A, M Ayub andAhmad A A R 1999 Weed-crop competition in maize relation to row spacing and duration *Pakistan Journal of Biological Sci.* **2**(2):363-364.
[6] Fadhly A F and Tabri F 2007 *Pengendalian Gulma Pada Pertanaman Jagung* http://balitsereal.litbang.pertanian.go.id. (Accessed 30 August 2017)

# Physiology response of fourth generation saline resistant soybean (*Glycine max* (L.) Merrill) with application of several types of antioxidants

**I R Manurung, Rosmayati and N Rahmawati**

Faculty of Agriculture, Universitas Sumatera Utara, Padang Bulan, Medan 20155, Indonesia
Email : nini@usu.ac.id

**Abstract.** Antioxidant applications are expected to reduce the adverse effects of soil saline. This research was conducted in plastic house, Plant Tissue Laboratory Faculty of Agriculture and Plant Physiology Laboratory Faculty of Mathematic and Natural Science, Universitas Sumatera Utara, Medan also in Research Centers and Industry Standardization, Medan from July-December 2016. The objective of the research was to know the effect of various antioxidant treatments with different concentrations (control, ascorbic acid 250, 500 and 750 ppm; salicylic acid 250, 500 and 750 ppm; α-tocopherol 250, 500 and 750 ppm) on fourth generation soybean physiology in saline condition (Electric Conductivity 5-6 dS/m). The results of this research showed that the antioxidant type and concentration affected not significantly to physiology of fourth generation soybean. Descriptively the highest average of superoxide dismutase and peroxide dismutase was showed on ascorbic acid 250 ppm. The highest average of ascorbate peroxidase was showed on α-tocopherol 750 ppm. The highest average of carotenoid content was showed on ascorbic acid 500 ppm. The highest average of chlorophyll content was showed on α-tocopherol 250 ppm. The highest average of ratio of K/Na was showed on salicylic acid 250 ppm.

## 1. Introduction

Soybean is one of the kind of plant nuts used as food source of energy and protein [1]. Efforts to increase soybean production can be done with intensification and extensification. Efforts that can be taken in the intensification is to plant high yielding varieties and apply the optimal package of technology. As for extensification efforts taken is to open a new agricultural area [2]. The saline soil is one of the largely untapped land for cultivation, due to toxic effects and increased root osmotic pressure resulting in disruption of plant growth [3].

one of the causes of plant damage in stressful conditions is occurrence of oxidative stress caused by accumulation of reactive oxygen species (ros) such as oxygen ($o_2$), hydrogen peroxide ($h_2o_2$), superoxide ($o_2^-$) and hydroxyl radicals ($oh^-$), due to inhibition of the process photosynthesis by closing stomata causing damage oxidative properties of photosynthetic [4].

some of the antioxidants that can be applied to plants are ascorbic acid, salicylic acid and α-tocopherol. application of ascorbic acid 500 ppm can help morphology and physiology of soybean to grow and produce better in the soil [5]. salicylic acid applied to soybean in saline soils can increase chlorophyll and superoxide dismutase (sod) [6]. the application of α-tocopherol may increase the activity of wheat leaf enzymes in saline soils [7].

the objective of the research was to know the effect of various antioxidant treatments with different concentrations on fourth generation soybean (*glycine max* (l.) merrill) physiology in saline condition (ec 5-6 ds/m).

## 2. Materials and methods

Field experiment was conducted on July-December 2016 in plastic house Faculty of Agriculture, Universitas Sumatera Utara. The experiment was arranged in a non factorial randomized block design with ten treatments (control, ascorbic acid 250, 500 and 750 ppm; salicylic acid 250, 500 and 750 ppm; α-tocopherol 250, 500 and 750 ppm) with four replications for each treatment. The research started from land preparation, planting, application of antioxidant, maintenance, fertilizing and physiology parameters analysis.

The antioxidants were applied at 2 weeks after planting (stadium $V_1$) to pod filling period (stadium $R_3$), with intervals of 1 week. To determine the volume of spraying, every week spray volume calibration was done on plant leaves by using hand sprayer. The application time was at 07.00 AM. The antioxidants solvent was dissolved by stirrer with aquades according to predetermined concentrations.

The observation on SOD, POD, APX, chlorophyll content, carotenoid content and ratio of K/Na were measured when the plants at stadium $V_4$. The physiology parameters analysis include SOD, peroxide dismutase (POD), ascorbate peroxide (APX) and carotenoid content was held in Plant Tissue Laboratory Faculty of Agriculture, Universitas Sumatera Utara. Chlorophyll content observation was held in Plant Physiology Laboratory Faculty of Mathematic and Natural Science, Universitas Sumatera Utara. Ratio of K/Na observation was held in Research Centers and Industry Standardization, Medan.

The data were analyzed statistically using F-test and then following by orthogonal contrast test at 5 % level.

## 3. Results and Discuss

Data presented in Table 1 shows that application of several types of antioxidants affected not significantly increased SOD, POD and APX compared with untreated plants.

**Table 1.** Effect of application of several types of antioxidants on SOD, POD and APX in soybean in saline soil

| Treatment | SOD (unit/mg protein) | POD (unit/mg protein) | APX (μM/min/mg protein) |
|---|---|---|---|
| $A_0$ (Control) | 4860.69 | 0.17 | 0.50 |
| $A_1$ (Ascorbic Acid 250 ppm) | 5213.31 | 1.16 | 0.50 |
| $A_2$ (Ascorbic Acid 500 ppm) | 4626.46 | 0.33 | 0.56 |
| $A_3$ (Ascorbic Acid 750 ppm) | 2312.50 | 0.16 | 0.35 |
| $A_4$ (Salicylic Acid 250 ppm) | 1775.86 | 0.07 | 0.46 |
| $A_5$ (Salicylic Acid 500 ppm) | 2558.40 | 0.09 | 0.43 |
| $A_6$ (Salicylic Acid 750 ppm) | 1323.05 | 0.10 | 0.24 |
| $A_7$ (α-tocopherol 250 ppm) | 4888.42 | 0.11 | 0.80 |
| $A_8$ (α-tocopherol 500 ppm) | 1695.73 | 0.10 | 0.74 |
| $A_9$ (α-tocopherol 750 ppm) | 3201.41 | 0.10 | 0.84 |

Data in Table 1 indicated treatment of ascorbic acid 250 ppm increased the activity of SOD and POD. This suggests that ascorbic acid is capable enhancing plant physiology against salinity stress. Ascorbic acid is an antioxidant with preventing and reducing the toxic effects of ROS by donating electrons to ROS, such as single oxygen, radical superoxide, radical hydroxyl; regenerate radical tocopherol for membrane protection; as well as electron donors for APX activities [8]. Ascorbic acid has many benefits to plant physiology in protein synthesis, as an antioxidant, enzyme cofactor and as a signal cell modulator in a variety of important physiological processes, including cell wall biosynthesis, secondary metabolites and phytohormone, stress physiology, photoprotection, cell division and growth [9].

Antioxidant treatment had no significant effect on POD enzyme activity. The application of salicylic acid in *P. sativum* in saline soil did not have a significant effect on POD enzyme activity. This is caused by the hydrogen peroxide ($H_2O_2$) produced by plants not only dispart by peroxidase enzyme but also catalase enzyme (CAT) and APX [10].

In APX observation, α-tocopherol 750 ppm can increase the activity of APX compared with other antioxidants. Decrease in APX activity occurs on the antioxidant salicylic acid 750 ppm. The decrease in APX activity is relate to the abundance of $H_2O_2$ accumulation as substrate enzyme APX, so the plant is unable to reduce the negative effects caused by $H_2O_2$. The APX enzyme uses ascorbic acid as electron receptors to reduce $H_2O_2$ into water and monodehydroascorbate (MDHA) [11].

**Table 2.** Effect of application of several types of antioxidants on chlorophyll content and carotenoid content in soybean in saline soil

| Treatment | Chlorophyll Content (mg/g) | Carotenoid Content (mg/g) |
|---|---|---|
| $A_0$ (Control) | 4.91 | 0.31 |
| $A_1$ (Ascorbic Acid 250 ppm) | 4.45 | 0.34 |
| $A_2$ (Ascorbic Acid 500 ppm) | 4.07 | 0.41 |
| $A_3$ (Ascorbic Acid 750 ppm) | 4.09 | 0.37 |
| $A_4$ (Salicylic Acid 250 ppm) | 4.90 | 0.32 |
| $A_5$ (Salicylic Acid 500 ppm) | 4.21 | 0.32 |
| $A_6$ (Salicylic Acid 750 ppm) | 4.38 | 0.30 |
| $A_7$ (α-tocopherol 250 ppm) | 4.95 | 0.29 |
| $A_8$ (α-tocopherol 500 ppm) | 4.40 | 0.31 |
| $A_9$ (α-tocopherol 750 ppm) | 3.95 | 0.33 |

On observation of chlorophyll content, the application of α-tocopherol 250 ppm was more effective in increasing total chlorophyll compared with α-tocopherol 750 ppm. Decrease in chlorophyll in the saline soil occurs due to disruption of chlorophyll formation in chloroplasts. Growth of plants in environments with high salt levels causes plants to hyperosmotic effects. As a result, interference with membrane function, metabolic toxicity, disturbance of the photosynthesis process, may even lead to crop death [12]. Salinity in soybeans can cause a decrease in green color on the leaves.

α-tocopherol in resolve the harmful effects of salt stress relate with the stability and protection of photosynthetic pigments from oxidative damage. α-tocopherol 250 ppm is more effective in increasing the amount of chlorophyll in soybeans. Increased chlorophyll indicates more active photosynthetic activity and will increase crop production. Salinity through increased osmotic pressure causes reduction of water uptake and metabolic disorders and physiological processes is under its influence. This causes decreased production by decreasing the amount of leaf chlorophyll [13].

In the observation of carotenoids, ascorbic acid 500 ppm can increase the physiology of soybean to salinity with high carotenoid content. The effect of ascorbic acid on reducing the harmful effects of salt stress is thought to result from activating several enzyme reactions in the metabolic process. Ascorbic acid can trigger the action of enzymes in the metabolism process of plants which will further increase the rate of photosynthesis where if the process of photosynthesis increases then the resulting assimilate is sufficient to be distributed to the seeds so the quality of the beans produced also increases [11].

**Table 3.** Effect of application of several types of antioxidants on ratio of K/Na in soybean in saline soil

| Treatment | Ratio of K/Na |
|---|---|
| $A_0$ (Control) | 0.21 |
| $A_1$ (Ascorbic Acid 250 ppm) | 0.20 |
| $A_2$ (Ascorbic Acid 500 ppm) | 0.27 |
| $A_3$ (Ascorbic Acid 750 ppm) | 0.29 |
| $A_4$ (Salicylic Acid 250 ppm) | 0.34 |
| $A_5$ (Salicylic Acid 500 ppm) | 0.23 |
| $A_6$ (Salicylic Acid 750 ppm) | 0.30 |
| $A_7$ (α-tocopherol 250 ppm) | 0.21 |
| $A_8$ (α-tocopherol 500 ppm) | 0.22 |
| $A_9$ (α-tocopherol 750 ppm) | 0.31 |

In observation of ratio of K/Na, salicylic acid 250 ppm can increase ratio of K/Na. Whereas, low ratio of K/Na indicates that antioxidant treatment has not been effective in improving soy physiology in saline soils. The physiology of soybean in saline soils can be seen from decreasing salt levels through the soil's ability to exchange Na ions to K. High salt levels in the soil affect plant growth through four mechanisms i.e. (1) elevated levels of salt cause osmotic stress, (2) inhibition in $K^+$ absorption which is the main nutrient of the plant, (3) $Na^+$ ions at elevated levels are toxic to cytosolic enzymes and (4) high salt concentration stimulates oxidative stress and cell death [14].

The soil's ability to convert K ions to Na will decrease the Na content to the plants. Physiological responses of soy genotype to salinity stress can be (1) prevention of ion transfer from roots to other parts of the plant, (2) not accumulate many salts in leaves and stems, and (3) have better osmotic adjustability in plant cells. The physiology of soybeans on salinity is also related to the regulation of stable moisture content in the canopy and the accumulation of soluble saccharides, dissolved proteins, amino acids, and $K^+$ and $Ca^+$ ions for osmotic adjustment [15].

## 4. Conclusions

The results of this research showed that the treatment of antioxidant type and concentration affected not significantly to physiology of fourth generation soybean. The best ascorbic acid concentration was obtained at 250 ppm to improve soybean physiology in saline soil.

## References

[1] Cahyadi, W 2007 *Soybean: Efficacy and Technology* Bumi Aksara, Jakarta
[2] Sunarto 2001 *Soybean Tolerance to Saline Soil.* Bul. Agron. 29 (1):27-30
[3] Slinger, D and K Tenison 2005 *Salinity Glove Box Guide - NSW Murray and Murrumbidgee Catchments* An initiative of the Southern Salt Action Team, NSW Department of Primary Industries
[4] Prohazkova D, RK Sairam, GC Srivastava and DV Singh 2001 Oxidative stress and antioxidant activity as the basis of senescence in maize leaves *Plant Sci.* 161:765-771
[5] Rahmawati N 2014 Increasing of Soybean Yields on Saline Soil with the Utilization of Resistance Genotypes, Ascorbic Acid and Arbuscular Mycorrhiza Fungi. Doctoral Program of Agricultural Science Agriculture Faculty Universitas Sumatera Utara, Medan
[6] Jaiswal, AV Pandurangam and SK Sharma 2014 Effect of salicylic acid in soybean (*Glycine max* (L.) Merrill) under salinity stress *The Bioscan* **9**(2): 671-676

[7] Farouk S 2011 Ascorbic Acid and α-tocopherol Minimize Salt-induced Wheat Leaf Senescence. J. of Stress Physiology and Biochemistry 7(3)

[8] Ahmad P, Jaleel CA, Azooz MM and Nabi G 2009 Generation of ROS and Non-Enzymatic Antioxidants During Abiotic Stress in Plants. Botany Research International 2(1):11-20

[9] Wolucka BA, A Goossens and D Inze 2005 Methyl Jasmonate Stimulates The de novo Biosynthesis of Vitamin C in Plant Cell Suspensions. J. Exp. Botany 56: 2527-2538

[10] Sajid ZA, M Safdar and SA Khilji 2016 Amelioration of Salinity Stress Tolerance in Pea (*Pisum sativum* L.) by Exogenous Application of Salicylic Acid. J. Biologia 62(1):69-78.

[11] Shigeoka S, Ishikawa T, Tamoi M, Miyagawa Y, Takeda T, Yabuta Y and Yoshimura K 2002 Regulation and function of Ascorbate peroxidase isoenzymes. Journal of Experimental botany 53: 1305–1319

[12] Malhotra RS and T Blake 2004 Breeding for Salinity Tolerance. *In*: M. Ashraf and P.J.C. Harris (*Eds*.). Abiotic Stresses, Plant Resistance Through Breeding and Molecular Approaches. Food Products Press, Binghamton, NY. p.125-143

[13] Kurniasari AM, Adisyahputra and R Rosman 2010 The Effect of Drought on NaCl Saline Soil on The Growth of Patchouli. Research Bulletin of Herbs and Medicine 21(1):18-27.

[14] Xiong L and Zhu JK 2002 Molecular and Genetic Aspects of Plant Responses to Osmotic Stress. Plant, Cell and Environment 25(2):131-139

[15] Pathan SMD, J Lee, JG Shannon, and HT Nguyen 2007 Recent Advances in Breeding for Drought and Salt Stress Tolerance in Soybean (Chapter 30). *In*: M.A. Jenks *et al.* (*Eds*.). Advances in Molecular Breeding Toward Drought and Salt Tolerant Crops. p.739-773

**Acknowledgments**

The authors wish to express sincere thankfulness to Directorate General of Higher Education, Ministry of Research and Higher Education, Indonesia for financial support on the Competitive Grant program with contract number : 017/SP2H/LH/DRPM/II/2016 in this study.

# Population dynamics of weeds in oil palm (*Elaeis guineensis* Jacq.) circle weeding area affected by herbicide application

**S Sidik[1], E Purba[1*] and E N Yakub[1]**

[1]Department of Agrotechnology, Faculty of Agriculture, Universitas Sumatera Utara, Medan, Indonesia 20155
*Email: epurba@yahoo.com

**Abstract.** Weed problems in oil palm field were mainly overcomed by herbicide application. The application certain herbicides may lead to rapid population dynamic of certain species due to their different response to herbicides. Some species may less susceptible to certain herbicide whereas other species more susceptible. The objective of this study was to determine the population dynamic of weed species in circle weeding of oil palm in Serdang Bedagai, North Sumatra. Six treatments using glyphosate singly and mixture compared with manual weeding were evaluated for weed control. The treatments were arranged in a randomized block design with four replicates. Each treatment consisted of four circle weedings. The results showed that glyphosate 720 g a.i/ha + indaziflam 50 g a.i/hareduced seedbank and regrowth of weeds. Up to 12 weeks after application glyphosate 720 g a.i/ha + indaziflam 50 g a.i/ha is 29.46% total weeds dry weight compared to manual weeding. The effect of herbicide application on changes on the weed composition and weed seedbank are affected by the characteristic of herbicides and weed response to herbicide application.

## 1. Introduction

One of the biggest challenges in increasing the potential for palm oil production in Indonesia is the management of weeds. Weeds are undesirable plants because they can compete with the crops. Weed growth can reduce the production of fresh fruit bunches (FFB) by 20% not onlydue to competition against nutrients, but also allelopathic substances that are toxic to plants.

Weeds compete with cultivated plants against limited resources such as water, nutrition, and light [1]. In addition, weed infestations also encourage disease problems, serve as host alternatives to destructive insects and diseases, slow down harvest operations, increase production costs, reduce the market value of crops and increase the risk of fires in annual crops, plantations and protected forests [1]. The quality and quantity of weed seeds in the seedbank is very important in the weeds cycle that reproduce through the seed itself [2] which is the source of the weed population in the future.

Knowledge of the weed seedbank composition in soil contributes to the predictive and historical understanding of the composition of a plant community. Understanding the mechanism of weed seed distribution is very important as an effort to weed management strategy. In this study will be revealed about the dynamics of weed populations in terms of the ability of weed seeds to grow to the soil surface, the pattern of distribution of weed seedbank, and the effect of herbicide use on weed seedbank.

## 2. Materials and methods

This study consisted of two units. First unit was to study the effect of herbicides on soil seedbank composition in the circle weedings and the second unit was to investigate the effect of indaziflam This research was conducted in the field at Adolina Estate PT. Perkebunan Nusantara IV from July to December 2016. Weed vegetation was recorded prior to herbicide application and four weeks after application on circle weeding. Observations were carried out using quadratic method. Data were analyzed descriptivelydescribing the existing conditions of the site of study. The treatments on circle weedings consisted of seven treatments (glyphosate 720 g ai/ha + 2,4 dimethyl ammonium 186.43 g ai/ha, glyphosate 720 g ai/ha + methyl metsulfuron 20 g ai/ha, glyphosate 720 g ai/ha + fluroxypir 40 g ai/ha , glyphosate 720 g ai/ha + indaziflam 50 g ai/ha, glyphosate 720 g ai/ha + diuron 800 g ai/ha, and glyphosate 720 g ai/ha + oxyfluorfen 240 g ai/ha and manual weeding without herbicide application for a comparison. All treatments were arranged in a randomized block design with 4 replicates. Each treatments consisted of four circle weedings.

Weed seedbank compositition in the circles was analized by germinating the seeds in the soil samples collected from the circles using soil core. Soil samples were collected both prior to and post herbicide application. Soil samples were divided into four depth categories (0to1, 1to3, 3to5 and 5to10cm). Seeds in the soil samples were germinated in polybags containing 20cm deth of sterilized soil. Soil from each depth category was distributed on the sterilized soil media and watered regularly in order to have seedbank germinating and emerging to be observed. Number of seedlings were recorded regularly every two days until no new seedling emerged.

The weed seed introduction test was carried out by taking a soil sample that has been applied with herbicide to a depth of 20 cm by 2 points per treatment to measure and observe weeds that grow at various depths (0 to1, 1 to3, 3 to5, 5 to10, 10 to20 cm). *Eleusine indica* weed seeds taken one week before introduction are planted in paralon pipes containing soil from different depths of 100 seeds per paralon. To obtain weed to free soil weeded sterilized pipe that already contains the soil from the experimental garden using sterilizer furnace with temperature 60to70 C for 30 minutes. bservation of germination done since one week after weed seed sowing. While on the seed germination test in the layer zone after herbicide application, it is done by taking soil samples from each treatment to measuring 20 cm x 20 x 5 cm. The soil samples were taken two points north of the tree at each treatment. Soil sampling after one month of herbicidal application at ground depth zones 0to1, 1to3, 3to5, 5to10 cm at each treatment. Furthermore, descriptive observation is done by evaluating some zone into the soil after herbicide application and its effect on the existence of weed seed inside the depth zone. The soil taken from the disk is mixed and stirred evenly into one sample per treatment. The stirred soil sample is fed into a growth box measuring 20 cm x 20 cm x 5 cm thick 2 cm above the sterile soil present in the box. We observed the number of seed weed seeds that appeared at twotoweek intervals.

## 3. Results and discussion

### *3.1. Analysis of weed vegetation*

The summed dominance ratio (SDR) value of the vegetation analysis on the circle weeding of oil palm illustrates the degree of dominance and the distribution of weeds. Table 1 shows the SDR value at the time of herbicide application which *Asystasia gangetica* was the predominant weed species in the treatment plots of H0, H1, H2, H3, and H4 whereas the species in the treatment plots of H5 and H6 were predominated by *Cleome rutidosperma* species.

SDR values were recalculated at four weeks after application (WAA) to determine the changes in the weed species dominance level in each treatment. Table 2 shows that *A. gangetica* was the dominant weed in the circles sprayed withH1, H2, H3, and H4and H0 (glyphosate 540 g a.e./Ha + 215 g a.i/2,4-D; glyphosate 540 g a.e./Ha + methyl metsulfuron 20 g a.i/Ha; glyphosate 540 g a.e./Ha + fluroxypir 50 g a.i/Ha; glyphosate 540 g a.e./Ha + indaziflam 50 g ai/Ha) and on manual weeding.There was a shift in the dominance of weeds in the treatment of H5 and H6 which was originally predominated by *C. rutidosperm* changed to bepredominated by *A. gangetica*. This suggests that the application of glyphosate with various mixtures is not able to control *A. gangetica* growth. Application of glyphosate herbicide mixed with diuron and oxyfluorfen able to control the growth of *C. rutidosperma*satisfactorily. While the result of vegetation analysis on manual weeding treatment did not show any change of weed dominance level.

**Table 1.** Weed dominance level at the time of herbicide application

| Weed species | Summed Dominance Ratio† | | | | | | |
|---|---|---|---|---|---|---|---|
| | H0 | H1 | H2 | H3 | H4 | H5 | H6 |
| | % | | | | | | |
| *Asystasia gangetica* | 27.53 (1) | 52.18 (1) | 52.78 (1) | 49.17 (1) | 54.42 (1) | 25.73 (2) | 19.51 (3) |
| *Eleusine indica* | 21.04 (3) | 28.85 (2) | 19.44 (3) | 28.04 (2) | 0.00 (4) | 19.96 (3) | 23.22 (2) |
| *Cynodon dactylon* | 16.49 (4) | 13.09 (3) | 0.00 (4) | 6.66 (4) | 0.00 (4) | 9.56 (5) | 12.37 (4) |
| *Cleome rutidosperma* | 13.25 (5) | 0.00 (5) | 0.00 (4) | 0.00 (5) | 34.61 (2) | 32.28 (1) | 34.82 (1) |
| *Cyperus kyllingia* | 21.69 (2) | 5.88 (4) | 27.78 (2) | 16.13 (3) | 10.97 (3) | 12.46 (4) | 10.09 (5) |
| SDR | 100.00 | 100.00 | 100.00 | 100.00 | 100.0 0 | 100.0 0 | 100.0 0 |

†Numbers in the brackets are the sequence of weed dominance value levels.
H0= control (manual weeding)
H1= glyphosate 540 g ae/Ha + 215 g a.i/Ha 2,4-D
H2= glyphosate 540 g ae/Ha + methyl metsulfuron 20g ai/Ha
H3= glyphosate 540 g ae/Ha + fluroxypir 50 g a.i/Ha
H4= glyphosate 540 g ae/Ha + indaziflam 50 g ai/Ha
H5= glyphosate 540 g ae/Ha + diuron 80 g ai/Ha
H6= glyphosate 540 g ae/Ha + oxyfluorfen 240g a.i/Ha

**Table 2.** Summed dominance ratio level of weed species at four weeks after application (WAA)

| Weed species | Summed Dominance Ratio† | | | | | | |
|---|---|---|---|---|---|---|---|
| | H0 | H1 | H2 | H3 | H4 | H5 | H6 |
| | % | | | | | | |
| *Asystasia gangetica* | 27.01 (1) | 39.90 (1) | 41.71 (1) | 41.87 (1) | 21.76 (1) | 42.66 (1) | 36.77 (1) |
| *Eleusine indica* | 19.72 (3) | 24.27 (2) | 18.92 (3) | 25.67 (2) | 12.94 (2) | 20.10 (2) | 21.72 (2) |
| *Cynodon dactylon* | 18.68 (4) | 6.46 (5) | 11.53 (4) | 5.30 (5) | 0.00 (4) | 15.26 (4) | 18.98 (3) |
| *Cleome rutidosperma* | 12.08 (5) | 9.17 (4) | 7.75 (5) | 15.89 (3) | 0.00 (4) | 4.46 (5) | 8.23 (5) |
| *Cyperus sp* | 22.50 (2) | 20.21 (3) | 20.09 (2) | 11.28 (4) | 5.29 (3) | 17.52 (3) | 14.30 (4) |
| SDR | 100.00 | 100.00 | 100.00 | 100.00 | 100.00 | 100.00 | 100.00 |

The numbers in the brackets are the sequence of weed dominance value levels.
H0= control (manual weeding)
H1= glyphosate 540 g ae/Ha + 215 g a.i/Ha 2,4-D
H2= glyphosate 540 g ae/Ha + methyl metsulfuron 20g ai/Ha
H3= glyphosate 540 g ae/Ha + fluroxypir 50 g a.i/Ha
H4= glyphosate 540 g ae/Ha + indaziflam 50 g ai/Ha

H5= glyphosate 540 g ae/Ha + diuron 80g ai/Ha
H6= glyphosate 540 g ae/Ha + oxyfluorfen 240g a.i/Ha

*3.2. Weed seed germination collected from circle weeding*

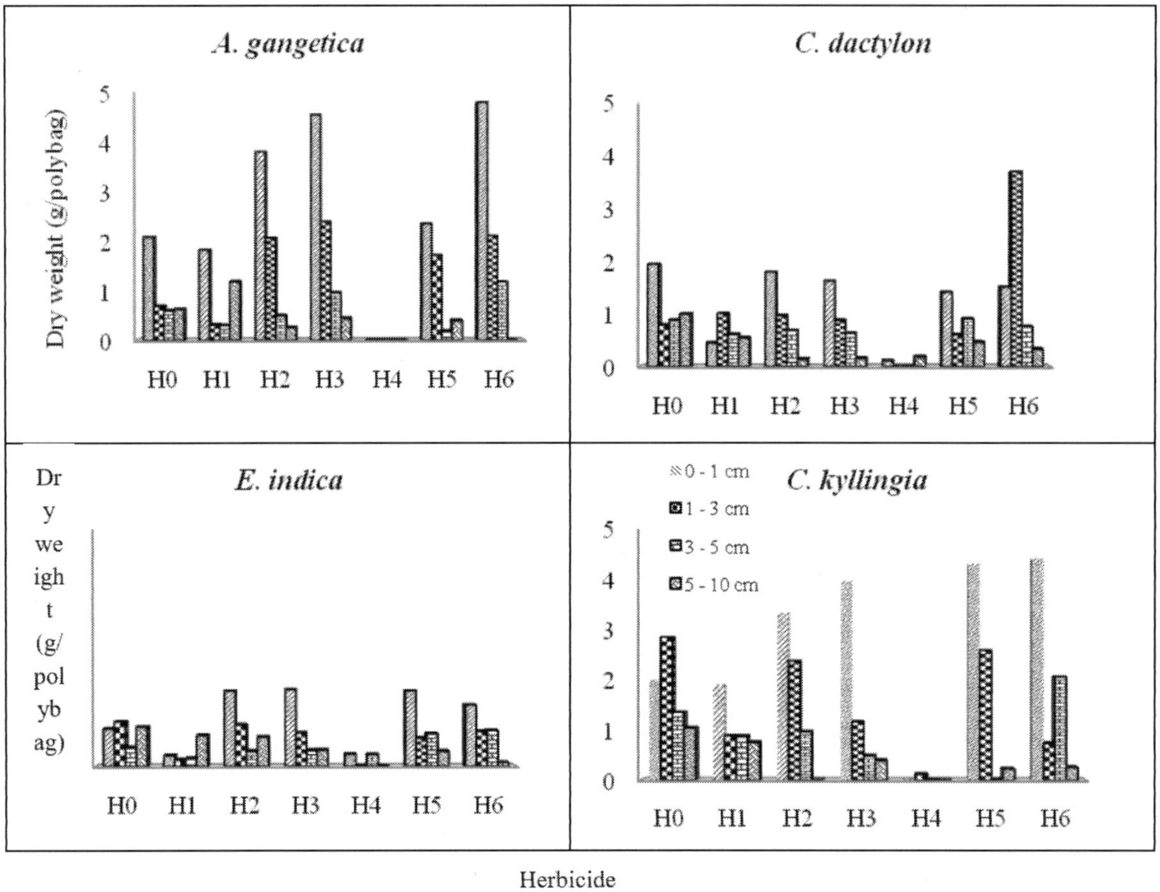


H0= control (manual weeding); H1= glyphosate 540 g ae/Ha + 215 g a.i/Ha 2,4-D; H2= glyphosate 540 g ae/Ha + methyl metsulfuron 20g ai/Ha; H3= glyphosate 540 g ae/Ha + fluroxypir 50 g a.i/Ha; H4= glyphosate 540 g ae/Ha + indaziflam 50 g ai/Ha; H5= glyphosate 540 g ae/Ha + diuron 80 g ai/Ha; H6= glyphosate 540 g ae/Ha + oxyfluorfen 240g a.i/Ha

**Figure 1.** Seedlingsemerged from different soil depth collected from circles which have been previously treated with herbicides.

A. gangetica methods seeds can grow well at a depth of 0 to 1 cm of soil with the most amount on H6 and decreased at the next depth. *C. dactylon* seeds also grow a lot in H6 treatment at a depth of 1 to 3 cm. *E. indica* seeds are grown on each treatment less than the other weeds. *E. indica* grow most at H3 treatment at 0 to 1 cm depth. While *C. kyllingia* seeds can grow well on H6 treatment at 0 to 1 cm depth. Based on these results, it is confirmed that weed seeds were able to grow well at a soil depth of 0 to 3 cm but decreased in deeper soil layer. In addition, the highest number of seeds germinated and emerged was those in soil collected from H6 treatment. This suggests that the oxyfluorfen herbicide soil residue is low in the treatment. Whereas in the H4 treatment, only a few weed seedsgerminated indicating that the indaziflam residue was higher in the soil compared to other treatments. The number of *A. gangetica* seedlings also indicated that the weed seedbank is plentiful in the soil. The soil contained a large number of seeds for several weed species [3] and the determination of weed seed banks is very important in the observation of population dynamics as well as in planning appropriate weed control.

The absence of seedlings on the soil treated glyphosate treatment combined with indaziflam indicated that the combination indaziflam was able to suppress overall weed growth. Indaziflam herbicide is a broad spectrum herbicide which controls broadleaves and grass with the mechanism of action inhibiting cellulosic biosynthesis [4]. Cellulosic biosynthesis is common in land plants. Therefore, herbicides with a working mechanism inhibiting cellulosic synthesis have a broad spectrum of weed control. Indaziflam has the ability to inhibit the growth of monocotyl and dicoty weeds. The large number of weed seeds grown on soils derived from the H6 treatment indicates that the application of glyphosate + oxyfluorfen herbicides is not able to kill the already mature weeds that will be the source of the seedbank. Thus the application of oxyfluorfen herbicide is preferable in the early phase of weed growth [5].

### *3.3. Seed germinationafter herbicide application*

The germination of *E. indica* seeds introduced in various herbicide applied soil depths is shown in Figure 2. Herbicide treatments of H1, H2, H3, and H4, the majority of weed seeds grew in soil layer of 0 to 3 cm depth and decrease at the deeper depth. At the H5 treatment, no weeds grew on the soil from a depth of 0 to 5 cm but there are weeds of *E. indica* emerged on soil from a depth of 10to20 cm. While in the treatment of H6 and H7, weed seeds that grew on soil that emerged from a depth of 0 to 1 cm were less than in the deeper depth. The ability of weed seeds to emerge indicates the pattern of herbicide distribution in each herbicide treatment. Herbicide 2,4-D, methyl metsulfuron, and fluroxypir are widely dispersed at a depth of 0 to 3 cm and decrease their residue at a depth of 3 to 20 cm. Indaziflam still remains active in the soil and accumulates at a depth of 0 to 5 cm. While diuron and oxyfluorfen herbicides also accumulate a lot in the soil surface with a depth of 0 to 3 cm which then declined at the next depth.

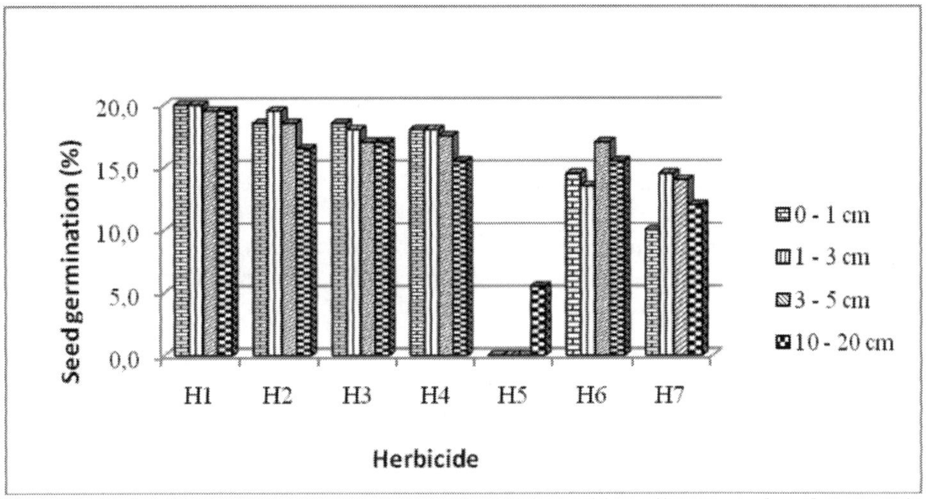


**Figure 2.** Percentage of *Eleusine indica* seed germinated in soil two weeks after spraying

The percentage of weed seeds growing on the soil of H6 and H7 treatments at 0 to 1 cm depth is lower than that of seedlings growing on H1, H2, H3, and H4treatments. This is probably due to oxyfluorfen tends to persist longer in the soil [5]. Diuron applied over the surface of soil is not readily soluble in water which lead to its ability to withstand the washing and high absorption rates by soil colloids [6].

### *3.4. Germination of seed in various soil layer zone after herbicide application*

One month after application, the soil sampling was carried out with a depth of 0 to 10 cm. Then the soil sample was placed in a growth plastic-box with size of 20 cm x 20 cm x 5 cm. The soil sample was distributed as tick of 2cm sterilized soil media in the growth box. The ability of seeds to germinate and emerge was observed at 3, 7 and 15 days after planting (Table 3).

**Table 3.** Germinating seed of selected weed species (*A.ganetica* and *E. indica*) planted at different depth of soil 3, 7, and 15 days after application (DAA)

| No. | Depth of seed burial | 3 DAA | | | | 7 DAA | | | | 15 DAA | | | |
|---|---|---|---|---|---|---|---|---|---|---|---|---|---|
| | | *A. gangetica* | | *E. indica* | | *A. gangetica* | | *E. indica* | | *A. gangetica* | | *E. indica* | |
| | | ................(biji)................ | | | | | | | | | | | |
| 1 | 0 – 1 cm | 6.00 | A | 1.67 | a | 18.33 | A | 18.33 | a | 20.00 | a | 19.67 | a |
| 2 | 1 – 2 cm | 2.67 | B | 0.67 | ab | 18.67 | A | 12.33 | b | 19.67 | a | 17.00 | a |
| 3 | 2 – 3 cm | 0.00 | C | 0.00 | b | 8.33 | B | 0.00 | c | 9.67 | b | 3.00 | b |
| 4 | 3 – 4 cm | 0.00 | C | 0.00 | b | 1.33 | C | 0.00 | c | 2.67 | c | 0.00 | b |
| 5 | 4 – 5 cm | 0.00 | C | 0.00 | b | 0.00 | C | 0.00 | c | 0.00 | d | 0.00 | b |
| 6 | 5 – 10 cm | 0.00 | C | 0.00 | b | 0.00 | C | 0.00 | c | 0.00 | d | 0.00 | b |

Table 3 shows that weed seeds could emerge from soil at a depth ranging from 0 to 4 cm. On the other hand no seedling emerged from soil depth of 4 to 10 cm. Therefore the potential weed problems where *A. gangetica* and *E. indica* predominant species comes from seedbank in the soil layers 0 to 4 cm.

## 4. Conclusions

Glyphosate 720 g a.i/ha + indaziflam 50 g a.i/hareduced seedbank and regrowth of weeds. Up to 12 weeks after application glyphosate 720 g a.i/ha + indaziflam 50 g a.i/ha is 29.46% total weeds dry weight compared to manual weeding. The effect of herbicide application on changes on the weed composition and weed seedbank are affected by the characteristic of herbicides and weed response to herbicide application.

## References

[1] Oudhia P 2004 Phytotosociological studies of rainy wasterlands weed with special reference to *Parthenrium hysterophous* L. IN Repair District http://www.iprng.org/IPRNG to and w 15. Htm.
[2] Gulden R H and Shirtliffe B M2009 Weed Seed Banks: Biology and Management.
[3] Roham R,Yoghubi, M, Noghabi, M A,Akbari N, Nazarian, F 2012. Spatial distribution of broadleaf weeds seedbank in sugar beet field *International Journal of Agriculture and Crop Sciences* **4(5)** 251 – 254
[4] Brabham C, Lei L, Gu Y, Stork J, Barrett M, DeBolt S 2014 Indaziflam herbicidal action: a potent cellulose biosynthesis inhibitor *Plant Physiology American Society of Plant Biologist* **166** 1177-1185.

[5] Jursik M, Andr J, Holec J, Soukup J 2011 Efficacy and selectivity of posttoemergent application of Flumioxazin and Oxyfluorfen in sunflower *Plant Soil Environ.* **57(11)** 532-539
[6] Agustanti V M F 2006 Studi keefektifan herbisida Diuron dan Ametrin untuk mengendalikan gulma pada pertanaman tebu (*Saccharum officinarum* L.). Skripsi. Departemen Agronomi. Institut Pertanian Bogor. Bogor. 55 halaman.

# Determination quercetin content, antioxidant and antimicrobial activity of genotype mutant Samosir shallots irradiated by gamma rays

**M Sinuraya[1], D S Hanafiah[1], A Romulo[2] and A Barus[1]**

[1]Faculty of Agriculture, Universitas Sumatera Utara, Medan 20155, Indonesia
[2]Department of crop sciences and Agroforestry, Faculty of Tropical Agrisciences, Czech Universuty of Life Sciences Prague 6 Suchdol Czech Republic
E-mail: mariati@usu.ac.id

**Abstract.** The aim of the research was to study the variation in antioxidant and antimicrobial activity as well as the total quercetin content of the fifth generation genotypes mutant Samosir shallot irradiated by gamma rays. The studies conducted included the assessment of quercetin content, antioxidant and antimicrobial activity in shallot bulbs after long- term storage (6 months in the room temperature). Quercetin content of 20 selected genotype mutants of irradiated shallot bulbs along with untreated populations were calculated using quercetin (QU) as a standard. Antioxidant activities of 8 genotype mutant were determined using DPPH. Antimicrobial activity of bulb extracts were tested against six bacteria including *Staphylococcus aurous*, *Enterococcus faecalis*, *Bacillus subtilis*, *Escherichia coli*, *Pseudomonas aeruginosa*, and *Klebsiella pneumoniae* and oneyeast*Candida albicans*. The results showed that population of genotype mutants irradiated with dosage 2Gy, 4 Gy, 5 Gy and 6 Gy have higher quercetin content than control samples. None of the genotype mutants exhibited antibacterial inhibitory against all microorganism tested except for the sample number 2 and 6 (bulbs generated from the plants irradiated by gamma rays with dosage at 2 Gy and 6 Gy). There was also none of the genotypes observed exhibited significant antioxidant efficacy.

## 1. Introduction

Shallot (*Allium cepa* L. var. ascalonicum Backer) belongs to the Liliaceae family, is a common food spice, used widely in many parts of the world. For many centuries various species of genus Allium have been used as flavoring vegetable and as medicines for curing various diseases.

Shallot is abundant with biologically active compounds of antioxidant properties. These include, among others, flavonoids and phenolic acids [1-5]. Flavonol levels in the edible portion of Allium vegetables including leeks, shallots, green onions, garlic, and onions range from less than 0.03 to 1 g/Kg of vegetables [6]. Analysis of shallot extracts has confirmed the presence of flavone and polyphenolic derivatives such as quercetin, quercetin 4=-glucoside, quercetin 7,4=-diglucoside, quercetin 3,4=-diglucoside, and quercetin mono-D-glucose, suggesting that it also may have antioxidant properties [7]. It is well known that quercetin has a pronounced effect to allergies, asthma, arthritis, cancer, diabetic complications, goat, neurodegenerative disorder and osteoporosis [8].Allium species are a rich source of phytonutrients, useful for the treatment or prevention of a number of diseases, including cancer, coronary heart disease, obesity, hypercholesterolemia, diabetes type 2, hypertension, cataract and disturbances of the gastrointestinal tract (e.g. colic pain, flatulent colic and dyspepsia) [9].

In in-vitro experiments, onion showed antibacterial and antifungal activity against both gram-positive and gram-negative bacteria (including enteropathogens), pathogenic yeast (*Candida* spp.) and some skin-pathogenic fungi. Alliums were important in warding off plague and other microbial infections over the centuries [10]. Allium crops are well known for their biological activity resulting from the presence sulphur compound in the bulbs, however other compounds, namely flavonoids and other polyphenols, are also in the area of interest. Flavonoid compounds have received considerable attention because of their potential health-promoting properties for human consumers. Flavonoids show wide range of biological properties, mainly connected with beneficial effect on cardiovascular system and with their antioxidant activity [11]. Allium vegetables are rich in flavonoles, primarily quercetin, and among the onion varieties the most abundant in quercetin are shallots and red onion [12]. It contains sulphur compound like allyl propyl disulphide, chromium and vitamin B6 known as compound that could reduce homocysteine, a factor cause heart attack and stroke. *Allyl propyl disulphide* in shallots could also increase insuline content and decrease glucose content in the blood.Chromium could decrease triglyceride and cholesterol in the blood. It also contains high flavonoids like quercetin that can reduce the risk of cancer [13].

Flavonoids and phenolic acids are the most important groups of secondary metabolism in plants that consider as good sources of natural antioxidant in human diets. In the last decade, gamma irradiation has been drawn the attention as a new and rapid method to improve the qualitative and quantitative characters of many crops. Irradiation with gamma rays is currently used as a tool in mutation breeding technology for enhancing the production of plant secondary metabolites like alkaloids or to increase biomass production in medicinally valuable plants [14]. Gamma rays were reported to be the most efficient ionizing radiation of creating mutants in plants as they can induce high mutation numbers in plants. It could also modify physiological characteristics to create new mutants with improved properties that can produce higher amounts of commercially essential metabolites, developing varieties that are agriculturally and economically significant, and contain high productivity potential [15-16].

A lot of work has been carried out on the beneficial effects of gamma irradiation in improving the crop, however for my knowledge, there is no report informing about impact of gamma irradiation on enhancement antioxidant and antimicrobial activity of shallot mutant.

The aim of the research was to investigate the variation in antioxidant and antimicrobial activity as well as the total quercetin content of the fifth generation genotype mutant Samosir shallot irradiated by gamma rays.

## 2. Materials and Methods

### *2.1. Preparing plant materials*

Uniform and healthy dry bulbs 2,5 months after harvest collected from Samosir farm, selected with the weight ranging from 1,3 g to1,7 g, by doses of irradiated along with non-irradiated local Samosir shallot were packaged in 0.1 mm thick paper bags of 10 cm x 22 cm dimension and sealed. The bags were subjected to gamma rays irradiation in radiator Chamber 4000 A using $^{60}$Co gamma source at Patir Batan, Jakarta, by exposing them to gamma irradiation with doses ranging from 1 to 9 Gy. Subsequently, the irradiated bulbs along with unirradiated bulbs (control) were planted in experimental field for generation advance. Bulbs of the $M_1V_1$ generation were planted in 2014 for generation advance in an experimental field. Bulbs harvested individually gave rise to the $M_1V_2$ population. In the following year, $M_1V_2$ were planted again in experimental field, aiming at the generation advance. Each plant was harvested individually, giving rise to $M_1V_3$genotype. Selected mutants by dosage of irradiation $M_1V_3$ were planted giving rise to $M_1V_4$ and $M_1V_5$. Six months after harvesting, populations of $M_1V_5$ were used as the samples to be analysed in antioxidant and antimicrobial activity and five months shallot bulbs storaged in the content of quercetin.

*2.2. Assay of Quercetin content*

The methanol extracts (250 ml each) of 19 genotype mutant of Samosir shallot along with control plant bulbs (after storage for 5 months in the room temperature) were mixed with 1.25 ml of distilled H2O and 75 ml of a 5% NaNO2 solution. After 5 minute, 150 ml of a 10% AlCl3 H2O solution was added and filtered for 6 minute. About 500 ml of 1 M NaOH and 275 ml of distilled H2O were added to the mixture, mixed well and the intensity of pink color was measured at 510 nm. *The level of total Quercetin concentration was calculated using quercetin (QU) as a standard* [17]. The results were expressed as mg of quercetin /g fresh weight of bulb tissues.

*2.3. Evaluation antimicrobial and antioxidant activity of $M_1V_5$ treated and untreated bulb*

Preparation of plant extracts

Plant materials were ground using an electric mill (GM100 Retsch, Germany). Each sample of 8 (eight) genotype mutant along with control plant bulbs (5 g) was macerated with 150 ml of 80 % ethanol and placed on a shaker (200 rpm) (GFL3005, Germany) for 24 h. All procedures, stated above, were carried out at room temperature. Extracts from each specimen were subsequently filtered and concentrated (evaporated to dryness) using a rotary vacuum evaporator (R-200 Buchi, Switzerland) at 40 °C. Dried residues were dissolved in 100 % dimethyl sulfoxide (DMSO) to obtain a stock concentration of 51.2 mg/ml, which was kept at −20 °C until used.

*2.3.1. Assay of antimicrobial activity*

In this study, four bacterial strains and one yeast were tested. The following American Type Culture Collection (ATCC) standard strains were purchased from Oxoid (United Kingdom) for analysis: *Enterococcus faecalis* ATCC 29212, *Escherichia coli* ATCC 25922, *Staphylococcus aureus* ATCC 25923, *Pseudomonas aeruginosa* ATCC 27853 and *Candida albicans* ATCC 10231. Six clinical isolates of antibiotic-sensitive as well as antibiotic-resistant *S. aureus* strains (SA1, SA2, SA3, SA4, SA5, SA6, SA7, SA8, SA9, SA10) were provided by University Hospital in Motol (Prague, Czech Republic). Microorganism cultures were stored in Mueller-Hinton broth (MHB) (Oxoid, United Kingdom) at 4 °C until used. Prior to antimicrobial tests, microorganisms were re-cultured at 37 °C for 24 h (48 h for C. albinos).

In vitro antimicrobial activity ethanol extracts derived from bulb of 8 genotype mutant along with control plant bulbs were determined by broth micro dilution method using 96-well microtiter plates as described by the Clinical and Laboratory Standards Institute and Cos *et al* (2006) in [18]. Minimum inhibitory concentration (MIC) is expressed as the lowest concentrations that inhibited bacterial growth by≥80% compared with that of the agent-free growth control. Antibiotics ciprofloxacin, oxacillin, tetracylline and tioconazole were used as positive control. Oxacillin and tetracylline were used as markers for methicillin and tetracycline resistance.

*2.3.2.DPPH radical-scavenging assay (assay of antioxidant activity)*

Slightly modified method previously described by [19] was used. Initially, two-fold serial dilutions of each sample were prepared in methanol (175 L) in 96-well microtiter plates. Subsequently, 25L of freshly prepared 1mM DPPH (2, 2-Diphenyl-1-picrylhydrazyl) in methanol solution was mixed with the sample in each well plates creating a range of concentrations from 2 to 5120 g/mL (final volume of 200 L) to start the radical-antioxidant reaction. The mixture was kept in the dark at room temperature. After a 30 min incubation period, the absorbance was read against blank at 517 nm using Infinite 200 reader (Tecan, Switzerland). Results were expressed as half maximal inhibitory concentration (g/mL IC50).

**3. Result and discussions**

In this study, 19 genotype mutants along with control bulbs(20 weeks after harvested) for the total of quercetin content and 8 genotype mutant (24 weeks after harvested) of irradiated shallots by gamma rays along with control bulbs (unirradiated) , were assayed for in vitro anti-microbial and antioxidant activity.

### *3.1. Total Quercetin content*

Biochemical differentiation based on quercetin content revealed that shallot bulbs irradiated at 6 Gy, 5Gy, 4 Gy and 2 Gy with the plant code number 6(3-4-4) and 6(2-5-6), 5(4-5-1), 4 (6-8-5) and 2 (6-7-9) exhibited quercetin content (QC) of 1.7 mg/g, 1.6 mg/g and 1.6 mg/g, 1,6 mg/g and 1.6 mg/g respectively,higher than unirradiated plant (control) (Table 1). However, the study also revealed that among genotypes themselves which were irradiated at the same dosage, exhibited different QC. For example between 5(4-5-1)and 5(5-1-2) that irradiated with 5 Gy exhibited QC 1.6 g/mg and 1.2 g/mg and among different dosage 4(6-8-5), 4(7-2-4), 3(13-2-3), 3(11-4-6) and 2(19-6-3) exhibited QC 1.6, 1.2, 1.2, 1.3, and 0.9mg/g respectively.

**Table 1**. The total quercetin concentration of Samosir Shallot irradiated by gamma Rays 5 months after harvest

| No | Plant code number | Dosage of irradiation | Total Quercetin concentration (mg/g) |
|---|---|---|---|
| 1 | 1(4-2-4) | 1 Gy | 1.3 |
| 2 | 1(5-4-2) | 1 Gy | 1.0 |
| 3 | 2(6-7-9) | 2 Gy | 1.6 |
| 4 | 2(9-7-5) | 2 Gy | 1.0 |
| 5 | 2(13-6-4) | 2Gy | 1.2 |
| 6 | 2(9-7-5) | 2Gy | 1.2 |
| 7 | 2(19-6-3) | 2Gy | 0,9 |
| 8 | 3(11-4-6) | 3Gy | 1.3 |
| 9 | 3(13-2-3) | 3Gy | 1.2 |
| 10 | 4(6-8-1) | 4Gy | 1.2 |
| 11 | 4(6-8-5) | 4Gy | 1,6 |
| 12 | 4(7-2-4) | 4Gy | 1.2 |
| 13 | 4(7-8-4) | 4Gy | 0.7 |
| 14 | 5(4-5-1) | 5Gy | 1.6 |
| 15 | 5(5-1-2) | 5Gy | 1.2 |
| 16 | 6(2-5-6) | 6Gy | 1.6 |
| 17 | 6(3-4-4), | 6Gy | 1.7 |
| 18 | 8(1-3-3) | 8Gy | 0.6 |
| 19 | 9(1-3-2) | 9Gy | 0.9 |
| 20 | Control | Unirradiated | 1.1 |

Gamma iradiation was reported to induce oxidative stress with over production of reactive oxygen species (ROS) such as superoxide radicals, hydroxyl radicals and hydrogen peroxide, which react rapidly with almost all structural and functional organic molecules, including proteins, lipids and nucleic acids causing disturbance of cellular metabolism [20]. This indirect effect of irradiation is important in vegetative cells, the cytoplasm of which contains about 80% water. These free radicals create alterations in plastid ultra-structure and mitochondria [21] and fragmentation of the endoplasmic reticulum, golgi apparatus, DNA and proteins by breaking chemical bonds [22].These radicals can induce damage in cellular membranes [23] and damage or modify important components of plant cells and have been reported to affect differentially the morphology, anatomy, biochemistry and physiology of plants depending on the radiation dose [24]. These effects include changes in the plant cellular structure and metabolism e.g., dilation of thylakoid membranes, alteration in photosynthesis, modulation of the anti-oxidative system, and accumulation of phenolic compounds [25].

To avoid oxidative damage, plants have evolved various protective mechanisms to counteract the effects of reactive oxygen species in cellular compartments [26]. This defense was brought about by alteration in the pattern of gene expression. This led modulation of certain metabolic and defensive pathways. The results of this study revealed that variation in quercetin content was observed in irradiated plants. Table 1 showed that gamma rays irradiation caused the variation in QC of bulbs, however the total QC of the sample was not determined by dosage of gamma irradiation, since genotype mutants irradiated with the same dosage also exhibited the different in total QC.

### *3.2. Antimicrobial activity*

Antimicrobial activity of nine (9) genotype mutants along with control plants had been evaluated in vitro against four bacterial species and one yeast. Complete results for antimicrobial activity of tested bulbs extracts are summarized in Table 2. It can be seen that none of the genotype mutants and control plant exhibited antibacterial inhibitory against all microorganism tested except for the sample number 2 and 6 (bulb generated from the plants irradiated by gamma rays with dosage at 2 and 6 Gy) that could only exhibited *Staphylococcus aureus* at high concentration. This results contrast with the statement mentioned before. This probably caused by the bulbs of all genotypes mutant along with the control plant had been kept too long (6 months) in improperly condition (at room temperature) which can affect their quality and biological characteristics. Prefously, the extract of *A. cepa* has been described to possess antimicrobial effect against *Vibrio cholera* [27] and *S. aurensus* [28]. The essential oil of this plant also exhibited strong antimicrobial activity against *Salmonella enteridis*, *Aspergillus Niger* and *Penicillium cyclopium* [29], complementing the results found in this study.

**Table 2.** Minimum inhibitory concentration (MIC) of the fresh shallot extract with different dosage of gamma ray irradiation

| Treatment Dosage of irradiation | MIC (μg/mL) | | | | |
|---|---|---|---|---|---|
| | *Staphylococcus aureus* ATCC 25923 | *Enterococcus faecalis* ATCC 29212 | *Escherichia coli* ATCC 25922 | *Pseudomonas aeruginosa* ATCC 27853 | *Candida albicans* ATCC 10231 |
| Control | >512 | >512 | >512 | >512 | >512 |
| 1 | >512 | >512 | >512 | >512 | >512 |
| 2 | **512** | >512 | >512 | >512 | >512 |
| 3 | >512 | >512 | >512 | >512 | >512 |
| 4 | >512 | >512 | >512 | >512 | >512 |
| 5 | >512 | >512 | >512 | >512 | >512 |
| 6 | **512** | >512 | >512 | >512 | >512 |
| 8 | >512 | >512 | >512 | >512 | >512 |
| 9 | >512 | >512 | >512 | >512 | >512 |
| ATB | 1 | 16 | 1 | 16 | 0.5 |

ATB Antibiotics used as positive control

### *3.3. Antioxidant activity*

Results for antioxidant activity of tested plants extracts are summarized in Table 3. It can be seen that none of these genotypes exhibited significant antioxidant efficacy. As mention before, this might be caused by the bulbs of all genotypes mutant along with the control plant had been kept too long (6 months) in improperly condition (at room temperature) which can affect their quality and biological characteristics.

**Table 3**. Antioxidant activity of fresh shallot extracts with Different irradiation dosage.

| Sample (Gy) | Antioxidant activity ($IC_{50}$ µg/mL) |
|---|---|
| Control | > 537.375 |
| 1 | > 537.375 |
| 2 | > 537.375 |
| 3 | > 537.375 |
| 4 | > 537.375 |
| 5 | > 537.375 |
| 6 | > 537.375 |
| 8 | > 537.375 |
| 9 | > 537.375 |
| Trolox (Vit.E) | 52.59 |

## 4. Conclusions

Population of genotype mutants irradiated with dosage 2Gy, 4 Gy, 5 Gy and 6 Gy have more benefit such as higher quercetin content, compared to control samples. None of the genotype mutants exhibited antibacterial inhibitory against all microorganism tested except for the sample number 2 and 6 (bulb generated from the plants irradiated by gamma rays with dosage at 2 and 6 Gy) that could only exhibited *Staphylococcus aureus* at high concentration. In addition, there was also none of the genotypes observed exhibited significant antioxidant efficacy.

## References

[1] Yang J, Meyers K J, Heide J and Liu R H 2004 Varietal differences in phenolic content and anti-oxidant and anti-proliferative activities of onion. J. Agric. Food Chem. Vol **52** pp. 6787-6793.

[2] Ly T N, Hazama C, Shimoyamada M, Ando H, Kato K and Yamauch R 2005Antioxidative compounds from the outer scales of onion. J. Agric. Food Chem. Vol**53** pp. 8183-8189.

[3] Jiang H, Ji B, Liang J, Zhou F, Yang Z and Zhang H 2006 Comparison on the antioxidant capacity of selected fruits and vegetables and their separations. Chem. Natur. Compound. Vol **42**pp. 410-414.

[4] Štajner D, Milić N, Čanadanović-Bruntel J, Kapor A, Štajner M and Popović 2006 Exploring Allium species as a source of potential medicinal agents. Phytother. Res. Vol **20** pp. 581-584.

[5] Stratil P, Kleidus B and Kubán V 2006 Determination of total content of phenolic compounds and their antioxidant activity in vegetables - evaluation of spectrophotometric methods. J. Agric. Food Chem. Vol**54** pp. 607-616.

[6] Leighton T, Glinther , Fluss L, Harte WK, Cansado J, Notario V 1992 Molecular characterization of quercetin and quercetin glycosides in Allium vegetables: their effects on cell transformation. In: Huang, M.T., Lee, C.Y., Ho C.T. (eds.) pp.221 -238

[7] Fattorusso E, Iorizzi M, Lanzotti V, Taglialatela-Scafati O 2002 Chemical composition of shallot (*Allium ascalonicum* Hort.) J Agric Food Chem. Vol **50** pp.5686 -5690.

[8] Kempuraj D, Castellani M, Petrarca C, Frydas, Conti P, Theoharides T, Vecchiet J 2006 Inhibitory effect of quercetin on tryptase and interleukin-6 release, and histidine decarboxylase mRNA transcription by human mast cell-1 cell line. Clin. Exp. Med. Vol **6** pp.150-156. [Pub Med].

[9] Capasso F, Gaginella TS, Grandolini G, and Izzo AA 2003 Phytotherapy.A Quick Reference of Herbal Medicine, Springer-Verlag, Heidelberg, Germany,.

[10]Lawson L D 1998 "Garlic: A Review of Its Medicinal Effects and Indicated Active Compounds." In Phytomedicines of Europe, edited by L. D. Lawson and R. Bauer. American Chemical

Society Symposium Series, no. 691. Washington, D.C. American Chemical Society pp.176–209

[11]Holman P C H, Katan M B 1999 Dietary flavonoids Intake, health effects and bioavailability. Food Chem. Toxicol Vol **37** pp. 937-942.

[12]Kyle JAM, Duthie G G 2006 Flavonoids in foods. In Flavonoids, Chemistry, Biochemistry and Health Implications (Andersen O.M., Markha, New York; Taylor and Francis Group. 219-263

[13]White K and Zellner J 2008 Hamilton College Seminar 235 Food for thought: The Science, Culture and Politics of Food Spring. Clinton New York. United States.

[14]Sharma J R, Lal R K , Misra HO, Gupta M M and Ram RS 1989 Euphytica. Vol **40** pp. 253 -258.

[15]Eroglu Y, Eroglu H E, and Ilbas A I, 2007. Gamma ray reduces mitotic index in embryonic roots of Hordeum vulgare L. Adv. Biol. Res. Vol **1** pp. 26-28

[16]Sato Y, Shirasawa K, Takahashi Y, Nishimura M and Nishio T 2006 Mutant selection from progeny of gamma-ray-irradiated rice by DNA heteroduplex cleavage using brassica petiole extract. Breeding Sci. Vol**56** pp. 179-183.

[17]Jia Z, Tang M and Wu J 1999 The determination of flavonoid contents in mulberry and their scavenging effects on superoxide radicals. Food Chem. Vol **64** pp. 555-559.

[18]Teka A, Rondevaldova J, Asfaw Z, Demissew S, Van Damme P, Kokoska L and Vanhove W 2015 In vitro antimicrobial activity of plants used in traditional medicine in Gurage and Silti Zones, south central Ethiopia. BMC Complementary and Alternative Medicine 15.

[19]Sharma O P and Bhat T K 2009 DPPH antioxidant assay revisited. Food Chem. Vol **113** pp. 1202-1205.

[20]Noreen Z and Ashraf M 2009 Changes in antioxidant enzymes and some key metabolites in some genetically diverse cultivars of radish (*Raphanus sativus* L.). Environ. Exp. Bot. Vol**67** pp. 395-402

[21]Kovacs E and Keresztes A 2002 Effect of gamma and UV-B/C radiation on plant cells. Micron.Vol **33** pp.199-210.

[22]Somosy Z 2000 Radiation response of cell organelles. Micron. Vol **31** pp. 165-181 [PumMed].

[23]Shuryak I and Brenner DJ 2009 A model of interactions between radiation-induced oxidative stress, protein and DNA damage in *Deinococcus radiodurans*. J Theor Biol.Vol 261 pp. 305-317.

[24]Ashraf M, Cheema A A, Rashid M and Qamar Z 2003 Effect of gamma rays on M1 generation in Basmati rice. Pak. J. Bot. Vol **35(5)** pp. 791-795.

[25]Wi S G, Chung B Y and Kim J S 2007 Effects of gamma irradiation on morphological changes and biological responses in plants. Micron. Vol **38** pp. 553-564.

[26]Kiong A, Ling Pick, Grace Lai S H and HarunAR 2008 Physiological responses of *Orthosiphon stamineus* plantlets to gamma irradiation. Am-Eurasian J. Sustain. Agric. Vol **2(2)** pp. 135-149.

[27]Hannan, Abdul, Tabish Humayun, Muhammad Barkaat Hussain, Muhammad Yasir, and Sumayya Sikandar 2010 "In Vitro Antimicrobial Activity of Onion (*Allium cepa*) against Clinical Isolates of Vibrio Cholerae," Journal of Ayub Medical College, Abbottabad. Vol**22(2)** pp.160-163.

[28]Eltaweel M 2013 "Assessment of Antimicrobial Activity of Onion Extract (*Allium cepa*) on *Staphylococcus aureus*; in vitro study," International Conf. Chem. Agric. Med. Sci. Dec. 29-30. Kuala Lumpur

[29]Benkeblia N 2004 " Antimicrobial Activity ofEssential oil Extracts of various Onions (*Allium cepa*) and Garlic (*Allium sativum*)." LWT-Food Science and Technology. Vol**37 (2)** pp. 263-68.

**Acknowledgments**
This research was supported by the Grant of Menristek Directorate General of Higher Education and Prof. Kokoska's Laboratory at Prague. We respectfully thank Menristek Directorate General of Higher Education (Republic of Indonesia) for providing financial assistance and greatly appreciate Prof. Kokoska for providing laboratory facilities in assaying antimicrobial and antioxidant activity of genotype mutant Samosir shallots.

IOP Conf. Series: Earth and Environmental Science **122** (2018) 012071 doi:10.1088/1755-1315/122/1/012071

# Maturity analysis of the innovation system in the livestock industries of West Sumatra, Indonesia

**A Suresti[1], U G S Dinata[2] and R Wati[3]**

[1,3]Animal Husbundry Faculty, Universitas Andalas
[2]Faculty of Engineering, Universitas Andalas, West Sumatera, Indonesia

**Abstract.** The main objective of this study was to analyse the ANIS of Livestock industry in West Sumatera according to its main determinants. This research aims to know and understand the maturity level of the regional innovationsystem of West Sumatra livestock industries , including recommendation for improvements. The maturity level of the regional innovation system was analyzed with the ANIS method by using an opinion survey to some experts from businessmen, government and universities. This expert opinion survey was conducted to measure expert perceptions on the implementations of determinants of west Sumatera regional innovations systems on livestock industries on macro level (innovations policy), mezzo level (innovation institutions and programs) and micro level (innovation capacity). By using Likert method, the result showed maturity level for the West Sumatra livestock industries innovation system was still develop. This is caused by low maturity in the macro and meso determinants which are at the central and regional government side. This research result may be useful as recommendations for the government for improving the maturity level and may be a basic for incoming researches to identify regional innovation system determinants with low maturity to improve.

## 1.Introduction

One of the great potential that is owned by the Province of West Sumatra in the development of the regional economy is the livestock sector, where the vision of the future development of livestock breeding is to realize advanced, efficient and resilient, competitive, independent and sustainable economic and simultaneously able to empower people, especially in the countryside therefore, directed livestock development in order to compete in national and international markets, strengthen national food security to food self-sufficiency, and increase the contribution of livestock in the Gross Regional Domestic Product (GDP). To establishfood security, attention tofood self-sufficiency program to meet the needs of animal protein per capitashould be increased to improve.Increasing the competitiveness of the livestock industry products can boost the local economy and, ultimately, improvepublic welfare western Sumatrahave almost all kindsof farm products such as meat, eggs, milk, leather and processed products.However, the production capacity andthe added valueoft hefarmcommodities have not been as expected. This plus there are many imported mainly meat and milk and derivative products nationally.

It is necessary to develop a strategy for the development of animal husbandry accurate. The development of the livestock industry has good prospects caused by the demand for materials derived from cattle will continue to increase in line with increases in population, income and awareness to consume foods high in nutrients. Innovation in West Sumatera existing farm and running, but has not been able to improve the competitiveness and economic improve. It is believed due to the lack of an integrated system. One strategy that can be put forward is to combine various concepts, namely the concept of the Regional Innovation System , the system of agribusiness, farm-based integrated region with attention to environmental sustainability. With this model expected above problems can be overcome, so that self-sufficiency in livestock products can be realized in order to build a strong food security and sustainable.

The important question is the extent to which the level of maturity Regional Innovation System in West Sumatra to improve the competitiveness of the livestock industry. In it, the extent to which there is completeness and the involvement of actors and facilitation support at each level of the macro, meso and micro regional innovation system sumatera. .Measurement of the maturity level using ANIS (Analysis of Nations Innovation Systems in the livestock industry in West Sumatra.

To answer the second question above, has made the identification of activities that can map the degree of maturity of innovation system livestock industry in west). The maturity level can be obtained from the perception of businessmen, academics and local government officials who are involved as actors and observers of regional innovation systems. Based on the formulation of the problem mentioned above, the proposed research with a common purpose, The main objective of this study was to analyse the ANIS of Livestock industry in West Sumatera according to its main determinants namely the identification of the level of maturity of the livestock industry innovation system of West Sumatera in increasing production capacity and added value to the competitiveness of livestock products . This study has several output targets as the specific purpose of knowing the perceptions of actor and observer of regional innovation systems consisting of businessmen, academics, and Local Government, determine the level of maturity of regional innovation systems west of livestock industry at every level of the macro, meso and micro, and issued a recommendation intervention to increase the maturity of innovation by building and / or strengthening actors- actor and facilitation of regional innovation systems.If the goal is reached, it will obtain various benefits from the results of this study, among others, the results of this study can determine the order of priority the establishment and / or strengthening factors of maturity of the innovation system based on minimal effort with maximum impact on industrial innovation system maturity of livestock.output this research can determine the roadmap plan of action to strengthen the regional innovation system in west sumatera. this research could create regional innovation systems that can improve the competitiveness of industrial products farms with increased production capacity and value-added.

## 2. Materials and Methods

The model used draws on a wide range of data from the Expert Opinion Survey (EOS). The EOS meets the need for up-to-date and far-reaching data, providing valuable qualitatively information for which hard data sources are scarce or nonexistent. The survey was completed by at least 20 national experts for each country. We asked the experts to provide their opinions on various aspects of innovation and the innovation environment in which they operate. The data gathered thus provide a unique source of insight and a qualitative portrait of each nation's innovation concept and how it compares with the situation in the other country.

ANIS approach is based on the assumption that innovation systems is strongly influenced by the opinions of the actors determinant of government, universities and the private sector each reflecting aspects of the innovation system. The determining factors can be grouped into three levels [1], namely:

a. Levelmakro (policy), which includes the national innovation policy, innovation policy areas, master plan, training and education, R & Dforesight, cluster policy, regulatory innovation.
b. Meso level which includes:1. Institutional innovation consists of a technology transfer center, technopark, technology incubators, clusters, business promotion agencies, service providers innovate and innovation funding agencies.2. The supporting program inovation consists of STI financing scheme, a program of basic research, applied research program, co-financing schemes STI assistance efforts, support entrepreneurship, cluster development programs, and international support.

Micro level (innovation capacity) including universicity, basic research institutions, private research institutions, innovators, private investors, entrepreneurial, technology-based SMEs and large industry-based technology.

The questions in the study follow a structure that asks the interviewees to evaluate, on a scale of 1 to 4, the current conditions of their particular operating innovation environment. At one end of the scale, 1 represents the worst possible operating condition or situation and at the other end of the scale, 4 represents the best (see figure 4). Indicator values above 3 are characteristic for well developed industrial countries where all determinants are established and well functioning, some slightly better than the others. A value between 1.5 and 3 means the determinant already exists and is in the phase of development. Values below 1.5 mean that a specific determinant may exist, but is not really operational in practice or it needs improvement.

## 3.Results And Discussion

By ANIS method can be seen the results of the maturity level of the Regional innovation system for livestock industry of west Sumatera in figure 1below.

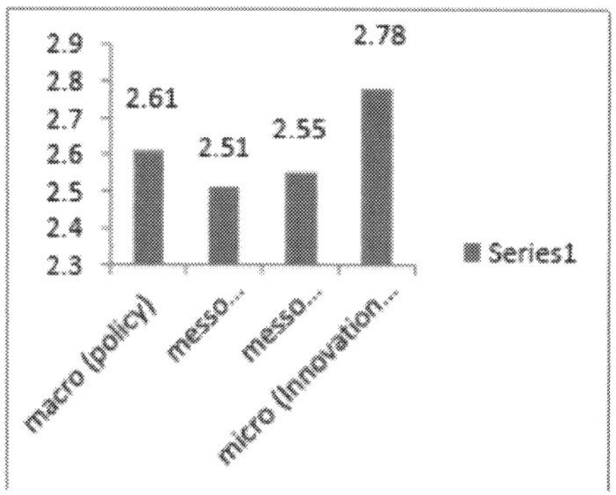


**Figure 1**. The average value of each level (top) and the levels below and above the overall average value (bottom)

The average value of each level and the levels below and above the overall average. The average value of the entire questionnaire is 2.61 on a scale from 1 to 4. This value indicates above 2.5 as the mean scale Lickert. This means that the respondents argued among disagree and agree that the Regional Innovation System for the livestock industry has been implemented in West Sumatra.

The figure 1 above also shows the average value of each level group comprising the macro level, meso-level institutional, meso-level programs, and micro level. Macro level are in the same position with the average value of the whole is 2.61. Meso level of both institutional and program is under the average value is the mean level of the weakest, especially on the institutional aspects. The highest level value is at the micro level, namely the innovation capacity for the livestock industry which shows the value above the overall average, which means that this level is the most mature in the innovation system ranch in West Sumatra.

Macro level were in the same position with the overall average value was 2.61.The lowest value occurred in the National Innovation Policy was 2.50, while the highest values were in theEducationand Training Policy 2,86.

The meso level for institutional support aspects of innovation was at 2.51 which is the lowest position in average group level. At the messo level, the lowest value occurred in the institutional Technopark with valueof 2.28, while the highest value was at the Institute of Business Promotion with a valueof 2.69.

The values determinant group messo leve lof innovation support programs. The messo level was in the low position was 2.55. At this level, the lowest value of the determinantis on Facilitation/Support

International with a valueof 2.35, while the highest score was on Facilitation/Support Entrepreneurship with a value of 2.85. The values of the micro-level determinants of the group that has the highest value was 2.78. At this level, the lowest value of the industry was in engagement with a value of 2.62, while the highest value was at university with a valueof 2.89 which is a determinant for the overall highest level.

At the meso level institutional innovation average value of 2.51, there was a determinant which is under the value that Technopark, Incubator, and livestock Cluster .Institute the form of Technology Transfer Center, Institute of Business Promotion, Innovation Service Providers and Institutions Financing Innovation in West Sumatra for the livestock industry, including the determinants of mature or above average on the meso level institutional innovation

Respondents assess determinants on the meso level to aspects of the innovation program, on average, at a value of 2.55 is not much different from the meso level institutional innovation. At the meso level, there are five programs were rated below average ie Financing Scheme of Science, Technology and Innovation (STI), Basic Research Program, the Joint Funding Scheme, Cluster Development Program, and facilitation following Internasional.Determinan, Applied Research Program, Assistance efforts of Science, Technology and Innovation (STI), and Support Entrepreneurship rated mature which is above the average value of the group.

Based on the recapitulation at the micro-level group, the average value of this level was 2.78 which is the overall highest level group. Determinant considered below average innovation capacity in the livestock industry is the Basic Research Institute, Applied Research Institute, and Large Industries. There is determinant that has a value equal to the average value of the micro-level group that innovators and entrepreneurs. Determinants above average is the University, Private Investors, and Small and Medium Enterprises (SMEs).

ANIS study results with 30 respondents from businesses, government officials and academics shows that there are a variety of strengths and weaknesses Regional Innovation Systems for the livestock industry of West Sumatra were evaluated from three levels, namely the level of macro, meso and mikro.Based on ANIS method, strength Regional Innovation Systems for the livestock industry found in West Sumatra these levels:

a. Macro Level Innovation Policy, namely the Master Plan, Training and Foresight R and D Agenda.
b. Messo level namely Innovation Technology Transfer Center, Institute of Business Promotion, Innovation Service Providers, and the Institute for Innovation Funding
c. Meso-level Innovation Program is Applied Research Program, Assistance Efforts Science, Technology and Innovation (STI), and Enterprise Support
d. Micro level which shows have the capacity of innovation that includes the University, Private Investors, and Small and Medium Enterprises (SMEs)

Regional Innovation Systems weakness for the livestock industry of West Sumatra are at these levels:

a. Macro Level Innovation Policy is the National Policy, Regional Innovation Policies, Policy and Regulatory Pro Cluster Innovation
b. Level Meso institutional namely Innovation Technopark, Incubator, and Cluster in the livestock industry
c. Level Meso Innovation Program Funding Scheme namely Science, Technology and Innovation (STI), Basic Research Program, the Joint Funding Scheme, Cluster Development Program, and the International Facilitation
d. Level Micro Innovation Capacity Research Institute which includes Basic, Applied Research Institute, and Large Industries.

## 4.Conclusion

The results showed that the average value of the innovation system maturity level livestock in West Sumatera was 2.61 with an average value At the macro level of 2.61 such as policy, for institutional meso level of 2.51, the meso level of 2.55 and the value to the micro level of 2.78 also as the highest value. It is mean that the maturity level of the innovation system livestock in West Sumatera still at the developing position (1.5 - 3) has still not been established yet ( 3-4). Determinants Reginal Innovation System based ANIS immature in West Sumatra and need to be prioritized to be empowered is the implementation of the National Policy, Regional Innovation Policies, Policy and Regulatory Pro Innovation Cluster (macro level), the implementation of Technopark, Incubator, and Cluster (meso-level institutional innovations ), and the implementation of the Scheme Funding of Science, Technology and Innovation (STI), Basic Research Program, the Joint Funding Scheme, Cluster Development Program, and the International Facilitation (at the meso level innovation program), as well as the Basic Research Institute, Institute for Applied Research and Industrial Large (at the micro level of innovation capacity).

Based on the conclusions above, intervention needs to be submitted recommendations to the Central Government and Local Government to increase the maturity of determinants Regional Innovation System supporting the livestock industry in west Sumatera must be prepared few determinant for determining priorities which are enhanced where effort is not too big but the effect it is large enough to maturity Regional Innovation Systemfor livestock industri inWest Sumatra.

**References**

[1] Koswara D V 2012s *Analysis of Nations Innovation System* Bahan Pelatihan Sistem Inovasi Daerah di Bengkulu

[2] Asheim B T and Isaksen A 2002 Regional innovation systems; the integration of local "sticky" and global "ubiquitous" knowledge *The Journal of Technology Transfer* Vol. **27** No. 1 pp 77-86

[3] Breschi S and Malerba F 1997 "Sectoral innovation systems: technological regimes, Schumpeterian dynamics and spatial boundaries". In: Edquist, C. (ed.): Systems of Innovation: Technologies, Institutions and Organisations ( London: Pinter)

[4] Chung S Building a National Innovation System through regional innovation systems, *Technovation* **22** pp 485-491

[5] Etzkowitz H 2002 The Triple Helix of university - industry - government, implications for policy and evaluation (Science Policy Institute Working paper 2002•11). Retrieved from http://www.sister.nu/pdf/wp_11.pdf

[6] Edquist C 1997 *Systems of Innovation: Technologies, Institutions and Organizations* London, Pinter

[7] Freman C 1995 The national system of innovation in historical perspective *Cambridge Journal of Economics* No 19 pp 5–24

[8] Freeman C 1987 *Technology Policy and Economic Performance: Lessons from Japan* (London: Pinter Publishers)

[9] Freeman C 1988 "Introduction". In: Giovanni, D.; Freeman, C.; Nelson, R.; Gerald, S. and Soete L. (eds.): Technical change and economic theorya Pinter

[10] Lundvall B Å.1992 *National Innovation Systems: Towards a Theory of Innovation and Interactive Learning* (London: Pinter)

[11] Malerba F 2002 Sectoral systems of innovation and production *Research Policy* No 31 Vol **2** pp 247-264

[12] Nelson R R 2002 Bringing institutions into evolutionary growth theory *Journal of Evolutionary Economics* **vol. 12**: 17-28

IOP Conf. Series: Earth and Environmental Science **122** (2018) 012072 doi:10.1088/1755-1315/122/1/012072

# Swamp land optimization in supporting food security and enhancing farmers welfare in South Sumatra Indonesia

**Herwenita and Y Hutapea**

South Sumatra Assessment Institute for Agricultural Technology (AIAT), Indonesian Agency for Agricultural Research and Development (IAARD), Ministry of Agriculture, Indonesia
E-mail: herwenita@gmail.com

**Abstract:** Swamp land in Indonesia spread in Sumatra, Kalimantan and West Papua. In Sumatra the largest swamp land area is located in South Sumatera Province. Unfortunately only few of the areas have been utilized due to its fragility, in which farmers could only cultivate rice on it once a year. The purpose of this paper is to develop a feasible farming pattern in swamp land to help farmers and practitioners in optimizing it by managing its water level. Shallow and mid swamp land can be cultivated using rotation model of crops (rice, corn, cassava), horticulture (cucumber, long beans, watermelon etc), fish farming (catfish, snake head fish, tilapia), and duck farming, whereas submergence tolerant rice varieties can be cultivated alternating with fish farming in deep swamp land. This study shows that such swamp land management is financially feasible showing by its positive NPV value, BCR value is above 1.00, and IRR value is greater than the interest rate. Therefore, implementation of this farming pattern is expected to increase farmers' income and household food supply as well as village food supply.

## 1. Introduction

Land, as mention in the first strategy of the Indonesian Ministry of Agriculture to strengthen agricultural development [1] is important is necessary to properly utilize effectively and efficiently. Due to its limited resources, land conservation, protection and expansion through land reclamation and optimization of marginal land and or low productive land are applied [2].

Swamp land as marginal land can be utilized for agriculture since it is quite large in Indonesia. It reaches 13.3 million hectares spreads in Sumatra, Kalimantan and West Papua [3]. In Sumatera, the largest swamp land area is located in South Sumatera Province (2.98 million ha). Unfortunately only 368,690 hectares have been utilized for rice which is divided into 70,908 hectares of shallow swamp land, 129,103 hectares of mid swamp land and 168,670 hectares of deep swamp land [4]. Low optimization of swamp land is happened due to its physical condition which considered as a fragile land. According to Djamhari [5], obstacles found in swamp land were physic-chemical condition with high acidity, low soil fertility, as well as sudden and erratic water logging and flooding in rainy season but drought in dry season. Moreover, biological problem of high weeds, pests and disease also occurs. All these problems lead to low productivity in swamp land.

Several projects and studies have been done to increase productivity in swamp land as well as to increase farmers' welfare. A study conducted by Djamhari [5] revealed with proper technology and proper water management system, swamp land can be cultivated three times a year using rice-rice-crops pattern which can increase productivity up to 7.00 tons of dry grain harvest. The weakness of this study was lower farmers' level of adoption due to higher technology intake which leads to higher cost. However, Galib [6] shows using relatively low input along with integrated crop management approach, corn cultivation in swamp land still yield 4.5 tons per hectares and even more.

In line with agriculture development strategic plan by Ministry of Agriculture to optimize and increase the use of land, this study is design to visualize utilization of swamp land optimally based on its natural characteristics. This feasibility study is different from existing studies and projects because it emphasize on optimal land utilization throughout the year using propose farming pattern based on swamp land characteristics. Moreover, simple technology is used in order to be easily adopted by farmers. Hence, the specific objectives of the study are: 1) to develop a feasible farming pattern for swamp land; and 2) to give an overview of the feasibility of the project. The expected result may provide greater insight for the farmers and government regarding the effort of swamp land optimization.

## 2. Methods

This study uses both primary and secondary data. Primary data was gathered from interview with existing rice farmers in Kota Daro II Village, Rantau Panjang Sub-district, Ogan Ilir in mid swamp land during April 2017. The random sample consisted of 15 farmers from Harapan Maju II farmer group (managers and members). Due to the study purpose to learn the farming feasibility through financial analysis, the interviews were not conducted individually but in groups to obtain an actual and valid data of input and output items so that this method is expected to represent the farming business in swamp area. However, since the farming pattern of rice-corn-fish has not yet found in this location, the benefit cost analysis of corn and fish are derived from the secondary data which is financial analysis of corn and fish farming in other swamp area outside the study location that compiled and analyzed for the purpose of this study.

According to Boslaugh [7], secondary data analysis research can be done by collecting secondary data, then examines it to look at the variables (aspects) of what is in the data to then raised the research question (problem formulation) by connecting various aspects (variables). Using Amirin [8] steps of secondary research methods, the main secondary data used in this study is derived from Statistics Indonesia and journals. In terms of data analyzing, production-consumption analysis is done by comparing data from 2010 to 2015 to see the pattern of production and consumption development and use the gap in projecting surplus or scarcity of each commodities in the future. Furthermore, to determine business feasibility, some of methods are used which are Net Present Value (NPV) where $C_t$ are net cash inflow during the period t, $C_0$ are the total initial investment costs, $r$ is a discount rate, and $t$ is the number of time periods; Benefit Cost Ratio (BCR); Internal Rate of Return (IRR) where $r_1$ is the lower discount rate, $r_2$ is the higher discount rate, $NPV_1$ is the NPV at $r_1$, and $NPV_2$ is the NPV at $r_2$; and Payback Period (PP) where $I_0$ is the initial investment amount, and $R$ is the amount of return on investment [9].

$$NPV = \sum_{t=1}^{T} \frac{C_t}{(1+r)^t} - C_o \quad (1)$$

$$BCR = \frac{Discounted\ Value\ of\ Benefit}{Discounted\ Value\ of\ Cost} \quad (2)$$

$$IRR = r_1 + \frac{NPV_1}{NPV_1 - NPV_2}(r_2 - r_1) \quad (3)$$

$$PP = \frac{Io}{R} \quad (4)$$

## 3. Results and discussion

This study proposes a feasible farming pattern in swamp area in South Sumatra Province. Ogan Ilir Regency is chosen as the study case location because it is the largest swamp area in South Sumatra where 55.2 percent (266,607 ha) of swamp land in South Sumatra is located in this regency [10]. This regency has shallow, mid, and deep swamp area which currently used for farming that makes it a suitable location for this study. Ogan Ilir has tropical wet climate (Type B) with dry season on May until October and rainy season from November to April [13].

In terms of planting time, the water logged in the land which affected by rainfall intense and river rise may cause differs in crops planting time. Most of the field is cultivated once a year and only small part of land is utilized for vegetables and crops grown in the bunds. Using climate data and planting time of the area above, thus the proposed farming pattern is formulated as shown in Table 1.

**Table 1.** The Proposed farming pattern in swamp land

| Swamp Land | Commodities | Month | | | | | | | | | | | |
|---|---|---|---|---|---|---|---|---|---|---|---|---|---|
| | | Jan | Feb | Mar | Apr | May | Jun | Jul | Agst | Sep | Oct | Nov | Dec |
| Shallow | Rice | | | ■ | ■ | ■ | ■ | | | | | | |
| | Crops | | | | | | | ■ | ■ | ■ | ■ | | |
| | Crops | ■ | ■ | | | | | | | | | ■ | ■ |
| Mid | Rice | | | | ■ | ■ | ■ | ■ | | | | | |
| | Corn | | | | | | | | ■ | ■ | ■ | ■ | |
| | Fish | ■ | ■ | ■ | | | | | | | | | ■ |
| Deep | Rice | | | | | | ■ | ■ | ■ | ■ | | | |
| | Fish | | ■ | ■ | ■ | ■ | | | | | | | |

In shallow swamp land when the water recede begins, farmers start to cultivate rice in March to June and plant crops such as corn from June/July to October followed by another crops in November to February. Mid swamp land can be cultivated started in April followed by corn in August and fish farming during rainy season in December to March. Furthermore in deep swamp land, where most of the times are flooded by high tide, cultivation can only be done once in June to September for rice while the rest of the year can be utilize for fish raising. Thus this study propose a throughout year optimization. However, mid swamp land is the only agro ecosystem that estimated for this feasibility study since it is the largest swamp area compare to shallow and deep swamp land. Furthermore, the mid swamp land has greater chance to be planted twice a year due to its water availability.

Rice is planted in April to July using early maturing submergence rice variety because it has higher level of resistance considering more problems occur in swamp land with expected yield per hectare per year is 5 tons rice. The leftover straw from rice production will be rotten and become organic media for the next crops. Corn is cultivated after the rice production in August to October. Sweet corn variety is used considering it short day plant period compare to other corn variety which has 75 days plant to harvest period so it may harvest in October with expected yield as much as 20,000 corncobs.

For fish farming, tilapia or catfish species is used since it has higher level of resistance to high acidity water and its relative easier to be raised. Fish farming is conducted in $48m^3$ ponds considering farmers' ability both capital and labor. It started to culture in December and harvested in May.

As mention above, rice is the existing commodity in swamp area. Based on literature, swamp land optimization is possible to be done either by a second rice production or other commodities. Corn and fish are chosen as additional commodities to optimize swamp land in this study because it is relatively easier to conduct. Production-consumption analysis of corn and fish is done in order to predict its market share in the future. The gap between the two will indicate whether a particular commodity is on its surplus or scarcity. It is important because by analyzing the production-consumption, the product that are needed and desired by consumers will be known to fill the expected target market share.

Corn is a major commodity crop after rice which contributes as feed. Corn production was obtained from corn production in Ogan Ilir Regency from 2010 to 2015 while corn consumption was compiled from livestock and poultry corn feed consumption which shown in Figure 1. The average increase of corn consumption was 49.47% per year during 2010-2013 while corn supply has decrease in average of 37.79% per year. It can be seen that corn production is not sufficient to cover corn feed consumption in the future.

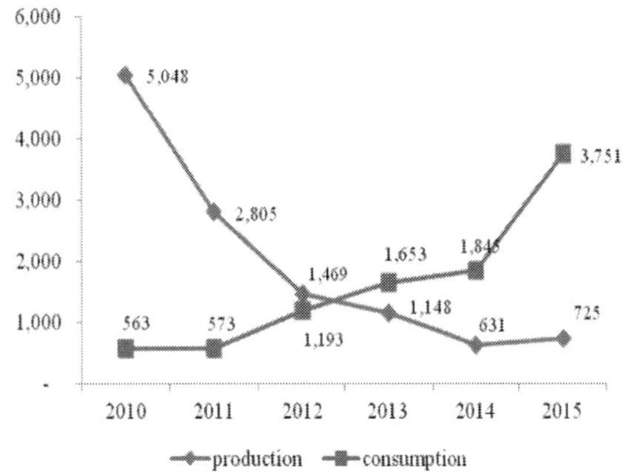


*Source: Compiled from Statistics Ogan Ilir, 2011-2016*

**Figure 1**. Production-Consumption of Corn in Ogan Ilir Regency (in tons)

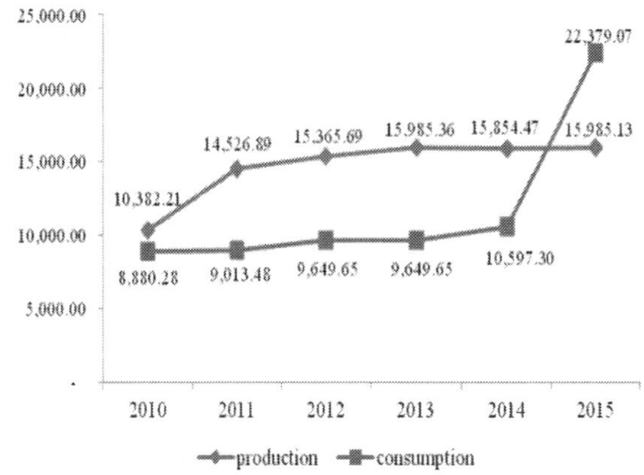


*Source: Compiled from Statistics Ogan Ilir, 2011-2016*

**Figure 2.** Production-Consumption Analysis of Fish in Ogan Ilir District (in tons)

Like other South Sumatrans, the people in Ogan Ilir also fond of fish compare to meat. The Ogan Ilir fish supply is coming from river catch and fish farming. The fish supply demand analysis in Ogan Ilir District is shown in Figure 2. During 2010 to 2015, fish supply has the highest increased in 2011 while slightly increase shown during 2012 to 2015. Contrary, fish consumption has tremendous increased in 2015 that has doubled its number. Therefore, any efforts to increase fish production to increase fish supply are necessary to meet the increasing demand of fish.

**Table 2.** The Proposed farming pattern in swamp land (in IDR)

| Item | | | Year – 1 | Year – 2 |
|---|---|---|---|---|
| 1. | Investment Costs | a. Land (1 ha) | 60,000,000 | 0 |
| 2. | Operating/Maintenance Costs | | | |
| | • Materials | a. Rice | 1,650,000 | 1,701,150 |
| | | b. Corn | 12,300,000 | 12,681,300 |
| | | c. Fish | 9,312,500 | 9,601,188 |
| | • Tools | a. Rice | 247,150 | 247,150 |
| | | b. Corn | 265,900 | 265,900 |
| | | c. Fish | 1,803,000 | 1,803,000 |
| | • Labors | a. Rice | 9,803,929 | 10,294,125 |
| | | b. Corn | 5,883,750 | 6,177,938 |
| | | c. Fish | 5,750,000 | 6,037,500 |
| Total Costs | | | 107,016,229 | 48,809,250 |

Several assumptions were used to provide a proper projection of the project feasibility. The assumptions used are increase in material costs as much as 3.1 percent per year and increase in labor cost as much as 5 percent per year while tools depreciation is assumed similar. With regards to the revenue, assumption used is increase in output price as much as 5 percent. These were based on inflation rate "year on year" (January 2015 to January 2016) [11]. Production cost of the project is shown in Table 2 while revenue of the project per hectare is shown in Table 3.

The largest production cost of this project is investment expenditure which is buying the land in the first year or the beginning of the project that reach IDR 60 millions. The purchase of land is chosen because this study wants to see the feasibility of the project with all the expected costs to see its breakeven point.

Operating and maintenance cost of the project are divided into three part namely materials, tools and labors. The highest material cost is found in corn production because of higher seed price that reach IDR 9 million for 12 kg sweet corn seeds for one hectare farm. Fish farming also has higher material cost due to high feed consumption which takes more than 80 percent of its total material cost. Moreover, it also has the highest tools cost compare to rice and corn production. In terms of labors, rice production has the highest labor input since it has more differ work activities from land processing to post harvesting compare to corn and fish farming so that requires higher labor intake.

Rice production yields 5 tons with a selling price of IDR 4,300 per kilograms. It also has side product in form of bran at a price of 5 percent of rice selling price. The corn yields 20,000 corncobs at IDR 1,500 selling price per cobs while fish farming yields 1.5 tons of fish at IDR 16,000 selling price per kilogram. Thus, the total revenue in the first year of the project reaches IDR 79,800,000. In the second year, the revenue increase using 5 percent increase in price output assumption, so that the expected total revenue is IDR 83,790,000 (Table. 3).

**Table 3.** Revenue of the project per hectare in mid swamp land (in IDR)

| Item | Year – 1 | Year - 2 |
|---|---|---|
| Rice | 25,800,000 | 27,090,000 |
| Corn | 30,000,000 | 31,500,000 |
| Fish | 24,000,000 | 25,200,000 |
| Total Revenue | 79,800,000 | 83,790,000 |

*3.1 Financial statement and analysis*

Financial statement and analysis will show the feasibility of project in term of financial approach. BCR, NPV, IRR and payback period were used to see the project feasibility.

**Table 4.** Profitability of the project (in IDR)

| | Year – 1 | Year - 2 |
|---|---|---|
| Total Cost | 107,016,229 | 48,809,250 |
| Total Revenue | 79,800,000 | 83,790,000 |
| Benefit | (27,216,229) | 34,980,750 |
| Discount Factor (25%) | 0.800 | 0.640 |
| PV of 1 (25%) | (21,772,983) | 22,387,680 |
| Discount Factor (30%) | 0.769 | 0.592 |
| PV of 1 (30%) | (20,929,280) | 20,708,604 |
| *Financial statement:* | | |
| NPV | 614,697 | |
| BCR | 1,01 | |
| IRR | 28.68 % | |
| Payback Period | 1 year 9 months | |

Net Present Value (NPV) is the difference between the present value of cash inflows and the present value of cash outflows. NPV is used in capital budgeting to analyze the profitability of a projected investment or project. The discount factor used in this analysis is 25 percent per year and 30 percent to obtain both positive and negative NPV value. Based on table 4, the NPV value is IDR 614,697 which shows that this project is feasible since NPV value is greater than zero.

Benefit-Cost Ratio (BCR) is used in cost-benefit analysis as an indicator to summarize the value for money of a project by determining the relationship between cost and benefit. The decision rule is to accept investment projects with B/C ratio greater than 1. The BCR value of this project is 1.01 or greater than 1 so that it is feasible.

The Internal Rate of Return (IRR) method is a method that calculates the interest rate equalizing the value of the initial investment. If the IRR is greater than interest rate, then investment can be considered feasible and if the IRR is smaller, then the investment is not feasible. Based on the latest data released by The Bank Indonesia [12] in The Basic Interest Rate of Credit applied in all operated Banks in Indonesia, the average interest rate is 12.03 which used in this study. Table 4 shows that IRR value is greater than interest rate (28.68% > 12.03%) meaning this project is feasible.

The last financial statement to see project feasibility is Payback Period. Payback Period is the number of years it takes to recover all the capital invested. Payback Period computation shows that this project needs 1 year and 9 months to be able to recover all costs. Therefore, based on financial analysis and all financial statement above, this project is feasible.

## 4. Conclusion

Swamp land is potential as one of alternative production land to support food security. Unfortunately swamp land has not optimally utilized since farmers only cultivate once a year. Research studies show that using appropriate treatment and integrated farming system, swamp land can be utilized thorough out the year. This study shows that swamp land can be cultivated twice a year with additional fish farming during wet season. Using actual input and low projected output, this project is financially feasible with the NPV value is IDR 614,697; BCR value is 1.01; IRR value is 28.68 percent; and Payback Period of 1 year 9 months. Since farming is very crucial for this community, government fully support is needed in protecting the input prices and regulate commodity prices particularly rice, corn, and fish to maintain the feasibility of this project.

## References

[1] Ministry of Agriculture 2011 *Strategic Plan of Ministry of Agriculture 2010-2014* Revised edition Indonesian Ministry of Agriculture

[2] Ministry of Agriculture 2015 *Agricultural and Region Development Policy 2015-2019* Indonesian Ministry of Agriculture

[3] Subagjo H and I P G Widjaja Adhi 1998 *Opportunities and Constraints of The Wetlands for Agricultural Development in Indonesia, Case: South Sumatra and Central Kalimantan* Main Papers on Discussion Meeting and Communication of Soil and Agro-climate Research Findings Bogor February 10 1998 p 1-22

[4] Indonesian Center for Agricultural Land Resources Research and Development (ICALRRD) 2002 *Climate Anomaly Impact Evaluation Forecasting and Technology Anticipation To Reduce The Risk of Decline in Production* Research Report ICALRR

[5] Djamhari S 2009 *Increasing Rice Production in Swamp Land as an Alternative of Agricultural Land Development Beyond Java* Indonesian Journal of Science and Technology Vol 11 No 1: 64-69

[6] Galib R 2010 *The Potential of Corn Farming in South Kalimantan's Swamp Land* The national cereal week proceeding 2010 ISBN: 978-979-8940-29-3

[7] Boslaugh S 2007 *Secondary Data Sources for Health: APractical Guide* Cambridge: Cambridge University Press

[8] Amirin T M 2015 *Secondary Research Methods* Retrieved from https://tatangmanguny.wordpress.com in June 15 2017

[9] van Kooten G C 2017 *Applied Welfare Economic, Trade and Agricultural Policy Analysis* University of Victoria Retrieved from http://web.uvic.ca in October 17 2017

[10] Ratmini S and Herwenita 2014 *Increasing Rice Farmers' Income Through Integrated Crop Management Approach of Swamp Land in Ogan Ilir Regency South Sumatra* Research Journal of Jambi University: Sains Series Vol 16 No 1 (2014)
[11] Statistics South Sumatra 2016 *Official Statistics News* Retrieved from https://sumsel.bps.go.id in June 18, 2017
[12] Bank Indonesia 2016 *The Basic Interest Rate of Credit* The Latest Data of November 2016 Retrieved from http://www.bi.go.id/ in October 17 2017
[13] Statistics Ogan Ilir 2016 *Ogan Ilir in Figures 2016* Statistics Indonesia

# Changing of Sumatra backswamp peat properties by seawater and zeolite application

**Sarifuddin[1], Z Nasution[1], A Rauf[1] and B Mulyanto[2]**

[1]Faculty of Agriculture, Universitas Sumatera Utara, Jl. Dr. Mansur no. 9 Kampus USU Medan (20155) Indonesia
[2]Faculty Of Agriculture, Institut Pertanian Bogor
E-mail address : sarifuddin@usu.ac.id

**Abstract**. This research attempts to improve the properties of backswamp peatsoil originated from Asahan District, North Sumatra Indonesia by adding sea water and zeolite using factorial randomized block design with volume of sea water as first factor, consisting of without seawater, 500 ml, 1000 ml and 1500 ml and second factor are dosages of zeolite consisting of without zeolite, 100 g, 200 g each 10 kgs of wet peat soil. at green house in faculty of agriculture University of Sumatra Utara (USU) Medan, Indonesia. The result showed that the application of seawater decreased pH, C/N and Cation Exchange Capacity and increased of base saturation of peat soil. Adding of zeolite minerals can buffered the increasing of acidity and Electric Conductivity caused by sea water application. Interaction seawater + zeolite decreased of C/N and increased of percent of base saturation.

## 1. Introduction

Indonesia's peatlands cover about 21 million hectares [1], unfortunately, to use for agriculture purpose there are some problems of the physical constraints such as excess of water and low bulk density, chemical constraints include high acidity, carbon-nitrogen high in C/N ratio and cation exchange capacity, availability of macro and micro nutrients are low and organic acids which are poison for the plants. High cation exchange capacity and low of bases exchangeable resulted lower in the availability of the base cations . The base saturation should be increase to 20-30% , to fulfill the plants requirement [2]. In fact the technology has not gained enough in a comprehensive and environmentally to manage of peatlands as a medium to plant growth.

Some researches had been done to improve peat soil fertility, among others, by mixing peat with mineral soils, liming and iron on soybean [3], giving Na and some micro elements on rice yield [4], ameliorant with high abundance of iron mineral soil [5], a high abundance of iron mineral soil material enriched high-grade iron [6], by liming [7],, application of steel slag in combination with a compound fertilizer and manure [8] in rice plants and the applied of fly ash in Acasia plant [9] and soil with pyrite combined with washing of soil in rice [10].

In terms of numbers, with area of 5.8 million km2 of marine waters, or 70% of the territory Indoneia, seawater available in abundance and easy to obtain [11]. Sea water also has a content of dissolved elements are quite high as 1,272 ppm Mg, S 884 ppm, 400 ppm Ca, 380 ppm K, although the content of Na and Cl are also quite high at 10,561 ppm and 18,980 ppm Na Cl [12]. Therefore sea water is potential as an alternative ameliorant to poor soil nutrients such as peat. Beside of that generally lowland peat are close to coastal areas even directly adjacent to the beach. However, the use of seawater as an ameliorant have a challenge that is high salinity. Furthermore to decrease the level of salinity, Na and Cl dissolved to an extent that does not interfere with the growth and production of crops is an effort that must be done. Research of [4] showed that Na can replace some of the functions of Cu in the binding of organic acids that are poisoned (phenolic and carboxylic) of peat material. The production of food crops (rice, pulses and sweet potato) was higher in tidal peatland than inland peat [13]. IR 64 rice variety grown on peat tidal (marine peat) and given a high abundance of iron ameliorant also gave the highest results than those grown on brackish peat and fresh water peat in Central Kalimantan [14]. Losing of carbon (C) through the emission of $CH_4$ and $CO_2$ released is also the lowest in the wetland peat affected sea water (1, 97 tons C / ha / yr) than in peat transition (1.99 tons C / ha / yr and peat influenced freshwater only (2.09 tons C / ha / yr) [14].

Zeolite is a mineral of aluminosilicate hydrate with tetrahedron crystalline structure in which the silicate or silicate was substituted aluminum as the core and four oxygen atoms on all four sides to form a tetrahedron and bind to each other to form a series of tertrahedron tectosilicate [15]. The use of zeolite in agriculture, such as to improve soil acidity, increasing the content of ammonium in the rice soil as well as a soil conditioner to increase crop yields [15]. The use of zeolite as ameliorant on peat has not been known including sorption properties (ion exchange), dehydration-rehydration and its ability to increase the buffering capacity of the soil, especially in soil with acidity and salinity problems.

This trial aims to determine the effect of sea water which is combined with the mineral zeolite to improved physical-chemical properties and soil buffering capacity to reduce fluctuations in salinity and acidity of peat soil.

## 2. Materials and Methods

The experiment was conducted in the glasshouse, Laboratory of Research and Technology at Faculty of Agriculture USU Medan on april to september 2011.

The peatsoil samples had taken from Rawasari, Aek Kuasan, Asahan Distric North Sumatra Indonesia (02°42'33,9" N dan 99°46'34,3" E). Sea water taken from Pantai Cermin, Serdang Bedagai Distric and zeolite obtained from PT. Dunia Windu Medan.

The tools used are plastic as container pots for plant media, scales, measuring cups, pH meter, EC meter and atomic absorption Spectrophotometer.

This study used factorial randomized block design consist of two treatments with the addition of sea water each 4 dosage and zeolite mineral with 3 dosage with 3 replications.: Factor I: Volume of seawater which consists of: A0 = 0 ml of sea water + 2000 ml fresh water; A1 = 500 ml of seawater + 1500 ml of fresh water; A2 = 1000 ml seawater + 1000 ml of fresh water; A3 = 1500 ml of sea water + 500 ml of fresh water each pots Factor II: Dosage of zeolite consists of: Z0 = 0 g / pot; Z1 = 100 g ; Z2 = 200 g each pot. Analysis of mean separation using Duncan's Multiple Range Test 5% [16][17].

The numbers of 10 kgs of wet peatsoil (equivalent to 10 liters) with bulk density (BD) 0.13 g / $cm^3$ put in a black plastic pot as plant media, then gave zeolite mineral powder and sea water then stirred to be paste conditions. During the incubation period, paste condition is maintained by adding free ion water and then soil sample carried out to laboratory for soil analysis.

The parameters measured are: pH (pH meter) 1: 2.5 [22] and electric conductivity (EC) (EC meter) 1: 2.5 [18] were measured every week; ratio of C/N (C-Walkley and Black; N-Kjedahl) and cation exchange capacity (CEC) (extract Ammonium Chloride 1 N) [19], [20] was measured every month and percent of base saturation by calculated until 3 months observation respectively.

## 3. Results and Discussion

Soil properties of back swamp peat soil before application seawater and zeolite mineral are listed in this following table.

**Table 1**. The nature of the soil before application

| No. | Parameters | Unit | Result |
|---|---|---|---|
| 1 | pH $H_2O$ | - | 3.79 |
| 2 | EC | (dS/m) | 0.56 |
| 3 | K-exc | (me/100 g) | 0.008 |
| 4 | Na-exc | (me/100 g) | 0.190 |
| 5 | Ca-exc | (me/100 g) | 0.140 |
| 6 | Mg-exc | (me/100 g) | 0.250 |
| 7 | CEC | (me/100 g) | 111.040 |
| 8 | Base Saturation | (%) | 0.530 |
| 9 | C-org. | (%) | 20.580 |
| 10 | N-tot. | (%) | 0.550 |
| 11 | C/N | - | 37.420 |

### *3.1. Soil Acidity (pH)*

The higher concentration of sea water progressively decrease soil pH, especially between A0 (without seawater) to A1 (sea water 500 ml / pot) and A1 to A2 (sea water 1000 ml / pot) and A2 to A3 (sea water 1500 ml / pot) although the decline was not as big as between treatment A0 to A1. The existence of zeolite without the addition of sea water raised the pH of the soil, but with sea water, the addition of zeolite particularly Z1 (100 g / pot) is able to buffering of decline in soil pH zeolite treatment, even 200 g / pot zeolite (Z2) is able to raise the soil pH compared Z0 (without zeolite) although the pH value is still lower than the initial pH of peat.

The period of incubation for each sea water treatment and zeolite lowered the pH of peat soil gradually (**Fig. 1**).

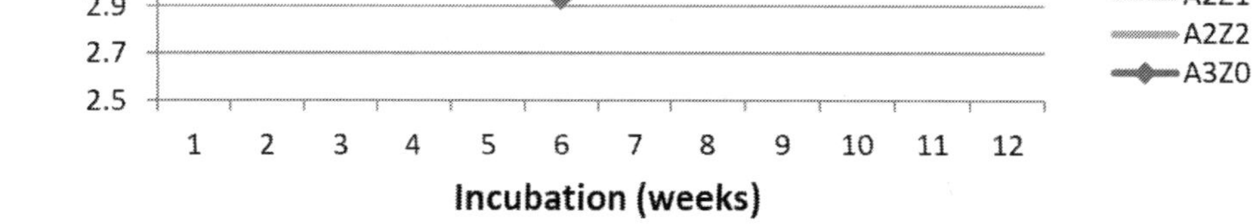


**Figure 1.** Changing of pH after sea water (A) and zeolite (Z) application

### *3.2. Electrical Conductivity (EC)*

The higher concentration of sea water increased EC and the period of incubation period also effected it. Highest EC increasing at 11$^{th}$ week and then decreased at 12 week of incubation.. The addition of zeolite particularly in the treatment Z2 (200 g / pot) is able to reduce the value of EC especially at 3$^{rd}$ month of incubation (**Fig. 2**).

IOP Conf. Series: Earth and Environmental Science **122** (2018) 012073 doi:10.1088/1755-1315/122/1/012073

The experimental results showed that the application of sea water with higher concentrations due to soil acidity, increasing as well as the value of the electric conductivity. The addition of zeolite able to slightly increase the soil pH, and reduce of soil EC increasing due to sea water application. This shows the role of the zeolite as ion exchanger material capable of supporting the changes in pH and EC of peat soil.

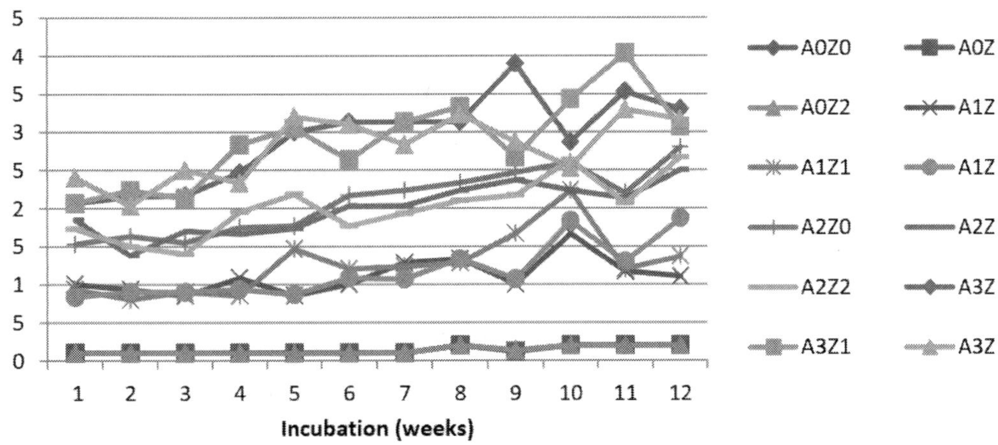


**Figure 2.** Changing of EC after sea water (A) and zeolite (Z) application

*3.3. The ratio of C-org and N-total (C N)*

Application of sea water with concentration of A3 (150 ml / pot) significantly decreased the ratio of C/N since the 1st to 3rd months incubation to control. While giving the zeolite also significantly decreases the ratio of C / N peat at 1st and 3rd months of incubation (Fig. 3)

Seawater and zeolite, significantly increasing the C / N peat because of a decline in the amount of C in the peat. These results indicate that both seawater and zeolite does not play a role in increasing levels of N in the soil due to both the material does not contain significant amounts of N, but the sea water and zeolite role in lowering organic C, especially in the treatment of A3 (1500 ml sea water) and Z2 (200 g zeolite). Respiration rate as an indicator of the rate of decomposition of the peat soil indicate that the low water level and complex bond carbon chains are the most important factors slowing down the rate of decomposition of peat but polyvalent metal-containing material is not significant [21]. Zeolite was role to decline of C/N peat ratio by binding water molecules in the peat material.

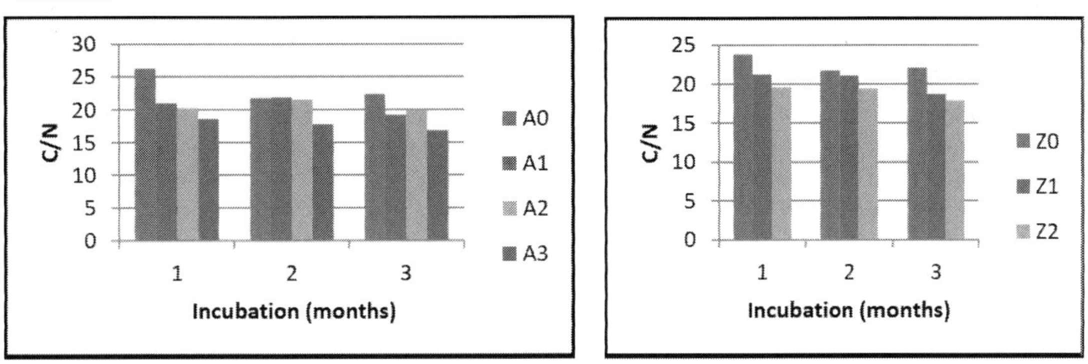


**Figure 3 .** Changing of C / N ratio after seawater (A) and zeolite (Z) application

*3.4. Cation Exchange Capacity (CEC)*

Application of sea water significantly decreased of soil CEC for 1st to 3rd months incubation, and not significant for zeolite but interaction both sea water and zeolite significant effect lowered CEC during 3 months incubation (Fig. 4).

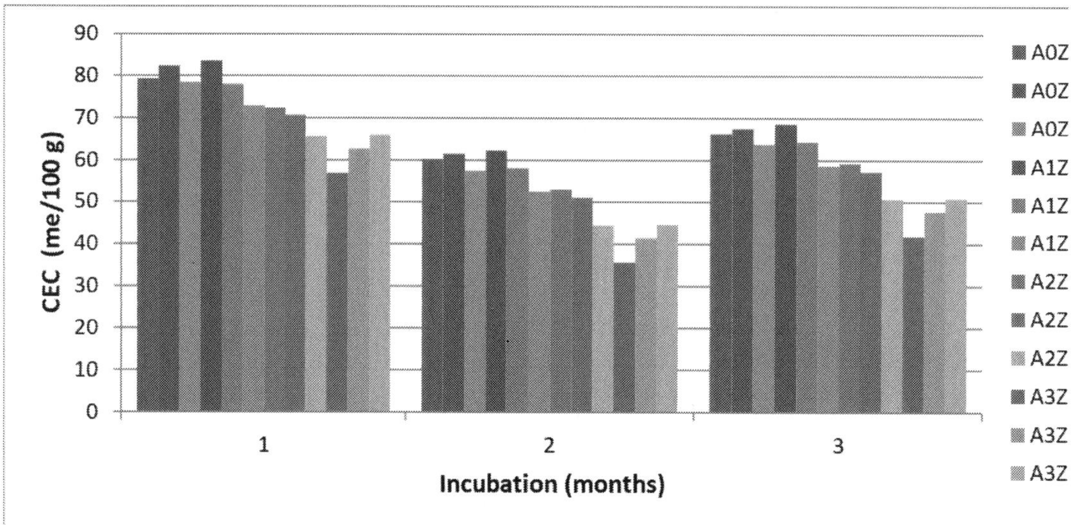


**Figure 4.** Changing of cation exchange capacity after sea water (A) and zeolite (Z) Application

Application of sea water significantly decrease soil CEC with lowest CEC at 2nd month incubation. The combination of the highest concentration of 1500 ml sea water/pot without zeolite (A3Z0) produce lowest CEC because its treatment also resulted in the lowest pH. Peat's colloid is entirely organic and variable charged. Changed highly acidity will decreasing the amount of negative charge on the surface of colloid so CEC will be lower also [22]. The correlation between EC versus CEC at 1st , 2nd and 3rd month had close correlation value (r) - 0, 875; -0.862 and - 0.877 respectively. This proved that the cations are present in sea water is capable of forming a complex bond with organic colloids and consequently the amount of negative charge that is active to be reduced and lowered the soil CEC. The combination of sea water 500 ml (A1) and 100 ml (A2) with the zeolite (Z1 and Z2) showed decreasing of soil CEC but the treatment of sea water 1500 ml (A3) with zeolite (Z1 and Z2) showed an increasing. This seems contradictory when viewed as a function of zeolite as ion exchanger material that has a large high CEC. This is thought to occur because of the influence of zeolite in influencing soil CEC offset by changes in a negative charge on the colloidal organic peat soils due to changes in pH. Tan [15] states that 50 % of organic acid's CEC contributed by the carboxyl group and began to dissociate to release of $H^+$ ions at pH 3 to 5.5. Declining of CEC occurs by reaction of bases in sea water with the COOH group to form salts of COONa R- and R-COOK and polyvalent cation forming coordination bonds with organic molecules such as humic [23].

*3.5. Base Saturation (BS)*

Application of sea water significantly increase percent of base saturation (BS) for 3rd month of incubation and sea water concentration concomitant with the base saturation (Fig. 5). Giving of zeolite does not affect to base saturation for 3 months incubation.

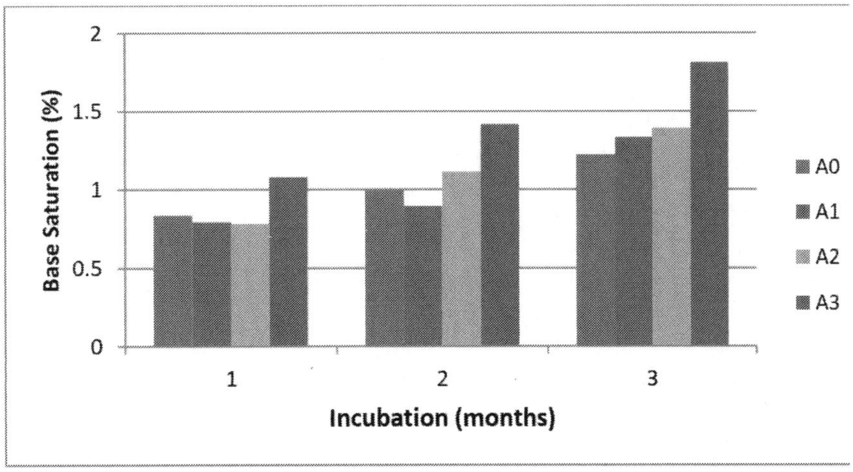


**Figure 5.** Changing of base saturation after sea water (A) application

Sea water, especially at the highest concentration, increasing of base saturation. The highest in base saturation occurs at 3 rd month incubation. The increasing in base saturation is in line with the decrease in soil CEC as well due to the increase in the number of bases the exchange itself.

## 4. Conclusion

Application seawater as ameliorant able to improve some properties of peat soils to be more suited to the needs of plant growths such as lowering the C / N ratio and the CEC were too high and to increase the percent of bases saturation as essential plant nutrients. On the other hand application of sea water is also gives the effect of a less favourable by soil pH decreasing and electrical conductivity increasing whom would interfere plant growth.

Application of zeolite as a material that functions as an ion exchanger capable to buffering a decrease in soil pH and increase electrical conductivity due to the giving of sea water,

A combination of seawater and zeolite provide better conditions on the properties of physic-chemical properties of peat soil in accordance with the requirements of crops growth unless the value of electrical conductivity is still high for most crops.

## References

[1] Wahyunto and Mulyani A 2011 *Pengelolaan Lahan Gambut Berkelanjutan (Sustainable Management of Peatlands*) Balai Penelitian Tanah Bogor

[2] Sagiman S 2007 Pemanfaatan Lahan Gambut Dengan Perspektip Pertanian Berkelanjutan *(Utilization of Peat Land With Sustainable Agriculture Perspective)* Orasi Ilmiah Guru Besar Tetap Kesuburan Tanah FP Untan. *http://www.untan.ac.id/wp-content/uploads/pengukuhan_guru_besar_23 juli 2007 Saeri_Sagiman.pdf.* (Diakses 20 Okt 2007).

[3] Halim A P K S 1989 Perbaikan *tanah gambut pedalaman melalui peningkatan kejenuhan basa untuk budidaya tanaman kedelai (improvement of inland peatland through increased saturation bases for soybean cultivation)* Prosiding Seminar Tanah Gambut Untuk Perluasan Pertanian FP UISU Medan

[4] Prasetyo T B 1996 Perilaku Asam-Asam Organik Meracun Pada Tanah Gambut Yang Diberi Garam Na Dan Beberapa Unsur Mikro Dalam Kaitannya Dengan Hasil Padi ( Disertasi. PPS IPB Bogor.

[5] Salampak 1999 Peningkatan produktivitas tanah gambut yang disawahkan dengan pemberian bahan ameliorant tanah mineral berkadar besi tinggi (productivity increasing of peat land

assisted by provision of ameliorant material of high iron mineral land) *Disertasi* Program Pasca Sarjana. IPB Bogor.
[6] Mario D M 2002 Peningkatan produktivitas dan stabilitas tanah gambut dengan pemberian tanah mineral yang diperkaya oleh bahan berkadar besi tinggi (productivity increasing and stability of peat land by providing mineral land enriched by high iron material) *Disertasi* Program Pasca Sarjana IPB Bogor
[7] Rumbang Nyahu 2003 Polyphenol in peat soil and growth and yield of rice (*Oriza sativa* L.) *Proc. Of Int. Symposium on Land Management and Biodiversity in South East Asia*. Bali, Indonesia 17 – 20 September 2002
[8] Suwarno 2003 Utilization of steel salg in wetland rice cultivation on peat soil *Proc. Of Int. Symposium on Land Management and Biodiversity in South East Asia* Bali, Indonesia 17 – 20 September 2002
[9] Iskandar, Djajakirana G and Marolop R 2003 The use of fly ash as ameliorant to improve the chemical properties of peat soil in pulau muda, Riau province *Proc. Of Int. Symposium on Land Management and Biodiversity in South East Asia* Bali, Indonesia 17 – 20 September 2002
[10] Suastika I W, S Sabiham and D A Suriadikarta 2006 Pengaruh pencampuran tanah mineral berpirit pada tanah gambut terhadap pertumbuhan dan hasil tanaman padi (Effect of mineral mixing on peatlands on growth and rice crops results) *Jurnal Ilmu-ilmu Pertanian Indonesia* Vol. **8** (2)
[11] Resosudarmo B P, Dj Hartono, T Ahmad, N I L Subiman, Olivia and A Noegroho 2002 Analisis penentuan sektor prioritas di kelautan dan perikanan Indonesia (Analysis of determining priority sectors in Indonesian marine and fisheries) *Jurnal Pesisir & Kelautan* Vol. **4**(3): 17-28
[12] Ilahude A G 1999 *Pengantar ke Oseanologi Fisika (Introduction to Oceanology Physics)* Pusat Penelitian dan Pengembangan Oseanologi LIPI
[13] Noor M 2001 *Pertanian Lahan Gambut-Potensi dan Kendala (Peatland Agriculture-Potential and Constraints)* (Yogyakarta: Kanisius)
[14] Sabiham S 2010 Properties of Indonesia peat in relation to the chemistry of carbon emission *Proc. Of Int. Workshop on Evaluation and Sustainable Management of Soil Carbon Sequestration in Asian Countries* Bogor Indonesia
[15] Mumpton F A 1985 Using Zeolit for Agriculture. www. Princeton.edu/~ota/disc 2/1985/8512/851210/pdf.
[16] Litle T M and Hills F J 1978 *Agricultural Experientation. Design and Analysis* (New York: John Wiley and Sons)
[17] Gomez K A and Gomez A A 1984 *Statistical Procedures for Agricultural reseach. 2nd. Edition* (Yogyakarta: John Wiley & Sons)
[18] U S Salinity Laboratory Staff 1954 *Diagnosis and Improvement of Saline and Alkali Soils. Agriculture Hand Book 60* (U.S. Department of Agriculture. Oxfoed and IBH Publishing Co., New Delhi)
[19] ISRIC 1987 *Procedures for Soil Analysis. Second Edition.* Wageningen Netherlands
[20] Agus F and Subiksa I G M 2008 Lahan Gambut: Potensi untuk Pertanian dan Aspek Lingkungan. Balai Penelitian Tanah dan ICRAF, Bogor.
[21] Edi H and Agus F 2011 Microbial activities as affected by peat dryness and ameliorant *Am. J of Environ. Sci.* Vol **7**(4)
[22] Uehara G and Gillman G 1981 The Mineralogy, Chemistry and Physics of Tropicals Soils with Variable charge Clays. WestviewPress Colorado.
[23] Stevenson F J 1994 *Humus Chemistry. Genesis, Composition, Reaction* (New York: John Wiley & Sons)

# Quality evaluation of probiotic capsule prepared from alginate, carrageenan and tofu waste flour based on bacterial activity and organoleptic test

**V Muhardina[1], D Ermaya[1], Y Aisyah[2] and S Haryani[2]**

[1]Serambi Mekkah University, Teungku Imum Lueng Bata, Banda Aceh, 23245, Indonesia
[2]Syiah Kuala University, Kopelma Darussalam, Syiah Kuala, Banda Aceh, 24411, Indonesia
E-mail: virna.muhardina@serambimekkah.ac.id

**Abstract.** Probiotic capsule is an innovation in functional food sector. It is used to preserve the living cells of probiotic bacteria during processing and storage. In this research, the improvement of probiotic viability is studied by using two kinds of encapsulating biomaterials and different concentration of tofu waste flour. Extrusion is selected method for encapsulation process. The purpose of this study is to examine the quality of probiotic capsule by evaluating the lactic acid bacteria performance and its physical characteristic. The article provides the data of probiotic bacteria activity related to their living cells present in capsule, activity in fermentation media compare to uncapsulated bacteria, and panelists' preferences of capsule's physical properties. The data is analyzed statistically by using ANOVA. The result shows that variables in this study affect the number of bacteria, their metabolic activity in producing acid during fermentation, and physical appearance of the capsule. Combination of alginate and tofu waste flour allows the multiplication of bacteria to a high number, and forms elastic, yellow and cloudy capsule, while with carrageenan, it causes the growth of a few numbers of bacteria which affects to a moderate pH and produces elastic, creamy and transparent capsule.

## 1. Introduction

Probiotic is essential live bacteria that deliver beneficial effect to human's health if they present in sufficient number [1]. The positive effect of this bacterial group, especially lactic acid bacteria, confirms it as one of functional food. A living condition of probiotic needs to preserve during processing and storage time, until the bacteria are released into gastrointestinal tract [2]. One of effective protection methods for probiotic is encapsulation. It is a technical procedure to protect probiotic bacteria from harmful condition for their growth by immobilizing the cells inside capsule and entrapping them within the gel matrix [3]. This physicochemical process produces tiny particles, ranging from nanometers to millimeters. The formation of these capsules can sustain the life of probiotic bacteria in the product and improve the effectiveness of its life in reaching the digestive tract [4, 5].

In encapsulation, the quality and physicochemical characteristic of capsule are determined by organic macromolecules used as encapsulating biomaterials. The common biomaterials derived from polysaccharide group, such as alginate and carrageenan. Alginate is a polymer of carbohydrate group extracted from seaweed. Alginate has a good capability as a capsule, but its gel is unstable under acidic conditions [6, 7]. The ratio of its chemical composition between guluronic and mannuronic acid defines the gel strength of its capsule [8, 9]. Carrageenan is also a class of polysaccharides obtained from red algae that contains sulphate and hydrophilic molecules [10]. Chemical binding of D-galactose, its molecular weight, and sulphate along the polymer chain determines hydrophilic characteristic and gel properties [11, 12]. Both of these biomaterials have different characteristics leading to different physicochemical properties of probiotic capsule.

In addition, probiotic bacteria also require source of energy (prebiotic) in their living media for the growth and metabolism activity. Natural prebiotic can be derived from plants, such as soybean. Soybean is a kind of bean commonly used as main ingredient in making tofu. This bean contains of oligosaccharide (raffinose and stachyose) considered as one of prebiotic sources [13]. Oligosaccharide still can be found in its byproduct, which is tofu waste. This waste also contains other nutrition, especially protein and fiber. The high nutritional value of tofu waste brings it as a prospective reusable material to be applied in several innovations, namely as prebiotic in probiotic encapsulation. The tofu waste performs as an important ingredient in encouraging living capability and multiplication of probiotic bacteria. Probiotic capsule made from alginate, carrageenan and tofu waste flour combination shows specific quality. Therefore, this study aims to examine through laboratorial tests regarding bacterial activity and panelists' preferences and description in terms of its physical characteristic prior to the capsule apply into food product.

## 2. Materials and Methods

Materials used in this research were probiotic bacteria; *Lactobacillus bulgaricus*, encapsulating materials; carrageenan and alginate, and prebiotic source; tofu waste flour. Regarding research analysis, bacteria enumeration [17] used MRS agar and broth, and pH calculation used milk as fermentation media following yogurt production method [18, 19]. Organoleptic test followed the procedure [20]. The tools used were syringe 2.5 cc, digital scales, laminar flow cabinet, incubator, water bath, thermometer, centrifuge, hot plate, and glassware.

This study used two factors, which were the variety of encapsulating biomaterials (alginate and carrageenan) and the percentage of tofu waste flour (1.5%, 2%, 2.5%, and 3%). The data comparison was statistically calculated by using ANOVA, in order to examine the significant level of variables to the tested product.

This research used extrusion method for encapsulation technique as it is easy and commonly used in laboratorial scale. *L. bulgaricus* in pellet form was dissolved into sterile distilled water for a culture solution. The encapsulating solution, either carrageenan or alginate, was homogenized and sterilized. This solution was mixed with tofu waste flour. The *Lactobacillus bulgaricus* culture solution and encapsulating solution were mixed with a ratio of 1: 4. This mixture was dripped by using syringe into the hardener solution [14, 15]. The tofu waste flour used in the encapsulation process had been prepared before making capsule. Tofu waste was taken from one of tofu industries in Banda Aceh, Indonesia. The tofu waste was pressed and steamed for ± 15 minutes before drying outdoor for 8 hours per day. Drying process took about 3 days. The dried waste was crushed and sieved with 140 mesh size [16].

Analysis performed in this study involved of laboratorial test; bacteria enumeration and pH analysis of fermentation media, and panelists' perception. According to bacteria enumeration, probiotic capsule was added with phosphate buffer and stirred for 2 hours for 1.5%, 2% and 5 hours for 2.5%, 3 % of additional tofu waste flour. After dilution, 100 μl injected to MRS agar and incubated in 37°C for 72 hours [17]. Fermentation followed the procedure of yogurt fermentation. Skim milk and sugar were added to milk then homogenized and pasteurized at 80°C for 15-30 minutes. The mixture was cooling down until 40-45°C. *Lactobacillus bulgaricus* as starter culture and probiotic capsule were added into the milk. It was incubated at 37°C for 24 hours [18, 19]. In terms of organoleptic test, it used 15 panelists that assessed all probiotic capsules and presented their observation in score [20].

## 3. Results and Discussion

### *3.1. Bacterial Activity (Enumeration of total LAB and pH of fermentation)*

The enumeration of total lactic acid bacteria (LAB) was conducted by calculating the number of living cells inside 1 gram of probiotic capsule. Based on Table 1, a high number of bacteria are discovered in alginate capsule rather than carrageenan capsule. It correlates to the quantity of capsule in 1 gram. Due to low weight of alginate capsule, so it requires more capsules to reach 1 gram compare to carrageenan capsule which has more weight and only requires a few capsules to obtain 1 gram as a total. In each capsule of alginate, there might be some living bacteria, therefore the more number of capsules, and the more bacteria are examined ($19.61x10^9$ CFU/gram).

**Table 1.** Total of Lactic Acid Bacteria (LAB)

| Variables | Total of Lactic Acid Bacteria ($1x10^9$ CFU/gram) |
|---|---|
| Encapsulating Biomaterials : | |
| Alginate ($K_1$) | $19.61^b$ |
| Carrageenan ($K_2$) | $11.32^a$ |
| Concentration of Tofu Waste Flour : | |
| 1.5% ($P_1$) | $10.85^a$ |
| 2.0% ($P_2$) | $12.40^b$ |
| 2.5% ($P_3$) | $15.07^c$ |
| 3.0% ($P_4$) | $23.53^d$ |

Moreover, the characteristic of biomaterials also determine the growth of bacteria. Alginate capsule has a liquid center bordered by semipermeable membrane [21, 22]. Semipermeable with thin layer and small diameter facilitates the movement of cells and nutrition, so that the bacteria could grow well [23]. The high viscosity and thick gel of carrageenan capsule could interrupt and restrict the diffusion and dispersion of nutrition for bacteria. Also, cell of bacteria could not move and pass through the membrane to absorb the nutrition due to its thickness. It may cause mortality to the bacteria and affect to the low number of LAB ($11.32x10^9$ CFU/gram). Additional prebiotic in capsule could increase the viability, endurance capability and growth of probiotic bacteria [24]. Moreover, prebiotic also could improve the quality and robustness of capsule by repairing the gel's porosity. Table 1 shows clearly the improvement of LAB along with the increase of tofu waste flour concentration. The highest amount of LAB is represented by $P_4$ or 3% ($23.53x10^9$ CFU/gram). This number fulfills the requirement of probiotics dose in a day, at least $10^6 - 10^7$ CFU/gram [25].

Bacterial activity inside capsule also could be evaluated from pH reached during fermentation. In this test, the capability of encapsulated bacteria in metabolic system was compared with starter culture *Lactobacillus bulgaricus* without capsule. *L. bulgaricus* without capsule produces the lowest pH, 4.0, due to its direct interaction with the media, while alginate and carrageenan capsule have a higher pH; 4.68 and 4.88 respectively (Figure 1a). Encapsulation could inhibit proliferation cells of probiotic in food product; therefore the sensory characteristic of product might not deteriorate and could preserve its quality [26]. In making capsule, alginate has porous layer properties that causes cells moved unrestrictedly through membrane [6], then release acid to media and lowering pH (4.68). Moreover, its higher number of total LAB (Table 1) compare to carrageenan is responsible to the lower pH in fermentation.

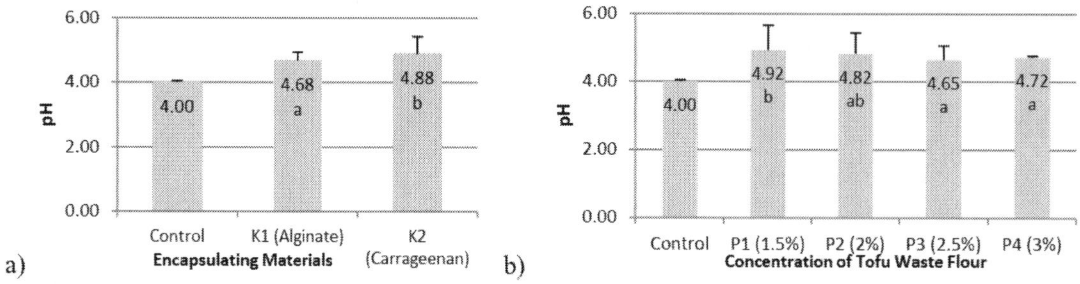


**Figure 1.** pH during fermentation; a) Encapsulating biomaterials; b) Tofu waste flour.

All capsules with tofu waste flour have a higher pH rather than sample without the flour (control). It confirms the function of prebiotic is not only as energy source for bacteria (Table 1), but also as supporting materials to build up good capsule by repairing porosity of biomaterials [27, 28]. Figure 1b shows that pH around 4.65 – 4.92 occurs due to the bacteria could not across easily from capsule to media. Moreover, a complex chemical structure of this prebiotic makes it is not able to be degraded optimally by metabolic reaction during fermentation time (24 hours), so that it produces low acid level. Therefore, pH reached is moderately high.

### *3.2. Organoleptic Test*

**Table 2.** Preference Level of Capsule Physical Characteristics

| Encapsulating Biomaterials | Preference Level of Organoleptic Parameters | | |
|---|---|---|---|
| | Texture | Color | Odor |
| Alginate ($K_1$) | $2.83^a$ | $2.55^a$ | $2.78^a$ |
| Carrageenan ($K_2$) | $3.81^b$ | $3.58^b$ | $3.12^b$ |

Scale: 1=very dislike; 2=dislike; 3=neutral; 4=like; 5=very like

Organoleptic test used in this research were hedonic and description tests. Hedonic test is aimed to assess the preference level of panelists in term of specific parameter. Regarding Table 1, it shows that panelists prefer capsule made from carrageenan, according to texture (3.81 or like) and color (3.58 or like) than alginate (2.83 or neutral) and (2.55 or neutral) respectively. The texture of carrageenan capsule is tenderer due to its capability in trapping and absorbing more water [12, 29]. Carrageenan flour has creamy to yellowish color, while alginate has yellow to brown color, which relates to the color of their raw material [30]. Moreover, hydrophilic properties of carrageenan makes it binds with more water, and produces capsule with more shiny and white. Panelists assume that both of encapsulating biomaterials produce almost none odor (neutral) (Table 2). The preference level of both capsules either alginate or carrageenan occupies approximately 2.78 – 3.12 (neutral).

Description test is intended to evaluate the specific physical characteristic of capsule based on panelists' perspective. From Figure 2a, it illustrates that, overall, the alginate and carrageenan capsules have similar elastic texture (3.8 – 4.4 or elastic). Based on color description, panelists express different perspective between alginate capsule and carrageenan capsule, in which alginate capsule ($K_1P_1$ – $K_1P_4$) has yellow color (3.9 – 4.5), while carrageenan capsule ($K_2P_1$ – $K_2P_4$) reflects creamy color (1.8 – 2.3). As overall appearance, panelists think that alginate ($K_1P_1$ – $K_1P_4$) produces cloudy or turbid capsule, while carrageenan ($K_2P_1$ – $K_2P_4$) is responsible to semitransparent appearance of the capsule. It is also related to the amount of water binding and its natural characteristics [30].

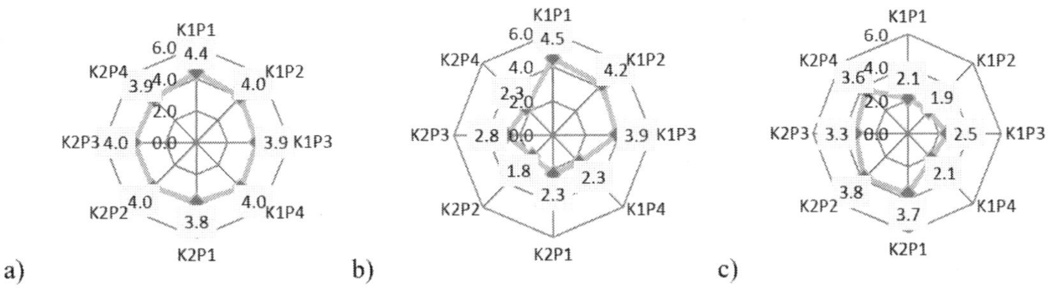


**Figure 2.** The Description of Capsule's Quality; a) Elasticity; b) Color; c) Appearance. Elasticity scale: 1=very soft; 2=soft; 3=slightly elastic; 4=elastic; 5=very elastic; 6=rigid. Color scale: 1=broken white; 2=cream; 3=light yellow; 4=yellow; 5=dark yellow; 6=brown yellow. Appearance scale: 1=opaque; 2=cloudy; 3=slightly cloudy; 4=semitransparent; 5=transparent; 6=clear

## 4. Conclusions

Encapsulating biomaterials and prebiotic addition could affect the probiotic living cells and capsule physical characteristic related to panelists' perspective. Alginate could provide a better growth condition for bacteria in capsule with higher total LAB ($19.6161x10^9$ CFU/gram) than carrageenan ($11.32x10^9$ CFU/gram). The porosity of alginate makes it reaches lower pH in 24 hours than carrageenan. Additional tofu waste flour prebiotic in encapsulation process can improve the growth of bacteria and their survival condition in capsule; 3% has $23.53x10^9$ CFU/gram. Its chemical compounds affect to the longer fermentation time in reaching proper pH; (pH 4.72 of 3% prebiotic). Panelists prefer carrageenan capsule than alginate (texture and color). Alginate capsule illustrates similar elasticity to carrageenan, yellow and cloudy appearance, while carrageenan capsule has creamy and transparent performance. Overall, even though carrageenan confirms lower total LAB, its combination with high percentage of tofu waste flour (3%) could produce a good quality of probiotic capsule.

## References

[1] FAO/WHO 2006 Probiotics in Food. Health and Nutritional Properties and Guidelines for Evaluation FAO Food and Nutrition Paper No. 85 *World Health Organization and Food and Agriculture Organization of the United Nations*, Rome.

[2] Burgain J, Gaiani C, Linder M and Scher J 2011 Encapsulation Of Probiotic Living Cells: From Laboratory Scale to Industrial Applications. *Journal of Food Engineering*. Vol. 104 (4) pp. 467-483.

[3] Vidhyalakshmi R, Bhakyaraj R and Subhasree R S 2009 Encapsulation "The Future of Probiotics": A Review. *Adv. Biol. Res*. Vol. 39. pp. 96–103.

[4] Ozyurt H and Otles S 2014 Properties of Probiotics and Encapsulated Probiotics in Food. *Acta Sci. Pol. Technol. Aliment*. Vol. 13 (4). pp. 413-424.

[5] Haffner F B, Diab R and Pasc A 2016 Review Encapsulation of Probiotics: Insights into Academic and Industrial Approaches. *AIMS Material Science*. Vol 3 (1) pp. 114-136.

[6] Gouin S 2004 Microencapsulation: Industrial Appraisal of Existing Technologies and Trends. *Trends in Food Science and Technology*. Vol. 15 (7-8) pp. 330-347.

[7] Gudmund S B 2006 Alginates, in Food Polysaccharides and Their Applications. *CRC Press*. 289-334.

[8] Dong Z, Wang Q and Du Y 2006 Alginate/Gelatin Blend Films and Their Properties for Drug Controlled Release. *J. Membrane Sci*. Vol. 280 pp. 37–44.

[9] Draget K I, Steinsvag K, Onsoyen E and Smidsrod O 1998 $Na^+$ and $K^+$ Alginate Effect on Ca Gelation. *Carbohydr. Polym*. Vol. 35, pp. 1–6.

[10] Chavarri M, Maranon I and Villaran MC 2012 Encapsulation Technology to Protect Probiotic Bacteria. *Intech*

[11] Gaaloul S, Turgeon S L and Corredig M 2009 Influence of Shearing on The Physical Characteristics and Rheological Behaviour of An Aqueous Whey Protein Isolate-Kappa-Carrageenan Mixture. *Food Hydrocolloids*. Vol. 23 pp. 1243–1252.
[12] Winarno F G 1996 Seaweed Processing Technology. Pustaka Sinar Harapan, Jakarta.
[13] Mongngram A, Trachoo N and Siringungwan N 2011 Low Molecular Weight Carbohydrates, Prebiotic, Content and Prebiotic Activity of Selected Food Plants In Thailand. *Adv Journal Food Science Technology*. Vol. 3 pp. 269-274.
[14] Le-Tien C, Millette M, Mateescu M A and Lacroix M 2004 Modified Alginate and Chitosan for Lactic Acid Bacteria Immobilization. *Biotechnology Applied Biochemistry*. Vol. 39 pp. 347-354.
[15] Lian W C, Hsio H C and Chou C 2002 Survival of *Bifidobacterium longum* After Spray Drying. *Journal Food Microbiology*. Vol. 74 pp. 79-86.
[16] Sudaryati H P, Mulyani T and Setiawan E B 2012 Study of Tofu Waste Substitution and Natrium Bicarbonate in Tortilla Production. EJournal UPN Jatim. http://ejournal.upnjatim.ac.id/index.php/rekapangan/article/view/421 [February 22nd, 2016].
[17] Ivanovska T P, Petrusevska-Tozi L and Kostoska M D 2012 Microencapsulation of *Lactobacillus casei* in Chitosan-Ca-Alginate Microparticles Using Spray-Drying Method. *Macedonian Journal of Chemistry and Chemical Engineering*. Vol. 31 (1) pp. 115-123.
[18] Koswara S 2009 Tchnology of Yogurt Production. eBookPangan.com
[19] Hendrati P M 2014 Yogurt Production Using *Lactobacillus bulgaricus* and *Streptococcus thermophilus*. Biology Faculty, University of Jenderal Sudirman.
[20] Soekarto 2003 Organoleptic Evaluation for Food and Agricultural Products. Bhrata Karya Aksara, Jakarta.
[21] Yoo I K, Seong G H, Chang H N and Park J K 1996 Encapsulation of *Lactobacillus casei* Cells in Liquid-Core Alginate Capsules for Lactic Acid Production. *Enzyme Microbial Technol.* Vol. 19, pp. 428-433.
[22] Hsu Y L and Chu I M 1992 Biotechnol. *Bioeng*. 40: 13001308.
[23] Kailasapathy K 2002 Microencapsulation of Probiotic Bacteria: Technology and Potential Applications. *Curr. Intest. Microbiol.* Vol. 3, pp. 39-48.
[24] Chen K N, Chen M J, Liu J R, Lin C W and Chiu H Y 2005 Optimization of Incorporated Prebiotics As Coating Materials for Probiotic Microencapsulation. *Journal Food Science.* Vol. 70 pp. M260-M266.
[25] Krasaekoopt W, Bhandari B and Deeth H 2003 Evaluation of Encapsulation Techniques of Probiotics for Yoghurt. *International Dairy Journal*. Vol. 13 (1) pp. 3–13.
[26] Favaro-Trindade C S and Grosso C R F 2002 Microencapsulation of *L. acidophilus* (La-05) and *B. lactis* (Bb-12) and Evaluation of Their Survival at the pH Values of the Stomach and in Bile. *Journal of Microencapsulation*. Vol. 19, pp. 485–494
[27] Anal A K and Singh H 2007 Recent Advances in Microencapsulation of Probiotics for Industrial Applications and Targeted Delivery. *Trends In Food Sci Tech*. Vol. 18, pp. 240-251.
[28] Mortazavian A, Razavi S H, Ehsani M R and Sohrabvandi S 2007 Principle and Methods of Microencapsulation of Probiotic Microorganisms. *Iranian J of Biotechnol.* Vol.5, pp. 1-18.
[29] William R B and Harper A R 2010 Carragenan. In Imerson A (Ed). Food Stabilizer, Thickener, and Gelling Agent. Willey-Blackwell Publishing Ltd, Oxford, England.
[30] Mtolera M S P and Buriyo A S 2004 Studies of Tanzanian Hypneaceae: Seasonal Variation In Content and Quality Of Kappa-Carrageenan From Hypnea Musciformis (Gigartinales : Rhodophyta). *J Mar Sci*. Vol. 3 (1) pp. 43-49.

# Some climatological factors of pine in the lake toba catchment area

**Z Nasution**

Faculty of Agriculture, Universitas Sumatera Utara; E-mail: zulnasution@usu.ac.id

**Abstract.** The article deals with climatological factors of Pine at the Lake Toba Catchment Area also called drained basin, *Pinus merkusii* is a plant endemic in Sumatra. A central population of Pine in North Sumatra is located in the Tapanuli region to south of Lake Toba. Junghuhn discovered the species in the mountains range of Sipirok. He provisionally named the species as *Pinus sumatrana*. The article presents a detail analysis of approaches to climate factors, considers rainfall, air temperature, humidity, stemflow, throughfall and Interception following calculation of regression to determine relationship between precipitation with stemflow and interception. Stemflow, it is highly significant with significance of difference between correlation coefficients and z normal distribution. Temperature and relative humidity are the important components in the climate. These components influence the evaporation process and rainfall in the catchment. *Pinus merkusii* has the big crown interception. Stemflow and Interception has an opposite relation. Increasing of interception capacity will decrease stemflow. This type of Pine also has rough bark however significant channels so that, it flows water even during the wet season and caused the stemflow in *Pinus merkusii* relatively bigger.

## 1. Introduction

The genus *Pinus* is one of the most widely distributed genera of trees in the extra-tropical Northern Hemisphere [1] In the evolutionary term the genus has evolved into 105 species with two main centers of speciation which are globally recognized. These are in southeastern Eurasia and southern North America. Perhaps from these sites *Pinus* populations have migrated into the tropics. The range of extension of pines in tropical South Asia has been described by [2] and [3] Schweinfurth [4] has compiled additional information and references on the distribution of pines in South East Asia which are provided by vegetation maps of the Himalayas, China, Cambodia, Vietnam, Sumatra, and Thailand.

A central population of Pine in North Sumatra is located in the Tapanuli region to south of Lake Toba. Junghuhn discovered the species in the mountains range of Sipirok. He provisionally named the species as *Pinus sumatrana*, The herbarium material was sent to Governor Merkus who in turn passed it on to Professor de Vriese who published the first botanical description in 1845, renaming the species *Pinus merkusii* Junghuhn et de Vriese in honour of the Governor who had died in the interim [3]

An indigenous species is the one that grows naturally in a country at certain site only. With these native species, there are not political or quarantine problems to obstruct its use in a plantation program, and in fact, there are some important biological advantages [5]: Growth of natural stands provides some indication of possible performance in plantation. The species is adapted to the environment and already filling an ecological niche. This may render it less susceptible to serious damage from diseases and pests since controlling agents (predators, viruses, and climatic factors) are already present. Indigenous species, even in monoculture, are generally considered more ecologically valuable than exotics for conservation of native flora and fauna. The timber is likely to be known to local wood-using industries.

For these reasons if an indigenous species is grown well in a plantation on the sites for afforestation, there is no compelling reason to widen the choice especially in using other introduced species [5] Buys *et al.* (1928) in [3] mentioned that several localities in the region from which *Pinus merkusii* has been reported, mostly in small stands, groups or even as scattered trees. The most extensive occurrence, estimated at about 1200 ha in the 1920s, is associated with two mountains, Dolok (mount) Tusam and Dolok Pardumahon. This stated region is the Dolok Saut Nature Reserve, which contains some 40 ha of *Pinus merkusii* associated with hardwood forest. Most of the reported occurrences are observed at an altitude of 1000 to 1500 m a.s.l.

Mirov [6] stated that the area around Lake Toba (1000 m a.s.l.) was once covered to a large extent by natural stands of pine as is evidenced by scattered remnants and the place name. For example Tusam for pine forest in Parapat which is in the vicinity of Girsang sampling station. In the Tapanuli region, 17,000 ha of pine are recorded but this which also includes large plantations in the vicinity of Lake Toba. No exact recent figures are available for the natural pine areas but these are probably in the order of less than 3,500 ha.

## 2. Research Methods

In this article the author study the typical cilmatological factors of Pine as an endemic plant in the catchment are of Lake Toba.

### *2.1. Rainfall*

The data of rain days, rain months and rain year were obtained from direct measurements with rain gauges (observatory type) at some stations: Bahal Batu station (02° 08' 893'' NL; 098° 59' 712'' EL); Girsang Simpangan Bolon station (02° 39' 044" NL; 098° 57' 853" EL);"); The diameter of gauge was 20 cm. The measured rainfall is calculated based on arithmetic means follows:

$$\sum_{i=1}^{n} R_{m,i} / n \quad \text{The monthly rainfall average at m month from all gauges} \tag{2.1}$$

$$\sum_{m=1}^{12} R_m \quad \text{The yearly rainfall average as a result of sum of monthly rainfall} \tag{2.2}$$

### *2.2. Air temperature*

All averages for a month, season, year or even a long period of years are built upon the mean daily temperatures as the basic unit [7]. The daily mean is thus the individual brick out of which the general temperature structure is composed. In some countries this basic unit of daily temperature is derived by computing the average of temperatures observed at certain definite hours during the day, as for instance, (0700 hours + 1400 hours + 1700 hours +1700 hours)/4. The United States Weather Bureau at present uses the formula (Maximum + Minimum)/2, or, in other words, the average of the highest and the lowest temperature recorded during the 24-hr period. According to Tjasyono [8], in Indonesia the average temperatures are derived by computing the average temperatures observed at:

$$T_{\square} = (2T_7 + T_{13} + T_{18})/4 \tag{2.3}$$

$T_{\square}$ = Daily average temperature

$T_7 + T_{13} + T_{18}$ are observed temperature at 0700 hrs; 1300 hrs and 1800 hrs.

### *2.3 Humidity*

The relative humidity RH is the relative of actual to saturation vapor pressure at the same temperature. In percentage terms

$$RH = 100\ ea/es \tag{2.4}$$

Saturation vapor pressure increases exponentially with increasing temperature [9]

$$es = 0.61078 \exp (17.269\ T/T + 237.30) \quad (2.5)$$

Where es, is the saturation vapor pressure in kPa and T is temperature in $^0$C

$$ea = es - AP\ (T - Tw) \quad (2.6)$$

Where T and Tw are dry and - wet bulb temperature ($^0$C)
P is the actual vapor pressure and es is the saturation vapor pressure at Tw.
A is a constant of proportionally and is slightly dependent on T. A, according to [10] has a value of:

$$A = 6.6 \times 10^{-4} (1 + 1.15 \times 10^{-3}\ Tw) \quad (2.7)$$

When the respiration rate is sufficiently rapid, say 3 – 4 $ms^{-1}$ or more.

*2.4. Factors Affecting Runoff (stemflow, throughfall and Interception)*

Part of the precipitation reaching a stream is evaporated or transpired. The remainder become runoff but takes many different routes over a long time span to reach a stream. The route and timing depend on character of the drainage basin and precipitation intensity. Each path affects the rate and amount of runoff in a different way.

- *Stemflow*

Some precipitation falls on the trunks of trees or stems of plants and runs to the ground. The water reaching the ground, called stemflow, which is largely dependent on the roughness of the trunks and stems. Stemflow from smooth surfaces is greater than compose the rough surfaces.

Measurement of stemflow can be done by string around stem of plastic tubing to collect water flow from the tree branch and stem, than flows to the catcher (bottle). The volume of is calculated with the equation:

$$Sf = (Vs/L) \times 10 \quad (2.8)$$

Where, Sf : Stem flow (mm)
Vs : Volume of flow ($cm^3$)
L : wide of crown ($cm^2$)
10 : conversion cm to mm

- *Throughfall*

If rain continues after the leaves can hold no more water, additional precipitation accumulates in droplets and falls to the next lower leaves. Eventually, the storage capacity of all leaves is reached and drops fall to the earth. The amount reaching the earth is called throughfall.

Measurement of throughfall can be done with gauges (in this study, place 10 observatory type gauges) under tree canopies. Volume of throughfall which is caught would be calculated with the equation

$$Th = (Vt/A) \times 10 \quad (2.9)$$

Where, Th : throughfall (mm)
Vt : volume of water catch ($cm^3$)
A : wide of longitudinal section ($Cm^2$)
10 : conversion cm to mm

- *Interception*

If the ground is covered with vegetation, the first drops of precipitation striking the plant leaves are retained, spreading over the leaf surface in a thin layer. This process and the water stored are called interception [11].

Measurement of interception at canopy scale would be carrying out with the volume balance approach i.e., with measure gross rainfall, stemflow and throughfall. The procedure of this method can be done with gauges, place stemflow and throughfall equipment at sampling site of study area with specific vegetation. The interception of the site would be calculated with the equation:

$$Ic = Pg - (Th + Sf) \quad (2.10)$$

Where, Ic : canopy interception (mm)
Pg : gross precipitation (mm)

Th : througfall trough canopy (mm)
Sf : (mm)
This equation should be developed to find out the total interception:

$$It = Ic + If \quad (2.11)$$

Where, If : litter interception

$$If = Pn - Pe \quad (2.12)$$

Where, Pn : net precipitation
Pe : effective precipitation

$$Pn = Th + Sf \quad (2.13)$$

$$Pe = Th + (Sf - If) \quad (2.14)$$

- *Evaporation*

The evaporation of the study site was calculated based on the evaporation from bare land and forest area. For bare land and under canopy, the evaporation was measured by using microlysimeters (tubing with 20 cm long and 8 cm diameter). In this study 10 lysimeters were placed in each station. The lysimeter is planted to the land with a hammer up to 0.5 cm left above soil surface and filled with soil. After that the tubing was lifted up by digging the surrounding soils and cleaned for the first balanced. After balancing the tubing was closed with the plastic below and replaced to the previous hole for the second balanced. The first balancing was done at 0600 hours and the second at 1800 hours over a one- year period.
Evaporation was calculated with the equation:

$$E = (Ws - Wf)/A \quad (2.15)$$

Where, E : evaporation
Wf : first weigh (gr)
Ws : second weigh (gr)
A : surface wide ($cm^2$)

*2.5. Regression model*

The independent variables are noted by $X_1$, $X_2$.......$X_m$ on the horizontal axis and the dependent variable by Y on the vertical axis. A hat on the $\hat{Y}$ is used when it value is estimated instead of actually measured. Regression analysis is appropriate for determining the coefficient (b's) of any prediction equation (or model) that takes the form [12]

$$\hat{Y} = b_0 + b_1X_1 + b_2X_2 + \ldots + BmXm \quad (2.16)$$

*3. Results and discussion*

*3.1. Rainfall*

Rainfall is one of important components of climate in the Lake Toba catchment area. The primary data that are measured since 1999-2002 in three stations are representing in Figure 1.

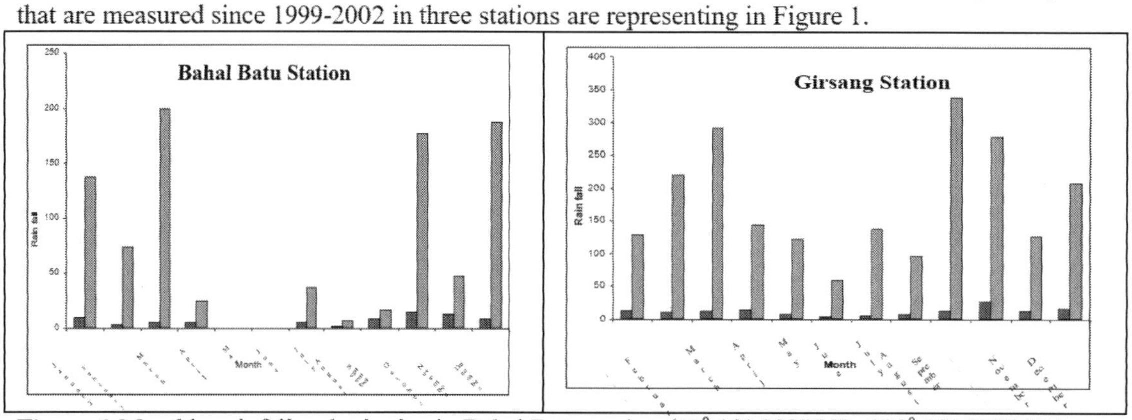


**Figure 1** Monthly rainfall and rain day in Bahal Batu station ($02^0$ 08' 893" NL; $098^0$ 59' 712" EL). Girsang station ($02^0$ 39' 044" NL; $098^0$ 57" 853" EL) Rainfall in mm, and Rainday in day

According to the graph of rainfall and rain day, the pattern is the same with the rain season in October to November and dry season in May to July. March to April is the transition season. The variability of rainfall is almost the same throughout the year.

### *3.2. Temperature and Relative humidity*

Temperature and relative humidity are the important components in the climate. These components influence the evaporation process and rainfall in the catchment. The fluctuation of daily temperature and daily relative humidity are showed in Figures 3.2

Based on those figures, the fluctuation of average daily temperature was relatively stable, whereas the temperature fluctuations between day and night are distinct as shown in Bahal Batu station. This situation most probably due to the area has more open bare ground compared with others. At Rianiate area (representative for Samosir Island), the fluctuation of temperature and relative humidity are small. This is probably due to the sites are very close to the lake.

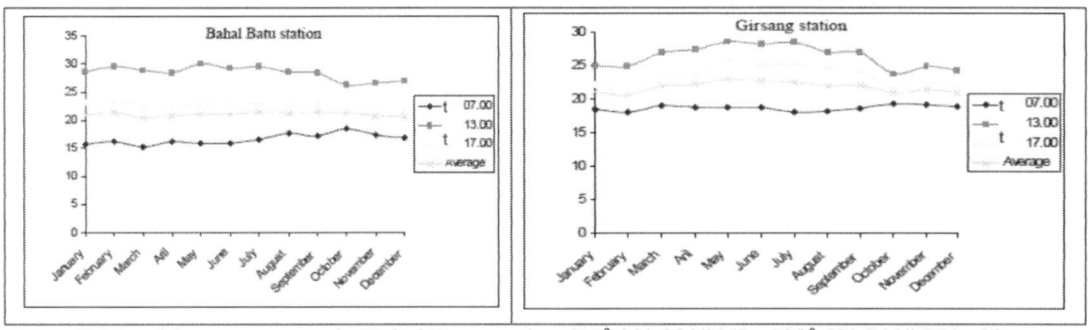


**Figure 2.** Monthly temperature in Bahal Batu station ($02^0$ 08' 893" NL; $098^0$ 59' 712" EL). Girsang station ($02^0$ 39' 044" NL; 0980 57" 853" EL)

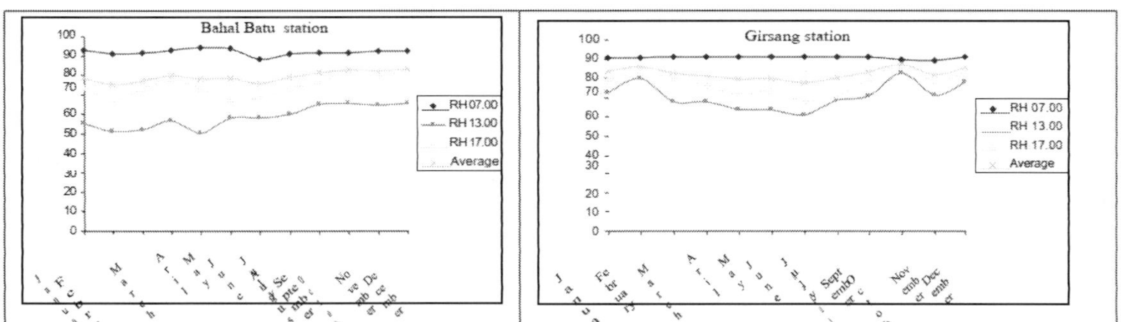


**Figure 3.** Relative humidity in Bahal Batu station ($02^0$ 08' 893" NL; $098^0$ 59' 712" EL). Girsang station ($02^0$ 39' 044" NL; $098^0$ 57" 853" EL)

### *3.3. Stemflow, Throughfall and Interception*

Stemflow, throughfall and interception are the important component in hydrologic cycle. Although small value, however these parameters are very important in the scale of a catchment area, particularly to find out the value of water balance. In a short time interception was not important but it can influence the water yield and net precipitation in a long time. Intensive works have done at the three stations however one addition station was added during the sampling period.

- *Girsang station*

Table 1 shows that from 149 days precipitation the value of T/P varies between 78.88 to 88.31 % with an average 83.95 %, S/P 0.101 to 0.378 % and an average 0.206 to 0.259 %. I/P are varied between 11.437 to 20.810 % with an average 15.786 %. The net precipitation in this area is 84%

- *Bahal Batu station*

Table 2 shows that from 129 days precipitation, the value of T/P vary between 73.62 to 88.52 % with an average 83.81 %. S/P 0.07 to 0.519 % and average 0.237 to 0.374 %. I/P are varied between 11.908 to 36.472 % with an average 19.45 %. The net precipitation in this area is 83%. Tables 1 and 2 also show the comparison between throughfall, and interception to rainfall. From these tables the readings can be obtained indicating that the parameters of throughfall, stemflow and interception are various.

**Table 1**. The comparison between Troughfall, Stemflow and Interception to Rainfall at Girsang Station

| Months | Rain days | (P) | Through Fall (T) (mm) | Stremflow (s) (mm) | | Interception(I) (mm) | | T/P (%) | S/P (%) | | I/P (%) | |
|---|---|---|---|---|---|---|---|---|---|---|---|---|
| | | | | 1 | 2 | 1 | 2 | | 1 | 2 | 1 | 2 |
| January | 13 | 129.28 | 110.29 | 0.405 | 0.382 | 18.588 | 18.611 | 85.31 | 0.314 | 0.296 | 14.378 | 14.396 |
| February | 11 | 219.86 | 190.65 | 0.484 | 0.411 | 28.729 | 28.802 | 86.71 | 0.22 | 0.817 | 13.067 | 13.100 |
| March | 12 | 291.22 | 243.06 | 0.539 | 0.561 | 47.616 | 47.594 | 83.46 | 0.185 | 0.193 | 16.351 | 16.343 |
| April | 15 | 143.73 | 117.64 | 0.367 | 0.447 | 25.718 | 25.639 | 81.85 | 0.255 | 0.311 | 17.893 | 17.838 |
| May | 8 | 121.96 | 97.13 | 0.147 | 0.214 | 24.681 | 24.614 | 79.64 | 0.121 | 0.176 | 20.236 | 20.182 |
| June | 4 | 59.20 | 47.79 | 0.067 | 0.080 | 11.339 | 11.326 | 80.73 | 0.113 | 0.135 | 19.155 | 19.132 |
| July | 6 | 137.36 | 120.80 | 0.259 | 0.311 | 16.310 | 16.258 | 87.94 | 0.189 | 0.227 | 11.873 | 11.835 |
| August | 8 | 95.80 | 75.57 | 0.294 | 0.294 | 19.936 | 19.936 | 78.88 | 0.307 | 0.307 | 20.81 | 20.81 |
| September | 14 | 337.24 | 297.83 | 0.69 | 0.715 | 38.717 | 38.692 | 88.31 | 0.205 | 0.212 | 11.481 | 11.473 |
| October | 27 | 278.17 | 238.48 | 0.282 | 0.697 | 39.4 | 38.98 | 85.73 | 0.101 | 0.251 | 14.164 | 14.015 |
| November | 14 | 126.32 | 108.66 | 0.299 | 0.556 | 17.358 | 17.101 | 86.02 | 0.237 | 0.440 | 13.740 | 13.538 |
| December | 17 | 207.70 | 172.08 | 0.467 | 0.784 | 35.151 | 34.834 | 82.85 | 0.225 | 0.378 | 16.924 | 16.770 |
| Total | 149 | 2147.8 | 1819.98 | 4.301 | 5.452 | 323.543 | 322.39 | 83.95 | 0.206 | 0.259 | 15.839 | 15.786 |

**Table 2.** The comparison between Troughfall, Stemflow, and Interception to the Rainfall at Bahal Batu Station

| Months | Rain days | Precipi tation (P) | Trough fall (T) | Stemflow (s) (mm) | | | Interception (I) (mm) | | | T/P (%) | S/P (%) | | | (I/P) (%) | | |
|---|---|---|---|---|---|---|---|---|---|---|---|---|---|---|---|---|
| | | | | 1 | 2 | 3 | 1 | 2 | 3 | | 1 | 2 | 3 | 1 | 2 | 3 |
| January | 11 | 116.55 | 188.33 | 0.448 | 0.73 | 0.655 | 27.77 | 27.46 | 27.56 | 86.97 | 0.23 | 0.40 | 0.34 | 14.74 | 14.58 | 14.63 |
| Feb | 4 | 110.40 | 91.03 | 0.229 | 0.327 | 0.239 | 19.14 | 19.04 | 19.13 | 82.46 | 0.25 | 0.36 | 0.26 | 21.03 | 20.92 | 21.02 |
| March | 13 | 153.12 | 128.01 | 0.331 | 0.644 | 0.601 | 23.98 | 23.67 | 23.71 | 84.12 | 0.26 | 0.50 | 0.47 | 18.62 | 18.38 | 18.41 |
| April | 17 | 157.21 | 127.24 | 0.325 | 0.581 | 0.512 | 29.64 | 29.39 | 29.46 | 80.95 | 0.26 | 0.46 | 0.40 | 23.28 | 23.08 | 23.13 |
| May | 8 | 113.65 | 99.80 | 0.219 | 0.318 | 0.240 | 13.63 | 13.53 | 13.61 | 87.81 | 0.22 | 0.32 | 0.24 | 13.66 | 13.56 | 13.64 |
| June | 2 | 22.38 | 16.39 | 0.013 | 0.031 | 0.022 | 5.97 | 5.95 | 5.96 | 73.26 | 0.08 | 0.19 | 0.13 | 36.42 | 36.32 | 36.37 |
| July | 6 | 95.09 | 84.68 | 0.214 | 0.321 | 0.293 | 10.19 | 10.80 | 10.11 | 89.06 | 0.25 | 0.38 | 0.35 | 12.04 | 11.91 | 11.94 |
| August | 7 | 70.94 | 53.07 | 0.118 | 0.167 | 0.133 | 17.75 | 17.70 | 17.74 | 74.80 | 0.22 | 0.32 | 0.25 | 33.46 | 33.37 | 33.43 |
| Sept | 12 | 258.48 | 223.77 | 0.433 | 0.649 | 0.587 | 34.28 | 34.07 | 34.13 | 86.57 | 0.19 | 0.29 | 0.26 | 15.32 | 15.23 | 15.25 |
| Oct | 23 | 239.49 | 326.25 | 0.811 | 0.996 | 0.907 | 66.42 | 66.24 | 66.33 | 82.91 | 0.25 | 0.31 | 0.28 | 20.36 | 20.20 | 20.33 |
| Nov | 14 | 129.93 | 114.80 | 0.349 | 0.596 | 0.509 | 14.77 | 14.53 | 14.62 | 88.36 | 0.30 | 0.52 | 0.44 | 12.87 | 12.66 | 12.73 |
| Dec | 12 | 223.14 | 197.52 | 0.631 | 0.898 | 0.916 | 24.89 | 27.72 | 24.7 | 88.52 | 0.32 | 0.46 | 0.46 | 12.65 | 12.52 | 12.51 |
| Total | 129 | 1944.49 | 1651.79 | 5.121 | 8.280 | 8.616 | 289.6 | 288.4 | 290.1 | 88.81 | 0.24 | 0.37 | 0.33 | 19.54 | 19.40 | 19.45 |

The relationship between interception and stemflow to precipitation is varied; even with the same species of plant in the same location the value may be different, due to a lot of influencing factors like rainfall characteristic, variety, age and density of plant, also growth period.

Morphological characteristic of plant like stem, canopy, leaf types and plant age gave big variation to stemflow and interception. The variation of plants with different type of canopy, leaf and stem is shown in Table 3. At Bahal Batu station, stemflow it is highly significant with significance of difference between correlation coefficients and z normal distribution (two-tailed test of the normal distribution) was found z = 3.07 (z table 1.96 at level 0.05). Morphologically there is the bark different among *Pinus merkusii*

At Girsang station, for stemflow is highly significant with significant difference between correlation coefficients and z normal distribution (two-tailed test of the normal distribution) is found z = 2.570 (z table 1.96 at level 0.05).

**Table 3.** Relationship between precipitation with stemflow and interception.

| Station | Relationship | Diameter (cm) | Regression equation | R |
|---|---|---|---|---|
| Bahal Batu | Precipitation and Stemflow | 43 | Y = 0.003x + 0.0011 | 0.918 |
| | | 23 | Y = 0.0022x + 0.0005 | 0.937 |
| | | 39 | Y = 0.0025x – 0.002 | 0.869 |
| | Precipitation and interception | 43 | Y = 0.1056x + 0.2214 | 0.794 |
| | | 23 | Y = 0.1064x + 0.223 | 0.752 |
| Girsang | Precipitation and Stemflow | 40 | Y = 0.0031x + 0.0015 | 0.787 |
| | Precipitation and interception | 40 | Y = 0.0418x + 0.7101 | 0.422 |

The range from 0.40 – 11 canopy interception capacity. Based on this analysis, the conclusions are as follow: *Pinus merkusii* has the big crown interception. Stemflow and Interception has an opposite relation. Increasing of interception capacity will decrease stemflow. *Pinus merkusii* has rough bark however significant channels so that, it flows water even during the wet season and caused the stemflow in *Pinus merkusii* relatively bigger

## References

[1] Critchfield, W.B & Little, E.L.1966. *Geographic distribution of the pines in the world.* Washington: USDA for Serv Misc Publ. 97 pp
[2] Kowal, N.E. 1966. Shifting cultivation, fire and pine forest in the Cordillera Central, Luzon, Philippines. *Ecol. Monogr.* 36: 389-419.
[3] Cooling, E.N.G. 1968. *Pinus merkusii. Fast-growing timber trees of the lowland tropics.* Oxford. No 4 Com For Inst. 168p.
[4] Schweinfurth, U. 1988. *Pinus in Southeast Asia.* Beitr Biol Pflanzen 63, 252-269.
[5] Evans, J. 1986. *Plantation forestry in the tropics.* Oxford. Clarendon Press. 472p.
[6] Mirov, N. T. 1967. *The genus Pinus.* New York. The Ronald Press Company. 602 p
[7] Trewartha, G.T. 1954. *An Introduction to climate.* Third edition. New York. Toronto. London. McGraw-Hill Book Company, Inc. 402 p.
[8] Tjasyono Bayong, H.K. 1987. *Iklim dan lingkungan.* Bandung. PT. Cendekia Jaya Utama. 190 p.
[9] Rosenberg, N.J., Blad, B.L. & Verma, S.B. (1983). *Microclimate. The Biological Environment.*
[10] Second Edition. New York. Brisbane. Toronto. Singapore. John Wiley & Sons. 495 p.
[11] Harrison, L.D. (1965). Fundamental concepts and definitions relating to humidity. *Humidity and moisture* (Waxler, A. ed). Meas. Contr. Sci. Ind. 3, 3-70.
[12] Hammer, M.J & Mac Kichan, K.A. 1981. *Hydrology and quality of water resources.* John Willey & Sons. New York. Chichester. Brisbane. Toronto. Singapore. 485p
[13] Myers, W. L & Shelton, R. L. 1980. *Survey Methods for ecosystems management.* New York. Chichester. Brisbane. Toronto. John Willey & Sons . 403 pp.

# Strengthening community participation in reducing GHG emission from forest and peatland fire

**A S Thoha[1*], B H Saharjo[2], R Boer[3] and M Ardiansyah [4]**

[1] Faculty of Forestry, Universitas Sumatera Utara (USU), Medan Indonesia
[2] Departement of Silviculture Faculty of Forestry, Bogor Agricultural University (IPB), Indonesia
[3] Departement Geophysic and Meteorologic, Mathematic and Natural Science, Bogor Agriculture University (IPB), Indonesia
[4] Departement of Soil and Land Resources Faculty of Agriculture, Bogor Agriculture University (IPB), Indonesia
E-mail: siddikthoha@gmail.com

**Abstract.** Strengthening community participation is needed to find solutions to encourage community more participate in reducing Green House Gas (GHG) from forest and peatland fire. This research aimed to identify stakeholders that have the role in forest and peatland fire control and to formulate strengthening model of community participation through community-based early warning fire. Stakeholder mapping and action research were used to determine stakeholders that had potential influence and interest and to formulate strengthening model of community participation in reducing GHG from forest and peatland fire. There was found that position of key players in the mapping of stakeholders came from the government institution. The existence of community-based fire control group can strengthen government institution through collaborating with stakeholders having strong interest and influence. Moreover, it was found several local knowledge in Kapuas District about how communities predict drought that have potential value for developing the community-based early warning fire system. Formulated institutional model in this research also can be further developed as a model institution in the preservation of natural resources based on local knowledge. In conclusion, local knowledge and community-based fire groups can be integrated within strengthening model of community participation in reducing GHG from forest and peatland fire.

## 1. Introduction

Land and forest fire and the impact had extended to various areas including Kapuas Regency in Central Kalimantan Province. Moreover, the impact of land and forest fire will get worse if taking place in peatland. Peatland fire in 1997/1998 as El Nino occurred in Indonesia, contributed emission 13-40% of global emission. [1-3]. The huge land and forest fire repeated in 2015. In Kapuas Regency, there are over four hundred peatlands locating at the former of One Million Hectare Peatland Development (Pengembangan Lahan Gambut/PLG)) since 1995. The PLG project opened hugely peatland including thick peat swamp forests with thousands of kilometers of canals construction that impact on peatland drainage ecosystems. Peatland drainage ecosystems initiated the prone area of land and forest fire occurring almost every year and causing massive carbon emission [4].

Early Warning System of Forest and Land Fire (EWSFLF)which is part of Forest and Land Fire Control still can not solve the problem of greenhouse gas emission from land and forest fire.

Currently, EWSFLF implementation may involve less the participation of the community. Some detection information and warning fire were used to provide information about fire activity from several government agencies. Currently, prediction information of fire period is managed by Indonesia Meteorological, Climatological and Geophysics Agency (BMKG). Moreover, the hot spot data were used as an indication of forest and land fire that was managed by The Ministry of Environment and Forestry (KLHK). Warning information of forest and land fire that has been used in national level has not been socialized to the community which are close to the location of fire-prone. In addition, the data and information of the forest and land fires probability are still centralized and dissemination of information is also impressed bureaucratic.

Currently, community institution has not been seen actively involved in forest and peatland fire control. Communities who live in fire-prone have the probability to organize themselves and contribute greatly in controlling forest and land fire by collaborating with the government and private sector. Institutional aspects and the involvement of community participation actively is to be important in reducing the spread of fire [5,6]. Consequently, it is needed to develop early warning system of forest and land fire to find solutions of the centralized and expensive current system as well as encouraging people to play more active the role in their own protection. Therefore, to achieve the reduction in greenhouse gas emissions from forest and land, it would require the development of early warning system involving community participation in Kapuas Regency of Central Kalimantan Province. This research aims to identify stakeholders who play role in controlling forest and peatland fire and to formulate strengthening model of community participation through the community-based fire early warning system.

## 2. Materials and Method

### *2.1. Study Area*

Primary and secondary data collected in Kapuas, Bogor and Palangkaraya from April 2012 - September 2013. The informants were from the government agency and non- governmental organizations which associated with the activities of forest and land fire control as well as the community from fire-prone villages in Central Kalimantan and Kapuas Regency. Focused group discussion was done in Kuala Kapuas Kapuas Regency and Bogor City.

### *2.2. Data Analysis*

Secondary data was legal documents and reports activities related to forest and land fire control in Central Kalimantan Province and Kapuas Regency which were derived from government agency and non-government organization. In addition, primary data collection was completed by structured interviews, in-depth interviews and focus group discussions (FGD). Structured interview conducted on government agency staff to explore the basic information related to the duties and functions of the institution, forest and land fires control program, the facilities of institutions and coordination with the other parties. The informants selection of stakeholder in this study were conducted in two ways: (1) snowballing sampling and (2) key person. Researchers began to collect information by finding the first gatekeeper or the first person who welcomed at place of the research that could give clues about the stakeholders and who could be interviewed in order to obtain the data research [7]. There were 15 informants of stakeholder which each institution chose a representative to be key person.

The FGD with various stakeholders at the community level and the institution level was done separately. FGD activities was conducted to reveal their knowledge related to forest and land fires, early warning systems and their expectations for the future. The FGD at the community level was followed by 10 participants [8]. Meanwhile, the FGD participant from government institution consisted of six agencies from the Regional Disaster Management Agency (BPBD) of Central Kalimantan Province, BPBD of Kapuas Regency, Forestry and Plantation Agency of Kapuas (Disbunhut Kabupaten Kapuas), Meteorology and Geophysics Agency of Central Kalimantan Province.

IOP Conf. Series: Earth and Environmental Science **122** (2018) 012076 doi:10.1088/1755-1315/122/1/012076

Stakeholder Analysis was used to determine the key actor involving in the implementation of policy and institutional of forest and land fire prevention. Interview result became a reference to conduct the classification and assessment of the level of interest and influence [9] and stakeholder mapping [10]. In 2 x 2 matrix, each stakeholder was mapped based on two criteria, namely the influence and interest. This mapping indicates a potential relationship that may be built namely a potential conflict, complementarity and cooperation [10].

The formulation of early warning system of forest and land fire comes from various stakeholders reflection on the incidence of forest and land fires and early warning systems that happen. The reflection according to the criteria in action research is the effort to understand the problem and criticize what has happened [11]. The formulation of early warning system of forest and land fire is the result of community and key stakeholder reflection who answered some aspects in terms of based community early warning systems from UNISDR [12], consisting of four key elements : (i) knowledge of risk, (ii) monitoring, analysis and threatening prediction, (iii) the deployment of warning and alert message and ( iv ) the ability of community to respond the received warning.

## 3. Results and Discussion

It was arranged key actor level based on ownership of the task, function and program that directly leads to action forest and land fire control. Moreover, the key actor were also assessed on its role in supporting the financial aspect and policy of forest and land fire control. The arrangement of key actor level was made in classification form of a role of the activities, interest and influence of stakeholder into the success of the program. Identification result of the interest and influence of the parties related to the reduction of GHG from forest and land fire was classified as can be seen in Table 1 .

**Table 1.** Classification of the parties according to the influence and interest in reducing of GHG emission from forest and peatland fires in Kapuas

| No. | Stakeholder Groups | Role in program | Stakeholder interests on the success of the program | Stakeholder influence on the success of the program |
|---|---|---|---|---|
| 1 | Regional Disaster Management Agencies of Kapuas (BPBD) | Decision maker and organizer | 4 | 5 |
| 2 | Environmental Agency of Kapuas (BLHD) | Executor | 3 | 3 |
| 3 | Forestry and Plantation Agency of Kapuas (Dishutbun) | Executor | 5 | 3 |
| 4 | Agriculture, Horticulture and Food Crops Agency of Kapuas | Executor | 3 | 2 |
| 5 | District Development Planning Agency of Kapuas (Bappeda) | Supporter | 2 | 4 |
| 6 | Manggala Agni | Executor and Supporter | 5 | 2 |
| 7 | District Government | Utilizer and Supporter | 2 | 2 |
| 8 | Village Government | Supporter | 5 | 3 |
| 9 | Fire Brigade Community (Community Group for Forest Fire Management/ Community Care Fire / Fire Extinguishing Team) | Utilizer and Supporter | 5 | 2 |
| 10 | Indigenous Institute | Supporter | 2 | 3 |

| | | | | |
|---|---|---|---|---|
| | (Mantir) | | | |
| 11 | Non-Governmental Organization (Indigenous Peoples Alliance of the Archipelago/AMAN) | Supporter / Opposition | 2 | 2 |
| 12 | Cooperation Project Local Governments and Donors (KFCP) | Supporter | 3 | 1 |
| 13 | Police | Executor | 3 | 2 |
| 14 | Provincial government (POSKO of Province) | Supporter | 3 | 2 |
| 15 | Meteorological, Climatological and Geophysics Agency of Central Kalimantan Province (BMKG) | Supporter | 2 | 3 |

format sources: (RTI , 2002)
Note: 1 = little / not important , 2 = quite important , 3 = moderate , 4 = very important ,
5 = a key player.

Institution as stakeholder continuously receives the impact of the activities or policies, where the institution also has an interest into the process and results of the activities and policies that affect [13]. Then, [14] added that the stakeholder plays as the key actor who has strong influence and interest related to the problems, needs and attention to success of the program. Institution that takes place in quadrant with high influence and interest is the institution that has highest participation towards the success of the action of forest and land fire control. These institutions have the technical participation through programs and political participation through duties and policies affecting capability in taking measures and supervision of certain situations and improve the independent action in strengthening and development of institutional and program [15].

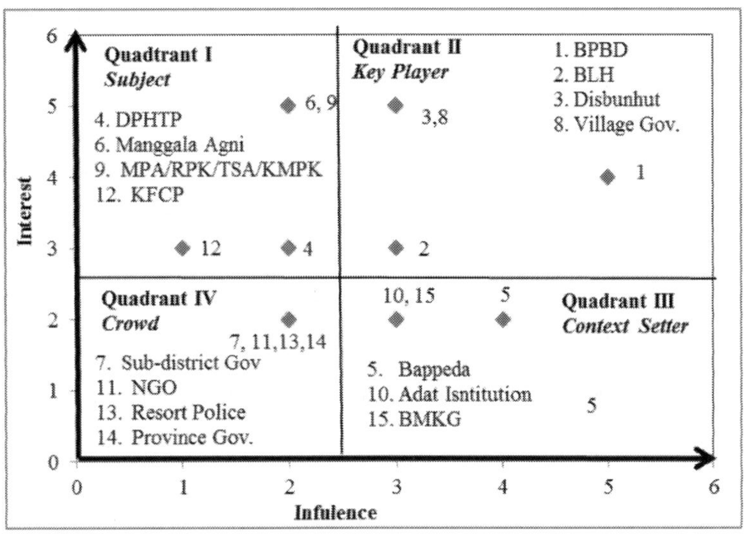


**Figure. 1**. Matrix of stakeholder's interest and influence in forest and land fire early warning system in Kapuas Regency, Central Kalimantan Province, Indonesia

According to [10] classified stakeholder based on the influence and interest as subjects, key players, crowd, and context setters. Subject has high interest but low influence. Although supporting the activity, subject capacity into the impact may be nothing. The stakeholder can be the influence when form the alliance with the other stakeholder. Key players are active stakeholders because they have high interests and influence to the development of a project. Crowd is stakeholder who has low interest but high influence. These are the reasons why crowd is involved in decision making. Context setter has high influence but low interest causing significant risk to be monitored. The mapping result of influence level and interest of the various stakeholders related to the forest and land fire control program in Kapuas, Central Kalimantan Province is shown in Figure 1.

Quadrant I position (subject) at Figure 1 is derived from government agencies, non-governmental organization and community. Stakeholders in Quadrant I is Department of Agriculture Food Crops and Horticulture (DPHTP) Kapuas, Mangala Agni Regional Operations II (Daops II) Kapuas, Community Development of Fires Control in Forest and Land (MPA / RPK / CPC / KMPK) and project KFCP (Kalimantan Forest and Climate Partnership). Stakeholders of community such as MPA / RPK / CPC / KMPK have high interest in forest fire control as they relate to the land that they have to protect. However, the ability of stakeholder is inadequate related to fund, facilities and human resources. This condition is similar to the research of Herawati et al. [16] on the program of in Community Plantation Forest (HTR), research of [17] and research of [18] when the high interest of community into the natural resources, especially related to economic and social interests. In Kapuas Regency, there are community based of forest and land fire of fire control institutions that formed and receive the guidance in Kapuas Regency. These institutions are KMPK, MPA, RPK and CPC. KMPK and CPC were formed and nurtured by local Kapuas governments namely BPBD and Social Service consisting of 23 and 15 groups. MPA is under the guidance of Agni Mangala Kapuas, while RPK was formed and receive guidance from the KFCP project (The Kalimantan Forest Carbon Partnership) in which each of them is formed of four and seven groups. Number of community-based fire control in Kapuas is 49 groups.

Stakeholder group in Quadrant I has plus point that most of the community has high interest in controlling forest and land fire. This is a valuable potential in strengthening community participation in forest and land fire control. However, the weakness of Quadrant I is most of the stakeholders coming from formal institutions (DPHTP and Manggala Agni) and the International cooperative institution (KFCP). Although those three institutions have programs related to community involvement in early warning fire, but the portion of the program and budget are minimum so they do not have strong influence in strengthening communities participation in each program.

Quadrant II position (key players) are occupied by BPBD (Regional Disaster Management Agency) of Kapuas, Environmental Agency of Kapuas (BLH), Department of Forestry Plantations of Kapuas (Disbunhut) and the local government. Those four stakeholders have important role in preventing fire, extinguishing, handling post fire, controlling land fire, developing community capacity, and empowering communities. These are common actions at any natural resource management project in which the manager who gets the power legally is always in main stakeholder position [19].

All stakeholders in Quadrant II come from the government as a formal institution. It has plus point where the institution would have strong legal and sustainable program. However, the weakness of the composition of stakeholders in Quadrant II is the absence of institution that comes from community so that forest and peat land fire control program potentially does not get strong response and participation from the community.

Quadrant III position (contex setter) are occupied by Regional Development Planning Agency of Kapuas (Bappeda), the Traditional Institute in village level and Meteorological, Climatological and Geophysics Agency (BMKG). These stakeholders can influence the development of community-based early warning system as these have high influence. Bappeda plays in the planning and development authorities including planning of disaster management program. Indigenous Mantri becomes a reference in determining forest fire and land danger warning information that can be trusted. BMKG has data to be a reference in determining the preparedness status of local government disaster related to climate and weather.

The plus point in quadrant III is derived from the group of stakeholders which are traditional institutions that have high impact, especially as resources of drought predictions based on local indicators. The weakness of the stakeholders are BMKG has no high interest so that warning information response dissemination of forest fire hazard is not up to the community level and the response of community is low .

Position in quadrant IV (crowd) are took place by the four stakeholders namely District Government, NGO (Non Governmental Organization) AMAN, Police of Central Kalimantan Province Government. The stakeholders also give attention to the development of early warning system for forest and land fire. However, the activities are only temporary with weak effect.

The lack in the composition of stakeholders in Quadrant IV is low role of the Provincial Government (POSKO of Central Kalimantan) in supporting the strengthening participation of community in forest and land fire control. As an institution that makes laws, provincial government has to increase its role throughout the district or town in Central Kalimantan Province for strengthening the institutional and participation of community for the control of forest and land fire.

According to all stakeholders which are mapped, it can be identified three relationships that occur between stakeholders which engages in the development of early warning system for forest and land fire namely a potential conflict, complementarity and cooperation. Potential interest conflict occurs between community or group of communities with the institutions that have the duty and function in the prevention of forest and land fires. Government institutions are less in socialization so many people do not know the regulations on land clearing and yard guidance and fire danger warning information. Potential conflicts also occur between KFCP and community. KFCP is not entirely successful in the dissemination and expansion of participation causing resistance from some local group of communities. Complementary occurs between Agni Mangala Kapuas, KFCP, Disbunbut, BLH and based community forest and land fire control group. These institutions are often in an activity that is in increasing the capacity of community for forest and land fire prevention. The potential cooperation can be established for all stakeholders, especially among providers of information and the dissemination of information to the community and the local government, BMKG, Manggal DOPS Agni II Kapuas, BLH, DPHTP and BPBD. They can cooperate to integrate the data of early warning forest and land fire that can be understood and responded well by the community. The local government and community of forest and land fire control could help the government to provide actual data on their villages to government agencies. In addition, the local government and community can help disseminate information of the dangers of forest and land fire from the government quickly and utilize effective communication in community.

Reflection from the community and government agencies regarding the early warning system of forest and land fire which has been conducting currently is formulated by a model dam strengthening community participation lowers GHGs through the development of early warning systems in Kapuas as presented at Figure 2. This model on the potential relationship between stakeholders are dynamic and result reflection of government agencies and the public that refers to the four key elements of early warning system according to UNISDR (2009), namely (i) knowledge of risk, (ii), analysis and peramalam hazards, (iii) information and communication warning message and (iv) response.

Reflections on the knowledge of risk will answer questions about local knowledge community concerning long drought in the Kapuas Regency, Central Kalimantan Province. One form of local knowledge of the tribes in Kapuas is how they predict the coming drought. There are some local wisdoms in Kapuas that how the community predict as presented in Table 2.

Based on Table 2, there are four public knowledge as indicators of fire-prone period in the Kapuas Regency that are easy to use and understand namely Beje, fallen leaves, animal collect on the riverbanks and Bali Calendar. These local indikator are easy to understand and even be able to provide clearly a period of time especially in Bali Calendar

According to monitoring, analysis and prediction hazard aspect, community have an understanding that most of fire occur on land that is unmaintained, especially in peatlands. Based on ground check and interview, the location that almost burnt in every drought are shrubs, former purun grass shrubs, secondary swamp forest, reeds and former agricultural land. The local knowledge also corresponding to study by [20] in Kapuas District that found the fire prone area located in grassland, shrubs and secondary swamp forest.

**Table 2.** List of Local Knowledge of Community in Disaster Risk in Kapuas, Central Kalimantan Province.

| No | Knowledge of Community | Application Level | | |
|---|---|---|---|---|
| | | Easy | Medium | Difficult |
| 1. | Beje (fish trap pool ) is receding | v | | |
| 2. | Many Fish go down to the river mouth | | | v |
| 3. | Snakeskin fish clot in the air | | | v |
| 4. | The umbrella of the sun | | | v |
| 5. | The fall of the leaves of trees | v | | |
| 6. | The animals collect on the riverbank | v | | |
| 7. | Season Kapat or severe drought | | V | |
| 8. | Use Farming Calendar (Bali Calendar) | v | | |

Source : interviews with community leader

Based on aspect of fire warning message, it is obtained information that early warning message of forest and land fires to community mostly come from village officials (Table 3). The message was conveyed by village officials on socialization of regents or government agency regarding drought and awareness status of forest and land fires. Both village official and community leader or activist can be a reference of information for the community regarding the arrival of long drought causing forest and land fire. Traditional leaders, especially at the village level, both religious figures and paramedics custom generally have the local knowledge of local indicators or signs of drought on the observation of nature. The tool that is used to disseminate information of early warning of forest fire is an informal meeting (Table 3). Informal meeting is usually able to collect a lot of people and has sufficient time to convey the message in attractive way and appropriate understanding.

**Table 3.** The tools of dissemination of messages regarding the dangers of forest fires and land in Kapuas District

| Institution | Tool of Message Dissemination | | | | | | |
|---|---|---|---|---|---|---|---|
| | Radio Commu-nications | Phone | Cellular phone | Letter | Home visit | Informal meeting (Mosque, Church, Informal meetings, etc. ) | Visual Aids (Warning boards, banners, billboards ) |
| BPBD | | v | v | v | | | v |
| BLH | | | | | v | v | v |
| Disbunhut | v | v | | | | | v |
| Manggala Agni | | v | v | | v | v | v |
| MPA | | | v | | | v | |
| BPK | v | | | | | | |
| RPK | | | | | v | v | |
| Local government | | | | v | v | v | |
| Total | 2 | 3 | 3 | 2 | 4 | 5 | 4 |
| Persentage (%) | 25.0 | 37.5 | 37.5 | 25.0 | 50 | 62.5 | 50 |

Source : interview with leader and village government April-September 2012

The ability of communities to respond warning fire message was found that they got hazard warning information from several sources that were Agni Mangala, banners / bulletin boards of governors, circulars regent, village officials and information from the mass media. For those who provided information regarding fire warning, community got it from the village head and staff, Chairman of RT, community leaders and BPD (the village assembly). If wildfire happens, the community responds by distributing information about hazards and land clearing activity through community-based Group Fire Control. Study by [21, 22] also found that community-based fire control group in Mantangai District Kapuas Regency ie Fire Brigade (RPK) had recognized the existence and benefit by most villagers. They also asked for help on the parties that were considered capable to control namely the Head of Village and staff, other rural communities, other land owners, Mangala Agni, Disbunhut and BPBD.

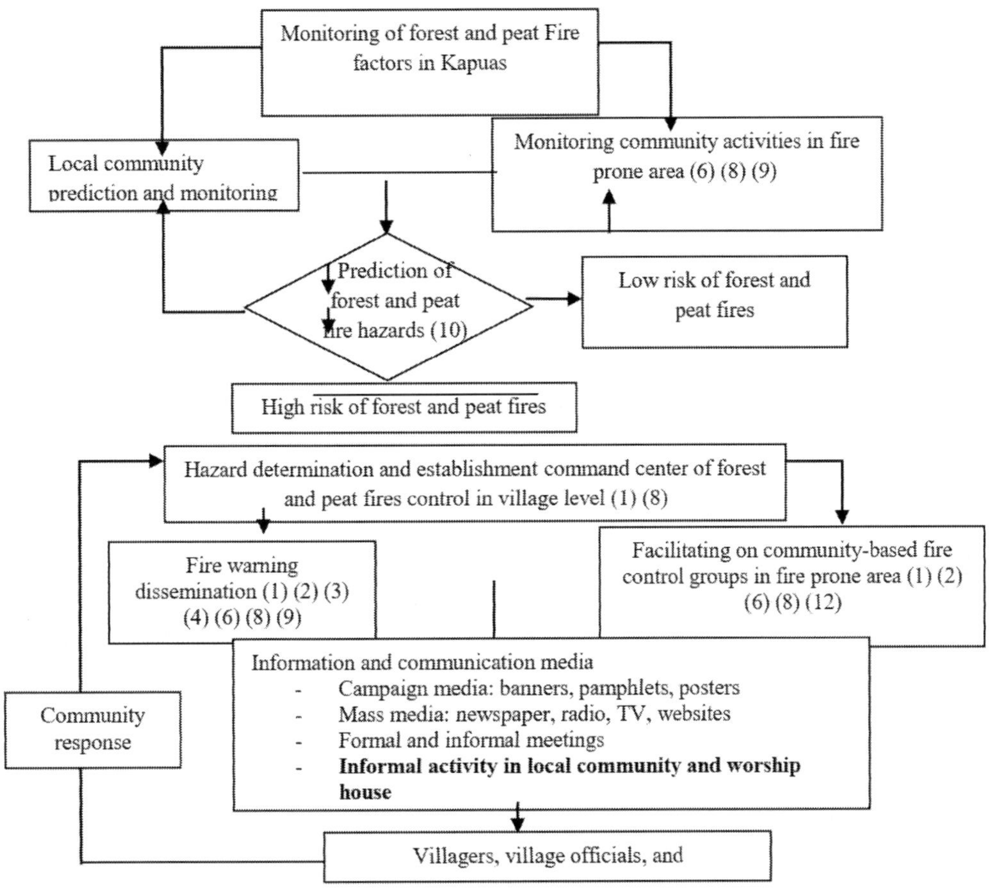


**Figure. 2**. Model of strengthening community participation of forest and peat fire through community-based early warning system in Kapuas District, Central Kalimantan Province, Indonesia (The numbers represent stakeholders such as at Figure 1).

Strengthening model of community participation can be flexible and prospective. Flexible means that each stakeholder role can change according to the capacity of the institution and the intervention of outside party to strengthen these institutions. Prospective means that this institutional model could be relied upon not only overcome the problem of forest and land fires, but also a wider role in improving community capacity to participate in disaster management in general as well as protecting the citizens, property and land in their respective villages. Therefore, this institutional model is also prospective to be developed into a model institution in the preservation of natural resources based on local knowledge to the village level. This step is conducted by strengthening the presence and role of community based forest and land fire control group into one Village Community Institution (LKD) as a partner of the village government.

## 4. Conclusion

Public knowledge regarding how to predict the fire season has a potential value for the development of community-based forest and peatland fire early warning system. Community-based fire control group and traditional institution require strengthening of influence and importance to increase participation in controlling forest and peatland fire till the village level. Institutional model and strengthening of participation in this research could also be developed for the further research into a model institution in the preservation of natural resources based on local knowledge till the village level by strengthening of community-based fire control group as a partner of the village government. It is recommended that the local and institutional knowledge can be integrated in the strengthening participation model in reducing GHG from forests and peatland fire.

## References

[1] Page S E, Siegert F, Rieley F O, Boehm H V, Jayak A and Limin S H 2002 The amount of carbon released from peat and forest fires in Indonesia during 1997 *Nature* **420**: 61-65.
[2] Harrison M E, Page S E and Limin S H 2009 The global impact of Indonesian forest fires *Biologist* **56** (3): 156-163
[3] Langmann B, Duncan B, Textor C, Trenmann J and van der Werf G E 2009 Vegetation fire emissions and their impact on air pollution and climate *Atmospheric Environment* **43** : 107-116
[4] Hooijer A, Van der Vat M, Prinsen G, Vernimmen R, Brinkman J, and Zijl F 2008 Hydrology of the EMRP Area – Water Management Implications for the Peatlands. Technical Report Number 2, Master Plan for the Rehabilitation and Revitalization of the Ex-Mega Rice Project Area in Central Kalimantan. Euroconsult Mott MacDonald and Deltares and Delft Hydraulics
[5] Akbar A 2008 Pengendalian Kebakaran Hutan Berbasis Masyarakat sebagai Suatu Upaya Mengatasi Risiko dalam REDD *Tekno Hutan Tanaman* **1**(1): 11-22
[6] Sunanto 2008 Peran serta masyarakat dalam pencegahan dan penanggulangan kebakaran lahan. Tesis Universitas Diponegoro Semarang Indonesia. URL : eprints.undip.ac.id/18422/1/SUNANTO.pdf
[7] Bungin B 2010 Penelitian Kualitatif (Jakarta: Kencana Prenada Media Group)
[8] Kriyantono R 2009 Teknis Praktis Riset Komunikasi (Jakarta : Kencana Press)
[9] [RTI] Research Triangel Internatiobal 2002 Citizen participation in decision making (Trainig Manual. Bucharest Romania)
[10] Reed MS, Graves A, Dandy N, Posthumus H, Huback K, Morris J, Prell C H, Quin C H and Stringer L C 2009 Who'sin and why? A typology of stakeholder analysis methodsfor natural resources management. *Journal ofEnvironmental Management* **30**:1-17
[11] Dick B 1997 Action learning and action research. URL : http://www.scu.edu.au/schools/gcm/ar/arp/actlearn.tml
[12] [UNISDR] United Nation InternationalStrategy for Disaster Reduction 2009 Disaster Risk Reduction Terminology (Bangkok : UNISDR Asia and the Pacific Office)
[13] Iqbal M 2007 Analisis peran pemangku kepentingan dan implementasinya dalam pembangunan pertanian *Jurnal Litbang Pertanian* ***26(3)*** : 89 – 99
[14] Bryson J and Crosby B 1992. Leadership for the Common Good: Tacklina Public Problemsin a Shared Power World (San Francisco CA: Jossey-Bass)
[15] Selener D 1997 Participatory action research and social change (New York : Cornell University)

[16] Herawati T, Widjayanto N, Saharuddin and Eriyatno 2010 Analisis respon pemangku kepentingan di daerah terhadap kebijakan hutan tanaman rakyat *Jurnal Analisis Kebijakan Kehutanan* **7**(1):13-25.

[17] Kusumedi P and Rizal A 2010 Analisis *stakeholder* dan kebijakan pembangunan KPH model Maros di ProvinsiSulawesi Selatan *Jurnal Analisis Kebijakan Kehutanan* **7**(3):179-193

[18] Rastogi A, Badola R, Hussain S A and Hickey G M 2010 Assessing the utility of stakeholder analysis to protectedareas management: the case of Corbett National Park, India *Biological Conservation* **143**:2956-2964.

[19] Sembiring E, Basuni S and Soekmadi R 2010 Resolusi konflik pengelolaan Taman Nasional Teluk Cenderawasih di Kabupaten Teluk Wondama *Jurnal Manajemen Hutan Tropika* **16**(2):84-91

[20] Thoha AS, Saharjo BH, Boer R and Ardiansyah M 2014 Spatiotemporal distribution of peatland fires in Kapuas District, Central Kalimantan Province, Indonesia *Agriculture, Forestry and Fisheries* **3**(3): 163-170

[21] Akbar A, Sumardi RH, Purwanto dan Sabarudin MS 2011 Studi sumber penyebab terjadinya kebakaran dan respon masyarakat dalam rangka pengendalian kebakaran hutan gambut di areal Mawas Kalimantan Tengah. *Jurnal Penelitian Hutan Tanaman* **5**: 287 – 300

[22] Thoha A S, Saharjo B H, Boer R and Ardiansyah M 2017 Forest and Land Fires Hazard Level Modeling : Case study of Kapuas, Central Kalimantan *Disaster Risk Reduction in Indonesia (Electronic Material vol IV)* ed Djalante R et al (New York: Springer International Publishing) chapter 22 pp 539-560

**Acknowledgement**

This study was funded by Columbia University and Bogor Agricultural University Partnership to Build Capacity for Adaptation to Climate Risks In Indonesia, USAID Grant No.AID-497-A-11-00011. We also thank to Regional Operation Manggala Agni II of Kapuas Natural Resources Conservation Center (BKSDA) Central Kalimantan for the support and assistance provided to the researcher during the data collection process. Sincerely appreciation is also extended to anonymous reviewer for correction and comments.

# Phylogenetic relationship of asam gelugur (*Garcinia atroviridis* Griff. ex T. Anders) based on morphological characters in Langkat and Serdang Berdagai, Sumatera Utara

**E S Bayu, A Lestami, E H Kardhinata and Rosemary**

Faculty of Agriculture, Universitas Sumatera Utara, Jl. Prof. A. Sofyan No.3, Padang Bulan, Medan, 20155, Indonesia
E-mail: tinigirsang@yahoo.com / 08116004596

**Abstract.** *Garcinia atroviridis* plant is a commodity that has the potential to become an export commodity in Sumatera Utara. Characterization is an activity in germplasm to determine the morphological properties that can be utilized in differentiating between accessions and assessing the magnitude of genetic diversity. The magnitude of genetic diversity based on morphological properties can support breeding programs. The research was conducted in several areas of *G. atroviridis* plant, i.e. Gebang, Bahorok, Padang Tualang, Pegajahan, Sei Rampah, Pantai Cermin, and Perbaungan sub-districts in September 2016 with direct observation method. The aim of this research is to explore and characterization between *G. atroviridis* plant in Langkat and Serdang Berdagai. Parameters morphology characters were observed based on IPGRI ,and observation data were analyzed by using SPSS version 21 to obtain the dendogram. The dendogram results showed that there are four groups of kinship relationships on the scale of the spacing (euclidean distance scale) 17. The lowest unequal value (closest kinship) in accessions G3(Bahorok) and G7(Sei Rampah) is 16.328 while the highest value of inequality (farthest kinship) in G5( Padang Tualang) and G6 (Pegajahan) is equal to 54.187. It needs the conservation of *G. atroviridis* although must be done next observation.

## 1. Introduction

Germplasm is one of the most important natural resources and is the basic capital needed to develop the agricultural industry. Germplasm is a necessary genetic resource, one of which is to develop new varieties or cultivars. The high diversity of germplasm we have, open the opportunity to search for, utilize, discover, and optimize uncharted genetic potential [1]. One of Indonesia's biological riches is *Asam Gelugur* (*Garcinia atroviridis* Griff. ex T. Anders.). *Asam Gelugur* is a fruit belonging to the *Guttiferae* family and the *Garcinia* genus. This fruit grows a lot in the area of Sumatera Utara.

*Garcinia* clan plants spread in Asia tropical region. The well-known species, *Garcinia cambogia*, are commonly found in southern India, while other species, namely *Garcinia atroviridis* (*Asam Gelugur*) commonly found in the Malay Peninsula region [2]. The *Asam Gelugur* or Gelugor is widely used as a flavoring dish by the Malay community. This plant is still a family with Mangosteen and *Asam Kandis* that spread in Southeast Asia. This plant belongs to the *Guttiferales* ordo and *Guttiferae* family whose height can reach 20 m [3].

In Indonesia *Garcinia* is a widely distributed plant and an important part of forest composition. In the forest, it is often encountered as a plant in the second storey based on the height of the tree. Based on existing data in Herbarium Bogoriense in Indonesia there are about 100 types of *Garcinia*. In the world, the number is estimated to reach 400 species. This means about a quarter of the world's *Garcinia* species are in the Indonesian region [4].

*G. atroviridis* is widely used as a flavoring cuisine by the Malay community, But also proved beneficial for lowering cholesterol. It is also an antioxidant and can reduce body weight and cholesterol. Mackeen [5] investigated the bioactivity of water ethanol extracts from this plant which gave the result that the extract had antibacterial, antifungal, antioxidant, antitumor, and antimalarial activity.

Commercially *G. atroviridis* has been exploited by the international industry as a health food to burn highly potent fats (slimming), lowering cholesterol, hypertension, rheumatism, processed in capsules and syrups. *G. atroviridis* fruit now has a high economic value and began commercialized with a fairly high selling price of 60000 IDR / kilogram in dry processing mode. This product has been marketed to Medan, Teluk Kuantan and overwhelmed to meet demand due to shortage of raw materials [6].

New varieties may arise due to environmental factors and genetic variation, such as due to cross-pollination. The differences and similarities of morphological emergence outside the species of a plant can be used to know the proximity of kinship relationships [7].

Information on diversity is essential in plant breeding programs, because with the availability of such information, it is easier in determining the position or kinship between varieties that can be used as the basis of plant selection. Renwain et al [8] stated that the success of plant breeding programs to improve the character of a plant is largely determined by the availability of genetic resources.

The collection of diversity information is carried out through enrichment activities such as exploration. With this activity the opportunity for the emergence of the desired genetic potential may be available [9]. Therefore, it is necessary to identify a plant. Characterization of collections (accessions) performed, aimed at obtaining data on the nature or character of agronomic morphology (basic morphological description) so as to distinguish the phenotype of each accession quickly and easily, by estimating the scale of genetic diversity it possesses [10].

## 2. Methode

The research was conducted by exploring Langkat and Serdang Bedagai regencies with direct observation method in September 2016. The materials used in this research are 9 accessions of asam gelugur in seven sub-districts in Deli Serdang regency, i.e. Gebang (G1, G2), Bahorok (G3, G4), Padang Tualang (G5), Pegajahan (G6), Sei Rampah G7), Pantai Cermin (G8), and Perbaungan (G9). Observations were performed on 34 morphological characters including stems, leaves, flowers, fruits and seeds. Morphological characters on stems, leaves, flowers, fruits and seeds were observed based on IPGRI (2003). Documentation is done using a digital camera (Nikon D5300). The color is measured according to the Royal Horticultural Society (5th Edition) standard color chart. Data is processed and analyzed using hierarchical cluster analysis method with Euclidean distance concept. Parameters observed include plant height, bar length, stem circle, stem surface appearance, crown shape, young leaf color, old leaf color, leaf size, leaf shape, leaf base shape, leaf tip shape, leaf margin, leaf bone, flower size, flower position, flower petal color, flower crown color, fruit shape, number of fruit segment, fruit diameter, fruit weight, fruit size, fruit skin thickness, ripe fruit color, seed length, seed width, number of seeds per fruit, seed shape and seed color.

## 3. Result and Discussions

The analysis results from the part of Garcinia atroviridis plant covering the tee height, trunk height, trunk circumference, trunk surface and crown shape are made in a dendogram (Figure 1). From the analysis of the trunk, nine accessions of G. atroviridis in Langkat and Serdang Berdagai district can be divided into three groups at eucledien distance scale 16. The first group is G1, G2, G3, G4, G5, G8 and G9. The second group is G6. The third group is G7.

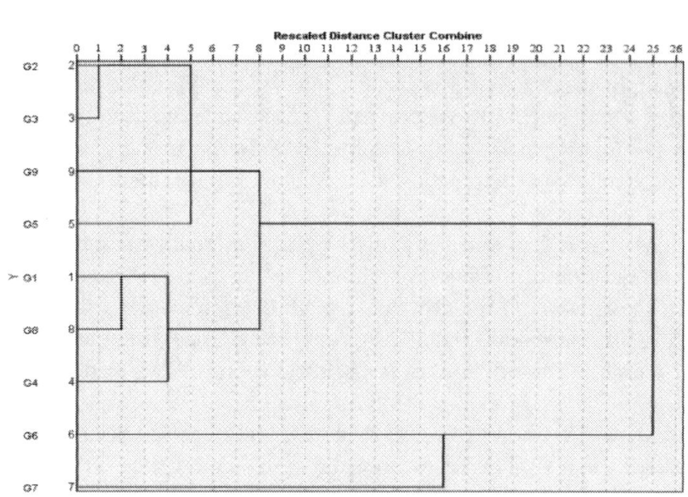


**Figure 1.** Results of trunk analysis

**Table 1**. Trunk characteristics of Garcinia atroviridis

| Accession Code | Characteristic | | | | |
|---|---|---|---|---|---|
| | Tree Height | Trunk Height | Trunk Circumference | Trunk Surface | Crown Shape |
| G1 | 11.5 | 2.7 | 105 | Rough | Pyramid |
| G2 | 11.5 | 2.4 | 115 | Rough | Spherical long |
| G3 | 12.5 | 3.5 | 110 | Rough | Spherical long |
| G4 | 7.5 | 2.1 | 90 | Rough | Pyramid |
| G5 | 9.5 | 2.1 | 85 | Rough | Spherical long |
| G6 | 27.5 | 12.0 | 330 | Very Rough | Spherical long |
| G7 | 21.5 | 3.7 | 310 | Very Rough | Pyramid |
| G8 | 12.5 | 2.5 | 97 | Rough | Pyramid |
| G9 | 15.5 | 6.3 | 94 | Rough | Spherical long |

Based on tree height. 9 observed accessions were grouped into 3 groups. i.e. short (5-10 m). moderate (11-16 m) and high (> 16 m). G4 and G5 include the short type. G1. G2. G3. G8 and G9 are of moderate type. while the high types are G6 and G7.(Tabel 1)

From 9 accessions of Garcinia atroviridis observed. G4 and G5 is the accession that has the shortest trunk height with 2.1 meters while the longest trunk height is G6. In the trunk circumference parameter. it is known that the smallest trunk is in G5. which is 85 cm. while the largest trunk is in G6 of 330 cm.

The analysis result from the leaf of Garcinia atroviridis included young leaf color. mature leaf color. leaf size. leaf shape. leaf base shape. leaf apex shape. leaf blade margin and leaf venation are made in a dendogram (Figure 2). From the results of leaf analysis. nine accessions of Garcinia atroviridis in Langkat and Serdang Berdagai district can be divided into three groups at eucledien distance scale 17. The first group is G3. G6 and G7. The second group is G8 and G9 while the third group consists of G1. G2. G4 and G5.

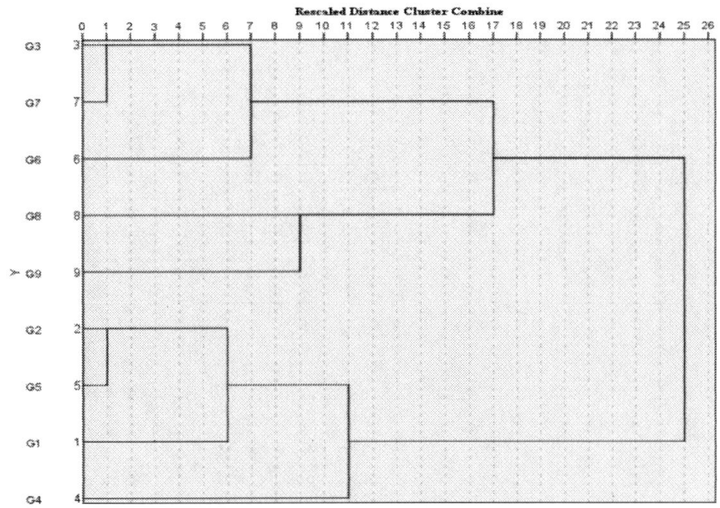


**Figure 2**. Result of leaf analysis

**Table 2**. Leaf characteristics of Garcinia atroviridis

| Accession Code | Characteristic | | | | | | | | |
|---|---|---|---|---|---|---|---|---|---|
| | Young Leaf Color | Mature Leaf Color | Leaf Length (cm) | Leaf Width (cm) | Leaf Shape | Leaf Base | Leaf apex | Leaf margin | Leaf venation |
| G1 | Light green | Dark green | 17.5 | 5.6 | Lanceolatus | Oblique | Acute | Entire | Prominent |
| G2 | Light green | Dark green | 19.5 | 6.2 | Oblong | Oblique | Acute | Entire | Prominent |
| G3 | Light green | Dark green | 23.1 | 8.4 | Oblong | Rounded | Obtuse | Entire | Prominent |
| G4 | Light green | Dark green | 22.5 | 7.2 | Oblong | Oblique | Acute | Entire | Prominent |
| G5 | Light green | Dark green | 20.2 | 6.8 | Oblong | Oblique | Acute | Entire | Prominent |
| G6 | Light green | Dark green | 20.5 | 8.2 | Oblong | Rounded | Obtuse | Entire | Prominent |
| G7 | Light green | Dark green | 21.3 | 8.5 | Oblong | Rounded | Obtuse | Entire | Prominent |
| G8 | Light green | Dark green | 21.6 | 8.4 | Elliptic | Oblique | Obtuse | Entire | Prominent |
| G9 | Light green | Dark green | 19.2 | 7.1 | Elliptic | Oblique | Obtuse | Entire | Prominent |

Based on the data obtained. it is known that from the nine accessions observed; the longest leaf length parameter found in accession G3 with 23.1 cm and the shortest leaf length is at accession G9. As for the largest leaf width parameter found in accession G7 with 8.5 cm and the smallest leaf width is in accession G1 with 5.6 cm.( Tabel 2)

The cluster analysis result for the flower size was made into dendogram (Figure 3). The size of the flowers can be divided into three groups at eucledien distance scale 9. The first (large) groups are G2. G4. G6. G8 and G9. The second (medium) group is G1. G5 and G7. The third (small) group is G3.

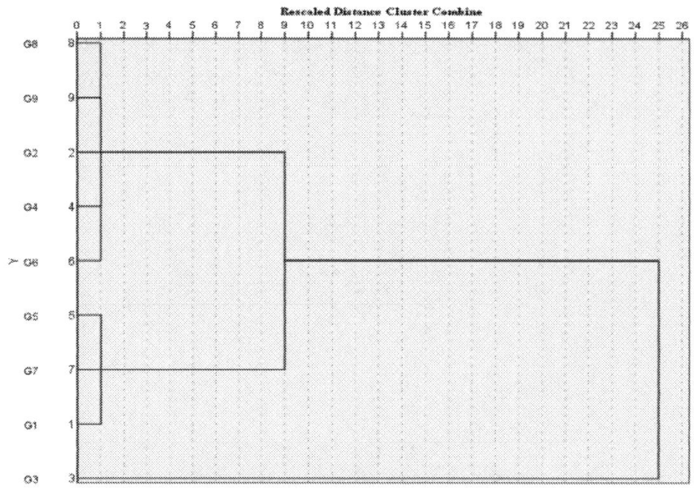


**Figure 3**. Flower size

**Table 3**. Flower characteristics of Garcinia atroviridis

| Accession Code | Characteristic | | | |
|---|---|---|---|---|
| | Flower Size | Position of Flower | Sepal Color | Petal Color |
| G1 | Medium | Terminal | Yellow with red margin | Red |
| G2 | Large | Terminal | Yellow with red margin | Red |
| G3 | Small | Terminal | Yellow with red margin | Red |
| G4 | Large | Terminal | Yellow with red margin | Red |
| G5 | Medium | Terminal | Yellow with red margin | Red |
| G6 | Large | Terminal | Yellow with red margin | Red |
| G7 | Medium | Terminal | Yellow with red margin | Red |
| G8 | Large | Terminal | Yellow with red margin | Red |
| G9 | Large | Terminal | Yellow with red margin | Red |

Primary parts or construction of flowers in this accession are very similar. They are composed of four sepal and four petals. Petal and color of this accession red and yellow with red margin. (Tabel 3) The analysis results of fruit from nine accessions of G. atroviridis in Langkat and Serdang Berdagai districts can be grouped into three groups at eucledien distance scale 24 (Figure 4). The first group is G1. G3. G5. G7. G8 and G9. The second group is G4 and G6. The third group consists of only one accession that is G2.

Observation on the morphological character of number of fruit segments on the G. atroviridis is discovered that the highest number of fruit segments are in G6 accession which is 15 segments. meanwhile the smallest number of fruit segments is in the G2 with 9 segments. For the parameter of fruit diameter the largest one found in G4 accession that is equal to 102.20 mm while the smallest is in G5 accession that is 75.30 mm. In the parameter of fruit weight the largest one is found in G4 accession with 420 g in accordance with the wide diameter. while the smallest fruit weight is in G7 accession with 200 g.(Tabel 4).

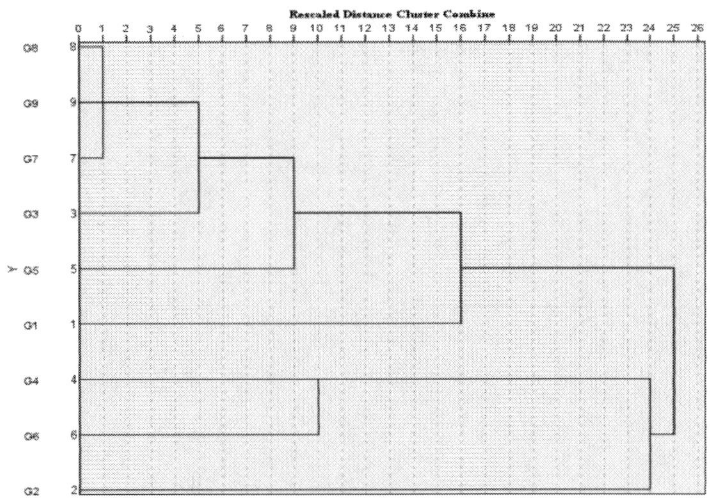


**Figure 4**. Result of fruits analysis

**Table 4.** Fruit characteristics of Garcinia atroviridis

| Accession Code | Characteristic | | | | | | |
|---|---|---|---|---|---|---|---|
| | Shape | Numberof Fruit Segmen | Fruit Diameter (mm) | Fruit Weight (g) | Fruit Size | Fruit Skin Thickness | Mature Fruit Color |
| G1 | Flattened | 16 | 93.30 | 370 | Big (>140g/fruit) | Thin | Bright yellow |
| G2 | Round | 9 | 83.25 | 320 | Big (>140g/fruit) | Thin | Bright yellow |
| G3 | Flattened | 14 | 85.20 | 280 | Big (>140g/fruit) | Thin | Bright yellow |
| G4 | Round | 12 | 102.20 | 420 | Big (>140g/fruit) | Thin | Bright yellow |
| G5 | Ovoid | 13 | 75.30 | 210 | Big (>140g/fruit) | Thin | Bright yellow |
| G6 | Round | 15 | 90.25 | 350 | Big (>140g/fruit) | Thin | Bright yellow |
| G7 | Flattened | 14 | 86.30 | 200 | Big (>140g/fruit) | Thin | Bright yellow |
| G8 | Flattened | 14 | 90.15 | 280 | Big (>140g/fruit) | Thin | Bright yellow |
| G9 | Flattened | 11 | 87.25 | 260 | Big (>140g/fruit) | Thin | Bright yellow |

**Table 5**. Seed characteristics of Garcinia atroviridis

| Accession Code | Characteristic | | | | |
|---|---|---|---|---|---|
| | Seed Length | Seed Width | Number of Mature Seed / Fruit | Seed Form | Seed Color |
| G1 | 10.5 | 8.4 | 5 | Reniform | Light brown |
| G2 | 11.0 | 7.5 | 1 | Reniform | Light brown |
| G3 | - | - | - | - | - |
| G4 | - | - | - | - | - |
| G5 | - | - | - | - | - |
| G6 | - | - | - | - | - |
| G7 | - | - | - | - | - |
| G8 | - | - | - | - | - |
| G9 | 11.5 | 8.5 | 5 | Reniform | Light brown |

The seed analysis results of Garcinia atroviridis (Table 5) showed that from nine accessions identified only three accessions that found have seeds. This is because the fruits of G. atroviridis generally only have a few seeds and some do not have seeds.

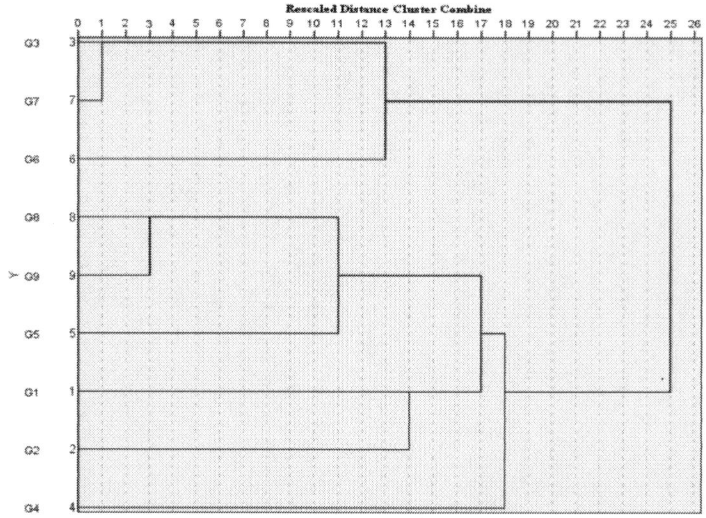


**Figure 5**. The result of all morphological characters analysis

*3.1. Phylogenetic Relationship of Garcinia atroviridis in Langkat and Serdang Bedagai*

The kinship relation analysis is done using hierarchical cluster analysis technique. which is grouping the objects based on the similarity of characteristics among the objects. Those objects are classified into one or more clusters (groups) so that objects that are in one cluster will have a resemblance to one another [11]. The cluster analysis technique was applied to the observation data of 27 morphological characters from 9 accessions of Garcinia atroviridis. so that the relationship of nine accessions of the Garcinia atroviridis was obtained.

Based on the dendogram formed (Figure 5). it is obtained two groups. three groups. and four groups of phylogenetic relationship on the scale of the spacing (euclidean distance scale) 17. 18. and 25. The smaller the euclidean distance between the objects analyzed. the closer the phylogenetic relationship of the object and the more similar character it has [11]. An analysis of phylogenetic relationship based on morphological characters on a distance scale of 25 indicates the presence of two plant kinship groups (Figure 5). The first group consists of three accessions of Garcinia atroviridis. namely G3. G6 and G7. The second group consists of 6 accessions. namely G1. G2. G4. G5. G8 and G9.

The analysis of phylogenetic relationship based on morphological characters of eucledien distance scale 18 showed there are three groups (Figure 5). The first group consists of three accessions of Garcinia atroviridis. namely G3. G6 and G7. The second group consists of 5 accessions. namely G1. G2. G5. G8 and G9. The third group consists of only one accession. G4.

The analysis of phylogenetic relationship based on morphological characters of eucledien distance scale 17 showed there are four groups (Figure 5). The first group consists of 3 accessions of Garcinia atroviridis that is G3. G6 and G7. The second group consists of 3 accessions: G5. G8 and G9. The third group consists of 2 accessions. namely G1 and G2. The fourth group consisted of only one accession. G4.

The lowest dissimilarity or closest relationship were obtained in G3 from Langkat and G7 from Serdang Bedagai district with 16.328 by some character difference from 27 characters observed. This is in accordance with the literature Suskendriyati et al [7] which states that the emergence of some differences and similarities within the species outer morphology of a plant can be used to determine the proximity of relationship.

The highest dissimilarity or the farthest relationship is obtained in G5 from Langkat and G6 from Serdang Bedagai regencies which is 54.186 with some character difference from 27 characters observed that is form of canopy. leaf shape. leaf base shape. leaf tip shape. size of flowers and fruit shape. This generally indicates that the morphological similarity level of the 9 identified accessions of Garcinia atroviridis is relatively close. This is in accordance with the literature of Cahyarini et al (2004) which states that the distance of similarity is said to be far if less than 0.6 or 60%. So from the grouping it can be said that the nine accessions observed have very little relationship.

## 4. Conclusion

Based on cluster analysis on morphological characters of nine accessions of *Garcinia atroviridis* in Langkat and Serdang Bedagai regencies. there are four groups of phylogenetic relationship on the euclidean distance scale 17. The closest relationship is G3 and G7 with a coefecient dissimilarity 16.328 and farthest relationship is G5 and G6 with a coefecient dissimilarity 54.187.

## References

[1] Prasetyo A 2006 Mempelajari Proses Pembuatan dan Daya Simpan Koktail Asam Gelugur (*Garcinia atroviridis* Griff. ex T. Anders.). Institut Pertanian Bogor. Bogor.
[2] Rittirut W. Siripatana C 2007 Diffusion properties of Garcinia fruit Acids (*Garcinia atroviridis*). Walailak J Sci and Tech Vol. 4(2):187-202.
[3] Jansen P C M 1992 Plant Resourches Sout-East Asia No 2: Edible fruits and nuts. Prosea. Hal 175-177.
[4] Sari R dan Hanan A 2000 Garcinia (Clusiaceae) di Kebun Raya Bogor: Fisiognomi. Keragaman dan Potensi. Diakses dari http://elib.pdii.lipi.go.id. pada 1 Maret 2016.
[5] Mackeen M M 1998 Bioassay-guided isolation and identification of bioactive compounds from *Garcinia atroviridis* (Asam gelugor). Tesis. Faculty of Food Science and Biotechnology. University Putra Malaysia.
[6] Mansyah E dan Edison H S 2012 Pendugaan Keragaman *Garcinia* sp dan *Nephelium* sp di Provinsi Sumatera Barat dan Jambi serta Potensi Pemanfatannya dalam Pertanian Bioindustri. Balai Penelitian Tanaman Buah Tropika. Solok.
[7] Suskendriyati H. Wijayati A. Hidayah N dan Cahyuningdari D 2000 Studi Morfologi dan Hubungan Kekerabatan Varietas Salak Pondoh (Salacca zalacca (Gaert.) Voss.) di Dataran Tinggi Sleman. UNS. Surakarta.
[8] Renwain J. Hartana A. Hambali G G dan Rumawas F 1994 Ubi Jalar Tetraploid dan Prospeknya Sebagai Sumber Genetik dalam Program Pemuliaan Ubi Jalar Pentaploid. Zuriat. 5(2) : 8-15.
[9] Puslitbangbun 2007 Petunjuk Pelaksanaan Pengelolaan Plasma Nutfah Tanaman Perkebunan. Pusat Penelitian Dan Pengembangan Perkebunan. Bogor.
[10] Bermawie N 2005 Karakterisasi Plasma Nutfah Tanaman. Buku Pedoman Pengelolaan Plasma Nutfah Perkebunan. Pusat Penelitian dan Pengembangan Perkebunan. Bogor. hal. 38-52.
[11] Santoso S 2002 Buku Latihan SPSS Statistik Multivariat. PT Elex Media Komputindo. Jakarta.

# Molecular performance of commercial MTG variety oil palm based on RAPD markers

**L A P Putri[*1], I E Setyo[2], M Basyuni[3], E S Bayu[1], H Setiado[1], N F Reynaldi[1], H Laia[1], S A K Puteri[4], D Arifiyanto[5] and I Syahputra[5]**

[1] Department of Agrotechnology, Faculty of Agriculture, Universitas Sumatera Utara, Jln. Prof. A. Sofyan no.3 Medan, North Sumatera 20155
[2] Pusat Penelitian Kelapa Sawit, Jl. B. Katamso 51 Kp. Baru Medan
[3] Department of Forestry, Faculty of Forestry, Universitas Sumatera Utara, Medan 20155
[4] PT. Rajawali Nusantara Indonesia (PT. RNI), Gedung RNI, Jl. Denpasar Raya, Kav D III, Kuningan, Jakarta 12950
[5] PT. Socfin Indonesia, Jl. Kl Yos Sudarso No. 106 Medan 20115, Sumatera Utara, Indonesia
[*]E-mail: lollie.agustina@usu.ac.id

**Abstract**. The oil palm, an economically important tree in Indonesia, has been one of the world's major sources of edible oil and a significant precursor of biodiesel fuel. This research is conducted by taking individual tree sample of commercial MTG variety germplasm oil palm one years old. The purpose of this research is to analyse molecular performance of some oil palm MTG variety based on RAPD markers. In this experiment, the DNA profile diversity was assessed using markers of oil palm's random RAPD markers (OPD-20, SB-19, OPM-01 and OPO-11). A total of 15 trees commercial MTG oil palm variety were used for analysis. The results of the experiment indicated out of 4 RAPD markers (OPD-20, SB-19, OPM-01 and OPO-11) showed polymorphic of PCR product. These preliminary results demonstrated RAPD marker can be used to evaluate genetic relatedness among trees of commercial MTG variety oil palm and detecting either genetic variants or mislabelled.

## 1. Introduction

African oil palm is a prominent export commodity and contributes significantly to Indonesia's current economy and Indonesia is the world's largest producer and exporter of palm oil (*Elaeis guineensis* Jacq.). North Sumatra Province is one of the oil palm plantation centers in Indonesia, with CPO production in 2016 (crude palm oil) reaching 5,440,594 tons and is the second largest province after Riau for CPO production [1]. The high productivity is not only in the production parameters of FFB (Fresh Fruit Bunch), but also the high oil to extraction rate (OER), the quality and composition of the material contained in the oil, such as the content of unsaturated fatty acids and carotenoids.

Optimal yield production of oil palm in Malaysia and other countries of South East Asia are hampered by the presence of devastating Basal Stem Rot (BSR) disease caused by Ganoderma boninense. For more than 40 years, Ganoderma remained to be the most serious problem in many areas in Malaysia and Indonesia. But unfortunately, there is no single reliable application in suppressing or controlling either the disease severity or a promising resistant variety of oil palm to this pathogen [2]. Many researchers claim that Ganoderma is an important disease in oil palm that causes significant losses in palm oil production. This disease is able to attack at various stages of oil palm development, ranging from germination stage to adult stadia. The symptoms are slow, but plants that have been infected will be death. The attack always starts from the root zone [3]. Genetic material (germplasm) is a key element in the use of molecular diagnostics to improve seed quality and productivity of oil palm. Currently, oil palm germplasm with resilience characteristic of ganoderma is already available from Indonesian producers of seedlings (eg, MTG varieties of Socfindo and PPKS) and can be commercially obtained, but this is of moderate resistance. Therefore, this applicative research is focused on the development and production of oil palm genotype with superior resistance to stem rot disease with molecular diagnostic approach.

The use of plant material resistant / resistant to diseases such as the use of plant material varieties MTG (Moderate Tahan Ganoderma) is one of the efforts to increase the productivity of oil palm plantations. For the long term, the use of resistant plant material is a great opportunity to control stem rot disease. Breeding for resistance is an obvious approach and a long-term solution for Ganoderma disease. It can also be used for controlling disease among oil palms by developing a variety of palm trees with high resistance to the disease. The most important factor in breeding for resistance is the source of resistance (the genes),which is available in the same gene pool that contains genes of all other inherited traits [4]. Breeding for Ganoderma disease tolerance is limited due to the lack of an effective screening method [5]. The use of DNA as a genetic material has been widely applied to know the diversity of a particular character in oil palm at the genomic level;[6-8]. Thus far, resistance to Ganoderma has not been clearly found yet in any oil palm germplasm. Furthermore, even when tolerance is present it may represent a relative level of tolerance, in the sense that the all palms get the disease, but the tolerant genotypes last longer.

## 2. Materials and methods

### *2.1. Plant material and DNA isolation*

The fresh young leaves plant material used is palm oil from commercial MTG (Moderate Tahan Ganoderma) derived from PT. Socfindo. The extraction and isolation of genomic DNA was performed by genomic DNA isolation procedures adapted from the CTAB method by Oroczo[9] with some modifications to the concentrations of polyvinylpolypirillidone (PVPP) and 2- mercaptoethanol [10-11]. The quantity of each isolated DNA was measured with a nanodrop spectrophotometer while its quality was electrophoresed on 0.8% electrophoresis gel. The material genetics (genomic DNA) were stored at -20 °C.

### *2.2. DNA amplification*

For DNA amplification, the method is based on [11]. Furthermore, PCR was performed on a total volume of 25 µl, 2 µl of DNA extract added to 12.5 µl reaction mix (Go Green Taq Promega), 9.5 µl nuclease free water and 1 µl random primer. Amplification using 6 primers ie OPD-20, SB-19, OPM-01, and OPO-11. PCR reaction using AB Biosystem thermocycler is programmed as follows: after 2 minutes of heating at 94 $^{0}$C, DNA amplification was performed at 45 cycles from 1 minute denaturation at 94 0C, 1 minute at 36-37 $^{0}$C, and 2 minutes extension at 72 $^{0}$C. Forty-five cycles are terminated after 4 minutes of extension at 72 $^{0}$C and cooled to 4 $^{0}$C. The amplified products were mixed with 4 µl of 6X loading dye and resolved in 2 % agarose gel containing ethidium bromide in a horizontal electrophoresis tank. The amplified RAPD patterns were analysed with the Gel Doc *UVITEC Cambridge* ( USA) and alleles were identified according to their base pair size. Molecular sizes of amplified products were estimated using a 1kb DNA ladder marker (Promega).

Each RAPD band is considered a single locus, only locus showing clearly defined used for scoring: there are (1) and empty (0). The distance matrix or genetic inequality for all combinations of individual pairs can be performed with two types of descriptive analysis of diversity: (1) Principal Coordinates Analysis (PCoA), a factorial analysis type in the inequality table for the main group origin and (ii) Neighbor-Joining Tree (NJtree) to obtain a picture of kinship among individuals. This calculation and descriptive analysis uses DARwin 6.05 software [12].

## 3. Results and Discussions

The use of molecular markers, revealing polymorphism at the DNA level, has been playing an increasing part in plant biotechnology and their genetic studies. In this work, the utility of RAPD markers for variation and identification of oil palm germplasm was studied.

### *3.1. Polymorphism detected using RAPD markers*

Out of 4 primers tested, all primers were chosen for further studies based on clarity of the banding patterns (table 1). The number of bands generated among the 15 oil palm MTG accessions using these 4 primers was 20, which ranged from 2 (OPO-11) to 7 (SB-19), with an average of 5 bands/primer. Out of these fragments scored, 14 (70 %) were polymorphic. A total of 20 amplicons (amplified fragment) were generated by four RAPD primers in which all of the primers were polymorphic, all as shown in Table 1.These preliminary results demonstrated RAPD markers can be used to evaluate genetic relatedness among trees of MTG variety oil palm genotype.

The results of the present study, using 15 MTG oil palm accessions and 4 RAPD primers, indicated that 70 % of the scored fragments were polymorphic, which was relatively lower compared to earlier studies utilizing dominant marker systems like RAPD [13-15]. The number of amplicon per locus is affected by the number of markers and sample size analyzed

**Table 1.** Sequence of RAPD primers, the number of scorable polymorphic bands and percentage of polymorphism of each primer

| Primer Name | Primer sequence | No. of amplified bands | No. of polymorphic bands | Polymorphism % |
|---|---|---|---|---|
| OPD-20 | ACCCGGTCAC | 5 | 4 | 80 |
| SB-19 | CAGCACCCAC | 7 | 4 | 57.14 |
| OPM-01 | GTTGGTGGCT | 6 | 5 | 83.33 |
| OPO-11 | GACAGGAGGT | 2 | 1 | 50 |
| | Average | 5 | 3.5 | 70 |
| | Total | 20 | 14 | |

### *3.2. Genetic diversity analysis among the MTG oil palm accessions*

The scored bands were used to calculate the genetic diversity among the 15 MTG oil palm accessions.The genetic similarity coefficient between the pair samples was evaluated by calculating the Jaccard's dissimilarity coefficient based on the proportion of shared bands. The pairwise dissimilarity coefficient was lowest (0.00) between MTG-12 and MTG-11; MTG-10 and MTG-5; MTG- 8 and MTG-17 (with yellow colour). The maximum genetic dissimilarity (1.00) was observed between MTG-5 and MTG-18; MTG-5 and MTG-6; MTG-6 and MTG-7; MTG-6 and MTG-10; MTG-10 and MTG-18 (with green colour) (table 2). Pair wise dissimilarity coefficient's ranged from 0.01 to 1.00 indicating the capability of RAPD markers to detect high levels of genetic diversity among the MTG oil palm accessions analyzed.

A dendrogram constructed using corresponding genetic dissimilarity coefficients obtained from Neighbor Joinining UPGMA analysis was used to determine the clustering pattern among the MTG oil palm accessions analysis (figure 1). Three main clusters (designated 'A' 'B' and 'C') were formed diverging at the dissimilarity coefficient of 0.28. The cluster ('A') was a large one comprising 9 accessions. Within this cluster (A), there were two sub-clusters (designated 'a1' and 'a2'). Sub-cluster 'a1' had a single cluster—comprising one accessions. The second cluster within sub-cluster 'a2' had 8 accessions. The second main cluster ('B'), had four MTG oil palm accessions. And the third main

IOP Conf. Series: Earth and Environmental Science **122** (2018) 012078 doi:10.1088/1755-1315/122/1/012078

cluster ('C') had two MTG oil palm accessions. The clades were supported by reliable bootstrap values.

**Table 2.** Dissimilarity matrix among 15 MTG oil palm accessions based on Jaccard's coefficient using RAPD markers

| | MTG 4 | MTG 5 | MTG 6 | MTG 7 | MTG 8 | MTG 10 | MTG 11 | MTG 12 | MTG 13 | MTG 14 | MTG 15 | MTG 16 | MTG 17 | MTG 18 |
|---|---|---|---|---|---|---|---|---|---|---|---|---|---|---|
| MTG5 | 0.63 | | | | | | | | | | | | | |
| MTG6 | 0.61 | 1.00 | | | | | | | | | | | | |
| MTG7 | 0.63 | 0.67 | 1.00 | | | | | | | | | | | |
| MTG8 | 0.35 | 0.85 | 0.47 | 0.65 | | | | | | | | | | |
| MTG10 | 0.63 | 0.00 | 1.00 | 0.67 | 0.85 | | | | | | | | | |
| MTG11 | 0.20 | 0.70 | 0.55 | 0.70 | 0.15 | 0.70 | | | | | | | | |
| MTG12 | 0.20 | 0.70 | 0.55 | 0.70 | 0.15 | 0.70 | 0.00 | | | | | | | |
| MTG13 | 0.45 | 0.60 | 0.59 | 0.83 | 0.40 | 0.60 | 0.25 | 0.25 | | | | | | |
| MTG14 | 0.55 | 0.54 | 0.71 | 0.81 | 0.50 | 0.54 | 0.35 | 0.35 | 0.13 | | | | | |
| MTG15 | 0.22 | 0.63 | 0.75 | 0.63 | 0.35 | 0.63 | 0.20 | 0.20 | 0.45 | 0.39 | | | | |
| MTG16 | 0.28 | 0.83 | 0.40 | 0.60 | 0.12 | 0.83 | 0.25 | 0.25 | 0.50 | 0.60 | 0.45 | | | |
| MTG17 | 0.35 | 0.85 | 0.47 | 0.65 | 0.00 | 0.85 | 0.15 | 0.15 | 0.40 | 0.50 | 0.35 | 0.12 | | |
| MTG18 | 0.85 | 1.00 | 0.86 | 0.70 | 0.59 | 1.00 | 0.65 | 0.65 | 0.78 | 0.75 | 0.72 | 0.71 | 0.59 | |
| MTG19 | 0.42 | 0.75 | 0.65 | 0.75 | 0.37 | 0.75 | 0.30 | 0.30 | 0.39 | 0.50 | 0.42 | 0.47 | 0.37 | 0.69 |

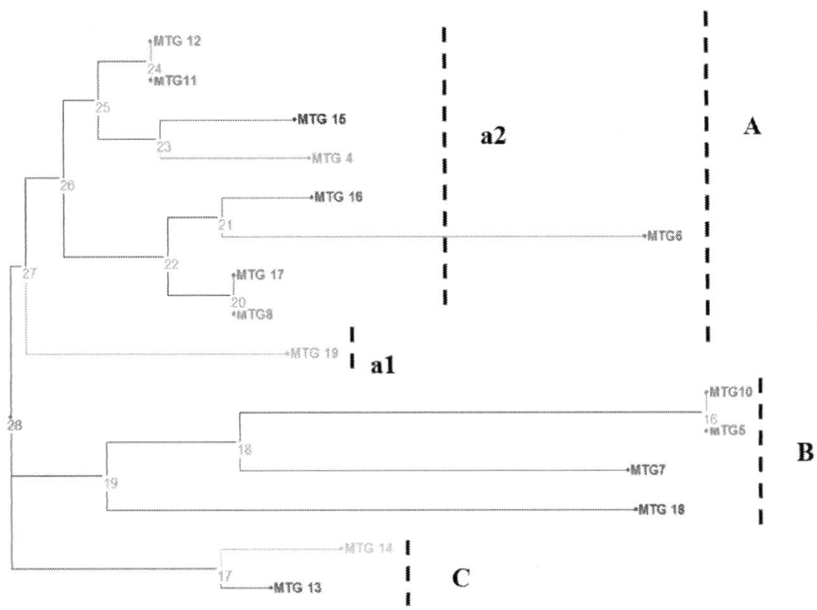


**Figure** 1. Dendrogram generated from genetic dissimilarity, using Jaccard's coefficient and UPGMA clustering method of 15 MTG oil palm accessions based on RAPD analysis

When compared to the previous report of Putri[16], we get the three main clustering of 10 MTG variety individual with 2 SSR markers. Ten sampels in this report is the same as previous research[16]. From that, we preliminary indicated there are three profiles of this commercial MTG variety germplasm collection derived from different crossing or different to MTG resistance characters.

## 4. Conclusions

This preliminary study indicates that RAPD markers are informative, could be used to detect polymorphism among MTG commercial oil palm germplasm accessions and were able to classify the genetic background of MTG variety oil palm and represent an essential tool for genetic analysis of this main oil palm germplasm. Based on molecular results, some of 15 individual MTG oil palm type have different crossing or derived from different parent (different MTG resistance trait) and detecting either genetic variants.

## References

[1] Dirjenbun 2017 Buku Statistik Perkebunan Tahun 2013-2017

[2] Chong K P, Markus A and Rossall S 2012 The Susceptibility of different varieties of oil palm seedlings to Ganoderma Boninense infection *Pak. J. Bot* **44**(6): 2001-2004.

[3] Naher, L, Yusuf U K, Ismail A, Soon Guan Tan and Mondal M M A 2013 Ecological status of Ganoderma and basal stem rot disease of oil palms(*Elaeis guineensis*. Jacq.) *Australian J Crop Sci* 7(11):1723-1727

[4] Hama-Ali E O, Panandam J M, Tan S G, Syed Alwee S S R, Sheong T J, Ling H C, Namasivayam P and Peng H B 2015 Association between basal stem rot disease and simple sequence repeat markers in oil palm, *Elaeis guineensis* Jacq *Euphytica* **202**:199–206 doi 10.1007/s10681-014-1270-6

[5] Rivas M, Barbieri R L and Maia L Cd 2012 Plant breeding and in situ utilization of palm trees *Cie ˆncia Rural* **42**(2):261–269

[6] Hetharie H, Wattimena G A, Thenawidjaya M S, Aswidinnoor H, Toruan-Mathius N and Ginting G 2007 Karakterisasi morfologi bunga dan buah abnormal kelapa sawit (*Elaeis guineensis* Jacq) hasil kultur jaringan *Bul. Agron.* **35**(1):50-57

[7] Budiani A 2010 Analysis of DNA sequences of the 5'-flsnking EGAD1 from normal and abnormal fruit from tissue culture derived oil palm *Menara Perkebunan* **78**(2):43-51

[8] Jaligot E, Rival A, Beule´ T, Dussert S, Verdeil J L 2000 Somaclonal Variation in Oil Palm (*Elaeis guineensis* Jacq.): the DNA Methylation Hypothesis *Plant Cell Report* **19**:684-690

[9] Oroczo-Castillo C, Chalmers K J, Waugh R, Powell W 1994 Detection of genetic diversity and selective gene introgression in coffee using RAPD markers *Theor. Appl. Genet.* **87**: 934-940

[10] Toruan-Mathius N dan Hutabarat T 1997 Pemanfaatan teknik penanda molekuler dalam usaha meningkatkan produktivitas tanaman perkebunan *Warta Puslit Biotek Perkebunan II* (**1**):2-9

[11] Asmono D, Toruan-Mathius N and Subronto 2000 Pemetaan Genom Pengendali Produktivitas Minyak pada Kelapa Sawit Laporan Riset Unggulan Terpadu (RUT) VII *Lembaga Ilmu Pengetahuan Indonesia*

[12] Perrier X and Jacquemoud-Colled J P 2016 DARwin Software. http://darwin.cirad.fr/darwin

[13] Upadhyay A, Jayadev K, Manimekalai R and Parthasarathy V A 2004 Genetic relationship and diversity in Indian coconut accessions based on RAPD markers *Sci Hort* **99**:353–362

[14] Manimekalai R and Nagarajan P 2006 Interrelationships among coconut (*Cocos nucifera* L.) accessions using RAPD technique. *Genet Resour Crop Evol* **53**:1137–1144

[15] A Manimekalai R and N agarajan P 2006 Assessing genetic relationships among coconut (*Cocos nucifera* L.) accessions using inter simple sequence repeat markers. *Sci Hort* **108**:49–

[16] Putri L A P, Setiado H and Hardianti R 2017 DNA Profiles of MTG (Moderat Tahan Gano) Oil Palm Variety Based on SSR Markers *IOP Conf. Series: Materials Science and Engineering* 180:0120144

## Acknowledgement

The authors gratefully acknowledge that the present research is supported by University of Sumatera Utara, Ministry of Research and Technology and Higher Education Republic of Indonesia. The support is under the research grant TALENTA USU of year 2017 Contract Number 5338/UN5.1.R/PPM/2017.

# Microwave-assisted extraction of pectin from cocoa peel

**M Sarah*, F Hanum, M Rizkyand M F Hisham**

Departement of Chemical Engineering, Universitas Sumatera Utara, Padang Bulan, Medan 20155, North Sumatera, Indonesia
*Corresponding Author Email: mayasharid@yahoo.com

**Abstract:** Pectin is a polymer of d-galacturonate acids linked by β-1,4 glycosidic bond. This study isolates pectin from cocoa peel (*Theobroma cacao*) using citric acid as solvent by microwave-assisted extraction method. Cocoa peels (moisture content of 10%) with citric acid solution (pH of 1.5) irradiated by microwave energy at various microwave power (180, 300, 450 and 600 W) for 10, 15, 20, 25 and 30 minutes respectively. Pectin obtained from this study was collected and filtrated by adding 96% ethanol to precipitate the pectin. The best results obtained from extraction process using microwave power of 180 Watt for 30 minutes. This combination of power and time yielded 42.3% pectin with moisture content, ash content, weight equivalent, methoxyl content and galacturonate levels were 8.08%, 5%, 833.33 mg, 6.51% and 58,08%, respectively. The result finding suggested that microwave-assisted extraction method has a great potency on the commercial pectin production.

## 1. Introduction

Cocoa plant (*Theobroma cacao*), native plant in Indonesia, contributes to economic growth from industrial sector. Chocolate industry utilizes cocoa bean to produce chocolate and disposes cocoa peel as solid waste. Increasing the demand of cocoa bean lead to accumulation of cocoa peel in the environment and disposal of cocoa bean become serious problem. Previous study found that cocoa peel is a source of pectin, so that the peel has a benefit to for pectin production. Pectin may be found as a fiber component in the lamella layer and primary cell wall of cocoa plant [1]. Pectin is popular polysaccharide used as gelling, stabilizing and thickening agent in food industry [2].It consists of d-galacturonic acid backbone linked by α-(1.4) glycosidic bond. Some carboxyl groups of pectin polymer esterified into methoxyl groups [3].

Pectin isolation commonly carries out by extraction of strong or weak acids [4]. Extraction may proceeds by various methods such as soxhlet extraction, maceration and stirring extraction. Recently, Wang S, et al. (2007) and Shkodina, O.G, et al. (1998) reported microwave-assisted extraction [5] and enzymatic method [6] to extract pectin from plants. Acid method carries out at high temperature using sulfuric, phosphoric, nitric, hydrochloric and citric acids [7]. In addition, microwave-assisted extraction (MAE), a technique to extract dissolved materials in plant by microwave energy, is introduced to improve the extraction method. The technology is suitable for the removal of thermolable dielectric materials such as cocoa. This method proved effectively reduces extraction timeless energy and solvent consumption, higher yields, higher accuracy and precision [8]. Lefsih, K et al.(2017) and Swamy, G.J and Muthukumarappan, K (2017) reported laboratory scale application of microwave-assisted extraction to extract pectin from *Opuntia ficus indica* [3] and banana peels [5].

Instead of extraction at high temperature, others reported utilization of microwave energy for heating and sterilization process [9, 10, 11]. Sarah et al. (2014) evaluated heating performance method for conventional heating and microwave heating using palm fruit sample [11]. Microwave heating proved more effective as compare to conventional heating, because microwave energy generates heat instantly from water molecule re-orientation inside plant during heating process [12]. Microwave energy creates instant and rapid heating to facilitate temperature required for extraction process.

In this study, microwave-assisted extraction method used to extract pectin from cocoa peel. Qiu et al. (2010) reported parameters associated with yield and pectin quality during extraction process such as pH, temperature, and extraction time, agitation and solid to liquid ratio [7]. In the case extraction process by microwave irradiation, level of microwave power during extraction process also taken into consideration instead of previous parameters

## 2. Materials and Methods

Material used in this study was *Theobroma cacao* from Pangkalan Brandan and Galang, Sumatera Utara. The chemicals used were citric acid, ethanol, silver nitrate solution, distill water, sodium chloride, sodium hydroxide and phenolphthalein. Experimental rig to extract pectin from cocoa peel is an erlenmeyer in a domestic microwave oven Samsung model ME731K with frequency of 2450 MHz and 800 Watt.

### *2.1. Pretreatment*

Prior to the experiment, the cocoa fruit was cleaned upon arrival from plantation by putting the fruits inside basin filled full with lime ash solution to remove the gums. The cocoa fruit was cut into cocoa slices with dimension of 2x2x0.5 cm and sun dried for approximately 8 hours, and another 4 hours using conventional oven at temperature of 40°C. Then, the dried cocoa slices was blended (Miyako model BL-151PFAP No 34060052217) to reduce cocoa size into smaller particles size (60 mesh).

### *2.2. Extraction process*

Cocoa peels powder (moisture content of 10%) diluted with citric acid solution (pH 1.5). The powder was placed inside a microwave cavity and irradiated by microwave energy at various microwave power (180, 300, 450 and 600 W) for 10, 15, 20, 25 and 30 minutes respectively. The extract was then filtered by using filter paper (Whatman no 2) to obtain the filtrate. Furthermore, the filtrate was stirred and heated at temperature of 95°C until the volume was reduced up to 50% of filtrate. The filtrate was cooled and pectin precipitation was carried out by adding acidified ethanol. The precipitated pectin was washed using ethanol to obtain pectin and remove chloride, and dried in an oven at temperature of 40°C for approximately 8 hours. The dried pectin was ready for further quality assessment such as moisture and ash content [13], equivalent weight [14], methoxyl content and galacturonate content [15].

## 3. Results and Discussion

The performance of microwave-assisted extraction method for pectin production was indicated by yield and quality of pectin in accordance to International Pectin Producers Association (IPPA) Quality Standard [16] at various extraction time and microwave power.

### *3.1. Effect of extraction time and microwave power on yield of pectin*

Figure 1 shows relationship between extraction time and yield of pectin. In general, prolonged extraction time, associated with longer heating period, and increment of microwave power increased yield of pectin. Prior to the extraction by microwave irradiation, the cocoa suspension was heated to maximum temperature before attaining constant temperature. Prolonged heating in this case occurred at constant temperature. During heating period,extraction of pectin from cocoa peel occurred due to the increase of temperature. The presence of acid facilitated the release of pectin from tissue cell wall of cocoa powder. Similar phenomena reported by Nurdjanah and Usmiati for pectin extraction from pumpkin [...]. Elevating extraction temperature increased kinetic energy of sample solution and facilitated solvent diffusion into the tissue cells of pumkin. In other words, elongated heating period would attain high temperature, the greater time and temperature extraction process, so that greater pectin was produced [17].

Figure 1 shows increment of microwave power from 180 to 450 Watt increase the yield of pectin. Additional microwave power will reduce the yield of pectin due to excessive time and temperature of extraction process destroyed the pectin [17]. The best extraction time to obtain 42.3% yield of pectin was 30 min at microwave power of 300 Watt . Increment of microwave power indicates more heat delivered into cocoa suspension, thus causing more pectin substances to be extracted.

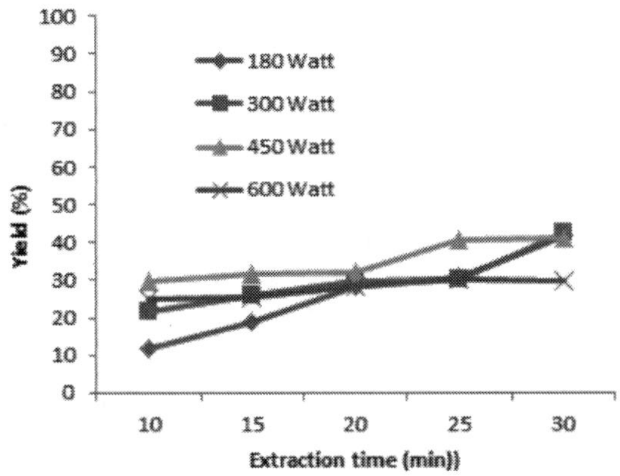


**Figure 1.** Effect of extraction time to yield of pectin at various microwave power

*3.2 Effect of extraction time and microwave power on quality of pectin*

Figure 2, Figure 3 and Figure 4 show relationship between extraction time and microwave power on the quality of pectin. Figure 2 illustrate that moisture content reduced with increment of microwave power and extraction time. Prolonged extraction process from 10 to 30 min at constant microwave power along with the increment of microwave power from 180 to 600 Watt reduced the moisture content. Moisture affects shelf life of pectin, in which higher water content causes susceptibility to microbial activity and reduces the quality of pectin. Moisture content in pectin observed ranged between 8.08 to 11.88% which was still below the allowed maximum moisture on IPPA standard [16].

Ash content represents mineral content of pectin, such as potassium, sodium, magnesium and iron. Ash content indicates pectin purity level, in which high pectin purity indicated by low ash content. The level of ash in pectin flour is influenced by the presence of inorganic material residues contained in cocoa peel, extraction methods and pectin insulation [19]. Figure 2b indicates that ash content increased with the extended of extraction period. Higher ash content was obtained at increment of microwave power from 180 to 450 Watt. Elevated power which also illustrated higher temperature and extraction period resulted in higher pectin ash content. The prolonged extraction time promoted hydrolysis of protopectin producing more calcium and magnesium contents of pectin [20]. In general, ash content in this study ranged from 1-5% or below IPPA standard. Maximum allowance ash content in pectin according to IPPA standard is 10% [16].

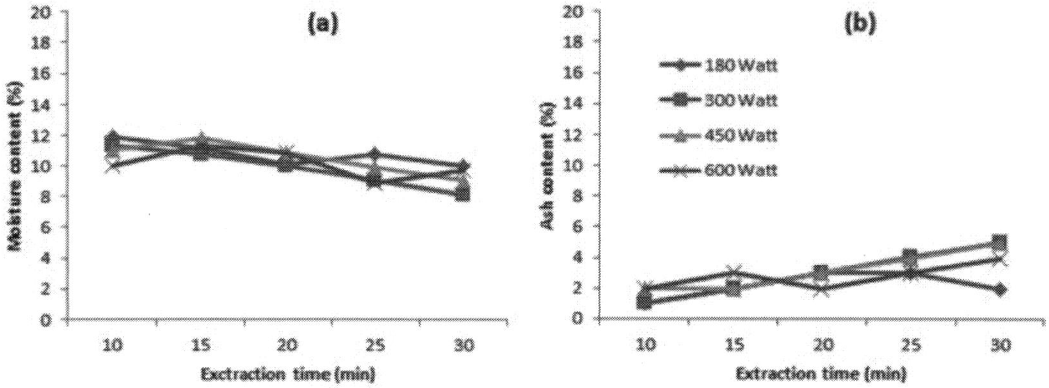


**Figure 2.** Effect of extraction time to moisture (a) and ash content(b) at various microwave power

Figure 3 shows relationship between equivalent weight and extraction time. Pectin equivalent weight represents anhydrouronic acid content and degree of esterification. This study observed extraction of pectin from cocoa peel assisted by microwave irradiation should occurred less than 21 min. The increase of extraction time yielded more pectin equivalent weight, indicating that greater pectin released from cocoa peel was resulted at the first 20 min of extraction time. However, lengthened extraction duration from 20 to 30 min reduced the pectin equivalent weight. This study suggests that pectin extraction period more than 20 min responsible for pectin degradation or pectin de-polymerization into pectic acid which was denoted by greater un-esterified galacturonate acid. In this microwave treatment, increment of microwave power from 180 to 200 Watt increased equivalent weight. However, increasing microwave power greater than 200 Watt resulted in lower equivalent weight. The equivalent weight of pectin ranged between 800 to 3000 mg, above IPPA standard (600-800 mg) [16]. Plant characteristic, raw material quality and extraction method are suggested for high pectin equivalent weight in this study as compared to IPPA standard. The results suggested that extraction should carried out at 300 Watt for 30 min to obtain pectin with equivalent weight meet the IPPA standard. Pectin equivalent weight resulted from this process combination was 833.33 mg.

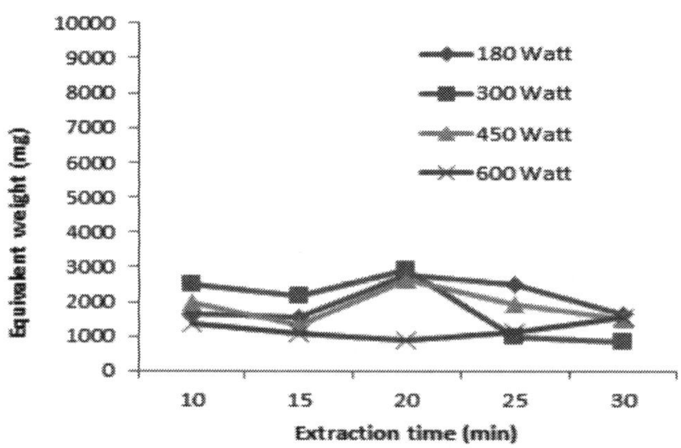


**Figure 3.** Effect of extraction time to equivalent weight of pectin at various microwave power

Figure 4.a shows relationship between methoxyl number and extraction time. Methoxyl number represents free esterified carboxyl group in the pectin and influences the gel formation. High Methoxyl (HM) pectin can form gels in the presence of sugars and acids in certain comparisons, whereas Low Methoxyl (LM) pectin can form gel in the absence of sugars [21]. Methoxyl number increased at longer extraction time at constant microwave power due to hydrolysis of methyl ester group with the present of acid. The increase of microwave power from 180 to 300 Watt improved the methoxyl number, while prolonging extraction duration from 300 to 450 and 600 Watt decreased the methoxyl content. However, all methoxyl content obtained in this study ranged between 2.5 to 7.12% , which meet the IPPA standard for LM content (2.5 to 7.2%) [16].

Galacturonate levels as well as the charge of pectin molecules play an important role in determining the functional properties of the pectin solution and affecting the structure and texture of the gel formation [22]. In addition to the methoxyl content, gel strength is also associated with the levels of galacturonate acid in the pectin. Galacturonate acid levels represent the amount of pectin present in pectin flour. High galacturonate acid indicates strong bonds in the pectin during gel formation [23]. Figure 4.b shows galacturonate levels increased along with the longer of extraction time at constant microwave power. According to the Food Chemical Codex [24], the acceptable amount of galacturonate content is not less than 65%. In this study, galacturonate content ranged between 22 to 58.08%. To obtain pectin with galacturonate specification meets pectin quality requirements, pectin extraction should be carried out more than 20 min using power of 450 and 600 Watt.

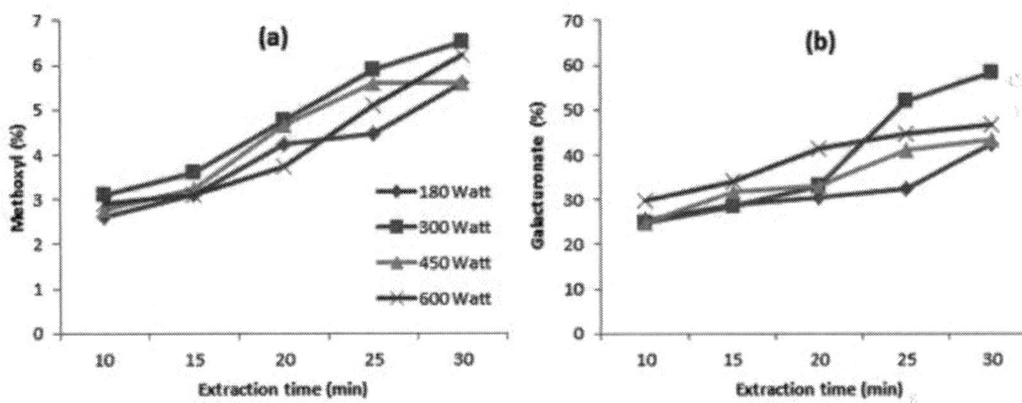


**Figure 4.** Effect of extraction time to galacturonate and methoxyl content at various microwave power

Overall pectin quality resulted from this study is tabulated in Table 1.

**Table 1.** Pectin quality as compared to IPPA standard

| Quality factor | This study | IPPA [16] |
|---|---|---|
| Methoxyl content (low) (%) | 2.5 - 7.12 | 2.5 – 7.2 |
| Levels of galactic acid (%) | 22 - 58.08 | Min 35 |
| Drying losses (moisture content) (%) | 8.08 - 11.8 | Max 12 |
| Ash content (%) | 1-5 | Max 10 |
| Equivalent Weight (mg) | 800 - 3000 | 600 - 800 |

## 4. Conclusions

Pectin extraction from cocoa peel using microwave-assisted method was influenced by microwave power and extraction time. In addition, microwave power was associated with extraction temperature. This study concludes that extraction time and microwave power affect yield of pectin and pectin quality, in which the increment of microwave power and extraction time result in greater yield, methoxyl content and galacturonate content in pectin. On the contrary, it reduced moisture content, while the ash content and equivalent weight was fluctuated. The best pectin extraction by microwave irradiation can be conducted at microwave power of 300 Watt for 30 min. This combination of power and time produce pectin (yield of 42.3%) with moisture content of 8.08%, ash content of 5%, equivalent weight of 833.33 mg, methoxyl content of 6.51% and galacturonate level of 58.08%. Overall, quality of pectin resulted in this study meet criteria established by IPPA [16], except for the equivalent weight.

**References**

[1] Sirotek, K., L. Slovakova, J. Kopecny and M. Marounek. 2004. Fermentation of Pectin and Glucose, and Activity of Pectin Degrading Enzymes in The Rabbit Caecal Bacterium Bacteroides Caccae. *Letters in Applied Microbiology*. 38, 327-332.

[2] Oliveira, C.F, Giordani, D, Lutckemier, R, Gurak, P.D, Cladera-Olivera, F and Marczak, L.D.F. (2016). Extraction of pectin from passion fruit peel assisted by ultrasound. *LWT-Food Science and Technology,*71, 110-115.

[3] Lefsih, K, Giacomazza, D, Dahmoune, F, Mangione, M.R, Bulone, D, San Biagio, P.L, Passantino, R, Costa, M.A, Guarrasi, V and Madani, K.(2017). Pectin from *Opuntiaficus indica*: Optimization of microwave-assisted extraction and preliminary characterization. *Food Chemistry,* 221, 91-99.

[4] May,C.(1990). Industrial pectins: Sources, production and applications. *Carbohydrate polymers*, 12, 79-99.

[5] Wang, S, Chen, F, Wu, J, Wang, Z, Liao, X and Hu, X.(2017). Optimization of pectin extraction assisted by microwave from apple pomace using response surface methodology. Journal of Food Engineering, 78, 693-700.

[6] Shkodina, O.G, Zeltser, O.A, Selivanov, N.Y and Ignatov, V.V.(1998). Enzymic extraction of pectin preparation from pumpkin. *Food Hydrocolloids*, 12, 313-316

[7] Qiu, L.P, Zhao, G.I, Hui, W, Lu, J, Li, X.F, Liu, J.J.(2010). Investigation of combined effects of independent variables on extraction of pectin of banana peel using response surface methodology. *Carbohydrate Polymers,* 80, 326-331

[8] Castillo-Israel, K.A.T. Baguio, S.F., Diasanta, M.D.B., Lizardo, R.C.M., Dizon, E.I. And Mejico, M.I.F. 2015. Extraction and Characterization of Pectin from Saba Banana [Moses 'Saba' (Musa acuminata x Musa Balbisiana)] Peels Wastes: A Preliminary Study. *International Food Research Journal* 22 (1): 202-207

[9] Cheng, S. F., Nor L, M. and Chuah, C. H. (2011). Microwave pretreatment: A clean and dry method for palm oil production. *Industrial Crops and Products, 34*(1), 967-971.

[10] Sukaribin, N. and Khalid, K. (2009). Effectiveness of sterilisation of oil palm bunch using microwave technology. *Industrial Crops and Products, 30*(2), 179-183.

[11] Sarah, M and Taib, M.R. (2013). Enzymatic Destruction Kinetics of Oil Palm Fruit by Microwave Sterilization. *International Journal of Chemical Engineering and Application,*4:3, 129-133

[12] Sarah, M, Taib, M.R and Adamu, A. (2014) Enzymatic Inactivation of Oil Palm Fruit: Comparison of Microwave Irradiation and Steam Batch Process. *Jurnal Teknologi*, 69:2,55-60

[13] British Standard. 1998. *Methods for Analysis and Testing Coal and Coke. Proximate Analysis*. British Standard Institution

[14] Ranggana, S. 2000. *Handbook of Analysis and Quality Control for Fruit and Vegetable Product.*2$^{nd}$Edition. New Delhi: McGraw-Hill Publishing Company Limited: 35.

[15] Devi, Elizabeth, W, Shukla, R.N, Bala, K.L, Kumar, A, Mishra, A and Yadav, K.C. 2014. Extraction of Pectin from Citrus Fruit Peel and Its Utilization in Preparation of Jelly. *International Journal of Engineering Research & Technology (IJERT),* 3(5).

[16] International Pectin Producers Association (IPPA). 2014. Pectin Commercial Production and Pectin in Organic Food Products.

[17] Nurdjanah, N, and Usmiati, S. 2006. Extraction and Pectin Characterization of Yellow Pumpkin. *Journal of Postharvest Agriculture Research* 3 (1),13-23.

[18] Yujaroen, P, Supjaroenkul, U and Rungrodnimitchai, S. 2008. Extraction of Pectin from Sugar Palm Meat. *Thammasat International Journal Science Technology*, 13, 44-47.

[19] Kalapathy, U and Proctor, A. 2001. Effect of Acid Extraction and Alcohol Precipitation Conditions on the Yield and Purity of Soy Hull Pectin. *Journal Food Chemistry*, 73: 393-396

[20] Budiyanto, Agus and Yulianingsih. 2008. Effect of Temperature and Extraction Time on Pectin Character from Siam Citrus (*Citrus nobilis L*). *Journal of Postharvest Agricultural Research,* 5 (2), 37-44

[21] Cruess, W.V. 1958. Commercially Fruits and Vegetable Products. New York: McGraw Hill Book

[22] Wachida, N and Yunianta. 2005. Pectin Extraction from Sweet Orange Skin Review ofMaturity Level and Type of Settler. Department of Agricultural Products Technology. BrawijayaUniversity. Malang

[23] Constenla, D and Lozano, J.E. 2003. Kinetic Model of Pectin Demethylation. *Latin American Applied Research* 33, Page 91-96

[24] Food Chemical Codex. 1996. IV monographs. Washington DC: National Press. Pp 283

# The combination of activated natural zeolite-bentonite to reduce Fe and Cu in refined bleached palm oil (RBPO) by using atomic absorption spectrophotometer method

**Zakwan, PM Raja and Giyanto**

Sekolah Tinggi Ilmu Pertanian Agrobisnis Perkebunan (STIPAP) / The College of Agriculture Agribusiness Plantation Science, Willem Iskandar Street (Pancing) 20222, Medan, Indonesia.

E-mail: zakwanzulkarnain11@gmail.com

**Abstract.** Indonesia is one of the crude palm oil (CPO) production country in the world. As many products are derivated from the CPO, the quality must be increased continuously. One of the things that influence the quality of palm oil is the Fe and Cu content. The objective of this research was to reduce Fe and Cu content in Refined Bleached Palm Oil (RBPO). In processing CPO or Refined Bleachead Palm Oil (RBPO) may be contaminated by Fe and Cu from metal tank and pipe in the factory. The zeolite and bentonite was activated by maceration method using hydrochloric acid (0,1 N). Four batch reactions consisting of refined palm oil (RPO), activated natural zeolite-bentonite (ANZB) was bleached by heating and stirring them at about 105°C and 1200 rpm for 30 minutes. The results showed that all combinations of ANZB can reduce the Fe content. Thereafter, the optimal combination of ANZB was obtained in K1, K2 and K4 with Cu content 0.02 ppm. In the future, it is needed to study on the reduction of the Fe and Cu content in palm oil with the other adsorbent.

## 1. Introduction

Indonesia is one of crude palm oil (CPO) producer country in the world. The production reached 26 million tons in 2012 and 28 million tons in 2013, by which about 75% CPO is exported and only 25% for domestic requirement [1]. So, the prospect of palm oil industry will be potential by processing CPO to the high value products.

Crude Palm Oil (CPO) is the extracting product of the mesocarp of palm fruit. The derivated products of CPO, i.e. cooking oil and margarine, are obtained by further processing steps such as refining, bleaching and deodorizing. The CPO that has been refined and bleached is known as refined bleached palm oil (RBPO). The purpose of refining process is to produce a clean, attractive looking and palatable oil or fat, by removing the impurities content that cause the oil to have an unattractive colour and taste. After refining, the crude oil still contains other impurities, odour, pigment, Fe or Cu from the metal tank or pipe in the factory, that need to be lowered. They are reduced by using bleaching earth or activated clay, as the bleaching earth adsorbs the undesirable impurities such as trace metal, moisture, and insoluble matter [2]. Natural zeolite or bentonite is usually used and the adsorbent.

The reduction iron Fe by using natural bentonite has been studied. The adsorption of Fe increase sharply by elevating natural bentonite dosage [3]. The study of adsorption method using activated zeolite as adsorbent to remove oil in water has been studied [4]. The heavy metals, Fe and Cu, still important to be researched, because they can contaminate to food product and make serious problems in healthy, as they can generate toxic derivatives in the body [5]. For instance, an excessive amounts of iron in specific tissues and cells (iron-loading) promote development of infection, neoplasia, cardiomyopathy and possibly neurodegenerative disorders [6-9].

Previous study used natural bentonite or activated zeolite as single adsorbent to reduce Fe and Cu in RBPO. This study, was aimed to study a combination of activated natural zeolite and bentonite as adsorbent for reducing Fe and Cu in refined bleached palm oil (RBPO).

## 2. Material and Methods

### *2.1 Materials and Equipment*

Materials used in this study are crude palm oil (CPO), natural zeolite activated, natural bentonite activated, HCl 0.1 N (Merck), $HNO_3$ 0.5 N (Merck), $HClO_4$ Pa (Merck), and aquades. While, the tools used are hammer mill (Van Aarsen), sieve 250 mesh (Patraproduk), hotplate-stirrer (Thermo Scientific), magnetic stirrer, oven (Jouan), vaccum (Value), desicator, whatmant 1, chemical equipment laboratory and atomic adsorption spectrophotometer (Hitachi Model Z-2000 Series Flame).

### *2.2 Zeolit and Bentonite Adsorbent Preparation*

Natural zeolite, originated from Sarulla area of North Tapanuli, was crused and sieved to obtain the appropriate size of 250 Mesh. Zeolite was dried in an oven at temperature of 200$^0$C for 2 hours, cooled to room temperature and stored in a desicator. Then, zeolite was activated with 0.5 N hydrochloric acid for two hours, 1200 rpm at temperature 75°C using hotplate stirrer (Thermoscientific Cimarec), washed with distilled water until the pH was neutral. Zeolite was dried in an oven at temperature of 105$^0$C for two hours, so that it can be used as an adsorbent.

Similarly, natural bentonite, originated from Bener Meriah area of Aceh Singkil, was crused and sieved to obtain the appropriate size of 250 Mesh. Bentonite was dried in an oven at temperature 200$^0$C for 2 hours, cooled to room temperature and stored in a desicator. Then, bentonite was activated with 0.5 N hydrochloric acid for two hours, 1200 rpm at temperature 75°C using hotplate stirrer (Thermoscientific Cimarec), washed with distilled water until the pH was neutral. Bentonite was dried in an oven at temperature 105$^0$C for two hours and ready to be used as an adsorbent.

### *2.3 Experimental Procedure*

The combination of activated natural zeolite:bentonite are K1 (30%:70%), K2 (40%:60%), K (60%:40%) and K4 (70%:30%) each one was mixed to CPO. It was stirred using 1200 rpm at temperature 105°C for 30 minute using hotplate stirrer (Thermoscientific Cimarec) and vacummed. CPO was filtrated with whatman 1. Then, it was dried in oven at temperature 105°C for 24 hours, cooled to room temperature and stored in a desicator. Two grams of sample was mixed with 1.5 ml of $HClO_4$ and 3.5 mL $HNO_3$, then the container was closed and left for 24 hours.

Furthermore, the obtained solution was heated over a water bath at 60-70°C for 2-3 hours (until the solution is clear). If the sample is not soluble, 5 mL $HClO_4$, 5 mL $HNO_3$, and 3 mL of aquades were added, then the mixture was reheated until the solution is almost dry. RBPO was cooled at room temperature and added 1 mL of $HNO_3$ and stirred slowly, then added 9 ml of aquadest. The samples are ready to be measured with AAS by using triplo replication. Further, the data will be analyzed by analysis of variance method that is run by using software microsoft office excel 2007.

## 3. Result and Discussion

### *3.1. The characteristic of crude palm oil (CPO)*

Based on the study conducted in the Laboratory of Mutu dan Kimia STIPAP, the characteristic of CPO of CPO is shown in Table 1.

**Table 1.** Characteristic of crude palm oil (before processing)

| Parameter | Unit | Value | Spesification | Standard |
|---|---|---|---|---|
| FFA content | % | 1.13 | Max. 5 | |
| Moisture content | % | 0.09 | Max. 0.25 | SNI 01-2901-2006 |
| Impurities content | % | 0.01 | Max. 0.25 | |

According to Table1, the characteristic of raw material (CPO) like free fatty acid (FFA), moisture and impurities, were still within standard range. Moisture could affect FFA content, because the higher moisture enhance hydrolysis reaction that effect (more than 5%) breaking and increase FFA in oil. It means that the quality of raw material in this research was good and beyond the maximum limit of the standard quality of CPO.

*3.2. The Fe and Cu content in RBPO with ANZB*

The results of Fe and Cu analysis in refined bleached palm oil (RBPO) with activated natural zeolite-bentonite (ANZB) are shown in Table 2.

**Table 2.** The Fe and Cu analysis in RBPO with ANZB

| | Activated natural zeolite (%) | Activated natural bentonite (%) | Replication (ppm) | | | Mean (ppm) | Stdev |
|---|---|---|---|---|---|---|---|
| | | | I | II | III | | |
| | | | Fe | | | | |
| CPO | 0 | 0 | 5.39 | 4.81 | 4.59 | 4.93 | 0.41 |
| K1 | 30 | 70 | 0.21 | 0.23 | 0.20 | 0.21 | 0.02 |
| K2 | 40 | 60 | 0.13 | 0.12 | 0.10 | 0.12 | 0.02 |
| K3 | 60 | 40 | 0.13 | 0.12 | 0.14 | 0.13 | 0.01 |
| K4 | 70 | 30 | 0.14 | 0.15 | 0.14 | 0.14 | 0.01 |
| | | | Cu | | | | |
| CPO | 0 | 0 | 0.03 | 0.02 | 0.03 | 0.03 | 0.01 |
| K1 | 30 | 70 | 0.02 | 0.02 | 0.03 | 0.02 | 0.01 |
| K2 | 40 | 60 | 0.02 | 0.03 | 0.02 | 0.02 | 0.01 |
| K3 | 60 | 40 | 0.02 | 0.03 | 0.03 | 0.03 | 0.01 |
| K4 | 70 | 30 | 0.02 | 0.03 | 0.02 | 0.02 | 0.01 |

The initial Fe content in CPO was 4.93 ppm, which is almost reached the maximum standard level of Fe content in CPO (5.0 ppm) [10]. The increase in Fe content migt be resulted by contamination of the corrosion equipment to CPO that by the suboptimal maintenance of metal tank or pipe in the factory. In addition, Cu content in CPO was 0.03 ppm, which was lower than the standard of allowed Cu content (0.4 ppm) [10]. According to the analysis of variance method was known that in probability $\alpha$=0,01 F-count > F-table values (Fe : 668,93 > 6,93 and Cu : 41,20 > 6,93). It means that all combinations of ANZB have signifficant effect to Fe and Cu content.

The analysis indicated that the initial Fe in CPO was very high, so that the process to decrease the Fe content by refining and bleaching is necessary. The reduction of Fe content could be reached by mixing the CPO with the activated natural zeolite-bentonite (ANZB), by which the zeolite and bentonite was acting as adsorbents.The chemical formula of natural zeolite and bentonite has negatively charged $Al^-$ and $(Si\text{-}O)^-$. These negative entities might bind metal ions especially at lower pH. Therefore, the HCl solution may achieve the optimum adsorption. The binding of metal ions might be influenced on the surface of natural zeolite and bentonite [3].

The aim of the activation with chemical method by using hydrochloric acid is to exchange $Ca^{2+}$ that exist in zeolite and bentonite to become $H^+$ and release ion Fe, Al and Mg and other impurities in the zeolite and bentonite structures, so that the zeolite and bentonite become more active. Hydrochloric acid (HCl) are used as the chemical compound and HCl are also used as catalyst in zeolit and bentonite activation. By dissolving in acid solution, the surface area of zeolite and bentonite become larger. So, the absorption was more effectively [12].

This research has two steps of treatment for zeolite and bentonite, there are the activated physically through the heating process in 75°C and chemically by using hydrochloric acid as activator, to give better results. The heating process on hot plate was conducted to keep the form of bentonite stable during the adsorption process. In thermal activation, high temperature can omit water molecule and other impurities. The change of structure and composition during the heating process can vary based on the chemical composition of natural zeolite and bentonite and time of heating.

In this current study, the reseachers do not use too high temperature because the over heating can make the short space in structure and interlayer of zeolite and bentonite, the small interlayer space can cause the particle diffusion closer to another atom that can affect the space surface. Moreover, the high temperature can make degradation the natural component of palm oil.

*3.3. The trending of ANZB to the Fe and Cu content in RBPO*

The trending of the combination activated natural zeolite-bentonite to Fe and Cu contents are shown in the Figure 1 and 2.

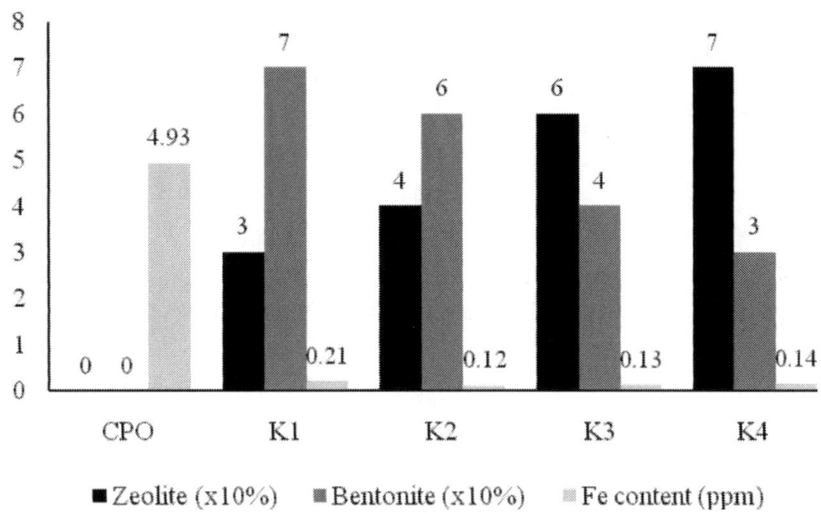


**Figure 1.** The histogram chart of combination ANZB to Fe content

The Figure 1 showed the histogram of the activated natural zeolite-bentonite (ANZB) combination to the Fe content in the CPO after refining and bleaching. It was found that the highest Fe content was noted at CPO that without mixing with ANZB and the lowest Fe content was found at K2 (activated natural zeolite : activated natural bentonite = 40 : 60). The Fe content of palm oil decreased slightly with the combination of ANZB (Table 2). The lower Fe content was achieved after mixing with combination ANZB because natural zeolite and bentonite as the adsorbents have negative ions that might bind the Fe in palm oil [3].

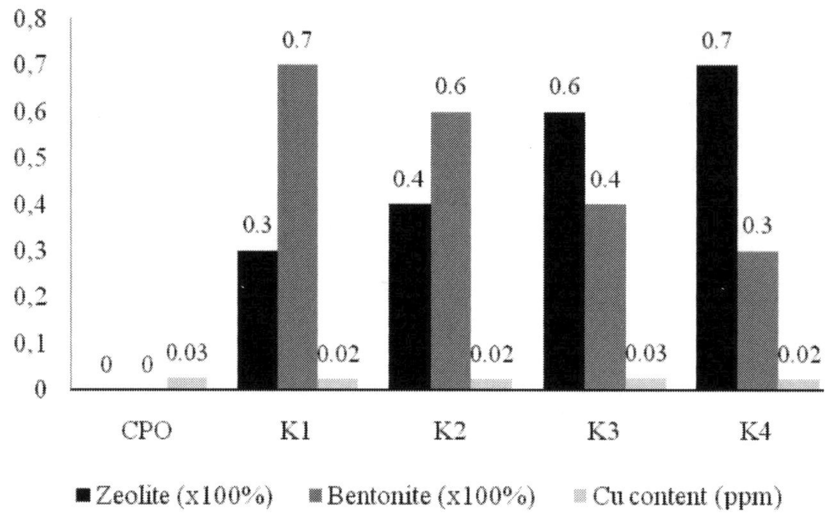


**Figure 2.** The histogram chart of combination ANZB to Cu content.

The Figure 2 showed the histogram of combination activated natural zeolite-bentonite (ANZB) to Cu content in RBPO. The highest activated zeolite and bentonite in K4 and K1 was 70%. Therefore, the lowest activated zeolite and bentonite in CPO was 0%. The Cu content in CPO, K1 and K4 was 0.03 ppm, 0.02 and 0.02 ppm, respectively. It means that the highest activated zeolite and bentonite is able to decrease the Cu content in RBPO. Because activated zeolite and bentonite as the adsorbents that might bind the Cu in palm oil [3].

## 4. Conclusion

The activated natural zeolite and bentonite was used as bleaching earth to reduce Fe and Cu contents in refined bleached palm oil. The Fe content in RBPO with combination ANZB K1; K2; K3 and K4 (ppm) was 0.21; 0.12; 0.13 and 0.14 respectively. The Cu content in RBPO with combination ANZB K1; K2; K3 and K4 (ppm) was 0.02; 0.02; 0.03 and 0.02. The result showed that the combination of ANZB can reduce the Fe content more effectively than the Cu content.

## Reference

[1] Direktorat Jenderal Perkebunan 2012. *Produksi kelapa sawit menurut provinsi di indonesia 2009-2012 (Province palm production in Indonesia 2009-2012)* (Jakarta)
[2] Yusoff M S A 2000 Advances in oil palm research *Refining and Modification of Palm and Palm Kernel Oil* vol 2 ed Y Basiron Jalani B S and Chan K W (Malaysia : MPOB) chapter 20 pp 783-790
[3] Al-Anber M A 2010 *Elsevier : Desalination* 250 885-891
[4] S Zaidi et al 2015 *Arabian Journal of Chemistry* 8 1-12
[5] M Wessling-Resnick 1999 Biochemistry of iron uptake *Crit. Rev. Biochem. Mol. Biol.* 34 285-314
[6] E D Weinberg and G A Weinberg 1995 The role of iron in infection *Curr. Opin. Infect. Dis.* 8 164-169
[7] T P Tuomainen K Punnonen K Nyyssonen and J T Salonen 1998 Association between body iron stores and the risk of acute myocardial infraction in men *Circulation* 97 1461-66
[8] J M McCord 1996 Effecs of positive iron status at a cellular level *Nutr. Rev.* 54 85

[9] P Sarin V L Snoeyink J Bebee K K Jim M A Beckett W M Kriven and J A Clement 2004 Iron release from corroded iron pipes in drinking water distribution system *Water Res.* 38 1259-1269

*[10]* Codex Standard 1999 Codex standard for named vegetable oils *CodexStan 210*

[11] Badan Standardisasi Nasional 2006 Minyak kelapa sawit mentah (crude palm oil) *SNI 01-0018-2006* Jakarta

[12] Dewi R Ratna S and Syarifuddin 2016 The effect of the natural bentonite to reduce COD in palm oil mill effluent by using a hybrid adsorption-flotation method *IOP Conf. Seies : Materials Science and Engineering* 210 1-8

**Acknowledgements**
The authors gratefully acknowledges STIPAP for funding and supporting this research.

# The tocol content of crude palm oil based on the level ripeness and their relationship to the quality and their stability

**I U P Rangkuti [1], E Julianti[2] and J Elisabeth[3]**

[1] Department of Plantation Technology, STIPER- Agrobisnis Perkebunan, Medan
[2]Department of Food Science and Technology, Universitas Sumatera Utara
[3]Department of Research and Development Oil/Fat, Wilmar Group Indonesia

[1] uchastipap@gmail.com, [2] elisayulianti@yahoo.com, [3]jenny.elisabeth@id.wilmar-intl.com

**Abstract**. Palm oil contain tocol and betacarotene which has an antioxidant function to protect the oil from oxidation processPalm oil comes from different maturity have not been specifically determined their tocol and betacarotene content, and their relation to quality and stability. The purpose of this research was to find the content of tokol and betacarotene in palm oil which comes from different maturity and to determine its stability of quality. The level of maturity was fraction 1 (unripe), fraction 2 and fraction 3 (ripe), fraction 4 (over-ripe).The result showed that palm oil which comes from different level of maturity from fraction 1, 2, 3 and 4 contained different amount of tocol and betacarotene. The quality of crude palm oil which were analysed by the value deteoriation of bleachability index (DOBI) had closed relationship to the number of beta-carotene in the palm oil. The color stability and oxidation were different in every level of maturity. The level of maturity which had the best quality and its stability was fraction 3. The tokol and betacarotene compound were 346 ppm and 727 ppm respectively, while the DOBI value was 2,75.

## 1. Introduction

Crude palm oil (CPO) which produced from palm oil fruit (*Elaeis guineensis*) is the main vegetable oil in the world market today. Crude palm oil is rich in carotenoids with concentrations between 500-700 ppm, where the content of beta carotene as retinol equivalent was 15 times more than carrots and 50 times more than tomatoes. Beside carotenoids, which consist of alpha- and beta carotene, crude palm oil is also rich in other minor components that provide unique nutritional properties, especially the tocopherol and tocotrienol components, both of which are also called tocol. Crude palm oil contains 600-1000 ppm of tocol. Tocol was nutritional substance and has antioxidant activity that contribute to the oxidative stability of crude palm oil. Tocopherol may prevent oxidation of fat by blocking the formation of hydroperoxides in the propagation chains as well as the process of decomposition and formation of aldehyde compounds [1]. The fruit maturity level is very influential on the productivity and quality of crude palm oil produced. Change in the color of the palm fruit flesh to yellowish green after three months indicates that the oil has formed the formation of carotene. In addition, the change of fruit color from green to yellow because the loss of chlorophyll content in the fruit is replaced by the formation of carotenoids and will increase along with the fruit [2]. Fat-soluble carotenoids are very sensitive towards heat and also susceptible to enzymatic, chemical, and photochemical oxidation. During the storage process these carotenoids is oxidized and form oxidation compounds [3].The effect of fruit maturity level on the quality and stability of crude palm oil is estimated to be related with the minor component content which is found in the palm fruit, especially the antioxidant component such as beta carotene and tocol. Due to its easily oxidized characteristics, the beta carotene component issuspected of not contributing greatly to crude palm oil quality's stability.

## 2. Method

### *2.1. Materials*
The materials used in this research were the DxP crossbreeding palm oil fruit derived from Kebun Sampali owned by Perkebunan Nusantara (PTPN) II Ltd. The chemicals materials used are; 3:2 (v/v) acetic acid solution-chloroform, 0.01 N sodium thiosulfate solution, saturated KI solution, starch indicator solution, N-hexane pa, methanol pa, methanol licrosolv, standard tocopherol with isomer A-T1, δ-T1, α-T3, γ-T3, and δ-T3.

### *2.2. Palm Oil Extraction*
Palm oil fruits were peeled using machetes, then boiled using autoclave with a temperature of 120°C for 90 minutes and extracted the oil using a hydraulic press. The resulting crude palm oil is separated by centrifuges to obtain crude palm oil that is free from impurities. The oil extraction results are stored in the refrigerator (LG brand freezer) at -5°C until the extraction of all samples is completed to proceed with the sample analysis.

### *2.3. Analysis of Tocopherol and Tocotrienol Content in Palm Oil (In house method PPKS, 2011)*
The crude palm oil sample was weighed 2 g in a 10 ml measuring flask (measuring flask 1), then hexane was added to the marker boundary marker. The crude palm oil sample is homogenized, then taken 2 ml with a volume pipette, put into a 10 ml measuring flask (measuring flask 2), then added with pa methanol to the marking margin and homogenized. The sample solution on the second measuring flask is inserted into the centrifuge tube, then centrifuged for 10 minutes at 2500 rpm. The solution at the top layer was separated, then analyzed the content of tocopherol and tocotrienol by using Shimadzu LC-6A High Performance Liquid Chromatography (HPLC) with Fluoresence Detector.

### *2.4. Beta Carotene Content (MPOB, 2004)*
The crude palm oil sample is melted at 60-70°C and shaken until homogeneous. Crude palm oil is filtered with Whatman No.1 paper, and then crude palm oil is weighed as much as 0.04 g and put into a 10 ml measuring flask. In the sample solution the crude oil is added N-hexane solvent up to the mark line. Absorbance of crude palm oil is read by spectrophotometer at $\lambda = 446$ nm.

### *2.5. DeteriorationofBleachabilityIndex/DOBI (MPOB, 2004)*
The crude palm oil sample is melted at 60-70°C. Samples were filtered with Whatman no. 1 paper then weighed 0.04 g. Put into a 10 ml measuring flask and add n-hexane solvent up to the mark line. Absorbance of crude palm oil sample is read by spectrophotometer at $\lambda = 446$ nm and 269 nm

### *2.6. Color (AOCS Official Method Cc 13e-92)*
Measurement of oil sample color is done with Lovibond Tintometer in soft light atmosphere and not exposed to direct sunlight. The crude palm oil sample is inserted into a 1" size glass cell and the readings are corresponding on the glass cell limit containing the crude palm oil sample placed in the light cabinet close to the observation tube. The color of crude palm oil is measured using a color rack with ratio of 10 for yellow and 1 for red. Correction till the accurate color matching were obtained using the minimum number of blue or neutral, where the color obtained is no more than blue 9.0 and neutral 3.0. The results of the red and yellow readings of the oil samples are then recorded.

## 3. Result and discussion

### *3.1. The Content of Tocol Components in Crude Palm Oil*

Crude palm oil contains 600-1000 ppm tocopherol and tocotrienol, where these two compounds can be called tokol compounds. Tocopherol and tocotrienols are composed of 4 different isomers i.e. as α-, β-, γ- or δ-, which corresponds to the number and position of the methyl group attached to the chroman ring [1]. Vitamin E has alpha, gamma and beta forms, and delta-tocotrienols serves as a natural antioxidant in protecting cell membranes from oxidative damage [4]. The analysis results of tocol components which is contained in crude palm oil was derived from the palm fruit with different levels of maturity showed different of tocol total content, as shown in Figure 1. Analysis of variance indicated that fruit maturity level had significantly different effect ($P < 0.05$) on tocol content. Further test results using Duncan test showed a difference of tocol component content of crude palm oil from each level of palm fruit maturity.

Figure 1 shows trend of tocol total content that has increased along with the increasing maturity level of the palm fruit. This indicates that palm fruit maturity level have an important role in the formation of tocopherols and tocotrienols in palm oil. The amounts of Vitamin E in fruits and vegetables are affected by species, varieties, maturity, growth conditions (weather, growing season, solar intensity,and soil), distribution of unbalanced tocopherol, and harvesting time/process [6].

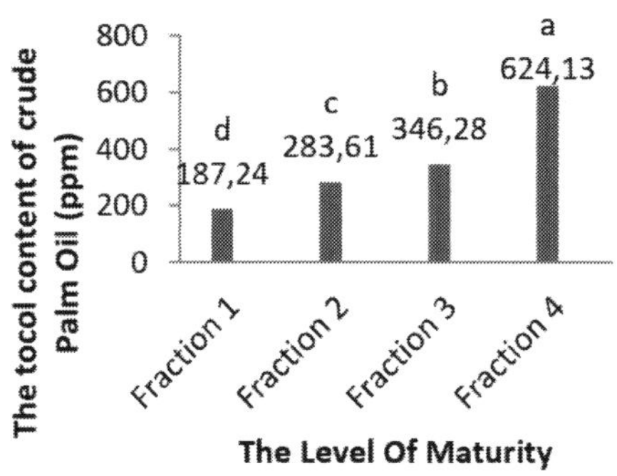


**Figure 1.** The Content of Tocol Component in Crude Palm Oil

In the early stages of fruit maturation (fraction 1 and fraction 2) the composition of tocol compounds in crude palm oil consists of γ-tocotrienol (90.04 and 146.94 ppm), δ-tokotrienol (66.45 and 49.57 ppm), and α-tocotrienol (25.35 and 68.54 ppm), also α-tocopherol (5.40 and 18.56 ppm). In the final stages of fruit maturation (fractions 3 and 4) the composition of tocol compounds contained in crude palm oil consists of γ-tocotrienol (104.56 and 313.75 ppm), α-tocotrienol (154.36 and 196.06 ppm), δ-tocotrienol (20.48 dan 44.19 ppm) and α-tocopherol (66,88 dan 70,13 ppm), the number of tocol compounds in the form of their respective isomers increases with the age of the palm fruit. Tocotrienols contained in fruits and vegetables, usually consist of α-tocotrienol and γ-tocotrienol which is quite low in concentration compared to tocopherol. Palm oil contains higher α-tocotrienol and δ-tocotrienol than fruit and vegetable [7].

### *3.2. The Content of Betacaroten in Crude Palm Oil*

The content of betacarotene has increased in fraction 1 and fraction 2. The content of beta-carotene reaches the optimum limit at fraction 3 or at the level if the fruit ripening process has been completed and then at fraction 4, i.e the expected phase of the fruit has entered the level of passing through, then the content of beta-carotene in palm oil is decreased. The decrease of beta carotene content after reaching this optimum point is related to the research result of carrot, where it is stated that the carotene will accumulate and reach the maximum point after the carrot tuber is about 90-120 days old and will gradually decrease [8].

Figure 2 shows the graphic trend of beta-carotene content that increases with the increasing maturity level of palm fruit, but the content of beta carotene decreases at the last fruit maturity level (fraction 4). The increase in beta carotene content in palm oil which derived from different fruit maturity levels is also not the same, where at the beginning of fruit maturity, the increase in the amount of beta carotene is greater, and the amount of beta carotene decreases thereafter.

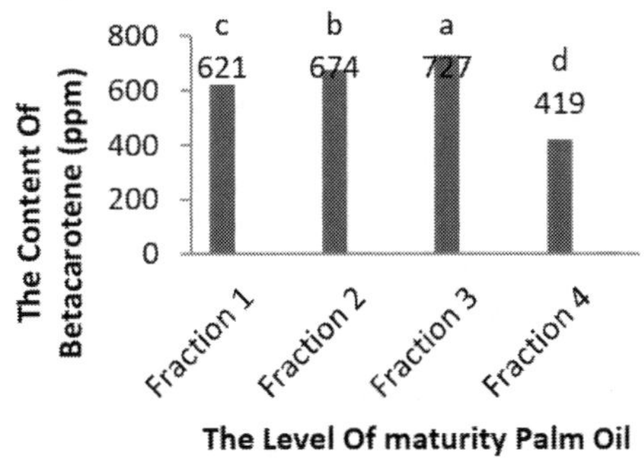


**Figure 2.** The Content of Betacarotene in Crude palm Oil

Carotene content in crude palm oil from raw fruit is significantly smaller than ripe fruit and ripe fruit [9]. Carotene also has relatively less stable properties, which can be degraded by environmental factors [10].

*3.3. The Deterioration of Bleachability Index (DOBI Index) in Crude Palm Oil*
Crude palm oil from palm fruits derived from different degrees of maturity in Figure 3 can explain that the value of DOBI obtained is still within the standard number of $\geq 1.5$. A good DOBI value is $\geq 1.5$. A good DOBI value for the purification process is $\geq 1.5$ so that the resulting oil color does not darken [11]. Figure 3 shows the graphical trend of decreasing DOBI value in the final stages of the fruit maturity level. The analysis of variance shows that the maturity level of the palm fruit gives a significant effect ($P < 0{,}05$) to the value of DOBI. The maturity level in immature palm fruit i.e. fraction 1 and fraction 2 is having enough high of DOBI value compared to maturity level of ripe and over ripe palm fruit i.e fraction 3 and fraction 4.

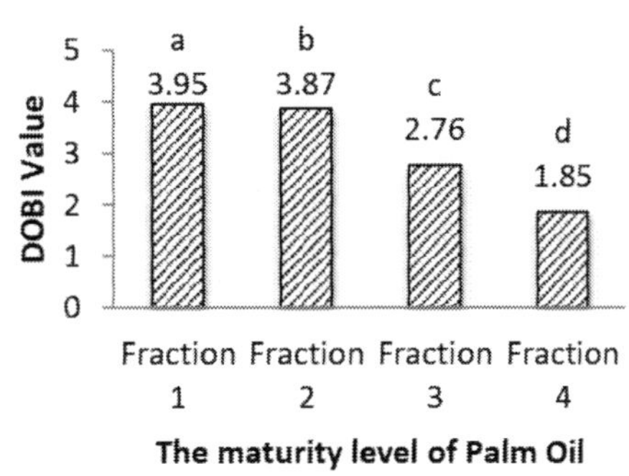


**Figure 3.** The DOBI Value in Crude Palm Oil

*3.4. The oxidative stability of crude palm oil as measured by color*
Red color changes in crude palm oil is due to heating process at 100 ° C for 24 hours as can be seen in Figure 4. Red color stability is reduced after 12 hours of heating on crude palm oil which derived from

fruit with maturity levels of 1, 2 and 4. In crude palm oil which derived from fraction 3, a decrease in the stability of the red color occurs in 18 hours of heating. This is because the content of beta carotene in the fruit with maturity level 3 higher than the level of maturity 1, 2 and 4 (Figure 2). Although fraction 4 has higher total tocol content than fraction 3 but its beta carotene content is much lower than fraction 3 so that its oxidative stability is lower.

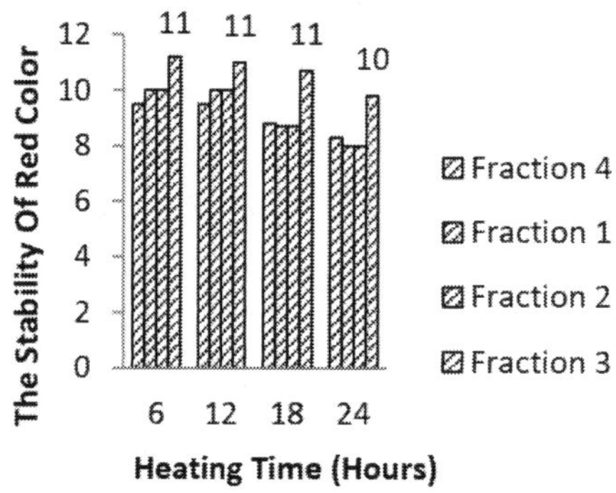


**Figure 4.** The Oxidative Stability Of Red Color in Crude Palm Oil

*3.5. The Oxidative Stability of crude palm oil as measured by peroxide index*

The oxidative stability of crude palm oil for each different maturity level at the end of heating (24 hours) of fractions 3 and 4 has better stability than fraction 1 and fraction 2. This is because the content of beta carotene and tocol contained in those fractions is high enough to maintain the stability of crude palm oil. Beta carotene becomes an antioxidant that first works to prevent oxidation damage, then followed by the performance of tocopherol and tocotrienols as antioxidants. When linked to previous studies on antioxidant properties derived from beta carotene, red palm oil has the potential to slow down oil damage or slow the formation of free fatty acids and peroxides during heating [12].

## 4. Conclusion

Tocol content in crude palm oil is influenced by the fruit maturity level, where the higher the fruit maturity level then the content of tocol will increase. The content of beta carotene in crude palm oil is influenced by the fruit maturity level. The higher the fruit maturity level, the beta carotene content will increase, but in fraction 4 beta carotene content decreased again. DOBI value content in crude palm oil is influenced by the maturity level of the palm fruit. This is related to the content of beta carotene and the condition of the fruits to be processed. The highest of beta carotene amount is oxidized, the lower the DOBI value and the resulting oil will become darker. The content of tocol and beta carotene in crude palm oil acts as an antioxidant that protects crude palm oil from oxidation damage. Beta carotene becomes an antioxidant that first works to prevent oxidative damage to palm oil then followed by the performance of tocopherol and tocotrienol as antioxidants.

## References

[1] Sundram K, Sambanthamurthi R, dan Tan Y 2003 *Palm Fruit Chemistry and Nutrition*, (Malaysia)
[2] Anonim 2008 *Technological Culture and Processing Palm Oil , Indonesian Oil Palm Research Institute* (Medan)
[3] Tagoe S, Dickinson M and Apetorgbor M M 2012 *Factor influencing quality of palm oil produced at cottage industry level in Ghana* ( International food research journal) vol 19(1).271-278
[4] Malaysia Palm Oil Board. 2004. *MPOB Test Methods* (Bangi) 395 p
[5] Food and Nutrition Board 2000 *Dietary references intakes for vitamin C vitamin E selenium*

*and carotenoids* (Washington: Washington National Academy press)

[6] Bauernfeind, J 1980 Tocopherols in foods In: Machlin, L.J. (Ed.) Vitamin E: A Comprehensive Treatise Marcel Dekker (New York) pp 99–167

[7] Jiyeon Chun Junsoo Lee Lin Yee Jacob Exler Ronald Eitnmiller 2005 *Tocopherol and tocotrienol contents of raw and processed fruits and vegetables in the united states diet* (Journal of food composition and analysis vol 19) 196-204

[8] Rubatzky, S and Yamaguchi M 1998 The Vegetable World 2nd Edition Institut Teknologi Bandung 166 Bandung

[9] Donald Siahaan Nuryanto Manik dan Tampubolon 2006 *Study on carotene content of palm oil in different varieties,maturity and unit process in palm oil mill International oil Palm Conference Juni 19-23* (Bali)

[10] Syahputra M.Rio Ferry F Karwur and Leenawaty limantara 2008 *The Analyze Of The Composition and Total Carotenoid Content and Vitamins A Solid and Liquid Fraction Of Crude Palm Oil (CPO) Using KCTK Detector* (Journal Of Natur Indonesia Vol 10)

[11] Syakirah Rosnah and Robiah 2014 *The effect of storage time of chopped oil palm fruit bunches on the palm oil quality 2nd International Conference on Agricultural and Food Engineering CAFEi 2014* (Agricultural Science Procedia 2)

[12] Budiyanto Silsia Efendi and Janika 2010 *The Changing of Beta-carotene, Free Fatty Acids And Peroxide Value in Red Palm Oil During Heating* (Bengkulu : Journal Of Food Processing Vol 30 ) Universitas Bengkulu

IOP Conf. Series: Earth and Environmental Science **122** (2018) 012082 doi:10.1088/1755-1315/122/1/012082

# Estimation of shelf life of wikau maombo brownies cake using Accelerated Shelf Life Testing (ASLT) method with Arrhenius model

**S Wahyuni, Holilah, Asranudin and Noviyanti**

Department of Food Science and Technology, Faculty of Technology and Agricultural Industry, Universitas Halu Oleo, Indonesia
Email : sriwahyuni_aan@yahoo.com

**Abstract**. The shelf life of brownies cake made from wikau maombo flour was predicted by ASLT method through the Arrhenius model. The aim of this study was to estimate the shelf life of brownies cake made from wikau maombo flour. The storage temperature of brownies cake was carried out at 20°C, 30°C and 45°C. The results showed that TBA (*Thio Barbaturic Acid*) number of brownies cake decreased as the storage temperature increase._Brownies stored at 20°C and 30°C were overgrown with mold on the storage time of six days. Brownies product (WT0 and WT1) had shelf life at 40°C approximately six and fourteen days, respectively. Brownies made from wikau maombo and wheat flour (WT1) was the best product with had the longest of shelf life about fourteen days.

## 1. Introduction

Brownies are one of the most popular cake-based wheat flour, a chocolate cake with attractive colors, a delicious aroma, and a less-fluffy texture [1]. During this time, production of brownies cake using wheat flour. To reduce the dependence on wheat flour and the food development based local resource, substitution of wheat flour is needed in production of brownies cake. One example of local food that can be used to substitute wheat flour is wikau maombo flour. Wikau maombo flour produced from varieties of cassava ie. *Manihot utillisima* and *Manihot esculenta* Crantz, respectively [2-3].

Production of wikau maombo flour is through the process of immersion in sea water, fermentation, drying and milling [2]. Wikau maombo flour has similar characteristics of wheat flour can be seen from the color and texture of flour. Compared with ordinary cassava flour, wikau maombo flour has a better appearance ie. whiter, soft and odorless. Wikau maombo flour has a different content with wheat flour [3]. The fundamental difference in content is that wikau maombo flour does not contain gluten. One of the requirements for the industrialization of brownies cake should have nutritional value information and have information of the shelf life of product.

Based on the description, it is necessary to study the proximate analysis of brownies cake from wikau maombo flour and to estimation of the product shelf life using ASLT (Accelerated Shelf Life testing) with Arrhenius model. So it is known deadline brownies cake that safe to be consumed by society. The ASLT method is a method of estimating shelf life by conditioning food products above normal storage conditions [4]. In the ASLT method the temperature acts as a key parameter determining food damage, because the higher the temperature, the food damage will be faster [5]. The correlation between temperature and speed of deterioration can be seen using the Arrhenius equation.

## 2. Materials and Methods

### *2.1 Materials*

Cassava (*Manihot esculenta* Crantz) roots were obtained from Buton island, Southeast Sulawesi, Indonesia. Wheat flour, sugar, egg, margarine, vanilla, chocolate powder, dark chocolate and skim milk were purchased from traditional market, Kendari, Southeast Sulawesi. All materials used in this work were analytical grade. Hydrogen chloride (HCl, ≥37%) was purchased from Merck, Germany. Acetic acid ($CH_3COOH$, ≥99%) and Thiobarbituric Acid ($C_4H_4N_2O_2S$, ≥98%) were purchased from Sigma-Aldrich, Singapore.

### *2.2 Preparation of Wikau maombo flour*

Preparation of Wikau maombo flour used is according to the procedure that have been conducted by Wahyuni *et al.* [2]. Cassava roots were manually peeled, washed with tap water and chipped using a manual chipper. The pieces then immersed in seawater for 3 h. In order to remove mucus formed during immersion, cassava was washed with fresh water. Cassava was fermented for 3 days. Fermentation of cassava was conducted by storing of cassava in closed condition. After fermentation was completed and then dried in oven at 60°C for 24 h. And then the dried cassava millded into powder, sieved (70 mesh) and dried at 60°C for 2 h.

### *2.3 Preparation of brownies cake*

Preparation of brownies cake was carried out by shaking the mixture of egg and sugar until expands for 15 minutes. And then little by little wikau maombo flour that sieved and roasted added. Vanilla, TBM, skim milk, cocoa powder and salt were also added. The dough is stirred until blended well. The melted margarine and chocolate are added to the dough and stir until blended. The dough is then poured into the mold. After that, baked at 160°C for 45 minutes. Production of brownies cake with the addition of wheat flour of 0% and 25% were denoted with WT0 and WT1 samples, respectively.

### *2.4 Estimation of shelf life of brownies cake*

The estimation of shelf life of brownies cake using ASLT method wuth Arrhenis model. Storage of the product was carried out at 20°C, 30°C, 40°C. The test at various temperatures were used to estimate the shelf life from rancidity rates of product through the test of TBA content. Determination of the initial TBA number is the TBA number of product after the production, the initial TBA content of the product is measured before it is placed at various temperature. Determination of critical of TBA number was conducted over a period of time which determined in accordance with Table 1.

**Table 1**. Parameters for determination of critical TBA number.

| Temperature (°C) | Time (day) |
|---|---|
| 20 | 20 |
| 30 | 10 |
| 40 | 5 |

## 3. Results and Discussion

### *3.1 Determination of k parameter*

From observation of brownie cake to time can be plotted and got 3 regression equation obtained from three temperature (20, 30, 40°C) of different storage. From each equation obtained the value of the slope (b) and the constant (k). Determination of the reaction order to be used is the order reaction of zero which is the relationship between the value of k with the time of storage and reaction order of one which the relationship between ln k with the time of storage. From the two equations obtained the largest $R^2$ selected as the reaction order. For Arrhenius approach the value of k is plotted by 1/T through the linear regression equation ln k = ln k - (E/R) (1/T) with ln ko is intercept and E/R is slope.

Ea is the activation energy and R is the ideal gas constant of 1.986 cal/mol.K (Figure 1). After coefficient values obtained are the pre exponential factor and the activation energy value of the change reaction on the characteristic of brownies where EA = E, Arrhenius equation is obtained which the equation of reaction rate of characteristic change in brownies with the equation k -ko.e-E / RT with T is the storage temperature. From the Arrhenius equation is obtained, it can be calculated Arrhenius constant value in each storage temperature. The parameters that have the lowest activation energy value are the key parameters [8]. The shelf life was calculated using the reaction equation based on the reaction order. To estimate the shelf life of brownies by entering the temperature value into the equation ln k = ln ko - (E/R) (1/T). The obtained k values are included in equation of the reaction order to obtain the shelf life of brownies. Summary of shelf life prediction of brownies (WT0 and WT1) is shown in Table 2.

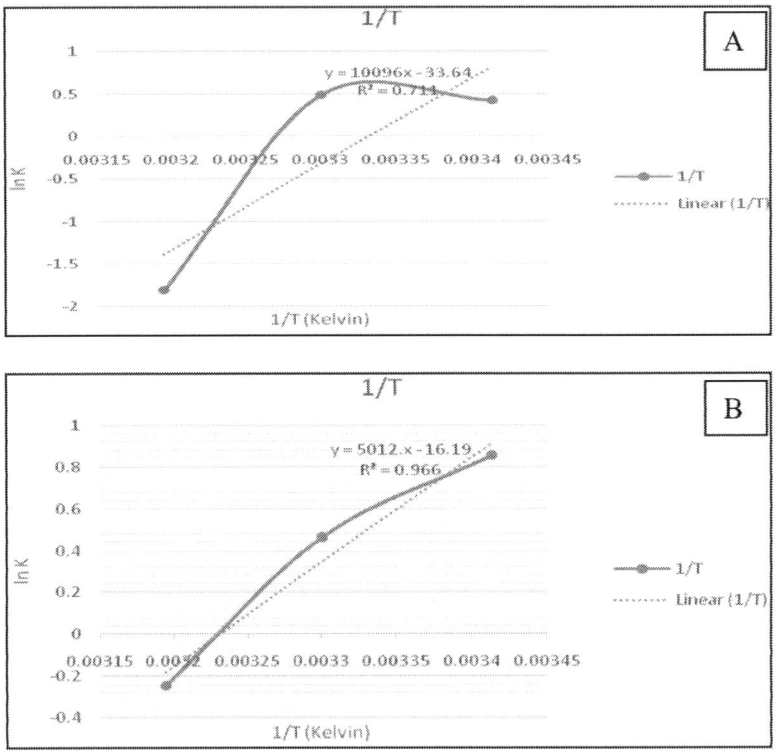


**Fig. 1**. The relationship of storage temperature (1/T) and ln K on WT1 (A) and WT0 (B) samples.

**Table 2**. Parameters of shelf life prediction of brownies (WT0 and WT1).

| Sample | T (K) | K | Initial TBA number | ct/co | Ln ct/co | Shelf life (day) |
|---|---|---|---|---|---|---|
| WT1 | 313 | 0.24 | 0.08 | 34.96 | 3.55 | 14.27 |
| | 303 | 0.72 | 0.08 | 34.96 | 3.55 | 4.92 |
| | 293 | 2.25 | 0.08 | 34.96 | 3.55 | 1.57 |
| WT0 | 313 | 0.83 | 0.01 | 192.30 | 5.25 | 6.30 |
| | 303 | 1.41 | 0.01 | 192.30 | 5.25 | 3.71 |
| | 293 | 2.48 | 0.01 | 192.30 | 5.25 | 2.11 |

*3.2 Estimatioon of shelf life of brownies using Arrhenius model*

Arrhenius model was applied of food products that easily damaged by the result of chemical reaction, such as fat oxidation, Maillard reaction, denaturation of proteins, and so on. In general, the rate of chemical reactions will accelerate at higher temperatures, which means the decrease in product quality is faster [6]. Storage by acceleration method is done to accelerate the process of damage including oxidation reactions that cause rancidity in the product. Acceleration method can be done by raising the storage temperature. In this study, variations of three storage temperatures are used ie temperature 20°C, 30°C and 40°C. During storage, measurement of TBA number was performed. The graph of TBA number for 30 days storage for brownies cake is presented in Figure 2.

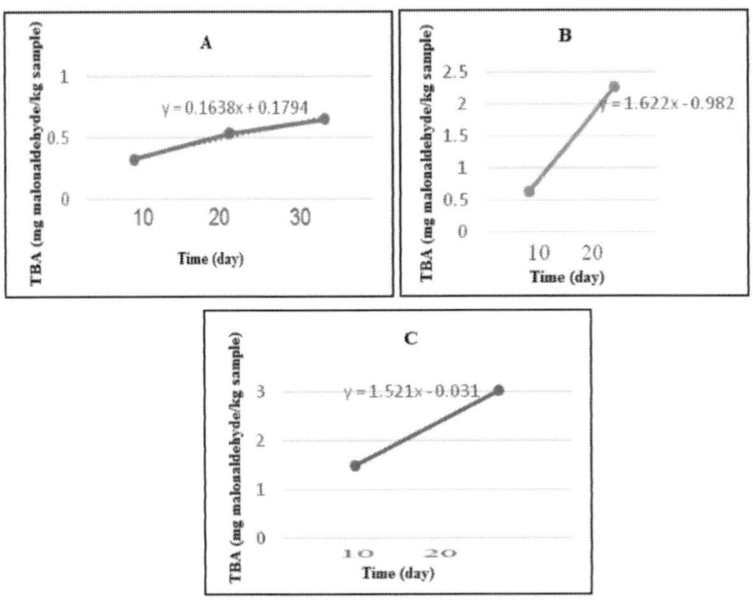


**Fig. 2**. TBA number of WT1 sample at 40°C (A), 30°C (B) and 20°C (C).

Based on the number of TBA in WT1 sample showed that the value of TBA decreased based on the increase of storage temperature. The storage at 40°C (days 5) showed a TBA number of 0.32 mg/kg sample. While storage days 10 and 15 showed TBA number of 0.53 mg/kg and 0.65 mg/kg sample, respectively. The number of TBA in the sample increases as storage time increases. TBA number at 30°C on days 10 and 20 ie. 0.71 and 2.30 mg/kg smple, respectively. TBA number at 20°C on days 15 and 30 are 1.48 and 3.01 mg/kg samples, respectively. Determination of shelf life of brownies can be estimated through the Arhennius equation at each storage temperature. Temperature increases (20, 30 and 40°C) showed an increase of shelf life of 1.57; 4.9 and 14 days.

WT0 sample (composition of wikau maombo flour 100% and wheat flour 0%) showed that the number of TBA at 40°C with the storage of 5; 10 and 15 days ie. 0.28; 0.58 and 1.84 mg/kg samples, respectively. TBA number increases with the increasing of storage time. Based on the equation of y = 0.7839x-0.663, the estimated of shelf life of WT0 is 6 days. The number of TBA at 30°C with storage time of 10 and 20 days are 0.71 mg/kg and 2.30 mg/kg sample, respectively. While TBA number at 20°C on storage time of 15 and 30 days are 2.32 mg/kg sample and 4.67 mg/kg sample, respectively. The shelf life of brownies on storage temperature of 30 and 20°C ie 4 and 2 days. Determination of shelf life of both brownies formulas through the reaction order of one approach. Figure 3 shows the number of TBA in WT0 sample at different storage temperatures.

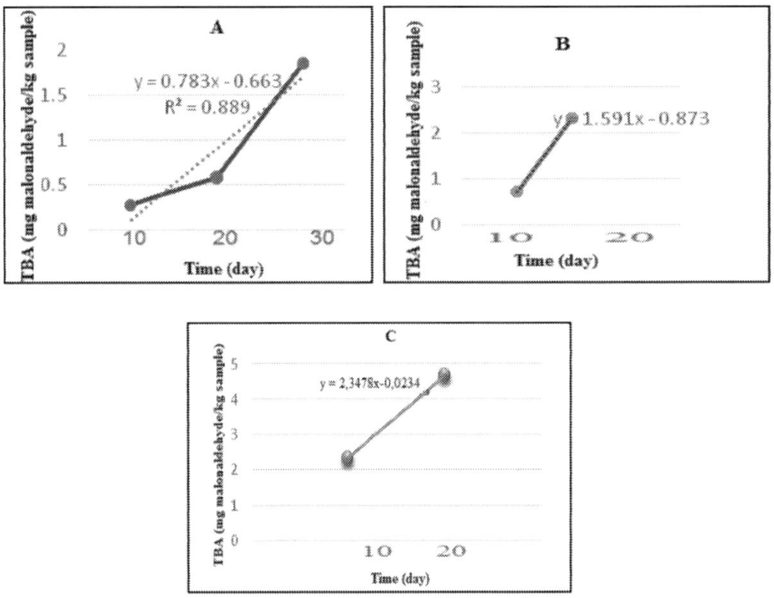


**Fig. 3**. TBA number of WT0 sample at 40°C (A), 30°C (B) and 20°C (C).

Figures 2 and 3 showed that the increase of brownie storage temperature causes an increase in the number of TBA. TBA number indicates the amount of malonal aldehyde produced due to the fat oxidation reaction. Fat oxidation will gradually produce aldehyde compounds such as (E/Z) hexenal. The shelf life is relatively short because the product is a very rich product of fat and sugar so it is susceptible to damage due to thermal treatment and yeast contamination. To increase the resistance of food products from physical, chemical or biological damage can use various antioxidants and antimicrobials of food grade standard. The use of an additive in food can be a natural or synthetic substance that safe for humans [10-11].

*3.3 Parameter of shelf life of selected brownies sample*

Characteristics of quality brownies stored at different storage temperatures are significantly altered when storage time of 20 days. Basically, brownie cake products are dark brown as they are influenced by the addition of dark chocolate and cocoa powder. Brownies were stored at room temperature and at 20°C overgrown with mold after storage time of 6 days. Decrease in quality occurs due to oxidation reactions that can cause changes in color, aroma, and texture of the product. In addition, the oxidation process that occurs is also caused by the growth of microbes in large numbers resulting in damage to food characteristics. While the product storage at 40°C until 20 days has not been covered with mold. The storage results of brownies are shown in Figure 4.

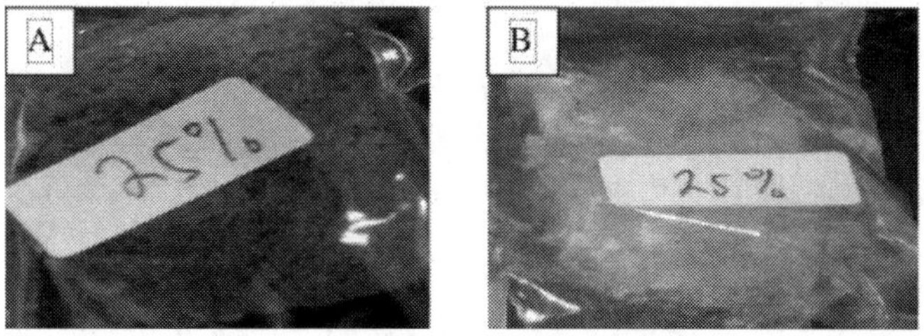


**Fig. 4**. Brownies before storage process (A) and brownies overgrown with white mold after 6 days storage at 20°C (B).

The extreme temperatures will accelerate the occurrence of product degradation and are often identified as the testing temperature of product shelf life [7]. Temperature control, humidity, and poor physical handling can be categorized as an abnormal food distribution condition. Distribution and temperature conditions will determine the shelf life of food products [8].

Based on the research results of Kusumo [5] has been reported that the shelf life of corn cookies product by Accelerated Shelf Life Test (ASLT) method, Arrhenius model with several parameters ie TBA, hardness, water content and Aw. Dry corn cookies was stored for 6 weeks, the sample is packaged in two different packages (metalized plastic and OPP-PP plastic with a thickness of 60 μm) and stored at 25°C, 35°C, and 45°C. The results showed that the estimation of shelf life, corn cookies packaged in plasticized metal has longer shelf life (2.4 months) than packaged in OPP-PP plastic (2 months). The TBA number (rancidity) is a major indicator in the destruction of corn cake quality. In addition, the shelf life of spirulina biscuit products by Arrhenius method, storage is done at 25°C and 35°C with observation interval is 5 days for 30 days. The parameters observed were fat content and peroxide number. The results showed that based on the parameters of packaging fat content that can maintain the quality of spirulina biscuits is packed aluminum foil when stored at 35°C that can last for 52 days 4 hours 49 minutes [9].

The use of room temperature can lead to greater reaction rate with increasing TBA numbers. If the reaction rate gets faster, then the product will be more quickly damaged which affects of the shelf life. Thus the good shelf life used from the results of this study is storage of WT1 sample at 40°C because the shelf life of the product the longer the 14 days.

The change of nutritional value such as fat during storage can form rancidity in foodstuffs. According to Ketaren [10], the change of peroxide numbers that occur during storage can suppress the rate of increase in the TBA number. The number of TBA is the most important for determining the degree of damage to fats and oils. The results showed that the number of TBA in brownies increased during storage of seven days at 30°C and 20°C. Bith temperature are suitable for mesophilic microbial growth that can accelerate the damage of brownies. The storage at higher temperature is suitable for WT0 and WT1 samples.

## 4. Conclusions

Rancidity is a critical quality parameter that affects the damage of brownies. In this study showed that the shelf life of brownies with wheat flour substitution has a longer shelf life at 40°C.

## References

[1] Edma AV, Pratjojo W and Budisusatyo E 2015 Indo. Uji proksimat dan organoleptik brownies dengan subtitusi tepung *MOCAF* (Modified Cassava Flour) (Proximate analyses and organoleptic of brownies with *MOCAF* substitution) *Indo Journal of Chemical Science* **4(3)** 168-171.

[2] Wahyuni S, Ansharullah, Saefuddin, Asranudin and Holilah 2017 Physico-chemical properties of Wikau maombo flour from cassava (*Manihot esculenta* Crantz) *Journal of Food Measurement and Characterization*. **11** 329–336.

[3] Wahyuni S, Ansharullah, Saefuddin, Asranudin and Holilah 2016 Characterization of Wikau maombo Flour from Fermented Cassava (*Manihot utilissima*) *International Journal of Chemical, Environmental and Biological Sciences (IJCEBS)*. **4(2)** 134-137.

[4] Labuza TP and Riboh D 1982 Theory and Aplication Of Arrhenius Kinetics to The Prediction of Nutrien Losses in Food *Journal of Food Technology*. **36** 66-74.

[5] Kusumo ME 2013 Shelf Life Estimation Of Corn Cookies Packed In OPP-PP And Metalized Plastic Using Accelerated Shelf-Life Testing Method-Arrhenius Equation. [Skripsi]. Program Studi Teknologi Pangan. Fakultas Teknologi Pertanian. Universitas katolik Soegijapranata Semarang.

[6] Hariyadi P dan Andarwulan N 2006 Perubahan Mutu (Fisik, Kimia dan Mikrobiologi) Produk Pangan Selama Pengolahan dan Penyimpanan (Quality changes (Physical, Chemical, Microbiological) in Food Products during Processing and Storage). Di dalam: Modul Pelatihan Pendugaan dan Pengendalian Masa Kadaluarsa Bahan dan Produk Pangan. Departemen Ilmu dan Teknologi Pangan dan Seafast Center IPB, Bogor.

[7] Hariyadi P 2004 Prinsip-prinsip pendugaan masa kedaluwarsa dengan metode Accelerated Shelf Life Test (Principles in determining life-shelf using Accelerated Shelf Life Test method). Pelatihan Pendugaan Waktu Kedaluwarsa (Self Life). Bogor, 1−2 Desember 2004. Pusat Studi Pangan dan Gizi, Institut Pertanian Bogor.

[8] Dewi R 2010 Pendugaan Umur Simpan Keripik Cakar Ayam (ceker) Menggunakan Metode Accelerated Shelf Life Testing Dalam Berbagai Bahan Pengemas [Skripsi] (Determining chicken feet chips shelf-life using Accelerated Shelf Life Testing in various packing materials [Essay]). Fakultas Teknologi Pertanian Universitas Brawijaya Malang.

[9] Astari MD 2015 Pendugaan Umur Simpan Biskuit Spirulina dengan Jenis Kemasan yang Berbeda (Determining Spirulina Shelf-life with different packing materials) *Jurnal Online Mahasiswa (JOM) Bidang Perikanan dan Ilmu Kelautan*. **2(2)** 1-7.

[10] Ketaren S 1986 *Pengantar Teknologi Minyak dan Lemak Pangan (Introduction to Oil and Fat)*. Edisi 1. Jakarta: UI-Press.

[11] Sunaryo E 2002 Pengolahan produk dan biji-bijian. [Skripsi] (Product and cereals processing [Essay]. Jurusan Teknologi Pangan dan Gizi, FakultasTeknologi Pertanian. IPB. Bogor.

**Acknowledgments**

The authors would like to acknowledge the Directorate General of Strengthening Research and Development, Ministry of Research, Technology and Higher Education, Indonesia for financial support under "MP3EI" grant research.

# Effect of sodium azide addition and aging storage on casein micelle size

**H Sinaga[1], H Deeth[2], B Bhandari[2]**
[1]Food Science and Technology, Universitas Sumatera Utara, Medan, Indonesia.
[2]School of Agriculture and Food Sciences, The University of Queensland, Brisbane, Australia
E-mail: hotnida.sinaga@uq.net.au

**Abstract.** Casein micelles affected most of milk properties, therefore the use sodium azide as milk preservation is not expected to alter milk properties during storage, including the casein micelle size. The aim of this study was to analyse casein micelle size after the addition of sodium azide during storage. The experiment was performed as a complete block randomised design with three replications. The addition of 0.02-0.10% Na-azide do not lead to any noticeable differences in average casein size at the same day and show similar trend after 14 day-storage. At concentration of 0.02% sodium azide (Na-azide), the size of pasteurised milk did not change up to 12 days, while the size of raw skim milk slightly increased by ageing time at day 5. The treated concentration did not affect the size distribution, except for milk with 0.02% Na-azide which had narrower distribution compared to other treated and control milk. The finding from this study suggests that the role of Na-azide in this experiments during storage at 4°C is only for preventing the microbial growth.

## 1. Introduction

Milk contains high amount of proteins, lipids and minerals, along with carbohydrates and vitamins. However, it is believed that the change in protein state is a basis for creating new dairy ingredients or products [1], because milk proteins mainly affect the properties of many dairy products [2]. Consequently, the properties of casein, comprises nearly 80% of the milk protein, have great impact on the functional behaviour of dairy products.

Based on the milk composition, it can be assumed that the main structural components of milk are fat globules and casein micelles (CMs). The white colour of skim milk is caused by the colloidal casein micelles scattering significant amount of light. Casein micelles exist in milk as aggregates of the casein proteins and a proportion of mineral calcium phosphate [3]. The casein micelles have an extremely variability of size and are sterically stabilised by a hairy layer of κ-casein molecules and perhaps β-casein [4, 5].

There are some components that contribute to the stability of casein micelles, including the integration of κ-casein, colloidal calcium phosphate and hydrophobic interactions [6] (Roach and Harte 2008). The collapse of surface protein κ-casein will affect the micellar solubility and generate aggregation [7]. On the other hand, the removal of calcium phosphate induces the dissociation of the micelles [8]. Hydrophobic interaction is important in the link between sub-micelles, so the distraction of the interactions might cause the micellar dissociation or aggregation [7]. Therefore, modification or disruption of the components might initiate either micelle aggregation or dissociation into submicelles and lead to the change in casein micelle size.

Various study reported that sodium azide (Na-azide), in various concentrations, was added to prevent microbial growth at experiments on casein micelle size [9,10,11]. However, it was not reported whether the addition would have an impact on the parameter measured. Sodium azide, also known as sodium trinitride, is the inorganic compound with the formula $NaN_3$. Hydrazoic acid or azoimide is a colourless liquid, which boils and solidifies at 37°C and -80°C, respectively. The azomide salt of sodium is stable and dissociate in solution [12]. In laboratories, Na-azide is useful for a bacteriostatic and also as a preservative. It is also a metabolic inhibitor, which inhibits oxidative phosphorylation. The role of sodium azide has been studied many years ago, not only in milk but also other products. Other research found that sodium azide can inhibit yeast and mold activities, caused by genus Candida in tomato juice. However, the effect significantly decreased, if the tomato juice is refrigerated for more than 5 days [13].

Limited research has been performed which relates to casein micelle size on milk during experiment. Therefore, the research findings are expected to provide more information on casein micelle size after the addition of sodium azide and how long the preservative can prolong shelf life of the milk.

## 2. Materials and methods

### *2.1. Materials*

Fresh pasteurised skim milk was purchased from local supermarket. For comparison, raw skim milk from local farm was used.

### *2.2. Methods*

In order to understand effects that might relate to the size measurements, we carried out the experiments on the addition of Na-azide (antimicrobial agent) with various concentrations and also ageing of the skim milk. Then, the samples were stored cold at 4 °C until further treatments. The initial pH was around 6.6 and the average size and size distribution of the casein micelles were determined. All experiments were repeated three times on different batches of milk samples.

### *2.3. Sodium azide*

Pasteurised milk was divided into 4 parts of 100 ml sample. Na-azide was added to the each sample at the concentrations of 0, 0.02, 0.05 and 0.1%. The sub-samples were stored in refrigerator at 4°C then subjected to analysis for casein micelle size at 0, 3, 5, 7, 10, 12 and 14 days of storage. Check batches were milk sub-sample without preservative and were analysed only until 7 days.

### *2.4. Effect of storage*

Pasteurised and raw milk were stored in refrigerator at 4°C for 14 days. Sodium azide (w/v) was added at day 0. The casein micelle size was measured every 2 or 3 days for 14 days of storage.

### *2.5. Experimental Design and Statistics*

All experiments were performed as a complete block randomised design with two or three replications. Statistical analyses to determine any treatment effects were done using Minitab 16.0. The corresponding data were determined by analysis of variance (ANOVA) using the General Linear Model. Tukey simultaneous test at the level of $P = 0.05$ was carried out to assess whether different treatments resulted in statistically significant differences.

## 3. Results and discussion

### *3.1. Effect of sodium azide on casein micelle size*

Casein micelles size at different concentration of Na-azide are presented in Figure 1. Although the CM size ranged between 251-258 nm, analysis result of the size among all samples showed no statistical

IOP Conf. Series: Earth and Environmental Science **122** (2018) 012083 doi:10.1088/1755-1315/122/1/012083

difference until day 7 of storage. Moreover, the addition of 0.02 - 0.1% Na-azide do not lead to any noticeable differences in average casein size at the same day. It was clear that the milk without Na-azide addition (0% Na-azide) became more concentrated, so that CM measurements after day 7 was terminated for this samples. Milk with various Na-azide concentration showed similar trend after 14-day of storage.

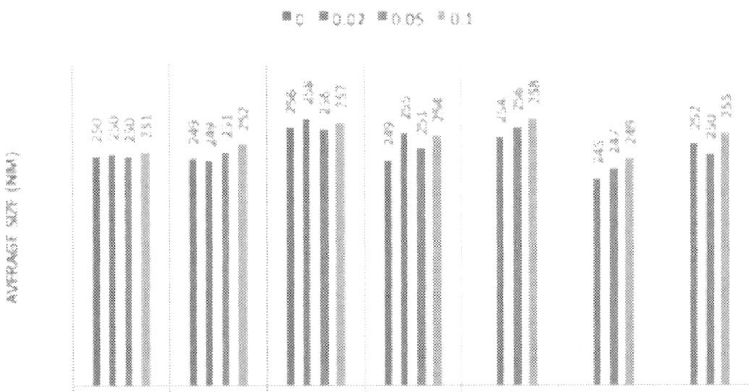


| | 0 | 3 | 5 | 7 | 10 | 12 | 14 |
|---|---|---|---|---|---|---|---|
| 0 | 250 | 249 | 256 | 249 | | | |
| 0.02 | 250 | 249 | 258 | 255 | 254 | 245 | 252 |
| 0.05 | 250 | 251 | 256 | 251 | 256 | 247 | 250 |
| 0.1 | 251 | 252 | 257 | 254 | 258 | 249 | 255 |

**Figure 1.** The particle size of skim milk with Na-azide addition after 14-day storage at 4°C

Figure 2 reported the size distribution of casein micelles at day 0, 7 and 14 days. At day 0, all untreated and treated milk has similar size distribution. In agreement with the identical trend amongst the Na-azide addition, only little differences were observed between micelle size distributions in the treated samples, especially after 14 day of storage. However, at day 7, although the addition of 0.02, 0.05, 0.1% Na-azide did not affect the average size of casein (255, 251, 254 nm, respectively), the 0.02% Na-azide had narrower distribution compared to other treated and control milk.

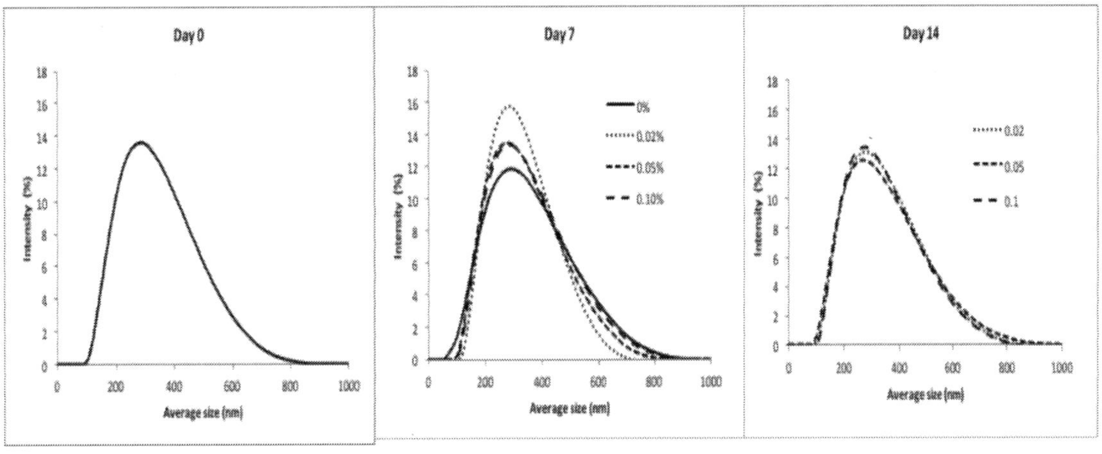

**Figure 2.** The size distribution of skim milk with Na-azide addition at day 0, 7 and 14 stored at 4°C

Na-azide is an ionic solid, soluble in water with the formula $NaN_3$. This colourless azide has been used as a common preservative agent in stock solution in laboratory and also applied in milk samples [14, 15, 16, 17, 18, 19]. Although some papers did not mention the type and the concentration of preservative being employed in the milk samples, the amount of Na-azide being used is noted in the range of 0.01% [20, 21], 0.02% [22], 0.04%, 0.05% [23] and up to 0.1%.

Result of this study indicated that the preservation concentrations mentioned can, therefore, be classified to be equally good, as they did not influence the parameter measured. Therefore, throughout the experiments, 0.02% Na-azide will be used to prevent the microbial growth.

*3.2. Effect of ageing of skim milk*

The average casein size of skim milk (SM) and raw skim milk (RSM) during 14-day storage are presented in Table 1. Storage time had significant effect on the parameter measured, either for SM and RSM. For both milk samples, the size tended to increase during the storage, rise up to 20 nm in SM and 30 nm in RSM.

For skim milk, the size after 2-day storage was similar with the control milk. However, there was a noticeable increase in the size at prolong storage up to 14 days, with the maximum size at day 5, increase from 243 nm to 266 nm. Although a milder, linear-like decrease occurred after 7 days, the size increased again at day 14.

Similarly, the size of raw skim milk increased over the storage time, with the maximum size was after day 5. The Na-azide addition has no significant effect on CM size until 14-day of storage. This study indicated that the trend for CM size of SM (skim milk) and RSM (raw skim milk) was not affected by the application of pasteurisation for skim milk.

**Table 1.** Casein micelle size of skim milk and raw skim milk during ageing storage for 14 days

| **Day** | **0** | **2** | **5** | **7** | **9** | **12** | **14** |
|---|---|---|---|---|---|---|---|
| **SM** | 243±2.0*d* | 245±2.0*d* | 266±2.4*a* | 257±3.3*abc* | 250±1.7*bcd* | 249±1.9*cd* | 260±2.0*ab* |
| **RSM** | 242±2.0*b* | 245±2.0*ab* | 262±2.0*ab* | 255±2.0*ab* | 258±2.0*ab* | 261±2.0*ab* | 270±2.0*a* |

SM: skim milk RSM: raw skim milk

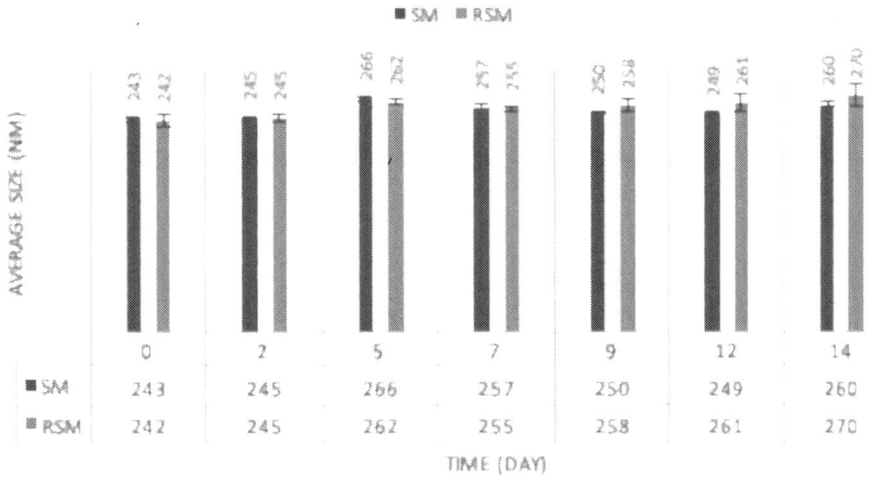


**Figure 3.** The particle size of skim milk (SM) and raw skim milk (RSM) after 14-day of storage at 4°C

The size distribution of skim milk and raw skim milk are presented in Figure 4 and 5, respectively. Raw skim milk was used without any pasteurisation treatment after milk separation from cream. However, only little differences were observed between size distributions in skim milk and raw skim milk under the same day of storage. Both of milk types had similar size and size distribution.

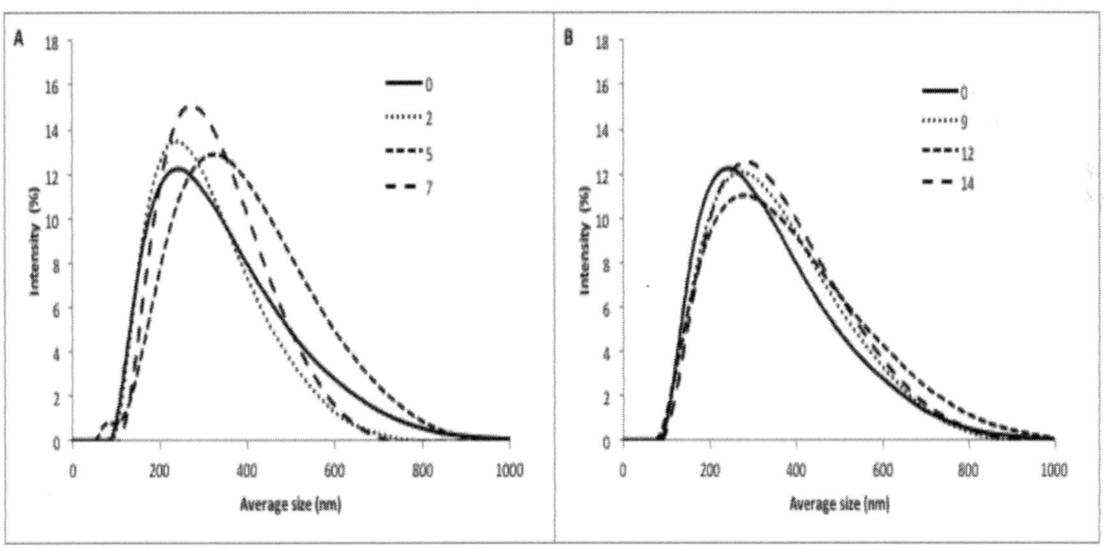


**Figure 4.** The size distribution of skim milk after 9-day (A) and 14-day (B) storage at 4°C

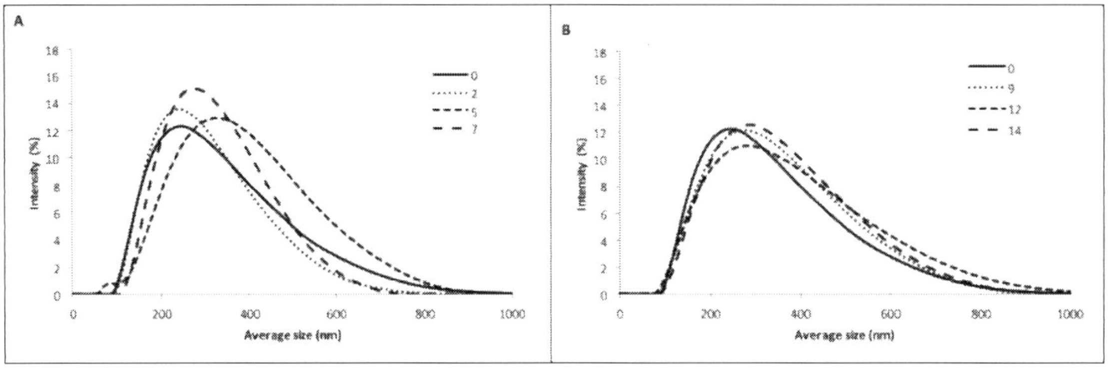


**Figure 5.** The size distribution of raw skim milk after 9-day (A) and 14-day (B) storage at 4°C

The size of casein micelles in fresh milk is well established, however, it seemed that the casein behaviour during storage of milk cannot be neglected. This result is in agreement with result from previous study, which noted the slight increase in particle size after 8-day storage [24]. He found that the aging process leads to an increase in particle size by 20 nm at day 7 and 30 nm at day 8. Other research reported that on routine determination of urea in milk, the addition of Na-azide at ten times higher level than recommended did not affect the added urea content [25].

## 4. Conclusions

It is concluded that Na-azide did not change the CM size during ageing storage. The addition of Na-azide up to 1% was only used as a preservative for preventing the microbial growth. It is also suggested that preserved milk samples should be kept in the refrigerator and treated or analysed within less 5 days, to minimise the effect of storage time.

## References

[1] Onwulata C, Phillips J, et al. 2010 Texturized dairy proteins." *Journal of Food Science* 75(2): E100.

[2] Phadungath C 2005 "Casein Micelle Structure: A concise Review." *Songklanakarin J. Sci. Technol.* 27(1): 201-212.

[3] Walstra P 1999 "Casein sub-micelles: do they exist?" International Dairy Journal 9(3-6): 189-192.

[4] Slattery CW 1976 Review: Casein micelle structure - Examination of models." *Journal of Dairy Science* 59(9): 1547-1556.

[5] Muller-Buschbaum P, Gebhardt R, et al. 2007 Effect of calcium concentration on the structure of casein micelles in thin films." *Biophysical Journal* 93(3): 960-968.

[6] Roach A. and Harte F 2008 Disruption and sedimentation of casein micelles and casein micelle isolates under high-pressure homogenization." *Innovative Food Science & Emerging Technologies* 9(1): 1-8.

[7] de Kruif CG 1999 Casein micelle interactions." *International Dairy Journal* 9(3-6): 183-188.

[8] Panouillé M, Nicolai T and Durand D 2004, 'Heat induced aggregation and gelation of casein submicelles', *International Dairy Journal,* vol. 14, no. 4, pp. 297-303.

[9] Sinaga H, Bansal N and Bhandari, B 2016. Partial renneting of pasteurized bovine milk: Casein micelle size, heat and storage stability. *Food Research International*. Vol 84 pp 52-60

[10] Sinaga H, Bansal N and Bhandari, B. 2017. Gelation properties of partially renneted milk. *International Journal of Food Properties*, 20 (8): 1700-14

[11] Sinaga H, Bansal N and Bhandari B 2016 Effects of milk pH alteration on casein micelle size and gelation properties of milk. *International Journal of Food Properties*. DOI 10.1080/10942912.2016.1152480
[12] Slawson MS, Snyder ML 1952 Mechanism of action sodium azide on the genus Candida. *Journal of Dental Research* vol 31 (1): 47
[13] Graham JDP 1949 Actions of sodium azide *Brit. J. Pharmacol.* 4: 1
[14] Hamann J, Gyodi P, et al. 1991. "On the Influence of Varying Preserving Agents and Storage Times on Somatic-Cell Counting in Milk Using the Fossomatic-360 and Fossomatic-180." Kieler Milchwirtschaftliche Forschungsberichte 43(4): 273-295.
[15] Hanus O, Gencurova V, et al. 1992 Tests of Milkofix, a New Preservative Substance for Milk Samples Used for the Purposes of an Infrared-Analysis of Basic Milk-Composition .2. Checks of Preservative Effects in Relation to the Infrared-Analysis." *Veterinarni Medicina* 37(1): 33-43.
[16] Chalermsan N, Vijchullata P, et al. 2003 Influence of preservatives on raw milk components and somatic cell counts analysis. *Proceedings of 41st Kasetsart University Annual Conference, 3-7 February, 2003.* Subject: Animals and Veterinary Medicine. Kasetsart University Annual Conference: 127-135.
[17] Gallaher JJ, Hollender R, et al. 2005 Effect of composition and antioxidants on the oxidative stability of fluid milk supplemented with an algae oil emulsion *International Dairy Journal* 15(4): 333-341.
[18] Haddadi K, Moussaoui F, et al. (2005). "E-coli proteolytic activity in milk and caseinbreakdown." *Reproduction Nutrition Development* 45(4): 485-496.
[19] Besse NG, Cauquil A, et al. (2008). Comparative Study of Different Milk Samples Preservation Procedures for Bacteriologic Examination *Food Analytical Methods* 1(1): 36-42.
[20] Madadlou A, Mousavi ME, et al. 2009 Alkaline pH does not disrupt re-assembled casein micelles *Food Chemistry* 116(4): 929-932.
[21] Knudsen JC and Skibsted LH 2010 High pressure effects on the structure of casein micelles in milk as studied by cryo-transmission electron microscopy." *Food Chemistry* 119(1): 202-208.
[22] Sandra S and Dalgleish DG 2007 The effect of ultra high-pressure homogenization (UHPH) on rennet coagulation properties of unheated and heated fresh skimmed milk." *International Dairy Journal* 17(9): 1043-1052.
[23] Huppertz, T., P. F. Fox, et al. (2004a). "Properties of casein micelles in high pressure-treated bovine milk." Food Chemistry 87(1): 103-110.
[24] Lieske B 1998 Effects of ageing of raw milk on some structural properties of native casein micelles." *Milchwissenschaft-Milk Science International* 53(10): 562-565.
[25] Luzzana M and Giardino R 1999 Urea determination in milk by differential pH technique. *Lait* vol 79 (2): 261-67.

**Acknowledgements**
The authors would like to thank Dr. John M. Schiller (Honorary Senior Fellow, School of Agriculture and Food Sciences, The University of Queensland), for his assistance in the proof-reading of the Abstract of this manuscript.

# Effect of microfluidization on casein micelle size of bovine milk

**H Sinaga[1], H Deeth[2], B Bhandari[2]**

[1]Food Science and Technology, Universitas Sumatera Utara, Medan, Indonesia
[2]School of Agriculture and Food Sciences, The University of Queensland, Brisbane, Australia
E-mail: hotnida.sinaga@uq.net.au

**Abstract.** The properties of milk are likely to be dependent on the casein micelle size, and various processing technologies produce particular change in the average size of casein micelles. The main objective of this study was to manipulate casein micelle size by subjecting milk to microfluidizer. The experiment was performed as a complete block randomised design with three replications. The sample was passed through the microfluidizer at the set pressure of 83, 97, 112 and 126 MPa for one, two, three, four, five and six cycles, except for the 112 MPa. The results showed that microfluidized milk has smaller size by 3% with pressure up to 126 MPa. However, at each pressure, no further reduction was observed after increasing the passed up to 6 cycles. Although the average casein micelle size was similar, elevating pressure resulted in narrower size distribution. In contrast, increasing the number of cycles had little effect on casein micelle distribution. The finding from this study can be applied for future work to characterize the fundamental and functional properties of the treated milk.

## 1. Introduction

Milk has so many benefits in human life as it contains the most nutrition food in nature. Different types of milk will have wide variation in nutrient proportions, however, generally milk containing high amount of proteins, lipids and minerals, along with carbohydrates and vitamins. Whole milk is composed of 87.42% water and 12.58% total solids, which compromised 3.86% lipid components and 8.73% of solids non fat. Lactose, nitrogenous matter, salts and other solids account for 4.67%, 3.28%, 0.66% and 0.17%, respectively, of the solids non fat [1].

Caseins as a dominant protein are a family of phosphorylated proteins. Caseins are about 80% of the protein in bovine milk, represents 4 gene products: αs1-, αs2-, β- and κ- caseins in the molar ratio 4:1:4:1 [2] and a trace amount of gamma casein [3]. Caseins also contain a large proportion of minerals (calcium and phosphate), which are required for bone growth for the newborns [4]. Most of caseins exist as casein micelles; they tend to associate because of their high hydrophobic bonding. The smallest aggregates can contain 10-20 moles casein [5].

Various processing technologies produce particular change in the average micelles sizes (CMs) [6, 7, 8]. As the properties of milk are likely to be dependent on the micelle size, the micelle size modification might affect the behaviour of protein during processing [9, 10]. In the past, most of the works are focused on the casein micelles model and the composition. Only little is known about the main effect of casein micelle size and on how the manipulation in their micelle structural properties contribute to the functionality of protein in general.

Homogenization in milk processing is generally applied to produce more consistent raw milk. Naturally fat milk will separate and form the top layer of cream, as the fat has a less dense than the milk serum. Initially homogenisation refers to break down the fat globules, and allow the smaller particles distribute consistently in milk [11]. Other research suggested that high pressure homogenisation can also be used to produce particular change in the molecular structure of protein, that can alter the milk constituent and properties [11, 12]. Along with homogenisation, microfluidisation is the most common method used in high shear processing. The samples are processed in an inline homogeniser and micro channels of a Microfluidizer, for homogenisation [13, 14], dispersion [15], emulsification [16, 17, 18] and particle size reduction [19, 20].

The equipments were originally used only for pharmaceutical emulsions, however nowadays there are widely used in dairy production [20]. When compared to conventional homogenised milk, microfluidized milk have smaller particle size and narrow size distribution [19], but resulting in larger particles in yoghurt [21]. Although Olson et al [21] observed that the particles size only decreased at the pressure < 100 MPa, others found that the decrease still continued at 150 MPa [21]. Other study had analysed functions to predict average diameter and size distribution of the microfluidized fat globules in dairy model emulsions [20].

Fat globules and emulsions prepared by microfluidization have been repeatedly documented in literatures. The structure of protein differed from those produced by valve-homogeniser. The change in particle size distribution was best measured by dynamic light scattering (DLS), as the instrument can measure the size smaller than 100 nm [22]. The results provided good overviews of using the device for the reduction of casein micelles size.

In this study, we will attempt to develop methods to manipulate the casein micelle size. Although various study reported the casein micelle size of by applying pH [23], partial renneting [24, 25], limited research has been performed which relates to the application of a Microfluidizer. The research findings are expected to solve many unknown size-property relationship, thus can provide enormous opportunities to develop new generation dairy products with variable functionality.

## 2. Materials and methods

In attempt to manipulate the size of casein micelles, a Microfluidizer was employed, with operating pressure ranged from 0 - 126 MPa. Further treatments were conducted by recirculating the sample through the homogeniser for up to six cycles. Then, the average size and size distribution of the casein micelles were determined.

### *2.1. Microfluidizer*

Pasteurised skim milk was homogenised using a Microfluidizer (Model B12-04DJCM3, Watts Fluidair Inc, Kittery, Maine) operating in continuously or recycles. The sample was passed through the microfluidizer at the set 12, 14, 16 and 18 KPsi (83, 97, 112 and 126 MPa, respectively) pressure for one cycle. The portion of the microfluidized sample was taken for size analysis, and the remaining was passed through a Microfluidizer again for second pass, the third and the fourth pass at 83, 97 and 126 MPa.

### *2.2. Experimental Design and Statistics*

All experiments were performed as a complete block randomised design with three replications. Statistical analyses to determine any treatment effects were done using Minitab 16.0. The corresponding data were determined by analysis of variance (ANOVA) using the General Linear Model. Tukey simultaneous test at the level of $P = 0.05$ was carried out to assess whether different treatments resulted in statistically significant differences.

## 3. Results and discussion

Microfluidized treatment significantly reduced the particles size of milk from 246 $\pm$ 0.1 nm to 239 $\pm$ 1.5 nm (Figure 1). However, elevated pressure from 83 to 126 MPa for one pass had no significant effect on the average size. This meant that casein micelle size did not change if the milk was subjected to a Microfluidizer with pressure less than 130 MPa. The average size for all treated milk was around 240 nm, only < 10 nm different with the untreated sample.

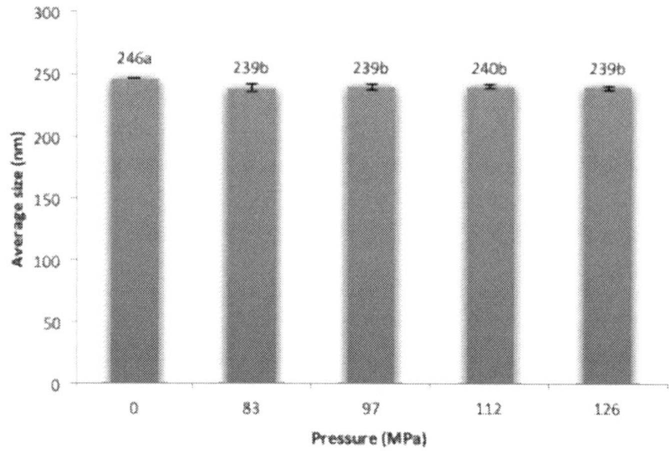


**Figure 1.** The particle size of microfluidized milk at pressure of 0 - 126 MPa

Although the average size of casein after treatment 83 - 126 MPa was similar, the size distribution in intensity (%) at lower and higher pressure treatment (112 and 126 MPa), was different (Figure 2). The micelle size distribution in milk treated at higher pressure became narrower than that of the untreated milk. The distribution was shifted towards smaller size.

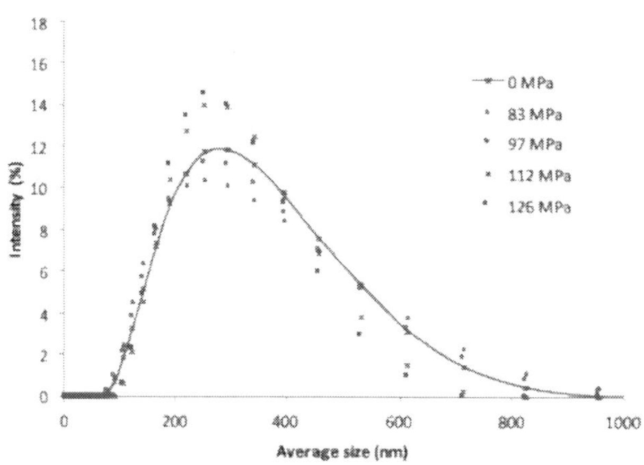


**Figure 2.** The size distribution of microfluidised milk at pressure of 0 - 126 MPa

In the multi-passes treatment, the results revealed that more cycle have no greater impact on the reduction of casein micelles size of microfluidized milk (Figure 3). The size decreased only up to 15 nm at 126 MPa at single pass. Further increasing the number of passes did not significantly change the size.

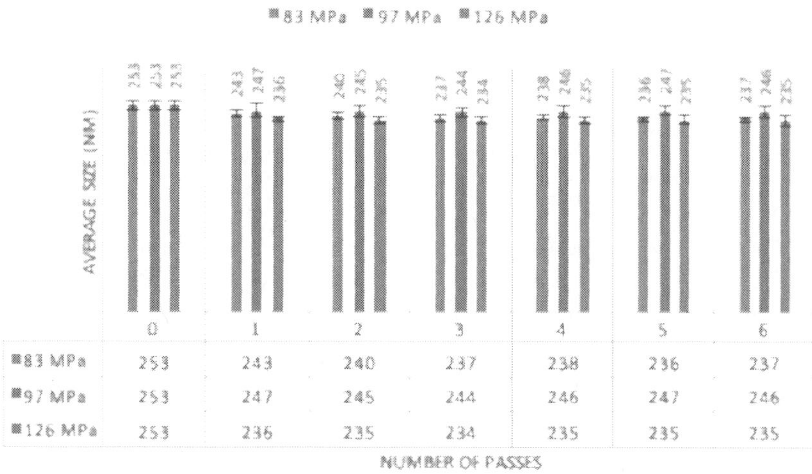


**Figure 3.** The particle size of microfluidized milk at pressure of 83 - 126 MPa as a factor of number of passes

Treatment of skim milk at 83 - 126 MPa reduced casein micelles in all 6 passes, however, only little differences was observed between micelle size distribution of treated and untreated milk (Figure 4). At all milk samples the size of micelles being present is < 800 nm.

These data suggest that elevating the pressure could lessen the size, however, no further reduction was observed after increasing up to 6 passes. Treatment up to 150 MPa had little effect on casein micelles size [8, 26, 27]. Other study found that only large micelles (300 - 600 nm) might dissociate at this pressure treatment [28]. As the maximum pressure of the Microfluidizer is 126 MPa, other method such as high pressure homogenisation (HPH) application can be used for higher pressure treatment.

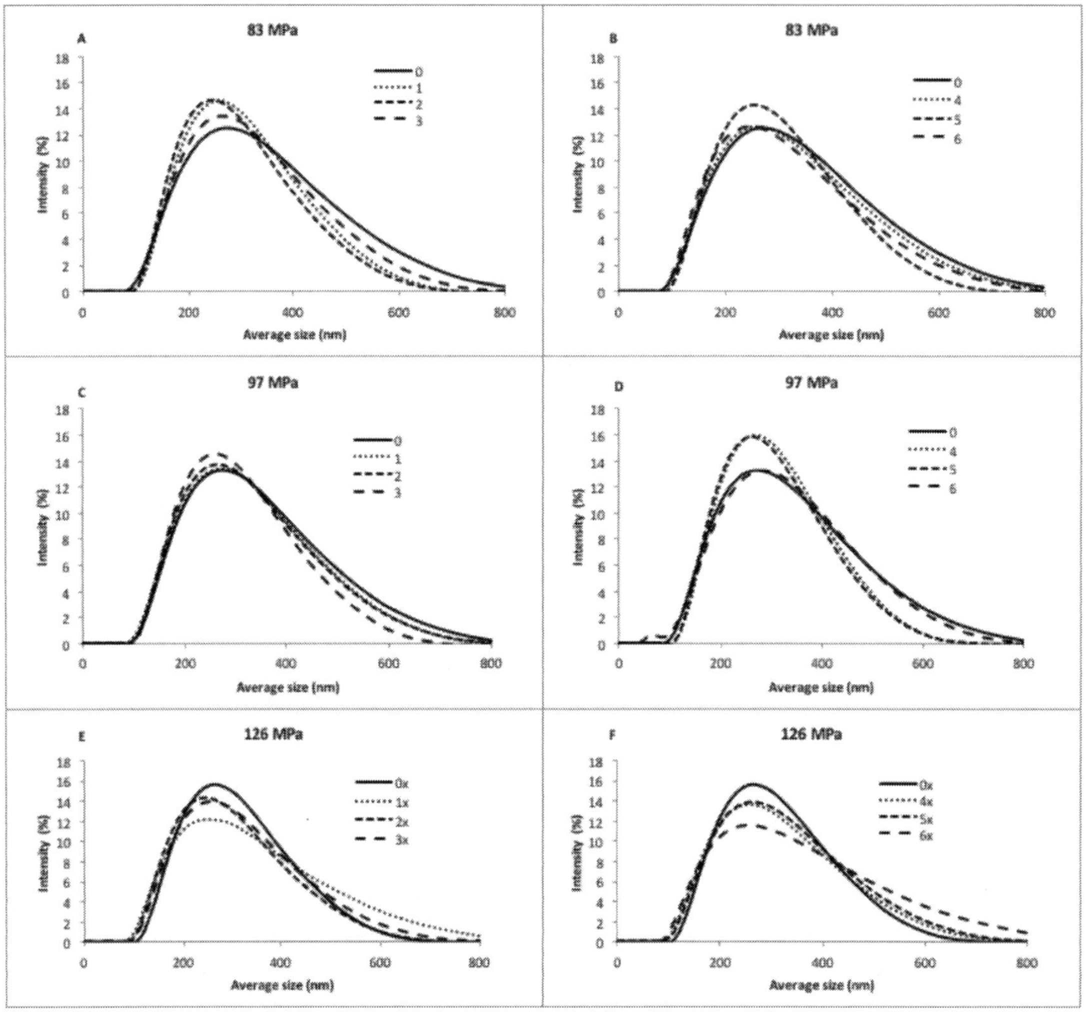


**Figure 4.** The size distribution of microfluidized milk at pressure of 83 - 126 MPa as a factor of number of passes

## 4. Conclusions

It is concluded that casein micelle size can be manipulated by the application of Microfluidizer. The CM size decreased at pressure of 83 MPa. However, there was no significant different amongst the size, even though the pressure was increased up to 126 MPa. Similarly, the number of passes up to 3 cycles did not affect the CM size. The findings suggested that applying Microfluidizer at 83 MPa for one cycle was enough to manipulate the CM size.

## References

[1] Modler HW, Ed. 2000 *Milk processing*. Food proteins: Processing Applications. New York, Wiley-VCH, Inc.

[2] Chowdhury PB and Luckham PF 1995 Interaction forces between kappa-casein adsorbed on mica', *Colloids and Surfaces B-Biointerfaces* Vol. 4(6): 327-334.

[3] Swaisgood, HE, Ed. 2003 *Chemistry of the caseins*. Advanced dairy chemistry - I: Proteins. Part A. New York, Kluwer Academic/ Plenum Publishers.

[4] Erkkola M, Kronberg-Kippila CK, et al. 2005 Maternal consumption of dairy products during pregnancy and lactation, and the development of cow's milk antibodies in the offspring. *Acta Paediatrica* Vol. **94** (6): 696-704.

[5] Fox PF and Brodkorb A 2008 The casein micelle: Historical aspects, current concepts and significance. *International Dairy Journal* Vol. **18** (7): 677-684.

[6] Ono T, Murayama T, Kaketa S and Satoshi 1990, 'Changes in the protein composition and size distribution of bovine casein micelles induced by cooling. *Agricultural and Biological Chemistry,* Vol. **54** (6): 1385-92.

[7] Needs EC, Stenning RA, Gill AL, Ferragut V & Rich GT 2000 High-pressure treatment of milk: effects on casein micelle structure and on enzymic coagulation. *Journal of Dairy Research* Vol. **67** (1): 31-42.

[8] Orlien V, Boserup L and Olsen K 2010 Casein micelle dissociation in skim milk during high-pressure treatment: Effects of pressure, pH, and temperature. *Journal of Dairy Science* Vol. **93** (1): 12-18.

[9] McMahon DJ and Brown RJ 1984 Composition, structure, and integrity of casein micelles - A review. *Journal of Dairy Science* Vol. **67** (3): 499-512.

[10] De Kruif CG, Tuinier R, Holt C, Timmins PA and Rollema HS 2002 Physicochemical study of kappa- and beta-casein dispersions and the effect of cross-linking by transglutaminase. *Langmuir,* Vol. **18** (12): 4885-91.

[11] Lopez-Fandino R 2006 Functional improvement of milk whey proteins induced by high hydrostatic pressure. *Critical Reviews in Food Science and Nutrition* Vol. **46**(4): 351-363.

[12] Garcia-Risco MR, Ramos M and Lopez-Fandino 2002 Modifications in milk proteins induced by heat treatment and homogenization and their influence on susceptibility to proteolysis. *International Dairy Journal* Vol. **12**(8): 679-688.

[13] Whiteley AJ and Muir DD 1996 Heat stability of homogenised concentrated milk .1. Comparison of microfluidizer with a valve homogeniser *Milchwissenschaft-Milk Science Internasional* Vol. **51** (6): 320-323.

[14] Hardham JF, Imison BW and French HM 2000 Effect of homogenisation and microfluidisation on the extent of fat separation during storage of UHT milk. *Australian Journal of Dairy Technology* Vol. **55** (1): 16-22.

[15] Jafari SM, He Y & Bhandari B 2007 Optimization of nano-emulsions production by microfluidization. *European Food Research and Technology* Vol. **225** (5-6): 733-741.

[16] McCrae CH 1994 Homogenization of milk emulsions - Use of Microfluidizer. *Journal of the Society of Dairy Technolog* Vol. **47** (1): 28-31.

[17] Jafari SM, He YH and Bhandari B 2007 Production of sub-micron emulsions by ultrasound and microfluidization techniques. *Journal of Food Engineering* Vol. **82** (4): 478-488.

[18] Mao LK, Yang J, Xu DX, Yuan F and Gao YX 2010 Effects of homogenization models and emulsifiers on the physicochemical properties of beta-carotene nanoemulsions. *Journal of Dispersion Science and Technology* Vol. **31** (7): 986-993.

[19] Strawbridge KB, Ray E, Hallet FR, Tosh SM and Dalgleish DG 1995 Measurement of particle size distributions in milk homogenized by a Microfluidizer: estimation of populations of particles with radii less than 100 nm. *Journal of Colloid and Interface Science* Vol. **171** (2): 392-398.

[20] Robin O, Blanchot V and Paquin P 1992 Microfluidization of dairy model emulsions. 1. Preparation of emulsions and influence of processing and formulation on the size distribution of milk fat globules. *Lait* Vol. **72** (6): 511-531.

[21] Ciron CIE, Gee VL, Kelly AL and Auty MAE 2010 Comparison of the effects of high-pressure microfluidization and conventional homogenization of milk on particle size, water retention and texture of non-fat and low-fat yoghurts. *International Dairy Journal* Vol. 20 (5): 314-320.

[22] West SJ 1996 Characterization of particles in emulsion systems formed by microfluidization. University of Guelph (Canada). Canada, University of Guelph (Canada), 1996; AAT MM14488.
[23] Sinaga H, Bansal N and Bhandari B 2016 Effects of milk pH alteration on casein micelle size and gelation properties of milk. *International Journal of Food Properties.* DOI 10.1080/10942912.2016.1152480
[24] Sinaga H, Bansal N and Bhandari, B 2016. Partial renneting of pasteurized bovine milk: Casein micelle size, heat and storage stability. *Food Research International.* Vol 84: 52-60
[25] Sinaga H, Bansal N and Bhandari B 2017 Gelation properties of partially renneted milk. *International Journal of Food Properties*, Vol 20 (8):1700-14
[26] Huppertz T, Fox PF and Kelly AL 2004 Properties of casein micelles in high pressure-treated bovine milk. *Food Chemistry.* Vol. **87** (1): 103-110.
[27] Anema SG, Lowe EK and Stockmann R 2005 Particle size changes and casein solubilisation in high-pressure-treated skim milk. *Food Hydrocolloids.* Vol. **19** (2): 257-267.
[28] Regnault S, Thiebaud M, Dumay E and Cheftel JC 2004 Pressurisation of raw skim milk and of a dispersion of phosphocaseinate at 9 degrees C or 20 degrees C: effects on casein micelle size distribution. *International Dairy Journal.* Vol. **14** (1): 55-68.

**Acknowledgements**
The authors would like to thank Dr. John M. Schiller (Honorary Senior Fellow, School of Agriculture and Food Sciences, The University of Queensland), for his assistance in the proofreading of the Abstract of this manuscript.

IOP Conf. Series: Earth and Environmental Science **122** (2018) 012085 doi:10.1088/1755-1315/122/1/012085

# Removal efficiency of methylene blue using activated carbon from waste banana stem: Study on pH influence

**E Misran*, O Bani, E M Situmeang and A S Purba**

Department of Chemical Engineering, Universitas Sumatera Utara,
Jl. Almamater Kampus USU, Padang Bulan, Medan 20155, Indonesia

erni_misran@yahoo.com

**Abstract**. The effort to remove methylene blue in artificial solution had been conducted using adsorption process. The abundant banana stem waste was utilized as activated carbon precursor. This study aimed to analyse the influence of solution pH to removal efficiency of methylene blue using activated carbon from banana stem as adsorbent. Activated carbon from banana stem was obtained by chemical activation using $H_3PO_4$ solution. Proximate analysis result showed that the activated carbon has 47.22% of fixed carbon. This value exhibited that banana stem was a potential adsorbent precursor. Methylene blue solutions were prepared at initial concentration of 50 ppm. The influence of solution pH was investigated with the use of 0.2 g adsorbent for 100 mL dye solution. The adsorption was conducted using shaker with at a constant rate of 100 rpm at room temperature for 90 minutes. The results showed that solution pH influenced the adsorption. The activated carbon from banana stem demonstrated satisfying performance since removal efficiencies of methylene blue were higher than 99%.

## 1. Introduction

The occurrence of synthetic dyes gives negative effects to aquatic organism and human health when discharge directly to environment. The dyes consist of poisonous and complex components with slow degradation rate. Furthermore, the presence of dyes components affects the undesirable color change in water. The unfavorable impact is not only from esthetics point of view but also from the decline of sun light penetration, thus reducing photosynthetic activity [1, 2].

Dye is used widely in industries such as textile, food, paper and textile. The annual production of synthetic dye is about 70,000 tons with 100,000 types of dye. During the production process, about 15-20% are lost and carried away in wastewater [3]. Methylene blue ($C_{16}H_{18}N_3SCl$, C. I. No. 52015) is one of cationic dyes that most frequently used in industry [4]. Concerning the quantity and its harmful impacts, it is necessary to make efforts to reduce the dye components before discharging to environment.

IOP Conf. Series: Earth and Environmental Science **122** (2018) 012085 doi:10.1088/1755-1315/122/1/012085

The removal of methylene blue from wastewater were reported by using numerous methods such as liquid-liquid extraction, reverse osmosis, advanced oxidation process, electrocoagulation, electrochemical oxidation, ozonation, and membrane filtration . However, adsorption method offers some advantages due to its simple design, high efficiency and low costs with unhazardous byproducts [2, 5].

The utilization of activated carbon obtained from agricultural wastes as adsorbent has stimulated the interest of researchers, because it provides both low cost adsorbent and solve agricultural waste problem altogether.

Banana stem is one of lignocellulose waste and exist abundantly in Indonesia. Banana production in Indonesia increases annually and reached 6.862 ton in 2014 [6]. From the total production, banana stem takes the largest portion of 60%, whereas 30% is fruit and the rest is banana leaf [7]. The composition of cellulose, hemicellulose and lignin in banana stem is 43.3%, 20.6% and 27.8% respectively [8]. At this composition, the utilization of banana stem as the source of activated carbon becomes potential.

In this study, the analysis of removal efficiency of methylene blue was analyzed using activated carbon from banana stem as adsorbent. The activated carbon was prepared by chemical activation with $H_3PO_4$ solution. The adsorption was conducted in various pH of dye solution in order to investigate the influence of pH on removal efficiency of methylene blue.

## 2. Experimental Procedure

### *2.1. Activated Carbon Preparation and Characterizations*

The activated carbon from banana stem was prepared with the same method as previously published [9, 10]. The first outer layer and the white middle part of banana stem were omitted from raw material. The sample was cut and washed several times. Afterwards, the sample was dried under sun light for 5 days and followed by oven drying at 105 °C until the constant mass was achieved. The samples were grinded and then sieved to 32 mesh. Sieved samples were impregnated in $H_3PO_4$solution at sample to solvent ratio of 1:1 and left overnight. Pyrolysis reactor equipped with nitrogen ($N_2$) flow was operated at 400 °C for 15 minutes. The produced activated carbon was washed with warm distilled water until solution reached neutral pH and then dried. Proximate analysis was carried out to obtain the moisture content, ash content and volatile matter content.

### *2.2. Adsorption of Methylene Blue*

The adsorption of methylene blue was studied in batch mode. The solution was prepared at initial concentration ($C_0$) of 50 ppm. At this concentration, the pH of solution was 7 (neutral). The influence of solution pH was investigated in the range of 4-9. The pH was adjusted using 0.1 N HCl and 0.1 N NaOH. The dose of adsorbent was 0.2 g [2] for each 100 mL dye solution. The adsorption processes were conducted using shaker at 100 rpm at room temperature for 90 minutes. The concentration of methylene blue after adsorption ($C_e$) was measured using UV-VIS spectrophotometer in triplicate. Removal efficiency ($R$) of methylene blue was calculated using Eq. 1.

$$R=\frac{C_0 - C_e}{C_0} \quad (1)$$

## 3. Result and Discussion

### *3.1. Characteristics of Activated Carbon*

Based on proximate analysis, the dried activated carbon had of 7.5% moisture, of 10.64% ash and of 34.64% volatile matter. The total amount of moisture, ash, volatile matter and fixed carbon content should be 100%. Therefore, the fixed carbon content in the activated carbon was 47.22%.

Based on SNI (Indonesian National Standard) for activated carbon [11], maximum value of moisture, ash and volatile matter content are 5%, 10% and 25% respectively, and minimum value for fixed carbon is 65%. Thus, only moisture and ash content of the activated carbon that met the requirements of SNI. However, previous study reported that proximate analysis of activated carbon from apple pulp was 15.11%, 15.35%, 62.69% and 6.56% respectively for moisture, ash, volatile matter and fixed carbon content. Whereas the value of moisture, ash, volatile matter and fixed carbon content of apple peel activated carbon were 12.27%, 7.43%, 72.89% and 7.40% respectively [12]. Both activated carbons of apple pulp and apple peel were applied to adsorb methylene blue with removal efficiency of up to 94.6%. Therefore, the potency of activated carbon from banana stem would merit further investigation.

### *3.2. Standard Curve*

The measurement of methylene blue concentration in final solution was initiated with searching the maximum wavelength. From UV-VIS spectrophotometer result, it was observed that maximum peak was acquired at 664 nm. Subsequently, methylene blue solutions at various concentrations were prepared and analyzed at 664 nm. The standard curve was obtained by plotting the recorded absorbance from spectrophotometer versus the prepared concentration of solution. The accuracy was excellent with coefficient correlation ($R^2$) of 0.9975 (Fig. 1).

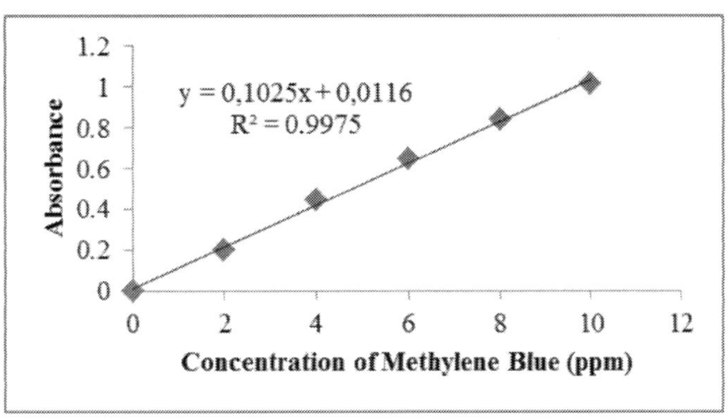


**Figure 1.** Standard Curve for Methylene Blue Solution

### *3.3. Solution pH Influence on Removal Efficiency*

Fig. 2 showed the influence of solution pH on final concentration of methylene blue. After 90 minutes of adsorption process, the concentration of methylene blue in solution decreased dramatically in the range of pH 4-9 (Fig. 2a). Thus, the removal efficiency of methylene blue could be calculated using Eq. 1 and showed excellent results that were higher than 99% (Fig. 2b).This was also confirmed with the change of solution color from dark blue to bright and clear color (Fig. 3). Besides that, the solution pH decreased 1 level after adsorption process.

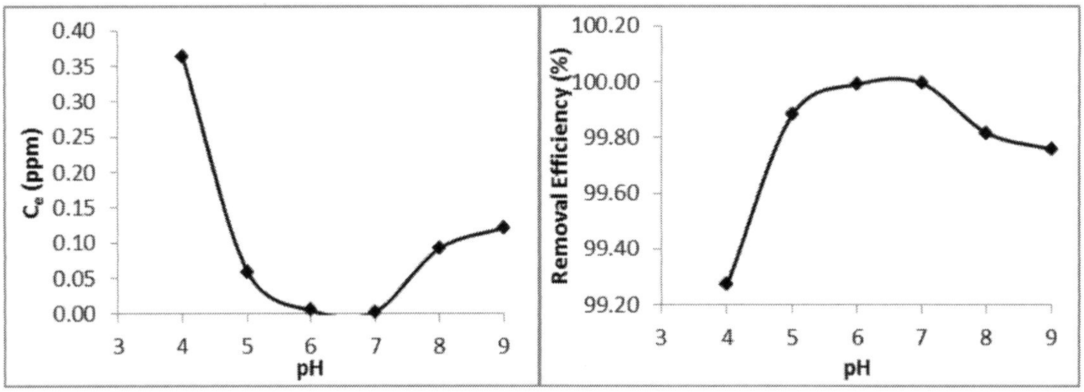


**Figure 2.** Influence of pH on (a) final concentration of methylene blue and (b) removal efficiency of methylene blue in adsorption process using activated carbon from banana stem

**Figure 3.** Color of methylene blue solution (a) before and (b) after adsorption process at pH = 7

The removal efficiency increased with the increasing pH of solution from 4 to 7, then decreased at pH 8 and 9. Methylene blue is cationic dye that generally demonstrates the positively charged ions when dissolved in water. The increasing of removal efficiency in the range of pH 4 to 7 might be because of the presence of acidic medium. The adsorbent surface may contain positive charge that does not attract the positively charged methylene blue in the solution [13]. As the pH of methylene blue solution increases, the adsorbent surface attains the negative charge that increases the adsorption capacity of activated carbon from banana stem. This is due to the formation of electrostatic interaction between the positively charged methylene blue and the negatively charged activated carbon [2].

Hence, the maximum removal of methylene blue using activated carbon from banana stem in this study was obtained at pH 7. However, the removal efficiency increased slightly with the rise of pH. Previous study reported that the best removal efficiency was also obtain at solution pH 7 for methylene blue adsorption using microwave assisted corncob activated carbon [2]. Whereas use of *Ficus carica* bast activated carbon showed that maximum methylene blue removal was obtained at pH 7.8 [14]. Both studies showed that adsorption of methylene blue were pH dependent.

In this study, the use of adsorbent dose of 0.2 g showed excellent performance of banana stem as activated carbon precursor in removing methylene blue in 100 mL solution with initial concentration of 50 ppm. The highest methylene blue removal in this study was 99.99%. A former study that used the same adsorbent dose showed almost the same removal efficiency. However, the treated initial concentration of methylene blue was smaller i.e. 25 ppm [2].The use of 0.5 g *Ficus carica* bast activated carbon in 100 mL methylene blue solution with initial concentration of 50 ppm gave about 85% removal efficiency [14]. These results showed that banana stem activated carbon is a promising adsorbent that can remove methylene blue effectively.

## 4. Conclusion

The increase of pH of the solution improved the removal efficiency of methylene blue using adsorbent of banana stem activated carbon. The maximum removal efficiency of 99.94% was achieved at pH 7. Methylene blue concentration dramatically decreased after contacting for 90 minutes with the adsorbent. This was also confirmed by significant color change of solution. Excellent performance was obtained when applying only a little amount of adsorbent mass in this study. Hence, banana stem activated carbon is a promising adsorbent in removing methylene blue.

## Acknowledgements

The authors would like to acknowledge Lembaga Penelitian of Universitas Sumatera Utara for the financial support under research grant TALENTA Research of Universitas Sumatera Utara, Fiscal Year of 2017, Contract No. 5338/UN5.1.R/PPM/2017, 22nd May 2017.

## References

[1] Spagnoli A A, Giannakoudakis D A and Bashkova S 2017 Adsorption of methylene blue on cashew nut shell based carbons activated with zinc chloride: The role of surface and structural parameters, *Journal of Molecular Liquids* 229 465–71

[2] Tharaneedhar V, Kumar P S, Saravanan A, Ravikumar C and Jaikumar V 2017 Prediction and interpretation of adsorption parameters for the sequestration of methylene blue dye from aqueous solution using microwave assisted corncob activated carbon *Sustainable Materials and Technologies* 11 1–11

[3] Duran-Jimenez G, Hernandez-Montoya V, Montes-Moran M A, Bonilla-Petriciolet A and Rangel-Vazquez N A 2014 Adsorption of dyes with different molecular properties on activatedcarbons prepared from lignocellulosic wastes by Taguchi method *Microporous and Mesoporous Materials* 19999–107

[4] Fu J, Xu Z, Li Q S, Chen S, An S Q and Zeng Q F 2010 Treatment of simulated wastewatercontaining Reactive Red 195 by zero-valent iron/activated carbon combinedwith microwave discharge electrodeless lamp/sodium hypochlorite *J. Environ. Sci.* 22512–8

[5] Kyzas G Z, Fu J and Matis K A 2013 The change from past to future for adsorbent materials in

treatment of dyeing wastewaters *Materials (Basel)* 6 5131–58
[6] Direktorat Jenderal Hortikultura Kementerian Pertanian Republik Indonesia (Directorate General of Horticulture Ministry of Agriculture of the Republic of Indonesia) 2015 *Horticultural Statistics 2014* Ir. Anastasia Promosiana, M S Jakarta 95
[7] Advena D 2014 Fermentasi batang pisang menggunakan probiotik dan lama inkubasi berbeda terhadap perubahan kandungan kering, protein kasar, dan serat kasar (Banana stem fermentation using probiotics and different incubation times for changes in dry, crude protein, and crude fiber content), Program Studi Peternakan, Fakultas Pertanian (Faculty of Agriculture), Universitas Taman Siswa, Padang
[8] Noeline B F, Manohar D M and Anirudhan T S 2005 Kinetic and equilibrium modelling of lead(II) sorption from water and wastewater by polymerized banana stem in a batch reactor *Separation and Purification Technology* 45 131–40
[9] Misran E, Maulina S, Dina S F, Harahap S A A and Nazar A 2016 Characterizations of activated carbon produced from bagasse and banana stem using $H_3PO_4$ as activating agent *Journal of Engineering and Applied Sciences 12(15) 3839-3842*
[10] Misran E, Maulina S, Dina S F, Harahap S A A and Nazar A 2017Activated carbon production from bagasse and banana stem at various times of carbonization *TALENTA CEST (Conference on Engineering, Science and Technology) 2017* Medan – Indonesia September, 7-8 (Accepted).
[11] Badan Standarisasi Nasional National Standardization Agency of Indonesia 1995 Karbon aktif (Activated Carbon): SNI No. 06–3730-1995 Jakarta
[12] Hesas R H, Arami-Niya A, Wan Daud W M A and Sahu J N 2013 Preparation and characterization of activated carbon from apple waste by microwave-assisted phosphoric acid activation: Application in methylene blue adsorption *BioResources* 8 2950-66
[13] Peydayesh M and Kelishami A R 2015 Adsorption of methylene blue onto Platanus orientalis leaf powder: Kinetic, equilibrium and thermodynamic studies *J. Ind. Eng. Chem.* 211014–9
[14] Pathania D, Sharma S and Singh P 2017 Removal of methylene blue by adsorption ontoactivated carbon developed from Ficus carica bast *Arabian Journal of Chemistry* 10 S1445–51

IOP Conf. Series: Earth and Environmental Science **122** (2018) 012086 doi:10.1088/1755-1315/122/1/012086

# Effect of pretreatment on purple-fleshed sweet potato flour for cake making

**M S Hutasoit, E Julianti and Z Lubis**

Department of Food Science and Technology, Faculty of Agriculture, Universitas Sumatera Utara
E-mail: minahutasoit@gmail.com

**Abstract.** The purple-fleshed sweet-potato (PFSP) flour was produced by varying pretreatment of washed chips: dipping in 0.5 and 1.0% (w/v) citric acid solution for 30 min, dipping in 0.5 and 1.0% (w/v) citric acid solution for 30 min and followed by steam blanching for 5 min. The pretreatment effect on cake quality was investigated. The results showed that PFSP flour produced from pretreatment with dipping in 0.5% citric acid for 30 min followed by steam blanching for 5 min had higher lightness (L*) value and lower browning index, higher hedonic value of color and aroma and baking expansion. The specific volume of cake from pretreated flour, untreated flour and wheat flour were 44.87, 43.83, and 50.43$cm^3/g$, respectively. The sensory evaluation of cake indicated that cake from pretreated PFSP flour was acceptable compare to those of cake from wheat flour.

## 1. Introduction

Sweet potato (*Ipomoea batatas* L. Lam) has been grown for domestic consumption in Indonesia. They are four common local varieties: the purple skin-white fleshed, yellow skin-yellow-fleshed, yellow skin-orange fleshed, and purple skin-purple fleshed. Sweet potato plays an important role in improving national food security, health, and livehoods of poor family, due to its wide range of agronomical and nutritional advantages such as high yield even in marginal soil conditions, wide ecological adaptability, low input requirements, and shorter growing period than other crops [1]. Sweet potato contains high carbohydrate content, low glycemic index [2] and dietary fiber [3]. In particular, purple-fleshed sweet potato has stronger antioxidant activity than many other vegetables due to its higher content of anthocyanin [4].

Even though they supply carbohydrate and anthocyanin that essential for proper functioning of the human body, their utilization is still limited. Most sweet potatoes are commonly consumed fresh (boiled, fried or make it into cake). Therefore, the use of sweet potato flour for bakery production would help lessen the need of imported wheat flour. The use of sweet potato increases depend on the processing technology to produce flour or starch with desirable physicochemical and functional properties along with the understanding of processing effect on their properties [5]. Therefore, the objective of this research was to study the effect of washing pre-treatment on sweet potato flour for cakes production.

## 2. Material and methods

The deep purpled fleshed sweet potato (PFSP) was purchased from farmers at Berastagi Village North Sumatera Province Indonesia. Roots were washed with tap water to remove dirt and soil, and after drying its surface, the washed sweetpotato was stored at 14 °C without curing before being processed. Other ingredients, such as eggs, sugar, skim milk, margarine, vanilla flavoring, xanthan gum, and baking powder were purchased from a market in Medan.

*2.1. Sample preparation and treatment*
The cleaned roots were peeled manually with stainless steel kitchen knife and sliced into 2 mm-thickness with the use of an electrical chipping machine. The chips were prepared into four flour samples. For the first and second group the chips were dipped in aqueous 0.5 and 1.0% (w/v) citric acid at room temperature for 30 min. For the third and fourth group, the chips were dipped in aqueous 0.5 and 1.0% citric acid followed by steam blanching for 30 minutes.

*2.2. Preparation of PFSP flour*
The treated chips were dried using a convection drying oven at 55 °C for 12 hours. The flour (moisture content about 10%) was obtained by milling the dried slices using a milling machine into flour with particle size of about 200 μm and the resulting flour was sieved to obtain a fine flour. The flours were then packaged in polyethylene bags, sealed and stored at ambient temperature (28 ± 2°C) before being used.

*2.3. Analysis of the physical and functional properties of PFSP flour samples*
The physical properties of flour analyzed were color (L* value), browning index, and hedonic value of aroma and color. L* value was determined by using a Minolta Chromameter CR-400 (Minolta Camera Co., Ltd., Tokyo, Japan). Browning index was determined using the method described by Youn and Choi [6]. Hedonic value of aroma and color were tested by 15 semi trained panelist using 5 point hedonic scale with 5 = very like and 1 = very dislike.
The functional properties of PFSP flour such as water absorption index (WAI), oil absorption index (OAI) and swelling power were determined by standard methods. WAI and OAI were determined according to the methods of Niba *et al* [7]. The swelling power and solubility of flour was determined based on a method of Leach *et al.*[8] and Anderson [9], respectively. The measurement of baking expansion uses the method of Demiate *et al* [10] with minor modification.

***2.4. Cakes production***
PFSP flour with the higher L* and hedonic value and the lower browning index was used for cake production. Wheat flour and PFSP flour without pretreatment were used as control. The formulation for cake according to Bennion and Pamford [11] was followed for the production of cake with modification: flour (100 g), baking powder (3 g), egg (125 g), sugar (100 g), skim milk (50 g), margarine (10 g) and vanilla flavoring (2 g).

Flour and baking powder were blended manually to get a uniform distribution. Egg, sugar, skim milk, margarine and vanilla flavoring were mixed by using handled mixer (HR 1538 Phillips) at high speed for 6 minutes. The flour mix was added into the cream and dough was mixed for 3 minutes at low speed. The dough was poured into baking pans, then placed in a preheated oven and baked at 180 °C for 35 minutes. Cakes were allowed to cool for 30 minutes at the pans at room temperature.

*2.5. Analysis of cake quality*
Resulting cakes were analyzed for their color (L* value) by using a Minolta Chromameter CR-400 (Minolta Camera Co.,Ltd., Tokyo, Japan). The volume of the cake was measured using seed replacement test according to the approve method by AACC [13].

A panel of 15 semi-trained judges of both genders aged 18-21 years evaluated the cakes on a 5 point hedonic scale (1=dislike extremely, 2=dislike, 3= neither like nor dislike, 4= like, 5= like extremely). Cakes were sliced into quarter and identified by a three-digit random number. The sampels were offered to the judges on a white plate at room temperature. The plates were cleansed with water before they are used for the next samples.

*2.6. Data analysis*

Data using completely randomized design were analyzed using SAS Version 9.2 for windows. The data reported in all tables are an average of triplicate observations subjected to one-way analysis of variance (ANOVA). Differences between the ranges of the properties were determined using the method of Least Significant Difference (LSD) tests at 95% confidence level (P< 0.05).

## 3. Results and discussions

*3.1. Effect of pretreatment on physical and functional properties of PFSP flour*

Table 1 shows that there was a significant difference at the 5% level in the lightness (L* value), browning index and hedonic value of color, but there is no significant difference in the hedonic value of aroma among the samples. The PFSP flour made from chips that dipped in citric acid 0.5% or 1.0% had the higher L* value (lightness) and lower browning index, but their acceptability in aroma and color are lower, this was due to the pale appearance of its color. PFSP flour made from chips that dipped in in citric acid 1.0% followed by steam blanching for 30 minutes had the higher acceptability in aroma and color but its lightness (L* value) was lowest and its browning index is higher than other pretreated flour, so its color appearance was darker than other flour. This was due to the structure of anthocyanin was depended on medium pH [14]. The PFSP flour made from chips that dipped in citric acid 0.5% followed by steam blanching 30 minutes was more acceptable in its color appearance than other flour.

**Table 1.** Effect of pretreatment on physical properties of PFSP flour

| Pretreatment | Parameter | | | |
|---|---|---|---|---|
| | Color (L* value) | Browning Index | Hedonic Value | |
| | | | Aroma | Color |
| CA 0.5 | 44.93 ±0.37b [b] | 0.78±0.09 [b] | 3.55±0.06 | 2.97±0.28 bc |
| CA 1.0 | 46.64±0.33a [a] | 0.76±0.03 [b] | 3.54±0.0 | 2.60±0.17c |
| BCA 0.5 | 45.62±0.94b [b] | 0.84±0.06 [b] | 3.48±0.27 | 4.00±0.53a |
| BCA 1.0 | 40.66±0.03d[d] | 0.86±0.07 [b] | 3.62±0.16 | 4.35±0.10a |
| Control | 43.23±0.16c [c] | 1.12±0.09 [a] | 3.50±0.19 | 3.16±0.06b |

*) CA 0.5 and CA 1.0 = the chips were dipped in aqueous 0.5 and 1.0% (w/v) citric acid at room temperature for 30 min, respectively. BCA 0.5 and BCA 1.0 = the chips were dipped in aqueous 0.5 and 1.0% citric acid followed by steam blanching for 30 minutes, respectively. Control was PFSP flour without washing pretreatment
The values are expressed as the mean of three replications.
Means followed by different letter in the same column are significantly different among drying temperature (p<0.05)

Table 2 shows the results of functional properties of pretreated and untreated flour. At the 5% level there were significance difference in WAI, OAI, swelling power, and baking expansion of pretreated PFSP flour. The washing pre-treatment with higher concentration of citric acid and blanching caused the water absorption index and swelling power increase, while the oil absorption index was decreased. The similar result was reported by Jangchud *et al.*[5]. Blanching can cause the gelatinization of starch and consequently will disrupt the starch granule and increase the starch-water interaction, resulting in an increase of WAI and swelling power [15]. Blanching also cause an increased of baking expansion due to increasing viscosity and internal bubble pressure which resulted the disruption of cell at the end of oven rise [16]. There is no significance difference in solubility of pretreated and untreated PFSP flour.

**Table 2.** Effect of pretreatment on functional properties of PFSP flour

| Preatreatment | Parameter | | | | |
|---|---|---|---|---|---|
| | WAI (g/g) | OAI (g/g) | Swelling Power (%) | Solubility (%) | Baking Expansion (ml/g) |
| CA 0.5 | 2.33±0.20[c] | 1.44±0.14[b] | 5.61±0.74[bc] | 3.55±0.35 | 0.92±0.46[ab] |
| CA 1.0 | 2.55±0.65[c] | 1.43±0.19[b] | 6.35±0.56a[b] | 3.58±0.12 | 0.89±0.00[a] |
| BCA 0.5 | 2.93±0.15[b] | 1.38±0.09[b] | 6.51±0.39[ab] | 3.56±0.63 | 0.98±0.07[a] |
| BCA 1.0 | 3.83±0.56[a] | 1.26±0.00[b] | 6.69±0.60[a] | 3.61±0.22 | 0.89±0.04[b] |
| Control | 1.90±0.11[d] | 1.73±0.13[a] | 5.19±0.22[c] | 3.43±0.13 | 0.90±0.01[ab] |

*) CA 0.5 and CA 1.0 = the chips were dipped in aqueous 0.5 and 1.0% (w/v) citric acid at room temperature for 30 min, respectively. BCA 0.5 and BCA 1.0 = the chips were dipped in aqueous 0.5 and 1.0% citric acid followed by steam blanching for 30 minutes, respectively. Control was PFSP flour without pretreatment
The values are expressed as the mean of three replications.
Means followed by different in each column are significantly different among drying temperature ($p<0.05$)

*3.2. Quality of cake from pretreated PFSP flour compared with untreated PFSP flour and wheat flour*

The quality of cake made from pretreated PFSP flour compared with untreated PFSP flour and wheat flour are presented in Table 3. Cake made from wheat flour had a higher L* and b* value. This result showed that cake made from wheat flour had a higher lightness than cake from PFSP flour. There is no significant difference ($P>0.05$) in specific volume among the cakes, but cake made from untreated PFSP flour had the lowest specific volume. There is no significant difference in aroma of cake among the cakes, but there are significant differences ($P<0.05$) in color, taste and texture. Cake made from wheat flour still has a higher hedonic value in color, taste and texture than cake made from pretretaed and untreated PFSP flour. There are no significant difference in color, and taste of pretreated and untreated PFSP flour, but untreated PFSP flour had a higher hedonic value in texture than pretreated flour. Respondent were neither like nor dislike cake made pretreated PFSP flour.

**Table 3.** Quality of cake from pretreated OFSP flour compared with untreated OFSP and wheat flour

| Parameter | BCA 0.5 | Untreated OFSP Flour | Wheat Flour |
|---|---|---|---|
| Color | | | |
| - L* | 21;08±3.23[b] | 22.81±1.17[b] | 69.50±0.62[a] |
| - a* | 10.65±1.40[a] | 8.65±1.12[c] | 9.73±1.22[b] |
| - b* | 1.07±0.40[c] | 5.83±1.03[b] | 39.76±1.03[a] |
| Specific Volume (ml/g) | 44.87±2.30[a] | 43.89±4.54[a] | 50.43±5.24[a] |
| Hedonic Value | | | |
| - Color | 3.30±0.20[b] | 3.30±0.17[b] | 3.93±0.35[a] |
| - Aroma | 3.30±0.45 | 3.40±0.20 | 3.76±0.20 |
| - Taste | 3.16±0.28[b] | 3.13±0.1[b] | 3.86±0.30[a] |
| - Texture | 2.61±0.30[c] | 3.08±0.20[b] | 3.58±0.20[a] |

*) BCA 0.5 was the flour made from chips that dipped in aqueous 0.5 citric acid followed by steam blanching for 30 minutes.
The values are expressed as the mean of three replications.
Means followed by different in the same raw are significantly different among drying temperature ($p<0.05$)

## 4. Conclusion

The study has shown that different pretreatments have an effect on the quality of flour produced. PFSP flour made from chips that pretreated with soaking in citric acif 0.5% followed by steam blanching 30 minutes has a higher hedonic value in color and baking expansion so it then was used in cake production. Cake made from pretreated PFSP flour has similar specific volume with cake made from untreated PFSP flour and wheat flour, but in general still has a lower acceptability in sensory properties. Pretreated PFSP flour is an alternative flour in the production of cakes.

**References**

[1] Horton D, Prain G, and Fregory P 1989 High-level investment return for global sweet potato research and development CIP Cir 17 1-11

[2] International Life Science Institute (ILSI) 2008 Nutritionally improved sweet potato. *In* : Assessment of foods and feeds. Comprehensive Review in Food Science and Food Safety 7 81-91

[3] Zhang ZE, Fan SH, Zheng YL, Lu J, Wu DM, Shan Q, and Hu B 2009 Purple sweet potato color attenuates oxidative stress and inflammantory response induced by D-galactose in mouse liver. Food and Chemical Toxicology 47 (2) 496-501

[4] Van Hall M 2000 Quality of sweet potato flour during processing and storage J Food Reviews Int 16 1-37

[5] Jangchud K, Phumolsiripol Y, and Haruthaithananasan V 2003. Physicochemical properties of sweet potato flour and starch as affected by blanching and processing Starch/Starke 55 258-264

[6] Youn KS and Choi YH 1996 Drying characteristics of osmotically pre-treatedcarrot. Korean Journal of Food Science and Technology28 11-28.

[7] Niba LL, Bokonga MM, Jackson EL, Schlimme DS, and Li BW 2001 Physicochemical properties and starch granular characteristics of flour from various *Manihot esculenta* (cassava) genotypes J Food Sci 67 5 1701-1705.

[8] Leach HW, McCowan LD, and Schoch TJ 1957 Structure of the starch granule : Swelling power and solubility patterns of different starches. Cereal Chem 36 534-544.

[9] Demiate IM, Dupuy N, Huvenne JP, Cereda MP, Wosiacki G 2000. Relationship between baking behavior of modified cassava starches and starch chemical structure determined by FTIR spectroscopy. Carbohyd Polym 42 149-158

[10] Anderson RA 1982 Absorption and solubility and amylograph characteristics of roll-cooked small grain products. Cereal Chemistry 59 265-269.

[11] Bennion EB and Bamford GST 1997 The Technology of Cake Making 6th Ed. pp 112-120, 277 and 285-288 Chapman& Hall London

[12] AOAC 1995 Official Methods of Analysis of The Association of Official Analytical Chemists Washington.

[13] AACC 2000. Approved Methods of the American Association of Cereal Chemists.10th Edn., American Association of Cereal Chemists Press, St. Paul, MN. .

[14] Xiu-Li HE, Xiu-Li LI, Yuang-Ping LV, and Qiang HE 2015 Composition and color stability of anthocyanian-based extract from purple sweet potato Food Science and Technology,

[15] Eliasson AC and Gudmundsson M 1996 Starch: Physicochemical and functional aspects. In Eliasson AC (Ed) Carbohydrates in food pp 431–503 New York Marcel Dekker.

[16] Vatanasuchart N, Naivikul O, Charoenrein S, and Sriroth K 2005 Molecular properties of cassava starch modified with different UV irradiation to enhance baking expansion Carbohydrate Polymers 61 80-87.

**Acknowledgments**

We want to thank Lembaga Penelitian Universitas Sumatera Utara for funding this research according to the research contract TALENTA Universitas Sumatera Utara 2017 No. 5338/UN5.1.R/PPM/2017 dd. 22 May 2017.

# Soy-yamgurt probiotic drink as a natural potential of antioxidant

**H Rusmarilin[1*], Nurhasanah[2], R Y Andayani[3].**

[1]Postgraduate Lecturer, Departement of Food Science, Faculty of Agriculture, Universitas Sumatera Utara, Medan. 20155, Indonesia.
[2]Post Graduate student, Departement of Food Science, Faculty of Agriculture, Universitas Sumatera Utara, Medan, 20155, Indonesia
[3]Postgraduate Lecturer, Departement of Economy Study, Program Study of Management, STIE Pelita Bangsa , Bekasi, West Java, Indonesia.
[*]***Coresponding author:*** herla_surabaya@yahoo.com

**ABSTRACT.** Yogurt is a popular healthy food, consumed by many people. Probiotic are used for better growth and survival of probiotic bacteria as well as to improve organoleptic, rheological and technological properties of soy-yamgurt probiotic. The aims of this study were to determine physicochemical changes and survival of probiotic bacteria due to the effect of ratio of yam-bean with soy-bean extract on the quality of soy-yamgurt probiotic drink. The quality examined were total bacteria, antioxidant activity, lactic acid content, and acceptability including color, texture, flavor, and overall acceptance. This research had been performed using Completely Randomized Design two factorials, consist of five levels i.e.: ratio of yam-bean with soy-bean extract (100:0; 75:25; 50:50; 25:75; 0:100) and the fermentation time (4,6 and 8 hours), the process was repeated three times. The results showed that ratio of yam-bean with soy-bean extract of 50:50 and fermentation for 6 hours incubation produced the best quality. Acceptable probiotic drink containing 1.44 x $10^9$ CFU/mL lactic acid bacteria, $IC_{50}$ of soy-yamgurt in the attenuation of free radical DPPH ranged from 58.718-18.112 mg/L in 5 minutes incubation and 39.7204-11.9925 mg/L in 60 minutes, and 0.48% lactic acid. This appearance of soy-yamgurt had yellow greenish color, desired texture and flavor, and received the highest score of overall acceptance.

## 1. Introduction

Among all the fermented dairy products, yogurt is one of the most popular and had more acceptability worldwide. This product is produced through lactic acid fermentation by *Lactobacillus delbruekii* subsp. *bulgaricus* and *Streptococcus salivarus* subsp. *thermophillus*. Probiotics have introduced as microorganisms which in sufficient amounts had health benefits for the host [2,4]. The most important health benefit of probiotics include reducing serum cholesterol level [21] , improving nutritional value, improving calcium absorption, proteins and vitamin synthesis, improving lactose digestion [4], promoting immune system function [1], preventing different type of cancer especially colon, preventing growth and activity of pathogenic microbes [20] and synthesizing bacteriocines [7].

Soy-yamgurt was a product of fermented milk produced from the activity of lactic acid bacteria and contain microorganisms alive, active and abundant (>$10^7$ CFU/mL). Raw material of soy-yamgurt not only from fresh milk, but its can also come from plant-based product diversification that mixed with skim milk powder as a source of lactose and lactic acid bacteria growth media. One of them is the raw material from soy-bean and yam-bean (*Pachyrhizus erosus L.*) extract.

It has been proven that the yam-bean had many benefits and a strong antioxidant activity, it also contains water-soluble polysaccharides (WSP) which proved to decrease diabetes mellitus [12]. Probiotic soy-yamgurt drink (a blend of soybean and yam-bean extract) turned out to have the high antioxidant activity and content of nutrients that are easy to digest due to the fermentation process.

Soy-yamgurt relatively more durable than fresh milk or milk powder dissolved in water or yam-bean extract, but its storage should be cold for a few days.

To enhance durability, increasing flavor and marketing development needed a further treatment. The treatment could also be expected to maintain or just reduce the nutritional value decline as little as possible, physico-chemical and organoleptic value. To create a more varian in its aroma and taste as well as increasing nutrition value, have been soy-yamgurt form. Among others: total bacteria $10^7$ CFU/mL, total lactic acid 0.5-2% and minimum total solid 12%, good appearance, taste, and overall acceptance [14].

## 2. Material and Methods

### *2.1 Materials*

Milk powder contains 1.0% fat was purchased from Supermarket in Medan, Indonesia. Microbial strains derived from plain yogurt commercial contaning *L. bulgaricus*, *S. thermophillus* and *L. acidophilus*. Yam bean *Pachyrhizus erosus* L. (PeL)) were identified as varieties "elephant" and collected from Binjai city, the province of North Sumatera, Indonesia.

Material analysis i.e.: NaOH 0, 1N, $H_2SO_4$ p.a, 96% alcohol, absolute alcohol, n-hexan, $K_2SO_4$, $CuSO_4$, phenolptalein, DPPH (2.2-3-diphenyl pikril hidrazil), aquadest, 0,9% physiological NaCl solution and MRS broth.

### *2.2 Primary culture preparation*

To prepare the primary culture , 250 mL of crude milk heated 80-85°C for 20 min, it is transferred to 500 mL erlenmeyer flasks, then culture solid (plain yogurt commercial) containing yogurt starter added to Erlenmeyer flasks (3%). It is incubated at 40-42°C for 6 hours. The process was repeated 3 times, to form a solid structure and used for better growth and survival of probiotic bacteria. At the end they were refrigerated.

### *2.3 Sample preparation.*

The making of yam-bean extract from yam-bean tubers (5 months harvesting time) was selected, trimmed, and shelled. Yam-bean were sliced and washed, then blanched for 7-10 minutes, blended and added water (yam-bean: water = 1:1), filtered.

The soybeans used were first sorted, then soaked in a solution of $NaHCO_3$ (baking soda) of 0.5% for 30 minutes and boiled for 30 minutes. The soybean shell was separated by squeezing and washed with water until the shell easy to separate. Soybean was milled with hot water as much as 6 parts. Soybeans was filtered by filter cloth that has been blanched and soybean extract was heated to boiling. After boiling, it is left for 20 minutes at 80°C [5].

### *2.4 Symbiotic yogurt production*

The materials used to make soy-yamgurt were 250 mL sterile containers containing the pasteurized of soybean extract and yam-bean extract (80°C for 20 minutes), 10% dried skim milk (1% fat) were inoculated simultaneously with 3% of the starter (*L. bulgaricus*, *S. thermophillus*, and *L. acidophillus*) derived from the commercial culture, 1% sugar, and 0,6% gum Arabic and then incubated at 40-42°C for 4, 6 and 8 hours.

### *2.5 Observation*

The observations were analyzed by total lactic acid [10]), determination of water content [18], sensory evaluation [15], activity antioxidant by DPPH method [8], determination of LAB colonies [3]. Soy-yamgurt were evaluated using 4 score Hedonic test. The samples were evaluated by 15 panelists regarding the organoleptic attributes including, color, flavor, taste and texture.

### *2.6 Microbial test*

Microbial test consisted of sample culture in MRS broth using pour plate method according to the standard. In order to provide this, proper dilution of the samples were made in sterile ringer solution and the plates were incubated at 37°C following the culture preparation. Colony counts were measured following 72h incubation period [3].

*2.7 Test of antioxidant activity by DPPH method*

The antioxidant activity was determined by the free radical method of DPPH [16]. 0.004% DPPH solution was prepared by dissolving 100% of the DPPH crystals in 100 mL ethanol p.a. A total of 200 μL of ethanol (pro analyzed) was transferred into the cuvette and then added 3 mL DPPH solution, stir well and immediately made visible light spectra (400-600 nm). For test of sample, 200 μL sample was transferred into the cuvette, then added 3 mL DPPH solution, subsequently made visible light spectra (400-600 nm). At the 5th minute, read absorbance at 497-517-537 nm and done in 60th minute. The calculation of DPPH absorbance and solution sampling determined using Equation 1, whereas non antiradical activity determined using equation 2.

The Absorbance value of DPPH and sample solution = A517 - ((A497+A537)/2) ........(1)

The % of DPPH Attenuation = (1-(A value of test sample/A value of the DPPH)) x 100% .....(2).

## 3. Results and Discussion

*3.1 Probiotic bacterium counts (PBC)*

Figure 1 shows the higher of yam-bean extract of the sample had the higher PBC, it means sample solution containing water soluble polysaccharide (WSP) from yam-bean [, there was significantly different between the sample that contains soy-bean extract. The addition of various amount of yambean and soybean extract of the soy-yamghurt probiotic drink gave significantly different effect to the total bacteria ($p<0.01$). Figure 1 shows that the higher soybean extract, the total bacteria of soy-yamgurt were declined, but it was still in a standardized Standard National Indonesia.

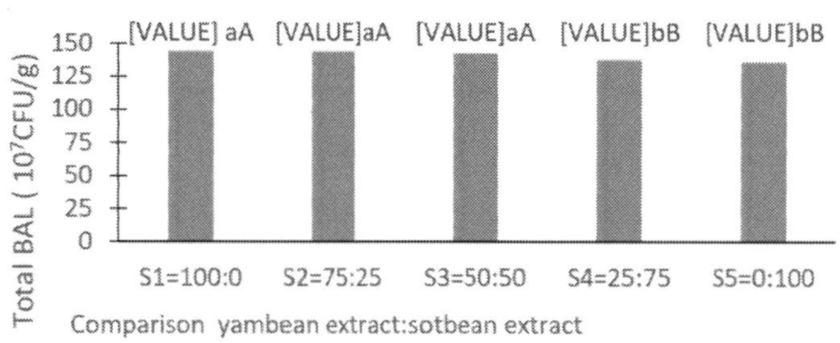


**Figure 1.** Total plate count of lactic acid bacteria in each treatment.

The total of bacteria on soy-yamgurt products is *L. bulgaricus*, *S. thermophillus* and *L. acidophilus* without any other microbial impurities, it can be seen from the growth of bacterial colonies in MRSA medium. The high number of total LAB soy-yamgurt is caused by the relatively higher sugar content of yam-bean extract, so it is suspected that all the glucose is fermented as a source of energy to form lactic acid. This indicates that the presence of FOS (fructo-olygosaccharide) deposits in yam bean and soy bean extract in the form of GOS (galactooligosaccharide) were not entirely used by microbes for growth. In this research using a banana yogurt products for the growth of *Lactobacillus acidophillus* CCRC 10695, pointed out that the source of FOS on bananas did not change significantly [6]

The starter bacteria growth is influenced by various factors, among them are nutrients that are closely related to the composition of the raw materials, such as lactose for growth and amino acids, vitamins and other supporting factors such as environmental conditions (pH, oxygen, and water activity), and the type of bacteria [13]. The addition of skim milk powder on this research serves as a source of lactose and protein for the growth and activity of the starter bacteria, because both soy-bean and yam-bean extract did not contain lactose or casein. This is supported by the opinion of [13] which states that the availability of the nutrients that are appropriate to the needs of the bacteria produce a high increasing number of bacterial cells. At the beginning of fermentation, the pH of the initial raw material is relatively high around 5.6-6.8 which is.

At that point, the *S. thermophilus* grew earlier and caused a decrease in pH to 5.0-5.5, then the *L. bulgaricus* will multiply and changed the lactose into lactic acid and caused the pH decreased to 4.4-3.8 [9]. When *L. bulgaricus* began to grow, it will release amino acids, such as valin, histidin and glycine that are needed for the growth of *S. thermophilus*. In contrast, S. *thermophilus* will help lower the pH (increasing levels of acid), synthesis the formic acid and folic acid that stimulate the growth of *L. bulgaricus* [9]. The *L. bulgaricus* growth at the beginning of fermentation was slow, but the growing activity of proteolytic *L. bulgaricus* released a number of peptides and amino acids that can stimulate the growth of *L. acidophilus*. In mutual, *L. acidophilus* produced folic acid that can stimulate the growth of *L. bulgaricus* [17].

### *3.2 Total dry solids*

Total dry solids of soy-yamgurt probiotic drink have a very important role, beside as a source of nutrition for the activity and bacterial starter culture developments, as well as to form a texture and flavor components in the produced fermented product. Total dry solids is influenced by the composition of the raw material used. Skim milk powder has a role as the starter bacteria growth media, because the powdered milk contains a high dry ingredients consisting of relatively high protein, fat and lactose as the main energy source to speed up the bacteria starter activities and to form lactic acid and flavor components. The bacteria are able to use the nutrients in the ingredients to form secondary metabolites and a mass of cells, so the total dry solids has increased with the increasing amount of soy-bean extract and the increasing of fermentation temperature [16], it is presented in Figure 2.

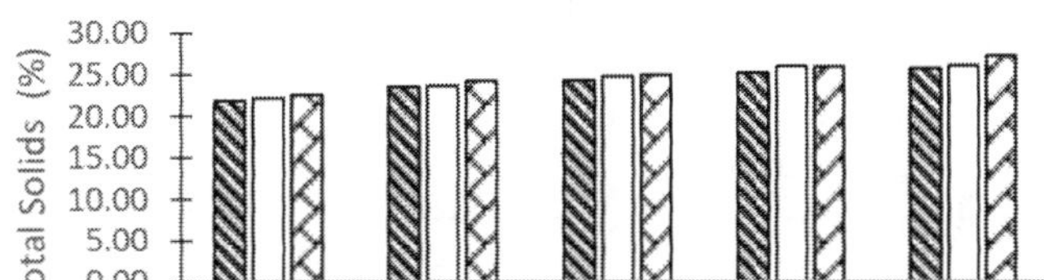


| | S1=100:0 | S2=75:25 | S3=50:50 | S4=25:75 | S5=0:100 |
|---|---|---|---|---|---|
| S1 | 21.8643 | 23.5197 | 24.2813 | 25.1867 | 25.6960 |
| S2 | 22.1317 | 23.6303 | 24.7183 | 25.9667 | 26.0220 |
| S3 | 22.5483 | 24.2180 | 24.9343 | 25.9350 | 27.1910 |

Comparison of yam-bean and soy-bean extract

**Figure 2.** The chart of average total solids in each treatment.

### *3.3 Lactic acid levels*

The lactic acid content are obtained by titration which was assumed as the levels of acid dominant formed during fermentation. The addition of yam-bean extract was significantly effect from 0.682 to 0.424%, the higher the amount of yam-bean extract the higher lactic acid value. It means that all the treatments of the addition of soy-bean and yam-bean extract produced beverage with the lactic acid levels were 0.5-2.0% [13], except for the addition of 100% soy-bean extract, to be more clear it can be seen in Figure 3.

Lactic acid is the main metabolite (85%) that are produced from lactose by the LAB starter. The process of lactose into lactic acid are influenced by several factors i.e.: the number and type of starter,

starter condition, temperature, incubation time and nutritional content of lactose in milk [13]. Soy-yamgurt had a specific sour taste derived from lactic acid bacteria activity. Lactose as a source of energy utilized by the bacteria starter to support its growth through the glycolysis; lactose is hydrolyzed by the β-galactosidase into glucose and galactose.

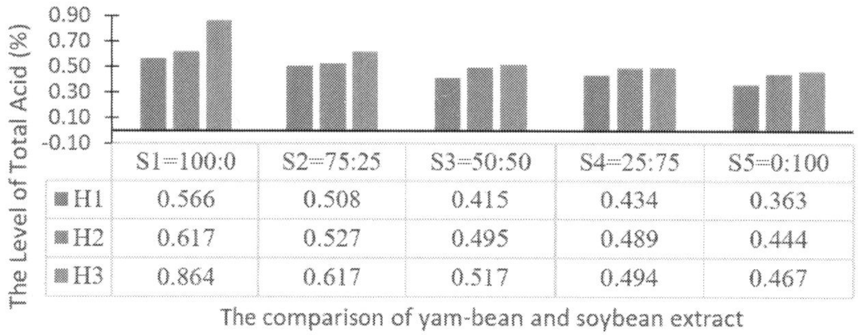


**Figure 3.** The chart of the average total acid in each treatment.

Furthermore, glucose is converted into pyruvate and then pyruvic acid is converted to lactate acid by the lactate dehydrogenase enzyme [19]. The overhaul process of lactose into lactic acid is influenced by several factors were the number and type of starter, starter conditions, temperature, incubation time and nutritional content of lactose in milk, the added initial raw material [13]. In the fermentation process, the growth of *S. thermophillus* is faster than *L. bulgaricus* and *L. acidophilus*. The temperature and incubation time effected the levels of lactic acid, as it caused the difference in the growth and composition of microbes contained in the starter, so it will produce lactic acid which varies on the final fermented product [9]

*3.4 Influence of the substrate variation and fermentation time on the antioxidant activity*

The antioxidant activity is calculated based on the DPPH free radical attenuation by soy-yamgurt antioxidant components. The presence of antioxidant activity in the raw materials is expected to add functional value on the soy-yamgurt, beside the probiotic bacteria content. The antioxidant activity of soy-yamgurt can be seen in Figure 4 and 5.

$IC_{50}$ value is the concentration of sample that gives 50% DPPH attenuation [17]. The result showed that the $IC_{50}$ value were obtained on the $5^{th}$ minutes or $60^{th}$ minutes, where the value ranged between middle to low, except in the treatment of the S1H1 and S2H1 on the $5^{th}$ minutes, the value of $IC_{50}$ is still above 100 mg/L. According to [8] that the compound had to be a powerful antioxidant activity if the $IC_{50}$ value is less than 50 ppm. This indicates that soy-yamgurt has a powerful to strong activity to reduce free radicals.

*3.5 The Characteristics of Soy-yamgurt Sensory Evaluation Quality of Substrate Variation with Fermentation Time.*

The characteristics of soy-yamgurt sensory evaluation were including the hedonic values of color, flavor, taste and texture. The quality of sensory evaluation is an important parameter associated with the panelist acceptance of a product. For more details can be seen in Tables 1 and 2. The hedonic value of soy-yamgurt on taste gave highly significant interactions ($p<0.01$). In general the acceptance of panelists on taste and texture qualities tended to increase in longer fermentation time, but all the statistical results showed that 6 hours fermentation gave the best results. It is showed that the substrate variation with higher soy-bean extract, the consistency and firmness of the soy-yamgurt tended to be more viscous.

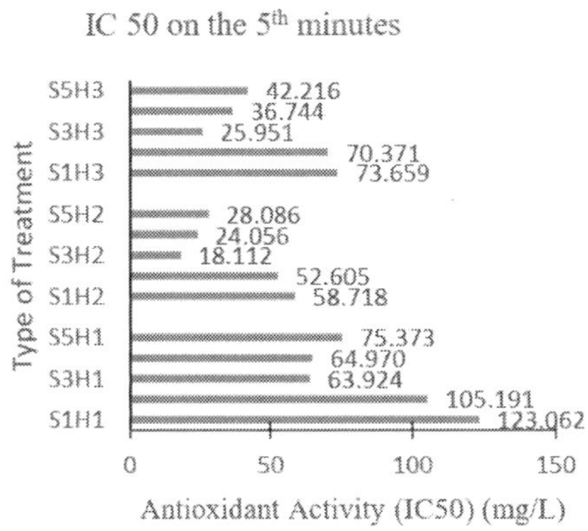


**Figure 4.** IC 50 antioxidant activity in various treatment on the 5th minute.

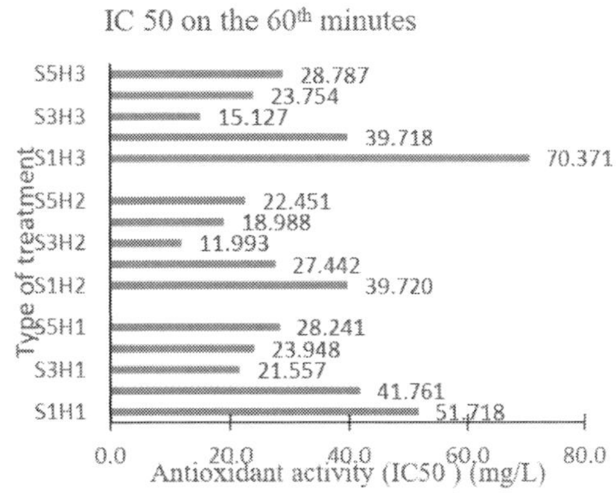


**Figure 5.** IC 50 antioxidant activity in various treatment on the 60th minute.

The increasing of soy-bean extract, tend to increase the organoleptic value, more acceptable taste. The best quality of sensory was S4 ( yam-bean extract with soy-bean extract = 25:75).

**Table 1.** The effect of the substrate variation for the sensory evaluation quality characteristic of soy-yamgurt.

| The Comparison of Yam-bean extract with soy-bean extract | The hedonic value (numeric) | | | |
|---|---|---|---|---|
| | Color | Flavor | Taste | Texture |
| S1=100:0 | 3,682bB | 3,061cC | 2,991cC | 2,782dC |
| S2=75:25 | 3,790aA | 3,417aA | 3,373bB | 3,447bcB |
| S3=50:50 | 3,380dD | 3,101cC | 3,399bB | 3,581aA |
| S4=25:75 | 3,600cC | 3,282bB | 3,380bB | 3,383cB |
| S5=0:100 | 3,412dD | 3,449aA | 3,561aA | 3,474bB |

Description: the data were composed of 3 replication, the value that are followed by different small letters in one column showed significant ($p < 0.05$) and highly significant effect ($p < 0.01$) (Capital letters).

**Table 2.** The effect of fermentation time on the sensory quality characteristic of soy-yamgurt

| Fermentation time | Hedonic value (numeric) | | | |
|---|---|---|---|---|
| | Color | Flavor | Taste | Texture |
| H1= 4 hours | 3,485aA | 3,383bB | 3,432bB | 3,289cB |
| H2= 6 hours | 3,521aA | 3,449aA | 3,711aA | 3,453aA |
| H3= 8 hours | 3,340bB | 3,361bB | 3,317cC | 3,387bA |

Description: The data were composed of 3 replication, the value that are followed by different small letters in one column showed significant ($p < 0.05$) and highly significant effect ($p < 0.01$) (capital letters)

## 4. Conclusions and Suggestions

The soy-yamgurt from yam-bean and soy-bean extract was analyzed in the quality characteristic including total solids, total acid titration, organoleptic test and antioxidant activity with the DPPH free radical attenuation method.

It can be concluded that the using of yam-bean and soy-bean had resulted in improving sensory and rheological attributes of soy-yamgurt. The best treatment in this study was the S3H2 (the comparison of yam-bean and soy-bean extract 50:50) and 6 hours fermentation time.

## REFERENCES

[1] Cross, M.L. 2002. Microbes versus microbes, immune signals generated by probioticLactobacilli and their role in protection against microbial pathogens, FEMS Immunlogy and Medical Microbiology, 34: 345-253.
[2] Donkor, O.N., Nilmini, S.L.I., Stolic., P., Vasilgevic, T. And Shah, N.P. 2007. Survival and activity of selected probiotic organism in set-type yoghurt during cold storage. International Dairy Journal, 17:92-151.
[3] Fardiaz, S. 1992. Petunjuk Laboratorium Mikrobiologi Pengolahan Pangan. IPB-press, Bogor.
[4] Heenan, C.M. Adams, M.C. Hosken, R.W. and Fleet, G.H. 2004. Survival and sensory acceptability of probiotic microorganism in a nonfermented frozen vegetarian dessert. Lebensmittel Wissenschaft and Technology. 17: 461-466.
[5] Koswara, S. 2006. Isoflavon Senyawa Multi Manfaat in Kedelai. www.ebookpangan.com (15Agustus 2015).
[6] Kusumaningrum, A.P. 2011. Kajian Total Bakteri Probiotik Dan Aktivitas ntioksidan Yoghurt Tempe dengan Variasi Substrat. Skripsi. Universitas Sebelas Maret. Fakultas Pertanian. Surakarta
[7] Majeed, M and Prakash, L. 2007. Probiotics for Health and Wellbeing. Sabinsa Corporation.1-12.
[8] Molyneux, P., 2004, The Use of Stable Free Radical diphenylpicrylhidraxyl (DPPH) for estimating antioxidant activity, Songklanakarin. J. Sci. Technol., 26 (2): 211-219.
[9] Rahman A, S. Fardiaz, W.P. Rahayu, Suliantari, C.C. Nurwitri. 1992. Teknologi Fermentasi Susu. Pusat Antar Universitas Pangan dan Gizi. Institut Pertanian Bogor. Bogor.
[10] Ranganna, S. 1978. Manual of Analysis for Fruit and Vegetables Product. Tata McGraw Hill. New Delhi.
[11] Robinson, T. 1991. Kandungan Organik Tumbuhan Tinggi (Terjemahan Kosasih T.). ITB, Bandung.
[12] Rusmarilin, H., E. Julianti, M. Nurminah. 2014. Peranan Senyawa Antioksidan Ekstrak Umbi Bengkoang (*Pachyrrhizus erozus* L.) Dalam Meredam Aktivitas 2,2-Diphenyl-2-Picrylhidrazil (DPPH). Prosidinng Seminar Nasional Bilogi,Fakultas MIPA USU, Medan.
[13] Salminen, S. and A. von Wright. 1993. Lactic Acid Bacteria. Marcel Dekker, Inc. New York.
[14] Standar Nasional Indonesia (SNI) No. 01.2981. 1992. Yogurt. Pusat Standarisasi Industri Departement Perindustrian, Jakarta. Indonesia
[15] Soekarto, T.S. 1985. Penilaian Organoleptik Untuk Industri Pangan dan Hasil Pertanian. Bharatara Karya Aksara. Jakarta. 79-80.
[16] Usmiati, S. dan T. Utami. 2008. Pengaruh bakteri probiotik terhadap mutu sari kacang tanah fermentasi. Jurnal Pasca Panen. 5(2): 27-36.
[17] Surono, I.S. 2004. Probiotik Susu Fermentasi dan Kesehatan. Yayasan Pengusaha Makanan dan Minuman Seluruh Indonesia (YAPMMI). PT. Tri Cipta Karya (TRICK). Jakarta.
[18] Sudarmadji, S., Haryono dan Suhardi. 1989. Prosedur Analisis untuk Bahan Makanan dan Pertanian. Liberty, Yogyakarta.
[19] Tamime, A.Y, R.K. Robinson. 1989. Yogurt Science and Technology. Pergamon Press. Britain.
[20] Tamime, A.Y. 2005. Probiotic Dairy Products. Blackwell Publishing. Oxford, 1-216.
[21] Yeganehzad, S., Mazaheri-Tehrani, M. and Shahidi, F. 2007. Studying microbial, physicochemical and sensory properties of directly concentration probiotic yoghurt. African Journalof Agricultural Research, 2 (8): 366-369.

## ACKNOWLEDGEMENTS

We thank to Directorate General of Higher Education, The Ministry of Research, Technology and Higher Education the Republic of Indonesia and Universitas Sumatera Utara for supporting this work through Competence Research Grant.

IOP Conf. Series: Earth and Environmental Science **122** (2018) 012088 doi:10.1088/1755-1315/122/1/012088

# Hypoglicemic activity of gambier (*Uncaria gambir robx.*) drinks in alloxan-induced mice

**E A Zebua[1,4], J Silalahi[2] and E Julianti[1,3]**

[1]Departement of Food Science, Faculty of Agriculture, Universitas Sumatera Utara, Medan, Indonesia
[2]Department of Pharmaceutical Chemistry, Faculty of Pharmacy, Universitas Sumatera Utara, Medan, Indonesia
[3]Center of Excellence for Marketing Technology of North Sumatera Indigenous Agricultural Product, Universitas Sumatera Utara, Medan 20155 Indonesia
[4]Email: amuritazebua@gmail.com

**Abstract.** Diabetes may cause complications in various parts of the body and increase the risk of early death. Gambier contains bioactive compounds and can be used as raw material in preparing gambier drink. The purpose of this study was to extract gambier, to determine the hypoglycemic activity of gambier drinks in alloxan-induced mice. In this study, gambier were extracted by 3 types of solvents with distilled water, ethyl acetate, and ethanol. The gambier drinks was tested on 5 groups of mice wich consisting of 5 male mice. Group 1 was given 0.5% CMC 1%/bw, group 2 was given metformin 65 mg/kg bw, groups 3, 4 and 5 were given gambier drinks with dosage of 100, 200, and 300 mg/kg bw. The results showed that extraction using distiled water produced the best gambier drinks. Gambier drinks with dosage 100, 200 and 300 mg/kg bb given for 15 days decreased the blood glucose level 27.69%, 38.75%, and 50.62% respectively, and increased the body weight 13.14%, 10, 91%, and 10.18% respectively. Treatment with gambier drinks was found to improve the condition of pancreatic langerhans island of alloxan-induced mice.

## 1. Introduction

Diabetes is a condition signalized by blood glucose level above normal called hyperglicemia due to dysfunction of blood glucose regulation[1]. There are two hormones which are responsible and play important roles to regulate glucose level in blood including insulin and glucagon hormones[2]. Metabolism abnormalities caused by β cell pancreas breakdown resulting in decreased insulin secretion and also insulin appeared to be resistent resulting in hyperglycemic as one of the diabetes symptoms causing complications in body parts and increasing the risk of early death[3],[4]. To prevent diabetes epidemics, it needs strategy to consider and prevent the complexibility of disease progression, and medications cost so increased access and knowledge on prevention strategy and medications on the disease is necessary by using components of traditional medicine for centuries[5]. Indonesia is known as main producer of gambier in international market[6]. Gambier is mainly used as tanner, dye substances and beauty product components, as a medicine mixture to treat burns, headache, diarrhea, dysentery, mouthwash, sprue and skin medicine[7,8]. Gambier which obtained from Indonesian market has very good antihyperglichemic activity[9], and it has been consumed to treat diabetics as traditional medicines. Public interest in gambier consumption decreases because of the astringent taste and less attractive appearance[10]. The astringent taste can be reduced by decreasing tannin content and other impurities by extraction. Gambier can be processed into functional drinks with better flavour and easy to consume[10]. Moreover, the processed food is expected to gain public interest in consuming gambier, as it also functions as anti diabetes. The purpose of this study was to extract gambier, and determine the hypoglycemic activity of gambier drinks in alloxan-induced mice.

## 2. Materials and Methods

### *2.1.Materials*

Gambier plants obtained from Mado Laoli Village, Gunung Sitoli City, North Sumatra. Other material used in this experiment were cinnamon, gom arabic, and synthetic sugar (Tropicana Slim DIABTX®). The chemichals used in this study were distiled water, ethyl acetate and ethanol technical grade, cathechin standart (Sigma), gallic-acid (Sigma), methanol, phenol, $H_2SO_4$, KMNO4, $Na_2CO_3$ and Folin-Ciocalteu pro-analyzed (Merck), indigocarmin, DPPH, glucose, alloxan tetrahydrat (Sigma), metformin, formalin, alcohol, paraffin, gliserin, xylene, xylol, eoisin and hematoxyline mayers. The equipments used in this study were analytical balance (AND GR-200), micrometer pipettes (Eppendorf), Whatman filter papers No. 41, spectrophotometer (Shimadzu UV-1800), glucose kit (GlucoDr), automatic tissue processor (Lens TP 1020), mikroturn (Leica), light microscope (Primo Star Zeiss).

### *2.2.Gambier Production*

Gambier leaves were boiled for 30 minutes on boiling water, crushed by using mortar, squeezed until all juice out from the leaves then saturated for 24 hours. After that, gambier's gum was molded and dried in oven at 50°C for 60 hours.

### *2.3.Gambier's Extractions*

Gambier's extractions were done by maceration method, using three different solvents: distiled water ($E_1$), ethyl acetate ($E_2$), and ethanol ($E_3$). Gambier was mashed and sieved with 60 mesh filter. 100 g gambier powder were dissolved in 500 ml solvent, stirred using shaker for an hour and maceration for 24 hours at 30°C then filtered using filter paper Whatman No. 41. The obtained residue was extracted again twice with 500 ml solvent. The extract was evaporated by using rotary evaporator to obtain dry gambier extract .

### *2.4.Gambier drinks production*

The mixture of gambier drinks consisted of 2.67% gambier extract, 0.67% cinnamon, 0.33% gom arabic, 2.33% synthesis glucose (Tropicana Slim DIATBX®) and 94% of water. The mixture was shaken to homogenous, boiled for three minutes, cooled, and packed in a bottle.

### *2.5.Catechin content determination*

Catechin content was determined with procedure as described in Indonesian National Standard[11]. 50 milligrams dry catechin standard was transferred into a 50 ml flask, dissolved and diluted with ethyl acetate to the volume of 50 ml (solution A), put on ultrasonic bath for 5 min. Transferred 2 ml of solution A into a 100 ml Erlenmeyer and add 50 ml of ethyl acetate (solution B) and placed on ultrasonic bath for 5 min, and the solution can be used for measurement. One (1) gram sample transferred into 50 ml flasks, dissolved and diluted with ethyl acetate until it reached the calibration mark (solution C), put on ultrasonic bath for 5 min, filtered. The first 15 ml filtrate was discarded and the next filtrate was collected. Put 2 ml of filtrate from solution C into 100 ml Erlenmeyer and added 50 ml of ethyl acetate solvent (solution D) and put on ultrasonic bath for 5 min before measurement. Then, the absorbance of the standard solution and the sample solution using UV spectrophotometer at 279 nm wavelength were measured. Catechin content was counted by the formula:

Catechin (%) = (Et 279 / Ec 279) x (Ws/ W) x 100% (1)

Note: Et 279 is absorbance of sample solution, Ec 279 is absorbance of catechin standard solution, Ws is catechin standard weight (g) and W is weight of sample (g).

*2.6.Tannin content determination*

Tannin content was determined by AOAC method[12]. Fife (5) gram of sample was added 400 ml of distilled water and boiled for 30 min. After cooling, the solution was added distilled water to a volume of 500 ml, and filtered (filtrarte 1). Then, 10 ml of filtrate 1 was mixed up with 25 ml of indigocarmin solution and 750 ml of distilled water then titrated with a 0.1 N KMnO4 solution until the color turned to golden color (titration A). Filtrate 2 was resulted from the mixture of 100 ml of filtrate 1, 50 ml of gelatin solution, 100 ml of acid salt solution and 10 g caolin powder. The mixture was then shaken for 5 minutes and filtered. Twenty five (25) ml of filtrate 2 was mixed with 25 ml indigocarmin solution and 750 ml of distilled water and titrated with a 0.1 N $KMnO_4$ solution (titration B). Tannin content was calculated by the following equation:

Tannin (%) = ((50A-50B) x N x 0,0416 ) / g (2)

Note: A is titration volume of sample (ml) and B is titration volume of blank (ml), N is normality of standard $KMnO_4$ solution, g is weight of sample (g). 1 ml of $KMnO_4$ 0.1 N is equivalent to 0.00416 gram of tannin

*2.7.Total phenols content determination*

Total phenols content was determinated by using Folin Ciocalteu with gallic acid as standard[13]. The standard solutions of gallic acid was prepared with concentration of 250, 500, dan 1000 μg/ml and measured the absorbance at 765 nm wavelenght using a spectrophotometer (Shimadzu UV-1800). 0.5 ml of sample dissolved in distilled water (1:100 b/v) added 2.5 ml Folin-ciocalteu 50%, 2.5 ml distiled water, and 2 mL $Na_2CO_3$ 15%, homogenized and kept in a dark place for an hour. Then measured the absorbance at 765 nm wavelength. The concentration was calculated based on equation obtained from the standard curve. Total phenolic content of the sample were calculated by the formula:

TPC = (C.V.DF) / g x 100% (3)

Note: TPC is total phenolic content (μg GAE/mg), V is volume of sample (ml), C is concentration (μgGAE/ml), DF is dilution factors, g is weight of sample (mg)

*2.8.Antioxidant activity determination*

Antioxidant activity was assessed using the radical capture method using DPPH[14]. Briefly 1 ml of sample solution (diluted with methanol) with concentration of 10, 30, 50, 70 dan 90 μg/ml, added 2 ml solution of methanol and DPPH respectively. The solution was shaken until homogeneous and allowed to stand for 30 minutes, then absorbance was measured at a wavelength (λ) 517 nm using a spectrophotometer (Shimadzu UV-1800). The percentage of inhibition were calculated with the following formula:

%inhibition = (Absorbance of control – Absorbance of sample) / Absorbance of control x 100% (4)

By using the value of the inhibition percentage as the ordinate (y) and concentration of the extract as the absciss (x) using the linear regression method, a linear equation was obtained. IC50 values was determined based on inhibition concentration at 50%.

*2.9.Total sugar content determination*

Total sugar content was determined by using phenol method[15]. Standard glucose solution was prepared with concentrations of 40, 50, 60, 70, 80 μg / ml. 2 ml of each solution was mixed with 1 ml of 5% phenol solution and shaken. Then, 5 ml of concentrated sulfuric acid was added, left for 10 minutes into a shaken water bath. The absorbance was measured using a spectrophotometer at 490 nm wavelength. 2 ml of sample solution (1: 100 m/v) was mixed with 1 ml of 5% phenol solution and shaken. Added 5 ml of concentrated sulfuric acid, left for 10 minutes, shaken then placed in a waterbath for 15 minutes. The absorbance was measured using a spectrophotometer at a wavelength of 490 nm. The concentration of sugar in the sample solution by the equation obtained from the standard curve. Then calculated percentage of total sugar sample with equation formula:

$$\text{Total sugar } (\%) = (C.V.DF)/g \times 100\% \quad (5)$$

Note : C is concentration of sugar (μg/ml), V is volume of sample (ml), DF is dilution factors and g is weight of sample (μg).

*2.10.Organoleptic test of color, flavor and taste*

Organoleptic tests of color, flavor and taste were determined using a hedonic value scale[16] and randomly tested by 20 panelists. Testing was done on a sensory basis based on the hedonic scale.

*2.11.Selection of the best treatment*

The best treatment of gambier drinks was chosen based on weighting technique[17] with variables including weight of organoleptic value of color 1, organoleptic value of taste 0.9, organoleptic value of flavor 0.8, total glucose 0.7, tannin content 0.6, catechin content 0.5, total phenolic content 0.4 and antioxidant activity 0.3.

*2.12.Hypoglycemic Effect Assay*

The animal used in this study were male Wistar mice weighing 28-32 g. Before the experiment, mice were acclimatizated for a week in a good cage for environmental adaptation, i.e. the reception of light, 12 hours dark and 12 hours light and all the mice were subjected to normal mice chow and distilated water ad libitum.

Alloxan solution was prepared by dissolving alloxan in distilled water. The mice were induced with aloxan solution of 125 mg/kgBW administated intraperitoneally. The blood glucose level (BGL) of mice were measured on the third day. The mice with glucose levels greater than 200 mg/dl were separated and used as experimental animals. The mice were divided randomly into 5 groups consisting of 5 mice of each group.

Group I : Diabetic mice were given a 0.5% CMC suspension, dose of 1% of body weight (BW)
Group II : Diabetic mice were given a Metformin suspension, dose of 65 mg/kgBW
Group III : Diabetic mice were given gambier drink 100 mg/kgBW
Group IV : Diabetic mice were given gambier drink 200 mg/kgBW
Group V : Diabetic mice were given gambier drink 300 mg/kgBW

The suspension for the tested material was administered orally for 15 consecutive days and the BGL and body weight were measured on the day 1, day 3, day 6, day 9, day 12, and day15 after administration of the tested material.

*2.13 Histolopathologycal of the Pancreas Mice*

The mice from each group was killed by cervical dislocation after completion of treatment, and pancreas were dissected out quickly, washed in saline (0.09 NaCl), and fixed in formaldehyde solution (10%). Tissue were processed and embedded in paraffin wax. Sections were cut at 5μ thickness and stained with Hematoxyline-Eosin (HE). After completion of staining sections were observed under microscope for histogical change.

A study approval was obtained from Animal Research Ethics Committees (AREC) of Faculty of Math and Science, University of Sumatera Utara.

*2.14.Statistical analysis*
Data were analyzed using SPSS software (22.0). Differences in mean values were tested by analysis of variance, and significance levels were obtained with Duncan's test. A significance level of $P<0.05$ was used.

## 3. Result and Discussion

*3.1.The Quality of Gambier Drinks*
Quality parameters based on the bioactive components present in gambier drinks by different solvent extractions shown in Table 1.

**Table 1.** The effect of gambier extraction method on gambier drinks' quality parameters observed

| Quality parameters | Methods of extraction of gambier | | |
|---|---|---|---|
| | Extraction using distiled water ($E_1$) | Extraction using ethyl acetate ($E_2$) | Extraction using ethanol ($E_3$) |
| Catechin content (%) | $1.85 \pm 0.05^{b}$ | $1.98 \pm 0.05^{a}$ | $1.93 \pm 0.54^{ab}$ |
| Tannin content (%) | $0.21 \pm 0.04^{b}$ | $0.29 \pm 0.04^{a}$ | $0.19 \pm 0.02^{b}$ |
| Total phenolic content (μgGAE/mg) | $639.75 \pm 5.70^{c}$ | $732.34 \pm 4.86^{a}$ | $697.34 \pm 4.84^{b}$ |
| Antioxidant activity ($IC_{50}$) (μg/ml) | $37.41 \pm 1.37^{a}$ | $35.96 \pm 0.50^{a}$ | $36.23 \pm 0.87^{a}$ |
| Total sugar (%) | $1.92 \pm 0.10^{a}$ | $1.96 \pm 0.09^{a}$ | $1.95 \pm 0.09^{a}$ |
| Organoleptic of color value (numeric) | $4.02 \pm 0.10^{a}$ | $3.73 \pm 0.16^{b}$ | $3.72 \pm 0.10^{b}$ |
| Organoleptic of flavor value (numeric) | $3.83 \pm 0.10^{a}$ | $3.60 \pm 0.25^{b}$ | $3.55 \pm 0.13^{b}$ |
| Organoleptic of taste value (numeric) | $3.70 \pm 0.13^{a}$ | $3.42 \pm 0.10^{b}$ | $3.49 \pm 0.05^{b}$ |

Note : Data is the means of three replicates

Table 1 shows that gambier drinks $E_1$ has low tannin level and higher organoleptics color and taste value. Therefore, the $E_1$ is more preferable for consumers because it has low tannin level lead to lower astringent taste. Tannin is a compound which cause astringent taste in gambier[10]. In addition, gambier drinks $E_2$ has the highest of catechin because polarities of ethyl acetate value similar to the polarities substance extracted so it is more effective[18]. Based on weighting technique, the best gambier drinks is a drink which used gambier as its raw material extracted with distilled water ($E_1$) and it is used for in vivo test on research.

*3.2.Hypoglycemic activity of gambier drinks*
Hypoglycemic activity of tested material was observed in interval time of 3 days or 15 days and the results presented in Table 2 and Figure 1.

**Table 2.** The effect of gambier drinks on blood glucose levels of mice induced with alloxan

| Groups | Gluose blood levels (mg/dl) with interval time (days) | | | | | | |
|---|---|---|---|---|---|---|---|
| | D0 | D1 | D3 | D6 | D9 | D12 | D15 |
| Group I | $120.0 \pm 8.22^{e}$ | $393.2 \pm 6.91^{d}$ | $402.6 \pm 9.13^{c}$ | $412.8 \pm 4.38^{b}$ | $423.0 \pm 3.39^{a}$ | $416.2 \pm 3.03^{ab}$ | $410.0 \pm 3.61^{bc}$ |
| Group II | $135.2 \pm 6.76^{g}$ | $395.6 \pm 9.76^{a}$ | $352.2 \pm 4.21^{b}$ | $302.0 \pm 8.69^{c}$ | $240.0 \pm 4.30^{d}$ | $207.0 \pm 4.74^{e}$ | $181.0 \pm 7.28^{f}$ |
| Group III | $130.4 \pm 9.24^{g}$ | $390.0 \pm 9.35^{a}$ | $355.4 \pm 4.51^{b}$ | $329.6 \pm 10.74^{c}$ | $307.6 \pm 9.40^{d}$ | $295.6 \pm 7.30^{e}$ | $282.0 \pm 7.07^{f}$ |
| Group IV | $141.2 \pm 7.50^{g}$ | $349.4 \pm 7.50^{a}$ | $319.6 \pm 6.50^{b}$ | $293.4 \pm 9.61^{c}$ | $264.2 \pm 6.76^{d}$ | $238.6 \pm 5.86^{e}$ | $214.0 \pm 4.74^{f}$ |
| Group V | $148.6 \pm 9.45^{g}$ | $356.4 \pm 9.76^{a}$ | $309.8 \pm 6.80^{b}$ | $271.0 \pm 6.96^{c}$ | $234.0 \pm 4.74^{d}$ | $201.6 \pm 8.41^{e}$ | $176.0 \pm 7.65^{f}$ |

Note : Data is the means of 5 replicates

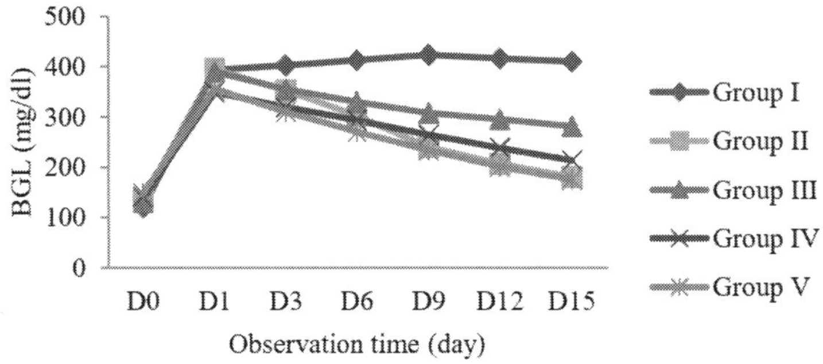


**Figure 1.** The hypoglycemic activity of material treatmens in mice

Table 2 and Figure 1 show that the BGL in the control group (group I) increased with the time, but the levels were decreased after 15 days of storage. Metformin (group II) was able to reduce blood glucose levels from 395.6 mg/dl to 181 mg/dl (~54%). Gambier drinks with dose of 100 mg/kgBW (group III) reduced BGL from 390 mg/dl to 282 mg/dl (~28%), dose of 200 mg/kgBW (group IV) reduced BGL from 349.4 mg/dl to 214 mg/dl (~39%), and a dose of 300 mg/kg BW (group V) reduced BGL from 356.4 mg/dl to 176 mg/dl (~51%). The gambier drinks' ability in reducting BGL was caused by phenolics compound in gambier with strong antioxidant activities. The antioxidant activity for gambier is 6.4 μg/ml and phenolics compounds are 1142.5 μg GAE/mg[19]. Gambier also had inhibitory activity on α-glucoside of 40.65-89.87 μg/ml[9]. The main bioactive compound which play role to inhibit α-glucoside is catechin[20]. BGL decrease is also caused by coumarin compound in cinnamon which can fix glicemics control by inhibiting mechanism on α-glucoside enzyme in intestine, organizing the balance of enzym Pyruvate Kinase (PK) and Pyruvate Carboxikinase (PEPCK) which play role to arrange glucose metabolism in the liver[21].

### *3.3.Body weight of mice*

The effect of gambier drinks on body weight of mice can be seen in Table 3. Statistically, gambier beverages did not give significant different effect (P> 0,05) on body weight of mice.

**Table 3.** The effect of gambier drinks on body weight of mice induced with alloxan

| Groups | Body weight (g) with interval time (days) | | | | | | |
|---|---|---|---|---|---|---|---|
| | D0 | D1 | D3 | D6 | D9 | D12 | D15 |
| Group I | 29.8±2.05ª | 29.8±2.05ª | 30.4±2.51ª | 31.4±3.51ª | 32.6±4.28ª | 34.2±5.31ª | 35.2±5.31ª |
| Group II | 29.2±0.45ª | 29.2±0.45ª | 29.4±1.14ª | 30.0±1.87ª | 30.6±2.88ª | 31.2±3.27ª | 31.8±3.77ª |
| Group III | 30.4±1.67ª | 30.4±1.67ª | 31.0±2.12ª | 31.4±2.88ª | 33.2±3.77ª | 34.2±4.76ª | 35.0±5.10ª |
| Group IV | 29.4±0.55ª | 29.4±0.55ª | 29.8±1.92ª | 30.4±3.51ª | 31.2±4.44ª | 32.0±4.85ª | 33.0±6.04ª |
| Group V | 30.0±1.41 ª | 30.0±1.41ª | 30.2±1.48ª | 31.0±2.92ª | 31.8±4.66ª | 33.0±5.70ª | 33.4±6.07ª |

Note : Data is the means of 5 replicates

### *3.4.Histology of langerhans islet of mice*

The histology of langerhans islet of mice can be seen in Figure 2.

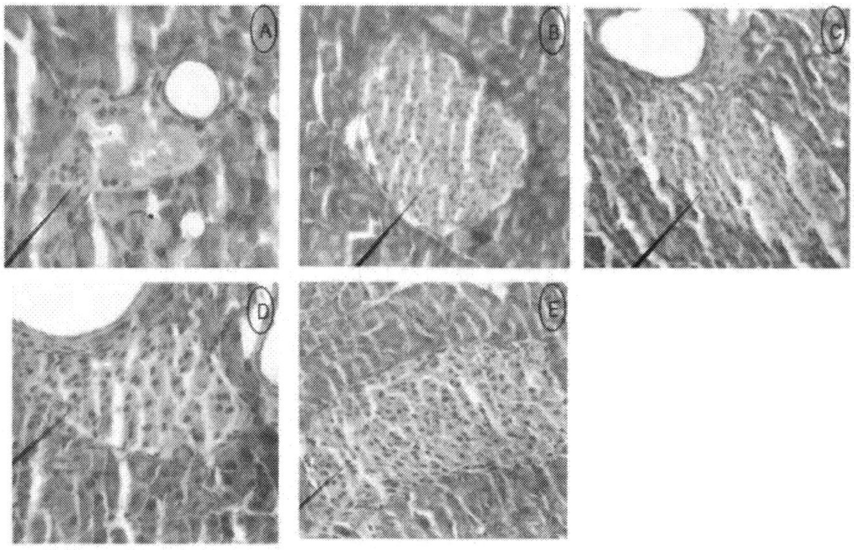


Figure 2. Islets of Langerhans of pancreas stained with HE. A is pancreas of group I mice, B is pancreas of group II mice at 40X. C is pancreas of group III mice, D is pancreas of group IV mice and E is pancreas of group V mice at 40X.

The group of control mice (K1) had langerhans islet impairment caused by alloxan which induced damage to the DNA of β cells located in langerhans islet and induced by high blood glucose levels[22]. Metformin as an oral hypoglycemic agent showed improvement of β cells in which β cell density increases because metformin provides protection against pancreatic β-cell apoptosis by a mechanism interference with the production of ROS and inhibits the influx of free fatty acids and CD36 (cluster determinant 36)[23] and prevent cell death due to oxidative stress in non-neuronal cells[24].

Gambier drinks with various doses seem to improve β cell structure of mice's pancreas, the higher dose of gambier drinks repair β-cell density of langerhans islet better because it can restore pancreatic β cells to secrete insulin, resulted in the decrease of BGL. Gambier drinks contain high phytochemical compounds, which play a role in repairing the β cells, improve the condition of blood glucose by preventing oxidative stress, improve damage conditions and prevent the occurrence of lesions on the islet of langerhans[25].

## 4. Conclusions

The gambier extraction method using distiled water produces the best quality gambier. This study indicates that gambier drink has the hypoglicemic effect to decrease blood glucose levels. The dose of 300 mg/kgBW has hypoglycemic activity which is similar to antidiabetic agent metformin, and can repair the damage of β-cells of langerhans islet.

## References

[1] Masharani U 2008 *Diabetes Demystified* (New York: MCGraw Hill Companies) p 3, 77
[2] Chee F and Fernando T 2007 *Closed-Loop Control of Blood Glucose* (Berlin: Springer) p 1-2.
[3] A D Association 2014 *Diagnosis and classification of diabetes mellitus* J. Diabetes Care **37** 1
[4] WHO 2016 *Global Report on Diabetes* WHO Library Cataloguing in Pub. Data (France: WHO Press) p 6
[5] Ballali S and Lanciai F 2012 Int. J. Food Science and Nutrition **63** S1
[6] Dhalimi A 2015 *Gambier (Uncaria gambir L.) problems in Sumatera Barat and its alternatives* J. Prespektif **5** 1

[7] Hidayat R S and Napitupulu R M 2015 *Book of Medicinal Plants* (Jakarta: AgriFlo) p 124

[8] Arbain D, Putra D P and Bachtiar A 2009. *Recent study on the chemistry of sumatran medicinal plants* Planta Medica J. **75**

[9] Widiyarti G, Sundowo A and Hanafi M 2011 *The free radical scavenging and anti-hyperglycemic activities of various gambiers avaiable in indonesia market* J. Makara Sains **15** 2

[10] Muchtar H, Yeni G, Hermianti W and Diza Y H 2010 *Preparation of polyphenol gambier (Uncaria gambir Robx) concentrate as a food antioxidant* J. Riset Industri **4** 2

[11] Indonesian National Standard 2000 *Gambier SNI 01-3391-2000* (Jakarta: Badan Standarisasi Nasional) p 2-4

[12] Sudarmadji S, Haryono B, and Suhardi 1984 *Analytical Procedures for Food and Agricultural Materials* (Yogyakarta: UGM-Press) p 108

[13] Singleton V L, Orthofer R, and Raventos R M L 1999 *Analysis of total phenols and other oxidation substrates and antioxidants by means of folin–ciocalteu reagent* Methods in Enzymology **299** p 152–178

[14] Mosquera O M, Correa Y M, Buitrago D C, and Nino J 2007 *Antioxidant activity of twenty five plants from colombian biodiversity*. Mem Inst Oswaldo Cruz **102** 5

[15] Apriyantono A, Fardiaz D, Puspitasari N L, Sedarnawati and Budiyanto S 1989 *Food Analysis* (Bogor: IPB-Press) p 50-51

[16] Setyaningsih D, Apriyantono A, Sari M P 2010 *Sensory Analysis for Food and Agro Industry* (Bogor: IPB Press) p 59-67

[17] De Garmo E P, Sullivan W G and Canada J R 1984 *Engineering Economy* (New York: Macmillan Publishing Company) p 265

[18] Yeni G, Syamsu K, Suparno O, Mardliyati E, and Muchtar H 2014 *Repeated extraction process of raw gambiers (Uncaria gambir Robx.) for the catechin production as an antioxidant.* Int. J. of Applied Engineering Research **9** 4.

[19] Sazwi N N, Nalina T and Rahim Z H A 2013 *Antioxidant and cytoprotective activities of Piper betle, Areca catechu, Uncaria gambir and betel quid with and without calcium hydroxide* J. BioMed Central Complementary and Alternative Medicine **13** 351.

[20] Apea-Bah F B, Hanafi M, Dewi R T, Fajriah S, Dawarman A, Artanti N, Lotulung P, Ngadymang P and Minarti B 2009 *Assessment of the DPPH and α-glucosidase inhibitory of gambier and qualitative identification compound* J. of Medicinal Plants Research **3** 10.

[21] Medagama A 2015 *The glicaemic outcomes of cinnamon: a review of the experimental evidence and clinical trials* Nutrition J. **14** 108.

[22] Walvekar M V, Potphode N D, Desai S S, dan Desmukh V M 2016 *Histological studies of islet of langerhans of pancreas in diabetic mice after curcumin administration* Int J of Pharmaceutical and Clinical Research **8** 9

[23] Moon J S, Karunakaran U, Elumalai S, Lee I K, Lee H W, Kim Y W and Won K C 2017 *Metformin prevents glucotoxicity by alleviating oxidative and ER stress-induced CD 36 expression in pancreatic beta cell.* J. of Diabetes and Its Complications **31**

[24] El-Mir, M Y, Detaile D, Villanueva G R, Esteban M D, Guigas B, Attia S, Fontaine E, Almeida A, and Leverve X 2008 *Neuroprotective role of antidiabetic drug metformin against apoptotic cell death in primary cortical neurons.* J. Molecular Neuroscience **34**

[25] Hosseni A, Nick R S and Ghorbani A 2016 *Pancreatic beta cell protection/ regeneration with phytotheraphy.* Brazilian J. of Pharmaceutical Science **51** 1

# The extraction of antimicrobials component of andaliman (*Zanthoxylum acanthopodium* DC.) and its application on catfish (*Pangasius sutchi*) fillet

**A Muzafri, E Julianti and H Rusmarilin**

Department of Food Science Technology, Faculty of Agriculture, Universitas Sumatera Utara, Indonesia

E-mail: al.outsiders@yahoo.com

**Abstract.** Andaliman *(Zanthoxylum acanthopodium* DC.*)* is a well known wild species in North Sumatera and used for seasoning in Batak's traditional cuisine. This study was aimed to examine the phytochemical constituents of andaliman fruit extracts after simple macerated in water, methanol, ethyl acetate and hexana using qualitative phytochemical analysis, and to determine its potential antimicrobial activity against *Staphylococus aureus*, *Escherichia coli* and *Salmonella sp* by using agar well difussion method and minimum inhibitory concentration (MIC). Phytochemicals such as alkaloids, flavonoid, glycosides, saponins, tannins, triterpene/steroid and glycoside anthroquinones were detected in the methanol extracts, but steroids and glycisode antraquinones were absent in the ethyl acetate extract. The ethyl acetate extracts showed maximum zone of inhibition and minimum inhibitory concentration against all the experimental microorganisms. The minimum zone of inhibition was determined in hexane extracts showing less antimicrobial activity against all the experimental microorganisms. The MIC of the ethyl acetate extracts was 0,5% w/v for all tested bacteria. Apllication of ethyl acetate extracts of andaliman fruits showed effective for catfish *(Pangasius Sutchi)* fillet stored in refrigerator (5 °C) for 3 days.

## 1. Introduction

Andaliman (*Zanthoxylum acanthopodium* DC.) fruits is known as the Batak society spicy and used to eliminate the smell of fish and raw meat. They are wildly grown in Tapanuli, North Sumatera at 1500 meters above sea level [1] at a temperature of 15-18 °C [2,3]. The fruits shape are like pepper, small round, green, but becomes black when they are dry. The fruits contain aromatic compounds with a bitter taste and produce the effect of thrilling sensation on taste buds and cause the tongue felt numb. Wijaya [4] found that citronellan and limonen were the most impacting compounds on the aroma of andaliman, but β-myrcene, (z)-β-ocimene, linalool, β-citronellol,neral, geraniol, geranial, geranyl acetate, unknown compound, and a sesquiterpene were also contributing to andaliman's fresh citrus and warm sweet-peppery aroma. It had been reported to has an anti-inflammatory activity [5] and antioxidant activity [6]. Based on Cahyana and Mardiana [7], andaliman has an antimicrobial activity. Siswadi [8] has evaluated the antibacterial activity of andaliman extract against *Bacillus stearothermophilus*, *Pseudomonas aeruginosa* and *Vibrio cholera*.

Fish product has more spoilage than other foods with high protein [9]. It should be noted that psychrophilic bactera spoilage occurs in cold storage [10]. The most factors for determining the quality of fish product are the smell, taste, freshness, absence of spesific microorganism, size and composition [11]. The enzymatic and chemical deterioration occur in fish prior to microbial spoilage, because of the high content of saturated fatty acid, free amino acid and othe reactive compound [9].

Nowadays consumers demand minimally preserved foods for maximum nutrient retention, without the addition of chemical preservatives. On the other hand, foods need to be safely preserved, with prolonged shelf-life [12]. An increasing interest in the use of natural antimicrobials as food preservatives has been recorded. Fortunately, antimicrobials from plants can be used as an alternative to chemical preservatives in order to satisfy consumers' demand for safe, convenient and wholesome food (13,14). Therefore, the aim of this research is to study the antimicrobial component of andaliman extracts and to evaluate its potency on catfish fillet quality during storage in refrigerator.

## 2. Material and methods

### *2.1. Preparation of andaliman fruits*

Andaliman fruits was collected from farmer at Brastagi Village North Sumatera. All fruits were washed with fresh water and dried in the oven at 40°C of temperature for 48 hours, The dried fruits were powdered by using electrical blender and stored in sterile container for further use.

### *2.2. Preparation of andaliman extracts*

The dried fruit powder were divided into 4 groups, and kept in contact with the solvent (water, methanol, ethyl acetate and hexana) in a stoppered container for 72 hours with occasional shaking. The solvent from the extract was filtered, and concentrated by using vacum evaporator at 50 °C to yield a crude extract and then weighed.

### *2.3. Qualitative analysis of phytochemical compound*

The individual extract was subjected to the yield and qualitative phytochemical analysis for the presence of some chemical constituents as per standard methods described by Brain and Turner [15] and Evans [16] included alkaloids, flavonoid, glycosides, saponins, tannins, triterpene/steroid and glycoside anthroquinones.

### *2.4 Evaluation of Antimicrobial Activity*

About 2 g of each extract was dissolved in 2 ml DMSO to obtain a stock solution (100% extract) and then the stock solution was diluted to obtain 75%, 50% and 25% of extract concentration. *Escherichia coli, Staphylococcus aureus* and *Salmonella typhimurium* were obtained from Biology Laboratory University of Sumatera Utara for the test and were performed by disc diffusion method according to the method described by Hussain *et al* [17]. Disc saturated with andaliman fruits extracts from different solvents with 25-100% concentration were placed on medium surfaces (Mueller-Hinton agar). Plates inoculated with tested organisms were incubated at 37 °C for 24 hours. Antimicrobial susceptibility testing was performed by the determination of the bacterial inhibition zone around the disc [17].

The minimum inhibitory concentrations (MIC) was performed by a serial dilution technique (0.25, 0.5, 0.75, 1.0, 2.0, 3.0, 4.0 and 5.0%) of concentrate according to the method used by Kubo *et al*.,[18, 19]) with some modifications. The extracts were mixed with 10 μL of bacteria test culture in a shaker incubator at speed 150 rpm for 24 hours. MIC value is the minimum extract concentration that inhibits 90% of the growth of bacteria during a 24 hours incubation.

### *2.5 Preparation of catfish fillet*

Fresh catfish (*Pangasius sutchi)* were bought from catfish pond in Medan. The live cathfish of similar weight and size were selected for each experiment. The fishes were cleaned and filleted. They were transported in isothermal ice boxes to the laboratory. The catfis fillet then dipped in andaliman extract at its MIC (according to the result of MIC determination) for 15 min, removed, placed in plastic bags, stored in refrigerator (5 °C) and taken for analysis on 1, 2 and 3 days. The quality of catfish fillet observed were total plate count and total volatile base nitrogen (TVBN) according to method of AOAC [19]. Sensory analysis was conducted by a panel 05 15 non-tranined taste panelist using 5-point hedonic scale [20]. The panelists scored the sample based on the characterization and differention of the various sensory characters such as odor, color and texture [21]. The experiment were performed with three replications.

The statistical analysis was used to carry out mean values, standard deviasion in addition to an overall analysis of variance (ANOVA) and Least Significant Differences (LSD) at 0.05 level.

## 3. Results and discussions

### *3.1. Phytochemical compunds of andaliman extracts*

The single stage extractions was performed in this research and the yield of extraction of andaliman fruit is shown in Table 1. The highest yield of andaliman fruit extract was found in methanol extract and followed by ethyl acetate and water. Results of the analysis of the phytochemcial compounds of andaliman fruit extract (Table 2) extracted with methanol has more phytochemical compounds and hexana extracts only have alkaloids and triterpens/steroids. The results revealed the presence of phytochemicals compound such as alkaloids, flavonoids, glycoside, saponins, tanins, triterpen/steroids, and glycoside anthraquinones [22]. Successful determination of biologically active compounds from plant material is largely dependent on the type of solvent used in the extraction procedure [23]. Phenolic compounds such as saponins and tannins are generally soluble in organic solvents that are polar such as methanol. Phenolic compounds are substances which have aromatic rings with one or more hydroxyl groups so it is easy soluble in polar solvents [24].

**Table 1.** The yield of andaliman fruits extract

| Solvent | Yield (%) |
|---|---|
| Water | 3.23 |
| Methanol | 4.15 |
| Ethyl Acetate | 3.97 |
| Hexana | 3.02 |

**Table 2.** Qualitative phytochemicals analysis of andaliman fruits extract

| Phytochemical | Solvent Extract [(+) means present, (-) means absent] | | | |
|---|---|---|---|---|
| | Water | Methanol | Ethyl Acetate | Hexana |
| Alkaloids | - | + | + | + |
| Flavoniods | + | + | + | - |
| Glycoside | + | + | + | - |
| Saponins | - | + | + | - |
| Tanins | + | + | + | - |
| Triterpens/steroids | - | + | - | + |
| Glycoside anthroquinones | - | + | - | - |

The presence of alkaloids, flavonoids, and tanins have been confirmed to have antimicrobial activity [25,26]. In general, plant phenolic compounds have been shown to have antibacterial activity. For example, the antimicrobial activity in germinated fenugreek seeds which may be due partly to the presence of flavonoids and polyphenols has been reported [27]. Scazzocchio *et al.* [28] reported that alkaloids (berberine, beta-hydrastine, canadine and canadaline) of *Hydrastis canadensis* L. (Ranunculaceae) have antibacterial activity against 6 strains of microorganism: *Staphylococcus aureus, Streptococcus sanguis, Escherichia coli* and *Pseudomonas aeruginosa.* Steroids of Alstonia macrophylla plants have been reported to possessed antibacterial activity against *E. coli, Salmonella*

Typhi, *Staphylococcus aureus* and *B. subtilis* [29]. Triterpenoids compounds that have antimicrobial activity include merediol, linalool, indole and kadinen that effective to inhibit the growth of *B. subtilis, S. aureus* and *E. coli* [19].

### *3.2. Antimicrobial activity of andaliman extracts*

The results of the antibacterial activity of an andaliman fruits extracts are tabulated in (Table 3). Minimum inhibitory concentrations of extracts against the tested organisms of andaliman fruit extracts exhibited MIC (Table 4) against the tested organisms. Table 2 shows that ethyl acetate extract has the higher antibacterial activity against *S.a*ureus and *S.typhimurium* at the concentration 100% while methanol extract has the higher antibacterial activity against *E.coli* at the 100% concentration. In the methanol extract, phenolic is more polar than flavonoid. Phenolic has higher antimicrobial activities than flavonid [30].

**Table 3.** Antimicrobial activity (Inhibition zone in mm) from andaliman fruit extracts

| **Inhibition zone (mm) of andaliman fruit extracts** | | | | |
|---|---|---|---|---|
| Test Bacteria | Concentration of andaliman fruit extracts (%) | | | |
| | 25 | 50 | 75 | 100 |
| **Solvent : Water** | | | | |
| *Escherichia coli,* | 8.25 | 8.75 | 9.00 | 8.88 |
| *Staphylococcus aureus* | 9.77 | 11.26 | 13.76 | 15.40 |
| *Salmonella typhimurium* | 8.49 | 8.72 | 9.14 | 15.46 |
| **Solvent : Methanol** | | | | |
| *Escherichia coli,* | 10.75 | 10.38 | 10.75 | 11.75 |
| *Staphylococcus aureus* | 11.34 | 13.88 | 15.81 | 17.78 |
| *Salmonella typhimurium* | 10.86 | 12.86 | 15.31 | 17.76 |
| **Solvent : Ethyl Acetate** | | | | |
| *Escherichia coli,* | 10.30 | 10.48 | 11.28 | 10.73 |
| *Staphylococcus aureus* | 12.23 | 14.97 | 17.22 | 20.06 |
| *Salmonella typhimurium* | 14.41 | 16.15 | 18.52 | 20.19 |
| **Solvent : Hexana** | | | | |
| *Escherichia coli,* | 10.38 | 10.13 | 10.00 | 9.63 |
| *Staphylococcus aureus* | 7.70 | 8.69 | 10.28 | 13.13 |
| *Salmonella typhimurium* | 9.42 | 10.30 | 10.86 | 13.67 |

**Table 4.** MIC determination of ethyl acetate extract of andaliman fruit against *Staphylococus aureus, Escherichia coli* and *Salmonella typhimurium*

| Extract Concentration (%) | *Staphylococus aureus* | *Escherichia coli* | *Salmonella typhimurium* |
|---|---|---|---|
| 5 | + | + | + |
| 4 | + | + | + |
| 3 | + | + | + |
| 2 | + | + | + |
| 1 | + | + | + |
| 0.5 | + | + | + |
| 0.25 | + | - | - |
| 0.1 | - | - | - |

(+) means that andaliman fruit extract still be able to inhibit the bacteria growth
(-) means that andaliman fruit extract is unable to inhibit the bacteria growth

Further testing of the extract of ethyl acetate was conducted to determine the value of MIC against *E. coli, S.aureus* and *S.typhimurium* with the direct method on nutrient broth media. MIC values of extracts of andaliman fruit ranged from 0.1 – 0.25%, with *E. coli* and *S.typhimurium* showing MIC of 0.25% and *S.aureus* showing MIC of 0.1%. The differences observed in the susceptibilities of the bacteria tested against the ethy acetate extracts could due to their cell wall component, such as the porin protein [31]. For the application of andaliman fruit extract in catfish fillet, the extract concentration used was 0.5%.

### *3.3. Application of andaliman fruit extract on catfish fillet*

The addition of 0.5% (volume per mass) andaliman fruit extract to catfish fillet can reduce the population of spoilage microflora as shown Tabel 4. Total plate count (TPC) reached the level 4.50 x $10^6$ in treated catfish sample within 3 days of storage at 5°C. Unacceptable TPC (almost $10^6$) were found in control catfish fillet (day 3).

The TVBN value at the beginning of experiment was 16.66 mg N/100 g and reached the level of 26.18 mg N/100 g and 30.62 mg N/100 g for catfish with and without addition of andaliman fruit extract, respectively. The limit of acceptability of TVBN in fish and fishery product is 30-35 mg N/100 G [32], therefore, the value of TVBN for untreated catfish fillet stored in refrigerator was unacceptable after 3 days of storage. TVBN is an indicator of freshness of fish [33].

**Table 5.** The TPC and TVBN value of catfish fillet with and without the addition of andaliman fruit extract for 3 days of storage in 5°C.

| Days | Samples * | Total Plate Count (CFU/ml) | TVBN (mg N/100g) | Sensory Value | | |
|---|---|---|---|---|---|---|
| | | | | Odor | Color | Texture |
| 1 | Treated catfish fillet | 4.18 x $10^5$ | 18.10 | 7.5 | 7.3 | 7.3 |
| | Control | 4.25 x $10^5$ | 22.14 | 5.0 | 5.0 | 5.0 |
| 2 | Treated catfish fillet | 4.35 x $10^5$ | 21.53 | 6.5 | 6.5 | 6.5 |
| | Control | 3.26 x $10^6$ | 25.01 | 5.0 | 5.0 | 5.0 |
| 3 | Treated catfish fillet | 4.50 x $10^6$ | 26.18 | 6.5 | 6.5 | 6.3 |
| | Control | 5.60 x $10^6$ | 30.62 | 5.0 | 3.0 | 1.0 |

* Treated catfish was catfish fillet with the addition of ethyl acetate andaliman extract with the 0.5% concentratiom, and control was catfish fillet without the addition of andaliman fruit extract

Table 4 showed the changes in sensory evaluation (odor, color and texture) for 3 days of catfish storage. The sensory scores of fillets in both groups decreased with the storage time, and the sensory scores of the control catfish fillet was lower than the treated one. It can indicate that the ethyl acetate extract of andaliman fruit extract with 0.5% concentration could be used as natural source of antimicrobial since it has better consumer acceptance. Fan *et al* [32] reported that sensory scores in common carp treated with tea polyphenols had a higher score than the control.

## 4. Conclusions

Andaliman fruit extracts contain various bioactive compounds such as alkaloids, flavonoids, glicoside, saponins, tanins, triterpene/steroids and glicoside anthraquinone. This bioactive compounds have antimicrobial activity against the *E.coli, S.aureus* and *S.typhimurium*. The higher antimicrobial activity was found in the ethyl acetate extract with the MIC value of 0.5%. The andaliman fruit extract can be used to inhibit the growth of spoilage microbial in the catfish fillet during 3 days of storage at 5°C. Consequently, the application of grape extract improved the quality of fillets which is quite promising for the food industry.

**References**

[1] Siregar B L 2002 Andaliman *(Zanthoxylum acanthopodium DC.)* di Sumatera Utara : Deskripsi dan Perkecambahan (Andaliman *(Zanthoxylum acanthopodium DC.)* in North Sumatera : Description and Germination) J Hayati 10 (1) 38-40

[2] Wijaya C H 1999 Andaliman, rempah tradisional Sumatera Utara dengan aktivitas antioksidan dan antimikroba (Andaliman, North Sumatera spice with antioxidant and anti-microbial activity), Buletin Teknologi Industri Pangan 10 59-61

[3] Hasairin A 1994 Etnobotani rempah dan makanan adat masyarakat Batak Angkola dan Mandailing (Spice and traditional food ethnobotany in Batak Angkola and Mandailing) Tesis Program Pascasarjana Institut Pertanian Bogor

[4] Wijaya CH, Triyanti I, and Apriyantono A 2002 Identification of volatile copmpunds and key aroma compounds of andaliman fruit (*Zanthoxylum acanthopodium* DC). J Food Sci Biotechnol 11 (6) 680-683

[5] Yanti, Pramudito TE, Nuriasari N and Juliana K 2011 Lemon pepper fruit extract (*Zanthoxylum acanthopodium* DC.) suppresses the expression of inflammatory mediators in lipopolysaccharide induced macrophages in vitro Am.J. Biochem. and Biotec. 7 (4) 176-186

[6] Suryanto E, Sastrohamidjojo H, Raharjo S. and Tranggono., Antiradical activity of andaliman (*Zanthoxylum acanth*opodium DC.) fruit extract, Indonesian Food and Nutrition Progress., 2004, 2,1, 15-19.

[7] Cahyana H, Mardiana L 2003 Senyawa kimiawi minyak atsiri andaliman (Zanthoxylum *acanthopodium* DC.*)* dan kemampuan sebagai antioksidan alami (Chemical compounds in Andaliman (Zanthoxylum *acanthopodium* DC.*)* essential oil and its ability as natural antioxidant). Jurnal Ilmu dan Teknologi Pangan. 1 (1). 105-111

[8] Siswadi I 2002 Mempelajari aktivitas antimikroba ekstrak buah andaliman (*Zanthoxylum acanthopodium* DC.) terhadap mikroba patogen perusak makanan (A study on antimicrobial activity in andaliman (Zanthoxylum *acanthopodium* DC.*)* extract towards foodborne pathogen) Fakultas MIPA Universitas Sumatera Utara , Medan.

[9] Huss HH 1998 Fresh Fish Quality and Quality Changes Danish International Development Agency, FAO, Rome, pp. 43-45

[10] Connell JJ 2002 Quality control in fish industry. Torry Advisory Note No 58

[11] Gram L and Huss HH 1996 Microbiological spoilage of fish and fish products Int J Food Microbiol 33 121-137

[12] Gould G W 1996 Industry perspectives on the use of natural antimicrobials and inhibitors for food applications J Food Prot (Suppl.) 59 82–86.

[13] Nychas GJE, Skandamis P, Tassou CC 2003 Antimicrobials from Herbs and Spices. In: Natural Antimicrobials for the Minimal Processing of Foods, Roller S (Ed.) CRC Press, WoodheadPublishing Limited, Cambridge, UK pp. 176– 200.

[14] Smid EJ and Gorris LGM 1999 Natural Antimicrobials for Food Preservation. In: Handbook of Food Preservation, Rahman MS (Ed.), Marcel Dekker, New York, NY, USA pp. 285–308.

[15] Brain KR and Turner TD 1975 The Practical Evaluation of Phytopharmaceuticals, Wrightscience Technical, 1st Ed , Bristol Britain., 144.

[16] Evans WC 1966 Trease Evans Pharmacognosy 14th Ed, London: WB Saunders Ltd, 119- 159.

[17] Hussain AI, Anwar F, Hussain Sherazi ST, and Przybylski R 2008 Chemical composition, antioxidant and antimicrobial activities of basil *(Ocimum basilicum)* essential oils depend on seasonal variations. Food Chemistry 108 (3) 986–995.

[18] Kubo I, Muroi H, and Himejima M 1992 Antimicrobial activity of green tea flavor components and their combination effects. Journal of Agricultural and Food Chemistry 40(2) 245-248

[19] Kubo A, Lunde, CS, and Kubo I 1993 Antimicrobial activity of olive oil flavour compounds. Journal of Agricultural and Food Chemical 40(6) 999-1003.

[20] AOAC 2002 Official Methods of Analysis of AOAC International 17th eds. Association of Official Analytical Chemistry, USA.
[21] ASTM 1969 Manual on Sensory Testing Methods. American Society for Testing and Materials, Philadelphia, PA pp 34-42
[22] Ruth EK, Junie S 2014 Cytotoxic and Antioxidant activity of Petroleum Extract of Andaliman Fruits (*Zanthoxylum acanthopodium* DC.). Int. J Pharm Tech Res. 6(3) 1064-1069.
[23] Tiwari K, Mandeep K, Gurpreet K and Harleem K 2011 Phytochemical screening and extraction : A review. Int Pharmac Sci 1 (1)
[24] Houghton PJ and Raman. 1998. Laboratory Handbook for The Fractionation of Natural Extract. Chapman and Hall. London.
[25] Alagesaboopathi C and Sivakumar R 2011 Antimicrobial properties of various extracts of *Andrographis neesiana* Wight-an endemic medicinal species from India", J. Pharm. Res. 3(1), 27-31.
[26] Kavit M, Patel BN, Jian BK 2012 Phytochemical analysis of leaf extract of *Phllanthus fratenus*", Res J Recent Sci 2 12-15
[27] Norziah MH, Fezea FA, Bhat R, and Ahmad M 2015 Effect of extraction solvents on antioxidant and antimicrobial properties of fenugreek seed (*Trigonella foenum-graecum* L.) Int Food Res J 22 (3) 1261-1271
[28] Scazzocchio F, Cometa MF, Tomassini L, and Palmery M. 2001. Antibacterial activity of Hydrastis canadensis extract and its major isolated alkaloids. Planta Medica 67(6) 561-564.
[29] Chattopadhyay D, Maiti K, Kundu, A P, Chakraborty M S, Bhadra R, Mandal S C. and Mandal AB 2001. Antimicrobial activity of *Alstonia macrophylla*: a folklore of bay islands. Journal of Ethnopharmacology 77(1) 49-55.
[30] Nikaido H 2003. Molecular basis of bacterial outer membrane permeability revisited. Microbiology and Molecular Biology Reviews 67(4) 593-656.
[31] Hasani S and Hasani M 2014 Antimicrobial properties of grape extract on common carp (Cyprinus carpio) fillet during storage in 4°C. J Food Technol 12 (2) 48-53
[32] Fan W, Chi Y, and Zhang S 2008 The use of a tea polyphenol dip to extend the shelf life of silver carp (Hypophthalmicthys molitrix) during storage in ice. Food Chem 108 148-153

# Revamping of entisol soil physical characteristics with compost treatment

**Sumono, S P Loka and D L S Nasution**

Department of Agricultural Engineering, Faculty of Agriculture, Universitas Sumatera Utara. Jl. Prof. Dr. A. Sofyan No.3 Kampus USU Medan 20155
E-mail: sumono@usu.ac.id

**Abstract.** Physical characteristic of Entisol soil is an important factor for the growth of plant. The aim of this research was to know the effect of compost application on physical characteristics of Entisol soil. The research method used was experimental method with 6 (six) treatments and 3 replications of which K1= 10 kg Entisol soil without compost, K2 = 9 Kg Entisol soil with 1 kg compost, K3= 8 kg Entisol soil with 2 kg compost, K4= 7 kg Entisol soilwith3 kg compost, K5= 6 kg Entisol soil with 4 kg compost and K6= 5 kg Entisol soil with 5 kg compost. The observed parameters were soil texture, soil organic matter, soil thickness, porosity, soil pore size, soil permeability and water availability.The results showed that the Entisol soil texture was loamy sand texture, the value of soil organic matter ranged from 0.74% to 4.69%, soil thickness ranged from 13.83 to 20.16 cm, porosity ranged from16% to 37%, soil pore size ranged from 2.859 to 5.493 μm, permeability ranged from 1.24 to 5.64 cm/hour and water availability ranged from 6.67% to 9.12% by each treatment.

## 1. Introduction

Agricultural development in Indonesia has many constraints due to the conversion of productive land into non-agricultural activities. In an effort to expand agricultural land, land use leads to marginal land which is less fertile and not supporting in matter of soil physical properties .Therefore, it is necessary to improve the physical, chemical and biological properties of the soil with the addition of organic materials and adequate water supply so that marginal land can be used for agricultural enterprises. Entisol soilis oneof marginal land that characterized by loose soil consistency, low level of agregation, easy to eroded, and low nutrient and organic matter content [1]

Organic Matter has the main role in maintaining soil quality. In the condition of low soil organic matter, the addition of compost is very recommended to increase soil organic matter and maintain soil quality. The aplication of compost influences soil quality and will reducing chemical ferlitilizer needs, increasing plant growth, increasing soil biomass and microbes activity [2]

Response of compost proportion on soil will depend on composition of compost, soil type, and compost ratio to soil, which also lead to different maximum enhancementin improving soil physical properties. It is necessary for the study to determine the exact proportion of composting Entisol soil by some compost treatment to soil in improving soil physical properties and the suitability for plant growth.

## 2. Materials and Methods

The material used in this study is sample of Entisol soil from PTPN II Tanjung Garbus. Compost is used as a addition of soil in hoping to increase soil fertility. This study uses compost biotic product Ipteks Bagi Inovasi dan Kreativitas Kampus (IBIKK) Compost Center Universitas Sumatera Utara. Polybag is used as the container for compost and soil, the labels are used to mark on both the ring samples and polybag, and water is used to stabilize the soil.

### *2.1 Research Methods*

The research method used is an experimental method in Greenhouse and soil analysis in the Laboratory of Research and Technology Faculty of Agriculture, Universitas Sumatera Utara with 6 treatments and 3 replications on each treatment, where 10 kg Entisol soil (control / K1) as first traetment, 9 kg Entisol soil and 1 kg compost (K2) as second treatment, 8 kg Entisol soil and 2 kg compost (K3) as third treatment, 7 kg Entisol soil and 3 kg compost (K4)as fourth treatment, 6 kg Entisol soil and 4 kgcompost (K5)as fifth treatment and 5 kg Entisol soil and 5 kgcompost (K6) as sixth treatment.

Entisol soil sample and compost are dried in room temperature. Each dried soil and dried compost was crushed and sieved using 10-mesh-siever. Then, the sieved soil and compost were mixed and stirred evenly. Soil treatment and compost by each treatment were poured into 10 kg polybag. The mixture of soil and compost were sprinkled in polybags for soil stabilization. The soil was watered continuously until it was steady. The soil criterion was stable when there was no reduction in soil thickness and constant drained water. Soil samples were taken after the soil reached field capacity using the ring sample to determine the soil physical characteristics in the laboratory.

Test in the Laboratory of Research and Technology Faculty of Agriculture includes measuring soil texture using hydrometer method and analyzed using USDA triangle, analyzing C-organic material by Walkley & Black method, analyzing bulk density, particle density, soil porosity, soil thickness, and soil permeability by Constant Head Test method. Moisture content of field capacity and permanent wilting point and water availability are analyzes by pF test 2.54 (field capacity) and pF 4.2 (permanent wilt point) in Palm Oil Research Centre. Soil pore size analyzed with SEM (Scanning Electron Microscope) and Research parameters were soil texture, soil organic matter, bulk density, particle density, porosity, permeability, the moisture content of field capacity, water available, thickness of the soil and pore size.

## 3 Results and discussion

### *3.1 Soil texture*

From the analysis of soil texture (Table 1), it is known that Entisol soil texture is loamy sand, which means sand fraction is more dominant than the fraction of silt and clay. The soil texture was determined using USDA. Knowing the texture of the soil will facilitate in the improvement of physical properties and soil fertility.

**Table 1.** Results of the analysis of soil texture

| Soil Entisol | Fraction | | | Soil texture |
|---|---|---|---|---|
| | Sand(%) | Silt (%) | Clay (%) | |
| $K_1$ | 85.84 | 5.89 | 8.26 | Loamy sand |
| $K_2$ | 85.84 | 6.56 | 7.60 | Loamy sand |
| $K_3$ | 84.56 | 7.94 | 7.49 | Loamy sand |
| $K_4$ | 83.89 | 7.94 | 8.16 | Loamy sand |
| $K_5$ | 80.56 | 11.94 | 7.94 | Loamy sand |
| $K_6$ | 78.56 | 15.28 | 6.16 | Loamy sand |

Table 1 shows that addition of compost treatment decreases the percentage of sand fraction and increase the amount of silt fractions. More porous a soil (the higher fraction of sand) will easier roots to penetrate and easier water and air to circulate, but water will easily lost from the soil, and otherwise. The sand textured soils have small surface area making it difficult to hold water and nutrients

*3.2 Soil Organic Matter*

Organic matter analysis results are presented in Table 2, the treatment K6 had the biggest organic matter content of 4.69% with moderate criteria, whereas K1 treatment had the lowest organic matter content of 0.74% with very low criteria, from [12] criteria carbon organic <1% (very low), 1-2% (low), 2.01-3 % (medium), 3.01-5% (high) and >5% (very high). This is caused by the higher ratio of compost in each treatment, will higher organic content of the soil and it increases soil fertility and improve soil physical and chemical properties, especially in soils with low nutrient such as Entisol

**Table 2.** Results of analysis of soil organic matter content

| Treatment | Levels of C-organic(%) | Organic matter content(%) | Criterion |
|---|---|---|---|
| $K_1$ | 0.43 | 0.74 | Very low |
| $K_2$ | 0.93 | 1.60 | Very low |
| $K_3$ | 1.29 | 2.22 | Low |
| $K_4$ | 1.88 | 3.24 | Low |
| $K_5$ | 2.31 | 3.98 | Medium |
| $K_6$ | 2.71 | 4.69 | Medium |

Neata, et al [2] reported that in areas where the soil organic matter content is low, the use of compost in agriculture is highly recommended to increase soil organic matter content and to improve and maintain soil quality.

*3.3 Soil thickness*

Soil thickness measurement results in a 40 cm x 50 cm and 23 cm diameter polybag size in wet conditions (field capacity of measuring drainage) free (Table 3).

**Table 3.** Results of the soil thick with revamping Entisol compost

| Treatment | Soil thickness (cm ) |
|---|---|
| K1 (Control) | 13.83 |
| K2 | 15.25 |
| K3 | 16.67 |
| K4 | 17.29 |
| K5 | 18.70 |
| K6 | 20.16 |

From Table 3, it is known that soil without compost had the lowest soil thickness which was 13.83 cm thickness and the highest is the treatment K6 which was 20.16 cm thickness. This is because the K1 are denser than the other treatment, where compost will friabling the soil and improve the ability of soil to retain water. Solids will reduce the water holding capacity, reducing the air content, providing major physical barriers on root penetration, reducing its ability to harvest water, air and nutrients. Israelsen and Hansen [3] stated that soil thickness will inhibit the regulation of air from soil pores that retarding water movement and it will reduce permeability rate.

*3.4 Bulk density*

The measurement results soil bulk density, soil particle density, and porosity can be seen from this table below

**Table 5.** The soil bulk density, soil particle density, and porosity

| Treatment | Bulk density(g/cm$^3$) | Particle density(g/cm$^3$) | Porosity(%) |
|---|---|---|---|
| K1 | 1.44 | 1.70 | 16 |
| K2 | 1.24 | 1.52 | 19 |
| K3 | 1.15 | 1.48 | 23 |
| K4 | 1.07 | 1.43 | 26 |
| K5 | 0.97 | 1.35 | 29 |
| K6 | 0.83 | 1.31 | 37 |

Table 5 showed that soil bulk density measurement results are different in each treatment. The value of the highest Entisol soil density was 1.44 g/cm$^3$(K1) and the lowest was 0.83 g/cm$^3$(K6).The variation in bulk density is due to relative proportions of organic particles and inorganic solids and soil porosity. It's suitable to Xin statement that compost aplication will decrase soil bulk density significantly and increase porosity [4]

Hossain et al stated that varity of bulk density value due to the relative proportion of anorganic solids dan soil porosity. Most of mineral soil bulk density soil is about 1,0 and 2,0 g/cm [5]

*3.5 Particle density*

It can be seen from Table 5 that the treatment K6 has the lowest density soil particles value of 1.31 g/cm$^3$. This result is directly proportional to the mass density of Entisol soil. The using of high compost comparison value will lower bulk density as well as the particle density. The more organic content contained in the soil, the smaller the value of particle density

Treatment K1 (control) particle density value is 1.70 g/cm$^3$ which below the average value of the density of mineral soil particles in the field that is 2.65 g/cm$^3$. This is caused by the texture of the Entisol is loamy sand which has a lot of pore texture macro or coarse pores, causing a small volume of soil density. Besides, at the beginning of soil in a state trial disrupted due to be crushed and sieved to obtain a uniform grain before it was poured in a polybag, then land needs through a process of consolidation and density is not the same as the conditions on the ground (soil structure is different)

*3.6 Soil porosity*

Results from Table 5 showed the highest soil porosity at 37% (K6) and the lowest (K1) at 16%. Porosity of K6 has a higher porosity for the K6 has a ratio of bulk density and particle density that is lower than other treatments, where the high and low value bulk density and particle density is influenced by organic matter soil so that the soil containing organic matter has a high porosity, too. The increase in soil organic matter content can increase the porosity of the soil so that it will further strengthen the structure and texture of the soil as well as the development of surface soil biota. Such conditions lead to improvement soil physical properties including improved infiltration capacity. Benefit from the presence of organic matter in the soil is to reduce the bulk density of the soil so that the soil minerals dissolve. Low bulk density is usually associated with increasing porosity caused by the fractions of organic and inorganic soil. It can be concluse that porosity is parameter that influences the most to absorption coefficient [6]

*3.7 Pore size*

The result given the size of the Entisol soil pores ranging from 2.859 to 5.493 μm belong to the class of medium-coarse. The soil with more silt content have the more macro pores than the clay soil. Soil witth more macro pores will more difficult to retain water that make the plant easy to dry.

The Entisol soil pore size measurement results with SEM test, from Table 6.

Table 6. Data pore size Entisol

| Treatment | Pore Size (μm) |
|---|---|
| KI | 4.125 |
| K2 | 2,859 |
| K3 | 3.845 |
| K4 | 4.125 |
| K5 | 5.070 |
| K6 | 5.493 |

*3.8 Soil Permeability*

Permeability measurement results are presented in Table 7.

**Table 7.** Results of analysis of soil permeability

| Treatment | Permeability (cm/hour) | Criteria |
|---|---|---|
| K1 | 3.81 | Average |
| K2 | 5.64 | Average |
| K3 | 4.10 | Average |
| K4 | 3.17 | Average |
| K5 | 2.30 | Average |
| K6 | 1.24 | A Bit slow |

Soil permeability measurement in a saturated condition shows that the rate of permeability of K1to K5 has average criteria while K6 criteria are a bit slow. Wherein, the highest permeability rate K2 indicated 5.64 cm/hour and the lowest is K6 which is 1.24 cm/hour. Permeability is affected by the porosity of the soil, Soil pores which initially measurement was macro will be changed to the size of meso because most have been filled with compost. Compost has the same properties as the clay, the more compost, the stronger the bond to nutrients and water, so this is what causes the permeability of the soil with a high content of the compost becomes slower. Soil permeabilty is also affected by soil thickness, that thicker the soil thickness will increase soil permeability [3]

*3.9 Field Moisture Capacity*

Measurement results of field capacity moisture content is presented in Table 8.

**Table 8.** Data on soil moisture field capacity and permanent wilting point

| Treatment | 2.54 pF(%)(field capacity) | 4.2 pF (%)(permanen wilting point) | The water is available(%) |
|---|---|---|---|
| K1 | 13.16 | 4.83 | 8.33 |
| K2 | 15.03 | 5.91 | 9.12 |
| K3 | 16.40 | 9.31 | 7.09 |
| K4 | 29.62 | 22.75 | 6.87 |
| K5 | 33.09 | 26.42 | 6.67 |
| K6 | 33.25 | 26.18 | 7.07 |

Table 8 shows the test value of 2.54 pF (field capacity). The treatment K1 has the lowest soil moisture content of 13.16% and the highest K6 at 33.25%. In 4.2 pF (permanent wilting point), K1 value is the result of low soil moisture content which is 4.83% and the highest K5 at 26.42%. Ability of soil to hold water in the conditions of field capacity and permanent wilting point with more compost ratio will be higher compared to Entisol soil without compost. There is a close relationship between the increase in organic matter and available water capacity and the ability of soil to survive in drought soil by increasing soil water content with increasing organic carbon.

*3.10 Water Available*

From Table 8 it can be seen that the highest amount of water available is K2 at 9.12% and the lowest is at 6.67%(K5). From the amount of water available, the highest for Entisol soil is treatment K2, with a big use of compost on the soil compost binding to water the more powerful it in accordance with the nature of the same compost with clay. Humus acts together with clay in retaining nutrients in the form provided to leaching and retain nutrients in a form available to plants and microorganisms. Water available affected of teh differences of soil texture and structure which affect field capacity and wilting point because the soil texture and structure determine the content of water soil in certain matrix in retaining water [7]

This suggests that the availability of water for plants on soil with compost improvement will depend on the type of soil and the soil texture. The big difference in soil type and soil texture can affect the suitability of the amount of compost different comparison. In Entisol soil that contains more sand ratio of good compost for water availability is K2 (9 kg mineral soil and 1 kg compost)

## 4. Conclusion

Entisol soil texture is loamy sand, where soil organic matter, soil thickness, and porosity was increase in each treatment by the increasing of compost using, whereas soil pore size obtain the best size in K6 treatment (5kg soil with 5 kg compost), permeability obtain the best performance in K2 treatment (9 kg soil with 1 kg compost) and water availability obtain the best result in K2 treatment (9 kg soil with 1 kg compost).The availability of water for plants on soil with compost improvement will depend on the type of soil and the soil texture.

## Reference

[1] Tan and Kim H 1986 *Degradation of Soil Minerals by Organic Acid* SSSA Publ. **17** pp 1-25
[2] Neata G, Teodorescu R, Dinca L and Basaraba 2015 Physico-chemical and microbiological composition of compost from Bucharest municipal waste *Agriculture and Agricultural Science Procedia* pp 486 – 491
[3] Israelsen O and Hansen V E 1962 *Irrigation Principles and Practices* (New York: Wiley)
[4] Xin, Zhang J, Zhu A and Zhang C 2015 Effect of long term (23 years) minerals fertilizer and compost application on physical properties of fluvo – aquic soil in the North China plain Soil and Tillage Research **156** pp 166 – 172
[5] Hossain M F, Chen W and Zhang Y 2015 Bulk density of mineral and organic soils in the Canada's arctic and sub – arctic Information *Processing in Agriculture* **2** pp 183 – 190
[6] Barros J L, O Thiers and F Torres 2015 Feasibility study of estimating the porosity of soils from sounds absorption measurements *Measurement* **77**(2016) pp 213 – 221
[7] Campos I, Piqueras J G, Carrara A, Villodre J and Calera A 2016 Estimation of total available water in soil layer by integrating actual evapotranspiration data in a remote sensing – driven soil water balance *Journal of Hydrology* **534**(2016) pp 427 - 439

IOP Conf. Series: Earth and Environmental Science **122** (2018) 012091 doi:10.1088/1755-1315/122/1/012091

# Ball mill tool for crushing coffee and cocoa beans base on fraction size sieving results

**B Haryanto, M Sirait, M Azalea, Alvin, S E Cahyani**

Universitas Sumatera Utara

bode.haryanto@usu.ac.id

**Abstract.** Crushing is one of the operation units that aimed to convert the size of solid material to be smoother particle's size. The operation unit that can be used in this crushing is ball mill. The purpose of this study is to foresee the effect of raw material mass, grinding time, and the number of balls that are used in the ball mill tool related to the amount of raw material of coffee and cocoa beans. Solid material that has become smooth is then sieved with sieve mesh with size number: 50, 70, 100, and 140. It is in order to obtain the mass fraction that escaped from each sieve mesh. From the experiment, it can be concluded that mass percentage fraction of coffee powder is bigger than cocoa powder that escaped from the mesh. Hardness and humidity of coffee beans and cocoa beans have been the important factors that made coffee beans is easier to be crushed than cocoa beans.

## 1. Introduction

Indonesia is a one of cocoa and coffee producer in the world. Cocoa and Coffee have a unique characteristics, that were developed in Indonesia's area. Indonesia has potential to improve the quality of cocoa beans into a product that can compete with other cocoa-producing countries [1]. Cocoa and coffee can contribute to increase the economic value to Indonesia. Coffee plants were introduced during the Dutch East Indies (now Indonesia) [2].

Cocoa have sorghum crystals in its components. Toughness of cocoa make more tasty when produces in indutrial operation processes and the toughness points of cocoa exists in conditional room temperature [1]. Meanwhile, Coffee have the skin contains the complex chemical substances. Physical changes in coffee during roasting include reduction in mass due to loss of moisture and decomposition of carbohydrates, increase the volume of coffee beans, lowering the density due to puffing and increase in brittleness. It is well known that during roasting, coffee beans lose their strength and toughness and become brittle and fragile [3].

Grinding in ball mills is an important technological process applied to reduce the size of particles, which may have different nature, and a wide diversity of physical, mechanical and chemical characteristics. However, ball mills have been widely applied in different fields such as the mining, chemical and pharmaceutical industries. One of the most popular ball mills is the tumbling ball mill, which consists of a rotating cylindrical drum with grinding media such as balls. The use of balls in the grinding of ores is a major item in the costs of milling operations [4]. Various factors like hardness, toughness, stickiness, slipperiness, moisture content, melting or softening point, abrasiveness and others (material structure, size, shape, flow, and bulk density of product) ratio of feed size to product size, affect the size reduction [5].

The objective of this investigation is to evaluate the crushing operation with ball mill to convert the size of coffee and cocoa beans to be smoother particle's size. The ability to reduce the sample size of diversity materials of coffee and cocoa beans were the interest of this study. Applying the variation of ball size, operation time and weight of material sample coffee and cocoa beans were completed as variables.

IOP Conf. Series: Earth and Environmental Science **122** (2018) 012091 doi:10.1088/1755-1315/122/1/012091

## 2. Material and Methods

The cocoa and coffee were washed by pure water then dried naturally in the hot sun temperature until really dry. Following the weight variation of sample, ingredients were prepared 30-50 grams. Then the sample put in the ball mill space. The certain number of balls as variation size was then put to the equipment ball mill space. The mill was then running at certain operation time 40-60 minutes. The crushing mill product was then taken and put in a container. The powdered products are weighed with laboratory balance to confirm with its initial weight. The product was then sieved using a mesh size of 50 70 100 140. Each mass that passes and is suspended is weighed as mass and recorded to obtain a fraction of mass escaped and retained. This procedure is repeated and done according to the variations in Tables 1 and 2.

**Table 1.** Ball mill sizes

| Size | Diameter | Weight |
|---|---|---|
| Small (ceramic) | 1.63 cm | 16.06 gram |
| Medium (metal) | 3.56 cm | 123.17 gram |
| Big (ceramic) | 5.97 cm | 160.88 gram |

**Table 2.** Variables measurement

| Variable | The range of variable | Run times/Operation Time |
|---|---|---|
| Weight of samples | 30 – 50 gram | 40-60 minutes |
| Type of ball mill | Small, medium, big | |

## 3. Results and Discussions

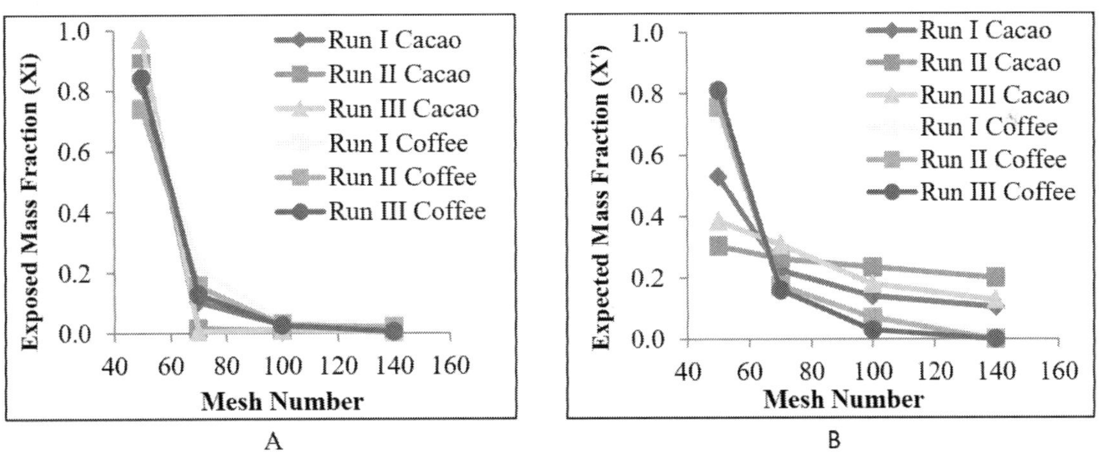


**Figure 1.** Graph of Exposed Mass Fraction versus Mesh Number with Feed Weight Variations

The effect of feed weight with the variation such as: run I : 30 grams with 10 medium balls, 8 small balls and operation time 40 minutes; run II : 40 grams with 10 medium balls, 8 small balls and operation time 40 minutes. Run III : 50 grams with 10 medium balls, 8 small balls and operation time 40 minutes. Figure 1A and B show the relationship of the retained mass fraction with the mesh number of the sieve used with the influence of the weight of the feed. From the results obtained that the highest fractured mass fraction occurred in a sample of cocoa beans with a weight of 50 gram and mesh number 50.

Based on the theory of the relationship between the mass fraction detained with the feed weight can be written by the equation:

$$A = \frac{6m}{\Phi_s \rho_p D_p} \quad (1)$$

$$A_w = \frac{6m}{\Phi_s \rho_p} \sum_{i=1}^{n} \frac{x_i}{\bar{D}_{pi}} \quad (2)$$

[6]

Where: A = total surface area of particle; Aw = specific surface area; m = mass; $\Phi_s$ = sphericity; Xi = retained mass fraction; $\bar{D}_{pi}$ = average particle diameter

From the equation can get the relationship of m (mass) and Xi (retained mass fraction). Increasing the mass sample number will increase the fraction of masses retained. However, the results for the cacao beans are not following the theories (equation 1 and 2). Increasing the sample number was not in accordance, where more weight should be retained greater mass fraction. It may a deviation on the sample of cacao bean for run II and run III.

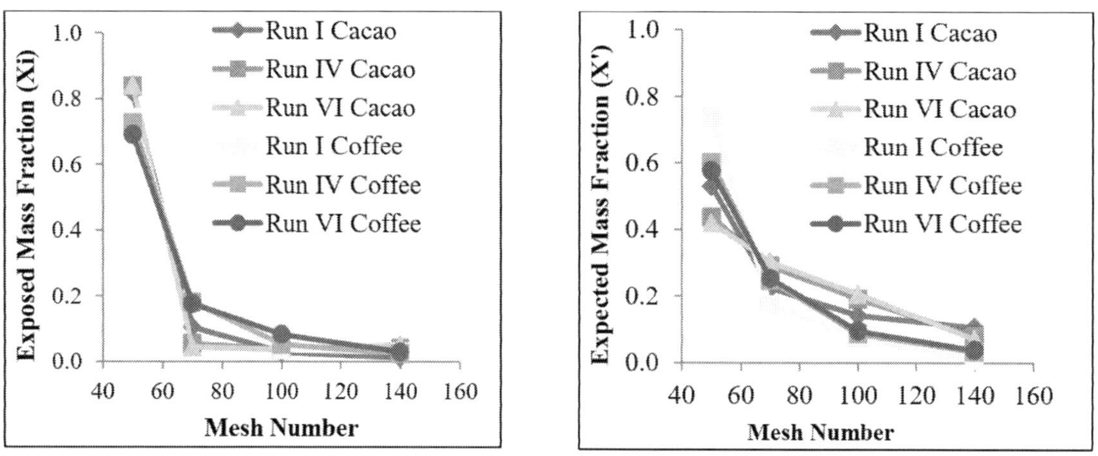


A B

**Figure 2**. Graph of Exposed Mass Fraction versus Mesh Number with Ball Sizes Variations

The effect of variation ball sizes such as: run I : 30 grams with 10 medium balls, 8 small balls for 40 minutes; run IV : 30 grams with 12 medium balls, 8 small balls for 40 minutes; run VI : 30 grams with 15 medium balls, 8 small balls for operation time 40 minutes. From Figure 2A and B, one can see the relationship of the retained mass fraction with the mesh number of the sieve used with the influence of the number of ball used. From the results obtained that the highest fractional mass fraction occurred in brown seed samples with the number of balls of 12 medium balls and 8 small balls and the mesh number was 50.

The relationship of mass fraction to the weight and size of the ball is:

$$\overline{S_i} = \sum S_{i,k}\, m_k \quad (3)$$

$$S_i = a x_i^{\alpha} \quad x_i \ll d, \; a>0 \quad [7,8]$$

Where: $S_i$ = Mass constant loss rate; d = Ball size; a and α = free parameters; $m_k$ = ball weight; $x_i$ = particle size

IOP Conf. Series: Earth and Environmental Science **122** (2018) 012091 doi:10.1088/1755-1315/122/1/012091

The result show that with greater the weight of sample and the larger the size of the ball can decrease the sample size become smoother. It impacts positively to the less of mass fraction retained. The results obtained are not in accordance with existing theories especially on the sample of cocoa seeds for run I and run III. It should by increasing the number of balls used, the retained fraction of the mass should be decreased. The deviation may accure during operation steps.

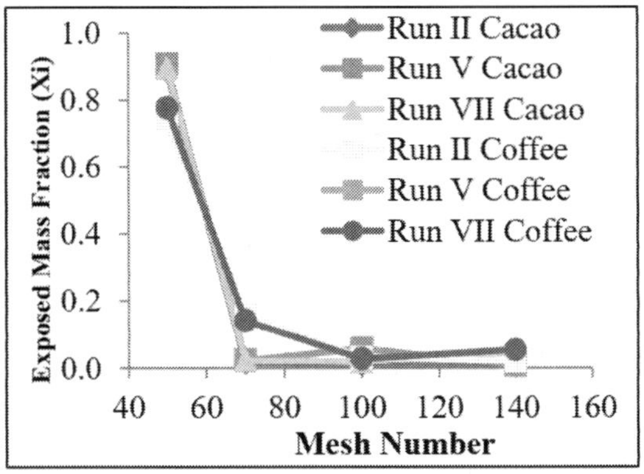


A

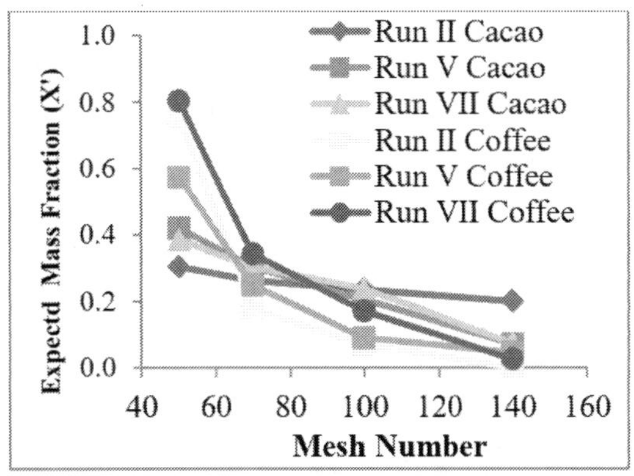


B

**Figure 3.** Graph of Exposed Mass Fraction versus Mesh Number with Operation Time Variations

The effect of variation operation time for run II : 40 grams sample with 10 medium balls and 8 small balls with variation operation time 40 minutes. Run V and VII with the same weight sample and ball mill size and the operation time for 50 and 60 minutes respectively. The relationship of the retained mass fraction with the mesh number of the sieve used with the influence of the milling time is shown in Figure 3 A and B. One can see from the results obtained that the highest fractured mass fraction occurred in cacao beans samples with 50 minutes grinding operation time for 50 mesh number.

In theory, the relationship between fractured mass fractions and milling time can be written by the equation:

$$\frac{dw_i(t)}{dt} = -S_i w_i(t) + \sum_{j-1}^{i-1} b_{i,i} S_j w_j(t) \tag{4}$$

[7]

Where, $S_i$ = the material selection function by size i; t = milling operation time; wi(t) = the mass fraction with i size of initial measure in mill at time t; $b_{i,j}$ = the next mass fraction at the crushing size j.

From the results, it can be concluded that the longer the grinding operation time could produce smoother the resulting powder. The fraction of the retained masses sample will be less. However, the results obtained are not all in accordance with the existing theory. The increasing crushing operation time should increase the smaller sample fraction then decrease the retained mass of sample. The deviation was havening on the sample of cocoa beans and coffee beans both at runs I, II and III.

The pretreatment of samples may impact the physical properties of the beans. The initial hardness and humidity of coffee and cocoa beans may be an important factor to crush to be smaller powder product [5]. The deviation results for this study may be infected by some aspect such as: the sample physical properties, operation control of the mill and human error. With different physical properties of both beans show that the coffee beans is easier to be crushed than cocoa beans. With tree variation in this investigation for crushing by ball mill type, weight of sample and operation time was possible to give the information about ability to decrease the sample size as the result after sieving the product.

## 4. Conclusion

Crushing by ball mill was possible to separate the materials base on the mess size after converting the size to be smoother particles. With the variation of sample weight was found that 30 gram sample is the best variation in comparing with coffee bean with sieve mesh with size number: 50 mesh. The balls type variation was shown that in using the 15 medium balls and 8 big balls size was possible to increase the fraction of coffee/ cocoa beans to be smoother size with sieve mesh with size number 70 mesh. Higher operation time was possible to convert the raw material of coffee/cocoa beans to become smaller in compare to coffee/cocoa beans with sieve mesh with size number 50 mesh. It was found that the most significant variation was 60 minutes. The physical properties of the hardness and humidity of coffee beans and cocoa beans have been the important factor that made coffee beans is easier to be crushed than cocoa beans.

**Acknowledgment**

The authors wish to express thankful to Unit Operation Laboratory of Chemical Engineering Department, Faculty of engineering USU in using the Ball Mill unit for this study.

## References

[1] Badan Pusat Statistik 2015 https://www.bps.go.id
[2] D Teketay 1999 *History, botany and ecological requirements of coffee* Walia **20** 28–50
[3] S Abrar; M Ali and S Endris 2014 *Sky Journal of Agricultural Research* **3(5)** 82–88
[4] Ashgar, S Ziaeddin, M Noaprast and M Karamoozian 2015 *Journal of Mining & Geophysics-Engineering* **49(1)** 83-91
[5] S Sushant and A Kamath. 2013 *International Research Journal of Pharmacy* **4**(8) 57-64
[6] W L McCabe, JC Smith and PE Harriot 1993 *Unit Operations of Chemical Engineering* Fifth Edition. McGraw-Hill
[7] M Vladimir, B Sokolov and S Stoenchev 2012 *Bulgarian Academy of Science Cybernetics and Information Technologies* **12(2)** 51- 68

# Delignification and Hydrolysis Lignocellulosic of Bagasse in Choline Chloride System

**R Manurung [1], A Syahputra, M A Alhamdi, W Satria, E M Barus, R Hasibuan and M Z Siswarni**

Chemical Engineering Department, Faculty of Engineering, University of Sumatera Utara
Jl. Almamater Kampus USU, Medan 20155, Indonesia.

[1] renitachem@yahoo.com

**Abstract.** Bagasse was the waste which has a fairly high content of lignocelluloses and has not been utilize optimally. With a cellulose content of up to 40%, bagasse then potentially be used as raw material for bioethanol. In this research, delignification process was carried out using sodium hydroxide (NaOH) in the ionic liquid system and without ionic liquids. The purpose of this research was to find out the highest content of cellulose which contained in the bagasse and the best hydrolysis conditions was obtained at the hydrolysis process in the choline chloride (ChCl) system. The hydrolysis stage in this research was carried out at temperature 105 °C, catalyst ($H_2SO_4$) 10% (w/w) cellulose, ChCl 10%, 15%, and 20% (w/w) cellulose and it was stirred at constant speed 120 rpm with reaction time of 30, 60 and 90 minutes. Delignification research results used ChCl obtained highest content of cellulose was 39.8%, hemicellulose 18.59%, and lignin 3.62% in cooking treatment 90 minutes and 20% ChCl. While delignification without ChCl obtained highest content of cellulose is 24.98%, hemicellulose 8.25%, and lignin 18.99% in cooking treatment 90 minutes. The maximum glucose yield of 39.4% was obtained at reaction time 90 minutes and 15% of ChCl.

## 1. Introduction

Sugarcane is a plant grown to be raw materials of sugar and vetsin. This plant can only grow in tropical climates. Sugarcane is of an herbage kind. The age of the plant between planting and harvesting is up to approximately 1 year. In Indonesia, the cultivation of sugar cane is done mostly in Java and Sumatra. The area of sugarcane plantations in Indonesia reaches 398,260 Ha. The process of sugarcane cultivation would produce 5% sugar, 90% bagasse and the remainder in the form of drops (molasses), and water [1]. Bagasse is one of the bioethanol raw materials that contain lignocellulose. The utilization of bagasse is still in the development process. Generally, bagasse is used for fuel, paper raw material, brake pad raw material, mushroom industry and others. Bagasse is possible to be utilized as an alternative energy source such as bioethanol or biogas. Bagasse contains 52.7% cellulose [2]. In the process of manufacturing bioethanol, lignocellulose delignification is the first phase to be done to break the bond between cellulose, hemicellulose, and lignin. Delignification process is a process of removing lignin from the raw material in order that the result of this process contains cellulose with adequate purity [3]. Cellulose delignification with Ionic Liquid media is more effective than the process without Ionic Liquid media. Ionic liquids are salts that are liquid at room temperature. This happens because the ionic liquid reduces the degree of crystallinity and increases the sample porosity so it becomes easier to delignify the bagasse [4]. Delignification of cellulose in Ionic Liquid media (ionic liquids) is more effective than without Ionic Liquid. Ionic liquids are salts that are at room temperature in the liquid form. This is because the ionic liquids have decreased the degree of crystallinity and increased the porosity of the samples making it easier to delegate fruit empty bunch

[5]. Hydrolysis plays an equally important role in bioethanol manufacture. Bagasse has the potential to be used as a source of glucose through a process of hydrolysis by using acids or enzymes. The resulting sugar solution can then be converted into various products such as alcohol, which owns a higher economic value. Lignocellulose hydrolysis with dilute acid is applied as the most common way to obtain sugar. Ionic liquids can generally be used at room temperature and does not produce toxic side reactions [4]. Choline chloride ionic liquid is a weak acid with low level toxicity and environmentally friendly. Choline chloride (ChCl) under the name IUPAC 2-hydroxy-N, N, N-trimethylethanaminium chloride or (2-hydroxyethyl) trimethylammonium chloride is one of the most widely used ammonium salts for DES formation because ChCl is cheap and can be easily extracted from biomass [6]. Initially, ChCl was used because when ChCl is mixed with most hydrogen bonding donors or metal halides, the physical properties of ChCl are almost always greatly increased compared to the use of other quaternary ammonium salts and the freezing depression tends to be one of the largest. An example is a mixture of ChCl: urea with freezing points of 303 ° C and 135 ° C, this mixture also shows lower viscosity than most other systems containing different ammonium salts and the conductivity is often higher when using ChCl [7].

## 2. Research Methodology

### *2.1. Materials*

Raw material used in the research is bagasse. The chemicals are aquades (distilled water), choline chloride, sodium peroxide 3%, sulfuric acid 72%, cellulose from sulfuric acid ($H_2SO_4$) delignification 97%, potassium iodide (KI 20%), sodium thiosulfate $(NH_4)_2HPO_4$, starch indicator, Luff solution, Aquadest ($H_2O$).

### *2.2. Delignification Procedure by Using NaOH without Ionic Liquids*

Bagasse powder is weighed 30 grams, and then put in a beaker, 150 ml of NaOH solution is added to the beaker containing powder bagasse. Then, the mixture is stirred equally until powder is soaked. The powder is soaked in NaOH 3% solution during a specified time. The powder that has been soaked with NaOH 3% is washed with water and dried at 105 °C for 16 hours before use.

### *2.3. Delignification Procedure by Using NaOH with Ionic Liquids*

The bagasse powder is weighed 30 grams, and then put into a beaker. Then, 150 ml of NaOH 3% solution is added to the beaker containing bagasse powder. Then, the mixture is stirred equally to soak the powder for 90 minutes. The powder that has been soaked with NaOH 3% is then washed with water and dried at 105 °C for 16 hours before use. The powder is sifted, mixed with choline chloride ionic liquid, heated to a temperature of 130 °C for a specified time. After that, the powder is washed with water to remove ionic liquid to obtain cellulose. Wash it again with water, and then put it into the oven until the weight is constant. Cellulose, hemicellulose and lignin analysis are done with *Chesson-Datta* method.

### *2.4. Hydrolysis Procedure*

An amount of 10 grams of cellulose from bagasse delignification is put into 250 ml Erlemmeyer. It is then added with concentrated sulfuric acid as much as 10% of the sample weight that has been diluted by distilled water and choline chloride ionic liquid in a determined amount into Erlemmeyer 250 ml. The erlenmeyer is then closed with cork and heated at 105 °C while stirred with a magnetic stirrer during a determined time. Then, pH of hydrolyzed sample is measured to achieve a pH of 7. After that, the hydrolysis is filtered using filter paper to obtain polysaccharide filtrate. Then, the glucose level is analyzed with Luff Schoorl method.

IOP Conf. Series: Earth and Environmental Science **122** (2018) 012092 doi:10.1088/1755-1315/122/1/012092

### *2.5. Analysis of Glucose*

Weigh as much as 2 grams of sample and put it into a 250 ml measuring flask, add water and homogenize. Add 5 ml of half alkaline Pb-acetate and shake. Add one drop of $(NH_4)_2HPO_4$ 10% solution. Shake and match the contents of flask up to the mark with distilled water boundary, shake 12 times and filter. Pipette the filtered solution and put in a 500 ml Erlenmeyer. Add 15 ml of distilled water and 25 ml of Luff solution and some boiling stones. Connect the erlenmeyer with a cooler, heat over an electric heater to boil for 10 minutes. Heat continuously for 10 minutes, and then remove and cool. Once being cool, add 10 ml of KI 20% solution and 25 ml of $H_2SO_4$ 25% solution. Titrate it with 0.0990 N of tio solution with 0.5% starch solution as an indicator. Make blank solution by using 25 ml of water and 25 ml of Luff solution in the same way without the use of sample solution.

## 3. Results and Discussions

### *3.1. Effect of Choline Chloride (ChCl) Amount and Delignification Time to the Remained Lignin Level in Holocellulose*

The process of bagasse delignification is conducted by using NaOH in choline chloride (ChCl) ionic liquid system in a wide variation of ChCl amounts: 10%, 15%, and 20% of bagasse weight. The result shows that the different number of ionic liquids affects the production of lignin. Figure 1 shows the relationship between levels of remained lignin in holocellulose with ChCl amount and delignification time. In the process of delignification, bagasse in ChCl system is expected to produce cellulose with low lignin level. Figure 1 shows that the lignin content decreases as the number of ChCl rises. Decreased level of lignin is caused by a long cooking time that triggers the base group degradation of cooker solution that attacks alpha and beta lignin, with the result that unstable degradation triggers condensation. Condensation causes the loss of lignin bond from hemicellulose and cellulose. Because ChCl is acid, acidic atmosphere tends to increase the molecular weight of lignin. This event causes the lignin to sediment and dissolve in water during the washing process. However, the ChCl number more than 15% makes remained lignin level in holocellulose constant on 90 minutes as a result of the addition reaction of hydroxyl group causing lignin not to dissolve in washing water [8]. Lignin produced is about 3.62%.

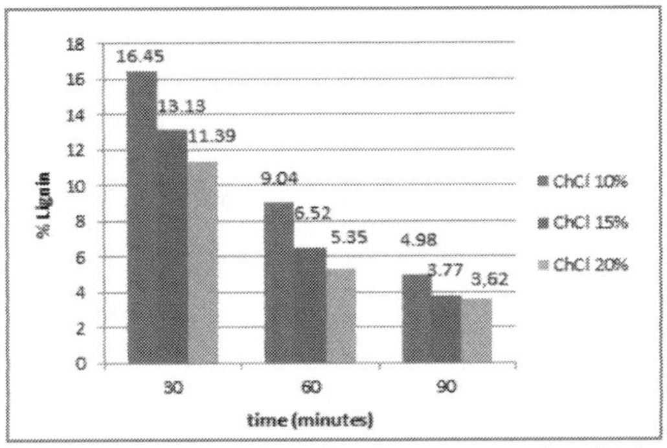


**Figure 1.** The Effect Between ChCl and Delignification to Remained Lignin Level in Holocellulose

### *3.2. Effect of Total Choline Chloride (ChCl) and Delignification Time towards Cellulose and Hemicellulose Levels*

The following is a graph showing the relationship between levels of cellulose and hemicellulose and CHCl amount and timing delignification.

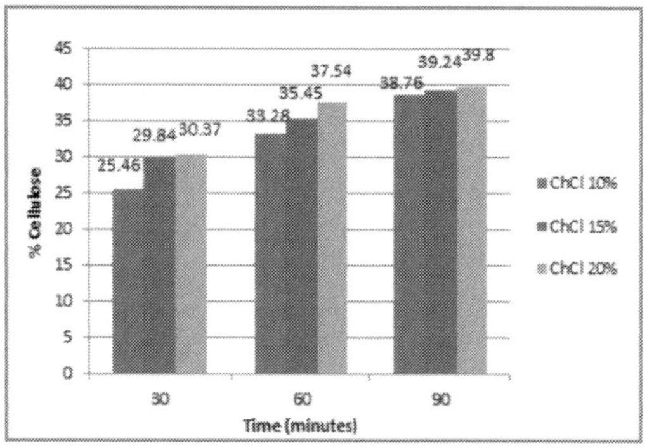


**Figure 2.** The effect of the amount of ChCl and Delignification Time Towards Cellulose Level

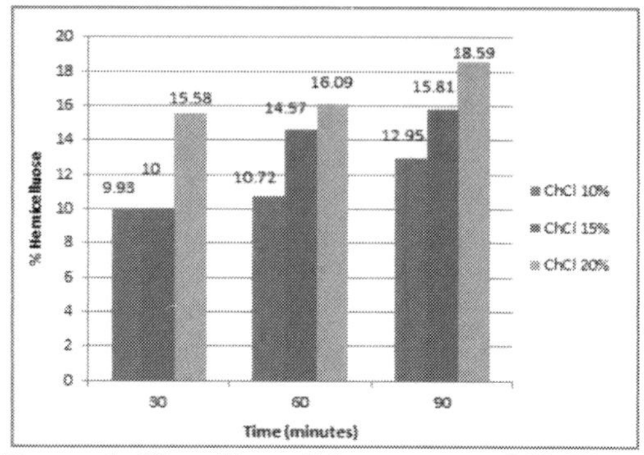


**Figure 3.** The effect of the amount of ChCl and Delignification Time Towards Hemicellulose

It can be seen in Figure 2 and Figure 3 that the levels of cellulose and hemicellulose increase with the rise of the number of ChCl and delignification time. Increased levels of cellulose and hemicellulose are caused by the lignin bond detachment from biomass due to a degradation of the alpha and beta lignin, so that cellulose and hemicellulose increase. However, the cellulose produced by ChCl amount is more than 15% due to degradation in polysaccharide groups, cellulose and hemicellulose, contained in the raw material. The highest cellulose level is 39.8% of the weight of bagasse, obtained in ChCl 20% and 90 minutes of time. While the highest hemicellulose level is 18.59%, obtained in 20% ChCl and 90 minutes.

### *3.3. Comparison of Remained Lignin levels in Delignified Holocelulose With and Without the Use of ChCl*

The lowest remained lignin level in holocellulose produced using ChCl is about 3.62%, while the lowest remained lignin content in holocellulose produced without using ChCl is up to 18.99%.

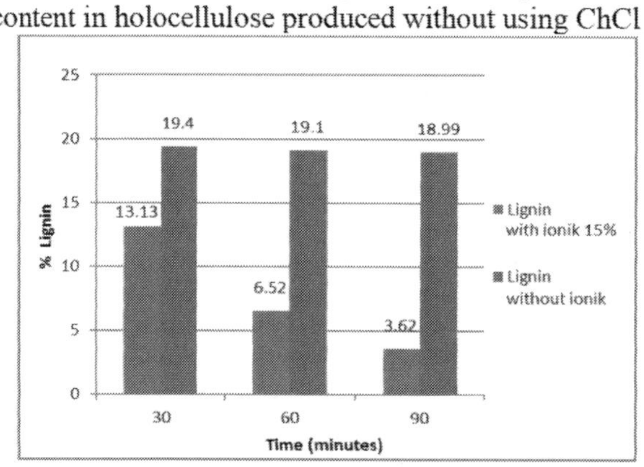


**Figure 4.** Comparison of Remained Lignin levels in Delignified Holocellulose With and Without the Use of ChCl

Lignin level decreases significantly with the increase of ChCl amount and delignification time, whereas lignin level without ChCl only slightly decreases that can be seen in Figure 4. The significant decrease of lignin is resulted by the double work between ChCl and NaOH that detach the lignin bond from raw materials to increase the wasted lignin compound. The remained lignin level in holocellulose is less. However, the remained lignin level in delignified holocellulose uses NaOH without bigger ChCl. This happens without the help of ChCl.

### *3.4. Comparison of Delignification Process With and Without ChCL Towards Cellulose and Hemicellulose Levels*

Figure 5 and 6 show the effect of the ratio between the use of ChCl 15% and without ChCl towards cellulose and hemicellulose levels. In the process of bagasse delignification, NaOH and ChCl work together as a cooking solution to detach lignin bond. ChCl helps NaOH increase levels of cellulose and hemicellulose.

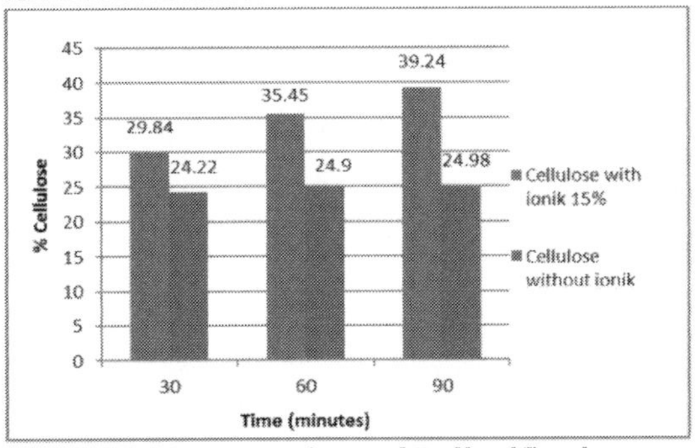


**Figure 5.** Comparison of Delignification Process With ChCl and Without ChCl Towards Cellulose Level

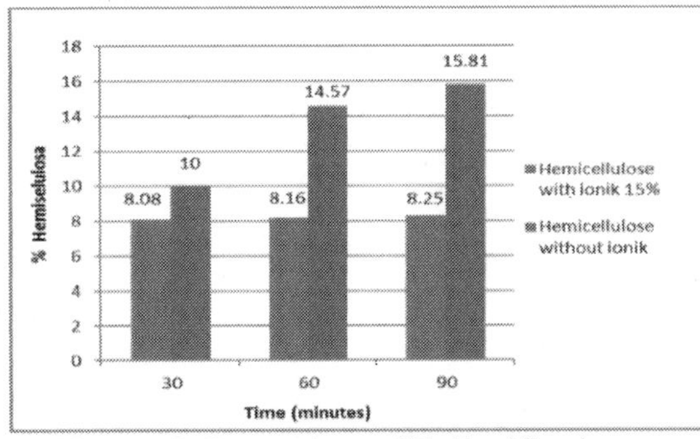


**Figure 6.** Comparison of Delignification Process With ChCl and Without ChCl Towards Hemicellulose Level

It can be seen in Figure 5 that the cellulose content increases significantly by using NaOH in CHCl ionic liquid systems, while the cellulose level obtained without ChCl does not increase significantly, even relatively constant. This is due to the acidic ChCl resulted in the degradation of the building compound of lignin that the cellulose level increases. While the level of cellulose obtained without ChCl increases slightly because NaOH is an alkaline base which has relatively small ability to dissolve organic and inorganic compounds. In the study, researcher obtains cellulose level of 39.8% at 90 minutes with ChCl 20%, while the highest content of hemicellulose is 18.59% at 90 minutes with ChCl 20%.

### *3.5. Effect of Reaction Time Change and Concentration of Choline Chloride Towards Glucose Level*

Bagasse hydrolysis process is conducted in the choline chloride ionic liquid system with ChCl variation of 10%, 15% and 20% of bagasse weight and variations hydrolysis time 30, 60 and 90 minutes. The highest number of glucose produced in the reaction time of 60 minutes with CHCl 20% by weight of bagasse is 39.4%. The analysis shows that the different number of ionic liquids affects the amount of glucose produced. Figure 7 shows the relationship between glucose levels with the amount of CHCl and hydrolysis time.

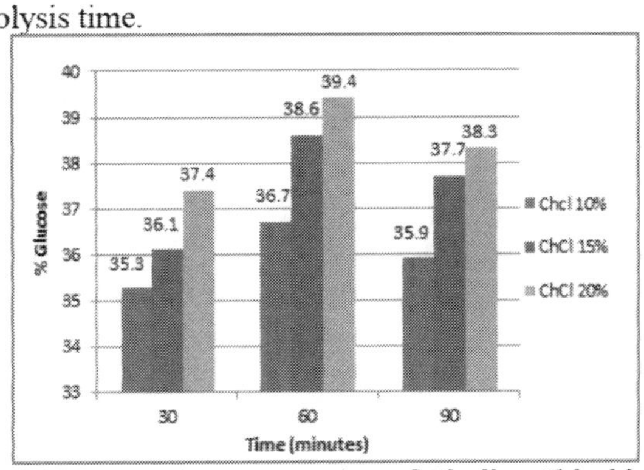


**Figure 7.** Effect of Reaction Time and Concentration of Choline Chloride Towards Glucose Level

As seen in Figure 7 that the glucose level increases with the increase of hydrolysis reaction time and concentration of choline chloride. In this study, for an increase of concentration of choline chloride from 10% to 20%, there is an increase in the percentage of glucose levels: 2.1% in 30 minutes, 2.7% at 60 minutes, and 2.4% at 90 minutes. As for the amount of choline chloride 20%, for an increase in hydrolysis time between 30 minutes to 90 minutes, there is a fluctuating glucose level. The more the increase of the reaction time, the more the glucose levels are generated and until at a certain time limit there will be obtained the maximum glucose level. This is due to the contact between the substances that can react much longer and when the time is extended, the increase of glucose levels becomes very small even decreases. If reaction time is longer, it will be insoluble in water and therefore the breaking of polysaccharide chains into glucose can not take place properly and can possibly damage the glucose produced as a result of continuous warming. ChCl is able to accelerate the hydrolysis reaction and increase the conversion of cellulose and lignocellulose into sugar. The ionic liquid reacts with water and can help bind hemicellulose in order that the hemicellulose is not taken away with cellulose to form good glucose compound and also because of its environmentally friendly fluid properties, it can reduce the concentration of sulfuric acid catalyst, making it safe for the environment and does not corrode the appliance.

*3.6. Comparison of Hydrolysis Process Using Ionic Liquids and Without Using Ionic Liquids*
Optimum glucose level is obtained at 60 minutes reaction time condition with the amount of glucose 39.4% in ChCl 20% by weight of bagasse. As for the hydrolysis process without the use of ionic liquids, the glucose level obtained at 90 minutes is 30.87% with a 10% concentration sulfuric acid. In this study, there is an increase in the percentage of glucose levels to be 6.4% at 30 minutes, 5.2% at 60 minutes, and 4.1% at 90 minutes with the use of 20% of ChCl bagasse weight as shown in Figure 7. It can be seen in Figure 8 that the glucose level increases with the increase of reaction time and the highest glucose level is obtained by using ionic liquids. Therefore, based on Figure 4, the optimum glucose level is obtained at 90 minutes of reaction time with glucose level of 39.4% and 15% of ChCl bagasse weight. As for the use of sulfuric acid 10%, glucose level is less in the same condition of 90 minutes that is 30.87%.

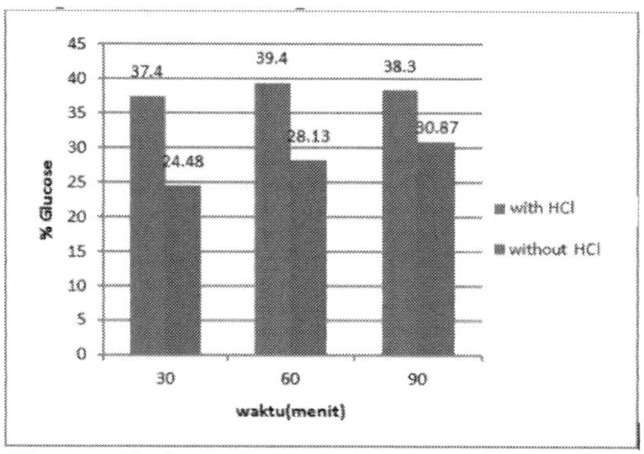


**Figure 8.** Comparison of Glucose Hydrolysis Process Using Ionic Liquids and Without Using Ionic Liquids

## 4. Conclusion
In bagasse delignification process with ChCl, cellulose level is best obtained when the ChCl use is 20% at 90 minutes where the result of cellulose level is 39.80%. The use of ChCl 20% of the bagasse weight is sufficient to the occurrence of degradation of the cooking solution base group that attacks alpha and beta lignin, which increases the cellulose level. The highest glucose level is 39.4%, obtained in 90 minutes hydrolysis time condition with 15% ionic liquid choline chloride concentration, whereas 30.87% glucose level is obtained without the use of choline chloride ionic liquid.

## 5. References
[1] Enny K A, Margaretha N and Vissia Widhie 2010 *Konstanta Kecepatan Reaksi Sebagai Fungsi Suhu Pada Hidrolisis Selulosa dari Ampas Tebu dengan Katalisator Asam Sulfat (Speed Reaction Constant As Temperature Function On Cellulose Hydrolysis from Sugar Cane with Sulfuric Acid Catalyst)* **Ekuilibrium IX** 1-4
[2] Samsuri M G, Baiquni M, Hermansyah A, Wijanarko A, Prasetya B, and Nasikin M 2007 *Pemanfaatan Selulosa Bagas untuk Produksi Etanol Melalui Sakarifikasi dan Fermentasi Serentak dengan Enzim Xylanase (Utilization of Cellulose Bags for Ethanol Production*

*Through Saccharification and Fermentation Simultaneously with Xylanase Enzyme*) **Makara Teknologi Vol. 11** 17-24.

[3] Puja I S 2012 *Produksi Glukosa Dari Tandan Kosong Kelapa Sawit Yang Didelignifikasi Dengan Ozonolysis Pretreatment Melalui Metode Hidrolisis Enzimatik (Glucose Production Of Empty Bunches Of Oil Palm In Delignification With Pretreatment Ozoneolysis Through Enzymatic Hydrolysis Methods)* Proceeding of National Seminar AvoER IV (Palembang: University of Sriwijaya, Indonesia)

[4] Ida B W G 2010 *Pengaruh Perlakuan Delignifikasi dengan Larutan NaOH dan Konsentrasi Substrat Jerami Padi Terhadap Produksi Enzim Selulase Dari Aspergillus niger NRRL A-II (Effect of Delignification Treatment with NaOH Solution and Rice Straw Substrate Concentration to Cellulase Enzyme Production From Aspergillus niger NRRL A-II)* **Jurnal Biologi XIV (1)** 55 – 61

[5] Hwa-Jeong Lee, Bernardi Sanyoto, Jae-Wook Choi, Jeong-Myeong Ha, Dong Jin Suh, Kwan-Young Lee 2013 *Effects of Lignin on The Ionic Liquid Assisted Catalytic Hydrolysis of Cellulose: Chemical Inhibition By Lignin, Springer Science Business*. Korea.

[6] Zhang, Qinghua., Karine De Oliveira Vigier, Sebastien Royer and Francois Jerome 2012 *Deep Eutectic Solvents: Syntheses, Properties and Applications* **Chem. Soc. Rev**. (41) 7108–7146

[7] Harris, Robert Christopher 2008 *Physical Properties of Alcohol Based Deep Eutectic Solvents*, Thesis. Department of Chemistry, University of Leicester

[8] Ahmad S Z, Sri H and Rafina J A P 2014 *Kajian Delignifikasi Pulp Formacell Dari Tandan Kosong Kelapa Sawit Menggunakan Hidrogen Peroksida ($H_2O_2$) Dalam Media Asam Asetat (Study of Formacell Pulp Delignification From Oil Palm Bunches Using Hydrogen Peroxide ($H_2O_2$) In Acetic Acid Media)* ***Jurnal Teknologi Industri dan Hasil Pertanian*. Volume 19 No.2** 194-204

**Acknowledgments**
Many thanks to Department of Chemical Engineering, Faculty of Engineering University of Sumatera Utara, Medan, Indonesia.

# Herbal dryer: drying of ginger (*zingiber officinale*) using tray dryer

**B Haryanto, R Hasibuan, Alexander, M Ashari and M Ridha**

Department of Chemical Engineering, Universitas Sumatera Utara, Padang Bulan, Medan 20155, Indonesia

E-mail: bode.haryanto@usu.ac.id

**Abstract**. Drying is widely used as a method to preserve food because of its convenience and affordability. Drying of ginger using tray dryer were carried out at various drying conditions, such as air-drying flow, air-drying temperature, and sample dimensions, to achieve the highest drying rate. Samples with various dimensions were placed in the tray dryer and dried using various air-drying flow and temperatures. The weights of samples were observed every 3 minutes interval. Drying was stopped after three times of constant weighing. Data of drying was collected to make the drying curves. Drying curves show that the highest drying rate is achieved using highest air flow and temperature.

## 1. Introduction

Ginger (*Zingiber officinale*) is a herb plant. Fresh ginger root is usually consumed as spice in the tropical countries and dried ginger is used as medicinal plant internationally. Dried ginger is usually used as spice or medicine. Quality specifications for export as medicinal herb required it to be properly cut into pieces, well dried and proper storage [1].

Drying process plays an important role in the preservation of agricultural products due to its energy efficiency. Drying has been used since ancient times. Hot air drying is the most frequently used in dehydration operation in the food and chemical industry. The wide variety of preserved foods, which today are available to the consumers and the interesting concern for meeting quality specifications and energy conservation, emphasize the need for a thorough understanding of the drying process [2,3].

Drying rate is calculated by measuring the weight of a drying material over function of time. Drying rate curves represented in three different types of plots that are drying rate versus sample dimension, drying rate versus drying-air temperature and drying rate versus drying-air flow.

The objective of this study is to determine the effects of the variables and to find which variable resulting the fastest drying rate of ginger samples.

## 2. Materials and Methods

### 2.1. Equipment

The tray dryer (Armfield UOP 8) is shown in Figure. 1. This is an equipment unit at The Unit Operation Laboratory of the Department of Chemical Engineering Universitas Sumatera Utara. This equipment consists of three trays. The tray dryer also equipped with air-drying temperature and air-drying flow control.

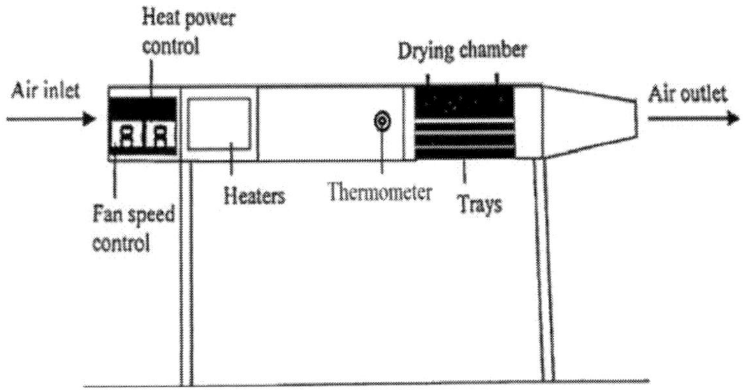


**Figure 1**. Tray Dryer (Armfield UOP 8)

### 2.2. *Materials*

Ginger used in all experiments were obtained from traditional market around Universitas Sumatera Utara. In order to perform all the experiments in identical conditions, ginger are selected manually. Then the gingers were skinned and cut into 2 different bar shape with different thickness.

**Table 1.** Dimension of Ginger Samples

| Sample | Length | Width | Thickness |
|---|---|---|---|
| Sample I | 4 cm | 2 cm | 0,1 cm |
| Sample II | 4 cm | 2 cm | 0,2 cm |

### 2.3. *Experimental Procedure*

In order to obtain the evaluation factors used for the research, measurement of air-drying temperature and moisture content were made. Drying time was determined with a stopwatch.
The drying air temperature was controlled with the aid of a thermostat attached to the heating element of the dryer, which was previously calibrated to read its maximum at the drying air temperature and turn off heating if maximum threshold is exceeded.

Two drying-air temperature (36°C, 39°C) and two drying-air flow (4,0 m/s, 4,5 m/s) were applied. The ginger samples were weighed to obtain the initial sample weight. Three of same dimension samples that had been cut were placed into each trays. The change of weight recorded every 3 minutes. Drying was stopped after three times of constant weighing. Data obtained from the weight measurement then used for calculating the moisture content and drying rate of the samples. Then the dry-weight basis moisture content ($M_d$) can be written as follows [4]

$$M_d = \frac{w-d}{d} \quad (1)$$

Where, w = initial sample weight, g & d = weight of dry product, g. Moisture content on dry-weight has been used for all calculation purpose in this paper. $M_d$ then used for calculating the drying rate (R). R can be written as follows [5]

$$R = \frac{M_d}{A}\left(\frac{M_{n+1}-M_n}{t_{n+1}-t}\right) \quad (2)$$

Where, R = drying rate, g/m$^2$.minute; A = surface area of samples, m$^2$; M = moisture content & t = time, minute.

**Table 2.** Experiment Design of Drying Process

| Run | Sample | Drying-Air Temperature | Drying-Air Flow |
|---|---|---|---|
| Run I | Sample I | 36°C | 4,0 m/s |
| Run II | Sample II | 36°C | 4,0 m/s |
| Run III | Sample I | 36°C | 4,0 m/s |
| Run IV | Sample I | 39°C | 4,0 m/s |
| Run V | Sample I | 39°C | 4,0 m/s |
| Run VI | Sample I | 39°C | 4,5 m/s |

## 3. Results and Discussion

### 3.1. *Effects of Sample Dimension on the Drying Rate*

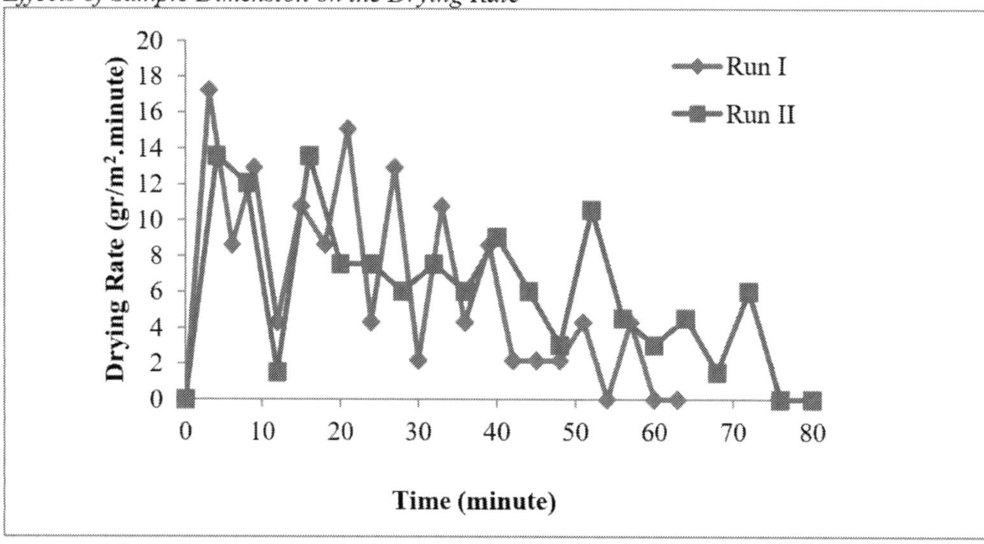


**Figure 2**. The Variation of Drying Rate for Different Sample Dimension

Figure 2. shows the drying rate when the ginger samples were dried at the respective dimension of (4 x 2 x 0,1) cm and (4 x 2 x 0,2) cm and drying-air temperature of 36°C and drying-air flow of 4,0 m/s. The drying rate reach constant value at minute 63 for run I and minute 81 for run II. Kim *et al.* [6] reported that with the increasing of surface area, the drying rate is decrease.

### 3.2. *Effects of Drying-Air Temperature on the Drying Rate*

Figure 3. shows the drying rate when the ginger samples were dried at the respective air-drying temperature of 36°C and 39°C and sample dimension of (4 x 2 x 0,1) cm and drying-air flow of 4,0 m/s. The drying rate reach constant value at minute 57 for run III and minute 63 for run IV. Tzempelikos *et al.* [7] state that drying-air temperature influencing the drying rate. The heat transfer seems to be most significant for higher drying-air temperature.

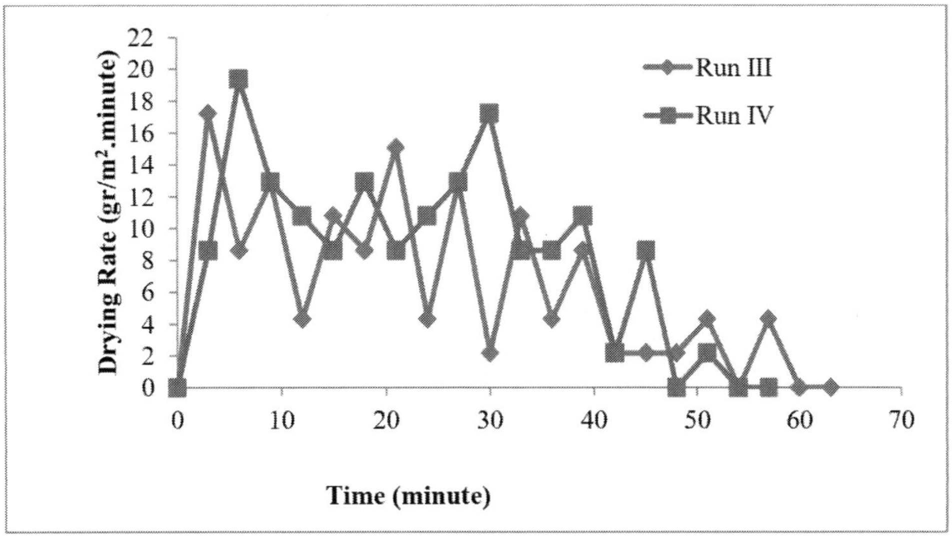


**Figure 3**. The Variation of Drying Rate for Different Drying-Air Temperature

*3.3. Effects of Drying-Air Flow on the Drying Rate*

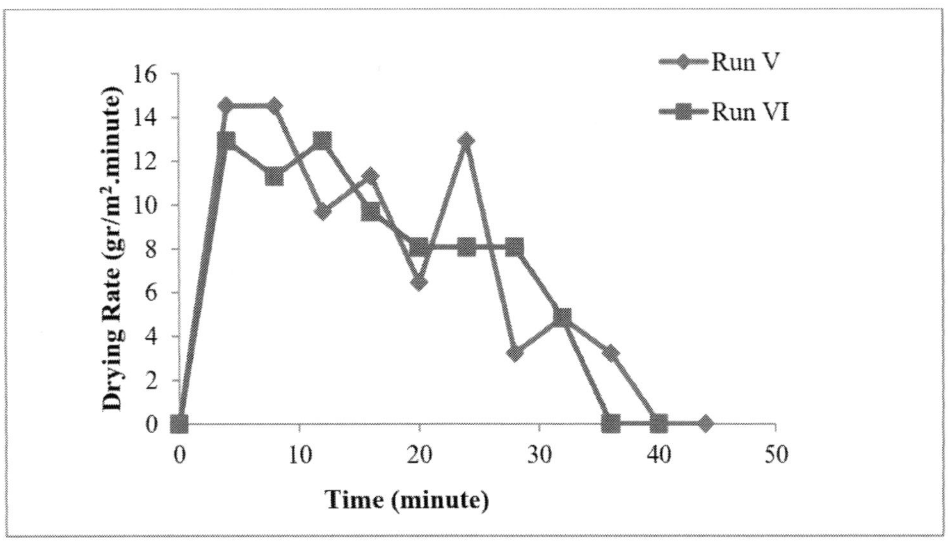


**Figure 4**. The Variation of Drying Rate for Drying-Air Flow

Figure 4. shows the drying rate when the ginger samples were dried at the respective air-drying flow of 4,0 m/s and 4,5 m/s and sample dimension of (4 x 2 x 0,1) cm and drying-air temperature of 39°C. The drying rate reach constant value at minute 39 for run V and minute 45 for run VI. Chinenye [8] state that drying rate increase with increasing drying-air flow but decrease with increasing time.

## 4. Conclusion

From the study it can be concluded that the drying rate is affects by variables, such as drying-air temperature, drying-air flow, and sample dimension. The best drying rate in this study is achieved by using the drying-air temperature of 39°C and drying-air flow of 4,5 m/s for the drying of ginger sample with dimension of (4 x 2 x 0,1) cm (Run VI).

## References

[1] Hoque M A, B K Bala, M A Hossain and M Borhan Uddin 2013 Drying Kinetics of Ginger Rhizome (*Zingiber officinale*) J. *Agril. Res.* **38(2)** 301-309

[2] Fudholi A, M Y Othman, M H Ruslan, M Yahya, A Zaharim and K Sopian 2011 The Effects of Drying Air Temperature and Humidity on the Drying Kinetics of Seaweed *Recent Researches in Geography, Geology, Energy, Enviroment and Biomedcine* 129-133

[3] Osueke C O, B O Akinnuili, A A Adediran, P P Ikubanni and O Agboola 2015 An Algorithm of a Convectional Factory Electric Tray Dryer *The International Journal Of Engineering and Science (IJES)* **4** 17-25

[4] Desmukh V V, Neeraj Chandraker, R S Thakur and A K Chandrakar 2015 Analysis of Moisture Content of Pulses Pellets using Fluidized Bed Dryer *Journal of Chemical and Pharmaceutical Research* **7(6)** 447-451

[5] Chin S K, Siew E S and Soon W L 2015 Drying Characteristics and Quality Evaluation of Kiwi Slices under Hot Air Natural Convective Drying Method *International Food Research Journal* **22(6)** 2188-2195

[6] Kim T H, J G Hampton, L U Opara, A K Hardacre and B R Mackay 2002 Effects of Maize Grain Size, Shape and Hardness on Drying Rate and the Occurence of Stress Cracks *Journal of the Science of Food and Agriculture* **82** 1232-1239

[7] Tzempelikos D A, A P Vouros, A V Bardakas, A E Filios and D P Margaris 2014 Case Studies on the Effect of the Air Drying Conditions on the Convective Drying of Quinces *Case Studies of Thermal Engineering*

[8] Chinenye N M 2009 Effect of Drying Temperature and Drying Air Velocity on the Drying Rate and Drying Constant of Cocoa Bean *Agricultural Engineering International: the CIGR Ejournal* **XI**

## Acknowledgements

We would like to thank the Unit Operation Laboratory of Department of Chemical Engineering Universitas Sumatera Utara for facilitating the equipment used in this study.

# Growth performance and yield stability of selected local upland rice genotypes in Buton Utara of Southeast Sulawesi

**G R Sadimantara[1], B Kadidaa [2], Suaib [1], L O Safuan [1] and Muhidin[1]**

[1]Department of Agrotechnology, Faculty of Agriculture, Halu Oleo University Kendari, Southeast Sulawesi, 93232 Indonesia

[2]Postgraduate Program in Agricultural sciences, Halu Oleo University Kendari, Southeast Sulawesi, 93232 Indonesia

Email : gusti5@yahoo.com

**Abstract.** To evaluate the growth performance and yield stability of seven selected local upland rice genotypes, field experiment was conducted in Buton Utara during rainy season 2015 and 2016. Seven selected upland rice genotypes were grown in six different locations and evaluated in a randomized complete block design with three replications. The variables observed were plant height, leaf area, number of tillers, number of productive tillers, panicle length, number of grains per panicle, number of filled grain per panicle, percentage of unfilled grain, 1000 grain weight, grain dry weight, and grain yield (t/ha). The results were analyzed using analysis of variance and combined analysis of variance for the interaction of environment x cultivars. Data on various growth characters and yield stability indicated different among genotypes and significant effect followed by the DMRT test at the significant level $\alpha = 0.05$. The genotypes of *Wakawondu, Wabalongka*, and *Mantebeka* have high yield stability, while the genotypes of *Waapolo, Warangka, Wabila Lambale*, and *Warara* have low yield stability. Hence, Wakawondu, Wabalongka, and Mantebeka could be recommended for cultivation by the farmers in Buton Utara and these cultivars should be popularized in larger scale to make use of its superiority.

## 1. Introduction

Rice is the second largest produce cereal in the world in 158.3 million hectare area with annual production of about 685.24 million metric tons [1] and the staple food for over one third of the world's population, but more than 90% of rice is produced and consumed in Asia. Rice grain contains 75 to 80% starch, 12% water and 7% protein [2-3].

Rice is a very important staple food in Indonesia. Indonesia has been major rice producing but in another side as the main rice consumer [4-5]. The rice demanding is expected to increase in line with the increase in population. The government of Indonesian has two police to reach rice self-sufficiency [6]. First is through the increasing rice production and second through the decreasing level rice consumption, and promoting local source staple foods such sago [7], cassava or corn [5]. One effort to improve rice production is through development of upland rice in dry land [8].

Upland rice comprises 11% of the total global rice production and is cultivated on about 14 million hectares [9]. It is also important in cropping systems, because of the lack of irrigation facilities and lower cost of production [10]. Genotype relationship is mainly based on information about plant characteristic. Agronomic and morphological characters are usually used as an initial tool to distinguish between varieties [11]. The sustainability of rice production depends on the development of new rice cultivars with high yield and stable performance across diverse environments [12]. Many tools related to grain yield analysis are available for this purpose. One of the effective factors to study of stability is to determine interaction between genotype and environment and it was studied by many researchers on the various genotypes of rice [13-14]. The G × E interaction can cause differences in grain yield performance and stability of a genotype developed in different environments [15].

The stability of yield in cultivars in different places can be due to cultivar performance that derived from a specific collection of genes (G), the characteristic that associated factors of the environment in which it is grown (E), and the interaction between genotype and location which are usually conducted in various years and locations to satisfactorily stand for spatio temporal variation [16].

Statistical analyzed using AMMI analysis and biplot facility, the promising rice yield data were analyzed to determine the nature and magnitude of G x E interaction effects on grain yield in diverse production environments to identify high yielding and stable genotypes adapted to diverse production environments, and to determine the areas where rice genotypes would be adapted and produce economically competitive yields [17]. Present study was carried out to evaluate the growth performance and yield stability of seven promising local upland rice genotypes in the various locations and agro-climatic regions of North Buton Regency, Indonesia.

## 2. Materials and Methods

Seven selected upland rice genotypes were grown in six different locations and evaluated in a randomized complete block design (RCBD) in each location. The field was divided into three blocks; representing three replications. The seeds were directly grown in the soil plots. The local upland rice cultivars consisted of *Wabalongka, Waapolo, Warangka, Wakawondu, Mantebeka, Wabila Lambale*, and *Warara*.

Land preparation for the planting of seeds involved slashing of the bush, plohwing and harrowing. The seeds were planted at the rate of three seeds per hole at a depth of 5 cm, with an inter and intra–row spacing of 25 x 25cm. The plot size was 3 x 4 $m^2$. Several agronomic traits, i.e: plant height, leaf width, leaf length, flag leaf length, tillers number, and productive tillers number were observed to characterize the cultivars phenotypically. Data on grain yield and some other yield components, i.e: harvesting date, the number of grains per panicle, filled grains, unfilled grains, 1000 grain weight, and total grain yield (t $ha^{-1}$) at 14% moisture level were collected. The results were analyzed using analysis of variance and combined analysis of variance for the interaction of environment x cultivars. Data on various growth characters and yield stability indicated different among genotypes and significant effect followed by the DMRT test at the significant level $\alpha = 0.05$.

Overall characterization results data were analyzed using analysis of variance (ANOVA) and displayed descriptively. The mean differences were adjudged with Duncan's Multiple Range Test (DMRT), Genotype-Environment Interaction (GEI) were estimated by the AMMI model [18]. All data were subjected to analysis using Cropstat version 7.2.

## 3. Results and Discussion

### *3.1. Growth components*

Plant height of local upland rice recorded different among seven rice genotypes at observation on 30, 60, 90, and 120 days after planted (DAP) (Fig. 1). All cultivars gained different plant height at maturity due to various internodal lengths. The cultivars having longer internodes produced taller plants.

The findings are in agreement with those of other research stated that longer internodes increased plant heights in rice [19]. Genetic differences between different varieties also caused variation in plant lengths. Similar results were described that plant height depends on genetic character of a plant and environmental conditions [20]. Montebeka cultivar gained the highest plant length at maturity (Fig. 1). From transplanting to maximum tillering stage it was similar to Warangka and Wabila lambale cultivars but afterwards length increased at much faster rate than others, so it became vulnerable to lodging. Other reported lodging in varieties having length of 4th and 5th internode equal to or more than 15 cm [21].

*Wakawondu* cultivar produced higher number of tillers as compared to other cultivars, except *Warangka* and *Montebeka* cultivars statistically the same (Fig. 2). It was due to their genetic variations. At maximum tillering stage *Warangka* achieved higher number of tillers than *Wakawondu* but this number decreased rapidly as compared to Wakawondu which retained significantly higher number at maturity.

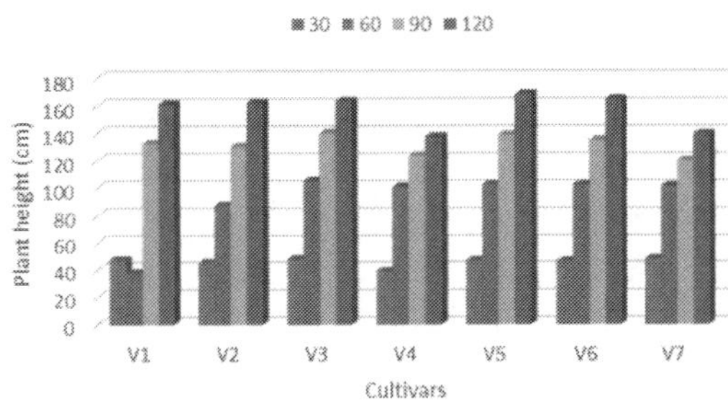


Fig. 1. Plant height of seven cultivars at 30, 60, 90 and 120 DAP
*V1= Wabalongka, V2= Waapolo, V3= Warangka, V4= Wakawondu, V5= Montebeka, V6= Wabila lambale, V7= Warara*

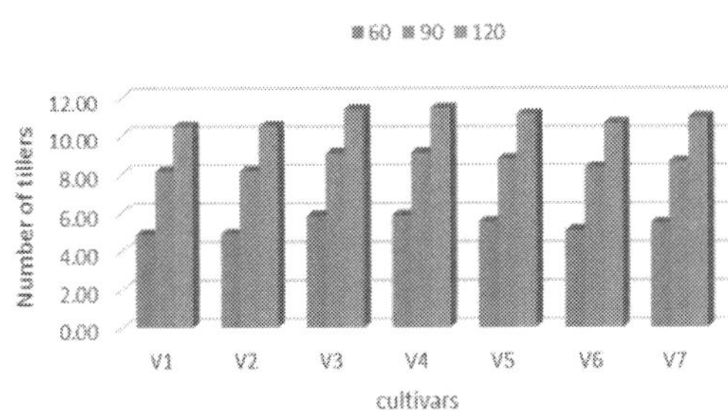


Fig. 2. Tillers number of local upland rice at 60, 90, and 120 DAP
*V1= Wabalongka, V2= Waapolo, V3= Warangka, V4= Wakawondu, V5= Montebeka, V6= Wabila lambale, V7= Warara*

### *3.2. Yield components and yield*

Yield components and yield of local upland rice cultivar obtained differed based on panicle length, grains number, filled grain number, fresh weight grain, dry weight grain, grain weight per clump, and yield t.ha$^{-1}$ (Table 1). The DMRT result shows that *Wabalongka* has longest panicle length compared with other cultivars, but was not significantly different with *Waapolo* and *Wabila Lambale* cultivars (Table 1). *Wa Apolo* found to be the cultivar with highest number of grains but was not significant different with *Wakawondu* cultivar.

The largest number of filled grain was obtained from *Wa Apolo* cultivar which was significantly different with other cultivars. The heaviest fresh and dry weight grains recorded on *Wakawondu* cultivar which was significantly different with other cultivars. *Warara* obtained as cultivar with the heaviest grain weight clump$^{-1}$, but was not significantly different with *Wakawondu*, *Wabalongka* and *Warangka* cultivars, respectively.

The highest yield (t.ha$^{-1}$) recorded on *Wakawondu* cultivar which was not significantly with *Wabalongka* and *Warangka* cultivars, but found significantly different from other cultivars. It can be considered that panicle number and grain number of almost local upland rice in the present study depends on tillers number and panicle length. Other reported that number of panicles depends on number of tillers and proportion of effective tillers [4].

**Table 1.** Yield components of local upland rice

| Varieties (V) | Panicle Length | Grains Number | Filled Grain Number | Fresh Weight Grain | Dry weight grain | Grain weight clump$^{-1}$ | Yield (t.ha$^{-1}$) |
|---|---|---|---|---|---|---|---|
| *Wabalongka* | 41.84 a | 141.76 b | 134.85 b | 41.62 e | 32.38 e | 53.43 ab | 4.23 d |
| *Wa Apolo* | 40.50 ab | 154.42 a | 147.00 a | 49.08 c | 38.33 c | 50.99 bc | 4.24 cd |
| *Warangka* | 39.89 bc | 125.73 ef | 117.39 d | 44.06 d | 31.86 ef | 53.07 ab | 4.53 bc |
| *Wakawondu* | 37.32 d | 149.37 ab | 141.31 a | 53.80 a | 44.88 a | 56.65 a | 4.81 a |
| *Mantebeka* | 39.00 bcd | 128.88 cde | 121.41 cd | 51.49 b | 35.49 d | 51.27 bc | 4.38 bc |
| *Wabila Lambale* | 40.69 ab | 130.68 cd | 118.09 d | 50.60 bc | 39.44 c | 52.07 bc | 4.59 bc |
| *Warara* | 39.55 bc | 132.50 c | 123.70 c | 51.31 b | 41.56 b | 56.80 a | 4.65 ab |

*Mean value followed by same letter at the same column are not significantly different at 5 % level by DMRT*

The grain weight character per clump for all cultivars tested were stable at all locations, there were *Wabalongka* cultivar with specifically adapted to several locations (Table 2). *Mantebeka* and *Warara* cultivars were most adaptable at Eelahaji and Ronta, while *Warangka* cultivar was most adaptable at Lauki. *Waapolo, Wakawondu* and *Wabila lambale* cultivars were most adapted at Peteta'a, West Bubu and Labajaya, respectively.

**Table 2.** Stability analysis for grain dry weight (clump$^{-1}$) of seven upland rice genotypes in six environments

| Cultivars | Average | IPCA 1 | IPCA 2 | ASV |
|---|---|---|---|---|
| Wabalongka | 31.433 | 2.954 | 3.718 | 6.407* |
| Waapolo | 32.329 | -4.370 | 5.427 | 9.436* |
| Warangka | 30.749 | -0.706 | -6.743 | 6.857* |
| Wakawondu | 23.172 | -8.381 | 2.762 | 15.061* |
| Mantebeka | 30.222 | 10.715 | 0.795 | 18.944* |
| Wabila lambale | 31.947 | -3.596 | -6.128 | 8.827* |
| Warara | 26.487 | 3.385 | 0.168 | 5.982* |
| **Location** | **Average** | **IPCA 1** | **IPCA 2** | |
| Lauki | 28.354 | 5.633 | -3.115 | |
| Peteta'a | 32.036 | 10.372 | -0.254 | |
| Eelahaji | 37.109 | -2.408 | -2.079 | |
| West Bubu | 22.582 | -0.348 | 4.730 | |
| Labajaya | 33.496 | -8.167 | -6.636 | |
| Ronta | 23.285 | -5.082 | 7.354 | |

*significant different to zero on ASV

Based on AMMI analysis with ASV value approach for the character of grain dry weight per panicle (Table 2) showed that all cultivars significantly different to zero. This means all the cultivars are not stable and only adaptable to the specific region. Lauki and Peteta'a location have large absolute scores of IPCA 1 and small IPCA 2 means high productivity environment, so genotypes that planted at Lauki and Peteta'a sites will reflect the true genetic ability due to small environmental influences.

According to the character of grain dry weight per clump indicated all tested genotypes were stable in all locations, there were also specific cultivars adapted very well at some locations. *Wabalongka* and *Warara* cultivars were most adaptable at West Bubu. Mantebeka was most adapted at Peteta'a location, *Wabila lambale* and *Warangka* were most adapted cultivars at Eelahaji and Labajaya locations, respectively. *Waapolo* and *Wakawondu* cultivars were most adapted at Ronta.

In Figure 3, the environments fell into three sections. Among the environments Eelahaji, West Bubu, Lebajaya and Ronta had short spokes and they do not exert strong interactive forces but the environments Lebajaya, Ronta and Peteta'a had long spokes therefore represent the most discriminating environments.

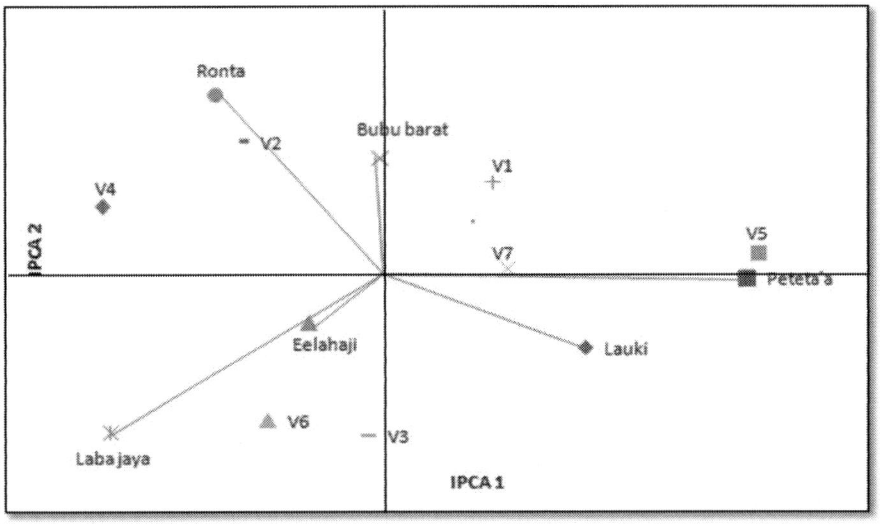


**Fig. 3.** AMMI 2 biplot for grain dry weight (clump$^{-1}$) showing the interaction of IPCA2 against IPCA1 scores of seven upland rice genotype (V) in six environments

Genotypes and environments that fall in the same sectors interact positively; in contrast if they fall in opposite sectors interact negatively. If they fall into adjacent sectors, interaction is somewhat more complex. In the present study, the best genotype with respect to sites Lebajaya and Eelahaji was *Wabila lambale* and *Warangka* genotypes but unsuitable for Peteta'a. Genotype *Waapolo* and *Wakawondu* were suitable for Ronta and West Bubu but unsuitable for Lauki and Peteta'a. The AMMI model is better for partitioning the G x E into the causes of variation, which ease identification of environments' potential and is used to identify superior genotypes either with specific adaptation or wide adaptation [26].

## 4. Conclusions

The genotypes of *Wakawondu, Wabalongka*, and *Mantebeka* have high yield stability, while the genotypes of *Wa Apolo, Warangka, Wabila Lambale*, and *Warara* have low yield stability. Hence, *Wakawondu, Wabalongka*, and *Mantebeka* could be recommended for cultivation by the farmers in Buton Utara and these cultivars should be popularized in larger scale to make use of its superiority.

## References

[1] FAOSTAT 2011 *Food and Agricultural commodities production / Countries by commodity* (Rome : Food and Agriculture Organization of the United Nations Statistics Division)

[2] Oko AO, Ubi BE, Efisue AA and Dambaba N 2012 Comparative analysis of the chemical nutrient composition of selected local and newly introduced rice varieties grown in Ebonyi State of Nigeria *Int. J. Agric.For.* **2(2)** 16-23

[3] Hossain MB, Islam MO and M. Hasanuzzaman 2008 Influence of different nitrogen levels on the performance of four aromatic rice varieties. *Int. J. Agric. Biol.* **10** 693-96

[4] Muhidin, Jusoff K, Syam'un E, Musa Y, Kaimuddin, Meisanti, Sadimantara GR and Baka LR 2013 The development of upland red rice under shade trees *World Applied Sciences Journal* **24(1)** 23-30

[5] Sutariati GAK, Khaeruni A, Pasolon YB, Muhidin and Mudi L 2016. The effect of seed bio-

invigoration using indigenous rhizobacteria to improve viability and vigor of upland rice (*oryza sativa* L.) Seeds *International Journal of PharmTech Research* **9(12)** 565-73

[6] Kadidaa B, Sadimantara GR, Suaib, Safuan LO, Muhidin 2017 Genetic diversity of local upland rice (*Oryza sativa* L) genotypes based on agronomic traits and yield potential in marginal land of North Buton Indonesia *Asian Journal of Crop Science* **9(4)** 109-117

[7] Muhidin, Leomo S, Alam S and Wijayanto T 2016 Comparative studies on different agroecosystem base on soil physicochemical properties to development of sago palm on dryland *International Journal of ChemTech Research* **9(8)** 511-18

[8] Syaiful SA, Syam'un E, Dachlan A, Jusoff K, Haerani N 2013 The effect of inoculating nitrogen fixing bacteria on production of rice *World Applied Sciences Journal* **26(26)** 94-99

[9] Sohrabi M, Rafii MY, Hanafi MM, Akmar ASN and Latif MA 2012 Genetic diversity of upland rice germplasm in Malaysia based on quantitative traits *The Scientific World Journal* **2012** Article ID416291 p9

[10] Fageria NK, Carvalho MCS and dos Santos FC 2014 Response of upland rice genotypes to nitrogen fertilization *Communications in Soil Science and Plant Analysis* **45(15)** 2058–66

[11] Sadimantara GR, Muhidin and Cahyono E 2014 Genetic analysis on some agro-morphological characters of hybrid progenies from cultivated paddy rice and local upland rice *Advanced Studies in Biology* **6(1)** 7–18

[12] Akter A, Jamil Hassan M, Umma Kulsum M, Islam MR, Hossain K and Rahman MM 2014 AMMI biplot analysis for stability of grain yield in hybrid rice (*Oryza sativa* L.) *J.Rice Res.* **2(2)** 1-4

[13] Lestari PA, Buang A, Ahmad J and Hajrial A 2010 Yield stability and adaptability of aromatic new plant type (NPT) rice lines *J. Agron. Indonesia* **38(3)** 199–204

[14] Bose LK, Jambhulkar NN, Pande K and Singh ON 2013 Use of AMMI and other stability statistics in the simultaneous selection of rice genotypes for yield and stability under direct-seeded conditions *Chilean Journal of Agricultural Research* **74** 3-9

[15] Colombari Filho JM, de Resende MDV, de Morais OP, Pereira A, Guimarães EP and Pereira JA 2013 Upland rice breeding in Brazil: a simultaneous genotypic evaluation of stability, adaptability and grain yield *Euphytica* **192** 117-29

[16] Karimizadeh R, Mohammadi M, Sabaghni N, Mahmoodi AA, Roustami B and Seyyedi F 2013 GGE biplot analysis of yield stability in multi-environment trials of lentil genotypes under rainfed condition *Notulae Scientia Biologicae* **5** 256-62

[17] Islam MR, Anisuzzaman M, Khatun H, Sharma N, Islam MZ, Akter A and Biswas PS 2014 AMMI analysis of yield performance and stability of rice genotypes across different haor areas. *Eco-friendly Agril. J* **7(02)** 20-24

[18] Zobel RW, Wright MJ and Gauch HG 1988 Statistical analysis of a yield trial *Agronomy Journal* **80** 388-93

[19] Ashrafuzzaman M, Islam MR, Ismail MR, Shahidullah SM and Hanafi MM 2009 Evaluation of six aromatic rice varieties for yield and yield contributing characters *Int. J. Agric. Bio.* **11** 616-620

[20] Kanegana MZ and Kargbo AM 2011 *Effect of nitrogen topdressing at different growth stages on rice plant growth, yield and yield components*. In: Report on experiments in rice cultivation techniques development course 3 Tsukuba Int Center Japan Int Coop Agency. pp-43-45

[21] Anandan A, Eswaran R, Sabesan T and Prakash M 2009 Additive main effects and multiplicative interactions analysis of yield performances in rice genotypes under coastal saline environments *Advances in Biological Research* **3(1-2)** 43-4

# The design of dapog rice seeder model for laboratory scale

**UI Purba[1], T Rizaldi[2], Sumono[2], R Sigalingging[2]**

[1] Student in Department of Agricultural Engineering, Universitas Sumatera Utara, Indonesia. Jl. Prof. A. Sofyan No. 3 Kampus USU, Medan 20155
[2] Department of Agricultural Engineering, Universitas Sumatera Utara, Indonesia. Jl. Prof. A. Sofyan No. 3 Kampus USU, Medan 20155
Co-authors e-mail: taufik.rizaldi@usu.ac.id

**Abstract.** The dapog system is seeding rice seeds using a special nursery tray. Rice seedings with dapog systems can produce seedlings in the form of higher quality and uniform seed rolls. This study aims to reduce the cost of making large-scale apparatus by designing models for small-scale and can be used for learning in the laboratory. Parameters observed were soil uniformity, seeds and fertilizers, soil looses, seeds and fertilizers, effective capacity of apparatus, and power requirements. The results showed a high uniformity in soil, seed and fertilizer respectively 92.8%, 1-3 seeds / $cm^2$ and 82%. The scattered materials for soil, seed and fertilizer were respectively 6.23%, 2.7% and 2.23%. The effective capacity of apparatus was 360 boxes / hour with 237.5 kWh of required power.

## 1. Introduction

Rice is consumed by nearly half of the world's population, with rice cultivation using 24-30% of the world's developed freshwater. A major challenge for food security in Asia in the coming decades is feeding the increasing population under water-scarce conditions. By 2025, an estimated 15-20 million ha of irrigated rice will suffer some degree of water scarcity include alternate wetting and drying, and the aerobic rice system. Another potential intervention is to reduce the growth duration of transplanted rice plants in the main field either by using short-duration varieties or by planting older but healthy seedlings [3]. Conventional way in rice breeding is usually done in rice fields so it requires a large area. The resulting seeds have poor roots and soil mixed. To support modern agriculture that uses rice transplanting as a mechanical planting apparatus makes this conventional method is considered less support. To avoid any confounding effect, sowing date in the nursery, seed rate and crop management in the main field were all the same [6]. Seeds with good root system and uniform plant height strongly support the performance of rice transplanting apparatus. The process of growing rice without the need for a long time to seeds the seeds back is one simple and advantageous method because it no longer will be limited supply of water [2]. Furthermore, for transplanted rice, seedlings can be produced on wet bed nurseries, dry bed nurseries, mat nurseries, or dapog nurseries for manual transplanting or in specialized seed trays for machine transplanting [1]. Therefore, appropriate seeding method can be used with dapog system. Dapog system can be operated with wet system in paddy field and dry system indoors can used with media tray. This research was conducted to make a model of dry dapog system with mechanical using media tray.

The modeling of these fixtures is carried out to ensure that the designs made are operable. This can save costs and simplify the manufacture of apparatus in large capacity because the constraints that occur can be overcome on the model. Model of small capacity seed breeding (laboratory scale) can be used by students as learning materials in the laboratory [5]. Enable students to learn about the rotor rationing mechanism as well as the use of electric motor power and transmits power through pulleys. The design of the seed sowing machine is based on the design of the three rotor metering device mechanisms and a conveyor shaft powered by an electric motor.

## 2. Materials and Methods

The materials used in the research are soil, rice seed, fertilizer and tray. The tools used are stopwatch, camera, tachometer and wattmeter.

### *2.1 Stages of research*

This research is based on design principles with several stages, (1) literature study, (2) problem analysis, (3) design of apparatus concept, (4) evaluation and design of apparatus, (5) manufacture of apparatus, (6) performance testing of apparatus

### *2.2 Functional design*

Firstly hopper is the functional design consists of three hoppers placed on the main frame for the entry of soil, seeds and fertilizers as material for rice breeding before being dropped on to the tray. The size for each hopper is 143 mm long, 103 mm wide and 228.64 mm high made of iron plate with a thickness of 1 mm. Secondly is rotor, consists of three rotor that will control the process of dropping materials onto a tray that is 290 mm long, 140 mm wide and 40 mm high. The rotor of soil metering device is designed to drop the soil into the tray with a thickness of 12 mm. Based on this thickness, the rotor has a diameter of 75 mm and a length of 140 mm with a flow width of 28.10 mm and 7.50 mm groove height. The rotor of seed metering device is designed to drop 3 mm (3-4 seed / $cm^2$) seed into the tray so that based on the rotor calculation it has a diameter of 75 mm and a length of 140 mm with a width of 43.56 mm and 5 mm groove. In the rotor of fertilizer metering device is designed to drop the fertilizer with a thickness of 5 mm to the top of the tray so that based on the calculation of the rotor has a diameter of 75 mm and length 140 mm with the width of the groove 28.10 mm and the height of the groove 2.50 mm.

### *2.3 Structural Design*

This rice seeding apparatus has several structural components, namely: firstly is beam, its function is to place all components of rice seeding with length 560 mm, width 253 mm and height 534,79 mm using material of iron plate and iron strip with 3 mm thick. Secondly is bearing, its function is to reduce friction on the rotating system on the apparatus. This model of rice seeding apparatus using bearing type UCP 204. Thirdly is gear box, its function is to reduce the rotational speed of the shaft. This paddy seeding apparatus uses a size 40 gear box. Fourthly is belt conveyor, its function is to bring the tray in a horizontal direction with a roller lever system. The belt is driven by an electromotor through a pulley. In this model of rice seedling use a conveyor with a width of 66 mm. Fifthly is sprocket, its function is to transmit power and rotation from gearbox to sprocket through chain to rotate rotor of metering device. Sprocket used has 42 teeth. Sixth is pulley, its function is to forward the power and rotation of the electromotor through v-belt to rotate the gear box and to the three rotor of metering device. In the model of rice seeding apparatus uses 2 inch, 5 inch, 6 inch, and 8 inch pulleys. Seventh is electromotor are machines that convert electrical energy into mechanical energy. Using an electric motor of 1420 rpm.

### *2.4 Apparatus mechanism*

This apparatus works by utilizing the rotary power of the electromotor that is transmitted through the pulley to rotate the rotor of metering device contained within the hopper and the conveyor roller. This apparatus has three hoppers each filled with soil, rice seed and fertilizer. When the electromotor is switched on, the conveyor roller will move the belt conveyor to bring the tray horizontally while the three rotors will rotate to drop the soil, seed and fertilizer according to the desired dosage.

The capacity of the soil hopper can be filled with soil as much as 750 gr, the capacity of the seed hopper can be filled with seeds as much as 500 gr. While the capacity of the fertilizer hopper can be filled as much as 750 gr of fertilizer. In the process of seeding required 363.3 gr of land, 40 gr of seed and 210 gr of fertilizer.

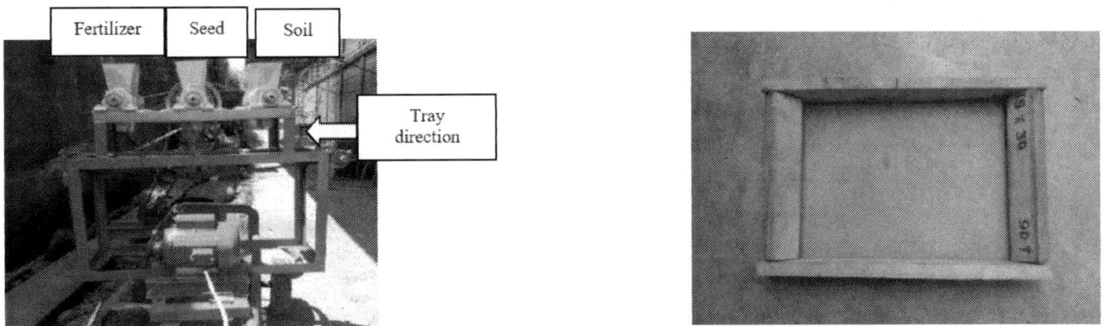


a) b)

**Fig. 1.** Model of seeding equippments (a), seeding tray (b)

*2.5 Testing*

Soil, seed and fertilizer are placed into each hopper while the tray position is placed as in Fig.1. Electromotor is switched then each shaft rotates, the tray is carried by the conveyor belt through the outlet of each hopper with rationing device starting from the soil, seeds and fertilizers respectively. The desired soil thickness for once seeding is 12 mm, seed density of 1-3 seeds / $cm^2$ and fertilizer with a thickness of 5 mm. This measure becomes the reference for determining the design accuracy made during the test.

*2.6 Parameters*

2.6.1. Percentage of uniformity of soil, seed and fertilizer distribution

Uniformity can be determined by calculating the standard deviation from the observed data on the targets to be achieved by equation 1.

$$S = \sqrt{\frac{\sum_{i-1}^{n} (xi-x)^2}{n-1}} \quad \ldots\ldots\ldots\ldots \quad 1)$$

Where S mean standard deviation with $n$ is the number of sample, $x_i$ is value of $x$ to-$i$ and $x$ is sample data.

2.6.2. Percentage loosing of soil, seed and fertilizer

The amount of soil, seed and fertilizer that falls outside the tray at the time of seeding is considered a lost material. The percentage can be determined by equation 2.

$$L = \frac{BBT}{BBA} \times 100\% \quad \ldots\ldots\ldots\ldots \quad 2)$$

Where $L$ mean percentage loosing of material (%), $BBT$ is weight of material lost (gr) and $BBA$ is total weight of material (gr).

2.6.3. Effective capacity of apparatus (KEA)

The effective capacity of the apparatus (tray/hour) can be determined by calculating the length of time required by the seeding tool to tray upon the capacity of the existing material with equation 3.

$$KEA = \frac{\text{Number of trays}}{\text{time required}} \quad \ldots\ldots\ldots\ldots \quad 3)$$

#### 2.6.4. Power Requiremet

The actual power requirements used for the operation of this apparatus can be measured using a digital wattmeter. Digital wattmeter is a measuring instrument readings of electric power in watt which is a combination of voltmeters and ammeters. With these tool of digital wattmeter, can know the actual power and will be displayed on the LCD display screen on digital wattmeter.

## 3 Results and Discussion

### *3.1 Percentage of uniformity of soil, seed and fertilizer distribution*

Figure 3 (a) explains that the measurements of the soil thickness have not fully reached the desired target. It can be seen that there are some measurement points whose values are not exactly on the target line (1.2 cm). Based on the calculation of percentage uniformity obtained an average deviation of 7.2%. However, this result can be expressed uniformly because of the great uniformity of 92.8%. Figure 3 (b) explains that the average seed density present in the tray is within the targeted range of 80%. The deviation occurs twice from 10 repetitions that is on the first repeat and the seventh repetition. If averaged, the seed density obtained is very good that is equal to 3.6 seeds / $cm^2$. Figure 3 (c) explains that the thickness of the fertilizer that falls into the tray is close to the desired target. From the experiment with 10 repetitions, the uniformity rate of fertilizer 82% was found or the average deviation of 18%. This can be caused by the rotation of the rotor of the fertilizer metering device is too fast, so further research is needed to determine the rotor speed of the appropriate metering device of fertilizers.

a) b) c)

**Fig. 2.** (a) the result of soil allotment, (b) the result of seed allotment, and (c) the result of fertilizer allotment

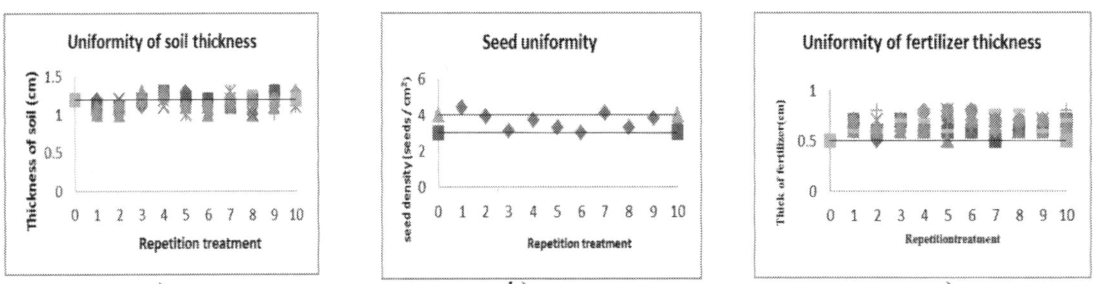


a) b) c)

**Fig. 3.** Graph of: (a) soil thickness uniformity, (b) seed uniformity,and (c) uniformity of fertilizer

Based on recommendation from BPPP East Java [4] stating that the requirement of seeding for thickness of soil material on dapog system is 1.2 cm Fig 2 (a). The result of comparison of soil thickness measurement on tray conducted at 10 point randomly with 10 repetition treatment can be presented in Fig. 3 (a).

According to Tang [7], seed treatments have been developed in an efficient, safe, and economical which allows farmers to control insect pest on various crops of rice worldwide. Seed treatment technology is relatively simple and inexpensive in controlling diseases and insect pest in agricultural crops. With seed treatment before seedling determines quality of the seed for the success a farm. On the measurement of the distribution of seed uniformity is done 10 repetitions Fig 2 (b). An analysis of the measurement results to determine how close to the expected target value. The results can be presented in Figure 3 (b).

BPPP East Java [4] states that the soil cover (soil mixed with organic fertilizer) should have a thickness of 0.5 cm then sprinkled with water. Measurements of soil height in the tray were performed at 10 random points with 10 repetitions Fig 2 (c). The final stage in the seeding is the closing of seeds with fertilizer. To measure the thickness of the fertilizer is done by determining the total targeted seeding thickness is 2 cm. The thickness measurement of the fertilizer is the total thickness of the material with the thickness of the seed and the thickness of the soil. Analysis of measurement results with expected target values is presented in Figure 3 (c).

### *3.2 Percentage loosing of soil, seed and fertilizer*

Percentage loosing of soil, seed and fertilizer of the experimental results are presented in Figure 4. Figure 4 explains that the average percentage loosing of soil, seed and fertilizer is 6.23%, 2.17% and 2.23%, respectively. The percentage loosing of soil is greater than seed and fertilizer because the volume of soil dropped by rotor of soil metering device is greater than the amount of soil and fertilizer dropped. The way to fix this is to modify the rotor of metering device channel material according to the size of the tray used.

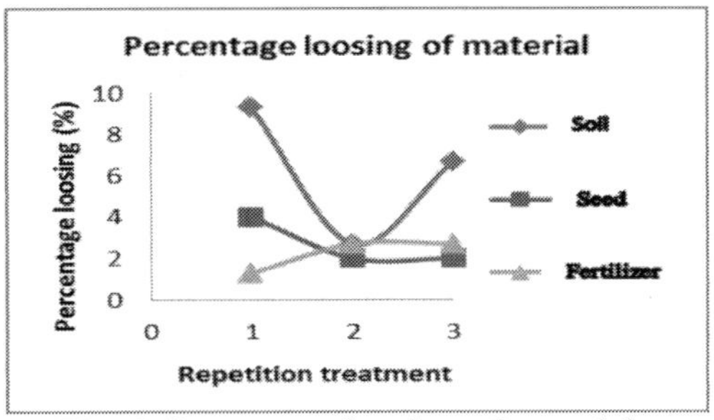


**Fig. 4.** Percentage loosing of material

### *3.3 Effective capacity of apparatus (KEA)*

The effective capacity of apparatus is the productivity apparatus during the rice seeding process. The capacity of the seeding tool is presented in Table 1. The effective capacity of apparatus is the productivity apparatus during the rice seeding process. Experiments conducted by three times repetitions obtained an average seeding of 2.5 trays for 24 seconds so that if converted per hour average can complete the seeding of 360 tray.

**Table 1.** Effective capacity

| Repetition | Number of tray | Time (second) | Effective capacity (tray/hr) |
|---|---|---|---|
| I | 2.5 | 24 | 360 |
| II | 2.5 | 24 | 360 |
| III | 2.5 | 24 | 360 |

### *3.4 Power Requirement*

The results of measuring the use of electrical power during the seedling process using a digital wattmeter are presented in Table 2. The results of measuring the use of electrical power during the seedling process using a digital wattmeter. Of the three repetitions performed, the highest electrical power requirement is at the first repeat of 240.0 watts, and the lowest electrical power requirement in the third repeat is 234.0 watts. The occurrence of the difference in power requirements can be caused by load conditions and unstable electrical current at the time of measurement.

**Table 2.** Power requirement

| Repetition | Electric current (Watt) |
|---|---|
| I | 240 |
| II | 238.5 |
| III | 234 |
| Average | 237.5 |

## 4. Conclusion

Design model of small-scale rice seeding tools can be used to produce seeding of dapog rice seed system using tray in laboratory. The average effective capacity of the apparatus in the laboratory-scale model of laboratory seedlings was is 360 tray / hour for the seedbed of rice seed. The uniformity of soil, seed and average fertilizer in the model of rice plant seeding system laboratory dapog were 92.8%, 3-4 seeds / $cm^2$ and 82% for seeding of rice seeds. The average percentage loosing of soil, seeds and fertilizers on the laboratory scale model of laboratory seeding were 6.23%, 2.7% and 2.23% for seeding of rice seeds.

## References

[1] Horgan, F., G. Figueroa, J., Y, and M. L. P. Almazan. *Seedling broadcasting as a potential method to reduce apple snail damage to rice*. Crop Protection 64 **168-176,** 2014.

[2] Hill, R. D. *The cultivation of perennial rice, an early phase in Southeast Asian?*. Journal of Historical Geography 36 **215-223**, 2010.

[3] Lampayan, R. M., J.E. Faronilo, T. P. Tuong, A. J. Espiritu, J. L. Dio, R. S. Bayot, C. S. Bueno, Y. Hosen. *Effects Of Seed bed management and delayed transplanting of rice seedlings on crop performance, grain yield, and water productivity*. Field Crops Research 183 **303-304**, 2015.

[4] Badan Penelitian dan Pengembangan Pertanian(Agency for Agricultural Research and Development). Pembibitan Padi Sistem Dapog *(Rice Breeding System Dapog)*. Serpong: Balai Besar Pengembangan Mekanisasi Pertanian (Serpong: Center for Agricultural Engineering Research and Development), 2016.

[5] Deininger, M., S.R. Daly, K.H. Sienko, J.C. Lee. *Novice designers' use of prototypes in engineering design*. Design Studies 002 **23-36**, 2017.

[6] Pasuquin, E., T. Lafarge and B. Tubana. *Transplanting young seedlings in irrigated rice fields: Early and high tiller production enhanced grain yield*. Field Crops Research 105 **141–155**, 2008.

[7] Tang, T., X. Liu, P. Wang, W. Fu, M. Ma. *Thiamethoxam seed treatment for control of rice thrips (Chloethripsoryzae) and its effects on the growth and yield of rice (Oryza sativa)*. Crop Protection 98 **136-142**, 2017.

## Acknowledgements

Thanks to the University of Sumatera Utara Research Institute for funding provided for research under the TALENTA research contract for year 2017 Number: 5338 / UN5.1.R / PPM / 2017 dated May 22, 2017.

# The utilization of ultisol soil for horticulture crops cultivation

**Sumono, SM Parinduri. N Huda, N Ichwan**

Agricultural Engineering, Faculty of Agriculture, University of Sumatera Utara
Jalan Prof. A. Sofyan No. 3 Kampus USU Padang Bulan, Medan, North Sumatera, Indonesia

Email: nazif.ichwan@gmail.com

**Abstract**. Ultisol soil is a marginal soil commonly used for palm oil cultivation in Indonesia, its very potential for cultivation of horticulture crops. The utilization of ultisol soil can be done with adding compost with certain proportions. The research aimed to know best proportion of ultisol soil and compost, and proportion of water concentration, and its relationship with fresh and dry weight of horticulture crops . The research was divided 3 steps. The first, mixed ultisol soil and compost with certain proportion and flooding until steady. The second, watering with different concentration to soil mixture. The last, studied its relationship with fresh and dry weight of crops. The result show that physical properties and nutrient content of ultisol soil was increasing with adding compost. SC4 (70% soil and 30% compost) is the best composition to soil mixture. Watering with different concentration show that trend decreased from reference and the bulk density and porosity decreased not significantly at the significant level $\alpha = 0.05$.Watering affect mass of pakcoynot significantly at the significant level $\alpha = 0.05$.Hence, ultisol soil was a potential marginal soil to utilizing as a media for cultivating horticulture crops.

## 1. Introduction

Ultisol soil (red soil) is a marginal soil, it's determined low soil content and high pH value[1][2]. It explains that the soil chemist content is poor and so does the soil physics. Bulk density and particle density affect the stability of soil aggregate, and itcan affect water movement and storage in soil[3]. The soil physics is the main properties, and the soil chemical and biological has been degradecaused by the use of chemical fertilizers[1]. It can be Increased by using compost as a natural fertilizer, which it can increase organic matter, pH, enzyme activities, and soil respiration[4]. Compost contains many micro nutrient such as Mg, Co, Fe, N, P, and K, and it can affect available water[5].

Correlation of both soil particle and structure is important in water holding capacity[6]. Land use, processing, fertilizing, and compaction can change porosity, soil particle, and pores function, that affect chemical, physical and biological properties of soil[7]. Distribution of aggregate, organic matter content, mineral composition, and water content affect fluid transportation[8], the easiness of water flow through pores, it's called permeability[9].

## 2. Materials and Method

This research was conducted at greenhouse laboratory, Agriculture Faculty, University of Sumatera Utara. The research was divided into three steps, they weremixed ultisol soil and compost with certain proportions, wateringthe soil mixture with different concentration, and studied its relationship between fresh and dry weight of pakcoy.

*2.1 Mixing ultisol soil and compost by certain proportions*
The research started by entering mixture soil and compost to polybag with 10 kg capacity. Each polybag consists of ultisol soil and compost with certain proportions. The proportions were SC1 (100% soil + 0% compost ) as control, SC2 (90% + 10%), SC3 (80% + 20%) and SC4 (70%+30%).Soil compaction was done by completely flooding theeach polybag.Texture, bulk density, particle density, porosity, permeability, C-organic, organic matter, pH, nitrogen, phosphor, potassium are observed in the end of season with soil sample from each polybag.

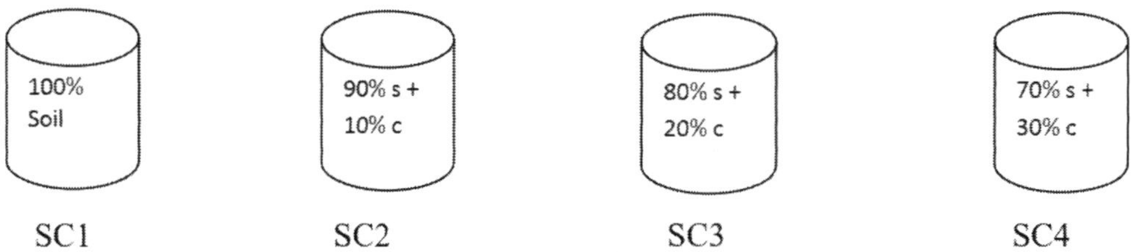


**Fig 1.** Mixture soil and compost inside of polybag

*2.2 Watering the soil mixture by different concentration*
The best proportion used as a media for horticulture cultivation. The experiment waswatering the sample by different concentration to lookat the best concentration. The watering concentrations were 100% available water (AW) from field capacity as W1, 80% AW as W2 and then 60% AW as W3. As soon as watering was done, horticulture cultivation was done with pakcoy (*Brassica rapa* L.) as sample plants. Watering was done until the harvest-age of pakcoy. Soil sample was done in the end of season, hence, texture, bulk density, particle density, porosity, permeability was observed.

**Fig 2.** Mixture soil and compost inside polybag with 70:30 composition

*2.3 Relationship both watering and weight of crops*
Pakcoy was harvested in 45 days after planting. Fresh and dry mass as a parameter observed in this research. Fresh and dry mass were measuredto obtain the relationship of mass and watering.

## 3. Results and discussion

*3.1 Mixing ultisol soil and compost*
Soil texture is one of main parameter to study about soil, hence, the proportion of fraction affected soil physics characteristic. The comparison of sandy, clay and loam fraction should be known.

**Table 1.** Soil Physics and chemist analysis from mixture of soil and compost

| No. | Parameters | | SC1 | SC2 | SC3 | SC4 |
|---|---|---|---|---|---|---|
| 1 | Texture | | Sandy loam | Sandy loam | Sandy loam | Sandy loam |
| 2 | Bulks Density | $g.cm^{-3}$ | 1.41 | 1.36 | 1.29 | 1.23 |
| 3 | Particle Density | $g.cm^{-3}$ | 2.98 | 2.93 | 2.86 | 2.79 |
| 4 | Porosity | % | 51.37 | 53.50 | 54.00 | 55.80 |
| 5 | Permeability | $cm.hour^{-1}$ | 5.62 | 6.56 | 6.75 | 7.97 |
| 6 | C-organic | % | 0.97 | 1.87 | 3.10 | 3.61 |
| 7 | Organic matter | % | 1.67 | 3.22 | 5.34 | 6.22 |
| 8 | PH | | 6.52 | 7.09 | 7.20 | 7.33 |
| 9 | Nitrogen | | 0.013 | 0.137 | 0.217 | 0.213 |
| 10 | Phosphor | | 20.90 | 29.13 | 35.48 | 42.25 |
| 11 | Potassium | | 0.491 | 0.669 | 0.702 | 0.761 |

Table 1 shows that sandy loam was the texture for each samples. The biggest proportion was sand (69.3%), followed by clay (16,3%) and loam (14.3%). Soil texture also affected bulk density and particle density. Adding compost caused the value of bulk density and particle density to decrease but not significantly, and increasing the value of porosity and permeability. If the value of bulk density decreasing, hence, the value of porosity will be increasing. There was negative correlation of bulk density and porosity.

Beside the soil physics, adding compost also affected the soil chemistry. C-organic, matter, pH, nitrogen, phosphor, potassium value was increasing which is caused by adding compost. The best experiment can be concluded that SC4 (70% soil and 30% compost) given the highest value than the other composition.

### *3.2. Watering with different concentration*

Table 1 shows that the best composition was 70% ultisol soil and 30% compost. Watering was done every day according the evapotranspiration value (ET) for each polybag (Table 2).

Sandy loam was soil texture of each samples with composition of sand (74.56%), clay (17.09%), and loam (8.35%). Table 2 showshow soonthe value of bulk density and particle density increases and particle density. Watering affected bulk density and particle density but not significantly. Permeability was categorized to rather quick as categories[10].

**Table 2.** Soil physics analysis from watering experiment

| No. | Parameters | | Control | W1 | W2 | W3 |
|---|---|---|---|---|---|---|
| 1 | Texture | | Sandy loam | Sandy loam | Sandy loam | Sandy loam |
| 2 | Bulk Density | $g.cm^{-3}$ | 1.41 | $0.79^{aA}$ | $0.84^{aA}$ | $0.85^{aA}$ |
| 3 | Particle Density | $g.cm^{-3}$ | 2.98 | 1.74 | 1.77 | 1.79 |
| 4 | Porosity | % | 51.37 | $54.39^{aA}$ | $52.34^{aA}$ | $52.72^{aA}$ |
| 5 | Permeability | $cm.hour^{-1}$ | 5.62 | 7.76 | 7.07 | 6.15 |

### *3.3. The relationship both watering and weight crops*

To obtain the relationship of watering and weight of crops, pakcoy used as experiment plant. Steam and leaves, and root weight also were scaled to obtain the effect of watering with different concentration (Table 3).

**Table 3.** Fresh weight and dry weight of pakcoy

| No. | Treatment | Fresh weight (gram) | | Dry Weight (gram) | |
|---|---|---|---|---|---|
| | | Stems and leaves | Root | Stems and leaves | Root |
| 1 | W1 | 57.50$^{aA}$ | 1.64$^{aA}$ | 4.04$^{aA}$ | 0.21$^{aA}$ |
| 2 | W2 | 58.75$^{aA}$ | 1.69$^{aA}$ | 3.61$^{aA}$ | 0.18$^{aA}$ |
| 3 | W3 | 70.00$^{aA}$ | 1.99$^{aA}$ | 5.09$^{aA}$ | 0.22$^{aA}$ |

Table 3 shows thatthere is negative correlation of watering concentration and fresh weight of pakcoy. The lower watering concentration, hence, the higher fresh weight of pakcoy. However, W2 was different and having positive trend on fresh weight.

Fresh weight was 57.50-70.00 grams for stems and leaves, and 1.64-1.99 grams for root. It's lower than standard mass of fresh.Dry weight was 3.61-5.09 grams for stems and leaves, and 0.18-0.22 grams for root. Overall, there was no significant difference all of them.

The data above can explain that adding compost to ultisol soil can increase soil physical and chemical. The proportion of soil fraction, affects bulk density and also porosity. The value of bulk density affected porosity as high. The focused concern of data was the different value both SC4 flooding and SC4 watering in soil physical properties. Flooding has not donedaily like watering. Aggregate stability can be movement parallel with water push from watering [3]. It can affect the value of bulk density to decrease from the reference.

There are significant difference between flooding and daily watering in bulk density and particle density. Flooding caused water to fill all of pores, and caused soil compaction which the bulk density and particle density value to increase. It's different with watering while water given according to ET value. Water is forced to fill in all of pores, and it's also given was according to field capacity of soil. Table 2 shows that there was no significant difference between W1, W2, and W3. It meansthe water can be givento match field capacity of available water.

Mass of pakcoy does not satisfythe standard mass of pakcoy which government had issued, KEMENTAN No. 253/kpt/TP.240/5/2000 which stated that standard mass of pakcoy is 250-300 g. Compost additionalso can increasewater holding capacity while increasing organic matter. Whereas water given has no significantly affected the mass of pakcoy, but it needs a big energy to water suction in ultisol soil. Mass of pakcoy not fulfillthe standard, furthermore, ultisol soil is a potential soil to be utilized as a media for cultivating the horticulture crops.

## 4. Conclusions

The research can be concluded that the best proportion is SC 4 that consist of 70% ultisol soil and 30% compost which can be use as a media of horticulture cultivation. Flooding and watering given affects bulk density and porosity, and watering affects bulk density and porosity but not significantly. Watering affects mass of pakchoy not significant also. Ultisol soil is a potential soil to utilizing as a media for cultivating the horticulture crops.

## References

[1] Anikwe M A N, Eze J C, Chima M C and Ikenganyia E E 2015Influence of lime and gypsum application on soil properties and yield of cassava (Manihot esculenta Crantz.) in a degraded Ultisol in Agbani, Enugu Southeastern Nigeria*Soil & Tillage Research* **158**, 32-38.

[2] Anikwe M A N, Eze J C, Chima M C and Ikenganyia E E 2016 Soil physicochemical quality in contrasting tillage systems and its effect on nodulation and nodulation effectivity of groundnut, Bambara groundnut and soybean in a degraded Ultisol in Agbani, Enugu Southeastern Nigeria*Rhizosphere***1**, 14-16.

[3] Wang J G, Yang W, Yu B, Li Z, Cai C and Ma R 2016 Estimating the Influence of Related Soil Properties on Macro- and Micro- aggregate Stability in Ultisols of South-Central China*Catena* **137**, 545-553.

[4] Oliva S R, Mingorance M D andPena A 2016Effect of two different composts on soil quality and on the growth of various plant species in a polymetallic acidic mine soil *Chemosphere***168**, 183-190.

[5] Donn S, Wheatley R E, McKenzie B M, Loades K W andHallett P D 2014 Improved soil fertility from compost amendment increases root growth and reinforcement of surface soil on slopes*Ecological Engineering***71**, 458-465.

[6] Ding D, Zhao Y, Feng H, Peng X, andSi B 2015 Using the double-exponential water retention equation to determinate how soil pore-size distribution is linked to soil texture*Soil & Tillage Research***156**, 119-130.

[7] Gao Lu S, Malik Z, Chen D PandWu C F 2014Porosity and pore size distribution of ultisols and correlations to soil iron oxides*Catena***123**, 79-87.

[8] Pla C, Cuezva S, Martinez J M, Cortes A F, Anton E G, Fusi N, Crosta G B, Gonzales J C, Cañaveras J C, Moral S S and Benavente D 2016 Role of soil pore structure in water infiltration and $CO_2$ exchange between the atmosphere and underground air in the vadose zone: A combined laboratory and field approach*Catena***149**, 402-416.

[9] Gupta P, Alam J and Muzzammil M 2016 Influence of thickness and position of the individual layer on the permeability of the stratified Soil*Perspectives in Science***8**, 757-759.

[10] Uhland R E and O'neal A M 1951Soil Permeability Determination For Use In Soil and Water ConservationUnited States of Agriculture, Washington.

# The effects of heating temperatures and time on deformation energy and oil yield of sunflower bulk seeds in compression loading

**A Kabutey[1], D Herak[1], R Sigalingging[2] and C Demirel[3]**

[1]Department of Mechanical Engineering, Faculty of Engineering, Czech University of Life Sciences Prague, Kamycka 129, 165 21 Prague, Suchdol, Prague 6, Czech Republic
[2]Department of Agricultural Engineering, Faculty of Agriculture, Universitas Sumatera Utara, Indonesia.
[3]Departement of Agricultural Machinery, Faculty of Agriculture, Ondokuz mayis University, Samsun, Turkey

E-mail: kabutey@tf.czu.cz

**Abstract.** The deformation energy (J) and percentage oil yield (%) of sunflower bulk seeds under the influence of heat treatment temperatures and heating time were examined in compression test using the universal compression testing machine and vessel diameter of 60 mm with a plunger. The heat treatment temperatures were between 40 and 100 °C and the heating time at specific temperatures of 40 and 100 °C ranged from 15 to 75 minutes. The bulk sunflower seeds were measured at a pressing height of 60 mm and pressed at a maximum force of 100 kN and speed of 5 mm/min. Based on the compression results, the deformation energy and oil yield increased along with increasing heat treatment temperatures. The results were statistically significant ($p < 0.05$). At a 40 °C heat treatment temperature in relation to varying heating time, deformation energy increased while the percentage oil yield decreased. However, at a 100 °C, the deformation energy almost showed no correlation but the oil yield similarly decreased. From the ANOVA statistical analysis, the effect of heating time at 40 and 100 °C heat treatment temperatures on deformation energy and oil yield was not significant ($p > 0.05$).

## 1. Introduction

The use of vegetable oils as part of renewable energy sources is rapidly growing over the concerns of fossil fuels or climate change [1]. Although renewable energy can provide a clean source of electricity, petroleum fuels still account for the vast majority of the world's energy generation [2]. Sunflower (*Helianthus annus* L.) produces edible vegetable oil with an annual production of 25.1 million tonnes being one of the potential feedstocks for biodiesel production [3]. The conventional method for sunflower oil extraction involves seed preparation, mechanical extraction and solvent extraction using n-hexane [4, 5]. Mechanical pressing is the most common method for oil extraction from oilseeds especially in the rural areas of developing countries [6]. The main purpose of oilseed post-harvest technology is the recovery of percentage oil yield and quality of oil which is dependent on several factors such as storage, pretreatment and processing [7, 8]. Studies have shown that applied pressure, pressing or heating temperature, heating time, pressing time, pressing speed and seed moisture content thus influence the oil recovery and energy requirement [9, 10, 11, 12]. To obtain a higher percentage oil yield and minimum residual oil in the seed cake, it is very important to control the above-mentioned processing parameters during oil extraction process [13]. The study was aimed to investigate the influence of heat treatment temperatures and heating time on deformation energy and percentage oil yield of sunflower bulk seeds in compression loading and also to describe analytically the dependency between the force and deformation curves as well as the deformation energy.

## 2. Materials and methods

### *2.1 Sample and compression test*

A sunflower bulk seed of moisture content of 12 w.b. % was used for the compression test. The moisture content was determined using the procedure specified by [14]. The equipment (MEMMERT GmbH + Co. KG, Germany) was used for the samples heat treatment temperatures from 40 to 100 °C and heating time from 15 to 75 minutes at specific heat treatment temperatures of 40 and 100 °C respectively. A universal testing machine of type ZDM 50, Czech Republic together with a pressing vessel of diameter 60 mm with a plunger were used for recording the data points of force (N) and deformation (mm) of sunflower bulk seed of 60 mm measured pressing height [15]. A maximum force of 100 kN and speed of 5 mm/min were applied. The test was repeated twice. The determined amounts namely the deformation energy (J) and oil yield (%) were statistically evaluated using the Statistica software (version 13) [16]. The deformation energy and oil yield were numerically determined based on the equations proposed by the authors [17, 18]. The theoretical relationships between the force and deformation curves as well as the deformation energy were described by the tangent curve mathematical model [19, 20] applicable in Mathcad software (version 14) [21].

## 3. Results and discussion

The measured amounts of deformation, oil yield and deformation energy in relation to heat treatment temperatures and heating time are presented in Tables 1 to 3 respectively.

**Table 1.** Measured data (mean±standard deviation) of sunflower bulk seeds at varying temperatures

| Temperatures (°C) | Deformation (mm) | Oil yield (%) | Deformation energy (J) *Numerical | **Analytical |
|---|---|---|---|---|
| *Control | 43.400±0.962 | 19.062±0.431 | 456.481±3.007 | 434.481±3.278 |
| 40 | 42.955±0.741 | 18.092±0.701 | 485.250±9.159 | 546.102±0.903 |
| 50 | 44.255±0.813 | 18.199±0.103 | 470.057±7.678 | 526.211±13.062 |
| 60 | 43.425±0.898 | 19.068±0.087 | 470.539±13.067 | 453.991±18.435 |
| 70 | 42.000±0.594 | 18.431±0.881 | 489.051±8.034 | 461.813±5.963 |
| 80 | 42.805±0.559 | 18.028±0.299 | 500.084±4.958 | 514.531±50.973 |
| 90 | 43.905±1.054 | 20.727±0.053 | 508.325±7.788 | 588.006±29.070 |
| 100 | 43.375±0.106 | 19.120±0.163 | 510.661±3.155 | 608.121±13.063 |

* Based on the area of a trapezoid [17], ** Based on the tangent curve model [19, 20]

**Table 2.** Measured data (mean±standard deviation) of sunflower bulk seeds at 40 °C

| Heating time (min) | Deformation (mm) | Oil yield (%) | Deformation energy (J) *Numerical | **Analytical |
|---|---|---|---|---|
| *Control | 42.955±0.741 | 18.092±0.701 | 485.250±9.159 | 546.102±0.903 |
| 15 | 45.375±0.375 | 18.979±0.749 | 485.288±9.509 | 478.305±9.673 |
| 30 | 44.975±0.686 | 18.176±0.103 | 489.252±17.171 | 500.232±2.137 |
| 45 | 43.990±1.655 | 16.697±1.641 | 495.644±9.134 | 477.881±25.710 |
| 60 | 43.365±3.429 | 16.831±0.959 | 506.362±6.164 | 533.771±1.740 |
| 75 | 43.920±0.467 | 18.024±0.462 | 509.229±12.489 | 554.103±24.579 |

Based on the ANOVA statistical analysis, the heat treatment temperatures significantly ($p < 0.05$) influenced the amounts of the deformation energy and percentage oil yield (Table 1). This means that the measured amounts mentioned above increased along with increasing heat treatment temperatures with a very high correlation of 0.935 and 0.941 respectively. The influence of heating time at particular heat treatment temperatures of 40 and 100 °C on deformation energy and oil yield was not statistically significant ($p > 0.05$) in terms of decreasing or increasing trends. However, from the experimental data, the deformation energy increased in relation to the heating time at 40 °C but slightly decreased at 100 °C. The oil yield, on the other hand, decreased at both heat treatment temperatures.

**Table 3.** Measured data (mean±standard deviation) of sunflower bulk seeds at 100 °C

| Heating time (min) | Deformation (mm) | Oil yield (%) | Deformation energy (J) *Numerical | **Analytical |
|---|---|---|---|---|
| *Control | 43.375±0.106 | 19.120±0.163 | 510.661±3.155 | 608.121±13.063 |
| 15 | 41.540±0.651 | 20.805±0.455 | 508.488±11.635 | 550.307±81.070 |
| 30 | 43.205±0.629 | 20.143±1.248 | 517.113±1.874 | 529.770±9.656 |
| 45 | 43.025±0.035 | 20.141±0.345 | 509.943±0.103 | 562.462±12.590 |
| 60 | 42.770±0.509 | 19.268±0.926 | 506.657±0.389 | 553.101±45.837 |
| 75 | 41.910±0.778 | 19.124±0.108 | 509.362±0.844 | 532.237±9.099 |

**Table 4.** One-way ANOVA analysis of measured data at varying heat treatment temperatures

| Measured data | $F_{ratio}$ | $F_{critical}$ | $p$ | $R^2$ |
|---|---|---|---|---|
| Deformation energy (J) | 8.193 | 3.866 | 0.006 | 0.875 |
| Oil yield (%) | 9.115 | 3.866 | 0.005 | 0.887 |

$p < 0.05$ level of significance or $F_{ratio} > F_{critical}$ is significant

**Table 5.** Factorial or Repeated measures ANOVA analysis of measured data at both 40 and 100 °C in relation to heating time

| Measured data | $F_{ratio}$ | $F_{critical}$ | $p$ | $R^2$ |
|---|---|---|---|---|
| Deformation energy (J) | 2.584 | 3.478 | 0.078 | 0.699 |
| Oil yield (%) | 5.344 | 3.478 | 0.008 | 0.828 |

$p > 0.05$ level of significance or $F_{ratio} < F_{critical}$ is non–significant

The theoretical description of the experimental dependency between the force and deformation curves of a sunflower bulk seed in relation to heat treatment temperatures is shown in Figure 1 similar to varying heating time at specific temperatures of 40 and 100 °C respectively. The area under the curve is the numerical deformation energy [17] (Tables 1 to 3) which can also be determined analytically using the tangent curve model [19, 20]. The determined coefficients of the tangent curve model using Mathcad 14 software are presented in Tables 6 to 8 respectively. The coefficients were statistically significant with high coefficients of determination ($R^2$) where the $F_{ratio}$ values were higher than the values of the $F_{critical}$. In addition, the $p$ values were higher than the significance level of 0.05. The tangent curve mathematical model considers the boundary conditions of the compression process being zero compressive force relate to zero deformation, deformation thus reaches a maximum limit as the force increases to infinity and the integral of the tangent curve function is the energy [19, 20]. The tangent curve mathematical model is described in equation 1 as follows:

$$F(x) = A \cdot (\tan(B \cdot x))^n \quad (1)$$

where $F$ is the pressing force (kN), $x$ is the deformation (mm), $A$ is the force coefficient of mechanical behaviour (kN), $B$ is the deformation coefficient of mechanical behaviour (mm$^{-1}$), $n$ is the fitting curve function exponent (-).

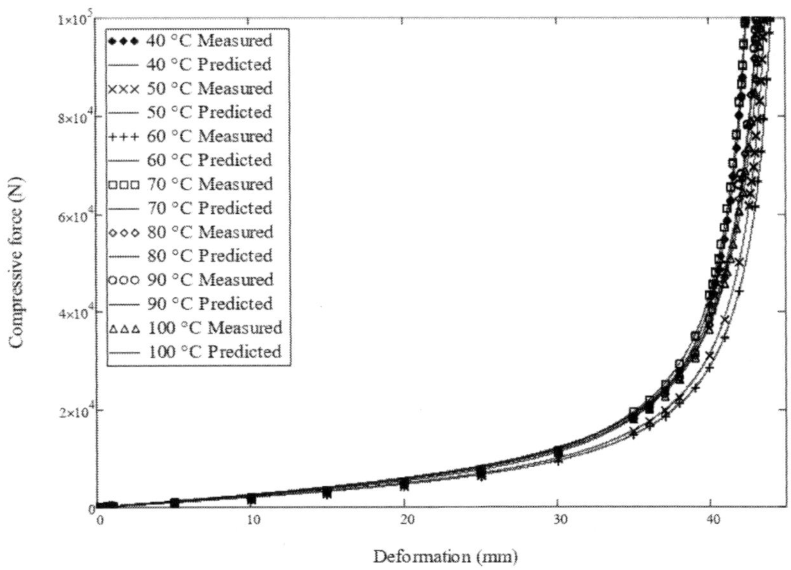


**Figure 1.** Force and deformation curves of sunflower seeds at heat treatment temperatures.

**Table 6.** Tangent curve model coefficients and statistical analysis (mean±standard deviation) at varying heat treatment temperatures.

| Temperatures (°C) | Tangent curve model coefficients | | | Statistical analysis | | | |
|---|---|---|---|---|---|---|---|
| | *A* (kN) | *B* (mm-1) | *n* | $F_{ratio}$ | $F_{critical}$ | *p* | $R^2$ |
| *Control | 5.752 ±0.235 | 0.035 ±0.001 | 1 | 0.0010 ±0.0002 | 3.870 ±0.004 | 0.975 ±0.003 | 0.999 ±0.001 |
| 40 | 6.298 ±0.087 | 0.036 ±0.001 | 1 | 0.0008 ±0.0009 | 3.867 ±0.004 | 0.979 ±0.013 | 0.999 ±0.001 |
| 50 | 5.821 ±0.127 | 0.035 ±0.001 | 1 | 0.0009 ±0.0007 | 3.864 ±0.004 | 0.976 ±0.009 | 0.999 ±0.001 |
| 60 | 5.984 ±0.374 | 0.035 ±0.001 | 1 | 0.0009 ±0.0004 | 3.864 ±0.003 | 0.977 ±0.006 | 0.999 ±0.001 |
| 70 | 6.485 ±0.231 | 0.036 ±0.001 | 1 | 0.0004 ±0.0006 | 3.865 ±0.004 | 0.988 ±0.017 | 0.999 ±0.001 |
| 80 | 6.818 ±0.027 | 0.035 ±0.001 | 1 | 0.0020 ±0.0003 | 3.866 ±0.001 | 0.965 ±0.002 | 0.999 ±0.001 |
| 90 | 7.048 ±0.026 | 0.035 ±0.001 | 1 | 0.0079 ±0.0029 | 3.865 ±0.001 | 0.929 ±0.014 | 0.999 ±0.001 |
| 100 | 7.227 ±0.018 | 0.035 ±0.001 | 1 | 0.0073 ±0.0025 | 3.863 ±0.001 | 0.933 ±0.012 | 0.999 ±0.001 |

* Sunflower bulk seeds at room temperature of 25 °C, $p > 0.05$ level of significance or $F_{critical} > F_{ratio}$ is significant.

IOP Conf. Series: Earth and Environmental Science **122** (2018) 012097 doi:10.1088/1755-1315/122/1/012097

**Table 7.** Tangent curve model coefficients and statistical analysis (mean±standard deviation) at 40 °C in relation to heating time.

| Heating time | Tangent curve model coefficients | | | Statistical analysis | | | |
|---|---|---|---|---|---|---|---|
| (min) | *A* (kN) | *B* (mm⁻1) | *n* | $F_{ratio}$ | $F_{critical}$ | *p* | $R^2$ |
| *Control | 5.752 ±0.235 | 0.035 ±0.001 | 1 | 0.0010 ±0.0002 | 3.870 ±0.004 | 0.975 ±0.003 | 0.999 ±0.001 |
| 15 | 6.034 ±0.267 | 0.033 ±0.001 | 1 | 0.0038 ±0.0006 | 3.863 ±0.001 | 0.951 ±0.004 | 0.999 ±0.001 |
| 30 | 6.102 ±0.202 | 0.034 ±0.001 | 1 | 0.0026 ±0.0002 | 3.864 ±0.001 | 0.960 ±0.002 | 0.999 ±0.001 |
| 45 | 6.310 ±0.411 | 0.034 ±0.001 | 1 | 0.0012 ±0.0011 | 3.865 ±0.002 | 0.974 ±0.013 | 0.999 ±0.001 |
| 60 | 6.551 ±0.447 | 0.035 ±0.003 | 1 | 0.0007 ±0.0012 | 3.866 ±0.002 | 0.983 ±0.023 | 0.999 ±0.001 |
| 75 | 6.590 ±0.189 | 0.035 ±0.001 | 1 | 0.0011 ±0.0006 | 3.865 ±0.002 | 0.974 ±0.007 | 0.999 ±0.001 |

p > 0.05 level of significance or $F_{critical} > F_{ratio}$ is significant.

**Table 8.** Tangent curve model coefficients and statistical analysis (mean±standard deviation) at 100 °C in relation to heating time.

| Heating time | Tangent curve model coefficients | | | Statistical analysis | | | |
|---|---|---|---|---|---|---|---|
| (min) | *A* (kN) | *B* (mm⁻1) | *n* | $F_{ratio}$ | $F_{critical}$ | *p* | $R^2$ |
| *Control | 5.752 ±0.235 | 0.036 ±0.001 | 1 | 0.0010 ±0.0002 | 3.870 ±0.004 | 0.975 ±0.003 | 0.999 ±0.001 |
| 15 | 7.575 ±0.203 | 0.035 ±0.001 | 1 | 0.0054 $\pm 4.6 \cdot 10^{-5}$ | 3.864 ±0.003 | 0.941 ±0.001 | 0.999 ±0.001 |
| 30 | 7.243 ±0.035 | 0.035 ±0.001 | 1 | 0.0087 ±0.0090 | 3.864 ±0.002 | 0.933 ±0.041 | 0.999 ±0.001 |
| 45 | 7.197 ±0.111 | 0.035 ±0.001 | 1 | 0.0081 ±0.0055 | 3.864 ±0.002 | 0.931 ±0.026 | 0.999 ±0.001 |
| 60 | 7.388 ±0.074 | 0.035 ±0.001 | 1 | 0.0113 ±0.0039 | 3.863 ±0.001 | 0.916 ±0.015 | 0.999 ±0.001 |
| 75 | 7.598 ±0.201 | 0.036 ±0.001 | 1 | 0.0077 ±0.0023 | 3.863 ±0.001 | 0.931 ±0.011 | 0.999 ±0.001 |

p > 0.05 level of significance or $F_{critical} > F_{ratio}$ is significant.

## 4. Conclusion

Heat treatment temperatures significantly ($p < 0.05$) increased the deformation energy and percentage oil yield of sunflower bulk seeds compressed at a maximum force of 100 kN and speed of 5 mm/min. The interaction effects of heating time and heat treatment temperatures of 40 and 100 °C did not significantly ($p > 0.05$) decrease or increase the deformation energy and percentage oil yield. The analytical deformation energy amounts were within the range of the numerical energy values.

The coefficients of determination of the theoretical description of the experimental relationships between the force and deformation curves in relation to heat treatment temperatures and heating time were 0.999. The theoretical model coefficients were statistically significant ($p > 0.05$). The exponent of the tangent model representing the value of the fitting curve was found to be 1 in all analyses. It is important to mention that the use of the Statistica software [16] interprets the statistical significance when the $p$ value is less than the significance level or the $F_{ratio}$ greater than the $F_{critical}$ (Table 4). This is, however, contrary to the Mathcad statistical package [21] where the $p$ value is greater than the significance level or $F_{critical}$ greater than the $F_{ratio}$ (Tables 6 to 8).

**References**

[1] YX Xu, Hanna M A and Josiah S J 2007 Hybrid hazelnut oil characteristics and its potential oleochemical application *Ind. Crop Prod* **26**:69-76.
[2] Lazkano I, Nostbakken L and Pelli Martino 2017 From fossil fuels to renewables: The role of electricity storage, https://doi.org/10.1016/j.euroecorev.2017.03.013.
[3] Saydut A, Erdogan S, Kafadar A B, Kaya C, Aydin F and Hamamci C 2016. Process optimization for production of biodiesel from hazelnut oil, sunflower oil and their hybrid feedstock *Fuel* **183**:512-517.
[4] Baumler E R, Carrin M E and Carelli A A 2016 Extraction of sunflower oil using ethanol as solvent *Journal of Food Engineering* **178**:190-197.
[5] Li Y, Fine F, Fabiano-Tixier A S, Abert-Vian M, Carre P, Pages X and Chemat F 2014 Evaluation of alternative solvents for improvement of oil extraction from rapeseeds *Comptes Rendus Chim* **17**(3):242-251.
[6] Achten W M J, Mathijs E, Verchot L, Singh V P, Aerts R, Muys B 2007 Jatropha biodiesel fueling sustainability? *Biofuels Bioprod. Biorefining* 1:283-291
[7] Nagaraj G 2009 Linseed. In: Oilseeds, Properties, Processing, Products and Procedures, Linseed, New India Publishing Agency, New Delhi, India, pp. 123.
[8] Kasote D M, Badhe Y S and Hegde M 2013 Effect of mechanical press oil extraction processing on quality of linseed oil *Industrial Crops and Products* **42**:10-13.
[9] Subroto E, Manurung R, Heeres H J and Broekhuis A A 2015 Mechanical extraction of oil from Jatropha curcas L. kernel: Effect of processing parameters *Industrial Crops and Products* **63**:303-310.
[10] Willems P, Kuipers N J M, de Haan A B 2008 Hydraulic pressing of oilseeds: experimental determination and modeling of yield and pressing rates *J. Food Eng* **89**:8-16.
[11] Acheheb H, Aliouane R, Ferradji A 2012 Optimization of oil extraction *Pistacia atlantica* Desf. Seeds using hydraulic press *Asian J. Agric. Res.* **6**:73-82.
[12] Karaj S. Muller J 2011 Optimizing mechanical oil extraction of *Jatropha curcas* L. seeds with respect to press capacity, oil recovery and energy efficiency *Industrial Crops and Products* **34**:1010-1016.
[13] Bamboye A I and Adejumo O I 2011 Effects of processing parameters of Roselle seed on its oil yield *Int. J Agric & Biol Eng* **4**(1):82-86
[14] ISI 1996 Indian standard methods for analysis of oilseeds IS:3579 New Delhi Indian Stand. Ins.
[15] Akangbe O.L and Herak D 2017 Mechanical behaviour of selected bulk oilseeds under compression loading *Agronomy Research* **15**(1), 947–951.
[16] Statsoft., Inc. Tulsa, OK74104, USA, 2010.
[17] Herak D, Kabutey A, Sedlacek A and Gurdil G 2012 Mechanical behaviour of several layers of selected plant seeds under compression loading *Res. in Agric. Eng.* **58**(1):24-29
[18] Deli S, Farah Masturah M, Tajul Aris Y and Wan Nadiah W A 2011 The effects of physical parameters of the screw press oil expeller on oil yield from *Nigella sativa* L. seeds *Int. Food Res.* **18**:1367–1373.
[19] Herak D, Kabutey A, Divisova M and Simanjuntak S 2013 Mathematical model of mechanical behaviour of Jatropha curcas L. seeds under compression loading *Bio. Eng.* **114**(3):279–288.

[20] Sigalingging R, Herak D, Kabutey A, Dajbych O, Hrabe P and Mizera C Application of a tangent curve mathematical model for analysis of the mechanical behaviour of sunflower bulk seeds *Int. Agrophys.* **29**(4):517–524.
[21] Mathsoft Engineering and Education, Inc. 101 Main Street, Cambridge, Ma 02142, USA, 2004.

IOP Conf. Series: Earth and Environmental Science **122** (2018) 012098 doi:10.1088/1755-1315/122/1/012098

# Utilization of FEM model for steel microstructure determination

**A Kešner[1], R Chotěborský[1], M Linda[2] and M Hromasová[2]**

[1]Department of Material Science and Manufacturing Technology, Faculty of Engineering, Czech University of Life Sciences Prague
[2]Department of Electrical Engineering and Automation, Faculty of Engineering, Czech University of Life Sciences Prague
E-mail: kesner@tf.czu.cz

**Abstract**. Agricultural tools which are used in soil processing, they are worn by abrasive wear mechanism cases by hard minerals particles in the soil. The wear rate is influenced by mechanical characterization of tools material and wear rate is influenced also by soil mineral particle contents. Mechanical properties of steel can be affected by a technology of heat treatment that it leads to a different microstructures. Experimental work how to do it is very expensive and thanks to numerical methods like FEM we can assumed microstructure at low cost but each of numerical model is necessary to be verified. The aim of this work has shown a procedure of prediction microstructure of steel for agricultural tools. The material characterizations of 51CrV4 grade steel were used for numerical simulation like TTT diagram, heat capacity, heat conduction and other physical properties of material. A relationship between predicted microstructure by FEM and real microstructure after heat treatment shows a good correlation.

## 1. Introduction

Chisels, plows, tines, discs are agricultural tools that are used for soil cultivation around the world[1, 2].Agricultural tools are in direct contact with the soil. Contacting tools with soil will cause the abrasive wear mechanism. Abrasive wear causes undesirable material loss until the time when the tool is unable to perform a required function or if its destruction[3–5].The resulting microstructure is important for the abrasion resistance but also other mechanical properties such as hardness, toughness, strength[6–8].For these reasons, manufacturers of agricultural tools focus on optimizing the heat treatment process[9, 10].

Optimizing heat treatment means accurate design and adherence to temperatures and times. Compliance times and temperatures is provided by adjusting the production line[11, 12].It is important to focus on the exact design of the heat treatment. Agricultural tools are produced by austempering[13, 14].Experimental generation of austempering data is time-consuming and requires high costs.[15]. FEM models are used for the exact design of austempering. FEM models allow prediction of microstructure after heat treatment[16, 17].The construction of FEM models must include the effects of heat treatment such as heat transfer, nonlinear material properties, material shape, phase transformation[18]. The kinetics of austenite decomposition are important for describing the heat treatment process.Johnson-Mehl-Avrami-Kolgomorovequations are used to solve this problem for

diffusion transformations[19]. Koistinen-Marburgerequations are used to solve this problem for non-diffusion transformations[20].
The aim of this work is to design an algorithm that describes microstructure phases after heat treatment of 51CrV4 steel.The mathematical model is verified by experimental measurement.

## 2. Material and methods

Steel 51CrV4 (C=0.53, Mn=0.89, Si=0.26, Cr=1.02, Cu=0.13, V=0.12, P=0.012, S=0.025, Ni=0.08, Mo=0.02 (wt. %)) was used for modelling and experimental verification. The chemical composition was subtracted from the material sheet. Samples were prepared from steel Ø 30 mm rod. Sample sizes were adjusted to 25 mm x 10 mm x 45 mm. All samples were heated to 800 °C for 30 min. The cooling overview is shown in Table 1. After that samples were cut, grinded, polished and etched in Nital solution for metallography. Optical metallography and electron scanning metallography were used for a determination of phases into steel specimens.

**Table 1.** Overview of sample cooling

| | cooling 1 | | | cooling 2 | | | cooling 3 | | |
|---|---|---|---|---|---|---|---|---|---|
| | temp. [°C] | Medium [-] | Time [s] | temp. [°C] | Medium [-] | time [s] | temp. [°C] | Medium [-] | Time [s] |
| **V1** | 300 | salt bath | 40 | 300 | air | 1000 | 20 | air | to 20°C |
| **V2** | 300 | salt bath | 40 | 20 | water | to 20°C | - | - | - |
| **V3** | 400 | salt bath | 40 | 400 | air | 1200 | 20 | air | to 20°C |
| **V4** | 400 | salt bath | 40 | 20 | water | to 20°C | - | - | - |

The simulation algorithm was designed to predict the final microstructure after heat treatment – see Figure 1. The ElmerFem program was used for algorithm calculations. The Flowchart is shown in Figure 1 for the algorithm.

Dimensions 25 mm x 10 mm were selected for both experiment and model. The mesh was formed in the core of the object 12.5 mm and under the surface 22 mm x 5 mm. The boundary conditions are taken from the experimental part for heat treatment – see Table 1. Specific heat capacity, thermal conductivity was taken from the article Chotěborský and Linda [21].

Step $\Delta t$ was chosen to solve the simulation. The **.vtu* file was created at each step $\Delta t$. The temperature change was stored in each part of the modeled sample at each step$\Delta t$. "*M*"files present the number of simulation steps.

The transformation of austenite to martensite is a non-diffusional process. The non-diffusional process is calculated according toKoistinen–Marburger law – see equation 1 [22]. The changes from austenite to bainite, ferite and perliteare diffusional processes. Diffusional processes are calculated according to Johnson–Mehl–Avrami–Koglomorovequation – see equation 2 [20].

$$V_M = 1 - e^{-\alpha \times (M_s - T)} \quad (1)$$

$$V_{P,B,F} = 1 - e^{-K \times t^n} \quad (2)$$

where $M_S$– martensite start [s], $T$ – temperature [K], $t$ – time [s], $K$ – overall rate constant [-],
$n$ – Avrami's exponent [-]

The calculations were made for the volume ratio of bainite$Vb$, martensite$Vm$, ferrite $Vf$, perlite $Vp$ in each step$\Delta t$according equations 3 to 6[21].

$$Vf = \sum_{i=1}^{n} -Kf \times Nf \times t^{Nf-1} \times e^{-Kf \times t^{Nf}} \times (1 - Vf_{max}) \quad (3)$$

$$Vp = \sum_{i=1}^{n} -Kp \times Np \times t^{Np-1} \times e^{-Kp \times t^{Np}} \times (1 - Vf) \quad (4)$$

$$Vb = \sum_{i=1}^{n} -Kb \times Nb \times t^{Nb-1} \times e^{-Kb \times t^{Nb}} \times (1 - Vf - Vp) \quad (5)$$

$$Vm = 1 - e^{-\beta \times (Tm_{start} - T)} \times (1 - Vf - Vp - Vb) \quad (6)$$

where *Kf, Kp, Kb* – overall rate constant of feritic, pearlitic and bainitic transformation that generally depends on temperature (-), *Nf, Np, Nb* – Avrami's exponent for feritic, pearlitic and bainitic transformation that depends on temperature (-)

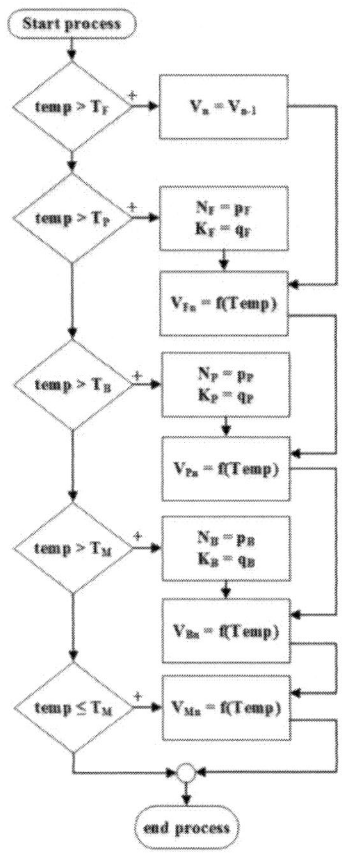


**Figure 1.** Flow chart for calculation of microstructure and hardness of steel after heat treatment

Conditions for formation of individual phases were taken from the TTT diagram for 51CrV4 steel. Parameters*Ni* and*Ki*are set forth in Kesneret al. [9]. The results were stored at a distance of 0.05 mm from the core to the edge of the sample.

## 3. Results and discussion

Bainite and martensite are only microstructures which have been identified in the results of experimental measurement and a mathematical model. The volume phases of perlite, ferrite, austenite were less representation than 0.01.For this reason, in this work only works with bainite and martensite in the next parts of this study.

Samples V1 and V3 wereaustempering in salt bath with a holding time in the air. Experimental measurements have shown bainitic structure with a very small amount of martensite for austempering– see Fig. 1.left. Sample V1 had a volume fraction 0.99 for bainite and 0.01 for martensite.Sample V3 had avolume fraction 0.98 for bainite and 0.02 for martensite.

Samples V2 and V4 were quenched in salt bath and then in water. The combination of bainite and martensite has been identified – see Fig. 1.right.A volume fraction of 0.67 martensite and 0.33 bainite was found for sample V2.A volume fraction of 0.76 martensite and 0.24 bainite was found for sample V4.The microstructure is shown in Figure 1 for the measured and calculated microstructure after heat treatment.

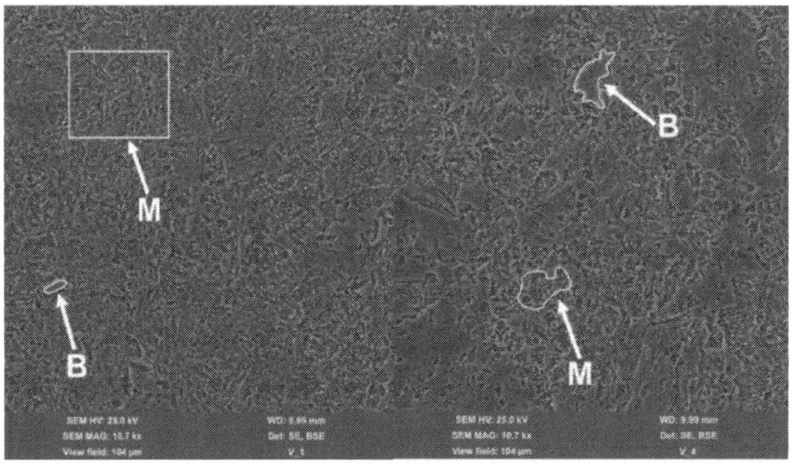


**Figure 2.** Microstructure with banite and martensite phases for steel51CrV4 –left for sample V1, rigth for sample V4

The comparison between the mathematical model and the experiment is shown in Fig. 3.for all samples. Samples V1 and V3 indicate good agreement between the mathematical model and experimental measurements. The results were displayed from the core to the edge of the sample at 0.2 mm for better clarity. The mathematical model indicates a volume fraction of 0.99 of bainite and 0.01 of martensite for samples V1 and V3. The difference between the mathematical model and the experimental measurement is the same for the sampleV1. The difference of 0.01 was found between the mathematical model and the experimental measurement.

Differences 0.26 for sample V2 and 0.16 for sample V4 were found between the mathematical model and the experimental measurement for the bainite and martensite phase. The cooling time was fast for these samples and this is one of the possible reasons for differences for the individual phases. Chemical composition can be another reason for high differences. The chemical composition has been taken from the material sheet into this work, and its batch may contain minor differences in some chemical elements. More heat treatment parameters could be used to refine the procedure in this work.

Difference was not found in the distribution of the individual phases from the core of the sample to its edge. Simsir and Gür[18]in their work dealt with the microstructure distribution. Their results show significant changes in the composition of the microstructure from the core of the sample to its edge during heat treatment in the salt baths and cooling in water. A larger sample size (ø60 mm) was used in their work. This may be one of the reasons why different microstructure across the sample is recorded in their work.

Chotěborský and Linda in their work were concerned with estimating the microstructure for the agricultural instrument. Differences are found between the mathematical model and the experimental measurement of phase representation in individual microstructures. Different values of specific heat capacity, heat flow, TTT diagram are given as possible causes. Different values may be due to different actual chemical composition of steel. This issue is the same for this work.

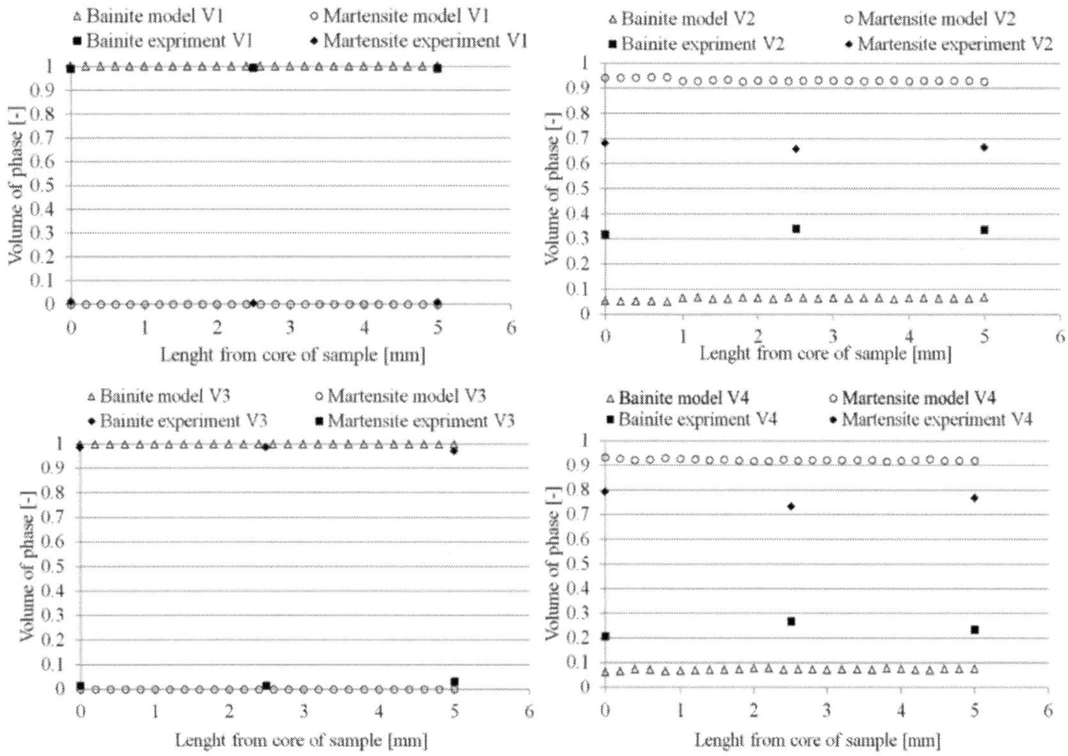


**Figure 3.** Dependency between volume of phase and length from core of sample

## 4. Conclusion

The mathematical model was assembled and experimentally verified for the detection of the microstructure of 51CrV4 steel. The results show a match for samples hardened in salt and then to the water. The resulting structure shows a bainitic structure with very small volume phase of martensite. The bainite structure is suitable for use in agricultural tools because it has good abrasion resistance. Samples hardened in a salt bath with a holding time in air showed 25% difference between the mathematical model and experimental measurements. These differences may be due to the different chemical composition of the steel. Chemical composition of steel affects boundary conditions such as heat capacity, heat flow and the values of the start and end of each phase during cooling. However, the above-mentioned procedure can be used to estimate the final microstructure composition after heat treatment.

## References

[1] Votava J 2014 Usage of abrasion-resistant materials in agriculture*Journal of Central European Agriculture***15(**2), pp. 119–128.

[2] Rose D C, Sutherland W J, Parker C, Lobley M, Winter M, Morris C, Twining S, Ffoulkes CH, Amano T and Dicks L V 2016 Decision support tools for agriculture: Towards effective design and delivery*Agricultural Systems***149**, pp. 165–174.

[3] Bansal A, Zafar S and Sharma A K 2015 Microstructure and Abrasive Wear Performance of Ni-Wc Composite Microwave Clad. *Journal of Materials Engineering and Performance***24**(10), pp. 3708–3716.

[4] Elalem K and Li D Y 1999 Dynamical simulation of an abrasive wear process. *Journal of Computer-Aided Materials Design***6**(2), pp. 185–193.

[5] Chotěborský R, Hrabě P, Müller M, Savková J, Jirka M and Navrátilová M 2009 Effect of abrasive particle size on abrasive wear of hardfacing alloys. *Research in Agricultural*

*Engineering***55**(3).

[6] Wei M, Wang S, Wang L and Chen K 2011 Effect of Microstructures on Elevated-Temperature Wear Resistance of a Hot Working Die Steel. *Journal of Iron and Steel Research, International***18**(10), pp. 47–53.

[7] Weber S, Li J R and Theisen W 2010 Microstructure and wear properties of novel sintered cold work steel and related particle reinforced composite materials. *Materials Science and Technology***26**(12), pp. 1494–1502.

[8] Buchely M F, Gutierrez J C, León L M and Toro A 2005 The effect of microstructure on abrasive wear of hardfacing alloys. *Wear***259**(1–6), pp. 52–61.

[9] Kešner A, Chotěborský R and Linda M 2016 A numerical simulation of steel quenching. In : *Proceeding of 6th international conference on trends in agricultural engineering.*

[10] Votava J, Kumbár V and Polcar A 2016 Optimisation of Heat Treatment for Steel Stressed by Abrasive Erosive Degradation. *Acta Universitatis Agriculturae et Silviculturae Mendelianae Brunensis***64**(4), pp. 1267–1277.

[11] Müller M, Chotěborský R, Valášek P and Hloch S 2013 Unusual possibility of wear resistance increase research in the sphere of soil cultivation | Posebna mogućnost istraživanja povećanja otpornosti na trošenje u području obrade tla. *Tehnicki Vjesnik***20**(4).

[12] Archer D W and Reicosky D C 2009 Economic performance of alternative tillage systems in the northern corn belt. *Agronomy Journal***101**(2).

[13] Šolić S, Godec M, Schauperl Z and Donik Č 2016 Improvement in Abrasion Wear Resistance and Microstructural Changes with Deep Cryogenic Treatment of Austempered Ductile Cast Iron (ADI). *Metallurgical and Materials Transactions A***47**(10), pp. 5058–5070.

[14] Erdogan M, Kilicli V and Demir B 2007 Transformation characteristics of ductile iron austempered from intercritical austenitizing temperature ranges. *Journal of Materials Science***44**(5), pp. 1394–1403.

[15] Sinha V K, Prasad R S, Mandal A and Maity J 2007 A Mathematical Model to Predict Microstructure of Heat-Treated Steel. *Journal of Materials Engineering and Performance***16**(4), pp. 461–469.

[16] Deng D 2009 FEM prediction of welding residual stress and distortion in carbon steel considering phase transformation effects. *Materials & Design***30**(2), pp. 359–366.

[17] Kešner A, Chotěborský R and Linda M 2017 Determining the specific heat capacity and thermal conductivity for adjusting boundary conditions of FEM model. *Agronomy Research.* **15**, Special Is.

[18] Şimşir C and Gür C H 2008 A FEM based framework for simulation of thermal treatments: Application to steel quenching. *Computational Materials Science***44**(2), pp. 588–600.

[19] Martin D 2010 Application of Kolmogorov–Johnson–Mehl–Avrami equations to non-isothermal conditions. *Computational Materials Science***47**(3), pp. 796–800.

[20] Babu K and Prasanna Kumar T S 2014 Comparison of Austenite Decomposition Models During Finite Element Simulation of Water Quenching and Air Cooling of AISI 4140 Steel. *Metallurgical and Materials Transactions B***45**(4), pp. 1530–1544.

[21] Chotěborský R and Linda M 2015 FEM based numerical simulation for heat treatment of the agricultural tools. *Agronomy Research***13**(3),pp. 629–638.

[22] Pietrzyk M, Kusiak J, Kuziak R, Madej Ł, Szeliga D and Gołąb R. Conventional and Multiscale Modeling of Microstructure Evolution During Laminar Cooling of DP Steel Strips. *Metallurgical and Materials Transactions A***45**(13), pp. 5835–5851.

## Acknowledgments

Supported by Internal grant 31140/1312/3114 agency of Faculty of Engineering, Czech University of Life Sciences in Prague.

# Evapotranspiration and crop coefficient of oil palm (*Elaeis guineensis* Jacq.) on the main nursery in a greenhouse

**R Sigalingging, Sumono and N Rahmansyah**

Department of Agricultural Engineering, Universitas Sumatera Utara, Prof. A. Sofyan No. 3 Kampus USU, Medan, Indonesia Republic-20155.

E-mail: riswanti@usu.ac.id or riswanti.rs@gmail.com

**Abstract.** The estimation of crop water requirement is an important part of oil palm plantation because fruit yield of oil palm can be affected by water stress. Evapotranspiration and crop coefficient of oil palm using Tenera variety at 7-12 months old was determined. Soil texture was sandy loam with 73.8 % sand, 10.8 % silt, 15.77 % clay and 1.41 % organic matter. The results showed that the oil palm getting older decreased significantly in bulk density, particle density and porosity of soil caused the root of oil palm enlarged (19.42 g to 53.37 g). This was indicated by increased the dry root weight. On the other hand, the value of evapotranspiration and crop coefficient increased significantly, that was 1.85 to 2.00 mm/day and 0.8 to 0.87 respectively.

## 1. Introduction

Nursery is the first step which determines for the successful planting of oil palm. The nursery stage requires big amounts of water and it is therefore important to choose a location with a sufficient water source for the nursery [1][2]. Crop water requirements are closely related to the rate of evapotranspiration. Evapotranspiration process is combination of two separate processes i.e evaporation and transpiration process. Water is lost from the soil surface by evaporation which mainly determined by the fraction of solar radiation approaching the soil surface. This fraction decreases over the growing period as the crop develops and the crop canopy shades more and more of the ground area. On the other hand, water is lost from the crop by transpiration which is the vaporization of liquid water contained in plant tissues and the vapour removal to the atmosphere through stomata. When the crop is small, water is predominately lost by soil evaporation, but once the crop is well developed and completely covers the soil, transpiration becomes the main process. During the early stages of the crop a high population planting would normally require somewhat more water than low density planting due to quicker development of full ground cover. In irrigated agriculture plant population has been considered of little importance in terms of total water needs.

The rate of evapotranspiration is not easy to measure. Specific devices and accurate measurements of various physical parameters or the soil water balance in lysimeters are required to determine evapotranspiration. The methods are often expensive, demanding in terms of accuracy of measurement and can only be fully exploited by well-trained research personnel. Although the methods are inappropriate for routine measurements, they remain important for the evaluation of evapotranspiration estimates obtained by more indirect methods. To account for the effect of the crop characteristics on crop water requirements, crop coefficients ($K_c$) are presented to relate $ET_o$ to crop evapotranspiration ($ET_c$). The $K_c$ value relates to evapotranspiration of a disease-free crop grown in large fields under optimum soil water and fertility conditions and achieving full production potential under the given growing environment. This study aims to determined crop evapotranspiration and crop coefficient of oil palm with Pan Method.

## 2. Materials and Methods

### *2.1 Experimental design*

The experiment was conducted in a greenhouse located in the experimental area of Agricultural Faculty, Universitas Sumatera Utara, Medan, Indonesia. Soil (Entisols type) was of sandy loam texture with percentages of sand, silt and clay of 73.8, 10.43 and 15.77 %, respectively, with 1.406 % soil organic matter, in polybag 21 cm diameter, 30 cm height and 22 cm soil height. Thirty-five polybag oil palm Tenera variety at 7 months old from Indonesian Oil Palm Research Institute (IOPRI) was used. The water requirements of the plants were filled by providing irrigation water on each plant manually with the same water volume that aims to fulfil the field capacity of the soil (36.79±1.54 % soil water content) and evapotranspiration. Provision of water is done periodically in accordance with the needs of water plants. Percolation (mm day$^{-1}$) was calculated by the ratio of water height (mm) to water moving time (day) for knowing water moving downward and through the soil horizons by the gravitational process for each month. Also, the wet roots weight (g) and dry roots weight (g) were determined as an indicator of water in the roots.

### *2.2 Calculation of crop evapotranspiration rate (ET)*

Crop evapotranspiration rate ($ET_c$, mm day$^{-1}$) was determined according to volumetric soil moisture (Eq. 1), where θ represent volumetric soil moisture (%), h is soil depth (cm) and t is time (day).

$$ET_c = \frac{1}{2}\frac{\theta h}{t} \quad (1)$$

Volumetric soil moisture (θ, %) was calculated using Eq. 2, where $M_c$ represent soil water content in dry basis (%), $\rho_s$ is bulk density of soil (g cm$^{-3}$) and $\rho_w$ is water density (1 gr cm$^{-3}$). The soil water content was determined by the conventional method using a standard hot air oven with a temperature setting 105 °C and a drying time of 24 h. On the other hand, soil bulk density ($\rho_s$, g cm$^{-3}$) was calculated by the ratio of dry soil weight (g) to soil volume (ring volume, 100.08 cm$^{-3}$). The experiment was repeated three times.

$$\theta = \frac{M_c \rho_s}{\rho_w} \quad (2)$$

### *2.3 Calculation of crop coefficient $K_c$*

Crop coefficient ($K_c$, -) is defined as the ratio of the crop evapotranspiration to the reference evapotranspiration ($ET_o$, mm day$^{-1}$) according to pan evaporation method with class A type as described by Eq. 3. The average of reference crop evapotranspiration ($ET_o$) was 2.3 mm day$^{-1}$. This value was determined according to potential evaporation with pan evaporation method which was measured every day under 29.29 to 29.87 °C as a daily temperature. Every month crop evapotranspiration and crop coefficient were calculated in six months.

$$K_c = \frac{ET_c}{ET_o} \quad (3)$$

## 3. Results and *discussion*

The average of soil water content ($M_c$,% in d.b), soil bulk density (g $cm^{-3}$), crop evapotranspiration ($ET_c$, mm $day^{-1}$) and crop coefficient ($K_c$, -) for each month were determined as shown at Table 1. Based on Table 1 that soil moisture content increase with growing age of oil palm due to increasing water requirement as shown by crop evapotranspiration. Oil palm. A plant has different water needs at different stages of growth, crop type and evaporative demand [3-4]. Young plant requires less water than when it is in the reproductive stage and water needs drops when approaches maturity. For good plant growth, a soil must have adequate room for water and air movement, and for root growth [4]. According to [5] the effect of soil water content on evapotranspiration varies with crop and is conditioned primarily by type of soils and water holding characteristics, crop rooting characteristics and the meteorological factors determining the level of transpiration. Entisols soil type with sandy loam texture was used in this study. Sandy loam soils are capable of quickly draining excess water but can not hold significant amounts of water or nutrients for crops. Crops grown in this type of soil will require more frequent irrigation and fertilization than soils with a higher concentration of clay and sediment.

**Table 1.** The values of soil water content, soil bulk density, crop evapotranspiration and crop coefficient in different age of oil palm

| Age of oil palm (Month) | Average of soil water content (%) | Average soil bulk density (g $cm^{-3}$) | Crop Evapotranspiration (mm $day^{-1}$) | Crop coefficient (-) |
|---|---|---|---|---|
| 7 | 11.82 | 0.998 | 1.85 | 0.80 |
| 8 | 11.98 | 0.994 | 1.87 | 0.81 |
| 9 | 12.32 | 0.989 | 1.01 | 0.83 |
| 10 | 12.62 | 0.978 | 1.94 | 0.84 |
| 11 | 12.95 | 0.968 | 1.96 | 0.85 |
| 12 | 13.33 | 0.959 | 2.00 | 0.87 |

$ET_c$ will be reduced if the rate of water supply to the roots is unable to cope with transpiration losses. This will be more pronounced in heavy textured than in light textured soils. Since reduction in evapotranspiration affects crop growth and/or crop yields, timing and magnitude of reduction in ET crop are important criteria for irrigation practices. Following an irrigation the crop will transpire at the predicted rate during the days immediately following irrigation. With time the soils become drier and the rate will decrease, more so under high as compared to low evaporative conditions. Whether or not the reduction in ETc is permissible during part or whole growing season can be determined only when the effect of soil water stress on yield during various stages of growth is known. The predicted ETc values should be applied unless specific objectives are pursued such as assuring that the greatest number of farmers benefit from irrigation or maximising yield per unit of water when available water supply is the limiting factor. However, available soil water has a balancing effect in meeting short duration, high ETc values; this effect is smaller for shallow, light soils than for deep, fine textured soils. Available soil water should therefore be considered. This calculation is usually done for months of peak water use.

Table 2 shows that percolation decreased with growing age of oil palm while wet and dry weight increased due to amount of water in root increased. Difference of percolation related to porosity of soil. Growing age of oil palm porosity decreases because of rootzone increases indicating with dry root weight increased (Table 2). This affects to the power of passing water to be smaller and the ability of soil to hold water stronger so that percolation is reduced. The big difference in percolation is also due to the water requirement of plants at the age of 12 months is greater due to larger plant size. So giving the same amount of water results in less unused water and is wasted through the percolation process. Water delivery should be given according to the needs of plant evapotranspiration, which is determined at the rate of growth and weather conditions.

**Table 2.** The values of percolation, wet root weight, dry root weight and amount of water in the root in different age oil palm

| Age of oil palm (Month) | Percolation (mm day$^{-1}$) | Average of wet root weight (g) | Average dry root weight (g) | Amount of water in the root (g) |
|---|---|---|---|---|
| 7 | 1.8 | 85.69 | 19.42 | 66.27 |
| 8 | 1.54 | 99.49 | 25.96 | 73.53 |
| 9 | 1.23 | 129.21 | 36.38 | 92.83 |
| 10 | 1.04 | 137.90 | 41.33 | 96.57 |
| 11 | 0.83 | 143.89 | 45.33 | 98.56 |
| 12 | 0.54 | 190.36 | 63.37 | 126.99 |

## 4. Conclusions
The oil palm getting older decreased significantly in bulk density, particle density and porosity of soil caused the root of oil palm enlarged (19.42 g to 53.37 g). This was indicated by increased the dry root weight. On the other hand, the value of evapotranspiration and crop coefficient increased significantly, that was 1.85 to 2.00 mm/day and 0.8 to 0.87 respectively. The predicted ETc values should be applied unless specific objectives are pursued such as assuring that the greatest number of farmers benefit from irrigation or maximising yield per unit of water when available water supply is the limiting factor.

## References
[1] Carr M K V 2011 *The Water Relations and Irrigation Requirements of Oil Palm (Elaseis guineensis):A Review* Expl Agric. **47** 629-52.
[3] Rogers D H, Aguilar J, KisekkaI, Barnes P L and Lamm F R 2014 *Irrigation Management Series: Soil, Water, and Plant Relationships* (Kansas State University Press) pp. 1-7 https://www.bookstore.ksre.ksu.edu/pubs/L904.pdf
[4] Simonne E H and Dukes M D 2010 *Vegetable Production Handbook:Principles and Practices of Irrigation Management for Vegetables* (University of Florida Press) chapter 3 pp. 17-23 http://hos.ufl.edu/vegetarian/09/Apr/VPH%202009-2010/Chap%203.pdf
[5] FAO 1992 *Crop Water Requirements* (Rome, Food and Agriculture Organization of United Nation) pp 1-144

## Acknowledgments
The research was funded by the Research Central of Universitas Sumatera Utara, Medan for Internal Grand Research (TALENTA 2017 Number: 5338/UN5.1.R/PPM/2017 at 22nd May 2017)

# Mechanical behaviour of Arabica coffee (*Coffea arabica*) beans under loading compression

**R Sigalingging[1], D Herak[2], A Kabutey[2] and C Sigalingging[3]**

[1]Department of Agricultural Engineering, Universitas Sumatera Utara, Prof. A. Sofyan No. 3 Kampus USU, Medan, Indonesia Republic-20155.
[2] Department of Mechanical Engineering, Czech University of Life Sciences Prague, Kamycka 129, Praha 6, Czech Republic.
[3]Department of Food Sciences, Universitas Sumatera Utara, Prof. A. Sofyan No. 3 Kampus USU, Medan, Indonesia Republic-20155.
E-mail: riswanti@usu.ac.id or riswanti.rs@gmail.com

**Abstract.** The uniformity of the product of the grinding process depends on various factors including the brittleness of the roasted coffee bean and it affects the extraction of soluble solids to obtain the coffee brew. Therefore, the reaching of a certain degree of brittleness is very important for the grinding to which coffee beans have to be subjected to before brewing. The aims of this study to show the mechanical behaviour of Arabica coffee beans from Tobasa (Indonesia) with roasted using different roasting time (40, 60 and 80 minutes at temperature 174 °C) under loading compression 225 kN. Universal compression testing machine was used with pressing vessel diameter 60 mm and compression speed 10 mm min$^{-1}$ with different initial pressing height ranging from 20 to 60 mm. The results showed that significant correlation between roasting time and the brittleness.

## 1. Introduction

The importance of economical production of crops seeds such as coffee and cocoa beans as base materials of their technological processing are ever increasing. The behaviour of most crops seeds deviates essentially from that of the generally known elastic materials. Therefore, the mechanics of agricultural materials, as a scientific discipline, is still being developed at present, and in many cases has no exact methods as yet. However, the methods developed so far can already be utilized successfully for designing and optimizing machines and technological processes. By knowing mechanical behaviour of coffee in different types and condition can be increased on energy efficiency during processing.

A good quality cup of coffee is depended on many factors, such as the quality of green beans, the roasting conditions, the time since the beans are roasted, and the type of water used for brewing. The pleasant aromas released during the grinding of roasted coffee beans are as attractive as the aromas of fresh brewed coffee [2]. More than 800 volatile compounds have been identified in roasted coffee, whereof around 30 compounds are responsible for the main impression of coffee aroma [1, 3]. Coffee flavour is directly related to the volatiles compounds produced such as acids, alcohols, aldehydes, esters, furan, ketones, lactones, phenolic compounds and pyrazines. These compounds are responsible for the coffee aroma and also called key odorants [4]. The development of flavour compounds is influenced by temperature and time of roasting.

However, the uniformity of the product of the grinding process depends on various factors including the brittleness of the roasted coffee bean and it affects the extraction of soluble solids to obtain the coffee brew (Clo and Voilley, 1983; Clarke and Macrae, 1987). The changes in the textural and mechanical characteristics to which the coffee bean undergo during roasting play a relevant role in the quality of the roasted beans, these properties have been scarcely studied until now. The grinding of the coffee represents a loading process which is dynamic rather than static. It means the mechanical properties like crushing strength should be evaluated at the dynamic loading [7]

The knowledge of the strength and toughness of the coffee beans is necessary in order to evaluate brittleness of coffee beans. The reaching of a certain degree of brittleness is very important for the grinding to which coffee beans have to be subjected to before brewing. This study aims to describe mechanical behaviour of Arabica coffee from Tobasa (Indonesia) with roasted using different roasting time.

## 2. Materials and Methods

Arabica coffee beans were obtained from Tobasa, North Sumatera, Indonesia. The roasting process was conventional process at different time (40, 60 and 80 minutes) under temperature 174 °C. The moisture content was determined using the procedure specified by [10]. The moisture content of beans for the compression test was determined by Equation (1) [11, 12]; where $M_C$ is the percentage moisture content on wet basis (% w.b.), $M_a$ is mass of coffee beans before heat-treatment (g) and $M_b$ is the mass of coffee beans after heat-treatment (g). A universal testing machine of type ZDM 50, Czech Republic together with a pressing vessel of diameter 60 mm with a plunger were used for recording the data points of force (N) and deformation (mm) of coffee beans of 20-60 mm measured pressing height [11]. A maximum force of 225 kN and speed of 10 mm/min were applied. The test was repeated three times. Minitab 16 statistical software (Minitab Inc, 2010) was used to analyse the statistical significance and correlation between parameters [13].

$$M_c = \left(\frac{M_a - M_b}{M_a}\right)100 \tag{1}$$

The deformation energy and oil yield were numerically determined based on the equations proposed by the authors [14].

## 3. Results and discussion

Initial and roasting moisture content of coffee beans and energy deformation were determined (Tables 1 and 2). Increase roasting time with moisture content decreased (Table 1) while energy deformation increased (Table 2). Deformation energy of coffee beans was affected by roasting time and biological characteristics of coffee beans due to the changing slope of deformation (Fig. 1). The values of these mechanical characteristics are dependent on the loading orientation. The orientation used to leads to the maximum of the breaking force.

**Table 1.** The values of moisture content of coffee beans

| | $M_a$ (g) | $M_b$ (g) | $M_c$ (w.b, %) |
|---|---|---|---|
| Initial Moisture Content | 100 | 87.23±0.46 | 12.77±0.46 |
| t1 = 40 min | 100 | 97.40±0.42 | 2.6±0.42 |
| t2 = 60 min | 100 | 97.70±0.14 | 2.30±0.14 |
| t3 = 80 min | 100 | 98.03±0.33 | 1.97±0.33 |

The dependence of the maximum of the loading (impact) force on the height h enables to evaluate the force at which the bean is fractured. The values of this static loading breaking force (strength) are plotted in the Fig. 1.

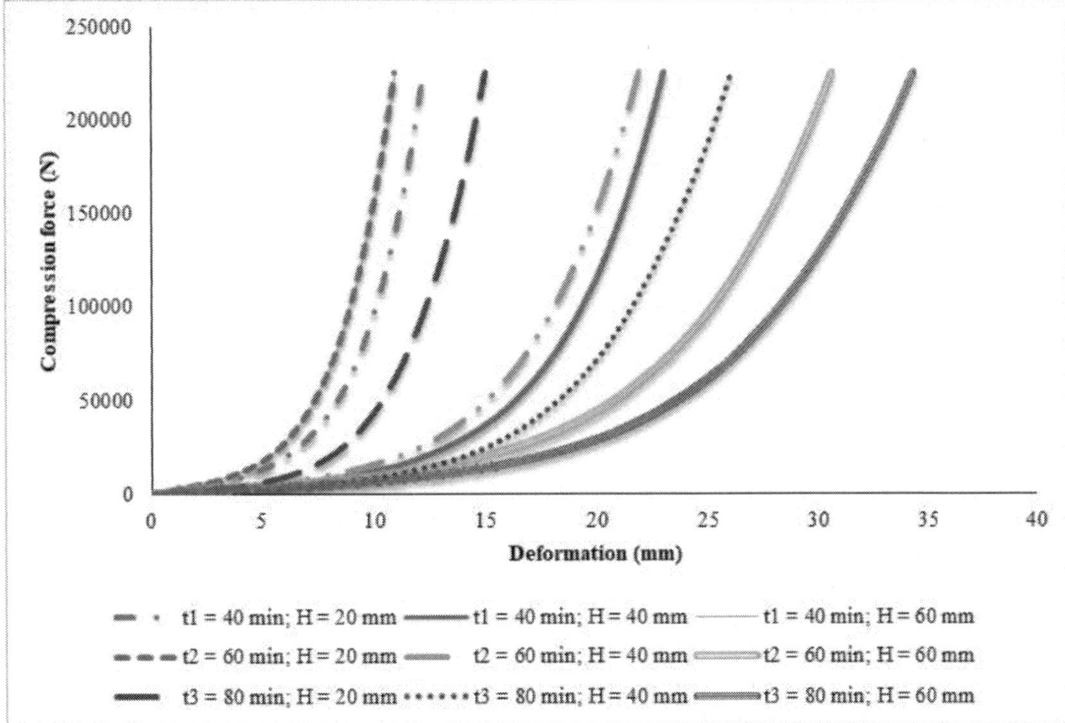


**Fig 1.** Mechanical behaviour of coffee at various height vessel press (H) with roasting time (t)

One can see that the coffee beans resistivity against to fracture is higher at the dynamic (impact) loading. The experimental points can be fitted by a linear function with a relatively good correlation. The increase the strength at transition from static to dynamic loading is observed for many technical materials including metals and their alloys, polymers etc [7]. This increase can be explained in terms of material structure. It means the explanation of this phenomenon in coffee beans must be based on the detail study of the microstructure of this material.

**Table 2.** The values of deformation and energy at various vessel press height and roasting time

| Time (min) | H (mm) | Deformation (mm) | Energy (J) |
|---|---|---|---|
| | 20 | 12.26 | 579.43 |
| t1=40 | 40 | 23.04 | 1007.87 |
| | 60 | 31.22 | 1418.87 |
| | 20 | 10.98 | 541.16 |
| t2=60 | 40 | 21.98 | 1032.10 |
| | 60 | 30.69 | 1477.33 |
| | 20 | 15.06 | 691.76 |
| t3=80 | 40 | 26.11 | 1169.99 |
| | 60 | 34.36 | 1606.62 |

## 4. Conclusions

Increase roasting time with moisture content decreased while energy deformation increased. Deformation energy of coffee beans was affected by roasting time and biological characteristics of coffee beans due to the changing slope of deformation. Correlation between roasting time and the brittleness was significantly

## References

[1] Baggenstoss J, Poisson L, Kaegi R, Perren R. And Escher F 2008 *Coffee Roasting and Aroma Formation: Application of Different Time temperature Conditions* Journal of Agricultural and Food Chemistry **56** (14) 5836-46

[2] Baggenstoss J, Thomann D, Perren R and Escher F 2010 *Aroma Recovery from Roasted Coffee by Wet Grinding* Journal of Food Science **75** (9) 697-702

[3] Ku Madihah K Y, Zaibunnisa A H, Norashikin S, Rozita O and Misnawi J 2013 *Optimization of roasting conditions for high-quality Arabica coffee* International Food Research Journal **20** (4) 1623-27

[4] Brohan M, Huybrighs T, Wouters, C. and Bruggen, B. V. 2009. Influence of storage conditions on aroma compounds in coffee pads using static headspace GC-MS. Food Chemistry 116: 480-483.

[5] Akimaya M, Murakami K, Ohtani N, Iwatsuki K, Sotoyama K, Wada A, Tokuno K, Iwabuchi H amd Tanaka K 2003 *Analysis of Volatile Compounds Released during the Grinding of Roasted Coffee Beans Using Solid-Phase Microextraction* J. Agric. Food Chem. **51** 1961-69

[6] Caprioli G, Cortese M, Sagratini G and Vittori S 2015 *The Influence of different types of preparation (espresso and brew) on coffee aroma and main bioactive constituents* Int. J. Food Sci. Nutr. 1-9

[7] Nedomova S, Trnka J, Stoklaso P and Buchar J 2013 *Strength of Coffee Beans Under Static and Dynamic Loading* ACTA University Agriculturae et Silviculturae Mendelianae Brunensis **LXI** (3) 743-49

[8] Uman E, Colonna-Dashwood M, Colonna-Dashwood L, Perger M, Klatt C, Leighton S, Miller B, Butler K T, Melot B C, Speirs R W and Hendon C H 2016 *The Effect of Bean Origin and Temperature on Grinding Roasted Coffee* Scientific Report **6** 1-7 https://www.nature.com/articles/srep24483

[9] Waters D M, Moroni A V and Arentd 2015 *Biocgemistry, Germination and Microflora Associated with Coffea Arabica and Coffea canephora Green Coffee Beans* Food Science and Nutrition

[10] ISI 1966 Indian standard methods for analysis of oilseeds IS:3579 New Delhi Indian Stand. Ins.

[11] Sigalingging R, Herak D, Kabutey A, Dajbych O, Hrabe P and Mizera C *Application of a tangent curve mathematical model for analysis of the mechanical behaviour of sunflower bulk seeds* Int. Agrophys **29**(4) 517–524.

[12] Kabutey A, Herák D, Chotěborský R, Sigalingging R, Mizera, Č 2015 *Effect of compression speed on energy requirement and oil yield of Jatropha curcas L. bulk seeds under linear compression* Biosystems Engineering **136** 8–13

[13] Minitab Inc (2010): Minitab® 16 Statistical Software, State College, Pa.

[14] Herak D, Kabutey A, Sedlacek A and Gurdil G 2012 Mechanical behaviour of several layers of selected plant seeds under compression loading Res. in Agric. Eng. **58**(1) 24-29

# The relation of sediment texture to macro- and microplastic abundance in intertidal zone

**H Wahyuningsih[1*], A P Bangun[2] and A Muhtadi[2]**

[1]Department of Biology, Faculty of Mathematics and Natural Sciences, Universitas Sumatera Utara, Jl. Bioteknologi No. 1, Medan 20155, North Sumatera, Indonesia, Tel./Fax. +661-821-3236,
[2]Department of Aquatic Resources Management, Faculty of Agriculture, Universitas Sumatera Utara, Jl. Prof A. Sogyan No. 3, Medan 20155, North Sumatera, Indonesia.
[*]Email: hestiw_ningsih@yahoo.co.id

**Abstract**. The intertidal zone is a waters area directly affected by the contamination of plastic debris from land and sea. The aim of this research were to analyze the relation of sediment texture to macro- and micro plastic abundance and also to determine appropriate management strategy. This research was conducted in intertidal zone Jaring Halus Village Langkat Regency North Sumatera Province on February-April 2017. Plastic debris was collected using quadrat transect. Sediment was collected with correct, up to a depth of least 30 cm. Abundance of micro plastic in Station 1 were positively tolerated with clay (0.509), and silt (0.787) and negatively correlations with sand (0.709) Station 2 were positively correlations with sand (0.645) and negatively correlations with clay (0.575), and silt (0.626) Station 3 were positively correlations with clay (0.435), and silt (0.466) and negatively correlations with sand (0.599). The abundance of microplastic was positively correlations with the abundance of microplastic (0.765). Microplastic density is directly proportional to the content of clay and dust. The higher the clay and dust content the higher the micro plastic density.

## 1. Introduction

Marine debris is any human waste, which is solid (state of matter, with fixed volume and shape) or matter that enters the marine environment either directly or indirectly [1, 2, 3]. Sea waste covers any manufactured form or processed material then discarded or abandoned in the marine environment. Sea waste consists of goods, food/snacks used by humans and then put into the sea, either intentionally or unintentionally, such as marine transport, drainage, and sewage or waste disposal systems by the wind [2, 3, 4]. The size category is used to classify marine debris, ie mega debris (> 100 mm), macro debris (> 20-100 mm), meso debris (> 5-20 mm), and micro debris (0.3-5 mm) [1, 2, 5].

One example of marine debris is plastic. Plastics is the largest contributor of marine waste in the world. Experts estimate more than 50% of marine waste comprises plastic [1, 2, 4, 6]. Chemically the plastic effect tends to increase the size of plastic particles (micro plastic) decreases, whereas the physical effect increases with increasing plastic size [5, 6].

Macroplastic can cover the surface of sediments and disrupt the movement of aquatic organisms [7, 8, 9] and can prevent the growth of mangrove seeds [8]. Macroplastic can also be a host for the growth of invasive species especially those attached to the substrate [7, 10]. Meanwhile, micro plastic provides a more dangerous threat than with micro plastic. This is because it can enter into the digestion and accumulate in the body of aquatic organisms [7, 9, 10, 11].

Plastic degradation occurs in all environments and is controlled by several environmental factors. The degraded plastic is due to photo-oxidation by ultraviolet (UV) light and is accelerated by high temperatures [12]. Thus, the plastics in the marine environment will be more stable than the soil surface due to the damping of temperature and light (UV) by sea water [13]. Additionally, the development of biofilms on plastic waste will minimize ultraviolet exposure on plastic surfaces that have the potential to slow down the degradation process [14]. Plastic wastes are found on the surface of the water and water columns as well as marine sediments in various places all over the world [4, 15, 16, 17].

Jaring Halus Village is one of the coastal settlements that have problems in waste management. Along with the increasing population in the area, many activities that occur in this village, such as fishing, fishery, agriculture, plantation, and household activities will all generate waste. The aim of this research were to analyze the relation of sediment texture to macro- and micro plastic abundance and also to determine appropriate management strategy.

## 2. Materials and Methods

### *2.1. Study site*

The research was conducted in JaringHalus Village, Langkat Regency, North Sumatra Province, Indonesia. Data were collected from February to April 2017. The tools used were Global Position System (GPS), meter, sample bottle, tool box, core, 1m × 1m board, filter, shovel, digital camera, and stationery.

### *2.2. Method of sediment sampling*

Substrate sampling was taken using a core. This process is done at low tide waters along with macro and microplastic sampling. Substrate sampling is done by immersing core based on depth stratification; 0-10, 10-20, 20-30 cm.

Examples of basic substrate waters obtained were analyzed to determine the composition (%) of clay, dust, sand, in the laboratory. The following are the substrate texture determining steps:

1. Determine the composition of each substrate fraction. For example 45% sand fraction, 30% dust and 25% clay.
2. Draw a straight line on the percentage of sand point 45% parallel to the percentage side of dust, then drawn a straight line on the percentage of dust at 30% parallel to the percentage of clay, and drag a straight line on the percentage of clay 25% parallel to the percentage of sand .
3. The point of intersection of the three lines will determine the type of substrate analyzed, for example, this is clay. For substrate analysis using Triangle The United States Department of Agriculture (USDA)

### *2.3. Method of macro- and microplastic survey*

The macro- and micro-plastic data used were taken from previous research [19].

### *2.4. Data analysis*

Correlation analysis between macro and microplastic to sediment texture using Pearson Correlation and Kruskal-Wallis with SPSS.

## 3. Results and Discussions

### *3.1. Quality of sediment*

The results of substrate texture analysis show that each station has a much different composition of the fraction of dust, clay, and sand. The highest sand fraction is in Depth 10 cm Plot 1 Station 2 of 93% and the lowest in Plot 2 Station 2 by 2%. Plot 2 Station 1 The depth of 10 cm has the highest dust fraction reaching 89% and the lowest at some observation point of 1%. The highest clay fraction composition was found in Plot 3 Station 3 The depth of 10 cm reached 50% and the lowest in Plot 2 Station 2 Depth 10 cm and Plot 3 Station 2 Depth 30 cm by 6%. The result of the measurement of the sediment quality parameter is the percentage of sediment fraction obtained from the observation station is presented in table 1.

The result of Kruskal-Wallis test shows that the percentage of clay, sand, and dust at depth of 10 cm, 20 cm and 30 cm in the intertidal zone is not significantly different. The average percentage of clay, sand, and dust at depths of 10 cm, 20 cm, and 30 cm were 36.59 - 44.83, 39.91 - 42.24 and 39.00 - 43.87. This proves that the percentage of the sedimentary fraction does not differ with increasing depth in each observation plot. The result of Kruskal-Wallis test shows that the percentage of clay, sand, and dust at the upper, middle and lower limit in the intertidal zone is significantly different. At the upper limit of clay, sand and dust have average percentage of 30.63%, 58.37%, and 19.37%, respectively. At the center boundary of clay, sand and dust have an average percentage of 51.80%, 31.19%, and 44.43% respectively. At the lower boundary of clay, sand and dust have an average percentage of 40.57%, 33.44%, and 59.20%, respectively. The average percentage of highest clay in the center (plot 2), upper sand (plot 1) and dust at the bottom (plot 3).

The difference of the sedimentary fraction in the intertidal zone is thought to be influenced by the presence of sediment particles carried by ocean currents and rivers (estuaries). The process of sediment formation in coastal waters other than influenced by wave forces is also determined by artificial activities (humans) in the mainland [20]. The artificial influence around the coast affects the distribution of sedimentary fractions because this activity suppliers poorly sorted sediment. The sedimentary fraction is affected by relatively strong tidal waves and currents [21].

**Table 1.** Quality of sediment

| Parameter | Station 1 | | | Station 2 | | | Station 3 | | |
|---|---|---|---|---|---|---|---|---|---|
| | 1* | 2* | 3* | 1* | 2* | 3* | 1* | 2* | 3* |
| % Clay | 21.31 | 8.87 | 20.82 | 7.28 | 24.33 | 15.12 | 35.44 | 24.91 | 32.28 |
| % Sand | 50.41 | 87.31 | 59.92 | 91.12 | 6.16 | 8.72 | 8.55 | 40.42 | 14.70 |
| % Silt | 28.50 | 3.87 | 19.26 | 1.57 | 69.36 | 76.09 | 57.46 | 36.05 | 53.24 |
| C-Organic | 1.57 | 2.13 | 1.88 | 1.56 | 2.79 | 1.88 | 3.12 | 2.99 | 2.71 |

***) Plot**

### 3.2. *Macro- and microplastic*

The total number of macroplastic collected from 3 stations divided into 27 observation plots is 308 types with total weight is 3689.87 g. The highest macroplastic density found in Station 2 ranged between 18.33-190.33 species $m^{-2}$ with weights ranging from 246.33 to 2103 gr. The lowest density found in Station 3 ranged from 3.33 to 11.67 species $m^{-2}$ with weights ranging from 13.46 to 117.67 gr. Density and macro and microplastic weights can be seen in table 2 and table 3 [19].

The highest microplastic density is found in Plot 1 of all observation stations which is the highest tidal boundary.Microplastic film type is the highest type found in all observation plots. The film comes from fragmentation of plastic waste such as packets of food and soft drinks, plastic bags and plastic wrapping fish commonly used in fishing activities. The film also has the lowest density so it is easily distributed by the presence of currents and tides [19]

**Table 2.** Abundance and Macroplastic Weights

| Station | Macroplastic Abundance (typem$^{-2}$) | | | Total | Macroplastic Weight (gr) | | | Total |
|---|---|---|---|---|---|---|---|---|
| | Plot 1 | Plot 2 | Plot 3 | | Plot 1 | Plot 2 | Plot 3 | |
| 1 | 32.67 | 9.67 | 5 | 47 | 393 | 115.67 | 65.67 | 574 |
| 2 | 190.33 | 31.67 | 18.33 | 240 | 2203 | 508 | 246.33 | 2957 |
| 3 | 11.67 | 5.33 | 3.33 | 20 | 113.67 | 31.08 | 13.46 | 158 |
| Total | 234.67 | 46.67 | 26.67 | 308 | 2709.67 | 654.75 | 325.46 | 3689.87 |

**Table 3.** The Percentage of Density and Average Density of Microplastic

| Station | Percentage of Microplastic Abundance | | | | Abundance Average (type/kg) |
|---|---|---|---|---|---|
| | Film | Fiber | Fragmen | Pellet | |
| 1 | 60.27 | 22.30 | 17.47 | 0 | 89 |
| 2 | 45.07 | 30.03 | 24.63 | 0.3 | 173.33 |
| 3 | 51.57 | 22.30 | 26.13 | 0 | 64 |
| Total | 52.30 | 24.88 | 22.74 | 0.1 | 326.33 |

*3.3 The Relation of Sediment Fraction with Microplastic*

The result of Pearson correlation test showed that the Stationplatz micro plastic density station 1 was positively correlated (0.509) with clay, negatively correlated (0.709) with sand and correlated positively (0.787) with dust. This suggests that microplastic density has a moderate relationship to clay and a strong relationship with dust. The higher percentage of clay and dust causes higher micro plastic density (table 4).

The microplastic density at Station 1 is strongly influenced by clay and dust proving that soft sediment is easier to become a micro plastic trap. In Station 2 the micro plastic density is negatively correlated (0.575) with clay, positively correlated (0.645) with sand and negatively correlated (0.625) with dust. The results of this test show differences with Station 1 where the micro plastic density has a negative correlation with sand. This happens because Station 2 is a community residential location where the highest plastic contamination pressure and micro plastic accumulate in the sediment layer.

In Station 3 the micro plastic density is positively correlated (0.435) with clay, negatively correlated (0.599) with sand and positively correlated (0.466) with dust. This suggests that microplastic density has a moderate relationship to clay and dust. The microplastic density of Stations 1 and 3 do not differ greatly in relation to clay, sand, and dust.

Microplastic density in the intertidal zone is influenced by two factors: The distance between the observation point and the percentage of the sedimentary fraction. The closer the distance to the main pollutant source causes higher micro plastic density. The higher percentage of clay and dust causes the sediment to soften and become a stronger trap against micro plastic. This is in accordance with the statement of [22] that soft sediments can more concurrently debris than rocky and gravel habitats [18, 23].

**Table 4.** Results of Pearson Correlation Test between Microplastic and Sediment Fraction

| Microplastic | Sediment Fraction | | |
|---|---|---|---|
| | Clay | Sand | Silt |
| 1 | 0.509 | -0.709 | 0.787 |
| 2 | -0.575 | 0.645 | -0.625 |
| 3 | 0.435 | -0.599 | 0.466 |

## 4. Conclusions

The coastal area of Jaring Halus Village, especially in the intertidal zone to obtain environmental pressure (pollution) of plastic in the form of macro- and micro plastic. The highest macro- and micro plastic densities are found in Station 2 which is a community residential area as well as the main source of plastic pollution. Microplastic density is directly proportional to the content of clay and dust. The higher the clay and dust content the higher the micro plastic density.

**References**

[1] United Nations Environment Programme 2011 UNEP Year Book Emerging Issues in Our Global Environment. UNEP. Nairobi.
[2] National Oceanic and Atmospheric Administration 2013 Programmatic Environmental Assessment (PEA) for the NOAA Marine Debris Program (MDP). NOAA. Maryland
[3] Engler 2012 The Complex Interaction between Marine Debris and Toxic Chemicals in Ocean. Office of Wetlands, Oceans, and Watersheds. U.S. Environmental Protection Agency. Pennsylvania Avenue. United States
[4] Galgani F*et al* 2015 Global Distribution, Composition and Abundance of Marine Litter. M. Bergmann et al. (eds.), Marine Anthropogenic Litter, DOI 10.1007/978-3-319-16510-3_2
[5] Hammer J et al 2012 Plastics in the Marine Environment: The Dark Side of a Modern Gift D.M. Whitacre (ed.), Reviews of Environmental Contamination and Toxicology. Springer Science + Business Media
[6] Thevenon F*et al* 2014 Situation Analysis Report. Gland, Switzerland: IUCN. 52 pp.
[7] Irwin Kimberly 2012 The Impacts of Marine Debris: A Review and Synthesis of Existing Research. Living Oceans Society December 2012
[8] Smith SDA and Markic A 2013 *Plos One*. **8** (12).
[9] National Oceanic and Atmospheric Administration Marine Debris Program 2016 Report on Marine Debris Impacts on Coastal and Benthic Habitats. Silver Spring, MD: NOAA
[10] Ogunola OS and Palanisami T 2016 *J Pollut Eff Cont* **4**: 161
[11] Eriksen M *et al* 2014 Plastic. *PLoS ONE*. **9** (12).
[12] Valadez-Gonzalez A *et al* 1999 *Polymer Degradation And Stability*. **63** (2): 253-260.
[13] Pegram JE and Andrady A L 1989 *Polymer Degradation and Stability*. **26** (1): 333–345.
[14] Artham T*et al* 2009 *International Biodeterioration & Biodegradation*. **63** (1): 884-890.
[15] Ryan PG*et al* 2009 *Philosophical Transactions of the Royal Society B*. **364** (1): 1999-2012.
[16] Browne MA 2011 *Environ. Sci. Technol*. **45** (1): 9175-9179.
[17] Claessens M et al 2011*Marine Pollution Bulletin*. **62** (1): 2199-2204.
[18] Coppock RL *et al* 2017 *Environmental Pollution*. **230** (2017) 829-837
[19] Bangun A P *et al* Impacts of Macro - and Microplastic on Macrozoobenthos Abundance in Intertidal Zone. International Confrence on Agriculture, Environtment,and Food security, Fakultas Pertanian USU 2017
[20] Efriyeldi 1999 *Natur Indonesia*. **2** (1): 85-92
[21] Rifardi 2008 Ekologi Sedimen Laut Modern UR-Press. Pekanbaru
[22] Mulyadi H *et al* 2015 *Jurnal Dinamika Lingkungan Indonesia* **2** (1): 26-31
[23] Watters DL *et al* 2010 *Marine Pollution Bulletin*. **60**: 131-138.

# Impacts of macro - and microplastic on macrozoobenthos abundance in intertidal zone

**A P Bangun[1]*, H Wahyuningsih[2], A Muhtadi[1]**

[1]Department of Aquatic Resources Management, Faculty of Agriculture, Universitas Sumatera Utara. Jl. Prof A. Sofyan No. 3, Medan 20155, North Sumatera, Indonesia. Tel./Fax. +661-821-3236,
[2]Department of Biology, Faculty of Mathematics and Natural Sciences, Universitas Sumatera Utara, Jl. Bioteknologi No. 1, Medan 20155, North Sumatera, Indonesia.
*E-mail: ariefpersadanta@ymail.com

**Abstract**. Plastics pollution in coastal areas is one of the topics that have received more attention over the past few years. The intertidal zone is a waters area that is directly affected by contamination of plastic waste from land and sea. The purpose of this study was to analyze the types and abundance of plastic waste in the intertidal zone and its impact on macrozoobenthos abundance. This research was conducted at Pesisir Desa Jaring Halus in February-April 2017. Macrozoobenthos and macro - micro plastic were collected by using quadratic transect. Sediments were collected with a core, to a depth of 30 cm. Microplastic and macroplastic abundances were analyzed using separation of sediment density and hand sorting. The dominant micro plastic types were film (52.30%), fiber (24.88%), fragments (22.74%), followed by pellets (0.1%). The total number of microplastics were 326,33 items and macro plastic were 308 items. Macroplastic abundance is positively correlated with microplastic (0.765). The abundance of macrozoobenthos is negatively correlated with microplastic abundance (-0.368) and with macro plastic abundance (-0.633). The management strategies were suggested clean up marine debris, decrease plastic using and built up the station of debris processing.

## 1. Introduction

Development is an inseparable part of human civilization. Life without development activity is a setback in human civilization. Often developments such as industrial and agricultural factories do not pay attention to the environmental aspect, thus destroying nature. The plastics industry is one of the industries that continue to grow along with the high use of plastics in society [1, 2]. The low public awareness of plastic waste management becomes one of the causes of the amount of plastic waste in the environment, especially in coastal waters.

Marine debris is a persistent solid, manufactured or processed by humans, directly or indirectly, intentionally or unintentionally, disposed of or abandoned in the marine environment. Type of marine waste including plastic, cloth, foam, styrofoam, glass, ceramics, metal, paper, rubber, and wood [3, 4]. The size category is used to classify marine debris, ie mega debris (> 100 mm), macro debris (> 20-100 mm), me so debris (> 5-20 mm), and micro debris (0.3-5 mm) [1, 3, 4].

Various problems arise due to marine debris, such as reduced coastal beauty, minims of various diseases, affecting food web, and reduced productivity of fish caught [3] [5] [6] [7] [8]. The potential effects of marine waste chemically tend to increase the size of plastic particles (micro plastic) decreases, while the physical effect increases with the increase in macro debris size [3]. Makrodebris provides physical impact such as closing the surface of the sediment and disrupting the movement of aquatic organisms [6, 9, 10] and can prevent the growth of mangrove seeds [9]. Macroplastic can also be a host for the growth of invasive species especially those attached to the substrate [6, 8].

Microplastic potentially threatens more seriously than large plastic materials as organisms that inhabit lower tropical levels, such as planktons that have particles susceptible to microplastic digestion processes, as a result, can affect high-level tropical organisms through bioaccumulation [6, 10, 11]. Microplastic is consumed by marine organisms when one of the micro plastic particles resembles food [8, 12, 13, 14].

The presence of macro and micro plastic in the environment especially the aquatic environment can have an impact on the existence of organisms. [7, 8] Macrozoobenthos is one of the several organisms that receive pressure from plastic waste contamination. Macrozoobenthos is often used to predict the imbalance of the physical, chemical, and biological environments of waters [15]. The polluted waters will affect the viability of macrozoobenthos organisms because macro-zoobenthos is a water biota that is easily affected by the presence of pollutants.

The phenomenon of marine waste in the form of plastic causes unrest in the community with the presence of garbage that has polluted coastal and marine areas, including the Village Jaring Halus Langkat. In addition, the absence of preliminary information about micro plastic in this region is one of the constraints to manage the potential of fisheries and marine-based environmentally friendly. Based on this, it is necessary to conduct a study to find out the macro plastic and microplastic distribution and its impact on macrozoobenthos in the Jaring Halus Village.

## 2. Materials and Methods

### 2.1. *Study site*

The research was conducted in Jaring Halus Village, Langkat Regency, North Sumatra Province, Indonesia. Data were collected from February to April 2017. The tools used were Global Position System (GPS), meter, sample bottle, tool box, core, 1m × 1m board, filter, shovel, digital camera, and stationery.

### 2.2. *Method of macroplastic survey*

The macrodebric samples (> 20mm) were collected with transects (1× 1 m) from each substation with the highest tidal boundaries and the lowest tidal boundaries divided into 3 sections [9]. The macrodebric composition is grouped into plastics, fabrics, foams, styrofoam, glass, metal, rubber, and wood. Samples are collected into sacks and labeled. The items (to further explain the flakes) in each macrodebric group are collected macroplastic, dried, calculated, and weighed. The parameters taken include the number of items (item $m^{-2}$) and weight (g $m^{-2}$) [16].

### 2.3. *Method of microplastic survey*

Sediment sampling (1L) was performed with core based on three stratified depths (0-30 cm). The core placement is performed on three sections (top, middle and bottom edge) at the substations at the highest tide and lows. The separation of microplastic particles (0.045-5 mm) from the sediments was carried out by several stages, namely (a) drying, (b) volume reduction, (c) separation of density, (d) filtration, and (e) visual sorting. Drying was done with 105 ° C oven for 72 hours. The dry-sediment volume reduction step was performed by filtration (5 mm in size) [17]. The density separation step is carried out by mixing the dry sediment sample (1 kg) and the saturated NaCl (3L) solution and the mixture are stirred for 2 minutes [18].

Floating plastics are polystyrene, polyethylene, and polypropylene. The filtration stage is carried out by filtering the supernatant (size 45 μm). Microplastic particles are visualized using a monocular microscope and grouped into four types, namely film, fiber, fragments, and pellets. The parameters taken were density (particle kg-1 dry sediment) [17].

*2.4. Sampling of macrozoobenthos*

A sampling of macrozoobenthos was done on the same plot/transect as macro and micro plastic retrieval. The macrozoobenthos samples were taken with the same transect as macro and micro plastic retrieval. Taking macro-zoobenthos is taken on each transect plot with a depth of 30 cm. All the substrate on the plot is removed with a shovel, then stored in a plastic bag for separation. The separation between macrozoobenthos and substrate is done in the field with the help of water and filter. Samples of macrozoobenthos that have been separated from the substrate are fed into a 70% alcohol-treated sample bottle to be identified in the laboratory.

*2.5. Data analysis*

The macroplastic abundance is calculated by the formula [4]:

$$K = \frac{n}{wxl} \tag{1}$$

Information:
K = Macroplastic Abundance
n = Number of Macroplastic (item / m3)
w = Width of Sampling Area
l = Length of Sampling Area

Microplastic abundance is calculated by the formula [4]:

$$K = \frac{n}{ax\,h} \tag{2}$$

Information:
K = Microplastic Abundance
N = Number of Microplastic (item / m3)
A = Sample Area of Sampling
H = Depth of Sampling

Biota Density of macrozoobenthos with formula [15]:

$$K = \frac{ni}{A} \tag{3}$$

Information :
K = Density
ni = Number of individuals of a type
A = Area Area

Correlation analysis between macro and microplastic to macrozoobenthos using Pearson Correlation SPSS.

## 3. Results and Discussions

*3.1. Macroplastic*

The total number of macro plastic collected from 3 stations divided into 27 observation plots were 308 items with total weight is 3689.87 gr. The highest macro plastic density found in Station 2 ranged between 18.33-190.33 species / $m^2$ with weights ranging from 246.33 to 2103 gr. The lowest density found in Station 3 ranged from 3.33 to 11.67 species / $m^2$ with weights ranging from 13.46 to 117.67 gr. Density and macro plastic weights can be seen in table 1.

Plastics is one of the most common types of waste found in various places both on land and in waters. The common marine waste category found in the Coastal Village of Jaring Halus is the macroplastic ie plastic waste with the size larger than 5 mm as many as 308 items (table 1). Hastuti (2014) found macro plastic density in Mangrove Ecosystem Pantai Indah Kapuk Jakarta as many as 6079 species. Some research results abroad have found that marine debris is dominated by macro plastic at 48% in Cassina Beach, Brazil [19], 67.6% Northeast Coast Brazil [20], 89% in Bootless Bay,

Papua New Guinea [21], 91 % In Midway, North Pacific [22], 68% in Monterey Beach, USA [23], 77% in Kaohsiung, Taiwan [24], 45.2% to 95% Eastern Mediterranean and Black Seas [25] and 60 5 to 80% In Gulf Coast of Guinea of Ghana [26]. The macro plastic proportion is dominant because its density is lower than glass, metal, and water density so it is easily transportable [1, 7, 25].

The highest macro plastic densities found at station 2 (near community settlements) ranged from 18.33 to 190.33 species $m^{-2}$. This is allegedly due to community activity that contributes most macro plastic compared to Station 1 (faced with Malacca Strait). Macroplastic density is influenced by the amount of marine debris carried from the sea such as fishing and shipping activities in the Malacca Strait and Station 3 (opposite the Ular River) where the macro plastic density is influenced by the amount of waste carried by the river. It also indicates that the station distance to the main pollutant source (settlement) effects macro plastic density.

**Table 1**. Abundance and Macroplastic Weights

| Station | Macroplastic Abundance ($item/m^2$) | | | Total | Macroplastic Weight (gr) | | | Total |
|---|---|---|---|---|---|---|---|---|
| | Plot 1 | Plot 2 | Plot 3 | | Plot 1 | Plot 2 | Plot 3 | |
| 1 | 32.67 | 9.67 | 5 | 47 | 393 | 115.67 | 65.67 | 574 |
| 2 | 190.33 | 31.67 | 18.33 | 240 | 2203 | 508 | 246.33 | 2957 |
| 3 | 11.67 | 5.33 | 3.33 | 20 | 113.67 | 31.08 | 13.46 | 158 |
| Total | 234.67 | 46.67 | 26.67 | 308 | 2709.67 | 654.75 | 325.46 | 3689.87 |

The result of Kruskal-Wallis test shows that macro plastic density between stations is significantly different with p value = 0.005. Mann-Whitney test results show that the macro plastic density of Station 1 is significantly different from Station 2 with p value = 0.031 and not significantly different from Station 3 with p value = 0,147. The station's macro plastic density 2 differs significantly from station 3 with p = 0.02. The result of Kruskal-Wallis test shows that macro plastic density between plot is significantly different with p = 0.004. Mann-Whitney test results showed that the macro plastic density of Plot 1 was significantly different from Plot 2 with p = 0.031 and significantly different from Plot 3 with p = 0.003. The macro plastic density of Plot 2 is not significantly different from Plot 3 with p = 0.63.

The result of Kruskal-Wallis test shows that macro plastic density between station plot of observation is not significantly different with p = 0.857. This indicates that the highest tidal distance and lows at each station do not affect macro plastic density. This is in accordance with the results of other studies [27] which obtain macro debris density in the mangrove ecosystem of Pantai Indah Kapuk is not significantly different between stations and indicates that the station distance from the sea does not affect macro debris density.

### 3.2. *Microplastic*

The highest micro plastic density is found in plot 1 of all observation stations which is the highest tidal boundary. Density decreases in the lowest tide and lows. This indicates that in the intertidal zone, the micro plastic decreases in density as the distance to the sea increases. Prove similarly that the micro plastic density in tidal zones at the highest tidal boundary is higher than at the lowest tidal boundary and there is a real difference between the two [14, 28]. The zone at the lowest tide is a very dynamic zone, deposition can occur constantly. Sediments in the upper layers in this zone are susceptible to runoff and become suspended again.

Microplastic film type is the highest type found in all observation plots. The film comes from fragmentation of plastic waste such as packets of food and soft drinks, plastic bags and plastic wrapping fish commonly used in fishing activities. The film also has the lowest density so it is easily distributed by the presence of currents and tides [18, 28]. Other research results on the Belgian Coast [28] gained micro plastic fiber type flavors (59%). While in Muara Badak, East Kalimantan [29] the highest abundance of microplastic fragments (58.97%) was found. This indicates the type of micro plastic depends on the macro plastic and the source of the plastic pollutant. The high fiber on the Belgian coast is due to the point of location close to the fishery port area. Fiber is derived from the activity of catching while the fragment is the result of a piece of plastic product with a very strong synthesis polymer [28].

**Table 2**. The Percentage of Density and Average Density of Microplastic

| Station | Percentage of Microplastic Abundance | | | | Abundance Average (item/kg) |
|---|---|---|---|---|---|
| | Film | Fiber | Fragment | Pellet | |
| 1 | 60.27 | 22.30 | 17.47 | 0 | 89 |
| 2 | 45.07 | 30.03 | 24.63 | 0.3 | 173.33 |
| 3 | 51.57 | 22.30 | 26.13 | 0 | 64 |
| Total | 52.30 | 24.88 | 22.74 | 0.1 | 326.33 |

The result of Kruskal-Wallis test shows that the density of micro plastic between stations is significantly different with the value of $p = 0$. This indicates that there is one station with a microplastic density that is significantly different from other stations. Mann-Whitney test results show that the stationary micro plastic density station 1 is significantly different from station 2 with p value = 0.031 and not significantly different from station 3 with $p = 0.094$. The station's micro plastic density station 2 differs significantly from station 3 with $p = 0.002$. This proves that the micro plastic density will be higher with reduced distance from the main pollutant source and with increasing distance from the main pollutant source the density is more affected by other pollutant sources such as river and sea currents [30]. The result of Kruskal-Wallis test showed that the micro plastic density between the observed plots was not significantly different with the value of $p = 0.984$.

### 3.3. *Macrozoobenthos*

The highest density of macrozoobenthos was found in station 3 with a density of 61 ind $m^{-2}$. In Plot 1 Station 2 no macro-zoobenthos is found in any depth stratification. The average density of macrozoobenthos at the observation point can be seen in Table 3.

The result of analysis using Pearson correlation shows the relationship between macro-zoobenthos density with macro plastic equal to 0.633 with the level of strong and micro plastic relationship of 0.386 with the low level of relationship. Macroplastic density with microplastic of 0.756 with strong relationship level. This suggests that macro plastic density affects macro-zoobenthos density. Increased macro plastic density causes decreased macro-zoobenthos density. The macro plastic can close the surface of the sediment and disrupt the movement of aquatic organisms [6, 9].

**Table 3.** The Average Density of Macrozoobenthos

| Stations | Macrozoobenthos Density | | | | | | | | | Total |
|---|---|---|---|---|---|---|---|---|---|---|
| | 1* | | | 2* | | | 3* | | | |
| | 10** | 20** | 30** | 10** | 20** | 30** | 10** | 20** | 30** | |
| 1 | 6 | 4 | 4 | 5 | 3 | 3 | 4 | 3 | 4 | 36 |
| 2 | 0 | 0 | 0 | 3 | 2 | 3 | 3 | 5 | 3 | 18 |
| 3 | 2 | 13 | 3 | 19 | 5 | 2 | 9 | 4 | 3 | 61 |
| Total | 8 | 17 | 7 | 26 | 10 | 8 | 16 | 12 | 10 | 115 |

*Plot | **Depth

Microplastic density has a low correlation to macro-zoobenthos it is presumably because micro plastic affects the digestive system of macrozoobenthos which can lead to accumulation in macrobenthos digestive organs [7, 8, 18, 30]. Macroplastic density with Microplastic of 0.756 with strong relationship level. This indicates that the micro plastic density is affected macro plastic. The higher the macro plastic density will lead to increased micro plastic density. Microplasticity results from macro plastic fragmentation [1, 7, 28, 29, 31].

## 4. Conclusions

An increase in the number of macroplastic densities has led to a decrease in macrozoobenthos density. Macroplastic density has a strong negative impact on macrozoobenthos density while microplastic density has a low negative impact on macrozoobenthos density. Increased macroplastic density leads to an increase in microplastic density.

## References

[1] Hammer J *et al* 2012 Plastics in the Marine Environment: The Dark Side of a Modern Gift D.M. Whitacre (ed.), Reviews of Environmental Contamination and Toxicology. Springer Science + Business Media.
[2] Thevenon F *et al* 2014 Plastic Debris in the Ocean: The Characterization of Marine Plastics and their Environmental Impacts, Situation Analysis Report. Gland, Switzerland:IUCN. 52 pp.
[3] United Nations Environment Programme 2011 UNEP Year Book Emerging Issues in Our Global Environment. UNEP. Nairobi.
[4] National Oceanic and Atmospheric Administration. 2013. Programmatic Environmental Assessment (PEA) for the NOAA Marine Debris Program (MDP). NOAA. Maryland
[5] Citrasari N *et al* 2012 *Journal of Biological Research*, **18** (1): 83–85.
[6] Irwin K2012 The Impacts of Marine Debris: A Review and Synthesis of Existing Research. Living Oceans Society December 2012
[7] Galgani F*et al*2015 Global Distribution, Composition and Abundance of Marine Litter. M. Bergmann et al. (eds.), Marine Anthropogenic Litter, DOI 10.1007/978-3-319-16510-3_2
[8] Ogunola OS and Palanisami T 2016 *J Pollut Eff Cont.* **4**: 161
[9] Smith SDA and Markic A 2013 PLoS *One*. **8** (12).
[10] National Oceanic and Atmospheric Administration Marine Debris Program 2016 Report on Marine Debris Impacts on Coastal and Benthic Habitats. Silver Spring, MD: NOAA
[11] Eriksen M *et al* 2014 Plastic. *PLoS ONE.* **9**(12):
[12] Browne MA *et al* 2008 *Environ. Sci. Technol.* **42** (1): 5026-5031.
[13] Boerger CM *et al* 2010 *Marine Pollution Bulletin*. **60** (12): 2275-2278
[14] Cauwenberghe LV *et al* 2013 *Marine Pollution Bulletin*. **73** (1): 161-169.
[15] Odum EP 1995 Dasar-Dasar Ekologi. Edisi Ketiga. Gajah Mada University Press. Yogayakarta.
[16] Peters K and Flaherty T 2011 Marine Debris in Gulf Saint Vincent Bioregion. Adelaide (AU): Government of South Australia. Australia.
[17] Hidalgo-Ruz V *et al* 2012 *Environmental Science and Technology*. **46** (1): 3060- 3075.
[18] Claessens M *et al* 2011 *Marine Pollution Bulletin*. **62** (1): 2199-2204.
[19] Tourinho P S and Fillmann G 2011 *Journal of Integrated Coastal Management*. **11**, 97–102
[20] Costa MF *et al* 2011 *Journal of Coastal Research*. **64** (1): 339-343.
[21] Smith SDA 2012 *Marine Pollution Bulletin*. **64** (9), 1880–1883
[22] Ribic C *et al* 2012 *Marine Pollution Bulletin*. **64** (8): 1726–1729
[23] Rosevelt C *et al* 2013 *Marine Pollution Bulletin*. **71**(1–2), 299–306
[24] Liu T *et al* 2013 *Marine Pollution Bulletin*. **72** : 99–106.
[25] Ioakeimidis C*et al* 2014 *Marine Pollution Bulletin.* **89** (1-2): 296-304.
[26] Van Dyck IP *et al* 2016 *Journal of Geoscience and Environment Protection*. **4** : 21-36
[27] Hastuti A R 2014 Distribusi Spasial Sampah Laut di Ekosistem Mangrove Pantai Indah Kapuk Jakarta. Skripsi. FPIK. Institut Pertanian Bogor. Bogor.

[28] Syakti *et al* 2017 Beach macro-litter monitoring and floating microplastic in a coastal area of Indonesia. *Marine Pollution Bulletin (in press)*
[29] Dewi IS 2015 *Depik.* **4**(3): 121-131.
[30] Andrady AL 2011 *Marine Pollution Bulletin.* **62** :1596–1605
[31] Cordova MR and Wahyudi AJ 2016 *Mar. Res. Indonesia.* **41** (1), 27–35.

IOP Conf. Series: Earth and Environmental Science **122** (2018) 012103 doi:10.1088/1755-1315/122/1/012103

# Structure of gastropod communities at mangrove ecosystem in Lubuk Kertang village, West Berandan District, Langkat Regency, North Sumatera Province

**T Manullang, D Bakti and R Leidonald**

Department of Aquatic Resources Management, Faculty of Agriculture, Universits Sumatera Utara, Medan 20155.
Email: dbakti06@yahoo.com

**Abstract.** Gastropod was one of the class from mollusca in mangrove ecosystem. Lubuk Kertang Village's mangrove forest was been converted into tourist areas, agricultural land and settlements. The purpose of this study was to analyze the structure of gastropods at Avicennia lanata, Rhizophora apiculata and Sonneratia alba. This research was conducted at Lubuk Kertang Village in February-March 2017. Gastropod were collected in 1 m × 1 m transect in mangrove. Examples of biota were taken by using shovel, then the biota was inserted into a plastic bag sample, wrote date of sampling and identified. The results showed there were 15 species Gastropods, namely Achatina fulica, Cerithidea alata, Cerithidea cingulata, Cerithidea obtusa, Chicoreus capucinus, Cymatium pileare, Ellobium aurimisdae, Ellobium aurisjudae, Littoraria melanostoma, Littoraria scabra, Murex tribulus, Nerita balteata, Nerita planospira, Pugilina cochlidium, Stramonita gradata, Telescopium telescopium and Terebralia sulcata. Diversity index ranged 1.702 to 2.165 was in medium category, Similarity index ranged 0.676 to 0.799 was in low category and Dominance index ranged 0.142 to 0.282 that categorized was low. The highest gastropod density was 31 individuals/m$^2$ in the Sonneratia alba. The conclusion of the research is the existing mangrove ecosystem in Lubuk Kertang Village in a stable state.

## 1. Introduction

Mangrove ecosystem is an ecosystem that is affected by water conditions that change at any time. It because effected by aquatic biota that associated with mangrove ecosystem. Potential mangroves as a source of nutrients for biota living in it as a place to live, feeding ground, nursery and spawning ground [1]. One of the fauna that capable to adapt and survive in such extreme environments is Gastropoda [2]. Abundancy and distribution of gastropods were influenced by local environmental factors, food availability, predation and competition. Environmental changes can affect the number of species and differences in the structure of the gastropod community. Gastropod meat is used as a source of animal protein and its shells are used as raw materials in industry and jewelry.

Utilization of mangrove forest to a plantation land resulted in decreased productivity of mangrove forest in Lubuk Kertang Village, the function destroys mangrove vegetation and disrupts the life of biota in it, one of them is Gastropoda. Gastropoda can be used as one of the biological indicators of a water. Gastropoda density is highly dependent on its tolerance to environmental change. Therefore, the linkage between mangrove species and Gastropoda located in Mangrove Ecosystem in Lubuk Kertang Village, Berandan Barat District Langkat, North Sumatera Province needs to be studied.

## 2. Materials and Method

*Study site*

The research was conducted in Lubuk Kertang Village, Langkat Regency, North Sumatra Province, Indonesia. Data were collected from February to March 2017. The materials used were Global Position System (GPS), meter, tool box, 1m x 1m board, digital camera, and Paper millimeter blok.

*Method of Gastropods*

Each observation station was set at 3 plots with an area along transects perpendicular line from the water to the mainland dominated mangrove species *Rhizophora apiculata, Avicennia lanata* and *Sonneratia alba* with transects squares of 10 m x 10 m , whereas in the plot are sub plot size of 1 m × 1 m placed to observe and identify the gastropod biota. The steps are as follows [3]:

A. Sampling of Gastropoda was done in transect observation of 10 m × 10 m mangrove vegetation.
B. In each plot of transect 10 m × 10 m, it is made sub plot with three randomly placed spots, where each point uses 1 m × 1 m transect.
C. Sampling of gastropods is done on each sub plot, ie at low tide, making it easier to calculate and identify the gastropods'species.
D. Biota retrieval is carried out on substrates, roots, stems and leaves to the highest tide limit.
E. Samples of biota were taken by using shovel (hand and picking) for the epifauna, then the biota is inserted into a plastic bag sample that has been given 70% alcohol solution and also wrote station location information, trasnsct, plot and date of sampling and further identified [4].

*Problem formulation*

Biota Density of Gastropods with formula [5]:

$$K = \frac{ni}{A} \quad (1)$$

Information:
K = Density
ni = Number of individuals of a type
A = Area

Diversity Index is calculated by the formula :

$$\mathbf{K} = -\sum_{i=1}^{s} \text{Pi ln Pi} \quad (2)$$

Information:
H' = Shannon-Wienner diversity index
Pi = Number The proportion of the number of individual species i to the number of individuals Total Pi = ni / N with ni: number of species i N: total sum species

Similarity Index is calculated by the formula:

$$K = \frac{H'}{H\ max} \quad (3)$$

Information :
K = Similarity index *(Evenness)*
H' = Shannon-Wienner diversity index
H max = Log $^2$S; S = Number of species or taxa

Dominance Index is calculated by the formula:

$$K = \sum_{i=1}^{s} \left(\frac{ni}{N}\right)^2 \quad (4)$$

Information :
K = Simpson's Dominance Index
ni = Number of individuals of a type of i
N = Total Number of individuals
S = Number of genera

## 3. Results and Discussions

*The diversity and composition of gastropods*

Gastropods found in *Avicennia lanata* was very diverse. Gastropods found mostly in the substrate and a bit attached to the mangrove stems. Gastropods composition highest species were *Cerithidea cingulata, Cymatium pileare, Nerita balteata* and *Cerithidea alata* 21%, 19% and 16%, while the value was lowest for the species *Ellobium aurismisdae* and *Nerita planospira* with a value of 1%. In mangroves *Avicennia lanata* found gastropod family potamididae with species *Cerithidea alata* and *Cerithidea cingulata* which has a high enough composition value among other types. This was due to Cerithidea was has a high adaptability and wide spread and substrate conditions in the form of sand and clay sand suitable to support the life of this genus. One of the gastropods that dominates the mangrove ecosystem is from the family Potamididae (*Cerithidae cingulata*). Potamididae conch is the original inhabitant of mangrove forests and lives in tidal areas and likes muddy areas [7].The composition of gastropods on *Avicennia lanata* can be showed on Table 1.

In *Rhizophora apiculata* vegetation found 13 species of gastropods. Of the 13 species of gastropods were the most dominant found attached to the roots of mangrove is the *Nerita balteata* by 49%, while the lowest value was found on substrate were *Achantina fulica, Chicoreus capucinus and Ellobium aurimisdae* with a value of 1%. *Nerita balteata* composition of 49% is dominant gastropods in *Rhizophora apiculata*, living on the mangrove root as their habitat and able to win the competition to get food and living place compared to other species. The highest density was found in the *Rhizophora apiculata* and the dominant gastropod was *Nerita balteata* species, the species likes mangrove forests as their habitat and were able to win the competition for food and livelihoods compared to other species [8]. The composition of gastropods on *Rhizophora apiculata* can be showed on Table 2.

**Table 1.** Composition species gastropods in *Avicennia lanata*

| Number. | Species Name | Family | Total | Percentage (%) |
|---|---|---|---|---|
| 1. | *Achantina fulica* | Achatinidae | 9 | 5 |
| 2. | *Ellobium aurisjudae* | Ellobidae | 1 | 0 |
| 3. | *Ellobium aurimisdae* | | 2 | 1 |
| 4. | *Littoraria melanostoma* | Littorinidae | 8 | 5 |
| 5. | *Littoraria scabra* | | 1 | 0 |
| 6. | *Pugilina cochlidium* | Melongenidae | 14 | 8 |
| 7. | *Stramonita gradate* | Muricidae | 7 | 4 |
| 8. | *Murex tribulus* | | 3 | 2 |
| 9. | *Nerita balteata* | Neritidae | 28 | 16 |
| 10. | *Nerita planospira* | | 2 | 1 |
| 11. | *Cerithidea alata* | Potamididae | 28 | 16 |
| 12. | *Cerithidea cingulata* | | 37 | 21 |
| 13. | *Terebralia sulcata* | | 1 | 0 |
| 14. | *Telescopium telescopium* | | 3 | 2 |
| 15. | *Cymatium pileare* | Rannelidae | 35 | 19 |

**Table 2.** Composition species gastropods in *Rhizophora apiculata*

| Number. | Species Name | Family | Total | Percentage (%) |
|---|---|---|---|---|
| 1. | *Achantina fulica* | Achatinidae | 3 | 1 |
| 2. | *Ellobium aurimisdae* | Ellobidae | 2 | 1 |
| 3. | *Littoraria melanostoma* | Littorinidae | 5 | 2 |
| 4. | *Littoraria scabra* | | 21 | 9 |
| 5. | *Pugilina cochlidium* | Melongenidae | 1 | 0 |
| 6. | *Chicoreus capucinus* | Muricidae | 2 | 1 |
| 7. | *Nerita balteata* | Neritidae | 121 | 49 |
| 8. | *Nerita planospira* | | 23 | 9 |
| 9. | *Cerithidea alata* | Potamididae | 15 | 6 |
| 10. | *Cerithidea cingulata* | | 35 | 14 |
| 11. | *Cerithidea obtuse* | | 4 | 2 |
| 12. | *Telescopium telescopium* | | 5 | 2 |
| 13. | *Cymatium pileare* | Rannelidae | 10 | 4 |

In the *Sonneratia alba* gastropods that most widely encountered was *Littoraria scabra* attached on mangrove's leaves and trunks by 44%, while the value was lowest for the *Terebralia sulcata* and *Telescopium telescopium* that found in the substrate with a value of 1%. The composition of gastropods on *Sonneratia alba* can be showed on Table 3.

**Table 3.** Composition species gastropods in *Sonneratia alba*

| Number. | Species Name | Family | Total | Percentage (%) |
|---|---|---|---|---|
| 1. | *Littoraria melanostoma* | Littorinidae | 12 | 4 |
| 2. | *Littoraria scabra* | | 123 | 44 |
| 3. | *Pugilina cochlidium* | Melongenidae | 1 | 0 |
| 4. | *Nerita balteata* | Neritidae | 21 | 7 |
| 5. | *Nerita planospira* | | 5 | 2 |
| 6. | *Cerithidea alata* | Potamididae | 38 | 14 |
| 7. | *Cerithidea cingulata* | | 52 | 19 |
| 8. | *Cerithidea obtuse* | | 7 | 3 |
| 9. | *Terebralia sulcata* | | 2 | 1 |
| 10. | *Telescopium telescopium* | | 4 | 1 |
| 11. | *Cymatium pileare* | Rannelidae | 13 | 5 |

In the mangrove vegetation, *Sonneratia alba* is found in seedlings category and gastropods of *Littoraria scabra* 44%, which lives attached to trunk and leaves by climbing trees. The dominant type of gastropod and found attached to the leaves of mangrove plants *is Littoraria scabra* and *Littoraria melanostoma*. Generally gastropods found in leaves took calcium carbonate contained in these plants by eating leaves [9]. Gastropods living in mangrove trees attached to roots or mangrove stems and some are climbing, for example in *Littoraria* and *Cassidula* [10].

*Value of Diversity, Similarity and Dominance Index of Gastropod on Mangrove*

Based on the data analysis, total population, Diversity Index (H'), Similarity Index (E) and Dominance (D) Gastropod at first, second and third sampling in each research station are presented in Table 4. Based on the result of research, the value of environmental chemical physics parameter and substrate type presented in Table 5.

**Table 4.** Value of Diversity, Similarity and Dominance Index of Gastropoda on Mangrove

| Stasion | Type of Mangrove | Σ Total | Σ Type Population | Index Value H' | Index Value E | Index value D |
|---|---|---|---|---|---|---|
| 1. | *Avicennia lanata* | 179 | 15 | 2. 165 | 0. 799 | 0. 142 |
| 2. | *Rhizophora apiculata* | 247 | 13 | 1. 735 | 0. 676 | 0. 282 |
| 3. | *Sonneratia alba* | 278 | 11 | 1. 702 | 0. 709 | 0. 260 |

**Table 5.** Physical Chemical Parameter and Substrate Type

| Station | Type of Mangrove | Water temperature (°C) | Substrate pH | Salinity (°/₀₀) | C-Organic (%) | Sediment type |
|---|---|---|---|---|---|---|
| I | *Avicennia lanata* | 29.3 | 6.2 | 24.3 | 1.89 | Sandy Clay Loam |
| II | *Rhizophora apiculata* | 29.7 | 5.4 | 24 | 4.12 | Sandy Clay Loam |
| III | *Sonneratia alba* | 29.7 | 5.8 | 25 | 7.22 | Sandy Clay Loam |

The value of Similarity index (E) ranges (0.676-0.799), Diversity index was 1.702 to 2.165 in the medium category, Dominance index from 0.142 to 0.282 belongs to the low category. A community was said to have high species diversity if the community was composed by many types Abundance of the same or almost the same kind. Conversely, if the community is made up of very few species and if only a few dominant species were of low biodiversity [11]. If the index of dominance is "low" then in the community being observed there is no extreme species dominating other species. This shows the condition of the community in a stable condition and environmental conditions were quite excellent [12].

The result of water temperature measurement at the three stations ranged from 29.3-29.7°C, the value of acidity degree (pH) of substrate ranged from 5.4-6.2, salinitas research location ranged from 24-25 ‰. The range of gastropod pH tolerance ranging from 5 to 9 is generally found in areas with more pH Greater than 7, High or low salinity will not affect the presence of gastropod species, because Gastropoda has adaptability or tolerance to salinity [13].

Average density of the gastropods on mangrove, the highest was found on *Sonneratia alba* 31 individuals / m $^2$ while the lowest found on *Avicennia lanata* 20 individuals / m $^2$. The difference in density is due to the selection of preferred feed sources depending on species of molluscs present, so that the availability of feed at the site is an important factor related to the density. In addition, the range of water depth is commonly associated with the way of life and foraging from some families of gastropods [14]. The density of Gasropoda on Mangrove can showed on figure 1.

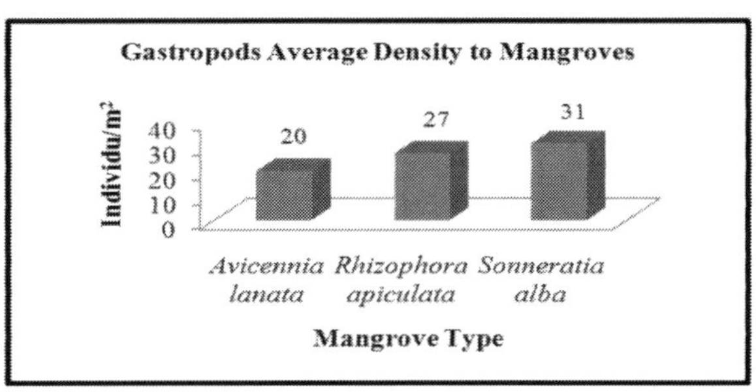


**Figure 1.** Gastropod Density of Mangrove Types

## 4. Conclusions

Gastropods that live on the mangrove ecosystem in Village Lubuk Kertang there were 8 Families and 15 species Gastropods found in three types different mangrove vegetation, results calculation diversity index in medium category, similarity index in low category and dominancy index was low. The highest gasropod density was 31 individuals/$m^2$ in the mangrove *Sonneratia alba.*

## 5. References

[1] Sari S Pratomo A and Yandri F 2014 Relation of mangrove density to pelecypoda abudance at costal area of tanjung pinang riau *Journal of University Maritime of Raja Ali Haji* Riau Kepulauan Tanjung Pinang

[2] Rusnaningsih 2012 Gastropoda community structure and population study cerithidea obtuse (lamarck 1822) in mangrove forest pangkal babu, tanjung jabung barat district, jambi. University of Indonesia Depok

[3] Lina F Lestari and Zulfikar A 2014 Gastropoda community in mangrove ecosystem nyirih river tanjungpinang district tanjung pinang municipality *Journal of University Maritime of Raja Ali Haji*, Riau Kepulauan Tanjung Pinang

[4] Carpenter K E and Niem V H 1998 The living marine resource of the western central pacificvol. 1 seaweeds, corals, bivalves and gastropods Italy

[5] Saptarini D I Trisnawati and Hadiputra M A 2011 Community structure gastropoda (moluska) mangrove forest sendang blue, South Malang *Journal of Surabaya Institute Technology* Surabaya

[6] Odum E P 1983 *Basic Ecology* (Canada: CBS College Publishing) p 613

[7] Afkar Djufri and Ali M S 2014 Makrozoobenthos association with mangrove ecosystem in reuleng leupung river, aceh besar distric *Journal of EduBio Tropika* **2**: 187-250

[8] Nasution N A Yunasfi and Rangkuti A M 2016 Relation of mangrove density to mollusc density in dusun ii of pulau sembilan village, pangkalan susu sub-district, langkat regency, north sumatera province *Journal of Aquacoastmarine* **14**: 11-18

[9] Muhsin Jamili and Hendra 2016 distribution of vertical gastropoda on mangrove rhizophora apiculata in kendari bay *Journal of Biological Research* p 2355-6404

[10] Pribadi R R Hartati and Suryono C A 2009 Composition of type and distribution of gastropoda in mangroveforest area segara anakan Cilacap *Journal of Marine Science* **14**: 102-111

[11] Ernanto R F Agustriana and Aryawati R 2010 Community structure of gastropoda on mangrove ecosystem in estuary of batang ogan komering ilir river South Sumatra *Journal of Maspari* p 73-78

[12] Ramses Notowinarto and Dewi I S 2016 Community structure mollusca penempel on mangrove vegetation in pulau kasu subdistrict padang city of Batam *University of Kepulauan Riau* Batam

[13] Romdhani A M Sukarsono and Susetyarini R E 2016 Diversity of gastropoda mangrove forest village baban subdistrict gapura sumenep regency as a source of biology learning *Journal of Biology Education Indonesia* pp 2442-3750

[14] Islami M M 2015 Spatial gastropoda and its relation to environmental characteristics on the coast of nusalaut island, central maluku *Journal of Tropical Marine Science and Technology* **7**: 365-378

# The growth and exploitation rate of yellowstripe scad (*selaroides leptolepis* cuvier, 1833) in the Malacca Strait, Medan Belawan Subdistrict, North Sumatera Province

**J Tambun, D Bakti and Desrita**

Program Study of Management of Aquatic Resources, Faculty of Agriculture, Universitas Sumatera Utara. Jl. Prof A. Sofyan No. 3, Medan 20155, North Sumatra, Indonesia. Tel./Fax. +661-821-3236.
Email: dbakti06@yahoo.com

**Abstract.** Yellowstripe scad included the one of commodity that has an important economic value in the Malacca Strait. Fish were found mostly in Indonesian of waters made this fish as one of the main target catch. But, it can had negative impact on the population of the fish. The study is done at Belawan Waters on March until May 2017 that which is purposed to study about the frequency distribution of length, determine the parameters of growth and, determine mortality rate and the rate of exploitation in order to provide appropriate management model for the fish resource. Yellowstripe scad was observed around 360 samples with the length range between 110 - 175 mm. The fish separated by bhattacarya method used the aid software FISAT II. A pattern of growth Yellowstripe scad alometrik negative with growth coefisien (K) 1.1 with length asimtotic ($L_\infty$) 181.65. The rate of mortality total ( Z) yellowstripe scad 4.34 per year at the rate of mortality natural ( M ) 1.204 per year and rate mortality by fishing (F) 3.136 per year in order to obtain the rate of exploitation 0.722. The value of this exploitation rate has exceeded the value of the optimum exploitation of 0.5.

## 1. Introduction

Yellowstripe scad is one type of important economic. Many people like it because it tastes good and the price is relatively. The fish is done by catching throughout the year so that the existence of behavior Yellowstripe scad is almost active like there every day.

Continuous capture without thinking the fish resources resulted in the threat of sustainability of this fish in the malacca strait. To know the extent of Yellowstripe scad exploitation, it is necessary to study the biological aspect in mean to growth pattern. Stated that the length measurement of fish weight aims to determine the variation of specific weights and lengths of individual fish or individual groups as guidance on obesity, health, productivity and visiologic conditions including gonadal development [7].

The economic and ecological potential of Yellowstripe scad is good enough so that it is necessary to review the basic information of fisheries biology to support the management of the resources of Yellowstripe scad in order to create a sustainable and environmentally friendly catch. Long-term frequency distribution and long-term relationship are important information to see growth and rate of exploitation which is one of consideration factor in setting fishery management strategy. Research purposes to determine the growth of fish in the malacca strait.

## 2. Material and Method

*Time and Place*

This research was conducted from March to May 2017. One time a month, of the length of fish weight. The fish samples were obtained from the catch of Yellowstripe scad in the Malacca Strait, Medan Belawan Subdistrict, North Sumatra Province.

*Tools and materials*

The tools used in this study is a *gill net* with a mesh size of 2 inches that serves to catch fish Selar Yellow, gauges fish with a precision of 1 mm to measure the length of the fish, digital scales to weigh the weight of fish, GPS to determine the coordinates of the *sampling station, refakto meter* to measure water salinity, thermo meter to measure water temperature, *Cool box* for the sample container, mengkukur DO meter for dissolved oxygen (DO) and the measuring vessel 5 GT.

Materials used in this study is Selar Yellowstripe scad *(Selaroides leptolepis)* as samples, ice for handling the fish so that the fish stay fresh after being arrested and wipes.

*Problem formulation*

Weights can be considered as a function of length. The length and weight relationship can follow the law of cubic where the weight of the fish is the rank of three of its length.But the actual relationship in fish is not so because the shape and length of the fish is different. Therefore, the long relationship weight of Yellowstripe scad using a common formula that is: [2]

$$W = aL^{b} \tag{1}$$

Information :
W = Weight (gram)
L = Length (cm)
A = Intersection of the long-weight relationship curve with the y-axis
B = Estimation of long-weight growth pattern

The condition factor is the state or the pledge of the fish expressed in the figures. The calculation of condition factors is based on length and weight. Calculation of this condition factor to look at the length and weight of some fish reach maximum or minimum conditions. The condition factor can be calculated by the formula: [1]

The value of $b \neq 3$ (allometric) then the condition is determined by the formula:

$$FK = \frac{W}{aL^{b}} \tag{2}$$

Information :
FK = Condition factor
W = Weight of fish (gram)
L = Total length of fish (mm)
A, b = Constants

Walford plot Ford is one of the simplest methods of predicting the growth parameters $L_{\infty}$ and K of the von Bertalanffy equation with interval sampling the same time [3]. The following is the von Bertalanffy growth equation:

$$Lt = L_{\infty} (1-e^{(-k(t-t_0))}) \tag{3}$$

Information :
Lt : Fish length at age t (time unit)
L∞: Theoretical maximum length (asymptotic length)

K : Growth coefficient (per unit time)
$T_0$ : Theoretical life at length is zero

The rate of mortality of capture (F) can be determined using the following formula:

$$F = Z - M \quad (4)$$

Determined by comparing the rate of exploitation of fishing mortality (F) of the total mortality (Z) [6] [9]

$$E = \frac{F}{F+M} = \frac{F}{Z} \quad (5)$$

The rate of fishing mortality (F) or the rate of exploitation *in the* optimum according to [9] are:

$$F_{optimum} = M$$
$$E_{optimum} = 0.5$$

Information :
E = Exploitation rate
F = Coefficient of capture mortality
M = Natural coefficient of death

Terms:
1. If E> 0.5 indicates a high rate of exploitation *(overfishing).*
2. If E <0.5 indicates a lack of utilization *(underfishing).*
3. If E = 0.5 indicates optimal utilization.

## 3. Results and Discussions

### *3.1 Results*

The From the analysis of heavy long relationship known that heavy long relationship equation Yellowstripe scad Selar from March to May was $W = 0.002L^{1,926}$ b Values obtained after t test (a = 0.05) on the value of b is known that fish Selar Yellow Has a negative allometric growth pattern, meaning that the length increase is faster than weight gain [1]. Can be seen in Figure 1.

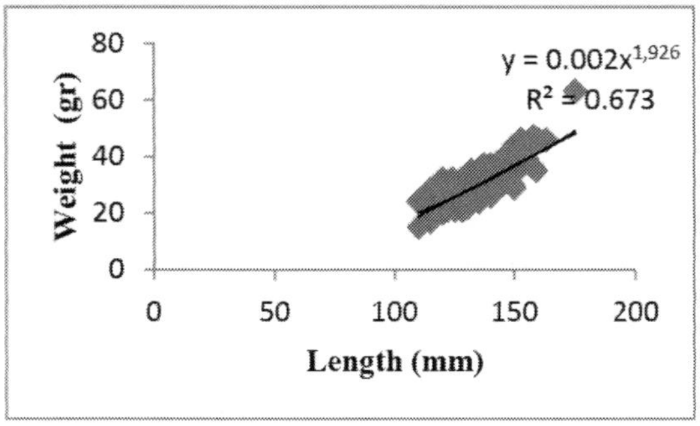


**Figure 1.** Long Relationship Relationship of Yellowstripe scad during March – May, n =360

The value of the condition of the female Yellowstripe scad in March was 0.865-1.276 while in the Yellowstripe scad Yellow male of 1.120-1.818. The value of Yellow female Selar Fish condition factor in April was 0.865-1.153 while the Yellowstripe scad Yellow male was 0.861-1.277. In May of 0.877-1.368 while in the Yellowstripe scad Yellow male of 0.836-1.201. The factor value of Yellowstripe scad condition in the Malacca Strait has range 0,836-1,818 (flat) with negative allometric growth pattern. Can be seen in Figure 2.

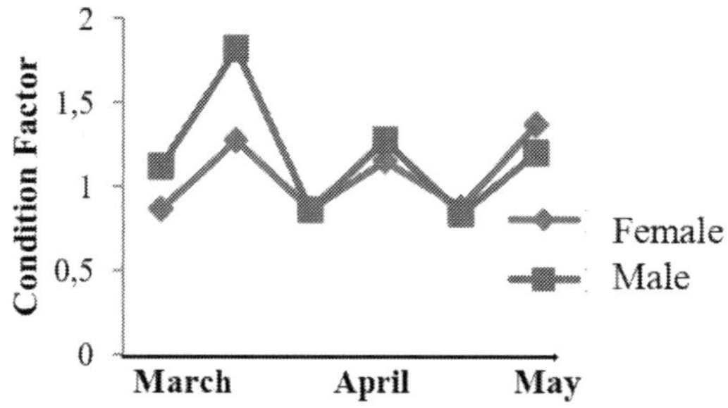


**Figure 2.** The Value of Yellowstripe scad Selection Factor Based on Observation Time.

The values of the growth parameters used as the basis to obtain Von Bertalanffy equation Yellowstripe scad Selar $L_t = = 181.65[1-e^{(-1,1(t+0.53394)}$ Based on the equation - the equation of Von Bertalanffy, growth charts Yellowstripe scad can be seen in Figure 3.

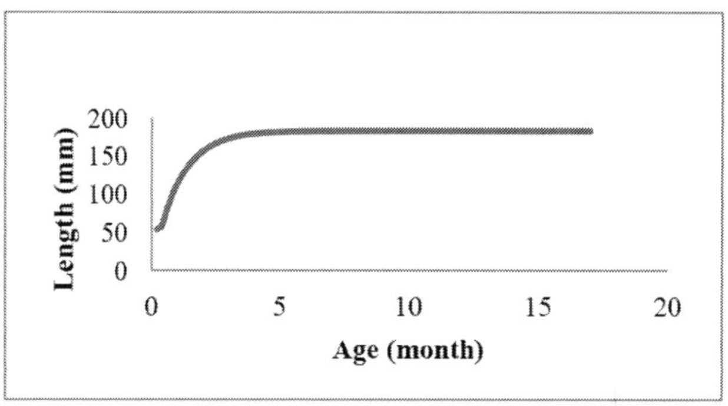


**Figure 3.** Growth Charts ($L_{t)}$ Fish Selar Yellow

The result of analysis of total mortality rate (Z) on Yellowstripe scad obtained 4.34 per year consisting of natural mortality (M) Yellowstripe scad is obtained 1.204 per year, and the fishing moortality (F) is 3.136 per year so that the rate of exploitation (E) 0.722 per year.

*3. 2. Discussion*

From the analysis of heavy long relationship known that heavy long relationship equation Yellowstripe scad Selar from March to May was $W = 0.002L^{1.928}$ . The value of b obtained after t test (a = 0.05) to the value of b is known that Yellowstripe scad has a negative allometric growth pattern, which means long increase faster than weight gain [1].

The pattern of Yellowstripe scad growth in Belawan waters, Malacca Strait is different from [8] study of long-term relationship analysis, obtaining a b value of 1.926. Differences in the value of b can be caused by several factors such as environmental conditions, different fishing locations and physiology of the fish itself. This explanation conform statement in [4] factors that cause different values of b in addition to species differences are environmental factors, different fish stocks in the same species, fish development stage, gender, maturity level of gonads, even differences Time of day due to changes in stomach contents. Add that the difference in the value of b can also be due to the difference in the number and variation of fish sizes observed [4].

Analysis of the relationship of length and weight of Yellowstripe scad in, Malacca Strait has negative allometric growth pattern with the calculation of condition factor (FK) ranged from 0.836 and

1.818 Value Factor of Yellowstripe scad condition shows that Yellowstripe scad is in flat condition. According to Effendie (2002) statement that the value of K in fish a slightly flat body ranged between 1 - 2, whereas in the less flat fish between 2 - 4. The results of this study Yellowstripe scad in Malacca Strait, Tanjung Beringin Subdistrict, Serdang Bedagai Regency has a condition factor value ranging from 0.711 - 1.551 [5]. The condition factors will also differ depending on the sex of the fish, the season or the location of the capture as well as the condition factor also influenced by the maturity level of gonad and the abundance of food [4].

Von Bertalanffy growth equation is formed for Fish Selar Yellow is the $L_t$= = 181.65[1-$e^{(-1.1(t+0,53394)}$] These results indicate that the length of fish caught is smaller than the length asymptote (Infiniti). Pauly (1983) states that the value of the maximum length obtained during sampling in the field generally do not have a considerable margin away with $L_\infty$ the results of this study are also in accordance with the fish yellowstripe scad caught in the Sea Natuna has a maximum length of 310 mm and has the asymptotic length ($L_\infty$) equal to 330 mm [3].

Curve of Yellowstripe scad growth by entering data of age (year) and data of theoretical length (mm) of fish, theoretically total length of fish is 182,7 mm and growth rate of Yellowstripe scad start stop at age 15.6 month and Selar fish Male yellow at the age of 13.8 months or less than 3 years. Azis (1989) in Sari, et al (2013) explains that the long-term growth curve of fish that occurs rapidly at a young age and increasingly slow along with age until reaching its asymptotic length where the fish does not grow longer.

The result of analysis of mortality rate of Yellowstripe scad obtained by total mortality rate (Z) Yellowstripe scad 4.34 per year with natural mortality rate (M) equal to 1.204 per year and mortality of catch (F) 3.136 per year. The estimated value of the exploitation rate (E) of this study is 0.722 per year, the value shows the difference of the estimated Z value obtained [5] in the Malacca Strait, Tanjung Beringin Subdistrict at 4.99 per year, Sapira (2013) at Landing Ikan Dusimas Malang Village Meeting is 0.784.Differences in total mortality values indicate that there is a difference in stock of Yellowstripe scad in waters caused not only fishing activity but also due to natural death.

## 4. Conclusions

Yellowstripe scad (*S. leptolepis*) in Malacca Strait, Medan Belawan Subdistrict, North Sumatera Province during the study has negative allometric growth pattern which means long increase faster than weight gain. The Bertonffy Von equation formed is $L_t$= = 181,65[1-$e^{(-1,1(t+0,53394)}$. The value of the exploitation rate of Yellowstripe scad in Malacca Strait, Medan Belawan Subdistrict, North Sumatera Province has overfishing status because the rate of exploitation exceeds the optimum value of 0.5 that is 0.722.

## References

[1] Effendie M I 1997 *Fisheries Biology* (Publisher of Pustaka Nusantara, Yogyakarta)
[2] Effendie M I 2002 *Fisheries Biology : Study Natural History* Bogor Agricultural Institute, Bogor
[3] Febrianti A, Efrizal T and Zulfikar A 2013 Study of Yellowstripe Scad Condition (*Selaroides leptolepis)* Based on The Relationship of Heavy Length and Condition Factors in the Natuna Sea which is Landed at KUD Tanjung Pinang. Faculty Marine and Fisheries Science. Maritime University of Raja Ali Haji, Riau.
[4] Harmiyati D 2009 Analysis of the Catch of Sumberaya Yellow Tail Fish that is Landing at PPI Pulau Pramuka, Kepulauan Seribu. Bogor Agricultural University, Bogor.
[5] King M 1995 *Fishery Biology, Assessment and Management* (Fishing News Books, London)
[6] Nasution F A 2015 *Growth and Exploitation Rate of Yellowstripe scad (Selaroides leptolepis) in The Waters of The Strait of Malacca Tanjung Beringin Subdistrict Serdang Bedagai North Sumatra* University Of North Sumatra, Medan
[7] Pauly D 1983 *Some Simple Methods For Assessment Of Tropical Fish Stock. Food and Agriculture Of The United Nations*, Roma

[8] Richter T J 2007 Development and evaluation of standard weight equations for bridgelip sucker and largescale sucker *North American Journal of Fisheries Management* **27**: 936-939
[9] Sharfina M, Boer M and Ernawati Y 2014 Potential sustainable yellowstripe scad (*Selaroides leptolepis*) in the Sunda Strait *Journal of Marine Fisheries 5* (1) ISSN: 2087-4235
[10] Sparre P and Venema S C 1999 *Introduction of Tropical Fish Stock Assessment* (FAO Fisheries Technical Paper, Roma)

**Acknowledgments**
Thanks to Faculty of Agriculture Universitas Sumatera Utara.

# Growth and exploitation rate of *Anadara gubernaculum* (reeve, 1844) Arcidae Family in Asahan Aquatic of North Sumatra

**M Fauzan, D Bakti*, I E Susetya and Desrita**

Department of Aquatic Resources Management, Faculty of Agriculture, Universitas Sumatera Utara, Medan 20155
*E-mail :dbakti06@yahoo.com

**Abstract.** High market demand for *A. gubernaculum*, tends to increase the greater catching capacity so that the decreasing population. The aims of this study were to determine the growth and rate of exploitation of *A. gubernaculum* in Asahan aquatic. It was conducted for 1 Month 14 Days from October to November 2016. Data analyzed by (Electronic Lenght Frequencys Assesment Tool) ELEFAN I method by using (FAO-ICLARM Fish Stock Assesment Tool) on FiSAT II software. Shells obtained 855 individual. The growth pattern of shells is negative allometric. The range of condition factor was 0.81 – 2.15. The frequency distribution of the *A. gubernaculum* ranges from 14 to 43 mm, the dominant size group was 20 - 22 mm. The prediction of growth parameter Von Bertalanfy showed that the asimptot length ($L\infty$) is 43.05 mm, the growth coefficient (K) is 1.2/year and the theoretical life ($t_0$) of the *A. gubernaculum* is -0.12. The total Mortality (Z) of *Anadara gubernaculum* was 2.121/year. Natural mortality estimation rate (M) was 1.9/year. The exploitation rate of *Anadara gubernaculum* is 0.1/year.

## 1. Introduction

Asahan aquatic is located in the east coast of North Sumatra. Asahan aquatic have a high potential and provide many marine products for the Province of North Sumatra. One of the main commodities of marine fisheries from Asahan aquatic is shells. The types of shells in Asahan aquatic are *Anadara granosa*, *A. gubernaculum*, *Perna viridis*, and *Meretrix meretrix*.

High market demand for *A. gubernaculum*, can lead to the greater catching and decrease populations. This situation will disrupt the resources in Asahan aquatic. Based on the above indication, basic information is needed to support its resources in sustainabling in exploitation. Assessment of stock is very important to be done in order to maintain the population of *Anadara gubernaculum* in Asahan aquatic of North Sumatra.

## 2. Material and Method

### *2.1 Study site*

The research was conducted in Asahan Aquatic of North Sumatera. Data were collected from October to November 2016. The Tools used were Analytical Scales, Calipers, Dark Bottle Sample, Stationery, Paper Label, Book identification, Black Plastic Bag, Coolbox. The materials used in this research are Shells

Samples, Microsoft Excel Software, FISAT Applications, Paper Label, Rubber, Aquadest, Alcohol, Tissue.

*2.2 Sampling of Anadara gubernaculum*
Shells sampling was done from October 2016 until November 2016 times with sampling time interval 2 weeks. Shells were taken randomly from 3 stations. Sampling using a scratching tool. It was made of iron. This tool has a size of 0.39 m x 0.24 m. How to operate this "garuk" by descending into the aquatic. After a few minutes, the "garuk" is lifted on to the boat.

*2.3 Problem formulation*
*Size Distribution*
The analysis of the length of shells data is :

a. Lengt-size data are grouped into long classes. Grouping shells into long classes is done by setting the first "range" or class region, class interval and lengt-term boundaries based on the number.
b. The data is plotted into a graph connecting the length of the shells (L) to certain length classes with the number of shells in that particular long class. The division of long-sized class hoses is done by 1 + 3.3 Log N, while for the width of the hose (Maximum length - Minimum length) divided by the number of class intervals that have been previously obtained.

*Length and weight relationships Anadara gubernaculum*
Long-weight relationships have practical value that allows converting long values into weight or vice versa. This Weight can be regarded as a function of its length, and this long-weight relationship follows the law of cubic expressed by the formula [1]:

$$W = aL^b \tag{1}$$

Information :
W = Weight (gram)
L = Length (mm)
A = Intercept
B = Prediction Lenght and weight growth patterns

*Condition Factor of A. gubernaculum*
The condition factor often referred to as K factor is applied from the analysis of length and weight relationships is an important derivative in the growth of bivalves.In this case, condition factors can illustrate both at least bivalve conditions seen in terms of physical capacity for survival and reproduction.
To see the condition factors in bivalves used formulas [1]:

$$Kn = \frac{W}{aL^b} \tag{2}$$

Information :
Kn = Relative condition factor
W = Body weight (gr)
L = Body length
Aa b = Constants

*Growth parameter of A. gubernaculum*
To know the growth parameters used growth model Von Bertalanffy namely [2]:

$$Lt = L\infty\,(1 - e^{-K(t-to)}) \quad (3)$$

Information:
Lt = shells length at time t (cm);
L = Asimtot length of shells (cm);
K = growth coefficient (per year);
$T_0$ = Theoretical age of the shells on the same length as zero (year)
T = Age of shells at Lt (year).
To estimate the theoretical age ($t_0$) when the shells length is equal to 0 (zero), as follows [3]:

$$\text{Log10}\,(-t_o) = -\,0.3922 - 0.2752\,\text{Log}_{10}\,L\infty\ -\ 1.038\,\text{Log}_{10}\,K \quad (4)$$

L∞ is the maximum length of the shells theoretically (asymptotic length), K is the growth rate coefficient (per unit time) and $t_0$ is the theoretical age of the shells when the total length of the shell is equal to zero.

*Mortality and Exploitation Rate*
The coefficient of total mortality is estimated by using the length-converted catch curve with the following equation [4]:

$$\ln M = -0.0152 - 0.279 * \ln L\infty + 0.6543 * \ln K + 0.463 * \ln T \quad (5)$$
$$M = e(\ln M)$$

The rate of mortality of catching (F) is determined by: F = Z - M. Furthermore, the rate of exploitation is determined by comparing the catch mortality (F) to total mortality (Z) [5]:

$$E = F/(F + M) \quad (6)$$

Information :
E = Status of exploitation;
F = Coefficient of catching mortality;
M = Natural coefficient of death;
If: E> 0.5 indicates a high level of exploitation (over fishing);
E <0.5 indicates a low level of exploitation (*underfishing*);
E = 0.5 indicates optimal utilization

## 3. Results And Discussion

### *3.1 Results*

#### *3.1.1 Biological A. gubernaculum*

*A. gubernaculum* have a left shell bigger than the right shell (*inequivalvis*). An elongated elliptical shell. The shell is thick, heavy, and white. The surface of the shell is decorated with very real radial ribs. Flat radial ribs, without bulge. The periostrakum layer is thick and there are modifications in the form of layers such as "hair". Layers of periostrakum is dark brown.

IOP Conf. Series: Earth and Environmental Science **122** (2018) 012105 doi:10.1088/1755-1315/122/1/012105

*3.1.2 Length frequency distribution*

Based on observations made during 1 month 14 days, the frequency distribution of *Anadara gubernaculum* showed different results at each station. The total length of the *A. gubernaculum* ranges from 14 to 43 mm. Groupings are differentiated based on each station and grouped by 10 class interval.

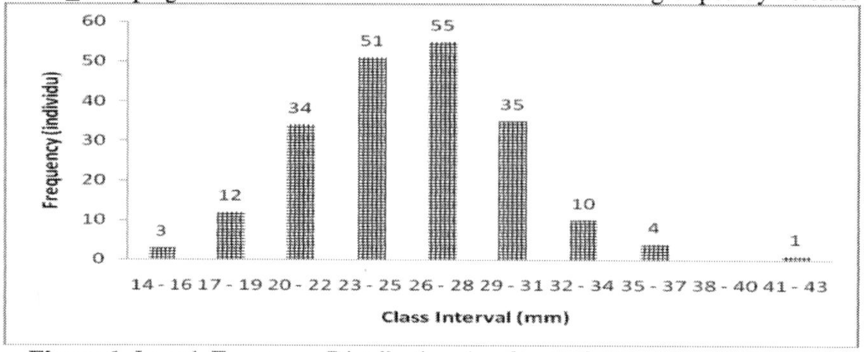


**Figure 1.** Length Frequency Distribution *Anadara gubernaculum* at Station 1

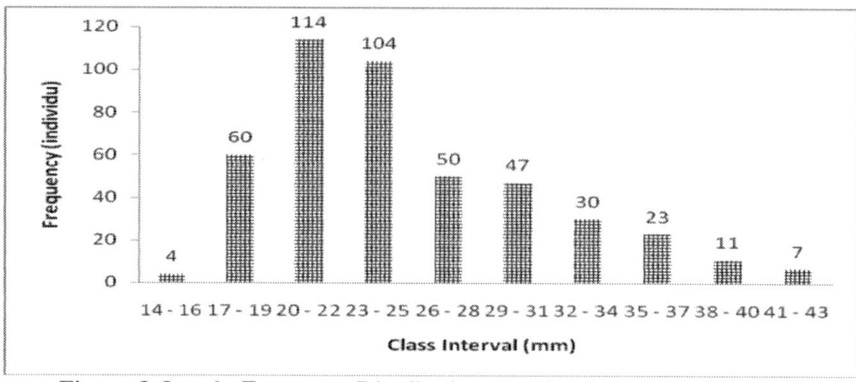


**Figure 2.** Lenght Frequency Distribution *A. gubernaculums* at Station 2

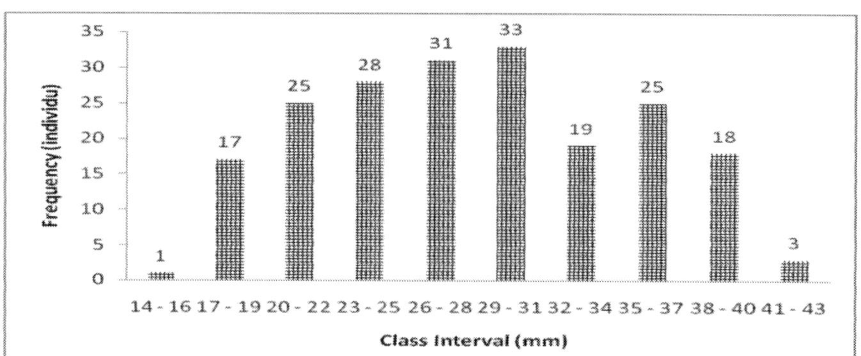


**Figure 3.** Length Frequency Distribution *Anadara gubernaculum* at Station 3

*3.1.3 Length and weight relationship of A. gubernaculum*

The result of measurement of length and weight of shells to get the coefficient of determination ($R^2$), intercept value, and regression coefficient based on the overall catch.

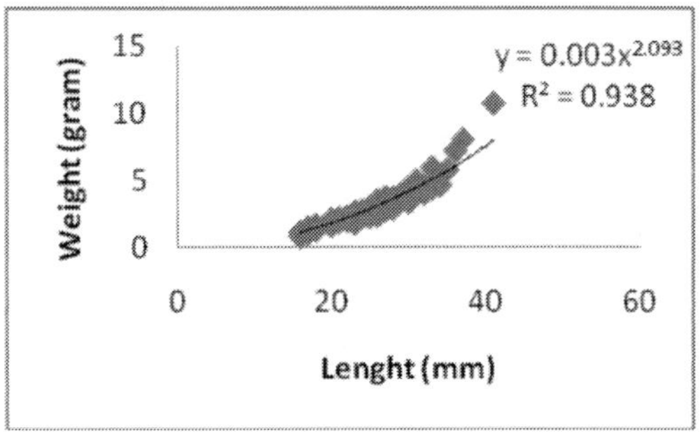


**Figure 4.** Length and weight relationship of *A.gubernaculum* at Station 1

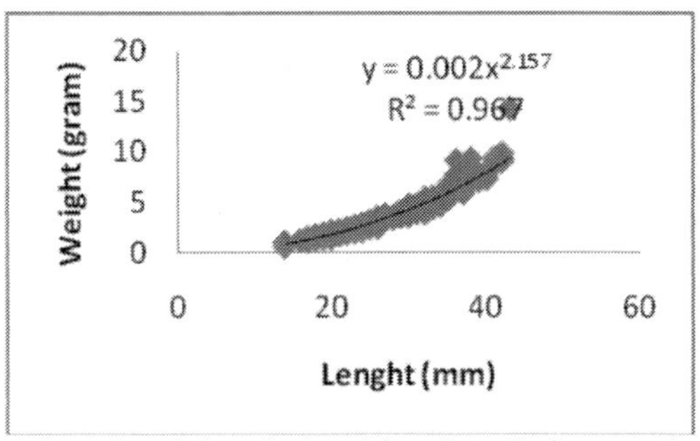


**Figure 5.** Length and weight relationship of *A. Gubernaculum* at Station 2

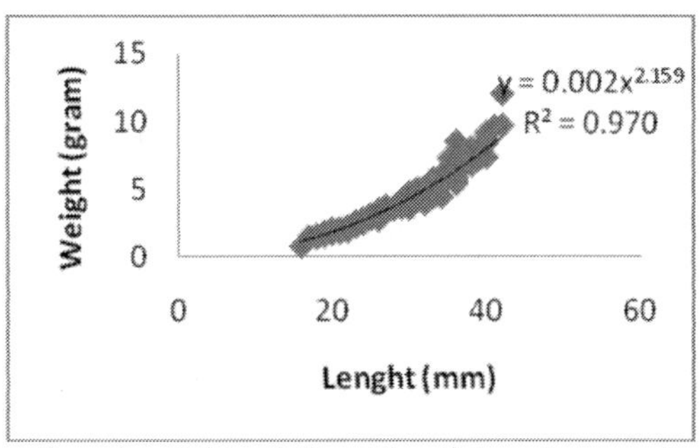


**Figure 6.** Length and weight relationship of *A. gubernaculum* at Station 3

*3.1.4 Condition Factor*

The result of calculation of the condition of the *A. gubernaculums* sampling in October - November 2016 where at station 1, it has value of condition factor ranged from 0.81 – 1.11. At station 2, it has a condition factor value ranged from 0.85 - 2.15. At station 3, it has a condition factor value ranged from 0.95 - 1.87.

*3.1.5 Growth parameter of A gubernaculum*

Result of analysis of growth parameters of *Anadara gubernaculum* consist of growth coefficient (K) is 1.2/year and infinitive length (L∞) is 43.05 mm and theoretical age of shells at the same length equal to zero ($t_0$) is -0.12 presented In Table 1 that was analyzed by ELEFAN method I in FiSAT II program.

**Table 1.** Growth parameter of *A. gubernaculum* in Asahan aquatic In October - November 2016

| | Growth parameter | | | |
|---|---|---|---|---|
| *A. gubernaculums* | $L_\infty$ (mm) | K (year$^{1}$) | $t_0$ (year) | Lt (mm) |
| | 43.05 | 1.2 | -0.12 | $43.05(1\text{-}e^{(-1.2(t+0.12)})$ |

*3.1.6 Mortality and Exploitation Rate*

Estimation of natural mortality rate of shells use empirical formula Pauly (Sparre and Venema, 1999) with average temperature of Asahan surface aquatic 31.17 °C. The results of the analysis of alleged mortality and the rate of exploitation of the *A. gubernaculum* consist of total mortality (Z) is 2.121/year, natural mortality (M) is 1.9/year, mortality of catching (F) is 0.21/year, and the rate of exploitation (E) is 0.1 (*Underfishing*)

*3.2 Discussion*

*3.2.1 Frequency Distribution of Length A. gubernaculum*

The results obtained can be seen that at station 1 is 205 individuals and most measuring 26 mm - 28 mm is 55 individuals, at station 2 is 450 individuals and most sized 20 mm - 22 mm is 114 individuals, and in station 3 is 200 individuals, and the largest size 29 mm - 31 mm were 33 individuals. The results of the three stations indicate that the amount of shells at station 2 more than at station 1 and at station 3 because *Anadara* did not occur in catches from this area at that time. So that its larger size as well as its great abundance at the time [6]

*3.2.2 Length and weight relationship of A. gubernaculum*

The obtained results at station 1 has growth pattern (b) of 2.093, at the station 2 is 2.157 while at the station 3 of 2.159. At each station has a negative allometric growth type (b<3) which is the increase in shells length is faster than the total weight gain. Therefore, thevalue of 'b' in the current study demonstrates that shells grows negative allometrically instead of isometrically and food availability can affect tissue growth, storage and utilization, which have the ability to change the ratio of total weight to shell length [7]. And negative allometric growth suggest that a negative growth may result in a rounded shell or body shape and may also influence the maximum shells length so that negative allometric growth may affect its maximum shells length and consequently their total weight [8]

*3.2.3 Condition Factor of Anadara gubernaculum*

The value of the condition of the shells *A. gubernaculum* station 1 was 0.81 - 1.11, station 2 of 0.85 - 2.15, and station 3 of 0.95 - 1.87. values of 1.0 and above indicate for better growth of the species, whereas values less than 1.0 indicate a stress condition or spent condition of the species [9]

*3.2.4 Growth parameter of A. gubernaculum*

Based on the analysis results obtained asymptotic length of L∞ *A. gubernaculum* is 43.05 mm with growth coefficient (K) is 1.2. There are big changes in growth between the same species in different populations. A variety of environmental factors such as the immersion period, turbidity, seawater temperature, salinity and water pollution have been confirmed to influence the growth rate of different *Anadara* spp populations [5]

*3.2.5 Mortality and Exploitation Rate of A. gubernaculum*

The value of mortality rate (Z) of *A. gubernaculum* is 2.121/year, natural mortality value (M) of *Anadara gubernaculum*is 1.9/year and the value of mortality of catching (F) of 0.21/year . Mortality that occurred in the study site was considered to be from natural causes (M). There were little empty shells and dead samples in the study site, which could be the reason for the existence of natural mortality In addition, the results of the theoretical calculation of total mortality in the current study showed that the shells mortality level was very low [8]. These results indicate that the analysis of the rate of exploitation (E) of shells in Asahan aquatic has a value of 0.1. The value of exploitation of *A.gubernaculum* is less than 0.5 indicated the 'under-fishing' condition (under-exploitation) which led to a decrease in biomass in the study area [7]

## 4. Conclusions

The conclusions obtained from this research are:

1. *Anadara gubernaculum* in the aquatic of Asahan has a negative allometric
2. The total mortality rate (Z) of *Anadara gubernaculum* of 2.121/year with natural mortality (M) of 1.9/year and the mortality of catching (F) of 0.21/year, and the rate of exploitation of 0.1/year. The value of exploitation rate indicates that this asahan aquatic is still under sustainable condition.

## 5. References

[1] Le Cren C D, 1951 The length-weight relationship and seasonal cycle in gonad weights and condition in perch, *perca jluviatilis. J. Anim. Ecol.* **20** 201- 209

[2] Pauly D Soriano-Bartz M Moreau J and Jarre-Teichmann A 1992 A new model accounting for seasonal cessation of growth in fishes. *J. Mar. Freshwater – Res.* **43** 1151-1156

[3] Pauly D 1983 Some simple methods for the assessment of tropical fish stocks. (Manila: Food and Agriculture Organization) pp 1-60

[4] Pauly D 1999 On the interrelationships between natural mortality, growth parameters and mean environmental temperature in 175 fish stocks. *J. Cons. CIEM* **39** 175-192.

[5] Gulland J A and Rosenberg A A (1992) A review of length-based approaches to assessing fish stocks. (Rome: Food & Agriculture Organization) ISBN 92-5-103121-5

[6] Tebano T and Paulay G 2000 Variable recruitment and changing environments create a fluctuating resource: the biology of *anadara uropigimelana* (bivalvia: arcidae) on tarawa atoll *Atoll Research Buletin* 1 - 15

[7] Mirzaei M R Yasin Z and Hwai A T S 2014 Length-weight relationship, growth and mortality of anadara granosa in penang island, malaysia: an approach using length-frequency data sets *Journal of the Marine Biological Association of the United Kingdom* 1 - 10.

[8] Albuquerque FS Peso-Aguilar MC Assuncao~Albuquerque MJT and Galves L 2009 Do climate variables and human density affect *achatina fulica* (bowditch) (gastropoda: *pulmonata*) shell length, total weight and condition factor? *Braz. J. Biol.* 879-885.

[9] Sharma R, Venkateshvaran K and Purushothaman C S 2005 Length-weight relationship and condition factor of *perna viridis* (linnaeus, 1758) and *meretrix meretrix* (linnaeus, 1758) from mumbai waters *J. Indian Fish. Assoc.* **32** 157-163

**Aknowledgment**

This researchs have financed by Non BNPB fund of University of Sumatera Utara on TALENTA research. We say thank you very much.

# Habitat characteristic of macrozoobenthos in Naborsahan River of Toba Lake, North Sumatra, Indonesia

**M Basyuni[1*], M S Lubis[2] and A Suryanti[3]**

[1] Department of Forestry, Faculty of Forestry, Universitas Sumatera Utara, Jl. Tri Dharma Ujung No. 1 Medan, North Sumatera 20155, Indonesia
[2] Department of Aquatic Resource Management, Faculty of Agriculture, Universitas Sumatera Utara, Medan, North Sumatera 20155, Indonesia
[3]Faculty of Marine Science and Fishery, Universitas Maritim Raja Ali Haji, Jl. Jl.Politeknik Senggarang, 29100 Tanjungpinang, Indonesia
[*]Email: m.basyuni@usu.ac.id

**Abstract.** This research described the relative abundance, dominance index, and index of macrozoobenthos equitability in Naborsahan River of Toba Lake, North Sumatra, Indonesia. The purposive random sampling at three stations was used to characterize the biological, chemical, and physical parameters of macrozoobenthos. The highest relative abundance of macrozoobenthos found at station 2 (99.96%). By contrast, the highest dominance index was at station 3 (0.31), and the maximum equitability index found at station 1 (0.94). The present results showed diversity parameters among the stations. A principal component analysis (PCA) was used to determine the habitat characteristics of macrozoobenthos. PCA analysis depicted that six parameters studied, brightness, turbidity, depth, temperature, dissolved oxygen (DO) and biochemical oxygen demand ($BOD_5$) play a significant role on the relative abundance, dominance index, and equitability index. PCA analysis suggested that station 3 was suitable habitat characteristic for the life of macro-zoobenthos indicating of the negative axis. The present study demonstrated the six parameters should be conserved to support the survival of macrozoobenthos.

## 1. Introduction

Toba Lake is the largest lake in Southeast Asia, and the deepest lake in the world which was formed 75,000 years ago after an earth splitting volcano eruption [1]. The Toba Lake located at seven Regencies of North Sumatra province. Patterns of streams in the inlet of Toba Lake is dominated by small rivers, with a total of 289 streams. Among them is the famous Naborsahan River which is one of the inlets of Toba Lake.

The river formed aquatic ecosystems to play a significant role in the hydrological cycle and serves as a water catchment area for the adjacent lake [2]. The environmental characteristics strongly influence the condition of a river. River waters have biotic and abiotic components interact to form ecosystems which mutually changed [3]. Macrozoobenthos was known as bioindicators to evaluate freshwater bodies receiving domestic and industrial waste matters [4]. Based on its size, macrozoobenthos is very easy to identify and sensitive toward the environment change. The size of macrozoobenthos (animals at least 2 mm in size) can reach at least 3 to 5 mm at the maximum growth [5]. To get more insight into ecosystem function of the Inlet of Toba Lake the habitat characteristics of macrozoobenthos surrounding of Toba Lake was studied. This present research aimed to characterize the biological, chemical, and physical parameters of macrozoobenthos in Naborsahan River of Toba Samosir, North Sumatra, Indonesia.

## 2. Materials and Method

### *2.1. Time and Location Research*

This study was conducted in April-June in Naborsahan River of Toba Samosir, North Sumatra. Sampling was carried out at three stations. Macrozoobenthos was sampled from the middle to downstream of the Naborsahan River to characterize the biological, chemical, physical parameters of macrozoobenthos. Three locations were determined with each distance 1-1.5 km. Station 1 located at 02° 39' 06.5" North latitudes and 098° 56' 12.7" East longitudes with sand dominated. Station 2 sited at 02° 39' 12.4" North latitudes and 098° 56' 06.6" East longitudes with sand substrate. Station 3 situated at 02° 39' 19.1" North latitudes and 098° 56' 03.6" East longitudes with mud substrate.

### *2.2. Measurement of the parameters*

Physical and chemical parameters were measured at each station. Measurement of the parameters was performed *in situ* namely temperature, currents, depth, brightness, pH, and DO (dissolved oxygen) as previously described [5]. Furthermore, C-organic content, turbidity, COD (chemical oxygen demand), and $BOD_5$ (biological oxygen demand) were conducted *ex situ* as previously reported [5].

### *2.3. Macrozoobenthos measurement*

Each station was divided into 4 points analysis as quintuplicate to identify and measure macrozoobenthos. Eikman Grab was used both to take macrozoobenthos (ind/$m^2$) samples and substrate. Samples were obtained at each point to be sorted accordingly to its morphology such as shape, colour, and size for data analysis as previously described [6].

## 3. Results and Discussion

### *3.1. Physical and chemical parameter of macrozoobenthos*

The measurement of temperature at all stations was 23-25 °C (Table 1). It has been reported the optimal temperature range for macrozoobenthos growth between 20-30 °C [6]. Naborsahan River, therefore, had good water quality due to the development of macrozoobenthos. Table 1 shows the values of current velocity was 0.23 to 0.44 m/s. The current speed affected the spread of macrozoobenthos [7]. Current velocity speed caused the spread of the genus macrozoobenthos each station was different. The entire depth ranged from 55-88.9 cm each site. Position 3 was the deepest position because it was at the mouth of the river and the inlet of Lake Toba with a height that reaches 905 m elevation above sea level. Water transparency depends on the turbidity [3]. The results of brightness measurements from 49.3-55.9 cm. The entire station indicated clear water as the result of depth and light did not show a significant difference (Table 1).

Turbidity in the river could limit the entry of sunlight into the water hence the concentration of dissolved oxygen (DO) was getting lower. The values of turbidity were 4.4-14.3 NTU. Naborsahan River could tolerate due to several genera exist in the entire station. The C-organic substrate was at each site range 2.36-2.72%. Location 3 was part of the downstream Naborsahan River, but C-organic content at the station 3 was lower than the position 2 (2.38%). This condition was because of the number of water hyacinth plants that grow on the station 3. The location that has lots of water hyacinth plants can absorb organic material so that the C-organic matter content is low [4].

The values of pH $<5$ or $> 9$ were incompatible for life of macrozoobenthos [5]. The results obtained by measuring the pH ranged 6-7. Based on the results of the measurement of pH in the habitat Naborsahan river still was suitable for the macrozoobenthos growth. According to the value of the DO was 3 mg/l, it was the minimum threshold value. The content of DO obtained ranged from 6.9 to 10.9mg/ l. The DO was the highest at the station three because the location 3 was an inlet of Lake

Toba and the lower reaches of the river to meet the Naborsahan waves and currents velocity caused the dissolved oxygen (DO) to be increased in the station. The $BOD_5$ was obtained for the entire sites *ex-situ* were equal to 1.8 to 7.1 mg/l. The different values in each research station were due to the difference amount of organic matter contained. Natural materials derived from domestic waste and agriculture which located on the Naborsahan river.

**Table 1.** Physical and chemical parameters of Naborsahan Rivers

| Parameter | Station | | |
|---|---|---|---|
| | 1 | 2 | 3 |
| Temperature (° C) | 23 | 23 | 25 |
| Current velocity (m/s) | 0.44 | 0.23 | 0.43 |
| Depth (cm) | 55.05 | 68.40 | 68.90 |
| Brightness (cm) | 49.30 | 50.40 | 55.90 |
| Turbidity (NTU) | 4.45 | 7.63 | 14.30 |
| C-organic content (%) | 2.36 | 2.72 | 2.38 |
| pH | 7.02 | 6.70 | 7.05 |
| DO (mg/l) | 6.90 | 7.20 | 10.90 |
| $BOD_5$ (mg/l) | 1.80 | 4.10 | 7.10 |
| COD (mg/l) | 5.56 | 5.70 | 5.75 |

DO: dissolved oxygen, $BOD_5$: biological oxygen demand, COD: chemical oxygen demand.
Data are the mean of quintuplicate analyses

### *3.2. Genus and distribution of macrozoobenthos*

Table 2 displays distribution macrozoobenthos genera in Naborsahan River. At all stations, 26 genera were found. Almost every point sampled Thiara dominated throughout stations. Not all macro-zoobenthos had an excellent tolerance to environmental changes such as changes in physical and chemical parameters which differ in each site. There were several genera that there were only at the station two such as Bezzia and there were only at the location 3 such as Anentome, Alasmidonta, Gillia, Gyraulus, Hydrophilus, Littoridinops, Planaria, Prostoma, Pseudosuccinea, Somatogyrus, and Uvarus.

Relative abundance of each station showed the number of genera was divided by the total number of individuals. Table 3 depicted the relative abundance of the highest value was 99.96% at the station 2. Genus Stenelmis had the highest relative abundance value by 25% at the station 2, and the genus that had the lowest value relative abundance at this station was Bezzia, Chaoborus, Elimia, and Subulina with 0.62%. It has been reported that a habitat corresponded to the development of an organism if the value of relative abundance was more than 10% [6]. We interpreted that habitat at station II according to the development of Stenelmis because the levels of dissolved oxygen (DO) at position II was 7.2 mg/l and station I has the lowest level of turbidity 4.45 NTU (Table 1).

Table 3 showed the highest dominance index value was 0.31 at the location 3. By contrast, the stations 1 and 2 had dominance index values were 0.13 and 0.11, respectively. This study revealed that no specific macrozoobenthos genus dominant at every position due to dominance index value was under 0.5, indicating no particular genera that dominated of macrozoobenthos at each station.

IOP Conf. Series: Earth and Environmental Science **122** (2018) 012106 doi:10.1088/1755-1315/122/1/012106

**Table 2**. Genus and distribution of macrozoobenthos at the Naborsahan River

| No | Genera | Station 1 | | | | Station 2 | | | | Station 3 | | | |
|---|---|---|---|---|---|---|---|---|---|---|---|---|---|
| | | 1 | 2 | 3 | 4 | 1 | 2 | 3 | 4 | 1 | 2 | 3 | 4 |
| 1 | Alasmidonta | - | - | - | - | - | - | - | - | - | - | + | + |
| 2 | Anentome | - | - | - | - | - | - | - | - | - | - | + | + |
| 3 | Bezzia | - | - | - | - | - | - | - | + | - | - | - | - |
| 4 | Branchiura | - | - | + | - | + | + | + | - | + | + | + | + |
| 5 | Chaoborus | - | - | + | - | + | - | - | - | - | - | - | - |
| 6 | Chrysops | - | - | - | - | + | - | + | + | - | - | - | - |
| 7 | Diamesa | - | + | + | - | + | - | - | + | - | - | - | - |
| 8 | Elimia | - | - | - | - | - | - | + | - | + | + | + | + |
| 9 | Gillia | - | - | - | - | - | - | - | - | + | + | + | - |
| 10 | Gyraulus | - | - | - | - | - | - | - | - | - | + | - | - |
| 11 | Hexatoma | - | + | + | + | - | + | + | + | - | - | - | - |
| 12 | Hydrophilus | - | - | - | - | - | - | - | - | - | - | + | - |
| 13 | Littoridinops | - | - | - | - | - | - | - | - | + | + | + | - |
| 14 | Melanoides | + | - | + | - | + | - | + | + | + | + | + | + |
| 15 | Physa | - | - | - | - | - | - | - | - | + | + | + | - |
| 16 | Pila | + | - | + | - | + | + | - | + | + | + | + | - |
| 17 | Planaria | - | - | - | - | - | - | - | - | + | - | - | - |
| 18 | Prostoma | - | - | - | - | - | - | - | - | + | - | - | - |
| 19 | Pseudosuccinea | - | - | - | - | - | - | - | - | - | - | + | - |
| 20 | Somatogyrus | - | - | - | - | - | - | - | - | - | + | - | - |
| 21 | Stenelmis | - | + | - | - | - | + | - | - | + | - | - | - |
| 22 | Subulina | - | - | - | - | - | - | + | - | - | + | - | - |
| 23 | Thiara | + | + | + | - | + | + | + | + | + | + | + | + |
| 24 | Tryonia | + | + | - | - | + | + | + | - | + | + | + | + |
| 25 | Uvarus | - | - | - | - | - | - | - | - | + | - | - | - |
| 26 | Viviparus | - | + | - | - | + | + | + | - | - | + | + | - |

Note :
"Number 1-4" : The time of samples are taken once in two week
"+" : Presence of genus; "–" : Absence of genus;

**Table 3**. Relative abundance (RA), dominance index (DI), dan equitability index (EI) of macrozoobenthos

| Biological parameters | Station | | |
|---|---|---|---|
| | 1 | 2 | 3 |
| Relative abundance | 88.52 | 99.96 | 99.92 |
| Dominance index | 0.11 | 0.13 | 0.31 |
| Equitability index | 0.94 | 0.82 | 0.52 |

Data are the mean of quintuplicate analyses

Table 3 displayed the station 1 had the highest equitability index. The value of equitability index closed to 1 then the spread of the same number of each genus and value of equitability index closed to 0 then the number was not the same deployment [8]. The range of the large of each genus macrozoobenthos was around the same station. Therefore it has shown that an area has a number of the same genus to another area then the area has a higher diversity. Macrozoobenthos habitat characteristics could be determined through physical and chemical parameters contained in macrozoobenthos habitat itself. Physical and chemical parameters would interact and influence each other so that the natural and artificial conditions would affect the status of macrozoobenthos [9].

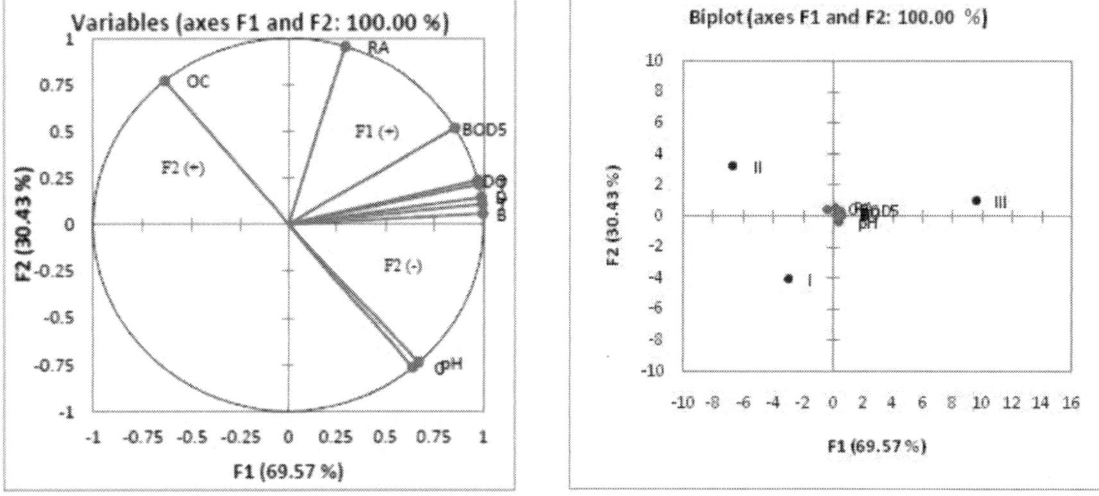


(a) (b)

Note:

a. F1(+): brightness (B), turbidity (Tr), depth (D), temperature (T), DO, $BOD_5$ and relative abundance (RA). F2(+): C-organic content (OC), F2(-): C- organic content and pH.

b. All physical and chemical parameters were clustered at the three stations.

**Figure 1**. a. PCA variables with physical and chemical parameters and relative abundance.
b. PCA Biplot with physical and chemical parameters and relative abundance at the research station.

Figure 1a showed the positive axis 1 (F1) had the characteristics of habitat could be interpreted that the correlation circles tend to cluster called brightness, turbidity, depth, temperature, DO and $BOD_5$. The relative abundance found on the same axis with intensity parameters, such as turbidity, depth, temperature, DO, and $BOD_5$, as clumped, approached the X axis. These parameters have been shown to have a considerable effect on the relative abundance of macrozoobenthos. On the other hand, other parameters such as flow, pH, and C-organic content were no correlation to RA.

Figure 2a illustrated the results of the analysis were represented using two principal axes that would account for 100% of the total variance. Data consists of the main components of 77.76% and 22.24% for the second element, respectively. The interpretation of PCA had been considered to represent a state that occurs without reducing the information from the data obtained [10].

Habitat characteristics are shown in the positive axis one clearly that contributed these parameters negatively to the shaft 2. Dominance index was on the same axis with brightness parameters, turbidity, depth, temperature, DO and $BOD_5$. The six parameters have an important role to the dominance index of macrozoobenthos. While other parameters such as the current, pH, and C-organic content were not correlated to the index due to the dominance index from the X axis.

Dominance index value was under 0,5 (Table 1) showed all macrozoobenthos had the adaptability and survival rate difference at every station on the physical and chemical parameters to characterize the habitat of each site [6]. To conserve the habitat characteristics of macrozoobenthos an ecological approach was required at Naborsahan River accordingly.

The measurements of physical and chemical parameters associated with equitability index using PCA analysis to determine the characteristics of macrozoobenthos habitat. Some parameters such as depth, brightness, turbidity, DO, and $BOD_5$ significantly affected the results of the study which are represented by two principal components they were a major part (76.43%) and the second element (23.57%). The result showed that almost all parameters strongly influenced the station 3. PCA analysis was presented in Figure 3a and 3b.

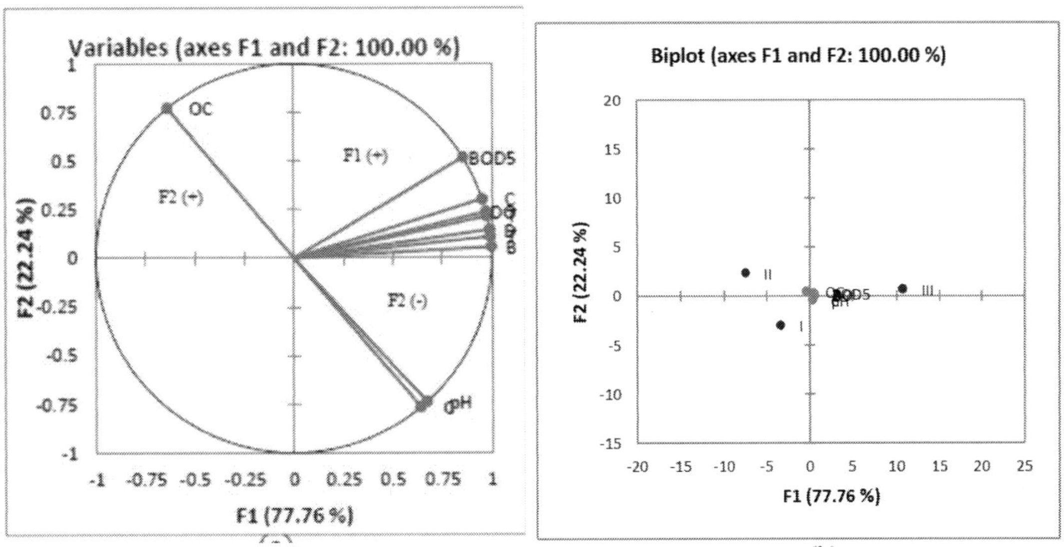


(a) (b)

Note:

a. F1(+): Brightness (B), turbidity (Tr), depth (D), temperature (T), DO, $BOD_5$ and dominance index (DI). F2(+): C-organic content (KC), F2(-): current velocity (C) and pH.
b. All physical and chemical parameters were clustered at the three stations.

**Figure 2.** a. PCA variables with physical and chemical parameters and dominance index,
b. PCA Biplot with physical and chemical parameters and dominance index at three research stations.

Figure 3a shows habitat characteristics could be illustrated that the correlation circle was at negative axis 2. Parameters such as brightness, turbidity, depth, temperature, DO and $BOD_5$ and equitability index or was clustered index uniformity approaching the X axis. Our current results had a considerable influence on the equitability index macrozoobenthos. While the flow, pH, and C-organic content showed no correlation to the abundance of individuals that not be approaching the X axis.

Figure 3b depicts PCA analysis of physical and chemical parameters of the research site. The results displayed the physical and chemical parameters had a significant role and an essential habitat for macrozoobenthos at the station three that was on the negative axis 2. The station 3 was the lowest equitability index macrozoobenthos (0.52). The physical parameters of the PCA analysis affected the chemical characteristics of the habitat at the station three which showed smaller value indicates uniformity low. This result showed that the density of each type could be same and tend to be dominated by a particular type.

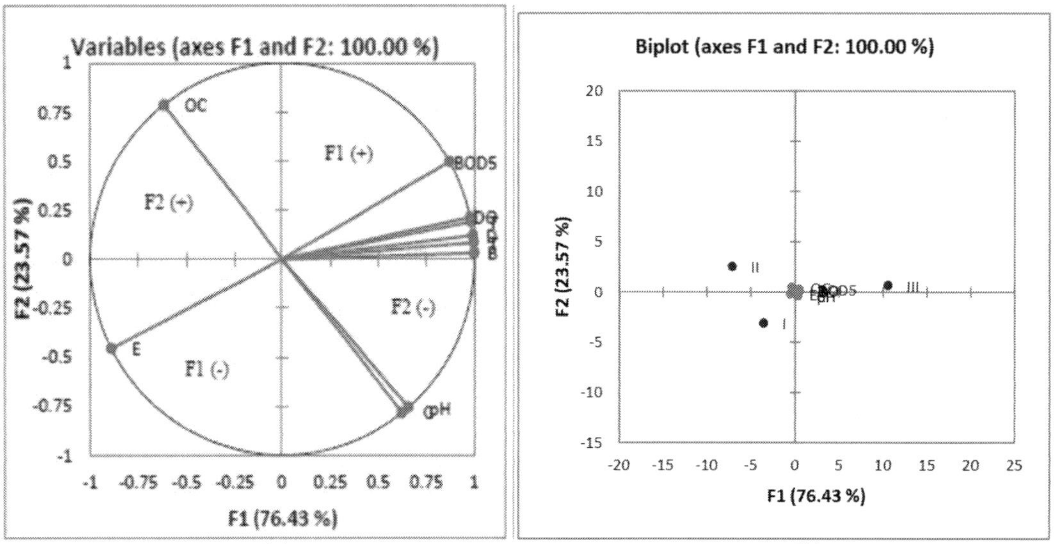


(a) (b)

Note:

a. F1 (+): Brightness (B), turbidity (Tr), depth (D), temperature (T), DO and $BOD_5$. F1 (-): Equitability index (E). F2 (+): C-organic substrate (OC). F2 (-): current velocity (C) and pH.

b. All physical and chemical parameters were clustered at the three station.

**Figure 3.** a. PCA variables with physical and chemical parameters and equitability index.
b. PCA Biplot with physical and chemical parameters and equitability index at the research station.

## 4. Conclusion

We have confirmed that physical and chemical parameters affected macro-zoobenthos habitat characteristics at each station. Six parameters play a significant role in Naborsahan river of Toba Samosir namely brightness, turbidity, depth, temperature, DO and $BOD_5$. The six parameters should be preserved to support the survival of macrozoobenthos in the sites.

## References

[1] Mark DF, Petraglia M, Smith VC, Morgan LE, Barfod DN, Ellis BS, Pearce NJ, Pal JN and Korisettar R 2014 A high-precision 40 Ar/39 Ar age for the Young Toba Tuff and dating of ultra-distal tephra: Forcing of Quaternary climate and implications for hominin occupation of India *Quat. Geochronol.* **21** 90–103.

[2] Jeppesen E, Brucet S, Naselli-Flores L, Papastergiadou E, Stefanidis K, Noges T, Noges P, Attayde JL, Zohary T, Coppens J and Bucak T 2015 Ecological impacts of global warming and water abstraction on lakes and reservoirs due to changes in water level and related changes in salinity *Hydrobiologia* **750** 201–27.

[3] Laverty C, Dick JT, Alexander ME and Lucy FE 2015 Differential ecological impacts of invader and native predatory freshwater amphipods under environmental change are revealed by comparative functional responses. *Biol. Invasions* **17** 1761–1770.

[4] Nesemann HF, Sharma G, Kumar R, Sheetal A and Roy S 2017 Do the rivers of Chota Nagpur Dry Forests (Jharkhand), Damodar and Subarnarekha differ in biodiversity of aquatic macrozoobenthos, functional feeding groups and biological water quality? *Aquat. Ecosyst. Health Manage.* **20** 116–129.

[5] Obolewski K 2011 Macrozoobenthos patterns along environmental gradients and hydrological connectivity of oxbow lakes *Ecol. Eng.* **37** 796–805.
[6] Wang Z, Zhang Z, Zhang J, Zhang Y, Liu H and Yan S 2012 Large-scale utilization of water hyacinth for nutrient removal in Lake Dianchi in China: the effects on the water quality, macrozoobenthos and zooplankton *Chemosphere* **89** 1255–1261.
[7] Maximov AA 2015 The long-term dynamics and current distribution of macrozoobenthos communities in the Eastern Gulf of Finland, Baltic Sea *Russ. J. Mar. Biol.* **41** 300–310.
[8] Browne HP, Forster SC, Anonye BO, Kumar N, Neville BA, Stares MD, Goulding D and Lawley TD 2016 Culturing of' unculturable' human microbiota reveals novel taxa and extensive sporulation *Nature* **533** 543–546.
[9] Bylak A, Kukuła K, Plesiński K and Radecki-Pawlik A 2017 Effect of a baffled chute on stream habitat conditions and biological communities *Ecol. Eng.* **106** 263–272.
[10] Rajsekhar D, Singh VP and Mishra AK 2015 Multivariate drought index: An information theory based approach for integrated drought assessment *J. Hydrology* **526** 164–182.

# Abundance of sea cucumbers on the ecosystem of seagrasses Inunggeh island, Tapanuli Tengah Regency North Sumatera Province

**M M Wisesa[1], D Bakti[2] and A Fadhilah[2]**

[1]Student of Water Management Resources, Faculty of Agriculture, Universitas Sumatera Utara

[2]Lecturer of Water Management Resources, Faculty of Agriculture, Universitas Sumatera Utara

E-mail: dbakti06@yahoo.com

**Abstract**. Unggeh Island is one area that has the potential of Sea Cucumber in the North Sumatra. Sea cucumbers have an important role in ecosystem waters, namely as a deposit feeder. Sea cucumbers can live in shallow waters, such as seagrass ecosystems. The purpose of this study is to knowing the abundance of sea cucumbers in the seagrass ecosystems on the island of Unggeh and to knowing the type of Sea Cucumber. The method used is a transect quadrant method with a size of 5x5 meters, on a transect line with a length of 100 meters. Sampling was done at three points observations, station 1 was at coordinate point 01°34'26,88 "LU and 098°45'40,25" BT, station 2 was at coordinate point 01°34'32,71 "LU and 098°45'37, 58 "BT, station 3 is at the coordinate point 01°34'24,22" LU and 098°45'38,06 "BT. The type of sea cucumber found in the seagrass ecosystem on the Unggeh island *Actinopyga ecinites, A. Miliaris, Holothuria scabra*. The density at station 1 was 0.16 ind / $m^2$, at station II a density was0.12 ind / $m^2$, at station III a density was 0.08 ind / $m^2$, and the total density at the research location was 0, 32 ind / $m^2$.

## 1. Introduction

The Indonesian sea has a very diverse wealth. In terms of ecosystems, biota-biota and biological resources contained therein. This natural wealth can be seen one of them from the biological resources of the waters. Various types of nekton, echinodermata, crustaceans, mollusca, plankton. Sea cucumbers are one of the marine biological resources that are quite abundant and scattered in Indonesian waters. These animals have many benefits for human life.

Sea cucumbers are spread over several marine waters, including in Indonesia. Habitat sea cucumbers in the form of seagrass ecosystems and coral reef ecosystems, ranging from the intertidal zone to a depth of 20 meters. In general, sea cucumbers require an environment that is not polluted. Some environmental parameters that affect the presence of sea cucumbers are temperature, pH, water turbidity, dissolved oxygen, ocean currents, light penetration, water salinity, substrate, nitrate, nitrite, orthophosphate, biological oxygen demand, and suspended solids [1]. Sea cucumbers include one of thorn-skinned animals or Echinodermata [2]. Sea cucumbers are commonly found in areas where algae or seagrass beds are found[3].

Unggeh Island is located in Sitardas village of Badiri district, Central Tapanuli district at coordinates between 1°34'23 "- 1°34'37" LU and 98°45'42 "BT. The unggeh island is located about 11 miles from the mainland of Sumatra and is an overgrown island of flora such as coconut trees bushes, ketaping trees and several other wood species [4].

## 2. Research methods

This research was conducted in the area of Pulau Unggeh Badiri District, Central Tapanuli Regency, North Sumatera Province. Observation abundance of sea cucumbers done in the field. Data analysis was conducted at Integrated Laboratory of Water Resource Management Faculty of Agriculture, University of Sumatera Utara. Water quality analysis was done at Balai Teknik Kesehatan Environments and Disease Control (BTKLPP) Class I Medan. Substrate analysis was conducted at the National Standardization Research Center. Research location can be seen in Fig.1

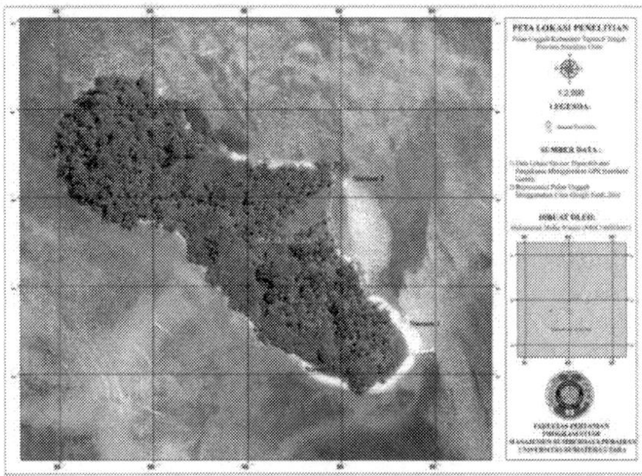

**Figure 1.** Research location of Pulau Unggeh Sub Badiri Kabupaten Tapanuli Tengah North Sumatra Province

The tool used is digital camera, Global Positioning System, identification book, stationery, calculator, secchi chip, tool box, stationery, ball, meter 100m, underwater camera, thermometer and pH meter, refractometer, 5x5 m , sample bottles, tracing paper, DO meter, cement ladle, General Manual of Identification and Monitoring of Sea Cucumber Population [5]. The materials used are Microsoft Excel software, plastic bags, sea cucumber samples, water samples, substrate samples and label paper.

The method used in collecting data of sea cucumber is observation (survey) field as shown in (Figure 1) to determine the location of sampling, plotting GPS. The area of the transect depends on the number of personnel, the density of sea cucumbers and the area of the area to be surveyed, 5 x 5 meters (distance between plots 5-10 meters) or 10 x 10 meters (distance between plots 5-10 meters). Distance between transects from 50 meters up to 10 meters. There are at least 3 plots in 1 transect, and the number of plots should be increased to represent the extent of the waters being monitored (reaching the edge) [5].

### *2.1 Water Quality Measurements*

Water quality data collection is done once before transect abundance of sea cucumber is done. Measurements are made by using the prepared equipment. Measurement of water quality parameters can be seen in Table.

**Table 1.** Measurement of water quality parameters

| Parameter | Pieace | Tool | Analysis |
|---|---|---|---|
| Physics | | | |
| Temperature | °C | Thermometer | *In situ* |
| Salinity | ppm | Refraktometer | *In situ* |
| Bright | % | Secchi disk | *In situ* |
| Depth | cm | Scaled sticks | *In situ* |
| Current Velocity | m/s | Ball guessed | *In situ* |
| Substrate | - | - | *Ex situ* |
| Chemical | | | |

| | | | |
|---|---|---|---|
| pH | - | pH meter | *In situ* |
| Nitrate | mg/L | - | *Ex situ* |
| Phospate | mg/L | - | *Ex situ* |
| Disolved Oxygen | ppm | DO meter | *In situ* |

*2.2 Data Analysis*

*The Density*

The density of Sea Cucumber (Holothuroidea) is calculated by using the formula [5], which are:

$$X = \frac{\Sigma Xi}{n} \qquad (1)$$

Information:

X : Average number of sea cucumbers widespread unity

Σx : Number of sea cucumbers in I unit

N : The total area of I unit

*Index of species diversity*

Species diversity can be regarded as an indication of the number of types of macrobenthos and how the spread of the number of individuals in each type and location of the sampling. To determine the diversity is calculated using the following Shannon-Wiener [5]formula:

$$H' = -\left(\sum \frac{ni}{N} \ln \frac{ni}{N}\right) \qquad (2)$$

Information:

H : Index of Species Diversity

Ni : Number of individual types of i

N : Total number of individuals

**Table 2.** The range of water stability based on the diversity index

| No | Stability Range | Diversity |
|---|---|---|
| 1 | $0 < H' \leq 1$ | Low (unstable) |
| 2 | $1 < H' \leq 3$ | Medium |
| 3 | $H' > 3$ | High (stable) |

*Index of Uniformity*

This index is used to determine the existence of the type that dominates the community and to know the spread of the number of individuals of each species. This index is obtained by comparing the index of diversity with its maximum value [5] using the following formula:

$$E = \frac{H'}{Hmax} \qquad (3)$$

Information:

E: Uniformity index

H ': Index of diversity

H max: Maximum species diversity (In S)

Values range from 0 and 1. The smaller the value of E, the smaller the value of uniformity. This shows the spread of the number of individuals of each species is not the same and there is a possibility that the population is dominated by a type of biota. Conversely, the greater the value of E, the higher the uniformity of the biota population. This shows that the number of individuals of each species is the same, where the population is not dominated by a biota type.

*Index of dominance*
To calculate the simpson dominance index [5] as follows:

$$D = \Sigma \left(\frac{n}{Ni}\right)^2 \quad (4)$$

Information :
D : The dominant index of simpson Ni
: Number of individual species i-i N : The
total number of species
S : Number of types

The D value is categorized as follows:
$0 < D \leq 0.5$ low dominance
$0.5 < D \leq 0.75$ moderate dominance
$0.75 < D \leq 1.00$ high dominance

## 3. Results and Discussions

### *3.1 Samples of sea cucumbers*
From research conducted in Pulau Unggeh Subdistrict of Badiri Regency of Central Tapanuli, got some kind of sea cucumber, Actinopyga milliaris, Actinopyga aechinites, Holothuria scabra.

### *3.2 Physical-Chemical Parameters*
The measurement of physics-chemical parameters of the waters in this study was conducted three times, ie morning, noon, and afternoon. Physical-chemical parameter values are taken before the sea cucumber transect is done. The parameters measured are temperature, depth, brightness, salinity, substrate, current, pH, DO, nitrate, phosphate. The value of physics-chemical parameter measurement can be seen in table 3, table 4, and table 5.

**Table 3.** Physical-chemical measurement results in the morning.

| Parameter | | Piece | Station I | Station II | Station III |
|---|---|---|---|---|---|
| Physics | Temperature | °C | 31 | 31 | 31 |
| | Depth | cm | 103 | 102 | 115 |
| | Bright | % | 100 | 100 | 100 |
| | Substrate | - | Sand | Sand | Sand |
| | Salinity | ppm | 32 | 33 | 29 |
| | Current Velocity | m/s | 0.58 | 0.46 | 0.55 |
| Chemical | pH | - | 8.05 | 7.92 | 7.99 |
| | DO | mg/l | 5.5 | 5.1 | 5.3 |
| | Nitrate | mg/l | 4.6 | 4.7 | 4.7 |
| | Phospate | mg/l | < 0.03 | < 0.03 | < 0.03 |

The following is data of physics-chemical measurement result of waters at Unggeh Island of Tapanuli Tengah Regency done in the morning.

**Table 4.** The results of physics-chemical measurements during the day

| Parameter | | Piece | Station I | Station II | Station III |
|---|---|---|---|---|---|
| Physics | Temperature | °C | 32 | 32 | 32 |
| | Depth | cm | 38 | 104 | 79 |
| | Bright | % | 100 | 100 | 100 |
| | Substrate | - | Sand | Sand | Sand |
| | Salinity | ppm | 27 | 29 | 30 |
| | Current Velocity | m/s | 0.05 | 0.067 | 0.025 |
| Chemical | pH | - | 7.91 | 7.91 | 7.99 |
| | DO | mg/l | 4.6 | 4.7 | 5.4 |
| | Nitrate | mg/l | 4.6 | 4.6 | 4.5 |
| | Phospate | mg/l | < 0.03 | < 0.03 | < 0.03 |

The following is the result of measurement of physics-chemical waters on Unggeh Island of Tapanuli Tengah Regency conducted at noon time.

**Table 5.** Physical-chemical measurement results in the afternoon.

| Parameter | | Piece | Station I | Station II | Station III |
|---|---|---|---|---|---|
| Physics | Temperature | °C | 30 | 30 | 30 |
| | Depth | cm | 105 | 112 | 118 |
| | Bright | % | 100 | 100 | 100 |
| | Substrate | - | Sand | Sand | Sand |
| | Salinity | ppm | 29 | 30 | 29 |
| | Current Velocity | m/s | 0.46 | 0.66 | 0.61 |
| Chemical | pH | - | 7.9 | 7.98 | 7.9 |
| | DO | mg/l | 5.3 | 5.6 | 5.9 |
| | Nitrate | mg/l | 4.6 | 4.6 | 4.8 |
| | Phospate | mg/l | < 0.03 | < 0.03 | < 0.03 |

The following is data of physics-chemical measurement result of waters on Unggeh Island of Tapanuli Tengah Regency conducted at afternoon time.

*Density of Sea Cucumber*

The density of sea cucumber at the location of research that is at station I has a density of 0.16 ind / $m^2$, on station II has a density value of 0.12 ind / $m^2$, while at station III has a density value of 0.08 ind / $m^2$, and total density of sea cucumber at research location that is 0,36 ind / $m^2$ can be seen in table 6.

**Table 6.** Density of sea cucumber (ind/$m^2$)

| Station | Number of individuals | Type of sea cucumbers | Density (ind/$m^2$) |
|---|---|---|---|
| Station I | 4 | *Actinopyga milliaris, A. Aechinites, Holothuria scabra* | 0.16 |
| Station II | 3 | *Actinopyga milliaris* | 0.12 |
| Station III | 2 | *Holothuria scabra, Actinopyga milliaris* | 0.08 |
| Total Density | | | 0.36 |

*Diversity Index (H '), Uniformity Index (E), Dominance Index (D)*
Based on the result of data processing of all stations, the value of the index of diversity of sea cucumber is 0.983 and has low category. For the value of uniformity is 0.895 and has a high category. While for the dominance value of 0.515 has medium category. The results can be seen in table 7.

**Table7.** Diversity Index (H '), Uniformity Index (E), Dominance Index (D)

| Index | H' | E | D |
|---|---|---|---|
| Value | 0.983 | 0.895 | 0.515 |

*3.3 Discussion*
*Physics-Chemical Parameters*
Based on the measurement of physics-chemical parameters of the waters of Unggeh Island, it is found that the temperature is around 30°C - 32°C. Based on the measurement of water quality parameters in the morning at 31°C, at day time measurement the temperature has a value of 32°C, and at the time of the afternoon measurement value of 30°C. According to [6], that the temperature is good for the life of sea cucumbers ranging from 22°C - 32°C. So from the results of these statements, the water temperature range at the study site is still ideal to support the life of sea cucumbers.

Based on the measurement of physics-chemical parameters of the waters of Unggeh Island, salinity value is obtained from 27-33 ppt. On the measurement of water quality parameters, the salinity is about 29 - 33 ppt in the morning, at the time of day 3 salinity value is 27 - 30 ppt, and salinity value at the time of afternoon measurement is 29 - 30 ppt. According to [7] in [8] states that salinity is one of the factors affecting the existence and life of biota in waters, including sea cucumbers. Sea cucumbers can live in the salinity range between 29 -34 ppt. In accordance with the above statement that the salinity in the waters of Unggeh Island is still good enough for the life of sea cucumbers.

The dissolved oxygen value of DO waters on Unggeh Island ranges from 4.6 to 5.9 which is obtained from measurements at 3 stations at 3 time. Measurement of dissolved oxygen in the morning has a value of 5.1 - 5.3 mg / l, dissolved oxygen niali at day time measurement is 4.6 - 5.4 mg / l, while the measurement at the afternoon is 5.3 – 5.9 mg / l. Dissolved Oxygen (DO) is a chemical variable that has a very important role for the life of aquatic biota as well as a limiting factor for biota life .. According to [9] the dissolved oxygen content is good for marine organisms is 5 - 7 mg / l. Based on the Decree of the Minister of Environment No.51 of 2004, the maximum threshold of dissolved oxygen for marine life is large from 5 mg / l. From the above statement that the value of dissolved oxygen in the waters of Unggeh Island is considered good for biota life.

*Density of Sea Cucumber*
The density of sea cucumber at the location of research is at station I has a density of 0.16 ind / $m^2$, on station II has a density value of 0.12 ind / $m^2$, while at station III has a density value of 0.08 ind / $m^2$, and total density of sea cucumber at research location that is 0,36 ind / $m^2$ can be seen in table 6. In this result the amount of cucumber density is low because the number of sea cucumber found at the time of research that is only 9 individual. This may be due to some environmental conditions that interfere with the existence of these sea cucumber organisms. Due to the existence of the tap fishing gear, as well as the location used for the manufacture of tap fishing gear that allegedly can reduce the habitat of sea cucumbers. This could be the possibility of a little sea cucumber found because few survive under these conditions. In addition, sea cucumbers are found at the time of sampling found on the sand subdivision at seagrass ecosystems. According to [10] sea cucumbers are organisms that occupy sandy substrate, because sea cucumbers are generally depositary feeder that is eating anything contained in the bottom of the water.

At the time of the research location also found the existence of other organisms such as sea urchins and starfish, it can also be one factor that affects the low density of sea cucumbers in the research location. [11] suggesting the presence of crabs, sea stars, and sea urchins can be one factor in decreasing the density of sea cucumbers. Crabs and starfish are predators and pests of sea cucumbers or sea cucumbers. The animal likes to stick to the body of sea cucumbers so it can cause injuries to the body of sea cucumbers. If the sea cucumber can not stand the wound will become larger and cause death on sea cucumbers.

The type of sea cucumber that is of significant economic value usually occupies a goba base (lagoon) with a depth of 5 to 30 meters. While the type of sea cucumber with medium and low economic value occupies shallow areas such as seagrass beds, algal growth areas and coral reefs with a depth of less than 2 meters. Usually sea cucumbers will appear on the surface of the water base at night, especially at the time of the tide, to find food. At noon the sea cucumbers prefer to immerse themselves in the sand. Sea cucumbers generally live in clusters, the type of Holothuria scabra usually live in groups consisting of three to five tails [12]. In accordance with the above statement, the results obtained for the smallest sea cucumber density can be caused by the living habits of sea cucumbers to feed and appear on the surface of the waters at night, while the sampling time is done at three times, morning, afternoon and evening .

In addition, according to local fishermen, the least number of sea cucumbers found on the island of Unggeh is due to over fishing exploitation some time later by fishermen outside Tapanuli Tengah area. Therefore, the number of sea cucumbers that are found very little and the type of commonly found fishermen on the island of Unggeh sandfish (swallow) (Holothuria scabra) and sea cucumber gamet (Actinopyga milliaris).

*Diversity indices, Uniformity Index, Dominance Index*

From the result of research known index value of diversity (H ') sea cucumber is 0.983. The value of this diversity is low in the waters of Unggeh Island. Because the number of species found at the time of sampling is 3 types. According to [13], diversity includes two important things: the number of species in the community and the abundance of each species, so that the smaller the number of species and variations in the number of individuals of each species have uneven distribution, the diversity will shrink and the low biodiversity occurs communities affected by seasonal or periodic disturbances by humans and nature.

The value of uniformity index (E) in the research is 0.895. From the value shows that uniformity of sea cucumbers in the waters of Unggeh Island high, this can be caused by the spread of uneven sea cucumber species in the waters. According to [14] the uniformity index ranges from 0-1. If the uniformity index is less than 0.4 then the ecosystem is in a depressed condition and has low uniformity. If the uniformity index is between 0.4 to 0.6 then the ecosystem is less stable and has moderate uniformity. If the uniformity index is more than 0.6 then the ecosystem is in stable condition and has high uniformity.

From the research result obtained value of Dominance Index (D) sea cucumber is 0.515. According to Simpson's dominant index [5] The dominant index value ranges from 0 to 1, the greater the index value the greater the tendency of one of the dominating types, the result can be concluded that there is no dominant type.

## 4. Conclusions

From the research results can be drawn the conclusion that the number of species of sea cucumber found in the study sites are 3 types, among others *Actinopyga millaris, A. Aechinites, Holothuria scabra*. The value of density of sea cucumber in research that is at station I has a density of 0.16 ind / $m^2$, on station II has a density value of 0.12 ind / $m^2$, while at station III has a density value of 0.08 ind / $m^2$, and total density of sea cucumber at research location that is 0.36 ind/$m^2$.

## References

[1] Aziz A 1997 Status Penelitian Teripang Komersial di Indonesia (Status of Commercial Sea Cucumber Research in Indonesia) *Oseana* **22**(1): 9-19

[2] Firth F E 1974 *The Encyclopedia ofa Marine Resources* (New York: Van Nostrand Reinhold Company)

[3] Palomares and Pauly 2011 www.Sealifebase.org. [terhubung berkala]. http://www.sealifebase.org/summary.php [03 Agustus 2011]
[4] CRITC-COREMAP II 2009 Pemantauan Perikanan Berbasis Masyarakat (Creel) Di Kabupaten Tapanuli Tengah Tahun 2008 *(Community-Based Fisheries Monitoring (Creel) In Kabupaten Tapanuli Tengah Year 2008)* LIPI Jakarta
[5] Direktorat Konservasi dan Keanekaragaman Hayati Laut, Kementerian Kelautan Perikanan. 2015 *Pedoman Umum Identifikasi Dan Monitoring Populasi Teripang (General Guidelines for Identification and Monitoring of Sea Cucumber Populatio)* Kementerian Kelautan dan Perikanan Jakarta
[6] Sutaman 1992 *Petunjuk Praktis Budidaya Teripang (Practical Guidance of Sea Cucumber Cultivation)* (Penerbit Kanisius Yogyakarta)
[7] Hyman L H 1955 *The invertebrate Echinodermata* Vol *IV (New York:* Mac Graw-Hill Book Co.)
*[8]* Saputra D A 2001 *Struktur Komunitas Teripang Holothuroidea Di Perairan Pantai Pulau Pramuka Dan Pulau Tikus Kepulauan Seribu (Community Structure Sea Cucumber Holothuroidea In Coastal Waters of Pramuka Island and Thousand Island of Pulau Seribu)* Institut Pertanian Bogor Bogor p 57
[9] Gufran M H, Kordi K and Tanjung A B 2007 *Pengelolaan Kualitas Air Dalam Budidaya Perairan (Water Quality Management in Aquaculture)* (Jakarta: Penerbit Rineka Cipta)
[10] Bakus G J 1973 *The Biologi and Ecology of Tropical Holothutians* (New York: Academic Press)
[11] Hana 2011 Evaluasi Pemacuan Stok Teripang pada Habitat Konservasi Lamun Pulau Pramuka, Kepulauan Seribu, Jakarta (Evaluation of Pemetuan Stok Teripang at Habitat Conservation of Seagrass Island Scout, Kepulauan Seribu, Jakarta) Program Studi Ilmu Kelautan IPB Bogor
[12] Direktorat Konservasi dan Keanekaragaman Hayati Laut, Kementerian Kelautan Perikanan 2015 *Pedoman Umum Identifikasi Dan Monitoring Populasi Teripang (General Guidelines for Identification and Monitoring of Sea Cucumber Populatio)* (Jakarta: Kementerian Kelautan dan Perikanan)
[13] Odum E P 1993 *Dasar-Dasar Ekologi (Fundamentals of Ecology)* (Yogyakarta: Gajah Mada University Press)
[14] Junianto, Irawan D, Yandri H 2014 Studi Ekologi Teripang (Holothuroidea) Di Perairan Desa Pengudang Kabupaten Bintan Universitas Maritim Raja Ali Haji Kepulauan Riau
[15] Direktorat Konservasi dan Keanekaragaman Hayati Laut, Kementerian Kelautan Perikanan. 2015 *Pedoman Umum Identifikasi Dan Monitoring Populasi Teripang (General Guidelines for Identification and Monitoring of Sea Cucumber Populatio)* Kementerian Kelautan dan Perikanan Jakarta

# Length-weight relationship and condition factor of white shrimp *Penaeus merguiensis* captured in ecosystem mangrove of Bagan Asahan, Tanjungbalai, Asahan, North Sumatra, Indonesia

**A Suryanti[1*], N Riza[2], and T S Raza'i[3]**

[1] Social Economic of Fisheries Departement, Marine science and Fisheries Faculty, Universitas Raja Ali Haji Maritim, Jl.Politeknik Senggarang, 29100 Tanjungpinang, Indonesia
[2] Department of Aquatic Resource Management, Faculty of Agriculture, Universitas Sumatera Utara, Jl. A. Sofyan No.3 Medan, North Sumatera 20155, Indonesia
[3]Aquaculture Departement, Marine science and Fisheries Faculty, Universitas Raja Ali Haji Maritim, Jl.Politeknik Senggarang, 29100 Tanjungpinang, Indonesia
[*]Email: yanti.ajb@gmail.com

**Abstract.** White Shrimp *Penaeus merguiensis* was commonly found in Mangrove Ecosystem of Bagan Asahan Village. The purpose of this research are to determine length-weight relationship and condition factor of white shrimp *Penaeus merguiensis* around ecosystem mangrove waters in Bagan Asahan Village. This research was conducted for 3 month in Maret until Mei 2017 with determination of research station used purposive sampling method. The shrimp samples were taken by shrimp trawl. The result showed that 98 shrimp which consists of 58 males and 40 female. The carapace length of female shrimp between 6,05 – 22,125 mm and total weight ranged from 0,12 – 6,95 g. Male shrimp had carapace length between 7.125 – 18.25 mm and total weigth ranged from 0.14 – 3.82 g. Female and male white shrimp had different growth pattern. Female shrimp had b = 2.984 included in negaive allometric and male shrimps with b = 3.187 included in positive allometric. The value of correlation coefficients was more than 90% for both male and female showed very strong relation between length carapace and body weight. The value of shrimp condition factor ranged from 0.570 – 1.773 and included to flat (thin) body shrimp.

## 1. Introduction

Mangroves are defined by the presence of trees that mainly occur in the intertidal zone, between land and sea, in the (sub) tropics. The aeral roots of mangroves partly stabilise this environment and provide a substratum on which many species of plants and animals live [1]. Mangrove is an important spawning and nursery ground for shrimp, fish and crabs [2]. Mangroves also serve as fish nurseries and spawning grounds for finfish, crabs, shrimps, mollusks, and other sea life [3]. [4] have demonstrated that there's positive correlations between mangrove extent and shrimp catch. Some types of commercial shrimp around of mangrove in Bagan Asahan village are white shrimp (*Penaeus* sp.) and giant prawns (*Macrobrachium* sp.).

White shrimp (*Penaeus merguiensis*) is a type of shrimp that lives on sand and mud, from coastline and river mouth to a depth of about 55 m, usually less than 20 m and likes cloudy water [5]. Many authors have reported on the length-weight relationship of *Penaeus merguiensis* at some place in the world. Persian Gulf and Oman Sea, Iran [6], Mumbai, Indian [7], and the Sonmiani Bay Lagoon, Balochistan, Pakistan [8]. However, the study length-weight relationship and condition factors of white shrimp in ecosystem mangrove of Bagan Asahan Village, Tanjungbalai Sub-district, Asahan Regency, North Sumatera Province, Indonesia never published.

The purpose of this research is to determine the relationship of length and weight and condition factor of White Shrimp *Penaeus merguiensis* that catch in ecosystem mangrove waters, Bagan Asahan Village, Tanjungbalai District, Asahan Regency. This research is expected to provide information related to length-weight relationship and the condition factor of white shrimp *Penaeus merguiensis* in the study area for sustainable management of shrimp resources.

## 2. Material And Method

### 2.1. Study site

This research was conducted for 3 months in March - May 2017 in ecosystem mangrove of Bagan Asahan Village, Tanjung Balai Sub-district, Asahan Regency, North Sumatera Province. Determination of sampling point by used purposive sampling method and determined 3 station sampling. The measurement of samples were conducted in Laboratory of Aquatic Resources Management, Faculty of Agriculture, Universitas Sumatera Utara. The tools were 0.01 mm vernier caliper, digital scales with 0.01 g accuracy, shrimp trwal, digital camera, stationery and magnifying glass. The material were white shrimp *Penaeus merguiensis* and alcohol 96%.

### 2.2. Sampling of shrimp

The sampling of shrimps in each station used shrimp trawl. The shrimp trawl had length 12 m and width / height of 2 m with 200 mesh with mesh size 0.5 cm. Shrimp samples obtained and measured the length of its carapace that measured from the front of base carapace to the limits of shrimp carapace with the body. The weight of the shrimp body was measured used a digital scales. Determination of the sex of the shrimp visually visible with using magnifying glass.

### 2.3. Problem formulation

The length-weight relationship was calculated by the formula [9]:

$$W = aL^b \tag{1}$$

Information :
W = weight (g)
L = length of carapace (mm)
a = the intercept on the Y-axis
b = exponent/ slope

Condition factor was calculated by the formula [9].

a. if isometric growth pattern (b = 3) use formula:

$$K = \frac{10^5}{L^3} W \tag{2}$$

b. if allometrik growth pattern (b ≠ 3) use formula:

$$K = \frac{W}{aL^b} \tag{3}$$

Information:
K = condition factor
W = weight (g)
L = length of carapace (mm)
a dan b = constanta

## 3. Results and Discussions

### *3.1. Length-weight relationship*

The number of white shrimp samples captured during the study was 40 female shrimps and 58 male shrimps. The result of white-shrimp body length-weight relationship analysis was presented in Table1.

**Table 1**. The Result of analysis of length-weight relationship of white shrimp *Penaeus merguiensis*

| Parameter | Female | Male |
|---|---|---|
| Number of shrimp sample | 40 | 58 |
| Range of carapace lenght (mm) | 6.05 – 22.125 | 7.125 – 18.25 |
| Range of weight (a) | 0.12 – 6.95 | 0.14 – 3.82 |
| b value | 2.984 | 3.187 |
| Coefficient correlation (r) | 0.967 | 0.959 |
| Regression equation | $0.0006x^{2.984}$ | $0.0004x^{3.187}$ |
| T-test | $T_{count} < T_{table}$ | $T_{count} > T_{table}$ |
| Type of growth pattern | Isometric | Allometric |

The result of t test in Table 1 showed that female white shrimps has equal to 3 (b = 3) with $T_{count}$ smaller than $T_{table}$ ($t_{count} < t_{table}$). It was significant and included to isometric growth pattern with means the length increase equals the weight gain. Male white shrimp has bigger than 3 (b> 3) with $T_{count}$ more than $T_{table}$ ($t_{count} > t_{table}$). It was not significant and included to allometric growth pattern that means the length increase was not same as the weight gain.

The value of b in Table 1 showed that growth pattern for female white shrimp with b = 2.984 belonging to negative allometric, while for male white shrimps with value of b = 3.187 belonging to positive allometric or weight gain faster than length increase. The value of b = 3 indicated that long growth was balanced with weight gain or isometric growth, whereas if the b value was greater or smaller it was called allometric growth [9]. The difference in growth pattern between male and female shrimp was similar to the research by [6] in Persian Gulf and Oman Sea with b value of 2.740 for male shrimp and b value of 2.843 for female shrimp, sama negative allometric growth. Different with the result by [7] in Sonmiani Bay Lagoon, Balochistan, Pakistan showed the growth patterns of both (male and female) shrimp included to allometric negative. The effect of length and body weight was very large on the value of b obtained, so the factors that influence the body size indirectly will be able to influence the variation pattern of the value b [9].

The coefficient of correlation (r) of long-weight relationship for female white shrimp was 0.967 and male white shrimp was 0.959. The correlation showed a very strong relationship between the increase of length carapace and the increase of body weight both male and female white shrimp.

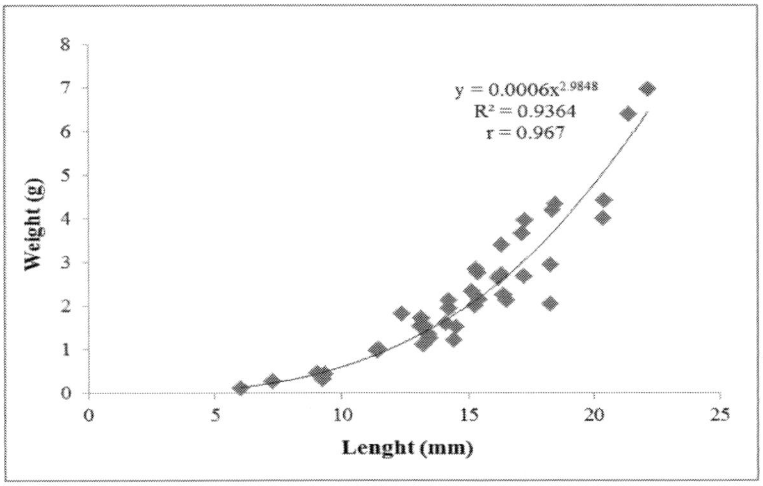


**Figure 1.** The Length-weight relationship of female shrimp

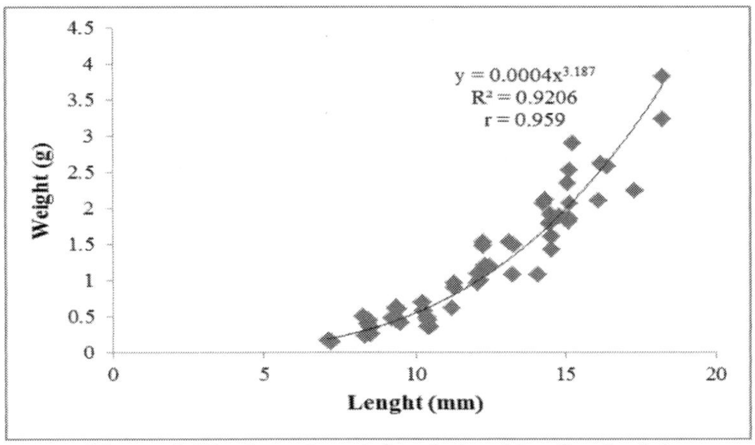


**Figure 2.** The Length-weight relationship of male shrimp

### *3.2. Condition Factor*

The result of Condition Factor of white shrimp *P. merguiensis* could be seen in Table 2.

**Table 2.** The value of condition factor of White Shrimp *P. merguiensis* in Ecosystem Mangrove Waters of Bagan Asahan Village

| Gender | Average L (mm) | Average W(gr) | Condition Factor Range (Kn) |
|---|---|---|---|
| Female | 14.858 | 2.326 | 0.588 – 1.676 |
| Malae | 12.132 | 1.232 | 0.490 – 1.485 |

Table 2 shows that condition factor of female white shrimp was 0.588 - 1.676 and the value of white male shrimp condition factor was 0.490 - 1.485. It showed the female white shrimp was more plump compared with male white prawns. The mature adult stage white shrimp as thicker, with higher weight in proportion to length, than younger organisms in the growth stage so affect the growth pattern and shrimp condition factor [10]. The value of condition factor showed that female white shrimp had better condition factor for reproduction process than male white shrimp. Condition factor was useful in expressing the wellbeing or fatness of the shrimp, it varies with sex, stage of maturity, and time of the year. The differences in the condition factor between males and females could be due to the presence of gravid females or due to higher weight of the females's gonads [11].

The value of condition factor for female and male white shrimp included to flat (thin) body. This is same with statement [9], that the range of the condition factor between 3 – 4 included to fat and the range of condition factor between 1 – 2 include to thin shrimp.

## 4. Conclusions

The length-weight relationship of female white shrimp included in isometric growth pattern or the increase of weight faster than length of carapace and male white shrimp included allometric possitive growth pattern or the increase of weight equal with length increase. White shrimp both male and female belonging to flat (thin) body shrimp.

## References

[1] Nagelkerken I, Blaber S J M, Bouillon S, Green P, Haywood M, Kirton L G, Meynacke J O, Pawlik J, Penrose H M, Sasekumar A and Somerfield P J 2008 *Aquatic Botany.* **89** 155-30

[2] Lee S Y, Primavera J H, Dahdouh-Guebas F, McKee K, Jared O, Bosire, Cannicci S, Diele K, Fromard F, Koedam N, Marchand C, Mendelssohn I, Mukherjee N and Record S 2014 *Ecology and Biogeography* **23** 726–19

[3] Okpiliya F I, Effiong E B, Eni I, Eja and Eja I 2013 *Journal of Sustainable Development.* **2** 149-13

[4] Jamizan A R and Chong V C 2017 *Sains Malaysia.* **46** 9-10

[5] FAO Species Identification Guide For Fishery Purposes 1998 *The Living Marine Resources of The Western Central Pacific, Vol 2: Cephalopods, Crustaceans, Holothurians, and Sharks. Food and Agriculture Organization of the United Nations*, ed Carpenter K E and Niem V H (Rome: FAO) pp. 687-709

[6] Safaie M 2015 *Tropical Zoology.* **28** 9-13

[7] Suman B and Biradar R S 2000 *Journal of the Indian Fisheries Association.* **27** 66-11

[8] Qureshi N A and Amanat Z 2014 *Journal of Anial and Plant Science.* **24** 503-8

[9] Ricker W E 1975 *Computation and interpretation of biological statistics of fish populations* Bulletin of the Fisheries Research Board of Canada Canada. pp 382

[10] Chow S and Sandifer P A 1991 *Aquaculture.* **92** 165–14

[11] Udoinyang E P, Amali O, Iheukwumere C C and Ukpatu J E 2016 *International Journal of Fisheries and Aquatic Studies.* **4** 109-5

IOP Conf. Series: Earth and Environmental Science 122 (2018) 012109 doi:10.1088/1755-1315/122/1/012109

# Evaluation of mangrove management through community-based silvofishery in North Sumatra, Indonesia

**M Basyuni[*], P Yani and K S Hartini**

Department of Forestry, Faculty of Forestry, Universitas Sumatera Utara, Jl. Tri Dharma Ujung No. 1 Medan, North Sumatera 20155, Indonesia

[*]Email: m.basyuni@usu.ac.id

**Abstract.** Aquaculture expansion has been reported as the primary driver of mangrove loss and a significant cause of mangrove deforestation in North Sumatra, Indonesia. Development of silvofishery based on creating balance condition between conserving mangrove forest and offering better livelihood for local communities surrounding mangrove. The present study evaluates of mangrove management through community-based silvofishery in three villages, namely Paluh Manan, Paluh Kurau, and Lama, Hamparan Perak of Deli Serdang Regency, North Sumatra, Indonesia. Three communities used the same ecological type-silvofishery, characterized by planted mangrove surrounded aquaculture. Results showed that in the Paluh Manan village, planted mangrove and aquaculture in the ratio of 75:25 with planting distance of mangrove 50x50 cm, containing 2,500 trees/ha, resulted in US$ 36.2/month/ha of fish and shrimp farming. In the Paluh Kurau village, a mixture mangrove and aquaculture in an 84:16 ratio, planting distance of 1x1 m, consists of 1,600 trees/ha, US$ 23.8 of generating revenue from crab farming. Furthermore, in the third village, Lama village, consists of mangrove and aquaculture in the proportions 90:10, with planting spacing 2x2 m, composing 1,000 trees/ha, led to US$ 45.8/month/ha from fish, shrimp and crab farming. The present study suggested the mangrove management through community-based mangrove-friendly aquaculture.

## 1. Introduction

Mangrove forests in North Sumatera, Indonesia covered approximately 50,369.8 ha in 2015 and existed in the eastern coast of Sumatra Island and commonly distributed in Asahan, Langkat, Deli Serdang, Batubara, Tapanuli, Nias, Labuhanbatu until Serdang Bedagai [1]. The significant ecology, economy, and environment of mangroves have been well documented; however, mangroves are currently the most threatened ecosystems due to indirect and direct deforestation and degradation [2]. Aquaculture expansion has been reported as leading driver of mangrove forest cover loss in Indonesia. Furthermore, conversion to shrimp aquaculture, oil palm plantation, and urban development has been the primary cause of mangrove deforestation in North Sumatra, Indonesia [1-4].

The livelihood of many people living in and adjacent to mangrove areas is strongly dependent on forestry and fisheries resources provided by the mangrove ecosystems. Mangrove ecosystems serve as vital spawning, nursery, and feeding grounds for aquatic animals such as shrimp, crabs, and fish, and provide habitats for resident and migratory birds and habitats for other animals [5]. Mangrove conservations as well as sustainable mangrove management, therefore, are needed to maintain the presence of mangrove forest and increased reforestation programs.

Development of silvofishery based on balance condition between conserving mangrove forest and fishery by offering better livelihood for local communities surrounding mangrove forest. Primavera

[6] has suggested that mangroves and aquaculture are not necessarily unharmonious. Such mangrove-friendly aquaculture is amenable to small-scale, family-based operation and may be adopted in mangrove conservation site [6]. The present study, therefore, aimed to evaluate on mangrove management through community-based silvofishery in three villages, namely Paluh Manan, Paluh Kurau, and Lama, Hamparan Perak of Deli Serdang Regency, North Sumatra, Indonesia.

## 2. Materials and Method

### *2.1. Study area*

The study was performed in three villages, namely Paluh Manan, Paluh Kurau, and Lama, belong to Hamparan Perak District, Deli Serdang Regency, North Sumatra, Indonesia. Three communities used the same ecological type-silvofishery, characterized by planted mangrove surrounded aquaculture. Some sample respondent in this study was 52 respondents from three villagers, which belongs farmer group, respectively as described previously [7]. One farmer group in Paluh Manan has 25 members, in Paluh Kurau Village has 14 members, and Lama Village has 13 members.

### *2.2. Data collection*

Figure 1 depicts the procedure for evaluation of silvofishery including the data collection as previously described [8]. Two scientific activities were done to obtain the data, performing research in the silvofishery ground and literature search to support this experiment. Interview, questioner, and observation were carried out in the silvofishery site to get primary data on the silvofishery system and its application. Furthermore, from literature search resulted in secondary data. Both data were combined through performing research and literature search.Secondary data was obtained from Mangrove Forest Management Agency (BPHM) North Sumatra and Forest Area Conservation Agency (BPKH) North Sumatra regarding with information of site and supporting data.

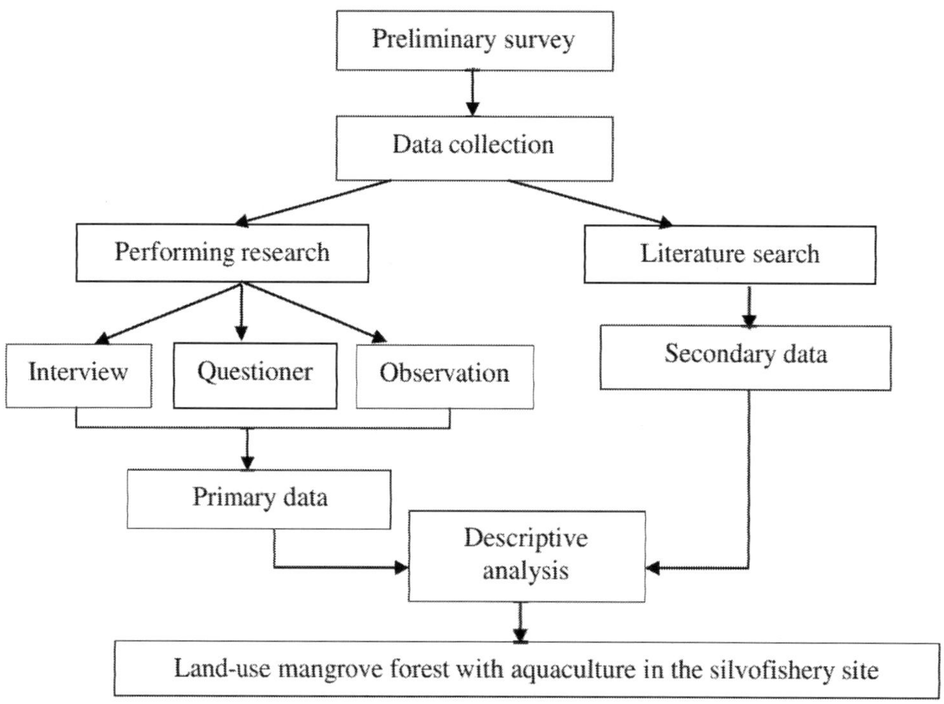


**Figure 1**. Research procedure including data collection.

## 3. Results and Discussion

### *3.1. Silvofishery in Paluh Manan Village*

The pond area is influenced by tidal sea process; therefore the community group managed the aquaculture by planting mangrove around the aquaculture. This activity was to strengthen the soil structure of beds as not to be crushed by the tides. The mangrove planted in the pond area is intended to restore fertility and give space to the fish or shrimp seeds that have been spread by farmers and neutralize toxins that exist in the previous fish farming [8].

Maintenance of ponds in the Paluh Manan village concerning water management for the production process is well organized as shown in Figure 2. Figure 2 also illustrates that water coming from the stream enters the pond through a pipe made for the opening of brackish water and regularly drains water. The planted mangroves species were *Rhizophora mucronata*, *R. apiculata* and *R. stylosa* along with intentionally oil palm plantations. Planted mangrove and aquaculture was in the ratio of 75%:25% with planting distance of mangrove 50x50 cm, containing 2,500 trees/ha (Figure 2).

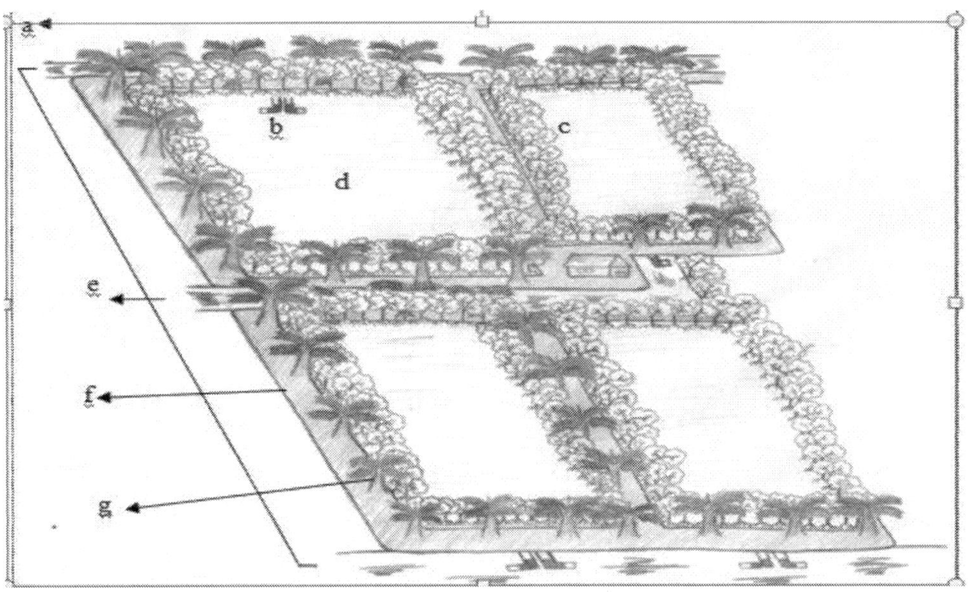


**Figure 2**. Silvofishery model in Paluh Manan Village. a. Primary road, b. Silvofishery gate for freshwater to brackish water, c. Mangrove species (*R. mucronata* and *R. stylosa*), d. Pond, e. Stream, f. Beds, g. Oil palm plantation

Mangrove planted with ponds in the silvofishery system of Paluh Manan Village consists of, *Rhizophora mucronata, R. apiculata*, and *R. stylosa*. These mangroves are planted on the edge of the ponds because they have an important role in triggering high phytoplankton growth. The type of pond dominated by *R. mucronata* is more suitable for plankton growth and was able to provide higher nutrient than other types of mangrove stands [9]. *Rhizophora* roots do not hamper the pond production process. *R. mucronata* and *R. stylosa* leaves can be used for animal feed such as cow and goat and provide shade for living creatures that exist around the pond [10].

Table 1 shows the cultivated species in the aquaculture are Nile tilapia and tiger shrimp but not the crab species. Crabs had a hole-like risk in the beds that increased the cost of land management. Silvofishery operational cost for six months US$ 186.5 (Table 1). Silvofishery production for one pond within six months is displayed in Table 2. This activity resulted in US$ 36.2/month/ha of fish and shrimp farming.

**Table 1**. Silvofishery operational in Paluh Manan Village for six months

| No | Activity | Cost (US$) |
|---|---|---|
| 1 | Pond activity | 38.5 |
| 2 | Feed | 23.1 |
| 3 | Nursery for Nile tilapia @ 1250 | 19.2 |
| 4 | Nursery for shrimp @ 2000 | 9.6 |
| 5 | Land rental | 38.5 |
| 6 | Employment | 57.7 |
| Total | | 186.5 |

1 US $: IDR 13,000

**Table 2**. Silvofishery production in one pond within six months

| No | Type | Price (US$) | Amount (Kg) | Cost (US$) |
|---|---|---|---|---|
| 1 | Nile Tilapia | 1.15 | 150 | 173.0 |
| 2 | Shrimp | 4.6 | 50 | 230.8 |
| Total | | | | 403.8 |

1 US $: IDR 13,000

*3.2. Silvofishery in Paluh Kurau Village*

Silvofishery in Paluh Kurau was initially mangrove production forest. Species composition between mangrove and fish pond in Paluh Kurau Village were *R. mucronata, R. apiculata, R. stylosa* and *Avicennia marina*. Species of *R. mucronata, R. apiculata, R. stylosa* were planted in ponds while *A. marina* planted in front of the sea because this species is expected to protect other species of waves from the sea and appropriate zoning as well [11].

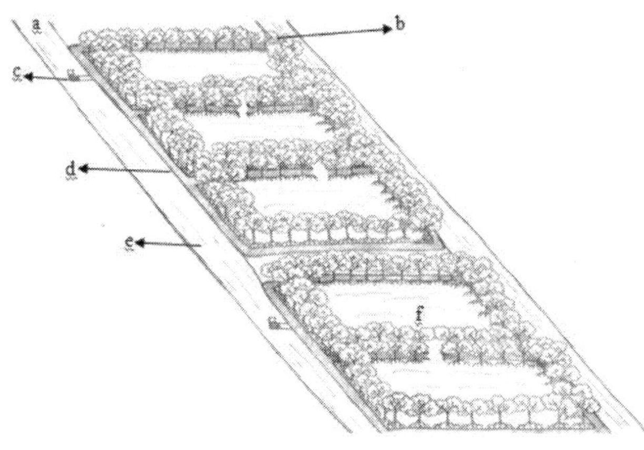


**Figure 3**. Silvofishery model in Paluh Kurau Village. a. Silvofishery gate for freshwater to brackish water b. Mangrove species (*R. apiculata, R. mucronata,* and *R. stylosa*), c. Pipe, d. Beds, e. Stream, f. Ponds

Community group planted 1,600 mangrove species in the pond area with a spacing of 1m x 1m in 1 ha. The composition of mangrove with ponds is 84%:16%. The silvofishery is believed to be able to combine the interests of mangrove conservation with improving the welfare of the surrounding community. The relationship is expected to form an ecological balance. Ponds that are ecologically lacking producer elements can be supplied through feeding, by the existence of producer subsidies (marine biota) from mangrove forests [9].

**Table 3**. Silvofishery operational in Paluh Kurau Village for six month

| No | Activity | Cost (US$) |
|---|---|---|
| 1 | Pond activity | 46.1 |
| 2 | Feed | 30.7 |
| 3 | Crab seeds @4000 | 153.8 |
| 4 | Land rental | 26.9 |
| 5 | Employment | 61.5 |
| | Total | 319.0 |

1 US $: IDR 13,000

The profit of silvofishery pond crab cultivation in Paluh Kurau Village was US$ 461.5/6 months. This activity resulted in US$ 23.8 of generating revenue from shellfish farming (Tables 3-4). Crab is a species that can only be cultivated in ponds because crabs live in salt water while shrimp and tilapia are not suitable grown using salt water but brackish water [12].

**Table 4**. Silvofishery production in one pond within four months

| No | Type | Price (US$) | Amount (Kg) | Cost (US$) |
|---|---|---|---|---|
| 1 | Crab | 2.3 | 200 | 461.5 |

1 US $: IDR 13,000

*3.3. Silvofishery in Lama Village*

In 2005 community group in Lama Village used intensive pond technology system without mangrove plants. Water did not depend entirely with the ups and downs of inclusion and discharge. This pond resulted from high production results in one-two years, but after three years later the production decreased dramatically.

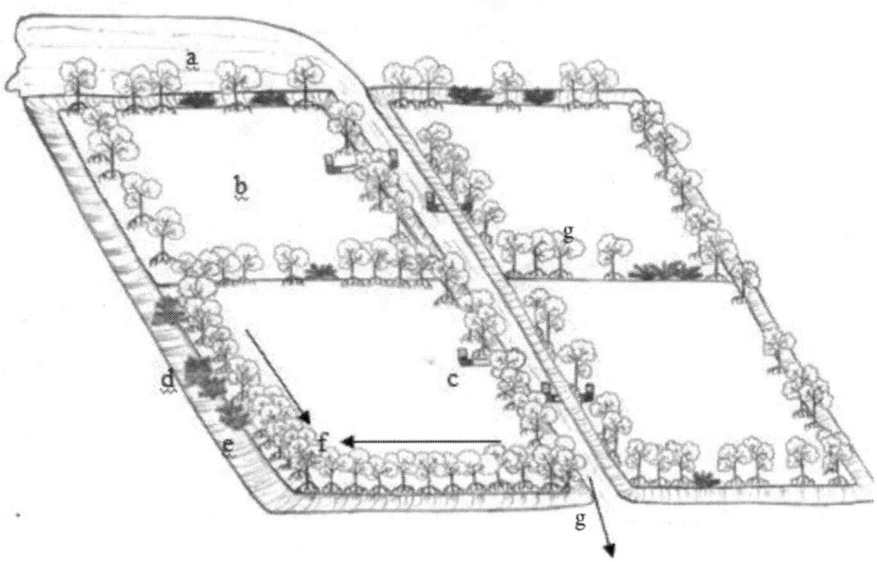


**Figure 4**. Silvofishery model in Lama Village. a. Reservoir, b. Pond, c. Pipe, d. *Acrostichum aureum*, e. Beds, f. Mangrove species, g. Stream

This aquaculture was destroyed because the results of ponds cultivated such as tilapia, shrimp, and crabs suffered many diseases consequently failed the production. In 2010 they planted in ponds with mangrove species such as *R. mucronata* and *R. stylosa* with total of 1,200 seedlings. In 2013, the existing mangrove added with *A. marina* and *A. aureum* around the pond area that unintentionally grown. Some crops are intentionally planted in the embankment of ponds like cassava.

As depicted in Figure 4 that the mangroves were planted on the underside or the edges of the ridges, such as *R. mucronata* and *R. stylosa* with a spacing of 2m x 2m with the 1,000 mangrove plants. The composition of mangrove (density) with pond per ha is 90%:10%, led to US$ 45.8/month/ha from fish, shrimp and crab farming (Tables 5-6).

**Table 5**. Silvofishery operational in Lama village for six months

| No | Activity | Cost (US$) |
|---|---|---|
| 1 | Pond activity | 38.5 |
| 2 | Feed and fertilizer | 38.5 |
| 3 | Nursery for Nile tilapia @ 1250 | 46.2 |
| 4 | Nursery for shrimp @ 2000 | 5.7 |
| 5 | Crab seeds @ 48 | 8.1 |
| 6 | Land rental | 38.5 |
| 7 | Employment | 57.7 |
| | Total | 233.2 |

1 US $: IDR 13,000

**Table 6**. Silvofishery production in one pond within six months

| No | Type | Price (US$) | Amount (Kg) | Cost (US$) |
|---|---|---|---|---|
| 1 | Nile Tilapia | 1.15 | 325 | 375.0 |
| 2 | Shrimp | 3.8 | 23 | 88.5 |
| 3 | Crab | 3.3 | 19.2 | 44.3 |
| Total | | | | 507.8 |

1 US $: IDR 13,000

In the mangrove-friendly aquaculture such as silvofishery, this site was planted with mangroves to provide firewood, fertilizers, and protection from the wave. The mixed mangrove-aquaculture system has various benefits such as low capital provision, livelihood diversification through the provision of regular income, and the recognition of an organic farming practice [3]. Thus, at present social induced-rapid developmental changes stipulate a better understanding of the dynamic of mangrove forest [13].

## 4. Conclusions

Conversion to aquaculture is responsible for deforestation in Indonesia, including in North Sumatran mangrove. This study confirmed that mangrove forest and fishery is compatible by developing silvofishery. The present study suggested the mangrove management through community-based mangrove-friendly aquaculture. Silvofishery benefited from feeding cows, chickens, and goats. Thus, the activity that originated from mangroves can be profitable in all aspects for others.

## References

[1] Basyuni M, Putri LAP and Murni MB 2015 Implication of land-use and land-cover changes into carbon dioxide emission in Karang Gading and Langkat Timur Laut Wildlife Reserve, North Sumatra, Indonesia *J. Man. Hut. Trop.* **21** 25–35

[2] Richards DR and Friess DA 2012 Rates and drivers of mangrove deforestation in Southeast Asia, 2000-2012 *Proc. Natl. Acad. Sci. USA* **113** 344–3349

[3] Bosma RH, Nguyen TH, Siahainenia AJ, Tran HT and Tran HN 2016 Shrimp-based livelihoods in mangrove silvo-aquaculture farming systems *Rev. Aquacult.* **8** 43–60

[4] Thomas N, Lucas R, Bunting P, Hardy A, Rosenqvist A and Simard M 2017 Distribution and drivers of global mangrove forest change, 1996–2010 *PloS one* **12** e0179302
[5] Barbier EB, Hacker SD, Kennedy C, Koch EW, Stier AC and Silliman BR 2011 The value of estuarine and coastal ecosystem services. *Ecol. Monogr.* 81 169–93
*[6]* Primavera JH 2006 Overcoming the impacts of aquaculture on coastal zone *Ocean Coast Manage* **49** 531–545
[7] Basyuni M, Rouf RA, Saragih M, Asbi AM and Yuriswan W 2017 Local wisdom and mitigation action to maintain secondary mangrove forest: a case study of Jaring Halus village in Langkat, North Sumatra, Indonesia *Adv. Soc. Sci. Educ. Hum. Res.* **81** 551–555
[8] Ha TT, van Dijk H and Visser L 2014 Impacts of changes in mangrove forest management practices on forest accessibility and livelihood: A case study in mangrove-shrimp farming system in Ca Mau Province, Mekong Delta, Vietnam *Land Use Policy* **36** 89–101
[9] Saifullah AS, Kamal AH, Idris MH, Rajaee AH and Bhuiyan MK 2016 Phytoplankton in tropical mangrove estuaries: role and interdependency *Forest Sci. Technol.* **12** 104–113.
[10]Baba S, Chan HT and Aksornkoae S 2013 Useful products from mangrove and other coastal plants. International Society for Mangrove Ecosystems, Okinawa, Japan
[11]Basyuni M, Baba S, Kinjo Y and Oku H 2012 Salinity increases the triterpenoid content of a salt secretor and a non-salt secretor mangrove *Aquat. Bot.* **97** 17–23
*[12]* De-León-Herrera R, Flores-Verdugo F, Flores-de-Santiago F, González-Farías F 2015 Nutrient removal in a closed silvofishery system using three mangrove species (*Avicennia germinans, Laguncularia racemosa,* and *Rhizophora mangle*) *Marine Poll. Bulletin.* 91 243–248
[13]Basyuni M, Oku H, Baba S, Takara K, Iwasaki H 2007 Isoprenoids of Okinawan mangroves as lipid input into estuarine ecosystem *J. Oceanography* **63** 601–608

**Acknowledgment**
A Grant-in-Aid for a Community Service Research-Based (No. 453C/UN5.2.3.2.1/PPM/2016 to MB) from the Non-PNBP University of Sumatera Utara 2016 supported this study partly.

# The prevalence of parasites in ornamental fish from fish market in Medan

**R R Dewi[1,2*], Desrita[2] and A Fadhilla[2]**

[1]Animal Science, Universitas Tjut Nyak Dhien, Indonesia
[2]Management of Water Resources, Universitas Sumatera Utara, Indonesia
E-mail : ritardewi@yahoo.com

**Abstract.** Parasites still become the major problem in ornamental fish as the fast grown of its trading in Indonesia. Parasites causes diseases in ornamental fish hence followed by death and reducing its appearence. In this study, the prevalence of parasites in 100 apparently healthy ornamental fishes namely Guppy (*Poecilia reticulate*) and Goldfish (*Carrasius auratus*) were determined. The method of this research used was survey in local fish market in Medan from March to May 2017 The aim of this study was to determine the parasite that infects aquarium fishes and find out its prevalence. For this purpose, ornamental fishes were examined for parasites from their skin, fin, gill and intestine using wet mount method under a light microscope. The survey result showed that parasites that infect Guppy and Goldfish were Monogenea (*Dactylogyrus* sp and *Gyrodactylus* sp) in their skin and gill, Protozoa (*piscinodinium* sp) in their gill and Nematoda (*Capillaria* sp) in abdominal cavity. Prevalence rate of parasites that attack Guppy in Medan is *Dactylogyrus* sp (8%), *Gyrodactylus* sp (14 %), *Piscinodinium* sp (6%) and *Capillaria* sp (8%). Then, prevalence rate of *Capilaria sp* that attack Goldfish is 4%. The conclusion of this work revealed that the prevalence rate of ectoparasite and endoparasites in ornamental fishes in Medan had the low rate as well as there was no pathological findings is observed. However, these parasites could be a harmful parasitic diseases in case there is the changes in aquarium environment and improper fish handling.Key word : Parasites, prevalence, ornamental fish

## 1. Introduction

Gold fish and Guppy is the main ornamental fish because its appearance and many hobbyist choose them as a pet. Despite of their appearance, an economic value of ornamental fish are the main reason for the remarkable grown in its trading both in Indonesia market and worldwide. The global wholesale value for aquarium fish trading estimated at US$ 900 million in 2000[1]. Asia provided more than 50 % of the worldwide total ornamental fish supply. Furthermore, Indonesia has been one of the main country in Asia which is import both marine and fresh water ornamental fish to Europe in 2009[1].

Parasitic diseases in ornamental fish is one of the main problem both for farmer and aquarist. Parasites make a great loss because it could reduce fish performance such as scale loss, colour change and many pathological aspects as well as mortality. Obviously, parasites possibly have the capability to directly lessen the appearance of their hosts, in terms of growth and reproduction, through their direct impact on fish health[2].High mortality rate by parasites due to the progress infection by parasites such as mechanical damages in gill [3], alterations in reproduction organ [4]and secondary infection by bacterial diseases[5].

The occurrence of parasites in ornamental fishes has been reported worldwide as well as Indonesia. Parasites in ornamental fishes that determined from Chumuci River in Mexico were from 3 taxas , namely Monogenean, Nematoda and Acantocephala[6]. In Korea, 7 imported fishes from Southeast Asia were examined for parasite infection and they found that there were 3 ciliates, 2 monogeneans, 1 nematode, and 1 copepod from 7 host species [7].Aquarium fishes that colected from pet shop in Sweden mostly infected by *Monogenea* parasites[8]. Parasites infection in imported fish also found in Pakistan [9]. In Iran, there were reported that parasites infection in fresh water ornamental fish farm also happened during 2011 -2012 [10]. Parasites infection in ornamental fishes also reported from Indonesia such as Jakarta [11] and Surabaya [12]. However, the occurance parasites of ornamental fishes in Medan still unknown for these years.

The distribution of parasites infection could by trading (imported fish), fish farm infection from natural resources or in fish market. The present study is aimed to investigate parasitic infection and to know the prevalence of parasites infection in guppy and goldfish species ornamental fishes that selling in fish market in Medan city.

## 2. Materials And Methods

One hundred ornamental fishes were collected from five pet shops in Medan city from March 2017 to May 2017 and were examined for the parasites infection. The fish samples were brought live in polyethylene bag and kept in aquarium aerated water in Management Water Resources laboratory, Faculty of Agriculture, Sumatera Utara University. The experiment methods was survey method.

The experimental fishes were measured and examined by standard parasitology examination [13].The external examination was observed to identified the pathological damage caused by parasites. The body, skin, fins, gill were throughly examined as well as intestine for external parasites and endoparasites infestations. The wet mount preparation or biopsy procedures were applied to observe the parasites from experimental fishes [13]. The surface of body and fns were scrapped and intestine were cut to observe the parasites infestation. The parasites examination and identification were done by using the micoscope examination [13][14]. Parasites, habitats, prevalence and intensity of parasites were analyzed by descriptive analysis.

## 3. Results and Discussions

The study of parasitic infection in ornamental fish both Guppy and goldfish from Medan were done then it found there were three group of parasites : monogenean( *Gyrodactylus* sp, *Dactylogyrus* sp), protozoa (*Piscinodinium* sp), and nematode (*Capillaria* sp) were observed.A total of 100 fishes both guppy and gold fish were examined for parasitic infection. The total length of the fish ranged 2,3 - 4,9 cm for guppy and 4,72 – 7,62 cm for gold fish. Out of the total 100 fishes, the number of nfected fishes varied for each type of parasites (Table 1). There were four species infected Guppy and one species infected goldfish. These were : *Piscinodinium* sp, *Gyrodactylus* sp, *Dactylogyrus* sp, and *Capillaria* sp. The parasites prevalence and intensity both Guppy and Goldfish are given in Table 1

**Table 1.** Prevalence and Intensity of Parasites that Infected Ornamental Fishes in Fish Market Medan

| Fish | Parasites | Prevalence (%) | Intensity |
|---|---|---|---|
| Guppy | Protozoa | | |
| | *Piscinodinium* sp | 6 | 10.67 |
| | Monogeneans | | |
| | *Gyrodactylus* sp | 14 | 1.43 |
| | *Dactylogyrus* sp | 8 | 1.25 |
| | Nematoda | | |
| | *Capillaria* sp | 8 | 1.25 |
| Gold Fish | Nematoda | | |
| | *Capillaria sp* | 2 | 1 |

The prevalence of Monogenean both *Gyrodactylus* sp and *Dactylogyrus* sp that infect the Guppy is the highest among all the parasites which is about 14% and 8 %. respectively. Then. the endoparasites ; *Capillaria* sp about 8 % . The lower prevalence of parasites in Guppy is *Piscinodinium* sp about 6 %. On the contrary. Goldfish have the fewest parasites which is shown by the lowest prevalence was about 2 %. Then. the parasites that infects goldfish was endoparasites namely nematodes (*capillaria* sp.).The intensity of parasites Piscinodinium sp is the highest among all the parasites infected the Guppy which is about 10.67. The intensity of monogenean. both *Gyrodactylus* sp and *Dactylogyrus* sp were 1.43 and 1.25. Hence. the intensity of *Capillaria* sp was 1.25. In goldfish. the intensity of nematode parasites was about 1.

Different parasite species were reported from various ornamental fish species around the world. Parasites infection in a variety of goldfish was reported from Pakistan. These parasites were; monogenean (*Dactylogyrus extensus. Gyrodactylus* sp.). protozoan (*Trichodina* sp. *Ichthyophthirius multifiliis*) and crustacean parasite(*Argulus foliaceus)*[9]. *Ambiphyra* spp. was reported from the skin of guppy [15]. The ectoparasites also observed from goldfish (*Carassius auratus*) in Iran. six were monogeneans (*Dactylogyrus vastator. D. baueri. D. formosus. Dactylogyrus* sp.. *Gyrodactylus chinensis* and *Gyrodactylus*. sp.) and three were protistans (*Ichthyophthirius multifiliis. Trichodina* sp. and *Cryptobia* sp)[16].

This study is the first report about parasitic infection in ornamental fish from market fish in Medan. North Sumatera. However. there were also many reports of parasite fauna from ornamental fishes in Indonesia market. Parasites that infected guppy fish were Trichodinid (Ciliophora). *Gyrodactylus* sp. (Platyhelminthes) dan *Lerneae* sp. (Crustacea)[3]. Types of ectoparasites that found in Koi fish (*Cyprinus carpio)* in East Java were *Trichodina* sp.. *Chilodonella* sp.. *Myxobollus* sp.. *Ichtyopthirius multifilis. Dactylogyrus* sp. *Gyrodactylus* sp.. and *Argulus japonicus*[12].

In this study. there were three group of parasites : monogenean( *Gyrodactylus* sp. *Dactylogyrus* sp). protozoa (*Piscinodinium* sp). and nematode (*Capillaria* sp) were observed. In guppy fish. The highest prevalence of parasite infection were monogeneans : *gyrodactylus* sp.(14%) and *dactylogyrus* sp (8%). Monogeneans are platyhelminths parasitic that commmon in fish and known to lead desease in all kind of fish culture [14]. Their existance in fish culture and aquarium probably they have the high rate of reproduction under poor sanitation and deterioting water quality such as over crowding. high ammonia or nitrite. organic pollution or low dissolve oxygen[13]. While this study found that the prevalence and intensity of monogeneans in Guppy was low. however it still became threatened for diseases outbreak because the epidemic study on monogenea infection in guppy result showed that in social hosts like guppies. the frequency of social contact largely rules disease epidemics independent of host density [17].

Protozoan parasites that infect gill of Guppy fish in this experiment was*Piscinodinium* sp. with prevalence rate was about 6 % and intensity was 10. 67. *Piscinodinium* sp. infection in ornamental fish also reported in Srilanka [18]. This parasites was one of genera in Dinoflagellates which are protozoa that are commonly found in aquatic ecosystems [14]. Hence. it is a a problem in tropical freshwater fishes and mostly were reported in aquarium fishes [18][19]. P*iscinodiinium* sp infest the gill and skin [14]. Furthermore. in this study. we found it infected the gill with pathological sign such as slightly deterioted in lamella.

Endoparasites findings in this study was nematodes infection in intestine both Guppy and Goldfish. The prevalence of this infection was 8% in Guppy and only 2% in goldfish. *Capillaria* sp infestation also reported in Guppy from Srilanka [18] and Iran [10]. Fish are either intermediate or final hosts for nematodes and neither adult or larval nematodes are usually a problem in most cultured fish[13]. *Capillaria* sp. could lead the high mortality in aquarium fish because it can cause chronic wasting if present in high number [13].

The site of infection varied for each species of parasites. The Protozoan infected gill and Monogenean infected skin and gill. Endoparasites also found in the intestine which is identified as nematode. The location of each parasites were given in Table 2.

**Table 2** . Location of Parasites in ornamental fish bodies

| Parasites | Host Species | Location |
|---|---|---|
| *Piscinodinium* sp | Guppy | Gill |
| *Gyrodactylus* sp | Guppy | Skin |
| *Dactylogyrus* sp | Guppy | Gill |
| *Capillaria sp* | Gold fish. Guppy | Intestine |

The location of parasites depend on their specifity and their feeding. Ectoparasites infect host in their external bodies such as skin. fin and gill.Piscinodiiniums infest the skin and gill and feed from epithelial cells [13]. Nematodes are endoparasites that had infestation in intestine mucus of fish.
The fish lenght also measured in this study. The existance of parasites and their relationship with length of fish were showed in figure 1. The chart represents that the prevalence of parasite infection decrease as long as the the increase of the fish size. The highest prevalence with the length between 2.0 – 2.9 cm then followed by fish size with 3.0 – 3.9 cm lenght both value were 50% and 35%. respectively. The fish with lenght between 4.0 – 4.9 cm have the prevalence about 25 %

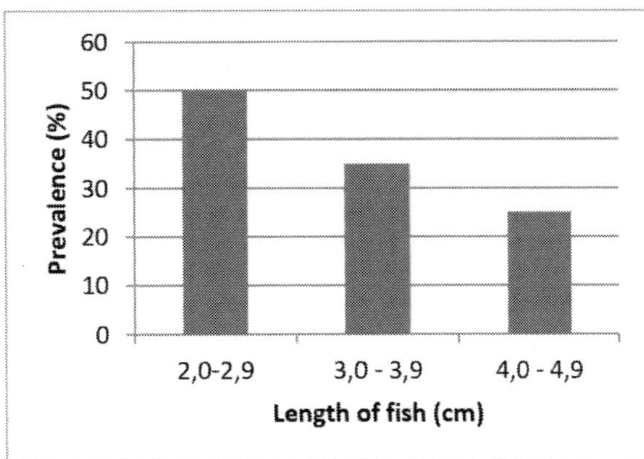


**Figure 1.** Parasites Prevalence and length of The Guppy

The diagram in figure 2 describes that the prevalence of parasitic infection gradually rise along with the increase of the lenght of the fish. The highest prevalence value is about 10% with the range of lenght between 7.0 – 7.9 cm then followed by the fish with lenght 6.0 – 6.9 cm with the prevalence 7.6%.

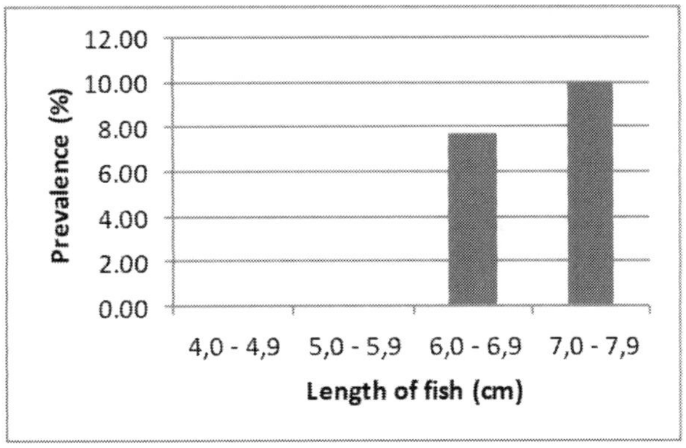


**Figure 2.** Parasites Prevalence and length of The Goldfish

The relation between host lenght and the existance of parasite in this study showed that there were difference pattern between guppy and goldfish prevalence related to the size of their body. The pattern in gold fish showed that large fish was more susceptible to parasites infection than the small fish and this work has similarities with the work of [20]. Hence. the given data showed that the smallest fishes were relatively less infected than the other length groups and the percentage of infection increases with enhancement of fish length. On the contrary. Guppy fish showed that the smallest fish have the higher prevalence of paraite infection. Some parasitic diseases namely piscinodiinium infestation most pathogenic in young fish [15].

The insidence of parasite infection in ornamental fish could be the main problem in ornamental fish industry in Indonesia particulary in Medan. The identified parasites in this study have not been reported as a parasitic problem in Medan. Though. the rate of infection in these aquarium fishes was low. the transmission of contamination among aquarium fishes and customer satisfaction taken into consideration.The prevention from severe infection should be done such as reducing stress condition through improved water quality a. reduction of organic matter. avoiding excessive density of fish and unnecessary manipulation and using appropriate disinfectants in farms can be benefit to avoid economic losses caused by parasitic disease in ornamental fishes

## 4. Conclusion

The parasites that infected ornamental both Guppy and Goldfish which was sold in fish market were four species infected Guppy and one species infected goldfish. These were : *Piscinodinium* sp. *Gyrodactylus* sp. *Dactylogyrus* sp. and *Capillaria* sp. The prevalence of Monogenean both *Gyrodactylus* sp and *Dactylogyrus* sp that infect the Guppy is the highest among all the parasites which is about 14% and 8 %. Goldfish have the fewest parasites which is shown by the lowest *Capillaria* sp infection prevalence in was about 2 %. The relation between host lenght and the existance of parasite in this study showed that there were difference pattern between guppy and goldfish prevalence related to the size of their body.

The information about advance research in parasitic infection of ornamental fishes is needed along with the economic value and the burden that caused by it. The routine monitoring in parasitic prevalence could prevent the huge economic loses caused by parasitic diseases.

## 5. References

[1] Andras. V.. 2012. Ornamental Fish-Product / Market Report. Thai Trade Center. Budapest. Hungary. Pp 1-13
[2] Barber I 2007*App Ani Behav Sci***104** 251
[3] Bennett S M. and Bennett M B 2001*J of Fish Dis***24**523
[4] Al-Jahdali M O and El-S Hassanine R M 2010*Acta Para.***55** 1
[5] Densmore C L. OttingerC A. BlazerV S. IwanowiczL R. and Smith D R 2004*J. of Aquatic Animal Health* **16** 73
[6] Fujimoto R. Nunes de BarosY. Marinho-FilhoZ M. DiniD G. and Eiras JC 2012. *Rev. Bras. Parasitol. Vet.*
[7] Kim JH. Hayward C J. Joh S J. and Heo G J 2002*Dis Aquat Org.* **52** 169
[8] Hongslo T and Jansson E 2009 *Bull. Eur. Ass. Fish Pathol..* **29(5)** 163
[9] Iqbal Z and Haroon F 2014*Pakistan J. Zool.* **46(3)** 651
[10] Adel M. Ghasempour F. Azizi H R. Shateri M H. and Safian A R 2015 *Vet Research. Forum.* **6** **(1)** 75
[11] Alifuddin M. Hadiroseyani Y and Ohoiulun I 2003*Jurnal Akuakultur Indonesia***2(2)** 93
[12] Prasetya N. Subekti S and Kismiyati 2013 *Jurnal Ilmiah Perikanan dan Kelautan***5** **(1)** 113
[13] Noga E J 2010*Fish Disease Diagnosis and Treatment*. 2nd ed. Wiley- Blackwell. USA.
[14] Woo P T K 1995 Fish Diesease and Disorders ; Protozoan and Metazoan Infections. CAB

International. UK.
[15] Kayis S. Ozcelep T. Capkin E 2009*Israel J Aquatic Bamid***61** 93
[16] Moussavi H E. Mood S M. Omrani B S. Mokhayer B. Ahmadi M. Soltani M. Mirzargar S S. Masoumian M and Pazooki J 2009 *Bull. Eur. Ass. Fish Pathol.* **29(5)** 175
[17] Johnson M B. Lafferty K D. van Oosterhout C. and Cable J 2001*PLosONE.***6**(**8**)
[18] Thilakaratne I D S I P. Rajapaksha G. Hewakopara A. Rajapakse R P V J. and Faizal A C M 2003. *Dis Aquat Org*. **4(54)** 157
[19] Piazza R S. Martins M L. Guiraldelli L.Yamashita M M 2006*B. Inst. Pesca. São Paulo*. **32**(**1**) 51
[20] Saha M. Bandyopadhyay P K. Roy A and Ghosh S 2015*IOSR J of Agr and Vet Sci***8** (**1**) 54

# Identification and stocking density of elements phosphate indigo system for fish farming cages nets floating in freshwater district Laut Tawar, Aceh Province

**N Eriyusni[1] and A Z Siregar[2]**

[1]Departement Fisheries, Universitas Sumatera Utara
[2]Departement Agrotechnology, Faculty Agriculture, Universitas Sumatera Utara
E-mail: eriyusni@hotmail.com and azsynati@gmail.com

**Abstract.** Lake Laut Tawar is located in Aceh district of Aceh province. The lake were used by the community as a livelihood by making a Cages. One of the elements that influence the fish growth is an element Phosphate ($PO_4$) and other such as, pH, nitrate-nitrite, temperature and others. The lake has undergone changes caused by the use of commercial feed. The sampling locations based on the environmental setting by using method "purposive sampling" from January to May. The results was found of 0.1407 mg/L phosphate with fish expenditure amounting to 23728.61 ton / year using cages of net with number as 92, stocking density at 48,950 of fish and marine area 19.77 Ha. However the standard classes showed is under class II Regulation with No. 82 Year 2001 3400, solid lowest stocking cages tail on the size of 2.5 x 2.5 x 3m (spacious, $181m^3$) with post-harvest fish expenditure amounted to 1,214 tons of fish/year, the highest 9,250 cages tail on the size e of 5 x 5 x 3m (spacious $123m^3$) with fish expenditure by 2,800 tones of fish/year. While the results of feed conversion and Phosphate content is 12.70% with the Feed Conversion Ratio (FCR) was 1,635 and 1,491.

## 1. Introduction

Laut Tawar is a lake formed from volcanic processes with an area of 5742.10 ha, with a depth of 84.23 and a length of 15.7 km, width 4.5 km, vast littoral zone with potential are the process of photosynthesis by aquatic 14.28 %. Lautt Tawar also has one outlet that river Batang Peusangan (the name of village) and there are as many as 25 small streams of water catchment areas and has a water flow varies between II L/ sec - 2554L / sec [1] so appropriate to be the location of freshwater fish farming. Water level in Laut Tawar in November to January very fluctuating then decreased until March and rose in April as written by Hygiene and Cropping (2014) . suitable for location of freshwater fish farming.

Fish farming in the system Cages net in Lautt Tawar has long been done by the people of Central Aceh. The system of fish farming every year cage net was continue increase, it is because knowledge of the fish farming cages net system is very good compared to the maintenance of fish naturally. The increasing public freshwater fish in the Laut Tawar, reduce impact of phosphate elements and oxygen elements due to feeding or household waste, that the occurrence of water pollution [2]

Feeding used by the public is in the form of feed floating nature of the feed drifting of stream direction of the wind, because the feed is floating not some favored fish. The feed will accumulate corner of the net so that there is contamination, thus the element Phosphate present in the water will be disrupted, this caused resulting in fish rearing is not achieved in accordance with what is expected. Death of fish is very high due to the high levels of Phosphate, the population levels of algae will grow so that oxygen will be reduced [3.4]. To avoid the occurrence of a mass mortality of fish, due to the increased element of phosphates in the Laut Tawar. It is necessary governance good governance by examining how much impact the influence Phosphate on fish cultivation system cages floating net by using additional feed. Feed Conversion Ratio (FCR) is one way of determining the amount of feed given in determining catches converted with elements phosphate contained in feed consumed fish.

Feeding is not appropriate trigger pollution because the feed is not eaten by the fish. Remaining in the other form, such as nitrogen and phosphorus pollution that would make the fishing mortality rate will be higher. Phosphorus was instrumental in the process of eutrophication in a water ecosystem, so increasing phosphorus in the water, the algae will increase, causing oxygen deficit leaching in water that eventually kill fish en masse.

To determine the level of fishing mortality caused by high Phosphate elements in water, and Phosphate elements contained in fish feed as well as the amount of fishing that takes place every year.

## 2. Materials and Methods

### *2.1. Samples*

Data collection was performed by means of purposive sampling, based on the point coordinates are generated. In this study was determined by 4 (four) observation station where the locations are considered in accordance with the purpose of research. Station 1, located in the village Kelitu, located at coordinates 4 ° 38 '12:45' 'N and 96 ° 54' 41.11 " BT. Water sampling is done at two points, with a distance of 50 and 100 meters from the shores of Lake. Station 2, the waters adjacent to agricultural activity and settlement. Station 3, is the only outlet Laut Tawar at coordinates 4 ° 37 '35.22' 'N and 96 ° 51' 32.45 " BT. Water sampling within the 100 and 150 meters from the shores of Lake. Station 3, is the only outlet Laut Tawar, located at coordinates 4 ° 37 '35.22' 'N and 96 ° 51' 32.45 " BT. Water sampling at a distance of 60-120 meters from the shore of Lake. At station four water sampling conducted on two groups of cages.

### *2.2 . Preserving Water Samples*

Pickling question is if the sample to be analyzed is not done directly at the time of data collection but were housed in some old dependent than the type of meter to be examined as an example of preservation under in:

*2.3. Measurement of Physical Chemistry Parameter Bodies* as follows:

1. The water temperature using secci Disk, and thermometer
2. Brightness, using Disk secci and count the number of dots on a string that has been given the scale.
3. The degree of acidity (pH). By dipping into the sample pH meter
4. Analysis Dissolved Oxygen (DO)
5. BOD (Biochemical Oxygen Demand)
6. Analysis Nitrate
7. Analysis of total phosphate.

*2.4. Calculation (Feed Conversion Ratio (FCR)* was calculated as follows:

FCR = Amount of feed given (tons)
Weight of fish produced (tons)

*2.5.Calculation of Morphology and Hydrology lake* with the formula:

Z = V / A
p = Q0 / V
Where:
Z = The average depth of the lake (m)
V = Volume of lake water (million m3)
A = area of the lake waters (m2)
p = rate of flushing water lakes (1 / year)
Q0 = Number of water flow out of the lake (Million m3 / year)

**Table 1.** Preserving water samples

| Type of Analysis | storage containers | Required Amount | Preservation | Limit of Storages |
|---|---|---|---|---|
| BOD | BOD Bottle | 100 | Refrigerant | 48 hours |
| Total Phosphate | G (A) | 100 | Refrigerant | 48 hours |
| Nitrate | P,G | 100 | H2SO4; pH≤ 2, Refrigerant | 48 hours |

Sources: [5]

*2.6. Pollution Load Calculation of Total P (La) with the formula:*

*La = Likan x A*
*Likan = Δ [P] Z.P / (1 / year)*
*Rikan = x + [(1-x) R]*
*R = 1 / (1 = 0,747 p0.507)*
*Where:*
*Likan = P-total waste of fish per unit large*
*A = area of the lake waters (Ha)*
*Z = The average depth of the lake (m)*
*p = rate of flushing water lake (1 year)*
*Δ [P] = value difference between the mean concentration Average phosphate in waters with a total maximum phosphate accepted farmed fish (mg / m3)*
*Rfish = The overall proportion of total phosphate lost to sediment*
*x = The proportion total phosphate missing permanently to precipitate in bottom waters was 0.45 to 0.55*
*R = Total phosphate dissolved stay together sediments*

*2.7. Calculated of Phosphorus in Fish Farming (PLP) by the formula:*
*PLP = FCR x Pfeed -Pfish*
*Where:*
*PPakan = levels of phosphate in the feed (Kg)*

*Where:*
P (La) = total phosphorus load carrying capacity. Where: The detail calculation in point 2.8 as following

below

*2.8. Calculation of Maximum Fish Production*

$$DD = \frac{P\ (La)}{PLP}$$

DD = Carrying capacity of water (tons / year) (Kg / year)
PLP = Number of phosphate released from
fish cultivation.

**3. Results and Discussions**
To determine the average stocking density, the amount of feed given and cages net production can be seen in Table 2 where the number was 920 cages net with indigenous fish stocking at 48.950. While the production of fish in the cages is 1,214 tons / year with size cages net is181 m3, stocking densities 3,400 fish, total fish/cycle amounted to 0.607 tones, the total feed of 1,986 tons/year and FCR amounting to 1,635, with a stocking density slightly but FCR owned larger, it shows that taking fish feed every uneven compared to the number of stocking density of fish and 9,250 cages, FCR was 1,491, but not all cages net which increased FCR. Fish expenditure average of the number of cages net is 2,135 tons / year, this indicates that the waters of Lut Tawar contributes greatly to the production of fish public waters [6] and if by Trix index, Laut Tawar categorized mesotrofik waters.
P / tons of feed) by the formula:

**Table 2.** Average Number of Total Solids Tebar and Fish Production in 181m$^3$ Cages and Total Feed

| Sizes of Cages net/m$^3$ | Solid stocking (total of fishes) (tails) | Total of fishes/cycles (year/tones) | Production of fishes (KG) Total of fishes /year (tones) | Feed/ cycles (tones) | Total of /year (tones) | FCR |
|---|---|---|---|---|---|---|
| 123 | 9,250 | 1,400 | 2,800 | 2,087 | 4,174 | 1.491 |
| 97 | 8,800 | 1,200 | 2,400 | 1,840 | 3,680 | 1.533 |
| 111 | 8,000 | 1,418 | 2,836 | 2,203 | 4,406 | 1.554 |
| 129 | 7,000 | 1,066 | 2,132 | 1,533 | 3,066 | 1.533 |
| 146 | 7,000 | 1,000 | 2,000 | 1,500 | 3,000 | 1.500 |
| 133 | 5,000 | 0,783 | 1,566 | 1,133 | 2,266 | 1.447 |
| 181 | 3,400 | 0,607 | 1,214 | 0,993 | 1,986 | 1.635 |
| Mean | | 1,068 | 2,135 | 1,613 | 3,225 | 1.527 |

Total phosphate in Laut Tawar experience likely fluctuations on a monthly basis as in January the state of phosphate was obtained 0.100 and 0.280 in the station 1 and 3, in February occurred for station 3 is higher than stations 1 such as phosphate is at 0.132 and 0.180 as shown in Table 3. Meanwhile, in April and Mei station 1 is lower compared with the phosphate content of 3 stations that are in the 0.108 to 0.133, and 0.559 to 0.164. For station 4 only occurs in March is 0.112 to 0.180 for Station 1 and Station 4.

Abundance of phosphorus can lead to eutrophication, thereby encouraging the growth and development of algae rapidly resulting in lower dissolved oxygen content will cause the death of the organism [7], While the value of total raw phosphate based on Government Regulation No. 82 of 2001 grade II for freshwater fish farming is 0.2 mg/l.

**Table 3.** Average Total Phosphate in January, February, March, April and May in Laut Tawar at Stations 1, 2, 3 and 4

| Number of Station | Month | | | | | |
|---|---|---|---|---|---|---|
| | January (mg/l) | February (mg/l) | March (mg/l) | April (mg/l) | Mei (mg/l) | FCR |
| Station 1 | 0.100 | 0.162 | 0.112 | 0.108 | 0.112 | 0.119 |
| Station 2 | 0.154 | 0.143 | 0.144 | 0.115 | 0.137 | 0.139 |
| Station 3 | 0.280 | 0.132 | 0.156 | 0.559 | 0.164 | 0.148 |
| Station 4 | 0.163 | 0.152 | 0.180 | 0.142 | 0.151 | 0.157 |
| Mean | 0.136 | 0.147 | 0.148 | 0.131 | 0.141 | 0.140 |

The results of the analysis of chemical properties of physics in the waters of Laut Tawar at each station experiencing significant difference even though there is the same as the state of the lake water brightness average ranges above 5 m, there are only four stations has decreased with a brightness level ranging from 4,94 m viewable in( Table 4). the whole station DO increase ranged 5.74 to 6.98 mg / l is at station 4 and 3. the resulting BOD is no low levels averaging 2.18mg / l. Nitrate-N lower at Station 1 is located at 0.8mg / l, while the quality standard of Grade II said 10mg / l content of nitrate-N in Laut Tawar that it demonstrates Laut Tawar is still best used with the cultivation of floating net system cages , as claimed by [8] in which the nitrite content in excess of 1 mg /l can make fish mortality.

**Table 4.** Average Analysis Chemical Properties Physical Bodies Laut Tawar at each station.

| Number Station | Parameter | | | | | | |
|---|---|---|---|---|---|---|---|
| | Tempe rature ($^0$C) | Brightness (m) | DO (mg/l) | BOD (mg/l) | pH | Nitrate-N (mg/l) | Total Phospate |
| The quality standard Class 11 Deviation 3 | | | $\geq 4$ | 3 | 6-9 | 10 | 0,2 |

The results of the analysis of chemical properties of physics in the waters of Laut Tawar at each station experiencing significant difference even though there is the same as the state of the lake water brightness average ranges above 5 m, there are only four stations has decreased with a brightness level ranging from 4,94m viewable in( Table 4). the whole station DO increase ranged 5.74 to 6.98 mg / l is at station 4 and 3. the resulting BOD is no low levels averaging 2,18mg / l. Nitrate-N lower at Station 1 is located at 0,8mg / l, while the quality standard of Grade II said 10mg / l content of nitrate-N in Laut Tawar that it demonstrates Laut Tawar is still best used with the cultivation of floating net system cages , as claimed by [8] in which the nitrite content in excess of 1 mg / l can make fish mortality.

From the analysis parameters Chemical Physics Water Laut Tawar in every station 1,2,3 and 4 (Table 5), the average temperature levels ranging from 23.5 up to 24.25$^0$C, Brightness was there 4.96 m to 5. 32 m, DO is at 6.05 womb up to 6.54mg / l, BOD is at 1.8 up to 2.38mg / l, pH between 7.17 up to 8.17. Nitrate-N ranged from 1.04 up to 1,11mg / l, and Total phosphate is at 0.13 up to 0.15mg/l.

**Table 5.** Average Analysis Chemical Properties Physical Bodies Laut Tawar at each station.

| Number Station | Parameter | | | | | | |
|---|---|---|---|---|---|---|---|
| | Tempe rature ($^0$C) | Brightness (m) | DO (mg/l) | BOD (mg/l) | pH | Nitrate-N (mg/l) | Total Phospate |
| The quality standard Class 11 Deviation 3 | | | ≥ 4 | 3 | 6-9 | 10 | 0.2 |
| Station 1 | 23.80 | 5.41 | 6.54 | 2.25 | 7.9 | 0.8 | 0.12 |
| Station 2 | 23.60 | 5.14 | 5.90 | 2.06 | 8.08 | 1.02 | 0.14 |
| Station 3 | 24.40 | 5.06 | 6.98 | 2.26 | 8.20 | 1.31 | 0.15 |
| Station 4 | 24.00 | 4.94 | 5.74 | 2.18 | 8.03 | 1.13 | 0.16 |

**Table 6.** Average analysis Physical and Chemical on January, February, March, April and May in\ Lake Laut Tawar at station1, 2, 3 and 4

| Parameter | The quality standard Class 1 | Month | | | | |
|---|---|---|---|---|---|---|
| | Deviation 3 | January | February | March | April | May |
| Temperature ($^0$C) | | 24.25 | 24.25 | 24 | 23.5 | 23.75 |
| Brightness (m) | | 5.32 | 5.01 | 5.23 | 4.96 | 5.15 |
| DO (mg/l) | 4 | 6.22 | 6.54 | 6.31 | 6.32 | 6.05 |
| BOD (mg/l) | 3 | 2.38 | 2.28 | 1.8 | 2.05 | 2.32 |
| pH | 6-9 | 8.1 | 8.03 | 7.99 | 7.91 | 8.17 |
| Nitrate-N (mg/l) | 10 | 1.04 | 1.06 | 1.11 | 1.05 | 1.09 |
| Total Phospate (mg/l) | 0.2 | 0.14 | 0.15 | 0.15 | 0.13 | 0.14 |

Results of average total phosphate Ra of 0.1407mg / l with fish production amounted 23,728,615 tons of fish / year means Laut Tawar can accommodate the expected carrying capacity, where the productivity in cages in Laut Tawar average of 41,577kg / m3 /year. To reach the carrying capacity of the water volume of Laut Tawar that can be utilized is 570,714,935m3, if at Laut Tawar has a depth of 3 meters, the water area that can be utilized in accordance with its carrying capacity is 19.024 hectares or an area of 0.331% of the area of the lake , as shown in table 6.

**Table 7.** Carrying fluctuations of Laut Tawar, Aceh

| Month | Phospate (mg/l) | Carrying Capacity (ton/year) |
|---|---|---|
| | 0.136 | 24,748,967 |
| February | 0.147 | 22,360,908 |
| March | 0.148 | 22,143,812 |
| April | 0.131 | 25,834,448 |
| May | 0.141 | 23,662,846 |
| Mean | 0.1407 | 23,728,615 |

## 4. Conclusions

The research result have was found of 0.1407 mg/L phosphate with fish expenditure amounting to 23,728,61 ton / year using cages of net with number as 92, stocking density at 48,950 of fish and marine area 19.77 Ha, and standard classes showed is under class II Regulation with No. 82 Year 2001. While the results of feed conversion and Phosphate content is 12.70% with the Feed Conversion Ratio (FCR) was 1.635 and 1.491.

**Refferences**

*[1]* Center for Environment, Hygiene and cropping central Aceh 2014 *Profile Lake Laut Tawar*
[2] Pontoh O 2014 Analysis of Developmen of Fish Growth in Aquaculture in Floating Net In Tandengan Minahasa village *Journal Aquaculture Fresh Water Fish* Vol **2** No 1
[3] Ridwansyah 2002 *Effect Of Consentration Hydrogen Peroxida and long Immersion Pengaruh Konsentrasi Hidrogen Peroksida dan lama perendaman For Quality of Kembung fish in pindang cooker* (USU Library: Universitas Sumatra Utara)
[4] Siagian M 2010 Carrying capasity reservoir PLTA Koto Panjang Kampar Provinsi Riau *Journal Fish Marrine* Vol **15** (1): 25 – 38
[5] Reguilation of marrine minister of life environment Nombor 28 Tahun 2009 And capasity water polution in lake/reservoir
[6] Hall of the life environment, cleanliness of the park Midle Aceh Distric 2014 Profile Lake Laut Tawar
[7] Wetzel R G 2001 *Limnology Lake and River Ecosystems* (California: Academic Press)
[8] Lesmana D S 2001 *Water Quality for Decorative of Fresh Fish* (Jakarta: Publisher Penebar Swadaya)

# Fast and simultaneous prediction of animal feed nutritive values using near infrared reflectance spectroscopy

**Samadi[1*], S Wajizah[1] and A A Munawar[2]**

[1]Department of Animal Science, Universitas Syiah Kuala, Banda Aceh – Indonesia.
[2]Department of Agricultural Engineering, Universitas Syiah Kuala, Banda Aceh – Indonesia.
E-mail: samadi77@unsyiah.ac.id

**Abstract.** Feed plays an important factor in animal production. The purpose of this study is to apply NIRS method in determining feed values. NIRS spectra data were acquired for feed samples in wavelength range of 1000 – 2500 nm with 32 scans and 0.2 nm wavelength. Spectral data were corrected by de-trending (DT) and standard normal variate (SNV) methods. Prediction of in vitro dry matter digestibility (IVDMD) and in vitro organic matter digestibility (IVOMD) were established as model by using principal component regression (PCR) and validated using leave one out cross validation (LOOCV). Prediction performance was quantified using coefficient correlation (r) and residual predictive deviation (RPD) index. The results showed that IVDMD and IVOMD can be predicted by using SNV spectra data with r and RPD index: 0.93 and 2.78 for IVDMD ; 0.90 and 2.35 for IVOMD respectively. In conclusion, NIRS technique appears feasible to predict animal feed nutritive values.

## 1. Introduction

In animal production and livestock enterprise, feed plays an important role and spends about 70% of production cost [1]. Sago starch can be utilized to become animal feed which are readily in an abundant number in Indonesia. As stated by [2], it contains around 28. 27% crude fibres, and approximately 1. 359 % crude protein. Moreover, according to [3] sago residues may contain starch (57.9%), cellulose (23.1%), hemicellulose (9.19%) and lignin (3.89%) after starch extraction.

Generally, crude fibre in high amount will cause slow and limited ruminal degradation of carbohydrates. Further, lower content nitrogen is one of the main factors deficiencies from residues causing the low value of residues as feed for ruminants [4]. Fermentation technique by using microbial seems to be a practical method to improve nutritional value of agro residues. It is crucial to evaluate nutritive value of animal feed before being fed to animals mainly feedstuff quality for feed formulation to obtain high production of animals. Recently, to determine feed quality, several methods were employed. Yet, most of them were used based on standard laboratory analysis with complicated procedures, time consuming and using of hazardous chemical solutions, from which may cause environmental pollutions.

Therefore, alternative methods should be found to alter chemical analysis at laboratories to evaluate the nutritive values of feed stuff. Near infrared reflectance spectroscopy (NIRS) has been developed as rapid, direct, non-destructive and non-invasive bioanalytical technique. By this technique, the quantity, composition, structure and distribution of chemical constituents and functional groups in a tissue can be revealed.

Based on advantages and excellence of NIRS as novelty method to measure feed quality, this study is truly required. Thus, the main purpose of this present study is to apply NIRS as a rapid and simultaneous method in determining feed nutritive values in form of digestive organic materials (KCBK and KCBO). The prediction models were developed using raw untreated spectra data and enhanced one.

## 2. Materials and Methods

### *2.1. Material*

Sago residues were used as a model feed to measure feed quality after fermentation. Sago from agro industry by product was fermtened with 40% water content. commercial fermentation products (Saus Burger Pakan® (SBP) was used to fermented with different day fermentation as follow control (without fermentation), 7d, 14d, 21d and 28d.

### *2.2. Spectra data acquisition*

Near infrared (NIR) spectra data of feed samples were obtained using Fourier transform infrared instrument. The basic measurement chosen for this spectra data measurement is high resolution with integrated sphere. Absorbance spectra data in form of diffuse reflectance in wavelength range from 1000 to 2500 nm were obtained 20 scans and averaged. Spectra data were recorded and saved in local computer with three different file formats (*.SPA, *.JDX and *.CSV) [5].

### *2.3. Measurement of feed nutritive values*

After collecting and recording infrared spectra data, all feed samples were analyzed in Animal Feed Laboratory, Syiah Kuala University to measure its relevance quality parameters: protein, digestive organic materials (KCBK and KCBO), crude fiber, crude lipid and minerals. The method used for these measurements were standard laboratory methods.

### *2.4. Spectra data correction*

Spectra data analysis were firstly performed to inspect spectra visualization and noise recognition. Noise, caused by light scattering were enhanced and corrected using de-trending (DT) and standard normal variate (SNV) methods. All spectra enhancement methods were processed using The Unscrambler X 10.3 software.

### *2.5. Prediction model*

Prediction models were established for each quality attributes (KCBK and KCBO) based on untreated raw and enhanced spectra data (DT and SNV). Prediction models were developed using principal component regression (PCR) with full leave one out cross validation (LOOCV).

### *2.6. Evaluation performance*

Prediction models performances were judged for their accuracies and robustness using these following indicators: the coefficient of correlation (r), root mean square error (RMSE) and the residual predictive deviation (RPD) defined as the ratio between the standard deviation and the RMSE. The higher the RPD, the greater and robust the model to predict KCBK and KCBO of animal feed samples [6].

## 3. Result and discussion

### *3.1. Spectra features*

Typical diffuse reflectance spectrum for feed samples in wavelength range from 1000 to 2500 nm is shown in Fig.1. this spectra data shows the absence o nutritive values of animal feed samples which is derived from the absorbance bands that result from the vibration of molecular bonds of O-H, C-H, C-O and N-H with the IR radiation.

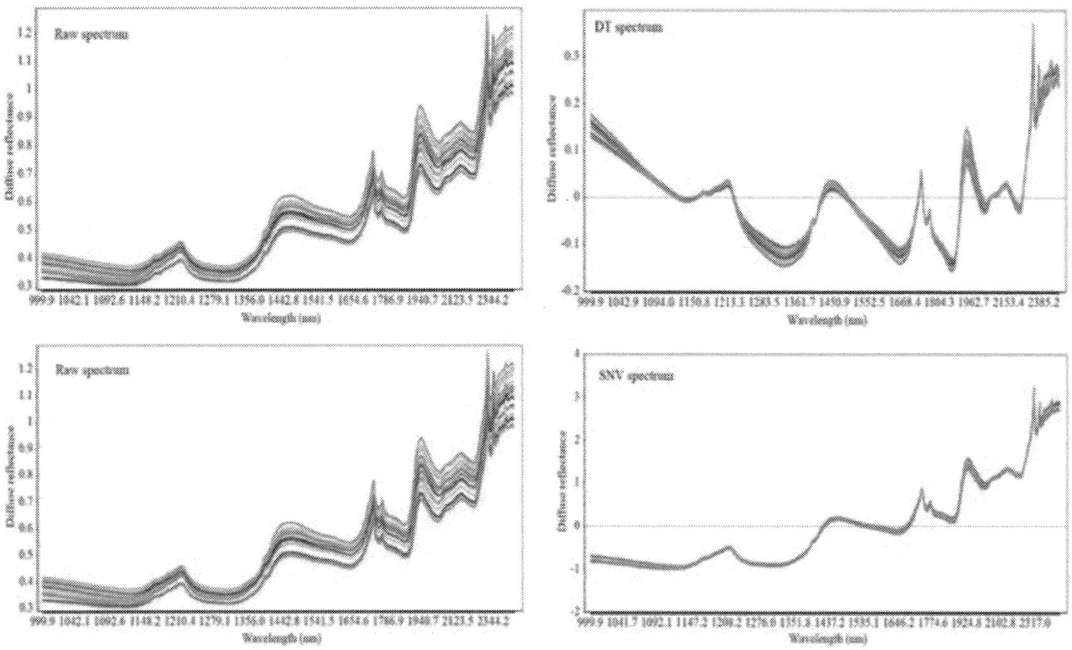

**Figure 1.** Spectra data of feed samples and corrected data using DT and SNV method.

Animal feed samples are scattered with the infrared radiation and their reaction including reflectance, absorbance and transmittance are captured and recorded. The different reaction of this specular data are depend on composition, cell structure and other chemical or physical properties of the feed samples. A captured near infrared spectra of biological object consists the response of the molecular bonds O-H, C-H, C-O and N-H. These bonds are subject to vibrational energy changes when irradiated by NIR frequencies [7].

As shown in Fig.1, spectra data were corrected using DT and SNV method. DT enhanced peak and valley by multiplying polynomial order while SNV spectra corrected spectra data by scaling spectra to its ideal one which normally mean spectra. These three spectra data (Raw, DT and SNV) were then used to develop prediction model used to predict feed nutritive values.

### *3.2. Prediction models*

The main part of near infrared spectroscopy application is building a calibration model used to predict quality attributes or chemical constituents of this feed samples to be observed. The models were established using principal component regression is a two-step procedure; first decomposes the X-variables by a principal component analysis (PCA) and then fits a MLR model, using a small number of principal components (PCs) or latent variables (LVs) instead of the original variables (NIR spectra) as predictors. The advantage with respect to MLR is that the X-variables (principal components) are uncorrelated, and that the noise is filtered. Also, usually a small number of principal components are preferable and sufficient for the models.

Prediction model consists of the relationship between the observed response variable y (Y-variables: nutritive values KCBK and KCBO) and the independent variable x (X-variables: NIR spectra data). The primary information that can be gathered from the interaction of the near-infrared radiation with the biological object is its physical, optical and chemical properties. Feed samples have shown to have identifiable C-H, N-H, C-H-O and O-H absorption bands in the near-infrared region whereas each have a specific vibrational frequency [8].

Prediction result for KCBK and KCBO parameters were shown in Table 1 and Table 2 respectively. It is obvious that even without spectra correction, NIRS methods seems feasible to predict KCBK and KCBO with coarse performance (RPD = 1.64 for KCBK and RPD = 1.60 for KCBO). Their prediction performances were significantly improved when spectra data were corrected by DT and SNV method. As shown in both Table.

**Table 1**. KCBK prediction performance of NIR spectra data

| Spectrum | Statistical Parameter | | | |
|---|---|---|---|---|
| | $R^2$ | r | RMSE | RPD |
| Raw | 0.61 | 0.78 | 2.03 | 1.64 |
| DT | 0.72 | 0.85 | 1.71 | 1.95 |
| SNV | 0.86 | 0.93 | 1.20 | 2.78 |

**Table 2**. KCBO prediction performance of NIR spectra data

| Spectrum | Statistical Parameter | | | |
|---|---|---|---|---|
| | $R^2$ | R | RMSE | RPD |
| Raw | 0.59 | 0.77 | 2.51 | 1.60 |
| DT | 0.69 | 0.83 | 2.18 | 1.84 |
| SNV | 0.81 | 0.90 | 1.71 | 2.35 |

Based on prediction results, SNV provide the most accurate and robust prediction results by achieving maximum correlation coefficient r = 0.93 and RPD = 2.78 for KCBK and maximum r = 0.90 and RPD = 2.35 for KCBO prediction. Scatter plot drawn for the best results is shown in Fig. 2.

In general, judging from prediction performance evaluation, NIRS can be used and improved by correcting spectra data. Regression approach probably can be enhanced using another method such as partial least square regression (PLSR) or support vector regression (SVR).

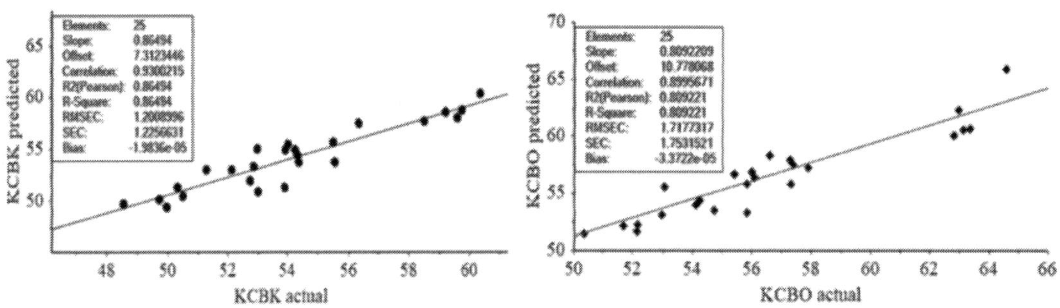


**Figure 2**. Scatter plot between actual and predicted feed nutritive values using SNV spectra data.

Based on regression and loading curve derived from SNV-PCR data, important and relevant wavelength for KCBK and KCBO prediction were observed around 1403 -1484 nm, 1908 - 2006 nm, and 2310 nm as shown in Figure. 3.

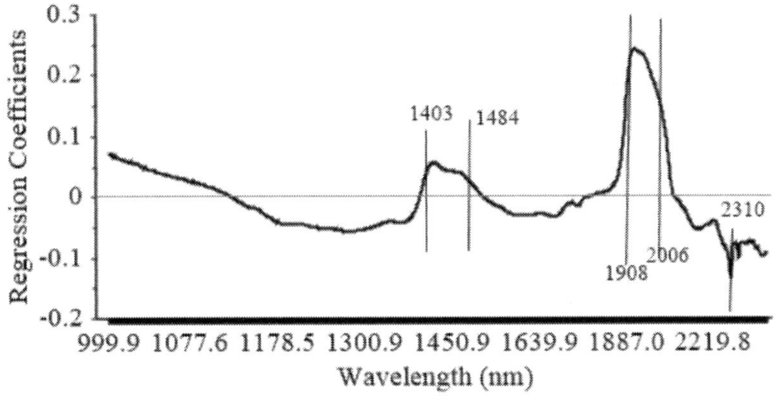


**Figure. 3**. Important wavelength in NIR region for KCBK and KCBO prediction.

## 4. Conclusion

Based on obtained results, it may conclude that NIRS can be used as an alternative fast and simultaneous method used to predict feed nutritive values (KCBK and KCBO). SNV spectra data provide the most accurate and robust prediction result with correlation coefficient r = 0.93 and RPD = 2.78 for KCBK and maximum r = 0.90 and RPD = 2.35 for KCBO prediction.

## References

[1] Samadi and Liebert F 2008 Modelling the optimal lysine to threonine ratio in growing chickens depending on age and efficiency of dietary amino acid utilisation *British poultry science* **49**(1), 45-54

[2] Tampoebolon B I M 2009 Kajian perbedaan aras dan lama pemeraman fermentasi ampas sagu dengan aspergillus niger terhadap kandungan protein kasar dan serat kasar (Study of different levels and duration of fermentation of sago waste by aspergillus niger to crude protein and crude fibre contents) *In Prosiding Seminar Nasional Kebangkitan Peternakan–Semarang* 20 Mei 2009 pp 235-243 (Fakultas Peternakan UNDIP Semarang)

[3] Linggang S, Phang L Y, Wasoh M H and Abd-Aziz S 2012 Sago pith residue as an alternative cheap substrate for fermentable sugars production.Applied biochemistry and biotechnology **167**(1): 122-131

[4] Van-Soest P J 2006 Rice straw, the role of silica and treatments to improve quality *Animal Feed Science and Technology* **130**(3): 137-171

[5] Munawar A A, Hörsten D V, Mörlein D, Pawelzik E and Wegener J K 2013 Rapid and non-destructive prediction of mango sweetness and acidity using near infrared spectroscopy. In Gesellschaft für Informatics lecture notes: Proceeding of the GIL Jahrestagung.

[6] Nicolai B M, Beullens K., Bobelyn E, Peirs A, Saeys W, Theron K I and Lamertyn J 2007 Nondestructive measurement of fruit and vegetable quality by means of NIR spectroscopy: a review *Postharvest Biology and Technology* **46**:99-118

[7] Chen L, Xue X, Ye Z, Zhou J, Chen F and Zhao J 2011 Determination of Chinese honey adulterated with high fructose corn syrup by near infrared spectroscopy *Food Chemistry* **128**: 1110–1114

[8] Cen H Y, Bao Y D, He Y and Sun D W 2007 Visible and near infrared spectroscopy for rapid detection of citric and tartaric acids in orange juice *Journal of Food Engineering* **82**: 253–260

**Acknowledgments**
Authors wishing to acknowledge to LPPM Syiah Kuala University for providing research grant through H-index scheme 2016.

# The role of lactic acid bacteria (*Lactobacillus sp yel133)* from beef in inhibiting of microbial contaminants on various fillers of starter culture

**Yunilas and E Mirwandhono**

Faculty of Agriculture, Universitas Sumatera Utara, Medan 20155, Indonesia
E-mail:yunilas11@yahoo.co.id

**Abstract**. The role of *Lactic Acid Bacteria* (LAB) on the starter culture can be seen from the ability to grow and suppress the growth of microbial contaminants (fungi). The research aimed to investigate the role of LAB (*Lactobacillus sp YEL133*) in inhibiting microbial contaminants (fungi) on starter cultures of various fillers. The materials used in this research was *Lactobacillus sp YEL133* from beef and various fillers (rice flour, corn starch and wheat flour). The research methods used completely randomized design (CRD) with 3 treatments and 4 replications. The treatments of this research was P1(rice flour), P2 (corn starch) and P3 (wheat flour) that inoculated with *Lactobacillus sp YEL133*. Parameters which is observed such as: growth of lactic acid bacteria, total microbes and total fungi as microbial contaminants. The results showed that the starter culture with a filler material of rice flour produce lactic acid bacteria and microbes were highly significant (P <0.01) for corn starch and wheat flour, as well as able to suppress the growth of microbial contaminants (fungi). The conclusion of the research is the use Lactobacillus sp YEL133 can suppress the growth of fungi on the starter culture using rice flour.

## 1. Introduction

Lactic acid bacteria that is use as starter cultures should be able to meet the criteria: a) be able to compete with other microorganisms, b) quickly produce lactic acid, c) be able to grow at a salt concentration of less than 6%, d) capable to react with a concentration of less than 100 mg / kg, e) capable to grow at temperatures between 15-40°C, f) include homofermentatif bacteria, g) does not generate peroxide in large quantities, h) can reduce nitrite and nitrate, i) can improve the flavor of the final product, j) does not produce amino acid compounds, k) can kill spoilage bacteria and pathogens, and l) synergistic with other starter compounds [1]. LAB (lactic acid bacteria) was able to convert sugars (including lactose) and other carbohydrates into lactic acid. These not only provides a unique sour taste of fermented dairy food such as fermented milk, but also acts as a provider by reducing the pH and create opportunities for harmful organisms to grow a little more.

Starter culture is a strain of microorganisms' that has been selected and is known to carry out metabolic activities that can improve the characteristics of the fermented material. Furthermore, starter culture is a selected strain of food-grade microorganism that is known to have metabolic activity and other stable property that is used in producing fermentation products with desired or attributes characteristics [2]. Starter functions include forming flavor (which is a product of organic acids, carbonyl component and partial hydrolysis of proteins and fats); inhibit unwanted microorganisms;

improve the texture of fermented products such as viscosity and cohesiveness also contribute as a functional food product with the use or addition of probiotic microorganisms [3], [4].

Pickling starter is generally done by drying [5], to maintain the viability and the particular traits of bacteria such as the acid-forming activity, produce aroma, texture and shape as the nature of the probiotics [4]. Heat stress that occurs during the process of drying and dry storage is the major cause of the loss of bacteria activity [6], also the stress of acid [7], nutritional deficiencies [8], osmotic stress and oxidative stress [9] affect to the performance of the starter.

The use of bacteriocin-producing bacterial strains as starter cultures or protective culture recently developed and is able to control the presence of pathogens and spoilage bacteria in ready food products [10]. Culture that is often used for the fermentation of meat and commercially available come from the *Streptococcus* [11], class *Micrococcus* [1], *Lactobacillus piantarum*, *Lactobacillus sake*, *L. Curvatus*, *Pediococcus iacidactici* and the right combination with *P. Pentosaceus* [12]. Naturally, there are species of lactic acid bacteria that grow on pure beef, one of which is *L. plantarum* [13].

Starter culture in the manufacture of sausages can be divided into two categories. The first generation contains lactic acid bacteria derived from plant material. The second generation comes from meat material that is specifically adapted to the ecology of fermented meat [14]. Starter cultures of lactic acid bacteria isolated from other than meat still less adaptive and less than optimal for the meat fermentation is characterized by fluctuating viability during the process [15] and [16].

Commercial dry cultures for fermented sausages have a total lactic acid bacteria at least $10^6$ CFU / g. Lactic acid bacteria *Lactobacillus sp YEL133* is one of lactic acid bacteria isolated from beef can be used as starter cultures in the making probiotic sausage [17]. However, the bacteria Lactobacillus sp YEL133 need to be tested role as starter cultures in inhibiting the growth of fungi, testing the viability on the various fillers.

## 2. Material and Methods

### *2.1. Materials and Equipment*

Probiotic sausages dry starter culture is made from pure cultures of lactic acid bacteria (*Lactobacillus sp YEL133*) isolated from beef purely based on research results of [17]. The medium used is MRSB (de Mann Rogose Sharpe Broth) and MRSA (de Mann Rogose Sharpe Agar) brand Merck, NB (Nutrient Broth), NA (Nutrient Agar), peptone for casein, yeast extract, CaCO3, buffered peptone water (BPW), ethanol, acetone, glycerol, grandula agar, 70% alcohol, 96% alcohol, 99% acetone, BCP (brome creasol purple), H2O2, violet crystal, lugol, acetone, safranin, cotton, distilled, spirits, vortex, hot plate, autoclave, incubator, vacuum cabinets, wire loop, laminar air flow, micropipette, analytical balance, oven, Bunsen burners, petridish, pH meter, a spatula, a 1.5 mL tube, vortex and glassware.

### *2.2. Methods*

*2.2.1. Purification:* Purification is done by growing lactic acid bacteria isolates from the results of previous research on medium de man Rogose Sharpe broth (MRSB) and de man Rogose Sharpe agar (MRSA) alternatively to seek a uniform pure colony. Purity test then followed by biochemical tests based on its ability to ferment several types *of sugars to ensure that the culture used is a pure culture. Testing is done by fermenting culture on sugar medium.*

*2.2.2. Refresher:* Refresher is done by growing isolates in MRSB media for 24 hours at a temperature of 37°C. Refresher is continue until the culture could adapt to medium that is characterized by clouding on the growth media. Refresher aims is to update and expand the culture of lactic acid bacteria before it is used as a starter culture of fermented sausage.

*2.2.3. Breeding:* Culture that has been freshened up, then inoculated in a medium containing 2% skim milk, which aims to reproduce and adapt to the culture medium. Medium then incubated at a temperature of 37°C for 48 hours. The result is called the main culture. This process is continued until it obtained the between culture and work culture.

2.2.4. *Manufacture Dry Culture and Viability Test:* Culture that has been bred will then added with fillers respectively 50% rice flour, corn flour and wheat flour. The aim of using filler material in the manufacture of dry culture is to preserve the activity of starter cultures during drying. Starter that has been added fillers will then dried using chemo reaction method with water content of culture almost ±10%.

2.2.5. *Methods:* The research methods used completely randomized design (CRD) with 3 treatments and 4 replications:
P1 = rice flour
P2 = corn starch
P3 = wheat flour

2.2.6. *Parameter:* Parameters measured include: the viability of lactic acid bacteria, total microbial and total molds (fungi) as microbial contaminants.

## 3. Results and Discussion

### 3.1. *Total Lactic Acid Bacteria (LAB) Dry Starter Culture Sausages*

The experimental results obtained total lactic acid bacteria (LAB) on the dry starter cultures varies with a range of $1.10 \times 10^6$ CFU / g (P2) - $1.70 \times 10^7$ CFU / g (P1) Table 1. The analysis showed that the treatment of various fillers was highly significant (P <0.01) to total lactic acid bacterial starter culture dried sausage.

**Table 1.** Total Lab In The Dry Starter Culture Of Sausages

| Filler material | Σ LAB (colony/g) |
|---|---|
| Rice flour | $1.70 \times 10^7$ A |
| Corn flour | $1.10 \times 10^6$ C |
| Wheat flour | $1.40 \times 10^7$ B |

Description: different super scripts letters in the same column indicate significantly different (P <0.01)

Duncan advanced test showed that the total LAB dry starter culture with rice flour filler treatment were significantly (P <0.01) higher than other treatments. This is because the starch content of rice flour was higher than corn and wheat flour. High starch content will increase the water holding capacity in the medium or cells, so it can protect microbial cells during the drying process. Consist to a statement from Chaplin [18] that one of the function of starch is a binder to water (water binder).

[19] states that the content of starch in wheat flour is about 70% while the starch content of rice flour is 90% of the dry weight. According [11], the use of heating medium that contains carbohydrates will increase the heat resistance of microbes contained in there. Components of the initial medium in microbes before being dried are mostly water. Rice flour has a higher starch content, so the amount of water and microbes that are bound and protected from direct exposure to heat during the drying starter using rice flour filler material will be greater.

Besides the starch content of rice flour, the properties of starch and protein in rice flour can withstand and protect *Lactobacillus sp YEL133* isolate to heat during the drying process, resulting in microbial viability higher than other treatments.

Differences in total LAB dry starter culture is also influenced by the levels of amylose and amylopectin starch constituent. Amylose is easily bound to water; this is because amylose consists of hydroxyl chain in large numbers so it could increase the water holding capacity in the material. Amylose prepare *amorphous* region and amylopectin prepare crystalline regions of starch granules. *Amorphous* region will absorb water earlier than crystalline regions because amylose is more hydrophilic. Amylose is hygroscopic; it can increase water absorption and gel formation easier because the straight chain could easily form a three-dimensional network. According to [20], amylose prepares *amorphous* regions and amylopectin prepare crystalline regions of starch granules. At the time of gelatinization, *amorphous* region absorbs water earlier because amylose is more hydrophilic, but amylose undergoes syneresis and crystallized faster.

Rice flour has amilopectin levels higher than corn and wheat flour so that the amount of water that is bound also greater after amylose. Therefore, the dry starter culture with higher levels amilopectin excipients capable of producing higher cell viability that is shown by the total LAB generated is higher in rice flour rather than corn and wheat flour.

### *3.2. Total Fungi Dry Starter Culture Sausage*

Mean total fungi dry starter culture sausage on various types of fillers can be seen in Table 2. The results show the average total fungi in dry starter cultures sausages ranging from 1.46 x $10^4$ to 3.00 x $10^4$ CFU / g, Table 2.

Total fungi obtained allegedly came from contamination that occurred during the manufacturing process of dry starter, especially in the drying process and also influenced by the type of flour used as filler.

Results of analysis of variance showed that the treatment of various fillers is highly significant (P <0.01) of the total fungi dry starter cultures sausage. Duncan advanced test showed that total fungi in dry starter culture made from wheat flour is not significantly different from rice flour, but highly significant with cornstarch. Allegedly due to the content of gluten in wheat flour is higher than corn flour. The interaction between water and wheat flour will produce gluten that has strong dough structure. This is thought to inhibit the penetration of fungi into the starter.

**Table 2.** Total Fungi In The Dry Starter Culture Sausages

| Filler Material | Σ fungi (colony/g) |
|---|---|
| Rice flour | 1.0 x $10^{3}$ |
| Corn flour | 3.00 x$10^{4}$ |
| Wheat flour | 1.46 x $10^{4}$ |

Description: different super scripts letters in the same column indicate significantly different (P <0.01)

Total fungi were lower in the treatment of dry starter culture in rice flour and wheat flour than corn flour. It is affected because the total LAB that is obtained is higher in that treatment. The high total LAB will produce a high lactic acid, causing the pH of the medium is low. Low pH medium will suppress the growth of pathogenic microbes, including fungi as a contaminant. LAB also produces other compounds, namely hydrogen peroxide, diacetyl, carbon dioxide, reuterin and bacteriocins which also serves as an antimicrobial.

### *3.3. Total Microbial Dry Starter Culture Sausages*

Total microbial in dry starter culture is the total number of microbes contained in dry starter culture. The mean total microbial in dry starter culture sausages ranged between 5.65 x $10^{5}$ (CFU / g) - 8.00 x $10^6$ (CFU / g), Table 3.

Based on the analysis of variance, the treatments of various types of fillers have a very significant effect (P <0.01) of the total microbial dry starter culture sausage. Duncan advanced test, indicate that the interaction of treatment using rice flour as a filler material, produces a number of microbes that is significantly higher than other treatments. This is because the total microbial dry starter culture is influenced by the total LAB and also influenced by the total fungi as microbial contaminants in the dry starter culture.

**Table 3.** Total microbes in the dry starter culture sausages

| Type of filler material | Σ Live microbe (colony/g) |
|---|---|
| Rice flour | $8.00 \times 10^6$ A |
| Corn flour | $5.65 \times 10^5$ C |
| Wheat flour | $7.00 \times 10^6$ B |

Description: different super scripts letters in the same column indicate significantly different (P <0.01)

The high total microbes on the treated filler rice flour is due to the high total lactic acid bacteria that being produced. Total high lactic acid bacteria are determined by high starch content. This is consistent with the statement from [18] that one of the functions of starch is as a water binder. According to [11], rice flour has a higher starch content, so the amount of water and microbes that bound and protected from direct exposure to heat during the drying starter using rice flour filler material will be even greater.

While the total microbial in treatment using corn flour is lower than other treatment. This is because the starch content of corn was lower than rice and wheat flour, so that the water-binding power is lower. Thus the ability to protect the microbes from directs exposure to heat during the drying is also low. This leads the viability power of microbes also low especially LAB.

## 4. Conclusions

Based on the results of this study concluded that *Lactobacillus sp YEL133* can suppress the growth of fungus in dry starter culture by using rice flour.

## References

[1] Varnam A N and Sutherland J P 1995 *Meat and Meat Products* Chapman and Hall, London
[2] Ray B 2004 *Fundamental Food Microbiology. 3rd Ed.* CRC Press, New York
[3] Tamime A Y, Skriver and Nilsson L E 2006 *Starter Culture* In: A.Y. Tamime (Ed) Fermented Milks Blackwell Science Ltd. Oxford
[4] Fonseca F, Beal C, Mihoub F, Marin M and Corrieu G 2003 Improvement of cryopreservation of Lactobacillus delbrueckii subsp. bulgaricus CFL1 with additives displaying different protective effects *Int.J. Food Microbiol.* **87**:1-8
[5] To C S B dan Etzel M R 1997 Spray drying, freeze drying or freezing of three different lactic acid bacteria species *J. Food Sci.* 62:576-578, 585
[6] Desmond C, Stanton C, Fitzgerald G F, Collins K dan Ross R 2002 Environmental adaptation of probiotic Lactobacilli towards improvement of performance during spray drying *Int. Dairy J.* 12:183-190
[7] De Angelis M, Bini L, Pallini V, Conconseli P S dan Gobbeti M 2001 The acid-stress response in Lactobacillus san fransciscessis CB1 Microbiology **147** pp 1863-1873
[8] Maus J E and Ingham S C 2003 Employment of stressful conditions during culture production to enhance subsequent cold and acid tolerance of bifidobacteria *J. Appl. Microbiol.* **95**:146-154
[9] Guerzoni M E, Lanciotti R and Cocconcelli P S 2001 Alteration in cellular fatty acid composition as a response to salt, acid, oxidative and thermal stresses in Lactobacillus helveticus *Microbiology* **147**: 2255-2264
[10] Hugas M, Garriga M, Aymerich M and Montfort J M 1995 In-hibition of *Listeria* in dry fermented sausages by the bacteriocinogenic *Lactobacillus sake* CTC494 *J. Appl. Bacteriol.* **79**:322– 330
[11] Fardiaz S 1992 *Analysis of Food Microbiology* Inter-University Center for Food and Nutrition, Bogor Agricultural University, Bogor
[12] Erdo™rul O T, Cetin O D dan ErgUn O 2002 A study on metabolic and antimicrobial activities of *Pediococcus pentosaceus* isolated from fermented sausage *Pakistan J. Biological Sciences* **5**(5):594-596

[13] Arief I I, Maheswari R R A and Suryati T 2005 *Isolation of some of Lactic Acid Bacteria of Beef. Papers Research Seminar Department of IPT* December 2005 The Faculty of Animal Husbandry, Bogor
[14] Hugas M dan Monfort J M 1997 Bacterial starter cultures for meat fermentation *Food Chemistry* **(59)** 4:547-554
[15] Arief I I, Maheswari R R A and Suryati T 2003 The processing of Beef Dark Firm Dry (DFD) Through Technology Fermentation by Lactic Acid Bacteria Lactobacillus plantarum. Basic research reports. Bogor Agricultural University, Bogor.
[16] Hapsari D, Irawati L, Irfan and Primary P 2003 *Beef Quality Improvement Microbiology Dark Firm Dry (DFD) Through Fermentation Process Using a starter culture of Lactobacillus plantarum* Research Report PKM Bogor Agricultural University Bogor
[17] Yunilas Warly L and Mirwandhono E 2014 Toward probiotict food product from meat through isolation and identification lactic acid bacteria as probiotic culture stater *International Journal On Advanced Science Engineering Information Technology*/ **Vol.4(2014)** No.2 ISSN:2088-5334/ INSIGHT Pub
[18] Chaplin M 2002 Starch.(http://www.sbu.ac.uk/water/hysta.html)
[19] Gaonkar A G 1995 *Inggredient Interactions Effects on Food Quality* (New York: Marcell Dekker, Inc.)
[20] Philpot K 2006 Environmental factors that affect the ability of amylose to contribute to retrogradation in gels made from rice flour *Journal of Agricultur and Food Chemistry* **54** : 5182 –5190

# Nutrition quality of extraction mannan residue from palm kernel cake on brolier chicken

**M Tafsin, N D Hanafi, E Kejora and E Yusraini**

Faculty of Agriculture, Universitas Sumatera Utara (USU), Padang Bulan, Medan 20155 Indonesia
Email: martafsin@yahoo.com

**Abstract.** This study aims to find out the nutrient residue of palm kernel cake from mannan extraction on broiler chicken by evaluating physical quality (specific gravity, bulk density and compacted bulk density), chemical quality (proximate analysis and Van Soest Test) and biological test (metabolizable energy). Treatment composed of T0 : palm kernel cake extracted aquadest (control), T1 : palm kernel cake extracted acetic acid ($CH_3COOH$) 1%, T2 : palm kernel cake extracted aquadest + mannanase enzyme 100 u/l and T3 : palm kernel cake extracted acetic acid ($CH_3COOH$) 1% + enzyme mannanase 100 u/l. The results showed that mannan extraction had significant effect ($P<0.05$) in improving the quality of physical and numerically increase the value of crude protein and decrease the value of NDF (Neutral Detergent Fiber). Treatments had highly significant influence ($P<0.01$) on the metabolizable energy value of palm kernel cake residue in broiler chickens. It can be concluded that extraction with aquadest + enzyme mannanase 100 u/l yields the best nutrient quality of palm kernel cake residue for broiler chicken.

## 1. Introduction

The palm kernel cake (PKC) high in coarse fiber, this is one of the limiting factors in its use as a source of monogastric animal feed especially in poultry. Poultry is a livestock that is intolerant to feed ingredients that contain high crude fiber because in the gastrointestinal tract there is no cellulase enzyme such as ruminants. The dominant component of crude fiber in PKC is mannose which reaches 56.4% of the total PKC and exists in the form of β-mannan bond [1]. Tafsin [2] reported that the detected sugar component of PKC is composed of mannose, glucose and galactose components with a ratio of approximately 3: 1: 1. High mannan content being the limiting factor for PKC digestibility in monogastric animals can also be considered as potential for feed like prebiotics that will improve the health of livestock. Sundu [3] suspect that there is a similarity between PKC and mannanoligosakarida (MOS) that will improve the health and immune system of poultry.

So far, PKC is only used as one source of feed, but see the potential can be added value added to feed raw material feed additives. In the process of increasing the added value of the use of PKC as a feed additive is done processing technology, in this case with the combination of extraction using acetic acid ($CH_3COOH$) which is a group of weak acids that have the ability to break the fiber and harmless when consumed by humans and animals with the right dose and with enzyme mannanase which has the ability to break the bond non starch polysaccharide by increasing the digestibility of PKC. From the result of PKC extraction with acetic acid ($CH_3COOH$) and mannanase enzyme will be produced supernatant (liquid) which is considered able to become immunostimulator in chicken [2] and residue (solid). The residue of PKC resulting from the extraction is considered waste and not utilized again. By looking at the potential of supernatant that can become immunostimulator, PKC residue is expected to be utilized as one source of feed material for poultry.

Research on residual extraction of mannan from PKC has not been done and there is no related information about the physical-chemical properties and the value of its metabolic energy. Starting from that point, this research needs to be done. The purpose of this research is to know the physical-chemical quality and energy value of residual metabolism of mannan extraction process from PKC.

## 2. Materials And Methods

The ingredients used are palm kernel cake (PKC), Acetic Acid ($CH_3COOH$) 1%, mannanase enzymes, aquades and chemicals for laboratory analysis obtained from commercial companies and 25 broiler chickens aged 5 weeks used to measure metabolizable energy .

Tools used include grinder, digital scales, oven, sieve, freezer, autoclave, water bath shaker, centrifuge, measuring cup, aluminum foil, vortex mixer, test tube, Erlenmeyer flask, gourd flask, micro pipette, heat resistant plastic, metabolic cage size 50 x 20 x 50 cm with a fecal container, feeding and water holder.

### *2.1. Mannan extraction process from palm kernel cake*

The palm kernel cake obtained from the industry is separated using a mesh (2 mm / 0.787 inches). The sifted palm kernel cake weighed 50 g and then poured into a 500 ml Erlenmeyer flask was added aquadest of 500 ml. The ratio of PKC to aquades is 1:10. The PKC solution with homogenized aquades is then wrapped in a heat resistant plastic. Sterilized with autoclave at 110 °C for 1 hour then cooled to 40 °C then mixed PKC with aquades dicentrifuge at 15 °C 4200 rpm for 15 minutes. The supernatant and residue are separated, collected and stored in a freezer at 50 C. The residue is sterilized at 60 °C for 24 hours. This dry residue becomes the treatment for T0. The same process was performed to obtain P1 treatment by substituting aquades with 1% acetic acid.

The P2 treatment material was obtained by sieving palm kernel cake weighed 50 g, then poured into a 500 ml Erlenmeyer flask with 500 ml added aquades and 83 µl mannanase enzyme. The PKC comparison with aquadest is 1:10. The solution was incubated with water bath shaker at 60 °C for 72 hours then cooled to 40 °C, then mixed BIS with aquadest + enzyme mannanase dicentrifuge at 15 °C 4200 rpm for 15 minutes. The supernatant and the residue were separated, collected and stored at a temperature of 5 $^0$C. The residue was washed at 60 °C for 24 hours. This residue becomes the treatment of P2. The same process was performed to obtain P3 treatment by substituting aquadest with 1% acetic acid.

### *2.2.Testing of physical and chemical quality*

Measurements were made by taking representative of PKC residue for five observations each of 200 g of 1 kg of PKC residue. The measurement of physical properties following the Khalil [4] method includes the density of the pile, the density of pile compaction and specific gravity. Meanwhile, chemical properties use proximate analysis [5] and fiber components [6].

### *2.3.Measurement of metabolizable energy*

Calculation of the value of metabolizable energy is calculated by the formula Sibbald [7]. Experiments started with broilers fasted for 24 hours in a metabolic cage. Then, the feed was given forcibly as much as 35 g with the help of a funnel (fed) on 20 broiler, while 5 other tailed back to measure endogenous energy. Drinking water is given ad libitum. Collection of excreta done for 24 hours. During excreta collection, every ± 2 hours of excreta was sprayed with aqueous $H_2SO_4$ solution (0.01N). The collected excreta is stored in the freezer. The excreta stored in the freezer is then removed, diluted and dried in a 60 $^0$C oven for ± 24 hours. The crushed dry excreta is measured in its bruton energy by using a calorimeter bomb.

### *2.4. Experiment design and data analysis*

The experimental method used was experimental using complete randomized design with 4 treatments and 5 replications with each chicken as experimental unit on biological test (metabolic energy). The ratio of palm kernel extract to the extract in each treatment is 1:10. The treatment in this research are: T0 = Palm Kernel Cake + Aquadest (Control), T1 = Palm Kernel Cake + Acetic Acid 1%, T2 = Palm Kernel Cake + Aquades + Enzyme Mannanase 100 U/l, T3 = Palm Kernel Cake + Acetic Acid 1% + Enzyme Mannanase 100 U/l. Metabolizable energy measurement data and physical properties data were analyzed variance (Anova). If there is a noticeable difference in the further tests of Duncan [8]. Data analysis was performed according to SAS procedure 9.1.3. Meanwhile, the chemical properties were processed using descriptive statistics.

### *2.5. The variables observed*

The variables observed in the research are physical test (pile density, pile compaction density, specific gravity), chemical test (moisture content, ash content, dry weight, crude protein, crude fiber, crude fat, gross energy, ADF and NDF) (Metabolizable energy).

## 3. Results And Discussion

### *3.1. The physical and chemical quality of residues from palm kernel cake*

The result of statistical analysis by using Anava (Varian Analysis) showed that extraction with mannanase enzyme 100 u/l gave a real effect ($P<0.05$) in increasing the physical properties value on the extracted PKC residue.

**Table 1.** The results of the physical and chemical quality of residues from palm kernel cake

| Information | Treatment | | | |
|---|---|---|---|---|
| | T0 | T1 | T2 | T3 |
| Chemical quality: | | | | |
| - *DW* (%) | 95.21 | 96.66 | 95.26 | 95.74 |
| - *AC* (%) | 2.76 | 2.07 | 2.42 | 2.15 |
| - *CP* (%) | 17.85 | 18.15 | 19.20 | 17.84 |
| - *CF* (%) | 22.74 | 23.38 | 23.92 | 24.72 |
| - *C.Fat* (%) | 8.57 | 8.47 | 7.89 | 7.36 |
| - *ADF* (%) | 55.74 | 57.67 | 51.55 | 36.45 |
| - *NDF* (%) | 80.59 | 78.76 | 75.70 | 77.75 |
| Physical quality: | | | | |
| - *SG* ($Kg/m^3$) | $162.64^{b} \pm 0.09$ | $162.68^{ab} \pm 0.08$ | $162.78^{a} \pm 0.09$ | $162.70^{ab} \pm 0.08$ |
| - *BD* ($Kg/m^3$) | $31.87^{b} \pm 0.18$ | $319.99^{ab} \pm 0.15$ | $320.13^{a} \pm 0.18$ | $320.07^{ab} \pm 0.18$ |
| - *CBD* ($Kg/m^3$) | $330.90^{b} \pm 0.19$ | $330.99^{ab} \pm 0.15$ | $331.17^{a} \pm 0.19$ | $331.10^{ab} \pm 0.15$ |
| Mannosa content (ppm) | $7.327 \pm 0.05$ | $2.433 \pm 0.31$ | $173.36 \pm 15.5$ | $629.14 \pm 0.96$ |

Explanation : T0 = PKC extracted with aquadest; T1 = PKC extracted with acetic acid 1%; T2 = PKC extracted with aquadest + enzyme mannanase 100U/l; T3 = PKC extracted with acetid acid 1 % + enzyme mannanase 100U/l; DW= Dry Weigth; AC= Ash Content; CP = Crude Protein; CF = Crude Fiber; C.Fat = Crude Fat; ADF = Acid Detergent Fiber; NDF= Neutral Detergent Fiber; SG= specific gravity; BD= bulk density; CBD= compacted bulk density.

The results using Duncan's multiple range test showed that the treatment for T2 (PKC extracted with aquades + 100U/l mannanase enzymes) had a physical quality value higher than or equal to the

T3 treatment (PKC extracted with 1% acetic acid + 100 U/l mannanase enzyme) and T1 treatment (PKC extracted with 1% acetic acid) between treatments.The treatment T0 (PKC extracted with aquades) showed a significant effect.

Calculation of physical quality correlation on PKC residue of treatment and NDF content shows that there is a very strong relationship between nutrient content contained by residue of PKC treatment result with value of physical quality possessed. The NDF content represents the chemical component of PKC residue treated by the effect of the value of the physical quality of the treated residue, the lower the NDF content the higher the physical quality of the PKC residue of treatment. This is in accordance with the opinion of Khalil [4] which states that the variation in the value of physical quality is influenced by the nutrient content of the material, particle size distribution and particle surface characteristics.

Based on the chemical test table on the residual PKC treatment results showed that the residue PKC extraction process using 100 U/l mannanase enzyme (T2) has the highest protein content among treatments. This may occur because of the addition of the mannanase enzyme in the PKC extraction process. This is in accordance with the opinion Jhonson [9] argues that mannanase is the enzyme decomposers mannan and galaktomanan become mannose and galactose. This enzyme randomly cuts the main chains of mannan and hetero ß-D-mannan into dissolved sugars ie manodextrins and mannose.

The table above shows that the use of the B-mannanase enzyme shows the highest mannosa component compared to other treatments. The lowest yield was shown on the treatment using 1% acetic acid. Measurements are made to the sample directly without using the hydrolysis process. These results indicate that the detected component of the mannosa is a relatively simple component (close to monosaccharides). The use of autoclave, centrifuge and mannanase enzyme activity results in increased protein molecules liberated from fiber bonds. So that the residue of PKC of T2 treatment had high crude protein content in this study.

The fiber component can be seen from contents ADF (Acid Detergent Fiber) with the Van Soest analysis. The ADF content in PKC reaches 50.74% while in residue the extraction process is ranged from 36.45% - 57.67%. The lowest ADF content was shown in a 1% acetic acid combination treatment with b-mannannase enzyme which showed a value of 36.45%. These results indicate that the use of acetic acid treatment combinations with enzymes is most effective compared to other treatments in overhauling the fiber component into a soluble material in the extraction process.

The higher the ADF, the higher the digestibility of feed ingredients. Judging from the value of ADF content of PKC residue of treatment result, T3 treatment has the lowest ADF content, but the crude protein content of T3 including the lowest and the highest crude fiber content among treatments. The result is assumed that the nutritional value of T3 is dissolved in the extraction process with the combination of acetic acid and mannanase enzyme 100 u / l and the ADF content possessed by T3 treatment is suspected that the fiber component already resistant is already in the form of lignin.

### *3.2. The result metabolizable energy of palm kernel cake residue*

The result of variance analysis showed a very real effect ($P<0.01$) from the extract treatment of palm kernel to the metabolic energy value of PKC residue of treatment. Further test results using Duncan multiple range test showed that T2 treatment (PKC extracted with aquades + 100 U/l mannanase enzyme) or T1 treatment (PKC extracted with 1% acetic acid) had higher metabolic energy values between treatments. Furthermore, T3 treatment (PKC extracted with 1% acetic acid plus 100 u/l enzyme mannanase) or T0 treatment (PKC extracted with aquadest) showed significant effect. Different metabolic energy values of PKC residue due to differences in nutrient content and physical quality of PKC residue of treatment.

Factors that affect the value of metabolic energy in poultry is the ability of livestock to metabolism rations in the body. This factor is influenced by the physical properties of feed, pH, gastrointestinal enzymes, food composition, environmental temperature and physiological livestock [5].

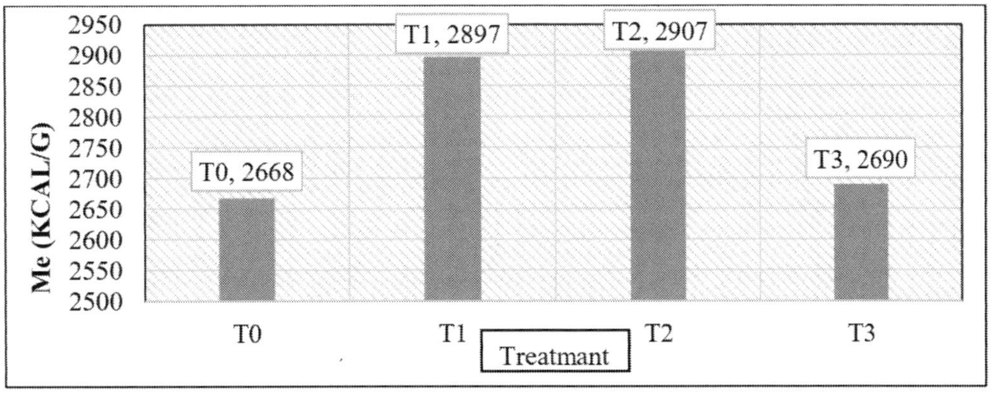


**Figure 1.** The Graph of result metabolizable energy residue from palm kernel cake

## 4. Conclusion

The extraction residue results in a significantly different value of physical quality. PKC extracted with aquadest + 100 u/l enzyme mannanase result specific gravity, bulk density and compacted bulk density greater than between treatments and can increase the value of crude protein and decrease the NDF value and produce different metabolic energy values. The PKC extracted with aquadest + 100 u/l mannanase enzyme or extracted with acetic acid is greater in the value of its metabolic energy than the PKC contracted with aquadest or extracted acetic acid + 100 u/l enzyme mannanase.

## References

[1] Daud M J, Jarvis M C and Rasidah A 1993 Fiber of PKC and its potential as poultry feed *Proc., 16th MSAP Annual. Conference* Kuala Lumpur, Malaysia

[2] Tafsin M 2007 Study of mannan polysaccharide from palm oil cake as controller of *Salmonella thypimurium* and immunostimulan in chicken [dissertation] (Bogor: Graduate School, Bogor Agricultural University)

[3] Sundu B, Kumar A and Dingle J 2005 Response of birds fed increasing levels of palm kernel cake supplemented with different enzymes *Aust. Poult. Sci. Symp.* **17**: 227 – 228.

[4] Khalil 1999 Effect of water content and particle size on changes in physical behavior of local feed ingredients: pile density, pile compaction density and specific gravity *Med. Pet.* Vo **22**:1(1-11)

[5] AOAC 1995 *Official Method of Analysis* (Association of Official Analytical Chemists Arlington, Virginia)

[6] Van Soest P J 1985 *Definition of Fibre in Animal Feeds*. In : Cole D J A and Hersign W (Ed.) Recent Advances in Animal Nutrion Butterworths. London. Cornell University. Ithaca, New York

[7] Sibbald I R 1989 Metabolizable energy evaluation of poultry diets. In: Cole, D. J.A. And W. Haresign (ed). *Recent Development in Poultry Nutrition*. University of Nottingham School of Agriculture. Butter Worths. London

[8] Steel R G D and Torrie J H 1995 *Statistics Principles and Procedures* (Jakarta : PT. Gramedia)

[9] Johnson J R 1994 The realities of bulk solid properties testing *Bulk Solid Handling* **14** (1): 129-132

IOP Conf. Series: Earth and Environmental Science **122** (2018) 012115 doi:10.1088/1755-1315/122/1/012115

# The effect of fermented cocoa pod (*Theobroma cacao*) husk supplemented with mineral on in vitro digestibility, rumen bacteria population and rumen liquid characteristics

**Nurhaita[1]*, N Definiati[1], U Santoso[2], S A Akbar[3] and Y L Henuk[4]**

[1]Department of Animal Science, Faculty of Agriculture, Universitas Muhammadiyah, Bengkulu 38119 Indonesia
[2]Department of Animal Science, Faculty of Agriculture, Universitas Bengkulu 38119 Indonesia
[3]Department of Animal Science, Faculty of Agriculture, Universitas Muhammad Yamin, Solok, West Sumatera 27321 Indonesia
[4] Department of Animal Science, Faculty of Agriculture, Universitas Sumatera Utara, 20155 Indonesia
E-mail: nurhaita@gmail.com

**Abstract**. This study aimed to determine the effect of mineral supplementation, such as S, P and Zn on the nutrients digestibility of fermented cocoa pod husk, the population of rumen bacteria and rumen liquid characteristics in vitro. The study used a randomized block design with 5 treatments and 4 replicates. The treatments tested were: T0 = without minerals; T1 = 0.2% S mineral; T2 = 0.27% P mineral; T3 = S and P; and T4 = S, P and Zn at 50 ppm. Parameters measured were: (1) digestibility of dry matter and organic matter; (2) rumen bacterial and cellulolytic bacterial populations; (3) characteristics of rumen liquid in vitro. The results of the study showed that mineral supplementation significantly ($P < 0.05$) improved dry matter and organic matter digestibility. Mineral supplementation had no effect on the total population of rumen bacteria and cellulolytic rumen bacterial populations. The characteristics of rumen liquid such pH, VFA and $NH_3$ were in optimal condition. In conclusion supplementation of S, P and Zn simultaneously gave the best results to improve the digestibility of dry matter and organic matter and to maintain rumen liquid characteristics under optimal conditions for growth and microbial activity

## 1. Introduction

Cocoa (*Theobroma cacao*) is primarily grown for its bean to be further processed by the chocolate industry. The bean itself is located inside the cocoa fruit. When the bean is removed from its fruit, the leftover is called the pod and may constitute about 75% of the weight of the whole fruit [1]. Although other by-products from cocoa processing such as cocoa shell and cocoa dust exist and may also be used as animal feeds, the pod is the predominant cocoa by-product [2].

Indonesia is the second largest producer of cocoa in terms of total production after the Ivory Coast. Therefore, quite an amount of cocoa by-product is available in the country, especially in the areas of cocoa plantations or where cocoa processing plants are located. Unfortunately, despite its abundant quantity, so far cocoa pod has not been optimally utilized by farmers. Usually the pod is left to rot on the plantation area or is used as mulching material. If the cocoa pod could be fed to animals, its economic value would be improved [3]. In general, a main obstacle of utilizing cocoa pod as an animal feed is its high fiber and low protein contents. It also contains a considerable amount of lignin, i.e. between 12% to 19% dry matter (DM), and such value is 2 to 3 times higher than that of rice straw [4]. Studies using high inclusion levels of untreated cocoa pod in diets have resulted in lower digestibility and animal performance [5], confirming its low nutritional quality. Two main nutritional strategies have been proposed to overcome such limitation of cocoa pod, i.e. either by mixing with a more fermentable or digestible feedstuff [6].

Pod cocoa fermentation with local microorganisms of rumen contents proved to increase dry matter digestibility of 30.07% [7], but still needed mineral supplementation to improve rumen bacterial populations to digest higher. S and P minerals are essential for rumen microbes digesting fiber. S mineral needs from 0.14 to 0.26% (average 0.2%) and P 0.16 to 0.38% (average 0.27%) of dry matter [8]. The mineral is often deficient in feed materials derived from agricultural waste, and become a limiting factor for the growth of rumen microbes. This is due to the feed in the tropics and feed derived from agricultural waste are often deficient in essential minerals for the growth of microbes, such as S and P [9]. In addition biolavalibility of mineral is also low. This will negatively affect the synthesis of microbial protein and digestibility of nutrients. S and P supplementation on palm leaves ammoniation improve in vitro digestibility of 36.68% from 34.67% to 47.39% [10]. Furthermore, the in vivo digestibility of ammoniation palm leaf supplemented with mineral S and P was increased from 51.5% to 56.0% in sheep [11]. In cattle, supplementation of S and P on ammoniation palm oil frond improve feed digestibility, and give similar body weight gain to natural grass [12]. The present study was conducted to investigate the effect of supplementation of S, P and Zn in fermented cocoa pod on dry matter, organic matter digestibility, rumen bacterial populations and characteristics of rumen fluid in vitro.

## 2. Materials and Methods

This research material was fermented cocoa pod, sulfur as a source of S, TSP fertilizer as a source of mineral P, and ZnCl2 as a source of Zn, Mc Doughall's solution as a buffer and as a donor microbial rumen fluid. The tools used are in-vitro devices, digital pH meter to measure the pH of rumen fluid, and a set of laboratory equipment for analysis Proximate Van Soest, VFA and NH3-N. The study used a randomized block design with 5 treatments and 4 replicates. The treatments were supplementation of S, P and Zn in fermented cocoa pod as follows: T0 = without mineral supplementation (control); T1 = 0.2% S mineral supplementation; T2 = supplementation of P at 0.27%; T3 = supplementation of S and P; and T4 = supplementation of S, P and Zn 50ppm. The amount of supplementation of S and P were calculated based on the content of S and P in the material so that the amount be the same as the standard [8], whereas Zn supplementation of 50 mg/kg ration [13]. Data were analyzed using analysis of variance according to R G D Steel, and J H Torrie [14]. The difference between treatments was tested by Duncan's Multiple Range Test (DMRT).

### 2.1. *Laboratory analysis*

This study used an in vitro method as described by J M ATilley and Terry[15]. Samples of fermented cocoa pod before used they were dried and finely ground, as well as sulfur and TSP. They were inserted into the erlenmeyer and added minerals in accordance with each treatment, and mineral and sample mix evenly. They then were given the buffer solution Mc Doughall's and rumen fluid with a ratio of 4: 1, the CO2 gas flowed for 30 seconds so that the anaerobic conditions and incubated for 2x24 hours in a water bath shaker. After fermentation ends, all the samples were centrifuged at 1200 rpm for 15 minutes to separate the filtrate and precipitate. The filtrate was used for the measurement of rumen fluid characteristics (pH, $NH_3$-N and VFA) and analysis of rumen bacteria, whereas in vitro fermented esidues were dried to analyze content of nutrients. Parameter observed was the digestibility of dry matter and organic matter, the total population of rumen bacteria and cellulolytic rumen bacterial populations, and the characteristics of rumen fluid such as pH, levels of VFA and steam distillation, and the content of $NH_3$-N.

## 3. Results and Discussion

Effect of mineral supplementation on dry matter (DM) and organic matter digestibility of fermented cocoa pod are presented in Table 1.

**Table 1.** Dry matter and organic matter digestibility

| Treatments | DM (%) | OM (%) |
|---|---|---|
| $T_0$ | 28.62$^{a}$ | 23.88$^{a}$ |
| $T_1$ | 30.23$^{ab}$ | 27.09$^{c}$ |
| $T_2$ | 29.29$^{a}$ | 25.11$^{ab}$ |
| $T_3$ | 29.06$^{a}$ | 26.26$^{bc}$ |
| $T_4$ | 30.52$^{b}$ | 27.04$^{c}$ |

Notes: Different superscipts in the same column shows significantly different ($P<0.05$)

Data on Table 1 showed that the mineral supplementation significantly affect digestibility of dry matter and organic matter of fermented cocoa pod. Dry matter ranged from 28.62% to 30.52% and organic matter digestibility ranged from 23.88% to 27.04%. Supplementation of S, P plus Zn also showed the highest dry matter and organic matter contents. S supplementation improved the digestibility of dry matter and organic matter cocoa pod. This may be due to the low S content (0.007%) of the cocoa pod, which is far below the needs of S (0.14 to 0.26%) [16]. Therefore, S supplementation as much as 0.2% gives a positive response to microbial activity resulting in better digestibility.P supplementation had no effect on dry matter and organic matter of fermented cocoa pod. This may be due to the P content of fermented cocoa pod at 0.67% has exceeded the standard requiremnets, namely 0.18% - 0.36% [16]. The higher digestibility of feed means greater benefit in supporting the productivity of livestock. Nutrient digestibility in this study is only around 28.62% - 30.52% which indicated that the fermented cocoa pod is a low quality feed. Nutrients contained in the material is not entirely available for ruminants. Feed quality is determined by the digestibility of nutrients of these materials, which describes what percentage of substances that are digested and what percent is excreted through feces [11].

### *3.1 Bacterial Population and Cellulolitic Bacteria*

The total population of bacteria ranged from (log) 4.66 to 6.15, while the population of cellulolytic bacteria ranged from (log) 4.81 to 5.79. S, P, and Zn supplementation had no significant effect ($P> 0.05$) on the total population of bacteria and cellulolytic rumen bacteria (Table 2).

Bacteria play an important role in the process of fermentative digestion of feed in the rumen. Bacteria population in the rumen ranged from $10^9$ to $10^{12}$ per ml of rumen fluid. To obtain a higher digestibility of low quality wastes, an increase in bacterial population should be pursued through the provision of nutrient precursor for rumen microbial growth. S is essential for fiber digestion in the rumen, and adequate sulfur supply will optimize the digestibility of cellulose through the stimulation of specific cellulolitic bacteria and the activity of ciliate protozoa and fungi in rumen [9].P needed by rumen microorganisms for cellulose digestion, but it is not easy to prove that P can stimulate the production of VFA [17]. P is specifically required for cell wall digestibility as main elements, especially for selulolisis requiring higher than for hemiselulolisis and amilolisis. In most in-vivo study, deficiency of P showed a negative effect on the digestibility of the fiber fraction and the ability to digest organic matter [9].

**Table 2**. Total Bacterial Population and Cellulolitic bacteria

| Treatments | Total Bacteria (log) | Cellulolitic Bacteria (log) |
|---|---|---|
| $T_0$ | 6.15 | 5.38 |
| $T_1$ | 5.84 | 5.58 |
| $T_2$ | 4.66 | 4.81 |
| $T_3$ | 5.97 | 5.48 |
| $T_4$ | 5.92 | 5.79 |
| SE | 0.038 | 0.035 |

*3.2. Rumen Fluid Characteristics*

Characteristics of rumen fluid is generally indicacated optimum rumen conditions for microbial growth and activity. This condition is a necessary condition that must be met fatherly support high livestock production. Three main factors can be used as criterions in assessing the condition of rumen i.e., pH, total VFA and NH3 consentration rumen fluid. The results of measurements of pH, total VFA and NH3 concentrations in this study can be seen in Table 3.

**Table 3**. Rumen Fluid Characteristics.

| Treatments | pH | VFA (mM) | $NH_3$-N (mg/100ml) |
|---|---|---|---|
| $T_0$ | $7.18^b$ | $85.01^a$ | $5.16^a$ |
| $T_1$ | $7.00^{ab}$ | $106.28^{ab}$ | $6.30^b$ |
| $T_2$ | $7.08^{ab}$ | $96.90^{ab}$ | $6.79^b$ |
| $T_3$ | $7.00^a$ | $99.23^{ab}$ | $6.15^b$ |
| $T_4$ | $6.78^a$ | $115.52^b$ | $6.50^b$ |

Notes: Different superscipts in the same column shows significantly different (P<0.05)

Results of analysis of variance in Table 3 showed that the mineral supplementation significantly (P <0.05) on the parameters of rumen fluid characteristics. Rumen fluid pH values obtained ranged from 6.78 to 7.18. Near-neutral pH value is obtained for the use of artificial saliva as a buffer capable of maintaining stability conditions of the rumen fermentation activity influence. This is in accordance with D C Church[17], who reported that saliva acts as a buffer to maintain the stability of rumen fluid. High or low pH of rumen fluid is determined by the type of food, the time available for the rumen microbes to ferment food ingredients, buffers and system capacity levels and total VFA NH3 rumen fluid, so that the rumen pH can vary in the time before and after meals. The pH value obtained in this study are in optimal conditions to support microbial growth and activities. Our findings in accordance with R A Erdman [18] and E R Orskov and M Ryle [19], they reported that rumen pH range of 6.3 to 7.0 can assure the growth and activity of rumen microbes. If the rumen pH below 6.2 then the cellulolytic microbial life will be disrupted and the digestibility of the fiber will decrease.Production of total VFA in this study ranged from 85.01 to 115.52 mM. Statistical analysis showed that the effect of different treatments were highly significant (P <0.01) of total VFA. DMRT highest VFA concentration 115.52 mM obtained at SP + Zn supplementation treatment. The high production VFA at SP + Zn supplementation treatment is in line with the high dry matter digestibility. It is easy to understand, because the VFA is the end product of carbohydrate fermentation by rumen microbes so that the increase in the digestibility will lead to an increase in the final product, i.e., VFA.

H L Davies [20] suggested that elevated levels of VFA reflects the increased solubility soluble carbohydrate feed. VFA ruminants have a dual role as a source of energy for livestock and a source of carbon skeletons for the growth of microbial protein [21]. VFA is a dynamic element that amount depends on the material fermentability, absorption in the rumen wall and utilization by rumen microbes [22].VFA production is highly dependent on a variety of carbohydrates contained in a material. Increased production meant that the total VFA also increase carbohydrate fermentation in the rumen [23].

VFA total production obtained in this study could support microbial growth and activity. This is in accordance with P J Van Soest [24], who reported that total VFA needed for the growth and activity of microbial range of 80-160 mM. NH3 is the result of the digestibility of protein or NPN compounds by rumen microbes. The concentration of NH3 obtained in this study ranged from 5.16 to 6.79mg/100 ml rumen fluid. Consentrationof NH3 rumen fluid is strongly influenced by the type of food, solubility of nitrogen and the level of digestibility of protein, concentration of nitrogen ration, time after feeding, rate of use of nitrogen for microbial biomass rumen, absorption of NH3-N or recycling of urea and nitrogen rumen bacteria [3]. The concentration level of NH3-N in this study was at S supplementation, supplementation of P, SP, and SP + Zn was not significantly different. Overall concentrations of NH3-N in this study have been indicated to sufficient for NH3-N for growth of rumen microbes. This findings is supported by L D Satter and L L Slyter [25], they reported that the concentration of $_{\text{NH3}}$ in the rumen varied between 0-130mg/100 ml rumen fluid, whereas the minimum levels for rumen microbial protein synthesis that is optimal is 5 mg / 100 ml rumen fluid.

## 4. Conclusions

Supplementation of S, P and Zn simultaneously on fermented cocoa pod improved the digestibility of nutrients in vitro, but it did not affect the population of bacteria and cellulolytic bacteria in rumen. The mineral supplementation maintained rumen fluid characteristics under optimal conditions for growth and microbial activity.

## References

[1] Owusu-Domfeh K 1972 The future of cocoa and its by-products in the feeding of livestock *Ghana J. Agric Sci.* **5**:57–64

[2] Aregheore E M 1972 Chemical evaluation and digestibility of cocoa (*Theobroma cacao*) byproducts fed to goats *Trop Anim Health Prod.* July, **34**(4):339-48

[3] Erika B L and Jayanegara A 2015 Improving Nutritional Quality of Cocoa Pod (*Theobroma cacao*) through Chemical and Biological Treatments for Ruminant Feeding: In vitro and In vivo Evaluation Asian-Australas *J. Anim. Sci. Mar.* **28**(3): 343–350

[4] Sutikno A I 1997 Cocoa pod for feeding of ruminant livestock *Wartazoa* **6**:38–43

[5] Smith O B and Adegbola A A 1982 The feeding value of cocoa-pod for cattle *Trop Anim Prod.* **7**:290–295

[6] Alemawor F, Dzogbefia, V P, Oddoye, E O K, and Oldham, J H 2009 Effect of Pleurotusostreatus fermentation on cocoa pod husk composition: Influence of fermentation period and Mn2+ supplementation on the fermentation process African J Biotech. **8**:1950–1958

[7] Nurhaita, Nurhaita D, Neli and Suliasih Fermentation with local microorganism to improve pod cacao quality as ruminants feed *In: Proc. ISEPROLOCAL. Badan Penerbitan Fakultas Pertanian Universitas Bengkulu, Bengkulu, Indonesia* pp 218-221

[8] National Research Council 1985 *Nutrient Requirements of Sheep : 6$^{th}$ Revised Edition* (Washington DC: National Academy Press X)

[9] Komizarczuk S and Durand M 1991 Effect of mineral on microbial metabolism. In. Rumen Microbial Metabolism and Ruminant Digestion. *J.P. Jouany (Ed) INRA Publ. Versailes, France*

[10] Nurhaita, Jamarun N, Saladin R, Warly L and Mardiati Z 2008 Efek suplementasi mineral sulfur dan phospor pada daun sawit amoniasi terhadap kecernaan zat makanan secara in-vitro dan karakteristik cairan rumen (The effects of sulfur and phosporal mineral supplementation on ammonium palm leaves on the digestibility of dietary substances in-vitro and rumen fluid characteristics) *J Peng Pet Trop.* **33**(1): 51-58

[11] Nurhaita, Jamarun N, Saladin R, Warly L and Mardiati Z 2010 Kecernaan ransum domba

berbasis daun sawit amoniasi yang disuplementasi S, P dan daun ubi kayu (Digestibility of sheep ration based on ammonium palm leaves supplemented by S, P and cassava leaves **33**(3): 144-149

[12] Nurhaita, Ruwendi, Wismalinda R, and Robiyanto 2014 Optimizing rumen bioprocess through supplementation of microbe precusor nutrient in ammoniation of palm oil frond-base cattle ration *J Adv Agric Tech.* **1**(1) Juni : 10-13

[13] Fariani A, Abrar A and Muslim G 2011 Peningkatan performa kambing kacang melalui suplementasi biomineral-zn (Performance improvement of goat beans through biomineral-zn supplementation) *Prosiding Seminar Nasional "Prospek dan Potensi Sumber daya Ternak Lokal dalam Menunjang Ketahanan Pangan Hewani"* Fakultas Peternakan Universitas Jenderal Soedirman, Purwokerto

[14] Steel R G D and Torrie J H 1995 *Principles and Procedures of Statistics : 2nd Edition* (London: McGraw- Hill International Book Company)

[15] Tilley, J M A and Terry 1963 A two stage technique for in-vitro digestion of forage crops *J Brit Grassl Soc.* **18** (2):104 – 111

[16] National Research Council 1995 *Nutrient Requirements of Laboratory Animals: 4th Revised Edition* (Washington DC: National Academy Press)

[17] Church D C 1979 *Gigestive Physiology and Nutrition of Ruminant Vol 1. Digestive Physiology. 2nd Edition* (New York: Jhon Wiley and Sons)

[18] Erdman R A 1988 Dietary buffering requirement of lactating dairy cow: A Review *J Dairy Sci.* **71**: 3246

[19] Orskov E R and Ryle M 1990 *Energy Nutrition in Ruminant* (London: Elsevier Applied Science)

[20] Davies H L 1982 *Nutrition and Growth Manual* (Published by Australian Universities International Development Program, Melbourne)

[21] Sutardi T, Sigit N A dan Toharmat 1983 Standardisasi mutu protein makanan ruminansia berdasarkan parameter metabolismenya oleh mikroba rumen (Protein quality standarizationof ruminant food based on metabolism parameter by rumen microbe) *Fakultas Peternakan IPB bekerja sama dengan Direktur Pembinaan dan Pengabdian pada Masyarakat Depdikbud* Jakarta

[22] Warly L 1998 Pemanfaatan hasil ikutan agro-industri sebagai bahan makanan ternak ruminansia (evaluasi secara in-vivo) (Utilization of agro-industrial follow-up results as ruminant animal feed (in-vivo evaluation)) *Laporan Penelitian Hibah Bersaing V/2 Perguruan Tinggi Tahun Anggaran 1997/1998* (Padang: Fakultas Peternakan Universitas Andalas)

[23] Baldwin R L and Denham S C 1979 Quantitative and dynamic aspects of Nitrogen metabolism in rumen: A modelling analysis *J Anim Sci.* **49**:1631-1637.

[24] Van Soest P J 1982 *Nutritional Ecology of Ruminant* (Virginia: O&B Books. Inc.)

[25] Satter L D and Slyter L L 1974 Effect of ammonia concentration on rumminal microbial protein production in Vitro *British Journal Nutrition and Animal Science* **32** : 1999

**Acknowledgements**

This study was supported by funding from the Directorate General of Higher Education, Ministry of Research, Technology and Higher Education under the Letter Agreement on Implementation of Competitive Grants Research Grants for Lecturers PTS Kopertis Region II number 3778/K2/KM/2015 dated April 7, 2015 and Letter Grant Implementation agreement of Competitive Research Grant for UniversitasMuhammadiyah Bengkulu number. 100 / LPPM-UMB / 2015, April 10, 2015, for which the authors are very grateful. We also thank to Prof. Yusuf L. Henuk, Ph.D (E-mail: yusufhenuk62@live.com) from Faculty of Agriculture, University of Sumatera Utara (USU-Medan) for his critical reading and editing of this manuscript.

# The influence of *Aspergillus niger* inoculum dosage on nutritive value and metabolizable energy of apu-apu meal (*Pistia stratiotes L.*) on broiler chicken

**J Gloria[1], M Tafsin[2], N D Hanafi[2] and A H Daulay[2]**

[1]Graduate Student, Faculty of Agriculture, Universitas Sumatera Utara, Medan, North Sumatera 20155, Indonesia
[2]Lectures, Faculty Agriculture, Universitas Sumatera Utara, Medan, North Sumatera 20155, Indonesia
E-mail: nevydiana@yahoo.co.id

**Abstract.** Apu-apu lives at tropical and subtropical fresh waterways. The apu-apu meals ultization as feed still limited. The problem of ultization apu-apu meals as ingredients is a high crude fiber and need a treatment to decrease crude fiber. This study aim to find out the influence of Aspergillus niger inoculums dosage on apu-apu meal (Pistia stratiotes L.) on metabolizable energy on broiler chicken. This research used completely randomize design (CRD). The treatments consists of Aspergillus niger inoculum dosage (CFU/g) such as P0 (0), P1 ($10^4$ CFU/g), P2 ($10^6$ CFU/g), and P3 ($10^8$ CFU/g). The variable were observed : apparent metabolizable energy (AME), true metabolizable energy (TME), apparent metabolizable energy nitrogen corrected (AMEn) and true metabolizable energy nitrogen corrected (TMEn).The results showed that the dosage of Aspergillus niger increase nutritive value of Aspergillus niger. Dosage of Aspergillus niger also influence ($P<0.05$) metabolizable energy of apu-apu meals. Dosage $10^8$ CFU/g had metabolizable energy significantly higher than other treatments. Conclusion of this research is the Aspergillus niger at the dosage $10^8$ CFU/g increased nutritive value and metabolizable energy of apu-apu meal.

## 1. Introduction

In composed an animal feed was always based on the balance of energy and protein. One of the origin of vegetable protein, namely plant apu-apu (Pistia stratiotes L.), which live on freshwater waterways, measuring approximately 5 cm-10 cm, shape slightly rounded of the color green and white roots are a bit dirty. The leaves on these plants with delicate feathery. This plant grows as a weed in rice field acreage mainly rice and many farmer not yet know that these plants can be used as animal feed, especially poultry. So after the rice harvest, usually these plants join the cleaned along with the processed of land for planted the next rice.

According to Sutama [1], plant the apu-apu is potentially as constituents of feed because it contains crude protein 19.5%, ash 25.6%, fat 1.3% crude fibre 11.7% and 37.0% NNE. Crude fiber of high will be hard to digest by broilers, because thedigestion do not secretion the enzyme cellulase. Utilizing of crude fibers that are high in food can decrease the components easy to digest and decrease enzyme activity and solve the food substances like enzyme that help digestion of carbohydrates, proteins and fats [2]. Therefore the processing needs to be done in order to make use of apu-apu (Pistia stratiotes L.)

as material feed poultry can be optimum. One of the feed processing technique that was doing the fermentation.

Fermentation is the process of chemical change of organic compounds in either aerobic or anaerobic circumstances, through the work of enzymes produced by the microbes [3].

One of the factors which affect the fermentation was inoculum dosage. The right of Inoculum dosage was important to know because too much inoculum in the substrate would cause competition in obtaining food and affect to the ability of microorganisms in fermentation. Too much density of inoculum can also cause the death of microbes.

The mold used for the fermentation of starch apu-apu was Aspergillus niger. This mold was used because it can grow quickly and already used commercially. The existence of the cellulase enzyme produced by Aspergillus niger were able to degrade cellulose is chemically reducing crude fiber. The process of fermentation apu-apu meals can decrease crude fiber and increase the nutritive of apu-apu meals.

According to Mc Donald et al[4], metabolizable energy is the energy that is ready to be utilized by livestock in a variety of activities such as physical activity, maintaining body temperature, metabolism, tissue formation, reproduction and production. Therefore important to knowed the nutritional feed ingredients to determine a ration needs energy metabolis which will be consumed by livestock.

## 2. Materials And Methods

This research has been carried out in the Microbiology Laboratory and Livestock Biology Laboratory, Animal Husbandry Department, Agriculture Faculty University of North Sumatra.

Broiler chickens, apu-apu (Pistia stratiotes L.), Aspergillus niger as fermentator, Potatoes Dextrose Agar (PDA) as a growth medium, water, hot plate, spiritus, pumpkin elemenyer, sterile cotton wool, aluminium foil, ose, test tubes, petri dish, clear plastic, electronic scales, ovens and stationery, cages metabolism sized 52 x 25 x 45 cm by as much as 23, each comprising 1 Coop chickens. This enclosure equipped place drinking water and excreta container.

The experimental design used in this study was experimentally by using random design complete with 4 treatments and five replicates. As for the treatment as follows:P0 = Apu-apu meals without fermentation of Aspergillus niger; P1 = Aspergillus niger Fermentation $10^4$ CFU/g; P2 = Aspergillus niger Fermentation $10^6$ CFU/g; P3 = Aspergillus niger Fermentation $10^8$ CFU/g.

Observations include proximate analysis (dry matter, ash, crude protein, fat and crude fiber) and metabolizable energy(Apparent metabolizable energy (AME), true metabolizable energy (TME), apparent metabolizable energy nitrogen corrected (AMEn), and true metabolizable energy nitrogen corrected (TMEn).

### *2.1. Implementation*

Implementation of the research included the processing of apu-apu meals that was collecting, sorting, washing, drying, milling and proximate analysis.

Then the apu-apu meals had been packaged in clear plastic and then sterilized at temperature $121^0$C for 15 minutes then cooled, after reaching room temperature, substrates were incubated with prepared inoculum suitable with the treatment of $10^4$ CFU/g, $10^6$ CFU/g and $10^8$ CFU/g then blended in a tray then covered with a cling wrap and give a small hole then put into the laminar air flow. Then incubated at $28^0$C for 4 days. Each combination is repeated 5 times. After incubation time is reached, the apu-apu meals was dried on oven at a temperature of 60°C for 24 hours and constant weight is obtained. Then done the proximate analysis to knowed a nutritive value.

In the forced feeding, used a 20 broiler chickens have been fasted for 24 hours. Feed given forced as many as 25 g with the help of a funnel. Water given ad-libitum. After 24 hours the giving rations, excreta was collected and subsequent handling similarly as in feeding techniques without force.

## 3. Result And Discussion

*3.1. Nutritive value of apu-apu meals*

Table 1 shows that the best quality of nutrients in P3 crude fiber as 14.83% and crude protein as 16.95%. This shows that the fermented apu-apu meals using Aspergillus niger was decreased the crude fibers and increased crude protein. This statement suitable to Mangisah et al [5] that the Aspergillus niger capable to producing the enzyme cellulase and hemicellulase that will degrade cellulose and hemicellulose so that it is able to decrease the crude fibre in feed raw material and according to Hassanudin and Mujnisa [6] that results of sago pulp inoculation of Aspergillus niger contains a higher crude protein and lower crude fiber than was not inoculated with Aspergillus niger.

**Table 1**. Nutritive value of apu-apu meals

| Treatment | % Dry matter | % Ash | % Crude protein | % Crude Fiber |
|---|---|---|---|---|
| P0 | 89.51 | 9.40 | 14.62 | 17.33 |
| P1 | 92.36 | 9.20 | 16.83 | 14.95 |
| P2 | 91.91 | 9.82 | 16.92 | 14.92 |
| P3 | 92.58 | 10.25 | 16.95 | 14.83 |

*3.2. Metabolizable energy*

Metabolizable energy is a different within feed gross energy and excreta gross energy. Table 2 shows the different value of metabolizable energy in parameters. Based on the analysis of diversity showed that the apu-apu meals at P3 ($10^8$ CFU/g)give a different real influence ($P < 0.05$) against the apparent metabolizable energy (AME), true metabolizable energy (TME), apparent metabolizable energy nitrogen corrected (AMEn) and true metabolizable energy nitrogen corrected (TMEn) value compared with others treatment of pseudo compared on the dose on other treatments. However, P0, P1 and P2 did not show the real difference.This is caused by high levels of crude fibers at the P3 than other treatments in P3 so increased the feed digestibility that effect to metabolizable energy. This is in accordance with the opinion of Tillman et al[7] that increased the quality of the feed causes a high digestibility values, and the effect to increase the value of metabolizable energy. Laras[8] that a mixture of fermented onggok by Aspergillus niger increased the value apparent metabolizable energy (AME), true metabolizable energy (TME), apparent metabolizable energy nitrogen corrected (AMEn) and true metabolizable energy nitrogen corrected (TMEn). It predicted due to the role of the cellulase, Aspergillus niger's products which able to degrade cellulose into glucose. This is in accordance with the opinion of Sibbald [9] and Mc Donald et al[4].

**Table2.** Apparent metabolizable energy (AME), true metabolizable energy (TME), apparentmetabolizable energy nitrogen corrected (AMEn) and true metabolizable energy nitrogencorrected (TMEn) value of apu-apu meals

| Treatments | Parameters | | | |
|---|---|---|---|---|
| | AME (kcal/kg) | TME (kcal/kg) | AMEn (kcal/kg) | TMEn (kcal/kg) |
| P0 | $2496.73^b \pm 238.26$ | $2663.23^b \pm 238.26$ | $2496.48^b \pm 238.25$ | $2662.99^b \pm 238.25$ |
| P1 | $2415.79^b \pm 48.98$ | $2582.30^b \pm 48.98$ | $2415.51^b \pm 48.98$ | $2582.02^b \pm 48.98$ |
| P2 | $2443.94^b \pm 137.33$ | $2610.44^b \pm 137.33$ | $2443.67^b \pm 137.32$ | $2610.14^b \pm 137.32$ |
| P3 | $2859.19^a \pm 447.42$ | $3025.69^a \pm 22.83$ | $2858.85^a \pm 22.82$ | $3025.36^a \pm 22.82$ |
| Average | 2553,91±111,85 | 10881,68±447,42 | 2553,62±111,85 | 2720,13±111,85 |

Description: Different Superscript on the same line shows the real difference ($P < 0.05$)

## 4. Conclusions

Apu-apu mealsfermented Aspergillus niger increase crude protein and decrease crude fiber of apu-apu meals. Inoculum dose of Aspergillus niger on apu-apu meals significantly increase the value of metabolizable energy on broiler chicken with the best dose of $10^8$ CFU/g than other treatments

IOP Conf. Series: Earth and Environmental Science **122** (2018) 012116 doi:10.1088/1755-1315/122/1/012116

**References**

[1] Sutama S 2005 *The influence of Supplementation Kapu-kapu (Pistoia stratiotes l.) in Rations Against cholesterol in Serum and meat is free-range chicken* Bali Peternakan scientific journal, **8** (2): 9 pp

[2] Lunar AM, Supratman H and Abun 2012 *The influence of Inoculum Dose and Long Fermentation of the fruit of Terminalia Catappa (Ficus lyrata) by Aspergillus niger against dry, Rough Fiber and Gross Energy* Laboratory Science Nutrition farmed poultry Ruminant Livestock and food industries Faculty Of Animal Husbandry Padjajaran University

[3] Fardiaz 1989 *Food Microbiology*(Jakarta: Gramedia Pustaka Utama)

[4] Mc Donald, Edwards P A and Haigh J F D G 1994 *Animal Nutrition* 4th Ed. Longman scientific and technical Copublishing in The USA with (New York: John Wilety and Sons. Inc.)

[5] Mangisah, Sukamto B and Nasution M H 2009 *Implementation of Fermented Eceng Gondok in Duck Ration* Journal of the Indonesian Tropical Animal Agriculture **34** (2): 127-133

[6] Hasanuddin dan Mujnisa 2011*The influence of Long Incubation Dregs Sago With Aspergillus niger against the Protein content is rough and Coarse Fiber Sagu Dregs* The newsletter nutrition and forage **11**(1) : 37-42

[7] Tillman AD, Hartadi H, Reksohadiprodja S, Prawirokusumo S and Lebdosoekojo S 1991 *The Basic Forage Science*(Yogyakarta: Gajah Mada University Press)

[8] Laras, R 2010*The value of Energy metabolism and retention of Nitrogen Mix Onggok, Urea, and Ammonium sulfate Zeolite Fermented By Aspergillus niger In Broiler Chickens*Institut Pertanian Bogor Bogor

[9] Sibbald IR 1980*A New Technique For Estimating The Energy Metabolizablecontent Of Feeds For Poultry In : Standarization of Analitical Methodology for Feeds* International Development Research Central. Canada

# The response of red ginger *(Zinggiber officinalle var rubra)* with various processing in broilers were infected by *Eimeria tenella*

**E Z J Nasution, M Tafsin and N D Hanafi**

Faculty of Agriculture, Universitas Sumatera Utara (USU), Padang Bulan, Medan, 20155, Indonesia
E-mail : eka.nasution86@gmail.com

**Abstract.** Red ginger contains high antioxidants and have anti –inflamatory properties. Ginger also has the ability to treat kimiatif, antiemetic, antinausea, and antiparasitik. The aim of this experiment was identified the response of red ginger in broilers were infected by *Eimeria tenella*. This research used Completely Randomized Design (CRD) with 5 treatments and 4 replications. *Eimeria tenella* were infected by 10.000 oocysts/ head and red ginger solution were aplicated with 1% concentration. The treatments consist of KP (positive control), KO (coccidiostat), K1 (red ginger powder), K2 (extracted red ginger by ethanol) and K3 (extracted red ginger by water) . The results showed that the treatment of red ginger was significant effect ($P<0.05$) to lower oocyst production in broilers were infected by *Eimeria tenella*. The comparison between extracted red ginger by ethanol is better than by water or in powder form to decreased. The utilization of red ginger showed the percentage of heterophile and eosinophile close to normal when compared with positive control. Assesment of caecum lesion score was not significant ($P>0.05$) different effect between all the treatments. It is concluded that the treatment by red ginger better than coccidiostat and positive control.

## 1. Introduction

The poultry breeding sector in Indonesia is experiencing rapid progress due to the increasing demand for animal protein. In order to maintain poultry production, it is necessary to conduct regular monitoring efforts to check emerging diseases in animals. One of the most common diseases in broilers is coccidiosis. Coccidiosis is a parasitic disease in broilers that cause many losses, with the result of decreasing the efficiency of feed use and growth restriction, to death [1]. The disease is easy to develop in Indonesia because it is in accordance with the optimum temperature for the development of *Eimeria* is $21^0$C- $32^0$C, and enough moisture so that oocyst can sporulate. Sporulated oocyst can infect host [2].

Since the monetary crisis that occurred in Indonesia until now the price of manufactured medicines (imports) is very expensive, so it is not affordable by the breeders, especially breeders in the middle to lower scale [3]. In addition, most of the Indonesian population resides in the countryside, so the problem of distribution, lack of communication makes it difficult for the area to be reached by modern medicine and veterinary power [4]. Based on this the breeders are trying to find another alternative by utilizing some medicinal plants as traditional medicines called animal herbs that can be given in the form of solution through drinking water and or in the form of simplicia (flour) mixed into the ration as "feed additive" or "feed supplement " [3].

One of these medicinal plants is ginger. Ginger (*Zingiber officinale*) is a herb and medicine that has long been known to the public. Besides being used as a seasoning and traditional herbs, this plant also became a commodity trade as a material industry of drugs, cosmetics, beverages, snacks and kitchen needs. Ginger rhizome is known to contain oleoresin and essential oils of 1-3%. Ginger in the form of powder of 10 mg / kg body weight of chicken can decrease the result of coccidiosis disease [5].

This study aims to determine the response of red ginger for broilers which infected by *Eimeria tenella*, includes the value of cytotoxic injury, oocyst production, and differentiation leucocyt in blood.

## 2. Materials and Methods

The study was conducted in January 2015. The preparatory phase was held in early January 2015 at Pharmacy Laboratory of Faculty of Pharmacy, University of Sumatera Utara, Medan. Furthermore, the implementation stage of the research was conducted for five weeks at the experimental cage of Livestock Biology Laboratory, Faculty of Agriculture, University of Sumatera Utara Medan and Parasitology Laboratory of Balai Veteriner Medan.

### *2.1 Phase I. Propagation of Isolate Eimeria tenella*

After obtaining *E. tenella* isolate, then it is proportionalized with addition of 2-2.5% potassium bichromate solution (5-10 times from sample) for 1-2 days in room temperature (26$^0$C - 28$^0$C). Cover the Petridis / glass cup slightly with open air to give chance. For propagation, *E.tenella* isolates were inoculated in 2-week-old coccus-free chickens, 6 days post-inoculation was harvested as previously worked and then stored in the refrigerator until use.

### *2.2 Phase II. Making Red Ginger Solution*

In making red ginger solution, this study uses 3 kinds of ginger processing which will be used as a solution. The form of ginger processing in question is ginger powder, ginger extract using ethanol, and ginger extract using water.

*2.2.1 Ginger Powder.* Red Ginger is obtained from Pasar Medan. The fresh red ginger is washed and then slashed thinly and dried in an oven at 37$^0$C for 48 hours until dry, then ground powder [1]. The solution was prepared in the manner indicated in the Indonesian Pharmacopoeia, to obtain a solution of 1 g of dried powder in 1 ml [1].

*2.2.2 Ginger Extraction Using Ethanol.* The freshly cleaned red ginger rhizome is dried with oven blower (40-60° C) for 30-36 hours to obtain dry ginger with 8-11% water content. The dried ginger is milled and then filtered to produce 30 mesh ginger powder. A total of 250 grams of ginger powder in the extract 4 times by using ethanol solvent (500 ml). The extract obtained was filtered with filter paper under vacuum. The obtained liquid is fed into a weighed rotary tube, then distilled with a rotary vacum-evaporator. The distillation is stopped after the solvent stops dripping, it is obtained oleoresin consisting of semi-solid colored light brown to dark brown. Further weighing the oleoresin produced in the rotavapor flask. To homogenize the results of ginger extract with water at the time of preparation of the solution it is necessary to add 1% solution of CMC (Carboxyl Methyl Cellulose). Comparison of powder dosage with extract result is 1: 5. So Red ginger extract solution using ethanol made with concentration of 1% (2mg ginger / ml).

*2.2.3 Ginger Extraction Using Water.* Red ginger extraction uses water as the extracting solution. Ginger extraction is done on ginger powder. Every 25 g of ginger powder requires 125 ml of water. Extraction is done 4 times. To obtain the ginger extract, the filtrate is dried so that the solvent and water evaporate. Red ginger extract solution using water is made with a concentration of 1% (2mg ginger / ml).

*2.3 Phase III. In Vivo Test*

In this study, 80 chickens aged one day (DOC) Cobb 500 were randomized into 5 treatments (KP, KO, K1, K2, K3) with 4 replications, and each replication consisted of 4 chickens. At the age of 23 days, five chicken treatments (KP, KO, K1, K2, K3) were infected with *Eimeria tenella* each of 10.000 oocysts / oral. Five (5) days post infected *Eimeria tenella*, given the treatment of ginger solution as much as 1ml / head per oral (K1, K2, K3), and aquadest (KP) for 3 days, rest for 2 days, then given again for 3 days (system 3-2-3), as well as Coxymas coccidiostat administration (as directed by PT.Mensana).

*2.4 Determination of the Number of Oocysts per Gram of Excreta*

Six (6) days post infected *Eimeria tenella*, chicken excreta accommodated every day. In each treatment was taken excreta as much as 2 sample of cage, then collected and separated based on each treatment for 1 week. Excreta weighed 1 gram which was then dissolved into 29 ml of saturated salt solution, centrifuged for 10 minutes at 1500 rpm. After a centrifuge, the top of the supernatant (clear) is taken with a Pasteur pipette then dripped on both sides of the McArm's count chamber. Further observed under a microscope with a magnification of 10 x 10 [6].

*2.5 Determination of SLS (Score Lesion Score)*

The caecum lesion score (SLS) was determined based on pathological-anatomical changes in the degree of damage from infected chicken according to the Johson and Reid (1970) method.

*2.6 Leucocyte Differentiation Examination*

Blood sampling was done 3 times (age 27 days, 30 days, and 35 days). In each treatment 2 samples were taken per plot. To make blood slides and blood sampling according to the Piatina (2001) method [7].

## 3. Results and Discussion

*3.1 Oocyst per Gram Excreta*

Oocyst is the result of fertilization of microgamet and macrogamet in the sexual stage. After zygote fertilization will form oocyst. Oocyst *E. tenella* will come out with the excreta in a sporulated state, and will sporulate within 1-2 days of obtaining oxygen, suitable temperature, and humid environment [8]. The average production of oocysts per gram of excreta that is stored for seven days in *E.tenella* infected chickens can be seen in Table 1.

**Table 1.** Average production of Oocyst Per Gram (OPG) excreta were infected by *E.tenella* and giving red ginger solution (*Zingiber officinale var Rubra*)

| Treatment | Observation on day-to-day (post infection) | | | | | | |
|---|---|---|---|---|---|---|---|
| | 6 | 7 | 8 | 9 | 10 | 11 | 12 |
| KP | $9425^{a}$ | 9463 | $12725^{a}$ | $12950^{a}$ | $11163^{a}$ | $8600^{a}$ | $6363^{a}$ |
| KO | $3925^{b}$ | 8138 | $9000^{a}$ | $7800^{bc}$ | $5175^{b}$ | $3350^{b}$ | $2988^{ab}$ |
| K1 | $7888^{ab}$ | 8475 | $12725^{a}$ | $7950^{bc}$ | $5400^{b}$ | $3675^{b}$ | $4338^{ab}$ |
| K2 | $4025^{b}$ | 4988 | $3838^{b}$ | $3488^{c}$ | $2875^{b}$ | $3250^{b}$ | $1625^{b}$ |
| K3 | $3725^{b}$ | 7575 | $9900^{a}$ | $9980^{ab}$ | $7188^{ab}$ | $4675^{b}$ | $3313^{ab}$ |

Description: Superscript with different letters in the same column states differ significantly at the level (P <0.05)

The result of statistical analysis using Anova (Analysis Varian) showed that giving 1% red ginger solution gave a significant effect (P <0.05) to lowering the production of oocysts in broiler were infected by *E. tenella*. This can be seen from the treatment of K1, K2, K3 < positive control (KP).

While the treatment of coccidiostat (KO) gave no significant difference (P> 0,05) with 1% red ginger solution (KO = K2 = K3).

Furthermore, at the 7th and 8th days post infection, there was no significant difference (P> 0.05) between treatment of KP, KO, K1, K3. The number of oocyst of these four treatments has increased. Maudya (1994) stated that on the 6th and 7th days after the oocyst production infection will increase again because on that day the bleeding has somewhat decreased. However, all four treatments (KP, KO, K1, K3) were significantly different (P <0.05) by treatment of K2 (on day 8). K2 treatment produces fewer oocysts than other treatments.

On the 9th and 10th day post-infection showed that there was a significant difference (P <0.05) between the KO, K1, K2 treatment with positive control (KP). On that day seen the treatment of K3 still continue to increase the number of ookista. And slowly decline until the 12th day. However, statistically, the treatment showed no significant difference (P> 0.05) with positive control treatment (KP).

On the 11th day post infection showed no significant difference (P> 0.05) among all treatments (KP, KO, K1, K2, K3). The peak of oocyst secretion / expulsion on day 8 and the number of oocysts will decrease on the 9th day after infection. It will gradually decline by the 11th day, the oocyst remains small but may still be present in the stool for several months after infection. Oocyst left in the chicken body will serve as a stimulant to form immunity. Conversely high doses of 50,000 oocysts *E.tenella* will cause an outbreak of disease [9].

On the 12th day post infection showed no significant difference (P> 0,05) between coccidiostat treatment (KO) with treatment 1% red ginger solution (K1, K2, K3) to decreasing oocyst production. While K2 treatment was significantly different (P <0,05) with positive control treatment (KP). The treatment of K2 (red ginger solution of ethanol extract 1%) tends to decrease oocyst production from day 8 to day 12. The decrease of oocyst production by giving the red ginger solution of ethanol extract of 1% indicates the existence of anticoxy activity. The anticoxy activity of ginger extract depends on the processing and chemical content [10].

Red ginger contains bioactive components such as oleoresin and gingerol. Gingerol is a phenol-derived compound that interacts with protozoan cells through an adsorption process involving hydrogen bonding. Low levels of phenol interact with proteins form the phenol protein complex. The bond between protein and phenol is a weak bond and immediately decomposes. The free phenol penetrates into the cells, causing precipitation and protein denaturation. At high levels of phenol causes coagulation of proteins so that cell membranes undergo lysis. The anticoxy power of *E.tenella* oocysts occurs through its lysis of the *E.tenella* cell. Oleoresin and gingerol are anti-inflammatory and anti-bacterial. The consistency of the presence of chemical compounds in an herbal origin preparation is influenced by many factors such as where to grow, harvest time, extraction and solvent used. Treatment given anticoxy will work faster to reduce oocyst production. Red ginger can improve digestion, the process that occurs is arousal intestinal stomach and intestinal stimuli by ginger oil. In addition, red ginger will stimulate and launch blood circulation.

*3. 2. SLS (Score Lesi Scores)*

Coccidiosis is an intracellular protozoal infection. Infection by *Eimeria tenella* as the cause of coccidiosis will always have implications for epithelial cell tissue damage, particularly in the mucosal and submucous parts of the caecum [5].

The result of statistical analysis using Anova showed that red ginger solution gave an unreal effect (P> 0,05) in decreasing the score of caecal lesions in broiler infected with *E.tenella*. The treatment that showed the highest score of the caecal lesion was the treatment of KP (positive control) and K1 (1% powdered ginger solution) of 0,56%, while the mean score of the lowest (mild) Treatment K2 (ginger solution ethanol extract 1%) that is equal to 0,37%. This means that the administration of red ginger is effective and respones in preventing tissue / bleeding damage to broiler caecum.

In addition, the way of processing herbs and active compounds oleoresin, essential oils, and gingerol contained in herbal raw materials of red ginger can work effectively in suppressing E.tenella infection. Compounds contained in red ginger has been widely known to streamline the absorption of food into the body with high antioxidant content and strong anti-inflammatory properties [9]. This is also in accordance with the opinion of Iskandar *et al* (2000) which states that the infusion of red ginger as much as 1% can be useful as coccidiostat in broiler [1].

Based on the degree of tissue injury and bleeding in chicken caecum, the lesion score for all treatments is still relatively mild. It's because the resulting lesion score still ranges from 0-1. If sorted by the score of the lowest (mild) to the highest, the following sequence is K2, KO, K3, K1, KP. K2 treatment is lighter when compared with the treatment of KO, K3, K1, KP.

This suggests that red ginger solution effectively treats lesions in chicken caecum. KP treatment showed no significant difference (P> 0,05) with K1, K2, K3 and KO treatment because the amount of oocyst was not enough to give effect to broiler. In addition, it is possible chicken body response immediately form antibodies that can prevent tissue damage and bleeding in broiler caecum. The severity of coccidiosis attack is influenced by the number of attacking parasites (dose of infection), immune power, and age of host.

### *3.3 Leukocyte Differentiation*

Leukocytes or white blood cells are cells that have nuclei and organelles. Leukocytes are active unit of the body's defense system. Leukocytes serve to protect the body from infection and cancer, and help the healing process [11]. The granulocyte group is characterized by the presence of granules in the cytoplasm, while the agranulocyte group has no granules [12]. The results obtained after the calculation of leukocyte differentiation are as follows:

**Table 2.** Mean percentage of heterophils, basophile, eosinophile, lymphocytes, and monocytes in broilers were infected by *E.tenella*

| Leucocyte Differentiation (%) | | Treatment | | | | |
|---|---|---|---|---|---|---|
| | | KP | KO | K1 | K2 | K3 |
| After Treatment : | | | | | | |
| Day (0) | Heterofil$^{ns}$ | 30.63 | 33.75 | 28.00 | 30.75 | 44.50 |
| | Basofil$^{ns}$ | 9.54 | 7.50 | 8.25 | 8.62 | 7.75 |
| | Eosinofil$^{ns}$ | 10.00 | 9.00 | 6.50 | 7.13 | 6.50 |
| | Limfosit$^{ns}$ | 47.34 | 47.13 | 55.00 | 50.13 | 39.12 |
| | Monosit$^{ns}$ | 3.12 | 2.62 | 2.25 | 3.37 | 2.13 |
| Day (3) | Heterofil | 16.17$^{b}$ | 22.13$^{ab}$ | 28$^{a}$ | 27.37$^{a}$ | 26.50$^{a}$ |
| | Basofil$^{ns}$ | 6.00 | 5.75 | 5.62 | 4.50 | 5.50 |
| | Eosinofil | 10.63$^{a}$ | 8.12$^{b}$ | 4.25$^{c}$ | 4.50$^{c}$ | 5.88$^{bc}$ |
| | Limfosit$^{ns}$ | 65.00 | 61.75 | 59.63 | 60.63 | 59.25 |
| | Monosit$^{ns}$ | 2.25 | 2.25 | 2.50 | 3.00 | 2.87 |
| Day (8) | Heterofil$^{ns}$ | 30.50 | 35.75 | 35.00 | 25.63 | 36.37 |
| | Basofil$^{ns}$ | 7.87 | 4.25 | 4.00 | 2.75 | 5.25 |
| | Eosinofil$^{ns}$ | 9.87 | 6.87 | 5.00 | 5.50 | 6.25 |
| | Limfosit$^{ns}$ | 49,88 | 50.38 | 54.50 | 64.50 | 50.38 |
| | Monosit$^{ns}$ | 1.88 | 2.75 | 1.50 | 1.62 | 1.75 |

Description : ns = not significant

**Table 3.** The percentage of normal differentiation leucocytes in blood cell

| Differentiation Leucocytes | (%) Normal in Blood Cell |
|---|---|
| Heterofil | 27.20 |
| Basofil | 1.7 |
| Eosinofil | 2-8 |
| Limfosit | 59.1–64.6 |
| Monosit | 8.9-10.2 |

Srurkie (1976)

After Treatment (Day 0), the results of statistical analysis using Anova showed that the mean percentage of heterophils, basophils, eosinophils, lymphocytes, and monocytes were not significantly different ($P > 0.05$) in each treatment (Table 2). After Treatment (Day 3), the results of statistical analysis using Anova showed that the mean percentage of basophils, lymphocytes, and monocytes were not significantly different ($P > 0.05$) in each treatment (Table 2). There was a noticeable change where the positive control decreased heterophils count. While in the treatment of KO, K1, K2, K3 has slightly decreased the number of heterophils with the lowest percentage at KO (22.13%). The amount of heterophils in the treatment of red ginger (K1, K2, K3) tends to approach the normal value. This occurs because the heterophils are mobilized to enter the bloodstream from the bone marrow. Another possibility is the presence of antioxidant content in ginger is efficacious as anti-inflammatory. The mean percentage of eosinophil was significantly different ($P < 0.05$) between KP treatment with KO, K1, K2, K3. Treatment treated with coccidiostat and red ginger solutions are still able to maintain normal eosinophil count in blood despite being infected by *E. tenella*. The content of antioxidants in ginger can stimulate the intestinal mucous membrane to reduce inflammation. After Treatment (Day 8), the results of statistical analysis using Anova showed that the mean percentage of heterophils, basophils, eosinophils, lymphocytes, and monocytes were not significantly different ($P > 0.05$) in each treatment (Table 2). The anticoxy activity of ginger extract depends on the way processing and chemical content [10]. Red ginger contains bioactive components such as oleoresin and gingerol. Gingerol is a phenol-derived compound that interacts with protozoan cells through an adsorption process involving hydrogen bonding. Then when the chickens are able to survive past the 8th day and the 9$^{th}$ where oocyst production reaches its peak, then the chickens will go to healing by itself [13]. Some leucocyte have started normal.

## 4. Conclusion

The treatment which given red ginger solution by extract ethanol 1% (K2) was effective in decreasing oocyst production and showed the best result compared to other treatment. The administration of coccidiostat and red ginger solution showed not significant difference in reducing the caecum lession value. The percentage of heterophils and eosinophils on the 3$^{rd}$ day post treatment tended to be more stable (normal) in the treatment of red ginger than the treatment of coccidiostat (KO) and positive control (KP).

## References

[1] Iskandar T, Murdiati T B, dan Subekti D T 2000 Pengaruh pemberian infus jahe merah (*Zingiber officinale var rubra*) terhadap koksidiosis sekum pada ayam pedaging (The influence of red ginger (*Zingiber officinale var rubra*) infusion against cecum causidosist at broiler chicken). Balai Penelitian Bogor Bogor

[2] Allen P C, Fettere R H 2002 Clinical microbiology reviews: Recent advanceins biology and immunobiology of eimeria species and in diagnosis and conffol of infection with these coccidian parasites of poultry 1 *Soc. Microbiol* **15** No. 1:58-65

[3] Zainuddin B, Bakrie B, Andayani D dan Yanis M 2006 Tanaman obat meningkatkan efisiensi pakan dan kesehatan ternak unggas (Medical plants improve the efficiency of animal feed and poultry healthy) *Prosiding Seminar Nasional Teknologi Peternakan dan Veteriner* September 2003 Puslitbang Peternakan Bogor
[4] Iskandar T dan Husein A 2003 Pemberian campuran serbuk jahe merah (*Zingiber officinale var Rubra*) pada ayam petelur untuk penanggulangan koksidiosis (Giving of red ginger (*Zingiber officinale var Rubra*) powder in laying chicken for coccidian prevention) *Pros. Seminar*
[5] Iskandar T, Didik T S, dan Toni S 2006 Pengaruh pemberian RBM5 terhadap koksidiosis pada ayam broiler (Influence of RBM5 on coccidian on broiler chicken) Balai Penelitian Veteriner Bogor Bogor
[6] Alamsari O S 2000 Pengaruh larutan lempuyang wangi (*Zingiber aromaticum val*) terhadap produksi ookista *Eimeria spp* pada ayam (Effect of lembayung wangi (*Zingiber aromaticum val*) to ookista *Eimeria spp* production in chicken) Institut Pertanian Bogor Bogor
[7] Piatina V Z 2001 Pengaruh pemberian berbagai konsentrasi larutan biji paria (*Momordica charantia Linnaeus*) terhadap differensiasi leukosit pada ayam yang terinfeksi *Eimeria spp.* (The effect of various concentration of paria seed solution to leukocytes differentiation in infected *Eimeria spp.* chicken) Institut Pertanian Bogor Bogor
[8] Tampubolon M P 1996 *Protozoologi (Protozoology)* (Bogor : Pusat Antar Universitas Ilmu Hayat Institut Pertanian Bogor) hlm 116 – 118
[9] Sanda A 1985 Immunoprophylaxis of coccidiosis in chicks *Monatshefie for veterinar medizin* 40 (5) : 165-167 *In protozoological Abstract* **9** No.6. pp .229
[10] Goto H, Sasaki Y, Fushimi H, Shibahara N, Shimada Y and Komatsu K 2005 Effect of curcuma herbs on vasomotion and hemorheology in spontaneously hypertensive rat *American J. of Chinese Medicine* **33** : 449-457
[11] Corwin J E 2000 *Patofisiologi* (Jakarta: EGC)
[12] Caceci T 1998 Formed elements of blood *J. Cancer* **11**(3) Pp : 1743-1826 Califonia Avian Laboratory USA
[13] Soulsby E J L 1972 *Immunity to Animal Parasities* (Academic Press New york and London) pp 336-382

IOP Conf. Series: Earth and Environmental Science 122 (2018) 012118 doi:10.1088/1755-1315/122/1/012118

# Contribution of local beef cattle production on farmer's income in the dryland farming of Kupang Regency, Indonesia

**M M J Kapa[1**], Y L Henuk[2*], Hasnudi[2] and Suyadi[3]**

[1]Department of Agribusiness, Faculty of Agriculture, Universitas Nusa Cendana (Undana) Kupang, , East Nusa Tenggara, Indonesia
[2]Department of Animal Science, Faculty of Agriculture, Universitas Sumatera Utara (USU), Medan, North Sumatera 20155, Indonesia
[3]Department of Animal Production, Faculty of Animal Husbandry, Universitas Brawijaya (UB), Malang, East Java, Indonesia
E-mail: yusufhenuk62@live.com

**Abstract**. Study on contribution of local beef cattle enterprise on income of dryland farmers in Kupang Regency was conducted from September to December 2016. The study aimed to: (1) determine composition of farm household income in the dry land area of Kupang Regency, Indonesia, (2) analyze contribution of income from local beef cattle enterprise to farm household income. A survey was done on 56 beef cattle farmers who were purposively selected as respondents. All respondents were interviewed using structured questioners with focus on farm household activities and their income. The results showed that total net income of farm household was Rp 14,854,550 per year, out of this Rp3,246,550 to Rp 5,404,750 equals to 21.85 to 36.38% was from local beef cattle enterprise. To enhance the role of beef cattle enterprise, the owner should improve cattle husbandry management through providing good quality and quantity of feed continuously, as well as empowering livestock extension workers to deliver proper information and technology on beef cattle husbandry to the farmer.

## 1. Introduction

The predominant role of traditional livestock production in the developing world like Asian and African countries is obvious. In Indonesia, livestock business is an un-separated component of the farming systems. Some African countries such as Zimbabwe considers cow as a sustainable means to improve rural community source of revenue [1]. Livestock production in East Nusa Tenggara (ENT) Province of Indonesia is an integral part of farming systems. Various types of livestock kept by farmers, namely small livestock (goats, sheep, and pigs), large livestock (cattle, buffaloes, and horses), and poultry (chicken and ducks). Among them, beef cattle are the most important livestock for farm households. Beef cattle provide economic, social and cultural roles to the farm households and they become a leading commodity for the province economy development [2]. They further explained that ENT province is one of the centers for Bali cattle development in Indonesia. This province is also the main supplier of beef cattle for other area of Indonesia including West Java and Jakarta, and supplier of breeding cattle to other provinces including transmigration areas in Indonesia. The dominant type of beef cattle in ENT is Bali cattle and cross bred Ongole known as Sumba Ongole (SO). Bali cattle is mainly concentrated in Timor and Flores, while SO are developed well in Sumba island [2][3] explained factors that encouraged the development of beef cattle business in ENT were due to (1) high demand for beef both locally and nationally, (2) the availability of pasture and leucaena forage that encourages the development of fattening systems of beef cattle mainly in the area Amarasi Sub-district, Regency of Kupang, and (3) increase demand for beef and buffalo from the island of Java, especially Jakarta. Like in other countries, beef cattle in Indonesia has very important roles in fulfilling the demand for meat, a source of breeding cattle and providing agricultural labor, source of income of farm households. In addition, the cattle trade between the islands is one of the backbone of rural economy [4].

In the traditional system as practiced by farm household in Kupang Regency, farmers generally pay less attention on their cattle husbandry. However, in the last decade beef cattle development in Kupang Regency received serious attention from the government because Kupang Regency is a production center for beef cattle and suppliers for national needs, especially for Jakarta Province. The latest data showed the population of beef cattle (cattle Bali) in Kupang Regency in 2014 was 149,243 head while in 2015 the number increased to 154,811 heads or there was an increase by 5.48 %. East Amarasi Sub-district is one of the sub-districts in Kupang Regency that has the largest number of beef cattle. It was reported that cattle population in this area increased by 25 % during the period of 2014-2015, i.e. from 5,202 heads in 2014 to 6,503 heads in 2015 [5]. However, the increase of beef cattle population in the sub-district of East Amarasi did not necessarily increase the welfare of farmers. Beef cattle business in this area is operated in an extensive traditional way. It is difficult to determine how much revenue and cost used to run the business, and how significant is the contribution of the enterprise to the farm household income. Thus, it is necessary to do analysis of farm household income in this area to describe the potential of beef cattle rearing by farmers in contributing the household farm income.

## 2. Materials and Methods

This research was carried out in August - September 2016 in Pakubaun Village, East Amarasi Sub-district, Kupang Regency, Indonesia. The location was determined purposively because this village has large number of beef cattle in East Amarasi. This research was done by using a survey to collect both primary and secondary data. The primary data collection was conducted by interviewing 56 respondent farm household. The main data collected were (1) respondent characteristics (age, level of education, and farming experience), (2) land tenure and cultivated land area; (3) number of cattle owned, (4) Total costs incurred; (5) revenues obtained from all farm and nonfarm; (6) number of cattle sold; (7) total household incomes including income derived from cattle. The secondary data was obtained from relevant institutes and agencies such as Statistical Central Board (SCB), Department of Animal Husbandry and data from previous studies, these include: information on natural condition of study area, and human resources, beef cattle population. Data analysis was done by descriptive analysis, income analysis of beef cattle and farm household income analysis and total farm household and beef cattle enterprise revenues farm household revenues consisted of revenues derived from farm production and nonfarm revenue including revenue obtained from beef cattle selling. Therefore, total revenues of farm household are the sum of values of all farm products and values of nonfarm product before deducting by all production costs incurred. The costs can be cash costs and in-kind (non-cash) costs. Theoretically, production costs consisted of fixed costs and variable costs.

## 3. Results and Discussion

### *3.1. Description of Research Location*

This research was conducted in the Pakubaun Village in East Amarasi Sub-district of Kupang Regency, Indonesia. East Amarasi Sub-district has an area of 162.9 km$^2$. East Amarasi Sub-district has boundaries as follows: North- Sub-district of Amabi Oefeto and Amabi Oefeto East; South by the Timor Sea and South Amarasi Sub-district, East by Timor Tengah Selatan Regency and West by the Amarasi Sub-district [5].

### *3.2. Climatic circumstances*

In general, the climate in the study area is a tropical climate with two seasons namely dry and rainy seasons. The dry season is relatively longer (approximately 7-8 months) due to the influence of wind flows from Australia which does not contain water vapour. Conversely, short rainy season (December to March) occurs because the wind flows from Asia and the Pacific Ocean that contains a lot of moisture of rain. Climate conditions in this area were 5 wet months and 7 dry months with monthly rainfall of 276.7 mm and daily rain of 10.63/day [6].

### *3.3. Population and Livelihoods*
Pakubaun Village had human population of 2,557 people. The population density in Pakubaun Village was 28 people per km$^2$. Most of the population in the study area are farmers. Agricultural activities are mainly carried out during the rainy season by relying upon field gardens as land for planting crops and only a small proportion of respondents who possess and cultivate rainfed land. Common crops planted in the fields are corn, cassava and beans. Meanwhile, annual crops such as coconut, jackfruit, banana, betel nut were planted in the home garden and agroforestry land. In addition to crops production, the farmers also raise livestock mainly cattle, pigs, goats, and chickens [6].

### *3.4. Characteristics of Respondent*
The average age of respondent farmers was 48.6 years with a range of 23-70 years 28-64 years in the village. Based on the age structure, it can be considered that the respondents in the study area were categorized as productive farmers. Productive age ranges from 15 years to 64 years of age. Result of the study showed the average farming experiences of respondent farmers in rearing cattle was 6.9 years with a range of 4.0-37 years. The data illustrated that all respondents have good experiences in beef cattle husbandry. The average level of formal education of farmer was relatively low, where 67 % of them did not complete primary school, 14 % of respondents were graduated from Junior High School, 9.0 % were graduated from High School, 4.0 % were university graduates. This study showed that none of respondents attended any informal education in agriculture fields such as training in the field of animal husbandry. The dominant farming patterns practiced in the study area is dryland farming. Preparation of land for cultivation of the dryland farming is different from the wetland farming. Farmers in the dryland farming started preparation of land by cutting existing shrubs on the land and allowing to dry and then burning it. This practice is well known as slash and burn systems. Average farm land owned by farmers was 1.42 ha with the arable land of 0.54 ha or about 35.21 %. The dominant plant species cultivated by dryland farmers were maize, *nasi* beans (*Vigna umbellata* (Thunb.), *turis* bean (*Cajanus cajan*), green beans (*Vigna radiata*), pumpkin and cassava. The planting practices of dryland farmers were to plant more than one seeds in one hole. The mixed seeds were especially maize mixed with green beans, *nasi* beans, *turis* bean and pumpkins. This practice is locally known as *salome* means putting together more than one types of seed in one hole. Due to limited rainfall and short of rainy season, hence planting of all crops was done only once a year. The average production of food crops in the study area for one planting season of maize was 1,800 kg per ha, *nasi* beans 455.4 kg per ha, green beans 150 kg per ha year$^{-1}$. The priority use of the products was to meet their own family requirements for food, which reached 80 % and the rest 20 % were stored as seeds for the next planting season. Selling food crop product was rarely done by farmers, farmers would sell their food crops product when there was an emergency need of the family.

### *3.5. Cattle Husbandry in the Dryland Area*
Raising cattle for the farmers in the rural areas of Pakubaun Village, Kupang Regency, is inseparable from farming activities. The culture of raising livestock, especially cattle has been considered as inheritance. There were diverse motivations and goals of respondents in rearing cattle, such as, (1) a source of income, which was expressed by 95.0 % of respondents, (2) customs purposes (66.7 % of respondents), (3) support the farm (90.5 % of respondents), and (d) tradition (22.2 %). The average number of cattle kept by farmers in the research area was presented in Table 1. Data in Table 1 showed that average number of cattle reared by respondents was same in proportion between males

and females both calf and adult cows. The cattle was still traditionally and extensively reared by farmers, where the cattle were grazed in pastures or forest lands during the day and at night the cattle were corralled. The owners had very low interference to their animals.

**Table 1**. Average number of cattle ownership by respondents based on sex and age group.

| Description | Calf | Young | Adult | Total |
|---|---|---|---|---|
| Male (AU) | 2.50 (±1.78) | 2.60 (±1.75) | 3.12 (±1.90) | 8.22 (±3.58) |
| Female (AU) | 2.24 (±1.26) | 3.09 (±2.00) | 4.12 (±1.26) | 9.45 (±4.19) |

(Source: Primary data, 2016).

However, rearing cattle management especially for male cattle has shifted in the last few years from extensive systems to semi-intensive systems in which cattle is tethered on a tree or placed in a very simple cage continuously in farmer's home garden and fed with cut and carried leucaena leaf, grass and other leaves by the owners (cut and carry system). Utilization of inputs in beef cattle business in the study area was very low. The only main input given by farmers in the beef cattle business was labor mainly for collecting fodders from home gardens, fields and surrounding areas. While attention for the health care of cattle was very low. Allocation of family labor in the cattle business was high, especially during the dry season. Farmers had to collect fodders from forest or other rangelands area which were relatively far from their homestead or from where their cattle being tethered. Time allocation for different activities in rearing beef cattle by farmers was presented in Table 2. Data on Table 2 explains that most of time allocation by farmers was collecting and providing feed and herding of the cattle. Of the total time available, more than half was used to collect fodders and herding the cattle. Farmers allocated only 12.56% of their time to provide water for the animal. It is understandable because water is very scarce and difficult to get in the area of study, then farmers prefer to give banana stem as a source of water for their cattle.

**Table 2.** Average labor utilization in cattle husbandry.

| Type of Activities | Time allocation (MD/month) |
|---|---|
| Collecting feeds or fodders | 17.14 |
| Feeding and watering cattle | 4.61 |
| Herding | 12.85 |
| Health care and maintaining animal house | 2.10 |
| Total | 36.70 |

(Source: Primary data, 2016).

### *3.6. Income analysis of beef cattle*

Average household income from beef cattle sales in research area was presented in Table 4. Data in Table 3 indicates that farmer operated small scale beef cattle production. The average number of selling cattle was only 2.75 AU annually with gross revenue of Rp 17.9 million and net income of Rp 5.4 million including labor cost or Rp 3.25 million excluding labor cost.

It should be noted that labor for beef cattle production was fully from family labor, therefore in the case of net income calculation, it was not separated from total net income. In some cases for the purposes of economic analysis, this labor cost has to be calculated by applying opportunity costs. In this analysis, there are two ways to calculate the net income of beef cattle production namely with and without incorporating labor cost. Average net income of farm households from food crops and nonfarm activities was presented in Table 4.

**Table 3.** Average Household Income from Beef Cattle Sales in Research Area.

| No. | Categories | Total income |
|---|---|---|
| 1. | Number of sale (AU) | 2.75 |
| 2. | Price per AU (Rp) | 6.500.000 |
| 3. | Gross Revenue (Rp) | 17.875.000 |
| 4. | Cost of buying stored bull (Rp) | 12.375.000 |
| 5 | Retribution Cost (village administration) (Rp) | 27.500 |
| 6. | Buying rope (5 m)(Rp) | 67.750 |
| 7. | Labor costs (Rp) | 2.158.200 |
| 8. | Net Annual Income without labor cost (Rp) | 5.404.750 |
| 9 | Net Annual Income with labor cost (Rp) | 3.246.550 |

(Source: Primary data, 2016)

**Table 4.** Average of Net Income of farm households from non beef cattle enterprise.

| No. | Categories | Total |
|---|---|---|
| 1. | Net income of from food crops (Rp) | 12.242.000 |
| 2. | Net Income from Nonfarm activities(Rp) | 2.612.500 |
| 3. | Total (Rp) | 14.854.550 |

(Source: Primary data, 2016).

Income from food crops derived from maize, beans, and cassava (tuber) products, while nonfarm income was derived from selling forest product such as tamarind pulp, and also selling lobster and fish. Collecting tamarind and caching lobster and fish is usually done during dry season.

### *3.7. Contribution of beef cattle enterprise on Household Income*

Income analysis of farm households in the dryland area of Kupang Regency revealed that the average yearly farm household income was Rp 14.9 million. Yearly farm household income derived from beef cattle enterprise was Rp 5.4 million (without labor costs) and Rp 3.25 million (with labor costs). Thus, the contribution of local beef cattle enterprise to the total farm household income in the dryland area of Kupang Regency was 36.38 % and 21.85 % with and without incorporating labor costs in the analysis, respectively. This finding showed that the contribution of beef cattle income to the farm household income in the dry land area, Kupang Regency was slightly higher than the results of the study from B Hartono and E S Rohaeni [7] about the contribution of beef cattle to the traditional household income in Tanah Laut Regency, South Kalimantan that is 15-26 %.

## 4. Conclusions

Farmer respondents operated small scale farming with an average arable land ownership less than 1 ha and beef cattle ownership of 1-6 heads per household. Total annual income of the farm household in the dryland area of Kupang Regency was low. The farm household income was only Rp 1.25 million per month. This total income was below the Regency Minimum Wage of Rp 1.55 million. The contribution of local beef cattle enterprise to the household income ranged from 21.85 to 36.38 %. We suggested that in order to improve the welfare of livestock farmers in dry land areas, the government and the various parties concerned with rural development need to make a plan of action in order to increase the productivity of dryland farming. These can be done through providing farmers access to the capital source, and empowering the role of animal extension worker to offer better services in beef cattle husbandry in general and especially in feed processing and fodder preservation.

**References**

[1] Tavirimirwa B, Mwembe R, Ngulube B, Banana N Y D, Nyamushamba G B, Ncube S and Nkomboni D 2013 Communal cattle production in Zimbabwe: A review *Livest. Res. Rur. Dev.* **25**: 24 – 28

[2] Kapa M M J, Soemarno, Yanuwiyadi B and Suyadi. 2017. Sustainability status of biology dimension of local beef cattle development in the dryland region, Indonesia. *J. Econ.Sustain. Dev.* **8**(6): 102-108

[3] Nulik J and Bamualim A 1998 Large ruminant feed in Nusa Tenggara. Agricultural Technology Assesment Board Naibonat in collaboration with Eastern Island Veterinary Services Project, Kupang.

[4] Prasetyo E, Sunarso, Santosa P B and Rianto E 2012 The influence of agribusiness subsystem on beef cattle fattening farm's profit in Central Java *J.Indon. Trop.Anim. Agric.* **37**(2):121-126

[5] Statistical Central Board of Kupang Regency 2016 *East Amarasi in figure 2016* (CBS, Kupang Regency)

[6] Statistical Central Board of ENT 2016 *Kupang Regency in figure 2016* ENT CBS

[7] Hartono B and Rohaeni E S 2014 Contribution to income of traditional beef cattle farmer households in Tanah Laut Regency, South Kalimantan, Indonesia *Livest. Res.Rur. Dev.* **26**(8): 1-9

IOP Conf. Series: Earth and Environmental Science 122 (2018) 012119 doi:10.1088/1755-1315/122/1/012119

# Analysis of cattle breeder's income in South Kualuh sub-district of Labuhan Batu Utara Regency

**Hasnudi, I S Berutu, A H Daulay, N Ginting and I Sembiring**

Animal Production Program, Faculty of Agriculture, Universitas Sumatera Utara, Medan 20155, Indonesia
E-mail: nurzainahginting@gmail.com

**Abstract**. Beef cattle breeding is a business of the majority people in South Kualuh Subdistrict among others business. Therefore it is necessary to know the people income of beef cattle farms in the district. This research was conducted in South Kualuh Subdistrict of Labuhan Batu Utara Regency of North Sumatera Province from October to December 2016. This research used survey method with family respondent unit that raising beef cattle. The sample was obtained through Proportional Stratified Random Sampling method and obtained by 97 farmers such as, from Tanjung Pasir village, 64 respondents, Simangalam village was 24 respondents and Lobu Huala village was 9 respondents. Data was analysed by multiple linear regression analysis. The results showed that scale of business, education of farmers positively affect the income of beef cattle farms while the age, breeding experience and the number of family dependents negatively affect to the income of beef cattle farms.

## 1. Introduction

The development of livestock sub-sector in Indonesia has a goal to increase livestock production. This increase in production is expected to have an impact on increasing farmer's income, improving environmental conditions, increasing business opportunities, opening up new jobs and expanding existing employment opportunities. The long-term goal of livestock sub-sector development is the achievement of nutritional adequacy standards from livestock products for the people of Indonesia. FAO[1].2001)

The livestock development strategy has good prospects in the future, as demand for livestock materials will continue to increase as population, income, and public awareness to consume nutritious food is high as a result of the rising education level of the average population. Development and development is one of the development in agriculture which includes development in the field of animal husbandry, where one of the many farms undertaken by the people in the countryside is raising beef cattle, which is in the form of community livestock business. FAO [2]

Livestock business is also a livestock activity where breeders and their families do livestock raising that aims to earn income from the sale of livestock. For farmers, cattle serve as a source of income, animal protein, and fertilizer producers. Other functions are as seeds and savings. The contribution of cattle to income depends on the type of cattle that is maintained, how to maintain, and the allocation of resources available in each region. Livestock business in Indonesia is dominated by small-scale community farms. Farming is not a rare thing to do. Only the scale of management is still a side that is not balanced with adequate capital and management. Almost all households (especially in rural areas) are cultivating cattle as part of daily activities. Zenihun Nigussie et al [3].

There are several factors that cause the amount of beef production is still low, among others, population and production of low cattle. It appears that in North Sumatra there are some very dense areas, some are moderate, but there are very rare or limited population distribution of beef cattle. Of course this greatly affects the amount of income or income society in the area so that differences arise in economic terms. IFAD [4].

In conducting cattle business, the farmer acts as a decision maker that seeks to make effective and efficient decisions in running and managing his business. The socio-economic characteristics of farmers (business scale, number of cattle, age of breeder, level of education, farming experience, number of dependents of the family can influence in making decisions that can give a profit for his business. Hartono, B, Rohaeni E.S [5].

Beef cattle business in North Sumatera spread in various regions with different population density. Southern Kualuh Sub-district is one of the districts where cow livestock population spread in 2016 reaching 531 heads Central Bureau of Statistics of Labuhan Batu Utara Regency[6].

## 2. Material and Methods

The research was conducted in Tanjung Pasir Village, Simangalam Village and Lobu Huala Village, South Kualuh Sub-district, Labuhan Batu Utara District, North Sumatera Province from November 10 to December 20, 2016.

The method of withdrawal of respondents used is as follows: in the first stage determine 3 village fruit from several villages in Southern Kualuh Subdistrict with the withdrawal of respondents by Proportional Stratified Random Sampling Soekartawi [7].

That is the village whose cattle population is high, Is determined by looking at data from the Central Bureau of Statistics Labuhan Batu Utara District in 2016 number selected three villages are:
- Village with high population density is Tanjung Pasir.
- The village whose population density is medium cattle is Simangalam.
- Villages whose population density is rare are Lobu Huala

In the second stage of simple random selection of respondents, 30% of all farmers from each sample village were taken. For research that will use statistical data the sample size of 30% is already representative of the population Prashant Palvia [8]. From each village in getting the number of farmers who become the sample as much as:
- Desa Tanjung sand 212 farmers, then the number of samples 64 breeders.
- Simangalam village 80 farmers, then the number of samples 24 breeders.
- Lobu Huala village 9orang farmers, then the number of samples 9 breeders.

Data obtained from the results of field interviews processed and tabulated then made an average. Then the average data entered into the balance sheet of each breeder and taken the average income farmers. Then the data is analyzed by using the method of income analysis and multiple regression analysis with the following formula

$$\mathbf{Pd = TR - TC} \quad (1)$$

Based on the results that have been obtained, then to see the factors that affect revenue can be seen using the approach model of econometric techniques by using multiple linear regression analysis with Statistical Package for Social Sciences (SPSS 19) software tools.

$$\hat{Y} = a + b_1X_1 + b_2X_2 + b_3X_3 + b_4X_4 + b_5X_5 + \mu \quad (2)$$

The variables in the hypothesis are tested simultaneously and partially to determine whether the variable has a dominant influence or not. If the variable is influential simultaneously then used the test F namely:

$$F = \frac{r^2/k}{(1-r^2)/(n-k-1)} \quad (3)$$

if the variables affect partially can be tested by T testie:

$$T\ hit = \frac{b_1}{Sb_1} \qquad S^2y_{123} = \frac{\sum(\hat{y} - y)^2}{n - k - 1} \qquad S^2b_1 = \sqrt{\frac{s2y123}{\sum X_i^2(1 - R_i^2)}} \tag{4}$$

Parameters of income analysis research include business scale, age of breeder, education level, experience (long) breeding, number of family dependents, receipts, expenses and income.

## 3. Result and discussion

**Table 1**. Revenue and expenditure of farmers in research areas in 2016

| Characteristics of sample breeders | Unit | Total | % |
|---|---|---|---|
| Income | | | |
| Livestock sales | Rp/years | 2,403,000,000 | 100 |
| Expenses | | | |
| Seeds | Rp/years | 1,339,000,000 | 84,197 |
| Drugs | Rp/years | 17,600,000 | 1,106 |
| Feed | Rp/years | 53,700,000 | 3,376 |
| Labor | Rp/years | 180,000,000 | 11,318 |
| B/C Ratio | | 1.51 | |

In the business of beef cattle in the research area obtained the total income from cattle business for 1 (one) year is ranged from Rp 10,000,000 to Rp 52,000,000 / year / breeders with an average of Rp, 24,773,196 / year / breeder,

The net income of each respondent from the business of beef cattle for 1 (one) year ranged from Rp, 28,400,000 up to Rp, 223,050,000 with average Rp 88,002,577 / year / breeder, From the average value of family income from the business of beef cattle this can illustrate that the respondents are motivated enough to do the development of their cattle business, they see very well that the beef cattle they earn can earn bigger income when done seriously,

The B / C Ratio analysis is used in an attempt to find out whether or not the effort is to be continued to the next period or otherwise the business is terminated because it is less feasible, From the above table obtained B / C Ratio 1.51, This indicates that the livestock business is feasible to continue, An effort can be said to be feasible or provide an advantage if the value of B / C ratio above 1 (> 1), While a business is declared not feasible or does not provide benefits if the value of B / C ratio is smaller 1 (<1), The greater the value of B / C ratio the more efficient the business and vice versa the smaller the B / C ratio the more inefficient the business, Andres Algabaa [9],

**Table 2**, Analysis of variant income and variable estimates

| Model | The Sum of Square | Free Degrees | Middle squares | F | Sig, |
|---|---|---|---|---|---|
| Regression | 1,132E17 | 5 | 2,264E16 | 281,947 | ,000[a] |
| Residual | 7,308E15 | 91 | 8,031E13 | | |
| Total | 1,205E17 | 96 | | | |

a. Predictors: (Constant), Number of dependents, Level of education, Breeding experience, Number of livestock, Age of breeder
b. Dependent Variable: income

Based on the above table, simultaneously the value of F (281,947) is greater than F-table (2,31), This indicates that simultaneously all variables are business scale, age of breeder, education level, length of livestock and the number of family dependent significantly influence to earnings of beef cattle farmer with level of significance 0,000a and at 95% confidence level, Jeff SauroJames [10],

**Table 3**. Coefficient of Determination ( $R^2$)

| Model | R | Coefficient of Determination | Adjusted R Square | Standard Error and Estimasion |
|---|---|---|---|---|
| 1 | .969[a] | .939 | .936 | 8961310.59200 |

Predictors: (Constant). Number of dependents. Level of education. Breeding experience. Number of livestock. Age of breederBased on the above table shows that the coefficient of determination (R Square) is worth 0.939 means that all free variables of cattle business scale (number of livestock). age of breeder. education level. breeding duration and number of dependents affect dependent variable 93.9% and the rest is equal to 6.1% was explained by another variable (μ) not investigated in this study.

So to see the factors that affect the income by using the Model of Econometric Approach by using multiple linear regression analysis of Statistical Package for Social Sciences (SPSS 19) software tools can be seen in Table 4.

**Table 4.** Multiple linear regression analysis of the scale of business / number of livestock. age of breeder. education level. farming experience and the number of dependent breeder to beef cattle rancher income

| Model | | Unstandardized Coefficients | | Standardized Coefficients | T | Sig. |
|---|---|---|---|---|---|---|
| | | B | Std. Error | Beta | | |
| 1 | (Constant) | 4411049.755 | 5235016.264 | | .843 | .402 |
| | Number of livestock | 9832525.557 | 302860.466 | .960 | 32.466 | .000 |
| | Age of breeder | -273444.655 | 155450.396 | -.067 | -1.759 | .082 |
| | Level of education | 2620840.973 | 1177024.737 | .058 | 2.227 | .028 |
| | Breeding experience | 588037.574 | 286683.363 | .075 | 2.051 | .043 |
| | The number of dependents | 135681.213 | 820654.806 | .005 | .165 | .869 |

Based on the above table we get the following equation:

$\hat{Y} = 4411049.755 + 9832525.557X1 - 273444.655X2 + 2620840.973X3 + 588037.574X4 + 135681.213X5 + \mu$

Information:

$\hat{Y}$: Income of beef cattle ranchers
X1: Business scale / number of cattle (tail)
X2: Age of breeder (year)
X3: Education level (year)
X4: Breeding experience (year)
X5: Number of dependents of the family (people)
μ : Uncompromised variables

Based on the above equation model can be interpreted that:

a. If the free number of livestock (X1) has increased by 1 tail. then there will be an increase in income (Y) of Rp. 9.832.525.557.

b. If the age-free variable of breeders (X2) has increased by 1 year. then there will be a decrease in income (Y) of Rp 273.444.655.
c. If the independent variable of education level (X3) has increased by 1 level. then there will be increase of farmer income (Y) equal to Rp 2.620.840.973.
d. If the free variable of breeding experience (X4) has increased by 1 year. then there will be increase of farmer income (Y) equal to Rp 588.037.574.
e. If the dependent variable of family dependent number (X5) has increased by 1 person. then there will be decrease of farmer income (Y) equal to Rp 135.681.213.
f. If the X1. X2. X3. X4 and X5 variables in the analysis are ignored (not doing the activity). the beef farmer will still receive income of Rp 4.411.049.755 / year

The influence of the number of livestock. age of breeder. level of education. experience and number of dependents partially to the income of breeder can be known as follows:

a. Variable of cattle counted have significant effect to beef cattle farmer income. if measured at 95% confidence level indicated by t-count value (X1) equal to 32.466 bigger than t-table ($\alpha = 0.05$) that is equal to 1.98. Pytrick Reidsma and Marie Helene Jenttroy [11]. that the income of cattle business is influenced by the number of livestock sold by the farmers themselves so that the more the number of cattle. the higher the net income earned.
b. The variable of age of the farmer does not significantly affect the income of beef cattle. if measured at 95% confidence level indicated by the value of t-count (X2) equal to -1.759 smaller than t-table ($\alpha = 0.05$) that is equal to 1. 98 this variable is negative because because the criteria of age breeders do not encourage farmers in developing a business of beef cattle in Southern Kualuh Subdistrict. Suggests that the younger age of breeders (productive age 20-45 years) is generally a sense of curiosity towards something higher and interest to adopt towards the introduction of higher technology.
c. The variable of education level has significant effect to beef cattle farmer income. if measured at 95% confidence level shown by t-count value (X3) 2.227 bigger than t table value ($\alpha = 0.05$) that is equal to 1.98. Irina and Helena [12] states that education has a bit more important role to the productivity of farmers in managing their livestock.
d. Variable of experience / duration of livestock have real effect to beef cattle farmer income. if measured at 95% confidence level indicated by t-count value (X3) equal to 2.051 bigger than t table value ($\alpha = 0.05$) that is equal to 1.98 . A long breeding experience will provide information on the purpose of breeding is to provide added value for their lives Hasan [13].2000).
e. The variable of number of family dependent have no significant effect to beef cattle farmer income. if measured at 95% confidence level shown by t-count (X5) 0.165 smaller than t-table ($\alpha = 0.05$) that is equal to 1.98 That the dependents of the child in the breeder's family can not provide a positive boost to the income increase of the farmer.

## 4. Conclusion

Business scale. farmer education level. breeding experience give positive influence to increase farmer's income. while the age of breeder and the number of family dependent do not give increase of beef cattle farmer income in Kecamatan Kualuh Selatan. Labuhan Batu Utara Regency.

## References

[1] FAO 2001 Mixed crop-livestock farming: A review of traditional technologies based on literature and field experience. Animal Production and Health Papers 152 Rome (Italy):FAO
[2] FAO 2010 Sete Lagons "consensus" on integrated crop-livestock-tree systems form sustainable development (IC;LSD). An international consultation on integrated crop-livestock system development-the way forward for sustainable production intensification. Integrated Crop Management13:2010
[3] Abanades J *et al.* 2014 *Wave Farm Impact on the Beach Profile* (UK: A Case Study)
[4] IFAD 2009 Integrated crop-livestock farming systems Rome (Italy) IFAD headquarters

[5] Hartono B and Rohaeni E S 2014 Contribution to income of traditional beef cattle farmer households in Tanah Laut Regency South Kalimantan Indonesia *Livest Res Rur Dev.* 26.2014
[6] Central Bureau of Statistics Kabupaten Labuhan Batu Utara 2016.
[7] Soekartawi A 1995 *Farming Analysis* (Jakarta: UI Press)
[8] Prashant P 2015 CIT for socio Economic development a citizens perspective USA
[9] Andres A 2016 *Generalized Financial Ratios to Predict the Equity Premium* Belgium
[10] James J S 2016 *Chapter 10 – An introduction to Correlation. Regression. and ANOVA* USA.
[11] Reidsma R and Jenttroy M H 2016 *Farming System Analysis and Design for Sustainable Intensification : New Methods and Assessments* France
[12] Malganova I and Zagladina H 2015 *Instrumental and Methodology Approaches to research of Socio Economic Development of the region* Kazan Rusia
[13] Hasan I 2000 *Analysis of Coffee Production in Mbenti Vill age. Minyambow District. Manokwari District*. Http: //www. Papuaweb. Org / unipa / dlib- s123 / hasan / s1. Pdf.

# Prey preference of *Myopopone castanea* (hymenoptera: formicidae) toward larvae *Oryctes rhinoceros* Linn (coleoptera: scarabidae)

**Widihastuty[1], M C Tobing[2], Marheni[2] and R A Kuswardani[3]**

[1]Postgraduate student in Ilmu Pertanian Dept Pasca Sarjana Universitas Sumatera Utara Medan
[2]Lecturers in Agroteknologi Dept Agriculture Faculty Universitas Sumatera Utara Medan
[3] Lecturer in Agroekoteknologi Dept Agriculture Faculty Universitas Medan Area Medan
Agriculture Faculty Universitas Sumatera Utara, Jl Prof A. Sofyan No. 3 Medan
E-mail: cyccutobing@gmail.com

**Abstract.** Myopoponecastanea (Hymenoptera: Formicidae) ant is a predator for larvae Oryctes rhinoceros (Coleoptera: Scarabidae) which is a pest on oil palm. These ants are able to prey on all stadia of O. rhinoceros larvae. This study was conducted to determine prey preference of M. castaneatoward its prey O. rhinoceros larvae.. The study was conducted using a Factorial Complete Random Design with two factors (using log and no log) and five replications. Preferences test was done by choice test and no choice test. The results of no choice preference test on the log treatment, M. castaneaprefer preyed on firstinstar larvae of O. rhinoceros ( $\overline{X}$ = 2.6 individual) with a preference index was 0.194 and on no log treatment, M. castanea prefer for both first instar larvae and second instarlarvae ( $\overline{X}$ = 4.6 individual) with a preference index 0.197. The results of the choice preference test using logs, showed that M. castanea prefer the firstinstar larvae of O. rhinoceros ( $\overline{X}$ = 2.6individual), with a preference index (0.35), and on no log treatment, M. castaneaprefer the second instar larvae( $\overline{X}$ =1.4 individual) with a preference index 0.189. Both first and secondinstarlarvaeof O. rhinoceros were preferred by predator M. castanea

## 1. Introduction

Ants have an important role in the ecosystem, which can act as pollinators, seed dispersal, and also as predators of herbivore insects [1]. Various species of the ants group have been widely used as biological agents for plant pest control, such as weaver ant or green ant (*Oecophylla smaragdina*) can prey on caterpillars (*Setora nitens*) with 83% high predation rate [2], *Dolichoderus thoracicus*, is able of suppressing *Helopeltis sp* attacks on cocoa plantations in Sulawesi [3]. Marheni [4] found that *Myopopone castanea* ant (Hymenoptera: *Formicidae*) is a predator of *Oryctes rhinoceros* pest (*Coleptera: Scarabidae*) which is one of the pests of oil palm plants. These ants usually live on the ground and decayed logs. In the palm oil plantations, these ants can be found in the fallen palm trunks and have been decayed due to old or because of stem rot disease. *O. rhinoceros* larvae in oil palm plantations usually live in decayed palm oil trunks and organic material piles. The similarity of living places between *M. castanea* and *O. rhinoceros* larvae reveal great opportunities to exploit these predatory ants as potential biological agents for *O. rhinoceros*. Based on the observations of Marheni [8] the ability to prey *M. castaneae* ants on *instar* larvae *O. rhinoceros* in the laboratory using 20 working ants could reach 4-5 prey/day. *M. castaneae* ant attacks its prey alive by biting and stinging it to death then eating its *hemolymph* liquid. The initial symptoms that were indicated on the larvae of *O. rhinoceros* were the larval cuticle which turns into brownish and gradually darkened. The body of the larvae will be blackened and damaged due to the bite and sting of *M. castaneae* ants so that remaining only the cuticle part. This ant is also able to eat the first *instar* larvae of *O. rhinoceros* until it runs out [5].

Predators are the generalist (have much prey). Despite it is generalist, predators will form selection against its prey and this selection takes place naturally. Rasool *et al.*, [6] suggest that preference is a selection against the proportions of available prey in a particular environment. Preferences could be predicted by using the equations developed by Strauss [7]. Linear index of prey preferences or preference index (*Li*) is the difference between the proportion of prey predator (*ri*) prey and the proportion of available prey *(pi)*. Components that may affect the prey preference are interest and conformity to the prey, prey recognition, decision of attack or not, and ability to capture and consume the prey.
This study aims to determine the preference of *M. castanea* ants against the prey *instar* of larvae *O. rhinoceros*.

## 2. Materials and methods

The study was conducted from December 2016 - March 2017 at the Plant Pest Laboratory of the Faculty of Agriculture, University of North Sumatra Medan.

### *2.1 Insect Test Collected*

The research was conducted by collecting predator ants *M. castanea* from decayed palm oil stems from palm oil plantations in the southern Binjai, Binjai City. *M. castanea* ants obtained from the field then cultivated in the laboratory of Pest Plant, Faculty of Agriculture, University of North Sumatra. The colony of the ants maintained in a glass box with a size of 70 x 30 x 30 cm. Inside the glass box was placed two pieces of decay palm stems with size 20 x 20 x 3 cm and symmetric forms as the nest of the ants. At the centre of the palm stem, there was a small hole to place the prey larvae *O. rhinoceros*. Every day the log was sprayed with water to moist the nest of the ants. The allocation of prey for *O. rhinoceros* larvae was administered according to the ant quartile requirement. When the prey is dead and begins to dry, soon it will be given the new prey of *O. rhinoceros* larvae.

### *2.2 Preference Test*

The preference test of *M. castanea* ant prey was done into two ways, the choice preference test and the no choice preference test. The design used in this test was a Completely Randomised Design of two factors (using logs and without logs) with five replications. This test had been done inside a glass box measuring 40cm x 20cm x 20cm. In the log-treated treatment, 2 pieces of logs from the decayed palm stem measuring 15 cm x 15 cm x 4 cm were made symmetrically and had been dredged slightly in the middle to place the *O. rhinoceros* larvae, whereas in a log-free treatment, inside the glass box provided a layer of soil and some flakes of decayed stem palm that has thickness of ± 1.5 cm.

In the choice preference test, inserted 40 worker ants of *M. castanea* and 20 larvae into each glass box. *M. castanea* ant was first unfed (starvation) for 24 hours. The next day, inoculated six larvae of *O. rhinoceros* consisting of two first-*instars* larvae, two second-*instars* larvae and two third-*instars* larvae. Observations were made a day later by counting the number of dead larvae.

In the preference test with no choice, in each box of glass inserted 40 worker ants *M. castanea* and 20 ants larvae which have been starved for 24 hours. Five boxes were given six individuals first-*instar* larvae prey *O. rhinoceros*, the next five boxes provided six individuals second- *instar* larvae prey *O. rhinoceros* and the last five boxes provided six individuals third-*instar* larvae. Observations were made a day later by counting the number of dead larvae. The data were analysed statistically with ANOVA and tested at 5% level. The preferred degree (preference index) of the predator is thought to be using the equation developed by Strauss [7] as follows:

$$Li = ri - pi \tag{1}$$

Information:

L = the linear index of prey selection
i = stadia of eaten prey
ri = prey proportion of preyed predator
(number of eaten prey stadia i/total predation)
pi = proportion of available prey
(number of available prey stadia i/total of available prey)

## 3. Results and Discussion

### *3.1 Prey Preference*

The results of the no-choice preference test revealed that 3 instars of prey were tested using the log, *M. castanea* ants preferred the first-*instar* larvae of *O. rhinoceros*, with the average number was 2,6 larvae. For the no-log treatment, *M. castanea* ants preferred both first-*instar* larvae and second-*instar* larvae with the ability to respectively an average of 4.6 larvae (figure 1). The result of the choice preference of the treatment using logs indicated that *M. castanea* ants preferred first-*instar* larvae (x = 1,2 larvae), whereas, for the no-log treatment preference test, *M. castanea* preferred second *instar* larvae with the average number ( $\overline{X}$ ) = 1.4 larvae (figure 2).

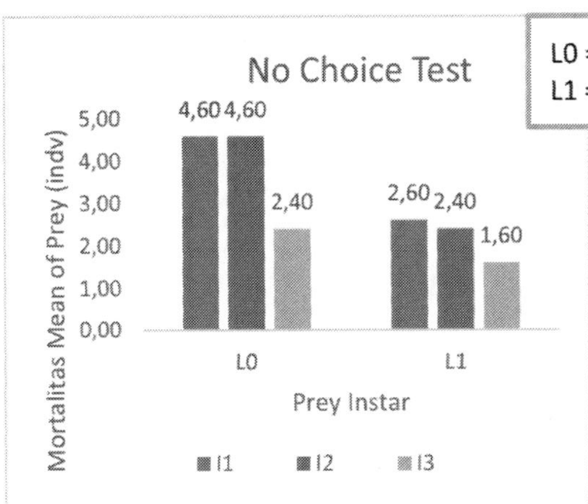


**Figure 1.** Prey Preference instar of *M. castanea* in *no choice test*

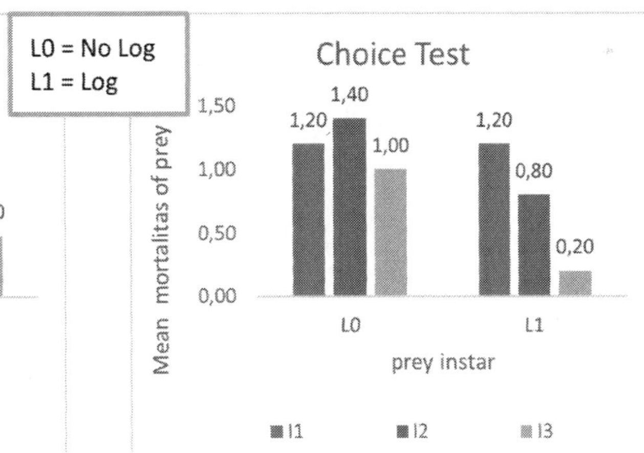


**Figure 2.** Prey Preference instar of *M. castanea* in *choice test*

On the no-choice preference test, both using log and no-log, *M. castanea* ants preferred both first-*instar* and second-*instar* larvae, as well as on the choice preference test. The ants preferred first and second *instar* since they have smaller form than the third *instar*. In addition, the prey morphology of the *instar* such as the cuticle of the prey is still thin, so it is easier to tear and suck the hemolymph liquid by the ants. As stated by Marheni [4] and Junaedi [5], the mortality level of the first and second *instar* that was preyed by *M. castanea* ants, was higher than the third *instar*. Due to its bigger form, the third prey *instar* larvae could fight more against the predator. During the observation of the research, within the prey process, there was five to seven ants that work together in paralyzing a prey. The third *instar* larvae was able to bite the *M. castanea* ants with its mandible. Thus, during its predation process, usually there was some predator ants died. Legaspi *et al.* [8] stated that even the acceptance rate for two types of prey is similar, but due to the ability to avoid both types of prey is different, the frequency meetings between predator and prey may differ. As a result, the prey preference of both prey is different. The preference by predators can also be affected by various physical factors such as light, color, shape, and prey size, as well as chemical factors in the form of odors produced by prey [9]. Legaspi *et al.* [8] stated that two important differentiators of the host or prey range are the host's taxonomy and ecology. In addition, the selection of hosts or prey by natural enemies can be influenced by physical and chemical factors of host or prey, generally these two factors can indeed determine the success of natural enemies in finding hosts or prey.

### *3.2 Preference Index*

Preferred linear index of prey or preference index (Li) is the difference between prey proportion of preyed predator (ri) and the proportion of available prey (pi). The symmetric value of preference index is linear. The selection number of prey varies from -1 to +1. Negative Li values indicated that the prey tends not to be chosen or disliked by predators. In contrast, positive Li values indicated that predators tend to prefer these prey to be consumed.

The value of preference index on the no-choice test in the no-log treatment for the first-*instar* prey was 0.197, for the second-*instar* = 0.197 and for the third-*instar* = 0.07. Meanwhile for treatment using log, its preferred index logs for first-*instar* was 0.194, for the second-*instar* = 0.164 and for the third-instar = 0.042 (figure 3). Preference index value of choice test on no-log treatment for the first-*instar* prey was 0.133, for the second-instar = 0,189 and for the third- *instar* = 0,078. Meanwhile for treatment using log, the preference index of the first-*instar* was 0,345, for the second-instar = 0.164 and for the third-*instar* prey = -0.109 (figure 4). Negative values for the third-*instar* prey indicate that these instars prey tend not to be selected by *M. castanea* ants to be preyed if all three types of *instar* prey are available as its prey.

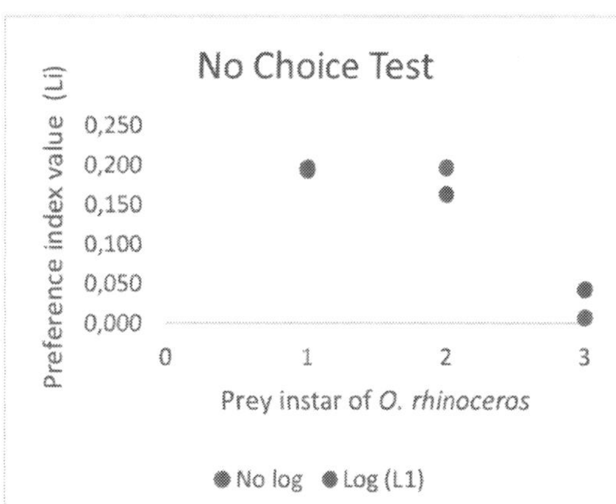


**Figure 3.** Preference index Value of *M. castanea* in no choice test

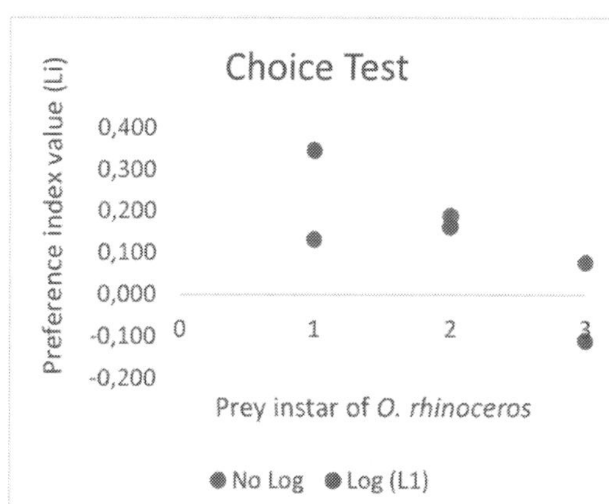


**Figure 4.** Preference index value of *M. castanea* in choice test

The preference index value on the no-choice with the log and no-log treatment, for each prey *instar O. rhinoceros* larvae, had positive value, meanwhile on the choice test, index preference of third-*instar* using log showed negative value. This suggests that the third-*instar O. rhinoceros* prey on log treatment is not favored by *M. castanea* ants. Third-*instar O. rhinoceros* larvae has a large form size and has the ability to avoid by dragging into the log to avoid the meeting of prey and predators. According to de Bach [10] natural enemies can select the match of host or prey and the selection takes place through a natural process. Jaworski *et al.* [11] stated that one of the factors that can influence the predator interaction with prey is the size of the prey form. Generally, predators consider the efficiency of predation when prey on its prey by choosing prey to be preyed. Bennett and Gratton [12] stated that in predating predators will use energy to search and then consume prey. This will cause predators to choose prey for consumption in order to maximize the ratio of energy and nutrient acceptance of the predating process [13].

## 4. Conclusion

In the preference test with no choice on the log treatment, *M. castanea* prefers to prey first-*instar* larvae of *O. rhinoceros* ( $\overline{X}$ = 2.6 larvae) with a preference index of 0.194, and in no log treatment preferred both first and second *instar* with the ability to prey on each, $\overline{X}$ = 4.6 larvae with a preference index value of 0.197.

In the choice preference test on log treatment, *M. castanea* prefers prey on first-*instar* larvae ( $\overline{X}$ = 1,2 larvae) with a preference index of 0.35, and in no log treatment, *M. castanea* prefers to prey second-*instar* larvae ( $\overline{X}$ = 1,4 ) with index of preference 0,189.

*M. castanea* ants were able to prey on all *instar* types of larvae of *O. rhinoceros*, but prefer prey first and second-instar of *O. rhinoceros* larvae if the three types of *instar* prey present as its prey.

## References

[1] Rizal S, Falahudin I dan Endarsih T 2011 Keanekaragaman semut predator permukaan tanah (Hymenoptera: Formicidae) di perkebunan kelapa sawit SPPN Sembawa Banyuasin (Diversity of ant land predator (Hymenoptera: Formicidae) in oil palm plantations SPPN Sembawa Banyuasin) *Sainmatika* **8(1)**:37-42

[2] Falahudin I 2013 Peranan semut rangrang (*Oecophylla smaradigna*) dalam pengendalian biologis pada perkebunan kelapa sawit (The role of rangrang ants in biological control of oil palm plantations) *Prosiding Konferensi AICIS **XII*** 2604-2618

[3] Anshary A dan Pasaru F 2008 Teknik perbanyakan dan aplikasi predator *Dolichoderus thoracicus* Smith (Hymenoptera: Formicidae) untuk pengendalian penggerek buah kakao *Conomorpha cramerella* (Snellen) di perkebunan rakyat (Propagation techniques and application of predators Dolichoderus thoracicus Smith (Hymenoptera: Formicidae) to control Conomorpha cramerella (Snellen) cocoa borer in smallholdings plantation *Journal Agroland*. **15(4)**:278-287

[4] Marheni 2012 *Karakteristik Ekologi dan Biologi Orytes rhinoceros pada Pertanaman Sawit di Sumatera Utara* (Ecological and Biological characteristic of *Oryctes rhinoceros* at palm oil plantation in North Sumatera [tesis] Universitas Gadjah Mada, Yogyakarta

[5] Junaedi D, Bakti D dan Zahara F 2014 Daya Predasi *Myopopone castaneae* (Hymenoptera: Formicidae) Terhadap Larva *Oryctes rhinoceros* L (Coleoptera: Scarabidae) di Laboratorium (Predation power of Myopopone castaneae (Hymenoptera: Formicidae) Against Larvae Oryctes rhinoceros L (Coleoptera: Scarabidae) in Laboratory). *J.Online Agroekoteknologi*. **3(1)**:112-117.

[6] Rasool S, Gulzar S, Javed S and Shahid M S 2015 Tritrophic response of *Coccinella septempunctata* towards *Aphis gossypii* in cotton. *Acad J. Entomol* **8(1)**:01-04

[7] Strauss R E 1979 Reliability estimates for Ivlev's electivity index, the forage ratio, and proposed linear index of food selection. *Trans American Fisheries Soc*. **108**:544-552.

[8] Legaspi, J C, Simmons, A M, & Legaspi Jr, B C 2006 Prey preference by *Delphastus catalinae* (Coleoptera: Coccinellidae) on *Bemisia argentifolii* (Homoptera:Aleyrodidae): effects of plant species and prey stages', *Bioone Online J.* **89**:218-222.
[9] Tarumingkeng R C 1994 *Dinamika Pertumbuhan Populasi Serangga (Insect Population Growth Dynamics)* IPB Press, Bogor
[10] de Bach P 1991 *Biologi Control by Natural Enemies*, Cambridge University Press, London.
[11] Jaworski C C, Bompard A, Genies L, Desneux E A, Desneux N 2013 Preference and Prey Switching in a Generalist Predator Attacking Local and Invasive Alien Pests *Plos one J* **8(12):**e82231
[12] Bennett A B, Gratton C 2012 Measuring Natural Pest Suppression at Different Spatial Scales Affects the Importance of Local Variables *Environ Entomol* **41(5):**1077-1085
[13] Roger C 1999 *Mechanisms of Prey Selection in the Ladybeetle Coleomegilla maculata lengi Timb. (Coleoptera: Coccinellidae)* [tesis] (Canada (US): McGill University)

# Carcass yields of two different strains of ducks raised in different altitude

**St Y F G Dillak, N G A Mulyantini, G M Sipahelut and U R Lole**

Faculty of Animal Science. Universitas Nusa Cendana Kupang – NTT Indonesia
E-mail: ngamulyantini@yahoo.com.au

**Abstract.**The objective of this research was to determine if there is a difference in performance and carcass yield between ducks of two different strains raised in different altitude. Ducks different strains (*Muscovy* vs *Pekin* ducks) and they raised either in high or low altitude (high altitude which was between 500 and 1000m vs low altitude which was below 500m). All ducks were given one of two different diets and provided water *ad libitum*. The diets were: 1) commercial diet, and 2) local diet. There were three replicate per treatment and there were 5 ducks per replication. Ducks from each strain were standardized to a similar weight. The results show that Pekin ducks carcass performance was significantly better than Muscovy ducks. Ducks given diet 2 had significantly ($P<0.001$) lower carcass percentage than those given diet 1. Pekin ducks had greater genetic potential for carcass performance in high altitude environment. However, abdominal fat percentage in Pekin ducks is significantly ($P<0.01$) higher than abdominal fat percentage in Muscovy ducks.

## 1. Introduction

The duck industry has considered duck strains in regard to carcass yields. Many factors influence duck performance, one of the factors that are less well understood is the geographical location, especially with respect to the altitude from sea level. The relationship between variable environmental parameters and duck performance has not been clearly described. The level of genetic variation within population decreased with the environmental altitudinal gradients increased. However, there are also reports showed opposite results, genetic variations was not by altitude at all. Thus, the relationship between the altitude and genetic diversities are needed to be more investigated. The objective of this research was to determine if there is a difference in performance and carcass yield between ducks of two different strains raised in different altitude. Ducks different strains (*Muscovy* vs *Pekin* ducks) and they raised either in high or low altitude (high altitude which was between 500 and 1000m vs low altitude which was below 500m). The objective of this research was to determine if there is a difference in performance and carcass yields between ducks of two different strains raised in different altitude.

## 2. Materials and Methods

The experiment method was used 2 x 2 x 2 factorial arrangement. 90 ducks were allocated to 6 treatments with 3 replicates of 5 ducks per replicate. The treatments were (2 different strains of ducks; 2 different diets; and 2 different altitudes). Ducks different strains (*Muscovy* vs *Pekin* ducks) obtained from traditional farmers. The ducks raised either in high or low altitude (high altitude which was between 500 and 1000m vs low altitude which was below 500m). Ducks were provided with regular pelleted starter diet until 3 weeks of age. From 3 to 6 weeks of age, all ducks were given one of two different finisher diets (commercial vs local). Diets and water were provided *ad libitum*. Composition of experimental finisher diets can be seen in Table 1. All of the finisher diets contain 16% crude protein and EM 3150 kkal/kg. At 6 weeks of age, ducks from each group were selected for carcass analysis. Ducks from each strain were standardized to a similar weight. Variables measured were : live weight, carcass weight, carcass percentage, breast meat yield and weight of abdominal fat. Data were subjected to One-Way ANOVA procedures of SAS software for analysis of variance. The significant level was set at $P<0.05$ by Duncan's multiple range test.

**Table 1.** Composition of experimental finisher diets (%) for two different strain of ducks in different altitude.

| Ingredients (%) | Diet 1 | Diet 2 |
|---|---|---|
| Corn | 68.00 | 68.00 |
| Soybean meal | 27.00 | 26.20 |
| Cornstarch | 0.50 | 0.50 |
| Dicalcium phosphate | - | 1.30 |
| Limestone, 38% Ca | 2.00 | 1.50 |
| Vitamin and mineral premix | 1.00 | 1.00 |
| Salt | 0.50 | 0.50 |
| Vegetable oil | 1.00 | 1.00 |
| Total | 100 | 100 |

### 3. Results and Discussion

It can be seen in Table 1 that live weight, carcass weight, carcass percentage and breast meat yields of two different strain of ducks were significantly ($P<0.05$) much better in high altitude compared with ducks raised in low altitude. However, abdominal fat percentages of 2 strains ducks were significantly higher in high altitude than in low altitude.

The results also show that Pekin ducks carcass performance whether raised in high and low altitude were significantly better than Muscovy ducks. The better carcass performance of Pekin ducks would be expected as it was the strain selected for higher body weight, faster growth and breast yield. Generally, studies comparing different strain of ducks found that strain affects breast weight. [1] reported that factors inherent to ducks that may affect product appearance and carcass yields are breed, selection within breed, and sex. Previous researchers found differences between duck breeds (Pekin and Muscovy) for breast meat color, cooking loss and shear value. Baeca *et al.* [2] stated that selection of Muscovy ducks improved body and breast meat weight. Witkiewiez et al. [3] found that two different strain of Pekin ducks were observed to have significantly different live weight, breast meat yield, and collagen content. Omojola found that breed (Pekin and Muscovy) affected live weight, carcass yield, shear value and color as evaluated by a sensory panel [4].

Numerically, ducks given diet 2 had lower carcass weight, lower carcass percentage and lower breast meat yield than those given diet 1. These results suggest that ducks given adequate phosphor in the diet 2 gave better carcass performance than ducks given diet inadequate P (diet 1). Omojola [4] stated that during finisher period, the requirement of avalaible Phosphorus for ducks is reduced from 0.45 to 0.30%. Pekin ducks had greater genetic potential for carcass performance in high altitude environment. However, abdominal fat percentage in Pekin ducks is significantly ($P<0.01$) higher than abdominal fat percentage in Muscovy ducks which has to be considered by consumers.

Table 2 shows live weight, carcass weight, carcass percentage, breast meat weight and abdominal fat weight of two different strains of ducks raised in different altitude.

**Table 2.** Live weight, carcass weight and carcass percentage of two strain of ducks given different diets in different altitude

| Variables measured | Low altitude | | | | | | | High altitude | | | | | | |
|---|---|---|---|---|---|---|---|---|---|---|---|---|---|---|
| | Muscovy | | | Pekin | | | Average in low altitude | Muscovy | | | Pekin | | | Average in high altitude |
| | Diet 1 | Diet 2 | Average for Muscovy | Diet 1 | Diet 2 | Average for Pekin | | Diet 1 | Diet 2 | Average for Muscovy | Diet 1 | Diet 2 | Average for Muscovy | |
| Live weight (g) | 1,461.00 ±32.34$^{a}$ | 1,508.00 ±32.35$^{b}$ | 1,484.50 ±32.34$^{a}$ | 1,736.22 ±32.36$^{b}$ | 1,801.23 ±32.35$^{b}$ | 1,768.73 ±32.34$^{b}$ | 1,626.61 ±32.34$^{b}$ | 1,904.00 ± 8.01$^{c}$ | 1,896.00 ± 8.01$^{c}$ | 1,900.00 ± 8.01$^{c}$ | 1987.01 ±7.36$^{c}$ | 1950.02 ±7.36$^{c}$ | 1,968.50 ±7.36$^{c}$ | 1,934.25 ±7.36$^{c}$ |
| Carcass weight (g) | 762.20 ±17.36$^{a}$ | 755.96 ±17.33$^{a}$ | 759.08 ±17.35$^{a}$ | 916.03 ±12.06$^{b}$ | 904.22 ±12.06$^{b}$ | 910.13 ±12.06$^{b}$ | 834.60 ±13.06$^{c}$ | 1077.76 ±39.06$^{b}$ | 1,048.49 ±39.08$^{b}$ | 1,068.13 ±39.06$^{b}$ | 1200.94 ±40.86$^{d}$ | 1160.06 ±40.86$^{d}$ | 1,180.50 ±40.86$^{d}$ | 1,124.31 ±40.86$^{d}$ |
| carcass percentage (%) | 52.17 ±3.08$^{a}$ | 50.13 ±3.06$^{a}$ | 51.15 ±3.08$^{a}$ | 52.76 ±2.88$^{a}$ | 50.20 ±2.88$^{a}$ | 51.48 ±2.88$^{a}$ | 51.32 ±3.08$^{a}$ | 57.13 ±2.76$^{b}$ | 55.30 ±2.76$^{b}$ | 56.22 ±2.76$^{b}$ | 60.44 ±3.16$^{b}$ | 59.49 ±3.16$^{b}$ | 59.97 ±2.76$^{b}$ | 58.09 ±2.76$^{b}$ |
| breast meat(g) | 231.21 ±12.26$^{a}$ | 221.48 ±12.26$^{a}$ | 226.35 ±12.26$^{a}$ | 207.12 ±11.34$^{a}$ | 219.26 ±11.34$^{a}$ | 213.19 ±11.34$^{a}$ | 219.77 ±11.34$^{a}$ | 251.17 ±8.14$^{b}$ | 247.23 ±8.14$^{b}$ | 249.20 ±8.14$^{b}$ | 243.00 ±9.15$^{b}$ | 239.01 ±9.15$^{b}$ | 241.01 ±9.14$^{b}$ | 245.10 ±8.14$^{b}$ |
| abdominal fat (g) | 1.48 ±0.14$^{a}$ | 1.55 ±0.14$^{a}$ | 1.52 ±0.15$^{a}$ | 2.34 ±0.13$^{b}$ | 2.30 ±0.13$^{b}$ | 2.32 ±0.13$^{b}$ | 2.17 ±0.13$^{b}$ | 2.25 ±0.13$^{b}$ | 2.18 ±0.13$^{b}$ | 2.22 ±0.13$^{b}$ | 2.30 ±0.16$^{b}$ | 2.89 ±0.16$^{c}$ | 2.60 ±0.13$^{c}$ | 2.41 ±0.13$^{c}$ |

*Values are means ± SEM*
*Means ± SEM followed by the same superscripts in each parameter are not significantly different at the 5% level*

## 4. Conclusion

In conclusion, Pekin ducks had greater genetic potential for growth carcass performance in high altitude environment. Live weight, carcass weight, carcass percentage and breast meat yields of two different strain of ducks given adequate phosphor diet had better carcass yields in high altitude compared with ducks raised in low altitude.

## References

[1] Fletcher D l 2002 Poultry meat quality *Worlds Poultry Science Journal* **58** : 131-145
[2] Baeza E, Dessay C, Wacenier N, Marche G and Listrat A 2002 Effect on selection for improved body weight and composition on muscle and meat characteristics in Muscovy ducks *British of Poultry Science Journal* **43** : 560-568
[3] Witkiewicz K, Kontecka H, Ksiazkiewicz J and Perz W 2004 Carcass composition and breast muscle microstructure in selected vs non selected ducks *Animal Science. Pap. Rep.* **22** : 65-73
[4] Omojola A B 2007 Carcass and organoleptic characteristics of duck meat as influenced by breed and sex *International Journal of Poultry Science* **6** : 329-334

# The utilization of endopower β in commercial feed which contains palm kernel cake on performance of broiler chicken

**S S A Purba[1] , M Tafsin[2], S P Ginting[2] and Y Khairani[3]**

[1]Graduate Student of Master Program of Livestock Science, Universitas Sumatera Utara
[2]Lecture of Master Program of Livestock Science, Universitas Sumatera Utara
[3]Student of Livestock, Science Faculty of Agriculture, Universitas Sumatera Utara
Indonesia
E-mail: purba_surya@yahoo.co.id

**Abstract.** Palm kernel cake is an agricultural waste that can be used as raw material in the preparation of poultry rations. The design used was Completely Randomized Design (CRD) with 5 treatments and 4 replications. Level endopower β used 0 % (R0), 0.02% (R1), 0.04% (R2) and 0.06% (R3). The results showed that R0a and R0b were significantly different from R3 in terms of diet consumption, body weight gain and the conversion ratio The utilization of endopower β in commercial diets containing palm kernel cake in broilers can increase body weight gain, feed consumption, improve feed use efficiency and even energy. It is concluded that utilization endpower β improve performances of broiler chicken fed by diet containing palm kernel cake.

## 1. Introduction

In the poultry industry, the ration cost component is the largest production cost component that can reach 70%. Therefore, many efforts are made to reduce the cost of ration so that the industry is more efficient and profitable for business. The high price of rations in the livestock business, then we need to find an alternative that can suppress and reduce the cost of a high ration. In this case with the utilization of Palm kernel cakewith a high level or reach 20% of the total ration can reduce the price of the ration. According [1], utilization of palm kernel cake (PKC) can be given up to 25% -30%.

Feed broiler chicken containing corn up to 40% this often causes expensive feed prices and compete with humans. To reach corn content of about 40% in feed, maize is often imported from other countries like from India this causes high feed prices. Along with the high price of feed more and more solutions are in can to reduce feed prices such as Palm kernel cake mixed to feed. The high nutrient content of Palm kernel cakeis also a factor used as an alternative to feed mixture of broiler where palm kernel cake contains metabolism energy 2.810 kcal / kg, crude protein 15.40% crude fat 6.49% crude fiber 15% this causes palm kernel cake Ogled as alternative feed to reduce high feed costs.

Feed industry still rarely uses palm kernel cake as in the ration mixture this is due to high coarse fiber, in dingding cell also contains mannosa of 56.4% [2], causing palm kernel cake is difficult to be broken and in Digested by broiler chickens, especially mannan found in gymnospermae. The purpose of this research is to know the influence of endopower β supplementation in feed containing palm kernel cake as much as 20% to consumption, body weight gain, feed conversion, income feed cost (IOFC) broiler.

## 2. Materials and methods

The research has been conducted in Animal Husbandry Laboratory of Animal Husbandry Study Program Faculty of Agriculture University of Sumatera Utara Medan. This study lasted for 35 days starting from November to December 2013. Materials used in the study is a day old chick (DOC) broiler as much as 160 heads. The diet material consists of commercial rations PT. Sabas Indonesia Feed Mill, palm kernel cake, fish meal and Endopower β. The tool used is a battery cage measuring 100cm x 100cm x 50cm, the number of cages as many as 20 units and each unit in the contents of 8 chicks aged one day and each unit is equipped with feed and drinking place. The research design used was a complete randomized design (CRD) consisting of 5 treatments and 4 replications and each replication consisted of 8 broiler. The treatments studied were:

R0a = Commercial diet (100%)
R0b = Commercial diet + 20% Palm kernel cake
R1 = Commercial diet + 20% Palm kernel cake+ 0.02% Endopower β
R2 = Commercial diet + 20% Palm kernel cake+ 0.04% Endopower β
R3 = Commercial diet + 20% Palm kernel cake+ 0.06% Endopower β

Parameters measured in the study are ration Consumption (g / head / week) ,the consumption of the ration is calculated on a weekly basis based on the difference between the amount of diet given and the amount of residual diets; Increase of Body Weight (g / head / week), measured by weighing the body every week minus the weight of the previous week; ration Conversion, ration conversion is calculated by comparing the amount of ration consumed by the weight gain achieved per week; Income Over Feed Cost (IOFC), obtained by calculating the difference in the income of farm business less the cost of ration. Revenue is the multiplication of weight gain due to the treatment (in Kg weight of life) to the selling price, while the ration cost is the cost incurred to produce the growth of the livestock body weight [3].

*Research phase: performance observation*

The cage used is a stage system, consisting of 27 units, each unit has 8 chicks (DOC). Before chicks are inserted, the cage is cleaned and disinfected with Rodalon and fumigation using 40% formalin and KMnO4. Before the chick is inserted into the cage, weighing first to determine the range of initial body weight to be used, then placed in the experimental unit.

The rationing material used consisted of commercial rations, palm kernel cake, fish meal and endopower β. The rationing material used was weighed first according to the composition of the ration structure specified in the formulation of each treatment. Chickens are kept in a heated and lighting treatment enclosure (incandescent 40 watt lamp). Rations and drinking water are administered in adlibitum. The data were collected every week during the study (7 weeks).

## 3. Results and discussion

### *3.1. Consumption of Rations*

The result of the study of endopower β in commercial ration containing palm kernel cake in broiler chicken to consumption can be seen in Figure 3.

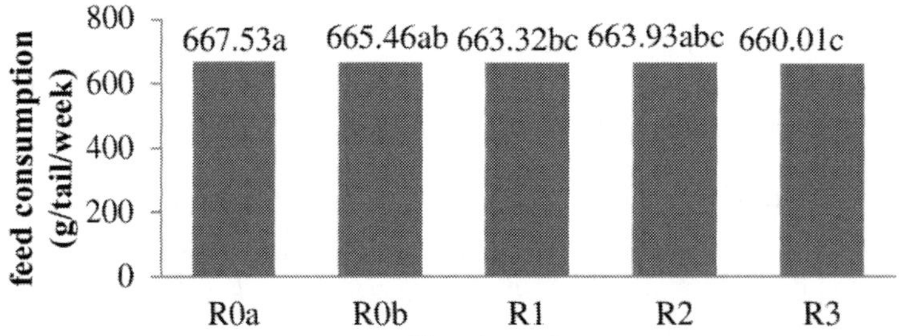


**Figure 1.** Average consumption of unit

The quality of ration in treatment of R0a was significantly higher than the treatment of R0b, R1, R2, R3 because the treatment of R0a was commercial feed having a standard whereas in the feed treatment R0b, R1, R2, R3 was a commercial feed mixture added by bus of 20 %. This affects the level of livestock consumption in this ration in accordance with the statement [4], which states that in consuming livestock ration is influenced by various factors, among others: age, palatability of ration, livestock activity, ration energy and protein level. Also determined by the quality and quantity of the given ration and its classification.

Feed consumption containing Palm kernel cakeis lower than commercial feed in the study [5], which uses 20% of palm kernel cake in basal feed produces lower consumption than commercial feed. The amount of difference in consumption of broiler chicken fed with a meal containing Palm kernel cakeis 32.61%. This is due to the high content of feed fiber in feed containing Palm kernel cakemay be the cause of low feed intake. According [6], the use of fiber needs to be limited, between 3% - 5%, because improper use can reduce the efficiency of feed.

*3.2. Body weight gain*

The addition of body weight can be calculated by weighing the body weight every week and reduced by the weight of the previous body.

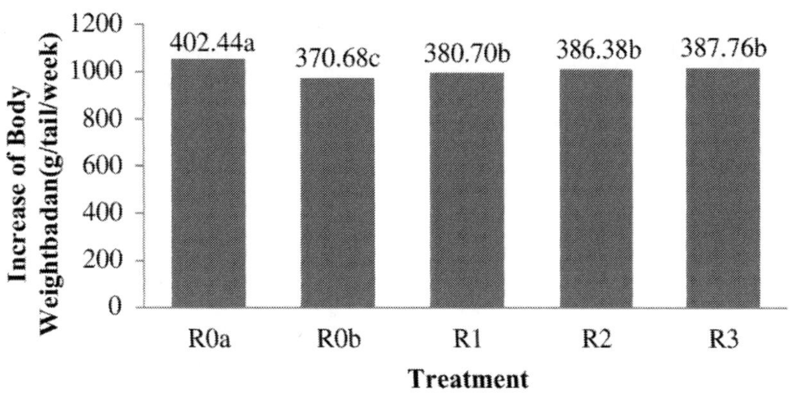


**Figure 2.** Average weight gain.

The addition of endopower β on treatment of R1, R2, R3 showed significant effect on weight gain. This suggests that the endopower β is also a kind of enzyme that contributes to the increase of feed digestion, so that the feed given endopower β can increase the weight of the broiler body according to the statement [7], Endopower β makes the productivity higher and more profitable . With Endo power β the digestibility of feed ingredients will increase. These results markedly lower feed costs, improving the appearance of livestock. The α-galactosidase enzyme contained in the endopower β will hydrolyze the complex sugars that can not be digested by the digestive system because broiler chickens can not produce α-galactosidase enzymes. The xylanase enzyme in the endopower β® will degenerate the polysaccharides, break down the cellulose and increase the nutrient content of the palm kernel meal contained in the feed. The xylanase enzyme breaks down the non-soluble nonpartic polysaccharides [8].

### *3.3. Feed conversion ratio*

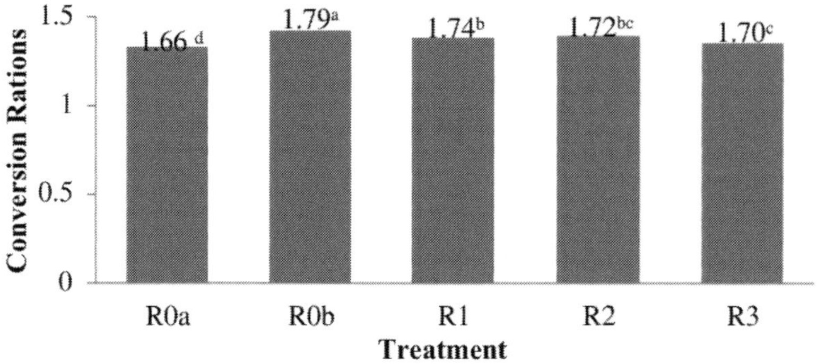


**Figure 3.** conversion of age ration 0-35 days

The lowest feed conversion is in the treatment of R0a by using commercial feed because of the difference in feed texture where in the treatment of R0b, R1, R2, R3 feed containing BIS where BIS contains high coarse fiber so that broiler can not efficiently absorb the nutrients contained in the feed And shells contained in the BIS mixture can damage the digestive system of the broiler. This is in accordance with the statement [9], which states a high crude fiber content (15%). Citrus cake meal is usually contaminated with shell pieces of palm with an amount of about 15-17%. This shell fragment has a very hard and sharp texture. This causes the substance to be less favorable to livestock and it is feared to damage the gastrointestinal wall in young cattle.

### *3.4. Income Over Feed Cost (IOFC)*

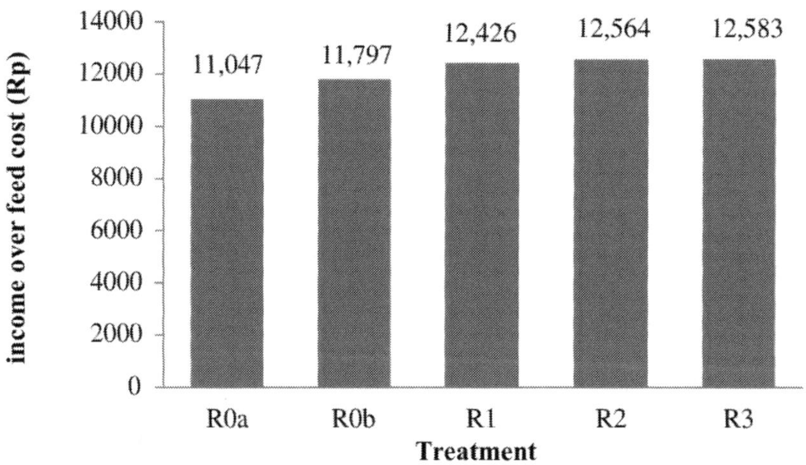


**Figure 4.** Average income feed cost (IOFC)

The results showed that increased doses of endopower β can increase the average income feed cost (IOFC) of broiler chicken ration. The average income feed cost (IOFC) in R1, R2 and R3 was higher than that of R0b without the addition of β endopower. The value of metabolism energy of R3 ration with dose of 0.06% endopower β higher than with R1 and R2 that is equal to Rp.12,583.

## 4. Conclusions

The use of endopower β in commercial rations containing palm kernel cake in broiler age 0-35 days can increase body weight gain, feed consumption and improve efficiency of use at level 0.06% endopower.

## References


*[1]* Sembiring P 2006 Bioconversion of palm oil factory waste with *Phanerochaete Chrysosporium* and its implication on broiler chicken performances *Dissertation* (Bandung: UNPAD)

*[2]* Daud M, Piliang W G and Kompiang I P 2007 Percentage and quality of carcass chicken broiler which is probiotic and prebiotic in rations *Jitv* **12**(3): 167-174

*[3]* Prawirokusumo S 1990 *Comparative Nutrition Science* (Yogyakarta: BPFE)

*[4]* Anggorodi R 1997 *General Food Science* (Jakarta: Gramedia Pustaka Utama)

*[5]* Vipassana 2013 Broiler chicken performed feeding contains powdery palm core flour or palm-core flour *Department of Nutrition Science and Feed Technology Faculty of Animal Husbandry* (Bogor: Institut Pertanian Bogor)

*[6]* Nurhayati 2006 Study of the effect of sugar and old levels of fermentation on quality of nata de soya *J. of Mathematics, Science, and Technology* **7**:40-47

*[7]* Easy Bio Systerm 2010 Endopower B. Union Center. Seoul. Korea

*[8]* Hughes B D 2003 Interaction of dietary calcium and protein in bone health in humans *Am. J. Clin. Nutr.* **133**:852s-854s

*[9]* Sundu B 2007 Proceedings of the National Seminar on Sustainable Use of Biological Resources Palu

# The potential and biological test on cloned cassava crop remains on local sheep

**R Ginting[1], S Umar[2] and C Hanum[2]**

[1] Postgraduate student of Livestock Science, Faculty of Agriculture, Universitas Sumatera Utara
[2]Lecturer of Postgraduate Program of Livestock Science, Faculty of Agriculture, Universitas Sumatera Utara
E-mail : ronistraginting@yahoo.co.id

**Abstract** This research aims at knowing the potential of cloned cassava crop remains dry matter and the impact of the feeding of the cloned cassava crop remains based complete feed on the consumption, the body weight gain, and the feed conversion of the local male sheep with the average of initial body weight of 7.75±1.75 kg. The design applied in the first stage research was random sampling method with two frames of tile and the second stage research applied Completely Randomized Design (CRD) with three (3) treatments and four (4) replicates. These treatments consisted of P1 (100% grass); P2 (50% grass, 50% complete feed pellet); P3 (100% complete feed from the raw material of cloned cassava crop remaining). Statistical tests showed that the feeding of complete feed whose raw material was from cloned cassava crop remains gave a highly significant impact on decreasing feed consumption, increasing body weight, lowering feed conversion, and increasing crude protein digestibility. The conclusion is that the cloned cassava crop remains can be used as complete sheep feed to replace green grass and can give the best result.

## 1. Introduction

Cassava plant (*Manihot utilisima*) is one kind of agricultural plants which has the potential to be grown in Indonesia. This plant is classified into *Euphorbiceace* family that is easy to grow even on dry and poor nutrient land and can resist either plant diseases or weeds. Cassava plant is easy to cultivate because the propagation of this plant usually applies branch cuttings and recently it even applies cloning technique. The production of cassava in Indonesia increases quite rapidly in the last five years from 19,321,183 tons in 2005 to 21,786,691 tons in 2009, or it increases 11.32% [1]. The increase of the production also increases the byproduct of cassava cultivation and its industry that it is potentially enough to be used as feed. The feed material comes from the cassava plant post crop products, such as cassava shoots, cassava stems, cassava tree bark and dried cassava. All are included in carbohydrate source feed that can be easily digested.

The cassava crop remains such as leaves, petioles, and young stems are much wasted at field. Some people have used the cassava crop remains as the greenery of cattle feed; for example local sheep. But the usage is limited because the crop remains are easily damaged and decayed on the field that an effort is needed to change the cassava crop remains to be the product for the raw material of cattle feed in the form of dry matter (DM). Then, it is processed to be complete feed in the form of pellet. The pellet has low content of water that it can be stored for a long time and it is practical to use for cattle.

Feed costs the most of all the production cost which takes around 70-80% of cost production [2]. The optimal use of local feed for agricultural products or the byproducts is hoped to decrease the cost for feed. Therefore, an effort is required to find the alternatives of feed material that are cheap, easy to get, good quality, and uncompetitive with human needs. The alternatives are leaves, cassava tree bark, and the byproducts of cassava agro-industry.

## 2. Material and Research Methodology

The first stage of research was conducted on a field of cloned cassava cultivation in Lau Tima village, Pancur Batu, Deli Serdang to measure the potential of dry matter and the second stage research was conducted at the laboratory of Livestock Biology, Livestock Department, the Faculty of Agriculture of the University of North Sumatera for biological test on sheep. The research was conducted for 4 months beginning from September until Desember 2013.

The sheep used in this research were 12 sheep with the average body weight was 7.75±1.75 kg, rations were cassava skin fruit, BIS, salt, cassava and mollasse. Drinking water was given in adlibitum way, drugs such as anthelmintic (*kalbazen*), anti bloating and vitamin.

12 units of individual sheepfold with its equipments, 12 feed and drinking water containers, scales of living weight which had the capacity of 50 kg with 50 gr sensitivity, scales which had the capacity of 2 kg with 10 gr sensitivity to weigh feed, feed chopper, cage cleaning tools like broom stick, shovel, knife and cutter, plastic pail, stationery, book of data and calculator.

### *2.1. Research Methodology*

This research was conducted in two stages. The first stage research was to get the potential of the dry matter of cloned cassava crop remains in one acre. The first stage research applied random sampling method by building two frames of tile in 1 acre field as the sample. Each frame consisted of 200 cloned cassava trees and after the data of each tile were taken then the average was measured and was multiplied with the number of cassava trees in one acre. The crop remains that were taken are leaves, leaf twigs and 30 cm long young stems from the shoots.

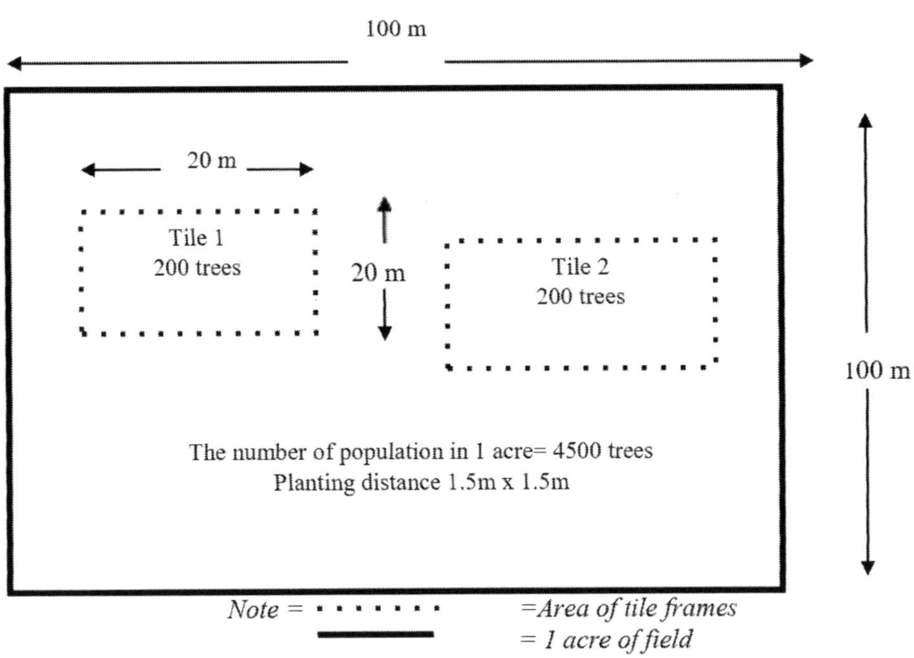


**Figure 1.** The way to decide tiles in 1 acre of cloned cassava field

**Table 1.** Feed formulation in research treatment

| Treatment | P0 | P1 | P2 |
|---|---|---|---|
| Grass | 100% | 50% | 0 |
| Complete Feed | 0 | 50% | 100% |

Table 2 contains the composition of nutrition of complete feed which was made of raw material mostly from cassava leaves and the young stems.

**Table 2.** The Composition of Complete Feed for the Second Stage Research

| Feed Material | Amount | CP | CF | TDN |
|---|---|---|---|---|
| Cassava Leaves | 50.00 | 8.56 | 9.73 | 38.25 |
| Young Stems | 20.00 | 1.23 | 7.59 | 12.95 |
| Dried Cassava | 5.00 | 0.10 | 0.82 | 3.85 |
| BIS | 20.00 | 3.08 | 2.10 | 16.20 |
| Molasses | 3.00 | 0.02 | 0.01 | 2.43 |
| Urea | 1.00 | 2.02 | 0.00 | 0.00 |
| Salt | 0.50 | 0.00 | 0.00 | 0.00 |
| Mineral | 0.50 | 0.00 | 0.00 | 0.00 |
| Total | 100.00 | 15.02 | 20.25 | 73.68 |

Table 3 below contains the comparison of crude protein (CP) and crude fiber (CF) of P0, P1 and P2 feed which were used as the feed for the sheep in biological test in the second stage research.

**Table 3.** Analysis of the Content of CP and CF of Feed Material of P0, P1, P2 for the Research

| Feed Material | Amount(%) | CP(%) | CF(%) |
|---|---|---|---|
| P0 | 100 | 10.62 | 23.25 |
| P1 | 100 | 12.82 | 21.75 |
| P2 | 100 | 15.02 | 20.25 |

(Cattle Feed Laboratory of PT. Sabas Indonesia feed mill Medan, 2014)

### *2.2. The Research Parameter*

The potential of the dry matter of cloned cassava crop remains in one acre is counted by building two frames of tile in 1 acre of field as the sample. Each tile contains 200 cassava tress and after that the data of dry matter are taken from the pieces of the crop remains such as the leaves, leaf twigs and 30 cm long young stems from the shoots in each tile then the average of the dry matter weight of each tree is taken in the form of dry matter and after that it is multiplied with the number of cassava trees in one acre that we can get the potential of dry matter in one acre in one period of 11 month-cloned cassava crop.

Body weight gain can be measured by dividing the difference of body weight (last weight – initial weight) with the length of weighing day. It is done once a week stated in gram per head per day.

$$DBWG = \frac{\text{last weigh t} - \text{initial weight (gr/head)}}{\text{length of keeping (day)}}$$

Feed consumption is achieved by counting the difference between the amount of feed given with the remains of the feed daily and is stated in gram per head per day.

Feed Consumption = feed which is given (in % of DM) – the remaining feed (in % of DM)

Feed conversion is measured by dividing the average number of dry matter consumption per head per day with the production of average number of body weight gain per head per day.

$$\text{Feed Conversion} = \frac{\text{Feed consumed (gr/head)}}{\text{DBWG (gr/day)}}$$

The crude fiber digestibility can be measured with the formula:

$$\text{CFD} = \frac{\text{CF Consumptio n - CF Feces}}{\text{CF Consumptio n}} \text{x100\%}$$

The consumption of feces excretion (CF) is achieved in the measuring time during the collection period within a week.

Crude protein digestibility can be measured with the formula:

$$\text{CPD} = \frac{\text{CP Consumptio n - CP Feces}}{\text{CP Consumptio n}} \text{x100\%}$$

CP Consumption and feces is achieved in the measuring time during the collection period within a week.

*2.3. The Research*

*2.3.1. The First Stage Research.* The crop remainings which are taken are leaves and leaf twigs and young shoot stems. The average length of the chopping from the shoots until the young stems is achieved by conducting repeated measuring. Then, all the leaves, leaf twigs and the length of the young stems which are taken is in accordance with the average of chopping that we get. Then the fresh weight is weighed then they are dried and then we get the dry weight. After the data of each tile are taken then the average potential of dry matter from each tree are taken in the form of dry matter and after that it is multiplied with the number of casava trees in one acre that we can get the potential of dry matter in one acre.

*2.3.2. The Second Stage Research.* The second stage research is done by doing the following steps: the preparation of sheepfold, sheep randomizing and the giving of feed and drinking water.

Before beginning the research, the sheepfold and all equipments are cleaned and washed. Then the liming is done on the floor and the wall of the sheepfold before the keeping process. Then the sheepfold and all the equipments are sprayed with *Rhodallon* (10 ml / 2.5 litre of water dosage).

The sheep used in this research are 12 sheep. The placement of the sheep is with random system which does not differ the sheep body weight. Before the weighing of the sheep initial body weight is done.

The feed given is fresh grass. The fresh grass is mixed with complete feed in the form of pellet and pelleted feed. Then feed and drinking water are given in adlibitum way. The water is changed everyday and the water container is cleanly washed. The crop remains are weighed to know the cattle consumption. Before the research is conducted, adaptation is done for a week.

*2.4. The Pathway of Complete Feed Processing*

The pathway of complete feed processing is shown in Figure 2.

IOP Conf. Series: Earth and Environmental Science **122** (2018) 012123 doi:10.1088/1755-1315/122/1/012123

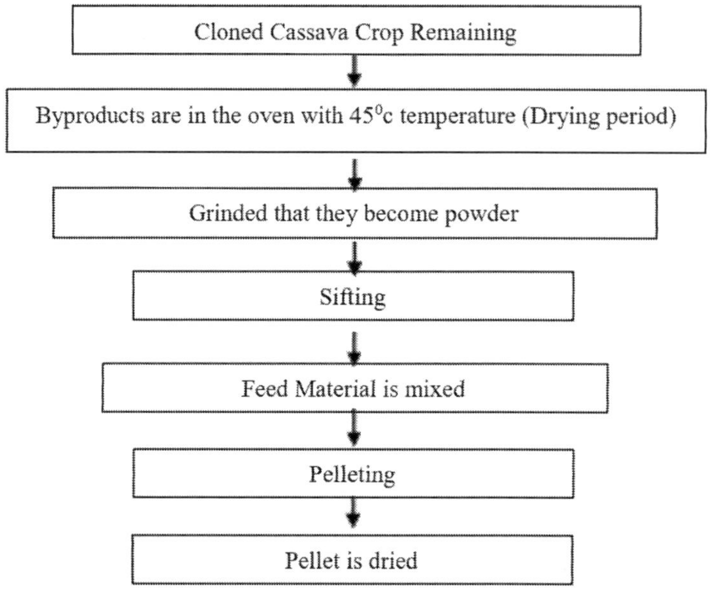


**Figure 2.** The pathway of complete feed proceesing

## 3. Result and Discussion

The result of the first stage research is measured in the form of dry matter which was taken from frames data of cloned cassava crop remaining. Table 4 is the data of the result of the first stage research to know the potential of the dry matter of cloned cassava crop remains in 1 acre of land.

**Table 4.** The list of recapitulation of result of the study of the potential of cloned cassava crop remains dry matter as the local male sheep complete feed

| Observed Parameter | Treatment | | |
|---|---|---|---|
| | P0 | P1 | P2 |
| The potential of crop remaining (kg/acre/crop) | | | 14265 |
| Feed Consumption (kg/head/3 months) | $43.01^{A}$ | $26.21^{B}$ | $22.37^{C}$ |
| Body Weight Gain (kg/3months) | $2.8^{A}$ | $3.26^{B}$ | $4.62^{B}$ |
| Feed Conversion | $15.45^{A}$ | $8.04^{B}$ | $4.90^{C}$ |
| Crude Fiber Digestibility (%) | $65.24^{a}$ | $64.68^{a}$ | $65.24^{a}$ |
| Crude Protein Digestibility (%) | $86^{A}$ | $88^{B}$ | $90.5^{C}$ |

Note : Superscript with different alphabet to the line shows highly significant difference (P<0.01)

**Table 5.** Cloned cassava crop remains

| Frames of Tile | Total of Fresh Weight(Kg) | Total of Dry Weight(kg) | The average of DM (kg / tree / crop) |
|---|---|---|---|
| I | 3161.6 | 632.32 | 3.2 |
| II | 3108.6 | 621.72 | 3.15 |
| Average | 3135.1 | 627.02 | 3.17 |

Source: Result of Study (2013)

### *3.1. The Potential of Cloned Cassava Crop Remaining*

Cloned cassava crop remains such as leaves, twigs and young stems as the raw material of complete feed are obtained with the average of 3.17 kg/tree/crop. With planting distance of 1.5m x 1.5m, the number of cloned cassava trees in 1 acre is 4500 trees. The number of cloned cassava population in 1 acre is multiplied by the average of dry matter of 1 tree that it is found the potential of cloned cassava crop remains as the raw material of complete feed per acre of cloned cassava field is 14265 kg/acre/crop. One Animal Unit (AU) needs 1140 kg of Dry Matter (DM) yearly that one acre of cloned cassava field is able to accommodate 12.5 Animal Unit (AU) or it is equivalent with 89 adult sheep (one adult male sheep is equivalent with 0.14 Animal Unit or one Animal Unit is equivalent with seven adult sheep). If we compare it with pastura grass in Indonesia, usually it can only support 1-2 Animal Unit/acre/year of cattle shepherding or it is equivalent with 7-14 sheep/goat/year. This shows that dry matter of cloned cassava crop remains has bigger potential than the dry matter of pastura grass [3].

Table 5 shows that the average of the dry matter of cloned cassava crop remains is 14,265 kg/acre/production. It means that cloned cassava crop remains can be used as the source of alternative feed to develop the integration of sheep farm with cassava that it can increase the additional value of cassava plantation. If it is compared to setaria green grass in a year, it can produce fresh green grass as much as 60-100 tons/acre/year (water content is around 70%) that cloned cassava crop remains has smaller potential compared to the special cultivation system of setaria grass which is made to be the source of sheep feed but the added value of the integration system is achieved.

### *3.2. Feed Consumption*

Feed consumption is cattle's capability to consume a certain amount of feed which is fed in a certain range of time. Feed consumption can be measured by the subtraction of the amount of the feed given with the feed remains. The feed given during the research is in the form of pellet and fresh green grass. The feed is given in *adlibitum* way. The consumption of weaning off local male sheep during the research can be seen in Table 4 and Figure 3.

Table 4 shows that the highest average of feed consumption is seen in P0 treatment as much as 43.01 kg/head/3 months and the lowest average of feed consumption is seen in P2 treatment as much as 22.37kg/head/3 months. The result of statistical analysis shows that Feed Treatment which is used is each of P0, P1, P2 feed treatments are significantly different ($P<0.05$) on the consumption of local male sheep feed.

Table 4 shows the lowest feed consumption is seen in P2. This is influenced by the form of feed that in P2, the feed is in the form of pellet. The feed in the form of pellet has lower level of consumption than the green grass feed. This is caused by the different level of density of feed where pelletized feed has higher level of density than green grass feed. The feed processing shows the decrease of feed consumption level. This is in accordance with Hofman [4] who says that the processing of pelletized feed will usually have effect of the increase of nutrition density in feed. The increase of feed density is one of the important factors in increasing feed efficiency.

### *3.3. Body Weight Gain*

Body weight gain is the difference of the last body weight with the initial body weight in a certain time (kg/head/3 months). Table 4 shows that the highest average of body weight gain can be seen in P2 treatment as much as 4.62kg/head/3 months or 51.33 gr./head/day and the lowest weight gain is seen P0 treatment as much as 2.8 kg/head/3 months or 31.11 gr./head/day. The feeding of feed by using pelletizing technology increases feed density and can remove the selection of certain feed components, therefore pelletization guarantees the achievement of certain nutritional intake more in accordance with what was calculated at the time of the making of ration formula. The increase of density and the removal of the selection of feed component affects nutritional intake that it is more efficient in body weight gain.

The feeding of feed using cassava byproducts based complete feed with 100% level shows significant effect on sheep body weight gain. This is because as a whole, cassava byproducts are easily degradated in rumen. Meanwhile P1 and P2 body weight gain does not show significant difference. This is because the comparison of nutritional content of P1 and P2 is not significantly different and with the existence of the mix of complete feed pellet in P1 feed. The composition of complete feed pellet which contains NPN (Non Protein Nitrogen) can trigger rumen microbe. Zinc content in cassava skin is really high compared to zinc content in the tuber. Cassava skin contains very high vitamin A which is as much as 11000 S.I and zinc content is as much as 2 mg each kg (DM), which indicates that cassava leaves can be used as quite good source of vitamin A and zinc [5], which is needed for sheep growth.

### *3.4. Feed Conversion*

Table 4 shows that the highest feed conversion average is seen in P0 treatment with as much as 15.451 and the lowest feed conversion is seen in P2 treatment with as much as 4.905.

Variance analysis shows that the treatment of giving P0, P1, P2 rations has different significant effects ($P<0.05$) on the conversion of local male sheep feed.

The smallest significant difference (SSD) test is conducted to find out the difference of P0,P1,P2 treatments on the conversion of local male sheep feed. The result of statistical analysis shows that Feed Treatment which is used is each of P0,P1,P2 feed treatments are significantly different ($P<0.05$) on the conversion of local male sheep feed.

Table 4 shows that the lowest feed conversion is seen in P2. This is affected by the form of feed where the feed in P2 is in the form of pellet. Feed in the form of pellet can increase fly fatty acid and carbohydrate digestibility. This thing causes the nutritional intake of cattle is higher than cattle that consume green roughage as much as 2-5 cm.

### *3.5. Crude Fiber Digestibility*

Table 4 shows that the digestibility of feed crude fiber in local male sheep does not show significant difference. High crude fiber digestibility is affected by the content of crude fiber in feed. Cassava byproducts based feed has no effect on the digestibility of feed crude fiber in local male sheep. This is because cassava byproducts based complete feed is easily degradated in the rumen that it can be similar with the digestibility of crude fiber which is used 100% green grass. After 24 hour incubation in sheep rumen, the loss of the dry matter of dried cassava skin is as high as 70 and 73% after ensiling process has been done. The lost of the dry matter of dried cassava becomes higher (83-84%) in 48 hour incubation.

The digestibility of crude fiber is affected by the content of crude fiber in feed and the microorganism capability in sheep rumen. The digestibility in P0, P1 and P2 treatments does not show significant difference. This is because the feed which is given in the form of complete feed is composed based on what sheep need.

### *3.6. Crude Protein Digestibility*

Table 4 shows that the highest crude protein digestibility is seen in P2 treatment and the lowest one is seen in P0 treatment. The high digestibility is affected by the quality of the feed where in P2 treatment, the feed is composed of 85% cloned cassava crop remains. Feed that contains cloned cassava crop remains such as dried cassava, dregs and cassava skin can increase carbohydrate digestibility level and increase the total of high digestibility. This is in accordance with Zinn and Depeters [6] who say that concentrate which contains dried cassava pellet can increase the total of digestibility as much as 91%.

The highest protein digestibility is seen in P2 treatment because the treatment uses the cassava byproducts complete feed. The cassava byproducts, especially the leaves, contain tannin. Tannin in cassava leaves and tropical plants is usually in the form of polyphenolics which is not easily dissolved in water but easily ties with protein in the form of tannin, high complex protein which is tied by hydrogen. But tannin can also increase the recycling of N in rumen that it increases microbe protein synthesis. This makes the crude protein digestibility higher. This is in accordance with Makkar [7] who says that toxic tannin for cattle can increase the recycling of N in rumen.

The content of protein in rumen also affects the digestibility level against crude protein. In P2 treatment the content of protein in feed is higher than the content of protein in feed in P0 and P1 treatment. This is in accordance with Tillman *et al* [8] who say that the crude protein digestibility depends on the content of protein in feed. Feed with low content protein, generally has low level of digestibility and feed with high content protein has high level of digestibility. The high and low level of protein digestibility depends on the content of protein in feed and the amount of protein which gets into digestive passage.

## 4. Conclusions

1. The potential of the dry matter of cloned cassava crop remains as the raw material of complete feed is as much as 3.17 kg/tree/crop.
2. The potential of the dry matter of cloned cassava crop remains as the raw material of complete feed in one acre is as much as 14,265 kg/acre/crop, one acre of cloned cassava field can accommodate 12.5 Animal Unit (AU) or similar with 89 adult sheep.
3. The crop remains such as leaves, leaf twigs and 30 cm long young stems of cloned cassava are processed to be the raw material of complete feed and can replace green grass.
4. The feeding of cloned cassava crop remains based complete feed has really significant effect on decreasing feed consumption, lowering feed conversion, increasing crude protein digestibility, increasing body weight gain and has no significant effect on crude fiber digestibility.
5. The result of the study shows that cloned cassava crop remains based pelletized feed can be fed 100% to local sheep and can give the best result.

## References

[1] Departemen Pertanian 2009 *Basis Data Pertanian* http://database.deptan. go.id/ bdsp
[2] Wahyu J 1988 *Ilmu Nutrisi Unggas* (Yogyakarta: Gadjah Mada University Press)
[3] Purnomo, Zulkifli A J, Prayitno S, Sudarjat S and Rahayu Y 2009 *Pelatihan Aplikatif Peternakan Domba Kambing* (Bandung: Agromania – Eka Agro Rama)
[4] Hofman R J 1995 *Trends.Ecol & Evol* The Changing Focus of Marine Mammal Conservation Vol 10 No.11:462-465
[5] Antari R dan Umiyasih U 2009 *Wartazoa* Pemanfaatan Tanaman Ubi Kayu dan Limbahnya secara Optimal sebagai Pakan Ternak Ruminansia. Vol. 19 No. 4 Th. 2009 192
[6] Zinn R A dan Depeters E J 1991 *Journal of Animal Science* Comparative Feeding Value of Tapioca Pellets for Feedlot Cattle **69**(12): 4726-4733
[7] Makkar H P S 2000 *Evaluation and Enhancement of Feeding Value of Tanniniferous Feed. In*: J.D. Brooker (Ed). Proc. Intern. Workshop on Tannins in Livestock and Human Nutrition. ACIAR Proc. No. 92: 71 –74
[8] Tillman A D, Hartadi H, Reksohadiprodjo S, Prawirokusumo S and Lebdosoekojo S 1998 *Ilmu Makanan Ternak Dasar* (Yogyakarta: Gadjah Mada University Press)

# Effectiveness duckweed *(Lemna minor)* as an alternative native chicken feed native chicken *(Gallus domesticus)*

**A Putra and M Z Ritonga**

Universitas Pembangunan Panca Budi, Jalan Jendral Gatot Subroto KM 4,5 Medan.
E-mail: mudhitaritongavet@gmail.com

**Abstract.** This study aimed to know the effectiveness duckweed as feed as native chicken (Gallus domesticus) on growth period (weight gain, feed intake and feed conversion). This research was conducted in Desa Telaga Jernih Kabupaten Langkat. The study was conducted in February 2017 until May 2017.This study use completely randomized design (CRD) with 4 treatments and 5 Replication, where each treatment consisting of 5 Native chickens unsexing. The treatment was used P0 = control (feed manufacturing), P1 = ration conventional with 10% duckweed, P2 = ration conventional with 20% duckweed, P3 = ration conventional with 30% duckweed. The parameters observed were weight gain, feed consumption and feed conversion. The results showed not significantly effect in body weight gain, feed consumption and feed conversion. Where the average of best weight gain on treatment P0 (control), P2 (20% duckweed), P3 (30% duckweed) and P1 (10% duckweed), average of best feed consumption in P0 (control), P2 (20% duckweed ) Of P1 (10% duckweed) and P3 (30% duckweed), P1 (10% duckweed) and P3 (30% duckweed), average of best feed conversion rate in P0 (control), P2 (20% duckweed) P1 (10% duckweed) and P3 (30% duckweed).

## 1. Introduction

In the need of protein needs, Indonesia's dependence on chicken meat is large, although, the need for chicken meat is dominated by chicken. While local chicken consumption is still low or only 6% of total national poultry production.

Association of Local Poultry Farmers of Indonesia (Himpuli), said that the government still considers chicken as complementary to the poultry industry, not as a commodity considered strategic for food security. The government until now still positioning this local chicken is just a complement, just there. There is no policy direction to make local chicken as a strategic source of animal protein such as beef or chicken (broiler). Though this chicken is genetic source of indigenous Indonesia and chicken is a native Indonesian Germplasm that should be utilized for national poultry.

Other countries such as China and Thailand, precisely prioritize the existence of their local chicken as a source of animal protein strategic country. In fact, he explained, Indonesia has many strains (species species) compared to other countries. The common problem faced by the chicken breeders generally is the availability of limited source of feed ingredients, the pattern of traditional village chicken maintenance system to reduce production costs in terms of feed availability.

The needs of the farmers for the fulfillment of animal feeds make the government attempt to import food ingredients such as corn, soybeans, fish meal and so on the gradually the price is more expensive and the farmers can not cover the greater output to be issued. It is reflected in the event. The economic recession of 1997 in which the large number of farmers who could not continue their livestock business and was forced to cover up the business due to soaring import food prices.

We can solve the problem of feed from alternative feed, we are encouraged to think creatively and try to dig, and utilize all the potential that exists to solve the problem such as by utilizing the potential of local as animal feed, the local feed must of course meet the criteria both in terms of aspects Nutrition, economics, social culture, and the sustainability. so that it can become a source of feed ingredients that are available, cheap, easy to obtain, not polluting, and still in accordance with the culture of society, so that later easy to be accepted among the community. One way to find alternative high-quality feed is easy to cultivate and does not require extensive land. It is expected that this feed material can substitute some of the use of imported feed ingredients.

Local feed is the source of feed ingredients whose existence is located in Indonesia and is located around the farmers communities whose amount of feed resources can meet as a source of feed ingredients. One of them is Lemna Minor Plant which its existence is often found in the community, especially found in rice field crops, this plant can be used as green manure , animal feed such as (Chicken, Duck, Rabbit, Horse, Goat, cow, Sheep, etc.), and can also be used as a fish feed (catfish, pomfret, etc.).

Lemna Minor is a species of water nail plant that floats mostly in stagnant waters, especially in paddy fields and in ponds, has a soft leaf surface easily develops and lives in symbiosis with Anabaena Lemna Minore that can fix Nitrogen (N2) from the air. Lemna Minor pinnata is a small plant that floats in water, looks triangular or rectangular. Lemna Minor measuring 2-4 cm x 1 cm, with branches, roots of rhizoma and floating leaves. The solitary root, suspended in water, hairy, 1-5 cm long, with a group forming 3-6 root hairs. The leaves are small, forming 2 rows, varying saturation, sitting attached, the lobe with the dorsal lobe holding on to the water surface and the ventral lobes float. In Indonesia Lemna Minor is known by the name of Mata catfish, while the local name Lemna Minor is "Mata Lele" (Java), "kakarewoan" (Sunda) its existence is Planteous, but not get good attention.

Lemna Minor plant has the characteristics of thick-green leafy green with a rather pale green edge, the growth of leaves overlap to form a cluster with a thickness of 1-3 cm, has a lot of spores. Lemna Minor is a type of water nail plant that floats mostly found in stagnant waters, especially in paddy fields and ponds, has a soft leaf surface easily

Lemna Minor plant has the characteristics of thick-green leafy green with a rather pale green edge, the growth of leaves overlap to form a cluster with a thickness of 1-3 cm, has a lot of spores. Lemna Minor is a type of water nail plant that floats mostly in stagnant waters, especially in paddy fields and ponds, has a soft leaf surface easily develops and lives symbiosis with Anabaena azolae that can fix Nitrogen (N2) from the air so it has the content N and high protein [1].

Lemna Minor is a nail-fern plant that floats on the water surface. Lemna Minor plant has a fairly high protein content of 28.12% dry weight [2], while [3] stated protein content in Lemna Minor sp for 23.42% dry weight with composition Complete essential amino acids. This plant can be used for livestock feed as well as ruminants because of its high protein content.

Several studies have found that the growth rate of Lemna Minor is 0.355 - 0.390 grams per day (in laboratory) and 0.144 - 0.860 grams per day (in the field). In general, maximum Lemna Minor biomass is achieved after 14 -28 days after inoculation.

From the various information it is increasingly clear that Lemna Minor has the potential to be utilized and developed in Indonesia as a source of feed ingredients when viewed from the nutrient content, the ability to breed and its usefulness as a feed ingredient can increase the rabbit cut weight, thus also needed further research on the utilization Lemna Minor as animal feed, for example a mixture of chicken, pork, sheep or goat Feed.

**Table 1.** Percentage of nutrients Lemma Minor of some sources

| Nutrient Lemna Minor | Source (%) | | | Average |
|---|---|---|---|---|
| | A | B | C | |
| Dry Matter | - | 6 | 4.93 | 5.42% |
| Crude Protein | - | 20 | 29.9 | 23.24% |
| Calcium | 1 | 1.1 | - | 1.3% |
| Fosfor | 1.4 | 0.5 | - | 1.3% |
| Crude Fat | 4 – 6 | 3.8 | 5.33 | 3.43% |
| Crude Fiber | 7 – 10 | 16 | 9.6 | 11.53% |
| Ash | 8 – 14 | 17 | 15 | 13.23% |

Source [$4^A$.$5^B$.$6^C$]

## 2. Materials and Method

The materials to be used in this research are:
- Lemna Minor
- Native chicken.
- Starter Feed and finisher feed.
- Trial Cage

### *2.1 Research design*

The research model used in the study was a complete randomized design (CRD) with 5 treatments with 4 replications. The treatment provided is as follows:
P0 = Control Feed (without addition of Lemna Minor)
P1 = Feed with addition of Lemna Minor 5%
P2 = Ration with the addition of Lemna Minor 10%
P3 = Ration with the addition of Lemna Minor 15%
P4 = Ration with the addition of Lemna Minor 20%

### *2.2 Research sites*

This research was conducted in Klambir Lima Village. Hamparan Perak Sub-district. Deli Serdang Regency. North Sumatera Province.
Parameters observed
1.Average Feed Consumption
2.Average Increase of Body Weight
3.Average Feed Conversion

## 3. Results and Discussion

### *3.1. Daily Gain*

Many factors influence the growth of livestock, especially native chicken. Native chicken is a local livestock with genetic ability to resist the surround environment. but relatively longer growth rate compared to other chicken livestock such as broiler Therefore, the Native chicken can utilize the available local feed sources to sustain their lives and to produce. The result of research that have been done by utilizing Lemna Minor on daily weight gain (g /head/day), ration consumption (g/head/day) and to feed conversion can seen in Based on Table 2.

Table 2 shows that the result of observation of average daily gain increase is not significant effect. It can also be said that the use of Lemna Minor with the use level 0% up to 20% in native chicken gave the same result. As it is known that Lemna Minor is a plant species that lives symbiosis with anabaena aziolae that can fix Nitrogen (N2) from air so it has high N and protein content [1] . Lemna Minor plant has a fairly high protein content of 28.12% dry weight [2] . while [3] indicated protein content in Lemna Minor sp of 23.42% Dry with a complete essential amino acid composition. This plant can be used for livestock feed as well as ruminants because of its high protein content. In addition to the high protein content as animal feed, but Lemna Minor also contains a high level of crude fiber is also high. according to the analysis of Biochemistry Nutrition Laboratory Faculty of Animal Husbandry University of Gajah Mada. the content of Lemna Minor flour is crude protein 17.95% . Crude fat 5.67% ash 28.95% crude fiber 18.74% and TDN 64.93%. This causes the addition of Lemna Minor to give the level of weight gain of different body is not real on the results of research observations.

**Table 2.** Recapitulation of Observation Results

| Treatment | Obsevation | | |
|---|---|---|---|
| | Average Daily Gain | Feed Consumption | Feed Converstion |
| P0 | 11.53[tn] | 47.45 [tn] | 4.12 [tn] |
| P1 | 11.56 [tn] | 47.07 [tn] | 4.08 [tn] |
| P2 | 11.94 [tn] | 47.10 [tn] | 3.96 [tn] |
| P3 | 11.30 [tn] | 46.82 [tn] | 4.15 [tn] |
| P4 | 11.19 [tn] | 46.62 [tn] | 4.18 [tn] |

Description: [tn]= Show different results not real (Source data is processed)

*3.2 Feed Consumption*

The result of average daily gain is not significant effect in the observation results also caused by the level of consumption of animal feed itself. In the observation results. it is also seen that the level of feed consumption of Native chickens using Lemna Minor shows not significant effect it is caused by the level of palatability of Lemna Minor which is similar to other food ingredients also derived from local food sources Such as rice bran corn flour and onggok.

Tables in general prefer feeds that contain low crude fiber. Factors that affect consumption is the nutritional quality and palatability of feed. Low palatability feed will be consumed in a limited way to meet the basic needs of livestock [7], fiber content in feed also affects feed consumption feed with high levels of high fiber have voluminous properties in livestock. Palatability affects the amount of feed consumption. High palatability feeds favored by livestock make increasing feed consumption, while low palatability feeds have relatively low feed intake [8].

Feed with high coarse fibers causes bulky properties, causing the rate of digestion in the slow rumen. The slow motion of digestion leads to lower feed intake because feed is in the rumen longer. Low-coarse grain feed material has a fast rate of digestion, so the feed can leave the rumen quickly and more feed in or consumed [9].

*3.3. Feed Conversion*

Feed conversion ratio (FCR) is a benchmark for assessing the efficiency of ration usage. The smaller the value of FCR, the higher the efficiency level of Feed use. The result of observation in this research is found that the feed conversion rate of native chicken shows not significant effect with addition lemna minor level 0% -20% feed conversion equal to 3.96 – 4.18.

Conversion of feed is the ratio between the amount of feed consumption and the increase of body weight [10]. Calculation of the conversion of native chicken ration can be seen in the Table 2. The smaller conversion value of the ration is more efficient to converting feed into the form of meat. The amount of feed consumed during the average maintenance period of 90 days from DOC (Day old Chick) to harvest is 4,231.10 grams per head for 90 days. [11] stated that the conversion of ration in 8 weeks old native chickens used rations with 17% and 20% protein content of 2.84 and 4.32.

[12]. states that some of the major factors affecting the conversion of rations are genetic quality of rations diseases temperature cage sanitation ventilation treatment and cage management. Factors of rationing lighting also take effect in influencing the feed conversion the feed process in gastrointestinal tract the physical form of ration and the composition of nutritional rations. [13] opinion that the smaller the conversion of ration means the feeding of rations is more efficient but if the conversion of rations is enlarged then there has been wastage. According [10] feed conversion affected several factors such as the age of strain the nutritional content of the ration the state of temperature and state of livestock. management and use of good seeds.

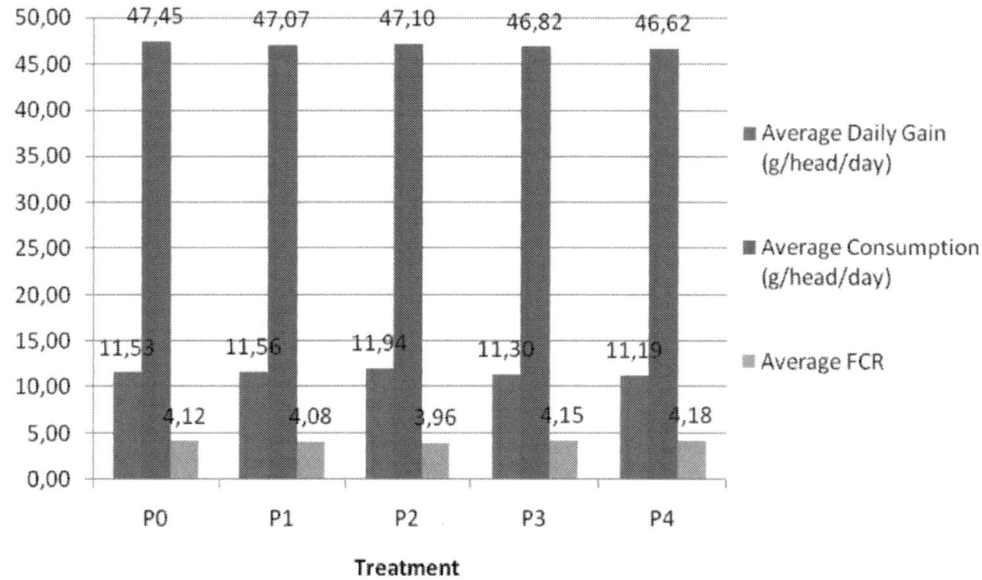


**Figure 1.** Recapitulation of research results

## 4. Conclusion

Based on the results of observations that have been carried out it can be concluded that the use of Lemna Minor as feed ingredients give a result that is not different from other conventional feed ingredients. The use of Lemna Minor can reduce /decrease cost of native chicken production due to the use of Lemna Minor does not require a large cost and easily obtained around paddy fields.

## References

[1] Imanuddin S 2007 *Pengaruh Pemberian Azolla terhadap Produksi Padi Sawah dan Jagung (Effects of Azolla Aplication to Rice Paddy and Corn Production* [Disertasi] Universitas Sumatera Utara Medan p 39
[2] Handajani H 2000 *Peningkatan kadar protein tanaman Azolla microphylla dengan mikrosimbion Anabaena azollae dalam berbagai konsentrasi N dan P yang berbeda pada media tumbuh* Tesis Program Pasca Sarjana Bogor
[3] Djojosuwito S 2000 *Azolla : Pertanian Organik dan Multiguna (Azolla : Organic Farming and Multifuction)* (Yogyakarta: Kanisius)
[4] Yılmaz E. Akyurt I and Günal G Use *of Duckweed. Lemna minor. as a Protein Feedstuff in Practical Diets for Common Carp. Cyprinus carpio. Fry.* Turkish Journal of Fisheries and Aquatic Sciences 4: 105-109 (2004)
[5] Anjana B and Matai S 1990 Composition of Indian aquatic plants in relation to utilization as animal forage *Journal of Aquatic Plant Management (USA)* **28**: 69-73
[6] Hassan M S and Edwards P 1992 Evaluation of duckweed (lemna perpusilla and spirodela polyrhiza) as feed for nile tilapia *(Oreochromis niloticus) Aquaculture* 104: 315-326
[7] Simanhuruk K dan Sirait J 2010 Silase kulit buah kopi sebagai pakan dasar pada kambing boerka sedang tumbuh *Jurnal Seminar Nasional Teknologi Peternakan dan Veteriner*

[8] Iskandar S. Resnawati H. Zainuddin D. Raharjo Y C dan Gunawan B 1998 Performance of pelung x kampung (pelung cross) chickens as influenced by dietary protein *Bulletin of Animal Science* Gadjah Mada University Yogyakarta
[9] Astuti P. Sampoerno and Ardian 2015 Uji beberapa konsentrasi pupuk cair azolla pinnata pada bibit kelapa sawit *(Elaeis Guineensis Jacq.)* di pembibitan awal Departement Of Agroteknologi Faculty Of Agriculture University Of Riau
[10] Anggorodi R 1985 *Ilmu Makanan Ternak Unggas* (Indonesia: Penerbit Universitas)
[11] Husmaini 2000 Pengaruh peningkatan level protein dan energi ransum saat refeeding terhadap performans ayam buras *Jurnal Peternakan dan Lingkungan* Vol.**6**(01)
[12] Lacy M dan Vest L R 2000 Improving feed conversion in broiler : a guide forgrowers.http://www.ces.uga.edu/pubed/c:793-W.html [6 Januari 2007]
[13] Rasyaf M 1994 *Beternak Itik Komersial* (Yogyakarta : Kanisus)

# Improved quality of frozen boer goat semen with the addition of sweet orange essential oil on tris yolk and gentamicin extender

**S A Sitepu[1], U Zaituni[2], Jaswandi[2] and Hendri[2]**

[1]Department of Animal Husbandry, Faculty of Agriculture, Pembangunan Panca Budi University, Indonesia
[2]Faculty of Animal Husbandry, Universitas Andalas, Indonesia.
E-mail : sukmaadityasitepu@yahoo.com

**Abstract**. This research aimed to determine the extent of frozen semen quality Boer Goat by essential oils of sweet orange peel in tris yolk and gentamicin extender. Research has been conducted at the Laboratory Loka Penelitian Kambing Potong Sei Putih, Deli Serdang, North Sumatra in February 2017. This study used a completely randomized design with 4 treatments and 5 replications. Treatments are 0.25; 0.5; 0.75 and 1% essential oils as additional diluent. The parameters were measured percentage Motility, membrane integrity, acrosome integrity and viability Boer Goat frozen semen. The results showed that the addition of essential oils as diluent semen was significant (P <0.01) in the percentage motility, Viability, membrane integrity and acrosome integrity Boer Goat frozen semen. Motility, membrane integrity, acrosome integrity and viability was significantly higher in all treated groups than the control group. The best results of all treatments In the study was the addition of essential oil as much as 1%.

## 1. Introduction

One effort to meet the needs of meat is to develop a broiler goat farm. The current constraints are the low population and genetic quality of local goats. The growth of local goat meat is still very low, but it has the advantage of easily adapt to the environment and the food was of low quality. It is necessary to consider ways to improve the population and genetic quality of local goats.

Efforts that can be made to increase the population and quality of livestock is to make artificial Insemination by using sperm from superior goats. One of the superior goats that can be used is the Boer Goat. Boer Goats have several advantages such as high body weight, high carcass, litter size 1.7 and rapid growth. It is expected that artificial insemination will increase the population and genetic quality of goats in Indonesia.

Currently the success of the artificial insemination program is still low. This is due to the low quality of frozen semen used, the low inseminator skills and the location of goat farms in rural communities is difficult to reach. The low quality of frozen semen is caused by damage to spermatozoa caused by improper handling of clotting processes. The most critical part of the process of freezing semen is when freezing and thawing. Besides the low quality of frozen semen is also caused by the growth of bacteria that can harm and kill spermatozoa.

The bacteria contamination in Boer goat cement is caused by many factors ranging from the holding process to the freezing. Sperm begins contaminated with bacteria ranging from male reproductive tract, cement holding process, dilution process, cooling and freezing. Bacterial contamination can also occur through the air, the materials and equipment used in the manufacture of frozen semen. By adding the antibacterial ingredient in frozen semen diluents boer goat expected to minimize the growth of bacteria.

Generally, to inhibit the growth of bacteria in frozen semen is to add antibiotics, such as gentamicin. Gentamicin can work on gram-negative bacteria. Gentamicin has been used on a commercial diluent is Andromed® [1]. In addition to antibiotics, sweet orange essential oil can also be used as an antibacterial on Boer goat frozen semen [2]. Essential oil of sweet orange peel contains the main components include limonene and linalool [3] that are toxic to bacteria [4] Research by the addition of essential oils of sweet orange peel in tris yolk and gentamicin extender on frozen semen boer goat's important to get quality frozen semen Boers better to minimize the growth of bacteria, able to maintain the quality and durability of spermatozoa.

## 2. Materials and methods

### *2.1. Place and Time Research*

Research has been conducted at the Laboratory Loka Penelitian Kambing Potong Sei Putih, Deli Serdang, North Sumatra in February 2017.

### *2.2. Materials*

Materials Research Used Boer goat bucks (approximately 3–4 years old) were used in the study, sweet orange essential oil, tris yolk and gentamicin as an ingredient diluent, citric acid, fructose, eosin 2% Media as observations of viability, formolsaline for observations acrosome integrity, akuabidestilata, Hypo osmotic Swelling Test (HOST) for observation membrane integrity, liquid N2 to freeze semen review, straw packing for a review of the cement and wipes for cleaning equipment.

### *2.3. Methods*

This study used a completely randomized design with 4 treatments and 5 replications. The treatments are:

$P_0$ = Tris Yolk + Gentamicin + Sweet Orange Oil 0.%
$P_1$ = Tris Yolk + Gentamicin + Sweet Orange Oil 0.25%
$P_2$ = Tris Yolk + Gentamicin + Sweet Orange Oil 0.5%
$P_3$ = Tris Yolk + Gentamicin + Sweet Orange Oil 0.75%
$P_4$ = Tris Yolk + Gentamicin + Sweet Orange Oil 1%

### *2.4. Evaluation of microscopic sperm parameters*

The observations made are:

Motility: the percentage of spermatozoa moving progressively forward. Evaluation is done by observing spermatozoa in eight different visual field with a light microscope magnification of 400 times [5].

Viability: evaluated using the eosin staining [5]. Live spermatozoa characterized by a head that do not absorb the dye. while the death was marked by a red head. The evaluation was done at a minimum of 200 spermatozoa were observed using a light microscope magnification of 400 times.

Membrane integrity: Evaluation of the integrity of the plasma membrane of the spermatozoa. The evaluation was done using methods hypoosmotic swelling test (HOST). Tests done by mixing 0.1 ml to 9.9 ml cement hypoosmotic medium. Once mixed. the preparations are incubated in a water bath temperature of 37 ° C for 30 minutes [6].

Acrosome integrity: Evaluation of the integrity of the sperm acrosome hood characterized by head spermatozoa black thick cement when exposed in physiological saline solution containing 1% formalin [7]. The evaluation was done at a minimum of 200 spermatozoa by using a light microscope magnification of 400 times.

## 3. Results and discussion

Evaluation of the quality of frozen semen Boers done before freezing and after freezing. Observations included the percentage of motility. Viability. Membrane Integrity and Acrosome Integrity. The results of the study can be seen in Table 1.

**Table 1**. Summary results of research quality frozen semen Boers.

| Parameters | Treatments | Observation | |
|---|---|---|---|
| | | Before Freezing | After Freezing |
| Motility | 0% | 71 | 47 |
| | 0.25% | 73 | 49 |
| | 0.5% | 75 | 51 |
| | 0.75% | 76 | 51 |
| | 1% | 77 | 53 |
| Viability | 0% | 79 | 67 |
| | 0.25% | 80 | 71 |
| | 0.5% | 83 | 73 |
| | 0.75% | 85 | 74 |
| | 1% | 86 | 75 |
| Membrane Integrity | 0% | 80 | 65 |
| | 0.25% | 82 | 67 |
| | 0.5% | 84 | 67 |
| | 0.75% | 86 | 69 |
| | 1% | 87 | 69 |
| Acrosome Integrity | 0% | 75 | 59 |
| | 0.25% | 78 | 61 |
| | 0.5% | 80 | 63 |
| | 0.75% | 81 | 64 |
| | 1% | 83 | 66 |

Explanation: *Different superscripts in the columns show very significant differences (P <0.01)*

Based on the results of the research. it is known that it is raising Boer goat frozen semen quality with the addition of sweet orange essential oil in tris diluents gentamicin yolk extender. The result of variance analysis showed that the influence of essential oil as diluent had a very real effect (P <0.01) on sperm motility before both freezing and after freezing. Further BNT test results show that the highest motility. viability. membrane integrity and acrosome integrity are found in the addition of 1% essential oil and the lowest without the addition of essential oil.

Motility is one determinant of the success of the sperm to reach the ovum in the fallopian tube channel and is a measure that is used as the ability of sperm to fertilize an egg. The entire treatment has the requirement to be stored and IB as stated by Evans and Maxwell [8]. a condition worthy of frozen semen motility IB have not less than 40% and have the appropriate national standards of Indonesia [9]. The higher the concentration of essential oil of sweet orange peel. the higher the percentage of sperm motility before and after freezing. Boer Goat Semen is very susceptible to damage during the freezing process for the formation of ice crystals that lead to the death of spermatozoa. During the freezing process ice crystals formed cement will cause the electrolyte concentration increases in the cell that would dissolve the cell wall sheathing lipoprotein spermatozoa and the thawing time will change so that the sperm plasma membrane permeability die [5]. Besides reduction in sperm motility was also caused by the treatment that causes damage and death of spermatozoa.

Spermatozoa are very susceptible to cell damage due to osmotic pressure changes suddenly due to the rapid melting during the thawing process. Only spermatozoa that have the power of the plasma membrane of the strong can survive. The decrease is also due to the reduced motility of energy supplies spermatozoa used to sustain life and support the movement of spermatozoa [10].

The big difference in sperm motility in each treatment may be due to the high level of use of essential oils of sweet orange in the diluent resulting in lipid peroxidation. According to [11] that hydroxynonenal is one of the lipid peroxidation can inhibit glycolysis and motility. In addition to the damage caused by lipid peroxidation. decreased motility can also occur as a result of several factors such as changes in medium pH. osmotic pressure and the effects of electrolytes and non electrolytes [5].

The decline in the average percentage of each evaluation time is a reasonable decline. According Toelihere [5] on the freezing cement will occur spermatozoa mortality up to 30% of the amount of fresh spermatozoa. Beconi *et al.* [12] states that the addition of cement with good quality antioxidants will maintain the vitality of frozen sperm. but not so in low-quality sperm because peroxidation process can not be treated with antioxidants. In the sweet orange essential oil also contains antioxidants because there are flavonoid compounds therein.

Cold shock is one of the main factors for the cryopreservation of sperm cells which can reduce cell viability. The level of cell sensitivity to cold shock is affected by the cooling rate and temperature interval [13]. Cold shock due to decreased body temperature to below 0 ° C. which would reduce cell viability. The phenomenon of cold shock on cells is not known clearly. but probably related to the phase transition of the membrane lipids that cause phase separation and inheritance - selective permeability of biological membranes of living cells [13].

The results showed a decrease in the quality of spermatozoa from cooling to freezing and thawing. Decrease in sperm motility before freezing decline caused by factors egg-yolk coagulating enzyme in seminal plasma of goats that are toxin and cold shock. The decrease is also due to the reduced motility of energy supplies spermatozoa used to sustain life and support the movement of spermatozoa. Decrease in sperm motility was also caused by the treatment that causes damage and death of spermatozoa. During the thawing process spermatozoa susceptible to cell damage due to osmotic pressure changes suddenly due to rapid melting. Only spermatozoa that have a strong ability of the plasma membrane that can withstand [10].

The percentage of spermatozoa with high life shows the percentage of intact plasma membranes are also high. Described Tambing [14] that the percentage of live spermatozoa with high indicates that the plasma membrane is still physically intact. so that the sperm cell organelles will be protected. the ions to the metabolic processes and the needs of food substances available. The concentration of lactose must cause a change in osmotic pressure on the diluent in the direction of hypertonic signaling molecules or particles outside cells more than in cells resulting in the expenditure of water from inside the cell to dilute molecules outside the cell that causes the cell will shrink [15].

Intact Membranes Plasma breakdown is usually accompanied by damage to the hood so that acrosome acrosome Whole Hood observation is important because damage can result in loss of the acrosome enzymes proteolytic enzymes and occur mostly during melting back that led to the failure of Artificial Insemination [16].

Another factor that determines the quality of spermatozoa when done thawing at the optimal temperature is glycerol intracellular diffusion occurs more rapidly and prevent the incidence of osmotic shock. According to Evans and Maxwell [8] thawing the frozen semen goat conducted at a temperature of 370C and not through a critical time limit. Thawing frozen semen are done properly will maintain osmotic balance and improve the configuration of the cell membrane lipid protein spermatozoa during the freezing process [17].

## 4. Conclusion

The addition of essential oils of sweet orange peel in tris diluents and gentamicin yolk extender on frozen semen boer goat's can improve quality frozen semen Boers better to minimize the growth of bacteria. able to maintain the quality and durability of Boers spermatozoa.

## References

[1] Minitub 2001 Certificate Andromed. Minitub Abfullund Labortechnik GmbH and Co KG. Germany.
[2] Sitepu S A. Zaituni U. Jaswandi and Hendri 2015 Pemanfaatan minyak atsiri kulit jeruk manis dalam pengencer tris kuning telur terhadap kualitas semen kambing boer setelah pengenceran dan eqiulibrasi *AGRIPET Jurnal Peternakan* Vol **11** No 2 Medan
[3] Agusta A 2010 *Minyak Atsiri Tumbuhan Tropika Indonesia* (Bandung: ITB)
[4] Fisher K and Phillips C A 2006 The effect of lemon. orange and bergamot essential oils and their components on the survival of Campylobacter jejuni. Escherichia coli O157. Listeria monocytogenes. Bacillus cereus and Staphylococcus aureus in vitro and in food systems *J Appl Microbiol.* 2006 Dec;**101**(6):1232-40
[5] Toelihere M R 1993 *Fisiologi Reproduksi pada Ternak* (Bogor: IPB Press)
[6] Rodriguezgil J E A Montserrat and Rigau T 1994 Effects of Hypoosmotic incubation on acrosome and tail structure on canine spermatozoa Theriogenology
[7] Saacke R G and White J M 1972 Semen quality tests and their relationship to fertility. Proceeding 4th Tech Conf on AI and Reprod NAAB.
[8] Evans G and Maxwell W M C 1987 *Salamon's Artificial Insemination of Sheeps and Goats* Butterworths London
[9] Badan Standardisasi Nasional 2014 *Standar Nasional Indonesia* (Jakarta: Badan Standardisasi Nasional) SNI 4869.**3**: 2014
[10] Maxwell W M C and Watson F P 1996 *Recent Progress in the Preservasion of Ram Semen* Animal Reproduction
[11] White I G 1993 Lipids and calcium uptake of sperm in relation to cold shock and preservation : a review *Reprod. Fertil. Dev.* **5**: 639-658
[12] Beconi M T. Francia C F. Mora N G and Afranchino M A 1993 Effect of natural antioksidant on frozen bovine semen preserpation *Therionology* **40**: 841-851
[13] Watson P F 2000 The caused of reduced fertility with cryopreserved semen *Anim Reprod Sci.* 60-61:481-492
[14] Tambing S N 1999 Efektivitas berbagai dosis gliserol dan waktu ekuilibrasi terhadap kualitas semen beku kambing Peranakan Etawah *Thesis Pascasarjana IPB*-Bogor
[15] Medeiros C M. Forell F. Oliveira A T and Rodrigues J L 2002 Current status of sperm cryopreservation: why in n't better *Theriogenology* **57**: 327-344.
[16] Valcarcel A. De Las Heras M A. Perez L. Moses D F and Baldassarre H 1997 Assessment ofthe acrosomal status of membrane-intact ram spermatozoa after freezing and thawing by simultaneous lectin/Hoechst 33258 staining *Anim. Reprod. Sci.* 45: 299-309
[17] Farstad W 1996 Semen Cryopreservation in dogs and foxes *Anim. Reprod. Sci.* 42 : 251-260

IOP Conf. Series: Earth and Environmental Science 122 (2018) 012126 doi:10.1088/1755-1315/122/1/012126

# Potential reproduction and response of selenium and zinc mineral supplementation on quality of goat samosir semen

**P Siswoyo[1], M Tafsin[2] and R Handarini[3]**

[1]Post-Graduate Student of Animal Science, Faculty of Agriculture, Universitas Sumatera Utara
[2]Animal Science, Universitas Sumatera Utara
[3]Animal Science, Universitas Djuanda
E-mail: woyo04@gmail.com

**Abstract.** The present study was conducted to investigate the effect of suppllementattion of selenium and zinc on semen quality and growth of samosir goat. The experimental design used latin square design (4x4). The treatment supplementation mineral on multi nutrient block (MNB) composed of without sipplementation (p0), +10ppm selenium (p1), +10ppm zinc (p2), +10ppm selenium and +10ppm zinc (p3). The result showed that supplementation mineral selenium and zinc increased significantly ($p<0.05$) average daily growth, feed consumtion, and lower feed convertion ratio. Semen quality of goat were supplemented by selenium and zinc influenced motility, viability, volume concentration, and responding hypo osmotic swelling (HOS). Combination supplementation selenium and zinc significanly had higher semen quality than ither treatment. It is concluded that supplementation selenium and zinc improve growth and semen quality of samosir goat.

## 1. Introduction

There are currently eigth local goat breeds in Indonesia that have been characterized based on the phenotype samosir, kacang, marica, muara, kosta, gembrong, bengali, and peranakan etawah (PE) [1]. Based on the historic information from local people samosir goat used for the material offering at relegius ceremony of one animinsm (parmalim). Goat must be white in color , so that the local residents selectively keep white colored goats as can adapt and breed on ecosystem on dry and rocky samosir island with hilly topograpy [1].

Genetic improvements should be supported by the improvement of nutrition, one of them with supplementation techniques so that can be seen the development and growth as well as livestock productivity is good as well. The succes of the supplementation technique is largely determited by the balance and completeness of nutrients added in the feed according to the needs of livestock. One of technique that can be used by farmers to overcome mineral defecieny is by mineral block separation with mineral blocks containing essential macro minerals such as selenium and zinc mainly in organic form can increase livestock productivity.

Selenium (Se) is present in the mid piece of spermatozoa and is associated with Cys-rich protein of the mitochondrial sheath [2], selenium deficiency resulted in abnormal spermatozoal mitochondria, and supplementation with nano-Se enhanced the testis Se content, testicular and semen GSH-Px activity, protected the membrane system integrity and the tight arrayment of the midpiece of the mitochondria [3]. Zn is an essential nutrient and indispensable element in growth and reproduction. Zn helps in testicular growth and development of seminiferous tubules, spermatogenesis, steroidogenesis in testes, synthesis and secretion of follicular stimulating hormone (FSH) and luteinizing hormone (LH) [4][5] and [6] and hence improves fertility [7]. Therefore, the aim of this experiment was to study the effect of dietary supplementation of ZN and Se on semen quality of samosir goat.

## 2. Material And Methods

Animal experimental design ; the present study was conducted on 4 samosir goat (1±0.2 years), weight 12.57 ±0.58kg, reared at laboratory animal science, Faculty Agrieculture, university of Sumatera Utara, Medan from june to december 2016. The animal were grazed in a block cage in laboratory animal science daily, and test composed with not supplement (p0), supplement with +10ppm selenium (p10, supplement +10ppm zinc (p2), and supplement +10ppm selenium +10ppm zinc (p3) concentrate MNB per animal daily with free access to water. All the experimental animal were regularly dewormed for parasites.

Animal experimental were diveded randomly into test containing four by four each. Analysis was conducted for 30 days, and the animal was rested for 7 days befor other test. The invetigation was under taken for the entire duration of the experiment. Assement was done by evaluating the goats body weight and collected the semen for further analysis in the end of each test.

The semen colected and analyzed from the donor samosit goat were trained and send semen was collected in the morning using artificial vagina. A nonoesterous due was used for the mounting of males, and semen was collected into the graduate cups with an accuracy attached to one end of the artifiacial vagina.

The semen quality was evaluated in terms of gross motility, progresif motility, sperm concntration (hemocytometer chamber method), hypo-osmotic swelling test and persentage live spermatozoa using microscope. The statistical analyzed using two way analysis of variance followed by tukey's post hoc test using spss/pc computer program(version 14.0 spss).

## 3. Result And Discussion

Effect of zinc and selenium supplementation on weight body samosir goat with supplemented by p3 is the higher ($p<0.05$) from the other test, consumtion was no diffrent from other and feed consomtion ratio p3 ($p<0.05$) is the less from the other. In test sperm quality motility p3 ($p<0.05$) is no diffrent from p1($p<0.05$) but significant higher diffrent with p2($<0.05$) and p0($p<0.05$). Viability p3 is the higher from the other but p2 is no diffrent from p1 and p0, volume semen p3 is the higher from the other but no diffrent from p1, concentration p3 no diffrent from p2 and p1 but p2 is no diffrent from p0, and HOST-test p3 and p1 the higher is no diffrent and p2 diffrent from p0 and less form p3 and p1. The benefit of zinc and selenium supplement on weight body were observed. Similiar result were also report Cole [8], Widhyari et al., [9] and kurnia et al., [10]. The effect benefit supplement on sperm was obseved and report Kendal et al., [11] and Lubis et al., [12]. Therefore, the increased higher weight body and semen quality could be due by supplement zinc and selenium.

**Table 1**. effect of zinc and selenium supplementation on weight body samosir goat

| Treatment | Liveweight gain (g/day) | consumtion (g/d) | convertion |
|---|---|---|---|
| p0 | 23.50 $^{d}$ | 1872.66 $^{a}$ | 79.73 $^{a}$ |
| p1 | 30.92 $^{c}$ | 1852.83 $^{a}$ | 59.19 $^{b}$ |
| p2 | 32.16 $^{b}$ | 1829.50 $^{a}$ | 57.61 $^{b}$ |
| p3 | 40.17 $^{a}$ | 1760.92 $^{a}$ | 43.86 $^{c}$ |

Superscript designate significant difference ($P> 0.05$) between rows and columns of each parameter respectively.

The sperm motility of individual spermatozoa on fresh cement were 65% for p0, 76,25% for p1 , 67,50% for p2 and 78,75% for p3. The individual motility was higher in p3 ($p<0.05$) treatment with supplement consist of + 10ppm zinc + 10ppm selenium. These results still fall within the normal range of between 60% - 80% (Hafez, 2000) and> 60% (Kartasudjana, 2001).

**Table 2**. effect of zinc and selenium supplementation on semen quality of samosir goat

| treatments | Motility (%) | Viability (%) | Volume Semen (ml) | Sperm concentration ($10^6$/ml) | HOST (%) |
|---|---|---|---|---|---|
| p0 | $65.00^b$ | $86.608^c$ | $0.125^b$ | $4150.0^b$ | $69.858^c$ |
| p1 | $76.25^a$ | $91.213^b$ | $0.300^{ab}$ | $4562.5^a$ | $80.033^a$ |
| p2 | $67.50^b$ | $89.743^{bc}$ | $0.200^b$ | $4450.0^{ab}$ | $73.795^b$ |
| p3 | $78.75^a$ | $96.515^a$ | $0.450^a$ | $4587.5^a$ | $82.430^a$ |

Superscript significant difference (P< 0.05) between rows and columns of each parameter respectively

Viabiility of the semen was also analyzed as it relates to the survival of spermatozoa [13]. The viability was 86.608% for p0. 91.213% for p1. 89.743% for p2. 96.515% for p3. The percentage percentage of spermatozoa viability obtained by percentage of spermatozoa viability at treatment of p3 better than treatment of p0. p1 and p2. The results of statistical tests showed that the different treatments gave a very real effect (P <0.05) on the viability of individual spermatozoa. according to Toelihere [14] normal cement ussualy has a minimum percentage of life of 50%. Therefore. the cement has good quality because the viability on this experiment were >50%.

Table 2 shows that the fresh semen volume of the average samosir goat is p0 0.125. p1 0.300. p2 0.200. p3 0.450 ml/ejaculation. The volume of samosir goat cement obtained in this study range between of 0.1-1.5 cc/ejaculation. the normal volume range which is in accordance to previous result [15]. The varying volume of cement at the time of shelter is influenced by differences in individual livestock. livestock. age. nutrition. ejaculation frequency. cement collection interval and cement collection [16].

The concentration of samosir goat cement for each treatment was $4150X10^6$ml for p0. 4562.5 $X10^6$/ml for p1. 4450.05 $X10^6$/ ml for p2. 4587.5 $X10^6$/ ml p3. Dally et al.. [17] reported that spermatozoa concentration of goat cement fall between 2.5-5.0x$10^9$ spermatozoa/mL. similiar to the results obtained on this assessment of the consentration is very important because it is used to reduce the quality of cement and loudspeaker [18].

HOST-Test has been proven as a good method to evaluated the integrity of cell membranes of spermatozoa of domestic animals by observing changes in the tail of the sperm [19]. In spermatozoa are alive or have a plasma membrane Which is good then the media were presented HOST enable active biochemistry contained in the membrane to balance the fluid inside and outside the sperm cells. so the solution hypoosmotik get into spermatozoa. media HOST cause expansion of the cell membrane of the tail of spermatozoa to Swast and peak force flagellum to roll Susilawati [20] says that intact plasma membrane characterized by screening the tail. because the plasma membrane of the spermatozoa still function well in absorbing water in an environment that is hypotonic. Instead of sperm with damaged or permeability of the plasma membrane increases. the solution can come out of the membrane hypoosmotik spermatozoa freely and not be trapped so that tail Looks straight [21]. The Host-test of the samosir goat was 69.858%. 80.033%. 73.795%. 82.430% for p0. p1. p2. p3 respectively.. The cell membrane is the outermost portion that confines the interior with the external environment of the cell and acts as a filter on the exchange of intracellular and extracellular substances maintained in the metabolic process [22].

## 4. Conclusion

It is concluded that the supplementation of zinc and selenium in samosir goat improved the body weight. consumtion. and quality semen but can reduce the Feed comsumtion ratio.

**Referances**

[1] Pamungkas F A, Batubara A, Doloksaribu M and Sihite E 2009 Technical Guidance for Local Goat Germplasm Potential Indonesia. Center for Research and Development of Animal Husbandry Agency of Agricultural Research and Development Department of Agriculture
[2] Kleene K C, Smith J, Bozorgzadeh A, Harris M, Hann L, Karimpour I and Gerstel J 1990 Sequence and developmental expression of the mRNA encoding the seleno-protein of the sperm mitochondrial capsule in the mouse *Develop Biol.* 137: 395-402
[3] Guang S L, Jie Y R and Bin Y W 2010 Animal Reproduction Science Volume 118 Issues 2–4. Pages 248-254
[4] Underwood E J 1977 Trace Elements in Human and Animal Nutrition 4th ed. Academic Press. New York
[5] Habib F K 1978 Zinc and the steroid endocrinology of human prostate. *J Steroid Biochem Mol Biol.* **9**: 403-7
[6] Bedwal R S and Bahugana A 1994 Zinc. copper and selenium in reproduction *Experientia* 50: 626-40
[7] Bray T M, Levy M A, Noaseworthy M D and Iles K 1997 The role of zinc in free radical mediated diseases. In: Fischer P W F, Abbe M R, Cockell K A and Gibson R S (eds.) Trace Elements in Man and Animals *Proceedings of ninth international symposium on trace elements in man and animals* NCR Research Press Ottawa Canada 3333-36
[8] Cole V G 1982 *Beef Cattle Production Guide* (New South wales: Mac. Arthur Press Parramata)
[9] Widhyari S D, Wientarsih I, Soehartono H, Kompiang I P and Winarsih W 2009 The Effectiveness of Combination of Zinc and Herbal Minerals as Immunomodulators, Jurnal Ilmu Pertanian Indonesia
[10] Kartasudjana 2001 Artificial insemination techniques in livestock Jakarta
[11] Kendall N R, Mc Mllen G, Reen S A and Rodway R G 2000 The effect of a zinc, cobalt and selenium soluble glass bolus on trace element status and semen quality of ram lambs *Anim Rep Sc.,* 62(4): 277-83
[12] Lubis F N L, Alfianty R and Sahara E 2015 Effect of Organic Selenium Supplementation (Se) and Vitamin E Against Performance of Pegagan Duck *Jurnal Peternakan Sriwijaya [page,29] Sriwijaya University* South Sumatra
[13] Setiadi B, Tiesnamurti B, Subandriyo, Sartika T, Adiati U, Yulistiani D and Sendow I 2002 *Collect and evaluate the characteristics of goat kitten and gembrong ex-situ, Collection of Research Results of APBN of Fiscal Year 2001 Book I Livestock Research Institute*, Ciawi-Bogor pp 59-73
[14] Toelihere M R 1993 *Artificial Insemination in Livestock* (Bandung: Angkasa)
[15] Hafez E S E 1993 *Semen evaluation In Hafez E S E (ed) reproduction in farm animals, 6th ed,* (Philadelphia: Lea and Febiger)
[16] Tambing, Toelihere, Yusuf, Purwantara and Sutama 2003 Quality of Sugar Frozen Cemented Goats in Different Kinds of Diluents *Hayati* 10:146-150
[17] Dally, Didion, Lenz, Love, Varner, Hafez and Bellin 2008 *Artificial Insenmination in Farm Animal Reproduction 7th ed by Hafez and Hafez* (Philadelphia: Lea and Febiger) pp 376-396
[18] Bearden H J and Fuquay 1984 *Applied Animal Reproduction* (Virginia: Reston Publishing Company Inc, Reston)
[19] Fonseca J F, Torres C A A, Maffilli V V, Borges A M, Santos A D F, Rodrigues M T and Oliviera R F M 2005 Hypoosmotic Swelling Test In Fresh Goat Spermatozoa, Anim, Reprod, 2(2): 139-144
[20] Susilawati T 2003 *Gender Determination and Arrangement, Faculty of Animal Husbandry* University Brawijaya, Malang
[21] Mafcruchati, Luqman M, Harjani M E, Widjiati N and Poernomo B 1997 *Inspection of Spermatozoa Membranes On Frozen Semen Sheep After Diluted* Research Faculty of Veterinary Medicine, Airlangga University 1-6

[22] Hafez E S E and Hafez B 2000 *Reproduction in Farm Animals* Seventh Edition (Baltimor: Lippincott Williams & Wilkins)

**Acknowledgement**
Authors acknowledge the Lab Biology animal facility, agriculture faculty, Universsity of North Sumatera, Medan, Indonesia,

IOP Conf. Series: Earth and Environmental Science **122** (2018) 012127 doi:10.1088/1755-1315/122/1/012127

# Nutrition quality test of fermented waste vegetables by bioactivator local microorganisms (MOL) and effective microorganism (EM4)

**E Mirwandono, M Sitepu, T H Wahyuni, Hasnudi, N Ginting, G AW Siregar, I Sembiring**

Animal Production Program, Faculty of Agriculture, Universitas Sumatera Utara, Medan 20155, Indonesia
E-mail: nurzainahginting@gmail.com

**Abstract.** Livestock feed mostly used waste which has low nutrition content and one way to improve feed content by fermentation. The objective of this study was to evaluate the effect of bioactifator types on fermented vegetables waste for animal feed.The research was conducted in Nutrition and Animal Feed Laboratory, Universitas Sumatera Utara from May until July 2016. The research was factorial completely randomized design of 3 x 3 with 3 replications. Factor I were bioactivator types which were control, local bioactivator and EM4 (Effective Microorganisms 4). Factor II were time of incubation 3, 5 and 7 days. Parameters were moisture content, ash, Nitrogen Free Extract (NFE) and Total Digestible Nutrient (TDN). The results showed that bioactivator types either local activator or EM4 has highly significantly different effect (P<0,01) on water content, NFE and TDN on vegetables waste while there was no different between local bioactifator with EM4 on all parameters. Time of incubation 7 days has highly significantly different effect (P<0,01) on NFE, TDN and significant different (P<0,05) on water content and ash. In conclusion local bioactifators could improve animal feed by fermenting vegetables waste and it is more available for livestockers.

## 1. Introduction

One of the biggest contributors in the environment is waste from traditional markets. Traditional market waste has slightly different characteristics than housing waste. The composition of market waste is more dominant organic waste rather than inorganic waste. Organic market waste consists of vegetable waste and fruit waste. As in Indonesia mostly waste is dumped into landfill which cause environmental problem, Ginting [1] suggested for any organic waste to be used which then will give enormous benefits.

Vegetable waste has potential as animal feed ingredients. This waste can be directly used as animal feed. But low levels of crude protein and high coarse fiber are usually the limiting factor in their use as feed. These wastes are also susceptible to decay and damage, so treatment is needed to prolong the shelf life and suppress the anti-nutritional effects generally in the form of alkaloids.

According to Susangka et al., [2] vegetable waste has a crude fiber of 5-38%. The coarse fiber is very high and can be derived by fermentation with local microorganisms (MOL) and Effective Microorganism 4 (EM4). According to Soccol et.al [3] fermentation was a very old way/thousand years ago which supported higher final concentration of products with environmental advantages.

Local microorganisms (MOL) contain microorganisms that have the potential to break down organic matter. EM4 is one of the microbes that can degrade crude fiber content because it has the ability to produce enzymes laccases and peroxidases that can break down and dissolve the lignin contained in feed ingredients that serve as a source of energy for livestock Mangisah et al. [4].

## 2. Materials and Methods

This research has been conducted in Nutrition and Feed Science Laboratory of Animal Husbandry Program of Faculty of Agriculture, University of Sumatera Utara from May 2016 to June 2016.

The materials for making MOL were 1 kg rotten papaya, bananas, pineapple, ½ kg rotten water spinach, spinach and long bean, 1 kg sugar, 60 g yeast tape, yeast tempe and youghurt. Araujo [5] found lactic acid bacteria (BAL) in youghurt. Other source of microorganism was EM4. Fermented vegetables waste were consisted of mustard, cabbage and corn chobot.

The research method used in this study was experimentally using a complete randomized design (RAL) with 3 x 3 and 3 replication factorial patterns. So that yielded 9 combination of treatment. The first factor is the type of bioactivator with 3 treatments, namely:
R0: No Fermentation (control)
R1: Local Microorganism (MOL)
R2: EM4 (Effective Microorganism 4)
The second factor is the duration of fermentation with 3 treatments, namely:
S1: 3 days
S2: 5 days
S3: 7 days
Parameters of research were water content, dry matter, ash content, TDN, BETN, crude fiber, crude fat, rough protein while there were 4 parameters already submit into local journals however it was not yet published ( dry matter, rough protein, crude fiber, crude lipid).

## 3. Result and Discussion

**Table 1.** Recapitulation of research results

| Treatment | Parameters (%) | | | | | | | |
|---|---|---|---|---|---|---|---|---|
| | Water content | Dry material | Rough fat | Crude protein | Rough fiber | ash content | BETN | TDN |
| R0 | $4.37^{B}$ | $95.62^{A}$ | $1.73^{A}$ | $4.30^{B}$ | $15.53^{A}$ | $21.76^{tn}$ | $56.65^{B}$ | $60.77^{B}$ |
| R1 | $8.41^{A}$ | $91.58^{B}$ | $1.23^{B}$ | $6.74^{A}$ | $11.41^{B}$ | $21.73^{tn}$ | $58.86^{A}$ | $64.58^{A}$ |
| R2 | $8.38^{A}$ | $91.61^{B}$ | $1.36^{B}$ | $6.90^{A}$ | $11.50^{B}$ | $21.79^{tn}$ | $58.43^{A}$ | $64.22^{A}$ |
| S1 | $7.16^{ab}$ | $92.83^{ab}$ | $1.48^{tn}$ | $5.04^{C}$ | $13.30^{A}$ | $21.73^{b}$ | $58.43^{A}$ | $62.88^{B}$ |
| S2 | $7.73^{a}$ | $92.26^{b}$ | $1.44^{tn}$ | $5.67^{B}$ | $12.93^{A}$ | $21.80^{a}$ | $58.14^{A}$ | $63.10^{B}$ |
| S3 | $6.27^{b}$ | $93.72^{a}$ | $1.40^{tn}$ | $7.23^{A}$ | $12.22^{B}$ | $21.76^{ab}$ | $57.36^{B}$ | $63.59^{A}$ |
| R0S1 | $4.49^{tn}$ | $95.50^{tn}$ | $1.73^{tn}$ | $4.30^{A}$ | $15.64^{tn}$ | $21.80^{b}$ | $56.51^{A}$ | $60.65^{tn}$ |
| R0S2 | $4.46^{tn}$ | $95.53^{tn}$ | $1.72^{tn}$ | $4.29^{A}$ | $15.67^{tn}$ | $21.77^{b}$ | $56.54^{A}$ | $60.69^{tn}$ |
| R0S3 | $4.17^{tn}$ | $95.82^{tn}$ | $1.75^{tn}$ | $4.33^{AB}$ | $15.28^{tn}$ | $21.73^{ab}$ | $56.89^{AB}$ | $60.97^{tn}$ |
| R1S1 | $8.58^{tn}$ | $91.43^{tn}$ | $1.28^{tn}$ | $5.32^{ABC}$ | $12.22^{tn}$ | $21.65^{a}$ | $59.51^{C}$ | $64.09^{tn}$ |
| R1S2 | $9.30^{tn}$ | $90.69^{tn}$ | $1.25^{tn}$ | $6.26^{CD}$ | $11.34^{tn}$ | $21.81^{b}$ | $59.32^{C}$ | $64.58^{tn}$ |
| R1S3 | $7.36^{tn}$ | $92.64^{tn}$ | $1.15^{tn}$ | $8.65^{E}$ | $10.68^{tn}$ | $21.75^{b}$ | $57.75^{ABC}$ | $65.08^{tn}$ |
| R2S1 | $8.42^{tn}$ | $91.58^{tn}$ | $1.42^{tn}$ | $5.50^{ABCD}$ | $12.06^{tn}$ | $21.73^{ab}$ | $59.27^{C}$ | $63.91^{tn}$ |
| R2S2 | $9.45^{tn}$ | $90.55^{tn}$ | $1.36^{tn}$ | $6.47^{DE}$ | $11.76^{tn}$ | $21.82^{b}$ | $58.57^{BC}$ | $64.02^{tn}$ |
| R2S3 | $7.29^{tn}$ | $92.70^{tn}$ | $1.31^{tn}$ | $8.73^{E}$ | $10.68^{tn}$ | $21.82^{b}$ | $57.44^{AB}$ | $64.74^{tn}$ |

Description: A. B. C. D. E : Very significant effect ( P<0.01)
a. b : Have real effect (P<0.01)
tn : Not real

The result of bioactivator type research gave no significant different effect (P>0,05) to water content, dry matter, crude fat, crude protein, BETN and TDN of vegetable waste. The duration of fermentation had significantly different effect (P <0.01) on crude protein, crude fiber, BETN, TDN and significantly different (P <0,05) to water content, dry matter, ash content. The interaction of type of bioactivator and fermentation duration gave a significantly different effect(P<0.05).

### *3.1. Effect of Treatment on Water Content of Fermented Vegetables Wastes*

The highest average moisture content was 9.45% in the R2S2 treatment (vegetable waste fermented with EM4 bioaktivator and fermentation time of 5 days) and the lowest was 4.17% in the non-fermentation treatment. The results showed that fermentation water content was higher than vegetable waste water content without fermentation. This indicates that the fermentation was already going well because the microbes on EM4 and MOL have started to grow and use water to ferment Buckle [6]. Vegetable waste water content was also a by-product of the fermentation process.

The result of variance analysis showed that bioactivator type treatment gave a very real effect (P <0.01) to fermented vegetable waste water content. The long treatment of fermentation also had significant effect (P <0.05) on the fermented vegetable water content. The interaction of type bioactivator treatment and fermentation time had no significant effect (P> 0.05) to water content. This means that the two treatment factors did not interact to increase the water content of Fermented vegetable wastewater.

The result of variance analysis showed that the influence of interaction factor between bioactivator type and fermentation time had no significant effect (P> 0.05) on fermented vegetable water content. so no further test was done.

### *3.2. Effect of Treatment on Dry Material Content of Fermented Vegetables*

The highest dry matter content was 95.82% in the non-fermentation treatment and the lowest drying material was 90.55% in the R2S2 treatment ((fermented vegetable waste with EM4 bioactivator and 5 day fermentation time) .The fermentation dry material was lower than the dry matter of waste vegetables without fermentation. Low dry matter due to microbes using carbohydrates. minerals and other substances for microbial growth Kamara et al. [7].

The result of variance analysis showed that bioactivator type treatment had very real effect (P <0.01) on dry matter of vegetable waste. The long treatment of fermentation also had a significant effect (P <0.05) on dry matter of vegetable waste. The interaction of bioactivator type treatment and fermentation time had no significant effect (P> 0.05) to the content of dry matter of vegetable waste. This means that the two treatment factors do not interact to reduce the dry matter content of fermented vegetable waste.

Compared to other treatments. possibly due to existing organic materials. such as carbohydrates. fats. and unused proteins as a source of nutrients for microbes. since no addition of EM4 was possible. Unlike the case in the addition of EM4 treatment. food substances. such as carbohydrates. fats and proteins used the source of nutrients for microbes contained therein.

The result of variance analysis showed that the influence of interaction factor between bioactivator type and fermentation time had no significant effect (P> 0.05) on dry matter of fermented vegetable waste. so no further test was done. The unreal effect is caused by the dry matter treatment is converted into energy (heat) which causes the decrease of dry matter substrates.

### *3.3. The Effect of Treatment on Crude Fat Content of Fermented Vegetable Waste.*

The highest rate of 1.73% was found in the treatment without fermentation and the lowest average was found at 1.15% in the R1S3 treatment ((vegetable waste fermented with MOL biocativator and 7 day fermentation time). From result of analysis of bioactivator type treatments have very real effect (P <0.01) to crude fat of vegetable waste. However. the long treatment of fermentation has no significant effect (P> 0.05) on the crude fat of vegetable waste. The result of variance analysis showed that there was no interaction between bioactivator type treatment and fermentation length on crude fat of fermented vegetable waste. This means that the two factors do not interact with each other to lower the raw fat of vegetable waste.

The result of variance analysis showed that the fermentation treatment had no significant effect (P> 0.05) to the crude fat of vegetable waste. It can be seen that the crude fat content decreases with increasing length of fermentation. This happens because the process of fermentation transforms complex compounds into simple compounds so that microbes can be utilized. This is in accordance with the statement of Mucra [8] which states that the fermentation aims to break down complex compound ds to be simpler so that can be utilized by microbes for its growth as an energy source in the form of VFA (Volatile Fatty Acid) in addition to energy from carbohydrates easily digested. Length of fermentation up to 7 days was not significantly different for the crude fat content in vegetable waste. This was because EM4 and MOL both contain the enzyme cellulase so there is not much of a rough fat change. This was consistent with the results of the Selviana [9] which states that each fermentor has the same enzyme ability to change the crude fat content. The absence of significant differences in crude fat content in each treatment was suspected because the content of fermentor was less dominant population of bacteria producing lipase enzymes that digest fat function but more dominant enzyme cellulase enzyme.

### *3.4. Effect of Treatment on Rough Protein Content of Fermented Vegetable Waste*

The highest mean rough protein was 8.73% in the R2S3 treatment ((vegetable waste fermented with EM4 bioactivator and fermentation time of 7 days) and the lowest was 4.29% in the fermented treatment.The fermented crude protein was higher than the crude protein of vegetable waste Without fermentation.

The result of variance analysis showed that bioactivator type treatment had highly significant effect (P <0.01) on crude protein of vegetable waste. Likewise in the long treatment of fermentation has a very significant effect (P <0.01) on crude protein of vegetable waste. The result of variance analysis showed that the interaction between bioactivator type and fermentation treatment had significant effect (P <0.01) on crude protein of vegetable waste. This means that the two treatment factors interact to raise the crude protein of vegetable waste.

This is consistent with the statement of Fajaruddin et al.. [10] which states that major microorganisms in EM4 solution consist of photosynthetic bacteria (bacteria fototropik). lactic acid bacteria. yeast. Actinomycetes and fermented mushrooms.

The addition of EM4 to the fermentation process serves to increase the growth of microorganisms found in the solid so that it can work optimally in breaking the unbroken cells and increasing the crude protein content due to the activity of microorganisms in the dry solids of organic sludge of bio gas units.

The result of variance analysis showed that the influence of interaction factor between bioactivator type and fermentation time had significant effect (P <0.05) to crude protein of fermented vegetable waste. so that further test was done. The influence of this type of biokativator increases the crude protein by increasing the length of fermentation resulting in interaction. This occurs because at the time of fermentation an increase in the number of microbial cell mass. This is consistent with the statement of Chandra et al.. [11] which states that the rice husk fermented with EM4 has a significantly higher effect than the unfermented ones. because at the time of fermentation there is an increase in the mass of microbial cell. The increase occurs because in the process of microbial fermentation produces microbial cells in the form of single cell proteins. microbial enzymes and microbial metabolism results are amino acids. nucleotides. and proteins.

### *3.5. Effect of Treatment on Rough Fiber Fertilizer Fermentation Wastes*

The highest rate of crude fiber was 15.67% in the non-fermented treatment and the lowest was 10.68% in the R2S3 treatment ((vegetable waste fermented with EM4 bioactivator and fermentation time of 7 days) .The coarse fermented fiber was lower than the crude fiber of the vegetable waste Without fermentation.This shows that the crude fiber of fermented vegetable waste is decreasing.

The result of variance analysis showed that bioactivator type treatment had highly significant effect (P <0.01) on crude fiber of vegetable waste. The long treatment of fermentation was very significant (P <0.01) to the crude fiber of fermented vegetable waste. The interaction between bioactivator type treatment and fermentation length did not give significant effect (P> 0.05) to crude fiber of vegetable waste. This means that the two treatment factors do not interact to reduce the crude fiber of vegetable waste.

The result of variance analysis showed that the influence of interaction factor between bioactivator type and fermentation time had no significant effect (P> 0.05) to crude fiber of fermented vegetable waste. so no further test was done. The effect is not significant is due to coarse fiber treatment decreases with the length of fermentation so that no interaction occurs. .

### *3.6. Effect of Treatment on Ash Content of Fermented Vegetables*

The highest rate of ash content was 21.82% in the R2S3 treatment (vegetable waste fermented with EM4 bioactivator and fermentation time of 7 days) and the lowest was 21.65% in the R1S1 treatment (vegetable waste fermented with MOL bioactivator and 3 day fermentation time) .

The result of variance analysis showed that bioactivator type treatment had no significant effect (P> 0.05) to ash content of fermented vegetable waste. The fermentation treatments had significant effect (P <0.05). The interaction between bioactivator type and fermentation time gave a significant effect (P <0.05) to ash content of vegetable waste. This means that the two treatment factors interact to reduce the ash content of vegetable waste.

### *3.7. Effect of Treatment on BETN Levels of Fermented Vegetable Waste*

The highest rate of BETN was 59.51% in the R1S1 treatment ((vegetable waste fermented with MOL bioactivator and 3 day fermentation time) and the lowest 56.51% in the non-fermented treatment.

The result of variance analysis showed that bioactivator type treatment had a very significant effect (P <0.01). The fermentation treatments had a very significant effect (P <0.01) on BETN of fermented vegetable waste. Interaction Treatment of bioactivator type and fermentation time had a significant effect on BETN content of fermented vegetable waste. This means that the two treatment factors interact with each other to lower vegetable waste BETN.

The result of variance analysis showed that the influence of interaction factor between bioactivator type and fermentation length significantly (P <0.05) to BETN of fermented vegetable waste. so that further test was done. BETN levels have increased indicating that many components of organic material in the digestible vegetable waste so that more energy will be generated.

### *3.8. Effect of Treatment on TDN Level of Fermented Vegetable Waste*

The highest rate of TDN was 65.08% in the R1S3 treatment ((vegetable waste fermented with MOL bioactivator with 7 day fermentation time) and the lowest 60.97% in the fermented treatment.

The result of variance analysis showed that the bioactivator type and fermentation treatment had a very significant effect (P <0.01) on TDN of vegetable waste. The interaction of bioactivator type treatment with fermentation time was not significantly different. This means that the two treatment factors are not mutually interacting to decrease TDN of vegetable waste.

The result of variance analysis showed that the influence of interaction factor between bioactivator type and fermentation time had no significant effect (P> 0.05) to TDN of fermented vegetable waste. so no further test was done. The unreal effect is due to TDN of fermented vegetable waste is the total of feed substance needed by the livestock which will increase with decreasing the raw fiber content of feed.

### *3.9. Description between MOL and EM4 on Price and Availability*

A description has been done to compare MOL and EM4 in terms of price and availability. MOL actually was already socialiced by Agricultural Department of Indonesia by extention program Agricultural Department [12] to local farmer thus many local farmers could done MOL. As MOL use rotten materials. any production cost was on tape yeast. sugar and youghurt. Meanwhile local farmers has to bought EM4 whenever they want to use EM4 as bioactivator. Local farmers anywhere even to a very remote area could have MOL while EM4 only could find in at least a regency or subregency. In conclusion. MOL was more feasible to be used as bioactivators.

**Table 2.** Description between MOL and EM4 on Price and Availability

| | Price/Liter (Rp) | Availability |
|---|---|---|
| MOL | 5,000 | Available everywhere |
| EM4 | 22,000 | Available only in regency and sub regency |

## 4. Conclusions

Either MOL or EM4 as bioactivator has an effect to increase water content. crude protein. BETN. TDN of vegetable waste and lowering dry material. coarse fat. crude fiber. The duration of fermentation / 7 days has an effect on increasing dry matter. crude protein. ash content. TDN of vegetable waste and lower water content. crude fiber and vegetable waste BETN. The interaction of bioactivator type and fermentation time have an effect on increasing crude protein of vegetable waste. In conclusions both MOL and EM4 was good as bioactivator on fermening vegetable waste. However. MOL was better as its price was cheaper than EM4 and it was also more avaiable for farmers.

## References

[1] Ginting N 2017 Benefits of using biogas technology in rural area: Karo District on supporting local action plan for greenhouse gas emission reduction of North Sumatera province 2010-2020 *In ternational Conference on Biomass: Technology Application and Sustainable Development IOP Publishing* IOP Conf. Series: Earth and Environmental Science **65 (2017)** 012007 doi: 10.1088/1755-1315/65/1/012007.

[2] Susangka I, Haetami K dan Andriani Y 2006 *Evaluasi Nilai Gizi Limbah Sayuran Produk Cara Pengolahan Berbeda dan Pengaruhnya Terhadap Pertumbuhan Ikan Nila (Evaluation Value of Waste Product Vegetables the Different Technique and the Influence of Parrotfish Growth)* UNPAD Bandung

[3] Soccol, Ricardo C, Ferreira da Costa E S, Letti L A J, Karp S G, Woiciechowski A L and Porto de Souza Vandenberghe L 2017 Recent developments and innovations in solid state fermentation. http://www.journals.elsevier.com/biotechnology-research-and-innovation/

[4] Mangisah I N, Suthama dan Wahyuni H I 2009 *Pengaruh Penambahan Starbio dalam Ransum Berserat Kasar Tinggi terhadap Performa Itik (the Influences of adding Starbio in high Crude Fiber Rations of duck Performance)* Universitas Diponegoro Semarang

[5] Araújo, Diego D, Alessandro B, Amorim, Mayra A D, Saleh, Curcelli F, Pedro L, Perdigon, Silvio J, Bicudo, Dirlei A and Berto 201 *J of Animal Nutrition* **2 (2016)** 149-153

[6] Buckle K A, Edwards R A, Fleet G H and Wooton M 1985 *Ilmu Pangan (Food Science)* Terjemahan: Purnomo H dan Adiono UI Press Jakarta

[7] Kamara D S, Rahman S D and Gaffar S 2008 Enzymatic degradation of cellulose from banana stalks for glucose production using cellulolytic activity of *Tricoderma viride In: Proceeding of the International Seminar on Chemistry* p 692-696

[8] Mucra D A 2007 *Pengaruh Fermentasi Serat Buah Kelapa Sawit terhadap Komposisi Kimia dan Kecernaan Nutrien secara In Vitro (Fermentation Influences of Palm Oil Fiber of Chemical Composition and Nutrient Digestibibity by In Vitro* Tesis Pascasarjana Peternakan. Universitas Gadjah Mada Yogyakarta

[9] Selviana E 2000 *Pengaruh Fermentasi Starbio Terhadap Kandungan Protein Kasar. Lemak Kasar dan Serat Kasar Kulit Buah Nanas (Starbio Fermentation Influences by*

*Containing Crude protein. Crude Fat and Crude Fiber from Pineapple Rind) Skripsi* Fakultas Peternakan Universitas Mataram Mataram

[10] Fajaruddin, Junus M dan Setyowati E 2013 Pengaruh lama fermentasi EM4 terhadap kandungan protein kasar padatan kering lumpur organik unit gas bio (the influences of lenght fermentation EM4 by containing crude dry solid protein of organic mud of gass bio) *Jurnal Ilmu-Ilmu Peternakan* ***23***(2) 14-18

[11] Andra, Kereh V G, Untu I M dan Rembet B W 2013 Pengayaan nilai nutritif sekam padi berbasis bioteknologi "*effective microorganism 4*" (EM4) sebagai bahan pakan organik (value nutritif enrichment of rich husk based in biotechnology "*effective microorganism 4*" (EM4) as organic material food) Universitas Sam Ratulangi Manado

[12] Indonesia Agricultural Department Annual Report 2011

# Financial analysis of biogas utilization : input cattle, pig feces and coffee waste in Karo, Indonesia

**N Ginting, F Zuhri , Hasnudi, E Mirwandhono , I Sembiring and A H Daulay**

Animal Production Program, Faculty of Agriculture, Universitas Sumatera Utara, Medan 20155
E-mail: nurzainahginting@gmail.com

**Abstract.** The community's need for renewable energy was very urgent. In addition, efforts to preserve the environment from waste caused biogas technology feasible to apply. This study aims to provide biogas technology with minimal cost and utilize agricultural waste that were coffee and livestock waste. The study was conducted from July to October 2016. The theoretical and empirical methods used in this study were included data from officials resources, field survey on 16 biogas locations, focus group discussion and interview with stake holders. Data were tabulated by Excel Program which then were analysed by SAS. Parameters were included Production Cost, Production Result, Profit Loss Analysis, Revenue Cost Ratio (R/C Ratio), Return On Investment (ROI), Net B/C, and IRR.The result of this research showed that the application of bioplastic gas with cow dung and coffee waste as bioplasticgas input cause the best results.

## 1. Introduction

Increased populations of the poor were also triggered by natural disasters, such as the eruption of volcanoes. In Karo District, since 2013, Mount Sinabung erupted, resulting in about 350 families had to move, which previously had run out of capital because the crops had failed several times as the impact of the eruption.

In Karo Regency, according to CBS [1] population of cattle and buffalo were 17,325 head. Most of cattle and buffalo feces were discharged into the environment on the puIDRoses to be compost naturally which in the processing would emit Green House Gases (GHG) and create an impact to the aesthetic of the environment Ginting [2]. According to Clark [3] GHG emissions from ruminant livestock comprised as much as 40% of total agricultural emissions.

**Table 1.** Cattle and buffalo feces production 2013 in Karo

| Number of Cows and buffalo | 17.325 head |
|---|---|
| Cow and Buffalo Weight/head | 450 Kg |
| Number of feces / head | 25 Kg |
| Total feces | 433.125 kg |

One way to mitigate GHG was by using biogas technology which according to Ginting [4] would gave benefits to farmers either energy for cooking, coffee roasting or fertilizer. Moreover, Ginting [2] mentioned that as Indonesia was one of the biggest GHG emiter in the world and committed to reduce its greenhouse gas (GHG) at the UNFCCC (Conference of the United Nation Framework Convention on Climate Change), Indonesia had to pay an effort to captured GHG than to freely discharged. By capturing GHG which, i.e CH4, it could used as renewable energy which would help Indonesia in saving its fossil fuel. Li et.al [5] found that in China, energy from livestock feces was equivalent to 20% of natural gas used.

Indonesia was under energy crisis. Utilization of organic waste as an alternative energy raw material was one of the answers to the energy crisis. People in Karo regency paid an average of IDR 22,000 per 3 kg LPG gas tube in one week just for cooking. Therefore, Karonese who use biogas would get cost savings on buying fuel. In addition, bio gas technology produces slurry which was ready made liquid fertilizer to cultivate agricultural commodities.

**Table 2.** Coffee Waste in Karo

| Year | Number of Coffee Production | Number of Coffee Waste |
|---|---|---|
| 2014 | 4823 Kg | 1688.05 Kg |

Source : CBS [6]

In Karo Regency CBS [6], there were 4,823 ton of coffee production. According to Siswati [7], the processing of coffee beans produced 35% of coffee waste which was a source of organic material cellulose content and was available abundantly in Indonesia.

## 2. Materials and Methods

The research was conducted in Karo District on 16 units of bio gas and this research was conducted from July to October 2016. The theoretical and empirical methods used in this study were included data from officials resources, field survey on 16 biogas locations, focus group discussion and interview with stake holders. There were 3 types of biodigesters, i.e. (A) plastic tube; capacity 4,950l, (B) fiber; capacity 7,500l and ( C) concrete; capacity 9,000l. Data were tabulated by Excel Program which then were analysed by SAS. Parameters were included Production Cost, Production Result, Profit and Loss, Revenue Cost Ratio (R/C Ratio), Return On Investment (ROI), NPV, Net B/C, and IRR. Arikunto [8] stated the survey was a critical observation or inquiry to get a good description of a particular problem in an area. Questionnaire was a way of collecting data by giving a list of questions or statements to the respondent. The execution of the research included preparation, preliminary survey, analysis of survey data and interviews, tabulation of data, concluding data.

## 3. Results and discussion

The results of production cost, profit and loss, R/C Ratio and ROI research on bio gas production using plastic, fiber and concrete biodigester per month were showed in Table 3.

**Table 3.** Recapitulation of biodigester research on Production Cost, Production Result, Profit Loss/month

| Biodigester type | Raw Materials | Production cost (IDR) | Profit (IDR) | Profit and loss (IDR) |
|---|---|---|---|---|
| (A) plastic tube; capacity 4,950l | Cow feces | 485,125 | 768,240 | 283,115 |
| | Cow and coffee waste | 485,125 | 854,670 | 369,545 |
| | Cow, pig and coffee waste | 485,125 | 825,210 | 340,085 |
| (B) fiber; capacity 7,500l | Cow | 839,575 | 1,161,780 | 322,205 |
| | Cow and coffee waste | 839,575 | 1,293,780 | 454,205 |
| | Cow, pig and coffee waste | 839,575 | 1,267,380 | 427,805 |
| (C) concrete; capacity 9,000l | Cow | 1,073,276 | 1,396,800 | 323,524 |
| | Cow and coffee waste | 1,073,276 | 1,555,170 | 481,894 |
| | Cow, pig and coffee waste | 1,073,276 | 1,522,170 | 448,894 |

**Table 4.** Production cost of biodigester/month (IDR)

| Biodigester type | Depreciation of biodigester | Raw material cost | Labour cost | Total cost |
|---|---|---|---|---|
| (A) plastic tube; capacity 4,950l | 65,125 | 330,000 | 90,000 | 485,125 |
| (B) fiber; capacity 7,500l | 158,575 | 501,000 | 180,000 | 839,575 |
| (C) concrete; capacity 9,000l | 203,276 | 600,000 | 270,000 | 1,073,276 |

**Table 5.** Biogas, Slurry and Total Production from every biodigester type

| Biodigester Type with Raw Materials | Biogas Production (m3)-LPG(kg)/ Day | Biogas Production (IDR/ day) | Slurry Production (l/day) | Slurry Production (IDR/day) | Total Production (IDR/month) | R/C | ROI (%) |
|---|---|---|---|---|---|---|---|
| (A) plastic tube; capacity 4,950l | | | | | | | |
| Cow feces | 1.65-0.79 | 5,808 | 165 | 19,800 | 768,240 | **1.58** | **58** |
| Cow feces and coffee waste | 2.47-1.19 | 8,689 | 165 | 19,800 | 854,670 | **1.76** | **76** |
| Cow. pig feces and coffee waste | 2.19-1.05 | 7,707 | 165 | 19,800 | 825,210 | **1.70** | **70** |
| (B) fiber; capacity 7.500l | | | | | | | |
| Cow feces | 2.49-1.19 | 8,726 | 250 | 30,000 | 1,161,780 | **1.38** | **38** |
| Cow feces and coffee waste | 3.74-1.79 | 13,126 | 250 | 30,000 | 1,293,780 | **1.54** | **54** |
| Cow. pig feces and coffee waste | 3.49-1.67 | 12,246 | 250 | 30,000 | 1,267,380 | **1.50** | **50** |
| (C) concrete; capacity 9.000l | | | | | | | |
| Cow feces | 3-1.44 | 10,560 | 300 | 36,000 | 1,396,800 | **1.30** | **30** |
| Cow feces and coffee waste | 4.5-2.16 | 15,839 | 300 | 36,000 | 1,555,170 | **1.44** | **44** |
| Cow. pig feces and coffee waste | 4.2-2.01 | 14,739 | 300 | 36,000 | 1,522,170 | **1.41** | **41** |

Note: LPG Price IDR. 7,333.33/kg ; Slurry Price based on NPK Price : IDR 12,000,-/kg : 250 ml slurry equal to 2.5 g NPK

In this research the production of gasbio was converted to LPG. MEMR [9] recommended the use of gasbio to replace LPG as Indonesia obtains much additional foreign exchange from exporting LPG. The use of biogas will be helpful in supporting more LPG exports.

Table 5 showed that plastic tube biodigester produced the best R/C and ROI. This was because the cost of making biodigester plastic was the cheapest. Also affecting R/C and ROI were gas and slurry production. Biodigester plastic tube, fiber and concrete have a balanced efficiency so that gas and slurry production of the three types was more influenced by the volume of stuffing. According to Pindyck et.al [10] R/C was used in an attempt to find out whether or not it was reasonable to proceed to the next period or otherwise discontinued because it was not feasible. R/C ratio was feasible if R/C Ratio was greater than 1. R/C Ratio was break even if R/C Ratio equal to 1. R/C ratio was not feasible if R/C Ratio less than 1. Ginting et .al [11] found that R/C on the use of goat feces and bagasse yielded R/C between 1.19 and 1.90. The use of biogas was feasible was associated with the cost of raw material which was cheap and the results. i.e gas and slurry were good. In this study all feasible treatments indicated as sustainable business. The highest biogas production was from the use cow feces combined with coffee waste. According to Ginting [4] which previously was mentioned by Zhang.et al [12] coffee waste contained a lot of glucose which triggered acetate to be formed as well as biogas/CH4.

The best ROI in this study was by biodigester plastic tube with an average ROI about 68%. This caused by the cost of plastic tube was cheaper. In this study almost all capital could be returned fastly. Pindyck et.al [10] said a faster attempt to return capital was a better undertaking. Of all the treatments. the used of cattle feces with coffee waste was the best. This was related to larger gas production.

**Table 6.** Recapitulation of biodigester research on Net B/C. IRR /Month

| Biodigester Type | Raw Materials | Parameter | Analysis Results | Mark as : |
|---|---|---|---|---|
| (A) plastic tube; capacity 4.950l | Cow feces | Net B/C | 6.07 | feasible |
| | | IRR | 127.04% | feasible |
| | Cow feces and coffee waste | Net B/C | 7.93 | feasible |
| | | IRR | 132.15% | feasible |
| | Cow.pig feces and coffee waste | Net B/C | 7.30 | feasible |
| | | IRR | 130.71% | feasible |
| (B) fiber; capacity 7.500l | Cow feces | Net B/C | 3.53 | feasible |
| | | IRR | 50.99% | feasible |
| | Cow feces and coffee waste | Net B/C | 3.53 | feasible |
| | | IRR | 51.00% | feasible |
| | Cow. pig feces and coffee waste | Net B/C | 3.53 | feasible |
| | | IRR | 50.99% | Feasible |
| (C) concrete; capacity 9.000l | Cow feces | Net B/C | 7.81 | Feasible |
| | | IRR | 38.46% | Feasible |
| | Cow feces and coffee waste | Net B/C | 6.84 | Feasible |
| | | IRR | 38.81% | Feasible |
| | Cow. pig feces and coffee waste | Net B/C | 6.97 | Feasible |
| | | IRR | 38.75% | Feasible |

According to Nachrowi [13] B/C and IRR indicated whether the business was useful to be implemented. B/C showed the amount of benefits compared to cost/investment. In this study. B/C and IRR values indicatd the utilization of biogas was very useful to be implemented. Osak et.al [14] found that utilization of biogas for farmers in remote area supported its economic value because farmers obtain savings by utilizing biogas and slurry. Slurry alone would lead to savings on Indonesia government because its fertilizers subsidizing to farmers Osorio et.al [15]. The fertilizer subsidy was huge in 2013 of IDR 25.9 trillion.

Based on the Money Market Information Center (PIPU) [16] it was known that the average deposit interest rate of all state banks was 7.5%. Therefore, the results of this study was very useful considering the IRR obtained all greater than the average deposit rate.

## 4. Conclusions

Plastic tube biodigester produced the best R/C and ROI. This was because the cost of making biodigester plastic was the cheapest while it also produced abundant biogas and slurry. B/C and IRR were feasible to be implemented by all type of biodigesters. The best gas production was by utilized cow feces and coffee waste.

## References

[1] North Sumatra Central Bureau of Statistic 2014

[2] Ginting N 2017 Benefits of using biogas technology in rural area: karo district on supporting local action plan for greenhouse gas emission reduction of north sumatera province 2010-2020 *In ternational Conference on Biomass : Technology. Application. and Sustainable Development IOP Publishing. IOP Conf. Series: Earth and Environmental Science* **65** (2017) 012007 doi: 10.1088/1755-1315/65/1/012007

[3] Clark H 2016 *The Estimation and Mitigation of Agricultural Greenhouse Gas Emissions from Livestock.* DOI: http://dx.doi.org/10.14334/Proc.Intsem.LPVT-2016-p5-13

[4] Ginting N 2017 Benefits of using biogas technology in rural area: karo district on supporting local action plan for greenhouse gas emission reduction of north sumatera province 2010-2020 *In ternational Conference on Biomass : Technology. Application. and Sustainable Development IOP Publishing. IOP Conf. Series: Earth and Environmental Science* **65** (2017) 012007 doi: 10.1088/1755-1315/65/1/012007

[5] Cheng L F S, Yu H and Dewei Y 2016 Waste from livestock and poultry breeding and its potential assessment of biogas energy in rural China *Journal of Cleaner Production* **126** (2016) 451 – 460

[6] North Sumatra Central Bureau of Statistic 2015

[7] Siswati and Dyah N 2014 *Bioetanol dari Limbah Kulit Kopi dengan Proses Fermentasi. Surabaya* Universitas Pembangunan Nasional

[8] Arikunt and Suharsimi 2010 *Prosedur Penelitian Suatu Pendekatan Praktek.* (Jakarta: Bhineka Cipta)

[9] MEMR 2007 Blueprint program pengalihan minyak tanah ke LPG Jakarta (Indonesia): *Ministry of Energy Mineral Resources of the Government of Indonesia*

[10] Pyndick, Robert S, Rubenfeld 2007 *Microeconomy. 6th Ed.* Jakarta (Indonesia); PT. Indeks

[11] Ginting N, Syahbana A F, Fadillah D M, Ginting S P 2016 *Biogas Productivity Financial Analysis, Livestock Mix Pasture Influenced by Biogas Input and Slurry DOI:*http://dx.doi.org/10.14334/Proc.Intsem.LPVT-2016-p.345-351

[12] Zhang, Zhikai, Zhang G. Li W, Li C and Xu G 2016 Enhanced biogas production from sorghum stem by co-digestion wih cow manure. international journal of hydrogen energy **41**(2016) 9153-9158

[13] Nachrowi, Usman H 2006 *Popular and Enonometric for Approaching Financial Analysis* Jakarta (Indonesia): Universitas Indonesia

[14] Osak REMF, Hartono B, Fanani Z, Utami HD 2011 *Biogas and Bioslurry utilization on dairy-horticulture integrated farming system in Tutur Nongkojajar. District of Pasuruan. East Java. Indonesia.* Livest Res Rur Dev. 27: 2015

[15] Osorio C G, Abriningrum D E, Armas E B, Firdaus M 2011 *Who is benefiting from fertilizer subsidies in Indonesia?* Policr research working paper 5758. East Asia and Pacific Region. Poverty Reduction and Economic Management Unit. Washington DC (US): The World Bank

[16] Money Market Information Center (PIPU) of Bank Indonesia 2016

# The utilization of coconut waste fermentated by *aspergillus niger* and *saccharomyces cerevisiae* on meat quality of weaning males rex rabbit

**T H Wahyuni, N Ginting, Yunilas, Hasnudi, E Mirwandono, G A Siregar, I G Sinaga and I Sembiring**

Animal Production Program, Faculty of Agriculture, Universitas Sumatera Utara, Medan 20155, Indonesia
E-mail : nurzainahginting@gmail.com

**Abstract.** Coconut waste (CW) could be applied for animal feed while its nutrition quality were low. This study aims to investigate fermented CW effect on meat quality of Rex rabbit which feed by fermented CW either by Aspergillus niger or Tape Yeast. This research was conducted in rabbit farm Brastagi, using 24 male Rex rabbits with initial weight 1012 ± 126.67 gram in July-October 2016. The design used was complete randomized design : 6 treatment 4 replications. Treatment were T1 (unfermented 10%); T2 (unfermented 20%); T3 (a.niger fermentation 10%); T4 (a niger fermentation 20%); T5 (tape yeast fermentation 10%) and T6 (tape yeast fermentation 20%). The parameters were pH, meat texture either raw or cooked, water content, fat content, protein content of meat and cooking loss. The results showed that effect of treatment was not significantly different (P>0.05) on pH and raw meat texture, but significantly different (P< 0.05) on texture of meat cooked and meat fat content and very significantly different effect ( P> 0,01) on cooking loss, water content and protein content of meat. The conclusion of this research was the utilization of fermented CW by *Aspergillius niger* and Tape Yeast improved the quality of Rex rabbit meat

## 1. Introduction

Coconut waste is usually only discharged to environment. Ginting [1] mentioned that as organic waste, it will decay and cause environmental problem include green housegases emission. Many coconut waste are found in markets where coconut dissolves, or in coconut processed industries. Coconut waste still has a fairly high nutritional value and need to be processed, i.e fermented first in order to increse its nutritional value. The results of analysis conducted by Miskiyah et al. [2] showed that an increased of protein content of coconut waste after fermentation from 11.35% to 26.09% or by 130% and decrease of fat content by 11.39%. Dry matter digestibility and organic matter increased from 78.99% and 98.19% to 95.1% and 98.82%, respectively. Usage in rations typically reaches 20% of the total ration feed ingredients of rabbits.

Low source of animal protein needs, especially meat is the causes of low consumption of animal protein in Indonesia. An alternative source of new animal protein with good, affordable and easy-to-obtain nutritional quality is needed to increase the consumption of animal protein in Indonesian. Rex rabbit, known as the fur-producing rabbit, also has the potential to produce meat, is one of the livestock sources of animal protein.

Meat is the main product of raising livestock. The availability of both quantity and quality of food is one factor that can affect the quality of meat, while other important factors are breed and maintenance management. Feed is very diverse can be a fresh forage, grain, or agricultural waste / industrial waste can affect the quality of meat. According to Kandeepan et al.,[3] the quality of feed can affect the quality of meat, which can affect the dressing yield, the comparison of bone meat, the ratio of fat protein, fatty acid composition, caloric value, color, physico-chemical, shelf life and sensory properties.

Fermented food is as a result of the activities of several types of microorganisms, including bacteria, yeasts, and molds. Microorganisms that ferment food can produce beneficial changes (fermentation products desired) and adverse changes (damage to foodstuffs). Of the food-fermenting microorganisms, the most important are lactic acid-forming bacteria, acetic acid, and some types of alcohol-producing yeasts Suprihatin [4].

*Aspergillus niger* is a member of the genus Arpergillus, the Eurotiaceae family, the order Eutiales, subclass Plectomycetetidae, ascomycetes class, ascomycotina subdivision and amastigmycota division. Aspergillus niger in its growth is directly related to the food substances contained in the medium. *Aspergillus niger* produces several extracellular enzymes such as amylase, amyloglucosidase, pectinase, cellulase, catalase and glucosidase Hardjo et al., [5]. Lehninger [6] added that *Aspergillus niger* produced urease enzymes that break urea into amino acids and $CO_2$ which are then used to form amino acid.

***Saccharomyces Cerevisiae*** (Yeast) is a facultative organism that has the ability to produce energy from organic compounds in aerobic or anaerobic conditions so that yeast can grow under different ecological conditions Winarno [7].

## 2. Material and Methods

This research was conducted at Jl. Air Gg. Rukun (Ranch Rabbit Rukun Farm) Berastagi and Food Technology Laboratory Faculty of Agriculture University of North Sumatra. This study will last for 3 months starting from July 2016 to October 2016.

The material used were 24 male rex rabbits, ration/pelleting ingredients consisting of fermented coconut waste, palm kernel cake, soybean meal, rice bran, fish meal, mineral mix and salt. Meat of rabbit chest and thigh as chemical test of meat. The experimental design used in this study was experimental using a completely randomized design with 6 treatments and 4 replications. The treatment is as follows:

P0a: Ration with 10% coconut waste without fermentation
P0b: Ration with 20% coconut waste without fermentation
P1: Ration with 10% coconut waste fermented by *Aspergillus niger*
P2: Ration with 20% coconut waste fermented by *Aspergillus niger*
P3: Ration with 10% coconut waste fermented by ***Saccharomyces cerevisiae***.
P4: Ration with 20% coconut waste fermented by ***Saccharomyces cerevisiae***.

Parameters were meat pH, cooking loss, tenderness, protein. Analysis on protein, water and fat contents was based on AOAC [8].

### *2.1. Protein levels (AOAC, 1995)*

The sample was weighed as much as 0.2 grams weight into a 30 ml Kjeldahl tube, then added 2 grams of $K_2SO_4$, 50 mg HgO and 2 ml of concentrated $H_2SO_4$. Samples boiled until clear. After that the liquid is cooled with water flowing slowly. The contents of the flask were then transferred to a distillation apparatus, added 8-10 ml of $NaOH\text{-}Na_2S_2O_3$ and then distilled. The destilate is accommodated in a 125 ml Erlenmeyer containing 5 ml of $H_3BO_3$ solution and 2 drops of indicator (mixture of red methyl and red blue) to about 15 ml of distillation. The distillation is diluted to about 50 ml and titrated with 0.02 N HCl until the gray color change.

### *2.2. Water levels (AOAC 1995)*
A sample of 2 grams is fed into a known aluminum cup. Then dried in a 100-105 oC oven oven to constant weight. After that it is cooled in the desiccator and weighed. Water content is obtained by using the formula:

$$\text{Content water (\%)} = \frac{\text{Initial weights} - \text{final weight}}{\text{Initial Weight}} \times 100\% \quad (1)$$

### *2.3. Fat Level (Soxhlet Extraction Method) (AOAC 1995)*
A total of 5 grams of stapled samples were wrapped in filter paper, loaded into soxhlet, and hexane added sufficiently and refluxed for 5-6 hours. Then, fat gourds containing extraction fat and solvent were heated at an oven at 105 ° C after it was cooled in a desiccot and weighed. Fat content is obtained by using the formula:

$$\text{Fat content (\%)} = \frac{\text{fat weight (g)}}{\text{sample weight (g)}} \times 100\% \quad (2)$$

## 3. Results and Discussion
Table 1. Recapitulation of research results of utilization of coconut fermented coconut flour on the quality of rabbit rex meat of male off weaning

| | Parameters | | | | | |
|---|---|---|---|---|---|---|
| Treatment | pH | Cooking Loss (%) | Tenderness (mm/g/dt) | Proteint content (%) | Water content(%) | Fat content (%) |
| P0A | 6.096 | 38.812$^{A}$ | 0.674$^{b}$ | 13.615$^{C}$ | 70.716$^{D}$ | 18.798$^{a}$ |
| P0B | 6.262 | 36.343$^{C}$ | 0.946$^{a}$ | 11.307$^{D}$ | 75.748$^{C}$ | 18.479$^{a}$ |
| P1 | 6.263 | 35.684$^{C}$ | 0.715$^{b}$ | 14.592$^{BC}$ | 79.721$^{AB}$ | 18.149$^{a}$ |
| P2 | 6.184 | 33.527$^{D}$ | 0.955$^{a}$ | 17.243$^{A}$ | 80.464$^{A}$ | 16.034$^{b}$ |
| P3 | 6.160 | 37.595$^{AB}$ | 0.733$^{b}$ | 13.975$^{BC}$ | 76.719$^{C}$ | 18.642$^{a}$ |
| P4 | 6.084 | 38.784$^{BC}$ | 0.584$^{b}$ | 15.193$^{B}$ | 77.792$^{BC}$ | 17.472$^{ab}$ |

Description: Superscript with different capital letters in the same column shows a very significant different effect ($P < 0.01$).

### *3.1. pH*
According to Lawrie [9]. pH value of meat needs to be known because the pH of the meat will determine the growth and development of bacteria. Virtually all bacteria grow optimally at a pH of about 7 and will not grow exactly below pH 4 or above 7. but the pH for optimal growth is determined by stimulant work of various other variables beyond the acidity factor itself. From the results of this study. pH value of rex rabbit meat showed that meat was feasible to be consumed and good quality because it showed pH value below 7.0 and not contaminated with bacteria with the largest meat pH ultimate is 6.1.

### *3.2. Cooking Loss*
The result of variance analysis in this study showed that the use of coconut waste fermented in the ration gave a very significant different effect ($P < 0.01$) on cooking loss of male rex rabbit meat. This related with nutrition contents of coconut waste after fermentation which improved thus improved meat quality. According to Warriss less cooking loss on meat meanes that meat has better quality than meat that has more cooking loss due to loss of nutrients during cooking loss. As the result. it also

related with better meat taste or juiceness. Based on the results of this study. the lowest cooking loss of rabbit meat was found in the P2 treatment of 33.527 ± 0.79934 which means that meat with this treatment has the best quality in terms of nutritional content.

### *3.3. Tenderness*

Analysis of rabbit meat texture variance during the study showed very different significant effect ($P < 0.01$) on meat tenderness. This is in accordance with the statement of Soeparno [10] which stated that antetorem factors affecting meat tenderness were species. physiological. age. management. gender and stress.

The mean value of tenderness in this study showed that the highest average was in treatment P2. while the lowest was in treatment P4. This means that the value of meat texture in P2 treatment is better than meat in treatment P4. This occured because the collagen levels present in P2 was higher than P4. The texture of meat is also influenced by the protein content of collagen in the meat. Collagen is the main protein in the connective tissue. where the connective tissue is present in almost every component of the body. Collagen connective tissue has an important role on the quality of meat. especially on the tenderness of meat. This is in accordance with the statement Soeparno [10] which states that the meat tenderness also influenced by the protein content of collagen in the meat This is also in accordance with Reny's [11] statement which states that factors affecting meat tenderness have something to do with the composition of the meat itself. which is in the form of woven binding. meat fibers. fat cells that exist between the fibers of meat.

This was consistent with the statement of Aberle et al. [12] which states that the regulation of rations for livestock directly affects the variation in the nature of the tendon after the fattening and the fattened cattle in the cage will produce more tender meat than the grazing animals. Bouton et al.. [13] stated that age under certain conditions does not affect the tenderness of meat produced. Older cattle but getting ration with good nutrition and handling can produce more tender meat compared to meat produced from young cattle but get ration nutrients and poor handling. Muscles can grow and develop well if they get good nutrition and handling. A good muscle has a smaller amount of collagen per unit area of muscle compared to the muscles of the livestock that get the nutrients and the poor handling. so that the meat will be more tender.

### *3.4. Protein levels*

The results of this study indicated that the provision of coconut waste fermented has very significant different effect ($P < 0.01$) on protein content of rabbit meat. The highest level of meat protein (17.243%) was found in the treatment of P2 while the lowest protein content (11.307%) was found in the treatment of P0B. This means that the type of fermentator and levels in the feed both have a very real effect ($P < 0.01$) on the protein content of meat.

The difference in protein levels was due to differences in levels in the feed ingredients. In addition. levels of fat and water content also affect protein levels in meat. In the results of this study. the fat content of meat in the treatment of P2 was the lowest. causing the meat has a high protein content. The content of fat and water content has a relationship that is inversely related to the protein content in the meat. This is in accordance with the statement Soeparno [10]. which stated that the chemical composition of meat is strongly influenced by the fat content. Increasing the fat content of meat and water content causes the protein content will decrease.

### *3.5. Water content*

The results of this study showed that coconut waste fermented very significantly different ($P < 0.01$) to the moisture content of rex rabbit meat weaning. The highest water content of meat was found in P2 treatment (80.464%) while the lowest water content was found in P0A treatment (70.716%). This means that the type and content of fermentor both have very significant different effect on the water content of meat ($P < 0.01$).

The average water content of rabbit meat in this study was still within the normal range (70-80%). in accordance with Lawrie's [14] statement which stated that generally the water content of meat ranges from 68 to 80%. The difference of water content of meat between treatments in this study was influenced by the fat content of the meat. this was because the high fat accumulation rate can loosen the bonding of the structure of the meat network so that much water is liberated. In this research. it was found that the fat content of meat in the treatment of P0A. P3. P0B. P1 and P4 was significantly different (P <0.05) to P2 treatment. so that water content in treatment P0A. P3. P0B. P1 and P4 tended to be higher compared to meat in treatment P2. This result is in accordance with Soeparno's [10] statement. which stated that the fat content has a negative relationship with moisture content. If the fat content of the meat increases the increase in the weight of life then the water content will decrease. so the increase in age will increase the fat content.

### *3.6. Meat Fat Level*

The results of this study indicated that the provision of coconu waste fermented has a significantly different effect (P <0.05) on the fat content of rex rabbit meat. Based on the data in Table 1 it can be seen that the highest content of fatty meat found in P0A treatment was equal to 18.798%. while the average fat content of the lowest meat found in P2 treatment is 16.034%.

Fat levels are influenced by the nation. age. species. location of muscles and feed. Rex rabbits that consume P0A pellet rations have higher levels of fat than rex rabbits that consume P2 rations. This was thought to occur because the feed on the treatment of P0A does not undergo fermentation so that it still has high levels of fat and coarse fiber and on feed treatment P2 has fermentation which causes the decrease in fat content and crude fiber in the feed ingredients. It was also suspected because the fermentation treatment feed done can increase the value of feed TDN. Differences in fat content are also suspected because the water content of rabbit meat treated P0A was lower than rabbit meat in P2 treatment. which causes fat content of P0A meat higher than fat content of meat in treatment P2. This was consistent with the statement by Browning et al. [15] which stated that the high level of fatty meat accumulation can loosen the bonding of meat tissue structures so that much water is liberated. therefore meat containing high levels of fat tends to contain low water content.

## 4. Conclusion

Coconut waste can be used as rabbit feed material. especially when processing by fermentation that can increase the content of waste nutrients as feed ingredients ration rabbit. Coconut waste fermented by Aspergillus niger at level 20% in pellet ration gives positive effect that can improve the quality of rabbit rex meat. especially in cooking loss. increase of protein content and water content. and decrease of fat content.

## References

[1] Ginting N 2017 Benefits of using biogas technology in rural area: Karo District on supporting local action plan for greenhouse gas emission reduction of north sumatera province 2010-2020 In ternational Conference on Biomass : Technology. Application. and Sustainable Development IOP Publishing. *IOP Conf. Series: Earth and Environmental Science 65* (2017) 012007 doi: 10.1088/1755-1315/65/1/012007

[2] Miskiyah I, Mulyawati dan Haliza W 2006 Pemanfaatan ampas kelapa limbah pengolahan limbah kelapa murni menjadi pakan (The Usage of Palm Waste from Waste Processing of Pure Palm to Food) *Prosiding. Seminar nasional teknologi peternakan dan venteriner*

[3] Kandeepan G, Anjaneyulu A S R, Rao V K, Pal U K, Mondal P and Das C K 2009 Feeding regimens affecting meat quality characteristics *Meso* **11** (4):240-249

[4] Suprihatin 2010 Teknologi Fermentasi (Surabaya: UNESA Press)

[5] Hardjo S S, Indrasti N S and Takjuddin B 1989 *Biokonveksi: Pemanfaatan Limbah Industri Pertanian (The Usage of Agriculture Waste Industry)* Pusat Antar Universitas Pangan dan Gizi IPB
[6] Lehninger W W 1991 *Dasar-dasar Biokimia (Biochemical Basis)* (Jakarta: Penerbit Airlangga)
[7] Winarno F G 2004 *Enzim Pangan (Food Enzim)* (Jakarta: PT. Gramedia Pustaka Utama)
[8] AOAC 1995 *Official Method of Analysis. 12th ed Association of Official Analytical Chemist* Washington DC
[9] Lawrie R A 2003 *Ilmu Daging ( The Science of Meat)* Penerjemah Aminuddin P. UI-Press Jakarta
[10] Soeparno 2005 *Ilmu dan Teknologi Daging (Science and Technology of Meat)* Cetakan III Gadjah Mada University Press Yogyakarta
[11] Reny D T 2009 *Kempukan Daging dan Faktor-faktor yang Mempengaruhuinya (The terderness of Meat and the Influences factors* Balai Pengkajian Teknologi Pertanian Lampung
[12] Aberle E D, Forrest J F, Gerrard D E, Mills E W, Hendrick H B, Judge M D dan Merkel R A 1981 *Principles of Meat Science. 1st Ed* (Iowa: Kendall/Hunt Publishing Company)
[13] Bouton P E, Harris P V, Shorthose W R dan Baxter R I 1978 *J. Food Sci.* **38** pp 932 – 939.
[14] Lawrie R A 2003 *Ilmu Daging (The Science of Meat)* Penerjemah Aminuddin P. UI-Press Jakarta
[15] Browning M A, Huffman D L, Egbert W R and Junst S B 1990 Physical and compositional characteristic of beef carcases selected for learnness *J. Food Science* **51**:51-56

# Genetic distance estimates and variable factors distinguishing between goat Kacang, Muara and Samosir

**Hamdan[1,2], H Saputra[1,2] E Mirwandhono[1,3], Hasnudi[1,2],I Sembiring[1,3], S Umar[1,2], N Ginting[1,2] and Alwiyah[1,2]**

[1]Animal Production Program Study, Faculty of Agriculture, Universitas Sumatera Utara, Indonesia
[2]Animal Production Laboratory, Faculty of Agriculture, Universitas Sumatera Utara, Indonesia
[3]Animal Social Economy Laboratory, Faculty of Agriculture, Universitas Sumatera Utara, Indonesia
Email: nurzainahginting@gmail.com

**Abstract.** The purpose of this research was to look the genetic distance and factors distinguishing variable betwen types of goats in North Sumatera. This research have been conducted in PayaBakung, Hamparan Perak and Klambir Lima village, Deli Serdang district, Batu Binumbun, Aritonang, HutaGinjang village, Muarasubdistrict, North Tapanuli district and ParbabaDolok, Siopat Sosor, Sinabulan village, Ronggur Nihuta Pangururan village, Sitonggi-tonggi village in the subdistrict RonggurNihuta, Samosir district of the month of July 2016. The data was analyzed using descriptive, discriminants, canonical, Principal Component Analysis, Distance genetic and Tree Phylogenetic. The result showed that the nearest genetic distance goat found in Kacang and Samosir (1.973), and the farthest genetic distnace find in Samosir and Muara (8.671). The variables made it difference was goat race Base Rim Horn (0.856) and Long Horn (0.878). Genetic distance values most far between Muaragoat with Samosir goat was (8.671). The conclude that the crossing superior result, must be cross between two goat types with value genetics most distance.It will have a better chance heterosis in cross result.

## 1. Introduction

Goats are the first animals to be domesticated by humans. Goats come from wild animals (Capra hircus aegagrus) that live in difficult areas and rocky areas. It is estimated that at the beginning the hunters bring back the goats from the prey. Goats are kept in the village as pets, then utilize for produce milk, meat and skin. In Indonesia, goat is found in rural areas.

Goat breeding has long been endeavored by farmers or the community as a side business or savings because the maintenance and marketing of their products (meat, dairy, dirt and leather) is relatively easy, although the traditional way of breeding is done.

Population of goats in Asia estimated about 225 million or 52% of the total population in the world. In Indonesia at 2014 have 19.2 million goats to be maintained. Livestock for small ruminants is mostly in the hands of small farmers, with the main business is farming, and the others was used as side business.

Many types of goats exist in Indonesia is known to have different qualities and quantitative depends the nation. To be able to know the qualities and quantitative should know about nation of goats.

The desirable properties of a breeder are beneficial or beneficial properties, such as high production power, number and weight of good birth puppies, rapid growth, low mortality and high food use efficiency Fahmy and Bernard [1]. Livestock productivity is enhanced through genetic improvement by selection and marriage and through environmental improvement[2].

One method of phenotypic differences that can be used to determine genetic distance is diversity analysis Komenes [3]. Phenotypic diversity can be determined by measuring body parts or morphometrics. Measurement of body size is used to distinguish the diversity of both size and body shape. Diversity and correlation analysis is widely used in characterizing the relationship of phenotypic and genetic traits [4].

To increase the production of livestock one of which is considered genetic factors, livestock with close kinship may be expected to be less likely to increase heterosis in its crosses. Therefore, before the marriage need know about genetic distance between types. It aims to obtain superior or profitable properties.

To increase goat production is not enough just feed and environmental factors but also genetic factors. Close kinship or low genetic distances have little chance of acquiring superior traits in crosses.

Therefore, the author is interested in do a research related to genetic distance factors from some goats in North Sumatra. That is the estimation of genetic distance and distinguishing factor of difference between goats Beans, Estuary and Samosir through Craniometric analysis.

## 2. Materials and Research Methods

The location of this research are:

1. Kacang goat in PayaBakung village, Hamparan Perak village and KlambirLima village, Hamparan Perak district, Deli Serdang district.
2. Muara goat in BatuBinumbun village, Aritonang village and HutaGinjang village, Muara sub-district, North Tapanuli district.
3. Samosir goat in the village of ParbabaDolok, SiopatSosor village and Sinabulan village in Pangururan sub-district and RonggurNihuta village, Sitonggi-tonggi village in RonggurNihutasubdistrict, Samosirdistrict.This study was conducted in July 2016.

The material used in this study was 100 head of Kacang goats with 34 males and 66 females, 100 head of Muara goats consists of 39 male and 50 female and 100 head Samosir goats with 37 males and 63 females who have adult genitals and adult body. The sample determination was done by census and the sample selection was done randomly.

The tool used in this study is the measuring tape or ruler in units (cm) and the sliding range in units (mm) as a gauge of the head part of the goat, the data book as a writing container of measurement results, the stationery to write the measurement results and the camera to document Measurement process, SAS (Statistical Analysis System) software, SPSS (Statistict Product and Service Solution) and MEGA are used to analyze data.

The research method used by census. The sampling technique was done randomly by taking the subject of goats Beans, Estuary and Samosir with age limit after adult body and adult sex. Parameters measured were skull length, skull width, horn length, horn circumference and inter-horn distance using unit gauge (cm) and unit slot (mm). Determination of the location of the research is selected in Hamparan Perak District, Muara District and Samosir District.

Parameters in this research are:

*1. Skull length (PTe)*

The length of the skull was measured from the base of the upper mouth to the upper part of the skull (X1).

*2. Skull width (LTe)*
The width of the skull was measured from the left cheekbone to the right cheek bone (X2).
*3. Long Horn (PT)*
The length of the goat's horn was measured from the base to the end of the horn (X3).
*4. Boundary of the Horn (LPT)*
The circumference of the goat horn was measured by encircling the meter at the base of the horn (X4).
*5. Between the Horns (JAT)*
The distance between the horns was measured horizontally from the base of the right horn to the base of the left horn (X5).
Measurements were made using a tape measure with units (cm) and with a unit sliding range (mm).

*2.1. Data Analysis*

*2.1.1Quantitative properties.* Descriptive statistical analysis was shown to obtain characterization of body sizes in goats. This analysis was done by calculating the value of mean (X), standard deviation (s) and coefficient of diversity (KK) with the following statistical procedure according to Wiley [5].

$$\overline{X} = \frac{\sum_{i-1}^{n} x^{i}}{n} \qquad S = \sqrt{\frac{\sum_{i-1}^{n}(xi-\bar{X})^2}{n-1}}$$

kk= (100 %)

*2.1.2. Genetic Distance.* The genetic distance of Mahalanobis as the minimum quadratic square distance measure used according to Mahalanobis [6]:

$$D^2_{(i,j)} = (\bar{X}_i - \bar{X}_j)\, C^{-1}(\bar{X}_i - \bar{X}_j)$$

*2.1.3. Canonical Analyzes.* Canonical analysis was performed to determine dispersion maps and similarity and mix values within and between groups of livestock Herrera [7].Canonical Discriminant Analysis were used to determine some variable to be used as discriminating variable or discriminator variable in the distribution of livestock based on the value of canonical structure or canonical correlation. This analysis was also used to estimate the genetic distance between livestock populations through the statistical distance of the mahalanobis and describes the relationship. Livestock in two dimensions based on the cumulative ratio of contributions from the canonical variablesGunawan and Sumantri [8].

*2.1.4. Discriminant Analysis.* When the test results of the two mean values of the properties used are different then the Fisher discriminant function was used to assess the differences in properties found among the goat groups. The discriminant function according to Wiley [9]:

Y = a 'X = (X1 - X2)' SG-1 X

*2.1.5. Main Component Analysis.* The size and shape equations are derived from the covariance matrix. The Main Component Analysis (MCA) used is based on Simanjuntak et.al [10]. The model size equation is as follows:

Y1 = a11X1 + a21X2 + a31X3 + ...... + a101X 10

*2.1.6. The Philogenetic Tree.* A common method used to design phylogenetic tree is to use the genetic distance matrix and maximum parsimony methods. The construction of phylogenetic tree with the genetic distance matrix method could be done more because the genetic distance can be obtained by observation and some parameters. The maximum parsimony method is generally more limited in phylogenetic tree construction because it uses the amino acid or nucleotide sequence data Mirabella [11].

## 3. Results and Discussion

From Table 1 it can be seen that the highest average of the three goats is Muara goat with the length of the skull (15.80 ± 1.58) with the KK (2.50%) and the lowest is the average of goat Beans from the intermediate distance variables with the value (2.66 ± 0.59) And KK (0.35%). The differences in body composition among livestock are mainly due to differences in adult body size. Environmental and genetic factors have a close relationship to express the individual's genetic capacity and an ideal environmental condition is required Saparto [12].

### *3.1. Differentiator of Goat*

From the results of the Main Component Analysis (Table 2)it can be seen thebody size variables that have a strong influence that distinguish the goat, and themost distinguishing variables are the length of the horn (0.878) and the hollow circumference (0.856). The estimate is based on the high eigenvalues of the hornsand circumference of the goat's horns. Prasetia [13] stated that the size of the phenotypic variables differ between types of cattle with one other.

### *3.2. Analysis nation clump of goat*

The results of the analysis show (Figure 1) that the type of goat Beans spread on the X axis clustered with goats Samosir while goats grouping itself on the axis Y. Analysis above can be seen the closest relationshipis Nuts and Samosir while the most distant is Estuary with Samosir. The plot of discriminant analysis data can be used to describe the maximum possible separation between the tested groups Noor [14].

### *3.3. Discriminant Analysis*

Based on the Table shows that the overall value of the observed variables has a comparison between goats species. The farthest value of comparison / variation is the value of Muara goat (-120.739) with Samosir goat (-94.387). The highest comparison value of the three types of goats was found in the Muara goat (11,814) from the skull length variant and the lowest was in Samosir goats (-2.091) from the skull length parameter. According toSaloko [15] size variations occur between different species, even in the same species. Several factors that limit the size and shape of animals, include genetic conditions, nutrient supply, toxicity, surface area to volume ratio, and structural limiting factors.

### *3.4. Determination of Estimation of Genetic Distance And Dendogram Between Goats*

From the results in the Table 3 above can be seen that the highest or most distant genetic distance was Muara goat with Samosir goat with value (8.671), from this result can be stated that goat and goat gaosSamosir possible no crossing outside the goat. The lowest genetic distance was Kacang goats with Samosir goats that is (1.937) while the genetic distance value of Kacang goat with Muara goat was (7.417).

### *3.5. Dendogram Analysis*

From the results of the Table 4 and the image of the dendogram analysis (Figure 2) above stated that the lowest distance value is in Kacang goat with Samosir goat that is(0.968) based on this value crossing between the two goats will not provide significant quantitative progress, if not accompanied by strict selection, it is possible Because of the small chance of heterosis (measured quantitative measurement of the superiority of children against the average parent) on the results of crossing between the two in the future. The goats that have the most distant kinship connections are Muara goats and Samosir goats and expected crossing between these two goats will provide significant quantitative development, this is possible because the greater the chances of heterosis in the cross between the two future.

## 4. Conclusion and Recommendation

### *4.1. Conclusion*

The highest or furthest genetic distance value from the three goats studied was Muara Goat with Samosir goat (8,671). From this result, it can be concluded that to get the superior crosses should be crossed between the two goats because the distance Themost distant genetic will have a greater chance of heterosis in the cross.

### *4.2. Recommendation*

We recommend that the purity of these three goats (Peanut, Estuary and Samosir) should be improved and maintained in order to be a germplasm for the region of North Sumatra and In carrying out this research is expected to carefully and accurately sample measurements to obtain accurate and precise data.

## References

[1] Lasley T J 1978 *Genetic of Livestock Improvement. 3rd Ed.* (New Delhi: Prentice Hall of India Private Ltd.)
[2] Komenes P A 1999 Alfa-casein and beta-lactoglobulin and growth hormone allelesfrequencis and genetics distances in Nelore, GYR, Guzera, Caracu, Charolais, Chanchim and Santa Getrudiscattles *Gen MolBiol* **22** : 539-541
[3] Salako A E and Ngere L O 2002 Application of multifactorial discriminant analysis in the morphometric structural diffrentiation of the WAD and Yankasa sheep in the humid southwest Nigeria. *Nig. J. Anim. Prod.* **29** (2) : 163-7
[4] Heridianto S 2015 Estimasi Jarak Genetik dan Faktor Perubah Pembeda Beberapa Bangsa Kambing di Sumatera Utara Melalui Analisis Morfometri (Medan: USU)
[5] Nei M 1987 *Moleculer Evolutionary Genetic* (USA:Columbia University Press)
[6] Rodero H M E, Gutierrez M J, Pena F and Rodero J M 1996 *J. Small Rum. Res.* **22**:39-47
[7] Gazpers V 1992 *Teknik Analisis dalam Perancangan Percobaan Penelitian* (Bandung: Penerbit Tarsito)
[8] Netro B 2014 Estimasi Jarak Genetik dan Faktor Perubah Pembeda Bangsa Babi (*Berkshire, Duroc, Landrace dan Yokshire* )Melalui Analisis Morfometrik Universitas Sumatera Utara Medan
[9] Wiley E O 1981 *Phylogenetics: The theory and Practice of Phylogenetic Systematic* (Canada: Jhon Wiley and Sons inc.)
[10] Afifi A A and Clark V 1996 *Computeraided Multivariate Analysis 3rd Edit* (New York: Chapman and Hall)
[11] Dani H 2005 *Kraniometri Beruk (Macaca Nemestrina)* Institut Pertanian Bogor Bogor
[12] Saparto 2004 Studi Kraniometri Sapi Jawa dari Beberapa Bangsa Sapi Potong di Indonesia UNDIP Semarang
[13] Arif P 2011 Studi Ukuran dan Bentuk Tubuh Sapi Pesisir, Sapi Bali, dan Sapi Peranakan Ongole Jantan InstitutPertanian Bogor Bogor
[14] Noor R R 2008 Genetika Ternak (Animal Genetic) Cet. Ke-4 (Jakarta: Pebar Swadaya)
[15] Salako A E dan Ngere L O 2002 Application of multifactorial discriminant analysis in the morphometric structural diffrentiation of the wad and yankasa sheep in the humid Southwet Nigeria *Nig. J. Anim. Prod.* **29** (2):163-7

**Table 1**. Average, standard deviation and coefficient of variation among goats (adult body)

| Observed Variable | Kacang | Muara | Samosir |
|---|---|---|---|
| Skull length | 14.45±1.06 | 15.80±1.58 | 13.40±0.78 |
| Average±SD, **CD (%)** | **1.12** | **2.50** | **0.61** |
| Skull Width | 11.22±1.2 | 10.55±1.34 | 10.63±1.08 |
| Average±SD, **CD (%)** | **13.75** | **6.72** | **8.16** |
| Long Horn | 10.82±3.7 | 8.36±2.59 | 8.09±2.85 |
| Average±SD, **CD (%)** | | | |
| Ring of the Horn | 10.69±2.30 | 8.00±1.80 | 9.33±1.79 |
| Average±SD, **CD (%)** | **5.31** | **3.24** | **3.22** |
| Distance Between Horns | 2.66±0.59 | 3.46±0.47 | 3.17±0.53 |
| Average±SD, **CD (%)** | **0.35** | **0.22** | **0.28** |
| Total (N) | 100 | 100 | 100 |

Inf : SD : Standard Deviation
CD : Coefficient Diversity

**Table 2**. Main Component Analysis Body Size Variable

| Variable Body Size | Factor 1 | Factor 2 |
|---|---|---|
| Skull length | 0.492 | 0.720 |
| Skull Width | 0.797 | 0.253 |
| Horn Length | 0.878 | -0.002 |
| Outer Horn Circle | 0.856 | -0.271 |
| Distance Between Horns | -0.403 | 0.798 |

**Table 3.** Discriminant Analysis

| | Kind | | |
|---|---|---|---|
| Parameter | Kacang | Muara | Samosir |
| Skull length | 9.595 | 11.814 | 8.839 |
| Skull Width | 4.970 | 4.556 | 5.038 |
| Long Horn | -1.899 | -2.030 | -2.091 |
| Ring of the Horn | 0.432 | -0.588 | 0.642 |
| Distance Between Horns | 6.833 | 8.172 | 8.715 |
| Y | -98.386 | -120.739 | -94.387 |

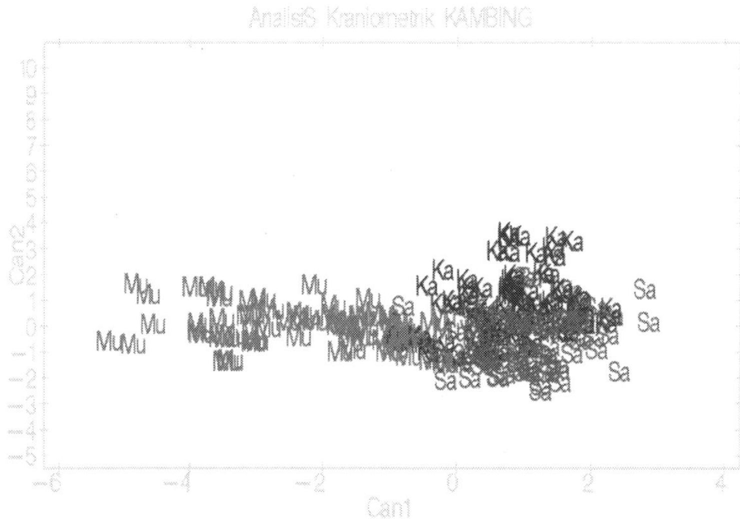


**Figure 1**. Goats Cariometric Analysis

**Table 4.** Matrix of goat's genetic distance

| Goat | Kacang | Muara | Samosir |
|---|---|---|---|
| Kacang | .000 | | |
| Muara | 7.417 | .000 | |
| Samosir | 1.937 | 8.671 | .000 |

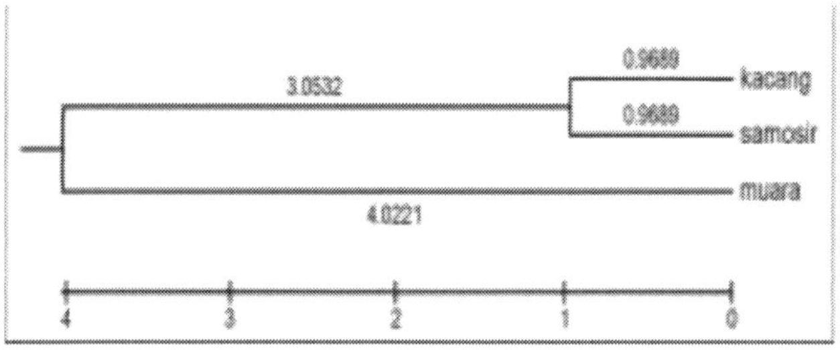


**Figure 2.** Dendogram analysis of goat

# Anylisis of botanical composition and nutrient content on natural pastures in Samosir Island of Samosir Regency

**N D Hanafi[1], M Tafsin[1], U Hutasuhut[2] and E Lubis[3]**

[1]Lectures, Faculty Agriculture, Universitas Sumatera Utara, Medan, North Sumatera 20155, Indonesia
[2]Postgraduate Student, Faculty of Agriculture, Universitas Sumatera Utara, Medan, North Sumatera 20155, Indonesia
[3]Graduate Student, Faculty of Agriculture, Universitas Sumatera Utara, Medan, North Sumatera 20155, Indonesia
E-mail : nevydiana@yahoo.co.id

**Abstract.** Samosir regency is one of the areas that have large grazing area. The potential of grazing production in the area plays an important role for the development of livestock, especially on ruminant livestock.The study aims to know the botanical composition and the nutritional content of forage on natural pasture at the Samosir Island. Animal feed assessment method on natural pastures in Samosir regency includes the determination of research location points based on the altitude through the survey method. Location of the study amounted to 15 locations. The result showed that at altitude 905 – 1200 meters above sea level had a botanical composition were 31 species with ratio of grass 80.58 %, legumes 9.14 % and weeds 9.63 % and the most dominant forage is *Imperata cylindrica l.* The botanical composition at altitude more than 1205 meters above sea level is 15 species with ratio of grass 92.72 %, legumes 2.87 % and weeds 4.39 % and the most dominant forage is *Axonopus compressus*. The forage which has the highest crude protein is *Starkuak* 15.13 %. The conclusion that the altitude in pastures give effect on the botanical composition of forages.

## 1. Introduction

Community livestock business in Samosir Regency is still very simple. Livestock breeding uses extensive systems (not stacked). Livestock only grazed above the hills of Lake Toba as well as on the slopes, hills especially on empty fields. If viewed the actual condition, the area of Samosir Regency is 1444.25 $km^2$ with an empty land area of 46036.39 ha and 21046 ha of shrubs (where the bushes and shrubs grow).

Based on altitude, Samosir Island is located at altitude 904–2157 meters above sea level. Condition of Samosir Island has hilly and bumpy contours with natural expanse of grassland that is wide enough spread in various places. The natural grasslands have the potential and deserve to be developed further as a provider of forage feed. Therefore, based on the altitudes of different places it is necessary to analyze the botanical composition and nutritional content of grasslands on Samosir Island.

This study aims to determine the botanical composition and nutrient content of forage on natural pastures contained in Samosir Island.

## 2. Materials and Methods

The research has been carried out in Samosir Island, Samosir District start from July to October 2016.

The materials used are forage samples that taken from Samosir Island. The tools used includes : 1x1 m square quadrant as a tool for measuring forage production, scissors for cutting forage, scales as a means of weighing fresh materials and dry materials, oven as a means of drying fresh material to obtain dry matter.

### *2.1. Determination of Sampling Site*

The selection of research location was carried out purposive in Samosir Island are:

a. At altitude 905–1200 meters above sea level consist of several villages. They are among others Simbolon, Unjur, Garoga, Marlumba, Suhi-Suhi Dolok, Parbaba Dolok, Lumban Pinggol, Onanrunggu, and Peanabolak.
b. At altitude more than 1205 meters above sea level or a maximum 1690 meters above sea level consist of several villages. They are among others Tanjungan, Sidihoni, Lintong Sunut, Lumban Simbolon and Sipira.

### *2.2. The Variables Observed*

The variables observed in this research are botanical composition and nutrient content analysis. The botanical composition is calculated on the basis of dry matter by the *Dry Weight Rank (DWR)* method. Nutrition Content Analysis using proximate analysis includes dry matter, ash content, crude protein, crude fiber, ether extract and beta-N.

## 3. Result and Discussion

### *3.1. Overview of Research Sites*

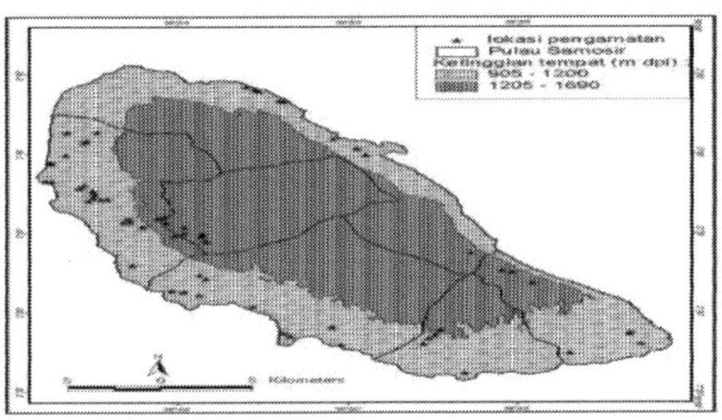

**Figure 1.** Altitude Map

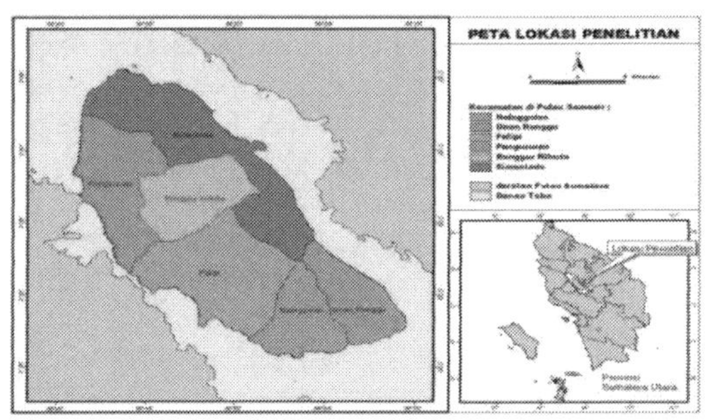

**Figure 2.** Samosir Island Map

Geographically Samosir Regency is located between 2°21'38" - 2°49'48" LU and 98°24'00" - 99°01'48" BT with altitude between 904–2157 meters above sea level. The total area is about 2069.05 km² and consists of land area of ±1444.25 km² or about 69.80%, that is all Samosir Island surrounded by Lake Toba and some areas of mainland of Sumatera Island and wide of Lake Toba ±624.80 km² (30.20%) [1].

Topography and soil contours are diverse among others flat, hilly, bumpy, tilted and steep. The soil structure is unstable and is in the path of tectonic and volcanic earthquakes. Topography and soil contours in Samosir District with slope composition: a) 0–20 (flat) ± 10%, b) 2–150 (ramps) ±20%, c) 15– 400 (inclined) ±55%, d) >400 (steep) ±15% [1].

### *3.2. Determination of Research Site*

Based on survey results on Table 1 it is known there are 63 points of natural pastures on the island of Samosir where the point is determined based on the interpretation of land area and a reasonable estimate to serve as a place of research. The survey results on the 63 points obtained are then mapped based on the height of the site, the class of land capability and land use to determine the points of the location where the sample will be taken.

**Table 1.** Natural Pasture Mapping Based on Altitude on Samosir Island

| Altitude | Land Of Capability Class | Land Use | | | | |
|---|---|---|---|---|---|---|
| | | Shrubs Land | Open Land | Dry Land Farming | Rice Fields | Swamp |
| 905–1200 meters above sea level | II | - | - | 4 | - | - |
| | III | - | - | - | - | - |
| | IV | - | 22 | 23 | - | - |
| More than 1205 meters above sea level | II | - | - | - | - | - |
| | III | - | - | - | - | - |
| | IV | 1 | 1 | 9 | 2 | 1 |

Sampling is then selected based on the analysis that has been done and taken a sample of 15 points that serve as a place for sampling research. Sampling at altitude 905–1200 meters above sea level amounted are 9 locations, while at an altitude above 1205 meters above sea level are 6 locations.

### *3.3. Botanical Composition*

At altitude 905–1200 meters above sea level the forage botanical composition highest is *Imperata cylindrica l* with fresh material production 21.34% and dry matter production 24.34%. Then followed by *Axonopus compressus* has fresh material production 17.40% and dry matter production 17.64%. Furthermore *Paspalum conjugatum* has fresh produce of 13.35% and dry matter production 13.77%.

The botanical composition of each type of forage varies according to the total production of fresh ingredients in each species. In addition, the highest difference is the cultivated *Imperata cylindrica l* because the grass can multiply rapidly due to the rapid development and spread and is easy to grow and resistant to stamping. This is consistent with the statement of Jayadi [2] which states that *Imperata cylindrica* can breed rapidly with its seeds spread rapidly with the wind, or through its rapid rhizomes through the loose soil. *Imperata cylindrica* usually grows on formerly damaged and open forest land, former fields, dry fields, roadsides and so on. To some extent, vegetation fires can stimulate the growth of *Imperata*.

The percentage of botanical composition based on forage weight on natural pastures on Samosir Island at altitude of 905–1200 meters above sea level is shown in Table 2 below.

**Table 2.** Botanical Composition and Number of Occurrence Frequencies on Natural Pasture in Samosir Regency at Altitude 905–1200 Meters above Sea Level

| No | Type of Forage | Botanical Composition (%) | | Frequency |
|---|---|---|---|---|
| | | Fresh Material | Dry Matter | |
| 1. | *Imperata cylindrica l* | 21.34 | 24.34 | 19 |
| 2. | *Axonopus compressus* | 17.40 | 17.64 | 28 |
| 3. | *Paspalum conjugatum* | 13.35 | 13.77 | 12 |
| 4. | *Zoysia matrella* | 10.37 | 12.75 | 13 |
| 5. | *Melastoma malabathricum* | 6.02 | 5.90 | 18 |
| 6. | *Cynodon plestosthacyrus* | 5.10 | 4.08 | 7 |
| 7. | *Stylosanthes capitata* | 4.34 | 3.71 | 16 |
| 8. | *Mimosa pudica l* | 2.92 | 2.87 | 14 |
| 9. | *Leptochloa chinensis l* | 2.91 | 1.91 | 2 |
| 10. | *Digitaria milanjiana* | 2.83 | 3.23 | 4 |

| | | | | |
|---|---|---|---|---|
| 11. | *Rabba-rabba* | 2.36 | 2.49 | 3 |
| 12. | *Stachytarpheta jamaicensis* | 2.31 | 1.43 | 3 |
| 13. | *Eupathorium adenophorum* | 1.77 | 1.34 | 3 |
| 14. | *Starkuak* | 1.37 | 0.38 | 2 |
| 15. | *Hiptis brevipus* | 0.90 | 0.54 | 2 |
| 16. | *Arachis pintoi* | 0.66 | 0.47 | 1 |
| 17. | *Desmodium triflorum* | 0.66 | 0.47 | 15 |
| 18. | *Vigna parkeri baker* | 0.62 | 0.62 | 9 |
| 19. | *Cassia sp* | 0.58 | 0.44 | 1 |
| 20. | *Leucena sp* | 0.40 | 0.35 | 4 |
| 21. | *Centella asiatica* | 0.33 | 0.18 | 4 |
| 22. | *Oryza sativa* | 0.29 | 0.19 | 1 |
| 23. | *Gleichenia linearis* | 0.28 | 0.27 | 4 |
| 24. | *Centrosema pubescens* | 0.26 | 0.22 | 2 |
| 25. | *Brachiaria mutica* | 0.24 | 0.28 | 1 |
| 26. | *Digitaria ciliaris* | 0.20 | 0.03 | 2 |
| 27. | *Sabi-sabi* | 0.12 | 0.04 | 1 |
| 28. | *Cynodon dactilon* | 0.04 | 0.03 | 1 |
| 29. | *Brachiaria decumbens* | 0.02 | 0.02 | 1 |
| 30. | *Chloris gayana* | 0.01 | 0.01 | 1 |
| 31. | *Ageratum conyzoides* | 0.01 | 0.01 | 1 |

At altitude more than 1205 meters above sea level the botanical composition highest is *Axonopus compressus* with fresh material production 66.36% and dry matter production 70.52%. Forage type with the second highest percentage is *Paspalum conjugatum* with fresh material production of 16.54% and dry matter production 15.81%. Then the third highest forage was *Centella asiatica* with fresh material production 4.04% and dry matter production 2.48%.

The percentage of botanical composition based on forage weight on natural pastures on Samosir Island at altitude more than 1205 meters above sea level is shown in Table 3 below.

**Table 3.** Botanical Composition and Number of Occurrence Frequencies on Natural Pastures in Samosir Island at Altitude More Than 1205 Meters Above Sea Level

| No | Type of Forage | Botanical Composition (%) | | Frequency |
|---|---|---|---|---|
| | | Fresh Material | Dry Matter | |
| 1. | *Axonopus compressus* | 66.36 | 70.52 | 13 |
| 2. | *Paspalum conjugatum* | 16.54 | 15.81 | 7 |
| 3. | *Centella asiatica* | 4.04 | 2.48 | 2 |
| 4. | *Flemingia macrofilia* | 3.10 | 2.27 | 1 |
| 5. | *Imperata cylindrica l* | 2.55 | 2.67 | 1 |
| 6. | *Eupathorium adenophorum* | 2.06 | 0.98 | 1 |
| 7. | *Zoysia matrella* | 1.78 | 2.01 | 1 |
| 8. | *Stylosanthes capitata* | 0.96 | 0.78 | 2 |
| 9. | *Amaranthus sp* | 0.89 | 0.69 | 1 |
| 10. | *Cynodon plestosthacyus* | 0.52 | 0.47 | 2 |
| 11. | *Mimosa pudica l* | 0.49 | 0.43 | 1 |
| 12. | *Chloris gayana* | 0.28 | 0.29 | 1 |
| 13. | *Lycopediella cernua* | 0.20 | 0.24 | 1 |
| 14. | *Paspalum notatum* | 0.13 | 0.18 | 1 |
| 15. | *Leucena sp* | 0.13 | 0.17 | 2 |

Comparison of botanical composition on Samosir Island at altitude of 905–1200 meters above sea level and more than 1205 meters above sea level can be seen through the classification of forage feeding on 3 cultivation that are grass, legumes and weeds. This is shown in Table 4 below.

**Table 4.** Comparison of Forage Species Number on Natural Pasture in Samosir Island At Altitude 905–1200 Meters Above Sea Level and More Than 1205 Meters Above Sea Level

| Type of Forage (%) | Altitude 905–1200 meters above sea level | Altitude >1205 meters above sea level |
|---|---|---|
| Grass | 80.58 | 92.72 |
| Legumes | 9.14 | 2.87 |
| Weeds | 9.63 | 4.39 |

Grass are more dominate grazing in Samosir Island than legumes. The most dominant forage botanical species at both altitude sites are *Axonopus compressus*, *Imperata cylindrica l* and *Paspalum conjugatum*. The result of botanical composition on grazing land in Samosir regency is grassland more dominate pasture land than legumes and even at altitude more than 1205 meters above sea level almost nothing. According to Yasin [3] opinion that during the rainy season, the potential grasses as forages grow well and the biomass production is relatively abundant. However, in the dry season, productivity is relatively small. That is, the limiting factor is the availability of water instead of rain. If water can be adequately supplied throughout the year, the production and availability of forage will not be an obstacle.

The high percentage of grass caused by grass is easy to grow and develop in almost all types of soil and in various types of climate. In addition, grass growth is faster than leguminous. This is because the grasses generally grow to form clumps, have a strong root system that is resistant to stalking and ripping cattle, its rapid regrowth, it is rhizomes creeping and forming new plants rapidly spreading if subjected to cuts by both livestock and defoliation. This is in accordance with the opinion of Hidayati [4] that the cutting period has an effect on the production and quality of grass and grazing production of pastureland.

### *3.4. Nutrition Content Analysis*

**Table 5.** Nutrient Content Analysis (based on Dry Matter)

| Type of Forage | Dry Matter | Ash Content | Crude Protein | Crude Fiber | Ether Extract | Beta-N |
|---|---|---|---|---|---|---|
| *Euphatorium adenophorum* | 82.90 | 9.44 | 8.65 | 15.67 | 0.73 | 48.43 |
| *Stylosantes capitata* | 90.11 | 9.09 | 9.89 | 27.64 | 1.37 | 42.12 |
| *Paspalum conjugatum* | 91.19 | 6.61 | 9.03 | 35.78 | 0.53 | 39.24 |
| *Cynodon dactylon* | 91.60 | 9.88 | 8.74 | 29.54 | 0.73 | 42.71 |
| *Flemingia macrofilia* | 90.44 | 11.69 | 14.21 | 32.45 | 1.56 | 30.53 |
| *Zoysia matrella* | 92.69 | 5.51 | 14.53 | 31.86 | 1.55 | 39.24 |
| *Axonopus compressus* | 91.88 | 10.79 | 10.44 | 28.17 | 1.41 | 41.07 |
| *Imperata cylindrica l* | 91.30 | 9.40 | 13.21 | 29.02 | 1.68 | 37.99 |
| *Rabba-rabba* | 90.25 | 8.27 | 9.17 | 26.82 | 1.63 | 44.36 |
| *Centella asiatica* | 90.37 | 10.05 | 9.07 | 18.24 | 1.33 | 51.68 |
| *Melastoma malabatricum* | 91.72 | 5.74 | 11.88 | 26.77 | 0.24 | 47.09 |
| *Mimosa pudica* | 90.92 | 4.74 | 9.42 | 30.11 | 1.05 | 45.60 |
| *Leptochloa chinensis l* | 89.54 | 10.56 | 13.88 | 27.06 | 1.22 | 36.82 |
| *Starkuak* | 91.12 | 11.08 | 15.13 | 22.83 | 1.32 | 40.76 |
| *Stachytarpheta jamaicencis* | 90.93 | 5.23 | 14.48 | 18.24 | 0.03 | 52.95 |

Based on analysis of dry matter content, the highest is *Zoysia matrella* 92.69%, followed by *Axonopus compressus* 91.88% and *Melastoma malabatricum* 91.72%. Dry matter highest content of a plant is influenced by soil conditions that experience drought so that forage that grows also experience water crisis so that the BK content of the plant has increased. At location of the study experienced a very long drought so that the crop on the pasture land experienced a water crisis or drought. According to Hidayati [4] that the quality of pastures represented by the amount of crude protein content and dry matter (almost entirely cellulose and hemi-cellulose) also displays large deviations throughout the year. Protein grass in the rainy season reaches 8.3% rapidly decreases to 5.5% at the beginning of the dry season and then falls to the lowest point, 2.2% at the end of the dry season. In the same period the leguminous protein also showed a decrease from 26.5% to 22.3% and so on 12.5%. Dry matter content also showed a large increase, from 20% in the rainy season for both grass and legumes increased sharply to 74% and 43% respectively at the end of the dry season.

## 4. Conclusions

Botanical composition at altitude of 905–1200 meters above sea level has more diversity that is 31 species with ratio of grass 80.58%, legumes 9.14%, weeds 9.63% and the most dominant is *Imperata cylindrica l.* While, botanical composition at altitude more than 1205 meters above sea level there are 15 species with ratio of grass 92.72%, legumes 2.87%, weeds 4.39% and the most dominant is *Axonopus compressus*. The result of nutrient content analysis which has the highest crude protein is *Starkuak* 15.13%.

## References

[1] BPS Samosir Distict 2016 *Samosir District In Figures 2016* ISSN: 2301-976X Catalog: 1102001.12.17

[2] Jayadi S 1991 *Tropical Livestock Feed Crops* Faculty of Animal Husbandry. Bogor Agricultural Institute

[3] Yasin S 2013 *Ruminant Eating Behavior As Bioindicator Phenology And Dynamic Pasture of Padang* Study Center for Sheep Shepherd's Production System and Tropical Pastoral Area, Faculty of Animal Husbandry of Mataram University, Mataram-NTB. ISSN: 2088-818X Pastura Vol. **3** No. **1**: 1-4

[4] Hidayati N, C Talib, A Pohan 2001 *Productivity of Native Pasture to Support Breeding Cattle Herd in East Kupang, East Nusa Tenggara* Bogor Livestock Research Center and Center for the Assessment and Application of Agricultural Technology of Kupang

# The effect of feeding bull Bali cattle kept in extensive husbandry system with concentrates contained gliricidia sepium leaf meal and banana strach tuber meal on their feed consumption and dried organic matter digestability

**S Fattah[1], Y U L Sobang[1], F D Samba[1], E Hartati[1], M M J. Kapa[2] and Y L Henuk[3]**

[1]Faculty of Animal Science, Universitas Nusa Cendana, Kupang, 85361 Indonesia
[2]Faculty of Agriculture, Universitas Nusa Cendana, Kupang 85361 Indonesia
[3]Faculty of Agriculture, Universitas Sumatera Utara, Medan, 20155 Indonesia
E-mail: yusufhenuk62@live.com

**Abstract.** This study aimed to evaluate the effect of feeding bull Bali Cattle kept in extensive husbnadry system with concentrates contained gliricidia sepium leaf meal and banana strach tuber meal in their feed consumptions and dried organic matter digestibility. Three bull Bali cattle aged 1 – 2 years old with an initial body weight of 135.5 kg – 168.0 kg were used in this study. The three treatments used were $T_0$ = local feeds (consisted of *Leucaena leucocephala, Acasia leochophloea*, and *Ficus sp.* leaves as commonly used by local farmers); $T_1 = T_0 + 1$ kg concentrate (contained banana strach tuber meal + *gliricidia sepium* leaf meal); $T_2 = T_1 + 2$ kg concentrate (contained banana strach tuber meal + *gliricidia sepium* leaf meal). The results showed that the dry matter intake were: 2.40, 3.52, and 4.14; organic matter intake were: 2.17, 3.32, and 3.62; dry matter digestible was 64.63%, 72.45%, 77.28% and organic matter digestible was 66.79%, 74.66%, 79.33% for $T_0$, $T_1$, and $T_2$, respectively. There was no effect ($P>0.05$) of treatments on the three parameters observed on bull Bali cattle kept in extensive husbandry system and fed with concentrates contained leaf gliricidia sepium meal and banana starch tuber meal.

## 1. Introduction

It is well known that beef cattle system in East Nusa Tenggara (ENT) Province is kept under extensive traditional regime. Knowing the characteristics of extensive management system of local cattle (Bali cattle) in ENT is helpful to identify the problem and the formulation of strategies for improving cattle productivity. Extensive traditional management system is defined as a way of husbandry where the cattle released during the day and stabled at night, and or released on grazing day and night. In this condition, the owner intervention is very limited. Population growth will normally followed by increasing demand for meat as a source of animal protein. The increase of demand on the one hand would boost economic growth but on the other hand will cause a negative impact on the sustainability of the local beef cattle reared under the extensive system. With the current husbandry system, the threats to the business sustainability of beef cattle farms in the dryland environment may occur [1]. ENT had been one of the major beef cattle suppliers under traditional management system in Indonesia. The beef cattle farming based on grazing native pasture and the introduction of shrub legumes (*Leucaena leucocephala*) may contribute to around 15 – 50% of the farmers' household income. In the last few years, supply of beef cattle tended to decline due to the decrease in cattle population in ENT [2]. The quality and quantity of feed in ENT is followed by the standard seasonal flow. In the dry season, the content of dry matter forage increases but in the rainy season the moisture content will be higher. In the rainy season, the amount of forage will be abundant while in the dry season is very little available so that the impact on the productivity of fattening cattle in the pattern of rearing breeders on the island of Timor is still low only around 0.25-0.30 kg/head/day. This is because the cattle fattening system conducted by farmers in ENT is still done without adequate technology input, especially in the aspect of feeding, where the livestock is dependent only on forage (grass and legume) without considering the aspect of nutrient adequacy [3]. Bali catlle farmers in ENT Province as a major supplier for domestic meat demand in Indonesia. However, beef cattle production in ENT province still facing constraints that caused low productivity of beef cattle. These constrains are incontinuity availability of forages, traditional breeding system, underutilized available local feed resources. *Gliricidia sepium* is one of legumes commonly found in ENT and well adapted to dry tropical environment and infertile soils such as most of neglected land in this region. *Gliricidia sepium* has not been widely used yet due to its specific smelt of coumarine a compound which lead to its low palatability [4]. This study aims to determine the effect of feeding bull Bali Cattle kept in extensive husbandry system with concentrates contained gliricida sepium meal and banana strach tuber meal on their feed consumption and dried organic matter digestibility.

## 2. Materials and Methods

The research was conducted in Desa Binaan Fapet Undana in Oelatsala Village, Taebenu Sub-district, Kupang District. Three bull Bali cattle aged 1 – 2 years old with an initial body weight of 135.5 kg – 168 kg were used in this study. The experimental design used was a Latin square design (3x3) consisted of 3 treatments and 3 replicates. The three treatment were $T_0$ = local feeds (consisted of leaves of *Leucaena leucocephala, Acasia leochophloea*, and *Ficus sp.* as commonly used by local farmers); $T_1 = T_0 + 1$ kg concentrate (contained banana strach tuber meal + *gliricidia sepium* leaf meal); $T_2 = T_1 + 2$ kg concentrate (contained banana strach tuber meal + *gliricidia sepium* leaf meal) (Table 1).

**Table 1**. Ingridients of concentrate as a suplement for fattening Bali cattle (DM).

| Ingredients | Treatments (%) | | |
|---|---|---|---|
| | $T_0$ | $T_1$ | $T_2$ |
| Rice bran | 50 | 50 | 40 |
| Corn meal | 24 | 24 | 20 |
| Banana Starch Tuber Meal | 0 | 15 | 15 |
| Gliricidia Sepium Leaf Meal | 15 | 0 | 15 |
| Fish Meal | 5 | 5 | 4 |
| Urea | 2.5 | 2.5 | 2.5 |
| Salt | 3 | 3 | 3 |
| Starbio | 0.5 | 0.5 | 0.5 |
| Total | 100 | 100 | 100 |

All cows were given *ad libitum* feed and water and used individual cages consisting of 3 plots measuring 1.20 m x 2.10 m cement floor, coconut-shaped roof and equipped with feeding and drinking water. Variables assessed were dry matter (DM) intake (kg/day) and organic matter (OM) intake (kg/day) of the experimental animals. Data were analyzed using analysis of variance according to R G D Steel, and J H Torrie [5]. The difference between treatments was tested by Duncan's Multiple Range Test (DMRT).

## 3. Results and Discussion

*3.1. The effect of treatments on DM and OM*

Consumption is a major factor affecting the supply of nutrients in cattle so that generally used as the main reference in assessing the capacity of livestock in utilizing feed with different quality. The effect of treatment on DM and OM is presented in Table 2. Average digestability of dry-matter and organic matter of Bali cattle is presented in Table 3.

**Table 2.** The average of DM and OM consumption of the Beef cattle (kg/head/day)

| Parameter | Treatments | | |
|---|---|---|---|
| | $T_0$ | $T_1$ | $T_2$ |
| DM intake | 3.40 | 3.22 | 4.14 |
| OM intake | 3.17 | 3.12 | 3.62 |

**Table 3.** Average Digestibility of DM and OM of Beef Cattle (%)

| Parameter | Treatments | | |
|---|---|---|---|
| | $T_0$ | $T_1$ | $T_2$ |
| DM intake | 69.22 | 63.15 | 76.00 |
| OM intake | 68.90 | 64.66 | 76.33 |

*3.2. The effect of Treatments to DM Consumption*

The result of analysis of variance showed that the treatment of $T_0$ and $T_1$ had no significant effect on DM consumption. However, $T_1 - T_2$ showed a significant effect ($P < 0.05$). This proved that the use of concentrate feed containing gamal leaf flour and banana cobs can increase the consumption of DM in the system of fattening patterns of breeders. The average consumption ability of ruminant livestock is 2-3% of body weight or 2.5 to 3.2%. Based on the mean score in table 3, the average DM consumption in this study was between 3.22-4.14 kg/ head/day. Mean $T_1$ (3.22 kg/head/day) is the lowest dry ingredient consumption. This is due to the consumption of feed is very dependent on the provision by farmers without regard to the aspect of nutritional adequacy characterized by low body weight so that the addition of constrates containing banana tubers as energy sources can not increase consumption of DM rations. DM consumption between $T_1$ - $T_2$ is not significant, this is due to the nutrient content of the feed is relatively the same between the treatments of DM feed pattern breeders that cause consumption is not significantly different. This is an accordance with N Silanikove, Z Nitsan and A Perevolotsky [6] they who reported that the ability of cows to consume DM is relative the same, cause no difference in feed consumption [7]. The unbalanced feed nutrient will affect feed consumption. The duration of feed ingredients in the rumen, will slow the rate of food so that the consumption level becomes low [7]. There was significant difference ($P < 0.05$) in $T_1$-$T_2$ with the addition of concentrate containing gamal leaf and banana flour in feed pattern of breeder characterized by DM consumption ($T_1$ = 3.22 kg/head/day $T_2$ = 4.14 kg/head/day), this is due to the level of concentrate and the concentrated nature of the palatability so that the cattle consume more and in the presence of the use of concentrate feed can improve DM digestibility in feed pattern breeders, so the rate of feed speed in the channel Digestion increases so that feed consumption increases. Accroding to P Faverdin, R Baumont, and K L Ingvartsen [8] the palatability of feed is a major factor explaining the difference in DM consumption between feed and low-yielding livestock. Consumption of DM between $T_1$ - $T_2$ is not significant, this is due to the nutrient content of the feed is relatively the same between the treatment of DM feed pattern that cause consumption is not significantly different. This findings is in line with N Silanikove, Z Nitsan and A Perevolotsky [6], they reported that the ability of cows to consume the same relative DM had given no difference in feed consumption. T R Preston and R A Leng [7] also described that the unbalanced feed nutrient will affect feed consumption, it further states that the duration of feed ingredients in the rumen, will slow the rate of food so that the consumption level becomes low. There was significant difference ($P < 0.05$) in $T_1$-$T_2$ with the addition of concentrate containing gamal and banana leaf banana on feed pattern of breeder characterized by consumption of DM ($T_1$ = 3.22 kg/head/day; $T_2$ = 4.14 kg/head/day), this is due to the level of concentrate and the concentrated nature of the palatability so that the cattle consume more and with the use of concentrate feed can improve DM digestibility in feed pattern breeders, so the rate of feed rate in the digestive tract increased, so the consumption of feed increases. Our findings supported by P Faverdin, R Baumont, and K L Ingvartsen [8], they reported the palatability of feed is a major factor explaining the difference in DM consumption between feed and low-yielding livestock. S D M Prawirodigdo dan D Andayani [9] stated that palatability affects the amount of feed consumed. R B Widyawati [2] also explained that with the use of supplement feed, microbial synthesis on the feed increases. With the increased synthesis of microbes in the rumen, the feed digestibility will increase, the feed rate in the rumen becomes faster and stomach will be empty quickly and feed consumption increases. A D Tillman, H Hartadi, S Reksohadiprojo, S Prawirokusumo. dan S Lebdosoekojo [10] reported that the easier the digestible feed in the gastrointestinal tract means that nutrient feed is absorbed faster so that feed flow leaves the digestive tract faster and causes more room to be available for feed growth. The faster the feed rate in the digestive tract, the consumption will increase. C Devendra [11] also reported that DM consumption has a strong relationship with undigested energy consumption. The stomach becomes empty faster and stimulates the cattle to consume the feed again. The standard DM requirement of livestock research is 4.04-4.85%/head/day from the mean body weight of 161.75kg which is characterized by daily weight gain of livestock receiving $T_0$ (0.38kg/head/day) and then $T_1$ (0.36kg/head/day) followed by $T_2$ (0.64kg/head/day). The results are slightly better than the findings of Y U L Sobang [12] that the addition of local concentrate feed can increase the consumption of ration DM and the productivity of Bali cattle fatten. The addition of local concentrate feed can increase the consumption of DM ration and cattle productivity Bali fattening pattern of breeders reaches 0.45-0.50 kg/head/day.

### *3.3. The Influence of Treatments on DM Consumption*

The result of the study showed that $T_1$ - $T_2$ had no significant effect but $T_2$ - $P_3$ treatment had significant effect ($P < 0.05$) on OM consumption. This is in line with the consumption of DM, the consumption of OM at low $T_1$ (3.12 kg/head/day) is characterized by a daily weight gain of 0.36 kg/head/day. This is due to the consumption of feed is highly dependent on provision by the owner and the existence of what feeding trends are met that are cut and given to livestock regardless of the nutritional adequacy aspect. $T_0$ = 3.17 kg/head/day with daily weight gain of 0.38 kg/head/day. The high OM consumption in animals treated with $T_2$ and $T_1$ is influenced by the level of palatability and concentrate, thus stimulating the animal to consume more feed. The increase of OM consumption of this ration along with the increase of consumption of DM. The organic material in $T_2$ consumption is high (3.62 kg/head/day) which is indicated by adding $T_2$ as much as 3.62 kg/head/day was followed by $T_0$ = 3.17 kg/head/day with daily weight gain of 0.38 kg/head/day. The high consumption of OM in animals treated with $T_2$ and $T_0$ was influenced by the level of palatability and concentrate, thus stimulating the livestock to consume more feed. It was found that the increase of OM of this ration along with the increase of consumption of OM ration. This result is supported by A A A Astuti dan P S B1 [13], they reported that consumption of OM is closely related to the consumption of DM, the higher the consumption of DM then the consumption of OM also increases. Nutrients contained in OM are the components of DM. The organic composition consists of fat, crude protein, crude fiber, and BETN. DM, plus ash [14]. So the amount of consumption of DM will affect the amount of consumption of OM. The amount of consumption of DM will affect the amount of nutrients consumed so that if the consumption of OM increases it will increase the consumption of nutrients. From this result that feeding pattern of breeder plus concentrate, give better effect to the productivity of cattle research.

### *3.4. The Effect of Treatments to DM Digestibility*

The result of variance analysis showed that the treatment had no significant effect ($P > 0,05$) to DM digestibility percentage. The DM digestibility of the rations using a forage counterpart - concentrates with different compositions other than those caused by the increase in DM digestibility is also due to the increased proportion of concentrates in the ration so as to give an indication that the feed is of good quality. Empirically the cattle that consume forage feed added to the concentrate containing banana herbs ($T_2$) has a lower DM digestibility than those added to the concentrate containing gamal leaf and banana flour, this proves that the use of concentrate feed can improve the digestibility of the ingredients Dry in the system of fattening pattern of breeders. There was a significant difference in $T_2$ and $T_3$ with the addition of concentrate containing gamal and banana leaf flour in feed pattern of breeder characterized by DM digestibility ($T_1$ = 63.15; $T_2$ = 76.00). This is because the addition of gamal and banana leaf powder in the concentrate and the concentrated nature of the palatability so that the cattle consume more and with the use of concentrate feed can improve DM digestibility in feed patterns of breeders, so the rate of feed speed in the digestive tract increases. According to P Faverdin, R Baumont, and K L Ingvartsen [8], palatability is a major factor explaining the difference in DM consumption between feed and low-yielding livestock. It is also said that the palatability of feed is generally associated with high digestibility of a feed. The high DM digestibility of ruminant livestock shows high levels of digested nutrients primarily by rumen microbes. This is in line with A D Tillman, H Hartadi, S Reksohadiprojo, S Prawirokusumo and S Lebdosoekojo [10] that the digestibility of a feed material depends on the harmony Food substances contained therein, feed ingredients with high digestible ingredients are generally high also the value of nutrients and digestibility can be the first measure of the high nutrient value of a feed material.

### *3.5. The effect of Treatment on OM Digestibility*

The results of analysis of variance showed that the treatment had no significant effect ($P > 0.05$) or in other words the treatment application did not give a significant effect on the percentage of OM digestibility in the fattening cattle breeder system. This is due to the nutrient content of feed consumed is relatively same, so that the digestibility result is not much different. The digestibility of OM was low at $T_1$ (64.6 %). It is well known that feed consumption is highly dependent on the feed provision method by the owner and feed types given to the animal. At $T_2$ the digestibility of OM is high (76.33 %) and it is caused by high OM intake by animal. The high digestibility of OM in animal given $T_2$ and $T_1$ is influenced by the addition of gamal leaf powder as a source of protein and banana cobs as a source of energy so that needs is sufficient and leads to increase the activity of rumen microorganisms in digesting feed material. The digestibility level of OM feed material has the same pattern with DM digestibility. The level of digestibility of OM is relatively higher than DM digestibility in all treatment rations. This is because the DM still contains ash, while the OM does not contain ash, so the material without ash content is relatively easier to digest. This result is supported by A D Tillman, H Hartadi, S Reksohadiprojo, S Prawirokusumo and S Lebdosoekojo [10] they stated that the DM digestibility value is closely related to the digestibility value of OM, as most of the DM component consist of organic material, the difference is only in the ash content. This result proven that the addition of concentrate into the ordinary ration gives good effect on animal performance. The digestibility value of OM is relatively the same among treatments. This happens due to the OM component and the extra materials without nitrogen as well as the crude fiber content of the ration treated are relatively similar. This is presumably because the microbes are unable to digest the crude fiber components contained in the feed optimally. The coarse fiber content in the feed will cause a low degradation value, since crude fibers in the form of cellulose and hemicellulose often bind to lignin and will be difficult to break down by digestive enzymes. The digestibility rate of OM is influenced by feed particle size, easy digestibility of carbohydrates, cellulose content, hemicellulose, lignin, and protein. The digestibility of OM in the digestive tract of livestock includes nutrients in the form of components of organic materials such as carbohydrates, proteins, fats and vitamins.

## 4. Conclusions

The feeding of concentrate containing gamal and banana flour concentrate is not significantly influenced ($P > 0.05$) the consumption and digestibility of Dry Matter, Organic Matter in Bali cattle male under farmer rearing systems. However, empirically there is an increase of consumption and Dry Matter, Organic Matter digestibility due to the addition of gamal flour and banana flour at a level of 15%.

## References

[1] Kapa M M J, Soemarno, Bagyoyanuwiyadi and Suyadi 2017 Sustainability status of biology dimension of local beef cattle development in the dryland region, Indonesia *J. Econ.Sustain. Dev.* **8**(6): 102-108

[2] Wirdahayati R B 2010 Penerapan teknologi dalam upaya meningkatkan produkitivitas sapi potong di Nusa Tenggara Timur (Technology application to increase productivity of beef cattle in Nusa Tenggara Timur *Wartazoa* **20**(1): 12-20.

[3] Fattah S 2012 *Produktivitas Ternak Sapi Bali (Bali cattle productivity)* (Yogyakarta: Penerbit The Phinisi Press)

[4] Fattah S, Sobang Y U L, Ratuwaloe J J A, and Henuk Y L 2014 The Effect of feeding concentrate containing *Gliricidia Sepium* leaves meal and addition vitamine B-Complex and worm medicine on dry matter intake and daily body weight gain of Bali Cattle raised based on local farmers' raising pattern. In: *Proceedings of the 16th AAAP Animal Science Congress*, 10-14 November 2014, Gadjah Mada University, Yogyakarta, Indonesia, II: 2291-2294.

[5] Steel R G D and Torrie J H 1995 *Principles and Procedures of Statistics. 2nd Edition* (London: McGraw-Hill International Book Company)

[6] Silanikove N, Nitsan Z and Perevolotsky A 1994 Effect of daily supplementation of polyethylen glycol on intake and digestion of tanin-containing leaves (*Cernatonia siliqua*) by sheep *J. Agric. Food Chem.* **42**: 2844 – 2847

[7] Preston T R and Leng R A 1984 Supplementation of diet based fibrous residues and by products. in: sundstol f and owen e (eds) *Straw and Other Fibrous by-Products as Feed. Elsevier*, Amsterdam pp 373-409

[8] Faverdin P, Baumont R, and Ingvartsen K L 1995 Control and prediction of feed intake in ruminants. in: m. journet, e. grenet, m-h. farce, m. theriez, and c. demarquilly (eds), proceedings of the 4th international symposium on the nutrition of herbivores. *Recent Development in the Nutrition of Herbivores* INRA Paris pp 95-120

[9] Prawirodigdo S D M dan Andayani D 1995 Subtitusi bungkil kedelai dengan bungkil biji kapok (ceip petandra) dalam ransum kelinci sedang tumbuh (Subtitution of soybean seed with cotton seed (Ceip petandra) into growing rabbit ration) *J. Ilm. Tern. Klepu* **1**(3): 26-31

[10] Tillman A D, Hartadi H, Reksohadiprojo S, Prawirokusumo S and Lebdosoekojo S 1991 *Ilmu Makanan Ternak Dasar (Basic Animal Food Science)* (Yogyakarta: Universitas Gadjah Mada Press)

[11] Devendra C 1983 The energy and protein requirements during pregnancy of katjang goats in Malaysia *Mardi Res. Bul.* **11**(2): 193-205

[12] Sobang Y U L 2005 Karakteristik sistem penggemukan sapi pola gaduhan menurut zona agroklimat dan dampaknya terhadap pendapatan petani di kabupaten Kupang NTT *Bulletin Nutrisi* Maret, **8**(2): 1-8

[13] Astuti A A A dan Subur P S B 2009 Pengaruh penggunaan high quality feed supplement terhadap konsumsi dan kecernaan nutrien sapi perah awal laktasi (The effect of high quality feed supplement addition on the nutrient consumption and digestibility of early lactating dairy cow *J. Pet.* **33**(2): 81-87

[14] Kamal M 1994 *Nutrisi Ternak (Animal Nutrition)* (Yogyakarta: Laboratorium Makanan Ternak Fakultas Peternakan)

IOP Conf. Series: Earth and Environmental Science **122** (2018) 012133 doi:10.1088/1755-1315/122/1/012133

# Income analysis of goat farmers on the farmers group in district of Serdang Bedagai

**J N Manurung, Hasnudi and T Supriana**
Faculty of Agriculture, Universitas Sumatera Utara, Indonesia
E-mail: tavihutasuhut@yahoo.co.id

**Abstract.** The farmers group are expected to reduce the production cost of goat breeding and improve the income of farmers which impact on the welfare of goat farmers. This research aim to analyze the factors that influence the income of farmers group, in sub-district Dolok Masihul Pegajahan, and Dolok Merawan, Serdang Bedagai. The method used is survey method with 90 respondents. Data was analysed by multiple linear regression. The result showed, simultaneously goat cost, sale price of goat, fixed cost and variable cost had significant effect on income of goat farmers. Partially, goat cost, variable cost and sale price of goat had significant effect on income of goat farmers, while fixed cost had no significant effect.

## 1. Introduction

Livestock can be one of the stimulator of regional economic growth, even in the national. One of livestock business which has promising prospect is goat livestock business. Some benefits of raising goats business are increasing the income of rural communities, obtaining side income from goat feces which can be used for making manure. Goats have more adaptability to the environment, easier to raise it, and the capital needed is relatively small.

Goat is one type of livestock familiar with farming systems in rural areas. This is because, the size of the body is not too large, easy treatment, fast breeding, the number of children per birth is often more than one tail, short birth spacing, and growing fast. In addition, goats have a high adaptability to the conditions of a place agroecosystem. Goat can still survive in the worst environments. Goats is farmed by farmers are generally a local goat. The local goat that reproduce by both in indonesia, namely goat kacang and goat etawah uterus (PE). Besides that, there are goat another local like a goat gembrong , costa , marica , jawarandu and bligon [1].

The farmers buy goat to start their business as well as to increase their income. Factors to be considered by farmers in the purchase of goats are certain superior nature such as body weight, number of children per birth, and disease endurance [2]. The quality of goat produced will determine the sale price of livestock. The better the quality then selling price will be the higher price with the higher price, the income to be obtained by farmers will also increase [3]. Income is the sum of money earned after all variable costs and fixed costs are covered. If the result of the deduction is positive, it means they will get some profit, while negative result of deduction means loss [4].

District of Serdang Bedagai is a suitable place for the development of goat because the human and natural resources are very supportive. This area has advantages as a livestock area in terms of climate and geography. This condition has been used by farmers to cultivate goats to date. The data from the Agriculture and Husbandry Office District of Serdang Bedagai which indicate the raising of the population of goats in the amount of 1,21% in 2014, from 77.161 heads become 78.106 in 2015.

Goat livestock conditions are characterized by low incomes, narrow land, lack of capital and knowledge of limited livestock breeding and farmers still tied to the main job and do the husbandry only as a side business (small-scale business i.e. an average of 3-5 goats per family). Many farmers do not have time to join groups with other farmers and still lack of farmers trust in the government, where small farmers generally prefer to raise potluck without any local government intervention. With farmers group, the production cost of goat breeding can be decreased, and the income of farmers can be increased which affect their wealth. The farmers group are farmers groups of livestock farmers. The Farmers group can be formed because of the unity of interest, especially regarding the mutual relationships that affect each other and awareness to help fellow members. The farmers group can also be formed because they want to get help from the government that can be either capital or other assistance to help increase the income of farmers. The Objective of this research is to analyze the factors that influence the income of farmers group, in sub-district Dolok Masihul Pegajahan, and Dolok Merawan, Serdang Bedagai.

## 2. Materials and methods

This research was conducted on July to September 2016. The selection of research area is in the District with high livestock population, that is Dolok Masihul District, District with the population of medium goat, and District Pegajahan, and District with low livestock population of goat namely Dolok Merawan District. The number of farmer respondents in the farmers group taken in this research were 30 respondents in each sub- district, so the number of respondents for the farmers group were 90 respondents.

The research method used is survey method with sampling technique which is called accidental sampling, that sample determination technique is based on spontaneity factor, meaning anyone who intentionally meet with the author and compatible with the characteristics (characteristics), then the farmers can be used as a sample (respondents). Parameters that become the object of observation in this research are the factors that influence the income of farmer like goat cost, fixed cost, variable cost and sale price of goat.

The model used in this this research is multiple linear regression model which is formulated as follows:

$$Y = a + b1X_1 + b2X_2 + b3X_3 + b4X_4 + \mu \quad (1)$$

**where:**

Y = Farmers Income (Rp/year)
a = Constants
$X_1$ = Goat Cost (Rp/Year)
$X_2$ = Fixed Cost (Rp/Year)
$X_3$ = Variable Cost (Rp/Year)
$X_4$ = Sales Price of Goat (Rp/Year)
$\mu$ = Error

## 3. Results and discussion

Based on the estimation results the factors that affect the income of farmers (Y) such as goat costs, fixed costs, variable costs and sale prices of goat are shown in Table 1 below.

IOP Conf. Series: Earth and Environmental Science **122** (2018) 012133 doi:10.1088/1755-1315/122/1/012133

**Table 1.** Factor Analysis Affecting Income on Farmers Group

| Model | Regression Coefficient | T-test | Sig. |
|---|---|---|---|
| Model | | | |
| (Constant) | 3331476 | 2.941 | 0.004 |
| Goat Cost | -0.614 | -4.815 | 0.000 |
| Fixed Cost | 0.420 | 1.129 | 0.262 |
| Variabel Cost | 0.429 | 1.148 | 0.035 |
| Sales Prices of Goat | 0.614 | 4.758 | 0.000 |
| $F_{test}$ | | | 9.028 |
| Sig | | | 0.000 |

Based on Table 1. the model of multiple linear regression equation in the income of farmers (Y) is:

$$Y = 3331428 - 0.614X_1 + 0.420X_2 + 0.429X_3 + 0.614X_4$$

Note:

$Y$ = Farmers Income (Rp/year)
A = Constants
$X_1$ = Goat Cost (Rp/Year)
$X_2$ = Fixed Cost (Rp/Year)
$X_3$ = Variable Cost (Rp/Year)
$X_4$ = Sales Price of Goat (Rp/Year)
$\mu$ = Error

### *3.1. Partial Variable Effect*

Partial significance test (T-test) is done to see partially (individual) effect of independent variables studied were goat cost ($X_1$). fixed cost ($X_2$). variable cost ($X_3$) and sale price of goat ($X_4$) to dependent variable which is called as income (Y).

### *3.2. Goat Cost ($X_1$)*

In Table 1 it can be seen that t-count on the variable of the goat cost ($X_1$) is 4.815 which is bigger than t-table which is equal to 1.991 with signification level of 0.000 ($P < 0.05$). so it can be concluded that $H_0$ is rejected and $H_1$ is accepted. Therefore. it can be concluded that the cost of goat partially has significant effect on farmers income. The negative coefficient value indicates that the goat cost ($X_1$) has negative effect (-0.614). which means that the increase of goat purchase cost of Rp 1,000. - will decrease the farmers income in the group by Rp. 614. -. In Table 2. it can be seen that the goat cost that farmers are issuing with average of Rp. 5,586,563. Goat purchased by local farmers in accordance with the quality of goat purchased. Therefore. the goat cost affects farmers income in the local area. The farmers buy goats to start their business as well as in increasing their income. Factors that farmers pay attention to in purchasing goats are certain superior nature such as body weight. number of children per birth. and disease endurance [2].

### *3.3. Fixed Cost ($X_2$)*

In Table 1. it can be seen that the t-count of fixed cost variable ($X_2$) consisting of the depreciation cost of the cage and the depreciation cost of the equipment is 1.129 which is smaller than the t-table 1.991 with the significance level of 0.262 ($P > 0.05$). So it can be concluded that $H_0$ is accepted $H_1$ is rejected. Thus it can be concluded that the fixed costs have no significant effect partially on the income of farmers. The positive value of coefficient indicates that the fixed cost ($X_2$) has a positive effect of 0.420 which means that with a fixed cost increase of Rp 1,000. - will increase the income of farmers of Rp. 420. -. It happens because the average of fixed cost that the farmer issued is large enough. that is Rp. 2,130,813 which can be seen in Table 2. The higher the cost incurred the lower received income. so the fixed cost does not affect the income of farmers. The cost of production (fixed costs) issued in the husbandry business is related to the income obtained by farmers [5].

### *3.4. Variable Cost ($X_3$)*

In Table 1. it can be seen that t-count on variable cost ($X_3$) consisting of feed cost. medicines. vehicle operation. electricity and labor wage is 1.148 which is smaller than t-table equals 1.991 with signification level equals 0.035 ($P< 0.05$). so it can be concluded $H_0$ rejected $H_1$ is accepted. Thus. it can be concluded that the variable costs have significant effect partially on farmer income. The positive value of coefficient indicates that the variable cost ($X_3$) has a positive effect of 0.429 which means that with the increase of the variable cost of Rp 1,000. - will increase the income of farmes Group of Rp. 429. -. This is because the average of variable cost issued by farmers is big enough that is Rp. 3,419,135 which can be seen in Table 2. The higher the cost incurred the higher received income. so the variable cost affected the income of farmers.

### *3.5. Sales Price of Goat ($X_4$)*

In Table 1. it can be seen that t-count on the sales price of goat ($X_4$) is 4.758 which is bigger than t-table equals 1.991 with significance level of 0.000 ($P <0.05$). so it can be concluded that $H_0$ is rejected and $H_1$ is accepted. Therefore. it can be concluded that the sales price of goat has significant effect partially on farmers income. The positive value of coefficient indicates that the variable cost ($X_4$) has a positive effect of 0.614. which means that with the increase of sales price of goat equals Rp 1,000. - will increase the income of farmers group of Rp. 614. -.

Farmers earnings are affected by Market demand and livestock prices. The price of livestock will increase when the demand for goat increases. Farmers generally sell their goats when they need cash for their family needs. The selling price of goat applicable in the study area is based on age and sex by considering the condition of livestock. too. Goat livestock sales system is estimated from the weight of the body and the market price in general. The quality of livestock produced will determine the sales price of the livestock. the better the quality then the hihger the sales price will be. With higher prices. the revenue to be gained will also increase [3].

### *3.6. Variable Influence Test Simultaneously (F - test)*

To see whether or not the influence of the independent variable on the dependent variable. can be seen from f-count value. In Table 1. it can be seen that the value of f-count in the group is 9.028 which is greater than the f-table value equals 2.49 with the significance level of 0.000 ($P <0.05$). It can be knwon that. simultaneously. goat cost ($X_1$). fixed cost ($X_2$). variable cost ($X_3$) and sales price of goat ($X_4$) influence the farmers income (Y). Every increase or decrease in goat costs. fixed costs. variable costs and prices of livestock sales will simultaneously lead to increase and decrease in farmers income.

### *3.7. Income*

Income is the difference between the income of livestock business per year with the total production cost per year. the amount of farmers income in the group. the average net income equals Rp. 5,943,555 per year can be obtained. Income is ithe difference between the receipt and all expenses incurred [6].

## 4. Conclusions

Based on the results of research obtained. simultaneously goat cost, sale price of goat. fixed cost and variable cost had significant effect on income of goat farmers. Partially. goat cost, variable cost and sale price of goat had significant effect on income of goat farmers. while fixed cost cost had no significant effect

## References

[1] Subandriyo 1996 The potential productivity of the livestock of sheep and goat in Indonesia

*Proceedings of the National Seminar on animal husbandry and Veterinary* The Department of Agriculture Bogor
[2] Rudiyanto 2000 *Analysis of Production Functions in Sheep and Goat Husbandry in Singasari Village, Jonggol District, Bogor Regency (Thesis)* Faculty of Animal Husbandry Bogor Agricultural University Bogor
[3] Suryadi U 2006 Influence of Number of Same Birth and Sex on the Performance of Lamb to Wean Department of Livestock State Polytechnic of Jember
[4] Moehar 2002 *Introduction to Agricultural Economics* (Jakarta: Bumi Aksara)
[5] Hendrawanto E 2008 *Analysis of Revenue and Production of Farming (Thesis)* Faculty of Animal Husbandry Bogor Agricultural University Bogor
[6] Soekartawi 1995 *Farming Analysis* University of Indonesia Jakarta

IOP Conf. Series: Earth and Environmental Science **122** (2018) 012134 doi:10.1088/1755-1315/122/1/012134

# In vitro digestibility of oil palm frond treated by local microorganism (MOL)

**M Tafsin , Y Khairani , N D Hanafi and Yunilas**

Animal sciences study program, Faculty of Agriculture, Universitas Sumatera Utara
E-mail : nevydiana@yahoo.co.id

**Abstract.** Oil palm frond is by product from oil palm plantation and were found in large quantity in Indonesia. This research aims to examine the ability of local microorganisms and buffalo rumen isolates in improving the digestibility of dry matter and organic matter *in vitro* of oil palm frond. The research used experimental method with four treatments and three replications. The treatments were given: Oil palms without treatment (P0); Starbio (P2); *Aspergillus niger* + *Saccharomyces cerevisiae* (P3); *Aspergillus niger* + *Saccharomyces cerevisiae* + Isolate of buffalo rumen bacteria (P4). The results showed that the fermented Oil Palm Frond had higher ($P<0.05$) DMD and OMD than control. The addition of *Aspergillus niger* and *Saccharomyces cerevisiae* plus buffalo rumen bacterial isolates had higher ($P<0.05$) DMD and OMD than other treatments. It can be concluded that the utilisation of MOL can improve the digestibility of oil palm frond in vitro.

## 1. Introduction

Oil palm plantation produces palm oil wastes that have potential prospective as feed providers of ruminant livestock. Each palm tree can produce 22 Oil palm frond (OPF) / year and the weight of OPF reaches 2.2 kg so that each hectare can produce oil palm frond of 9 ton / ha / year or equal to 1.64 ton / ha / year of dry matter [1] . Oil palm frond has nutrient content (% DM) equivalent to natural grass growing in grazing field. The content of nutrients of OPF had dry matter 48.78%, Crude Protein 5.3%, Hemicellulose 21.1%, Cellulose 27.9%, Crude Fiber 31.09%, Ash 4.48%, NFE 51.87%, Lignin 16.9% and Silica 0.6% [2]. Limiting the utilization of OPF as animal feed is the presence of high lignin content and protein content [3]. Provision of low quality feed with high lignin content, will affect yhe digestibility of nutrient, so the technology needed to fix it. Fermentation is one of the technologies to improve the quality of feed, due to the involvement of microorganisms in degrading crude fiber, reducing lignin levels and anti-nutrient compounds, so that the digestibility value of feed from waste can increase [4].

Fermentation using local microorganisms is simpler when compared with fermentation with commonly used commercial bacteria or fungi and can be directly used as inoculums in the substrate. It is hoped that the future of fermentation technology using this local microorganism can improve the quality of local feed continuously and replace commercial materials such as yeast tempe and EM4 [5].

Preliminary study was conducted using MOL containing *Aspergillus niger*, *Saccharomyces cerevisiae* and buffalo rumen isolates which showed that at a dose of 0.4% multi local microbial showed an increase in dry matter digestibility coefficient (DMD) and organic matter (OMD) in OPF than 0.2% , but the incubation time of 0-9 days did not affect the digestibility value [6]. Referring to the description, the authors are interested to examine the modification method on fermentation of OPF by local microorganisms and buffalo rumen isolates in Oil Palm Frond. This research aims to examine the ability of local microorganisms and buffalo rumen isolates in improving the digestibility of dry matter and the digestibility of organic matter on OPF *in vitro*.

## 2. Materials and methods

The research was conducted at the Laboratory of Animal Nutrition and Feed Science, Animal Husbandry Program, Faculty of Agriculture, University of Sumatera Utara Medan. This study was conducted from March to June 2017. Materials consisting of oil palm fronds, commercial probiotics (Starbio), MC Dougall solution, CO2 gas, 0.2% HCl pepsin solution, aquadest, saturated HgCl 2 solution, whatman no 41 filter paper, CMC1, XY5 and AT 4 isolates were obtained from the preliminary study of buffalo rumen isolation, *Aspergillus niger* and *Saccharomyces cerevisiae* were obtained from the collection of Animal Breeding and Reproduction Laboratory of Animal Husbandry Study Program

### *2.1. Research methods*

The research method used complete randomized design (CRD) with 4 treatments and 3 replications. The treatments were given composed of :
P0: Oil palm frond without treatment (control)
P1: Oil palm frond with addition of commercial product (Starbio)
P2: Fermented Oil Palm Frond with local microbe (*Aspergillus niger* + Saccharomyces cerevisiae)
P3: Fermented Oil Palm Frond with local microbe (*Aspergillus niger* + *Saccharomyces cerevisiae* + buffalo rumen isolate)

### *2.2. Isolate Bacteria from Rumen Buffalo*

Bacterial isolates were obtained from preliminary studies which obtained isolates by taking buffalo rumen fluid and then isolating buffalo rumen bacteria by growing them on selective medium. After isolation, there were 14 isolates consisting of three types, namely cellulose isolate, xylane isolate and lignin isolate. Then Isolate was characterized and measured the clearing zone to obtain the best isolate. The best 3 isolates for use in Oil Palm Frondpalm fermentation are CMC1, XY5 and AT4.

### *2.3. Fermentation Process of Oil Palm Frond*

600 grams of oil palm frond that has been chopped with Chopper were inserted in plastic and then sterilized (121$^0$C, 20 minutes), after which it was cooled and placed on the plastic tray, then treat according the experimental treatment. The treatment P3 using the *Aspergillus niger* $10^6$ CFU/g inoculum was added and stored at room temperature for 7 days. On day 7, the substrate were added *Saccharomyces cerevisiae* ($10^6$ CFU/g) and rumen buffalo bacteria isolate (0,4%) then incubated for 7 days.

### *2.4. In vitro digestibility [7]*

Samples of 0.5 grams of fermented oil palm frond were inserted into a 50ml fermentor tube added with a 30ml buffer solution (MgDougall), 15ml rumen fluid, then each fermentor tube was fed a CO2 gas for 30 seconds to create an anaerobic condition and covered with a rubber cover. The fermentor tube is inserted into a waterbath at 39°C, incubated for 48 hours. After 48 hours the incubation process is stopped and then drops 3 drops of saturated $HgCl_2$ to kill microbes Each fermentor tube was added 40ml 0.2% pepsin solution in 0.1% HCL then inserted into waterbath at 39°C for 48 hours, and centrifuge for 15 minutes at 2500 rpm. Then the residue is filtered with Whatman no. 41 filter paper. Each replication was done with duplo. The residue were collected and dry into the oven at 60°C for 24 hours then cooled in the desiccator for 15 minutes and then analized the dry matter content. Calculation DMD by using the formula as follows:

$$DMD\ (\%) = \frac{DM\ sample - (DM\ residue - DM\ residue\ blank)}{DM\ sample} \times 100\% \quad (1)$$

The OMD measurement was performed by the sample that had been tested on DMD measurements fed into the furnace for 6 hours at 600°C to calculated the ash content. . Calculation of OMD by using the following formula:

$$OMD\ (\%) = \frac{OM\ sample - (OM\ residue - OM\ residue\ blank)}{OM\ samples} \times 100\% \quad (2)$$

*Data analysis*

The analysis of treatment was calculated using the digestibility formula in vitro and the data obtained were analyzed using analysis of varians (ANOVA), and further analysis will be continued by using Duncan Multiple Range Test [8].

## 3. Results and discussion

### *3.1. The dry matter digestibility coefficient (DMD)*

Digestibility is an early indication of the availability of various nutrients contained in certain feed ingredients for livestock that consume them. High digestibility reflects the large contribution of certain nutrients to livestock, while feeds that have low digestibility indicate that the feed is less capable of supplying nutrients for basic living or for livestock production purposes [9]. The digestibility coefficient of dry matter of palm-oil based on in vitro analysis presented in Table 1.

**Table 1.** The average of In vitro Dry Matter Digestibility (%) of oil palm frond treated by local microorganism.

| Treatments | Replication | | | Average±SD |
|---|---|---|---|---|
| | U1 | U2 | U3 | |
| P0 | 36.89 | 33.45 | 36.08 | $35.47^{d}$ ±0.017 |
| P1 | 40.09 | 40.08 | 40.53 | $40.23^{c}$ ±0.003 |
| P2 | 45.67 | 47.08 | 45.29 | $46.01^{b}$ ±0.009 |
| P3 | 59.68 | 56.71 | 59.28 | $58.56^{a}$ ±0.016 |

Note: Different superscripts showed significant differences (P <0.01)

The result of statistical analysis showed that Oil Palm Frond with different microorganisms gave significant effect (P <0.01) on dry matter digestibility level. The highest DMD was found in P3 treatment with Oil Palm Frond with addition of *Aspergillus niger* and *Saccharomyces cerevisiae* plus rumen bacteria isolation (58.56%), while the lowest dry matter coefficient was found in treatment P0 ie Oil Palm Frond without treatment (35.47%). The result of analysis from Duncan Multiple Range Test showed that Oil Palm Frond without fermentation has DMD of 35.47% while fermented palm fronds using *Aspergillus niger* and *Saccharomyces cerevisiae* can increase DMD (46.01%) and fermentation of *Aspergillus niger*+*Saccharomyces cerevisiae*+buffalo rumen bacteria can increase DMD (58.56%). Fermentation by using commercial product (40.23%) had lower DMD compared with the utilisation of local microbial and rumen bacteria isolate..

Results of OPF using *Aspergillus niger* and *Saccharomyces cerevisiae* plus Isolate of buffalo rumen bacteria in this study were higher than MOL study containing *Aspergillus niger*, *Saccharomyces cerevisiae* and buffalo rumen bacterial isolates at dose 0.4% had dry matter digestibility coefficient value (DMD) of 45.37% [9]. This is because the isolates used in this study are the best rumen bacteria isolates that have passed characterization. The ability of the cellulolytic bacteria to dominate the bacterial population in the rumen and potential as a substitute for fresh rumen fluid in vitro studies [10].

The average percentage of dry matter digestibility coefficient showed that the fermented Oil Palm Frond (OPF) had higher digestibility than without fermentation, and indicated that microorganisms had ability to degrade OPF and could decrease lignin levels in OPF and the resulting increased the nutritional value of palm fronds. Fermentation degraded substrate by certain enzymes against non-digestible ingredients such as cellulose and hemicellulose into simple structure . During the fermentation process there is microbe growth produced by metabolic protein so that there is an increase in protein content [11].

Fermentation of OPF with rumen isolate showed the highest DMD caused by buffalo isolate containing bacteria capable of degrading fiber component in OPF. Isolate buffalo bacteria capable of degrading poor quality roughages. The rumen fluid from buffalo contains cellulolytic bacteria and is in good condition and ideal as a source of superior fiber digestive bacteria in utilizing fiber feed sources [12]. The presence of cellulolytic bacteria (fiber digesters) in the rumen fluid is *Butyrivibrio fibrisolvens, Bacteroides succinogenes* and *Ruminococcus albus* derived from rumen fluid of cow, buffalo and sheep [13]. The high ability of rumen bacteria isolates in degrading feedstocks of fiber source is possible because the tropical ruminant animals tend to consume the feed in the form of agricultural waste that has high lignocellulosic content [14].

*Aspergillus niger* plays a role in producing cellulase enzymes that function to convert cellulose into glucose so as to improve the digestibility of a feed ingredient. The addition of *Saccharomyces cerevisiae* is done to create anaerobic conditions for rumen bacteria to work more optimally. Yeast culture (*Saccharomyces cerevisiae*) used oxygen to feed particles metabolism into sugars and oligosaccharides in producing peptides and amino acids as the end product used by bacteria. Most rumen microorganisms are anaerobic, then oxygen utilization by yeast culture will improve the optimum conditions in the rumen. This condition, will protect the anaerobic rumen bacteria from damage by O2. Creates better conditions for cellulolytic bacteria growth, resulting in increased cellulolytic bacteria and increased digestion in the rumen [15].

Fermentation of OPF conducted on the research has been able to increase the nutritional value of OPF so it can increase DMD. Some things that affect the digestibility is the composition of the feed. Feed with a complete nutritional content will improve the digestibility of the feed itself. The higher the digestibility of the dry matter and the organic material of the high-fiber nutrient feed that can be used to meet the nutrient requirements of livestock.

### *3.2 Organic Components of Digestibility (OMD)*

The digestibility of organic matter is measured because the components of the organic material are needed by livestock for basic living and production. Organic materials produce energy for the growth and development of livestock. The higher the digestibility value of a feed ingredient, indicated that more nutrients for the animal body absorbs. The value of organic matter digestibility (OMD) can be seen in Table 2.

**Table 2.** The average of In vitro Organic Matter Digestibility (%) of oil palm frond treated by local microorganism.

| Treatment | Replications | | | Mean±SD |
|---|---|---|---|---|
| | U1 | U2 | U3 | |
| P0 | 56.19 | 50.31 | 54.55 | $53.68^{d}$ ±0.010 |
| P1 | 61.19 | 61.80 | 61.27 | $61.42^{c}$ ±0.003 |
| P2 | 64.22 | 66.70 | 64.38 | $65.10^{b}$ ±0.014 |
| P3 | 74.32 | 72.41 | 74.91 | $73.88^{a}$ ±0.013 |

Note: Different superscripts showed significant differences (P <0.01)

The result of statistical analysis showed that OPF from fermented sources with different microorganisms gave significant effect (P <0.01) on the digestibility level of organic matter. The highest OMD was found in P3 treatment of OPF with the addition of *Aspergillus niger* and *Saccharomyces cerevisiae* plus Buffalo rumen bacteria isolate (73.88%), while the lowest percentage of organic material low coefficient was found in the treatment of P0 (53.68%). The result of analysis from Duncan Multiple Range Test showed that OPF without fermentation (53.68%) significantly lower than other treaments, while fermented palm fronds using *Aspergillus niger* and *Saccharomyces cerevisiae* can increase OMD (65.10%) and fermentation of *Aspergillus niger* and *Saccharomyces cerevisiae* plus Rumen Buffalo bacteria isolate can increase OMD higher than other treatments. fermentation by using commercial product had lower OMD compared with the use of local microbial and rumen bacteria isolate.

The digestibility value of organic matter has the same scheme as dry matter digestibility. The digestibility of organic matter has a higher value than the dry matter digestibility. The digestibility value of organic matter is in line with dry matter digestibility value, this is because organic matter is part of dry matter. The high digestibility of organic matter is also caused by the presence of high crude protein content, which leads to increased development of microorganisms that digest these feed ingredients.

In this research, it can be seen that the ability of bacterial isolates from rumen buffalo fluid that is able to utilize fiber source feed is the main indication of the isolate as a bacteria isolate isolate. The high ability of rumen bacteria isolate in degrading feed material of fiber source is possible because of ruminant livestock in tropical area tend to consume food in the form of agricultural waste that has high lignocellulosic content. The group of extracellular enzymes from bacteria of fiber digesters can utilize them by breaking bonds from lignocellulose structures and increasing hydroxyl groups and carboxyl phenolics Increasing the digestibility of organic matter suspected of influence activity of local microorganisms and buffalo rumen bacterial isolates in degradation of lignocellulose and lignohemisellulose bonds on OPF. With the increasing number of rumen microbes, it can increase the activity of fermentatively diverting organic feed material into simple soluble compounds, consequently increasing the absorption of organic substances [16].

## 4. Conclusions

Fermented oil plam frond using *Aspergillus niger* and *Saccharomyces cerevisiae* had higher DMD and OMD than without fermentation. However, fermentation with the addition of *Aspergillus niger* and *Saccharomyces cerevisiae* plus buffalo rumen bacterial isolates had higher DMD and OMD than other treatment. It can be concluded that the utilisation of MOL can improve the digestibility of oil palm frond in vitro.

## References

[1] Diwyanto K D, I Manti, Mathius I W dan Soentoro 2003 Pengkajian pengembangan usaha sistem integrasi kelapa sawit- sapi (assessment of business development integration system of oil palm-cow) *Prosiding Lokakarya Nasional Bengkulu* 9 - 10 September 2003

[2] Imsya A 2007 Konsentrasi N-Amonia, kecernaan bahan kering dan kecernaan bahan organik pelepah sawit hasil amoniasi secara *in-vitro (*N-Ammonia concentrations, digestion of dry materials and digestion of organic ingredients of palm oil in-vitro ammonitation result) *Prosiding Seminar Nasional Teknologi Peternakan dan Veteriner* Bogor

[3] Prabowo A, Pramudyati Y S dan Susanti A E 2011 Potensi limbah pelepah dan daun kelapa sawit untuk pakan sapi potong di Sumatera Selatan (Potency of waste-midribs and palm oil leaves for beef feed in South Sumatera) *Prosiding Seminar Nasional Peternakan Berkelanjutan* Ke-3 Fakultas Peternakan Universitas Padjadjaran

[4] Wina E 2005 Teknologi pemanfaatan mikroorganisme dalam pakan untuk meningkatkan produktivitas ternak ruminansia di Indonesia (Technology utilization of microorganisms in

the feed to increase ruminant livestock productivity in Indonesia) Sebuah review *Wartazoa* **15** (4): 173-186

[5] Astuti T P, Yelni J G and Amir Y S 2015 The effect of a local biotechnological approach on rumen fluid characteristics (PH, NH3, VFA) of the oil palm fronds as ruminant feed International *Journal of Agriculture Innovations And Research* Vol **3**, Issue 6, ISSN (Online) 2319-1473

[6] Hanafi N D, Tafsin M dan Sujana W 2016 Penggunaan multi mikroba lokal dengan berbagai dosis dan lama inkubasi terhadap kecernaan bahan kering dan bahan organik pelepah kelapa sawit in vitro (Multiple local microbial utilization with various doses and length of incubation on dry food degradation and organic ingredients of palm oil in vitro) *Jurnal Peternakan Integratif* Vol **4** No 2

[7] Tilley J M A and Terry R A 1963 A two-stage technique for the *in vitro* digestion of forage crops *J. Brit. Grassland Soc.* **18**:104-111

[8] SAS Institute 1994 *SAS-STAT user's guide* (North Carolina; Cary)

[9] Rubianti A, Fernandez P T H, Marawali H H dan Budisantoso E 2010 Kecernaan Bahan Kering Dan Bahan Organik Hay *Clitoria Ternatea* Dan *Centrocema Pascuorum* Cv Cavalcade Pada Sapi Bali Lepas Sapih Balai Pengkajian Teknologi Pertanian Nusa Tenggara Timur

[10] Prihantoro I, Sari Y, Riyanti L, Sasmita T E, Evvyernie E, Suryani, Abdullah L and Toharmat 2012 Nutritive value of forages using a mixed bacteria isolated from the rumen liquor of buffalo *Proceeding of the 2nd International Seminar on Animal Industry* Jakarta

[11] Sembiring P 2006 Biokonversi Limbah Pabrik Minyak Inti Sawit Dengan Phanerochaete Chrisosporium *Disertasi* UNPAD Bandung

[12] Rifai A A 2010 Peran isolat bakteri selulolitik fakultatif asal rumen kerbau pada hijauan berbeda (The role of bacterial cellulolytic isolates from rumen buffalo origin in different forages) *Skripsi* Fakultas Peternakan. Institut Pertanian Bogor Bogor

[13] Thalib A 2002 Pengaruh imbuhan faktor pertumbuhan mikroba dengan dan tanpa sediaan mikroba terhadap performans kambing Peranakan Etawah (The effects of microbial growth factor with and without microbial preparation on Peranakan Etawah performance) *JITV* **7**:220-226

[14] Pandya P R, Singh K M, Parnerkar S, Tripathi A K, Mehta H H, Rank D N, Kothari R K and Joshi C G 2010 Bacterial diversity in the rumen of indian surti buffalo (Bubalus bubalis), assessed by 16S rDNA *analysis. J. Appl. Genet.* **51**: 395-402

[15] Jouany J P 2001 Twenty years of research and now more relevant than ever the coming of age of yeast cultures in ruminant diets. In : Responding to a Changing Agricultural Landscape. Alltech's European, Middle Eastern and African Lecture Tour.

[16] Murni S dan Putra S 2004 Manipulasi mikroba dalam fermentasi rumen salah satu alternatif untuk meningkatkan efisiensi penggunaan zat-zat makanan (Microbial manipulation in rumen fermentation as an alternative to increasing the efficiency of food substances utilization). Paper Jurusan Nutrisi dan Makanan Ternak Fakultas Peternakan, Universitas Udayana.

**Acknowledgement**
The author want to say thank you to Directorate General of Higher Education for supporting fund through DRPM 2017 for this research.

# The impact of sharing arrangement institution on beef cattle breeding performance in Kupang District, East Nusa Tenggara Province, Indonesia

**O H Nono[1], R Natawidjaja[2], B Arief[2], D Suryadi[2] and M M J Kapa[3]**

[1]Faculty of Animal Science; Universitas Nusa Cendana, Kupang Indonesia
[2]Faculty of Animal Science, Universitas Padjadjaran, Bandung Indonesia
[3]Faculty of Agriculture, Universitas Nusa Cendana, Kupang Indonesia
E-mail: yusufhenuk62@live.com

**Abstract**. The aim of this study was to analyse the impact of sharing arrangement systems to performance of beef cattle breeding. This research was conducted in Kupang Regency – East Nusa Tenggara Province, Indonesia. The study used multi stage cluster random sampling method to determine the sample area and respondents. The sample areas consisted of 2 sub-districts and 6 villages, while the total respondents were 117 people comprised 74 Participant Farmers (PF) of sharing arrangement systems (SAS) and 43 non-participant farmers (NPF). 23 investors were selected for the survey. The result of the study indicated that the performance of NPF in terms of revenue, net profit, and return on investment (ROI) was better than PF respondents. The value of ROI was between 16.69–32.23 %. This indicated that utilization of farm asset was not optimum yet. It was found that farm efficiency was 1.73 which indicated that SAS does not increase farm productivity.

## 1. Introduction

Livestock production is an important part of agricultural sector which contributes to Gross Domestic Product (GDP) of Indonesia, and to the Indonesian overall economy. It also plays a vital role in many different aspects of people daily life, such as ensuring food security, providing nutrition, generating income and savings, and in many other social and cultural functions. Beef cattle play a significant role in the livelihood of many smallholder farm people living in rural areas, particularly as their family source of income, nutrition and welfare. Indeed, most rural people rely on farming, in one form or another, as their main occupation [1]. Livestock subsector provides food in the form of animal protein for the people. As a source of animal protein, beef is expected to meet the nutritional needs of the people. Beef contains 10 kinds of amino acids and essential fatty acids especially conjugated linoleic acid which is beneficial for the growth of neurons in the brain that determines the level of human intelligence [2]. Currently, beef consumption in Indonesia continues to increase, but the increase was not offset by the addition of adequate production. This condition causes low contribution to the nationwide production of beef cattle and resulting in widening gap between demand and supply. One of the solutions to overcome this gap is to increase domestic beef production. It seems that these efforts will be constrained given that more than 90% of beef production in Indonesia is produced by small farmers with 2-3 cows per household farmers [3]. Downstream products of beef cattle have not been optimally processed to attain high added-value products. Therefore, strategic steps for beef agro-industries development using resource-based in East Nusa Tenggara (ENT) should be taken into consideration. Beef cattle industry in ENT is categorized as small and medium-scale enterprises and use simple technology [4]. This leads to low production and irregular product quality in ENT [5]. This study aimed to analyze the impact of sharing arrangements systems on performance of beef cattle breeding in Kupang District, ENT, Indonesia.

## 2. Materials and Methods

This study was conducted in Kupang Regency, ENT Province, Indonesia. Multistage cluster random sampling technique was used to determine the location and respondents. The Selection of the research area began from districts level (Kupang Regency), and the sub-district level (there are two sub-districts, namely Amarasi and Fatuleu) and village level (for each sub-districts, three villages were selected). The sample was determined using proportional allocation [6] and 74 participant farmers and 43 non-participant farmers were selected. In order to validate data of participant farmers' production performance, data from investors were used as a comparive data. Three types of investors were involved in this study. They are local government, external private insitution, and the internal private institution.There were 23 investors chosen to be interviewed. Comparison of the production performance of partipant farmers and non-participant farmers covered four performance variable,s namely income, revenue, profits, and return on investment from participant and non-participant farmers using two-tailed test [7].

## 3. Results and Discussion

### *3.1. Revenue Sharing Arrangment Institution*

The profit sharing arrangement institution among farmers and investors has been practiced for a long time. Majority (97.30 %) of respondents stated that the profit sharing has been done for more than 5 years. The sharing commitment is *in kind* pattern. That is, the farmer gets one calf from every 3-5 births. In general, the contract agreement between farmers and government is made in written form. Meanwhile, agreement between farmers and private investors is in the form of unwritten agreement. Both government and private investors require farmer participants to form a group before hand in order to get the breeding cattle aid. The group leader is responsible for the performance of its members.The contract contains the scope of work, rights and obligations, duration, execution, refund, sale, sanction, force majeure, cancellation of the agreement, dispute settlement. Breeders' obligations include feeding, health control, cattle safeguards and mating arrangement. The main thing to remember is every member of participants has to distribute a calf from every 3-5 births to another member of the group. The duration of sharingarrangement islimited to 3-5 years.The farmers are expected to have "sufficient" numbers of livestock so that they may develop their own business. They do not have any rights in terms of husbandry technology transfer in this case, feedingand reproduction technology aimed to increase productivity. This indicates that the existing rules of revenue sharing arrangement are not optimal yet. Institutional elements such as jurisdictional boundaries and authority between stakeholders are clearly define and revenue sharing arrangement is running well although the agreement between participant farmers and private investors are not in written agreement. According to participant farmers, the incentive of participant farmers does not compensate their efforts. The implementation of such institutional elements may be closely related to: (1) market profile, as one of the characterisitcs of revenue sharing commodities. This means that, the products can be sold anytime and related to socio-economic and cultural factors in the research area. This is confirmed by T Olviana, D P Darmawan, I G A A Ambarawati, and L Sulistyowati [5], who reported that market profile determines the pattern of institutional production, (2) low socio-economic level resulted in weak bargaining position. This is related to the elements of bounded rationality of participant farmers as a consequence of their capacity to access crucial information such as information on technology, final market and how to regulate these factors on agreement framework of revenue sharing with other parties, (3) scarcity or no other alternativesand productive economic activities.

### *3.2. Performance of Beef Cattle Breeding farms*

Performance indicators in beef cattle breeding consists of cash income (or inkind converted to cash value based on the price of cattle, revenue from value-added of capital (VAC), total revenues, net, Net profit (that's value obtained by substractnet income to transaction costs during a year and the value rate of return on investment (ROI). The data in Table 1 explained thatthe average net income (both cash and non cash)of participant farmer is lower than the average income earned by non participant farmers, and statistically there is a highly significant difference (P <0.01) in income earned between participant farmers and non-participant farmers. Data above also explain that revenue in sharing arrangement between farmers and government investors is more beneficial than revenue sharing arrangement between farmers and other investors. However, from institutional point of view, revenue sharing arrangement between farmers and government investors has several weaknesses such as the rule is loose, hence, the distribution of offspring (calf) to other group members is beyond the set schedule. This alsorelated to low abilityof government to monitoring and evaluation the systems.
Rules that have beeen regulated by both external and private investors are different. They are more assertive, hence, the revenue sharing is applied based on the agreement especially for breeding where 1 out of 3-5 births is belonged to participant farmers. On the other hand, non-particpant farmers' average revenues depends on the number of cattle that they have and this should be more than the average revenue recieved by the particpant farmers. Table 1 Summarisedthe analysis of the average varriance score of performan indicators of beef cattle breedingin Kupang District, ENT.

The statistcal test on performance indicator such as value-added capital (VAC) showed that only participant farmers funded by government that recieve VAC. In contrast, VAC of beef cattle breeding funded by private investors is not shared to the farmers. This significant differences (P <0.05) where average VAC recieved by participant farmers is lesser than independent farmers occurs because of the numbers of beef cattle that they have. Performance indicators such as total revenue (TR) of both participant and non-partcipant farmers is a cash income or its equivalent plus VAC. The results of variance test on mean value of total revenue (TR) shows that there is significant differences (P <0.01) between participant farmers and non-participant farmers. Based on Tukey's test, it reveals that TR of participant farmers funded by government is bigger than those who are funded by private investors (external and internal), but smaller than the TR of non-participant farmers.

The result of variance test on performance indicators such as net income is a difference between total income and production cost (especially calculated cost). It is clear that the differences is equal to the TR indicator. This also shows that the value of the calculated cost is relatively small and proportional, or does not differ by the type of farmers. In this calculation, the cost only includes health care fee (vaccine and medicince) which is partly covered by the participant farmers. For feeding cost and salary is not included as expenses because all workers are the farmers' family. This means that those two production factors are not considered as a part of sharing arrangement systems. This happens because the farmers want to maintain investors' trust in managing their assets. In addition, job or other business in this area is not available at all times so these two factors are economically invaluable.

**Table 1.** The analyisis of statistical test on average varriance of performance between participant farmers/and non-participant farmers

| Indicator | Code | Average (Rp000) | F-test | | Tukey-test | | T-test |
|---|---|---|---|---|---|---|---|
| | | | Among P1 | Between P1 &M | Among P1 | Between P1 & M | |
| Cash income(equal) | P1 | 2,787.60 | ** | ** | 1 vs 2 ** | 1 vs 2 ** | NS |
| | P2 | 1,733.75 | | | 1 vs 3 ** | 1 vs 3 ** | |
| | P3 | 1,949.20 | | | 2 vs 3 tn | 2 vs 3 | |

| | | | | | | | |
|---|---|---|---|---|---|---|---|
| | M | 2,380.20 | | | | 2 vs 4*<br>3 vs 4 | |
| Added value capital | P1 | 967.32 | | | 1 vs 2** | | ** |
| | P2,P3 | 0 | | | 1 vs 3** | | |
| | M | 2,595.35 | | | 1 vs 4* | | |
| Total Income | P1 | 3,774.93 | ** | ** | 1 vs 2 ** | 1 vs 2 ** | ** |
| | P2 | 1,744.29 | | | 1 vs 3 ** | 1 vs 3 ** | |
| | P3 | 1,972.80 | | | 2 vs 3 $^{tn}$ | 1 vs 4 ** | |
| | M | 4,962.56 | | | | 2 vs 3<br>2 vs 4*<br>3 vs 4** | |
| Net Income | P1 | 3,398.48 | ** | ** | 1 vs 2 ** | 1 vs 2 ** | ** |
| | P2 | 730.37 | | | 1 vs 3 ** | 1 vs 3 ** | |
| | P3 | 798.91 | | | 2 vs 3 $^{tn}$ | 1 vs 4 ** | |
| | M | 4,539.70 | | | | 2 vs 3<br>2 vs 4*<br>3 vs 4** | |
| Net Profit | P1 | 3,244.08 | ** | ** | 1 vs 2 ** | 1 vs 2 ** | ** |
| | P2 | 613.33 | | | 1 vs 3 ** | 1 vs 3 ** | |
| | P3 | 700.67 | | | 2 vs 3 $^{tn}$ | 1 vs 4 ** | |
| | M | 4,533.38 | | | | 2 vs 3<br>2 vs 4*<br>3 vs 4** | |
| ROI | P1 | 28.29 | ** | ** | 1 vs 2 ** | 1 vs 2 ** | ** |
| | P2 | 19.00 | | | 1 vs 3 ** | 1 vs 3 ** | |
| | P3 | 18.38 | | | 2 vs 3 $^{tn}$ | 1 vs 4 ** | |
| | M | 38.94 | | | | 2 vs 3<br>2 vs 4*<br>3 vs 4** | |

Description: P1 = participat farmers funded by government, P2 = participants farmers funded by external private institution, P3 = participant farmers funded by internal private institution, M = non-participant farmers. tn = no significant differences, ** = significant differences.

The results of mean variance test on performance indicators in a form of net profit is the difference between net income and transaction costs. Based on Table 2, the mean variance among participant farmers is equal to the variance of other performance indicators as explaind above, where the variance among participant farmers is P <0.01, between participant farmers and non-participant farmers is P <0.01. The result of Tukey test also indicates that the average net profit of farmers funded by government is bigger than the average net profit (NP) of participant farmers funded by external and internal private investors as well as the mean average NP of non participant farmers is bigger than NP of participant farmers (P <0.01). In terms of performance indicators, the result of mean varriance test on return of invesmentdescribes to what extent the investment can be returned in a particular production period. Table 2which shows the result of F-test indicates that there is a significant difference (P <0.01) in terms of return of investment among participant farmers, between participant farmers and non-participant farmers. Later on, Tukey test also shows similar result among participant farmers (P <0.01), as well as the comparison of the average ROI between participants farmers and average ROI of non-participant farmers (P <0.01).

The value of ROI on beef cattle breeding is high, which ranging between 18.38 to 38.94 %. Although this result is different with C Firmansayah, S Kuswaryan, and S Rahayu [8], who reported that ROI is 14.91 % only.This means that value of return investment is higher than annual commercial interest rates (16-18 %). Actually, ROI value on community beef cattle breeding in this study show minimal use of asset of beef cattle breeding, especially those owned by particpant farmers. In other words, the resources should be more optimized in order to increase the income of participant farmers. This is in accordance with the request of the majority of participant farmers that they need to increase the number of livestock so that the opportunity to develop their own business is greater. This is because it is supported by the availability of labor and land as a source of forage/feed. The net profit earned by participant farmers especially those who shared the revenue with external and internal private investors was very low. However, based on the interview with respondents, most of them prefer to continue this business with private investors. This is most likely related to the flexibility of the rules. This means that when the farmers have emergency need, they may get loan from investors that have equal value to the cattle that they are entitled to in the coming months. This kind of rules leads toproduce patron client relationship between investors and participant farmers. In addition, the sales activity is also flexible where the cattle can be sold at any time within a year, and at any age. Above explanation indicates that revenue sharing arrangements does not provesignificant impact on the producvity of beef cattle breeding. These results are similar to the finding of A Fujimoto [9], who reported about the role of contract farming. These two research show that there is no effect of tenure status as an elements of revenue sharing onrice productivity. This is different from the results obtained by O H Nono, M Yunus, and S A Adutae [10] who reported the performance indicators of agriculture on the irrigated area which include: total revenues, net income, net profit, R/C, and the value of return on invesment / ROI. In this case,the performance of pawner and tenantis better than the performance of owners and farmers. This facts do not indicate that the revenue sharing arrangements is not needed in business development. This occurs because farmers normally have several major limitations such as capital, acces on science and technology or access to markets (input and output). Thus, revenue sharing arrangments provides an opportunity for farmers to be able to optimize their resources so as to obtain additional income. It is asserted by O H Nono and S A Adutae [11] that the performance of agriculture businesses (variety of crops, livestock and plantations) is determined by the capacity of organizations and institutions in addition to other factors.

## 4. Conclusions

a. The performance of non-participant sharing arrangement business is more profitable than the profit earned by participant farmers.
b. Revenue sharing arrangements has not increased farms/business productivity yet. However, the existence of sharing arrangement of breeding cattle business provides an opportunity for farmers and their families to optimize resources (land and worker) usage which in turn increasing their income.

## References

[1] Gayatri S, Gasso-tortajada V and Vaarst M 2016 Assessing sustainability of smallholder beef cattle farming in Indonesia : a case study using the FAO SAFA framework *J SustainDev* **9**(3): 236 – 247
[2] Pighin D, Pazos A, Chamorro V, Paschetta F, Cunzolo S, Godoy F and Grigioni G 2016 A contribution of beef to human health: a review of the role of the animal production systems *The SciWrld J.* 1 – 10

[3] Isyanto A Y and Sugianto I 2016 Factors influencing population of beef cattle in Ciamis Regency, West Java Province, Indonesia *J Econ Sustain Dev.***7**(22), 34-38
[4] Kapa M M J, Soemarno, Bagyoyanuwiadi and Suyadi 2017 Sustainability status of biology dimension of local beef cattle development in the dryland region, Indonesia *JEcon Sustain Dev.***8**(6): 102-108
[5] Olviana T, Darmawan D P, Ambarawati I G A A and Sulistyowati L 2014 Development of beef cattle agroindustries performance model in East Nusa Tenggara, Indonesia *J Econ Sustain Dev.* **5**(21): 79 – 87
[6] Rasyid H 1994 Sampling technics and scaling. Postgraduate Programme Padjadjaran University, Bandung
[7] Siagian D and Sugiarto 2002 *Statistical Methods for Business and Economics* (Jakarta: Gramedia Pustaka Utama)
[8] Firmansayah C, Kuswaryan S and Rahayu S 2006 Financial benefit on local cattle breeding smallhoders sharing pattern *Anim Prod J.***6**(1) : 75-80
[9] Fujimoto A 1983 *Income Sharing Among Malay Peasants : A Study of Land Tenure and Rice Production* (Singapore: Singapore University Press)
[10] Nono O H, Yunus M and Adutae S A 2009 Analysis of land ownership status and performance of agricultural enterprises in irrigation areas in East Sumba regency. Research Report Cooperation Researsh Institution of University Nusa Cendana with Government of East Sumba Regency - East Nusa Tenggara Province
[11] Nono O H and Adutae S A 2010 Analysis of agricultural sector performance in east sumba regency *Research Report Cooperation Researsh Institution of University Nusa Cendana with Government of East Sumba Distric –East Nusa Tenggara Province*

**Acknowledgements**
The authors thanked to Prof. Yusuf L. Henuk, Ph.D (E-mail: yusufhenuk62@live.com) from Faculty of Agriculture, University of Sumatera Utara (USU)-Medan for his critical reading and editing of this manuscript.

# Factor affecting the whole beef prices in eight provinces in Sumatera, Indonesia

**D Yasin[1], A A Harahap[1], M A Syaputra[1], RMulianda[1], T J Adha[1], A S Harahap[1], A Rusdhi[1], M S Ibarahim[1], A R Siregar[1], BPurwadi[1], S P Ginting[2], Hasnudi[2]and Y L Henuk**

[1]Postgraduate Students, Faculty of Agriculture,Universitas Sumatera Utara (USU),Medan20155Indonesia
[2]Postgraduate Lecturers,Facultyof Agriculture, Universitas Sumatera Utara (USU),Medan 20155,Indonesia
E-mail: yusufhenuk62@live.com

**Abstract**.A survey has been conducted to study the fluctuation of whole beef prices in eight provinces in Sumatera, namely Lampung, South Sumatera, Jambi, Bengkulu, West Sumatera, Riau, North Sumatera, and NAD with the average cattle weight were 486kg, 492kg, 448kg, 464kg, 500kg, 480kg, 500kg, and 480kg, respectively. Their average prices and total price in these cities were IDR4,1000 & IDR19,926,000; IDR41,500 & IDR20,418,000; IDR43,000 & 19,264,000; IDR42,500 & IDR19,720,000; IDR43,000 & IDR21,500,000; IDR43,000 & IDR20,640,000; IDR42,500 & IDR21,250,000; IDR44,000 & IDR21,120,000,respectively. People West Sumatera, Riau and North Sumatra provinces prefer the final results of processed by products of beef which caused the high price of processed by products from cattle. Whereas, in NAD province, the meat selling price is very high compared to other provinces, while the selling price of processed beef by-products is much cheaper than the people outside NAD province. The selling price of processed beef by-product can not help to reduce the selling price of meat. Average carcass in Sumatera ranged from 49.00% (North Sumatera) – 51.00% (Jambi). Among the eigth provinces, the price of beef in Aceh is more expensive than other provinces not only in Sumatera, but also other provinces in Indonesia.

## 1. Introduction

The projection of national supply of consumption of beef in Indonesia relative stable from 2014 – 2024 (Figure 1 and Table 1), meanwhile the consumption of beef increases as population grows. It is estimated that if there is no significant technology advancement in producing the local beef and no significant increasing of cow population,gap between the local beef production and the number of beef demand will be widening. Therefore, it will affect the import volume of beef which will be higher in Indonesia [1]. It has been reported that there is no export on beef in Indonesia since 2013 and indeed Indonesia relied heavily on import of beef until now with the price of beef in Indonesia tend to increase[2].

IOP Conf. Series: Earth and Environmental Science 122 (2018) 012136 doi:10.1088/1755-1315/122/1/012136

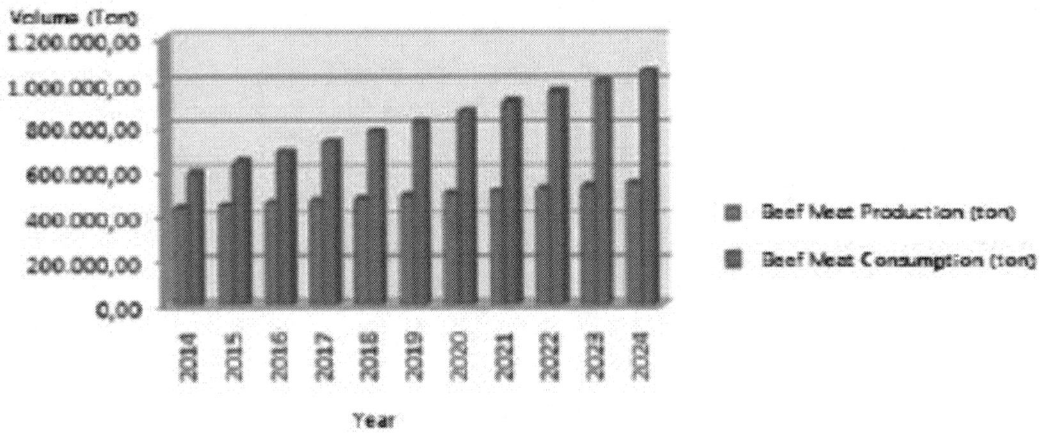


**Figure 1.** Projection of national supply and consumption of beef in Indonesia (2014 – 2024)[1].

**Table 1.** Volume of beef import, price of imported beef, GDP/capita and price of local beefin Indonesia (2010 – 2014) [1].

| Year | Volume of Beef Import (ton) | Price of Imported Beef (US$/kg) | GDP per capita (US$) | Price of Local Beef (IDR/kg) |
|---|---|---|---|---|
| 2010 | 84,508 | 3.2 | 3,125 | 57,944 |
| 2011 | 55,413 | 3.7 | 3,648 | 69,725 |
| 2012 | 30,377 | 4.1 | 3,701 | 76,925 |
| 2013 | 39,569 | 4.1 | 3,624 | 84,180 |
| 2014 | 57,052 | 4.9 | 3,492 | 99,056 |

[2]studied the effects of beef import restrictions policy on beef self-sufficiency in Indonesia. They aimed to analyse the impact of the restrictions of beef cattle industry and livestock sub-sector and the forecast of beef, self-sufficiency achievement in Indonesia. They suggested that restrictions on imports of feeder cattle and beef would increase domestic beef production and beef demand,but would reduce the population of cattle and livestock sub-sector performance. They also suggested that reduction in imports of beef and feeder cattle followed by technology improvement will accelerate the achievement of beef self-sufficiency in Indonesia. In general, the growing of population and the change in consumption pattern and the public demand of beef caused the national consumption of beef in Indonesia tends to rise. The need of beef in Indonesia is fulfilled by the three sources, which are: local cows, imported cows, imported beef. There are several efforts done by the Indonesian Government in stimulating the production of local livestock are: (1)enhancement of animal feed, (2) improvement of the quality of selective breeding through artificial insemination, and (3) disease eradication program [1]. Although Sumatera is the second producers (20.56%) of beef cattle (Figure 3) in Indonesia [4]. Among the eight provinces in Sumatera, West and North Sumatera is the highest beef meat production during the period of 2013 – 2017 [5]. This survey was conducted to analyse factors affecting the whole beef price in eight provinces in Sumatera as the highest population beef in Indonesia.

## 2. Materials and Methods

This survey analysed factor affecting which beef prices in eight provinces in Sumatera. It has been conducted during the period of January to April 2017. The data were collected by cutting reports on slaughterhouses or known in Indonesia as "Rumah Potong Hewan" (RPH) in eight large capital cities in Sumatera by collecting sampling from the average daily cut, the weight of live cattle, carcass weight, carcass, meat bone ratio of the cattle being cut, and the selling price of red meat and the follow-up products of cattle were cut.

## 3. Results and Discussion

There is a great variation in beef price factors in eight capital cities in Sumatera (Table 2). Table 2 shows that the average live cattle weight, live cattle price and total price of beef in eight cities from Lampung, South Sumatera, Jambi, Bengkulu, West Sumatera, Riau, North Sumatera, and NAD with the average cattle weight were 486kg, 492kg, 448kg, 464kg, 500kg, 480kg, 500kg, and 480kg, respectively.

**Table 2**. Average variation of live beef and meat characteristics in eight provinces in Sumatera.

| Beef Price Factors | Provinces | | | | | | | |
|---|---|---|---|---|---|---|---|---|
| | Lampung | South Sumatera | Jambi | Bengkulu | West Suamtera | Riau | North Sumatera | NAD |
| Live cattle weight (kg) | 486 | 492 | 448 | 464 | 500 | 480 | 500 | 480 |
| Live cattle price/kg (IDR) | 41,000 | 41,500 | 43,000 | 42,500 | 43,000 | 43,000 | 42,500 | 44,000 |
| Total price | 19,926,000 | 20,418,000 | 19,264,000 | 19,720,000 | 21,500,000 | 20,640,000 | 21,250,000 | 21,120,000 |
| Carcass (%) | 50.00 | 50.00 | 51.00 | 50.00 | 50.00 | 50.00 | 49.00 | 50.00 |
| Boning (%) | 35.40 | 35.00 | 36.00 | 35.00 | 35.60 | 35.00 | 35.50 | 35.50 |
| Total meat/ head (kg) | 172 | 172 | 161 | 162 | 178 | 168 | 178 | 170 |
| Meat price/kg (IDR) | 110,000 | 112,000 | 110,000 | 110,000 | 100,000 | 108,000 | 108,000 | 122,000 |
| Total meat price/head (IDR) | 18,924,840 | 19,286,400 | 17,740,800 | 17,864,000 | 17,800,000 | 18,144,000 | 19,170,000 | 20,788,800 |
| Head alone (IDR) | 2,000,000 | 2,100,000 | 2,500,000 | 2,500,000 | 4,500,000 | 3,500,000 | 3,200,000 | 1,500,000 |
| Total income (IDR) | 20,924,840 | 21,386,400 | 20,240,800 | 20,364,000 | 22,300,000 | 21,644,000 | 22,370,000 | 22,288,800 |
| Bruto (IDR) | 998,840 | 968,400 | 976,800 | 644,000 | 800,000 | 1,004,000 | 1,120,000 | 1,168,800 |
| Net profit buthcer/head (IDR) | 648,840 | 618,400 | 626,800 | 294,000 | 450,000 | 654,000 | 770,000 | 818,800 |

The average the cutting weight of beef in the eight cities in Sumatera (Table 2) within the average live cattle weight of 350- 500 kg/head which known as the period of innovation and efficiency toward a competitive beef cattle business (Tawaf,2016). Their average prices and total price in these cities were IDR4,1000 & IDR19,926,000; IDR41,500 & IDR20,418,000; IDR43,000 & IDR19,264,000; IDR42,500 & IDR19,720,000; IDR43,000 & IDR21,500,000; IDR43,000 & IDR20,640,000; IDR42,500 & IDR21,250,000; IDR44,000 & IDR21,120,000, respectively. There are some significant differences in the price of meat selling in each province which is caused by the consumption habits of beef and other processed products. People in West Sumatera, Riau and North Sumatra provinces prefer the final results of processed by-products of beef which caused the high price of processed by-products from cattle. While people in NAD province, the meat selling price is very high compared to other provinces, while the selling price of processed beef by-products is much cheaper than the people outside NAD province, so the selling price of processed beef by-product can not help to reduce the selling price of meat in the area. The average price of beef in these cities was higher than the government's recommended beef price of IDR80,000 /kg. Our survey supported by Susenas data showed that the participation of national beef consumption is only 16%, with income elasticity on demand greater than one (E> 1), meat that the higher the people's income, meat consumption will increase. In addition, this beef is also only enjoyed by 16% of Indonesians who have middle to upper income who live in urban areas [6]. M Sirod [7] conducted a study on precentage carcass of five types of cattle in 10 provinces in Indonesia, and reported the percentage of carcass of of the five types of cattle varied from 46.96% -- 53.26%; whereas the percentage of carcass in 10 provinces varied from 49.22% -- 53.72%. We found in our survey that average carcass in eight provinces in Sumatera ranged from 49.00% (North Sumatera) – 51.00% (Jambi; Table 2). The price of beef with the first quality in Aceh is still at IDR133,000 per kilogram (kg), according to the Indonesia Strategic Food Prices Information Center on January 18, 2017. The meat prices in NAD province beat the meat prices in Jakarta and South Kalimantan. In addition, the price of meat in Aceh is also the highest compared to beef prices in other provinces [8]. Our survey also proved that the price of beef in Aceh is more expensive than other provinces not only in Sumatera, but also other provinces in Indonesia.

## 4. Conclusions

There are some significant differences in the price of meat selling in each province which is caused by the consumption habits of beef and other processed products. People in West Sumatera, Riau and North Sumatra provinces prefer the final results of processed by-products of beef which caused the high price of processed byproducts from cattle. While people in NAD province, the meat selling price is very high compared to other provinces, while the selling price of processed beef by-products is much cheaper than the people outside NAD province, so the selling price of processed beef by-product can't help to reduce the selling price of meat in the area. Average carcass in eight provinces in Sumatera ranged from 49.00% (North Sumatera) – 51.00% (Jambi). Among the eight provinces, the price of beef in Aceh is more expensive than other provinces not only in Sumatera, but also other provinces in Indonesia.

**References**

[1] RudatinA 2016 Analysis on Indonesia's beef import*Econ.J.Emerg.Mark.***April**: 65-72
[2] Adam M 2016*Mendambakan Swasembada Daging.* www.ekonomi.metrotvnews.com/makro/ybD1GwZk-mendambakan-swasembada-daging.
[3] Kusriatmi O R, Syaukat Y, and Said A 2014 Analysis of the effect of beef import restriction policy on beef self-suffiency in Indonesia*J. ISSAAS***20**(1): 115 – 130
[4] Soehadji 2016*Analisa Perkembangan Peternakan di Indonesia.* http://wpsaindonesia.org/unduh/1suhadji analisa mengenai perkembangan peternakan indonesia.pdf
[5] Kementerian Pertanian RI 2017 *Beef Meat Production by Province*, 2013 – 2017. http://www.pertanian.go.id/NAK-2017fix/Prod_DagingSapi_Prop_2017.pdf.Accessed on 31October 2017
[6] Tawaf R 2016*RekonstruksiBisnisSapiPotong (Recovery of Beff Cattle Business)*http://koran-sindo.com/page/news/2016-08-03/1/1/Rekonstruksi_Bisnis_Sapi_Potong.Accessed on 31 October 2017
[7] SirodM 2012 *Survey KarkasPresentasi (Survey of Carcass Presentation)*https://www.slideshare.net/msirod/survey-karkas presentasi-2012.Accessed on 10 August 2017
[8] Cahyadi, G. 2017 HargaDagingSapi di Aceh Lebih Mahal dari Jakarta. http://databoks.katadata.co.id/datapublish/2017/01/18/harga-daging-sapi-di-aceh-lebih-mahal-dari-jakarta. Accessed on 31 October 2017

# Environmental performance, profitability, asset utilization, debt monitoring and firm value

**R Br Bukit[1], B Haryanto[2] and P Ginting[1*]**

[1]Faculty of Economic and Business, Universitas Sumatera Utara, 20155, Medan, Indonesia
[2]Faculty of Engineering, Universitas Sumatera Utara, 20155, Medan, Indonesia
E-mail: rina.bukit@usu.ac.id; rinabukit_dba@yahoo.co.id

**Abstract.** The growing issue on firm value shows that firm value is not only determined by the firm ability to increase financial profit, but also by the company's concern in maintaining the environmental condition. The industrial development produces waste that pollutes the environment that has potential to serious impact on the next life. In addition to provide financial benefits, companies are increasingly facing pressure to be socially responsible for the survival of the company. However, past findings demonstrate that the effect of environmental performance, profitability, and asset utilization to the firm's value are still unclear. This study aims to test whether environmental performance, firm profitability and asset utilization can effectively enhance firm value in two different conditions: intensive debt monitoring and less intensive debt monitoring. Sample of companies is taken from the list of Indonesia Stock Exchange during the period of 2013 to 2015. Using multiple regression analysis, discloses that: in intensive monitoring, managers tend to have high firm value when company has high environmental performance and or high profitability and high asset utilization. Monitoring system needs to be intensified especially for companies with the above characteristics.

## 1. Introduction

Firm value, which describes the company's ability to manage its assets, is important information to all parties concerned with the company. Generally, a firm has a goal to increase its value through increasing the wealth of owners or shareholders [1]. The firm's ability to create and deliver value to stakeholders may influence the perception of investors on the company, and also the firm share price.
Accounting literature has discussed many important factors that affect firm value such as environmental performance, profitability, asset utilization, and debt monitoring. However, past studies found the inconsistent results on the influence of the factors on firm value. For example, past studies showed the inconsistent results of the importance of environmental performance. Past studies documented that firms that practice the environmental performance have higher financial performance and profitability [2]; and high firm value [3]. However, past researchers have found that disclosure of environmental performance does not improve the quality of financial performance [4] and has no significant effect on firm value [5].

This study aims to examine the influence of environmental performance, profitability, asset utilization, and debt monitoring on firm value. This study also tests the moderating effect of debt monitoring on the relationship between environmental performance, profitability, asset utilization and firm value.

### *1.1. Environmental performance*
Environmental performance is the company's activities to create green practices and recover the environmental conservation. This study measures the environmental performance by using the PROPER rating value released by Ministry of Environment. The assessment of PROPER is based on the performance of the company in fulfilling the various requirements stipulated in the prevailing laws and regulations and the performance of the company in the implementation of various activities related to environmental management activities that have not become a compliance requirement.
Past studies documented that high environmental performance were related with financial performance and profitability [6]; and firm value [7]. This study argues the first hypothesis as follows:
H1: There is a positive relationship between environmental performance and firm value.

### *1.2. Profitability (return on assets, ROA)*
Return on asset is used to measure firm profitability. ROA means the ability of the company as a whole in generating profits with the total amount of assets owned. Past literatures documented that earning power from assets owned by the company will increase profit margin and firm value [8]. Yang et al.[9] confirm that firm profitability will result in more revenue, firm efficiency and thus increases the expected firm value. Thus this study argues that firm profitability influence firm value positively.
H2: There is a positive relationship between profitability and firm value.

### *1.3. Asset utilization*
Asset utilization means the total revenues earned divided by total assets owned by the company [10]. Past study showed that asset utilization influence firm value in a positive way. Firms with high asset utilization ratio tend to expand their current capacity to meet prospective demand in the market [11].This study argues that effective asset utilization will increase firm value, while ineffective asset utilization will decrease firm value. Thus, hypothesis 3 is stated as follows:
H3: There is a positive relationship between asset utilization and firm value

### *1.4. Debt monitoring (debt to total equity, DER)*
A company with higher DER ratio will get additional supervision from creditor or banking [12]. The company must be operated profitably in order to be able to pay the instalments of debt and the interest payable. In other words, companies with high debt ratios have high motivation and good effort that contribute to the improvement of company performance and value [13].
H4: There is a positive relationship between debt monitoring and firm value.

### *1.5. Interactions effect*
This study also argues that environmental performance, profitability, and asset utilization influence firm value more among firms with intensive debt monitoring compared to firms with low debt monitoring. Thus, it is hypothesized in the following manner.
H5A: There is a positive relationship between environmental performance/debt monitoring interaction with firm value.
H5B: There is a positive relationship between profitability/debt monitoring interaction with firm value.
H5C: There is a positive relationship between asset utilization/debt monitoring interaction with firm value.

## 2. Materials and Methods
This study constructs a model of multiple regression equation for the relationship between environmental performance, profitability, debt monitoring, asset utilization and firm value as follows:

$$Tobin\,Q = \alpha + \beta_1 PROPER + \beta_2 ROA + \beta_3 ASSETUT + \beta_4 DER + \beta_5 PROPER * DER + \beta_6 ROA * DER + \beta_7 ASSETUT * DER + \varepsilon \ldots \quad (1)$$

Where:

| | |
|---|---|
| Tobin Q | = Firm value |
| PROPER | = Environmental performance |
| ROA | = Profitability |
| DER | = Debt monitoring |
| ASSETUT | = Asset utilization |
| PROPER*DER | = the interaction between environmental performance and debt monitoring |
| ROA*DER | = the interaction between profitability and debt monitoring |
| ASSETUT*DER | = the interaction between asset utilization and debt monitoring |

### *2.1. Operationalization of variables*

*2.1.1. Firm value as dependent variable.* This study measures firm value using Tobin's Q that is the sum of market value of all outstanding stock and market value of all debt divided by the value of all capital placed in the production asset.

*2.1.2. Independent variable.* Environmental performance was measured by using the value of Corporate Performance Evaluation Rating Program in Environmental Management (PROPER) rating such as gold (5), green (4), blue (3), red (2), and black (1). Profitability is measured by using return on asset that is net income divided by total asset. Asset Utilization is the asset utilization ratio calculated from the total revenues earned for each Rupiah of assets owned by the company. Debt monitoring is measured using Debt to Total Equity ratio (DER). The operationalization of all variables is summarized in Table 1.

**Table 1**.Operationalization of variables

| Variable | Definition | Parameter | Scale |
|---|---|---|---|
| Firm value | Firm value is an economic measure reflecting the market value of a business. | Tobin Q = (MVA + Debt) / Total Asset | Ratio |
| Environmental performance | The company's performance in creating a green environment | PROPER; (1 = Gold, Green) and (0 = Blue, Red, and Black) | Dummy |
| Profitability | The effectiveness of the company in generating profits by utilizing its assets | ROA = net income / total asset | Ratio |
| Asset utilization | asset utilization ratio calculated from the total revenues earned for each Rupiah of assets owned by the company | ASSETUT = sales / total asset | Ratio |
| Debt monitoring | Additional firm' controlling from lenders | DER = total liability / total equity | Ratio |

### *2.2. Research model*

The final study sample is 16 companies in the period of 2012-2015 with a total of 60 observations from various sectors of companies. Details of the sample selection process are reported in Table 2.

### *2.3. Data Collection*

Data of environmental performance is obtained from the official website of the Indonesian Ministry of Environment. Data from profitability, asset utilization and debt monitoring variables are obtained from the annual report of firms listed in Indonesia Stock Exchange

**Table 2**. Sample selection criteria

| **Criteria** | **Number of firm-year** |
|---|---|
| Manufacturing companies consistently get PROPER ratings during the period of 2012 – 2015 | 18 |
| Total companies that are not delisted from Indonesia stock exchange | 16 |
| Total observation (16 x 4 tahun) | 64 |

## 3. Results and Disscussions

### *3.1. Descriptive statistic*

Table 3 presents the mean values of all variables. The mean score of the firm value is 3.38, the mean value of profitability is 11.62, the mean value of debt monitoring is 0.75, and the mean value of the asset utilization is 1.38. Table 4 shows that firm with high rating PROPER (1) is 37.5% vs low rating PROPER (0) is 62.5%.

**Table 3**. Descriptive statistics

| | N | Minimum | Maximum | Mean | Std. Deviation |
|---|---|---|---|---|---|
| ROA | 64 | -20.80 | 71.51 | 11.62 | 14.35 |
| ASSETUT | 64 | 0.35 | 10.21 | 1.38 | 1.31 |
| DER | 64 | -7.72 | 7.17 | 0.75 | 1.76 |
| TOBIN_Q | 64 | 0.23 | 27.31 | 3.38 | 5.30 |

**Table 4.** Frequency of PROPER

| | Frequency | Percent | Cumulative Percent |
|---|---|---|---|
| 0 | 40 | 62.5 | 62.5 |
| 1 | 24 | 37.5 | 100.0 |
| Total | 64 | 100.0 | |

### *3.2. Correlation analysis*

Correlations between independent variables were tested to see if there were any multicollinearity problems with the data. Table 5 indicates that only ROA have high correlation with DER with correlation level of 0.232 or about 23.2%. This study also tests the correlation between all independent variables includes the interaction variables (EP*DM, PRO*DM, and AU*DM). Because this correlation is still below 0.8 or 80%, it can be said that there is no serious multicollinearity problem. Past study states that the occurrence of multicollinearity when the correlation between independent variables is above 80% [14].

**Table 5**. Correlation coefficient

| | ASSETUT | PROPER | DER | ROA |
|---|---|---|---|---|
| ASSETUT | 1.000 | | | |
| PROPER | -0.007 | 1.000 | | |
| DER | 0.082 | 0.028 | 1.000 | |
| ROA | 0.207 | 0.227 | 0.232 | 1.000 |

### *3.3. Results of multiple regressions*

**Table 6.** Regression results model 1-4

| | Expected Sign | Model 1 | Model 2 | Model 3 | Model 4 |
|---|---|---|---|---|---|
| | | Dependent Variable: Firm Value | | | |
| (Constant) | ? | -2.526<br>(-4.845)*** | -2.175<br>(-4.234)*** | -2.769<br>(-6.025)*** | -.0538<br>(-0.939) |
| Environmental Performance (EP) | + | 0.601<br>(0.938) | -1.039<br>(-1.198) | 0.777<br>(1.381) | 0.109<br>(0.202) |
| Profitability (PRO) | + | 0.183<br>(8.003)*** | 0.170<br>(7.609)*** | 0.119<br>(4.796)*** | 0.136<br>(6.531)*** |
| Asset Utilization (AU) | + | 2.618<br>(11.042)*** | 2.489<br>(10.787)*** | 2.550<br>(12.261)*** | 0.452<br>(0.994) |
| Debt Monitoring (DM) | + | -0.088<br>(-0.496) | -0.163<br>(-0.950) | 0.336<br>(1.834)* | -1.552<br>(-4.944)*** |
| EP*DM | | | 2.119<br>(2.662)** | | |
| PRO*DM | | | | 0.048<br>(4.368)*** | |
| AU*DM | | | | | 2.304<br>(5.279)*** |
| F-value | | 61.113 | 55.353 | 67.687 | 76.733 |
| Sig | | 0.000 | 0.000 | 0.000 | 0.000 |

Notes: */ **/ *** is statistically significant at the 10% level/ 5% level/ 1% level

This research uses multiple linear regressions to analyze company value with several variables such as environmental performance, profitability, asset utilization and debt monitoring. The first step is to examine the direct relationship between independent variables and firm value. Table 6, Model 1 shows a significant positive relationship between profitability and firm value ($\beta$ = 0.183. sig = 0.00); and between asset utilization and firm value ($\beta$ = 2.618. sig = 0.00). It means that the higher the level of corporate profitability and asset utilization, the higher the value of the company. Thus, H2 and H3 are accepted. However, this study finds that the coefficients of environmental performance and debt monitoring are insignificant; It means H1 and H4 are rejected. This study also documents the positive coefficients of EP*DM, PRO*DM and AU*DM. The effect of environmental performance, profitability and asset utilization on firm value is moderated by debt monitoring. This study supports H5A, H5B and H5C.

## 4. Conclusions

The result shows that firm profitability has a positive and significant influence to company value, which is consistent with past researches [15]. Asset utilization also has a positive and significant effect on firm value; the result is consistent with past studies [10-11]. Thus, the higher the profitability ratio the more income that can be distributed to shareholders and thus the value of the company will be higher. This study also suggests that the more effective the utilization of assets of a company, the better the company's value. This study also proves the interaction effects of environmental performance and debt monitoring, of profitability and debt monitoring and of asset utilization and debt monitoring on firm value.

This study has several limitations. First, the sample data only retrieves company data that is ranked proper consistently during the study period so the data in this study is too small and is considered fewer representatives. Future research is expected not too tight in the sample selection. Second. the period of research is too short to see the effect of independent variables on firm value in each period. Thus, the results may be used with caution. Future research is needed to investigate the issue with a long span of time.

## References

[1] Brigham E F and Gapenski L C 2006 *Financial Management: Theory and Practice* 9. ed (Florida: Harcourt College Publisher)

[2] Al-Tuwaijri S A. Christensen T E and Hughes K 2004 The relations among environmental disclosure, environmental performance. and economic performance: A simultaneous equations approach. *Accounting, Organizations and Society* 29 issue 5-6

[3] Yadav P L. Han S H and Rho J J 2016 Impact of environmental performance on firm value for sustainable investment: Evidence from large US firms. *Business strategy and the environment* 25 issue 6

[4] Ofoegbu G N and Megbuluba A 2016 Corporate environmental accounting information disclosure in the Nigeria manufacturing firms *International journal of management sciences and business research* 5 issue 12

[5] Misani N and Pogutz S 2015 Unraveling the effects of environmental outcomes and processes on financial performance: A non-linear approach *Ecological economics* 109. No. 1

[6] Dasgupta S. Hettige H and Wheeler D 2000 What improves environmental compliance? Evidence from Mexican industry *Journal of environmental economics and management* 39 issue 1

[7] Jacobs B W, Singhal V R and Subramanian R 2010 An empirical investigation of environmental performance and the market value of the firm *Journal of Operations Management* 28. issue 5

[8] Modigliani F and Miller M 1963 Corporate income taxes and the cost of capital: A correction American *Economic Review* 53. Junepg. 433-443

[9] Yang CC. Lee CF. Gu Y X and Lee Y W 2010Co-determination of capital structure and stock returns – A LISREL approach: An empirical test of Taiwan stock market *The Quarterly Review of Economics and Finance* 50

[10] Ang J S, Cole R A and Lin W J 2000 Agency cost and ownership structure *Journal of Finance* 55

[11] Iskandar T M, Bukit R B and Sanusi Z M 2012 The moderating effect of ownership structure on the relationship between free cash flow and asset utilization *Asian Academy of Management Journal of Accounting & Finance* 8 (1)

[12] Gul F A and Tsui J S L 1998 A test of the free cash flow and debt monitoring hypotheses: Evidence from audit pricing *Journal of Accounting and Economics* 24

[13] Bukit R B and Nasution F N 2016 Voluntary disclosure, monitoring mechanism and firm value *IJABER* 14. no 11

[14] Gujarati D N 2003 *Basic Econometrics* 4 ed. (New York. USA: McGraw Hill).

[15] Agrawal A and Knoeber C R 1996 Firm performance and mechanisms to control agency problems between managers and shareholders *Journal of Financial and Quantitative Analysis* 31

## Acknowledgements

The authors would like to thank the editor and the anonymous reviewers for their very helpful comments and suggestions. They also be thankful for the comments and suggestions on earlier drafts of this paper from Prof. Robinson Tarigan. MRP (Universitas Sumatera Utara) and my accounting students especially Lisni Agnes.

# Passion fruit hulls particleboard: the effect of urea formaldehyde level on physical and mechanical properties

**A H Iswanto[1*], T Sucipto[1], E Adlina[1] and D S Prabuningrum[2]**

[1]Department of Forest Product, Faculty of Forestry, Universitas Sumatera Utara, Padang Bulan, Medan 20155, Indonesia
[2]Department of Forest Product, Faculty of Forestry, Institut Pertanian Bogor, Bogor, Indonesia
E-mail: apriheri@yahoo.com

**Abstract.** The purpose of this research was to explore the suitability of Passion Fruit Hulls (PFH) as a raw material particleboard with variants of urea formaldehyde adhesive content (UF). In this research, PFH particles filtered by sieve in size of 10 mesh to throw dust particles. Furthermore, the particles dried until reaches of 5% moisture content. Levels of UF adhesive was using comprise of 10%, 12% and 14%. Hot pressing conducted at 120°C temperature for 10 minutes at a pressure of 30 kg/cm$^2$. The results showed that in moisture content for 10% adhesive level, almost all the parameters such as thickness swelling, modulus of elasticity (MOE) and modulus of rupture (MOR) that produced did not fulfiled the standard. The 14% adhesive level produced of the best of PFH particleboard.

## 1. Introduction

Passion fruit hulls (PFH) are by product from syrup industries. In North Sumatra, the passion fruit hulls existence is abundant. It contains crude protein and lignin approximately 7.32 and 31.79% respectively [1]. Furthermore, PFH also contains lignocellulosic materials so it can be utilized as the raw material of particleboard. Recently, the data about properties of PFH's particleboard is not available.

The research about non-woods particleboard had been dominated by materials of bamboo, sugarcane bagasse, palm oil, banana fibre, rice hulls, etc. The research used the sorghum bagasse and jatropha fruit hulls had been reported in the previous research [2-3]. The results showed that these materials potentially could be developed for particleboard materials. The focused study in this research was to analyse the effect of adhesive level on particleboard properties. According to [4-6] that adhesive level is one of the parameters was affecting of quality particleboard.

## 2. Materials and Methods

### *2.1. Materials*

PFH waste from syrup industries in Berastagi-North Sumatra Indonesia, urea formaldehyde (UF) from PT. Palmolite Adhesive Industry (PT. PAI) Probolinggo, East Java Indonesia.

### *2.2. Methodss*

*2.2.1. Materials preparation.* PFH converted into a particle in size of 10 mesh. Furthermore, it dried in an oven at 103°C up to reach of 5% moisture content. The amount of UF adhesive prepared 10, 12, and

14% based on oven dry weight of the particle. Board density and thickness targets were set of 0.8 g/cm$^3$ and 1 cm respectively.

*2.2.2. Board producing.* PFH and UF mixed in a rotary blender. After that, it was placed into mat former in size of 25 by 25 cm$^2$. Furthermore, the mat was pressed using the hot pressing machine at 120°C temperature, 30kg/cm$^2$ pressure, and for 10 minutes. The next step, board conditioned at room temperature for seven days before testing.

*2.2.3. Board cutting.* After conditioning, board cutting refers to standard [7]. The parameter for evaluating included of physical (density, moisture content-MC, water absorption-WA, and thickness swelling-TS) and mechanical (modulus of rupture-MoR, modulus of elasticity-MoE, and internal bond-IB).

*2.2.4. Water absorption (WA) rate of particle.* The determination of WA rate of the particle was conducted by the weighting of 50 g PFH for initial weight (W0). Furthermore, it was immersed in water for one minute and weight after that (W1). This immersing continued every one minute until the saturated condition. The WA rate was calculated using the formula below:

$$V = \frac{W1 - W0}{t}$$

Note: V is WA rate (g/minute), W0 is Initial weight (g), W1 is Weight after immersing (g), and t is Time (minute).

## 3. Results and Discussions

### *3.1. Physical Properties*

The board density value in this research did not fulfiled the density target of 0.8 g/cm$^3$. This is due to the lose particles during the compression process as stated by Buffalino et al. [8]. The particle density values had met the standards required in standard ranging from 0.4 to 0.9 g/cm$^3$ [7].

**Table 1.** Physical properties parameters

| Adhesive Level (%) | Physical properties | | | | | |
|---|---|---|---|---|---|---|
| | Density (g/cm$^3$) | MC (%) | TS-2H (%) | TS-24H (%) | WA-2H (%) | WA-24H (%) |
| 10 | 0.83 | 14.70 | 60.99 | 83.69 | 118.64 | 168.34 |
| 12 | 0.75 | 12.73 | 37.04 | 60.99 | 76.38 | 107.91 |
| 14 | 0.78 | 13.58 | 37.83 | 61.10 | 90.90 | 107.99 |

*Note: H (Hours)*

Parameters of moisture content showed that the higher level of UF adhesive resulted in the decrease of moisture content particleboard. The high value of moisture content obtained in this study because the PFH's particle had hydrophilic properties where the particle can quickly absorb water in the environment.

The water absorption of the particleboard produces is still quite high because the UF adhesive had a low resistance to the water so that UF adhesive is more widely used for interior application. According to [9], UF adhesives were resulted from a condensation polymer of formaldehyde with urea, which has hydrolyzable on its hydrogen bond. When the particleboard immersed in the water, it will be easily to penetrate into the particleboard, as a consequence is the high value of water absorption. In addition, other factors that influence WA values are hygroscopic properties of particles. The hygroscopic is the ability of the materials that were easily passed by the water so that the material will be easy to expand after being immersed in the water. The rate of water absorption in the PFH can be seen in Figure 1.

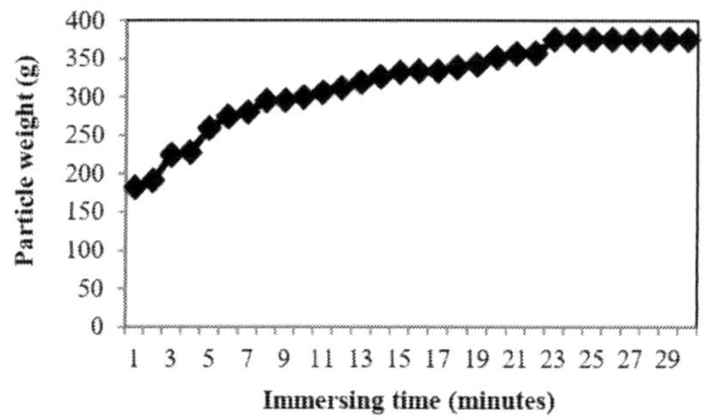


**Figure 1**. Water absorption rate of particle

Particle absorption rate test is performed by immersed the passion fruit hulls particles with a dry weight of 50 g. The results of the water absorption rate obtained on the PFH particles increase during the 30 minutes immersion until it reaches the water saturation point. After that, the final weight of PFH particles reached of 375.43 g. These results can prove that the PFH particles are porous materials. The rate of water absorption of PFH around 10.84 g/minute shows that PFH has hydrophilic properties. The high thickness swelling produced due to the use of UF adhesive, where the adhesive is used for the interior application. Bulky's property of non-wood materials is another reason that resulted in higher the thickness swelling value. Thickness swelling of the wood panel was influenced by adhesive quantity and distribution, MC of furnish, furnish compatibility, the chemical composition of furnish, etc [10]. The particles are resistant to MC will be able to produce of boards with good dimensional stabilization [11].

Increased of adhesive levels resulted in decrease the thickness swelling of the board. Iskandar and Supriadi [12] stated that the more adhesive resulted in better interparticles of bond in the particleboard produced. In addition, PFH particles have a high enough extractive substance that is equal to 37.16%. It can inhibit in glueing process on the particleboard manufacturing. According to [4], extractive substances affect to the adhesives consumption, the rate of adhesive curing and the durability of particleboard. Based on Standard, the thickness swelling value of PFH's particleboard produced did not meet the standard where the maximum thickness swelling is allowed less than 12% [7].

*3.2. Mechanical Properties*

The increasing of adhesive level resulted in higher MoE value, it was due to by higher of interparticle bonding that has resulted in improvement of strength to retain of the load. The MoE value did not meet standard, it is probably due to particle size. Small particle size resulted in lower mechanical properties, It is similar founding with [13]. Furthermore, particle geometry in form slenderness ratio is one of the important factors was affecting of modulus of elasticity properties [4-5].

**Table 2.** Mechanical properties parameters

| Adhesive Level | Mechanical properties | | |
|---|---|---|---|
| (%) | MoE (kg/cm$^2$) | MoR (kg/cm$^2$) | IB (kg/cm$^2$) |
| 10 | 6570.13 | 42.07 | 1.97 |
| 12 | 4523.30 | 31.72 | 2.60 |
| 14 | 7890.13 | 54.68 | 3.43 |

The similar trend with MoE also occurred in MoR value. The higher of adhesive content resulted of better MoR value. The MoR parameter also did not meet JIS A 5908-2003 that require of MoR value minimum of 82 kg/cm$^2$.

Trend IB value similar with MoE and MoR. It is supported by [12] research who reported that increasing of adhesive content resulted in higher of IB sugarcane bagasse particleboard. The increasing of adhesive level resulted in higher of covered the surface of the particle and its probably to improve the interparticle bond strength.

## 4. Conclusions

The physical and mechanical properties of particleboard were improved through to the higher of the adhesive level. Almost all parameters in this research did not meet standard execpted for density, moisture content and IB. The 14% adhesive level is the best resin content to produce passion fruit hulls particleboard, however, it is not an optimum condition.

## References

[1] Astuti T 2008 *Evaluasi Nilai Nutrisi Kulit Buah Markisa yang Difermentasi dengan Aspergillus niger dan Trichoderma harzianum sebagai Pakan Ternak Secara In-vitro*. Thesis of graduate school of Universitas Andalas Padang

[2] Iswanto A H, Sucipto T, Nadeak S S D, and Fatriasari W 2017 1st Annual Applied Science and Engineering Conference *IOP Conf. Series: Materials Science and Engineering* post-treatment effect of particleboard on dimensional stability and durability properties of particleboard made from sorghum bagasse. Vol **180** pp 012015. http://iopscience.iop.org/article/10.1088/1757-899X/180/1/012015

[3] Iswanto A H, Aritonang W, Azhar I, Supriyanto and Fatriasari W 2017 The physical, mechanical and durability properties of sorghum bagasse particleboard by layering surface treatment *Journal Indian Acad Wood Sci* **14**(1) 1–8

[4] Maloney T M 1993 Modern Particleboard And Dry-Process Fiberboard Manufacturing (USA: Miller Freeman Inc. San fransisco)

[5] Bowyer J L, Shmulsky and Haygreen J G 2003 *Forest Products and Wood Science - An Introduction* Fourth edition (USA: Iowa State University Pr.)

[6] Kelly MW 1977 *General Technical Report FPL-10-US Department of Agriculture Forest Service and Forest Products Laboratory University of Wisconsin*. Critical Literature Review of Relationship Between Processing Parameter and Physical Properties of particleboard

[7] Japanese Industrial Standard 2003 JIS A 5908 Japanese Standard Association Particleboard

[8] Bufalino L, Albino V C S, de Sá V A, Corrêa A A R, Mendes L M andAlmeida N A 2012 Particleboards made from australian red cedar: processing variables and evaluation of mixed-species *Journal. Tropical Forest Science* **24**(2) 162–172

[9] Pizzi A 1983 *Wood Adhesives, Chemistry of Technology*. National Timber Research Institute Council for Scientific and Industrial Research (South Africa: Pretoria)

[10] Guler C, Copur Y and Tascioglu C. 2008 The manufacture of particleboards using mixture of peanut hull (Arachis hypoqaea L.) and European Black pine (Pinus nigra Arnold) wood chips *Journal. Bioresource Technology* **99** 2893–2897

[11] Garay R M, MacDonald F, Acevedo M L, Calderón B and Araya J E 2009 Particleboard made with crop residues mixed with wood from Pinus radiata *Journal BioResources* **4**(4) 1396-1408

[12] Iskandar M I and Supriadi A 2013 The Effect of Adhesive Content on Properties Bagasse of Particleboard *Jurnal Penelitian Hasil Hutan* **31**(1) 19-26

[13] Gozdecki C, Zajchowski S, Kociszewski M, Wilczyñski A and Mirowski J 2011 Effect of wood particle size on mechanical properties of industrial wood particle-polyethylene composites *Polymery* **56**(5) 375-380

**Acknowledgement**
We would like to express my sincerely thanks to Directorate General of Higher Education-Ministry of Research, Technology and Higher Education Republic Indonesia for supporting research fund through to Competitive Research Grant (HIBAH BERSAING) years of 2016.

IOP Conf. Series: Earth and Environmental Science 122 (2018) 012139 doi:10.1088/1755-1315/122/1/012139

# The level of society's participation in critical land rehabilitation in DAS Randangan Pohuwato Regency

**M I Bahua[1*] and D O Suparwata[2]**

[1]Department of Agrotechnology, Faculty of Agriculture, Universitas Negeri Gorontalo, Jl. Jend. Sudirman No. 6 Kota Gorontalo Indonesia
[2]Department of Agribussiness, Faculty of Agriculture Sciences, Universitas Muhammadiyah Gorontalo, Jl. Jend. Sudirman Limboto Gorontalo Indonesia
Email: mohamad.bahua@ung.ac.id

**Abstract.** The aim of this study is to analyze the level of participation of the society and analyze the correlation of internal and external factor with the level of participation from the society. This research was done in Randangan watershed in Pohuwato district, Gorontalo Province from Februari to April 2017. The population of the research is 150 people with 60 respondents that is chosen through random sampling. The data were collected through structured interview with questionnaire. The data were analyzed with class interval analysis (low, average, high), with sperman rank correlatioon analysis, and with descriptive analysis. The result of the research showed that the participation of the society is in low category in terms of determining the rehabilitation location (65.0% is not participated) and evaluation for planting monitoring (68.3% does not participate), and average participation on implementing seeding (55,0% participated), seeding (58.3% participating). Among the internal and external factors of the society, the one that has a significant relation is $\alpha = 0.05$ is the level of program socialization is in implementing seeding (0.299). This indicates that the better the socialization the better the knowledge of the society in participating in implementing seeding

## 1. Introduction

Trustee act Number 41 year 1999 on forestry states that forests can be used for a wider audience fully for the welfare of the people of Indonesia, also contributed to enhance Watersheds (DAS) support and maintaining the adequacy of the forest at least 30% of the watershed with a vast proportion. However it does not mean the exclusion of environmental aspects, which can then perform logging freely towards natural resources. Because the main function of the forest is not only as the economic underpinning of society, but keeping ecology and maintaining social harmony in society.

Using agricultural-land without many conservation gave the negative impact on the rate of land degradation in the watershed area. The more the breadth of a degraded watershed area will form a critical land. The impact also occurs at the DAS in Gorontalo Province, one of which was DAS Randangan. DAS Randangan is located in the Regency of Pohuwato. The watershed has an area that is 268,078 Ha. The Randangan watershed conditions progressively worsen with increasing critical land-marked and its damage to the environment. Critical Land area to the DAS Randangan reached 91,494 Ha, comprising 18,832 Ha outside the forest area, and 72,662 Ha located in the forest area. This indicates that the high land degradation and soil erosion that occurred in the region of DAS Randangan [1].

Current conditions indicate that participation of the community to participate maintaining environment is still low. But when examined more deeply that maintaining the sustainability of nature provides benefits that will return to the community itself both directly and indirectly. It is less comprehensible and certainly needed an input policy strategy adopted by the Government to local communities.

## 2. Materials and Methods

This research was conducted in DAS Randangan, Pohuwato Regency, Gorontalo Province. The study was conducted from April to June 2017. Design the research using the survey approach. The population in this research is the community that rehabilitated critical land in DAS Randangan. The population in question is directly exposed to the impact of the farmers rehabilitation programs critical land. The population was 150 farmers. The sample is determined using the formula proportionately Taro Yamane cited in Riduwan (2013) with a precision of 10%, that is:

$$n = \frac{N}{N.d^2+1}$$

Where : n = number of sample
N = number of population
d = stipulated precision

So that the number of sample is 60 people that was chosen by systematic random sampling. This study uses two types of data namely, primary data and secondary data. While the primary data is data that is retrieved directly from the field, which is the results of the interviews with respondents/community. The technique of data collection performed is through structured interview, and questionnaires as a tool in the collection of the data. To find out the level of community participation in both stages of the planning, implementation and evaluation of rehabilitation activities on critical land in Randangan Watershed using scale analysis. The scale in question is to determine the interval class at each level of participation (low, medium, high). Determination of interval class categories low, medium and high that it can be calculated by using the formula [2], namely as follows:

$$k = \frac{R}{I} \quad (1)$$

Where: k = class interval
I = number of interval classes
R = range of relationship

Examines the correlation between internal and external factors against the level of public participation used Spearman rank correlation formula (Riduwan, 2013): (2)

$$r_s = 1 - \frac{6\sum d^2}{n\,(n^2-1)}$$

Where:
rs = Sperman Rank correlation Values, i.e., the correlation coefficient ranking by size for the relationship between the variable Y (the level of community participation) and the free variable X (internal and external factors)
$d^2$ = difference between the ranking of X and Y
n = number of data pairs
The trust level used is $\alpha = 0.05$

Analysis description used to describe the factors internal and external in the communities, describe the level of its public participation from the field data obtained, as well as to describe the depiction of relationships that occur between them.

## 3. Result and Discussion

Study of community participation is inseparable from related factors related to individual communities. These factors can be derived from the inside (internal factor) or outside (external factors) of each individual in society. In the study presented the community's internal factors such as: (a) the level of education, and (b) the level of people's income. While external factors are presented is the level of outreach/dissemination by the Government received by the community.

### *3.1. Internal and External Factors Of Critical Land Rehabilitator Community*

*3.1.1 The Level of Community Education.* The level of public education will have an effect on the level of participation in critical land rehabilitation. Qualified education level will have an impact both on the critical land rehabilitation activities, because of the level of education will determine the level of understanding, knowledge, admission information, intellect, participation, public perception against the critical land rehabilitation program. The reason is still the magnitude of public education at the primary school level (43%). As compared to Junior High/Secondary Schools (35%), Senior High School (20%). Especially on the college level, indicating that the level of education is still low (2%). Low levels of education are correlated at the low level of community participation in programmes planned by the Government. Soekartawi in Jariyah [3], those who are highly educated to be faster and instead they build low bit hard to receive the technology quickly.

*3.1.2. Income level of farmers.* The income of farmers is measured on the basis of income from farm and non-farm. Category income levels are divided into three categories, namely: <Rp.500,000/mths, Rp.500,000 – Rp.1,000,000/Mth, and > Rp. 1,000,000/Mths. The results showed that 66.7% of society have a range of income level of Rp.500,000 - Rp.1,000,000/month, compared to communities that have income >1,000,000/month (only 10%). The level of the income will affect the life of the community to better participate in critical land rehabilitation program. This indicates that between the people's income and participation in critical land rehabilitation will positively correlated to the success of the program. So does Erwiantono [4] says with a better income level/high can encourage someone to participate better/higher. Besides the internal factors external factors in the community, the community is also very influential towards the level of participation. External factors are defined as factors that affect the public who come from outside himself.

*3.1.3. The level of Outreach/socialization on critical land Rehabilitation.* Extension Program is the involvement of someone doing a conscious communication of information [5]. The success of the rehabilitation programs can be determined from the intensively extension is given. The level of the frequency extension is received by the community revolves around the 1-2 time extension (amounting to 75%). This means that the level of outreach done by the Government is in the medium category. The level of extention will be indicated on the abundance of information received by the community. Such information may include: (1) how to do a good soil for planting trees, (2) maintenance of measures including planting distances, (3) integration patterns that can be applied between the annuals crops, in utilizing critical land, and (4) include re-evaluating of activities that have been carried out by the community. The competence of the extension officers is dynamic. Competencies need to be adapted to the changes and development of science, technology and the changing profile of extension audiences.

*3.2. levels of community participation in critical land Rehabilitation*

*3.2.1. The form of community participation in the planning of the critical land rehabilitation.* Program of planning in other early indicators is not for the achievement of an objective of the program. Auliyani *et a.l* [6], are planning aims to prepare everything systematically so that the result is able to serve as a reference in the implementation of activities. The results of research in planning rehabilitation of critical land can be seen in Table 1.

Table 1 shows that the participation of the community in the rehabilitation of critical land classified as low (65.0%), this is included in category of nonparticipation. In this case the community is less or do not even play an active role in planning for the determination of the criteria for land rehabilitated. The community is on target with instant received information from the Government. Thus, the participation of the community in planning low indeed because of the dominance of the Government in the determination of the location and criteria can land rehabilitated.

**Table 1.** The form of planning public participation in the determination of the criteria and the location of critical land rehabilitation in DAS Randangan Pohuwato Regency

| Category (Value) | Form of the Community's participation on Planning Stage |
| --- | --- |

| | Number (People) | Percentage (%) |
|---|---|---|
| 1-2 (Low) | 39 | 65.0 |
| 3-4 (Middle) | 13 | 21.7 |
| >4 (High) | 8 | 13.3 |

*3.2.2. Forms of public participation in the implementation of the program of rehabilitation of critical land.* Community participation in the implementation phase of the rehabilitation of critical land includes several forms of activities like: nursery seed (Table 2) and planting the seeds of the plant (Table 3). More detailed explanations are presented as follows:

**Table 2.** The form of the implementation of the participation of the community in the nursery seed plants for critical land rehabilitation

| Category (Value) | Form of the Community's participation on Implementation Stage | |
|---|---|---|
| | Number (People) | Percentage (%) |
| 1-2 (Low) | 14 | 23.3 |
| 3-4 (Middle) | 33 | 55.0 |
| >4 (High) | 13 | 21.7 |

The level of participation of the Community belongs to the moderate level (55.0%); community participated in program implementation stages of critical land rehabilitation in DAS Randangan Pohuwato District. It means that the society should be its driving force of the implementation of the rehabilitation program. Thus the program launched by the Government will not be limited solely to project related service program. Furtermore, besides the shape of the seed of the plant breeding activities, community participation is also measured through their participation in the planting of the seed plants in the rehabilitation activities. Public participation on planting the seeds of the plant is presented in Table 3.

**Table 3.** Implementation of the public participation in the form of planting seedling plants for critical land rehabilitation

| Category (value) | Form of the Community's Participation on Implementation Stage | |
|---|---|---|
| | Number (People) | Percentage (%) |
| 1-2 (Low) | 9 | 15.0 |
| 3-4 (Middle) | 35 | 58.3 |
| >4 (High) | 16 | 26.7 |

The level of participation of the Community belongs to the medium (58.3%) community participation program implementation stages of critical land rehabilitation in DAS Randangan Pohuwato District. This indicates that the community could have been invited to negotiate and willing to get down in doing planting seedling plants, but still in the category of being. The wastewater treatment plant needs an extensive land, high investment and operational cost, technical trained persons, and a long time in the planning and implementation. The wastewater treatment plants are however inadequate to serve the need of urban sanitation as they only serve 10% of the Indonesian population . On the level of community participation, they heard and allowed to argue, but they don't have the ability to gain guarantee that their views will be considered by decision makers [7]. Some of the results of the study as reported that the level of public participation at the stage of implementation is generally classified as moderate (46.8%). Furthermore that in the implementation of most of the community participation program is active at the position (64.5%) [8].

*3.2.3. Form of public participation in the evaluation of the critical land Rehabilitation*

Evaluation of participation is systematically analysis with project management and a member of the group, assign the policy/institutional rearrangement, objectivity, or redevelopment of resources [9]. Characteristics of complex activities resulted in the rehabilitation of the evaluation process needs to be done carefully, systematic, and thorough [10]. In this form of research activity on evaluation of critical land rehabilitation that is seen from the activity monitoring/checking of planting that have been implemented. In more detail are described in Table 4.

**Table 4.** The form of the evaluation of the community participation in the activities of monitoring/checking planting on critical land rehabilitation

| Category (Value) | Form of the Community's Participation on Evaluation Stage | |
|---|---|---|
| | Number (People) | Percentage (%) |
| 1-2 (Low) | 41 | 68.3 |
| 3-4 (Middle) | 13 | 21.7 |
| >4 (High) | 6 | 10.0 |

The level of public participation in the evaluation of critical land rehabilitation program in the category of low (68.3%) community participated. This illustrates there are several factors why the society did not play an active role in the evaluation phase of the rehabilitation program, namely: (1) they are not be included in the Member's overall evaluation, the dominant moves in the evaluation is the core of the rehabilitators Group, (2) overflow duties and authorities are not evenly on each Member, the overflow concentrated on the task group Chairman the most widely received information from the Government (3) voice/opinion from members usually rarely found represented in decision-making, so that sometimes members reluctant to comment in the evaluation of the program, and (4) lack a sense of the community's own call to check continuously the plants that have been grown.

*3.3. The correlation of Internal and external factors to the level of public participation*

In the study of closeness relations it takes the variable X (internal and external factors), and the variable Y (the level of community participation). Research results about the relationship on the critical land rehabilitation consist of: (a) internal factors; the level of education and income, external factors; the level of extention, and (b) the level of participation of the stages of planning, implementation and evaluation. Described in more detail in Table 5.

**Table 5.** The correaltion of internal and external factors with the level of participation in the rehabilitation program of critical land in DAS Randangan Pohuwato Regency

| Factors | Form of the Community's Participation | | | |
|---|---|---|---|---|
| | Planning | Implementation | | Evaluation |
| | Criteria Location | Seed Plants | Planting Seedling | Monitoring |
| | (r) | (r) | (r) | (r) |
| Internal : | | | | |
| Education Level | 0,012 | 0,344** | 0,152 | -0,162 |
| Income Level | -0,090 | -0,192 | -0,219 | -0,213 |
| External: | | | | |
| Frequency Extension | 0,344** | 0,299* | 0,376** | 0,428** |

Description: r: Correlation's (CC)
** Extent significant at $\alpha = 0.01$
* Level significant at $\alpha = 0.05$

*3.3.1 The correlation between level of education with community participation in stages of planning, implementation and evaluation.* From the analysis results spearman rank (Table 5), the level of education with community participation in the determination of the location of the rehabilitation (r = 0.012), planting the seeds (r = 0,152), and monitoring (r =-0.162) does not indicate a real relationship. This indicates that there is a very weak correlation values between these elements. Whereas, in the activities of nursery seed (r = 0,344 ** level α = 0.01) shows a real correlation with the level of participation of the community, and the value of the correlation is positive. This shows that there is a strong link between educations with an understanding of farmers in sowing seeds in the nursery. Thus, the higher the education of a person then the knowledge and understanding of the community will be better in doing seed nursery in critical land rehabilitation.

*3.3.2. The correlation between the level of income with the public participation phase of planning, implementation and evaluation.* In Table 5, the results of the analysis spearman rank indicate that income level with the participation of the public in the determination of the location of the rehabilitation (r =-0.090), nurseries (r =-0.192), planting the seeds (r =-0.219), and monitoring (r =-0.213), the entire show is not a real touch of the participation elements of the land. This illustrates the existence of weak correlation values of the element, and the correlation coefficient is negative. This correlation indicates that not always the level of income of the community will always be affected in participating. On the other hand, that income is not a major factor in society participate. Because when communities have the time, opportunity, knowledge and ability then it is certain that society will want to participate.

*3.3.3. The correlation between level of outreach with the public participation phase of planning, implementation and evaluation.* From the analysis results of spearman rank (table 5) shows that the level of counseling at the stipulation of site rehabilitation (r = 0,344 **), planting the seeds (r = 0,376 **), and monitoring (r= 0,428 **) associated with the level of participation of the community at significant levels α = 0.01. While in the nursery of the seed (r = 0,299 *) shows a real relationship with the level of participation of the community at significant levels α = 0.05. This is indicated that the value of the correlation has a strong relationship between the participation elements, and is positive. It means that the better the outreach conducted by the Government and accepted by the community, it will also better public participation in critical land rehabilitation in DAS Randangan. Therefore, public awareness is a persuasive approach put forward measures and upholding the norms of participatory [11]. The existence of extension will be able to raise the curiosity and community participation are quite high, fueled by the presence of the Government as well as coaching from the activities of study appeals to some areas [4].

## 4. Conclusion

The level of community participation in rehabilitation programs in critical land DAS Randangan Pohuwato District for planning activities in the determination of the location of the rehabilitation (65.0% not participating), and evaluation for monitoring farming (68.3% did not participate) in the low category. On the implementation of the nursery (55.0% participated), planting the seeds (58.3% participating) so that it enters the category of participation are. Among the internal and external factors of the community, which has a real connection on a significant level α = 0.05 is the level of outreach program with the level of community participation in the activities of nursery seed plants (0.299 *). This indicates that the better the guidance provided will improve knowledge, understanding of the community to participate in the breeding of seed plants.

## References

[1] Balai Pengelolaan Daerah Aliran Sungai (BP-DAS) Bone Bolango 2013 *Buku Putih Sanitasi Kabupaten Gorontalo 2013* Departemen Kehutanan, Direktorat Jenderal Rehabilitasi Lahan dan Perhutanan Sosial Limboto Gorontalo

[2] Nasir M 2013 *Metode Penelitian (Research Methods) Cetakan Ke-8* (Bogor: Ghalia Indonesia) pp 77 – 82

[3] Jariyah N A 2014 Partisipasi masyarakat dalam rehabilitasi lahan dan konservasi tanah (RLKT) di sub DAS Keduang, Kabupaten Wonogiri, Jawa Tengah. Balai Penelitian Teknologi Kehutanan, Pengelolaan Daerah Aliran Sungai (Community participation in land rehabilitation and soil conservation in Keduang DAS, Wonogiri Regency, Central Java). Forest Technology Research Institute, Watershed Management, *Jurnal Penelitian Sosial dan Ekonomi Kehutanan*, **11**(3): 211 – 221

[4] Erwiantono 2006 Kajian tingkat partisipasi masyarakat dalam pengelolaan ekosistem mangrove di kawasan Teluk Pangpang-Banyuwangi (Community participation level assessment in mangrove ecosystem management in Pangpang-Banyuwangi Bay Area) *Jurnal EPP* **3**(1): 44-50

[5] Van den Ban AW dan Hawkins HS 2005 *Penyuluhan Pertanian* Edisi Revisi, Cetakan Ke-8. (Yogyakarta: Kanisius) pp 55-56

[6] Auliyani D, Hendrarto B dan Kismartini 2014 Partisipasi masyarakat dalam rehabilitasi mangrove di beberapa desa pesisir kabupaten Rembang: tinjauan berdasarkan tahap perencanaan (community participation in mangrove rehabilitation in several coastal villages of Rembang regency: review by stage of planning) *Maspari Journal* **6**(1): 13-19

[7] Ansori M 2012 *Pengelolaan Hutan Kemitraan untuk Menyejahterakan Rakyat Kasus Pola PHBM (Pengelolaan Hutan bersama Masyarakat) di Perum Perhutani BKPH Parung Panjang, KPH Bogor* Disertasi Sekolah Pascsarjana IPB Bogor p 132

[8] Dipokusumo B 2011 *Model Partisipatif Perhutanan Sosial Menuju Pengelolaan Hutan Berkelanjutan: (Kasus Pembangunan Hutan Kemasyarakatan pada Kawasan Hutan Lindung di Pulau Lombok*) Disertasi Sekolah Pascasarjana IPB Bogor p 115

[9] FAO 1988 *Participatory Monitoring and Evaluation (Handbook for Training Field Workers)* (Bangkok: Rapa Publication)

[10] Jatmiko A, Sadono R dan Faida L R W 2012 Evaluasi kegiatan rehabilitasi hutan dan lahan menggunakan analisis multikriteria (studi kasus di Desa Butuh Kidul Kecamatan Kalikajar Kabupaten Wonosobo, Jawa Tengah) *Jurnal Ilmu Kehutanan* **6**(1): 30-44

[11] Khoirul L 2009 Makna penyuluhan dalam pengelolaan kawasan konservasi *Majalah Penyuluhan Kehutanan (Kenari) komunikasi Edukasi Wana Lestari* Edisi **2**:2-6

# Analysis of community income on suren (*Toona sureni* (Blume) Merr.) and cacao crops (*Theobroma cacao* L.) in Simalungun, North Sumatera- Indonesia

**S Latifah[1], Y Afifuddin[1] and S Widya[1]**

Forestry Faculty, Universitas Sumatera Utara, Medan, Indonesia
E-mail: sitilatifah164@yahoo.co.id

**Abstract**. Agroforestry is the management and integration of trees, crops and/or livestock on the same plot of land and can be an integral component of productive agriculture. It may include existing native forests and forests established by landholders. The study was conducted in Mekar Sari Raya village, Panei sub, Simalungun regency, North Sumatera. This study aims to gain the ability to use agroforestry in suren crops and cocoa that provides benefits to farmers and the feasibility of the model farm. The study site has *Net Present Value* (NPV) is 2,670,306,905 ( IDR) for 15 of year , *Gross* B/C *Ratio* (BCR) is 2.3; *Internal Rate of Return* (IRR) is 28 %; and *Payback Period* (PP) for 5 years 4 months 24 days. Agroforestry using commodities cacao and suren crops are financially feasible to be cultivated and developed.

## 1. Introduction

Agroforestry is combines shrubs and trees in agricultural and forestry technologies to create more diverse, productive, profitable, healthy, ecologically sound, and sustainable land-use systems .Various efforts have been made for cocoa development. Improved cultivation techniques will ultimately bring benefits in the development effort. Efficient nursery techniques, attempts to obtain superior plant material through hybridization, pruning methods to establish good habitat, plant spacing, pest and disease protection efforts aimed at finding an efficient cocoa planting and maintenance period with maximum production target [1].

The reason the farmers do agroforestry pattern on suren and cocoa plants is: (1) increase revenue, (2) developing rural economic activity, (3) realizing food security system, (4) prevent erosion and landslides. Farmers in the area are the majority earning from farming rice, corn, coffee and cocoa. As for rice and maize with the ability of their land categorized fertile can produce or planted 2 times in 1 year. Developments in the Indonesian agricultural sector provide opportunities for farmers with their efforts. One of them is plantation crops, especially cocoa, which is one of the plantation commodities cultivated by many farmers. This commodity has a high economic value and strategic, both to provide increased income farmers can even increase foreign exchange for the country. This research is done to see public expenditure and earnings from their farm in Mekar Sari Raya Village Panei Sub district, Simalungun Regency, North Sumatera. Purposes of this study are: how is the feasibility level of farming by planting suren and cocoa in the research area.

## 2. Research Methods

The study was conducted from October to November 2016 in Mekar Sari Raya Village, Panei Sub district, Simalungun Regency of North Sumatra. The number of families in Mekar Sari Raya is 225 families. According to [2], if the subject is less than 100 people better taken all, if the number is greater than 100 people then taken between 10-25% or 20-25%. Thus, the data in this study used 34 head of families. In this study area, data collected in the field by interviewing the farmers directly, in the form of questionnaires on farmers' identity, land ownership, crop species, number of crops per ha and management methods, selling price of products (suren and cocoa ), Seed price, worker and others.
The revenue earned by the farmer is the product multiplied by the price of the product received by the farmer. While the income of farmers is the result of reduction of total receipts with the amount of costs incurred by farmers in one planting period [3]. Investment activity involves a relatively large amount of funds, which is invested in a certain time and is expected to benefit in the future to avoid losses. Some of the financial feasibility criteria to be used are given below [1] :

a. Net Present Value (NPV)
   Net Present Value, the current value of agroforestry activities with the following indicator :
   NPV = 0, when the agroforestry business value is at the prevailing interest rate in the bank.
   NPV > 0 when the agroforestry business is profitable.
   NPV < 0, when agroforestry business loses.
b. Benefit Cost Ratio (BCR)
   Benefit Cost Ratio, ccomparing the profit to the cost of an agroforestry activity or business with the following indicators:
   BCR > 1, when the agroforestry business is profitable.
   BCR < 1, when the agroforestry business is a loss.
   BCR = 1, neutral.
c. Internal Rate of Return (IRR)
   Internal Rate of Return, is a parameter at the interest rate of how the agroforestry business benefits with the following indicators:
   IRR $\geq i$,when agroforestry is feasible.
   IRR $< i$, when agroforestry is not feasible.
d. Payback Period (PP)
   Payback Period, is the period required to recover the investment expenditure using a net cash flow / proceed with a quick time so that funds invested in an investment can be recovered.

## 3. Result and Discussion

### *3.1 Cost of Suren and Cacao Farming*

#### 3.1.1 *Analysis of Agroforestry Revenue Suren and Cacao in Mekar Sari Raya Village.*

The form of suren planting is done on the sidelines of the cocoa tree line in the form of intercropping. Suren plants in the village of Mekar Sari Raya planted with spacing 2 x 3 m, 5 x 5 m, 10 x 10 m. While the distance of planting cocoa with a distance of 3 x 4 m (Figures 1, 2, 3 and 4). The planting of suren is done by people with different patterns such as planted as intercrops, as land boundaries. The shape of the cropping pattern undertaken by the farmers is adjusted to the interests of the farmers. This planting of suren is done with community initiative.

**Figure 1.** Suren and Cocoa Crops

**Figure 2.** Sketches of suren and cocoa crops patterns with 2 x 3 m spacing.

**Figure 3.** Sketch of suren and cocoa crops pattern with spacing 5 x 5 m.

**Figure 4.** Sketch of suren and cocoa crops pattern with spacing of 10 x 10 m.

*3.1.2 Investment Cost.*

Investment costs are the costs incurred for plant investment before there is a yield. Investment costs in suren and cocoa farming include the purchase of cocoa and cocoa seeds. Cacao plants still have to spend on investment until year 3. The first year, the investment cost includes the purchase of cocoa seeds and suren. The second and third year, investment costs include the purchase of cocoa seeds for the purpose of embroidery. The investment cost of cocoa per hectare is presented in Table 1.

**Table 1.** Average investment cost of suren and cocoa farming in Mekar Sari Raya village.

| Information | Amount (Units) | Price (IDR/Seeds) | Cost (IDR) |
|---|---|---|---|
| Cocoa seeds | 9,745 | 3,500 | 34,107,500 |
| Suren seeds | 1,448 | 3,000 | 4,344,000 |
| Amount | 11,193 | | 384,515,000 |

a. Labor costs

The labor used in cocoa farming comes from the family's own labor, neighbors and others. The cost of the family labor itself is a calculated cost where the cost is in cash not issued, while labor costs for other workers use the cash paid system and paid. Labor usage is calculated based on daily wages converted into working days (HK). The use of female labor is calculated as the average male laborer's wage, which is IDR 85,000.00 - IDR110,000.00 / day.

b. Fertilization and Pesticide Costs

The cost of fertilization is a considerable cost to farmers, the use of fertilizer enough each year to cause farmers to set aside income large enough to be used in purchasing fertilizer. The fertilizer costs incurred during farming can be shown in Table 2.

**Table 2.** Average cost of fertilizer done during suren and cocoa farming in Mekar Sari Raya village.

| Year | Manure spreader (IDR/ha) | fertilizer (IDR/ha) | | | Pesticide |
|---|---|---|---|---|---|
| | | Urea | TSP | KCL | Diazinon |
| 1-15 | 9,459,681 | 7,822,059 | 3,799,253 | 3,708,750 | 325,000 |

Fertilizers commonly used by farmers in the study sites are in the form of manure, Urea, and KCL. Manure is used at the time of land preparation, i.e. before the planting. Then for Urea and KCL fertilizer given the community at the time of planting is done. Urea and KCL fertilizers were administered two to four times a year by the community. In the maintenance activities, farmers also use pesticides. In addition to using pesticides, suren leaf extract is very helpful to prevent symptoms of damage caused by pest attacks. According to [4], states that suren leaf is often used as a vegetable pesticide and its pyrethropic active substances are capable of destroying the pest's nervous system. The substance works very quickly and cause symptoms of lethal paralysis. Suren extract can be an alternative in expelling aphis. According [5], states that the suren is rich in surenon, surenin and surenolakton which acts as a barrier to growth, insecticide, and antifeedant (inhibition of feeding power) against insect larvae.

*3.1.3 Suren and Cocoa Revenue.*

The results obtained from cocoa plants are dry cocoa beans in which the cocoa beans have been dried 1-3 days (if the weather is hot). Cocoa crops can be harvested at the age of 3 years with the production (dry beans) produced on average. The total production of dry cocoa beans and suren wood for each plant can be seen in Table 3 and Figure 5.

There are fluctuations in production from 3-15 years of suren and cocoa, because among farmers one with other farmers has a difference in the pattern of cocoa farming. For example, the difference in the use of fertilizers. The table above also shows that the largest cocoa seed production occurs at the time of the 13-year-old plant, which is 2,775 kg / ha / year.

According to [1] those cocoa plants have the largest production between 10 - 20 year-old plant, and the older age of cacao plants fruit production decreased. The lowest production occurs when the 3-year-old plant, which is 727 kg / ha / year. Because at the age of this plant, cocoa plants are still said to learn to bear fruit. According to [6], the average national productivity of cocoa reaches 845 kg / ha / year, while on private plantations slightly better with productivity of 907 kg / ha / year and 914 kg / ha / year respectively. The yield of cocoa production in Mekar Sari Village is below than the production potential. Based on Table 3, the average production of dry cocoa beans is about 1,960.42 kg / ha /year, so farmers get revenue of IDR 49,010,500 / year.

Suren trees are harvested in the 15th year of 168 trees, with an income of IDR 48,000,000, Suren trees can not be harvested together because the age of suren trees is not the same, Suren trees can be harvested with> 15 years and not all suren trees that are 15 years old can be harvested, due to the size of the tree diameter is different, Now, some farmers use suren trees only as intercrops, wind breakers, road shutters, and land boundaries, According to [7] (2016) states that suren tree can be sold which has a diameter of

≥ 30 cm, Suren diameter ≥ 30 cm has a price of IDR 1,000,000 - IDR2,000,000, the average market price of suren stands valued at IDR1,000,000 during the study.

**Table 3.** Average production of dry cocoa beans and suren wood

| Age of plants (Year) | Production of Dry Seed (Kg) | Price of dry cacao (Kg) | Suren Wood Producti on | Selling Price of Suren Stand (trunk) |
|---|---|---|---|---|
| 1 | - | - | - | - |
| 2 | - | - | - | - |
| 3 | 727 | 18,000 | - | - |
| 4 | 1,074 | 20,000 | - | - |
| 5 | 1,103 | 20,000 | - | - |
| 6 | 1,303 | 22,000 | - | - |
| 7 | 1,612 | 20,000 | - | - |
| 8 | 2,485 | 22,000 | - | - |
| 9 | 2,295 | 21,500 | - | - |
| 10 | 2,485 | 20,000 | - | - |
| 11 | 2,319,5 | 22,000 | - | - |
| 12 | 2,754 | 23,000 | - | - |
| 13 | 2,775 | 26,000 | - | - |
| 14 | 2,645 | 25,000 | - | - |
| 15 | 1,910 | 26,000 | 161 | 500,000 |
| Amount | 25,485,5 | 285,500 | 161 | 500,000 |
| Average | 1,960,42 | 21,961 | 21 | 500,000 |

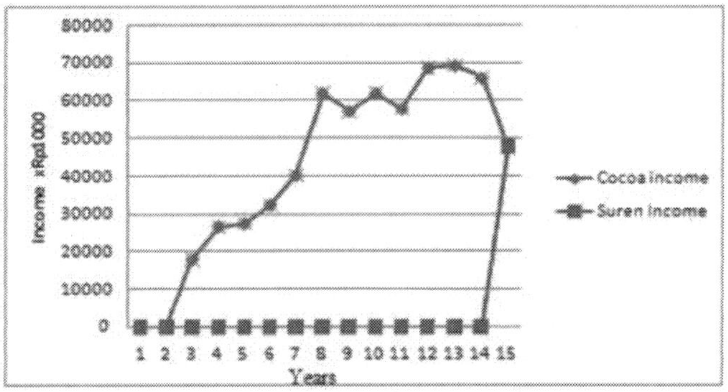


**Figure 5.** The average income of farmers in the village of Mekar Sari Raya for 15 years.

### *3.2 Revenue Analysis*

The income analysis is used to determine the ratio between the amount of costs incurred by the receipt of a production process, whether the production process is feasible to cultivate and can provide benefits. Table 4 shows the magnitude of NPV at 12% interest rate is IDR 2,670,306,905.84 for 15 years which means the value of NPV is positive or greater than zero, This indicates that net income of cocoa farm is bigger than total cost, so it can be said that cocoa farming in Mekar Sari Raya Village is feasible to be developed because of NPV> 0 value, The advantage gained by cocoa farmer causes the increase of farmer's cocoa welfare in Desa Mekar Sari is evident from the number of farmers who are able to increase the area of land, besides cocoa farmers are able to meet the needs of families and send their children to school.

**Table 4.** Analysis of cocoa and suren farm income per hectare at 12% (DF = 12%),

| Criteria | Results |
|---|---|
| Net Present Value (IDR) | 2,670,306,905.84 |
| Gross B/C | 2.3 |
| IRR (%) | 28% |
| Payback periode (year) | 5.4 (5 years, 4 months, 24 days) |

Table 4 shows that Gross B / C Ratio of 2.3> 1 is obtained which means that the benefits obtained are greater than the cost incurred, so that the farming business of suren and cocoa is feasible to continue, The IRR value obtained from the calculation is 28%. The IRR value obtained is greater than the prevailing market rate of 12%. It can be concluded that the agroforestry of cocoa and suren is feasible to be cultivated financially. Based on the calculation result, obtained payback period for 5.4 years. This means that the initial investment will be refunded within 5 years 4 months 12 days, so the investment payback period is shorter than the project's economical life of 15 year.

## 4. Conclusions

An income analysis conducted at the interest rate of 12% obtained Net Present Value (NPV) value of 2,670,306,905.84(IDR) for 15 years, Gross B / C Ratio (BCR) 2.3, Internal Rate of Return (IRR) 28% and Payback Period (PP) for 5 years 4 months 24 days. Based on the results of financial feasibility parameters namely NPV, BCR, IRR and PP obtained then agroforestry suren and cocoa in Mekar Sari Raya Village Panei Sub district financially feasible to be cultivated and developed.

## References

[1] Siregar T H, Slamet R and Laeli N 2005 *Cultivation, Processing and Marketing Chocolate* (Jakarta: Penebar Swadaya)
[2] Arikunto S 2006 *Research Procedures A Practice Approach* (Jakarta: Rineka Cipta)
[3] Mangdeska 2009 Analysis of Revenue of Cocoa Farming (*Theobroma cacao* L,) and factors influencing it in Nagari Lubuak Batikok Harau District, Lima Puluh Kota District, West Sumatra,
[4] Sesilia E P, Fidrianny L and Nawawi A 2006 Lean Suren Chemical Content Study (*Toona sinensis*), School of Pharmacy Bandung Institute of Technology, Bandung,
[5] Darwiati 2009 Test of Efficacy of Suren Plant Extract (*Tone sinensis* Merr) as Plant Insecticide in Leaf Pest Control (*Eurema* spp, And *Spodotera litura* F.), Sebelas Maret University, Surakarta,
[6] Ditjenbun 2013 Indonesian Plantation Statistics: Cocoa Commodities in 2012-2014, Secretariat General of Plantation, Jakarta,
[7] Situmeang H 2016 *Suren Agroforestry Financial Feasibility Analysis (Toona Sureni Merr,) And Arabica Coffee (Coffea Arabica L,)* (Medan: USU Press)

IOP Conf. Series: Earth and Environmental Science **122** (2018) 012141 doi:10.1088/1755-1315/122/1/012141

# Determination of wood wettability properties of oil palm trunk, *Shorea* sp. and *Paraserianthes falcataria* by contact angle method

**T Sucipto[1], R Hartono[1] and W Dwianto[2]**

[1]Faculty of Forestry, Universitas Sumatera Utara, Jl. Dr. Mansur Medan, 20135
[2]Research Center for Biomaterials, Indonesian Institute of Sciences, Jl. Raya Bogor Km.46, Cibinong Bogor 16911, Indonesia
E-mail: rudihartono@usu.ac.id

**Abstract.** The aim of this study was to determine the wettability of the inner part of oil palm trunk (OPT), the outer part of OPT, OPT that densified 50%, *Shorea* sp. and *Paraserianthes falcataria* wood, as raw material for laminated beams. The wettability of the wood was measured by using cosine-contact angle (CCA) method, which is measuring the angle between dripped resin liquid and the wood surface. The resins that used in this study is phenol formaldehyde (PF) and urea formaldehyde (UF). The results showed that the *Shorea* sp. and P. *falcataria* woods have the smallest contact angle or the best wettability properties than OPT. *Shorea* sp. has the best wettability on PF resin (83.00°), while P. *falcataria* on UF resin (90.89°), this is due to the levels of starch and extractive substances in *Shorea* sp. and P. *falcataria* wood are smaller than OPT. Furthermore, *Shorea* sp. and P. *falcataria* wood surfaces are flatter and smoother than OPT, so that the resin will flow easier and wetting the wood surface. In this condition, the liquid resin will flow easier and formed a smaller contact angle. The good wettability of wood will enhance the adhesion properties of laminated beams.

## 1. Introduction

Oil palm trunk (OPT) has a potential as as a raw material substitute for wood. In the other hand, OPT has many weaknesses, such as variations in the density, high shrinkage-swelling, high moisture content and low strength. The efforts to improve OPT quality has been done by compressed using close system compression [1] and by impregnated of phenol formaldehyde [2]. Through compression and impregnation of phenol formaldehyde into OPT can improve the physical and mechanical properties.

The other effort to improved OPT quality was make a laminated board. One of the properties that need before make the laminated board was to determine wettability properties. The wettability is a surface condition that determines the extent the fluid will be drawn by the surface, affecting absorption, adsorption, penetration and spread of adhesive [3]. By knowing the wettability properties of the wood, we can surmise any kind of wood that is suitable for done a gluing process, or what type of treatment can be done to increase the wettability properties of the wood so that the quality of gluing wood become better.

Wettability of the wood can be measured by the two methods, namely the method of Cosine-Contact Angle (CCA) and Corrected Water-Absorption Height (CWAH). CCA method and CWAH method has advantages and disadvantages of each. The weakness of the CWAH among other is the measurement results do not reflect the quality of the wood surface that will be glued, because the wood crushed into particles, so it can't be distinguished piece of wood surface to be glued or not to be glued. The advantage of CCA method among other its wettability measurement result is the wettability value of the wood surface that will be glued so that relatively more accurate if used to estimate the gluing strength.

In this study, wood wettability measurement was conducted by CCA. The wettability of wood depends on its wood species [4]. One of the easiest ways to determine wettability of the wood is a method of contact angle. Contact angle is done by measuring the angle formed between the surface of the wood with adhesive liquid that dripped in the board surface [5].

Smaller contact angle, indicating that the wood more easily wetted by adhesive.

The wood used is oil palm trunks (OPT). The OPT is comprised of three types, namely outer part of OPT, the inner part of OPT and the inner part of the OPT that is compressed to 50%. Meanwhile, as the comparison are meranti merah (*Shorea Sp.*) and Sengon (*Paraserianthes Falcataria*). The OPT included the monocots, while *Shorea* and *P. Falcataria* wood included the hardwood.

This study aims to analyze the nature of wettability of the inner and the outer of OPT and OPT that compressed, and also the wettablity of meranti merah and Sengon wood. The type of wood is the raw material that will be used for the manufacture of laminated beams.

## 2. Materials and Method

Materials used are OPT, *Shorea* sp, *P. Falcataria*, phenol formaldehyde (PF) resin and urea formaldehyde (UF) resin. Oil palm trunks used consisted of three types of oil palm trunk, there are outer part of OPT, inner part of OPT and compressed by 50% of OPT. Tools to be used are a saw, tape measure, calipers, oven, hot press machine, scales, moisture meter, digital camera and tripod, pipette, computer and Iconico Screen Protactor Software Version 4.0.

*The size of sample (OPT, Shorea* sp. and *P. falcatarial)* were 30 cm x 4 cm x 2 cm, in length, width, and thickness, respectively. The inner part of OPT for compressed sample was 30 cm x 4 cm x 4 cm and after compressed by 50% became 30 cm x 4 cm x 2 cm, using a hot press machine at a temperature of 120 °C and a pressure of 25 kgf for 30 minutes. Furthermore OPT board, *Shorea* sp. and *P. falcataria* wood dried in an oven at a temperature of 60 ° C until the moisture content of air dried. Sample test for the measurement of the contact angle is three boards per type of wood, with three measurements on each board. Total sample test is 30 boards (5 types of wood x 2 types of adhesive x 3 replications = 30 sample test).

Board was measured the moisture content and density. Measurements of water content using moister meter. The water content is denoted in percent (%). Density measurements were made by weighing the weight of the board (M) and measuring the average length, width and thickness to calculate the volume of the board (V). Density is calculated using dividing board's weight with board's volume, and it is denoted in units of $g/cm^3$.

Board wettability properties were measured using the Cosine-Contact Angle (CCA). Determination of board wettability is done by measuring the contact angle between the adhesive liquid that dripped in the board surface [5]. The trick is to do preparation beforehand in the form of a camera mounted on a tripod and eye lenses arranged parallel to the surface of the wood that will be dripped adhesive. The distance between camera lens with the wood is about 20 cm. The camera comes with a micro lens to produce an image that is clear and sharp. Dripping a liquid adhesive is done by using pipette and it is dripping on the surface of the wood that is flat and smooth. High of dripping adhesive is about 2 cm above the surface of the wood with a droplet volume of about 0.01 ml. Photo shoot is done 5 seconds after dripping. The magnitude of the contact angle was measured by the angle formed between the curved lines of the adhesive liquid to wood horizontal surface. Contact angle measurements using software Iconico Screen Protactor Version 4.0.

## 3. Results and Discussion

The value of the contact angle (CA), moisture content (MC) and the density (ρ) of OPT, *Shorea sp.* and *P. falcataria* wood are presented in Table 1. The results of these measurements show that the

OPT board, *Shorea sp.* and *P. falcataria* has a water content of 12.27~19.70%. Density board of OPT, *Shorea sp.* and *P. falcataria* is 0.26~0.78 g/cm$^3$. While, the contact angle OPT board, *Shorea sp.* and *P. falcataria* amounted to 83.00~108.78°.

**Table 1**. The contact angle value, moisture content and density of oil palm trunk, *Shorea sp.* and *P. falcataria*

| No. | The type of wood | PF Adhesives | | | UF Adhesive | | |
|---|---|---|---|---|---|---|---|
| | | CA (°) | MC (%) | ρ (g/cm$^3$) | CA (°) | MC (%) | ρ (g/cm$^3$) |
| 1. | Outer part of OPT | 105.44 | 17.90 | 0.39 | 98.11 | 17.13 | 0.38 |
| 2. | Inner part of OPT | 108.78 | 19.00 | 0.26 | 93.67 | 19.70 | 0.27 |
| 3. | OPT that compacted 50% | 98.89 | 12.27 | 0.38 | 107.33 | 13.83 | 0.38 |
| 4. | Meranti merah | 83.00 | 14.57 | 0.78 | 93.56 | 14.10 | 0.78 |
| 5. | Sengon | 90.11 | 16.20 | 0.46 | 90.89 | 15.67 | 0.49 |

Description: OPT = trunks of oil palm; PF = phenol formaldehyde; UF = urea formaldehyde; CA = contact angle; MC = moisture content; ρ= density

Wood moisture content is the content or the amount of water contained in the wood compared with the dry weight of the oven or the wood dry weight and expressed in percent. Wood is a hygroscopic material that has the ability to absorb and release water, either in the form of water or vapour. The greatest board water content is the board of the inner part of OPT is equal to 19.70%, while the water content of the smallest board is the OPT board that compressed 50% in the amount of 12.27%.

Wood moisture content was influenced by the wood itself and environmental factors. Wood moisture content is inversely proportional to its density. Smaller density of the wood, then the moisture content of the wood will be even greater, and vice versa. Board of the inner part of OPT that has a large water content, namely 19.70% is a board that has a small density is 0.27 g/cm$^3$. More decrease of wood density means that the wood is more porous and bulky or large volume. Low density wood is characterized by soft wood and large cell cavity. The magnitude of cell cavity can be a place to hold water in the wood. Water on the wood is contained in the wood cells. There are two types of water in the wood cells, the water in the cell walls of wood is called bound water, and water in the cell cavities is called free water. In the cell cavities are also saturated water vapour.

The contact angle was affected by wood moisture content [6]. All contact angles for all liquids increased with increasing wood moisture content. Accordingly that OPT have high moisture meter, especially the inner part and the outer part of OPT. The contact angle of inner part and outer part of OPT more higher from the other samples (Table 1).

Wood for contact angle measurements have been dried in an oven until the moisture content of air dried, the moisture content range of 15~20%. The water content of the dry air in wood occurs when an empty cell cavities and cell walls still contain some of the water. The OPT board that compacted 50% with a density of 0.38 g/cm$^3$ has the smallest water content, and it is equal to 12.27%. This is due to the OPT board has been experiencing a warming in hot press machine at a temperature of 120° C and a pressure of 25 kgf for 30 minutes. The warming has pulled out all the water from the cavity cells and some of the water from the cell walls of the wood. The compression process is able to improve the quality of the wood, lower water content and increase its density, such as compression of 50% on inner part of OPT being able to improve the physical and mechanical properties of OPT [7].

The highest density of the board is meranti merah board that is equal to 0.78 g/cm$^3$. The density of the lowest board is the board of the inner part of OPT is 0.26 g/cm$^3$. The results are consistent with other research. The density of meranti merah wood is 0.30-0.86 g/cm$^3$ with average 0. 52 g/cm$^3$ and density of sengon wood is 0.24-0.49 g/cm$^3$ with average 0.33 g/cm$^3$ [8]. In this type of OPT [9] suggested that a density value of OPT is 0.14~0.60 g/cm$^3$. In the oil palm trees aged 40 years, Hartono et al. [10] stated that OPT density is 0.23~0.74 g/cm$^3$. In this study of OPT was found that the density of the inner part of OPT lower than the density of the outer part of OPT. Then, all the physical and mechanical properties of the part near the bark decreases towards the central of trunk.

Compression of the inner part of OPT 50% able to improve OPT density of about 0.12 g/cm$^3$, so that the density of compressed OPT equivalent to the density of the outer part of OPT is 0.38 g/cm$^3$. OPT compression will be optimal if the known properties of the OPT, especially the radial distribution of its density value. The study of heat treatment [11] reported that heat treatment caused the wood more hydrophobic characteristics. Its mean that the contact angle values of the wood materials increased with the thermal modification. From data (Table 1) shown that contact angle of compressed OPT using FP resin was decreased if compare with the inner part and the outer part of OPT. The other hand, the contact angle of compressed OPT using UF was increased if compared with the inner part and the outer part of OPT.

The highest density of OPT is at the outer part near the bark, then decreased in the middle and the lowest in the central part of trunk. It is caused by distribution of vascular bundles, which is denser at the edges and tapers off towards the central [9], [10]. The other research [12] reported that horizontally the outer OPT density is higher than the inside. This is caused 1) on the outer is dominated by the vascular bundles (51%) which have a high density, while the inner part is dominated by the parenchyma tissue (70%) which having low density; and 2) the cell walls of parenchyma tissue of the outer is thicker than the inner.

Density of *Shorea* sp. include in the category of high density, due to its constituent cells has a thicker cell walls with a smaller cell cavity. Meranti merah has a robust class III~IV and suitable to use as a raw material of veneer, plywood, building materials, doors and windows, timber shipping, coffins, musical instruments, household furniture and crates wrapping [8].

Contact angle of OPT board, *Shorea* sp. and P. *falcataria* using phenol formaldehyde (PF) and urea formaldehyde (UF) resin is approximately 83.00~108.78°. The greatest contact angle is inner OPT by using PF resin, which amounted to 108.78°, whereas the smallest contact angle is *Shorea* sp. wood by using PF resin, which amounted to 83.00°. Contact angle of oil palm trunk, *Shorea* sp. and P. *falcataria* with PF and UF resin are presented in Figure 1.

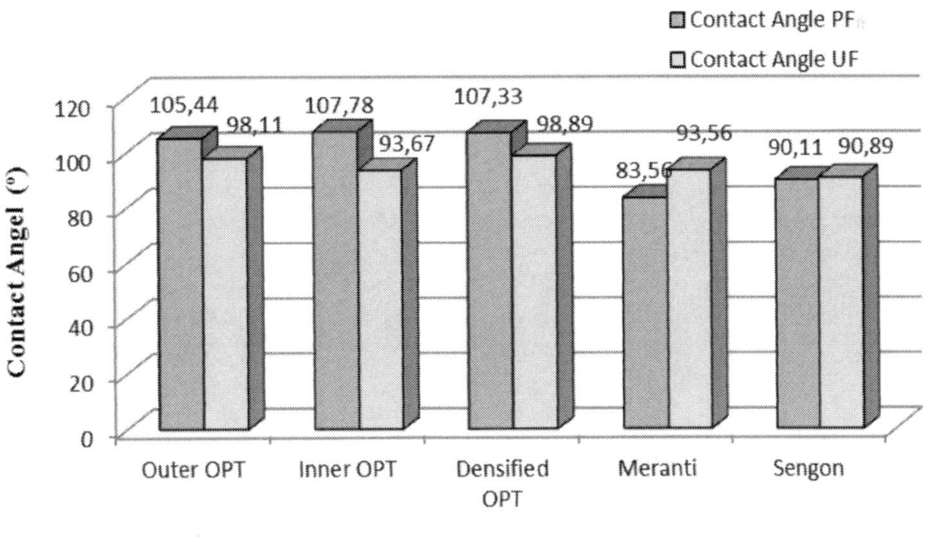


**Figure 1.** The contact angle of oil palm trunk, *Shorea* sp. and P. *falcataria*

The greater the contact angle formed between the wood and adhesive, then the lower the wettability properties of the wood. The results showed that *Shorea* sp. and P. *falcataria* wood have the most excellent wettability properties compared with OPT wettability. *Shorea* sp. and P. *falcataria* has the smallest contact angle or the best wettability properties than the wettability of oil palm trunk (OPT), both the outer part of OPT, the inner part of OPT, as well as OPT that compressed 50%. *Shorea* sp. wood has the best wettability for the adhesive properties of PF (the contact angle of 83.00°), while P. *falcataria* wood has the best wettability properties for UF resin (contact angle of 90.89°).

The results of wettablity several wood [13] showed that the wettability of P. *falcataria* wood higher than acacia and gmelina wood. The study also showed that the wettability of P. *falcataria* wood better than acacia wood, and acacia wood wettability is better than gmelina wood. This was caused by wood density and extractives deposit on the surface of the wood. The average density of acacia is 0.69 *(A. mangium)* and 0.61 *(A. auriculiformis)* and gmelina density is 0.42~0.61 [8]. In this study P. *falcataria* wood density is 0.46 ~ 0.49 (Table 1).

The study of contact angle [14] reported thath the contact angle dependant on wood species, resin type and on the location of the drop on the wood surface. Such as, the contact angle on sapwood is greater than that on heartwood for southern pine, but is smaller for Douglas-fir, and also for type resin, that contact angle difference between PF resin on southern pine and PMDI resin on Douglas-fir. From data (Fig 1), Contact angle of FP resin more higher than UF on inner part, outer part of OPT and compressed OPT. The otherwise, contact angle of UF is higher than PF on Sengon and Meranti wood.

The great of OPT contact angle or poor wettability properties allegedly caused by extractive substances that contained in it. In addition OPT has more rough wood surface and uneven than *Shorea* sp. and P. *falcataria*, so that the adhesive more difficult to flow and wetting the surface of the wood. In this condition, the liquid adhesive is more difficult to flow so that the contact angle was formed smaller. Good wettability properties of wood that will enhance the adhesion properties of the laminates beam to be made. The weakness of OPT are its starch contents, extractive and fatty acids are high, as well as hygroscopic properties is higher than wood [15]. Solubility of extractive substances of OPT in alcohol-benzene, cold water, hot water and NaOH 1% were 8.9%, 12.02%, 16.37% and 24.87%, respectively [16]. Then, solubility of extractive substances of P. *falcataria* wood in alcohol-benzene, cold water, hot water and NaOH 1% were 3.4%, 3.4%, 4.3% and 19.6% [17] and solubility of extractive substances of *Shorea* sp. wood in alcohol-benzene, cold water and hot water were 4.84%, 4.98% and 7.16%, respectively [18]

Solubility data of extractive substances on OPT, *Shorea* sp. and P. *falcataria* showed that the content of extractive substances in OPT is the greatest. The extractive agent has an enormous influence in decreasing hygroscopic, permeability and also increasing wood durability. Although its few in number, extractive influential in gluing of wood, which affect pH, contamination, and penetration [19]

The extractive substances occupy a place in the cell wall that it is usually occupied by water. Decreasing levels of extractive substances can raise the value of wettability due to the reduction of substances that block the absorption of fluid by the cell wall [20]. Also, the extractive substances composed of various substances differ in structure and chemical composition such as gum, fats, resins, sugar, oil, starch, and tannins the components are dissolved in cold water are tannin, gum, carbohydrate and pigment. Components that dissolved in hot water are the same as that dissolved in cold water plus starch component [21]

The presence of large extractive substances could interfere in the gluing process, especially from non-carbohydrate group (oils, fats, resins, salt, and etc) [22]. Therefore, if extractive substances that used quite a lot, wood should be treated to eliminate extractive substances by submersion, boiling or evaporation so that the wood was glued with satisfactory results.

Figure contact angle of outer part of OPT, inner part of OPT, OPT that compressed 50%, *Shorea* sp. and P. *falcataria* using PF and UF resin are presented in Figure 2 and Figure 3.

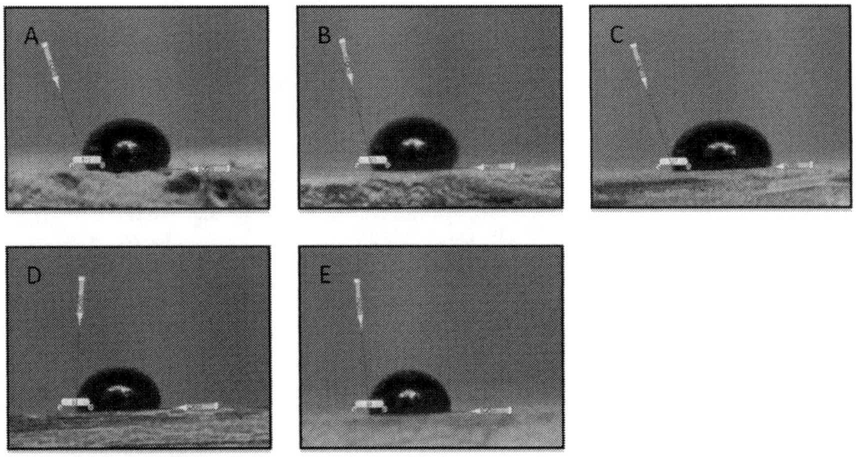


**Figure 2.** The contact angle of inner part of OPT (A), outer part of OPT (B), OPT that compressed 50% (C), *Shorea* sp. (D) and P. *falcataria* (E) using PF resin

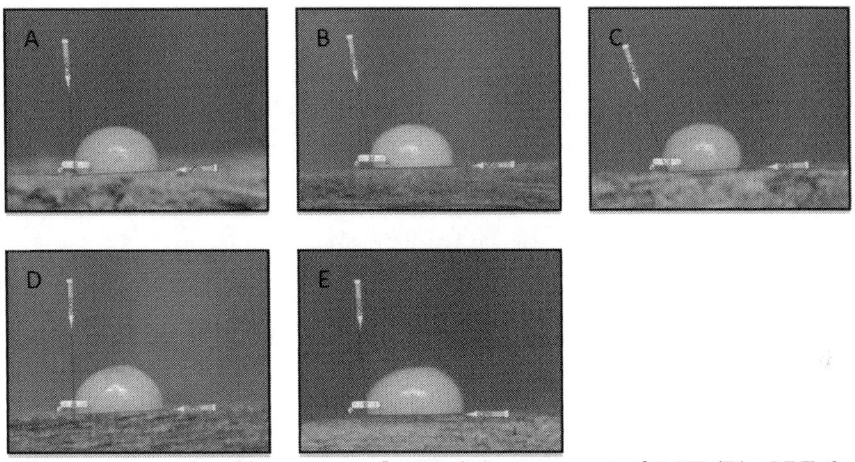


**Figure 3.** The contact angle of inner part of OPT (A), outer part of OPT (B), OPT that compressed 50% (C), *Shorea* sp. (D) and P. *falcataria* (E) using UF resin

The wettability is affected by the submersion treatment, either cold water or hot water submersion [23]. Wettability properties of wood that measured by the method of Corrected Water-Absorption Height (CWAH) rising after the levels of extractive substances reduced. Extractive substances out of the wood due to submersion treatments with cold water, hot water and steaming process. Reduced extractive substances cause a rise in the value of the absorption of water, which means also raise the value of wettability.

Then, wood extractives have correlation with the wettability of wood surfaces to be bonded by adhesives [24, 25]. Wettability is the ability of a material to absorb liquid (liquid system). Wettability properties can be used as an indication of the nature of its gluing. This is in line with the other research [26] that the wettability of wood by better adhesive will produce a stronger adhesiveness. This is shown with greater percentage of damage to the wood (not on the adhesion lines) and more resistant to delamination. At the time of adhesion process, the bond between the adhesive with adhered surface allows occurs because the first adhesive liquid wetting the surface.

Good wetting occurs when the contact angle (θ) between the adhesive and substrate is lower than $90^{\circ}$. Perfect wetting occurs when the molecular bond between liquids and solids (adherend) is greater than the molecular bonds in the liquid. Success or failure of a liquid wetting solid depends on the surface tension of both sustain, for example polymers and substrates. Increased surface energy which affected by water level, occurs due to the development of wooden structure that releases polar hydroxyl group will develop a liquid and the adhesive polymer. Liquid adhesive may cause the development of wood substantially. Thus, some of the wood adhesive has the capacity to convert wood surface with low energy become much higher. Sapwood transition process into heartwood will reduce surface energy, due to the non-polar extractives. Liquid adhesive with low pH such as UF has difficulty wetting and adhere to the surface. While the alkaline liquid adhesive (pH>8) as PF for plywood adhesive, can solve the extractive non-polar, and increase the surface energy which is sufficient to forming a good bond [27].

Wettability is affected by various factors, which are associated with an adhesive (surface tension, temperature, viscosity) and wood (density, porosity, extractive). Wood with low density (high porosity) has better wettability properties, while extractive in excessive amounts, like non-polar extractive such as terpene and fatty acids, having a bad influence. Wettability is also influenced by the cleanliness of the wood surface and the conditions of machining. For example, a blunt knife can cause the wood surface becomes overheating or compaction occurred (hardened surface). Veneer drying at high temperatures, above 160° C resulted in a decreased wettability [21].

The correlation between the glue ability index (adhesion strength) and wettability strengthen the theory of specific adhesion [28]. It is possible to predict the values of relative adhesion strength of a type of wood that is not recognized by measuring its wettability simply.

Based on these studies, it is possible to make laminated beams by using raw materials of OPT, *Shorea* sp. and P. *falcataria* wood. Laminated beams made of OPT with a combination of *Shorea* sp. and P. *falcataria* wood. On the one hand, it can utilize waste of oil palm plantations and on the other hand laminated beams that produced higher quality with better adhesive properties. To improve the gluing properties that will impact on the quality of laminated beams, it is necessary to pre-treatment in the form of submersion in hot water or in cold water. Most of extractive substances that can be disrupt the bonding process between adhesive and wood will be dissolved through the pre-treatment, so that the nature of its adhersion and the quality of the resulting laminated beams will increase.

## 4. Conclusion

*Shorea* sp. and P. *falcataria* wood have the smallest contact angle or the best wettability properties than the wettability of oil palm trunk (OPT), both outer part of OP, inner part of OPT, as well as OPT that is compressed 50%. *Shorea* sp. wood has the best wettability properties to phenol formaldehyde resin (contact angle of 83.00°), while P. *falcataria* wood has the best wettability properties for urea formaldehyde resin (contact angle of 90.89°).

## References.

[1] Hartono R, Wahyudi I, Febrianto F, Dwianto W, Hidayat W, Jang J H, Lee S H, Park S H and Kim N H 2016 *J. Korean Wood Sci. Technol* **44**(2): 172-183
[2] Hartono R, Hidayat W, Wahyudi I, Febrianto F, Dwianto W, Jang J H and Kim N H 2016 *J. Korean Wood Sci. Technol* **44**(6): 842-851
[3] Marra A A 1992 Technology of Wood Bonding:Principles in Practise (New York: Van Nostrand Reinhold)
[4] Petrie E M 2006 *Theories of Adhesives* ( http://www.specialchem4adhesives.com/editorial).
[5] Sucipto T 2009 *Determinasi Keterbasahan (Wettability) Kayu.* http://repository.usu.ac.id/bitstream/123456789/1035/1/10E00550.pdf
[6] Kúdela J, Wesserle F and Bakša J 2015 *Acta Facultatis Xylologiae Zvolen* 57(1): 25−35.
[7] Hartono R, Dwianto W, Wahyudi I, Febrianto F and Morooka T 2017 *IOP Conf. Ser.: Mater. Sci. Eng* **180** 012016
[8] Mandang Y I and Pandit I K N 2002 Pedoman Identifikasi Kayu di Lapangan (Indonesia: Yayasan Prosea dan Pusat Diklat Pegawai dan SDM Kehutanan, Bogor)
[9] Erwinsyah 2008 Improvment of Oil Palm Wood Properties Using Bioresin, Dissertation of graduated school of Universität Dresden.

[10] Hartono R, Wahyudi I, Febrianto F and Dwianto W 2011 *J. Ilmu dan Teknologi Kayu Tropis* **9**(1) 73-83
[11] Aytin A, Korkut S, Unsal O, Candan Z and Görgün H F 2015 *Proligno* 11(4): 373-376
[12] Bakar ES, Rachman O, Hermawan D, Karlinasari L and Rosdiana 1998 Jurnal Teknologi Hasil Hutan XI(1): 1-11
[13] Erniwati 2008 Pengembangan Papan Komposit Berlapis Anyaman Bambu dari Jenis Kayu Cepat Tumbuh dengan Perekat Poliuretan, Dissertation of graduated school of Bogor Agricultural University.
[14] Shi S Q and Gardner D J 2001 *Wood and Fiber Science* 33(1): 58-68
[15] Hartono R, Iswanto A H, Sucipto T and Maghfirah C Y 2012 Pengaruh Perendaman Partikel dalam Air Dingin dan Panas terhadap Sifat Fisis dan Mekanis Papan Partikel dari Limbah Batang Kelapa Sawit dengan Perekat Isosianat in: Prosiding Seminar Nasional Mapeki XV, Makassar, pp: 103-110
[16] Balfas J 2003 *Potensi Sawit sebagai Alternatif Bahan Baku Industri Perkayuan*. Seminar Nasional Himpunan Alumni IPB dan HAPKA Fakultas Kehutanan IPB Regional Sumatera. Medan.
[17] Martawijaya A, Kartasujana I, Mandang Y I, Prawira S A and Kadir K 1989 Atlas Kayu Indonesia Jilis II (Indonesia: Badan Penelitin dan Pengembangan Kehutanan Departemen Kehutanan, Bogor)
[18] Martawijaya A, Kartasujana I, Kadir K and Prawira S A 1985 Atlas Kayu Indonesia Jilis I (Indonesia: Badan Penelitin dan Pengembangan Kehutanan Departemen Kehutanan, Bogor).
[19] Ruhendi S, Koroh D N, Syamani F A, Yanti H, Nurhaida, Saad S and Sucipto T 2007 Analisis Perekatan Kayu (Indonesia: Fakultas Kehutanan IPB, Bogor)
[20] Bowyer JL, Shmulsky R and Haygreen J G 2003 Forest Products and Wood Science: An Introduction Fourth edition (USA: Lowa State University Pr.)
[21] Tsoumis G 1991 *Science and Technology of Wood: Structure, Properties, Utilization* (New York,Van Nostrand Reinhold)
[22] Kasmudjo 2010 Teknologi Hasil Hutan, Suatu Pengantar (Indonesia: Cakrawala Media. Yogyakarta)
[23] Pari G 1994 Pengaruh Keterbasahan Selumbar terhadap Keteguhan Rekat Papan Partikel, Undergraduate school of Faculty of Forestry, Bogor Agriculture University.
[24] Ruhendi S and Sucipto T 2007 *Wetabilitas Tandan Kosong Sawit* In: Prosiding Seminar Nasional Mapeki X, Pontianak, pp 17-23
[25] Rafael E 2016 *Appl Microbiol Biotechnol* **100**:1589–1596
[26] Alamsyah E M, Yamada M and Taki K 2005 *J Wood Science* **53***: 40-46*
[27] Wellons J D 1980 *Forest Products Journal* **30**(7): 53-55
[28] Warsa S R 1983 Gluability of Rotary-Cut Veneers of Some Indonesian Woods using Adhesives Extenden with Nami and Cassava Flours, Dissertation of the graduate school of University of the Philippines

**Acknowledgment**
The author would like to thank to the Ministry of Research, Technology and Higher Education on competitive grants for this research. And also to the Research Institute of the University of Sumatera Utara as a manager of a research grant at the university level.

# Pathotype profile of *Xanthomonas oryzae* pv. *oryzae* isolates from North Sumatera

**Z Noer[1,2*], Hasanuddin[2], Lisnawita[2] and D Suryanto[3]**

[1]Departemen of Agrotechnology, Universitas Medan Area, Jalan Kolam Nomor 1 Medan Estate Jl. Gedung PBSI, Medan 20223, North Sumatera, Indonesia
[2]Faculty of Agriculture, Universitas Sumatera Utara, Jl. Dr. A. Sofian No. 3, Padang Bulan, Universitas Sumatera Utara, Medan, 20155, North Sumatera, Indonesia.
[3]Department of Biology, Faculty of Mathematics and Natural Sciences, Universitas Sumatera Utara, Jl. Bioteknologi No. 1, Medan 20155, Indonesia
*Email: z_noer@yahoo.co.id

**Abstract**. The Bacterial blight disease caused by *Xanthomonas oryzae* pv. *oryzae* (Xoo) is one of the most important diseases and has caused crop failure in rice crops. This pathogen infects the leaves in all plant growth phases. The purpose of this study is to investigation 10 Xoo isolates pathotype obtained from North Sumatra based on their interactions with 10 near-isogenic rice lines (NIL) of IRRI. The results showed that there are 6 pathotypes of virulence in North Sumatra, they are; pathotype I with incompatible interaction to all Xa genes, pathotype II with compatible interaction to *Xa1* and *Xa3* genes, while it has incompatible interaction to other genes, pathotype III with compatible interaction to *Xa1, Xa5, Xa7, Xa8, Xa10* and *Xa11* genes, but it has incompatible interaction to other genes, pathotype IV with compatible interaction to all Xa genes, pathotype V with compatible interaction to *Xa1* gene and incompatible interaction to other genes, and pathotype VI with compatible interaction to *Xa3* gene and incompatible interaction to other genes. Based on the resistant genes in each individual*Xa2*, *Xa4*, and *Xa21* genes are the combination of Xa genes which are most suitable for use in the development of rice cultivars in North Sumatra.

## 1. Introduction

Rice is a staple food and is a worldwide model cereal [1]. Bacterial blight caused by *Xanthomonas oryzae* pv. *oryzae* (Xoo) is one of the most important diseases in rice-producing countries in Asia, including in Indonesia [2][3]. In Indonesia, this particular disease is found in various rice cultivation ecosystems and spread in the lowlands [4]. Bacterial blight is endemic and has been known to cause rice crop failures in irrigated and lowland areas across Asia [5]. Bacterial blight caused by Xoo appears in the form of epidemics in many parts of the world, resulting in 6-81% loss in some rice varieties. [6].

Bacterial blight generally infects leaves, but its distribution depends highly on its environmental conditions [3]. Infection usually starts from the tip of the leaf and spreads all the way to the base. Symptoms that occur in rice plants during its vegetative phase are called *kresek*, and in its generative phase they're called blight [7]. Pathogenic infections cause a disruption of photosynthetic functions, leading to infected plants producing much more empty grains compared to healthy plants [8]. If pathogen attacks the plant during the seed phase, its leaves will wither, curl, and become grayish-green in color. If the attack occurs on mature plants, the leaves will turn pale yellow [9].

The spread of Xoo-resistant rice varieties is a more sustainable solution for disease control than the use of chemicals [10]. The resistance of a rice variety to Xoo largely depends on the availability of resistance genes and information on virulence estimates in pathogen populations [11]. Information on the virulence of pathogenic populations is essential for efficient cultivation programs [12]. The population structure was determined through a series of tests using varieties with confirmed resistance genes, followed by classification of Xoo isolates into different pathotypes. Pathotypes are unique to one another because the researchers used various differential genotypes, both from Japan and the Philippines, as well as the newly selected differential genotypes in their research [13].

More than thirty bacterial leaf blight-resistance genes (called Xa genes) have been identified to provide resistance to a wide variety of race and pathotypes of Xoo [14]. Pathogens can easily damage a single-resistance gene [15]. One strategy that can be used to prolong the life of major gene resistance is the pyramid of several major resistant genes in resistant varieties [16]. This resistance gene shows specificity in terms of effectiveness against different pathogen races. As such, knowledge of diversity of pathotypes in target pathogenic population is essential to select resistant genes to be used in cultivation programs [10].

Rotation of resistant varieties to control bacterial leaf blight disease must be carefully designed, as they can last longer in the field. This strategy requires the support of various data, especially regarding the profile of Xoo pathotype in a population and the resistance history of he varieties to be planted. The purpose of this research is to identify the profile of 10 Xoo isolates obtained from the province of North Sumatera, Indonesia, using 10 differential genotypes of near-isogenic lines (NILs) from IRRI.

## 2. Material and Methods

### *2.1. Pathogen collection and artificial inoculation*

Ten pathogenic strains representing the North Sumatra region were collected in 2016 and 2017. Xoo isolates were grown on YDC (yeast extract-dextrose-CaCO3) or SPA (sucrose pepton agar) medium. The collected isolates were cultured and treated at the Pest and Disease Laboratory of the University of North Sumatra and the Laboratory of Indonesian Biotechnology and Genetic Resources Research and Development Center. Leaves with symptoms of bacterial blight underwent surface sterilization using 70% alcohol, then rinsed with sterile water three consecutives times, dried and cut to 5x5 mm pieces and eventually soaked in sterile water for 5 minutes in the test tube.The suspension was scratched on the medium, incubated for 48-72 hours until a single yellow colony is obtained and then stored in SPA medium at 4°C. The test plant was inoculated with Xoo at age 35 days after transplanting with scissors method[4]. The tip of rice leaves in three different plant varieties for every genotype was cut to a 10 cm piece using innoculation shear with 48 hours-old bacteria innoculum with a concentration of 108 cfu. Inoculation was carried out on three individual plants in each strain and repeated three times.

### *2.2. IRBB lines and their evaluation*

To identify the type of reaction of all isolates collected from North Sumatra, 10 different rice varieties with known resistant gene differenceswere employed. There were 10rice genotypes consisting of IRBB-1, IRBB-2, IRBB-3, IRBB-4, IRBB-5, IRBB-7, IRBB-8, IRBB-10, IRBB-11, and IRBB-21 from IRRI (Table 1).

### *2.3. Disease scoring, virulence analysis and morphological characterization*

Scoring is done by measuring the length of symptoms 15-30 days after inoculation. The severity of the disease was measured on three inoculated plant species by examining the five longest symptomatic leaves in each clump. The severity of the disease is determined by the ratio between the length of the symptoms and the length of the leaves. The resistance of varieties was grouped according to disease severity upon observation 30 days after the inoculation. The final grouping is based on the Standard Evaluation System (IRRI, 1996). The data were then used to calculate the disease intensity of each tested genotype. The crop endurance category was determined using terms based on lesion length (cm) with resistance (R; ≤ 10 mm) and susceptibility (S;> 10 mm)

## 3. Result and Discussion

### *3.1. Result*

A total of 10 Xoo isolates were collected from 10 different regions in North Sumatra, Indonesia, from March 2016 to March 2017. All isolates were tested for pathotype using 10 NILs with Xoo-resistant genes of *Xa1, Xa2, Xa3, Xa4, Xa5, Xa7, Xa8, Xa10, Xa11*, and *Xa21*. The results showed that there are 6 different pathotypes in this research (Table 1). Pathotype IV has the highest-frequency and accounts for 50% of all isolates. The Xoo strain included in this pathotype is widespread across North Sumatra. Compared with this pathotype, the other five are more rarely present, each accounting for 10% of the total batch (Figure 1).

**Table 1.** Pathotyping of North Sumatera, Indonesia of *X. oryzae* pv. *oryzae*

| Host differential | *X. oryzae* pv. *oryzae* Pathotype | | | | | |
|---|---|---|---|---|---|---|
| | I | II | III | IV | V | VI |
| IRBB-1 (*Xa1*) | R | S | S | S | S | R |
| IRBB-2 (*Xa2*) | R | R | R | S | R | R |
| IRBB-3 (*Xa3*) | R | S | R | S | R | S |
| IRBB-4 (*Xa4*) | R | R | R | S | R | R |
| IRBB-5 (*Xa5*) | R | R | S | S | R | R |
| IRBB-7 (*Xa7*) | R | R | S | S | R | R |
| IRBB-8 (*Xa8*) | R | R | S | S | R | R |
| IRBB-10 (*Xa10*) | R | R | S | S | R | R |
| IRBB-11 (*Xa11*) | R | R | S | S | R | R |
| IRBB-21 (*Xa21*) | R | R | R | S | R | R |

R; Resistance; S: Susceptible

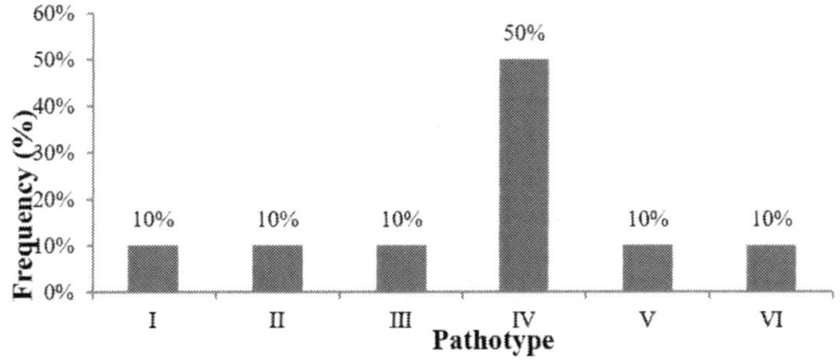


**Figure 1.** Frequency of *X. oryzae* pv *oryzae* pathotypes in North Sumatera, Indonesia Six different pathotypes, designated as I to VI, were identified in this study. The x-axis indicates pathotype and the y-axis indicates frequency amingst the 10 isolates that were collected and analyzed.

Pathotype I shows incompatibility with all differential plants, which indicates that this pathotype consists of a less virulent strain. This pathotype is present in 10% of the total sample of Xoo isolates. The possibility that pathotype I is a grouping of strains from another pathotype for unknown reasons has lost its virulence. Loss of virulence has been reported in the aging culture of Xoo [17]. The results of Xoo infection on NIL genotype are presented in Table 2. All the NIL genotypes of the differential rice varieties tested in this research show various reactions of resistance levels and susceptibilities to Xoo.

Average comparison of Xoo isolates shows that 5 isolates (Xo16-011, Xo16-222, Xo16-230, Xo16-254, and Xoo-7624) are significantly more virulent than other isolates. The dendogram analysis shows that the tested isolates are grouped into three main categories: resistant isolates (40%), moderately resistent isolates (10%), and susceptible isolates (50%) (Figure 2).

**Table 2.** Relationship between 10 NILs Genotypes and 10 Xoo isolates

IOP Conf. Series: Earth and Environmental Science **122** (2018) 012142 doi:10.1088/1755-1315/122/1/012142

| Isolate | IRBB-1 | IRBB-2 | IRBB-3 | IRBB-4 | IRBB-5 | IRBB-7 | IRBB-8 | IRBB-10 | IRBB-11 | IRBB-21 |
|---|---|---|---|---|---|---|---|---|---|---|
| Xo16-011 | S | S | S | S | S | S | S | S | S | S |
| Xo16-222 | S | S | S | S | S | S | S | S | S | S |
| Xo16-230 | S | S | S | S | S | S | S | S | S | S |
| Xo16-254 | S | S | S | S | S | S | S | S | S | S |
| Xo16-080 | R | R | R | R | R | R | R | R | R | R |
| Xo16-090 | S | R | S | R | R | R | R | R | R | R |
| Xo16-089 | S | R | R | R | S | S | S | S | S | R |
| Xo16-100 | S | R | R | R | R | R | R | R | R | R |
| Xo16-003 | R | R | S | R | R | R | R | R | R | R |
| Xoo-7624 | S | S | S | S | S | S | S | S | S | S |

R; Resistance; S: Susceptible

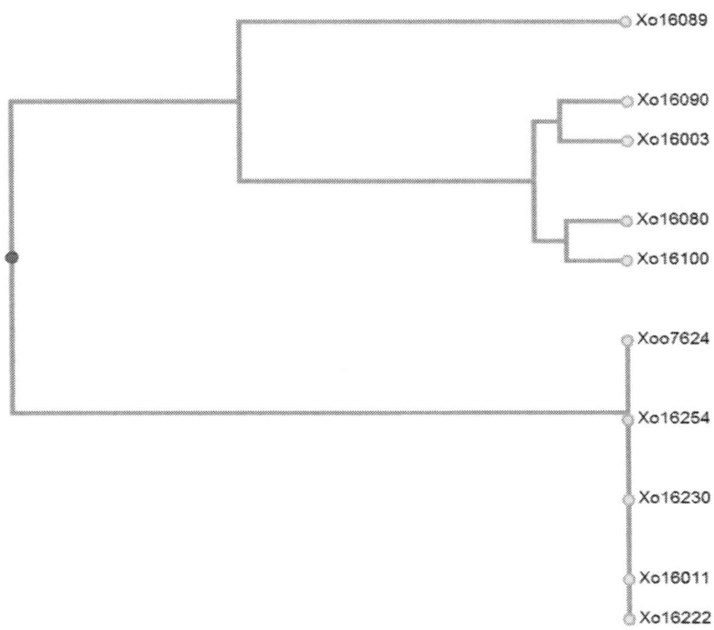


**Figure 2.** Dendogram of similarity among Xoo isolate based on their reaction to NILs

Based on the resistant genes in each individual, the genes *Xa2, Xa4*, and *Xa21* are the most effective genes that are resistant to 50% isolates, followed by *Xa5, Xa7, Xa8, Xa10* and *Xa11* (40%), *Xa3* (30%), and *Xa1* (20%). *Xa2, Xa4*, and *Xa21* genes are a combination of *Xa* genes that are most suitable for cultivating rice in North Sumatra (Figure 3).

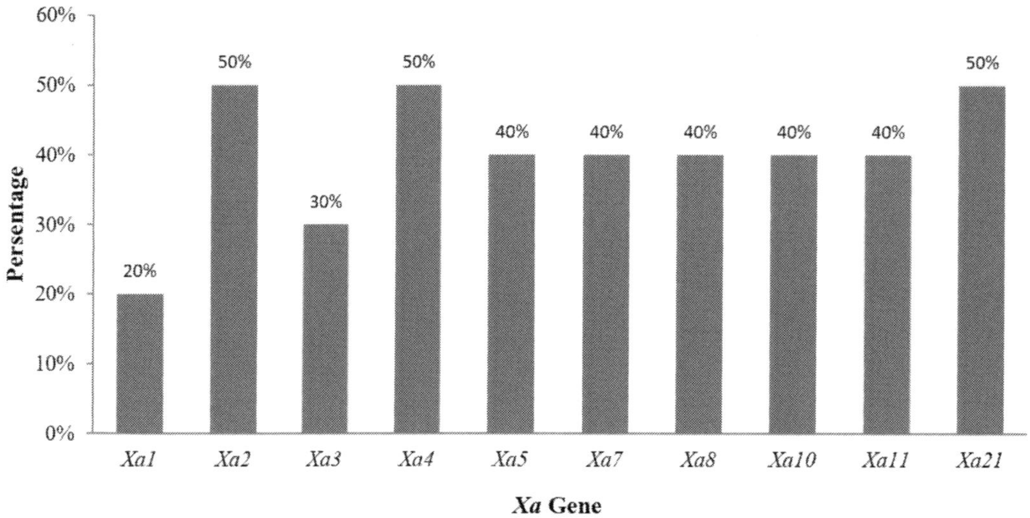


**Figure 3.** *Xa* Gene Effectiveness on 10 Isolates Xoo

### *3.2. Discussion*

Virulence of bacterial isolates in host genotype can be determined by the presence or absence of substantial differential interactions between the two [2]. Significant differential interactions between differences and Xoo isolates in the current research suggest that pathogen isolates have different in virulence responses, whereas host genotypes show vertical resistance [18]. Xoo isolates which are unique pathogens provide molecular mechanisms of plant pathogenic interactions. The diversity of Xoo isolates has been demonstrated in previous studies [13][19][20].

The reaction of different isolates on different rice varieties [21]. The grouping of isolates into different pathotypes follows a scheme provided by IRRI with slight modification [22]. Pathotype is numbered according to the pathotyping scheme designed by the Directorate of Rice Research. Numbering is based on virulence strength, from the least to the most virulent. The results show that isolates from different regions are classified as the same pathotype and at the same time, different isolates from the same region are grouped into different pathotypes. This shows that these pathogens are highly dynamic and their population in a particular region varies with different pathotypes [21]. Variations of virulence profiles between Xoo isolates from the same area have been reported previously [23] resulting in a strong selection pressure. The Xoo population produces variations of the virulence profile.

Bacterial blight is a major rice disease in the tropical and subtropical rice-growing countries. Cultivation for host resistance is the most effective, simple, economical, and environmentally-friendly method of controlling this disease [24]. To date, all Xa genes that provide resistance to these bacteria have been listed together with their source and country of origin [25]. In this study, the effective genes for all bacterial isolates were the genes *Xa2, Xa4* and *Xa21*, while the weakest against isolates were the genes *Xa1* and *Xa3* (Figure 3). Although the *Xa21* gene is resistant to most isolates, several isolates from Sri Lanka, Korea, Nepal, Bangladesh, and Pakistan have demonstrated virulence to *Xa21* [26]. Therefore, other *Xa* genes are indispensable to complement the *Xa21* gene in achieving longer lasting effects [27]. [28] conducted a screening study on 16 pyramid genes which carry the 2-5 resistance gene tested against 16 isolates and concluded that the genes *Xa13* and *Xa21* provide resistance to most of the Xoo isolates in Pakistan.

A better understanding of the population and its genetic structure is essential for planning and implementing better cultivation programs on the host plant's resistance. However, a study of the varieties themselves fails to show the true level of genetic diversity in the pathogenic population [29]. This study can be further developed using plants with Xa gene combinations in a pyramid, using molecular markers and transfer of plant resistance to bacterial blight through binary quarantine with local varieties and resistant varieties. The use of resistance genes found to be effective in this research can be carried out through dispersion under tropical climate conditions. This finding is also useful for pathologists to better understand the characteristics of bacterial isolates blight in North Sumatra in order to develop an efficient and long-lasting management strategy.

## 4. Conclusion

Analysis of rice plant patotypes in North Sumatra grouped into six types of patotype. A total of 50% of Xoo bacteria isolates belong to the type of patotype IV. Dendogram analysis showed that the isolates grouped into three categories: resistant isolates (40%), moderately resistent isolates (10%), and susceptible isolates (50%). Gene *Xa2*, *Xa4*, and *Xa21* are the most effective genes against isolate resistance (50%). Research can be continued by testing isolates using differential plants with pyramid Xa gene combinations to obtain which combinations of genes are most effective against Xoo bacteria.

## References

[1] Debnath S, Satya P and Saha B C 2013 Pathotype characterization of *Xanthomonas oryzae* pv *oryzae* isolates of West Bengal and evaluation of resistance genes of bacterial blight of rice (*Oryza sativa* L.) *J. Crop Weed.* vol 9 no 1 pp 198–200

[2] Nayak D, Bose L, Reddy P and Nayak P 2008 Host-pathogen interaction in rice-bacterial blight pathosystem *J. Plant Prot. Res.* vol 48 no 3 pp 371–384

[3] Sudir S and Yuliani D 2016 Composition and distribution of *Xanthomonas oryzae* pv. *oryzae* pathotypes the pathogen of rice bacterial leaf blight in Indonesia *Agrivita* vol 38 no 2 p 174

[4] Suparyono S, Sudir S and Suprihanto S 2003 Composition pathogenic pathotype of bacterial leaf blight in rice plants at grow different stadia (in Indonesian) *J. Penelit. Pertan. Tanam. Pangan* vol 22 no 1 pp 45–50

[5] Singh R A, Das B, Ahmed K M and Pal V 1980 Chemical control of bacterial leaf blight of rice *Int. J. Pest Manag.* vol 26 no 1 pp 21–25

[6] Shehzad F I D and Ahmad M 2012 Screening of local rice germplasm against bacterial leaf blight caused by *Xanthomonas oryzae* pv. *oryzae Sarhad J. Agric.* vol 28 no 4.

[7] Suparyono S, Sudir S and Suprihanto S 2004 Pathotype profile of *Xanthomonas oryzae* pv. *oryzae* isolates from the rice ecosystem in Java *Indones. J. Agric. Sci.* vol 5 no 2 pp 63–69

[8] Shanti M L, Shenoy V V, Devi G L, Kumar M, Premalatha P, Kumar G N, Shashidhar H E, Zehr U B & Freeman W H 2010 Marker-assisted breeding for resistance to bacterial leaf blight in popular cultivar and parental lines of hybrid rice *J. Plant Pathol.* vol 92 no 2 pp 495–501

[9] Lang J M, Hamilton J P, Diaz M G Q, Van Sluys M A, Burgos M R G, Vera C M, Cruz, Buell C R, Tisserat N A, Leach J E 2010 Genomics-based diagnostic marker development for *Xanthomonas oryzae* pv. *oryzae* and X. *oryzae* pv. oryzicola *Plant Dis.* vol 94 no 3 pp 311–319

[10] Mishra D, Vishnupriya M R, Anil M G, Konda K, Raj Y and Sonti R V 2013 Pathotype and genetic diversity amongst Indian isolates of *Xanthomonas oryzae* pv. *oryzae PLoS One* vol 8 no 11 p e81996

[11] Strange R N and Scott P R 2005 Plant disease: a threat to global food security *Annu. Rev. Phytopathol.* vol 43 no 1 pp 83–116

[12] Arshad H M I, Saleem K, Khan J A, Rashid M, Atiq M, Alam S S, Sahi S T 2017 Pathogenic diversity of *Xanthomonas oryzae* pv. *oryzae* isolates collected from Punjab Province of Pakistan *Eur. J. Plant Pathol.* vol 147 no 3 pp 639–651

[13] Khoshkdaman M, Ebadi A A, Majidi-Shilsar F and Dariush S 2014 Preliminary evaluation of resistance genes in rice against bacterial leaf blight in Guilan Province-Iran *Agric. Sci.* vol 5 no 2 pp 94–98

[14] Nino-Liu D O, Ronald P C and Bogdanove A J 2006 *Xanthomonas oryzae* pathovars: model pathogens of a model crop *Mol. Plant Pathol.* vol 7 no 5 pp 303–324

[15] Suryadi Y, Samudra I, Priyatno T P, Susilowati D N, Lestari P, Fatimah F, Kadir T S 2016 Determination of Pathotypes from Indonesian *Xanthomonas oryzae* Pv. *Oryzae* Population causing Bacterial Leaf Blight and their Reactions on Differential Rice *Makara J. Sci.* vol 20 no 3 pp 109–118

[16] Yugander A, Sundaram R M, Ladhalakshmi D, Kajira K, Prakasam V, Prasad M S, Madhav M S & Babu V R 2017 Virulence profiling of *Xanthomonas oryzae* pv. *oryzae* isolates causing bacterial blight of rice in India *Eur. J. Plant Pathol.* vol 149 no 1 pp 171–191

[17] Rajeshwari R and Sonti R V 2000 Stationary-phase variation due to transposition of novel insertion elements in *Xanthomonas oryzae* pv. *oryzae J. Bacteriol.* vol 182 no 17 pp 4797–4802

[18] Wang C, Su C, Zhai H and Wan J 2005 Identification of QTLs underlying resistance to a virulent strain of *Xanthomonas oryzae* pv. *oryzae* in rice cultivar DV85 *F. Crop. Res.* vol 91 no 2 p 337-343

[19] Salzberg S L, Sommer D D, Schatz M C, Phillippy A M, Rabinowicz P D, Tsuge S, Furutani A, Ochiai H, Delcher A L & Kelley D 2008 Genome sequence and rapid evolution of the rice pathogen *Xanthomonas oryzae* pv. *oryzae* PXO99 A *BMC Genomics* vol 9 no 1 p 204

[20] White F F and Yang B 2009 Host and pathogen factors controlling the rice-*Xanthomonas oryzae* interaction *Plant Physiol.* vol 150 no 4 pp 1677–1686

[21] Yugander A, Sundaram R M, Lakshmi D L, Shaik H, Madhav M S, Prasad M S, Viraktamath B C & Laha G S 2014 Pathogenic and genetic profile of *Xanthomonas oryzae* pv. *oryzae* isolates from Andhra Pradesh *Indian J. Plant Prot.* vol 42 no 2 pp 149–155

[22] Chaudhary R C 1996 *Standard evaluation system for rice*. Manila: International Rice Research Institute

[23] Shanti M L, George M L C, Cruz C V, Bernardo M A, Nelson J R, Leung H, Reddy J N & Sridhar R 2001 Identification of resistance genes effective against rice bacterial blight pathogen in eastern India *Plant Dis.* vol 85 no 5 pp 506–512

[24] Gautam R K, Singh P K, Sakthivel K, Srikumar M, Kumar N, Kumar K, Singh A K & Roy S D 2015 Analysis of pathogenic diversity of the rice bacterial blight pathogen (*Xanthomonas oryzae* pv. *oryzae*) in the Andaman Islands and identification of effective resistance genes *J. Phytopathol.* vol 163 no 6 pp 423–432

[25] Khan M A, Naeem M and Iqbal M 2014 Breeding approaches for bacterial leaf blight resistance in rice (*Oryza sativa* L.) current status and future directions *Eur. J. Plant Pathol.* vol 139 no 1 pp 27–37

[26] Lee S W, Choi S H, Han S S, Lee D G and Lee B Y 1999 Distribution of *Xanthomonas oryzae* pv. *oryzae* strains virulent to *Xa21* in Korea *Phytopathology* vol 89 no 10 pp 928–933

[27] Jeung J U, Heu S G, Shin M S, Vera C M, Cruz and Jena K K 2006 Dynamics of *Xanthomonas oryzae* pv. *oryzae* populations in Korea and their relationship to known bacterial blight resistance genes *Phytopathology* vol 96 no 8 pp 867–875

[28] Khan J A, Arshad H M I, Saleem K, Sandhu A F, Hasnain S and Babar M M 2012 Evaluation of resistance genes in rice against local isolates of *Xanthomonas oryzae* pv. *oryzae* in Punjab Province of Pakistan *Arch. Phytopathol. plant Prot.* vol 45 no 15 pp 1826–1839

[29] Yashitola J, Krishnaveni D, Reddy A P K and Sonti R V 1997 Genetic diversity within the population of *Xanthomonas oryzae* pv. *oryzae* in India *Phytopathology* vol 87 no 7 pp 760–765

**Acknowledgments**
This research was supported by KEMENRISTEK DIKTI through HIBAH DISERTASI DOKTOR 2017. Thanks to all parties involved in this study.

# AUTHOR INDEX

# AUTHOR INDEX

# AUTHOR INDEX

# AUTHOR INDEX